Small Air-Cooled Engines Service Manual

(16th Edition)

Only engines of less than 15 cu … *this manual.*

ACME
ADVANCED ENGINE PRODUCTS
BRIGGS & STRATTON
CHRYSLER
CLINTON
COX
CONTINENTAL
CRAFTSMAN
DECO-GRAND
HOMELITE
HONDA

…PRODUCTS
…OLMAR
…WA
…EH
…END
…SIN
…SIN ROBIN

Published by **INTERTEC PUBLISHIN**… **PARK, KS 66212**

All instructions and diagrams have been checked for accuracy and ease of application; however, success and safety in working with tools depend to a great extent upon individual accuracy, skill and caution. For this reason, the publishers are not able to guarantee the result of any procedure contained herein. Nor can they assume responsibility for any damage or injury to persons occasioned from the procedures. Persons engaging in the procedures do so entirely at their own risk.

CONTENTS

GENERAL

ENGINE SERVICE SECTIONS

CONTENTS CONT.

DUAL DIMENSIONS

This service manual provides specifications in both the Metric (SI) and U.S. Customary systems of measurement. The first specification is given in the measuring system used during manufacture, while the second specification (given in parenthesis) is the converted measurement. For instance, a specification of "0.28 mm (0.011 inch)" would indicate that the equipment was manufactured using the metric system of measurement and U.S. equivalent of 0.28 mm is 0.011 inch.

FUNDAMENTALS SECTION

ENGINE FUNDAMENTALS

OPERATING PRINCIPLES

The small engines used to power lawn mowers, garden tractors and many other items of power equipment in use today are basically similar. All are technically known as "Internal Combustion Reciprocating Engines."

The source of power is heat formed by the burning of a combustible mixture of petroleum products and air. In a reciprocating engine, this burning takes place in a closed cylinder containing a piston. Expansion resulting from the heat of combustion applies pressure on the piston to turn a shaft by means of a crank and connecting rod.

The fuel-air mixture may be ignited by means of an electric spark (Otto Cycle Engine) or by heat formed from compression of air in the engine cylinder (Diesel Cycle Engine). The complete series of events which must take place in order for the engine to run may occur in one revolution of the crankshaft (two strokes of the piston in cylinder) which is referred to as a "Two Stroke Cycle Engine," or in two revolutions of the crankshaft (four strokes of the piston in cylinder) which is referred to as a "Four-Stroke Cycle Engine."

OTTO CYCLE. In a spark ignited engine, a series of five events is required in order for the engine to provide power. This series of events is called the "Cycle" (or "Work Cycle") and is repeated in each cylinder of the engine as long as work is being done. This series of events which comprise the "Cycle" are as follows:

1. The mixture of fuel and air is pushed into the cylinder by atmospheric pressure when the pressure within the engine cylinder is reduced by the piston moving downward in the cylinder (or by applying pressure to the fuel-air mixture as by crankcase compression in the crankcase of a "Two-Stroke Cycle Engine" which is described in a later paragraph).
2. The mixture of fuel and air is compressed by the piston moving upward in the cylinder.
3. The compressed fuel-air mixture is ignited by a timed electric spark.
4. The burning fuel-air mixture expands, forcing the piston downward in the cylinder thus converting the chemical energy generated by combustion into mechanical power.
5. The gaseous products formed by the burned fuel-air mixture are exhausted from the cylinder so that a new "Cycle" can begin.

The above described five events which comprise the work cycle of an engine are commonly referred to as (1), INTAKE; (2), COMPRESSION; (3), IGNITION; (4), EXPANSION (POWER); and (5), EXHAUST.

DIESEL CYCLE. The Diesel Cycle differs from the Otto Cycle in that air alone if drawn into the cylinder during the intake period. The air is heated from being compressed by the piston moving upward in the cylinder, then a finely atomized charge of fuel is injected into the cylinder where it mixes with the air and is ignited by the heat of the compressed air. In order to create sufficient heat to ignite the injected fuel, an engine operating on the Diesel Cycle must compress the air to a much greater degree than an engine operating on the Otto Cycle where the fuel-air mixture is ignited by an electric spark. The power and exhaust events of the Diesel Cycle are similar to the power and exhaust events of the Otto Cycle.

TWO-STROKE CYCLE ENGINES. Two stroke cycle engines may be of the Otto Cycle (spark ignition) or Diesel Cycle (compression ignition) type. However, since the two-stroke cycle engines listed in the repair section of this manual are all of the Otto Cycle type, operation of two-stroke Diesel Cycle engines will not be discussed in this section.

In two-stroke cycle engines, the piston is used as a sliding valve for the cylinder intake and exhaust ports. The intake and exhaust ports are both open when the piston is at the bottom of its downward stroke (bottom dead center or "BDC"). The exhaust port is open to atmospheric pressure; therefore, the fuel-air mixture must be elevated to a higher than atmospheric pressure in order for the mixture to enter the cylinder. As the crankshaft is turned from BDC and the piston starts on its upward stroke, the intake and exhaust ports are closed and the fuel-air mixture in the cylinder is compressed. When piston is at or near the top of its upward stroke (top dead center or "TDC"), an electric spark across the electrode gap of the spark plug ignites the fuel-air mixture. As the crankshaft turns past TDC and the piston starts on its downward stroke, the rapidly burning fuel-air mixture expands and forces the piston downward. As the piston nears bottom of its downward stroke, the cylinder exhaust port is opened and the burned gaseous products from combustion of the fuel-air mixture flows out the open port. Slightly further downward travel of the piston opens the cylinder intake port and a fresh charge of fuel-air mixture is forced into the cylinder. Since the exhaust port remains open, the incoming flow of fuel-air mixture helps clean (scavenge) any remaining burned gaseous products from the cylinder. As the crankshaft turns past BDC and the piston starts on its upward stroke, the cylinder intake and exhaust ports are closed and a new cycle begins.

Since the fuel-air mixture must be elevated to a higher than atmospheric pressure to enter the cylinder of a two-stroke cycle engine, a compressor pump must be used. Coincidentally, downward movement of the piston decreases the volume of the engine crankcase. Thus, a compressor pump is made available by sealing the engine crankcase and connecting the carburetor to a port in the crankcase. When the piston moves upward, volume of the crankcase is increased which lowers pressure within the crankcase to below atmospheric. Air will then be forced through the carburetor, where fuel is mixed with the air, and on into the engine crankcase. In order for downward movement of the piston to compress the fuel-air mixture in the crankcase, a valve must be provided to close the carburetor to crankcase port. Three different types of valves are used. In Fig. 1-1, a reed type inlet valve is shown in the schematic diagram of the two-stroke cycle engine. Spring steel reeds (R) are forced open by atmospheric pressure as shown in view "B" when the piston is on its upward stroke and pressure in the crankcase is below atmospheric. When piston reaches TDC, the reeds close as shown in view "A" and fuel-air mixture is trapped in the crankcase to be compressed by downward movement of the piston. In Fig 1-2, a schematic diagram of a two-stroke cycle engine is shown in which the piston is utilized as a sliding carburetor-crankcase port (third port) valve. In Fig. 1-3, a schematic diagram of a two-stroke cycle engine is shown in which a slotted disc (rotary valve) attached to the engine crankshaft opens the carburetor-crankcase port when the piston is on its upward stroke. In each of the three basic designs shown, a transfer port (TP—Fig. 1-2) connects the crankcase compression chamber to the cylinder; the transfer port is the cylinder intake port through which the compressed fuel-air mixture in the crankcase is transferred to the cylinder when the piston is at bottom of stroke as shown in view "A."

Due to rapid movement of the fuel-air mixture through the crankcase, the crankcase cannot be used as a lubricating oil sump because the oil would be carried into the cylinder. Lubrication is accomplished by mixing a small amount of oil with the fuel; thus, lubricating oil for the engine moving parts is carried into the crankcase with the fuel-air mixture. Normal lubricating oil to fuel mixture ratios vary from one part of oil mixed with 16 to 20 parts of fuel by volume. In all instances, manufacturer's recommendations for fuel-oil mixture ratio should be observed.

FOUR-STROKE CYCLE. In a four-stroke engine operating on the Otto Cycle (spark ignition), the five events of the cycle take place in four strokes of the piston, or in two revolutions of the engine crankshaft. Thus, a power stroke occurs only on alternate downward strokes of the piston.

In view "A" of Fig. 1-4, the piston is on the first downward stroke of the cycle. The mechanically operated intake valve has opened the intake port and, as the downward movement of the piston has reduced the air pressure in the cylinder to below atmospheric pressure, air is forced through the carburetor, where fuel is mixed with the air, and into the cylinder through the open intake port. The intake valve remains open and the fuel-air mixture continues to flow into the cylinder until the piston reaches the bottom of its downward stroke. As the piston starts on its first upward stroke, the mechanically operated intake valve closes and, since the exhaust valve is closed, the fuel-air mixture is compressed as in view "B."

Just before the piston reaches the top of its first upward stroke, a spark at the spark plug electrodes ignites the compressed fuel-air mixture. As the engine crankshaft turns past top center, the burning fuel-air mixture expands rapidly and forces the piston downward on its power stroke as shown in view "C." As the piston reaches the bottom of the power stroke, the mechanically operated exhaust valve starts to open and as the pressure of the burned fuel-air mixture is higher than atmospheric pressure, it starts to flow out the open exhaust port. As the engine crankshaft

Fig. 1-1—Schematic diagram of a two-stroke engine operating on the Otto Cycle (spark ignition). View "B" shows piston near top of upward stroke and atmospheric pressure is forcing air through carburetor (C), where fuel is mixed with the air, and the fuel-air mixture enters crankcase through open reed valve (R). In view "A", piston is near bottom of downward stroke and has opened the cylinder exhaust and intake ports; fuel-air mixture in crankcase has been compressed by downward stroke of engine and flows into cylinder through open port. Incoming mixture helps clean burned exhaust gases from cylinder.

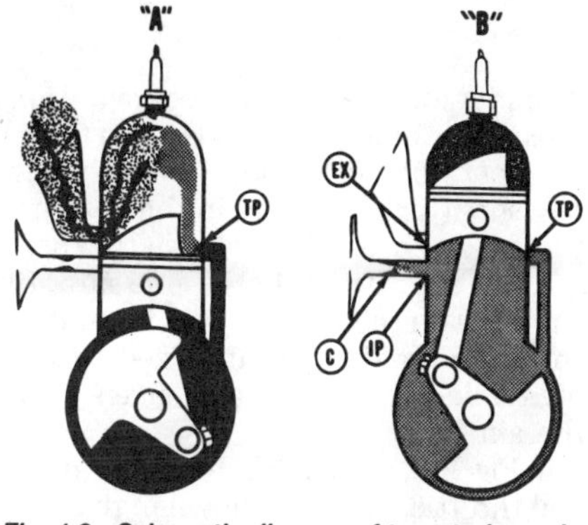

Fig. 1-2—Schematic diagram of two-stroke cycle engine operating on Otto Cycle. Engine differs from that shown in Fig. 1-1 in that piston is utilized as a sliding valve to open and close intake (carburetor to crankcase) port (IP) instead of using reed valve (R—Fig. 1-1).

C. Carburetor
EX. Exhaust port
IP. Intake port (carburetor to crankcase)
TP. Transfer port (crankcase to cylinder)

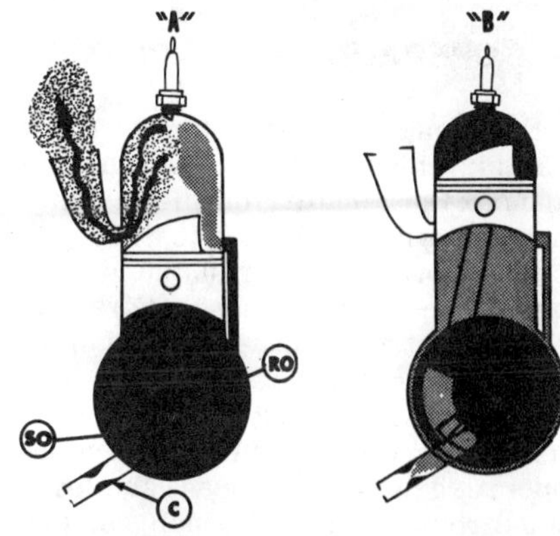

Fig. 1-3—Schematic diagram of two-stroke cycle engine similar to those shown in Figs. 1-1 and 1-2 except that a rotary carburetor to crankcase port valve is used. Disc driven by crankshaft has rotating opening (RO) which uncovers stationary opening (SO) in crankcase when piston is on upward stroke. Carburetor is (C).

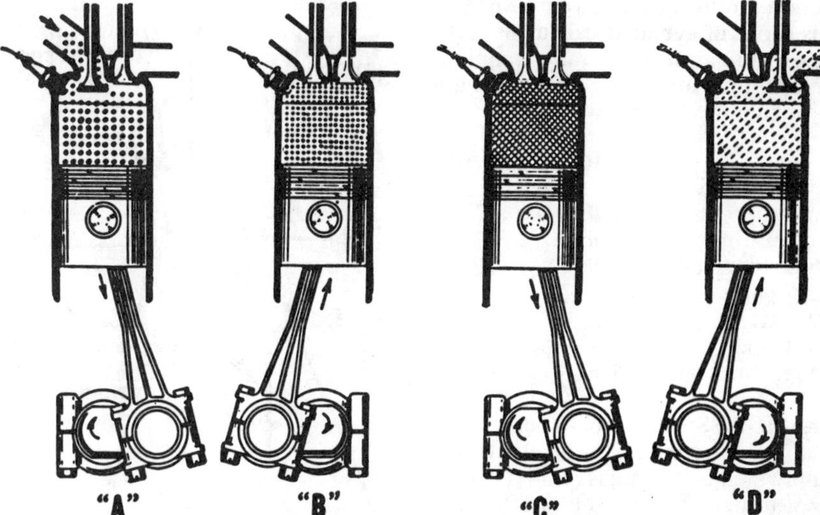

Fig. 1-4—Schematic diagram of four-stroke cycle engine operating on the Otto (spark ignition) cycle. In view "A", piston is on first downward (intake) stroke and atmospheric pressure is forcing fuel-air mixture from carburetor into cylinder through the open intake valve. In view "B", both valves are closed and piston is on its first upward stroke compressing the fuel-air mixture in cylinder. In view "C", spark across electrodes of spark plug has ignited fuel-air mixture and heat of combustion rapidly expands the burning gaseous mixture forcing the piston on its second downard (expansion or power) stroke. In view "D", exhaust valve is open and piston on its second upward (exhaust) stroke forces the burned mixture from cylinder. A new cycle then starts as in view "A."

turns past bottom center, the exhaust valve is almost completely open and remains open during the upward stroke of the piston as shown in view "D." Upward movement of the piston pushes the remaining burned fuel-air mixture out of the exhaust port. Just before the piston reaches the top of its second upward or exhaust stroke, the intake valve opens and the exhaust valve closes. The cycle is completed as the crankshaft turns past top center and a new cycle begins as the piston starts downward as shown in view "A."

In a four-stroke cycle engine operating on the Diesel Cycle, the sequence of events of the cycle is similar to that described for operation on the Otto Cycle, but with the following exceptions: On the intake stroke, air only is taken into the cylinder. On the compression stroke, the air is highly compressed which raises the temperature of the air. Just before the piston reaches top dead center, fuel is injected into the cylinder and is ignited by the heated, compressed air. The remainder of the cycle is similar to that of the Otto Cycle.

CARBURETOR FUNDAMENTALS

OPERATING PRINCIPLES

Function of the carburetor on a spark-ignition engine is to atomize the fuel and mix the atomized fuel in proper proportions with air flowing to the engine intake port or intake manifold. Carburetors used on engines that are to be operated at constant speeds and under even loads are of simple design since they only have to mix fuel and air in a relatively constant ratio. On engines operating at varying speeds and loads, the carburetors must be more complex because different fuel-air mixtures are required to meet the varying demands of the engine.

FUEL-AIR MIXTURE RATIO REQUIREMENTS. To meet the demands of an engine being operated at varying speeds and loads, the carburetor must mix fuel and air at different mixture ratios. Fuel-air mixture ratios required for different operating conditions are approximately as follows:

	Fuel	Air
Starting, cold weather	1 lb.	7 lbs.
Accelerating	1 lb.	9 lbs.
Idling (no load)	1 lb.	11 lbs.
Part open throttle	1 lb.	15 lbs.
Full load, open throttle	1 lb.	13 lbs.

BASIC DESIGN. Carburetor design is base on the venturi principle which simply means that a gas or liquid flowing through a necked-down section (venturi) in a passage undergoes an increase velocity (speed) and a decrease in pressure as compared to the velocity and pressure in full size sections of the passage. The principle is illustrated in Fig. 2-1, which shows air passing through carburetor venturi. The figures given for air speeds and vacuum are approximate for a typical wide-open throttle operating condition. Due to low pressure (high vacuum) in the venturi, fuel is forced out through the fuel nozzle by the atmospheric pressure (0 vacuum) on the fuel; as fuel is emitted from the nozzle, it is atomized by the high velocity air flow and mixes with the air.

In Fig. 2-2, the carburetor choke plate and throttle plate are shown in relation in the venturi. Downward pointing arrows indicate air flow through carburetor.

At cranking speeds, air flows through the carburetor venturi at a slow speed; thus, the pressure in the venturi does not usually decrease to the extent that atmospheric pressure on the fuel will force fuel from the nozzle. If the choke plate is closed as shown by dotted line in Fig. 2-2, air cannot enter into the carburetor and pressure in the carburetor decreases greatly as the engine is turned at cranking speed. Fuel can then flow from the fuel nozzle. In manufacturing the carburetor choke plate or disc, a small hole or notch is cut in the plate so that some air can flow through the plate when it is in closed position to provide air for the starting fuel-air mixture. In some instances after starting a cold engine, it is advantageous to leave the choke plate in a partly closed position as the restriction of air flow will decrease the air pressure in carburetor venturi, thus causing more flow fuel to flow from the nozzle, resulting in a richer fuel-air mixture. The choke plate or disc should be in full open position for normal engine operation.

If, after the engine has been started the throttle plate is in the wide-open position as shown by the solid line in Fig. 2-2, the engine can obtain enough fuel and air to run at dangerously high speeds. Thus, the throttle plate or disc must be partly closed as shown by the dotted lines to control engine speed. At no load, the engine requires very little air and fuel to run at its rated speed and the throttle must be moved on toward the closed position as shown by the dash lines. As more load is placed on the engine, more fuel and air are required for the engine to operate at its rated speed and the throttle must by moved

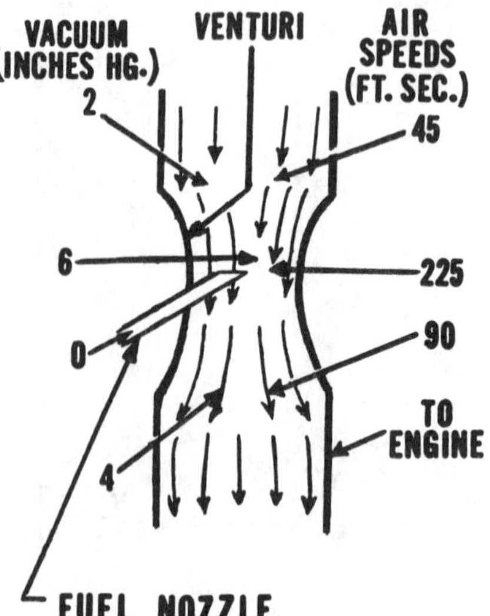

Fig. 2-1—Drawing illustrating the venturi principle upon which carburetor design is based. Figures at left are inches of mercury vacuum and those at right are air speeds in feet per second that are typical of conditions found in a carburetor operating at wide open throttle. Zero vacuum in fuel nozzle corresponds to atmospheric pressure.

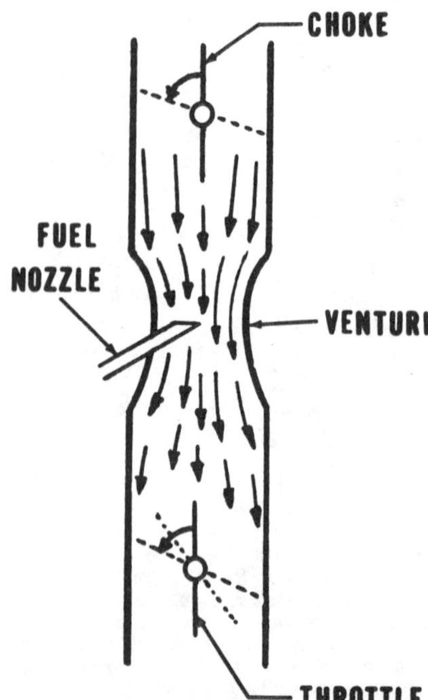

Fig. 2-2—Drawing showing basic carburetor design. Text explains operation of the choke and throttle valves. In some carburetors, a primer pump may be used instead of the choke valve to provide fuel for the starting fuel-air mixture.

closer to the wide open position as shown by the solid line. When the engine is required to develop maximum power or speed, the throttle must be in the wide open position.

Although some carburetors may be as simple as the basic design just described, most engines require more complex design features to provide variable fuel-air mixture ratios for different operating conditions. These design features will be described in the following paragraphs which outline the different carburetor types.

CARBURETOR TYPES

Carburetors used on small engines are usually classified by types as to method of delivery of fuel to the carburetor fuel nozzle. The following paragraphs describe the features and operating principles of the different type carburetors from the most simple suction lift type to the more complex float and diaphragm types.

SUCTION LIFT CARBURETOR. A cross-sectional drawing of a typical suction lift carburetor is shown in Fig. 2-4. Due to the low pressure at the orifice (O) of the fuel nozzle and to atmospheric pressure on the fuel in fuel supply tank, fuel is forced up through the fuel pipe and out of the nozzle into the carburetor venturi where it is mixed with the air flowing through the venturi. A check ball is located in the lower end of the fuel pipe to prevent pulsations of air pressure in the venturi from forcing fuel back down through the fuel pipe. The lower end of the full pipe has a fine mesh screen to prevent foreign material or dirt in fuel from entering the fuel nozzle. Fuel-air ratio can be adjusted by opening or closing the adjusting needle (N) slightly; turning the needle in will decrease flow of fuel out of nozzle orifice (O).

In Fig. 2-3, a cut-away view is shown of a suction type carburetor used on several models of a popular make small engine. This carburetor features an idle fuel passage, jet and adjustment screw. When carburetor throttle is nearly closed (engine is at low idle speed), air pressure is low (vacuum is high) at inner side of throttle plate. Therefore, atmospheric pressure in fuel tank will force fuel through the idle jet and adjusting screw orifice where it is emitted into the carburetor throat and mixes with air passing the throttle plate. The adjustment screw is turned in or out until an optimum fuel-air mixture is obtained and engine runs smoothly at idle speed. When the throttle is opened to increase engine speed, air velocity through the venturi increases, air pressure in the venturi decreases and fuel is emitted from the nozzle. Power adjustment screw (high speed fuel needle) is turned in or out to obtain proper fuel-air mixture for engine running under operating speed and load.

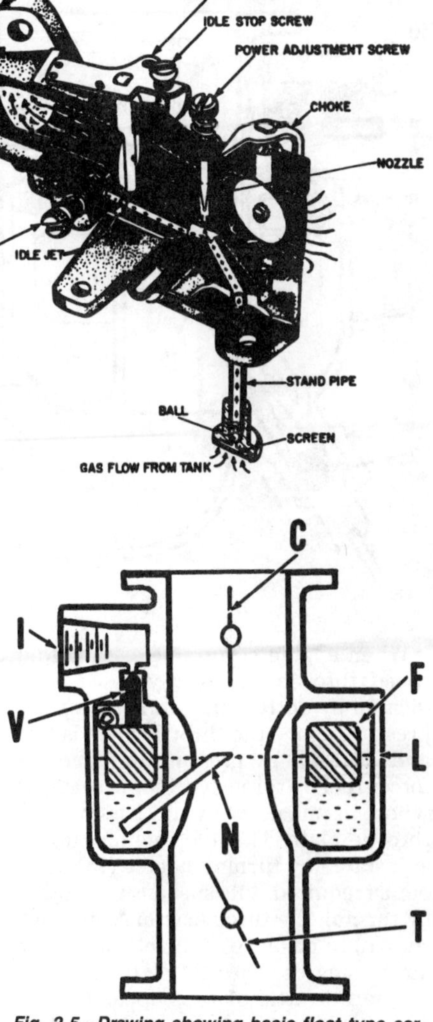

Fig. 2-3—Cut-away drawing of a suction lift carburetor used on one well known make engine. Ball in stand pipe prevents fuel in carburetor from flowing back into fuel tank.

Fig. 2-5—Drawing showing basic float type carburetor design. Fuel must be delivered under pressure either by gravity or by use of fuel pump, to the carburetor fuel inlet (I). Fuel level (L) operates float (F) to open and close inlet valve (V) to control amount of fuel entering carburetor. Also shown are the fuel nozzle (N), throttle (T) and choke (C).

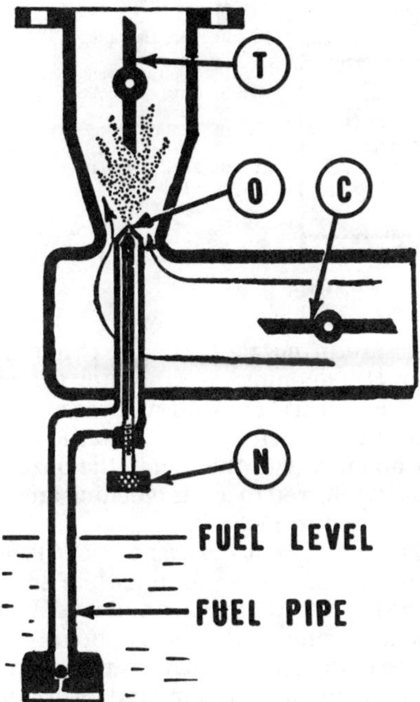

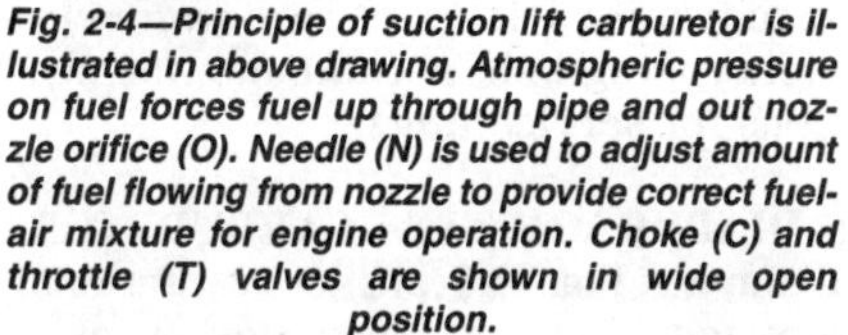

Fig. 2-4—Principle of suction lift carburetor is illustrated in above drawing. Atmospheric pressure on fuel forces fuel up through pipe and out nozzle orifice (O). Needle (N) is used to adjust amount of fuel flowing from nozzle to provide correct fuel-air mixture for engine operation. Choke (C) and throttle (T) valves are shown in wide open position.

FLOAT TYPE CARBURETOR. The principle of float type carburetor operation is illustrated in Fig. 2-5. Fuel is delivered at inlet (I) by gravity with fuel tank placed above carburetor, or by a fuel lift pump when tank is located below carburetor inlet. Fuel flows into the open inlet valve (V) until fuel level (L) in bowl lifts float against fuel valve needle and closes the valve. As fuel is emitted from the nozzle (N) when engine is running, fuel level will drop, lowering the float and allowing valve to open so that fuel will enter the carburetor to meet the requirements of the engine.

In Fig. 2-6, a cut-away view of a well known make of small engine float type carburetor is shown. Atmospheric pressure is maintained in fuel bowl through passage (20) which opens into carburetor air horn ahead of the choke plate (21). Fuel level is maintained at just below level of opening (O) in nozzle (22) by float (19) actuating inlet valve needle (8). Float height can be adjusted by bending float tang (5).

When starting a cold engine, it is necessary to close the choke plate (21) as shown by dotted lines so as to lower the air pressure in carburetor venturi (18) as engine is cranked. Then, fuel will flow up through nozzle (22) and will be emitted from openings (O) in nozzle. When an engine is hot, it will start on a leaner fuel-air mixture than when cold and may start without the choke plate being closed.

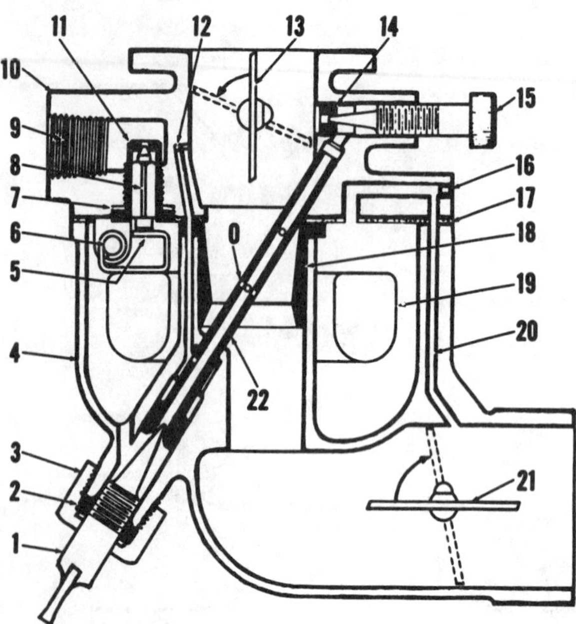

Fig. 2-6—Cross-sectional drawing of float type carburetor used on a popular small engine.

0. Orifice
1. Main fuel needle
2. Packing
3. Packing nut
4. Carburetor bowl
5. Float tang
6. Float hinge pin
7. Gasket
8. Inlet valve
9. Fuel inlet
10. Carburetor body
11. Inlet valve seat
12. Vent
13. Throttle plate
14. Idle orifice
15. Idle fuel needle
16. Plug
17. Gasket
18. Venturi
19. Float
20. Fuel bowl vent
21. Choke
22. Fuel nozzle

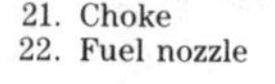

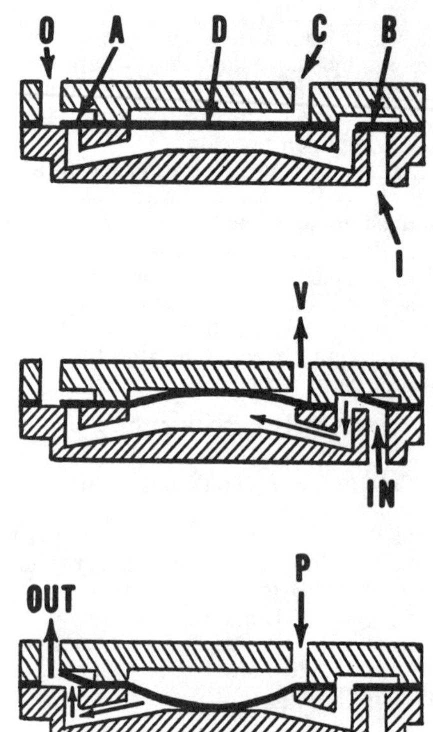

Fig. 2-9—Operating principle of diaphragm type fuel pump is illustrated in above drawings. Pump valves (A & B) are usually a part of diaphragm (D). Pump inlet is (I) and outlet is (O). Chamber above diaphragm is connected to engine crankcase by passage (C). When piston is on upward stroke, vacuum (V) at crankcase passage allows atmospheric pressure to force fuel into pump fuel chamber as shown in middle drawing. When piston is on downward stroke, pressure (P) expands diaphragm downward forcing fuel out of pump as shown in lower drawing.

When engine is running as slow idle speed (throttle plate nearly closed as indicated by dotted lines in Fig. 2-6), air pressure above the throttle plate is low and atmospheric pressure in fuel bowl forces fuel up through orifice in seat (14) where it mixes with air passing the throttle plate. The idle fuel mixture is adjustable by turning needle (15) in or out as required. Idle speed is adjustable by turning the throttle stop screw (not shown) in or out of control amount of air passing the throttle plate.

When throttle plate is opened to increase engine speed, velocity of air flow through venturi (18) increases, air pressure at venturi decreases and fuel will flow from openings (O) in nozzle instead of through orifice in idle seat (14). When engine is running at high speed, pressure in nozzle (22) is less than at vent (12) opening in carburetor throat above venturi. Thus, air will enter vent and travel down the vent into the nozzle and mix with the fuel in the nozzle. This is referred to as air bleeding and is illustrated in Fig. 2-7.

Many different designs of float type carburetors will be found when servicing the different makes and models of small engines. Reference should be made to the engine repair section of this manual for adjustment and overhaul specifications. Refer to carburetor servicing paragraphs in fundamentals sections for service hints.

DIAPHRAGM TYPE CARBURETOR. Refer to Fig. 2-8 for cross-sectional drawing showing basic design of a diaphragm type carburetor. Fuel is delivered to inlet (I) by gravity with fuel tank above carburetor, or under pressure from a fuel pump. Atmospheric pressure is maintained on lower side of diaphragm (D) through vent hole (V). When choke plate (C) is closed and engine is cranked, or when engine is running, pressure at orifice (O) is less than atmospheric pressure; this low

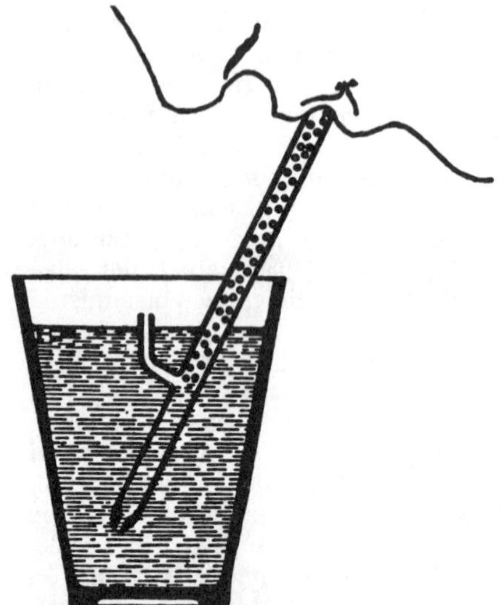

Fig. 2-7—Illustration of air bleed principle explained in text.

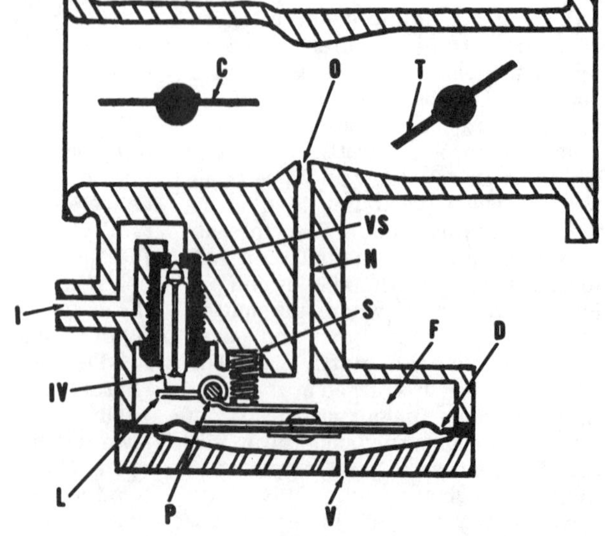

Fig. 2-8—Cross-section drawing of basic design diaphragm type carburetor. Atmospheric pressure actuates diaphragm (D).

C. Choke
D. Diaphragm
F. Fuel chamber
I. Fuel inlet
IV. Inlet valve needle
L. Lever
N. Nozzle
O. Orifice
P. Pivot pin
S. Spring
T. Throttle
V. Vent
VS. Valve seat

pressure, or vacuum, is transmitted to fuel chamber (F) above diaphragm through nozzle channel (N). The higher (atmospheric) pressure at lower side of diaphragm will then push the diaphragm upward compressing spring (S) and allowing inlet valve (IV) to open and fuel will flow into the fuel chamber.

Some diaphragm type carburetors are equipped with an integral fuel pump. Although design of the pump may vary as to type of check valves, etc., all operate on the principle shown in Fig. 2-9. A channel (C) (or pulsation passage) connects one side of the diaphragm to the engine crankcase. When engine piston is on upward stroke, vacuum (V) (lower than atmospheric pressure) is present in channel; thus atmospheric pressure on fuel forces inlet valve (B) open and fuel flows into chamber below the diaphragm as shown in middle view. When piston is on downward stroke, pressure (P) (higher than atmospheric pressure) is present in channel (C); thus, the pressure forces the diaphragm downward closing the inlet valve (B) and causes the fuel to flow out by the outlet valve (A) as shown in lower view.

In Fig. 2-10, a cross-sectional view of a popular make diaphragm type carburetor, with integral diaphragm type pump, is shown.

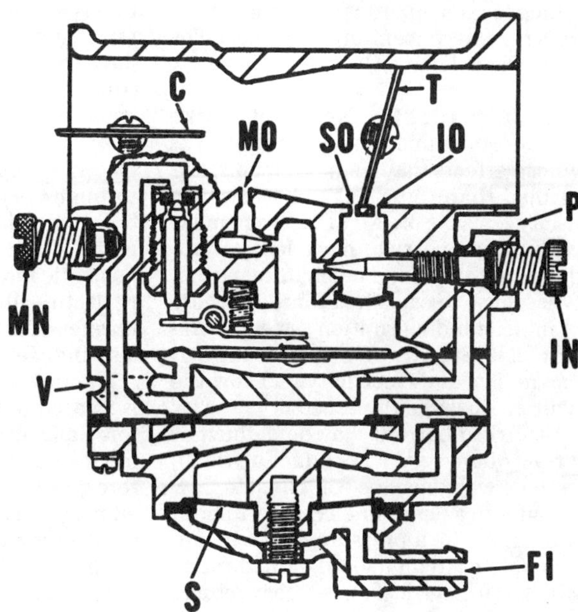

Fig. 2-10—Cross-sectional view of a popular make diaphragm type carburetor with integral fuel pump. Refer to Fig. 2-8 for view of basic diaphragm carburetor and to Fig. 2-9 for views showing operation of the fuel pump.

C. Choke
FI. Fuel inlet
IN. Idle fuel adjusting needle
IO. Idle orifice
MN. Main fuel adjusting needle
MO. Main orifice
P. Pulsation channel (fuel pump)
S. Screen
SO. Secondary orifice
T. Throttle
V. Vent (atmosphere to carburetor diaphragm)

IGNITION SYSTEM FUNDAMENTALS

The ignition system provides a properly timed surge of extremely high voltage electrical energy which flows across the spark plug electrode gap to create the ignition spark. Small engines may be equipped with either a magneto or battery ignition system. A magneto ignition system generates electrical energy, intensifies (transforms) this electrical energy to the extremely high voltage required and delivers this electrical energy at the proper time for the ignition spark. In a battery ignition system, a storage battery is used as a source of electrical energy and the system transforms the relatively low electrical voltage from the battery into the high voltage required and delivers the high voltage at proper time for the ignition spark. Thus, the function of the two systems is somewhat similar except for the basic source of electrical energy. The fundamental operating principles of ignition systems are explained in the following paragraphs.

MAGNETISM AND ELECTRICITY

The fundamental principles upon which ignition systems are designed are presented in this section. As the study of magnetism and electricity is an entire scientific field, it is beyond the scope of this manual to fully explore these subjects. However, the following information will impart a working knowledge of basic principles which should be of value in servicing small engines.

MAGNETISM. The effects of magnetism can be shown easily while the theory of magnetism is too complex to be presented here. The effects of magnetism were discovered many years ago when fragments of iron ore were found to attract each other and also attract other pieces of iron. Further, it was found that when suspended in air, one end of the iron ore fragment would always point in the direction of the North Star. The end of the iron ore fragment pointing north was called the "north pole" and the opposite end the "south pole." By stroking a piece of steel with a "natural magnet," as these iron ore fragments were called, it was found that the magnetic properties of the natural magnet could be transferred or "induced" into the steel.

Steel which will retain magnetic properties for an extended period of time after being subjected to a strong magnetic field are called "permanent magnets;" iron or steel that loses such magnetic properties soon after being subjected to a magnetic field are called "temporary magnets." Soft iron will lose magnetic properties almost immediately after being removed from a magnetic field, and so is used where this property is desirable.

The area affected by a magnet is called a "field of force." The extent of this field of force is related to the strength of the magnet and can be determined by use of a compass. In practice, it is common to illustrate the field of force surrounding a magnet by lines as shown in Fig. 3-1 and field of force is usually called "lines of force" or "flux." Actually, there are no "lines," however, this is a convenient method of illustrating the presence of the invisible magnetic forces and if a certain magnetic force is defined as a "line of force," then all magnetic forces may be measured by comparison. The number of "lines of force" making up a strong magnetic field is enormous.

Most materials when placed in a magnetic field are not attracted by the magnet, do not change the magnitude or direction of the magnetic field, and so are called "non-magnetic materials." Materials such as iron, cobalt, nickel or their alloys, when placed in a magnetic field will concentrate the field of force and hence are magnetic conductors or "magnetic materials." There are no materials known in which magnetic fields will not penetrate and magnetic lines of force can be deflected only by magnetic materials or by another magnetic field.

Alnico, an alloy containing aluminum, nickel and cobalt, retains magnetic properties for a very long period of time after being subjected to a strong magnetic field and is extensively used as a permanent magnet. Soft iron, which loses magnetic properties quickly, is used to concentrate magnetic fields as in Fig 3-1.

ELECTRICITY. Electricity, like magnetism, is an invisible physical force whose effects may be more readily explained than the theory of what electricity consists of. All of us are familiar with the property of electricity to produce light, heat and mechanical power. What must be explained for the purpose of understanding ignition system operation is the inter-relationship of magnetism and electricity and how the ignition spark is produced.

Electrical current may be defined as a flow of energy in a conductor which, in some ways, may be compared to flow of water in a pipe. For electricity to flow, there must be a pressure (voltage) and a complete circuit (closed path) through which the electrical energy may return, a comparison being a water pump and a pipe that receives water from the outlet (pressure) side of the pump and returns the water to the inlet side of the pump. An electrical circuit may be completed by electricity flowing through the earth (ground), or through the metal framework of an engine or other equipment ("grounded" or "ground" connections). Usually, air is an insulator through which electrical energy will not flow. However, if the force (voltage) becomes great, the resistance of air to the flow of electricity is broken down and a current will flow, releasing energy in the form of a spark. By high voltage electricity breaking down the resistance of the air gap between the spark plug electrodes, the ignition spark is formed.

ELECTROMAGNETIC INDUCTION. The principle of electro-magnetic induction is as follows: When a wire (conductor) is moved through a field of magnetic force so as to cut across the lines of force (flux), a potential voltage or electromotive force (emf) is induced in the wire. If the wire is part of a completed electrical circuit, current will flow through the circuit as illustrated in Fig. 3-2. It should be noted that the movement of the wire through the lines of magnetic force is a relative motion; that is, if the lines of force of a moving magnetic field cut across a wire, this will also induce an emf to the wire.

The direction of an induced current is related to the direction of magnetic force and also to the direction of movement of the wire through the lines of force, or flux. The voltage of an induced current is related to the strength, or concentration of lines of force, of the magnetic field and to the rate of speed at which the wire is moved through the flux. If a length of wire is wound into a coil and a section of the coil is moved through magnetic lines of force, the voltage induced will be proportional to the number of turns of wire in the coil.

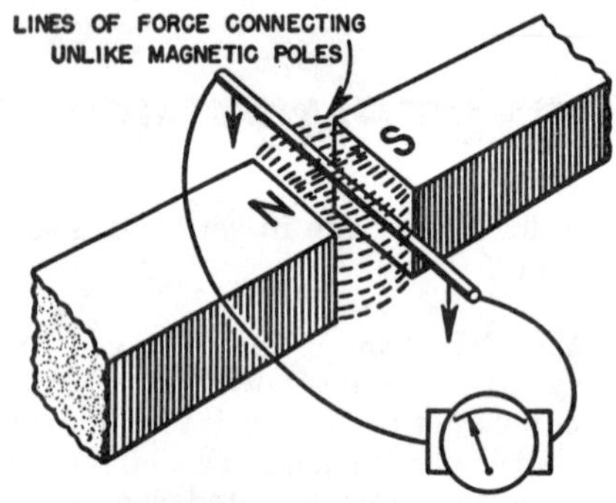

Fig. 3-2—When a conductor is moved through a magnetic field so as to cut across lines of force, a potential voltage will be induced in the conductor. If the conductor is a part of a completed electrical circuit, current will flow through the circuit as indicated by the gage.

ELECTRICAL MAGNETIC FIELDS. When current is flowing in a wire, a magnetic field is present around the wire as illustrated in Fig. 3-3. The direction of lines of force of this magnetic field is related to the direction of current in the wire. This is known as the left hand rule and is stated as follows: If a wire carrying a current is grasped in the left hand with thumb pointing in direction current is flowing, the curved fingers will point the direction of lines of magnetic force (flux) encircling the wire.

If a current is flowing in a wire that is wound into a coil, the magnetic flux surrounding the wire converge to form a stronger magnetic field as shown in Fig. 3-4. If the coils of wire are very close together, there is little tendency for magnetic flux to surround individual loops of the coil and a strong magnetic field will surround the entire coil. The strength of this field will vary with the current flowing through the coil.

STEP-UP TRANSFORMERS (IGNITION COILS). In both battery and magneto ignition systems, it is necessary to step-up, or transform, a relatively low primary voltage to the 15,000 to 20,000 volts required for the ignition spark. This is done by means of an ignition coil which utilizes the inter-relationship of magnetism and electricity as explained in preceding paragraphs.

Fig. 3-3—A magnetic field surrounds a wire carrying an electrical current. The direction of magnetic force is indicated by the "left hand rule"; that is, if thumb of left hand points in direction that electrical current is flowing in conductor, fingers of left hand will indicate direction of magnetic force.

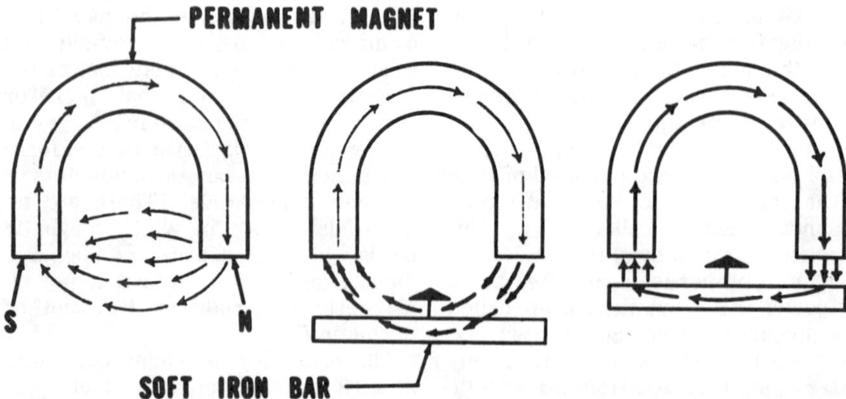

Fig. 3-1—In left view, field of force of permanent magnet is illustrated by arrows showing direction of magnetic force from north pole (N) to south pole (S). In center view, lines of magnetic force are being attracted by soft iron bar that is being moved into the magnetic field. In right view, the soft iron bar has been moved close to the magnet and the field of magnetic force is concentrated within the bar.

Fig. 3-4—When a wire is wound in a coil, the magnetic force created by a current in the wire will tend to converge in a single strong magnetic field as illustrated. If the loops of the coil are wound closely together, there is little tendency for lines of force to surround individual loops of the coil.

Basic ignition coil design is shown in Fig. 3-5. The coil consists of two separate coils of wire which are called the primary coil winding and the secondary coil winding, or simply the primary winding and secondary winding. The primary winding as indicated by the heavy, black line is of larger diameter wire and has a smaller number of turns when compared to the secondary winding indicated by the light line.

A current passing through the primary winding creates a magnetic field (as indicated by the "lines of force") and this field, concentrated by the soft iron core, surrounds both the primary and secondary windings. If the primary winding current is suddenly interrupted, the magnetic field will collapse and the lines of force will cut through the coil windings. The resulting induced voltage in the secondary winding is greater than the voltage of the current that was flowing in the primary winding and is related to the number of turns of wire in each winding. Thus:

$$\text{Induced secondary voltage} = \text{primary voltage} \times \frac{\text{No. of turns in secondary winding}}{\text{No. of turns in primary winding}}$$

For example, if the primary winding of an ignition coil contained 100 turns of wire and the secondary winding contained 10,000 turns of wire, a current having an emf of 200 volts flowing in the primary winding, when suddenly interrupted, would result in an emf of:

$$200 \text{ Volts} \times \frac{10{,}000 \text{ turns of wire}}{100 \text{ turns of wire}} = 20{,}000 \text{ volts}$$

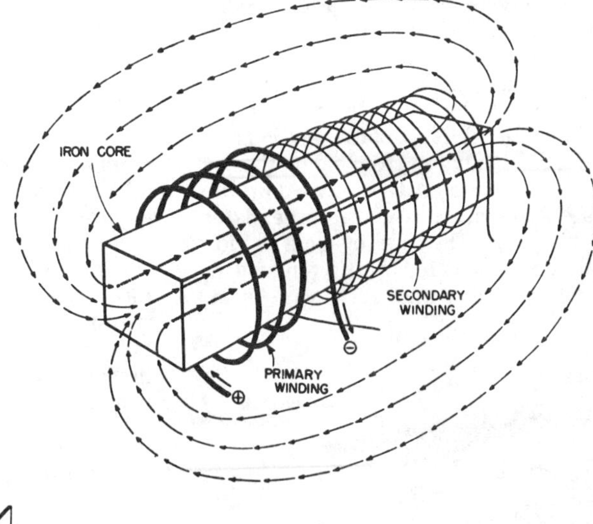

Fig. 3-5—Drawing showing principles of ignition coil operation. A current in primary winding will establish a magnetic field surrounding both the primary and secondary windings and the field will be concentrated by the iron core. When primary current is interrupted, the magnetic field will "collapse" and the lines of force will cut the coil windings inducing a very high voltage in the secondary winding.

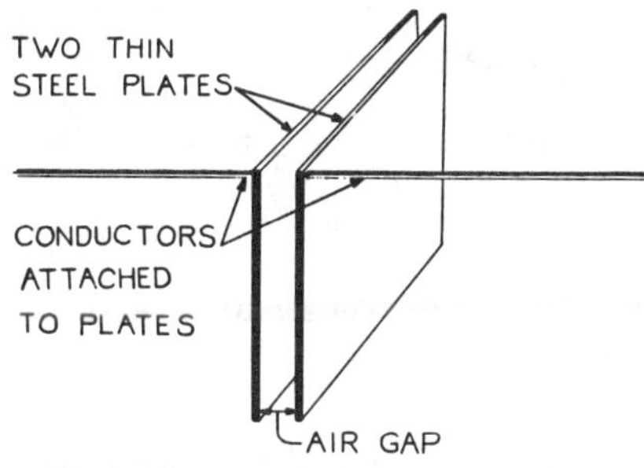

Fig. 3-6a—Drawing showing construction of a simple condenser. Capacity of such a condenser to absorb current is limited due to the relatively small surface area. Also, there is a tendency for current to arc across the air gap. Refer to Fig. 3-7 for construction of typical ignition system condenser.

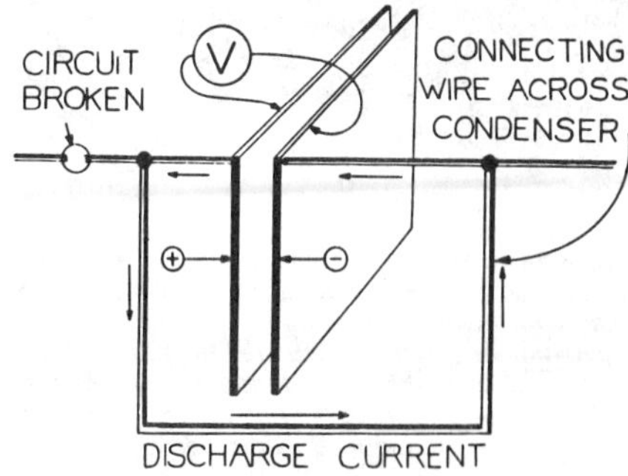

Fig. 3-6c—When flow of current is interrupted in circuit containing condenser (circuit broken), the condenser will retain a potential voltage (V). If a wire is connected across the condenser, a current will flow in reverse direction of charging current until condenser is discharged (voltage across condenser plates is zero).

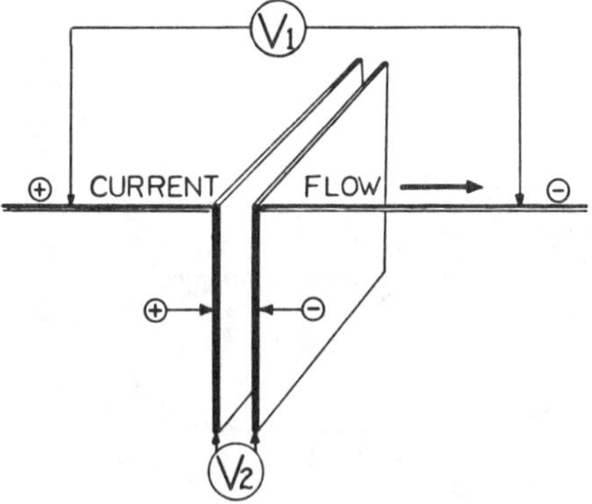

Fig. 3-6b—A condenser in an electrical circuit will absorb flow of current until an opposing voltage (V2) is built up across condenser plates which is equal to the voltage (V1) of the electrical current.

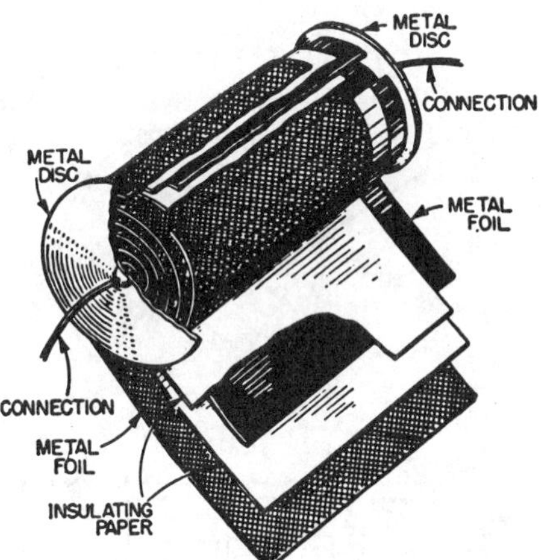

Fig. 3-7—Drawing showing construction of typical ignition system condenser. Two layers of metal foil, insulated from each other with paper, are rolled tightly together and a metal disc contacts each layer, or strip, of foil. Usually, one disc is grounded through the condenser shell.

SELF-INDUCTANCE. It should be noted that the collapsing magnetic field resulting from the interrupted current in the primary winding will also induce a current in the primary winding. This effect is termed "self-inductance." This self-induced current is such as to oppose any interruption of current in the primary winding, slowing the collapse of the magnetic field and reducing the efficiency of the coil. The self-induced primary current flowing across the slightly open breaker switch, or contact points, will damage the contact surfaces due to the resulting spark.

To momentarily absorb, then stop the flow of current across the contact points, a capacitor or, as commonly called, a condenser is connected in parallel with the contact points. A simple condenser is shown in Fig. 3-6a; however, the capacity of such a condenser to absorb current (capacitance) is limited by the small surface area of the plates. To increase capacity to absorb current, the condenser used in ignition systems is constructed as shown in Fig. 3-7.

EDDY CURRENTS. It has been found that when a solid soft iron bar is used as a core for an ignition coil, stray electrical currents are formed in the core. These stray, or "eddy currents," create opposing magnetic forces causing the core to become hot and also decrease the efficiency of the coil. As a means of preventing excessive formation of eddy currents within the core, or other magnetic field carrying parts of a magneto, a laminated plate construction as shown in Fig. 3-8 is used instead of solid material. The plates, or laminations, are insulated from each other by a natural oxide coating formed on the plate surfaces or by coating the plates with varnish. The cores of some ignition coils are constructed of soft iron wire instead of plates and each wire is insulated by a varnish coating. This type construction serves the same purpose as laminated plates.

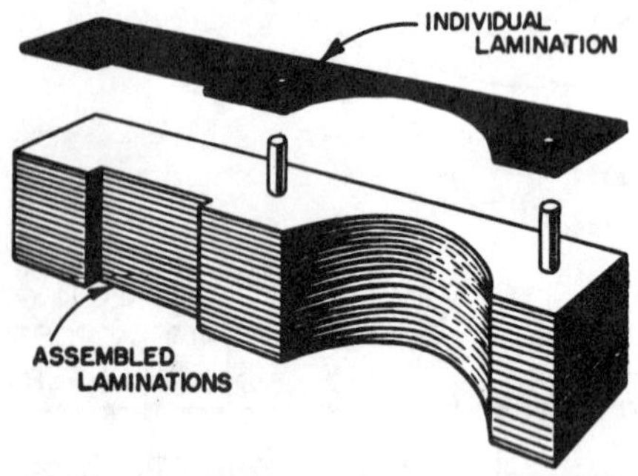

Fig. 3-8—To prevent formation of "eddy currents" within soft iron cores used to concentrate magnetic fields, core is assembled of plates or "laminations" that are insulated from each other. In a solid core, there is a tendency for counteracting magnetic forces to build up from stray currents induced in the core.

BATTERY IGNITION SYSTEMS

Some small engines are equipped with a battery ignition system. A schematic diagram of a typical battery ignition system for a single cylinder engine is shown in Fig. 3-9. Designs of battery ignition systems may vary, especially as to location of breaker points and method for actuating the points; however, all operate on the same basic principles.

BATTERY IGNITION SYSTEM PRINCIPLES. Refer to the schematic diagram on Fig. 3-9. When the timer cam is turned so that the contact points are closed, a current is established in the primary circuit by the emf of the battery. This current flowing through the primary winding of the ignition coil establishes a magnetic field concentrated in the core laminations and surrounding the windings. A cutaway view of a typical ignition coil is shown in Fig. 3-10. At the proper time for the ignition spark, the contact points are opened by the timer cam and the primary ignition circuit is interrupted. The condenser, wired in parallel with the breaker contact points between the timer terminal and ground, absorbs the self-induced current in the primary circuit for an instant and brings the flow of current to a quick, controlled stop. The magnetic field surrounding the coil rapidly cuts the primary and secondary windings creating an emf as high as 250 volts in the primary winding and up to 25,000 volts in the secondary winding. Current absorbed by the condenser is discharged as the cam closes the breaker points, grounding the condenser lead wire.

Due to resistance of the primary winding, a certain period of time is required for maximum primary current flow after the breaker contact points are closed. At high engine speeds, the points remain closed for a smaller interval of time, hence the primary current does not build up to maximum and secondary voltage is somewhat less than at low engine speed. However, coil design is such that the minimum voltage available at high engine speed exceeds the normal maximum voltage required for the ignition spark.

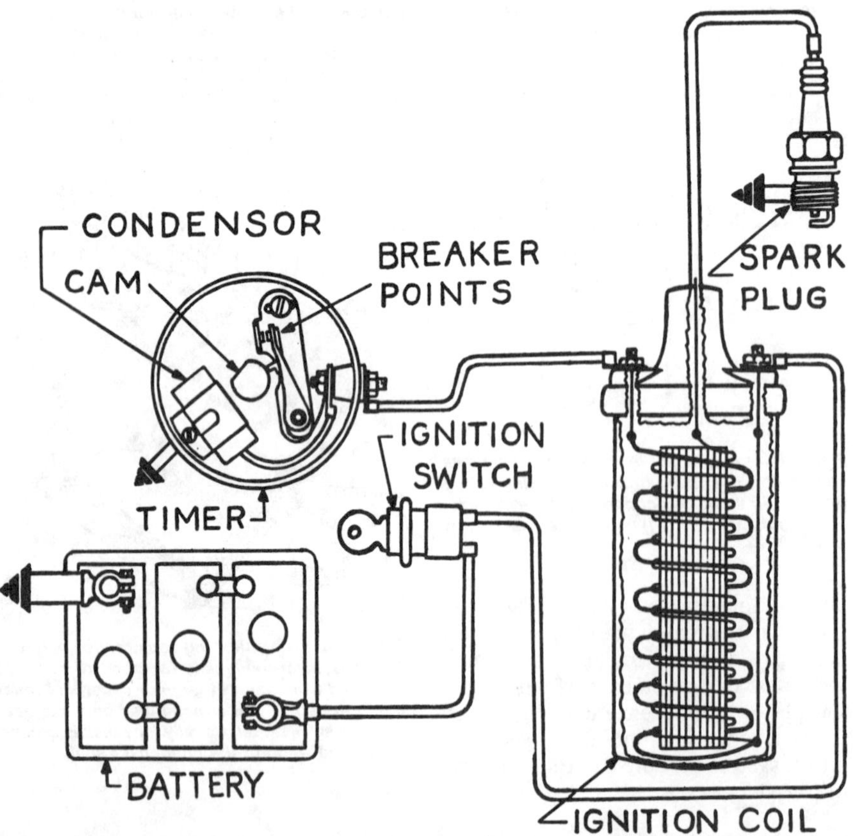

Fig. 3-9—Schematic diagram of typical battery ignition system. On unit shown, breaker points are actuated by timer cam; on some units, the points may be actuated by cam on engine camshaft. Refer to Fig. 3-10 for cutaway view of typical battery ignition coil. In view above, primary coil winding is shown as heavy black line (outside coil loops) and secondary winding is shown by lighter line (inside coil loops).

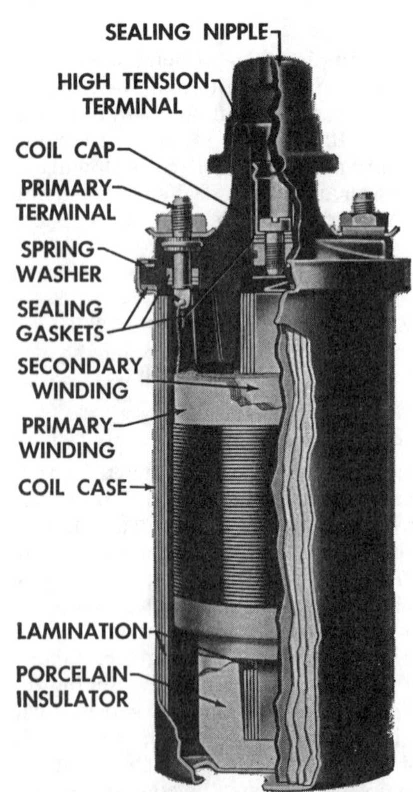

Fig. 3-10—Cutaway view of typical battery ignition system coil. Primary winding consists of approximately 200-250 turns (loops) of heavier wire; secondary winding consists of several thousand turns of fine wire. Laminations concentrate the magnetic lines of force and increase efficiency of the coil.

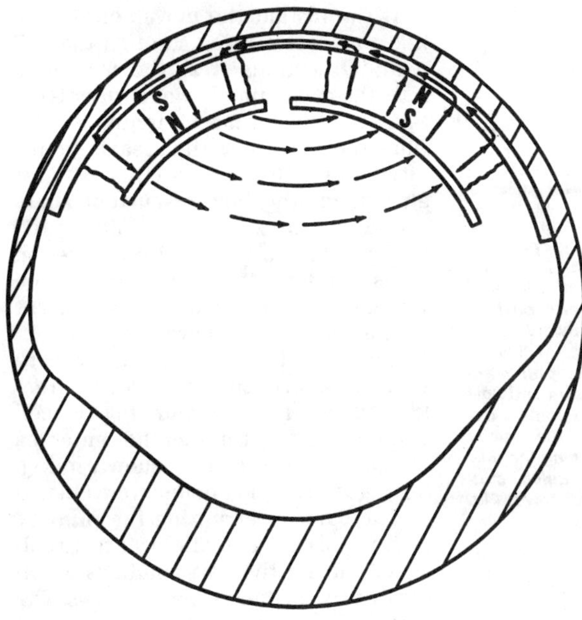

Fig. 3-11—Cutaway view of typical engine flywheel used with flywheel magneto type ignition system. The permanent magnets are usually cast into the flywheel. For flywheel type magnetos having the ignition coil and core mounted to outside of flywheel, magnets would be flush with outer diameter of flywheel.

MAGNETO IGNITION SYSTEMS

By utilizing the principles of magnetism and electricity as outlined in previous paragraphs, a magneto generates an electrical current of relatively low voltage, then transforms this voltage into the extremely high voltage necessary to produce the ignition spark. This surge of high voltage is timed to create the ignition spark and ignite the compressed fuel-air mixture in the engine cylinder at the proper time in the Otto cycle as described in the paragraphs on fundamentals of engine operation principles.

Two different types of magnetos are used on small engines and, for discussion in this section of the manual, will be classified as ''flywheel type magnetos'' and ''self-contained unit type magnetos.'' The most common type of ignition system found on small engines is the flywheel type magneto.

Flywheel Type Magnetos

The term ''flywheel type magneto'' is derived from the fact that the engine flywheel carries the permanent magnets and is the magneto rotor. In some similar systems, the magneto rotor is mounted on the engine crankshaft as is the flywheel, but is a part separate from the flywheel.

FLYWHEEL MAGNETO OPERATING PRINCIPLES. In Fig. 3-11, a cross-sectional view of a typical engine flywheel (magneto rotor) is shown. The arrows indicate lines of force (flux) of the permanent magnets carried by the flywheel. As indicated by the arrows, direction of force of the magnetic field is from the north pole (N) of the left magnet to the south pole (S) of the right magnet.

Figs. 3-12, 3-13, 3-14, and 3-15 illustrate the operational cycle of the flywheel type magneto. In Fig. 3-12, the flywheel magnets have moved to a position over the left and center legs of the armature (ignition coil) core. As the magnets moved into this position, their magnetic field was attracted by the armature core as illustrated in Fig. 3-1 and a potential voltage (emf) was induced in the coil windings. However, this emf was not sufficient to cause current to flow across the spark plug electrode gap in the high tension circuit and the points were open in the primary circuit.

In Fig. 3-13, the flywheel magnets have moved to a new position to where

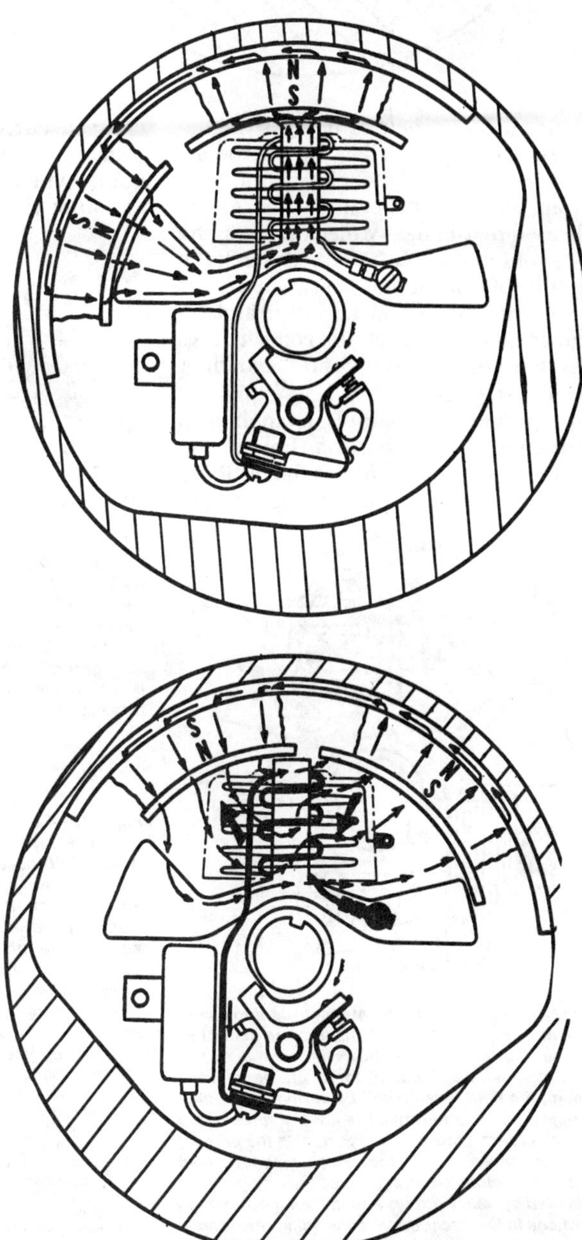

Fig. 3-12—View showing flywheel turned to a position so that lines of force of the permanent magnets are concentrated in the left and center core legs and are interlocking the coil windings.

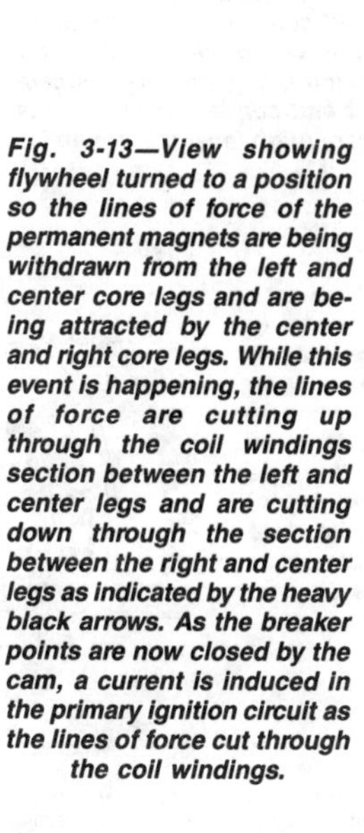

Fig. 3-13—View showing flywheel turned to a position so the lines of force of the permanent magnets are being withdrawn from the left and center core legs and are being attracted by the center and right core legs. While this event is happening, the lines of force are cutting up through the coil windings section between the left and center legs and are cutting down through the section between the right and center legs as indicated by the heavy black arrows. As the breaker points are now closed by the cam, a current is induced in the primary ignition circuit as the lines of force cut through the coil windings.

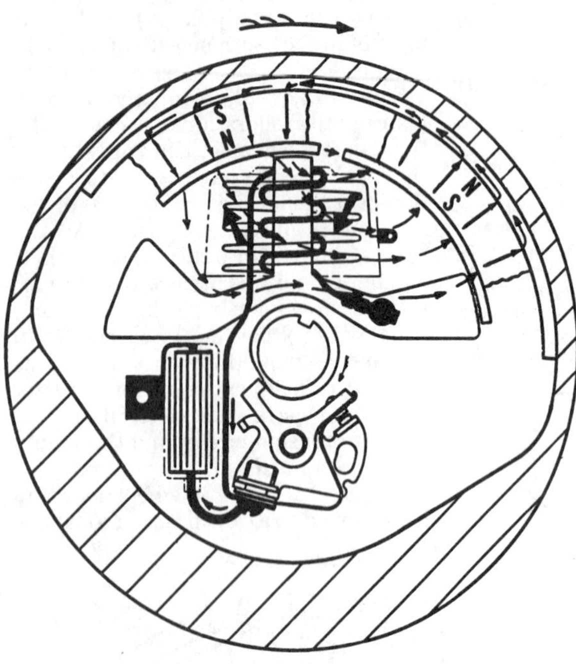

Fig. 3-14—The flywheel magnets have now turned slightly past the position shown in Fig. 3-13 and the rate of movement of lines of magnetic force cutting through the coil windings is at the maximum. At this instant, the breaker points are opened by the cam and flow of current in the primary circuit is being absorbed by the condenser, bringing the flow of current to a quick, controlled stop. Refer now to Fig. 3-15.

their magnetic field is being attracted by the center and right legs of the armature core, and is being withdrawn from the left and center legs. As indicated by the heavy black arrows, the lines of force are cutting up through the section of coil windings between the left and center legs of the armature and are cutting down through the coil windings section between the center and right legs. If the left hand rule, as explained in a previous paragraph, is applied to the lines of force cutting through the coil sections, it is seen that the resulting emf induced in the primary circuit will cause a current to flow through the primary coil windings and the breaker points which have now been closed by action of the cam.

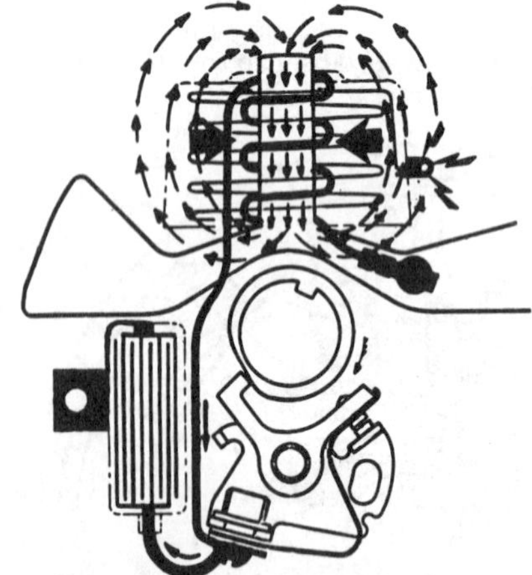

Fig. 3-15—View showing magneto ignition coil, condenser and breaker points at same instant as illustrated in Fig. 3-14; however, arrows shown above illustrate lines of force of the electromagnetic field established by current in primary coil windings rather than the lines of force of the permanent magnets. As the current in the primary circuit ceases to flow, the electro-magnetic field collapses rapidly, cutting the coil windings as indicated by heavy arrows and inducing a very high voltage in the secondary coil winding resulting in the ignition spark.

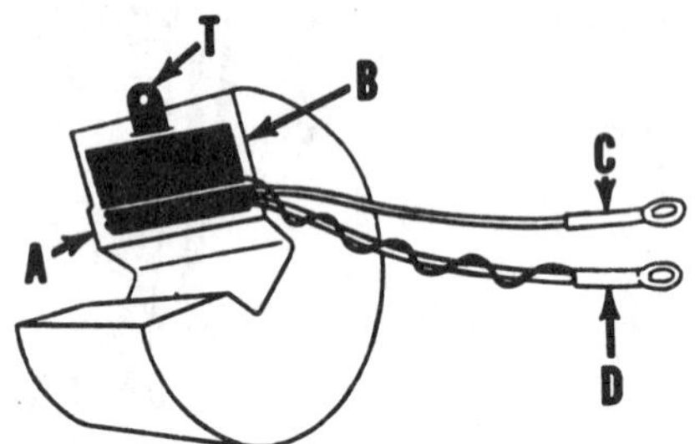

Fig. 3-16—Drawing showing construction of a typical flywheel magneto ignition coil. Primary winding (A) consists of about 200 turns of wire. Secondary winding (B) consists of several thousand turns of fine wire. Coil primary and secondary ground connection is (D); primary connection to breaker point and condenser terminal is (C); and coil secondary (high tension) terminal is (T).

At the instant the movement of the lines of force cutting through the coil winding sections is at the maximum rate, the maximum flow of current is obtained in the primary circuit. At this time, the cam opens the breaker points interrupting the primary circuit and for an instant, the flow of current is absorbed by the condenser as illustrated in Fig. 3-14. An emf is also induced in the secondary coil windings, but the voltage is not sufficient to cause current to flow across the spark plug gap.

The flow of current in the primary windings created a strong electromagnetic field surrounding the coil windings and up through the center leg of the armature core as shown in Fig. 3-15. As the breaker points were opened by the cam, interrupting the primary circuit, this magnetic field starts to collapse cutting the coil windings as indicated by the heavy black arrows. The emf induced in the primary circuit would be sufficient to cause a flow of current across the opening breaker points were it not for the condenser absorbing the flow of current and bringing it to a controlled stop. This allows the electro-magnetic field to collapse at such a rapid rate to induce a very high voltage in the coil high tension or secondary windings. This voltage, in the order of 15,000 to 25,000 volts, is sufficient to break down the resistance of the air gap between the spark plug electrodes and a current will flow across the gap. This creates the ignition spark which ignites the compressed fuel-air mixture in the engine cylinder.

Self-Contained Unit Type Magnetos

Some four-stroke cycle engines are equipped with a magneto which is a self-contained unit as shown in Fig. 3-20. This type magneto is driven from the engine timing gears via a gear or coupling. All components of the

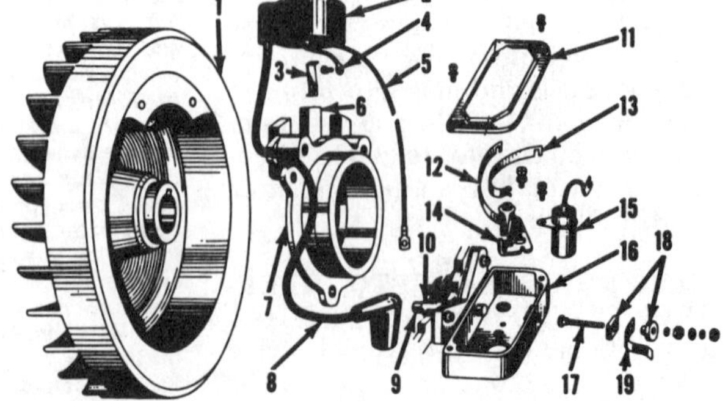

Fig. 3-17—Exploded view of a flywheel type magneto in which the breaker points (14) are actuated by a cam on engine camshaft. Push rod (9) rides against cam to open and close points. In this type unit, an ignition spark is produced only on alternate revolutions of the flywheel as the camshaft turns at one-half engine speed.

1. Flywheel
2. Ignition coil
3. Coil clamps
4. Coil ground lead
5. Breaker-point lead
6. Armature core (laminations)
7. Crankshaft bearing retainer
8. High tension lead
9. Push rod
10. Bushing
11. Breaker box cover
12. Point lead strap
13. Breaker-point spring
14. Breaker-point assy.
15. Condenser
16. Breaker box
17. Terminal bolt
18. Insulators
19. Ground (stop) spring

magneto are enclosed in one housing and the magneto can be removed from the engine as a unit.

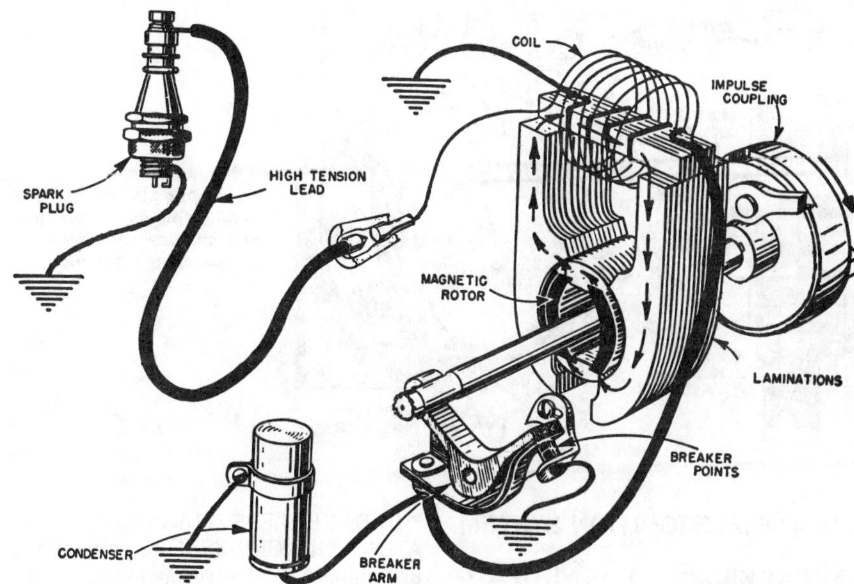

Fig. 3-21—Schematic diagram of typical unit type magneto for single cylinder engine. Refer to Figs. 3-22, 3-23 and 3-23A for views showing construction of impulse couplings.

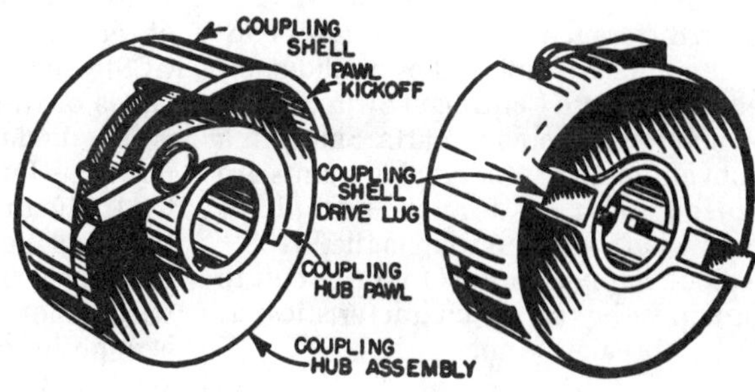

Fig. 3-22—Views of typical impulse coupling for magneto driven by engine shaft with slotted drive connection. Coupling drive spring is shown in Fig. 3-23. Refer to Fig. 3-23A for view of combination magneto drive gear and impulse coupling used on some magnetos.

UNIT TYPE MAGNETO OPERATING PRINCIPLES. In Fig. 3-21, a schematic diagram of a unit type magneto is shown. The magneto rotor is driven through an impulse coupling (shown at right side of illustration). The function of the impulse coupling is to increase the rotating speed of the rotor, thereby increasing magneto efficiency, at engine cranking speeds.

A typical impulse coupling for a single cylinder engine magneto is shown in Fig. 3-22. When the engine is turned at cranking speed, the coupling hub pawl engages a stop pin in the magneto housing as the engine piston is coming up on compression stroke. This stops rotation of the coupling hub assembly and magneto rotor. A spring within the coupling shell (See Fig. 3-23) connects the shell and coupling hub; as the engine continues to turn, the spring winds up until the pawl kickoff contacts the pawl and disengages it from the stop pin. This occurs at the time an ignition spark is required to ignite the compressed fuel-air mixture in the engine cylinder. As the pawl is released, the spring connecting the coupling shell and hub unwinds and rapidly spins the magneto rotor.

The magneto rotor (See Fig. 3-21) carries permanent magnets. As the rotor turns, alternating the position of the magnets, the lines of force of the magnets are attracted, then withdrawn from the laminations. In Fig. 3-21, arrows show the magnetic field concentrated within the laminations, or armature core. Slightly further rotation of the magnetic rotor will place the magnets to where the laminations will have greater attraction for opposite poles of the magnets. At this instant, the lines of force as indicated by the arrows will suddenly be withdrawn and an opposing field of force will be established in the laminations. Due to this rapid movement of the lines of force, a current will be induced in the primary magneto circuit as the coil windings are cut by the lines of force. At the instant the maximum current is induced in the primary windings, the breaker points are opened by a cam on the magnetic rotor shaft interrupting the primary circuit. The lines of magnetic force established by the primary current (Refer to Fig. 3-5) will cut through the secondary windings at such a rapid rate to induce a very high voltage in the secondary (or high tension) circuit. This voltage will break down the resistance of the spark plug electrode gap and a spark across the electrodes will result.

At engine operating speeds, centrifugal force will hold the impulse coupling hub pawl (See Fig. 3-22) in a position so that it cannot engage the stop pin in magneto housing and the magnetic rotor will be driven through the spring (Fig. 3-23) connecting the coupling shell to coupling hub. The impulse coupling retards the ignition spark, at cranking speeds, as the engine piston travels closer to top dead center while the magnetic rotor is held stationary by the pawl and stop pin. The difference in degrees of impulse coupling shell rotation between the position of retarded spark and normal running spark is known as the impulse coupling lag angle.

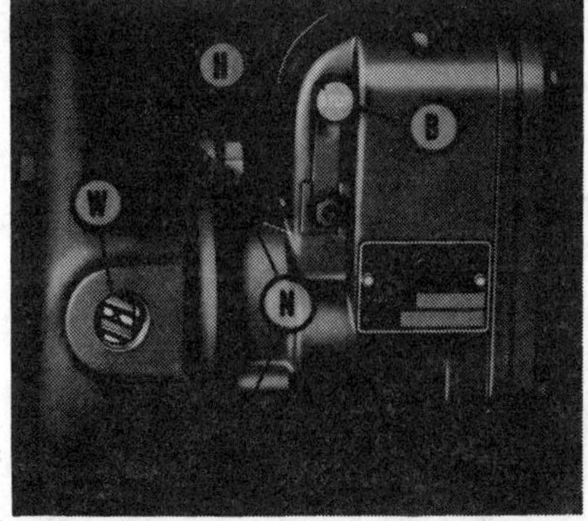

Fig. 3-20—Some engines are equipped with a unit type magneto having all components enclosed in a single housing (H). Magneto is removable as a unit after removing retaining nuts (N). Stop button (B) grounds out primary magneto circuit to stop engine. Timing window is (W).

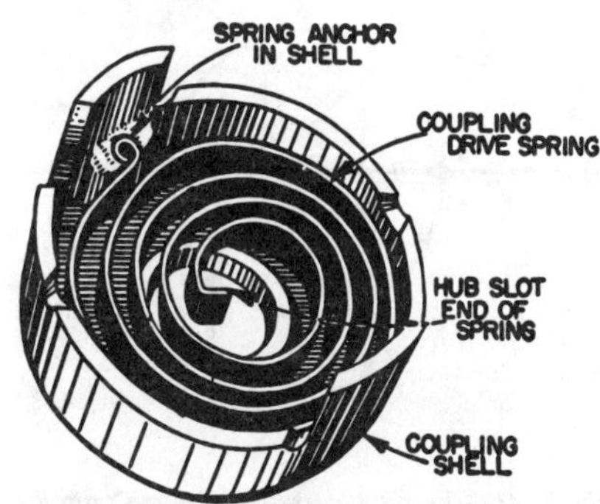

Fig. 3-23—View showing impulse coupling shell and drive spring removed from coupling hub assembly. Refer to Fig. 3-22 for views of assembled unit.

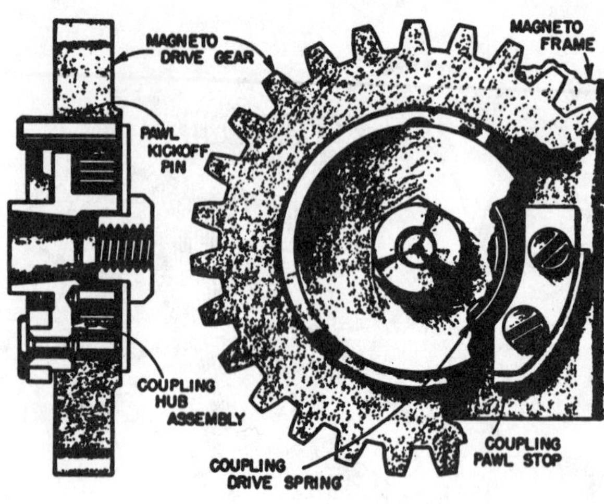

Fig. 3-23A—Views of combination magneto drive gear and impulse coupling used on some magnetos.

SOLID-STATE IGNITION SYSTEM

BREAKERLESS MAGNETO SYSTEM. The solid-state (breakerless) magneto ignition system operates somewhat on the same basic principles as the conventional type flywheel magneto previously described. The main difference is that the breaker contact points are replaced by a solid-state electronic Gate Controlled Switch (GCS) which has no moving parts. Since, in a conventional system breaker points are closed over a longer period of crankshaft rotation than is the "GCS", a diode has been added to the circuit to provide the same characteristics as closed breaker points.

BREAKERLESS MAGNETO OPERATING PRINCIPLES. The same basic principles for electro-magnetic induction of electricity and formation of magnetic fields by electrical current as outlined for the conventional flywheel type magneto also apply to the solid-state magneto. Therefore the principles of the different components (diode and GCS) will complete the operating principles of the solid-state magneto.

The diode is represented in wiring diagrams by the symbol shown in Fig. 3-24. The diode is an electronic device that will permit passage of electrical current in one direction only. In electrical schematic diagrams, current flow is opposite direction the arrow part of symbol is pointing.

The symbol shown in Fig. 3-24A is used to represent the gate controlled switch (GCS) in wiring diagrams. The GCS acts as a switch to permit passage of current from cathode (C) terminal to anode (A) terminal when in "ON" state and will not permit electric current to flow when in "OFF" state. The GCS can be turned "ON" by a positive surge of electricity at the gate (G) terminal and will remain "ON" as long as current remains positive at the gate terminal or as long as current is flowing through the GCS from the cathode (C) terminal to anode (A) terminal.

The basic components and wiring diagram for the solid-state breakerless magneto are shown schematically in Fig. 3-24B. In Fig. 3-24C, the magneto rotor (flywheel) is turning and the ignition coil magnets have just moved into position so that their lines of force are cutting the ignition coil windings and producing a negative surge of current in the primary windings. The diode allows current to flow opposite to the direction of diode symbol arrow and action is same as conventional magneto with breaker points closed. As rotor (flywheel) continues to turn as shown in Fig. 3-24D, direction of magnetic flux lines will reverse in the armature center leg. Direction of current will change in the primary coil circuit and the previously conducting diode will be shut off. At this point, neither diode is conducting. As voltage begins to build up as rotor continues to turn, the condenser acts as

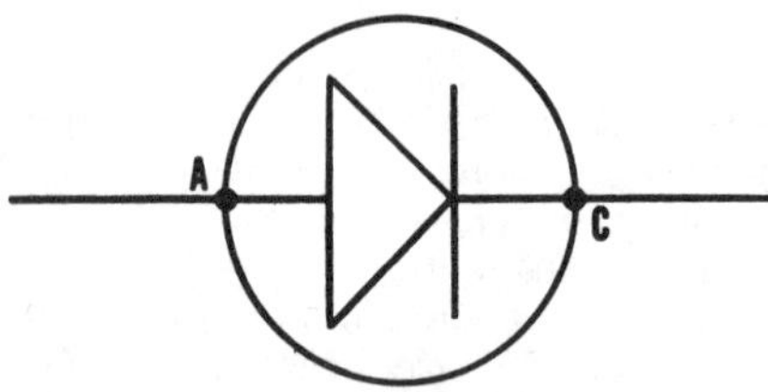

Fig. 3-24—In a diagram of an electrical circuit, the diode is represented by the symbol shown above. The diode will allow current to flow in one direction only, from cathode (C) to anode (A).

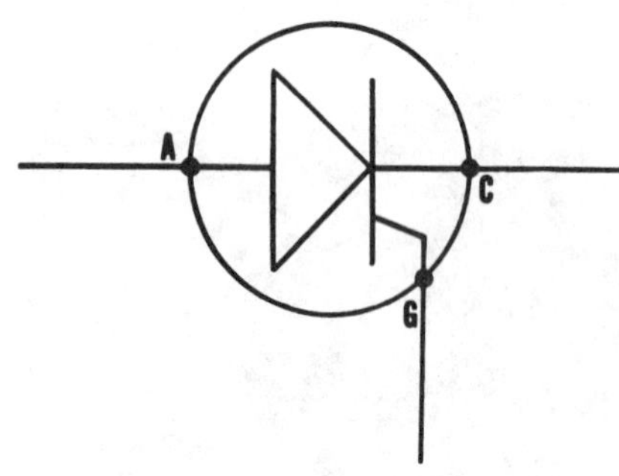

Fig. 3-24A—The symbol used for a Gate Controlled Switch (GCS) in an electrical diagram is shown above. The GCS will permit current to flow from cathode (C) to anode (A) when "turned on" by a positive electrical charge at gate (G) terminal.

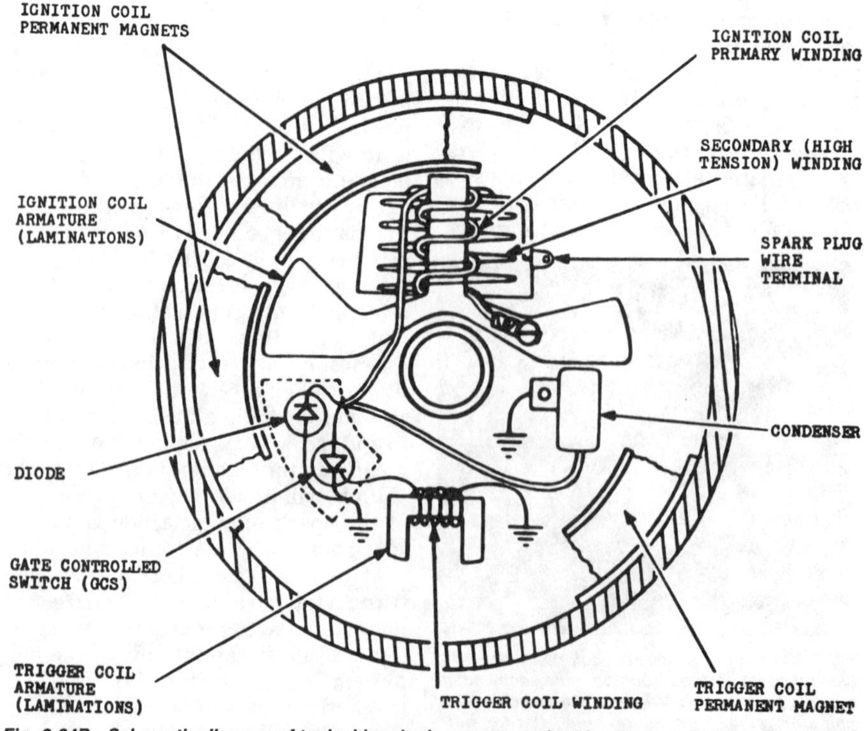

Fig. 3-24B—Schematic diagram of typical breakerless magneto ignition system. Refer to Figs. 3-24C, 3-24D and 3-24E for schematic views of operating cycle.

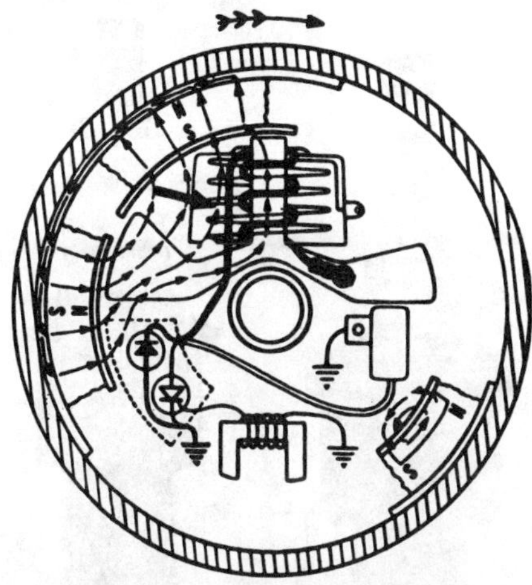

Fig. 3-24C—View showing flywheel of breakerless magneto system at instant of rotation where lines of force of ignition coil magnets are being drawn into left and center legs of magneto armature. The diode (See Fig. 3-24) acts as a closed set of breaker points in completing the primary ignition circuit at this time.

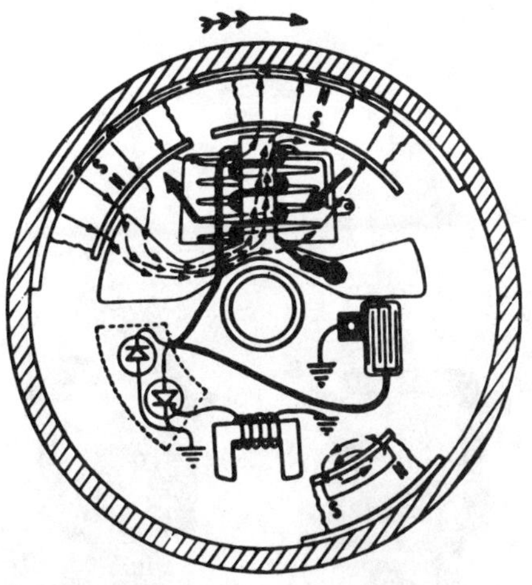

Fig. 3-24D—Flywheel is turning to point where magnetic flux lines through armature center leg will reverse direction and current through primary coil circuit will reverse. As current reverses, diode which was previously conducting will shut off and there will be no current. When magnetic flux lines have reversed in armature center leg, voltage potential will again build up, but since GCS is in "OFF" state, no current will flow. To prevent excessive voltage build up, the condenser acts as a buffer.

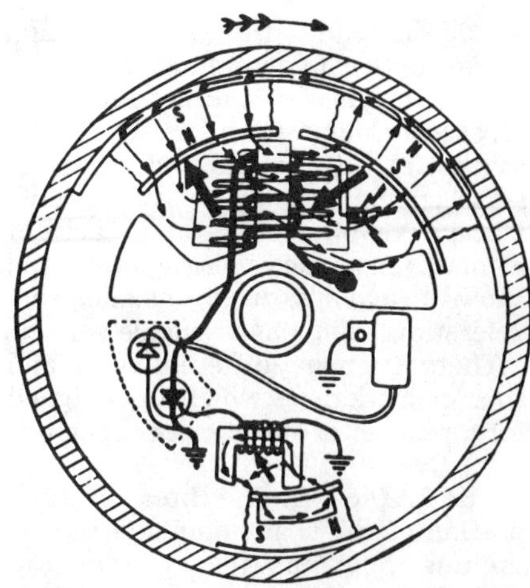

Fig. 3-24E—With flywheel in the approximate position shown, maximum voltage potential is present in windings of primary coil. At this time the triggering coil armature has moved into the field of a permanent magnet and a positive voltage is induced on the gate of the GCS. The GCS is triggered and primary coil current flows resulting in the formation of an electromagnetic field around the primary coil which inducts a voltage of sufficient potential in the secondary windings to "fire" the spark plug.

a buffer to prevent excessive voltage build up at the GCS before it is triggered.

When the rotor reaches the approximate position shown in Fig. 3-24E, maximum flux density has been achieved in the center leg of the armature. At this time the GCS is triggered. Triggering is accomplished by the triggering coil armature moving into the field of a permanent magnet which induces a positive voltage on the gate of the GCS. Primary coil current flow results in the formation of an electromagnetic field around primary coil which inducts a voltage of sufficient potential in the secondary coil windings to "fire" the spark plug.

When the rotor (flywheel) has moved the magnets past the armature, the GCS will cease to conduct and revert to the "OFF" state until it is triggered. The condenser will discharge during the time that the GCS was conducting.

CAPACITOR DISCHARGE SYSTEM. The capacitor discharge (CD) ignition system uses a permanent magnet rotor (flywheel) to induce a current in a coil, but unlike the conventional flywheel magneto and solid-state breakerless magneto described previously, the current is stored in a capacitor (condenser). Then the stored current is discharged through a transformer coil to create the ignition spark. Refer to Fig. 3-24F for a schematic of a typical capacitor discharge ignition system.

CAPACITOR DISCHARGE OPERATING PRINCIPLES. As the permanent flywheel magnets pass by the input generating coil (1—Fig. 3-24F), the current produced charges capacitor (6). Only half of the generated current passes through diode (3) to charge the capacitor. Reverse current is blocked by diode (3) but passes through Zener diode (2) to complete the reverse circuit. Zener diode (2) also limits maximum voltage of the forward current. As the flywheel continues to turn and magnets pass the trigger coil (4), a small amount of electrical current is generated. This current opens the gate controlled switch (5) allowing the capacitor to discharge through the pulse transformer (7). The rapid voltage rise in the transformer primary coil induces a high voltage secondary current which forms the ignition spark when it jumps the spark plug gap.

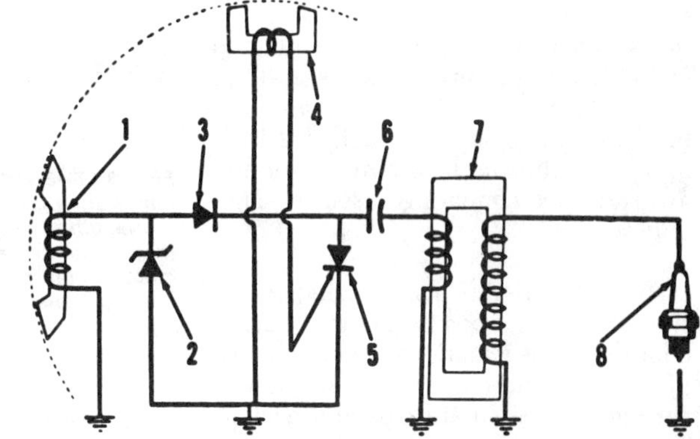

Fig. 3-24F—Schematic diagram of a typical capacitor discharge ignition system.

1. Generating coil
2. Zener diode
3. Diode
4. Trigger coil
5. Gate controlled switch
6. Capacitor
7. Pulse transformer (coil)
8. Spark plug

THE SPARK PLUG

In any spark ignition engine, the spark plug (See Fig. 3-25) provides the means for igniting the compressed fuel-air mixture in the cylinder. Before an electric charge can move across an air gap, the intervening air must be charged with electricity, or ionized. If the spark plug is properly gapped and the system is not shorted, not more than 7,000 volts may be required to initiate a spark. Higher

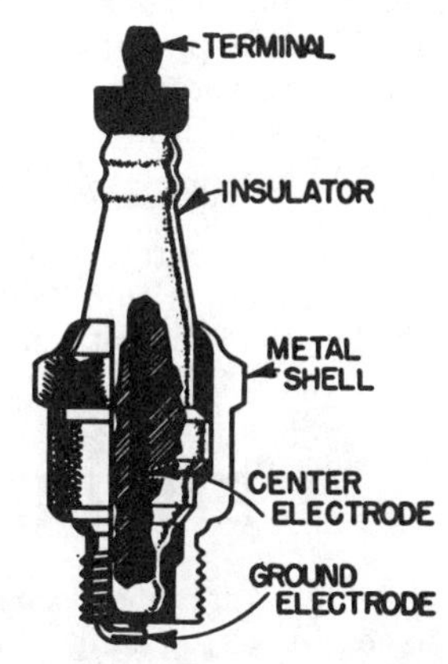

Fig. 3-25—Cross-sectional drawing of spark plug showing construction and nomenclature.

voltage is required as the spark plug warms up, or if compression pressures or the distance of the air gap is increased. Compression pressures are highest at full throttle and relatively slow engine speeds, therefore, high voltage requirements or a lack of available secondary voltage most often shows up as a miss during maximum acceleration from a slow engine speed.

There are many different types and sizes of spark plugs which are designed for a number of specific requirements.

THREAD SIZE. The threaded, shell portion of the spark plug and the attaching hole in the cylinder are manufactured to meet certain industry established standards. The diameter is referred to as "Thread Size." Those commonly used are: 10 mm, 14 mm, 18 mm, 7/8 inch and 1/2 inch pipe. The 14 mm plug is almost universal for small engine use.

REACH. The length of thread, and the thread depth in cylinder head or wall are also standardized throughout the industry. This dimension is measured from gasket seat of plug to cylinder end of thread. See Fig. 3-26. Four different reach plugs commonly used are 3/8 inch, 7/16 inch, 1/2 inch and 3/4 inch. The first two mentioned are the ones commonly used in small engines.

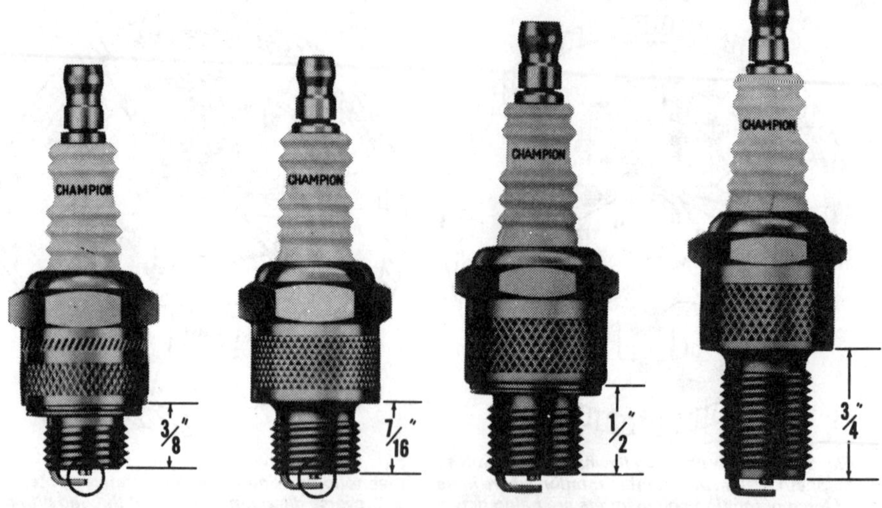

Fig. 3-26—Views showing spark plugs with various "reaches" available; small engines are usually equipped with a spark plug having a 3/8 inch reach. A 3/8 inch reach spark plug measures 3/8 inch from firing end of shell to gasket surface of shell. The two plugs at left side illustrate the difference in plugs normally used in two-stroke cycle and four-stroke cycle engines; refer to the circled electrodes. Spark plug at left has a shortened ground electrode and is specifically designed for two-stroke cycle engines. Second spark plug from left is normally used in four-stroke cycle engines although some two-stroke cycle engines may use this type plug.

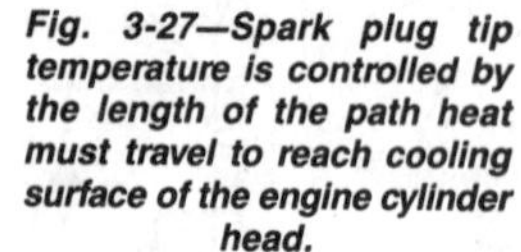

Fig. 3-27—Spark plug tip temperature is controlled by the length of the path heat must travel to reach cooling surface of the engine cylinder head.

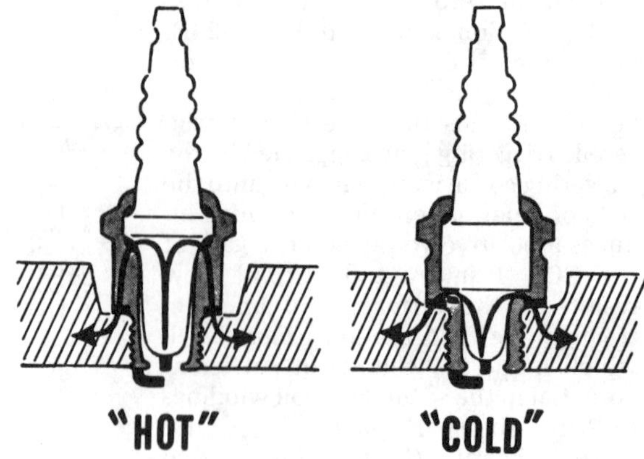

HEAT RANGE. During engine operation, part of the heat generated during combustion is transferred to the spark plug, and from the plug to the cylinder through the shell threads and gasket. The operating temperature of the spark plug plays an important part in engine operation. If too much heat is retained by the plug, the fuel-air mixture may be ignited by contact with heated surface before the ignition spark occurs. If not enough heat is retained, partially burned combustion products (soot, carbon and oil) may build up on the plug tip resulting in "fouling" or shorting out of the plug. If this happens, the secondary current is dissipated uselessly as it is generated instead of bridging the plug gap as a useful spark, and the engine will misfire.

The operating temperature of the plug tip can be controlled, within limits, by altering the length of the path the heat must follow to reach the threads and gasket of the plug. Thus, a plug with a short, stubby insulator around the center electrode will run cooler than one with a long, slim insulator. Refer to Fig. 3-27. Most plugs in the more popular sizes are available in a number of heat ranges which are interchangeable within the group. The proper heat range is determined by engine design and type of service. Refer to SPARK PLUG SERVICING FUNDAMENTALS for additional information on spark plug selection.

SPECIAL TYPES. Sometimes, engine design features or operating conditions call for special plug types designed for a particular purpose. Of special interest when dealing with two-cycle engines is the spark plug shown in the left hand view, Fig. 3-26. In the design of this plug, the ground electrode is shortened so that its end aligns with center of insulated electrode rather than completely overlapping as with the conventional plug. This feature reduces the possibility of the gap bridging over by carbon formations.

ENGINE POWER AND TORQUE RATINGS

The following paragraphs discuss the terms used in expressing engine horsepower and torque ratings and explains the methods for determining the different ratings. Some small engine repair shops are now equipped with a dynamometer for measuring engine torque and/or horsepower and the mechanic should be familiar with terms, methods of measurement and how actual power developed by an engine can vary under different conditions.

Fig. 4-1—A force, measured in pounds, is defined as an action tending to move an object or to accelerate movement of an object.

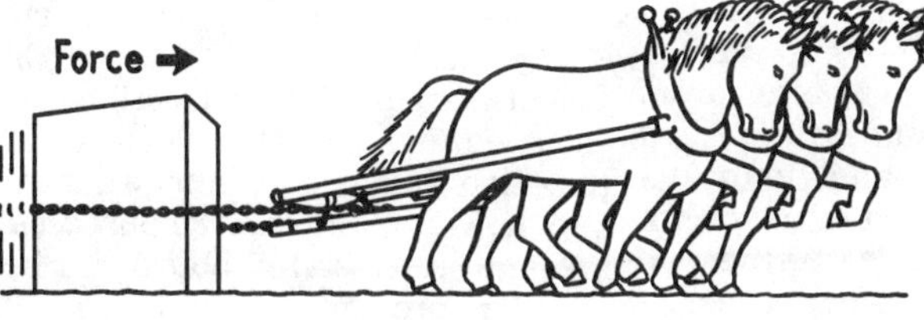

Fig. 4-2—If a force moves an object from a state of rest or accelerates movement of an object, then work is done.

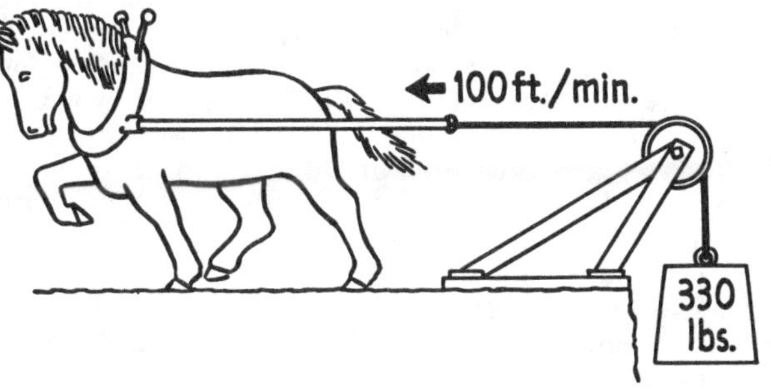

Fig. 4-3—This horse is doing 33,000 foot-pounds of work in one minute, or one horsepower.

GLOSSARY OF TERMS

FORCE. Force is an action against an object that tends to move the object from a state of rest, or to accelerate the movement of an object. For use in calculating torque or horsepower, force is measured in pounds.

WORK. When a force moves an object from a state of rest, or accelerates the movement of an object, work is done. Work is measured by multiplying the force applied by the distance the force moves the object, or:

$$\text{work} = \text{force} \times \text{distance.}$$

Thus, if a force of 50 pounds moved an object 50 feet, work done would equal 50 pounds times 50 feet, or 2500 pounds-feet(or as it is usually expressed, 2500 foot-pounds).

POWER. Power is the rate at which work is done; thus, if:

$$\text{work} = \text{force} \times \text{distance,}$$

then:

$$\text{power} = \frac{\text{force} \times \text{distance.}}{\text{time}}$$

From the above formula, it is seen that power must increase if the time in which work is done decreases.

HORSEPOWER. Horsepower is a unit of measurement of power. Many years ago, James Watt, a Scotsman noted as the inventor of the steam engine, evaluated one horsepower as being equal to doing 33,000 foot-pounds of work in one minute. This evaluation has been universally accepted since that time. Thus, the formula for determining horsepower is:

$$\text{horsepower} = \frac{\text{pounds} \times \text{feet}}{33{,}000 \times \text{minutes}}$$

When referring to engine horsepower ratings, one usually finds the rating expressed as brake horsepower or rated horsepower, or sometimes as both.

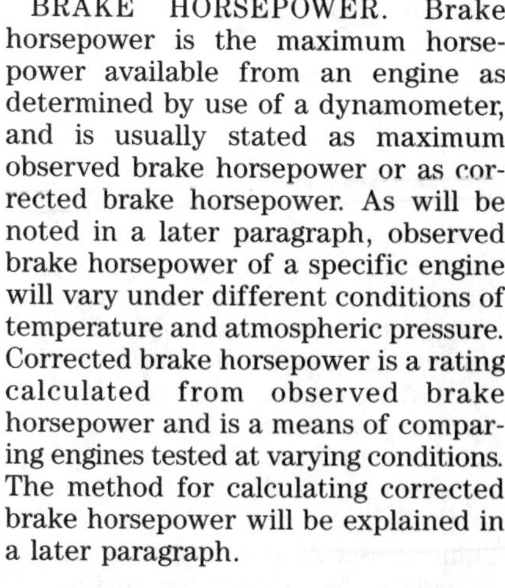
BRAKE HORSEPOWER. Brake horsepower is the maximum horsepower available from an engine as determined by use of a dynamometer, and is usually stated as maximum observed brake horsepower or as corrected brake horsepower. As will be noted in a later paragraph, observed brake horsepower of a specific engine will vary under different conditions of temperature and atmospheric pressure. Corrected brake horsepower is a rating calculated from observed brake horsepower and is a means of comparing engines tested at varying conditions. The method for calculating corrected brake horsepower will be explained in a later paragraph.

RATED HORSEPOWER. An engine being operated under a load equal to the maximum horsepower available (brake horsepower) will not have reserve power for overloads and is subject to damage from overheating and rapid wear. Therefore, when an engine is being selected for a particular load, the engine's brake horsepower rating should be in excess of the expected normal operating load. Usually, it is recommended that the engine not be operated in excess of 80 percent of the engine maximum brake horsepower rating; thus, the "rated horsepower" of an engine is usually equal to 80 percent of maximum horsepower that the engine will develop.

TORQUE. In many engine specifications, a "torque rating" is given. Engine torque can be defined simply as the turning effort exerted by the engine output shaft when under load. Thus, it is possible to calculate engine horsepower being developed by measuring torque being developed and engine output speed. Refer to the following paragraphs.

MEASURING ENGINE TORQUE AND HORSEPOWER

THE PRONY BRAKE. The prony brake is the most simple means of testing engine performance. Refer to diagram in Fig. 4-4. A torque arm is at-

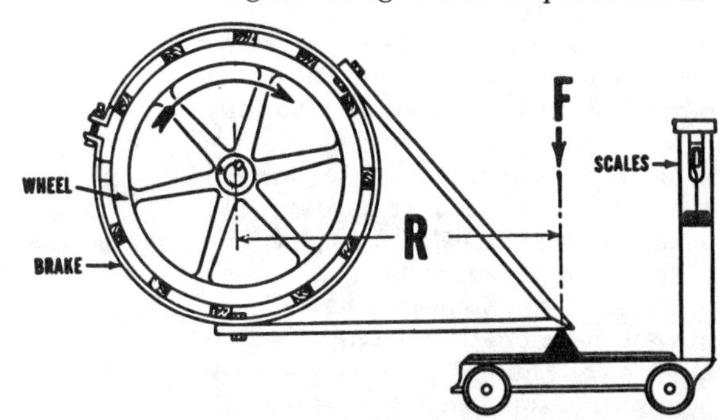

Fig. 4-4—Diagram showing a prony brake on which the torque being developed by an engine can be measured. By also knowing the RPM of the engine output shaft, engine horsepower can be calculated.

tached to a brake on wheel mounted on engine output shaft. The torque arm, as the brake is applied, exerts a force (F) on scales. Engine torque is computed by multiplying the force (F) times the length of the torque arm radius (R), or:

$$\text{engine torque} = F \times R.$$

If, for example, the torque arm radius (R) is 2 feet and the force (F) being exerted by the torque arm on the scales is 6 pounds, engine torque would be 2 feet x 6 pounds or 12 foot-pounds.

To calculate engine horsepower being developed by use of the prony brake, we must also count revolutions of the engine output shaft for a specific length of time. In the formula for calculating horsepower:

$$\text{horsepower} = \frac{\text{feet x pounds}}{33{,}000 \times \text{minutes}}$$

feet will equal the circumference transcribed by the torque arm radius multiplied by the number of engine output shaft revolutions. Thus:

$$\text{feet} = 2 \times 3.14 \times \text{radius} \times \text{revolutions}.$$

Pounds in the formula will equal the force (F) of the torque arm. If, for example, the force (F) is 6 pounds, torque arm radius is 2 feet and engine output shaft speed is 3300 revolutions per minute, then:

$$\text{horsepower} = \frac{2 \times 3.14 \times 2 \times 3300 \times 6}{33{,}000 \times 1}$$

or,

$$\text{horsepower} = 7.54$$

DYNAMOMETERS. Some commercial dynamometers for testing small engines are now available, although the cost may be prohibitive for all but the larger small engine repair shops. Usually, these dynamometers have a hydraulic loading device and scales indicating engine speed and load; horsepower is then calculated by use of a slide rule type instrument. For further information on commercial dynamometers, refer to manufacturers listed in special service tool section of this manual.

HOW ENGINE HORSEPOWER OUTPUT VARIES

Engine efficiency will vary with the amount of air taken into the cylinder on each intake stroke. Thus, air density has a considerable effect on the horsepower output of a specific engine. As air density varies with both temperature and atmospheric pressure, any change in air temperature, barometric pressure, or elevation will cause a variance in observed engine horsepower. As a general rule, engine horsepower will:

A. Decrease approximately 3 percent for each 1000 foot increase above 1000 ft. elevation;

B. Decrease approximately 3 percent for each 1 inch drop in barometric pressure; or,

C. Decrease approximately 1 percent for each 10° rise in temperature (Farenheit).

Thus, to fairly compare observed horsepower readings, the observed readings should be corrected to standard temperature and atmospheric pressure conditions of 60° F., and 29.92 inches of mercury. The correction formula specified by the Society of Automotive Engineers is somewhat involved; however, for practical purposes, the general rules stated above can be used to approximate the corrected brake horsepower of an engine when the observed maximum brake horsepower is known.

For example, suppose the engine horsepower of 7.54 as found by use of the prony brake was observed at an altitude of 3000 feet and at a temperature of 100 degrees. At standard atmospheric pressure and temperature conditions, we could expect an increase of 4 percent due to temperature (100°-60° x 1% per 10°) and an increase of 6 percent due to altitude (3000 ft.—1000 ft. x 3% per 1000 ft.) or a total increase of 10 percent. Thus, the corrected maximum horsepower from this engine would be approximately 7.54 + .75, or approximately 8.25 horsepower.

SERVICE SECTION

TROUBLE-SHOOTING

When servicing an engine to correct a specific complaint, such as engine will not start, is hard to start, etc., a logical step-by-step procedure should be followed to determine cause of trouble before performing any service work. This procedure is "TROUBLE-SHOOTING."

Of course, if an engine is received in your shop for a normal tune up or specific repair work is requested, trouble-shooting procedure is not required and the work should be performed as requested. It is wise, however, to fully check the engine before repairs are made and recommend any additional repairs or adjustments necessary to ensure proper engine performance.

The following procedures, as related to a specific complaint or trouble, have proven to be a satisfactory method for quickly determining cause of trouble in a number of small engine repair shops.

NOTE: It is not suggested that the trouble-shooting procedure as outlined in following paragraphs be strictly adhered to at all times. In many instances, customer's comments on when trouble was encountered will indicate cause of trouble. Also, the mechanic will soon develop a diagnostic technique that can only come with experience. In addition to the general trouble-shooting procedure, reader should also refer to special notes following this section and to the information included in the engine, carburetor and magneto servicing fundamentals sections.

If Engine Will Not Start—Or Is Hard To Start

1. If engine is equipped with a rope or crank starter, turn engine slowly. As the engine piston is coming up on compression stroke, a definite resistance to

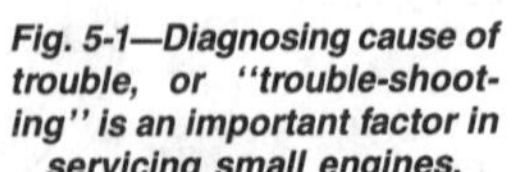
Fig. 5-1—Diagnosing cause of trouble, or "trouble-shooting" is an important factor in servicing small engines.

turning should be felt on rope or crank. See Fig. 5-2. This resistance should be noted every other crankshaft revolution on a single cylinder four-stroke cycle engine and on every revolution of a two-stroke cycle engine crankshaft. If alternate hard and easy turning is noted, the engine compression can be considered as not the cause of trouble at this time.

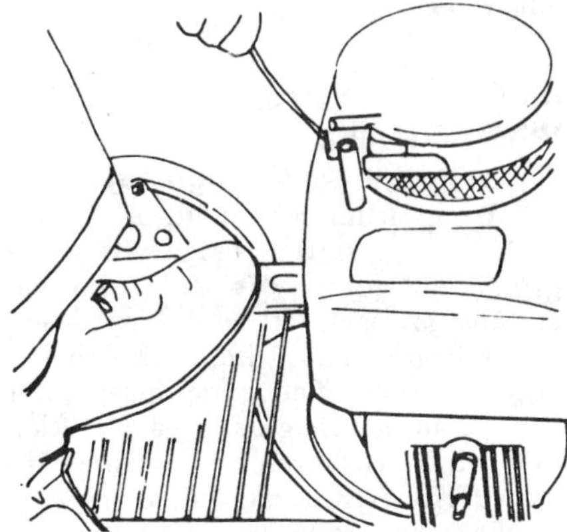

Fig. 5-2—Checking engine compression by slowly cranking engine; a definite resistance should be felt on starter rope each time piston comes up on compression stroke.

NOTE: Compression gages for small gasoline engines are available and are of value in trouble-shooting engine service problems.

Where available from engine manufacturer, specifications will be given for engine compression pressure in the engine service sections of this manual. On engines having electric or impulse starters, remove spark plug and check engine compression with gage; if gage is not available, hold thumb so that spark plug hole is partly covered. An alternating blowing and suction action should be noted as the engine is cranked.

If very little or no compression is noted, refer to appropriate engine repair section for repair of engine. If check indicates engine is developing compression, proceed to step 2.

2. Remove spark plug wire and hold wire terminal about 1/8 inch (3.18 mm) away from cylinder. (On wires having rubber spark plug boot, insert a small screw or bolt in terminal.

NOTE: If available, use of a test plug is recommended. See Fig. 5-3.

While cranking engine, a bright blue spark should snap across the 1/8 inch (3.18 mm) gap. If spark is weak or yellow, or if no spark occurs while cranking engine, refer to following IGNITION SYSTEM SERVICE section for information on appropriate type system.

Fig. 5-3—Checking ignition spark across gap of special test plug (TP).

NOTE: A test plug with 1/8 inch (3.18 mm) gap is available or a test plug can be made by adjusting the electrode gap of a new spark plug to 0.125 inch (3.18 mm).

If spark is satisfactory, remove and inspect spark plug. Refer to SPARK PLUG SERVICING under following IGNITION SYSTEM SERVICE section. If in doubt about spark plug condition install a new plug.

NOTE: Before installing plug, be sure to check electrode gap with proper gage and, if necessary, adjust to value given in engine repair section of this manual. DO NOT guess or check gap with a "thin dime"; a few thousandths variation from correct spark plug electrode gap can make an engine run unsatisfactorily, or under some conditions, not start at all. See Fig. 5-4.

If ignition spark is satisfactory and engine will not start with new plug, proceed with step 3.

3. If engine compression and ignition spark seem to be OK, trouble within the fuel system should be suspected. Remove and clean or renew air cleaner or cleaner element. Check fuel tank (See Fig. 5-5) and be sure it is full of fresh gasoline (four-stroke cycle engines) or fresh gasoline and lubricating oil mixture (two-stroke cycle engines) as prescribed by engine manufacturer. Refer to LUBRICATION paragraph in each two-stroke cycle engine service (engine repair) section for proper fuel:oil mixture for each make and model. If equipped with a fuel shut-off valve, be sure valve is open.

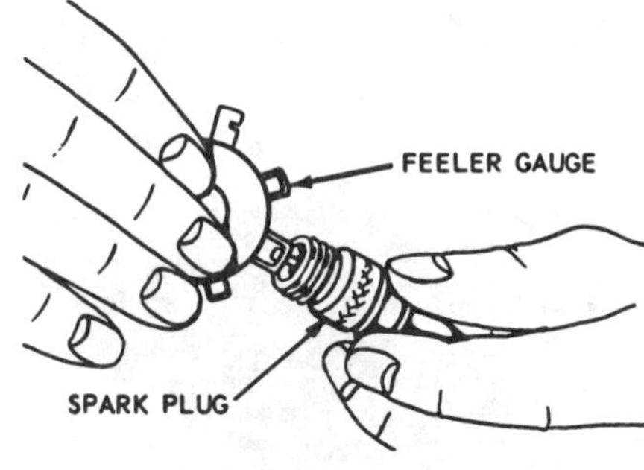

Fig. 5-4—Be sure to check spark plug electrode gap with proper size feeler gage and adjust gap to specification recommended by manufacturer.

If engine is equipped with remote throttle controls that also operate carburetor choke plate, check to be sure that when controls are placed in choke position, carburetor choke plate is fully closed. If not, adjust control linkage so that choke will fully close; then, try to start engine. If engine does not start after several turns, remove air cleaner assembly; carburetor throat should be wet with gasoline. If not, check for reason fuel is not getting to carburetor. On models with gravity feed from fuel tank to carburetor (fuel tank above carburetor), disconnect fuel line at carburetor to see that fuel is flowing through the line. If no fuel is flowing, remove and clean fuel tank, fuel line and any fuel filters or shut-off valve.

On models having a fuel pump separate from carburetor, remove fuel line at carburetor and crank engine through several turns; fuel should spurt from open line. If not, disconnect fuel line from tank to fuel pump at pump connection. If fuel will not run from open line, remove and clean the fuel tank, line and if so equipped, fuel filter and/or shut-off valve. If fuel runs from open line, remove and overhaul or renew the fuel pump.

After making sure that clean, fresh fuel is available at carburetor, again try to start engine. If engine will not start, refer to recommended initial adjustments for carburetor in appropriate engine repair section of this manual and adjust carburetor idle and/or main fuel needles.

If engine will not start when compression and ignition test OK and clean, fresh fuel is available to carburetor, remove and clean or overhaul carburetor as outlined in following CARBURETOR SERVICE section.

4. The preceding trouble-shooting techniques are based on the fact that to run, an engine must develop compression, have an ignition spark and receive the proper fuel:air mixture. In some instances, there are other factors involved. Refer to the special notes following this section for service hints on finding common causes of engine trouble that may not be discovered in normal trouble-shooting procedure.

If Engine Starts, Then Stops

This complaint is usually due to fuel starvation, but may be caused by a faulty ignition system. Recommended trouble-shooting procedure is as follows:

1. Remove and inspect fuel tank cap; on all except a few early two-stroke cycle engines, fuel tank is vented through breather in fuel tank cap so that air can enter the tank as fuel is used. If engine stops after running several minutes, a clogged breather should be suspected. On some engines, it is possible to let the engine run with fuel tank cap removed and if this permits engine to run without stopping, clean of renew the cap.

CAUTION: Be sure to observe safety precautions before attempting to run engine without fuel tank cap in place. If there is any danger of fuel being spilled on engine or spark entering open tank, DO NOT attempt to run engine without fuel tank cap in place. If in doubt, try a new cap.

2. If clogged breather in fuel tank cap is eliminated as cause of trouble, a partially clogged fuel filter or fuel line should be suspected. Remove and clean fuel tank and line and if so equipped, clean fuel shut-off valve and/or fuel tank filter. On some engines, a screen or felt type fuel filter is located in the carburetor fuel inlet; refer to engine repair section for appropriate engine make and model for carburetor construction.

3. After cleaning fuel tank, line, filters, etc., if trouble is still encountered, a sticking or faulty carburetor inlet needle valve, float or diaphragm may be cause of trouble. Remove, disassemble and clean carburetor using data in engine repair section and in following CARBURETOR SERVICE section as a guide.

4. If fuel system is eliminated as cause of trouble by performing procedure outlined in steps 1, 2 and 3, check magneto or battery ignition coil on tester if such equipment is available. If not, check for ignition spark immediately after engine stops. Renew coil, condenser and breaker points if no spark is noted. Also, on four-stroke cycle engines, check for engine compression immediately after engine stops; trouble may be caused by sticking intake or exhaust valve or cam followers (tappets). If no or little compression is noted immediately after engine stops, refer to ENGINE SERVICE section and to engine repair data in the appropriate engine repair section of this manual.

Engine Overheats

When air cooled engines overheat, check for:

1. "Winterized" engine operated in warm temperatures.

2. Remove blower housing and shields and check for dirt or debris accumulated on or between cooling fins on cylinder.

3. Missing or bent shields or blower housing. (Never attempt to operate an air cooled engine without all shields and blower housing in place.)

4. A too lean main fuel-air adjustment of carburetor.

5. Improper ignition spark timing. Check breaker-point gap, and on engine with unit type magneto, check magneto to engine timing. On battery ignition units with timer, check for breaker points opening at proper time.

6. Engines being operated under loads in excess of rated engine horsepower or at extremely high ambient (surrounding) air temperatures may overheat.

7. Two-stroke cycle engines being operated with an improper fuel-lubricating oil mixture may overheat due to lack of lubrication; refer to appropriate engine service section in this manual for recommended fuel-lubricating oil mixture.

Engine Surges When Running

Trouble with an engine surging is usually caused by improper carburetor adjustment or improper governor adjustment.

1. Refer to CARBURETOR paragraphs in the appropriate engine repair section and adjust carburetor as outlined.

2. If adjusting carburetor did not correct the surging condition, refer to GOVERNOR paragraph and adjust governor linkage.

Fig. 5-5—Condensation can cause water and rust to form in fuel tank even though only clean fuel has been poured into tank.

3. If any wear is noted in governor linkage and adjusting linkage did not correct problem, renew worn linkage parts.

4. If trouble is still not corrected, remove and clean or overhaul carburetor as necessary. Also check for any possible air leaks between the carburetor to engine gaskets or air inlet elbow gaskets.

Engine Misses Out When Running (Two-Stroke Cycle Engines)

1. If engine misses out only at no load high idle speed, first be sure that engine is not equipped with an ignition cut-out governor. (If so equipped, engine will miss out at high speed due to cut-out action.) If not so equipped, refer to appropriate engine repair section and adjust carburetor as outlined in CARBURETOR paragraph. Some two-stroke cycle engines will miss out (four-cycle) when not under load, even though carburetor is adjusted properly. If a two-cycle engine fires evenly under normal load, it can usually be considered OK.

Special Notes on Engine Trouble-shooting

SPECIAL APPLICATION ENGINES. An engine may be manufactured or modified to function as a powerplant for a unique type of equipment. Trouble shooting and servicing must be performed while noting any out-of-the-ordinary features of the engine.

TWO-STROKE CYCLE ENGINES WITH REED VALVE. On two-stroke cycle engines, the incoming fuel-air mixture must be compressed in engine crankcase in order for the mixture to properly reach the engine cylinder. On engines utilizing reed type carburetor to crankcase intake valve, a bent or broken reed will not allow compression build up in the crankcase. Thus, if such an engine seems otherwise OK, remove and inspect the reed valve unit. Refer to appropriate engine repair section in this manual for information on individual two-stroke cycle engine models.

TWO-STROKE CYCLE ENGINE EXHAUST PORTS. Two-stroke cycle engines, and especially those being operated on an overly rich fuel-air mixture or with too much lubricating oil mixed with the fuel, will tend to build up carbon in the cylinder exhaust ports. It is recommended that the muffler be removed on two-stroke cycle engines periodically and the carbon removed from the exhaust ports. Recommended procedure varies somewhat with different makes of engines; therefore, refer to CARBON paragraph of maintenance instructions in appropriate engine repair section of this manual.

On two-stroke cycle engines that are hard to start, or where complaint is loss of power, it is wise to remove the muffler and inspect the exhaust ports for carbon build up.

FOUR-STROKE CYCLE ENGINES WITH COMPRESSION RELEASE. Several different makes of four-stroke cycle engines now have a compression release that reduces compression pressure at cranking speeds, thus making it easier to crank the engine. Most models having this feature will develop full compression when turned in a reverse direction. Refer to the appropriate engine repair section in this manual for detailed information concerning the compression release used on different makes and models.

IGNITION SYSTEM SERVICE

The fundamentals of servicing ignition systems are outlined in the following paragraphs. Refer to appropriate heading for type ignition system being inspected or overhauled.

BATTERY IGNITION SERVICE FUNDAMENTALS

Service of battery ignition systems used on small engines is somewhat simplified due to the fact that no distribution system is required as on automotive type ignition systems. Usually all components are readily accessible and while use of test instruments is sometimes desirable, condition of the system can be determined by simple checks. Refer to following paragraphs.

GENERAL CONDITION CHECK. Remove spark plug wire and if terminal is rubber covered, insert small screw or bolt in terminal. Hold uncovered end of terminal, or bolt inserted in terminal about 1/8 inch (3.18 mm) away from engine or connect spark plug wire to test plug as shown in Fig. 5-3. Crank engine while observing gap between spark plug wire terminal and engine; if a bright blue spark snaps across the gap, condition of the system can be considered satisfactory. However, ignition timing may have to be adjusted. Refer to timing procedure in appropriate engine repair section.

VOLTAGE, WIRING AND SWITCH CHECK. If no spark, or a weak yellow-orange spark occurred when checking system as outlined in preceding paragraph, proceed with following checks:

Test battery condition with hydrometer or voltmeter. If check indicates a dead cell, renew the battery; recharge battery if a discharged condition is indicated.

NOTE: On models with electric starter or starter-generator unit, battery can be assumed in satisfactory condition if the starter cranks the engine freely.

If battery checks OK, but starter unit will not turn engine, a faulty starter unit is indicated and ignition trouble may be caused by excessive current draw of such a unit. If battery and starting unit, if so equipped, are in-satisfactory condition, proceed as follows:

Remove battery lead wire from ignition coil and connect a test light of same voltage as the battery between the disconnected lead wire and engine ground. Light should go on when ignition switch is in "on" position and go off when switch is in "off" position. If not, renew switch and/or wiring and recheck for satisfactory spark. If switch and wiring check OK, but no spark is obtained, proceed as follows:

BREAKER POINTS AND CONDENSER. Remove breaker box cover and, using small screwdriver, separate and inspect breaker points. If burned or deeply pitted, renew breaker points and condenser. If point contacts are clean to grayish in color and are only slightly pitted, proceed as follows: Disconnect condenser and ignition coil lead wires from breaker point terminal and connect a test light and battery between terminal and engine ground. Light should go on when points are closed and should go out when points are open. If light fails to go out when points are open, breaker arm insulation is defective and breaker points must be renewed. If light does not go on when points are in closed position, clean or renew the breaker points. In some instances, new breaker point contact surfaces may have an oily or wax coating or have foreign material between the surfaces so that proper contact is prevented. Check ignition timing and breaker point gap as outlined in appropriate engine repair section of this manual.

Connect test light and battery between condenser lead and engine ground; if light goes on, condenser is shorted out and should be renewed. Capacity of condenser can be checked if test instrument is available. It is usually good practice to renew the condenser whenever new breaker points are being installed if tester is not available.

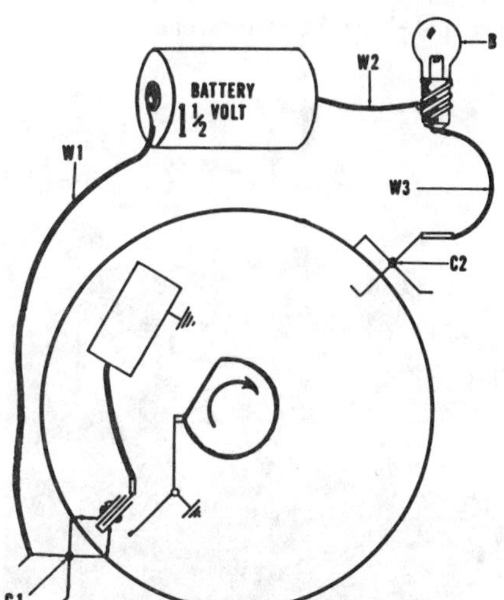

Fig. 5-6—Drawing showing a simple test lamp for checking ignition timing and/or breaker-point opening.

B. 1-1/2 volt bulb
C1. Spring clamp
C2. Spring clamp
W1. Wire
W2. Wire
W3. Wire

IGNITION COIL. If a coil tester is available, condition of coil can be checked. However, if tester is not available, a reasonably satisfactory performance test can be made as follows:

Disconnect high tension wire from spark plug. Turn engine so that cam has allowed breaker points to close. With ignition switch on, open and close points with small screwdriver while holding high tension lead about 1/8 to 1/4 inch (3.18 to 6.35 mm) away from engine ground. A bright blue spark should snap across the gap between spark plug wire and ground each time the points are opened. If no spark occurs, or spark is weak and yellow-orange, renewal of the ignition coil is indicated.

Sometimes, an ignition coil may perform satisfactorily when cold, but fail after engine has run for some time and coil is hot. Check coil when hot if this condition is indicated.

FLYWHEEL MAGNETO SERVICE FUNDAMENTALS

In servicing a flywheel magneto ignition system, the mechanic is concerned with trouble shooting, service adjustments and testing magneto components. The following paragraphs outline the basic steps in servicing a flywheel type magneto. Refer to the appropriate engine section for adjustment and test specifications for a particular engine.

Trouble-shooting

If the engine will not start and malfunction of the ignition system is suspected, make the following checks to find cause of trouble.

Check to be sure that the ignition switch (if so equipped) is in the "On" or "Run" position and that the insulation on the wire leading to the ignition switch is in good condition. The switch can be checked with the timing and test light as shown in Fig. 5-6. Disconnect the lead from the switch and attach one clip of the test light to the switch terminal and the other clip to the engine. The light should go on when the switch is in the "Off" or "Stop" position, and should go off when the switch is in the "On" or "Run" position.

Inspect the high tension (spark plug) wire for worn spots in the insulation or breaks in the wire. Frayed or worn insulation can be repaired temporarily with plastic electrician's tape.

If no defects are noted in the ignition switch or ignition wires, remove and inspect the spark plug as outlined under the SPARK PLUG SERVICING. If the spark plug is fouled or is in questionable condition, connect a spark plug of known quality to the high tension wire, ground the base of the spark plug to engine and turn engine rapidly with the starter. If the spark across the electrode gap of the spark plug is a bright blue, the magneto can be considered in satisfactory condition.

NOTE: Some engine manufacturers specify a certain type spark plug and a specific test gap. Refer to appropriate engine service section; if no specific spark plug type or electrode gap is recommended for test purposes, use spark plug type and electrode gap recommended for engine make and model.

If spark across the gap of the test plug is weak or orange colored, or no spark occurs as engine is cranked, magneto should be serviced as outlined in the following paragraphs.

Magneto Adjustments

BREAKER CONTACT POINTS. Adjustment of the breaker contact points affects both ignition timing and magneto edge gap. Therefore, the breaker contact point gap should be carefully adjusted according to engine manufacturer's specifications. Before adjusting the breaker contact gap, inspect contact points and renew if condition of contact surfaces is questionable. It is sometimes desirable to check the condition of points as follows: Disconnect the condenser and primary coil leads from the breaker-point terminal. Attach one clip of a test light (See Fig. 5-6) to the breaker-point terminal and the other clip of the test light to magneto ground. The light should be out when contact points are open and should go on when the engine is turned to close the breaker contact points. If the light stays on when points are open, insulation of breaker contact arm is defective. If light does not go on when points are closed, contact surfaces are dirty, oily or are burned.

Adjust breaker-point gap as follows unless manufacturer specifies adjusting breaker gap to obtain correct ignition timing. First, turn engine so that points are closed to be sure that the contact surfaces are in alignment and seat squarely. Then, turn engine so that breaker point opening is maximum and adjust breaker gap to manufacturer's specification. A wire type feeler gage is recommended for checking and adjusting the breaker contact gap. Be sure to recheck gap after tightening breaker point base retaining screws.

IGNITION TIMING. On some engines, ignition timing is nonadjustable and a certain breaker-point gap is specified. On other engines, timing is adjustable by changing the position of the magneto stator plate. (See Fig. 5-7) with a specified breaker-point gap or by simply varying the breaker-point gap to obtain correct timing, Ignition timing is usually specified either in degrees of engine (crankshaft) rotation or in piston travel before the piston reaches top dead center position. In some instances, a specification is given for ignition timing even though the timing may be nonadjustable; if a check reveals timing is incorrect on these engines, it is an indication of incorrect breaker-point adjustment or excessive wear of breaker cam. Also, on some engines, it may indicate that a wrong breaker cam has been installed or that the cam has been installed in a reversed position on engine crankshaft.

Some engines may have a timing mark or flywheel locating pin to locate the flywheel at proper position for the ignition spark to occur (breaker points begin to open). If not, it will be necessary to measure piston travel as illustrated in Fig. 5-8 or install a degree indicating device on the engine crankshaft.

A timing light as shown in Fig. 5-6 is a valuable aid in checking or adjusting engine timing. After disconnecting the ignition coil lead from the breaker-point terminal, connect the leads of the timing light as shown. If timing is adjustable by moving the magneto stator plate, be sure that the breaker point gap is adjusted as specified. Then, to check timing, slowly turn engine in normal direction of rotation past the point at which ignition spark should occur. The timing light should be on, then go out (breaker points open) just as the correct timing location is passed. If not, turn engine to proper timing location and adjust timing by relocating the magneto stator plate or varying the breaker contact gap as specified by engine manufacturer. Loosen the screws retaining the stator plate or breaker points and adjust position of stator plate or points so that points are closed (timing light is on). Then, slowly move adjustment until timing light goes out (points open) and tighten the retaining screws. Recheck timing to be sure adjustment is correct.

Fig. 5-7—On some engines, timing is adjustable by moving magneto stator plate in slotted mounting holes. Marks should be applied to stator plate and engine after engine is properly timed; marks usually appear on factory assembled stator plate and engine cylinder block.

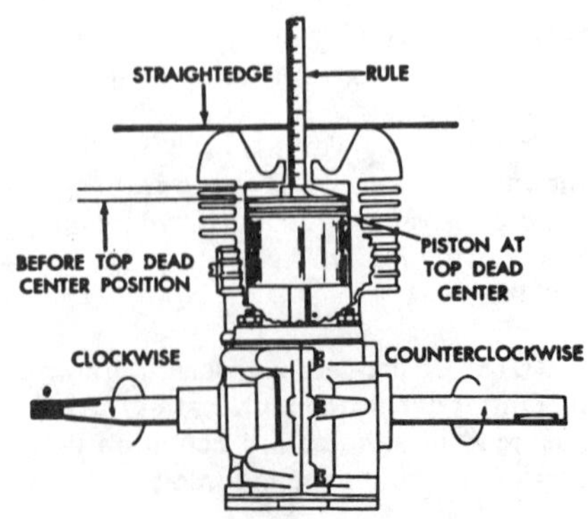

Fig. 5-8—On some engines, it will be necessary to measure piston travel with rule, dial indicator or special timing gage when adjusting or checking ignition timing.

ARMATURE AIR GAP. To fully concentrate the magnetic field of the flywheel magnets within the armature core, it is necessary that the flywheel magnets pass as closely to the armature core as possible without danger of metal to metal contact. The clearance between the flywheel magnets and the legs of the armature core is called the armature air gap.

On magnetos where the armature and high tension coil are located outside of the flywheel rim, adjustment of the armature air gap is made as follows: Turn the engine so that the flywheel magnets are located directly under the legs of the armature core and check the clearance between the armature core and flywheel magnets. If the measured clearance is not within manufacturers specifications, loosen the armature core mounting screws and place shims of thickness equal to minimum air gap specification between the magnets and armature core (Fig. 5-9). The magnets will pull the armature core against the shim stocks. Tighten the armature core mounting screws, remove the shim stock and turn the engine through several revolutions to be sure the flywheel does not contact the armature core.

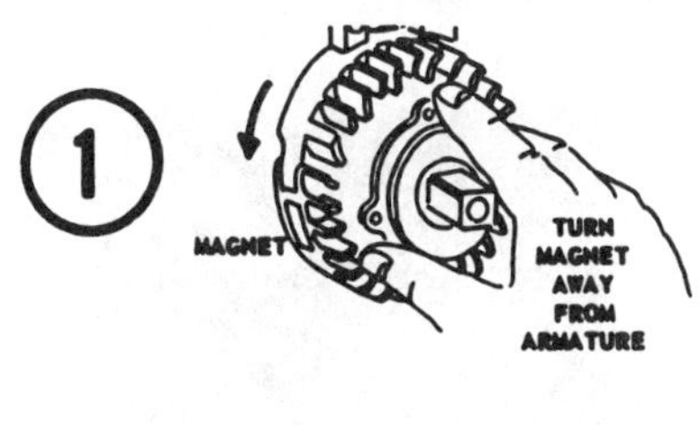

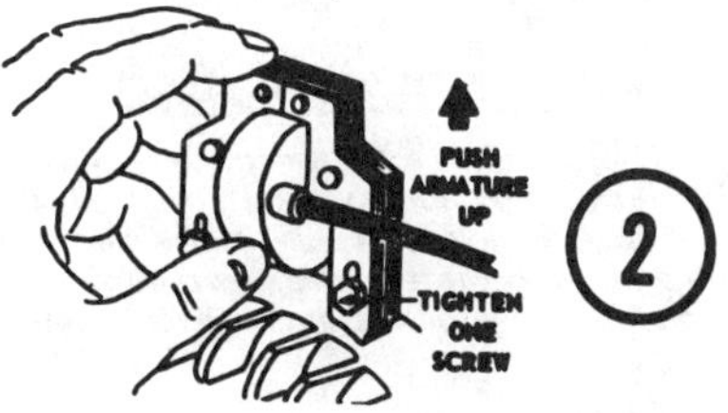

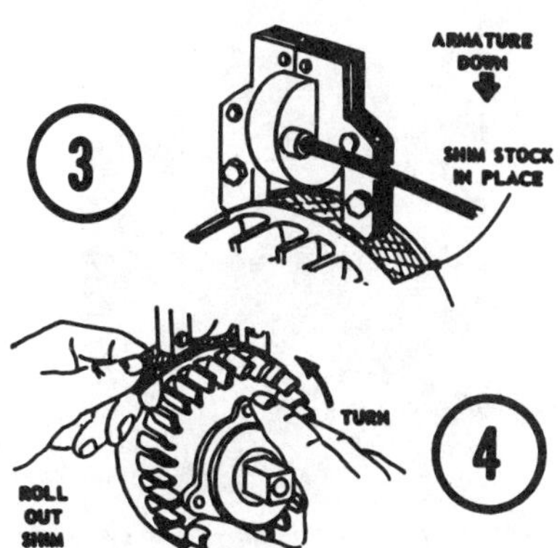

Fig. 5-9—Views showing adjustment of armature air gap when armature is located outside flywheel. Refer to Fig. 5-10 for engines having armature located inside flywheel.

Where the armature core is located under or behind the flywheel, the following methods may be used to check and adjust armature air gap: On some engines, slots or openings are provided in the flywheel through which the armature air gap can be checked. Some engine manufacturers provide a cutaway flywheel that can be installed temporarily for checking the armature air gap. A test flywheel can be made out of a discarded flywheel (See Fig. 5-10), or out of a new flywheel if service volume on a particular engine warrants such expenditure. Another method of checking the armature air gap is to remove the flywheel and place a layer of plastic tape equal to the minimum specified air gap over the legs of the armature core. Reinstall flywheel and turn engine through several revolutions and remove flywheel; no evidence of contact between the flywheel magnets and plastic tape should be noticed. Then cover the legs of the armature core with a layer of tape of thickness equal to the maximum specified air gap; then, reinstall flywheel and turn engine through several revolutions. Indication of the flywheel magnets contacting the plastic tape should be noticed after the flywheel is again removed. If the magnets contact the first thin layer of tape applied to the armature core legs, or if they do not contact the second thicker layer of tape, armature air gap is not within specifications and should be adjusted.

NOTE: Before loosening armature core mounting screws, scribe a mark on mounting plate against edge of armature core so that adjustment of air gap can be gaged.

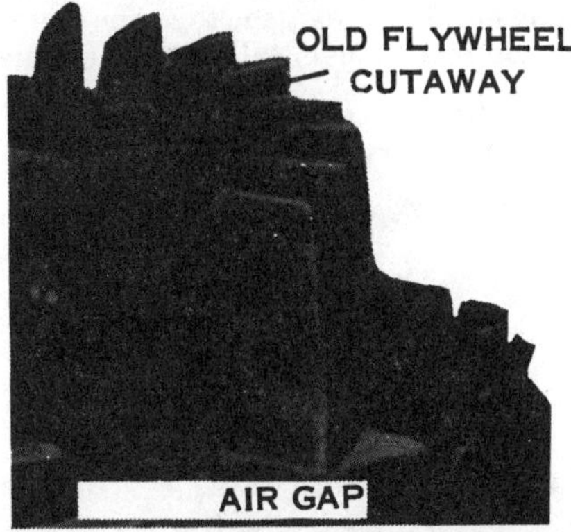

Fig. 5-10—Where armature core is located inside flywheel, check armature gap by using a cutaway flywheel unless other method is provided by manufacturer; refer to appropriate engine repair section. Where possible, an old discarded flywheel should be used to cutaway section for checking armature air gap.

In some instances, it may be necessary to slightly enlarge the armature core mounting holes before proper air gap adjustment can be made.

MAGNETO EDGE GAP. The point of maximum acceleration of the movement of the flywheel magnetic field through the high tension coil (and therefore, the point of maximum current induced in the primary coil windings) occurs when the trailing edge of the flywheel magnet is slightly past the left hand leg of the armature core. The exact point of maximum primary current is determined by using electrical measuring devices, the distance between the trailing edge of the flywheel magnet and the leg of the armature core at this point is measured and becomes a service specification. This distance, which is stated either in thousandths of an inch or in degrees of flywheel rotation, is called the Edge Gap or "E" Gap.

For maximum strength of the ignition spark, the breaker points should just start to open when the flywheel magnets are at the specified edge gap position. Usually, edge gap is non-adjustable and will be maintained at the proper dimension if the contact breaker points are adjusted to the recommended gap and the correct breaker cam is installed. However magneto edge gap can change (and spark intensity thereby reduced) due to the following:

a. Flywheel drive key sheared
b. Flywheel drive key worn (loose)
c. Keyway in flywheel or crankshaft worn (oversized)
d. Loose flywheel retaining nut which can also cause any above listed difficulty
e. Excessive wear on breaker cam
f. Breaker cam loose on crankshaft
g. Excessive wear on breaker point rubbing block or push rod so that points cannot be properly adjusted.

Unit Type Magneto Service Fundamentals

Improper functioning of the carburetor, spark plug or other components often causes difficulties that are thought to be an improperly functioning magneto. Since a brief inspection will often locate other causes for engine malfunction, it is recommended that one be certain the magneto is at fault before opening the magneto housing. Magneto malfunction can easily be determined by simple tests as outlined in following paragraph.

Trouble-shooting

With a properly adjusted spark plug in good condition, the ignition spark should be strong enough to bridge a short gap in addition to the actual spark plug gap. With engine running, hold end of spark plug wire not more than 1/16 inch (1.59 mm) away from spark plug terminal. Engine should not misfire.

To test the magneto spark if engine will not start, remove ignition wire from magneto end cap socket. Bend a short piece of wire so that when it is inserted in the end cap socket, other end is about 1/8 inch (3.18 mm) from engine casting. Crank engine slowly and observe gap between wire and engine; a strong blue spark should jump the gap the instant that the impulse coupling trips. If a strong spark is observed, it is recommended that the magneto be eliminated as the source of engine difficulty and that the spark plug, ignition wire and terminals be thoroughly inspected.

If, when cranking the engine, the impulse coupling does not trip, the magneto must be removed from the engine and the coupling overhauled or renewed. It should be noted that if the impulse coupling will not trip, a weak spark will occur.

Magneto Adjustments and Service

BREAKER POINTS. **Breaker points are accessible for service after removing the magneto housing end cap. Examine point contact surfaces for pitting or pyramiding (transfer of metal from one surface to the other); a** small tungsten file or fine stone may be used to resurface the points. Badly worn or badly pitted points should be renewed. After points are resurfaced or renewed, check breaker point gap with rotor turned so that points are opened maximum distance. Refer to MAGNETO paragraph in appropriate engine repair section for point gap specifications.

When replacing the magneto end cap, both the end cap and housing mating surfaces should be thoroughly cleaned and a new gasket be installed.

CONDENSER. Condenser used in unit type magneto is similar to that used in other ignition systems. Refer to MAGNETO paragraph in appropriate engine repair section for condenser test specifications. Usually, a new condenser should be installed whenever the breaker points are being renewed.

COIL. The ignition coil can be tested without removing the coil from the housing. The instructions provided with coil tester should have coil test specifications listed.

ROTOR. Usually, service on the magneto rotor is limited to renewal of bushings or bearings, if damaged. Check to be sure rotor turns freely and does not drag or have excessive end play.

MAGNETO INSTALLATION. When installing a unit type magneto on an engine, refer to MAGNETO paragraph in appropriate engine repair section for magneto to engine timing information.

SOLID-STATE IGNITION SERVICE FUNDAMENTALS

Because of differences in solid-state ignition construction, it is impractical to outline a general procedure for solid-state ignition service. Refer to the specific engine section for testing, overhaul notes and timing of solid-state ignition systems.

SPARK PLUG SERVICING

ELECTRODE GAP. The spark plug electrode gap should be adjusted by bending the ground electrode. Refer to Fig. 5-11. The recommended gap is listed in the SPARK PLUG paragraph in appropriate engine repair section of this manual.

CLEANING AND ELECTRODE CONDITIONING. Spark plugs are usually cleaned by an abrasive action commonly referred to as "sand blasting." Actually, ordinary sand is not used, but a special abrasive which is nonconductive to electricity even when melted, thus the abrasive cannot short out the plug current. Extreme care should be used in cleaning the plugs after sand blasting, however, as any particles of abrasive left on the plug may cause damage to piston rings, piston or cylinder walls. Some engine

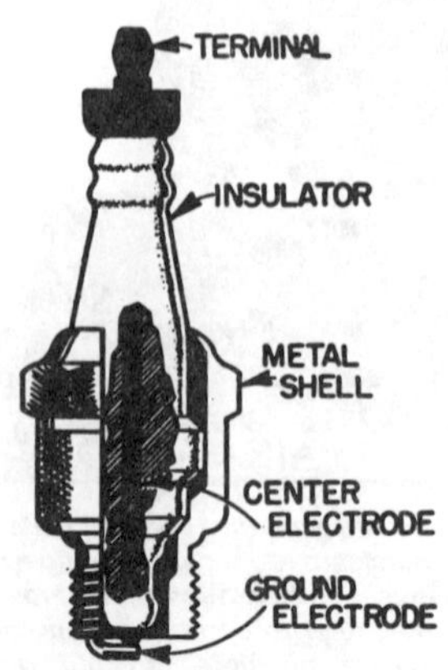

Fig. 5-11—Cross-sectional drawing of spark plug showing construction and nomenclature.

Fig. 5-12—Normal plug appearance in four-stroke cycle engine. Insulator is light tan to gray in color and electrodes are not burned. Renew plug at regular intervals as recommended by engine manufacturer.

Fig. 5-13—Appearance of four-stroke cycle spark plug indicating cold fouling. Cause of cold fouling may be use of a too-cold plug, excessive idling or light loads, carburetor choke out of adjustment, defective spark plug wire or boot, carburetor adjusted too "rich" or low engine compression.

Fig. 5-14—Appearance of four-stroke cycle spark plug indicating wet fouling; a wet, black oily film is over entire firing end of plug. Cause may be oil getting by worn valve guides, worn oil rings or plugged breather or breather valve in tappet chamber.

Fig. 5-15—Appearance of four-stroke cycle spark plug indicating overheating. Check for plugged cooling fins, bent or damaged blower housing, engine being operated without all shields in place or other causes of engine overheating. Also can be caused by too lean a fuel-air mixture or spark plug not tightened properly.

Fig. 5-16—Normal appearance of plug removed from a two-stroke cycle engine. Insulator is light tan to gray in color, few deposits are present and electrodes not burned.

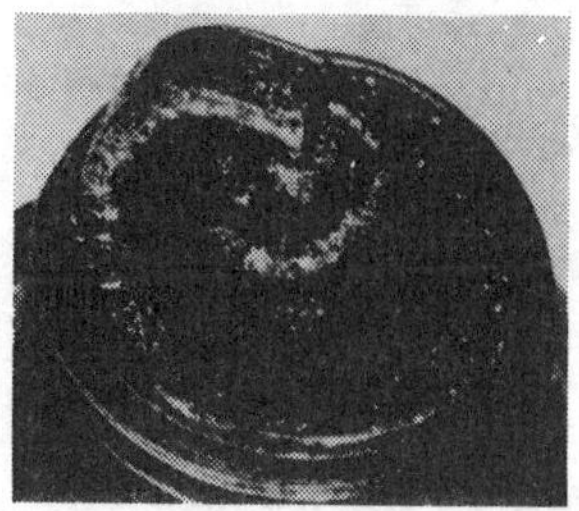

Fig. 5-17—Appearance of plug from two-stroke cycle engine indicating wet fouling. A damp or wet black carbon coating is formed over entire firing end. Could be caused by a too-cold plug, excessive idling, improper fuel-lubricating oil mixture or carburetor adjustment too rich.

manufacturers recommend that the spark plug be renewed rather than cleaned because of possible engine damage from cleaning abrasives.

After plug is cleaned by abrasive, and before gap is set, the electrode surfaces between the grounded and insulated electrodes should be cleaned and returned as nearly as possible to original shape by filing with a point file. Failure to properly dress the electrodes can result in high secondary voltage requirements, and misfire of the plug.

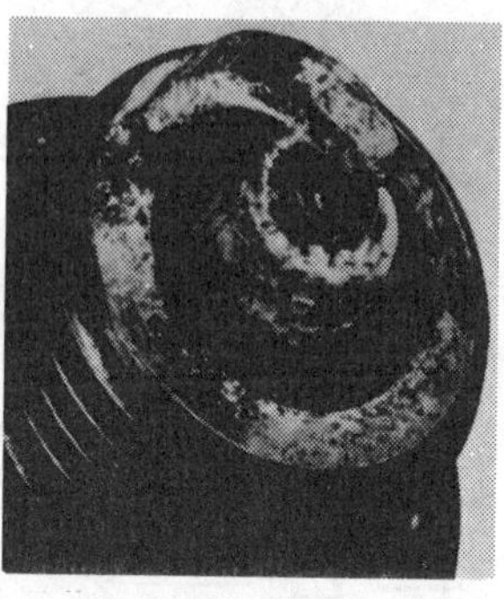

Fig. 5-18—Appearance of plug from two-stroke cycle engine indicating overheating. Insulator has gray or white blistered appearance and electrodes may be burned. Could be caused by use of a too-hot plug, carburetor adjustment too lean, "sticky" piston rings, engine overloaded, or cooling fins plugged causing engine to run too hot.

PLUG APPEARANCE DIAGNOSIS. The appearance of a spark plug will be altered by use, and an examination of the plug tip can contribute useful information which may assist in obtaining better spark plug life. It must be remembered that the contributing factors differ in two-stroke cycle and four-stroke cycle engine operations and although the appearance of two spark plugs may be similar, the corrective measures may depend on whether the engine is of two-stroke cycle or four-stroke cycle design. Figs. 5-12 through 5-18 are provided by Champion Spark Plug Company to illustrate typical observed conditions. Refer to Figs. 5-12 through 5-15 for four-stroke cycle engines and to Figs. 5-16 through 5-18 for two-stroke cycle engines. Listed in captions are the probable causes and suggested corrective measures.

CARBURETOR SERVICE

The bulk of carburetor service consists of cleaning, inspection and adjustment. After considerable service it may become necessary to overhaul the carburetor and renew worn parts to restore original operating efficiency. Although carburetor condition affects engine operating economy and power, ignition and engine compression must also be considered to determine and correct causes of poor performance.

Before dismantling carburetor for cleaning or overhaul, clean all external surfaces and remove accumulated dirt and grease. Refer to appropriate engine repair section for carburetor exploded or cross-sectional views. Dismantle carburetor and note any discrepancies to assure correction during overhaul. Thoroughly clean all parts and inspect for damage or wear. Wash jets and passages and blow clear with clean, dry compressed air. Do not use a drill or wire to clean jets as the possible enlargement of calibrated holes will disturb operating balance. The measurement of jets to determine the extent of wear is difficult and new parts are usually installed to assure satisfactory results.

Carburetor manufacturers provide for many of their models an assortment of gaskets and other parts usually needed to do a correct job of cleaning and overhaul. These assortments are usually catalogued as Gasket Kits and Overhaul Kits respectively.

On float type carburetors, inspect float pin and needle valve for wear and renew if necessary. Check metal floats for leaks and where a dual type float is installed, check alignment of float sections. Check cork floats for loss of protective coating and absorption of fuel.

NOTE: Do not attempt to recoat cork floats with shellac or varnish or to resolder leaky metal floats. Renew part if defective.

Check the fit of throttle and choke valve shafts. Excessive clearance will cause improper valve plate seating and will permit dust or grit to be drawn into the engine. Air leaks at throttle shaft bores due to wear will upset carburetor calibration and contribute to uneven engine operation. Rebush valve shaft holes where necessary and renew dust seals. If rebushing is not possible, renew the body part supporting the shaft. Inspect throttle and choke valve plates for proper installation and condition.

Power or idle adjustment needles must not be worn or grooved. Check condition of needle seal packing or "O" ring and renew packing or "O" ring if necessary.

Reinstall or renew jets, using correct size listed for specific model. Adjust power and idle settings as described for specific carburetors in engine service section of manual.

It is important that the carburetor bore at the idle discharge ports and in the vicinity of the throttle valve be free of deposits. A partially restricted idle port will produce a "flat spot" between idle and mid-range rpm. This is because the restriction makes it necessary to open the throttle wider than the designed opening to obtain proper idle speed. Opening the throttle wider than the design specified amount will uncover more of the port than was intended in the calibration of the carburetor. As a result an insufficient amount of the port will be available as a reserve to cover the transition period (idle to the mid-range rpm) when the high speed system begins to function.

Refer to Fig. 5-20 for service hints on diaphragm type carburetors.

When reassembling float type carburetors, be sure float position is properly adjusted. Refer to CARBURETOR paragraph in appropriate engine repair section for float level adjustment specifications.

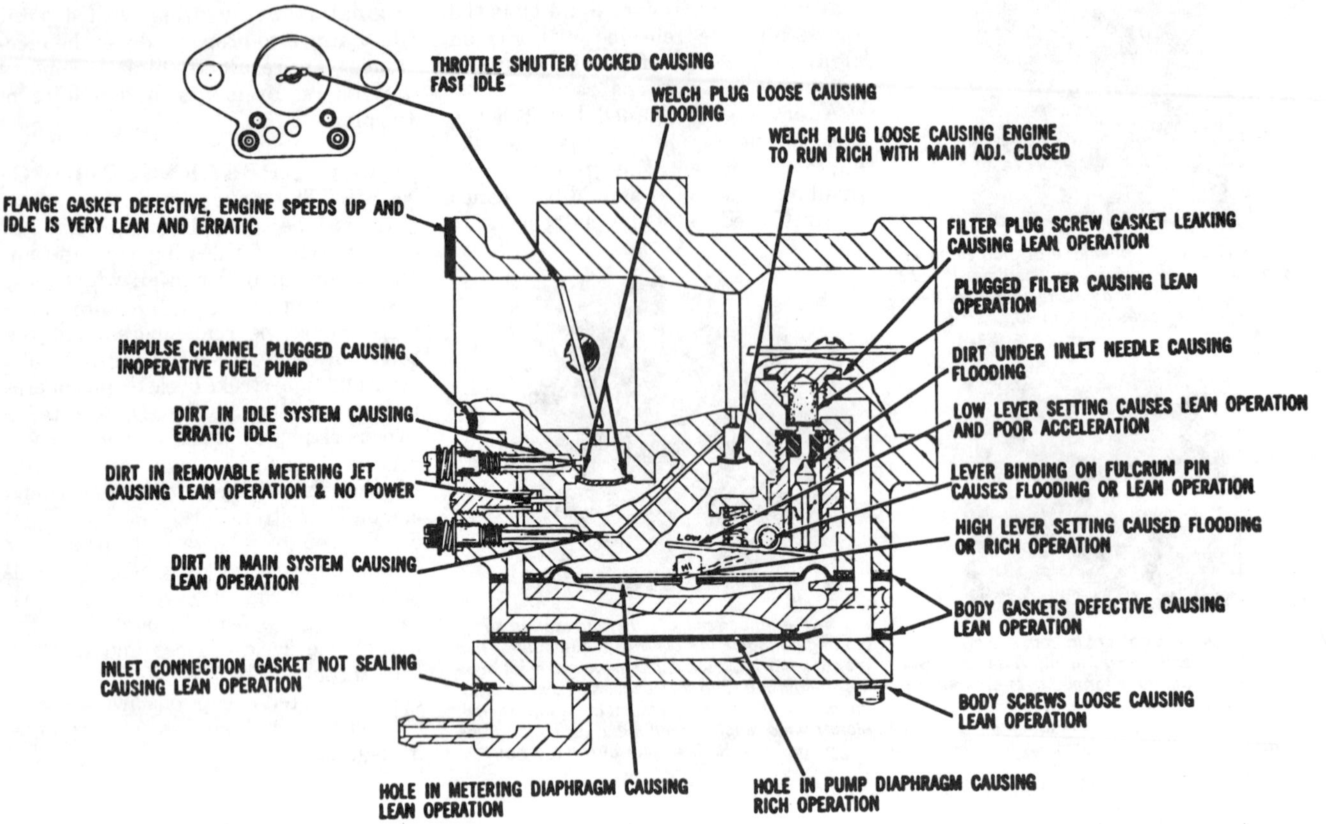

THROTTLE SHUTTER COCKED CAUSING FAST IDLE

WELCH PLUG LOOSE CAUSING FLOODING

WELCH PLUG LOOSE CAUSING ENGINE TO RUN RICH WITH MAIN ADJ. CLOSED

FLANGE GASKET DEFECTIVE, ENGINE SPEEDS UP AND IDLE IS VERY LEAN AND ERRATIC

FILTER PLUG SCREW GASKET LEAKING CAUSING LEAN OPERATION

PLUGGED FILTER CAUSING LEAN OPERATION

IMPULSE CHANNEL PLUGGED CAUSING INOPERATIVE FUEL PUMP

DIRT UNDER INLET NEEDLE CAUSING FLOODING

DIRT IN IDLE SYSTEM CAUSING ERRATIC IDLE

LOW LEVER SETTING CAUSES LEAN OPERATION AND POOR ACCELERATION

DIRT IN REMOVABLE METERING JET CAUSING LEAN OPERATION & NO POWER

LEVER BINDING ON FULCRUM PIN CAUSES FLOODING OR LEAN OPERATION

HIGH LEVER SETTING CAUSED FLOODING OR RICH OPERATION

DIRT IN MAIN SYSTEM CAUSING LEAN OPERATION

BODY GASKETS DEFECTIVE CAUSING LEAN OPERATION

INLET CONNECTION GASKET NOT SEALING CAUSING LEAN OPERATION

BODY SCREWS LOOSE CAUSING LEAN OPERATION

HOLE IN METERING DIAPHRAGM CAUSING LEAN OPERATION

HOLE IN PUMP DIAPHRAGM CAUSING RICH OPERATION

Fig. 5-20—Schematic cross-sectional view of a diaphragm type carburetor illustrating possible causes of malfunction. Refer to appropriate engine repair section for adjustment information and for exploded and/or cross-sectional view of actual carburetors used.

ENGINE SERVICE

DISASSEMBLY AND ASSEMBLY

Special techniques must be developed in repair of engines of aluminum alloy or magnesium alloy construction. Soft threads in aluminum or magnesium castings are often damaged by carelessness in overtightening fasteners or in attempting to loosen or remove seized fasteners. Manufacturer's recommended torque values for tightening screw fasteners should be followed closely.

NOTE: If damaged threads are encountered, refer to following paragraph, "REPAIRING DAMAGED THREADS."

A given amount of heat applied to aluminum or magnesium will cause it to expand a greater amount than will steel under similar conditions. Because of the different expansion characteristics, heat is usually recommended for easy installation of bearings, pins, etc., in aluminum or magnesium castings. Sometimes, heat can be used to free parts that are seized or where an interference fit is used. Heat, therefore, becomes a service tool and the application of heat is one of the required service techniques. An open flame is not usually advised because it destroys the paint and other protective coatings and because a uniform and controlled temperature with open flame is difficult to obtain. Methods commonly used are heating in oil or water, with a heat lamp, electric hot plate, or in an oven or kiln. See Fig. 5-21. The use of water or oil gives a fairly accurate temperature control but is somewhat limited as to the size and type of part than can be handled. Thermal crayons are available which can be used to determine the temperature of a heated part. These crayons melt when the part reaches a specified temperature, and a number of crayons for different temperatures are available. Temperature indicating crayons are usually available at welding equipment supply houses.

The crankcase and combustion chambers of a two-stroke cycle engine must be sealed against pressure and vacuum. To assure a perfect seal, nicks, scratches and warpage are to be avoided. Slight imperfections can be removed by using a fine-grit sandpaper. Flat surfaces can be lapped by using a surface plate or a smooth piece of plate glass, and a sheet of 120-grit sandpaper or lapping compound. Use a figure-eight motion with minimum pressure, and remove only enough

Fig. 5-21—In small engine repair, heat can be used efficiently as a disassembly and assembly tool. Heating crankcase halves on electric hot plate (above) will allow bearings to be easily removed.

metal to eliminate the imperfection. Bearing clearances, if any, must not be lessened by removing metal from the joint.

Use only the specified gaskets when re-assembling, and use an approved gasket cement or sealing compound unless the contrary is stated. Seal all exposed threads and repaint or retouch with an approved paint.

REPAIRING DAMAGED THREADS

Damaged threads in castings can be renewed by use of thread repair kits which are recommended by a number of equipment and engine manufacturers. Use of thread repair kits is not difficult, but instructions must be carefully followed. Refer to Figs. 5-22 through 5-25 which illustrate the use of Heli-Coil thread repair kits that are manufactured by the Heli-Coil Corporation, Danbury, Connecticut.

Heli-Coil thread repair kits are available through the parts departments of most engine and equipment manufacturers; the thread inserts are available in all National Coarse (USS) sizes from #8 to 1½ inch, National Fine (SAE) sizes from #10 to 1½ inch, Metric Coarse sizes from M3 to M20 and Metric Fine sizes form M8 to M20. Also, sizes for repairing M10, M12, M14, M18 and ⅞ inch spark plug ports are available.

Fig. 5-22—Damaged threads in casting before repair. Refer to Figs. 5-23, 5-24 and 5-25 for steps in installing thread insert. (Series of photos provided by Heli-Coil Corp., Danbury, Conn.)

Fig. 5-23—First step in repairing damaged threads is to drill out old threads using exact size drill recommended in instructions provided with thread repair kit. Drill all the way through an open hole or all the way to bottom of blind hole, making sure hole is straight and that centerline of hole is not moved in drilling process.

Fig. 5-24—Special drill taps are provided in thread repair kit for threading drilled hole to correct size for outside of thread insert. A standard tap cannot be used.

Fig. 5-25—A thread insert and a completed repair are shown above. Special tools are provided in thread repair kit for installation of thread insert.

VALVE SERVICE FUNDAMENTALS

(Four-Stroke Cycle Engines)

When overhauling small four-stroke cycle engines, obtaining proper valve sealing is of primary importance. The following paragraphs cover the fundamentals of servicing the intake and exhaust valves, valve seats and valve guides.

REMOVING AND INSTALLING VALVES. A valve spring compressor, one type of which is shown in Fig. 5-26, is a valuable aid in removing and installing the intake and exhaust valves. This tool is used to hold the spring compressed while removing or installing the pin, collars or retainer from the valve stem. Refer to Fig. 5-27 for views showing some of the different methods of retaining valve spring to valve stem.

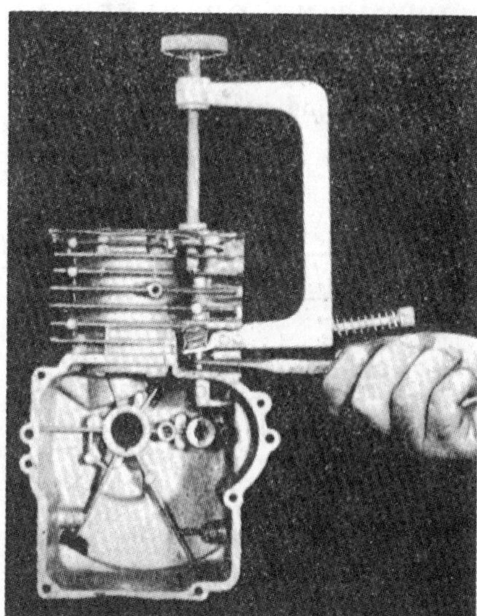

Fig. 5-26—View showing one type of valve spring compressor being used to remove keeper. (Block is cutaway to show valve spring.)

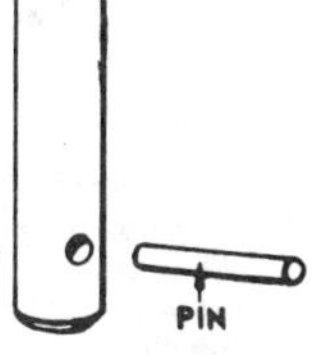

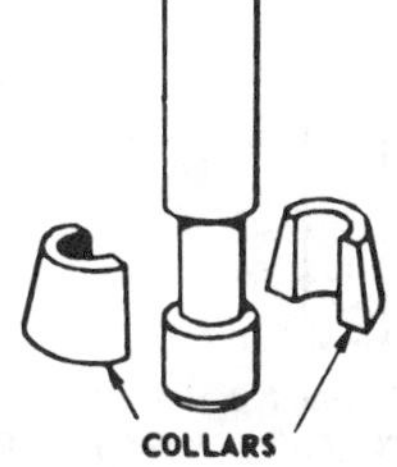

Fig. 5-27—Drawing showing three types of valve spring keepers used.

VALVE REFACING. If the valve face (See Fig. 5-28) is slightly worn, burned or pitted, the valve can usually be refaced providing proper equipment is available. Many small engine shops will usually renew the valves, however, rather to invest in somewhat costly valve refacing tools.

Before attempting to reface a valve, refer to specifications in appropriate engine repair section for valve face angle. On some engines, manufacturer recommends grinding the valve face to an angle of ½ to 1 degree less than that of the valve seat. Refer to Fig. 5-29. Also, nominal valve face angle may be either 30 or 45 degrees.

After valve is refaced, check thickness of valve "margin" (See Fig. 5-28). If margin is less than manufacturer's minimum specification (refer to specifications in appropriate engine repair section), or is less than one-half the margin of a new valve, renew the valve. Valves having excessive material removed in refacing operation will not give satisfactory service.

When refacing or renewing a valve, the seat should also be reconditioned, or in engines where valve seat is renewable, a new seat should be installed. Refer to following paragraph "RESEATING OR RENEWING VALVE SEATS." Then, the seating surfaces should be lapped in using a fine valve grinding compound.

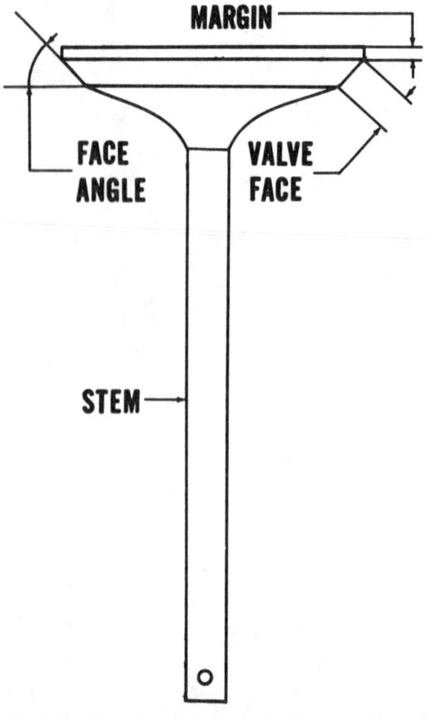

MARGIN
FACE ANGLE
VALVE FACE
STEM

Fig. 5-28—Drawing showing typical four-stroke cycle engine valve. Face angle is usually 30 or 45 degrees. On some engines, valve face is ground to an angle of ½ or 1 degree less than seat angle.

RESEATING OR RENEWING VALVE SEATS. On engines having the valve seat machined in the cylinder block casting, the seat can be reconditioned by using a correct angle seat grinding stone or valve seat cutter. When reconditioning valve seat, care should be taken that only enough material is removed to provide a good seating or valve contact surface. The width of the seat should then be measured (See Fig. 5-30) and if width exceeds manufacturer's maximum specifications, the seat should be narrowed by using one stone or cutter with an angle 15 degrees greater than valve seat angle and a second stone or cutter with an angle 15 degrees less than seat angle. When narrowing the seat, coat seat lightly with Prussian blue and check where seat contacts valve face by inserting valve in guide and rotating valve lightly against seat. Seat should contact approximate center of valve face. By using only the narrow angle seat narrowing stone or cutter, seat contact will be moved toward outer edge of valve face.

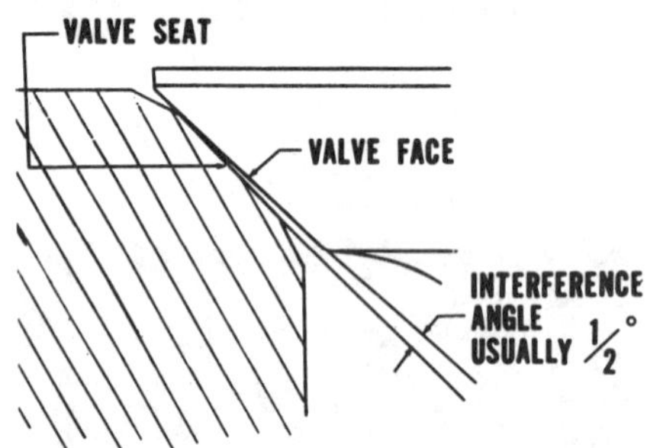

Fig. 5-29—Drawing showing line contact of valve face with valve seat when valve face is ground at smaller angle than valve seat; this is specified on some engines.

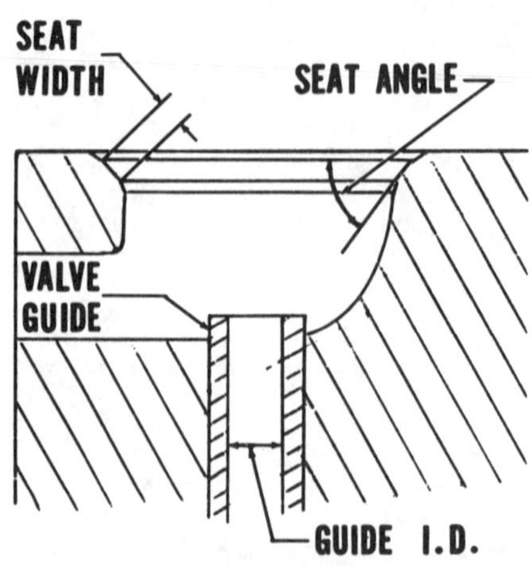

Fig. 5-30—Cross-sectional drawing of typical valve seat and valve guide as used on small engines. Valve guide may be integral part of cylinder block; on some models so constructed, valve guide ID may be reamed out and an oversize valve stem installed. On other models, a service guide may be installed after counter-boring cylinder block.

On engines having renewable valve seats, refer to appropriate engine repair section in this manual for recommended method of removing old seat and installing new seat. Refer to Fig. 5-31 for one method of installing new valve seats. Seats are retained in cylinder block bore by an interference fit; that is, seat is slightly larger than the bore in block. It sometimes occurs that the valve seat will become loose in the bore, especially on engines with aluminum crankcase. Some manufacturers provide oversize valve seat inserts (insert O.D. larger than standard part) so that if standard size insert fits loosely, the bore can be cut oversize and a new insert be tightly installed. After installing valve seat insert in engines of aluminum construction, the metal around the seat should be peened as shown in Fig. 5-32. Where a loose insert is encountered and an oversize insert is not available, the loose insert can usually be tightened by centerpunching the cylinder block material at three equally spaced points around the insert, then peening completely around the insert as shown in Fig. 5-32.

For some engines with cast iron cylinder blocks, a service valve seat insert is available for reconditioning the valve seat, and is installed by counterboring the cylinder block to specified dimensions, then driving insert into place. Refer to appropriate engine repair section in this manual for information on availability and installation of service valve seat inserts for cast iron engines.

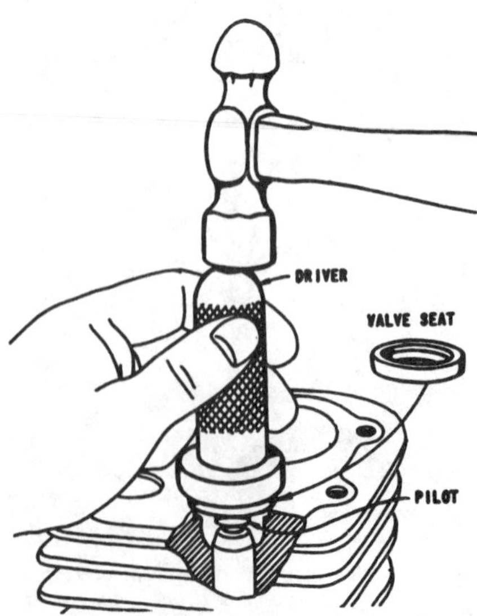

Fig. 5-31—View showing one method used to install valve seat insert. Refer to appropriate engine repair section for manufacturer's recommended method.

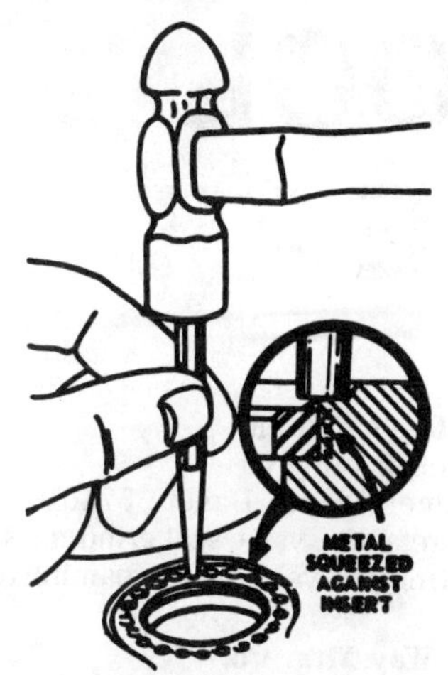

Fig. 5-32—It is usually recommended that on aluminum block engines, metal be peened around valve seat insert after insert is installed.

INSTALLING OVERSIZE PISTON AND RINGS

Some small engine manufacturers have oversize piston and ring sets available for use in repairing engines in which the cylinder bore is excessively worn and standard size piston and rings cannot be used. If care and approved procedure are used in oversizing the cylinder bore, installation of an oversize piston and ring set should result in a highly satisfactory overhaul.

The cylinder bore may be oversized by using either a boring bar or a hone; however, if a boring bar is used, it is usually recommended the cylinder bore be finished with a hone. Refer to Fig. 5-33.

Where oversize piston and rings are available, it will be noted in the appropriate engine repair section of this manual. Also, the standard bore diameter will be given. Before attempting to rebore or hone the cylinder to oversize, carefully measure the cylinder bore to be sure that standard size piston and rings will not fit within tolerance. Also, it may be possible that the cylinder is excessively worn or damaged and that reboring or honing to largest oversize will not clean up the worn or scored surface.

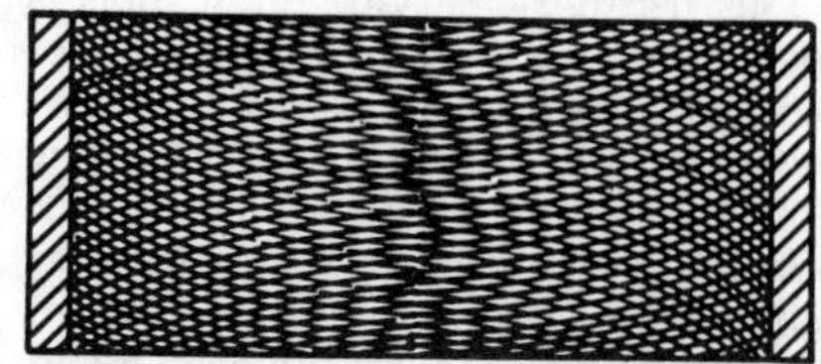

Fig. 5-33—A cross-hatch pattern as shown should be obtained when honing cylinder. Pattern is obtained by moving hone up and down cylinder bore as it is being turned by slow speed electric drill.

SERVICE SHOP TOOL BUYER'S GUIDE

This listing of Service Shop Tools is solely for the convenience of users of this manual and does not imply endorsement or approval by Intertec Publishing Corporation of the tools and equipment listed. The listing is in response to many requests for information on sources for purchasing special tools and equipment. Every attempt has been made to make the listing as complete as possible at time of publication and each entry is made from the latest material available.

Special engine service tools such as seal drivers, bearing drivers, etc., which are available from the engine manufacturer are not listed in this section of the manual. Where a special service tool is listed in the engine service section of this manual, the tool is available from the central parts or service distributors listed at the end of most engine service sections, or from the manufacturer.

NOTE TO MANUFACTURERS AND NATIONAL SALES DISTRIBUTORS OF ENGINE SERVICE TOOLS AND RELATED SERVICE EQUIPMENT. To obtain either a new listing for your products, or to change or add to an existing listing, write to Intertec Publishing Corporation, Book Division, P.O. Box 12901, Overland Park, Kansas 66212.

Engine Service Tools

Ammco Tools, Inc.
Wacker Park
North Chicago, Illinois 60064
Valve spring compressor, torque wrenches, cylinder hones, ridge reamers, piston ring compressors, piston ring expanders.

Black & Decker Mfg. Co.
701 East Joppa Road
Towson, Maryland 21204
Valve grinding equipment.

Bloom, Inc.
Route Four, Hiway 20 West
Independence, Iowa 50644
Engine repair stand with crankshaft straightening attachment.

Brush Research Mfg. Inc.
4642 East Floral Drive
Los Angeles, California 90022
Cylinder hones.

E-Z Lok
P.O. Box 2069
Gardena, California 90247
Thread repair insert kits for metal, wood and plastic.

Foley-Belsaw Company
Outdoor Power Equipment Parts Division
6301 Equitable Road
P.O. Box 419593
Kansas City, Missouri 64141
Crankshaft straightener and repair stand, valve refacers, valve seat grinders, parts washers, cylinder hones, gages and ridge reamers, piston ring expanders and compressor, flywheel pullers, torque wrenches, rotary mower blade balancers.

Frederick Mfg. Co., Inc.
1400 C., Agnes Avenue
Kansas City, Missouri 64127
Crankshaft straightener.

Heli-Coil Products Division
Heli-Coil Corporation
Shelter Rock Lane
Danbury Connecticut 06810
Thread repair kits, thread inserts, installation tools.

K-D Tools
3575 Hempland Road
Lancaster, Pennsylvania 17604
Thread repair kits, valve spring compressors, reamers, micrometers, dial indicators, calipers.

Keystone Reamer & Tool Co.
Post Office Box 310
Millersburg, Pennsylvania 17061
Valve seat cutter and pilots, adjustable reamers.

Ki-Sol Corporation
100 Larkin Williams Ind. Court
Fenton, Missouri 63026
Cylinder hone, ridge reamer, ring compressor, ring expander, ring groove cleaner, torque wrenches, valve spring compressor, valve refacing equipment.

K-Line Industries, Inc.
315 Garden Avenue
Holland, Michigan 49423
Cylinder hone, ridge reamer, ring compressor, valve guide tools, valve spring compressor, reamers.

K.O. Lee Company
101 South Congress
Aberdeen, South Dakota 57401
Valve refacing, valve seat grinding, valve reseating and valve guide reaming tools.

Kwik-Way Mfg. Co.
500 57th Street
Marion, Iowa 52302
Cylinder boring equipment, valve facing equipment, valve seat grinding equipment.

Lisle Corporation
807 East Main
Clarinda, Iowa 51632
Cylinder hones, ridge reamers, ring compressors, valve spring compressors.

Microdot, Inc.
P.O. Box 3001
800 So. State College Blvd.
Fullerton, California 92631
Thread repair insert kits.

Mighty Midget Mfg. Co.,
Div. of Kansas City Screw Thread Co.
2908 E. Truman Road
Kansas City, Missouri 64127
Crankshaft straightener.

Neway Manufacturing, Inc.
1013 No. Shiawassee
Corunna, Michigan 48817
Valve seat cutters.

Owatonna Tool Company
436 Eisenhower Drive
Owatonna, Minnesota 55060
Valve tools, spark plug tools, piston ring tools, cylinder hones.

Power Lawnmower Parts Inc.
1920 Lyell Avenue
P.O. Box 60860
Rochester, NY 14606-0860
Gasket cutter tool, gasket scraper tool, crankshaft cleaning tool, ridge reamer, valve spring compressor, valve seat cutters, thread repair kits, valve lifter, piston ring expander.

Precision Manufacturing
& Sales Co., Inc.
2140 Range Rd., P.O. Box 149
Clearwater, Florida 33517
Cylinder boring equipment, measuring instruments, valve equipment, hones,

porting, hand tools, test equipment, threading, presses, parts washers, milling machines, lathes, drill presses, glass beading machines, dynos, safety equipment.

Sunnen Product Company
7910 Manchester Avenue
Saint Louis, Missouri 63143
Cylinder hones, rod reconditioning, valve guide reconditioning.

Rexnord Specialty Fastener Division
3000 W. Lomita Blvd.
Torrance, California 90505
Thread repair insert kits (Keenserts) and installation tools.

Vulcan Tools
Div. of TRW, Inc.
2300 Kenmore Avenue
Buffalo, New York 14207
Cylinder hones, reamers, ridge removers, valve spring compressors, valve spring testers, ring compressor, ring groove cleaner.

Waters-Dove Manufacturing
Post Office Box 40
Skiatook, Oklahoma 74070
Crankshaft straightener, oil seal remover.

Test Equipment and Gages

Allen Test Products
2101 North Pitcher Street
Kalamazoo, Michigan 49007
Coil and condenser testers, compression gages.

Applied Power, Inc.,
Auto Division
P.O. Box 27207
Milwaukee, Wisconsin 53227
Compression gage, condenser tester, tachometer, timing light, ignition analyzer.

AW Dynamometer, Inc.
131-1/2 East Main Street
Colfax, Illinois 61728
Engine test dynamometer.

B.C. Ames Company
131 Lexington
Waltham, Massachusetts 02254
Micrometer dial gages and indicators.

The Bendix Corporation
Engine Products Division
Delaware Avenue
Sidney, New York 13838
Condenser tester, magneto test equipment, timing light.

Burco
208 Delaware Avenue
Delmar, New York 12054
Coil and condenser tester, compression gage, carburetor tester, grinders, Pow-R-Arms, chain breakers, rivet spinners.

Dixson, Inc.
Post Office Box 1449
Grand Junction, Colorado 81501
Tachometer, compression gage, timing light.

Foley-Belsaw Company
Outdoor Power Equipment
Parts Division
6301 Equitable Road
P.O. Box 419593
Kansas City, Missouri 64141
Cylinder gage, amp/volt testers, condenser and coil tester, magneto tester, ignition testers for conventional and solid-state systems, tachometers, spark testers, compression gages, timing lights and gages, micrometers and calipers, torque wrenches, carburetor testers, vacuum gages.

Fox Valley Instrument Co.
Route 5, Box 390
Cheboygan, Michigan 49721
Coil and condenser tester, ignition actuated tachometer.

Graham-Lee Electronics, Inc.
4200 Center Avenue NE
Minneapolis, Minnesota 55421
Coil and condenser tester.

K-D Tools
3575 Hempland Road
Lancaster, Pennsylvania 17604
Diode tester and installation tools, compression gage, timing light, timing gages.

Ki-Sol Corporation
100 Larkin Williams Ind. Court
Fenton, Missouri 63026
Micrometers, telescoping gages, compression gages, cylinder gages.

K-Line Industries, Inc.
315 Garden Avenue
Holland, Michigan 49423
Compression gage, tachometers.

Merc-O-Tronic Instruments Corporation
215 Branch Street
Almont, Michigan 48003
Ignition analyzers for conventional solid-state and magneto systems, electric tachometers, electronic tachometer and dwell meter, power timing lights, ohmmeters, compression gages, mechanical timing devices.

Owatonna Tool Company
436 Eisenhower Drive
Owatonna, Minnesota 55060
Feeler gages, hydraulic test gages.

Power Lawnmower Parts Inc.
1920 Lyell Avenue
P.O. Box 60860
Rochester, NY 14606-0860
Condenser and coil tester, compression gage, flywheel magneto tester, ignition tester.

Prestolite Electronics Div.
An Allied Company
Post Office Box 931
Toledo, Ohio 43694
Magneto test plug.

Simpson Electric Company
853 Dundee Avenue
Elgin, Illinois 60120
Electrical and electronic test equipment.

L.S. Starrett Company
1-165 Crescent Street
Athol, Massachusetts 01331
Micrometers, dial gages, bore gages, feeler gages.

Stevens Instrument Company, Inc.
Post Office Box 193
Waukegan, Illinois 60085
Ignition analyzers, timing lights, tachometers, volt-ohmmeters, spark checkers, CD ignition testers.

Stewart-Warner Corporation
1826 Diversey Parkway
Chicago, Illinois 60614
Compression gage, ignition tachometer, timing light ignition analyzer.

P.A. Sturtevant Co.,
Division Dresser Ind.
3201 North Wolf Road
Franklin Park, Illinois 60131
Torque wrenches, torque multipliers, torque analyzers.

Sun Electric Corporation
Instrument Products Div.
1560 Trimble Road
San Jose, California 95131
Compression gage, hydraulic test gages, coil and condenser tester, ignition tachometer, ignition analyzer.

Westberg Mfg. Inc.
3400 Westach Way
Sonoma, California 95476
Ignition tachometer, magneto test equipment, ignition system analyzer.

Shop Tools and Equipment

A-C Delco Division,
General Motors Corp.
400 Renaissance Center
Detroit, Michigan 48243
Spark plug cleaning and test equipment.

Applied Power, Inc.,
Auto. Division
Box 27207
Milwaukee, Wisconsin 53227
Arc and gas welding equipment, welding rods, accessories, battery service equipment.

Black & Decker Mfg. Co.
701 East Joppa Road
Towson, Maryland 21204
Air and electric powered tools.

Bloom, Inc.
Route 4, Hiway 20 West
Independence, Iowa 50644
Lawn mower repair bench with built-in engine cranking mechanism.

Campbell Chain Division
McGraw-Edison Co.
Post Office Box 3056
York, Pennsylvania 17402
Chain type engine and utility slings.

Champion Pneumatic Machinery Co.
1301 No. Euclid Avenue
Princeton, Illinois 61356
Air compressors.

Champion Spark Plug Co.
Post Office Box 910
Toledo, Ohio 43661
Spark plug cleaning and testing equipment, gap tools and wrenches.

Chicago Pneumatic Tool Co.
2200 Bleecker Street
Utica, New York 13503
Air impact wrenches, air hammers, air drills and grinders, nut runners, speed ratchets.

Clayton Manufacturing Company
415 North Temple City Boulevard
El Monte, California 91731
Steam cleaning equipment.

E-Z Lok
P.O. Box 2069
240 E. Rosecrans Avenue
Gardena, California 90247
Thread repair insert kits for metal, wood and plastic.

Foley-Belsaw Company
Outdoor Power Equipment
Parts Division
6301 Equitable Road
P.O. Box 419593
Kansas City, Missouri 64141
Torque wrenches, parts washers, micrometers and calipers.

G & H Products, Inc.
Post Office Box 770
St. Paris, Ohio 43027
Motorized lawn mower stand.

General Scientific Equipment Company
Limekiln Pike & Williams Avenue
Box 27309
Philadelphia, Pennsylvania 19118-0255
Safety equipment.

Graymills Corporation
3705 North Lincoln Avenue
Chicago, Illinois 60613
Parts washing stand.

Heli-Coil Products Div.,
Heli-Coil Corp.
Shelter Rock Lane
Danbury, Connecticut 06810
Thread repair kits, thread inserts and installation tools.

Ingersoll-Rand
253 E. Washington Avenue
Washington, New Jersey 07882
Air and electric impact wrenches, electric drills and screwdrivers.

Ingersoll-Rand Co.
Deerfield Industrial Park
South Deerfield, Massachusetts 01373
Impact wrenches, portable electric tools, battery powered ratchet wrench.

Jaw Manufacturing Co.
39 Mulberry Street, P.O. Box 213
Reading, Pennsylvania 19603
Files for renewal of damaged threads, hex nut size rethreader dies, flexible shaft drivers and extensions, screw extractors, impact drivers.

Jenny Division
of Homestead Ind., Inc.
Box 348
Carapolis, Pennsylvania 15108-0348
Steam cleaning equipment, pressure washing equipment.

Keystone Reamer & Tool Co.
Post Office Box 310
Millersburg, Pennsylvania 17061
Adjustable reamers, twist drills, tape, dies, etc.

K-Line Industries, Inc.
315 Garden Avenue
Holland, Michigan 49423
Air and electric impact wrenches.

Microdot, Inc.
P.O. Box 3001
800 So. State College Blvd.
Fullerton, California 92631
Thread repair insert kits.

Owatonna Tool Company
436 Eisenhower Drive
Owatonna, Minnesota 55060
Bearing and gear pullers, hydraulic shop presses.

Power Lawnmower Parts Inc.
1920 Lyell Avenue
P.O. Box 60860
Rochester, NY 14606-0860
Flywheel puller, starter wrench and flywheel holder, blade sharpener, chain breakers, rivet spinners, gear pullers.

Pronto Tool Division,
Ingersoll-Rand
2600 East Nutwood Avenue
Fullerton, California 92631
Torque wrenches, gear and bearing pullers.

Shure Manufacturing Corp.
1601 South Hanley Road
Saint Louis, Missouri 63144
Steel shop benches, desks, engine overhaul stand.

Sioux Tools, Inc.
2801-2999 Floyd Blvd.
Sioux City, Iowa 51102
Portable air and electric tools.

Rexnord Specialty
Fastener Division
3000 W. Lomita Blvd.
Torrance, California 90505
Thread repair insert kits (Keenserts) and installation tools.

Vulcan Tools
2300 Kenmore Avenue
Buffalo, New York 14207
Air and electric impact wrenches.

Sharpening and Maintenance Equipment for Small Engine Powered Implements

Bell Industries,
Saw & Machine Division
Post Office Box 2510
Eugene, Oregon 97402
Saw chain grinder.

Desa Industries, Inc.
25000 South Western Ave.
Park Forest, Illinois 60466
Saw chain breakers and rivet spinners, chain files, file holder and filing guide.

Foley-Belsaw Company
Outdoor Power Equipment
Parts Division
6301 Equitable Road
P.O. Box 419593
Kansas City, Missouri 64141
Circular, band and hand saw filers, heavy duty grinders, saw setters, retoothers, circular saw vises, lawn mower sharpener, saw chain sharpening and repair equipment.

Granberg Industries
200 S. Garrard Blvd.

Richmond, California 94804
Saw chain grinder, file guides, chain breakers and rivet spinners, chain saw lumber cutting attachments.

Ki-Sol Corporation
100 Larkin Williams Ind. Court
Fenton, Missouri 63026
Mower blade balancer.

Magna-Matic Div.,
A.J. Karrels Co.
Box 348
Port Washington, Wisconsin 53074
Rotary mower blade balancer, "track" checking tool.

Omark Industries, Inc.
4909 International Way
Portland, Oregon 97222
Saw chain and saw bars, maintenance equipment, to include file holders, filing vises, rivet spinners, chain breakers, filing depth gages, bar groove gages, electric chain saw sharpeners.

Power Lawnmower Parts Inc.
1920 Lyell Avenue
P.O. Box 60860
Rochester, NY 14606-0860
Grinding wheels, parts cleaning brush, tube repair kits, blade balancer.

S.I.P. Grinding Machines
American Marsh Pumps
P.O. Box 23038
722 Porter Street
Lansing, Michigan 48909
Lawn mower sharpeners, saw chain grinders, lapping machine.

Specialty Motors Mfg.
641 California Way,
P.O. Box 157
Longview, Washington 98632
Chain saw bar rebuilding equipment.

Mechanic's Hand Tools

Channellock, Inc.
1306 South Main Street
Meadville, Pennsylvania 16335

John H. Graham & Company, Inc.
617 Oradell Avenue
Oradell, New Jersey 07649

Jaw Manufacturing Company
39 Mulberry Street
Reading, Pennsylvania 19603

K-D Tools
3575 Hempland Road
Lancaster, Pennsylvania 17604

K-Line Industries, Inc.
315 Garden Avenue
Holland, Michigan 49423

Millers Falls Division
Ingersoll-Rand
Deerfield Industrial Park
South Deerfield, Massachusetts 01373

New Britain Tool Company
Division of Litton Industrial Products
P.O. Box 12198
Research Triangle Park, N.C. 27709

Owatonna Tool Company
436 Eisenhower Drive
Owatonna, Minnesota 55060

Power Lawnmower Parts Inc.
1920 Lyell Avenue
P.O. Box 60860
Rochester, NY 14606-0860
Crankshaft wrench, spark plug wrench, hose clamp pliers.

Proto Tool Division
Ingersoll-Rand
2600 East Nutwood Avenue
Fullerton, California 92631

Snap-On Tools
2801 80th Street
Kenosha, Wisconsin 53140

Triangle Corporation—Tool Division
Cameron Road
Orangeburg, South Carolina 29115

Vulcan Tools
Division of TRW, Inc.
2300 Kenmore Avenue
Buffalo, New York 14207

J.H. Williams
Division of TRW, Inc.
400 Vulcan Street
Buffalo, New York 14207

Shop Supplies (Chemicals, Metallurgy Products, Seals, Sealers, Common Parts Items, etc.)

ABEX Corp.
Amsco Welding Products
Fulton Industrial Park
3-9610-14
P.O. Box 258
Wauseon, Ohio 43567
Hardfacing alloys.

Atlas Tool & Manufacturing Co.
7100 S. Grand Avenue
Saint Louis, Missouri 63111
Rotary mower blades.

Bendix Automotive Aftermarket
1094 Bendix Drive, Box 1632
Jackson, Tennessee 38301
Cleaning chemicals.

CR Industries
900 North State Street
Elgin, Illinois 60120
Shaft and bearing seals.

Clayton Manufacturing Company
415 North Temple City Blvd.
El Monte, California 91731
Steam cleaning compounds and solvents.

E-Z Lok
P.O. Box 2069,
240 E. Rosecrans Ave.
Gardena, California 90247
Thread repair insert kits for metal, wood and plastic.

Eutectic Welding Alloys Corp.
40-40 172nd Street
Flushing, New York 11358
Specialized repair and maintenance welding alloys.

Foley-Belsaw Company
Outdoor Power Equipment
Parts Division
6301 Equitable Road
P.O. Box 419593
Kansas City, Missouri 64141
Parts washers, cylinder head rethreaders, micrometers, calipers, gasket material, nylon rope, universal lawnmower blades, oil and grease products, bolt, nut, washer and spring assortments.

Frederick Manufacturing Co., Inc.
1400 C Agnes Street
Kansas City, Missouri 64127
Throttle controls and parts.

Heli-Coil Products Division
Heli-Coil Corporation
Shelter Rock Lane
Danbury, Connecticut 06810
Thread repair kits, thread inserts and installation tools.

King Cotton Cordage
617 Oradell Avenue
Oradell, New Jersey 07649
Starter rope, nylon rod.

Loctite Corporation
705 North Mountain Road
Newington, Connecticut 06111
Threading locking compounds, bearing mounting compounds, retaining compounds and sealants, instant and structural adhesives.

McCord Gasket Division
Ex-Cell-O Corporation
2850 West Grand Boulevard
Detroit, Michigan 48202
Gaskets, seals.

Microdot, Inc.
P.O. Box 3001
800 So. State College Blvd.
Fullerton, California 92631
Thread repair insert kits.

Permatex Industrial
705 N. Mountain Road
Newington, Connecticut 06111
Cleaning chemicals, gasket sealers, pipe sealants, adhesives, lubricants.

Power Lawnmower Parts Inc.
1920 Lyell Avenue
P.O. Box 60860
Rochester, NY 14606-0860
Grinding compound paste, gas tank sealer stick, shop aprons, rotary mower blades, oil seals, gaskets, solder, throttle controls and parts, nylon starter rope.

Radiator Specialty Co.
Box 34689
Charlotte, North Carolina 28234
Cleaning chemicals (Gunk), and solder seal.

Rexnord Specialty Fastener Div.
3000 W. Lomita Blvd.
Torrance, California 90506
Thread repair insert kits (Keenserts) and installation tools.

Union Carbide Corporation Home & Auto Products Division
Old Ridgebury Road
Danbury, CT 06817
Cleaning chemicals, engine starting fluid.

ACME

ACME NORTH AMERICA CORP.
5209 W. 73rd St.
Minneapolis, Minnesota 55435

Model	Bore	Stroke	Displacement
A 180 B, A 180 P	65 mm (2.56 in.)	54 mm (2.13 in.)	179 cc (10.92 cu. in.)
ALN 215 WB, ALN 215 WP	65 mm (2.56 in.)	65 mm (2.56 in.)	215 cc (13.12 cu. in.)
A 220 B, A 220 P	72 mm (2.83 in.)	54 mm (2.13 in.)	220 cc (13.43 cu. in.)

ENGINE IDENTIFICATION

All models are four-stroke, air-cooled, single-cylinder engines.

All models are horizontal crankshaft with intake and exhaust valves located in the cylinder block.

Models with the suffix "B" are gasoline fuel engines and models with the suffix "P" are kerosene fuel engines.

Engine model number is located on a plate mounted on the right side of the shroud as viewed from flywheel side. The engine serial number is stamped into the engine block approximately 3 inches (76.2 mm) below model number plate, just above oil fill plug (Fig. AC1).

Always furnish engine model and serial number when ordering parts.

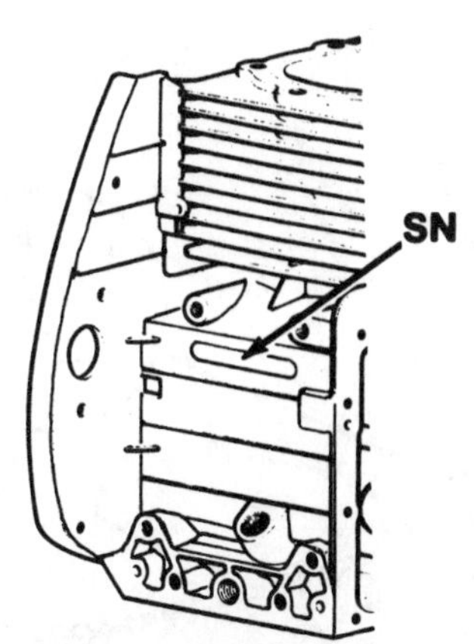

Fig. AC1—Engine serial number (SN) is stamped into engine block.

MAINTENANCE

SPARK PLUG. Recommended spark plug for Models A 180 B, A 180 P, A 220 B and A 220 P is a Bosch RO10846 or equivalent. Recommended spark plug for Model ALN 215 WB is a Bosch W95T1 or equivalent. Recommended spark plug for Model ALN 215 WP is a Bosch W45T1 or equivalent.

On all models, spark plug should be removed, cleaned and inspected after every 100 hours of operation. Spark plug electrode gap should be 0.6-0.8 mm (0.024-0.032 in.) for all models.

CARBURETOR. Models ALN 215 WB and ALN 215 WP are equipped with the float type carburetor shown in Fig. AC2. Models A 180 B, A 180 P, A 220 B and A 220 P are equipped with the float type carburetor shown in Fig. AC4. Refer to the appropriate following paragraphs for model being serviced.

Models ALN 215 WB and ALN 215 WP. Initial adjustment of idle speed mixture screw (3—Fig. AC2) from a lightly seated position is 1-1/4 turns open. Main fuel mixture is controlled by fixed main jet (13). Fixed main jet size is #75 for Model ALN 215 WB and #80 for Model ALN 215 WP. Idle jet (7) size is #50 for both models.

Final adjustments are made with engine at operating temperature and running. Adjust engine idle speed to 1100 rpm at throttle stop screw (5). Adjust idle mixture screw (3) to obtain smoothest engine idle and smooth acceleration.

When installing new fuel inlet needle and seat, install seat and measure from point (A) to edge of fuel inlet needle seat. Measurement should be 33-37 mm (1.30-1.46 in.). Vary number of fiber washers between fuel inlet seat and carburetor body to obtain correct measurement.

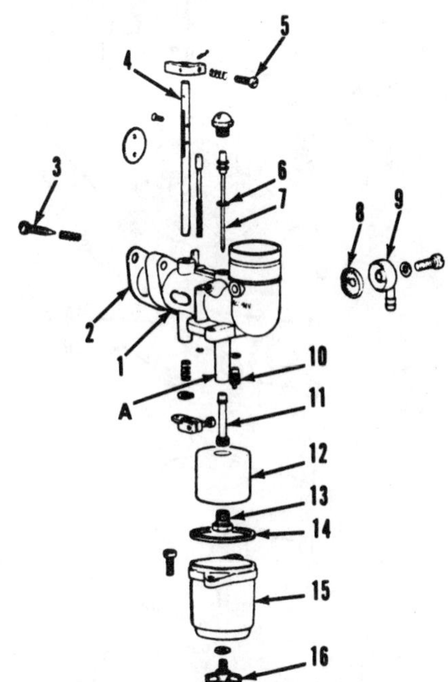

Fig. AC2—Exploded view of carburetor used on Models ALN 215 WB and ALN 215 WP.

1. Carburetor body
2. Gasket
3. Idle mixture screw
4. Throttle shaft
5. Throttle stop screw
6. Gasket
7. Idle jet
8. Screen (filter)
9. Filter housing
10. Fuel inlet needle valve
11. Nozzle
12. Float
13. Main jet
14. Gasket
15. Float bowl
16. Wing bolt

To check carburetor float (12) level, carefully remove float bowl (15) and measure distance from top of bowl (E—Fig. AC3) to fuel level (F) in bowl. Measurement should be 32-34 mm (1.26-1.34 in.). If float level is incorrect, fuel inlet needle seat must be shimmed as outlined in previous paragraph. Float weight should be 16.5 grams (0.6 oz.). If float is heavier than specified, it must be renewed.

Fuel filter screen (8—Fig. AC2) should be removed and cleaned after every 50 hours of operation.

Models A 180 B, A 180 P, A 220 B and A 220 P. Initial adjustment of idle speed mixture screw (3—Fig. AC4) from a lightly seated position is 1-1/2 turns open. Main fuel mixture is controlled by fixed main jet (13). Fixed main jet size is #95 for Model A 180 B, #98 for Models A 180 P and A 220 B and #100 for Model A 220 P. Idle jet (7) size for all models is #35.

Final adjustments are made with engine at operating temperature and running. Adjust engine idle speed to 1000-1100 rpm at throttle stop screw (5). Adjust idle mixture screw (3) to obtain smoothest engine idle and smooth acceleration.

To check float level, remove carburetor float bowl (15). Invert carburetor body (1). Float height should be 15 mm (19/32 in.) measured from carburetor body gasket surface to bottom of float (12). Float weight should be 8 grams (0.29 oz.). If float is heavier than specified, it must be renewed.

Fuel filter screen (8) should be removed and cleaned after every 50 hours of operation.

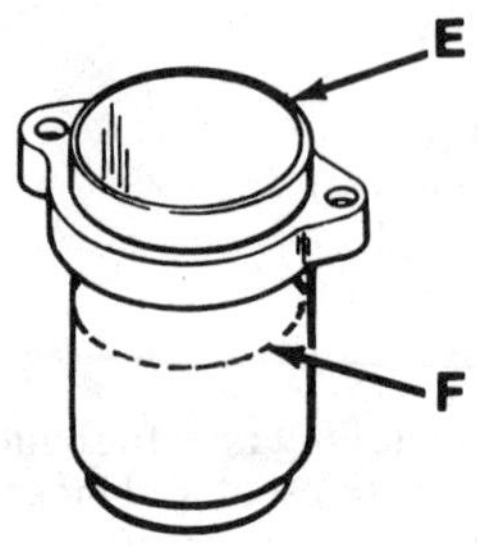

Fig. AC3—Measure distance from fuel level (F) to top of bowl (E). Distance should be 32-34 mm (1.26-1.34 in.).

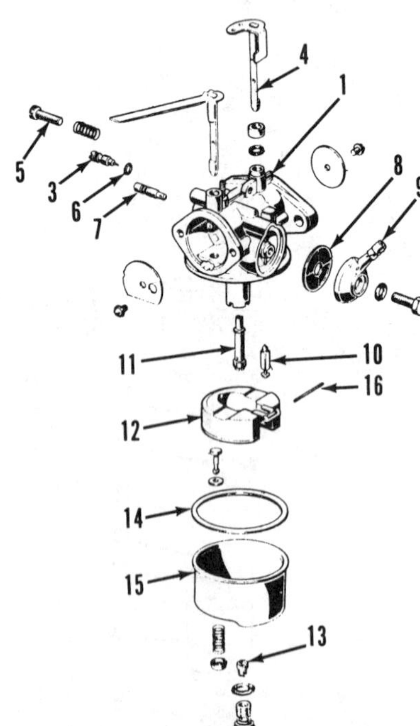

Fig. AC4—Exploded view of carburetor used on Models A 180 B, A 180 P, A 220 B and A 220 P.

1. Carburetor body
3. Idle mixture screw
4. Throttle shaft
5. Throttle stop screw
6. Gasket
7. Idle jet
8. Screen (filter)
9. Filter housing
10. Fuel inlet needle valve
11. Nozzle
12. Float
13. Main jet
14. Gasket
15. Float bowl
16. Float pin

AIR FILTER. Models ALN 215 WB and ALN 215 WP are equipped with an oil bath type air filter (Fig. AC5). Air filter element should be removed and cleaned after every 8 hours of operation. Discard old oil and refill with new engine oil to level indicated on oil reservoir (5).

Models A 180 B, A 180 P, A 220 B and A 220 P are equipped with a paper element type air filter. Filter should be checked daily. After removing element, clean element by gently tapping element to dislodge dust and dirt or use very low air pressure and blow from the inside of the air filter element toward the outside. Element should be renewed after 100 hours of operation when engine is operated under normal conditions. Shorten element renewal intervals when engine is operated under adverse conditions.

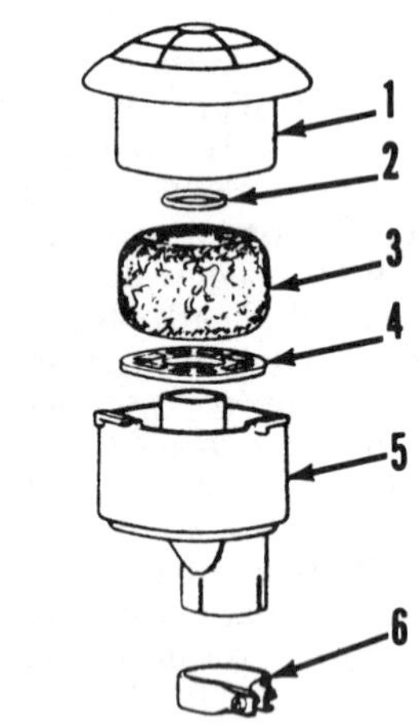

Fig. AC5—Oil bath air filter is used on Models ALN 215 WB and ALN 216 WP.

1. Cover
2. Gasket
3. Element
4. Plate
5. Oil reservoir
6. Clamp

GOVERNOR. Models ALN 215 WB and ALN 215 WP are equipped with a flyweight ball type centrifugal governor with governor assembly incorporated on camshaft gear. Models A 180 B, A 180 P, A 220 B, and A 220 P are equipped with a flyweight type governor. Governor gear and flyweight assembly is located on the crankcase cover and is driven by the crankshaft gear. On all models, the governor regulates engine speed via external linkage.

To adjust external linkage on Models ALN 215 WB and ALN 215 WP, first make certain all linkage moves freely with no binding or loose connections. Lock throttle control lever (5—Fig. AC6) in midposition with throttle lever locknut (6). Hook tension spring (3) in hole A for 3000 rpm engine speed, hole B for 3600 rpm engine speed and for special applications only, hole C for 4000 rpm engine speed. An alternate spring (part 551.107) is available. When tension spring 551.107 is hooked in Hole A, engine speed is governed at 2400 rpm. Place throttle control lever (5) in full speed position. Loosen carburetor-to-governor lever rod adjustment lock (7)

Fig. AC6—View showing relative position and relationship of external governor linkage and throttle parts on Models ALN 215 WB and ALN 215 WP.

1. Carburetor-to-governor lever rod
2. Governor lever
3. Tension spring
4. Spring
5. Throttle control lever
6. Throttle lever locknut
7. Carburetor-to-governor lever rod adjustment lock

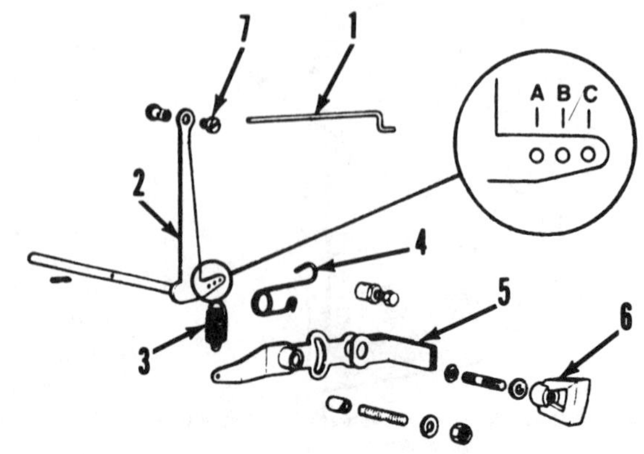

and push throttle fully open. With throttle held open, tighten rod adjustment lock (7). Governor and throttle linkage should work freely with no binding through entire operation range.

To adjust external linkage on Models A 180 B, A 180 P, A 220 B and A 220 P, first make certain all linkage moves freely with no binding or loose connections. Place throttle control lever (8—Fig. AC7) in full throttle position. Loosen clamp bolt (1). Insert screwdriver into slot of governor shaft (2) and rotate shaft clockwise as far as possible. Tighten clamp bolt (1). Place throttle control lever (8) in idle position, then start and run engine. Engine idle should be adjusted to 1000-1100 rpm. Place throttle control lever (8) in full throttle position and adjust throttle stop screw (9) to obtain 2400, 3000 or 3600 rpm. Do not exceed maximum engine speed of 3600 rpm.

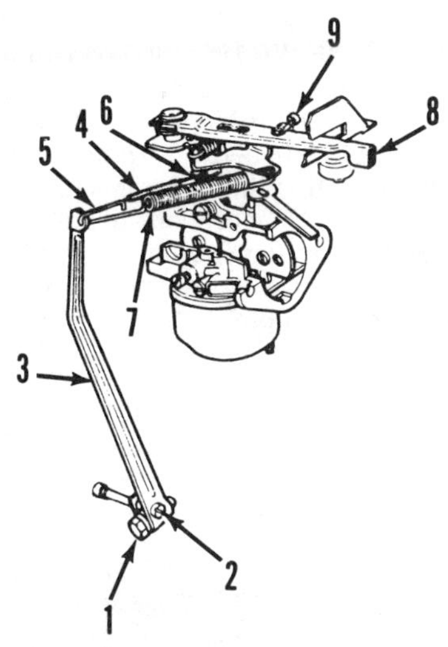

Fig. AC7—View showing relative position and relationship of external governor linkage and throttle parts on Models A 180 B, A 180 P, A 220 B and A 220 P.

1. Clamp bolt
2. Governor shaft
3. Governor lever
4. Spring
5. Governor lever-to-carburetor rod
6. Throttle lever
7. Spring
8. Throttle control lever
9. Throttle stop screw

IGNITION SYSTEM. A breaker point ignition system is standard on Models ALN 215 WB and ALN 215 WP and a solid-state electronic ignition system is standard on Models A 180 B, A 180 P, A 220 B and A 220 P.

Models ALN 215 WB and ALN 215 WP. The breaker-point set and condenser are located on the left-hand side of crankcase as viewed from flywheel side. Ignition coil is located behind flywheel. Breaker points are actuated by plunger (1—Fig. AC8).

Point gap should be checked and adjusted after every 400 hours of operation. To check point gap, remove cover (5) and use a suitable feeler gage. Point gap should be 0.4-0.5 mm (0.016-0.020 in.).

Ignition timing should be set at 21 degrees BTDC by aligning "AA" mark on flywheel with "PMS" mark on crankcase. Points should just begin to open at this position. Shift breaker point plate (3) if necessary. When "TDC" mark on flywheel is aligned with "PMS" mark on crankcase, piston is at top dead center.

Ignition coil is located behind the flywheel. Clearance between coil and flywheel magnets should be 0.6-0.8 mm (0.024-0.031 in.).

Models A 180 B, A 180 P, A 220 B and A 220 P. The solid-state electronic ignition system requires no regular maintenance. Use the correct feeler gage to check clearance between ignition coil and flywheel. Clearance should be 0.40-0.45 mm (0.016-0.018 in.).

LUBRICATION. Check engine oil level after every 8 hours of operation. Maintain oil level at lower edge of fill plug opening.

Change oil after every 50 hours of operation. Manufacturer recommends oil with an API service classification SC or CC. Use SAE 40 oil for ambient temperatures above 10° C (50° F); SAE 30 oil for a temperature range of 0° to 10° C (32° to 50° F); SAE 20W-20 oil for a temperature range of -10° to 0° C (14° to 32° F) and SAE 10W oil for temperatures below -10° C (14° F).

Crankcase oil capacity for Models ALN 215 WB and ALN 215 WP is 0.75 L (0.793 qt.). Crankcase oil capacity for Models A 180 B, A 180 P, A 220 B and A 220 P is 0.6 L (0.634 qt.).

VALVE ADJUSTMENT. Valves should be adjusted after every 200 hours of operation. Valve stem-to-tappet clearance (cold) should be 0.10-0.15 mm (0.004-0.006 in.) for intake and exhaust valves on all models. If clearance is not as specified, remove or install shims (7 and 8—Fig. AC9) in shim holder (9) as necessary. Shims are available in 0.1 mm (0.004 in.) and 0.2 mm (0.008 in.) thicknesses.

CRANKCASE BREATHER. Models ALN 215 WB and ALN 215 WP are equipped with a crankcase breather which must be removed and cleaned after every 50 hours of operation. Breather is located on tube attached to valve chamber cover. Rubber valve must be installed as shown in Fig. AC10 to ensure crankcase vacuum.

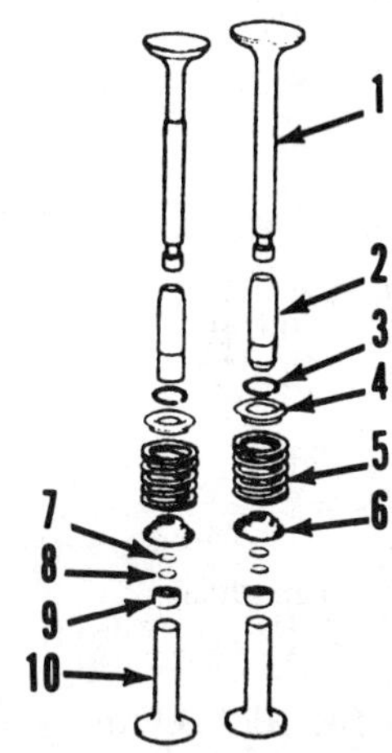

Fig. AC9—View of valve system components.

1. Valves
2. Guides
3. Seals
4. Seats
5. Springs
6. Retainers
7. Shims (valve adjustment)
8. Shims (valve adjustment)
9. Shim holder (cap)
10. Tappets

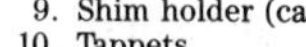

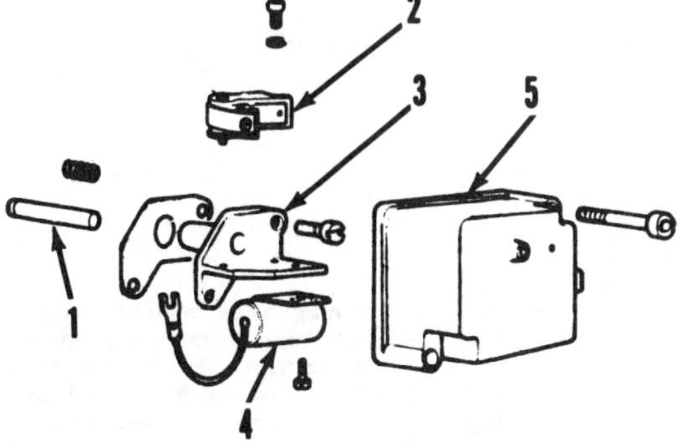

Fig. AC8—Breaker points and condenser on Models ALN 215 WB and ALN 215 WP are located on left-hand side of crankcase behind cover (5).

1. Plunger
2. Breaker points
3. Point plate
4. Condenser
5. Cover

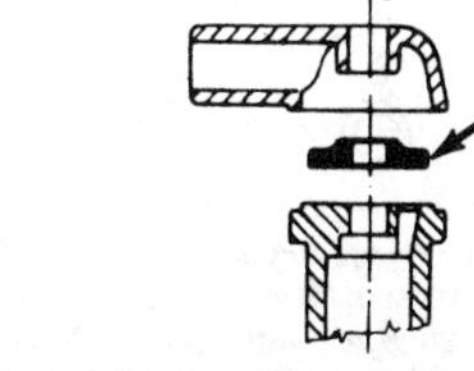

Fig. AC10—Breather valve must be installed as shown to ensure crankcase vacuum.

Models A 180 B, A 180 P, A 220 B and A 220 P are equipped with a crankcase breather which is an integral part of the valve chamber cover. No regular maintenance is required.

GENERAL MAINTENANCE. Check and tighten all loose bolts, nuts or clamps prior to each day of operation. Check for fuel or oil leakage and repair if necessary.

Clean dust, dirt, grease or any foreign material from cylinder head and cylinder block cooling fins after every 100 hours of operation. Inspect fins for damage and repair if necessary.

REPAIRS

TIGHTENING TORQUES. Recommended tightening torque specifications are as follows:

Cylinder head:
- A 180 B, A 180 P, A 220 B, A 220 P .... 24.5 N·m (18 ft.-lbs.)
- ALN 215 WB, ALN 215 WP .... 29 N·m (22 ft.-lbs.)

Connecting rod:
- A 180 B, A 180 P, A 220 B, A 220 P .... 11.8 N·m (9 ft.-lbs.)
- ALN 215 WB, ALN 215 WP .... 19 N·m (14 ft.-lbs.)

Crankcase cover:
- A 180 B, A 180 P, A 220 B, A 220 P .... 11.8 N·m (9 ft.-lbs.)
- ALN 215 WB, ALN 215 WP .... 15 N·m (11 ft.-lbs.)

Flywheel (all models) ...... 157 N·m (116 ft.-lbs.)

CYLINDER HEAD. All models are equipped with an aluminum alloy cylinder head which should not be removed when engine is hot.

To remove the cylinder head from all models, first allow engine to cool, then remove fuel tank and cooling shrouds.

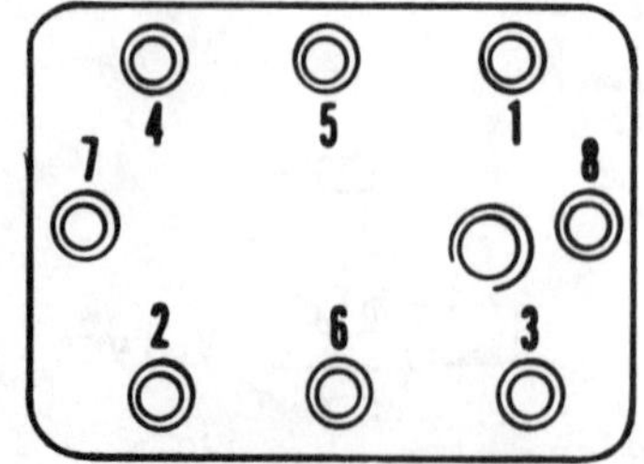

Fig. AC11—Loosen or tighten cylinder head bolts following sequence shown. Models A 180 B, A 180 P, A 220 B and A 220 P are equipped with two longer head bolts which must be installed in positions 1 and 3.

Loosen head bolts evenly following sequence shown in Fig. AC11. Remove cylinder head.

Check cylinder head for warpage by placing head on a flat surface and using a feeler gage to determine warpage. Warpage should not exceed 0.3-0.5 mm (0.012-0.020 in.).

Always install a new head gasket. Tighten cylinder head bolts to specification listed under TIGHTENING TORQUES following sequence shown in Fig. AC11. Note that Models A 180 B, A 180 P, A 220 B and A 220 P have two longer head bolts which must be installed in positions 1 and 3 shown in Fig. AC11.

CONNECTING ROD. On all models, an aluminum alloy connecting rod rides directly on the crankpin journal.

To remove connecting rod, remove cylinder head and crankcase cover. Use care when Models ALN 215 WB and ALN 215 WP crankcase cover is removed as governor flyweight balls will fall out of ramps in camshaft gear. On all models, remove connecting rod cap and push connecting rod and piston assembly out of cylinder head end of block. Remove piston pin retaining rings and separate piston from connecting rod. Camshaft and lifters may be removed at this time if required.

On all models, clearance between piston pin and connecting rod pin bore should be 0.006-0.022 mm (0.0002-0.0009 in.).

Clearance between crankpin and connecting rod bearing bore on Models A 180 B, A 180 P, A 220 B and A 220 P should be 0.030-0.049 mm (0.0012-0.0019 in.). Clearance between crankpin and connecting rod bearing bore on Models ALN 215 WB and ALN 215 WP should be 0.040-0.064 mm (0.0016-0.0025 in.). On all models, if clearance exceeds specified dimension, renew connecting rod and/or crankshaft.

On all models, connecting rod side play on crankpin should be 0.150-0.250 mm (0.0060-0.0100 in.).

On all models, renew connecting rod if crankpin bearing bore is excessively worn or out-of-round more than 0.10 mm (0.004 in.).

On all models, piston should be installed on the connecting rod with arrow (1—Fig. AC12), on top of piston, facing toward side of connecting rod with mark (2). Connecting rod and connecting rod cap marks (2) must align when components are assembled.

When installing piston assembly in cylinder block, connecting rod and cap match marks must be toward crankcase cover side of engine and arrow on top of piston must be on side opposite the valves on engines with clockwise rotation. On engines with counterclockwise rotation, arrow on top of piston must be toward valve side of engine and connecting rod and cap match marks will face toward flywheel side of engine. On all models, tighten connecting rod bolts to specification listed under TIGHTENING TORQUES. On Models ALN 215 WB and ALN 215 WP, use heavy grease to retain governor flyweight balls in camshaft ramps during assembly. On Models A 180 B, A 180 P, A 220 B and A 220 P, crankcase cover has two longer bolts which should be installed in the upper right and lower left positions. On all models, tighten crankcase cover bolts to specification listed under TIGHTENING TORQUES.

CYLINDER AND CRANKCASE. Cylinder and crankcase are an integral casting of aluminum alloy on all models. A high density perlite cylinder sleeve is cast as an integral part of the cylinder block.

Standard cylinder bore diameter for Models A 180 B, A 180 P, ALN 215 WB and ALN 215 WP is 65.000-65.013 mm (2.5591-2.5596 in.). Standard cylinder bore diameter for Models A 220 B and A 220 P is 72.000-72.013 mm (2.8300-2.8305 in.). If cylinder bore is 0.06 mm (0.0024 in.) or more out-of-round or tapered, cylinder should be bored to nearest oversize for which piston and rings are available.

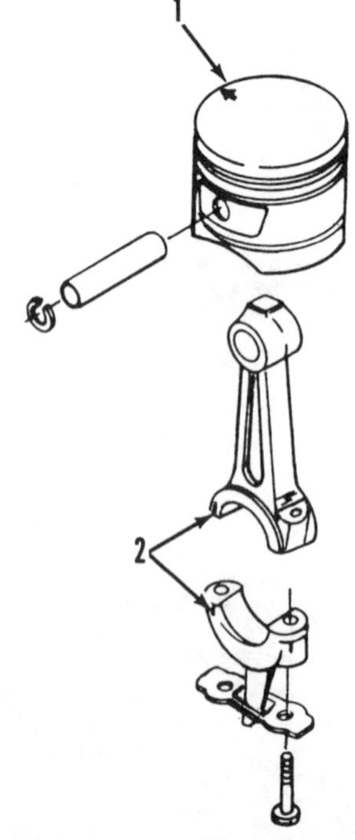

Fig. AC12—Piston must be installed on connecting rod so arrow (1) on top of piston faces toward side of connecting rod with mark (2). Connecting rod and connecting rod cap match marks (2) must align when components are assembled.

Crankshaft ball type main bearings should be a slight press fit in crankcase and crankcase cover. It may be necessary to slightly heat crankcase cover or crankcase assembly to remove or install main bearings. Renew bearings if they are loose, rough or damaged.

Models ALN 215 WB and ALN 215 WP are equipped with an oil slinger trough (2—Fig. AC13) installed in crankcase. If trough is removed, it must be securely repositioned prior to engine reassembly.

PISTON, PIN AND RINGS. Refer to previous CONNECTING ROD paragraphs for piston removal and installation procedure.

Standard piston diameter for Models A 180 B, A 180 P, ALN 215 WB and ALN 215 WP is 64.987-65.000 mm (2.5586-2.5591 in.). Standard piston diameter for Models A 220 B and A 220 P is 71.987-72.000 mm (2.8295-2.8300 in.). On all models, piston is a select fit at the factory.

Piston should be renewed and/or cylinder reconditioned if there is 0.013 mm (0.0005 in.) or more clearance between piston and cylinder bore.

Clearance between piston and piston rings in ring grooves should be 0.05 mm (0.002 in.) on all models.

Compression ring end gap on Models ALN 215 WB and ALN 215 WP should be 0.25-0.40 mm (0.010-0.016 in.). Compression ring end gap on Models A 180 B, A 180 P, A 220 B and A 220 P should be 0.25-0.45 mm (0.010-0.018 in.).

Oil control ring end gap on Models ALN 215 WB and ALN 215 WP should be 0.30-0.50 mm (0.012-0.020 in.). Oil control ring end gap on Models A 180 B, A 180 P, A 220 B and A 220 P should be 0.20-0.35 mm (0.008-0.014 in.).

Piston pin on all models should be a 0.004-0.012 mm (0.0002-0.0005 in.) interference fit in piston pin bore. It may be necessary to slightly heat piston to aid in pin removal and installation.

CRANKSHAFT. On all models, crankshaft is supported at each end in ball bearing type main bearings (11 and 16—Fig. AC14) and crankshaft timing gear (15) is a press fit on crankshaft (13). Refer to previous CONNECTING ROD paragraphs for crankshaft removal procedure and crankpin-to-connecting rod bearing bore clearance.

Standard crankshaft crankpin journal diameter on Models A 180 B, A 180 P, A 220 B and A 220 P is 25.989-26.000 mm (1.0232-1.0236 in.). Standard crankshaft crankpin journal diameter on Models ALN 215 WB and ALN 215 WP is 29.985-30.000 mm (1.1805-1.1811 in.).

Standard crankshaft main bearing journal diameter on Models A 180 B, A 180 P, A 220 B and A 220 P is 25.002-25.015 mm (0.98433-0.98484 in.). Standard crankshaft main bearing journal diameter on Models ALN 215 WB and ALN 215 WP is 30 mm (1.18 in.).

On all models, main bearings should be a slight press fit on crankshaft journal. Renew bearings if they are rough, loose or damaged. To prevent crankshaft damage, crankshaft should be supported on counterweights when pressing bearings or crankshaft timing gear onto crankshaft.

On all models, when installing crankshaft, make certain crankshaft and camshaft gear timing marks are aligned.

CAMSHAFT. Camshaft and camshaft gear on all models are an integral casting which rides in bearing bores in crankcase and crankcase cover.

On Models ALN 215 WB and ALN 215 WP, governor flyweight balls are located in ramps machined in face of camshaft gear.

On some models, a compression release mechanism is mounted on back side of camshaft gear. The spring-loaded compression release mechanism should snap back against camshaft when weighted lever is pulled against spring tension and released. Spring is in the correct position when dimension (A—Fig. AC15) of pin projection is 0.5-0.6 mm (0.020-0.024 in.).

Inspect camshaft journals and lobes on all models. Renew camshaft if worn, scored or damaged.

Standard camshaft bearing journal diameter for Models A 180 B, A 180 P, A 220 B and A 220 P is 14.973-14.984 mm (0.589-0.590 in.) at each end. Standard intake lobe height is 23.275-23.325 mm (0.916-0.918 in.). Standard exhaust lobe height is 17.575-17.625 mm (0.692-0.694 in.). Standard camshaft bearing journal

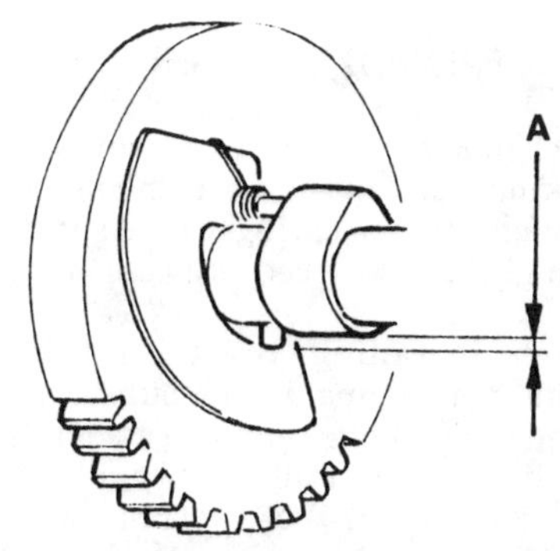

Fig. AC15—Dimension "A" of pin projection should be 0.5-0.6 mm (0.020-0.024 in.) for correct compression release mechanism operation.

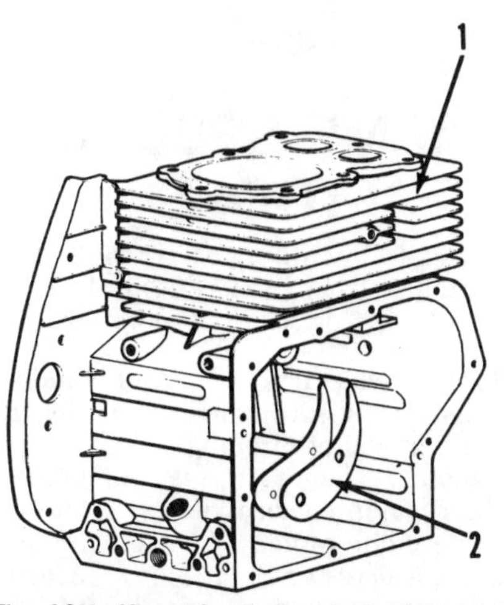

Fig. AC13—View of cylinder block (1) and oil slinger trough (2). If trough is removed, it must be securely repositioned prior to engine reassembly.

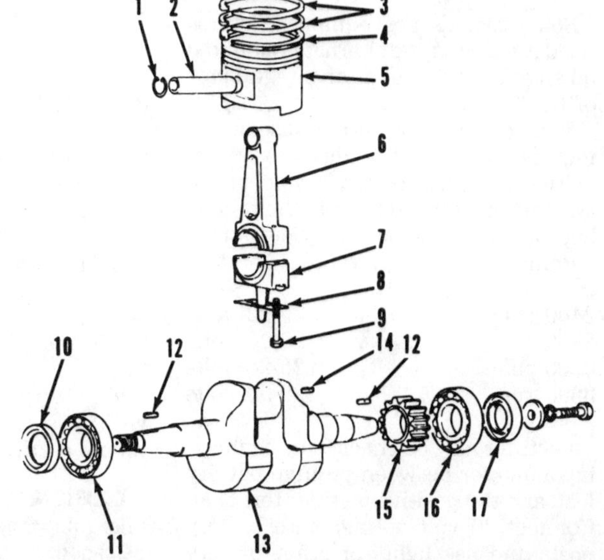

Fig. AC14—Exploded view of crankshaft, connecting rod and piston assembly.

1. Retaining rings
2. Piston pin
3. Compression rings
4. Oil control ring
5. Piston
6. Connecting rod
7. Connecting rod cap
8. Lockplate
9. Bolts
10. Seal
11. Main bearing
12. Key
13. Crankshaft
14. Crankshaft gear key
15. Crankshaft gear
16. Main bearing
17. Seal

diameter for Models ALN 215 WB and ALN 215 WP is 16 mm (0.6299 in.) at each end. Standard intake and exhaust lobe height is 26 mm (1.0236 in.).

When installing camshaft on Models ALN 215 WB and ALN 215 WP, retain governor flyweight balls in ramps with heavy grease. On all models, make certain camshaft and crankshaft gear timing marks are aligned.

VALVE SYSTEM. Refer to VALVE ADJUSTMENT paragraphs in MAINTENANCE section for valve clearance adjustment procedure.

Valve face and seat angles on all models is 45 degrees. Standard valve seat width is 1.2-1.3 mm (0.047-0.051 in.). If seat width is 2 mm (0.079 in.) or more, seat must be narrowed. If valve face margin is 0.5 mm (0.020 in.) or less, renew valve.

Standard exhaust valve stem diameter on all models is 6.955-6.970 mm (0.2738-0.2744 in.).

Standard intake valve stem diameter for Models A 180 B, A 180 P, A 220 B and A 220 P is 6.955-6.970 (0.2738-0.2744 in.). Standard intake valve stem diameter for Models ALN 215 WB and ALN 215 WP is 6.965-6.987 mm (0.2742-0.2751 in.).

Standard valve guide inside diameter on Models A 180 B, A 180 P, A 220 B and A 220 P is 7.015-7.025 mm (0.2762-0.2766 in.) for intake and exhaust valve guides. Standard valve guide inside diameter on Models ALN 215 WB and ALN 215 WP is 7.000-7.022 mm (0.2756-0.2764 in.). On all models, worn valve guides can be renewed using puller 365109.

Standard valve spring free length for Models A 180 B, A 180 P, A 220 B and A 220 P is 34 mm (1.34 in.). If spring free length is 31 mm (1.22 in.) or less, renew spring. Standard valve spring free length for Models ALN 215 WB and ALN 215 WP is 35 mm (1.38 in.). If spring free length is 32 mm (1.26 in.) or less, renew spring.

SERVICING ACME ACCESSORIES

REWIND STARTER

Refer to Fig. AC25 for exploded view of rewind starter used on all models so equipped. Rewind spring (5) and spring housing (4) are serviced as an assembly only.

When installing rewind starter assembly on engine, install but do not tighten the six bolts retaining assembly to cooling shroud. Pull cable handle (7) until 150 mm (6 in.) of cable has been pulled from housing and starter dogs (9) have centered assembly. Hold tension on cable while the six bolts are tightened.

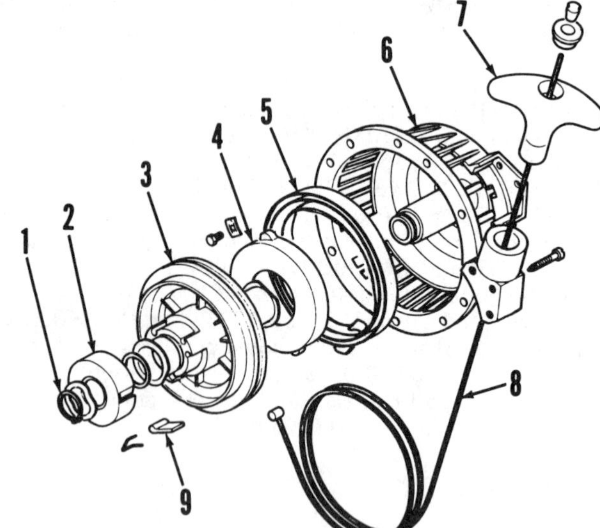

Fig. AC25—Exploded view of rewind starter assembly.
1. Snap ring
2. Starter dog housing
3. Cable pulley
4. Spring housing
5. Rewind spring
6. Housing
7. Handle
8. Cable
9. Starter dogs

ALTERNATOR

Some models are equipped with a fixed armature type alternator mounted on engine with rotor as an integral part of the flywheel.

To test alternator output, disconnect rectifier leads and connect to an AC voltmeter with at least a 30 volt capacity. Start engine and refer to the following specifications for voltage output according to engine operation.

2400 rpm . . . . 20-22 volts
2800 rpm . . . . 23-25 volts
3200 rpm . . . . 26-28 volts
3600 rpm . . . . 29-30 volts

Rectifier may be checked by connecting ammeter between positive battery lead and the positive rectifier terminal. Connect 20 volt voltmeter to battery posts and use lights or other battery drain method to lower battery voltage below 13 volts. Start engine and refer to the following specifications for amperage output according to engine speed.

1500 rpm . . . . 0.5 amp
2400 rpm . . . . 1.5 amp
3000 rpm . . . . 2.2 amp
3600 rpm . . . . 2.7 amp

If battery charge current is 0 amp with 12.5 volt or less battery voltage, renew rectifier.

CAUTION: Never operate engine with rectifier disconnected as rectifier will be damaged.

ACME SPECIAL TOOLS

The following special tools are available from Acme Central Parts Distributors or Acme Corporation.

Tool Description	Tool Number
Valve spring extractor	365110
Ignition coil positioning tool	365168
Valve guide check tool	365048
Valve guide puller	365109
Electrical tester	365180
Oil seal installation cone	365152
Engine flywheel and timing cover puller	365113

ACME CENTRAL PARTS DISTRIBUTORS

(Alphabetically by States)

These franchised firms carry extensive stocks of repair parts. Contact them for name of dealer in their area who will have replacement parts.

Fessler Equipment Company
2400 Commercial Drive
Anchorage, AK 99501

Southwest Continental
3040 North 27th Ave.
Phoenix, AZ 85017

Diesel Technology
4325 Hiway 60 West
Mulberry, FL 33860

I.C.L. Engine Sales & Repr.
1586 Iao Lane
Honolulu, HI 96817

Burlington Wholesale
1533 E. 3rd Road
Bremen, IN 46506

D & D Enterprises
Rte 3 Box 224
Salem, IN 47167

Acme Engines Gulf South
P.O. Box 46
Route 2 Box 58D
Mt. Hermon, LA 70450

Harris Castille, Inc.
104 E. Admiral Doyle
New Iberia, LA 70560

Mel's Repair & Supply
Rt. #1 Box 185
Utica, MN 55979

Phillips Diesel Corp.
Hwy. 6 & Interstate 25
Los Lunas, NM 87031

Moriches Agri. Supply
640 Montauk
E. Moriches, Long Island
New York 11940

Mainline of North America
Junction US 40 & Interstate Rte. 38
London, OH 43140

Bull Equipment Co.
9525 S.W. Commerce Cir.
Wilsonville, OR 97070

Shirk's Repair
Rd. 2 Box 102
Ephrata, PA 17522

Spak Distributing
3506 Babcock Blvd.
Pittsburgh, PA 15237

Middle Tennessee Assoc.
P.O. Box 90362
Nashville, TN 37209

Pioneer Industries
P.O. Box 6334
Beaumont, TX 77707

Pearson Equipment
3001 Romona
Fort Worth, TX 76116

Proctor Equipment
P.O. Box 6512 ATS
Midland, TX 79711

Proctor Equipment (Br)
4857 N. Chadburne
San Angelo, TX 76901

Proctor Equipment (Br)
6610 Topper Pkwy #102
San Antonio, TX 78255

ADVANCED ENGINE PRODUCTS

Model	Bore	Stroke	Displacement
Compact I	1.250 in. (31.8 mm)	1.032 in. (26.2 mm)	1.26 cu. in. (21 cc)
Compact II, Compact III, 13A, 13B	1.250 in. (31.8 mm)	1.096 in. (27.8 mm)	1.34 cu. in. (22 cc)

ENGINE IDENTIFICATION

Beginning mid-1967, a three-digit type number is stamped on the crankcase flange opposite spark coil and carburetor or on the large cylinder cooling baffle. Serial number indicates time of manufacture. The first digit denotes the year and the next two digits denote the month.

Engine may be equipped with direct drive, centrifugal clutch or centrifugal clutch and gear reduction. Rated engine speed is 6300 rpm and reduction units are available with output shaft speeds of 3300 rpm, 1700 rpm or 900 rpm.

MAINTENANCE

SPARK PLUG. On models using a 10 mm spark plug, use Champion UY-6, or equivalent. Models using a 14 mm spark plug use a Champion CJ14 or equivalent for light and normal use, or a Champion CKJ8 or equivalent for heavy duty applications.

Adjust spark plug electrode gap to 0.030 inch (0.76 mm) for normal operating conditions or to 0.025 inch (0.64 mm) for heavy duty applications.

CARBURETOR. A diaphragm type carburetor is used on all models. Model HU Tillotson carburetor (Fig. AE1) is used on late production 13B engines for constant speed operation. All other engines are equipped with the two-piece diaphragm carburetor shown in Fig. AE2. Refer to the appropriate paragraph for service procedure.

Tillotson Carburetor. Refer to Fig. AE1 for an exploded view of Tillotson HU model diaphragm type carburetor. This carburetor is a precalibrated unit designed for constant speed engine operation. It does not have throttle or speed and fuel flow adjustments. Starting enrichment is accomplished by pushing a primer button which moves the inlet needle off the seat allowing extra fuel flow. The diaphragm fuel pump assures a constant fuel supply to the metering chamber of carburetor. Operation of the pump diaphragm is caused by pulsations from engine crankcase. The rubber-tipped inlet needle valve seats directly on a machined orifice in body casting.

To disassemble the carburetor, unbolt and remove pump cover, gasket, pump diaphragm and inlet screen. Unbolt and remove metering diaphragm cover, diaphragm and gasket. Remove the pin retaining screw, fulcrum pin, control lever, spring and inlet needle.

Clean and inspect all parts and renew any showing excessive wear or other damage. Reassemble by reversing the disassembly procedure. Install pump diaphragm next to carburetor body. Install metering diaphragm next to diaphragm cover as shown in Fig. AE1.

Two-Piece Diaphragm Carburetor. Carburetor is equipped with only one

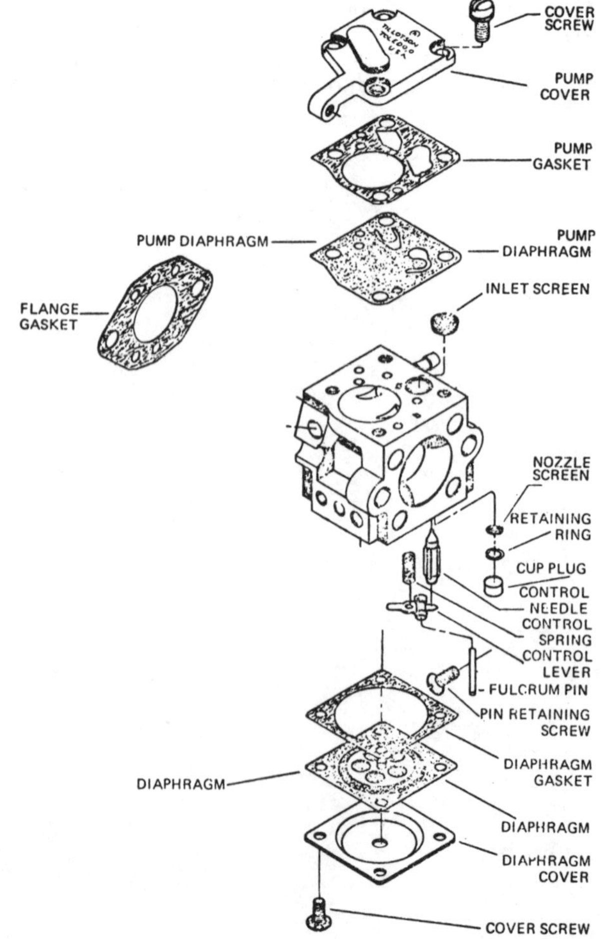

Fig. AE1—Exploded view of Tillotson HU carburetor designed for constant speed operation. Note absence of throttle, choke and adjusting needles. Primer button on metering diaphragm is used for starting enrichment.

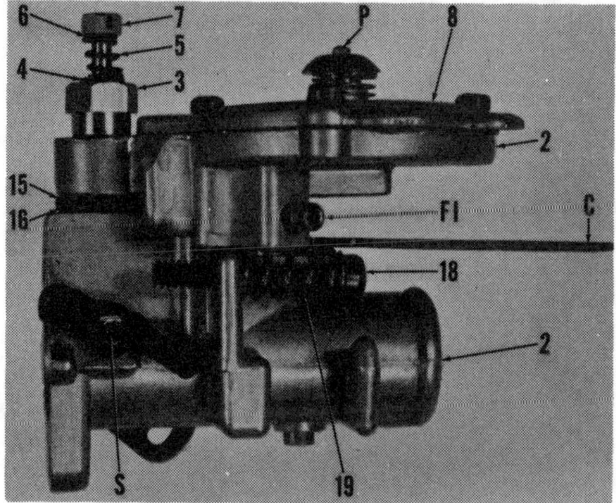

Fig. AE2—View of late production diaphragm carburetor. New cover (8) with primer unit (P) can be installed on some earlier carburetors. A conversion kit is available. Flat end of governor vane shaft fits in slot (S) in carburetor throttle shaft. Fuel inlet is (FI), chock lever is (C). Refer to Fig. AE3 for remainder of legend.

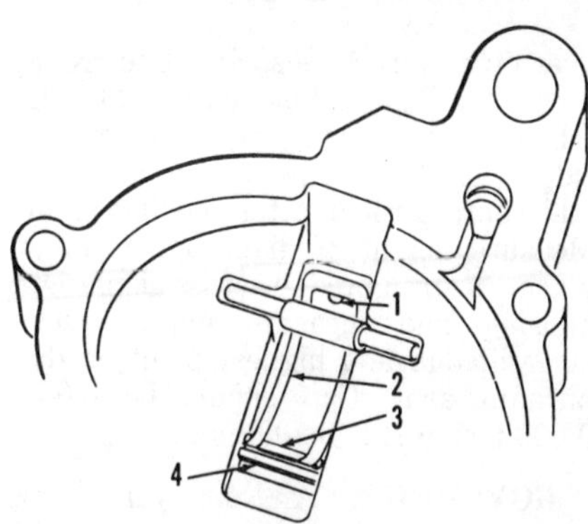

Fig. AE4—Install diaphragm arm as outlined in text.

1. Ball recess
2. Spring
3. Roller
4. Ridge

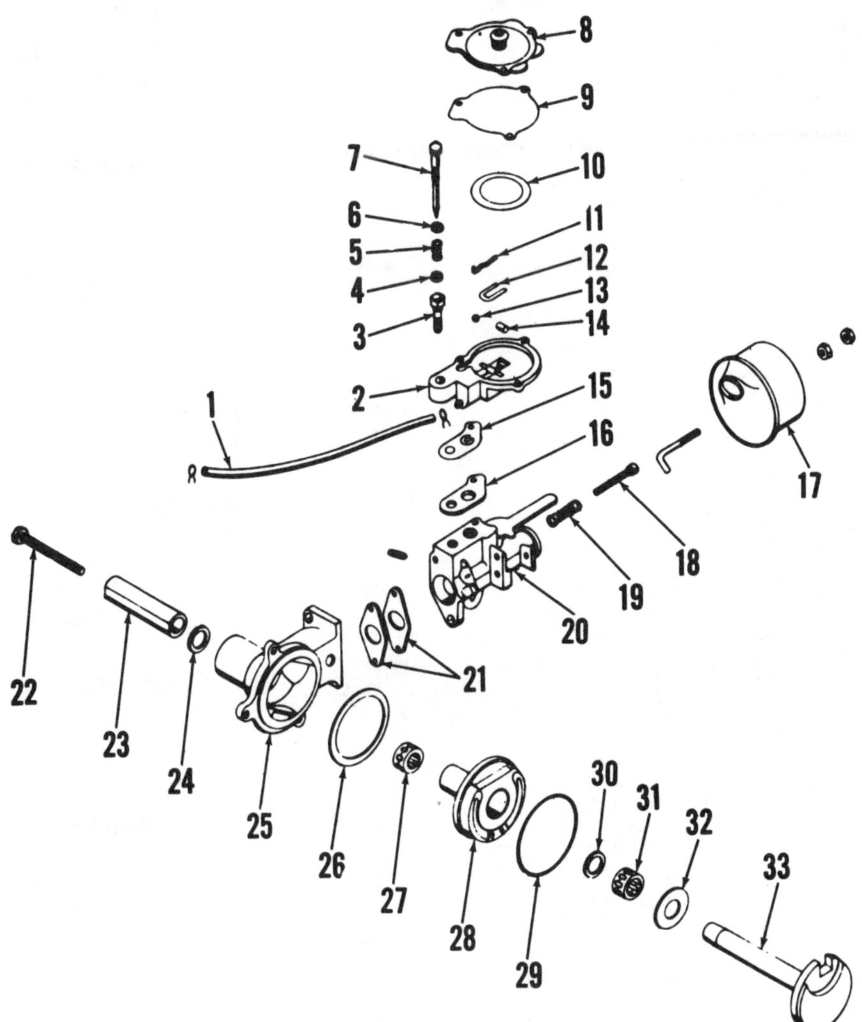

Fig. AE3—Exploded view of carburetor, induction housing and backshaft unit. Refer to Fig. AE2 for view of carburetor assembly. Refer also to Fig. AE7.

1. Fuel line
2. Diaphragm body
3. Needle valve housing
4. "O" ring
5. Spring
6. Washer
7. Needle valve
8. Primer diaphragm cover
9. Diaphragm
10. Diaphragm disc
11. Diaphragm arm
12. Arm spring
13. Diaphragm valve ball
14. Diaphragm roller
15. Diaphragm valve
16. Gasket
17. Air cleaner
18. Idle speed screw
19. Spring
20. Carburetor body
21. Gaskets
22. Extension shaft screw
23. Extension shaft
24. Sealing ring
25. Induction case
26. Gasket
27. Roller bearing assy.
28. Reed or "feather valve" assy.
29. "O" ring
30. Sealing ring
31. Roller bearing assy.
32. Thrust washer
33. Backshaft

mixture adjusting screw. On later carburetors, the adjusting screw has a spring to keep it from turning due to vibration. On some early models, the adjusting screw is held with a locknut.

Initial adjustment on early production models with screw head type jet instead of hex needle valve housing (3–Fig. AE2) is 1¼ to 1½ turns open from a lightly seated position.

Initial adjustment on later production models with mixture adjusting needle and hex needle housing (3–Fig. AE2) is ½ turn open from a lightly seated position.

Final adjustment on all models should be made with engine running under normal load. Adjust mixture screw in or out until engine is running smoothly at maximum rpm then turn screw out until engine slows down slightly. This will provide a fuel-air mixture rich enough for proper engine lubrication. If fuel mixture cannot be adjusted under load, adjusting screw should be turned to the position that will give smooth and rapid acceleration from low idle rpm to maximum governed speed. After adjusting carburetor, be sure adjusting screw locknut, if so equipped, is tightened securely. Idle speed should be 2200-2500 rpm.

When removing carburetor, first remove needle valve (7–Fig. AE3) and housing (3), and the nearest (long) screw retaining diaphragm cover (8) and lift off diaphragm body (2) as an assembly. Fuel line will not need to be disconnected. Carefully disconnect governor vane shaft from slot (S–Fig. AE2) in throttle shaft.

To reassemble diaphragm assembly, place diaphragm roller in housing as shown in Fig. AE4 and install spring (2) so bend in each end of spring is pointing down and closed end is adjacent to ball seat. Move spring away from roller (3) until diaphragm ball can be placed in seat while holding spring. Insert short end of diaphragm arm (11–Fig. AE3) under closed end of spring and hook arm over pivot pin. Move spring onto arm as far as it will go and move roller (3–Fig. AE4) to small cavity on other side of ridge (4). Be sure both ends of spring still contact roller after moving roller. Measure height of diaphragm arm by placing a straightedge across diaphragm housing and measure gap between straightedge and highest point on diaphragm arm. Gap should be 0.005-0.008 inch (0.127-0.203 mm).

GOVERNOR. An air vane type governor is used on all models not equipped with Tillotson HU constant speed carburetor. To remove or inspect air vane (44–Fig. AE5), remove screws retaining cylinder baffles to blower housing. Remove screws retaining blower housing to engine and lift blower housing from engine.

The governor air vane should be held against stop (43) by very light spring pressure from the coil spring on the carburetor throttle shaft. The vane and throttle shaft should work freely and return to position against the vane stop if released from any other position. Check for and remove any binding condition.

To renew governor spring, slide open end of new spring over throttle shaft and secure inner end of spring behind hook that is cast on the carburetor body. Slide a few coils of the spring outer end off the throttle shaft and twist spring clockwise only far enough to insert the straight outer end of the spring in the first hole of the throttle shaft. Be sure that none of the spring coils overlap when spring is in place.

The flat end of the governor vane shaft should engage the slot in the carburetor throttle shaft so that the throttle disc is wide open when the governor vane is against the stop.

MAGNETO AND TIMING. A flywheel type magneto is used. The breaker contact points are enclosed in a

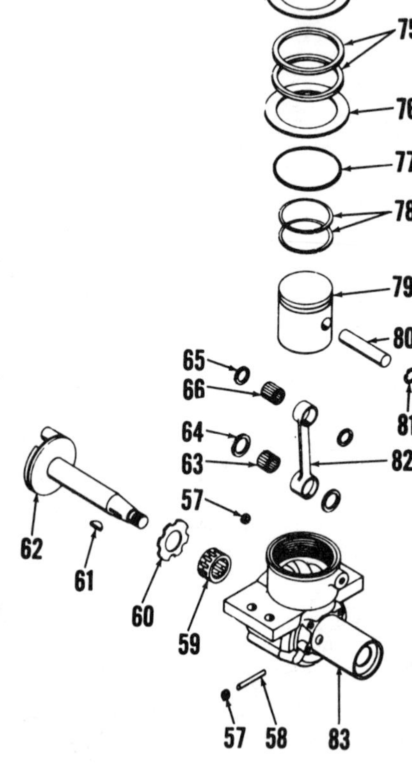

Fig. AE6—Exploded view of crankcase and cylinder assembly. Cylinder (71) is threaded into crankcase (83). Additional muffler sections (67) may be used on some engines. For exploded view of later production engine, refer to Fig. AE7.

57. "O" rings
58. Breaker pushrod
59. Roller bearing assy.
60. Thrust washer
61. Flywheel key
62. Crankshaft
63. Roller bearings
64. Retainer washers
65. Retainer washers
66. Roller bearing
68. Muffler cup
69. Muffler bolt
70. Gasket
71. Cylinder
72. Gasket
73. Exhaust collector
74. Gasket
75. Spacers
76. Spacer washer
77. "O" rings
78. Piston rings
79. Piston
80. Piston pin
81. Snap ring
82. Connecting rod
83. Crankcase

Fig. AE5—Exploded view of magneto, back plate, flywheel and governor vane unit. Governor vane guard tube (43) is not used on later models. Bearing and seal retainer (39) is pressed onto crankcase (83—Fig. AE6).

34. Flywheel nut
35. Washer
36. Starter dogs
37. Flywheel
38. Seal
39. Retainer
40. Roller bearing
41. Armature & coil assy.
42. Grommet (valve shaft)
43. Guard
44. Governor vane
45. Breaker cover
46. Breaker point arm
47. Insulators
48. Breaker point base
49. Breaker post
50. Insulator
51. Back plate
52. Insulated washer
53. Nut
54. Magneto short-out spring
55. Nut
56. Condenser

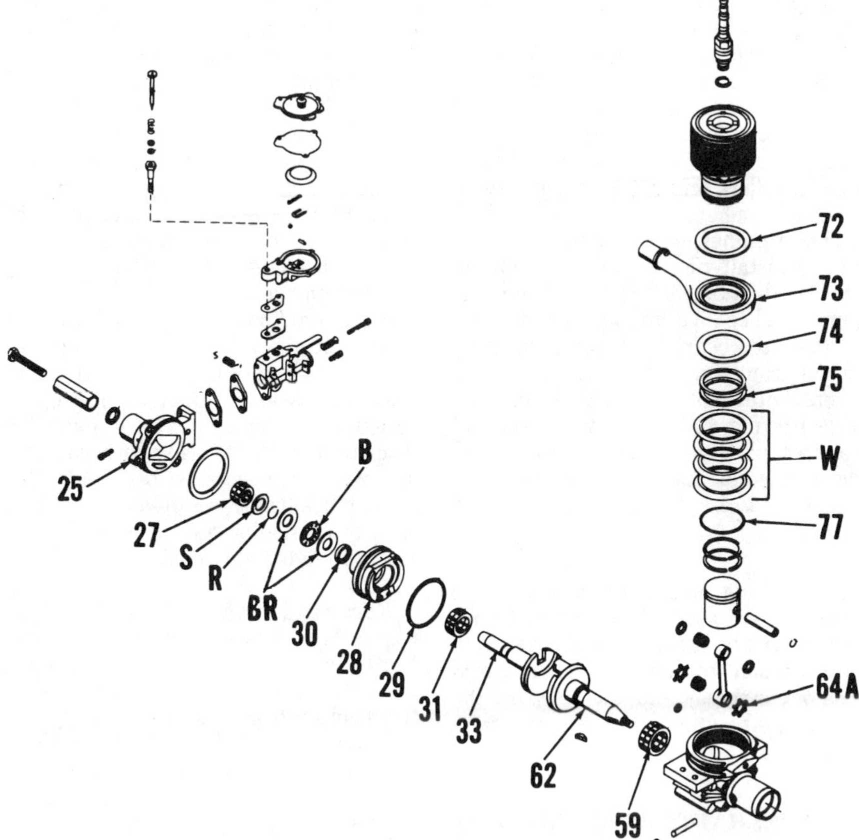

Fig. AE7—Exploded view of late production engine. Refer to Fig. AE3 and Fig. AE6 for earlier production engine, and for identification of parts not having call-out numbers in above view.

B. Thrust bearing
BR. Bearing races
R. Retaining ring
S. Spacer
W. Heat sink washers
25. Induction housing
27. Roller bearing
28. Reed (feather) valve assy.
29. "O" ring
31. Roller bearing
33. Backshaft
59. Roller bearing
62. Crankshaft
72. Gasket
73. Exhaust collector
74. Gasket
75. Spacers
77. "O" ring

box under the flywheel. The armature, coil and condenser are mounted outside the flywheel.

To gain access to the breaker contact points, remove the blower housing and governor vane if so equipped. Loosen flywheel retaining nut one turn and, while supporting engine with flywheel, tap the nut sharply with a small hammer to unseat the tapered fit between flywheel and shaft. A screwdriver may be used between flywheel and backplate to provide leverage against flywheel while tapping on nut. Remove nut and flywheel. Be careful not to lose the aluminum flywheel drive key. Remove breaker box cover (45–Fig. AE5).

Contact arm spring assembly can be removed by lifting arm from push rod and sliding assembly from pivot post. Stationary contact point can then be removed by removing adjustment lock screw. If breaker point actuating push rod is removed, reinstall with flat end towards crankshaft. Push rod length (new) is 0.837 inch (21.26 mm). Renew if worn to a length of 0.827 inch (21 mm) or less. An "O" ring (57–Fig. AE6) is used as a seal around push rod (58) and "O" ring should be renewed if there is any evidence of crankcase leakage.

On early production models, spring on breaker arm fits in slot of terminal post (49–Fig. AE5). On later production models, the post is not slotted and spring rests against the post. The post must be insulated from the magneto back plate as it carries the primary ignition current.

If necessary to renew the magneto back plate, disconnect governor linkage, pry flywheel shaft seal and bearing retainer (39) from crankcase and remove back plate retaining screws. Flywheel nut tightening torque is 100 in.-lbs. (11 N·m).

Adjust breaker contact gap to 0.020-0.022 inch (0.508-0.513 mm). Adjust armature air gap to 0.010 inch (0.254 mm).

LUBRICATION. Engine is lubricated by mixing SAE 30 API service classification SF motor oil with regular gasoline. Fuel:oil ratio is 16:1. Mix thoroughly in a separate container.

CARBON. The engine cylinder should be removed (see Fig. AE6), and any carbon present should be cleaned from exhaust ports, exhaust collector ring, piston, rings, ring grooves and muffler after each 30 to 100 hours of use. Use a dull tool to remove carbon and use care not to damage cylinder bore or piston.

REPAIRS

CAUTION: When disassembling engine, do not lose or intermix bearing rollers from the connecting rod, crankshaft, clutch or gearbox.

PISTON, PIN, RINGS AND CYLINDER. Lower end of cylinder is threaded into crankcase. To remove cylinder, first remove cylinder baffles, blower housing and governor vane shaft as equipped. Remove spark plug, hold exhaust collector from turning when unscrewing cylinder and lift cylinder straight up off of piston and rings when cylinder separates from crankcase.

Piston is equipped with two unpinned piston rings. Recommended piston ring end gap is 0.010 inch (0.254 mm) for early production wide rings and 0.0005 inch (0.0127 mm) for later production narrow rings. Narrow rings only are available for service and must be installed on new style piston.

Side clearance for either wide or narrow rings in piston grooves should not exceed 0.005 inch (0.13 mm). Parts area available in standard size only.

Piston rings should be installed with end gaps 180° apart. On early models with plain bearing piston pin, install pin with plugged end towards retaining snap ring.

New cylinder inside diameter is 1.2490 inch (31.725 mm) at the top and 1.2505 inch (31.763 mm) at the bottom. Renew cylinder if scored, cracked or if taper exceeds 0.005 inch (0.127 mm).

When installing cylinder, use new gaskets and "O" ring. Lubricate gasket and "O" ring, assemble exhaust collector on cylinder with gaskets, spacers and "O" ring as in Fig. AE6 or AE7. While holding piston at top dead center, push cylinder down over piston and rings. Carefully engage threads of cylinder into place. Be sure that exhaust collector ring is properly positioned and screw cylinder down snugly.

NOTE: The original cylinder used on COMPACT I series engines is no longer available for service. If cylinder must be renewed, the factory recommends installation of latest cylinder, exhaust collector ring and associated parts.

CONNECTING ROD. Piston and rod assembly may be removed after removing cylinder and induction housing. On models with a clutch or gearbox, remove extension shaft, gearbox cover or clutch cover and clutch. Remove screws retaining induction housing, clutch housing, radial mount or gearcase to engine crankcase. Complete assembly with carburetor attached may then be removed as a unit.

Use small needlenose pliers to remove outer half of connecting rod bearing retainer and bearing rollers (early production models). Slide connecting rod off of crankpin and remove rod and piston unit from engine. Remove inner half of bearing retainer (early model). Be sure to remove any loose rod bearing rollers from crankcase before proceeding.

To remove connecting rod from piston, remove snap ring at end of piston pin and push pin out of piston and connecting rod. On early production models with plain bearing piston pin, be careful not to push plug out of closed end of pin. On later engines with roller piston pin bearing, be careful not to lose any of the loose rollers and bearing retainer washers.

Renew any parts that are excessively worn or scored. Parts are available in standard size only.

To reassemble piston to connecting rod on early models with plain bearing in upper end of connecting rod, insert the pin, open end first, through piston and connecting rod and install snap ring.

To reassemble piston to connecting rod on later models, stick the loose needle rollers in upper end of connecting rod with a light coat of heavy grease. Slide piston pin slightly through pin bore in piston and place one retainer washer over inner end of pin. Insert connecting rod with roller bearings into piston and carefully push piston pin most of the way through rod. Slide remaining pin bearing retainer washer between rod and piston and push the pin fully into place. Install piston pin retaining snap ring.

To reinstall piston and rod unit on early production models, place half of rod bearing retainer on crankpin. Insert rod into crankcase and slide open end over bearing retainer and crankpin. Insert bearing rollers into retainer using small needlenose pliers. Place outer half of bearing retainer over crankpin.

To reinstall piston and rod unit on later production models, install one retainer washer on crankpin with flat side out (towards bearing). Place connecting rod on crankpin, lubricate rod, pin and rollers with motor oil and insert the loose bearing rollers between rod and crankpin with small needlenose pliers. Place second retaining washer on crankpin with flat side in towards bearing.

NOTE: New piston pin and crankpin roller sets are furnished in wax strips for easy assembly. DO NOT remove wax from rollers. Install the waxed roller strips in eyes of connecting rod and complete the assembly as outlined in preceding paragraphs.

CRANKSHAFT, BEARINGS AND SEALS. A two-piece crankshaft is used. Refer to Figs. AE3, AE6 and AE7. The flywheel crankshaft (62) with integral crankpin rides directly in two caged roller bearings located in the engine crankcase. The backshaft (33), or output shaft, rides directly in one caged roller bearing (31) located in the reed valve and in one caged roller bearing (27) located in the induction housing.

The flywheel end shaft is sealed with a lip type seal (38 – Fig. AE5) located directly behind the flywheel and seal is renewable after removing flywheel. The backshaft is sealed with an "O" ring (30 – Fig. AE3) located in the bore of the reed valve unit and "O" ring is renewable after removing the backshaft.

Remove Backshaft. Remove clutch or gearbox cover (as equipped) and remove clutch drive hub or drive gear hub from tapered outer end of backshaft. Remove induction housing retaining screws and work induction housing (complete with carburetor and clutch or gearbox housing) from engine crankcase. Backshaft can now be withdrawn from reed valve unit and induction housing. Reed valve unit may also be removed at this time.

Induction housing is sealed to reed valve unit and reed valve unit is sealed to engine crankcase with thin paper gaskets. Fuel-air induction passage is sealed with a lip type seal in induction housing and an "O" ring in reed valve unit which seals against backshaft. One induction housing retaining screw also has a small sealing "O" ring that fits into recess in induction housing.

Remove Flywheel End Shaft. To remove the flywheel end of the crankshaft, first remove flywheel, connecting rod and piston unit from engine. Turn flywheel shaft so ignition contact points are open (milled flat on shaft is away from pushrod) and withdraw shaft from engine. Reverse disassembly procedures and use all new gaskets and "O" rings to reassemble.

INTAKE VALVING. The reed valve assembly contains the inner bearing race for the engine backshaft and fits over the backshaft as shown in Fig. AE7. Protrusion on reed valve unit fits into induction housing. Eary models were fitted with metal reeds; later models have plastic reeds ("feather valve").

If the reed valve is not functioning properly, the complete valve assembly must be renewed as individual parts are not available. The reed valve assembly can be renewed after removing the engine backshaft. Refer to exploded views in Figs. AE3 and AE7.

ADVANCED ENGINE PRODUCTS

Model	Bore	Stroke	Displacement
20A	1.437 in. (36.5 mm)	1.250 in. (31.8 mm)	2.0 cu. in. (33.3 cc)

ENGINE IDENTIFICATION

The engine identification plate is attached to the magneto back plate and includes engine model number, engine type number and engine serial number. The engine model number indicates general engine features for servicing while the type number identifies specific engine features. Always furnish engine model number, serial number and type number when ordering parts or service information.

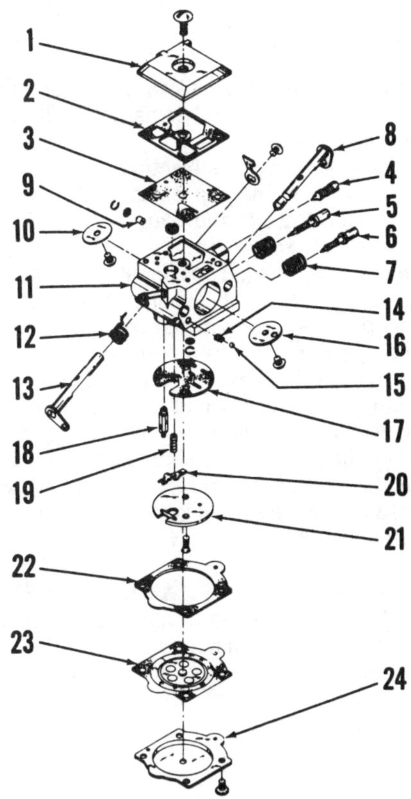

Fig. AE10—Exploded view of Walbro HDC carburetor.

1. Cover
2. Gasket
3. Fuel pump
4. Idle speed screw
5. Idle mixture screw
6. High speed mixture screw
7. Spring
8. Choke shaft
9. Limiting jet
10. Throttle plate
11. Body
12. Spring
13. Throttle shaft
14. Spring
15. Choke friction ball
16. Choke plate
17. Gasket
18. Fuel inlet valve
19. Spring
20. Metering lever
21. Circuit plate spring
22. Gasket
23. Metering diaphragm
24. Cover

MAINTENANCE

SPARK PLUG. Recommended spark plug is Champion CJ6 for normal use and CJ4 for heavy use. Spark plug electrode gap should be 0.025 in. (0.64 mm).

CARBURETOR. A Walbro HDC diaphragm carburetor is used on all models. Refer to Fig. AE10 for exploded view of carburetor. Initial adjustment for idle mixture screw is 1 turn open from a lightly seated position. Initial adjustment for main fuel mixture screw is 1¼ turns open from a lightly seated position. Make final adjustments with engine at normal operating temperature and running. Place engine under load and adjust main fuel mixture screw for leanest setting that will allow satisfactory acceleration and steady governor operation. Set engine at idle speed, no load and adjust idle mixture screw to obtain smoothest idle operation.

As each adjustment affects the other adjustment procedure may have to be repeated.

Engine idle speed is 2500 rpm and is adjusted by turning idle speed screw (4–Fig. AE10).

When installing carburetor, be sure crankcase pulse passages are aligned and clear in gaskets and reed valve housing.

GOVERNOR. An air vane governor is used on some engines. As engine speed varies the air vane turns the governor shaft. The governor link (10–Fig. AE11) between the governor shaft lever (15) and carburetor throttle will transmit variations in engine speed and open or close carburetor throttle. A governor spring (11) is connected between governor shaft lever and speed setting bracket (1). Spring may be connected to one of three holes to obtain desired governed engine speed.

Governor assembly should be inspected to be sure linkage does not bind or is restricted. Right angle end of governor link (10) is connected to carburetor throttle lever. Opposite end of governor link is connected to one of two holes in governor shaft (15) lever. Installing link in inner hole will increase governor sensitivity while installation in outer hole decreases sensitivity.

To adjust governed speed, refer to Fig. AE12 and connect governor spring in hole of desired governed rpm range. Loosen governor plate screw (S) and run engine under load. Move governor plate until desired governed rpm is obtained and retighten retaining screw. If engine "hunts", readjust carburetor.

If carburetor adjustment will not correct hunting, alter governor sensitivity by moving governor link (10–Fig.

Fig. AE11—Exploded view of governor and induction assemblies.

1. Governor plate
2. Governor bracket
3. Air filter housing
4. Gasket
5. Carburetor
6. Gasket
7. Reed valve seat
8. Reed petal
9. Gasket
10. Governor link
11. Governor spring
12. Bushing
13. Retainers
14. Pin
15. Governor shaft & lever
16. Crankcase
17. Magneto back plate

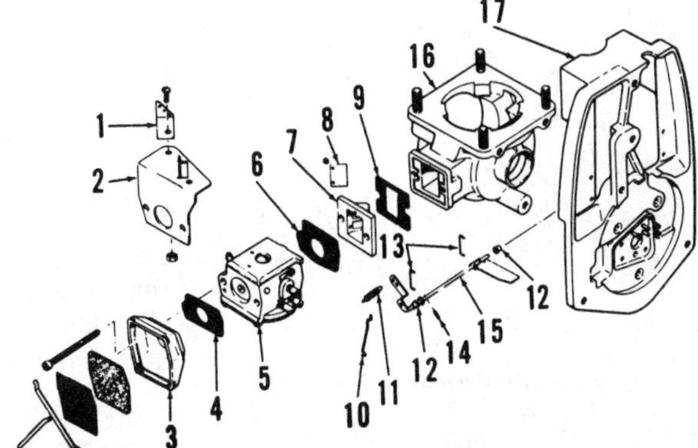

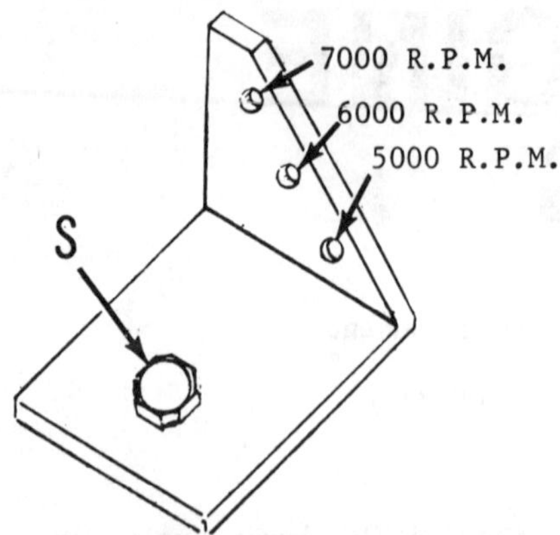

Fig. AE12—Governed speed is determined by location of governor spring in one of three holes in governor plate and by repositioning plate after loosening screw (S).

Fig. AE14—Exploded view of engine. Crankcase cover (16) is used with gear reduction unit. Cover (24) may be used with direct drive or clutch.

1. Spark plug
2. Compression release
3. Cylinder
4. Gasket
5. Piston rings
6. Piston
7. Piston pin
8. Retainer
9. Bearing
10. Connecting rod
11. Cap screw
12. Bearing rollers (28)
13. Rod cap
14. Crankcase
15. Seal
16. Crankcase cover
17. Gasket
18. Bearing
19. Snap ring
20. Crankshaft
21. Bearing
22. Seal
23. Snap ring
24. Crankcase cover
25. "O" ring

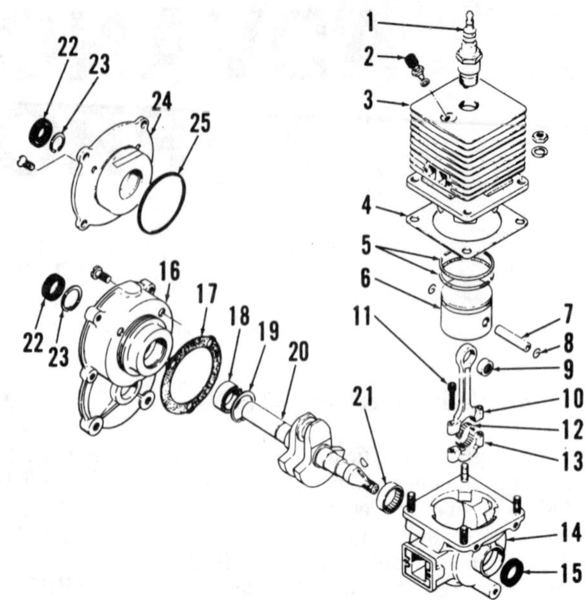

AE11) to alternate hole in governor shaft (15) lever.

MAGNETO AND TIMING. A flywheel type magneto is used. The breaker contact points are enclosed in a box under the flywheel and actuated by a cam on crankshaft. Armature and ignition coil are mounted outside flywheel. Refer to Fig. AE13.

To obtain access to breaker points, remove recoil starter assembly, blower housing and flywheel. On models with a kill switch mounted on blower housing, ground wire must be disconnected to remove blower housing. Tighten flywheel retaining nut to 120 in.-lbs. (13 N•m).

Ignition timing is fixed. Breaker point gap should be 0.015-0.018 inch (0.38-0.46 mm). Incorrect point gap will affect ignition timing. Armature air gap should be 0.010 inch (0.25 mm). Set air gap by loosening armature mounting screws and place 0.010 inch (0.25 mm) shim stock between armature and flywheel. Position armature against shim and tighten armature mounting screws. Remove shim stock.

Fig. AE13—Exploded view of magneto.

1. Magneto back plate
2. Armature & coil
3. Condenser
4. Breaker points
5. Point box
6. Flywheel

LUBRICATION. The engine is lubricated by mixing oil with the fuel. A good quality SAE 30 oil with API service classification SF should be mixed with fuel at ratio of 16:1. Oil designed for air-cooled engines should be used.

Gear reduction unit is lubricated with a Lubriplate type grease or equivalent. Oil should not be used as clutch action will be affected.

REPAIRS

PISTON, PIN, RINGS AND CYLINDER. Model 20A is equipped with a chrome plated cylinder. To remove cylinder, remove blower housing, flywheel and magneto back plate. Note position of exhaust so cylinder may be reinstalled correctly depending on engine application.

Inspect cylinder for excessive wear or damage to chrome bore. Excessive wear may be checked by placing a new piston ring in cylinder bore and measuring ring end gap. If gap exceeds 0.077 inch (1.96 mm), cylinder should be renewed. Remove, clean and inspect piston. Piston ring side clearance should be 0.0015-0.0035 inch (0.0381-0.0889 mm). Piston ring end gap should be 0.067-0.077 inch (1.70-1.95 mm). Piston rings are nondirectional. Install piston with piston ring locating pins toward flywheel side of engine. Tighten cylinder retaining nuts to 50 in.-lbs. (5.7 N•m).

CONNECTING ROD. To remove connecting rod, remove cylinder and piston as previously outlined. Remove carburetor and reed valve assembly. Hold connecting rod cap (13-Fig. AE14) by placing a finger through intake port and remove connecting rod bolts. Remove connecting rod being careful not to lose bearing rollers. Use a suitable press to remove and reinstall small end bearing.

Connecting rod is fractured and serrations of rod and cap must mesh during assembly. Rod and cap have match marks as shown in Fig. AE15 which must be aligned. To reinstall connecting rod, hold the 14 bearing rollers in connecting rod cap with heavy grease and hold cap and rollers on crankshaft. New bearing rollers are on wax strips which may be used in place of grease. Place remaining 14 rollers in connecting rod with heavy grease, align match marks and serrations on rod and cap and install rod on crankshaft. Tighten rod bolts to 40-50 in.-lbs. (4.5-5.7 N•m).

CRANKSHAFT AND CRANKCASE. To remove crankshaft, remove connecting rod as previously outlined. Remove gear reduction assembly and clutch if so equipped. Remove crankcase

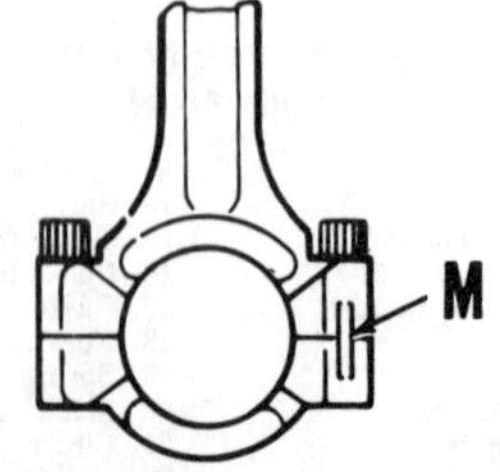

Fig. AE15—Connecting rod match marks (M) must be aligned during assembly.

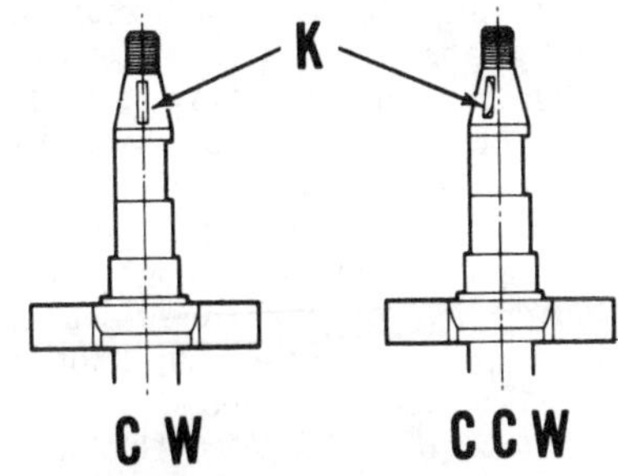

Fig. AE16—Crankshaft for clockwise (CW) or counterclockwise (CCW) rotation may be identified by keyway (K) location.

seal (22–Fig. AE14) and snap ring (23). Unscrew the three retaining screws and remove crankcase cover (16). Hold crankshaft while removing cover so crankshaft does not fall against crankcase. Withdraw crankshaft being careful not to lose loose bearing rollers in bearing (21). Remove seal (15), heat crankcase slightly and press bearing (21) race out of crankcase. Remove snap ring (19), heat crankcase cover slightly and press bearing (18) out of cover.

Note different locations of keyway on crankshaft used for clockwise or counterclockwise engine rotation. See Fig. AE16. Engine rotation is determined by rotation of output end of crankshaft and not gear reduction output shaft. Correct engine rotation is listed in parts book for each engine type number.

Inspect components for wear and damage. Heat crankcase cover slightly, press in bearing (18–Fig. AE14) and install snap ring (19). Heat crankcase and press in bearing (21) race so letters on race are to inside of crankcase. Install 18 bearing rollers in race with grease to hold rollers in place. Install crankshaft, gasket and crankcase cover. Install snap ring (23) and seal (22). Turn crankshaft and check for correct location of bearing (21) rollers and install seal (15). Reinstall remainder of engine components.

REED VALVE. Model 20A is equipped with a reed valve. Reed valve should be inspected for damage to reed valve petal or seat. Reed valve petal (8–Fig. AE11) should seat flat with an allowable gap of 0.010 inch (0.25 mm) between petal and seat. Reed petals are available separately. Install gaskets (6 and 9) and reed valve seat (7) so notches are aligned with notch in top of intake passage of crankcase (16). The notches form a passage for crankcase pulsations to the carburetor.

SERVICING ADVANCED ENGINE ACCESSORIES

RECOIL STARTER

Model 20A

To disassemble starter, remove rope handle and allow rope to rewind into starter. Remove cover (12–Fig. AE20). Rewind spring (11) should remain in cover but care should be taken when removing cover as spring may uncoil and could cause injury. Remove remainder of starter assembly from engine.

To reassemble starter, install starter dogs as shown in Fig. AE21. Note that crankshaft rotation is determined by viewing output end of crankshaft and NOT gear reduction output shaft. Place end of rope through hole in rope pulley and tie knot. Pass other end of rope through rope outlet in housing (9 or 9A–Fig. AE20). Assemble components (1 through 9). Pull rope out of starter so it is fully extended and attach rope handle so there is 28 inches (711 mm) from starter housing to bottom of rope handle. Wind rope on pulley by turning pulley in direction that does not engage starter dogs. Install rewind spring in cover as shown in Fig. AE21. Lubricate spring with a suitable lubricant. To preload starter rope, install spring and cover so inner hook of spring engages notch in rope pulley and turn cover 1 to 1½ turns in direction that will wind spring. Install cover screws.

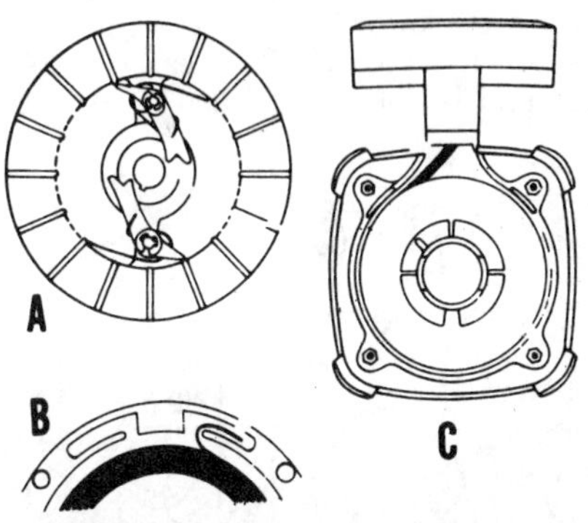

Fig. AE21—Views of location of starter dogs (A), rewind spring (B) and rope (C) on clockwise rotating engine. Reverse positions of starter dogs, rewind spring and rope shown above for counterclockwise rotating engine.

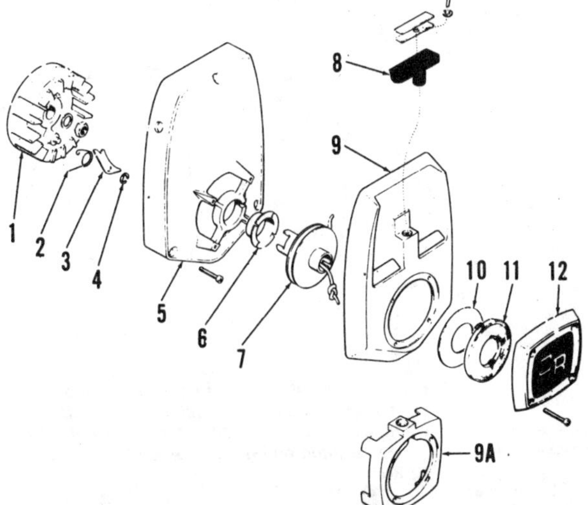

Fig. AE20—Exploded view of starter assembly. Housing (9) or (9A) may be used.

1. Flywheel
2. Spring
3. Starter dog
4. "E" ring
5. Blower housing
6. Bushing
7. Rope pulley & hub
8. Handle
9. Housing
9A. Housing
10. Washer
11. Rewind spring
12. Cover

Fig. AE22—On late models, except Model 20A, starter reel is retained to blower housing by a snap ring which must be removed as shown.

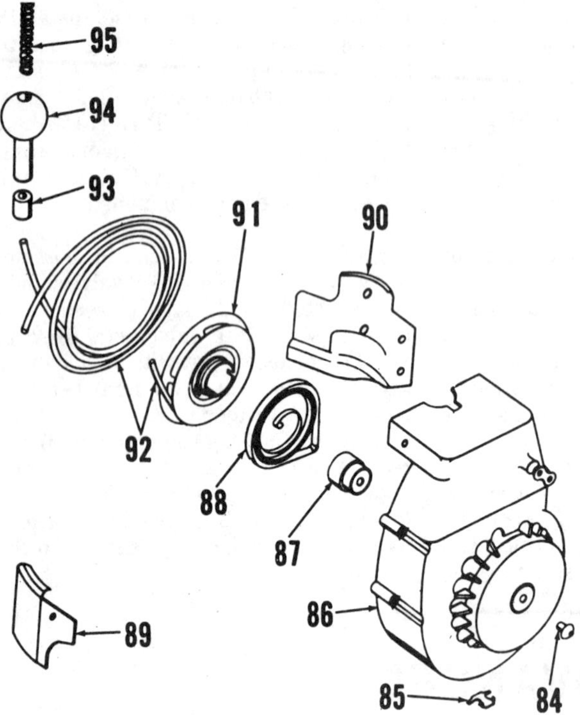

Fig. AE23—Exploded view of blower housing and recoil starter assembly. Screw (84) retains starter bearing (87) in blower housing.

84. Screw
85. Spring retainer (not used on late models)
86. Blower housing
87. Starter bearing
88. Starter spring
89. Cylinder baffle
90. Cylinder baffle
91. Starter reel assy
92. Starter cord
93. Cushion
94. Knob
95. Spring (not used on late models)

All Other Models

On late models, the starter reel is retained in blower housing by a snap ring as shown in Fig. AE22. On early models, the reel is not secured and is retained only by its close fit on bearing (87–Fig. AE23) and by rope and recoil spring.

To disassemble the starter, unseat the retaining snap ring if so equipped, and remove knob from rope. Allow spring to slowly unwind. Lift out the reel, being careful to disengage hooked inner end of recoil spring from reel hub. Spring should remain in recess in blower housing when reel is removed.

Early models use a 44-inch (1118 mm) nylon starter cord and hooked outer end of spring engages a slot in outside of blower housing. Late models use a 54-inch (1372 mm) cord and outer end of recoil spring contains a riveted loop. Cord and/or spring cannot be interchanged. On all models, rope is anchored to reel hub by a twisted soft iron wire.

When reinstalling the reel, unwind rope from reel sheave and pull entire rope out between spokes of reel. Install reel, making sure inner end of spring engages slot in reel hub. Install retaining snap ring if so equipped, making sure end gap is NOT positioned in wide slot where rope emerges. On all models, turn reel counterclockwise until recoil spring is completely wound, then clockwise one turn. Feed starter rope back through reel spokes and out through opening in blower housing. Pull rope tight through guide rollers and tie a slip knot in rope until knob is reinstalled.

Starter dogs (36–Fig. AE5) mount on pins on outer face of flywheel. On early models, dogs were retained by a machined slot in flywheel hub. On late models, a retaining washer is used. Washer is a press fit on flywheel and a new washer should be installed each time starter dogs do not bind or are not unduly worn. Stake new retaining washer in two places on models so equipped.

Fig. AE24—Remove threaded clutch hub by inserting screwdriver blade as shown and tapping with hammer.

DIRECT DRIVE CLUTCH

Model 20A

Clutch hub may be screwed on crankshaft if clutch sprocket is mounted inboard, or clutch may be retained on crankshaft with a snap ring. To unscrew hub, place blade of screwdriver in slot of clutch shoe as shown in Fig. AE24 and tap on screwdriver while holding flywheel. Disassemble clutch assembly and inspect components. Renew clutch shoes if grooves in rubbing surface are worn away or shallow. Renew clutch shoes in pairs. Inspect clutch drum for concentricity. If clutch engagement speed occurs at lower than desired rpm, renew clutch springs. Clutch springs must be renewed as a unit to obtain uniform clutch engagement.

Install clutch shoes on hub so relieved end of shoe is trailing surface. Install clutch springs in shoes with "V" of springs bent back against block on hub as shown in Fig. AE26. Secure retainer plate (11–Fig. AE25) screws with Loctite.

1. Seal
2. Gear cover
3. Bearing
4. Thrust washer
5. Output gear & shaft
6. Thrust washer
7. Bearing
8. Snap ring
9. Clutch housing
10. Washer
11. Retainer plate
12. Clutch springs
13. Clutch shoes
14. Clutch hub
15. Washer
16. Snap ring
17. Seal
18. Snap ring
19. Crankcase cover

Fig. AE25—Exploded view of gear reduction unit. Clutch components may be used without gear reduction as direct drive.

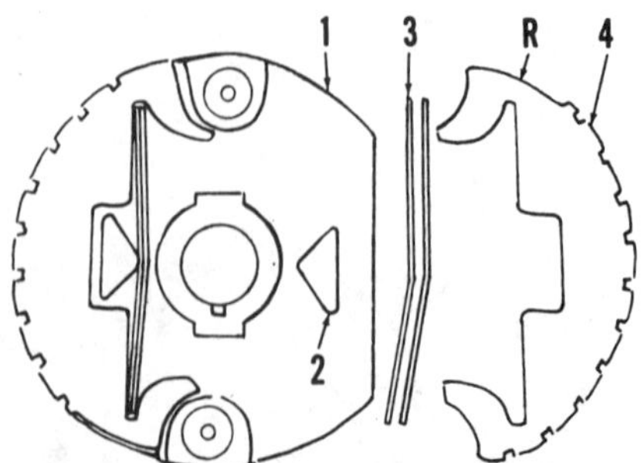

Fig. AE26—Install clutch springs (3) so arch of spring will be bent back against anchor block (2) when installed. Clutch shoe recess (R) should be on trailing end of shoe when installed.

R. Recess
1. Clutch hub
2. Anchor block
3. Clutch springs
4. Clutch shoe

All Other Models

EARLY PRODUCTION. To disassemble the early production direct drive clutch, loosen the extension shaft retaining cap screw (2–Fig. AE27) one turn. While supporting engine by holding extension shaft (4), tap loosened cap screw sharply with a small hammer to loosen taper fit between shafts. Remove extension shaft, clutch cover, thrust washer and roller bearing. If clutch drum (6) will not pull out of clutch bearing easily, reinstall extension shaft on clutch shaft and, while supporting engine with extension shaft, drive diecast clutch bearing (8) off of clutch drum with a screwdriver and small hammer.

CAUTION: Be careful that the engine does not drop when clutch bearing and drum separate. Hold engine close to bench surface.

Clutch shoes (7) can then be removed from clutch hub. Renew shoes if linings are excessively worn. To remove clutch hub from engine backshaft, loosen screw (10) one turn and support engine by holding clutch hub. Set screwdriver against loosened screw and tap screwdriver sharply with small hammer to loosen tapered fit between clutch hub and backshaft.

Lubricate clutch parts with SAE 30 oil and use new paper gasket between clutch housing and cover during reassembly. Reverse disassembly procedure to reassemble.

NOTE: Parts are no longer available for early production clutch. Install complete later production clutch if clutch renewal is necessary.

LATE PRODUCTION. To disassemble the late production direct drive clutch (Fig. AE28), loosen extension shaft retaining cap screw (1) one turn. While supporting engine by holding extension shaft (2), tap loosened cap screw sharply with small hammer to loosen taper fit between shafts. Remove extension shaft, clutch cover (3) and roller bearing retainer assembly (4). If clutch drum (5) will not pull out of clutch bearing easily, reinstall extension shaft and while supporting engine with extension shaft, lightly tap clutch bearing (7) off of clutch drum with a screwdriver and small hammer.

CAUTION: Be careful that engine does not drop when clutch bearing and drum separate. Hold engine close to bench surface.

Clutch shoes and hub assembly (6) can then be removed from engine backshaft. Loosen Phillips head screw (in end of backshaft) one turn and support engine by holding clutch bearing (7). Set screwdriver against loosened screw and tap screwdriver sharply with small hammer to loosen tapered fit between clutch hub and backshaft.

Lubricate clutch parts with grease and use new paper gasket between clutch housing and cover during reassembly. Reverse disassembly procedure to reassemble unit.

GEAR REDUCTION UNIT

Model 20A

Disassemble gear reduction assembly by removing gear cover (2–Fig. AE25) being careful not to lose bearing rollers in bearing (3). Withdraw output gear and shaft (5). Remove snap ring (8) and

Fig. AE27—View showing direct drive clutch drum and shoes removed.

1. Clutch cover
2. Cap screw
3. Clutch cover seal
4. Extension shaft
5. Outer clutch bearing
6. Clutch drum
7. Clutch shoes & spring assy.
8. Inner clutch bearing
9. Clutch hub
10. Clutch hub retaining screw

Fig. AE28—Exploded view of late production direct drive clutch assembly. Refer to Fig. AE27 for view of early type clutch.

1. Extension shaft screw
2. Extension shaft
3. Clutch cover
4. Roller bearing
5. Clutch drum
6. Clutch drum & shoe assy.
7. Clutch bearing
8. Engine backshaft

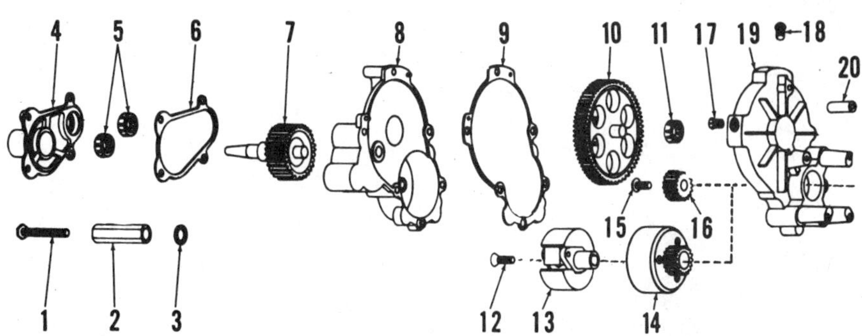

Fig. AE29—Exploded view of double reduction gear unit. On models without clutch (13 & 14), drive gear (16) is used.

1. Extension shaft screw
2. Extension shaft
3. Sealing ring
4. Reduction gear cover
5. Roller bearings
6. Gasket
7. Output gear
8. Gear housing cover
9. Gasket
10. Intermediate gear
11. Roller bearing
12. Phillips head screw
13. Clutch hub & shoe assy.
14. Clutch drum & drive gear
15. Phillips head screw
16. Drive gear
17. Plug
18. Plug
19. Gear housing
20. Spacer (reverse position gearcase models only)

remove clutch assembly. Refer to previous section to service clutch. To remove crankcase cover (19), remove snap rings (16 and 18) and seal (17). Unscrew retaining screws and remove cover while supporting crankshaft end. Refer to CRANKSHAFT section to service crankcase cover. When reassembling unit, lubricate bearings and seals with Lubriplate. Be sure bearing rollers are correctly installed. Note that gear reduction unit can be mounted on crankcase in three different positions.

All Other Models

Models with reduction gear drive (gearbox) may be equipped with or without centrifugal clutch. Models with 900 rpm output shaft speed have a compound gearbox.

To disassemble gear reduction drive, loosen cap screw retaining extension shaft (left hand threads on 1700 and 3300 rpm models, right hand thread on 900 rpm models) and, while supporting engine with extension shaft, tap loosened screw sharply with small hammer to loosen tapered fit of shaft. On 900 rpm models, remove compound gear cover, extension shaft and gear assembly. On 1700 and 3300 rpm models, remove extension shaft and gearbox cover.

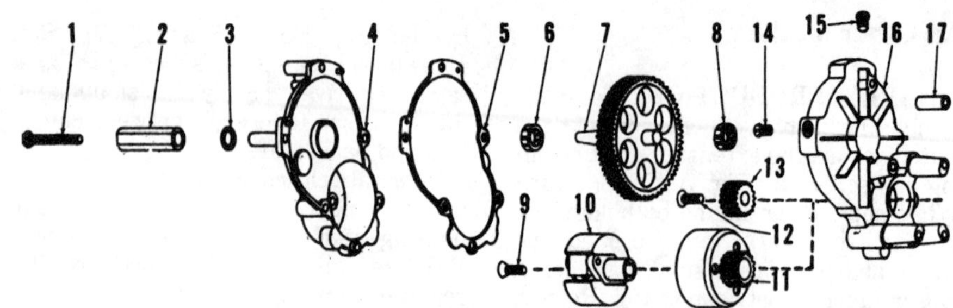

Fig. AE30—Exploded view of single reduction gear unit. Drive gear (13) is used on models without clutch (10 & 11). Spacer (17) is used on reverse position gearcase installation only.

1. Extension shaft screw
2. Extension shaft
3. Sealing ring
4. Gearcase cover
5. Gasket
6. Roller bearing
7. Output gear
8. Roller bearing
9. Phillips head screw
10. Clutch hub & shoe assy.
11. Clutch drum & drive gear
12. Phillips head screw
13. Drive gear
14. Plug
15. Plug
16. Gear housing
17. Spacer

Loosen the screw retaining clutch hub or pinion gear to engine backshaft and while supporting engine with clutch hub or pinion gear, tap loosened screw sharply with a small hammer and screwdriver to loosen taper fit of hub or gear to shaft. Remove pinion gear or clutch hub and large driven gear. Gear housing, induction housing and carburetor may now be removed from engine as a unit.

Reverse disassembly procedure to reassemble. Lubricate gearbox with steel gears with two tablespoons of heavy grease such as Union Oil Unoba F-1 or equivalent. One tablespoon of grease should be added any time gear cover is removed. On models without steel gears, lubricate unit with one tablespoon of SAE 30 motor oil. One teaspoon of oil should be added after each 10 hours of use.

CAUTION: Do not overfill gearbox as this will cause overheating. Gearbox is provided with a screw plug for lubricating gears.

BRIGGS & STRATTON

2-STROKE

BRIGGS & STRATTON CORPORATION
P.O. Box 702
Milwaukee, Wisconsin 53201

Model	Bore	Stroke	Displacement
62030, 62033	2.125 in. (53.975 mm)	1.70 in. (43.18 mm)	6.0 cu. in. (98.3 cc)
95700, 96700	2.362 in. (59.995 mm)	2.054 in. (52.172 mm)	9.0 cu. in. (147.5 cc)

BRIGGS & STRATTON NUMERICAL MODEL NUMBER SYSTEM

CUBIC INCH DISPLACEMENT	FIRST DIGIT AFTER DISPLACEMENT BASIC DESIGN SERIES	SECOND DIGIT AFTER DISPLACEMENT CRANKSHAFT, CARBURETOR GOVERNOR	THIRD DIGIT AFTER DISPLACEMENT BEARINGS, REDUCTION GEARS & AUXILIARY DRIVES	FOURTH DIGIT AFTER DISPLACEMENT TYPE OF STARTER
6 8 9 10 11 13 17 19 22 23 24 25 28 29 30 32 40 42	0 1 2 3 4 5 6 7 8 9	0 - Horizontal Diaphragm 1 - Horizontal Vacu-Jet 2 - Horizontal Pulsa-Jet 3 - Horizontal Flo-Jet Pneumatic Governor 4 - Horizontal Flo-Jet Mechanical Governor 5 - Vertical Vacu-Jet 6 - 7 - Vertical Flo-Jet 8 - 9 - Vertical Pulsa-Jet	0 - Plain Bearing 1 - Flange Mounting Plain Bearing 2 - Replaceable Bearing 3 - Flange Mounting Ball Bearing 4 - 5 - Gear Reduction (6 to 1) 6 - Gear Reduction (6 to 1) Reverse Rotation 7 - 8 - Auxiliary Drive Perpendicular to Crankshaft 9 - Auxiliary Drive Parallel to Crankshaft	0 - Without Starter 1 - Rope Starter 2 - Rewind Starter 3 - Electric - 120 Volt, Gear Drive 4 - Electric Starter - Generator - 12 Volt, Belt Drive 5 - Electric Starter Only - 12 Volt, Gear Drive 6 - Alternator Only 7 - Electric Starter - 6 or 12 Volt, Gear Drive, with Alternator 8 - Vertical Pull Starter *Digit 6 formerly used for "Wind Up" Starter on 60000, 80000 and 92000 Series.
9 9 Cubic Inch	5 Design Series 5	7 Vertical Flo-Jet	2 Replaceable Bearing	2 Rewind Starter

Fig. BS1—Explanation of engine model numerical code used by Briggs & Stratton to identify engine and optional equipment.

ENGINE INFORMATION

All models are two-stroke, single-cylinder, third port loop scavenged design engines. Models can be equipped with chrome plated cylinder bore or a cast iron cylinder bore. Refer to Fig. BS1 for interpretation of Briggs & Stratton engine model numbers.

MAINTENANCE

SPARK PLUG. Recommended spark plug for Models 62030 and 62033 is a Champion RCJ-8 or equivalent. Recommended spark plug for Models 95700 and 96700 is a Champion J19LM or equivalent. Specified spark plug gap for all models is 0.030 inch (0.76 mm).

CARBURETOR. Models 62030 and 62033 are equipped with a diaphragm type carburetor with an integral fuel pump and Models 95700 and 96700 are equipped with a float type carburetor. Refer to the appropriate paragraphs for model being serviced.

Diaphragm Type Carburetor. Refer to Fig. BS3 for an external view of carburetor.

Initial adjustment of carburetor mixture screw (3—Fig. BS4) is 1-1/2 turns counterclockwise from a lightly seated position. Make final adjustment with engine at operating temperature and running. Turn mixture screw (3) clockwise (leaner) until engine runs smoothly, then turn mixture screw counterclockwise (richer) until engine just starts to sputter. This position will provide the best engine performance under load. Mixture screw should never be less than 1-1/4 turns open from a lightly seated position or improper engine lubrication will result.

To disassemble carburetor, remove the five diaphragm cover screws and lift off cover. Be careful not to lose the pump valve springs (4—Fig. BS5) when removing diaphragm. Refer to Fig. BS6 and remove inlet needle assembly. Using a small hook, remove the C-ring holding needle seat to carburetor body. Using a 1/16 inch (1.6 mm) punch or Allen wrench, pry out check valve and seat. With fuel mixture screw (3—Fig. BS4) removed, remove "O" ring seal from shoulder on fuel mixture screw with a small hook. Clean carburetor body and diaphragm cover in carburetor cleaner.

Inspect for worn or damaged parts and renew as needed. Check carburetor body for warpage with a straightedge. Check with straightedge (1—Fig. BS7) placed from corner-to-corner as shown. Renew if more than 0.003 inch (0.076 mm) warpage is evident.

To reassemble, install fuel inlet needle seat (2—Fig. BS8) with grooved side facing toward bottom of cavity. Insert "C" ring (1) to a depth of 5/16 inch (7.92 mm) in cavity. Place spring (7—Fig. BS6) in spring pocket machined in carburetor. Assemble hinge pin (3), inlet needle lever (2) and inlet needle (5). Make certain inlet needle retaining clip (4) holds inlet needle against needle lever. Install assembly in carburetor body making certain spring is centered on lever dimple (6), then install retaining screw (1). Spring lever height should be 1.16 inch (1.57 mm). Measure as shown in Fig. BS10. Carefully bend lever to adjust. Install check valve (2—Fig. BS11) in pocket. Install Welch plug (1) and use a ¼ inch round punch to bend and seat Welch plug. Assemble fuel pump spring and cup, check valve springs, diaphragm, gasket and cover. Tighten diaphragm cover screws firmly follow-

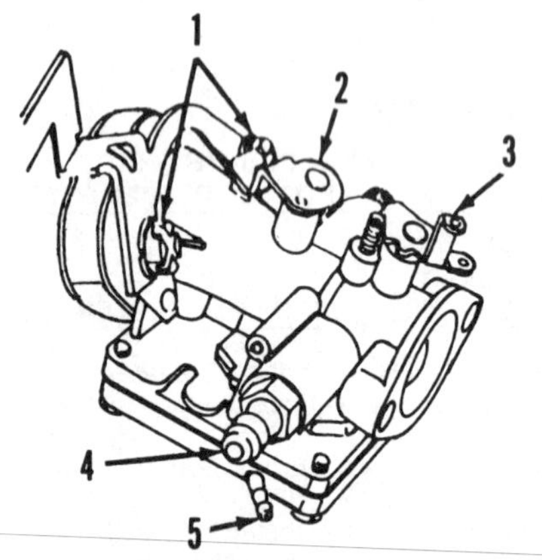

Fig. BS3—View of diaphragm type carburetor used on Models 62030 and 62033.

1. Mounting screws
2. Throttle lever
3. Choke lever
4. Fuel inlet
5. Primer inlet

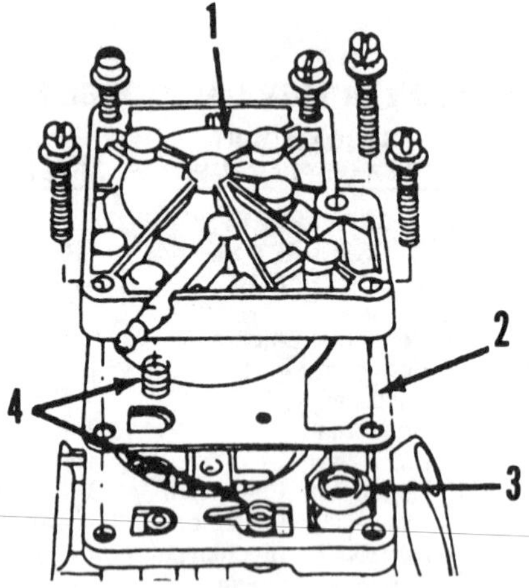

Fig. BS5—View of carburetor fuel pump cover (1), diaphragm (2), cup and spring (3) and valve springs (4).

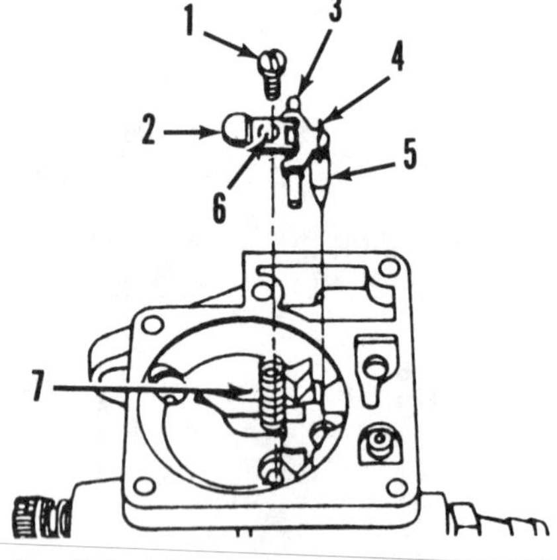

Fig. BS6—Internal view of diaphragm type carburetor.

1. Screw
2. Lever
3. Pin
4. Clip
5. Fuel inlet needle
6. Dimple
7. Spring

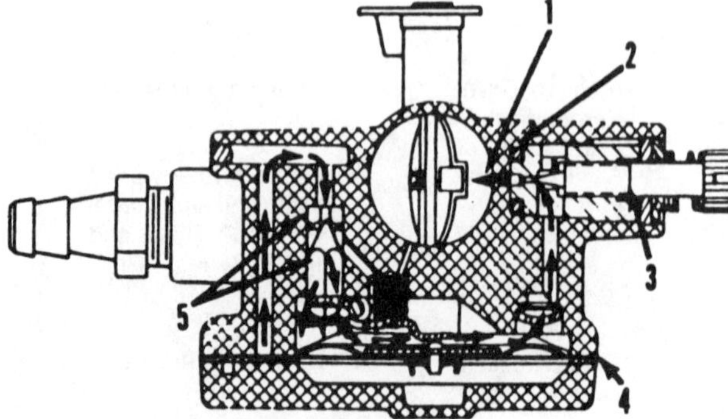

Fig. BS4—Cutaway view diaphragm type carburetor used on Models 62030 and 62033.

1. Metering hole
2. "O" ring
3. Mixture screw
4. Diaphragm
5. Inlet needle & seat

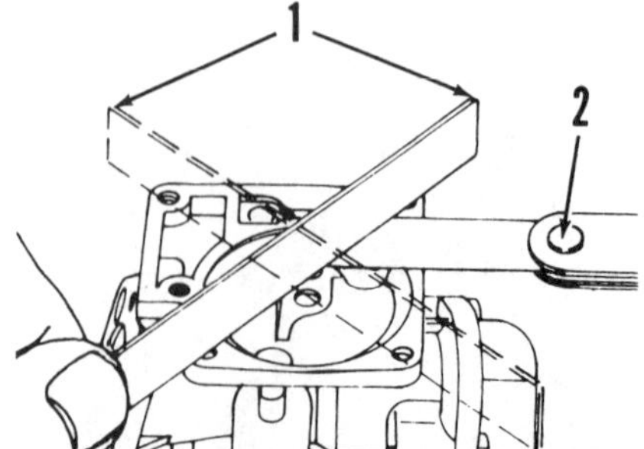

Fig. BS7—Check carburetor body for warpage with straightedge (1) and feeler gage (2). If warpage exceeds 0.003 inch (0.076 mm), renew carburetor.

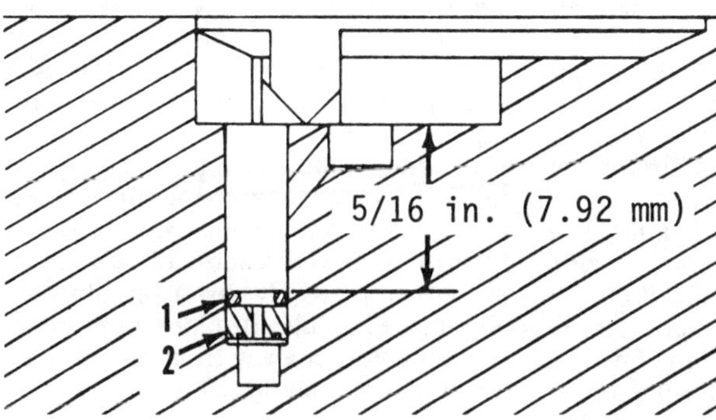

Fig. BS8—Refer to text for proper installation of "C" ring (1) and fuel inlet needle seat (2).

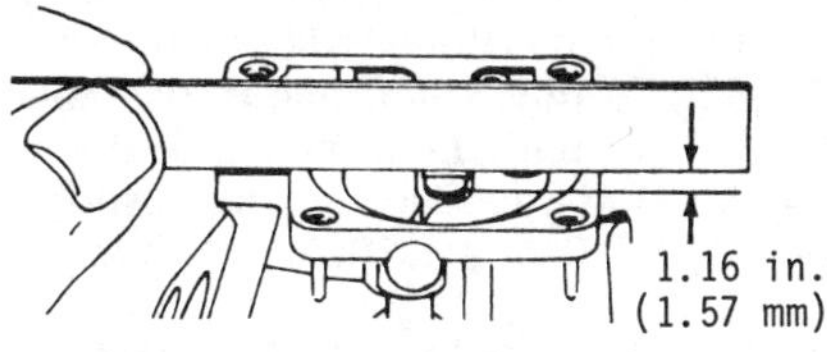

Fig. BS10—Lever height should be 1.16 inch (1.57 mm). Measure as shown.

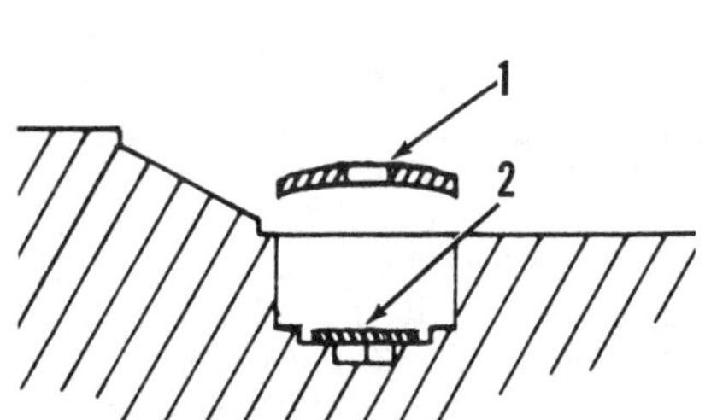

Fig. BS11—Install check valve (2) and Welch plug (1). Use a ¼ inch round punch to bend and seat Welch plug.

ing sequence shown in Fig. BS12. Assemble throttle plate and throttle shaft, choke shaft and choke plate. Install "O" ring (1—Fig. BS13) on shoulder of seat (2). Install spring (5), metal washer (4) and sealing washer (3) on fuel mixture screw (6). Align flat area (7) of seat (2) with flat area of seat bore and use oil fill tube (part 280131) to firmly seat assembly in bore. Screw fuel

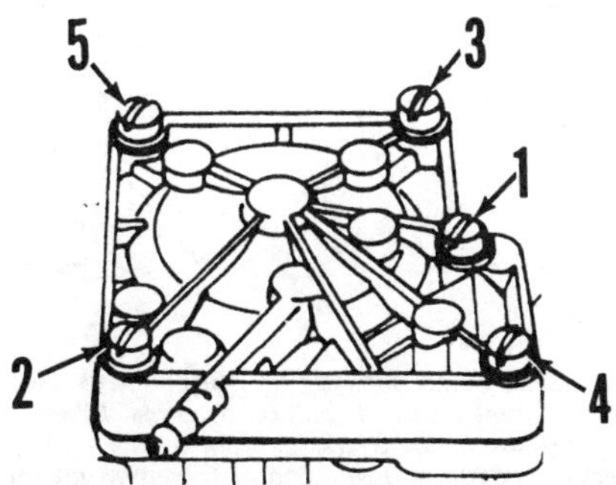

Fig. BS12—Firmly tighten diaphragm cover screws following the sequence shown.

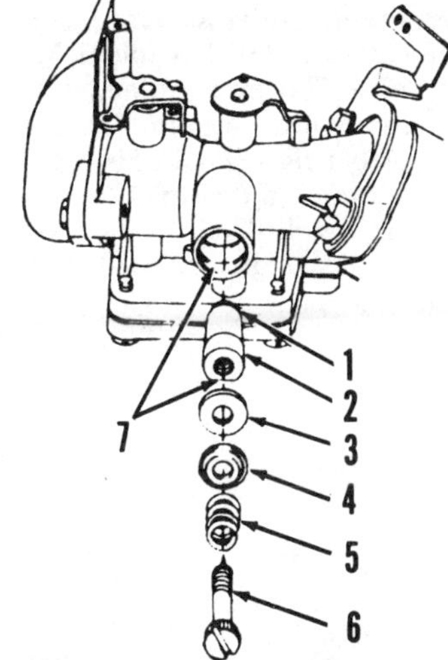

Fig. BS13—View of low speed mixture screw components on diaphragm type carburetor used on Models 62030 and 62033.

1. "O" ring
2. Seat
3. Sealing washer
4. Metal washer
5. Spring
6. Mixture screw
7. Flat area

mixture screw (6) in until lightly seated, then back out 1-1/2 turns.

Float Type Carburetor. The float type carburetor used on Models 95700 and 96700 is equipped with a fixed idle and main fuel jet. Standard fixed main jet size is #82.5. Optional #80 and #77.5 fixed main jets are available for high altitude operation. The only adjustment is engine idle rpm which is controlled by the throttle stop screw. Engine idle speed should be 1200 rpm with engine at operating temperature.

To disassemble carburetor, remove float bowl retaining screw (5—Fig. BS14), gasket (4) and float bowl (3). Float pin has one end which is flat on two sides (1—Fig. BS15) and one end that is round. Push pin from round end using a 5/64 inch (1.98 mm) punch to remove pin. Remove float, fuel inlet needle and float bowl gasket. Remove main jet (2) and idle jet (1—Fig. BS14).

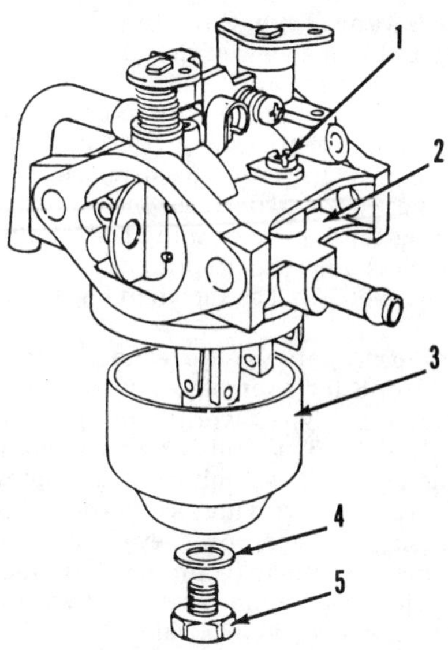

Fig. BS14—View of float type carburetor used on Models 95700 and 96700.

1. Idle jet
2. Carburetor body
3. Float bowl
4. Gasket
5. Screw

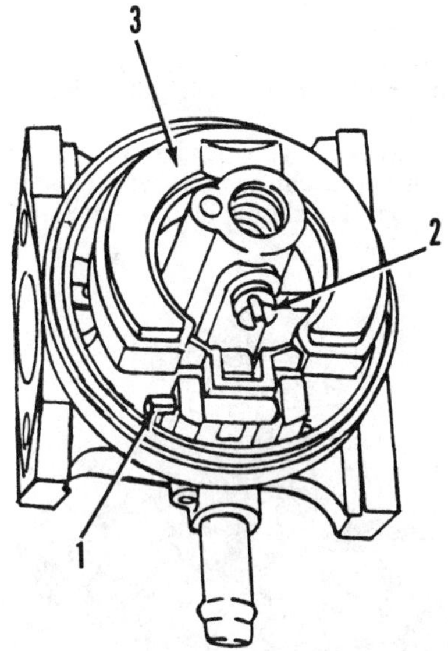

Fig. BS15—View of float chamber area on float type carburetor used on Models 95700 and 96700.

1. Squared side of float pin
2. Main jet
3. Float

Remove throttle plate, throttle shaft, choke plate, choke shaft and all shaft washers and seals.

To reassemble carburetor, reverse disassembly procedure. Float height of plastic float is nonadjustable.

GOVERNOR. Models 62030 and 62033 are equipped with an air-vane type governor and Models 95700 and 96700 are equipped with a centrifugal type governor. Refer to appropriate paragraph for service information.

Air-Vane Type Governor. Refer to Fig. BS16 for a view of the air-vane type governor used on Models 62030 and 62033. If adjustment of maximum no-load speed is necessary, tighten spring tension to increase speed and loosen spring tension to decrease speed. Adjust spring tension by bending the tang to which the governor spring is attached using tool 19229 as shown in Fig. BS17.

Centrifugal Type Governor. Refer to Fig. BS18 for a simplified view of the centrifugal type governor linkage used on Models 95700 and 96700. Maximum engine speed is limited by changing spring (4) and varying location of spring in holes located in speed lever (6). Three springs are available. Alternative springs provide maximum speed ranges of 2600-2900, 3000-3300 and 3400-3600 rpm depending upon spring selection and location.

IGNITION SYSTEM. Models 62030 and 62033 can be equipped with either a magneto type or a Magnetron ignition system. Models 95700 and 96700 are equipped with a magnetron ignition system. Refer to the appropriate paragraph for service information.

Magneto Type Ignition. Breaker-point gap on all models with magneto type ignition should be 0.020 inch (0.51 mm). Refer to Fig. BS19. One of the breaker contact points is an integral part of the ignition condenser. The breaker arm pivots on a knife edge retained by a slot in the pivot post. Breaker-point actuating plunger should be renewed if plunger is worn to a length of 0.931 inch (23.6 mm) or less. Specified air gap between armature and flywheel magnet is 0.006-0.010 inch (0.15-0.25 mm). The magneto type ignition uses a zinc flywheel key (1—Fig. BS20). Ignition coil can be tested using an ohmmeter. Resistance of ignition coil on the primary side should be 0.2-0.3 ohm. Resistance of ignition coil on secondary side should be 2500-3500 ohms.

Magnetron Ignition System. Some Models 62030 and 62033 and all Models 95700 and 96700 are equipped with the Magnetron ignition system. Installation of the correct flywheel key is required. Use the aluminum flywheel key (3—Fig. BS20) for Models 62030 and 62033 and aluminum key (2) for Models 95700 and 96700.

To check spark, remove spark plug. Connect spark plug cable to B&S tester 19051, then ground remaining tester lead to engine. Spin engine at 350 rpm or more. If spark jumps the 0.166 inch (4.2 mm) tester gap, system is functioning properly.

To remove armature and Magnetron module, remove flywheel shroud and armature retaining screws. Use a 3/16 inch (4.76 mm) diameter pin punch to release stop switch wire from module. To remove module, unsolder wires, push module retainer away from laminations and remove module.

Resolder wires for reinstallation and use Permatex or equivalent to hold ground wires in position. Ignition armature air gap for Models 62030 and 62033 should be 0.006-0.010 inch (0.15-0.25 mm). Ignition armature air gap for Models 95700 and 96700 should be 0.008-0.016 inch (0.20-0.40 mm).

CYLINDER HEAD AND COMBUSTION CHAMBER. Cylinder head on Models 62030 and 62033 is an integral part of one crankcase half and cylinder head on Models 95700 and 96700 is a separate casting from the crankcase halves. Cylinder head should be removed periodically and carbon and combustion deposits cleaned.

Specified compression for Models 62030 and 62033 without compression release is 90-110 psi (620-758 kPa). Specified compression for Models 62030 and 62033 with compression release mechanism is 80-90 psi (552-620 kPa). Specified compression for Models 95700 and 96700 is 90-120 psi (620-827 kPa).

LUBRICATION. Manufacturer recommends mixing a good quality BIA or NMMA oil certified for TC-W service with regular or low-lead gasoline at a 50:1 ratio.

FLYWHEEL BRAKE. Some 95700 and 96700 models are equipped with a band type flywheel brake which should stop the engine within three seconds, with remote control set at high speed

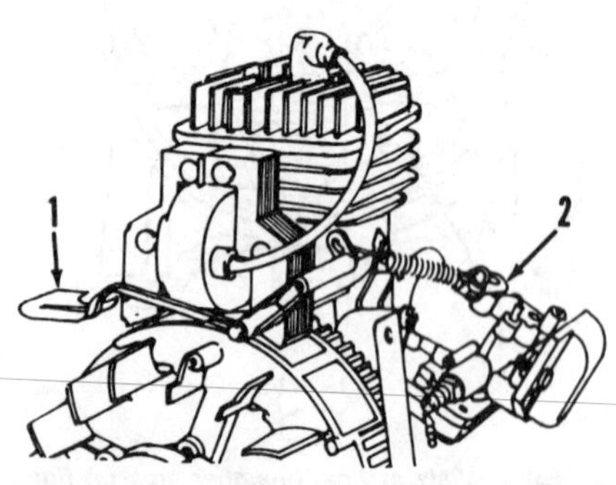

Fig. BS16—View showing air vane (1) and throttle lever (2) used on Models 62030 and 62033.

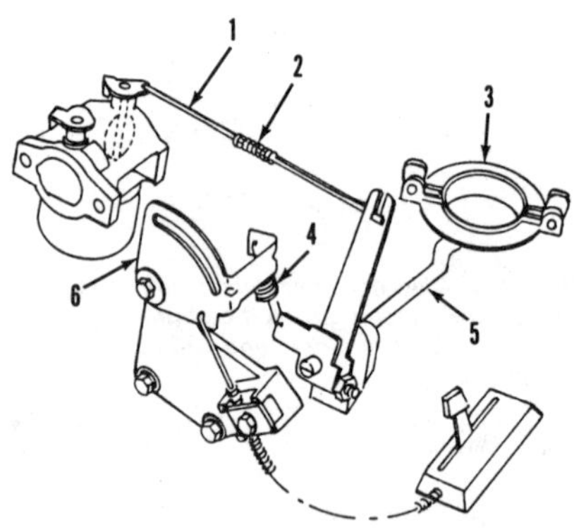

Fig. BS18—Simplified view of governor linkage used on Models 95700 and 96700.

1. Governor lever-to-carburetor rod
2. Spring
3. Flyweight assy.
4. Spring
5. Governor shaft
6. Speed lever

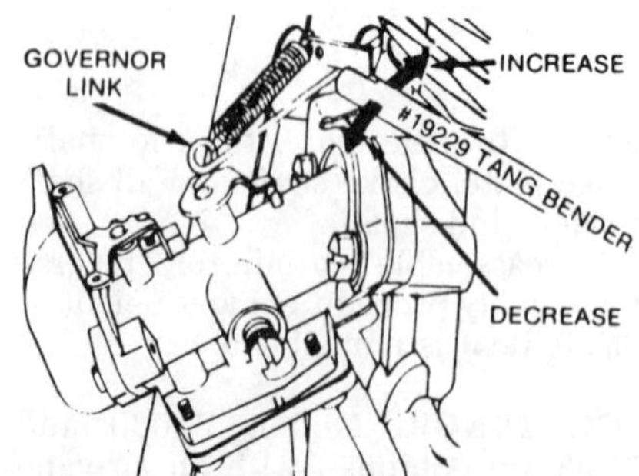

Fig. BS17—Use tool 19229 to adjust maximum engine speed on Models 62030 and 62033. Bend tang as needed to increase or decrease spring tension.

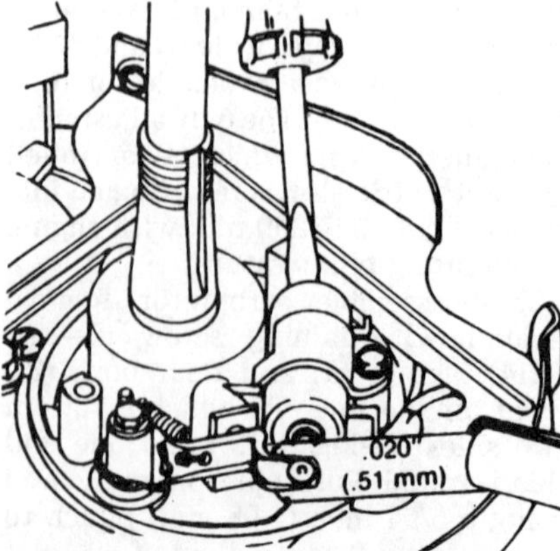

Fig. BS19—Breaker-point gap should be adjusted to 0.020 inch (0.51 mm) as shown.

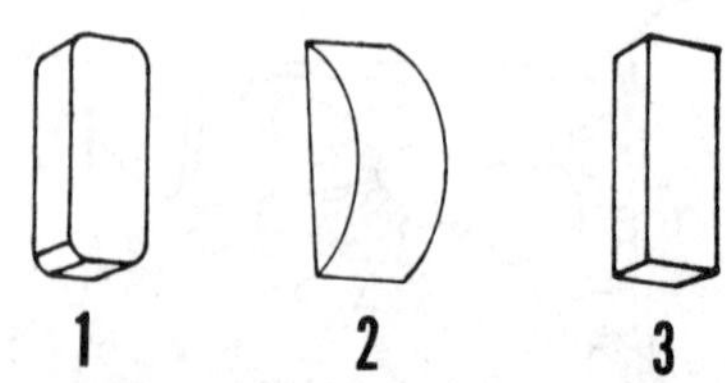

Fig. BS20—View showing the various keys used on different types of ignition systems. Use key style (1) for magneto type ignition, key style (2) for Models 95700 and 96700 and key style (3) for Models 62030 and 62033 with Magnetron ignition system.

position, when operator releases equipment safety control. Stopping time can be checked with tool 19255.

To adjust brake, first loosen both control bracket screws enough to allow movement of bracket. Place bayonet end of gage 19256 in control lever as shown in Fig. BS21. Push on control lever and install the other end of gage in cable clamp hole. Apply pressure to control bracket with a screwdriver and move it in direction shown by arrow until tension on gage is just eliminated. Hold control bracket in this position and tighten both screws to 25-30 in.-lbs. (2.82-3.39 N·m). As gage is removed a slight friction should be felt on gage and control lever should not move.

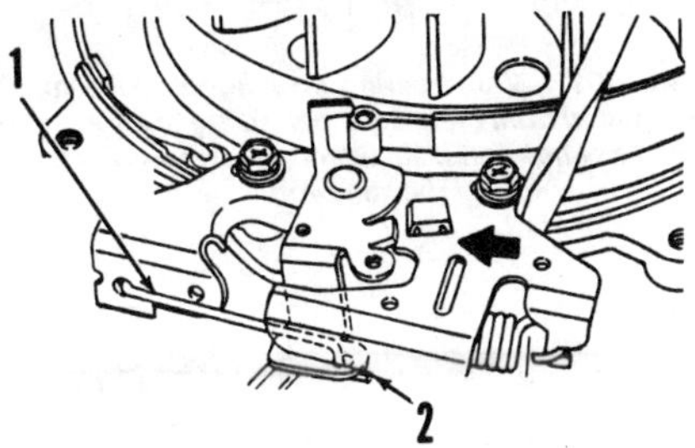

Fig. BS21—View showing flywheel brake adjustment on Models 95700 and 96700. Gage (1), part 19256, is inserted in hole (2). Refer to text.

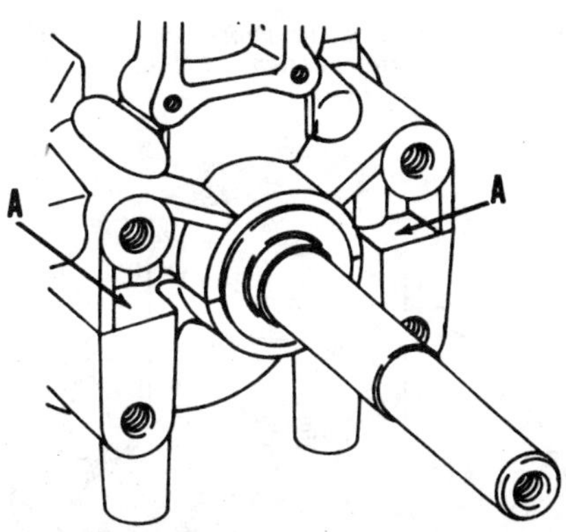

Fig. BS22—When separating crankcase halves, remove components as outlined in text, then tap locations (A) with a soft-faced mallet.

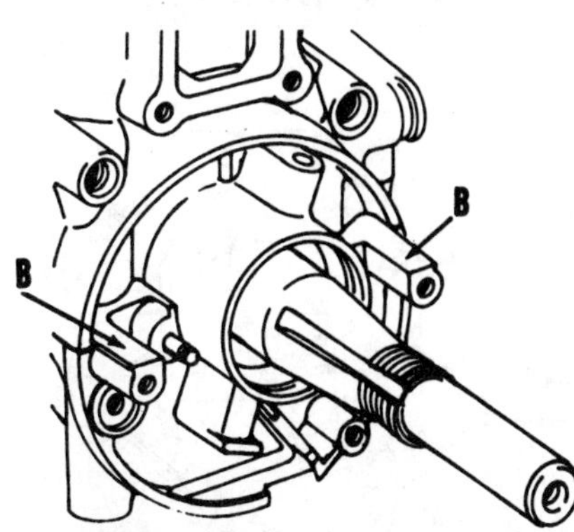

Fig. BS23—When separating crankcase halves, remove components as outlined in text, then tap locations (B) with a soft-faced mallet.

GENERAL MAINTENANCE. Check and tighten all loose bolts, nuts and clamps daily. Check for fuel leakage and repair if necessary. Clean cooling fins and external surfaces at 50 hour intervals.

REPAIRS

TIGHTENING TORQUES. Recommended tightening torque specifications are as follows:

Spark plug:
All Models . . . . 170 in.-lbs. (19.2 N·m)

Flywheel nut:
Models 95700 & 96700 . . . . 30 ft.-lbs. (40.7 N·m)

Connecting rod:
Models 62030 & 62033 . . . . 55 in.-lbs. (6.2 N·m)

Muffler:
Models 62030 & 62033 . . . . 115 in.-lbs. (13 N·m)
Models 95700 & 96700 . . . . 85 in.-lbs. (9.6 N·m)

Crankcase:
Models 62030 & 62033 . . . . 90 in.-lbs. (10.2 N·m)
Models 95700 & 96700 . . . . 60 in.-lbs. (6.8 N·m)

Cylinder:
Models 95700 & 96700 . . . . 110 in.-lbs. (12.4 N·m)

Carburetor:
Models 62030 & 62033 . . . . 100 in.-lbs. (11.3 N·m)
Models 95700 & 96700 . . . . 50 in.-lbs. (5.6 N·m)

CYLINDER AND CRANKCASE. On Models 62030 and 62033, the cylinder is an integral part of one crankcase half. On Models 95700 and 96700 the cylinder is a separate casting and can be separated from the crankcase halves.

To split crankcase halves on Models 62030 and 62033, remove the back plate, flywheel and crankcase cover screws. Using a soft-faced mallet, tap crankcase halves apart at locations (A—Fig. BS22) and (B—Fig. BS23). Lift out crankshaft, piston and rod assembly as shown in Fig. BS24.

Clean carbon from ports and inspect cylinder bore for scoring or other damage. Standard cylinder bore diameter is 2.124-2.125 inches (53.95-53.97 mm). If cylinder bore diameter is 2.128 inches (54.051 mm) or more, renew cylinder. Maximum allowable cylinder taper is 0.003 inch (0.076 mm) and maximum allowable out-of-round is 0.0025 inch (0.063 mm).

Refer to Fig. BS25 for a view of the compression release valve used on some models. Clean all carbon and foreign material from cavity and valve. When assembling valve on cylinder, seal and tighten the two screws to 30 in.-lbs. (3.4 N·m).

When reassembling crankcase halves, coat surfaces (A—Fig. BS26) with a silicone type sealer. Tighten all screws to specification listed under TIGHTENING TORQUES.

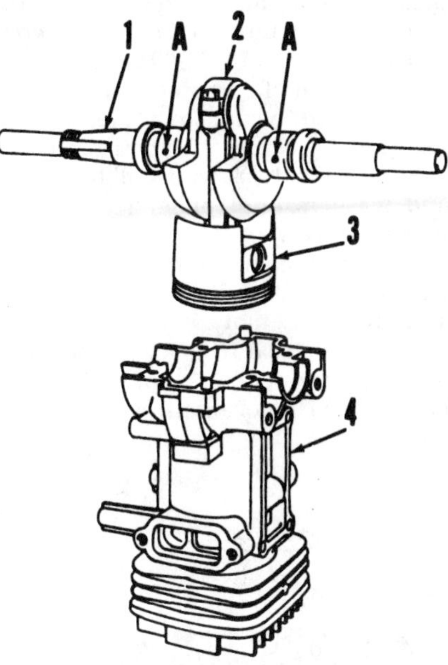

Fig. BS24—View showing crankshaft (1) connecting rod (2), piston (3), crankcase and cylinder (4) and main bearing locating tabs (A) used on Models 62030 and 62033.

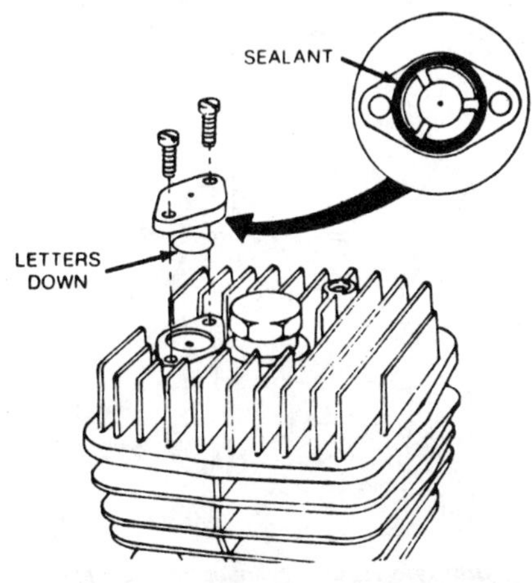

Fig. BS25—View showing compression release device used on cylinder head of some 62030 and 62033 models.

To split crankcase on Models 95700 and 96700, remove the four socket head screws retaining cylinder-to-crankcase. Using a soft-faced mallet, gently tap cylinder (Fig. BS27) to loosen cylinder. Carefully separate cylinder from crankcase. Remove the four cap screws retaining crankcase halves. Place flywheel side on a flat work surface and insert magneto end of crankshaft into flywheel taper. Refer to Fig. BS28 and insert a wooden hammer handle into crankcase to prevent crankshaft rotation. Place puller plate 19316, with "X" toward piston, and puller screw 19318 onto pto end of crankshaft. Thread puller studs 19317 into lower crankcase half. Turn puller screw until crankcase halves are separated, then withdraw puller assembly and lower crankcase. Insert puller studs 19317, with short threads down (Fig. BS29), into upper crankcase half. Protect crankshaft threads with flywheel nut. Place puller plate 19316, with "X" toward piston, onto crankshaft, then install four nuts and washers onto puller studs. Turn puller screw 19318 until upper crankcase is separated from crankshaft main bearing, then withdraw crankcase and puller assembly.

Remove oil seal from each crankcase half. Governor crank, on pto side, must be removed before ball bearing main bearings can be removed. Refer to Fig. BS30 for required tools and removal procedure for main bearing on pto side and to Fig. BS31 for required tools and removal procedure for main bearing on magneto side.

Remove snap rings retaining piston-to-connecting rod, then gently push piston pin from piston using tool shown in Fig. BS32. Separate piston from connecting rod. Remove piston rings from piston.

Connecting rod and crankshaft are considered an assembly and are not serviced separately.

Standard cylinder bore diameter is 2.362 inches (60 mm). Cylinder bore can be cast iron or chrome plated. If cylinder bore is chrome plated and cylinder diameter measures 2.368 inches (60.15 mm) or more, renew cylinder. If cylinder bore is cast iron and cylinder

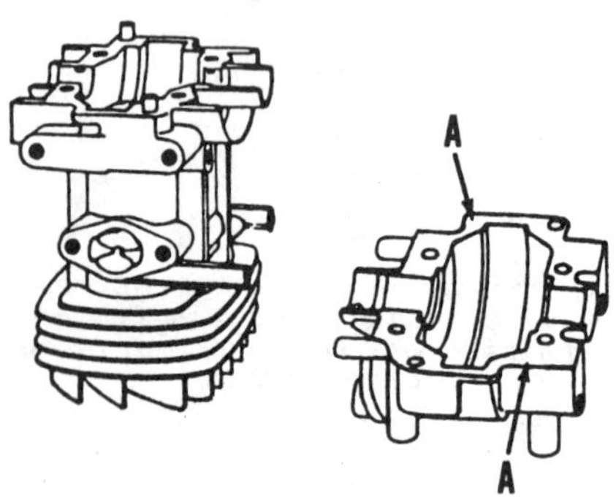

Fig. BS26—Apply sealer on surfaces (A) prior to reassembling crankcase.

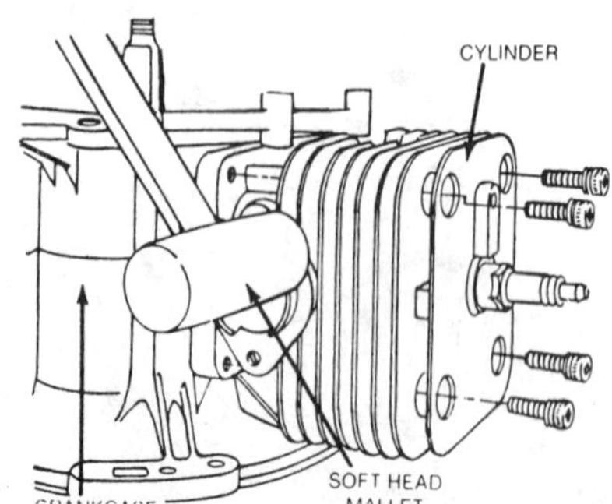

Fig. BS27—To remove cylinder head on Models 95700 and 96700, tap cylinder with a soft-faced mallet, then carefully lift cylinder off piston and connecting rod assembly.

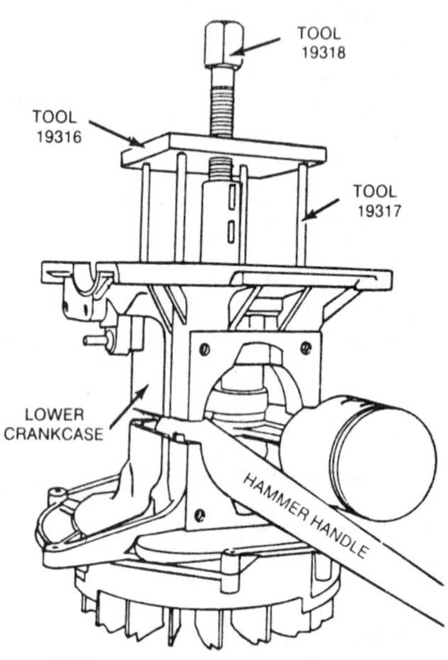

Fig. BS28—To separate crankcase halves, use puller assembly shown and insert a wooden hammer handle into crankcase to prevent crankshaft rotation. Refer to text.

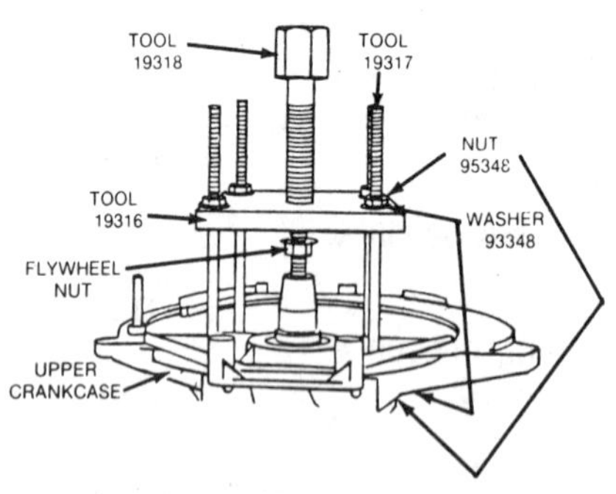

Fig. BS29—To remove crankshaft from crankcase, use puller assembly shown and refer to text.

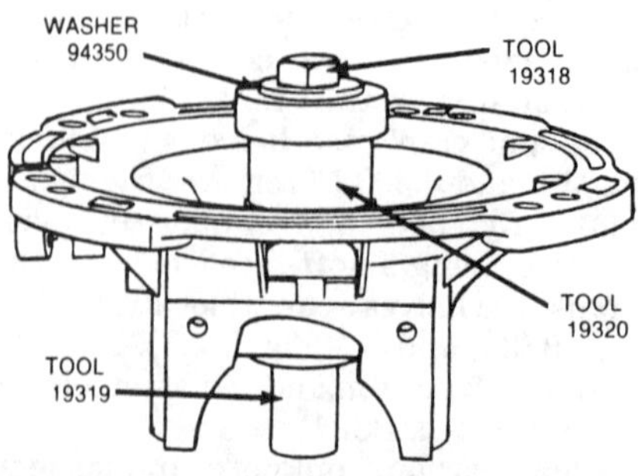

Fig. BS30—Main bearing on pto side must be pulled from crankcase half using tools assembled as shown. If main bearing is removed, it must be renewed.

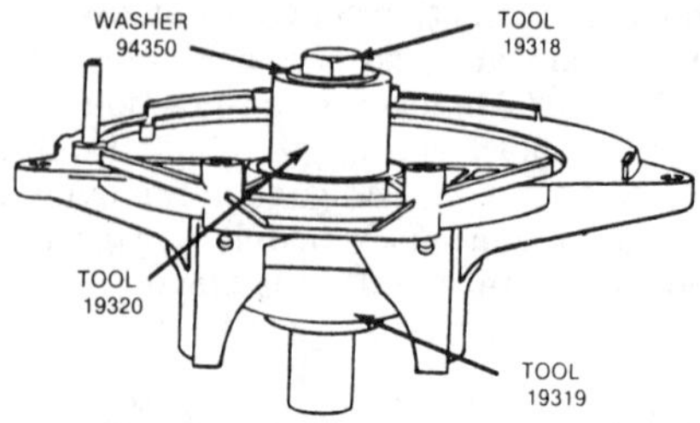

Fig. BS31—Main bearing on magneto side must be pulled from crankcase half using tools assembled as shown. If main bearing is removed, it must be renewed.

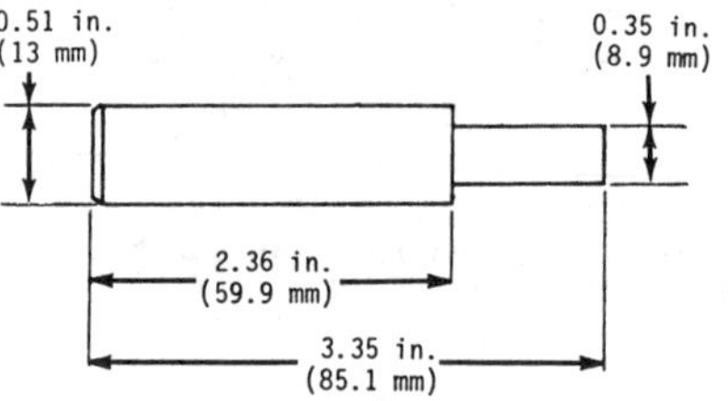

Fig. BS32—View shows dimensions for making piston pin removal tool.

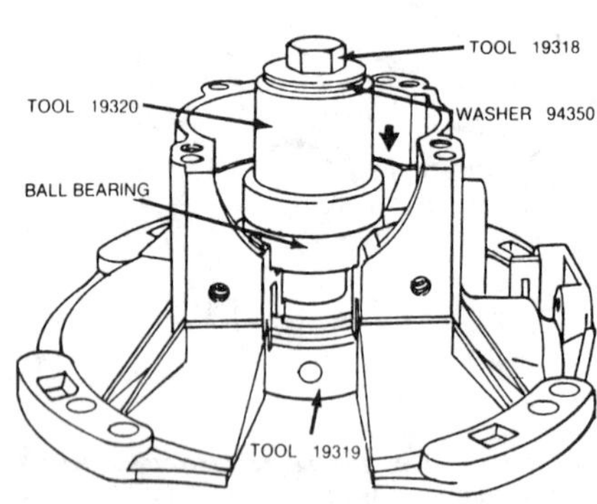

Fig. BS33—Assemble tools as shown to install ball type main bearing into pto side crankcase half on Models 95700 and 96700.

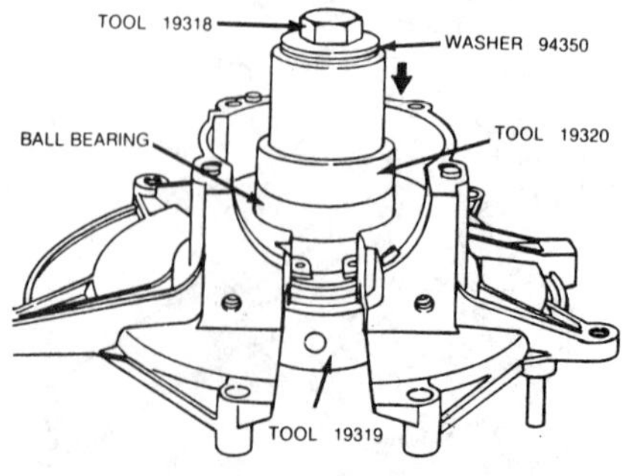

Fig. BS34—Assemble tools as shown to install ball type main bearing into magneto side crankcase half on Models 95700 and 96700.

diameter measures 2.369 inches (60.18 mm) or more, renew cylinder.

To install new main bearing into pto side crankcase half, assemble tools as shown in Fig. BS33. To install new main bearing into magneto side crankcase half, first install snap ring into bearing bore, then assemble tools as shown in Fig. BS34. Main bearing on magneto side seats against snap ring. To install new seals, refer to Fig. BS35 for magneto side seal and to Fig. BS36 for pto side seal. Pull seals into bores until they are seated.

To assemble magneto side crankcase half onto crankshaft, assemble tools as shown in Fig. BS37. Prevent crankshaft from turning, then turn nut to press crankcase half onto crankshaft until bearing seats. Remove tools and place crankshaft into flywheel taper on a suitable work bench. Install new gasket on crankcase, slide governor assembly down onto governor journal with weights as shown in Fig. BS38. Install new governor oil seal into crankcase until seal is flush with chamfer (Fig. BS39). Slide governor shaft through governor shaft bearing and place washer on shaft. Install "E" ring into groove cut in governor shaft. Position governor shaft as shown in Fig. BS40 and slide pto half of crankcase down onto crankshaft while holding governor crank in position. Assemble tools as shown in Fig. BS41. Prevent crankshaft from turning, then turn blade adapter bolt to pull crankcase onto crankshaft. Puller stud 19317 can be installed into crankcase halves to guide them together.

Refer to Fig. BS42 and install new rings on piston. Align ring end gaps with locator pins in ring grooves. Refer to following PISTON AND CONNECTING ROD paragraphs to assemble piston onto connecting rod. Install new gasket on cylinder. Make certain ring end gaps are aligned with locator pins, then slide cylinder and head assembly down over piston. Beveled portion on lower side of cylinder acts as a ring compressor. Install cylinder retaining bolts and tighten to specification listed under TIGHTENING TORQUES.

PISTON AND CONNECTING ROD. Refer to BS43 for a view of piston and

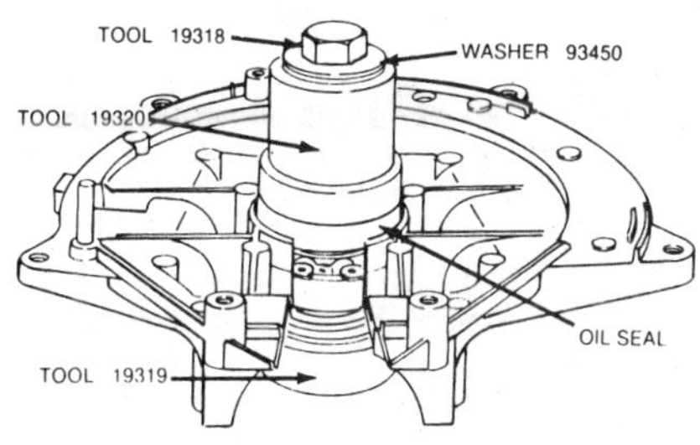

Fig. BS35—Assemble tools as shown to install seal into magneto side crankcase half on Models 95700 and 96700.

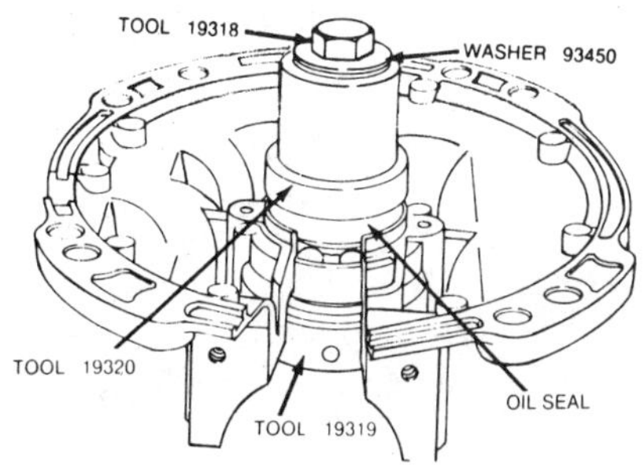

Fig. BS36—Assemble tools as shown to install seal into pto side crankcase half on Models 95700 and 96700.

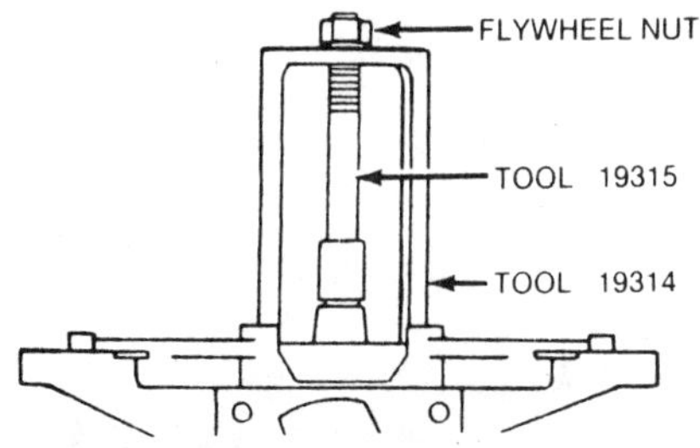

Fig. BS37—Assemble tools as shown to install magneto side crankcase half onto crankshaft. Make certain crankshaft does not turn during assembly procedure.

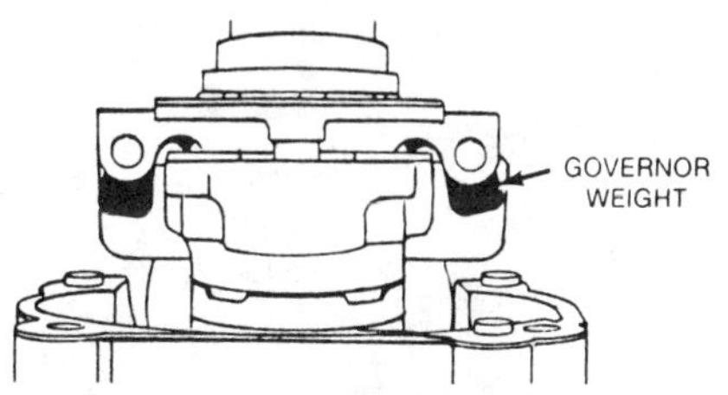

Fig. BS38—Governor weights must be held in position shown during crankcase assembly.

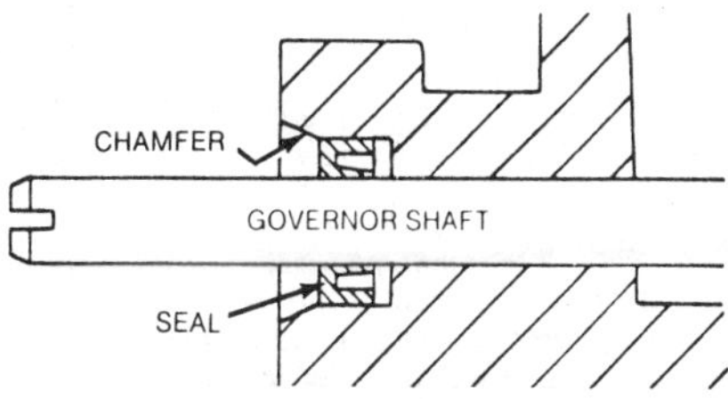

Fig. BS39—Governor shaft oil seal must be pressed into seal bore until outer seal edge is flush with chamfered edge of bore.

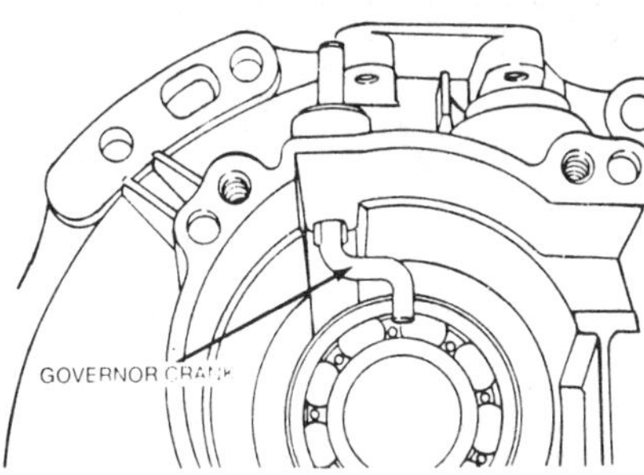

Fig. BS40—Governor crank must be in position shown during crankcase assembly.

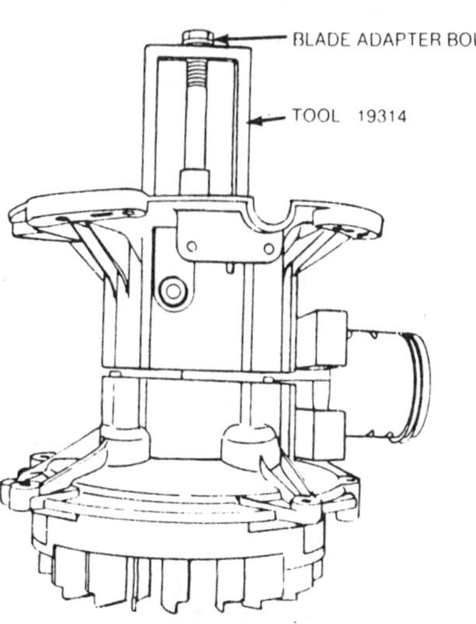

Fig. BS41—Assemble tools as shown to install pto side crankcase half onto crankshaft and magneto side crankcase half assembly.

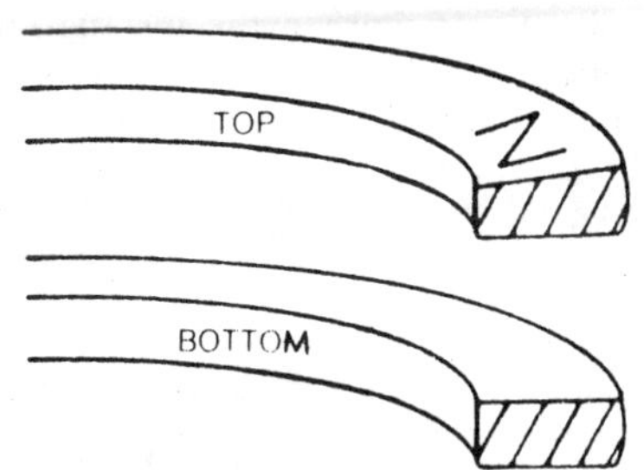

Fig. BS42—When installing piston rings on Models 95700 and 96700, install beveled ring into top ring groove of piston.

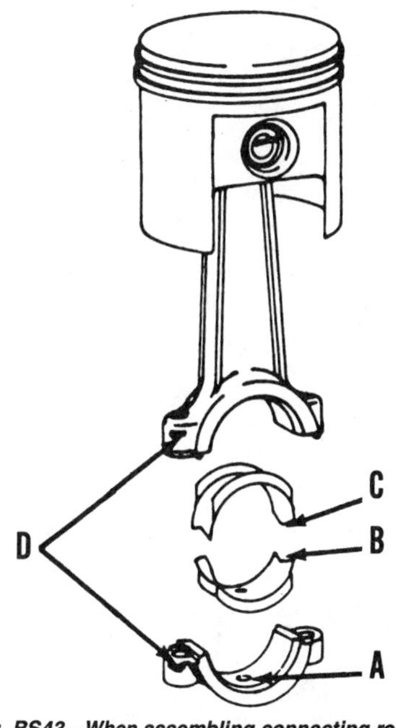

Fig. BS43—When assembling connecting rod on Model 62030 and 62033, align notch (B) and vee (C) in bearing liners. Match marks (D) on connecting rod cap and connecting rod must be aligned during assembly.

rod assembly used on Models 62030 and 62033. Piston pin floats in both piston pin bore and rod small end. Retaining rings are used to hold piston pin in place. Inspect piston for scuffing or excessive wear in ring lands and renew as needed. Connecting rod rides on needle bearings at crankshaft end. New needle bearings are retained with a wax coating on one side. Old needle bearings can be held in position with heavy grease. When assembling connecting rod, align vee (C) and notch (B) in connecting rod liners and match marks (D) on connecting rod and connecting rod cap. Tighten connecting rod bolts to specification listed under TIGHTENING TORQUES.

Refer to Fig. BS44 for a view of piston and rod assembly used on Models 95700 and 96700. Piston pin floats in piston pin bore and needle bearing which is installed in connecting rod small end. Retaining rings are used to hold piston pin in place. Open side of retaining rings should be located at the top or bottom of piston pin bore after installation. During assembly, arrow on top of piston crown must point toward exhaust port side of engine.

CRANKSHAFT. On Models 62030 and 62033, the crankshaft is supported at each end by encased needle bearings which are held in position by bearing locator tabs (Fig. BS45). Each bearing locator tab fits into a notch machined into respective crankcase half. Crankshaft end play should be 0.002-0.013 in. (0.05-0.33 mm). If crankshaft main bearing journals measure 0.7515 inch (19.088 mm) or less, renew crankshaft. If crankshaft crankpin journal measures 0.7420 inch (18.846 mm) or less, renew crankshaft.

On Models 95700 and 96700, the crankshaft is supported at each end by ball bearing type main bearings. Crankshaft and connecting rod are considered an assembly and are not serviced separately. If crankshaft main bearing journal diameters measure 0.982 inch (24.96 mm) or less, or if journals are out-of-round 0.0005 inch (0.0127 mm) or more, renew crankshaft.

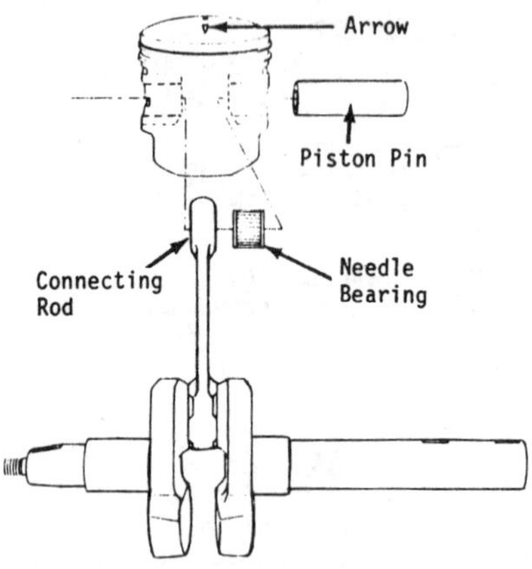

Fig. BS44—View showing connecting rod, piston and crankshaft assembly used on Models 95700 and 96700. Refer to text.

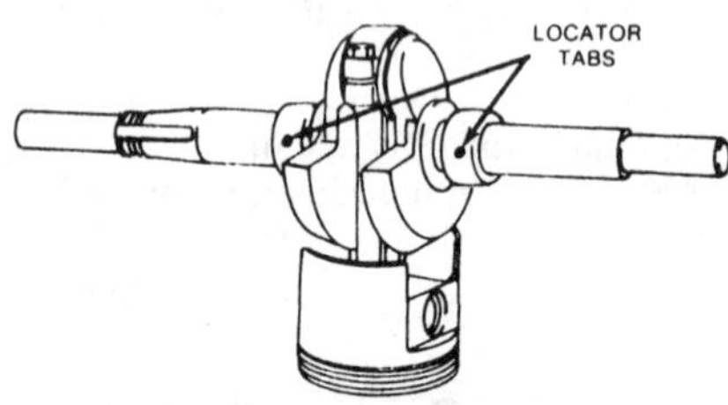

Fig. BS45—Locator tabs on encased needle bearings used on Models 62030 and 62033 must be aligned with tab notches machined into crankcase halves during assembly.

BRIGGS & STRATTON

4-STROKE

Model	Bore	Stroke	Displacement
5	2.000 in. (50.8 mm)	1.500 in. (38.1 mm)	4.7 cu. in. (77 cc)
6, N	2.000 in. (50.8 mm)	2.000 in. (50.8 mm)	6.3 cu. in. (103 cc)
8	2.250 in. (57.2 mm)	2.000 in. (50.8 mm)	8.0 cu. in. (131 cc)
9	2.250 in. (57.2 mm)	2.250 in. (57.2 mm)	9.0 cu. in. (147 cc)
14	2.625 in. (66.7 mm)	2.625 in. (66.7 mm)	14.2 cu. in. (233 cc)

ENGINE INFORMATION

All engines covered in this section are cast iron, one-cylinder, air-cooled engines. Engines with the suffix "S" after model number are equipped with a suction type carburetor. All other models are equipped with float type carburetor.

MAINTENANCE

SPARK PLUG. Recommended spark plug for all models is a Champion J8, or equivalent. If a resistor type plug is necessary to reduce radio interference, use Champion RJ8 or equivalent. Electrode gap for all models is 0.030 inch (0.76 mm).

CAUTION: Briggs & Stratton does not recommend using abrasive blasting method to clean spark plugs as this may introduce some abrasive material into the engine which could cause extensive damage.

FLOAT TYPE CARBURETOR. Float type carburetor is equipped with adjusting needles for both idle and power fuel mixtures. Counterclockwise rotation of the adjusting needles richens the fuel mixtures as shown in Fig. B3.

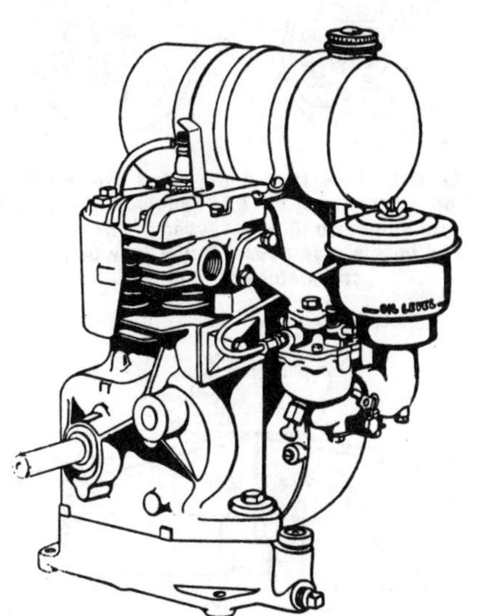

Fig. B1—Typical float carburetor installation on B&S engine. Fuel feed to carburetor may be either by gravity as shown or by fuel pump.

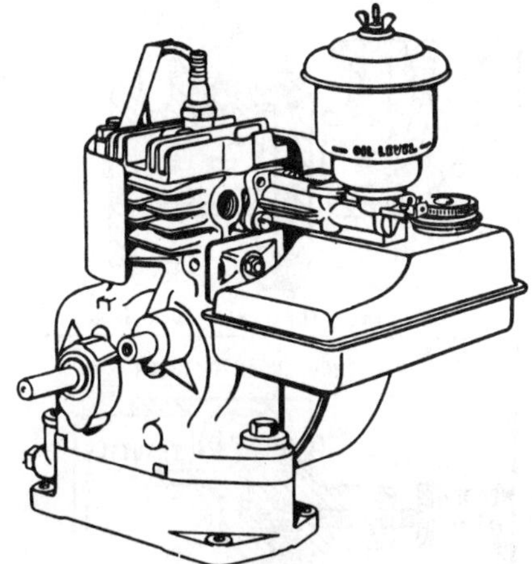

Fig. B2—Typical suction carburetor installation on B&S engine. Fuel is drawn from fuel tank by vacuum.

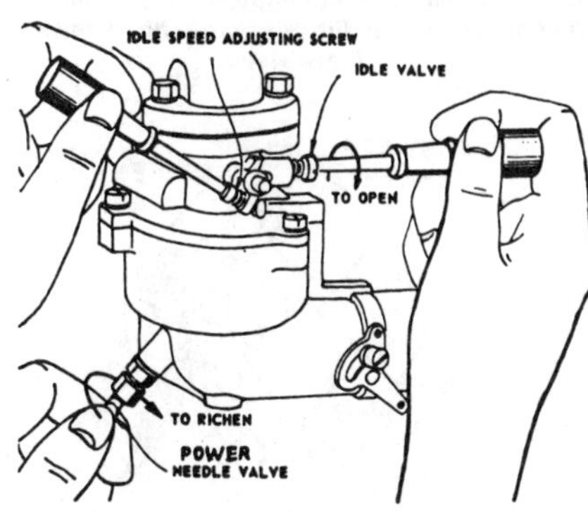

Fig. B3—Adjustment of fuel mixture and idle speed on float type carburetor.

Initial adjustment of carburetor mixture needles from a lightly seated position is ¾ turn open on the idle needle valve and 1½ turns open on the power needle valve. Make final adjustments with engine at operating temperature and running. Place engine under full load at operating speed and adjust power needle valve for smoothest engine operation. Set engine at idle speed, no-load, and adjust the idle speed screw on throttle shaft so engine is running at 1200 rpm on Models 9 and 14 engines and at 1750 rpm on all other models. Adjust idle needle valve for smoothest idle operation. As each adjustment affects the other, adjustment procedure may have to be repeated.

After making adjustments, engine should accelerate to full speed from idle position without hesitation. If it does not accelerate smoothly, turn the power needle valve slightly counterclockwise to provide a richer fuel mixture.

Float setting is correct when float is parallel with carburetor mating surface plus or minus 1/32 inch (0.794 mm). See Fig. B4. Carefully bend float tang that contacts float valve to bring the float setting within specifications.

The throttle shaft and bushings should be checked for wear, and if a diametral clearance of 0.010 inch (0.25 mm) or more can be found, installation of a new shaft and/or bushings is required. On carburetors not having bushings on the

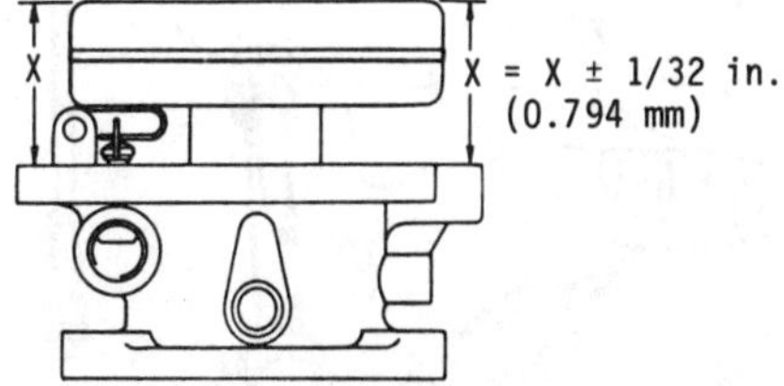

Fig. B4—Checking float setting on float type carburetor. Float should be parallel with carburetor mating surface plus or minus 1/32 inch (0.794 mm).

throttle shaft, the upper body must be renewed if installation of a new shaft does not reduce the clearance below wear limit.

Check the upper body for warpage using 0.002 inch (0.51 mm) feeler gage as shown in Fig. B5. If gage can be inserted between the upper and lower bodies as shown, a new upper body should be installed. Install new inlet float valve and seat if any wear is visible.

CAUTION: The upper and lower bodies of float type carburetors are locked together by the main nozzle. Be sure to remove the power needle valve and main nozzle before attempting to separate the upper body from the lower body. See Fig. B6.

SUCTION TYPE CARBURETORS. Suction type carburetor (Fig. B7) is much simpler than the float type and has only one fuel mixture adjusting needle. Initial adjustment is 1½ turns open from a lightly seated position.

Final adjustment is made with fuel tank half full and engine at operating temperature and running. Run engine at operating speed, no-load, and turn needle valve clockwise until the engine starts to lose speed. Turn the needle valve counterclockwise slowly until the engine begins to run unevenly. This should result in a fuel mixture rich enough for full load operation. If the engine does not pull its load satisfactorily, it may help to turn the needle valve counterclockwise a slight amount, but this richer mixture may cause the engine to idle unevenly. Adjust the idle speed screw until the engine idles at 1750 rpm.

FUEL TANK OUTLET. Figs. B8 and B9 illustrate two typical fuel tank outlets for engines with float type carburetors. The tank outlet shown in Fig. B8 is used on smaller engines and the fuel screen is within the fuel tank. The tank outlet shown in Fig. B9 is used on larger engines and incorporates a sediment bowl. The fuel screen is located in the bowl. Any dirt or lint should be brushed from the screen and the glass bowl cleaned. Varnish or other deposits may be removed with a suitable solvent.

If shut-off valve leaks, tighten the packing nut. If this does not stop the leak, remove the packing nut and install new packing.

The fuel tank outlet used with earlier suction type carburetors is shown in Fig. B10. Sectional views in this figure show two different types of check valves used in the fuel tank outlet. On later type suction carburetors, the fuel tank is mounted directly under the carburetor and the fuel tank outlet is a part of the carburetor. It is essential that the check valves operate to allow fuel to flow to the carburetor, but prevent fuel from returning to the tank. The valve should open when air is blown through the screen end of the outlet, and close tightly when air is blown through the connection end. If the check valve does not operate properly, it may be cleaned in a solvent. If cleaning does not correct valve operation, renew fuel tank outlet. Refer to Fig. B11 for method of remov-

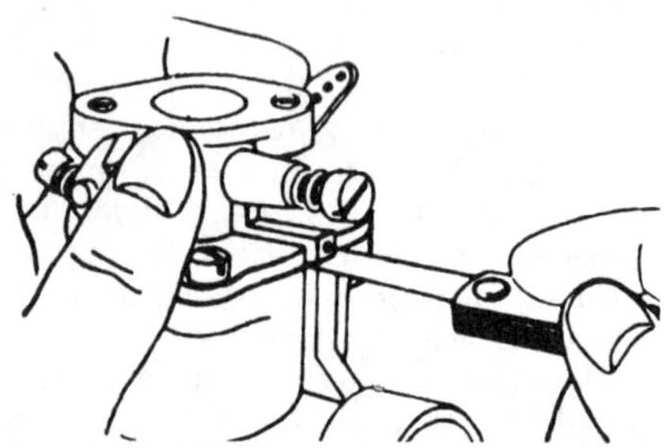
Fig. B5—Checking upper body for warpage on float type carburetor.

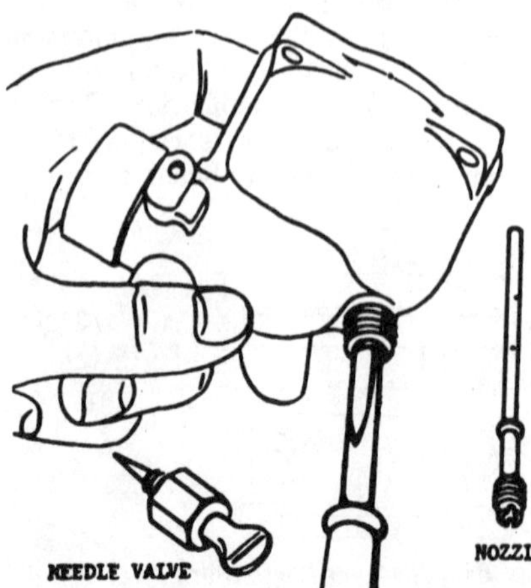

Fig. B6—Remove main nozzle before separating upper and lower bodies of float type carburetor.

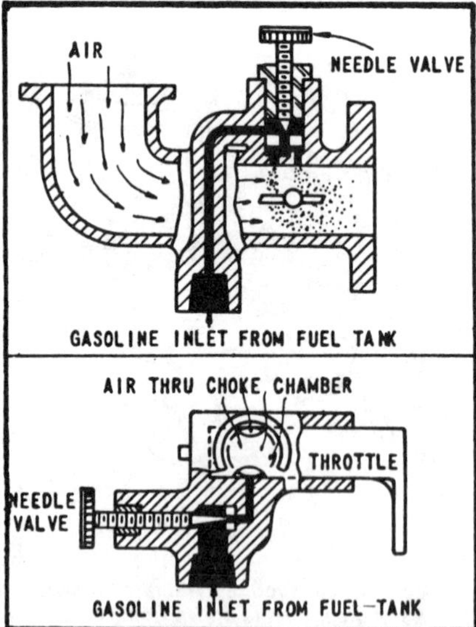

Fig. B7—Cross-sectional views of two typical B&S suction type carburetors. Idle speed adjusting screw on throttle shaft not shown in these views.

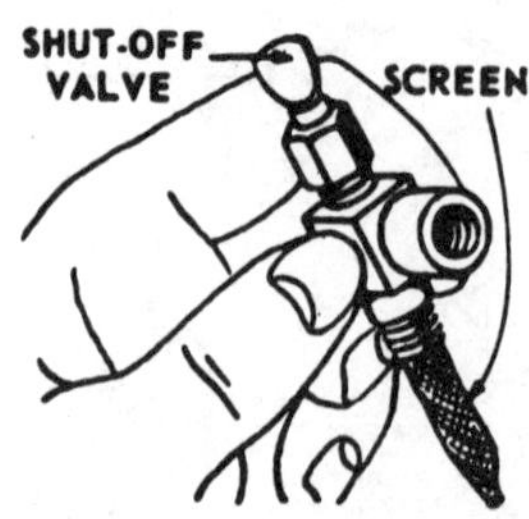

Fig. B8—Fuel tank outlet used on smaller B&S engines that are equipped with float type carburetors.

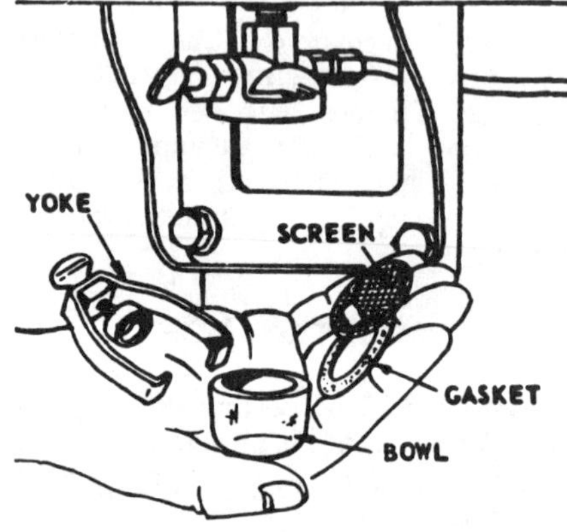

Fig. B9—Fuel tank outlet used on larger models of B&S engines with float type carburetors.

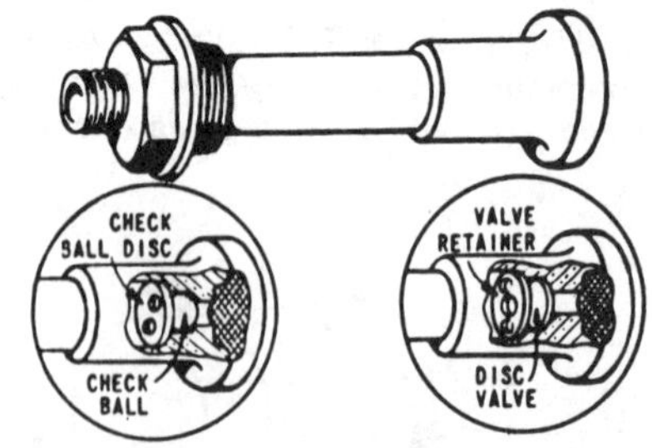

Fig. B10—Fuel tank outlet used with suction type carburetors on B&S engines. Sectional views show two different types of check valves used in this type outlet. Outlet may be part of carburetor as in Fig. B11.

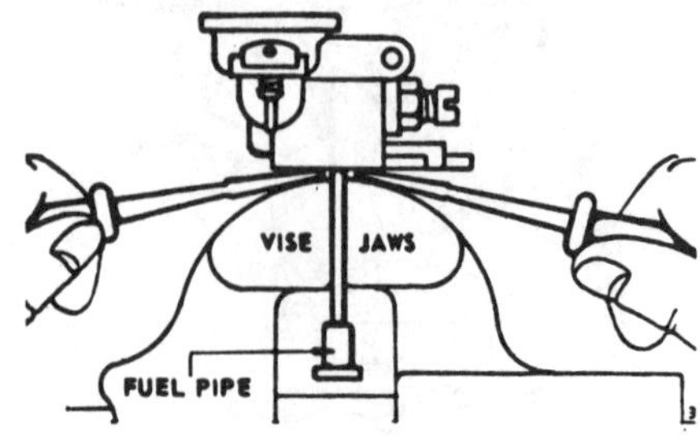

Fig. B11—Removing check valve and fuel pipe assembly from carburetor on models so constructed.

ing and installing fuel tank outlet on models where it is a part of the carburetor.

FUEL PUMP. The fuel pump that is used on some engines is actuated by a lever which rides in a groove on the crankshaft. The lever should be greased at the point shown in Fig. B12 before the fuel pump is installed. The fuel pump diaphragm can be renewed. Refer to FUEL PUMP paragraph in SERVICING BRIGGS & STRATTON ACCESSORIES section of this manual.

AUTOMATIC CHOKE. Exploded view of automatic choke is shown in Fig. B13. To adjust the automatic choke unit, refer to Fig. B14 and B15. Loosen setscrew (S) on the lever of the thermostat assembly and slide the lever on the shaft to ensure free movement of the choke unit. Turn the thermostat shaft clockwise until the stop screw strikes the tube as shown in Fig. B14. While holding the thermostat shaft in this position, move the shaft lever until the choke is open exactly 1/8 inch (3.17 mm) and tighten the lever setscrew. Turn the thermostat shaft counterclockwise until the stop screw strikes the tube as shown in Fig. B15. Manually open the choke valve until it stops against the choke link opening. At this time, the choke valve should be open at least 3/32 inch (2.38 mm), but not more than 5/32 inch (3.97 mm). Hold choke valve in horizontal position (wide-open) and check position of the counterweight lever. Lever should

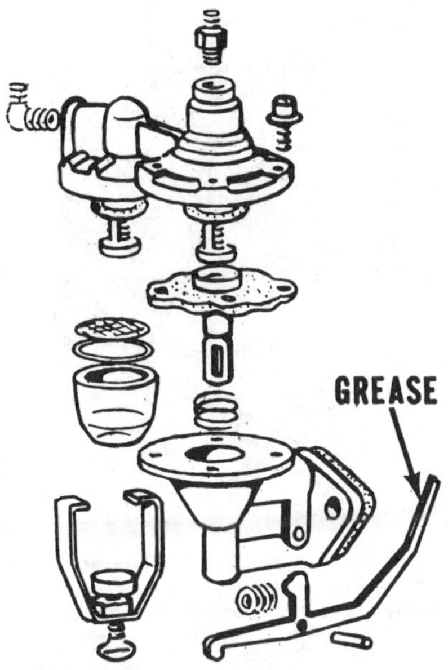

Fig. B12—Exploded view of fuel pump used on some models.

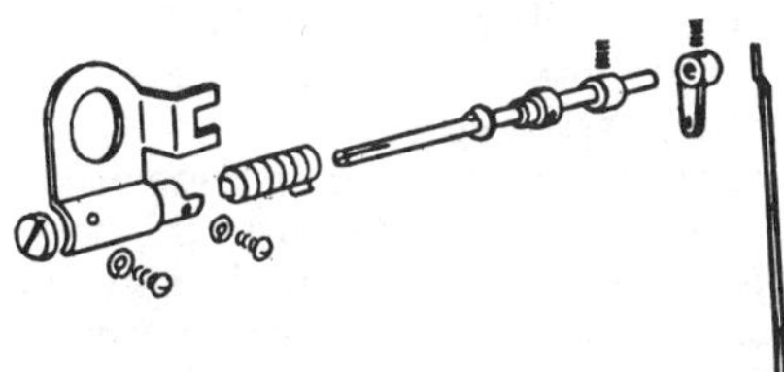

Fig. B13—Exploded view of automatic choke unit used on some models.

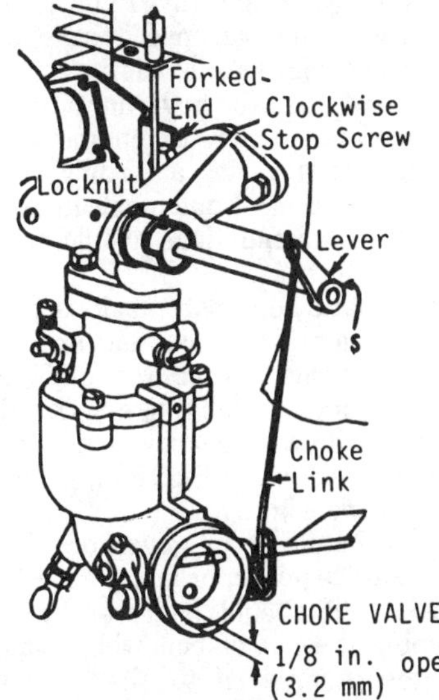

Fig. B14—Typical B&S automatic choke unit in "HOT" position.

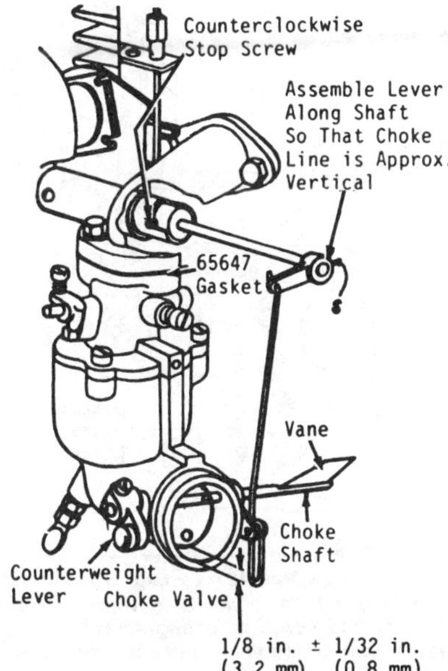

Fig. B15—Typical B&S automatic choke unit in "COLD" position.

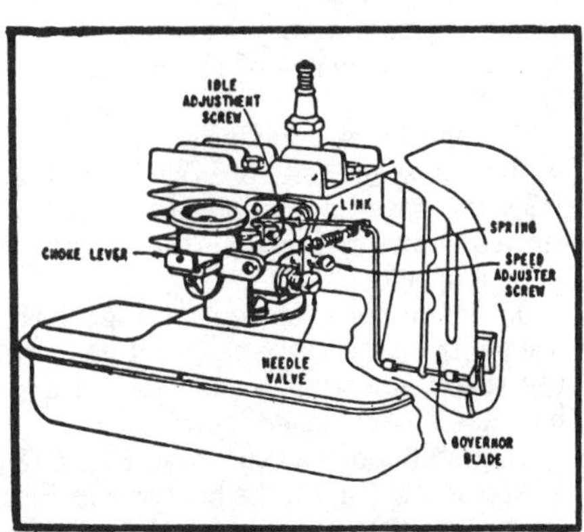

Fig. B16—Model 5S and 6S suction type carburetor. Model 6HS is similar. Air vane governor hook up shown. Turn the speed adjuster screw clockwise to increase engine speed.

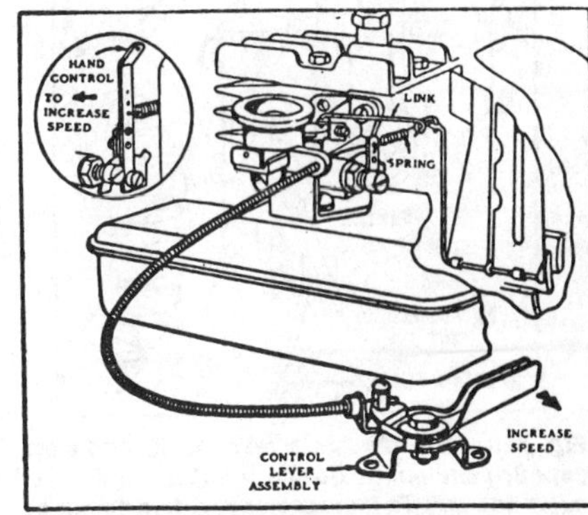

Fig. B17—Same as Fig. B16 except remote governor control is shown. Speed on some engines is changed by moving hand control lever. Refer to inset.

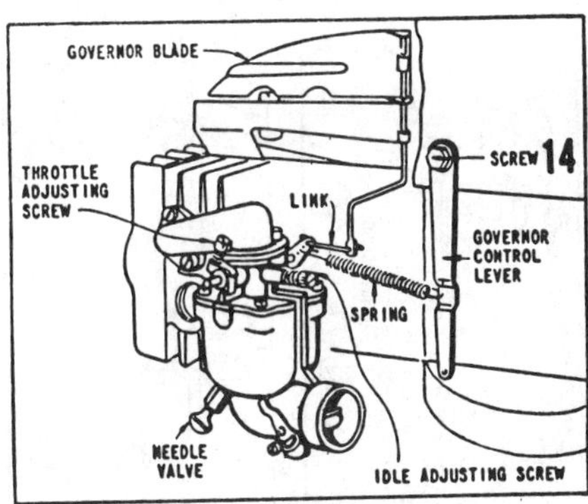

Fig. B18—B&S Models 6H (except 6HS) and 8H pneumatic (air vane) governor, showing hook up and clamp screw (14). To increase speed, loosen screw (14) and move control lever to right.

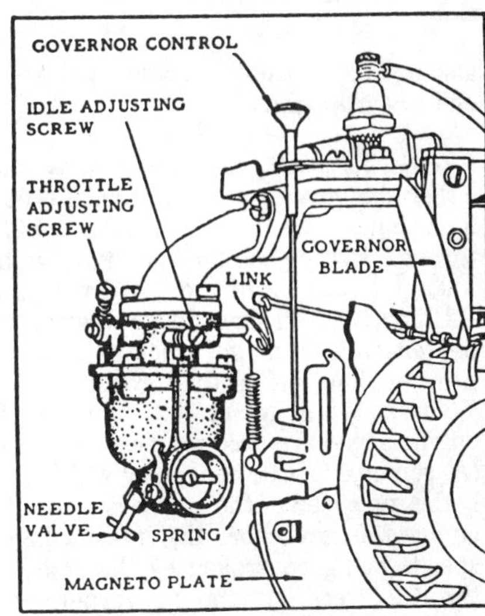

Fig. B19—B&S Model 5 pneumatic governor linkage and governor control. To increase speed, pull up on governor control.

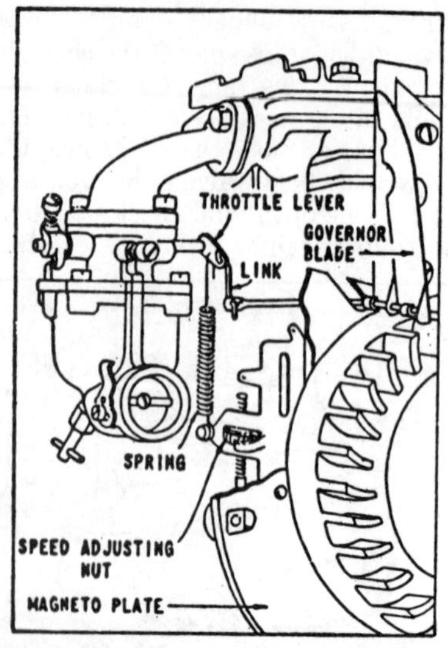

Fig. B20—B&S Models N (except NS) and 8 (except 8H) pneumatic governor linkage and speed adjusting nut. To increase speed, turn speed adjusting nut down.

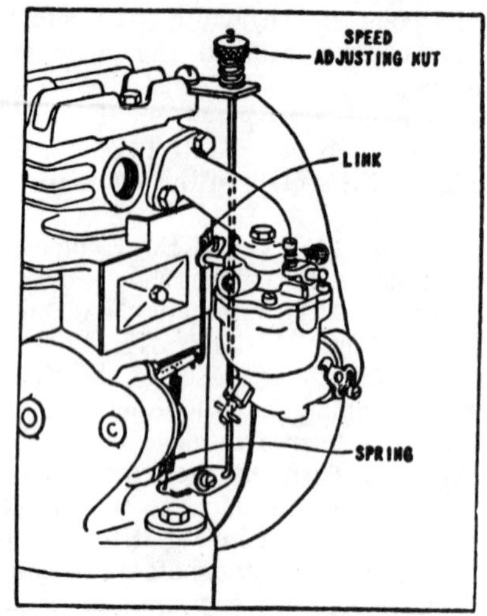

Fig. B22—B&S Models N (except NS) and 8 (except 8H) mechanical governor linkage, showing points of adjustment. To increase speed, turn adjusting nut down.

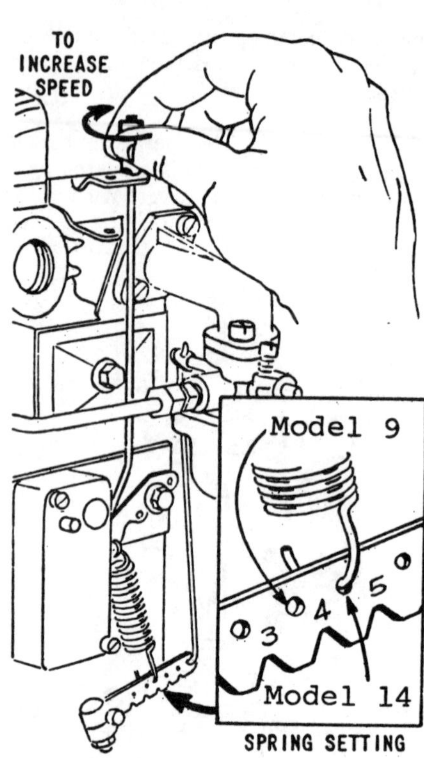

Fig. B24—On Models 9 and 14, connect governor spring as shown in inset. To increase speed on standard governor control, turn thumb nut as shown.

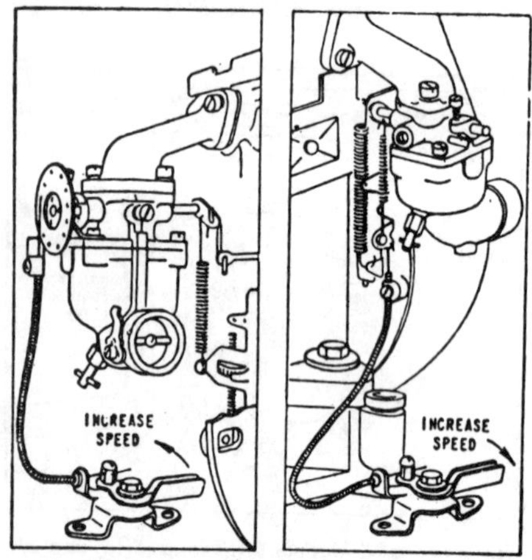

Fig. B21—Models N (except NS) and 8 (except 8H) pneumatic governor hook up with remote control. Speed is changed by moving the remote lever as shown.

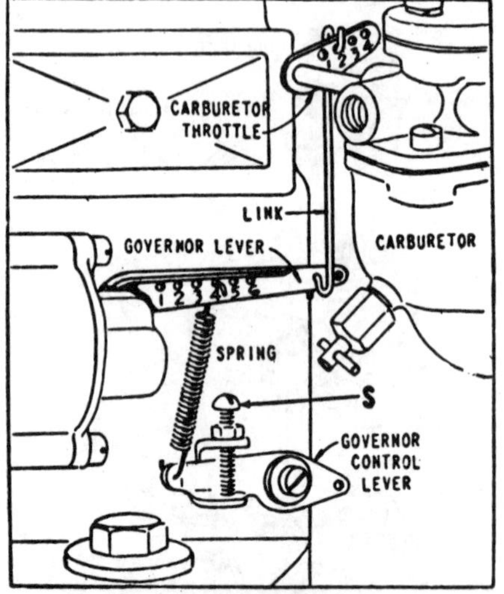

Fig. B23—B&S Models N (except NS) and 8 (except (8H) with mechanical governor showing hook up and speed adjusting screw (S). To increase speed, loosen locknut and turn screw (S) down.

also be in a horizontal position with free end toward right.

GOVERNORS. Governors may be either the gear driven mechanical type, with linkage as shown in Figs. B22 through B26 or the air vane type with linkage as shown in Figs. B16 through B21 and B27. All slack due to wear must be removed from governor linkage to prevent "hunting" of the throttle.

Three types of remote control of engine speed are used on these engines: (A) remote throttle control by which only the top speed of the engine is controlled by the governor, the intermediate speeds being controlled by the remote throttle lever; (B) remote governor control by which the governor speed is adjusted by movement of a remote control lever to provide (within the range of the governor) the effect of a variable speed governor; or (C) remote idle control by which the engine may be slowed to idle speed by movement of the remote control lever.

To adjust speed on air vane governors with fixed speed control as shown in Fig. B20, the speed adjusting nut should be turned down to increase the governed engine speed.

On installation shown in Fig. B16, used with suction type carburetors, turn the speed adjusting screw clockwise to increase engine speed.

On installations as shown in Fig. B18, loosen screw (14) on the blower housing, then move the governor control lever to the right to increase engine speed.

On installations shown in Fig. B27, turn speed adjusting screw down or move remote control lever to increase the engine speed.

On installation shown in Fig. B22, turn the speed adjusting nut down to increase engine speed.

The governor spring on these installations should be installed in the hole in the governor lever where the speed variation between load and no-load is smallest without producing hunting or erratic operation. Moving the springs closer to the governor shaft lowers no-load speed, but tends to produce hunting. Moving spring away from the shaft increases the no-load speed, but tends to reduce hunting. On installations shown in Fig. B24, turn the governor adjusting nut clockwise to increase engine speed.

Engine rpm varies according to engine application. Equipment manufacturers recommendations should be followed in all cases.

On air vane governors, make sure the air vane clears the magneto armature core and screws when the linkage is moved through its full range of travel. It can be made to clear by bending the bracket as shown in Fig. B28, or by filing the vane slightly.

On mechanical type governors, if the carburetor to governor linkage has been disturbed, it should be reset by first loosening the nut or bolt which clamps the governor lever to the governor shaft. After moving carburetor throttle to wide open position and turning gover-

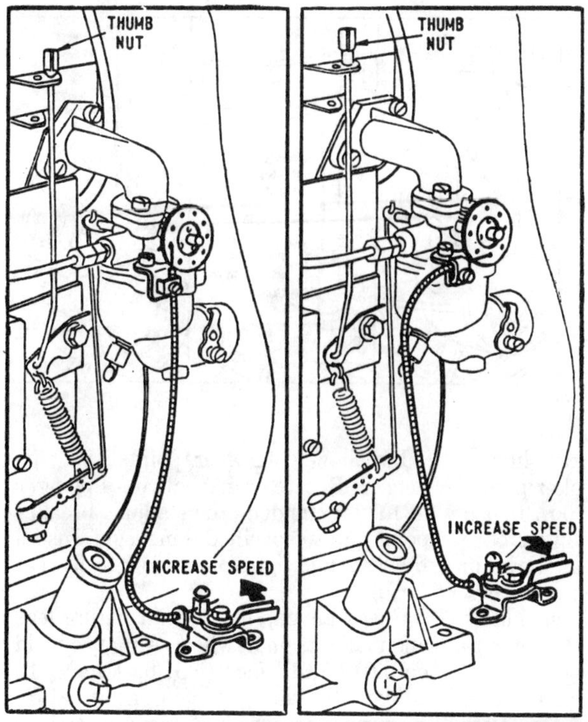

Fig. B25—Views showing methods of connecting different remote governor controls on Models 9 and 14. Moving the control lever will vary the governor spring tension, thus varying engine governed speeds. Refer also to Fig. B26.

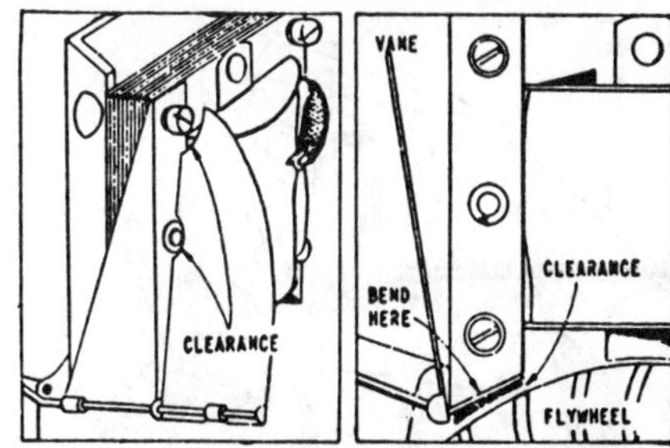

Fig. B28—B&S air vane governor adjustment.

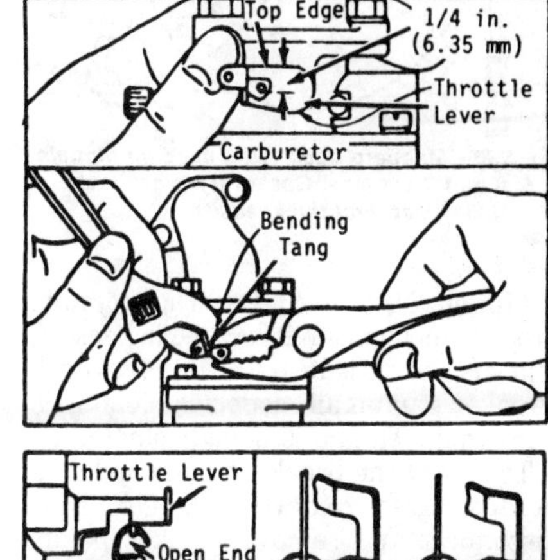

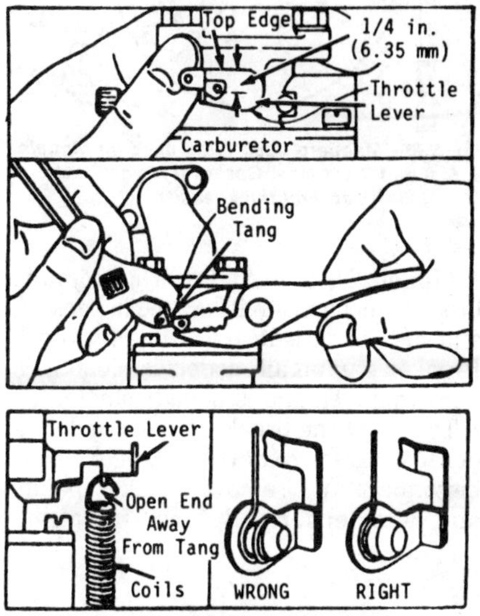

Fig. B29—Checking throttle spring and tang on early B&S Model N engines.

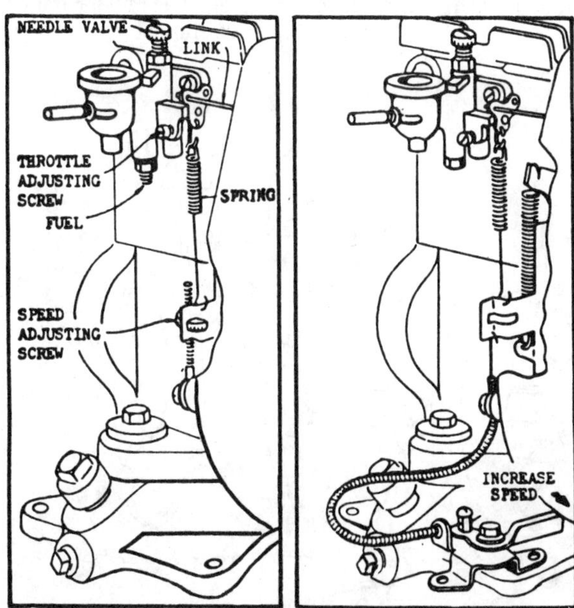

Fig. B26—Views showing methods of connecting remote throttle controls on Models 9 and 14. When control lever is in high speed position, the governor controls speed of engine and governed speed is adjusted by turning thumb nut. Moving control lever to slw speed position moves carburetor throttle shaft stop to slow down the engine.

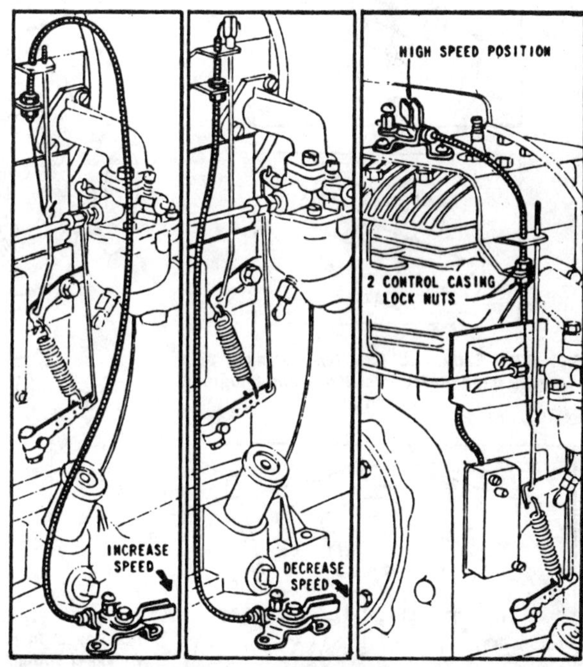

Fig. B27—B&S Model NS pneumatic governor hook up. Speed is increased by turning the speed adjusting screw down or moving the remote control lever as shown.

nor shaft so governor weights are in (as far clockwise as the shaft will turn), tighten the governor lever clamp bolt or nut.

On early N models engines, obtain 1/4 inch (6.35 mm) gap from the top of carburetor throttle lever (Fig. B29) to bottom of throttle spring hole by bending the lever tang. Solder the neck of tang if it cracks in bending. The loop at the lower end of the throttle spring on these engines should not be spread. Refer to right and wrong views in Fig. B29.

MAGNETO. Two different types of magnetos are used. Refer to appropriate following paragraph for model being serviced.

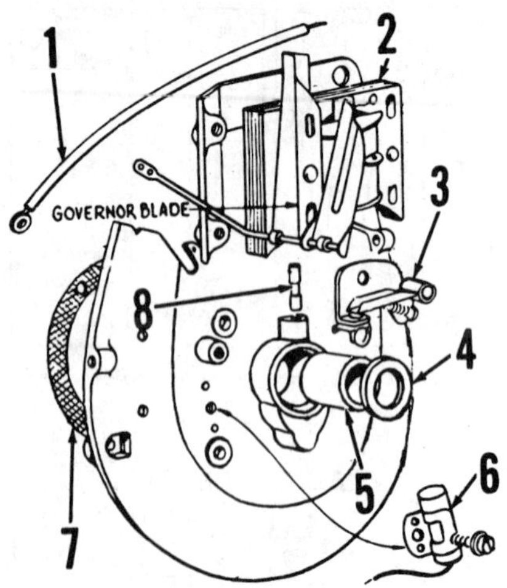

Fig. B30—Magneto back plate used on Models N, 5, 6 and 8 engines. Gasket (7) is available in several thicknesses.

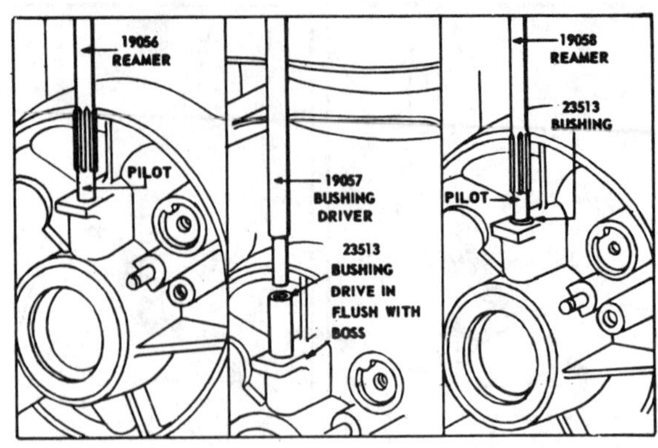

Fig. B33—Views showing reaming plunger bore to accept bushing (left view), installing bushing (center view) and finish reaming bore in bushing (right view).

Models N, 5, 6 And 8. The magneto breaker mechanism is located in back of, and is enclosed by, the flywheel. Breaker gap on all models is 0.020 inch (0.51 mm). Timing is nonadjustable.

To renew the breaker points and condenser, first remove flywheel, then disconnect and remove breaker points and condenser. Before installing breaker points, remove breaker plunger (8–Fig. B30) and renew breaker plunger if worn to a length of 0.870 inch (0.22 mm) or less. Check plunger bore with B&S plug gage #19055 as shown in Fig. B32. If plug gage will enter bore ¼ inch (6.35 mm) or more, bore should be reamed and bushing installed. Refer to Fig. B33 for steps in reaming bore and installing bushing. Early plunger (Fig. B31) can be installed either end up. Late style plunger must be installed with groove end at top.

Install new breaker points and condenser. Rotate crankshaft until plunger is fully extended, then adjust breaker point gap to specified dimension. Install flywheel and tighten nut to 57 ft.-lbs. (77 N•m).

Check air gap between armature legs and magnets on flywheel. Air gap should be 0.012-0.016 inch (0.31-0.41 mm). To adjust the air gap, loosen armature mounting screws and move armature away from flywheel. Place feeler gages or correct thickness shim stock between

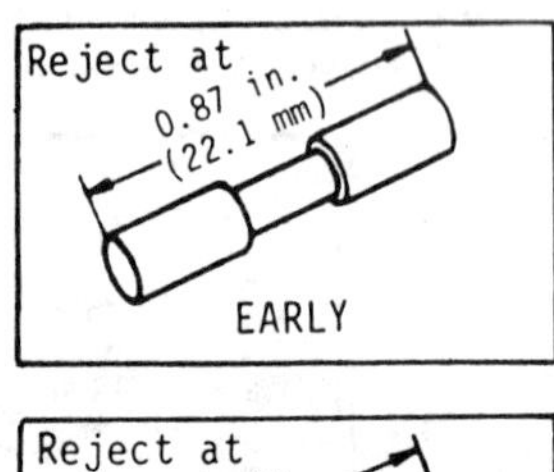

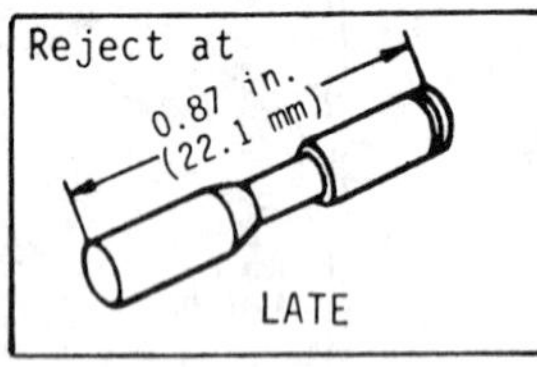

Fig. B31—Early and late type breaker point plungers used on Models N, 5, 6 and 8 engines. Refer to text.

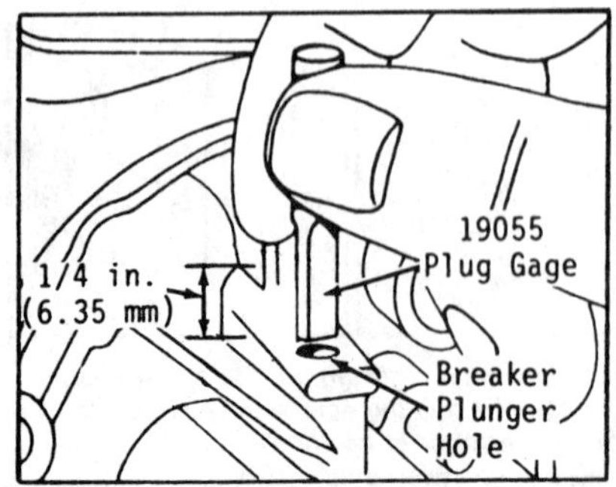

Fig. B32—If B&S plug gage 19055 can be inserted into breaker plunger bore a distance of ¼ inch (6.35 mm) or more, bore is worn and must be rebushed.

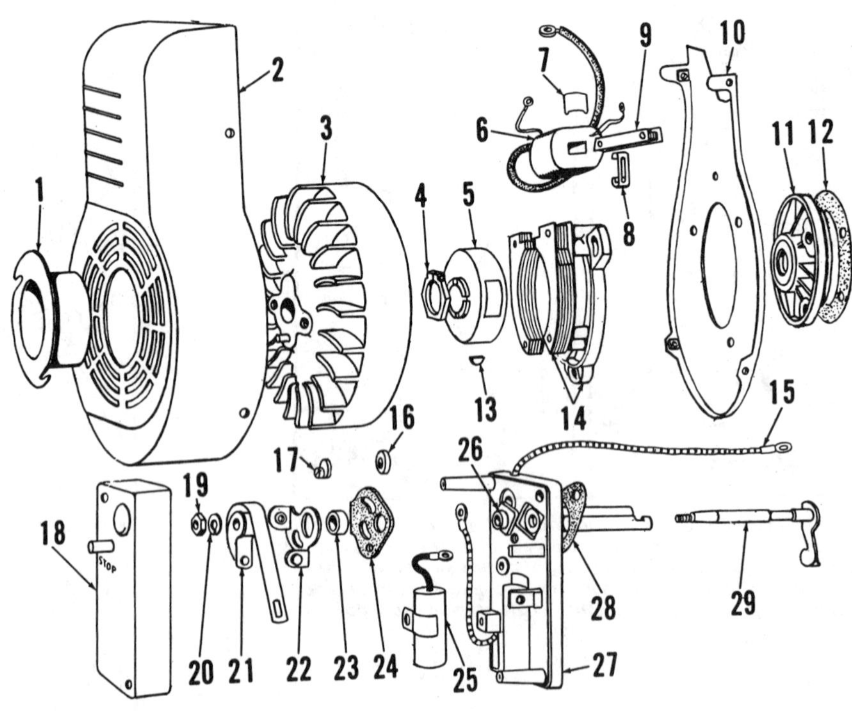

Fig. B34—Exploded view of magneto ignition system used on Models 9 and 14 engines. Flywheel is not keyed to crankshaft and may be installed in any position; however, on crank start models, flywheel should be installed as shown in Fig. B42. Breaker arm (21) is mounted on shaft (29) which is actuated by cam on engine cam gear. See Fig. B43. Two different methods of attaching magneto rotor (5) to engine crankshaft have been used. Refer to Figs. B38 and B39.

1. Starter pulley
2. Blower housing
3. Flywheel
4. Rotor clamp
5. Magneto rotor
6. Ignition coil
7. Coil clip
8. Core retainer
9. Coil core
10. Back plate
11. Bearing support
12. Shim gasket
13. Rotor key
14. Armature
15. Primary coil lead
16. Shaft seal
17. Eccentric
18. Breaker box cover
19. Nut
20. Washer
21. Breaker point arm
22. Breaker point base
23. Pivot
24. Insulator
25. Condenser
26. Seal retainer
27. Breaker box
28. Gasket
29. Shaft

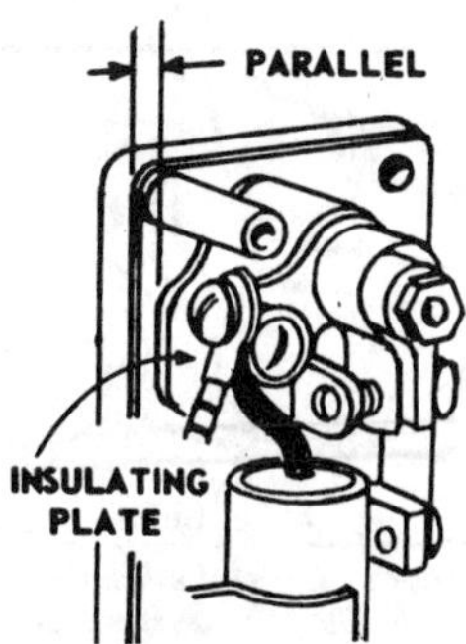

Fig. B35—When installing new breaker points on Models 9 and 14 engines, be sure dowel on breaker point base enters hole in insulator. Place sides of base and insulating plate parallel with edge of breaker box.

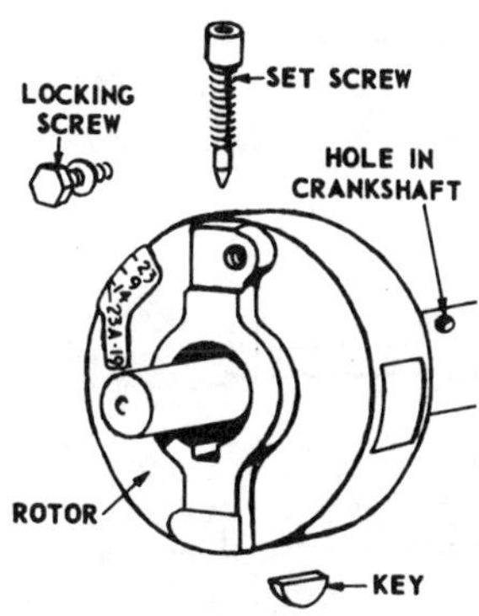

Fig. B38—On some models, magneto rotor is fastened to crankshaft by set screw which enters hole in shaft. Set screw is locked in place by a second screw. Refer to Fig. B39 for alternate method of fastening rotor to crankshaft.

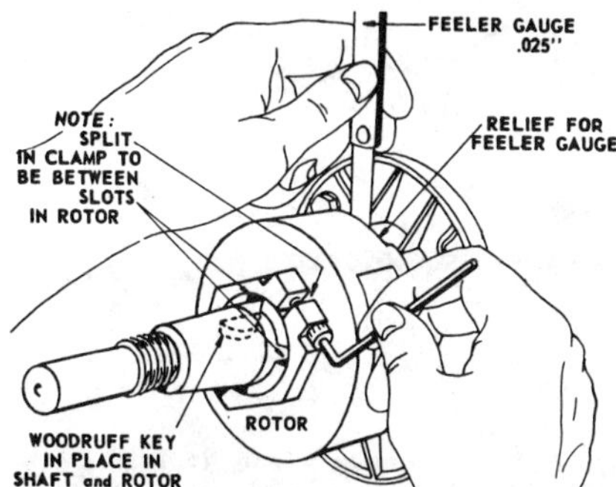

Fig. B40—On models having magneto rotor clamped to crankshaft, position rotor 0.025 inch (0.64 mm) from shoulder on shaft and split in clamp between slots of rotor, then tighten clamping screw.

armature legs and magnets on flywheel. Hold armature tight against feeler gages or shim stock, then tighten mounting screws.

Models 9 And 14. Refer to exploded view in Fig. B34 for typical magneto used on these models. Breaker points and condenser are mounted externally in a breaker box located on carburetor side of engine and are accessible after removing breaker box cover (18). Adjust breaker point gap to 0.020 inch (0.51 mm).

When renewing breaker points, or if oil leak is noted, the breaker shaft oil seal (16) should be renewed. To renew points and/or seal, turn engine so breaker point gap is at maximum. Remove terminal and breaker spring screws and loosen breaker arm retaining nut (19) so it is flush with end of shaft (29). Tap loosened nut lightly to free breaker arm (21) from taper on shaft, then remove breaker arm, breaker plate (22), pivot (23), insulating plate (24) and eccentric (17). Pry oil seal out and press new oil seal in with metal side towards breaker points. Place breaker plate on insulating plate with dowel on breaker plate entering hole in insulator, then install unit with edges of plates parallel with breaker box as shown in Fig. B35. Turn breaker shaft clockwise as far as possible and install breaker arm while holding shaft in this position. Adjust breaker points to specified gap.

Breaker box can be removed without removing points or condenser. Refer to Fig. B36. Disassembly of unit is evident after removal from engine.

To renew ignition coil, the engine flywheel must be removed. Disconnect coil ground and primary wires and pull spark plug wire from hole in back plate. Disengage clips (8–Fig. B34) retaining coil core (9) to armature (14), then push core from coil. Insert core in new coil with rounded side of core towards spark plug wire. On Model 9 engines, shorten spark plug wire by tying knot a distance (D) or 1½ inches (38 mm) from coil as shown in Fig. B37. Place coil and core on armature with retainer (7–Fig. B34) between coil and armature. Reinstall core retaining clips, connect coil ground and primary wires and insert spark plug wire through hole in back plate.

Two types of magnetic rotors have been used. Refer to Fig. B38 for view of rotor retained by set screw and to Fig. B39 for rotor retained by clamp ring. If rotor is as shown in Fig. B39, refer to Fig. B40 when installing rotor on crankshaft.

If armature core has been loosened or removed, rotor timing must be readjusted as follows: With point gap adjusted to 0.020 inch (0.51 mm), connect a static timing light from breaker point terminal to ground (coil primary wire must be disconnected) and turn engine in normal direction of rotation until light goes on. Then, turn engine very slowly in same direction until light just goes out (breaker points start to open). At that

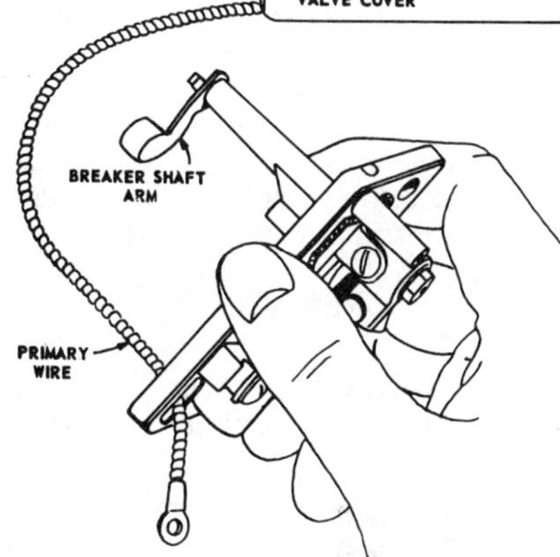

Fig. B36—Removing the breaker box on Models 9 and 14 engines. It is not necessary to remove points and condenser as the unit can be removed as an assembly.

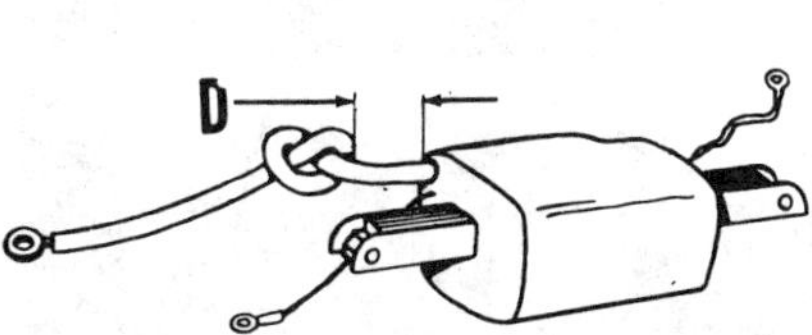

Fig. B37—When installing new ignition coil on Model 9 engine, tie a knot a distance (D) of 1½ inch (305 mm) from coil to shorten spark plug wire.

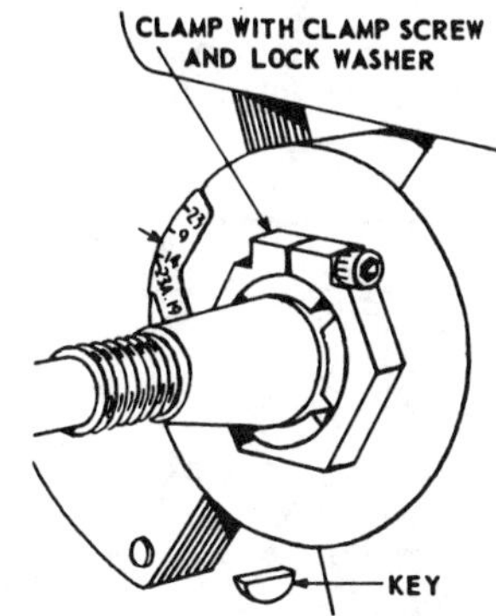

Fig. B39—View showing magneto rotor fastened to crankshaft with clamp. Be sure split in clamp is between two slots in rotor as shown and check clearance between rotor and shoulder on crankshaft as shown in Fig. B40. Refer to Fig. B38 for alternate method of fastening rotor to crankshaft.

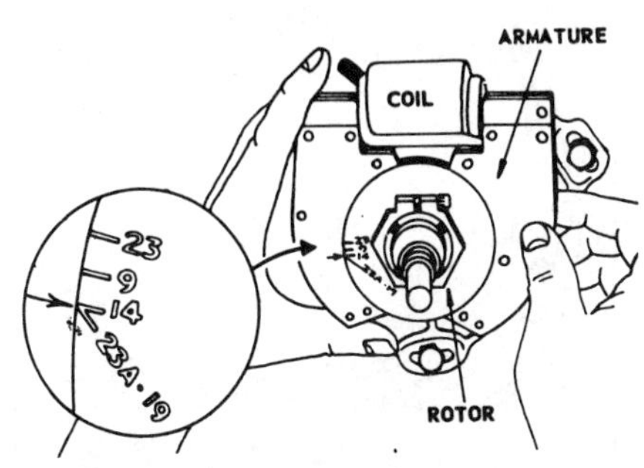

Fig. B41—With engine turned so points are just starting to open, align model number line on rotor with arrow on armature by rotating armature in slotted mounting holes.

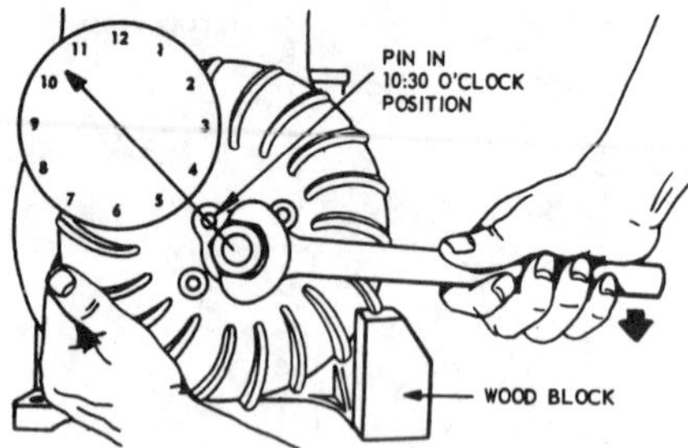

Fig. B42—On Models 9 and 14 with crank starter, mount flywheel with pin in position shown and with armature timing marks aligned as shown in Fig. B41.

time, engine model number on magneto rotor should be aligned with arrow on armature as shown in Fig. B41. If not, loosen armature core retaining cap screws and turn armature in slotted mounting holes so arrow is aligned with appropriate engine model number on rotor. Tighten the armature mounting screws.

When installing flywheel on models with crank starter, place flywheel on crankshaft as shown in Fig. B42 with magneto timing marks aligned as in previous paragraph. On models not having a crank starter, flywheel may be installed in any position.

Magneto breaker points are actuated by a cam on a centrigual weight mounted on engine camshaft. Refer to Fig. B43. When engine is being overhauled, or cam gear is removed, check action of the advance spring and centrifugal weight unit by holding cam gear in position shown and pressing weight down. When weight is released, spring should return weight to its original position. If weight does not return to its original position, check weight for binding and renew the spring.

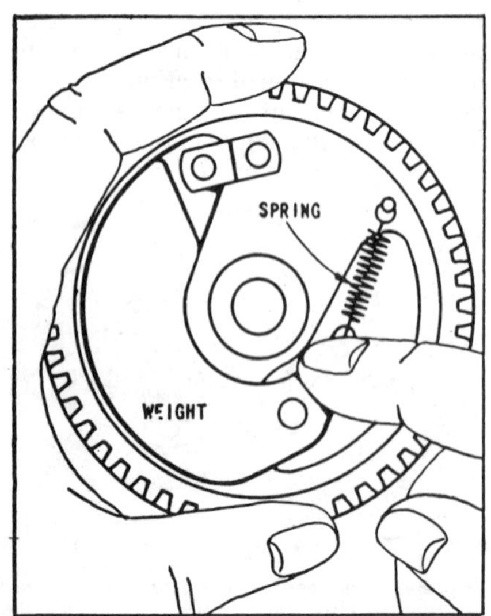

Fig. B43—Checking timing advance weight and spring on Models 9 and 14. Refer to text.

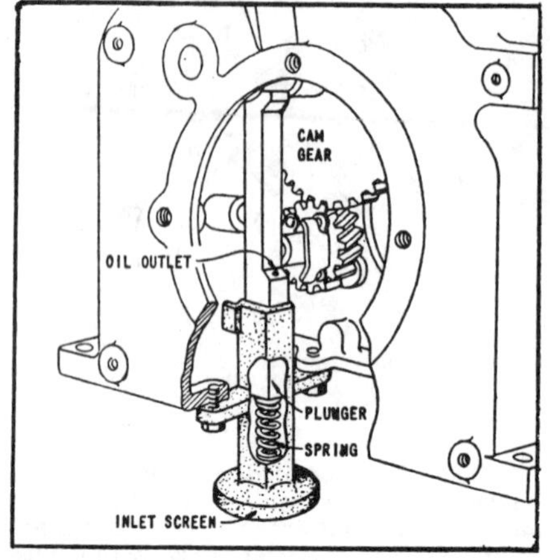

Fig. B44—View showing one type of oil pump used on some horizontal crankshaft models. Pump plunger is actuated by camshaft and oil sprays from outlet in plunger to lubricate engine. A second type of plunger type pump is similar, but oil is delivered via a tube connected to pump.

LUBRICATION. Model N engines from serial number 205000 to 205499 are lubricated by an ejector type pump (Fig. B44) wherein the pump squirts oil onto the internal moving parts. A similar system using a gear type pump is used on Model N engines from serial number 205500 to 205999 and all horizontal cylinder engines. Refer to Fig. B45. All other engines are lubricated by a plain splash system.

Manufacturer recommends using oil with an API service classification SF. Use SAE 30 oil when temperature is above 40°F (40°C) or SAE 10W-30 oil when temperature is below 40°F (4°C).

Fill crankcase to top of filler cap opening on vertical cylinder engines and to the "F" mark on sump dipstick for horizontal cylinder engines.

OIL RETURN VALVE. Some engines are equipped with an oil return valve located under the magneto end main bearing. See Fig. B53. If engine leaks oil at the main bearing, remove and clean or install a new oil return valve. Early production engines were equipped with a cast-in type of oil return valve and late production engines have a removable oil return valve. When overhauling early production engines, mill off the cast-in type and install the removable type. See Fig. B47.

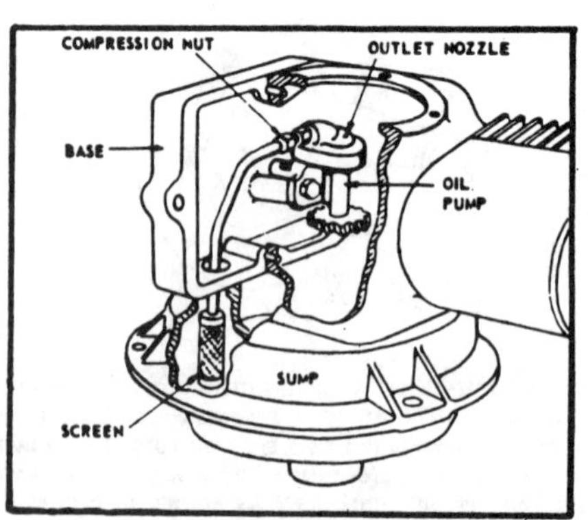

Fig. B45—Oil pump installation on horizontal cylinder engines.

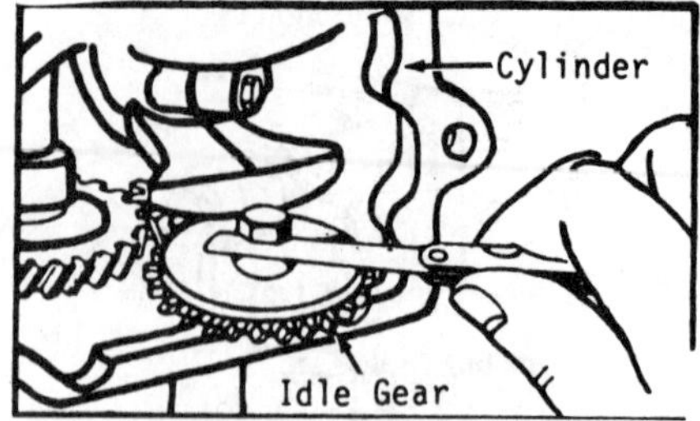

Fig. B46—Idler gear installation on Models 6AH, 8AH and 6AHS. The idler gear should have 0.010 inch (0.25 mm) end play on the shoulder screw when cold.

Fig. B47—Old style (cast-in) type oil return valve in B&S engines should be milled off and the new removable type installed.

CRANKCASE BREATHER. Outside type breathers should be cleaned and the filter element renewed at regular intervals. Some breathers are located inside the engine in the valve spring compartment as shown in Fig. B48. Disassemble, clean, and reassemble the parts in sequence shown.

REPAIRS

CYLINDER HEAD. When removing cylinder head, note position from which

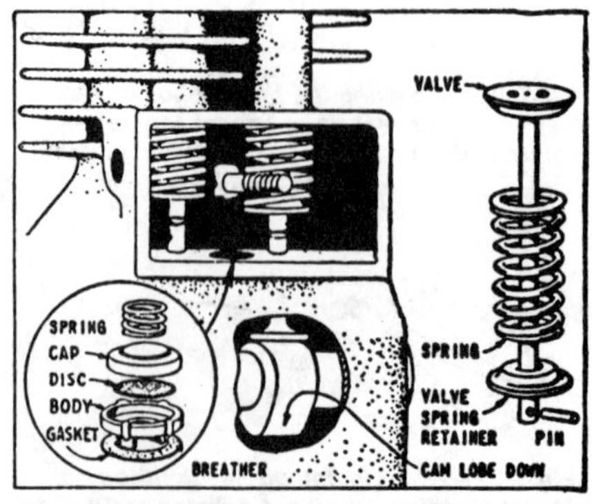

Fig. B48—Some B&S engines have internal breathers as shown. Insert shows assembly sequence.

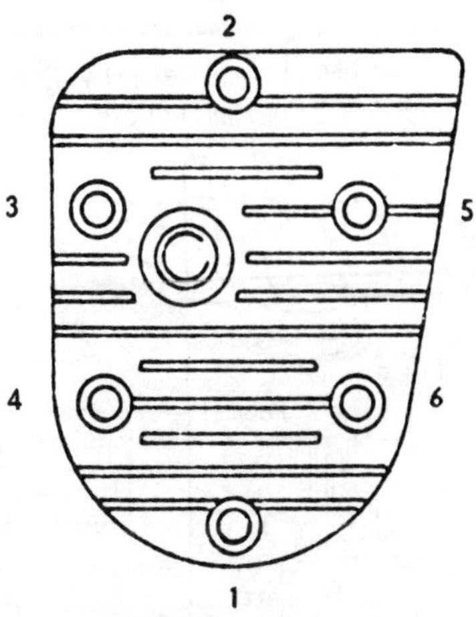

Fig. B49—Cylinder head cap screw tightening sequence on Models N, 5, 6 and 8 engines. Refer to text for tightening torque specifications.

different length cap screws were removed. If they are not used in same position when installing head, it will result in some screws bottoming in holes and not enough thread engagement on others. When installing cylinder head, tighten all screws lightly and then retighten them in sequence shown Fig. B49 or B50 to 165 in.-lbs. (19 N•m) on Model 14 and 140 in.-lbs. (16 N•m) on all other models.

Start engine and allow it to reach operating temperature. Stop engine and retighten cap screws using same sequence and specifications.

CONNECTING ROD. Connecting rod and piston assembly is removed from cylinder head end of block. The aluminum alloy connecting rod rides directly on the induction hardened crankpin. Connecting rod should be rejected if crankpin hole is out-of-round 0.0007 inch (0.018 mm) or more, or if piston pin hole is out-of-round 0.0005 inch (0.013 mm) or more. Discard connecting rod if worn to or beyond the following reject sizes.

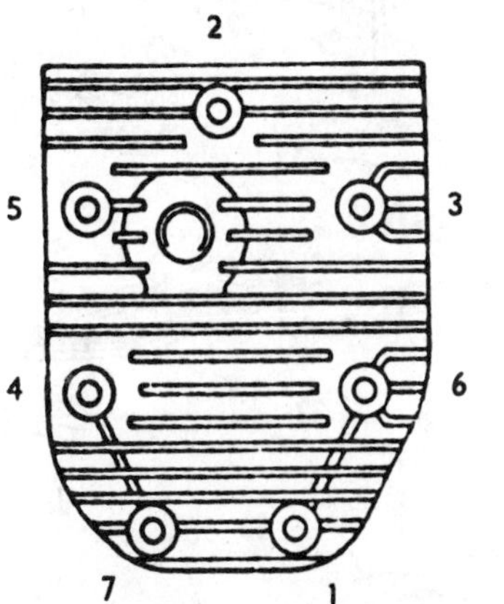

Fig. B50—Cylinder head cap screw tightening sequence on Models 9 and 14 engines. Refer to text for tightening torque specifications.

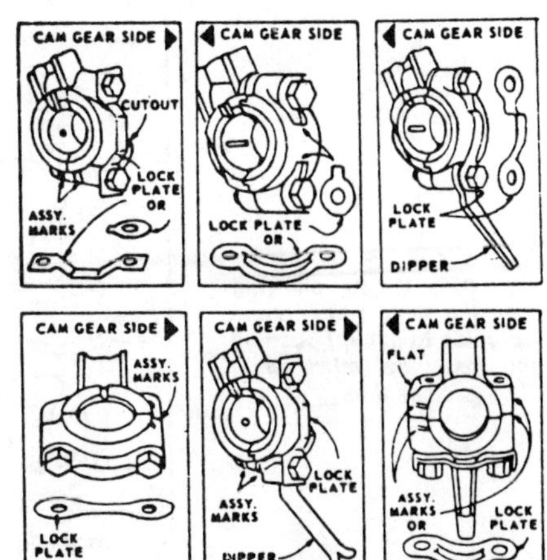

Fig. B51—B&S connecting rod position is determined by clearance flats or assembly marks as shown.

REJECT SIZES FOR CONNECTING ROD

Model	Bearing Bore	Pin Bore
5	0.752 in. (19.10 mm)	0.492 in. (12.50 mm)
N, 6, 8	0.751 in. (19.08 mm)	0.493 in. (12.52 mm)
9	0.876 in. (22.25 mm)	0.563 in. (12.30 mm)
14	1.001 in. (25.43 mm)	0.674 in. (17.12 mm)

If rod is otherwise servicable except for piston pin hole, rod and piston can be reamed and a 0.005 inch (0.13 mm) oversize piston pin installed.

Recommended clearance of rod to crankpin is 0.0015-0.0045 inch (0.038-0.114 mm) and recommended rod-to-piston pin clearance is 0.0005-0.0015 inch (0.013-0.038 mm).

Rod position in the crankcase is determined by the clearance flat, or assembly marks. See Fig. B51. Tighten connecting rod bolts to 90-110 in.-lbs. (11-12 N•m) for Models N, 5, 6 and 8; 130-150 in.-lbs. (15-18 N•m) for Model 9 and 175-200 in.-lbs. (20-23 N•m) for Model 14.

PISTON, PIN AND RINGS. Piston is made of aluminum alloy and is fitted with two compression rings and one oil control ring. If piston shows visible signs of wear, scoring or scuffing, it should be renewed. Also, renew piston if side clearance of new ring in top ring groove exceeds 0.005 inch (0.13 mm) on Models N, 5, 6 and 8 or 0.007 inch (0.18 mm) on Models 9 and 14. Reject piston or hone piston pin hole to 0.005 inch (0.13 mm) oversize if pin hole is 0.0005 inch (0.013 mm) or more out-of-round. Reject piston if pin hole is worn to a diameter of 0.563 inch (14.30 mm) on Model 9, 0.673 inch (17.09 mm) on Model 14 or 0.489 inch (12.42 mm) on Models N, 5, 6 and 8.

Renew piston pin if 0.0005 inch (0.013 mm) out-of-round, or if worn to a diameter of 0.561 inch (14.25 mm) on Model 9, 0.671 inch (17.04 mm) on Model 14 or 0.489 inch (12.42 mm) on Models N, 5, 6 and 8.

Pistons with an "X" mark on a pin boss are installed with "X" mark towards magneto side of engine. Pistons not having this mark may be installed either way.

Reject piston rings having an end gap of 0.030 inch (0.76 mm) for compression rings and 0.035 inch (0.89 mm) for oil control ring. If top ring has groove on inside, install with groove up. If second compression ring has groove on outside, install with groove down. Oil ring may be installed either side up.

Pistons and rings are available in a variety of oversizes as well as standard. A chrome ring set is available for slightly worn standard bore cylinders. Refer to note in CYLINDER section.

CYLINDER. Cylinder and crankcase are an integral cast iron casting. If cylinder is worn more than 0.003 inch (0.08 mm), or is more than 0.0015 inch (0.038 mm) out-of-round, it should be rebored to nearest oversize for which piston and rings are available.

Refer to the following table for standard cylinder bore diameter.

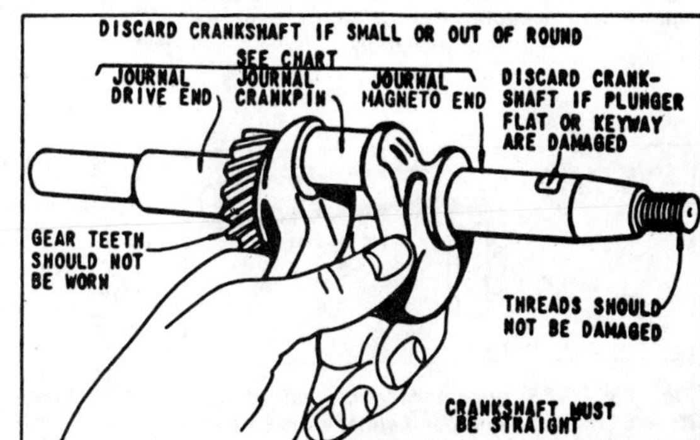

Fig. B52—Check points to determine if crankshaft should be reused or rejected.

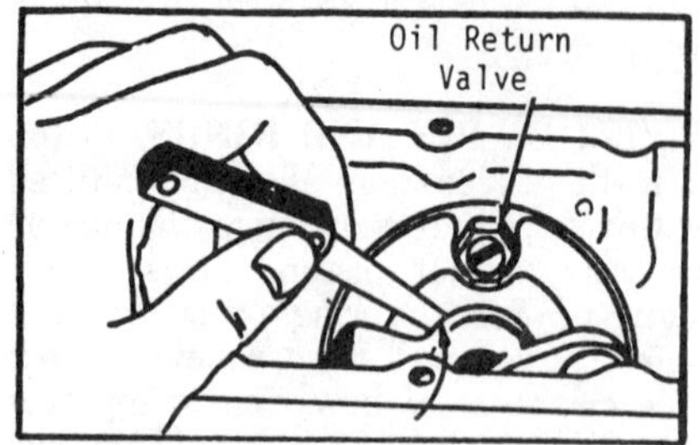

Fig. B53—Checking end play between bearing plate and crankshaft thrust face. Also note oil return valve used on some B&S engines.

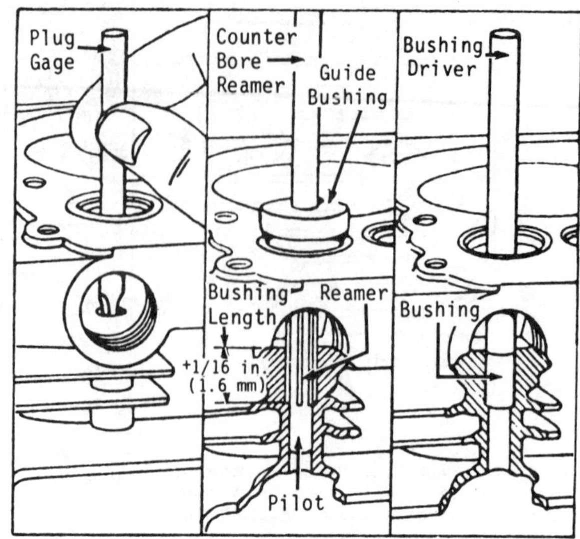

Fig. B56—Views showing checking valve guide, reaming guide to accept bushing and installing valve guide bushing.

STANDARD CYLINDER BORE SIZES

Model	Cylinder Diameter
5, 6, N	1.9990-2.0000 in. (50.77-50.80 mm)
8, 9	2.2490-2.2500 in. (57.12-57.15 mm)
14	2.6240-2.6250 in. (66.65-66.68 mm)

NOTE: A chrome piston ring set is available for slightly worn standard bore cylinders. No honing or cylinder deglazing is required for these rings. The cylinder bore can be a maximum of 0.005 inch (0.13 mm) oversize when using chrome rings.

CRANKSHAFT. The crankshaft is supported in two main bearings which may be either bushing or ball bearing type, or be one bushing and one ball bearing type, depending upon engine model or application.

On engines having bushing type main bearings, the running clearance should be 0.0015-0.0045 inch (0.038-0.114 mm). Wear on both the crankshaft and bushings will determine the running clearance; therefore, both the crankshaft and bushings should be checked. If the crankshaft is worn beyond rejection size listed in CRANKSHAFT REJECTION SIZE table, or if the journals are rough, scored, or more than 0.0007 inch (0.018 mm) out-of-round, install a new crankshaft. Refer to Fig. B52 for crankshaft check points.

CRANKSHAFT REJECTION SIZE

Model	Magneto Bearing	PTO Bearing
N,5,6,8	0.873 in. (22.17 mm)	0.873 in. (22.17 mm)
9	0.983 in. (24.97 mm)	0.983 in. (24.97 mm)
14	1.179 in. (29.95 mm)	1.179 in. (29.95 mm)

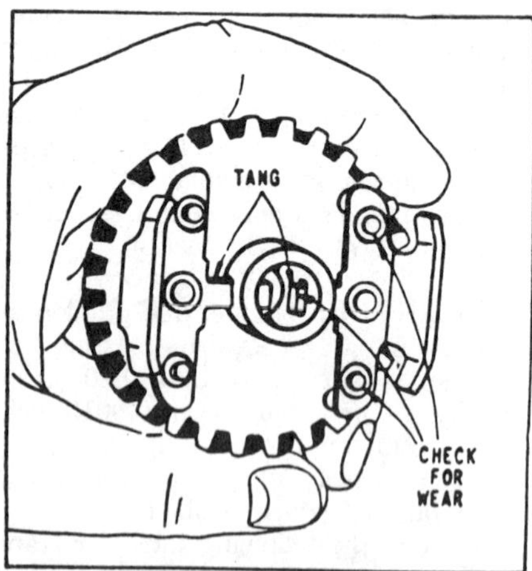

Fig. B54—B&S governor tangs. Tangs should be square and smooth. If not, renew gear and weight assembly.

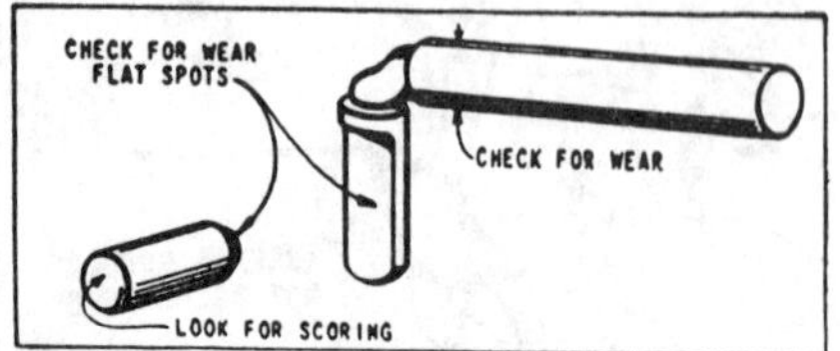

Fig. B55—B&S governor crank and plunger. Renew crank and plunger if either shows wear at points indicated.

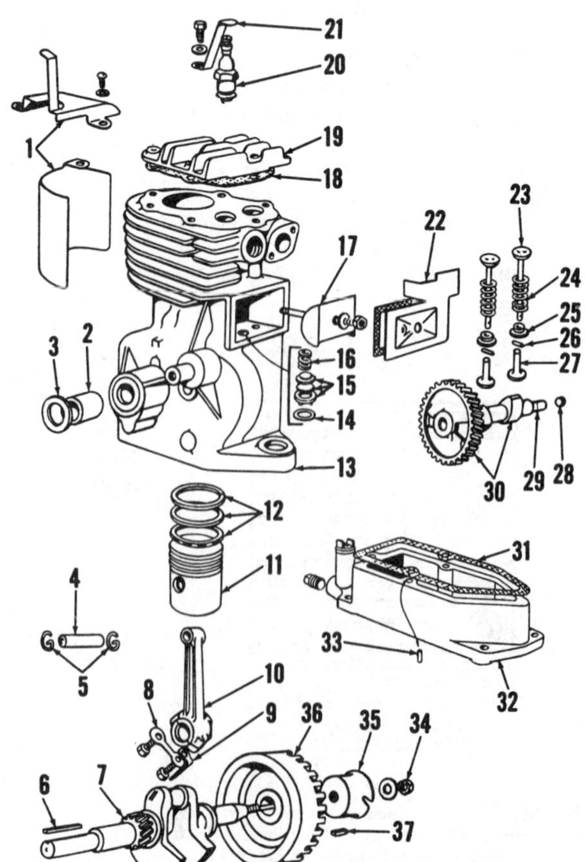

Fig. B57—Exploded view of Model 6 engine assembly. Models N, 5 and 8 are similar. Refer to Fig. B30 for view of magneto back plate and flywheel end crankshaft bearing. Some models are equipped with ball bearing mains.

1. Shields
2. Bushing
3. Oil seal
4. Piston pin
5. Snap rings
6. Output drive key
7. Crankshaft
8. Rod bolt lock
9. Oil dipper
10. Connecting rod
11. Piston
12. Piston rings
13. Crankcase
14. Gasket
15. Breather body & valve assy.
16. Spring
17. Oil spray shield
18. Cylinder head gasket
19. Cylinder head
20. Spark plug
21. Grounding spring
22. Tappet chamber cover
23. Valves
24. Valve springs
25. Spring retainers
26. Retainer pins
27. Tappets
28. Plug
29. Camshaft
30. Cam gear
31. Gasket
32. Engine base
33. Dowel
34. Flywheel nut
35. Starter cup
36. Flywheel
37. Flywheel key

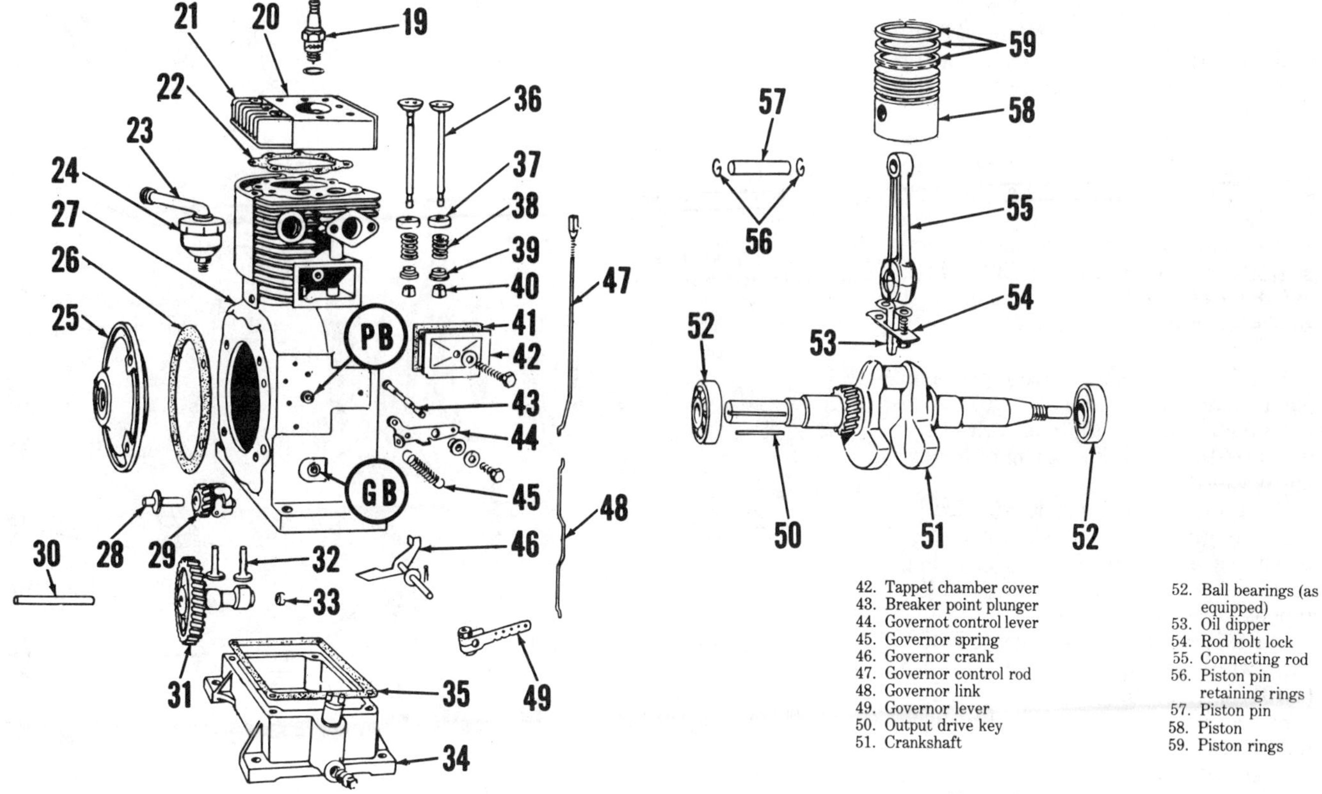

Fig. B58—Exploded view of Model 9 or 14 engine assembly not having "Magna-Matic" ignition system. For Magna-Matic ignition models, refer to Fig. B58A. Breaker plunger bushing (PB) and governor crank bushing (GB) in engine crankcase (27) are renewable. Breaker plunger (43) rides against a cam on cam gear (31).

19. Spark plug
20. Air baffle
21. Cylinder head
22. Gasket
23. Breather tube
24. Breather
25. Main bearing plate
26. Gasket
27. Engine crankcase & cylinder block
28. Governor shaft
29. Governor gear & weight unit
30. Camshaft
31. Cam gear
32. Tappets
33. Camshaft plug
34. Engine base
35. Gasket
36. Valves
37. Valve spring washers
38. Valve springs
39. Spring retainers
40. Keepers
41. Gasket

Crankpin journal rejection size is 0.7433 inch (18.87 mm) for Models N, 5, 6 and 8, 0.8726 inch (22.17 mm) for Model 9 and 0.9964 inch (25.30 mm) for Model 14.

On plain bushing equipped Models 9 and 14, main bearing bushings are not available separately from bearing plates. Renew bearing plates if bushing is 0.0007 inch (0.018 mm) or more out-of-round or if worn to a diameter of 0.9875 inch (25.09 mm) on Model 9 or 1.1850 inch (30.10 mm) on Model 14.

On Models N, 5, 6 and 8 equipped with plain bushings, renew bushings if they are rough, scored, out-of-round 0.0007 inch (0.018 mm) or are worn to a diameter of 0.878 inch (22.30 mm) or more.

When pressing or driving old main bearing bushings from block or crankcase, always support the crankcase wall with a Briggs & Stratton Crankcase Support or a pipe of suitable diameter to prevent damage to the casting.

When new main bushings are installed on Models N, 5, 6 and 8, ream bushings to an inside diameter of 0.8764-0.8770 inch (22.26-22.28 mm).

On engines equipped with ball bearing mains, check ball bearings for wear or roughness. If bearing is loose or noisy, renew the bearing. Ball bearings are a press fit on crankshaft and must be removed by pressing crankshaft out of bearings. Expand new bearings by heating them in oil to a maximum temperature of 325°F (163°C), and install them on crankshaft with shielded side towards crankpin journal.

Recommended crankshaft end play for all models is 0.002-0.008 inch (0.05-0.20 mm). End play is controlled by use of different thickness shim gaskets between the flywheel end main bearing plate and crankcase. End play can be checked by clamping dial indicator to crankshaft and resting indicator button against crankcase. On plain bushing models only, end play can be checked with engine base removed by using a feeler gage as shown in Fig. B53. Shim gaskets are available in a variety of thicknesses for all models.

CAM GEAR. On all models camshaft timing gear and camshaft lobes are an integral casting and is referred to as the "cam gear". The cam gear turns on a stationary shaft which is referred to as the "camshaft".

Reject cam gear if lobes are worn to a height of 0.875 inch (34.10 mm) on Models N, 5, 6 and 8, 1.124 inch (28.55 mm) on Model 9 or 1.115 inch (28.32 mm) on Model 14 or if gear teeth are worn or damaged.

Reject camshaft if worn to a diameter of 0.3719 inch (9.45 mm) on Models N, 5, 6, 8 and 9 engines or 0.4968 inch (12.62 mm) on Model 14.

On models with "Magna-Matic" ignition system, cam gear is equipped with an ignition advance weight (Fig. B43). A tang on the advance weight contacts breaker arm (29–Fig. B34) each camshaft revolution. On other models, breaker plunger rides against a cam on crankshaft.

On all models, align timing marks on crankshaft gear and cam gear when reassembling engine.

GOVERNOR (MECHANICAL). On mechanical type governors, the tangs on the governor weights should be square and smooth. If they are worn or do not operate freely, renew the gear and weight assembly. See Fig. B54.

Governor weight unit is removed from inside of crankcase on Models 9 and 14 and from outside of crankcase on Models N, 5, 6 and 8.

The governor crank should be a free fit in the bushing with only perceptible shake. If new crank fits the old bushing loosely, install a new bushing. On Models 9 and 14, ream bushing, after installation, to 0.2385-0.2390 inch (6.058-6.071 mm). Reject crank if it shows wear spots (Fig. B55).

VALVE SYSTEM. Valve tappet gap (cold) should be 0.007-0.009 inch (0.18-0.23 mm) for intake valve and 0.014-0.016 inch (0.36-0.41 mm) for exhaust valve.

Valve tappet clearance is adjusted by carefully grinding off end of valve stem to increase clearance or by grinding seats deeper or renewing valve and/or tappet to reduce clearance.

Valve face and valve seat angle is 45°. Seat width should be 3/64 to 1/16 inch (1.191 to 1.588 mm). Valve should be renewed if valve margin is 1/64 inch (0.397 mm) or less after refacing. All models are equipped with exhaust valve seat insert and an intake valve insert is available for service. Renewal of exhaust valve seat insert and installation of intake valve seat insert requires use of special tools. Refer to BRIGGS & STRATTON SPECIAL TOOL LIST following engine repair sections.

If Briggs & Stratton plug gage #19151 (Models 9 and 14) or #19122 (Models N, 5, 6 and 8) can be inserted a distance of 5/16 inch (7.938 mm) or more into valve guide, guide should be rebushed with a service bushing, Briggs & Stratton part #230655 (Models 9 and 14) or #63709 (Models N, 5, 6 and 8). On Models 9 and 14, ream (reamer #19192) guide to a depth of 1/16 inch (1.588 mm) more than length of bushing and press new bushing in until flush with top of guide. Bushings are presized at factory and should not require finish reaming.

On Models N, 5, 6 and 8, ream (reamer #19191) guide to a depth of 1/16 inch (1.588 mm) more than length of bushing and press new bushing in until flush with top of guide. Bushings are presized at factory and should not require finish reaming.

VALVE TIMING. When reassembling engine, align timing mark on cam gear with timing mark on crankshaft gear or crankshaft collar. Valve to piston timing will then be correct.

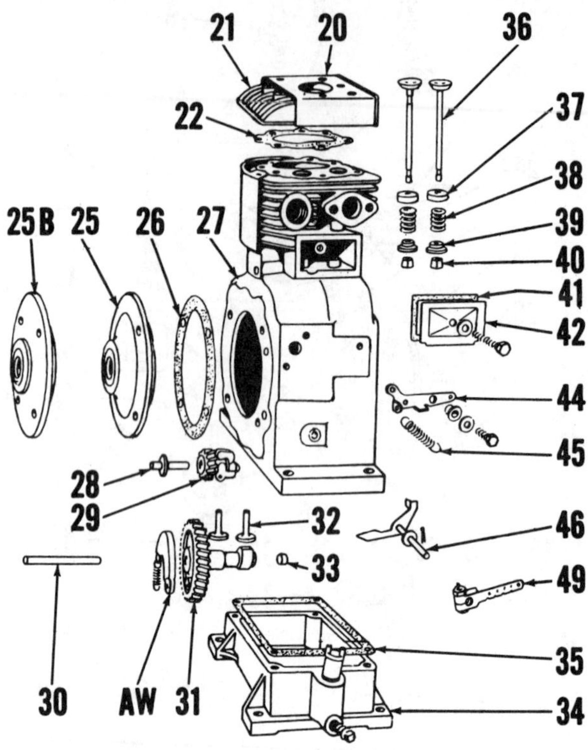

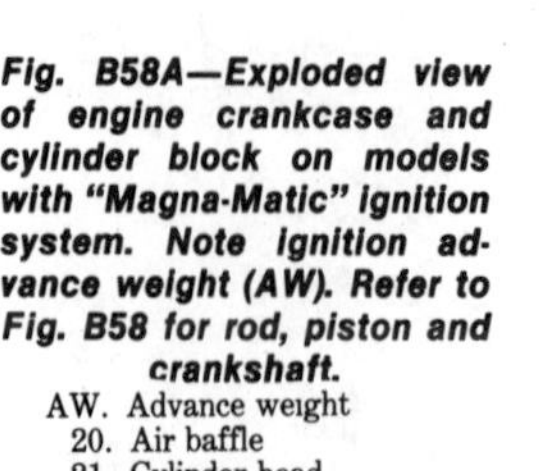

Fig. B58A—Exploded view of engine crankcase and cylinder block on models with "Magna-Matic" ignition system. Note ignition advance weight (AW). Refer to Fig. B58 for rod, piston and crankshaft.

AW. Advance weight
20. Air baffle
21. Cylinder head
22. Gasket
25. Bearing plate, plain bearing
25B. Bearing plate, ball bearing
26. Gasket
27. Crankcase & cylinder
28. Governor shaft
29. Governor gear & weight unit
30. Camshaft
31. Cam gear
32. Tappets
33. Camshaft plug
34. Engine base
35. Gasket
36. Valves
37. Valve spring washers
38. Valve spring
39. Spring retainers
40. Keepers
41. Gasket
42. Tappet chamber cover
44. Governor control lever
45. Governor spring
46. Governor crank
49. Governor lever

BRIGGS & STRATTON

4-STROKE

Model	Bore	Stroke	Displacement
6B, Early 60000	2.3125 in. (58.7 mm)	1.500 in. (38.1 mm)	6.3 cu. in. (103 cc)
Late 60000	2.3750 in. (60.3 mm)	1.500 in. (38.1 mm)	6.7 cu. in. (109 cc)
8B, 80000	2.3750 in. (60.3 mm)	1.750 in. (44.5 mm)	7.8 cu. in. (127 cc)
81000, 82000	2.3125 in. (60.3 mm)	1.750 in. (44.5 mm)	7.8 cu. in. (127 cc)
92000, 93500	2.5625 in. (65.1 mm)	1.750 in. (44.5 mm)	9.0 cu. in. (148 cc)
94000, 94500, 94900	2.5625 in. (65.1 mm)	1.750 in. (44.5 mm)	9.0 cu. in. (148 cc)
95500	2.5625 in. (65.1 mm)	1.750 in. (44.5 mm)	9.0 cu. in. (148 cc)

Model	Bore	Stroke	Displacement
100000	2.5000 in. (63.1 mm)	2.125 in. (54.0 mm)	10.4 cu. in. (169 cc)
110900, 111200, 111900	2.7813 in. (70.6 mm)	1.875 in. (47.6 mm)	11.4 cu. in. (186 cc)
112200	2.7813 in. (70.6 mm)	1.875 in. (47.6 mm)	11.4 cu. in. (186 cc)
113900	2.7813 in. (70.6 mm)	1.875 in. (47.6 mm)	11.4 cu. in. (186 cc)
13000	2.5625 in. (65.1 mm)	2.438 in. (60.9 mm)	12.6 cu .in. (203 cc)
131400, 131900	2.5625 in. (65.1 mm)	2.438 in. (60.9 mm)	12.6 cu. in. (203 cc)
140000	2.7500 in. (69.9 mm)	2.375 in. (60.3 mm)	14.1 cu. in. (231 cc)

CUBIC INCH DISPLACEMENT	FIRST DIGIT AFTER DISPLACEMENT BASIC DESIGN SERIES	SECOND DIGIT AFTER DISPLACEMENT CRANKSHAFT, CARBURETOR GOVERNOR	THIRD DIGIT AFTER DISPLACEMENT BEARINGS, REDUCTION GEARS & AUXILIARY DRIVES	FOURTH DIGIT AFTER DISPLACEMENT TYPE OF STARTER
6	0	0 -	0 - Plain Bearing	0 - Without Starter
8	1	1 - Horizontal Vacu-Jet	1 - Flange Mounting Plain Bearing	1 - Rope Starter
9	2	2 - Horizontal Pulsa-Jet	2 - Ball Bearing	2 - Rewind Starter
10	3	3 - Horizontal Flo-Jet (Pneumatic Governor)	3 - Flange Mounting Ball Bearing	3 - Electric - 110 Volt, Gear Drive
11	4	4 - Horizontal Flo-Jet (Mechanical Governor)	4 -	4 - Elec. Starter-Generator - 12 Volt, Belt Drive
13	5	5 - Vertical Vacu-Jet	5 - Gear Reduction (6 to 1)	5 - Electric Starter Only - 12 Volt, Gear Drive
14	6	6 -	6 - Gear Reduction (6 to 1) Reverse Rotation	6 - Alternator Only *
17	7	7 - Vertical Flo-Jet	7 -	7 - Electric Starter, 12 Volt Gear Drive, with Alternator
19	8	8 -	8 - Auxiliary Drive Perpendicular to Crankshaft	8 - Vertical-pull Starter
20	9	9 - Vertical Pulsa-Jet	9 - Auxiliary Drive Parallel to Crankshaft	
23				
24				
25				
30				
32				

* Digit 6 formerly used for "Wind-Up" Starter on 60000, 80000 and 92000 Series

EXAMPLES

To identify Model 100202:

10	0	2	0	2
10 Cubic Inch	Design Series 0	Horizontal Shaft - Pulsa-Jet Carburetor	Plain Bearing	Rewind Starter

Similarly, a Model 92998 is described as follows:

9	2	9	9	8
9 Cubic Inch	Design Series 2	Vertical Shaft - Pulsa-Jet Carburetor	Auxiliary Drive Parallel to Crankshaft	Vertical Pull Starter

Fig. B59—Explanation of numerical code used by Briggs & Stratton to identify engine and optional equipment from model number.

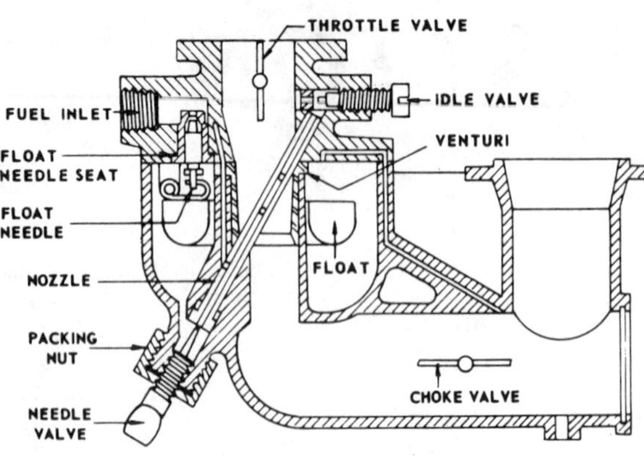

Fig. B60—Cross-sectional view of typical B&S "two-piece" carburetor. Before separating upper and lower body sections, loosen packing nut and unscrew nut and needle valve as a unit. Then, using special screwdriver, remove nozzle. Refer to text.

ENGINE IDENTIFICATION

Engines covered in this section have aluminum cylinder blocks with either plain aluminum cylinder bore or with a cast iron sleeve integrally cast into the block.

Early production of the 60000 model engine were of the same bore and stroke as the 6B model engine. The bore on 60000 engine was changed from 2.3125 inches (58.7 mm) to 2.3750 inches (60.3 mm) at serial number 5810060 on engines with plain aluminum bore, and at serial number 5810030 on engines with a cast iron sleeve.

Refer to Fig. B59 for chart explaining engine numerical code to identify engine model. Always furnish correct engine model and model number when ordering parts.

Fig. B61—Checking upper body of "two-piece" carburetor for warpage. Refer to text.

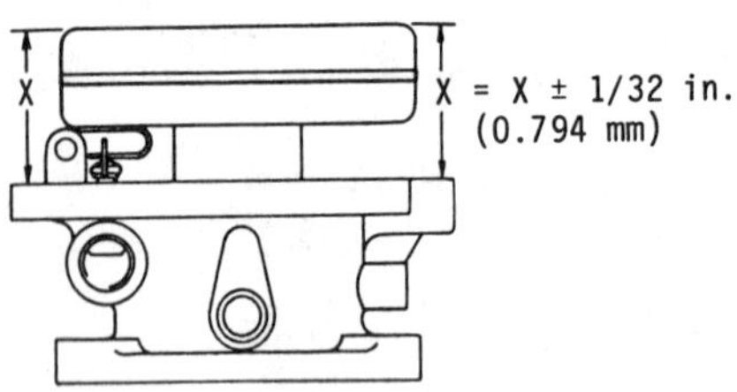

Fig. B62—Carburetor float setting should be within specifications shown. To adjust float setting, bend tang with needlenose pliers as shown in Fig. B63.

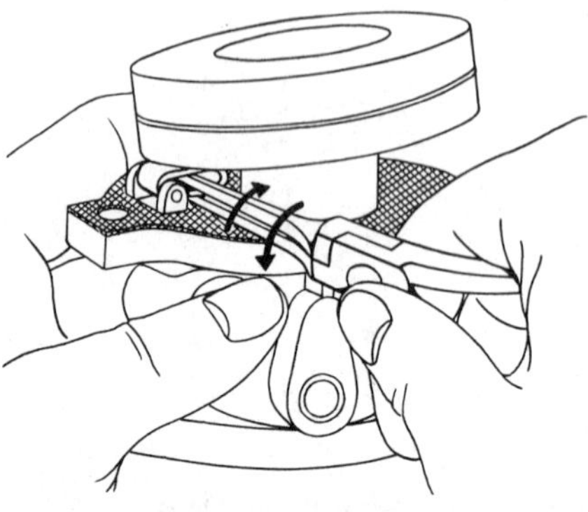

Fig. B63—Bending tang with needlenose pliers to adjust float setting. Refer to Fig. B62 for method of checking float setting.

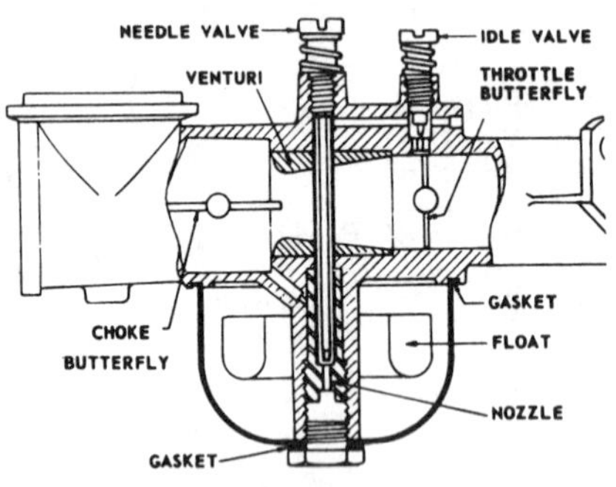

Fig. B64—Cross-sectional view of typical B&S small "one-piece" float type carburetor. Refer to Fig. B65 for disassembly views.

MAINTENANCE

SPARK PLUG. Recommended spark plug for all models is a Champion 8, or equivalent. If resistor type plugs are necessary to decrease radio interference, use Champion RJ8 or equivalent. Electrode gap for all models is 0.030 inch (0.76 mm).

CAUTION: Briggs & Stratton does not recommend using abrasive blasting method to clean spark plugs as this may introduce some abrasive material into the engine which could cause extensive damage.

FLOAT TYPE (FLO-JET) CARBURETORS. Three different float type carburetors are used. They are called a "two-piece" (Fig. B60), a small "one-piece" (Fig. B64) or a large "one-piece" (Fig. B66) carburetor depending upon the type of construction.

Float type carburetors are equipped with adjusting needles for both idle and power fuel mixtures. Counterclockwise rotation of the adjusting needles richens the mixture. For initial starting adjustment, open the main needle valve (power fuel mixture) 1½ turns on the two-piece carburetor and 2½ turns on the small one-piece carburetor. Open the idle needle ½ to ¾ turn on the two-piece carburetor and 1½ turns on the small one-piece carburetor. On the large one-piece carburetor, open both needle valves 1-1/8 turns.

Make final adjustments with engine at operating temperature and running. Set the speed control for desired operating speed, turn main needle clockwise until engine misses, and then turn it counterclockwise just past the smooth operating point until the engine begins to run unevenly. Return the speed control to idle position and adjust the idle speed

Fig. B65—Disassembling the small "one-piece" float type carburetor. Pry out welch plug, remove choke butterfly (disc), remove choke shaft and needle valve; venturi can then be removed as shown in left view.

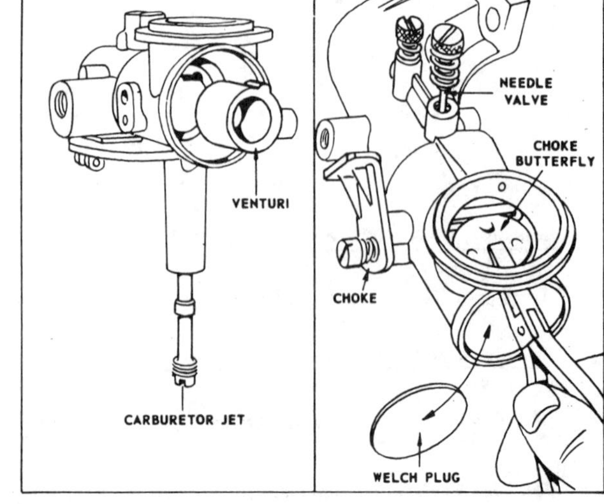

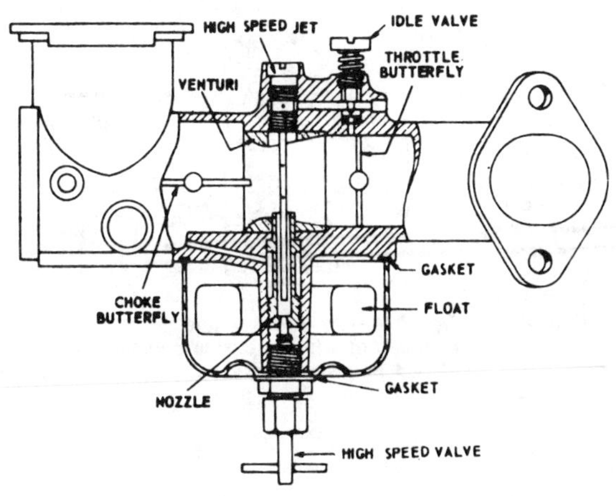

Fig. B66—Cross-sectional view of B&S large "one-piece" float type carburetor.

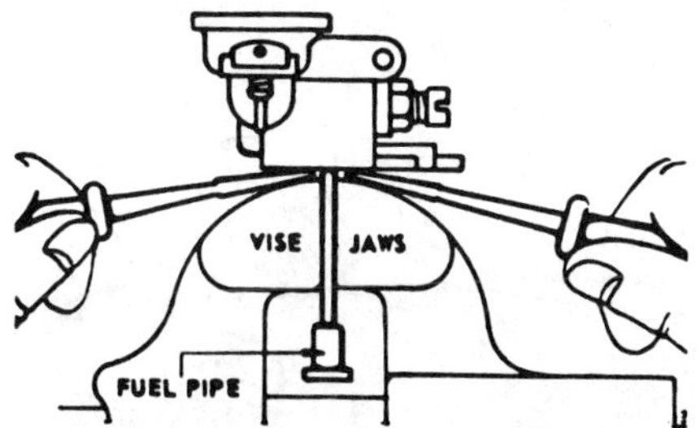

Fig. B68—Removing brass fuel feed pipe from suction type carburetor. Press new brass pipe into carburetor until it projects 2-9/32 to 2-5/16 inches (57.9-58.7 mm) from carburetor face. Nylon fuel feed pipe is threaded into carburetor.

stop screw until the engine idles at 1750 rpm. Adjust the idle needle valve until the engine runs smoothly. Reset the idle speed stop screw if necessary. The engine should then accelerate without hesitation. If engine does not accelerate properly, turn the main needle valve counterclockwise slightly to provide a richer fuel mixture.

The float setting on all float type carburetors should be within dimensions shown in Fig. B62. If not, bend the tang on float as shown in Fig. B63 to adjust float setting. If any wear is visible on the inlet valve or the inlet valve seat, install a new valve and seat assembly. On large one-piece carburetors, the renewable inlet valve seat is pressed into the carburetor body until flush with the body.

NOTE: The upper and lower bodies of the two-piece float type carburetor are locked together by the main nozzle. Refer to cross-sectional view of carburetor in Fig. B60. Before attempting to separate the upper body from the lower body, loosen packing nut and unscrew nut and needle valve. Then, using special screwdriver (B&S tool 19061 or 19062), remove nozzle.

If a 0.002 inch (0.05 mm) feeler gage can be inserted between upper and lower bodies of the two-piece carburetor as shown in Fig. B61, the upper body is warped and should be renewed.

Check the throttle shaft for wear on all float type carburetors. If 0.010 inch (0.25 mm) or more free play (shaft-to-bushing clearance) is noted, install new throttle shaft and/or throttle shaft bushings. To remove worn bushings, turn a 1/4 inch × 20 tap into bushing and pull bushing from body casting with the tap. Press new bushings into casting by using a vise and if necessary, ream bushings with a 7/32 inch drill bit.

SUCTION TYPE (VACU-JET) CARBURETORS. A typical suction type (Vacu-Jet) carburetor is shown in Fig. B67. This type carburetor has only one fuel mixture adjusting needle. Turning the needle clockwise leans the air:fuel mixture. Adjust suction type carburetors with fuel tank approximately one-half full and with the engine at operating temperature and running at approximately 3000 rpm, no-load. Turn needle valve clockwise until engine begins to run unevenly from a too rich air:fuel mixture. This should result in a correct adjustment for full load operation. Adjust idle speed to 1750 rpm.

To remove the suction type carburetor, first remove carburetor and fuel tank as an assembly, then remove carburetor from fuel tank. When reinstalling carburetor on fuel tank, use a new gasket and tighten retaining screws evenly.

The suction type carburetor has a fuel feed pipe extending into fuel tank. The pipe has a check valve to allow fuel to feed up into the carburetor but prevents fuel from flowing back into the tank. If check valve is inoperative and cleaning in alcohol or acetone will not free the check valve, renew the fuel feed pipe. If feed pipe is made of brass, remove as shown in Fig. B68. Using a vise, press new pipe into carburetor so it extends from 2-9/32 to 2-5/16 inches (57.9-58.7 mm) from carburetor body. If pipe is made of nylon (plastic), screw pipe out of carburetor body with wrench. When installing new nylon feed pipe, be careful not to overtighten.

NOTE: If soaking carburetor in cleaner for more than one-half hour, be sure to remove all nylon parts and "O" ring, if used, before placing the carburetor in cleaning solvent.

PUMP TYPE (PULSA-JET) CARBURETORS. The pump type (Pulsa-Jet) carburetor is basically a suction type carburetor incorporating a fuel

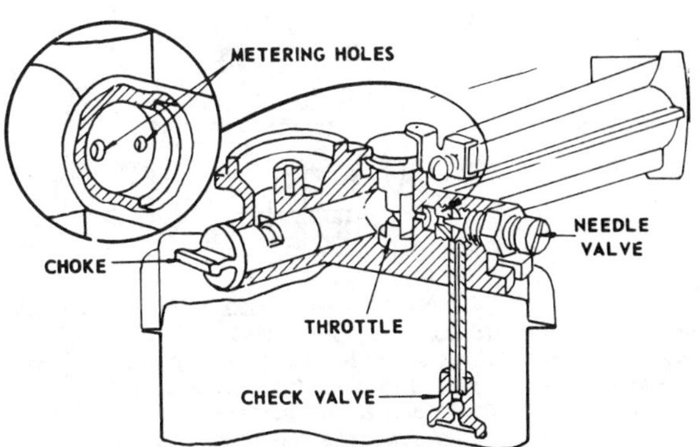

Fig. B67—Cutaway view of typical suction type (Vacu-Jet) carburetor. Inset shows fuel metering holes which are accessible for cleaning after removing needle valve. Be careful not to enlarge the holes when cleaning them.

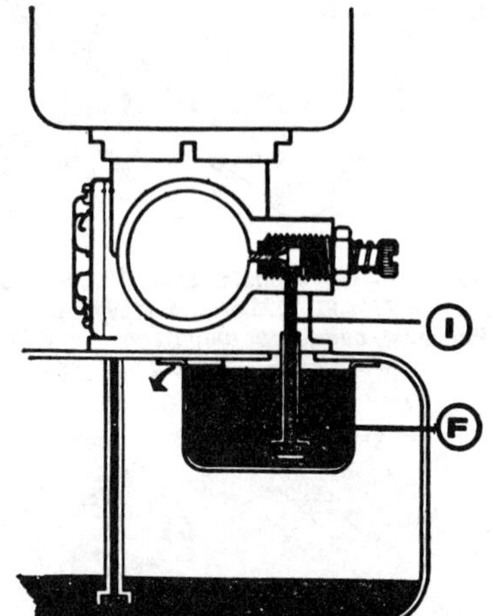

Fig. B69—Fuel flow in Pulsa-Jet carburetor. Fuel pump incorporated in carburetor fills constant level sump (F) below carburetor and excess fuel flows back into tank. Fuel is drawn from sump through inlet (I) past fuel mixture adjusting needle by vacuum in carburetor.

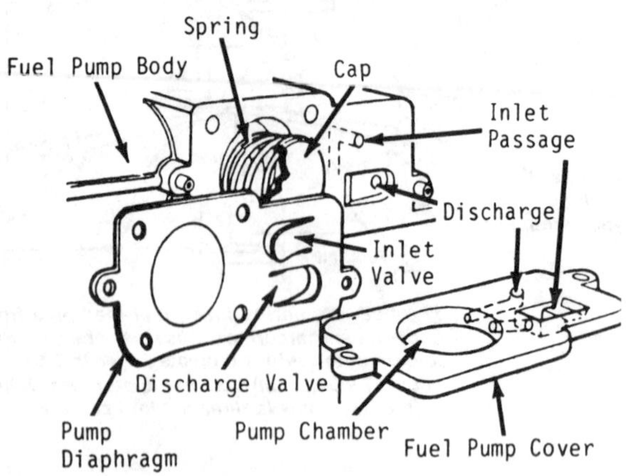

Fig. B70—Exploded view of fuel pump that is incorporated in Pulsa-Jet carburetor except those used on 82900, 92900, 94900, 110900, 111900, 112200 and 113900 models; refer to Fig. B71.

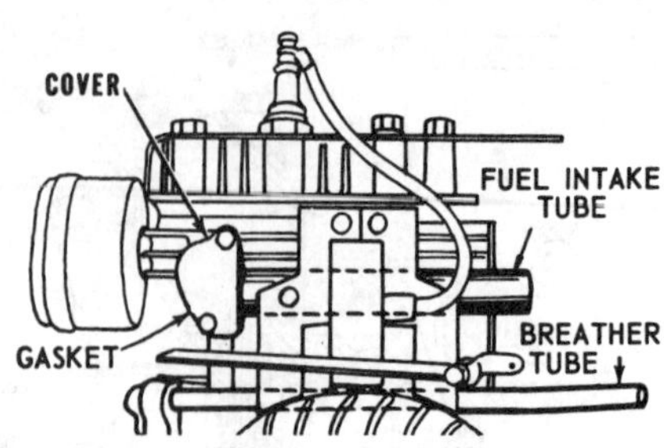

Fig. B74—On 82000 models, intake tube is threaded into intake port of engine; a gasket is placed between intake port cover and intake port.

pump to fill a constant level fuel sump in top of fuel tank. Refer to schematic view in Fig. B69. This makes a constant air:fuel mixture available to engine regardless of fuel level in tank. Adjustment of the pump type carburetor fuel mixture needle valve is the same as outlined for suction type carburetors in previous paragraph, except that fuel level in tank is not important.

To remove the pump type carburetor, first remove the carburetor and fuel tank as an assembly; then, remove carburetor from fuel tank. When reinstalling carburetor on fuel tank, use a new gasket or pump diaphragm as required and tighten retaining screws evenly.

Fig. B70 shows an exploded view of the pump unit used on all carburetors except those for Models 82900, 92900, 94900, 110900, 111900, 112200 and 113900 the pump diaphragm is placed between the carburetor and fuel tank as shown in Fig. B71.

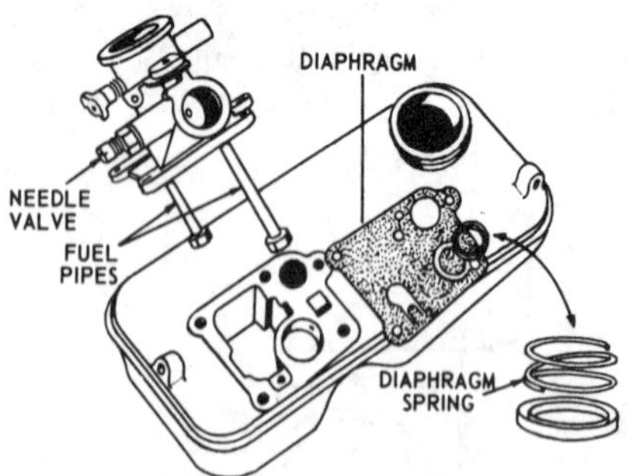

Fig. B71—On Models 82900, 92900, 94900, 110900, 111900, 112200 and 113900, pump type (Pulsa-Jet) carburetor diaphragm is installed between carburetor and fuel tank.

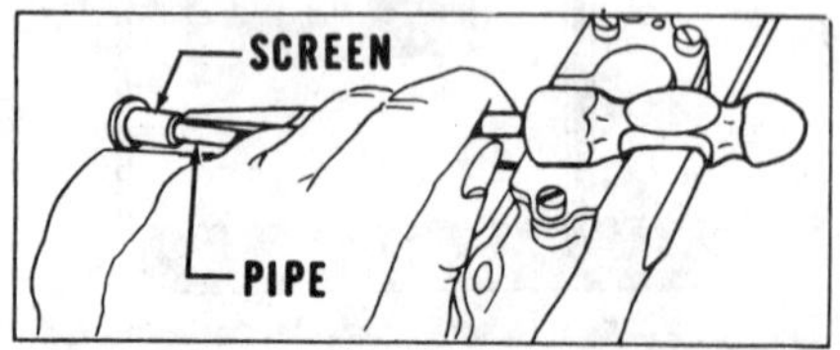

Fig. B72—To renew screen housing on pump type carburetor with brass feed pipes, drive old screen housing from pipe as shown. To hold pipe, clamp lightly in a vise.

The pump type carburetor has two fuel feed pipes. The long pipe feeds fuel into the pump portion of the carburetor from which fuel then flows to the constant level fuel sump. The short pipe extends into the constant level sump and feeds fuel into the carburetor venturi via fuel mixture needle valve.

As check valves are incorporated in the pump diaphragm, fuel feed pipes on pump type carburetors do not have a check valve. However, if the fuel screen in lower end of pipe is broken or clogged and cannot be cleaned, the pipe or screen housing can be renewed. If pipe is made of nylon, pipe snaps into place and considerable force is required to remove or install pipe. Be careful not to damage new pipe. If pipe is made of brass, clamp pipe lightly in a vise and drive old screen housing from pipe with a screwdriver or small chisel as shown in Fig. B72. Drive a new screen housing onto pipe with a soft faced hammer.

NOTE: If soaking carburetor in cleaner for more than one-half hour, be sure to remove all nylon parts and "O" ring, if used, before placing carburetor in cleaning solvent.

NOTE: On engine Models 82900, 92900, 94900, 95500, 110900, 111900 and 113900, be sure air cleaner retaining screw is in place if engine is being operated (during tests) without air cleaner installed. If screw is not in place, fuel will lift up through the screw hole and enter carburetor throat as the screw hole leads directly into the constant level fuel sump.

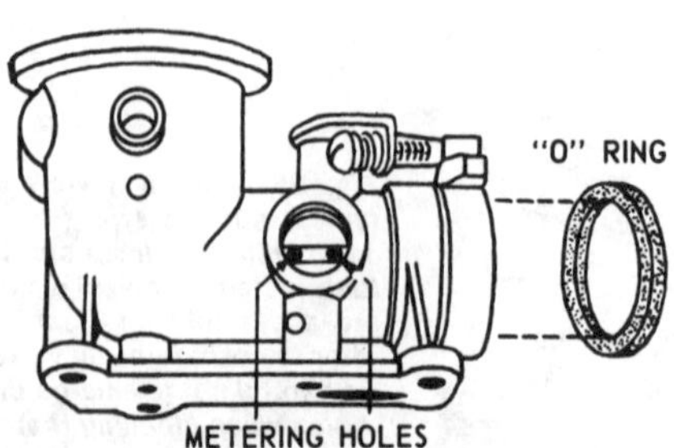

Fig. B73—Metering holes in pump type carburetors are accessible for cleaning after removing fuel mixture needle valve. On models with intake pipe, carburetor is sealed to pipe with "O" ring.

INTAKE TUBE. Models 82000, 92000, 93500, 94500, 94900, 95500, 100900, 110900, 111900, 113900, 130900 and 131900 have an intake tube between carburetor and engine intake port. Carburetor is sealed to intake tube with an "O" ring as shown in Fig. B73.

On Model 82000 engines, the intake tube is threaded into the engine intake port. A gasket is used between the engine intake port cover and engine casting. Refer to Fig. B74.

On Models 92000, 93500, 94500, 94900, 95500, 110900, 111900 and 113900 engines, the intake tube is bolted to the engine intake port and gasket is used between the intake tube and engine casting. Refer to Fig. B75. On Models 100900, 130900 and 131900 intake tubes are attached to engine in similar manner.

CHOKE-A-MATIC CARBURETOR CONTROLS. Engines equipped with float, suction or pump type carburetors may be equipped with a control unit with which the carburetor choke, throttle and magneto grounding switch are operated from a single lever (Choke-A-Matic carburetors). Refer to Figs. B76 through B82 for views showing the different

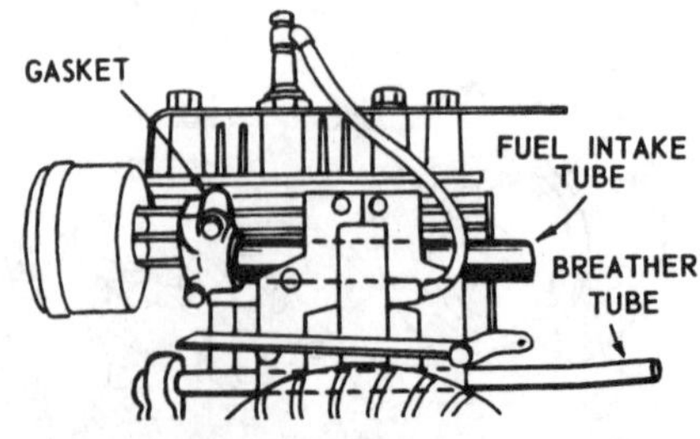

Fig. B75—On 92000, 93500, 94500, 94900, 95500, 110900, 111900 and 113900 models, fuel intake tube is bolted to engine intake port and a gasket is placed between tube and engine. On 100900, 130900 and 131900 vertical crankshaft models, intake tube and gasket are similar.

Fig. B76—Choke-A-Matic control on float type carburetor. Remote control can be attached to speed slide.

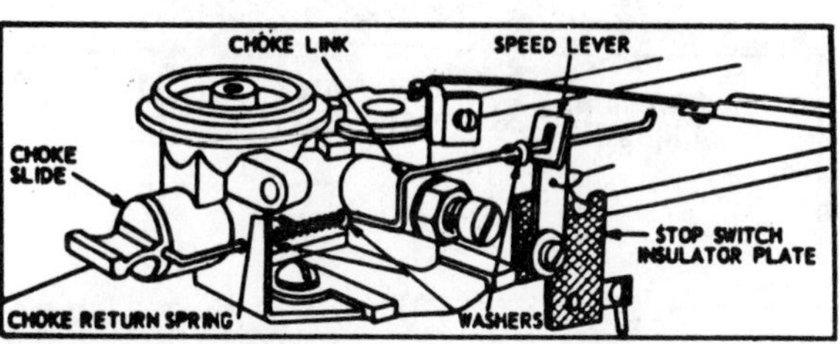

Fig. B77—Typical Choke-A-Matic control on suction type carburetor. Remote control can be attached to speed lever.

Fig. B78—Choke-A-Matic control in choke and stop positions on float carburetor.

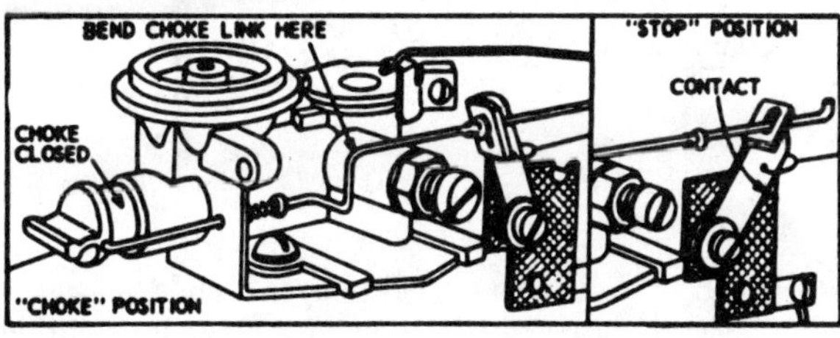

Fig. B79—Choke-A-Matic controls in choke and stop positions on suction carburetor. Bend choke link if necessary to adjust control.

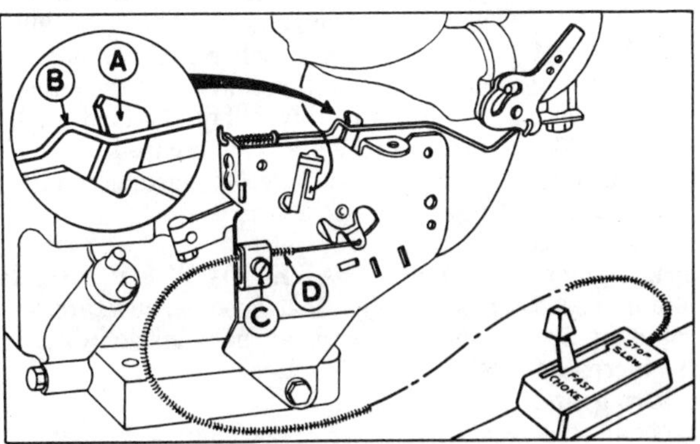

Fig. B80—On Choke-A-Matic control shown, choke actuating lever (A) should just contact choke link or shaft (B) when control is at "FAST" position. If not, loosen screw (C) and move control wire housing (D) as required.

Fig. B81—When Choke-A-Matic control is in "START" or "CHOKE" position, choke must be completely closed as shown in view A. When control is in "STOP" position, arm should contact stop switch (view B).

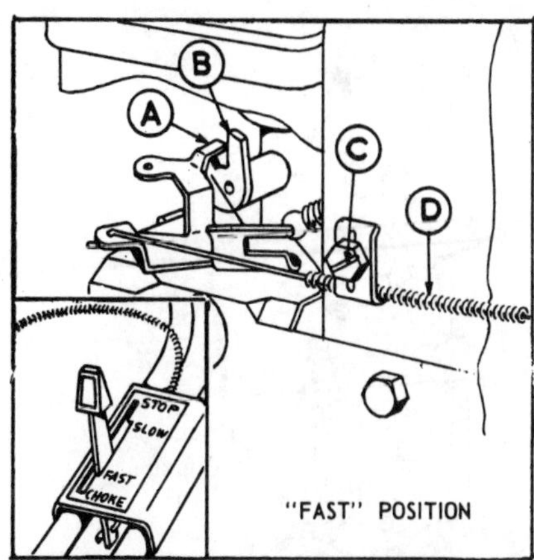

Fig. B82—On Choke-A-Matic controls shown, lever (A) should just contact choke shaft arm (B) when control is in "FAST" position. If not, loosen screw (C) and move control wire housing (D) as required, then tighten screw.

types of Choke-A-Matic carburetor controls.

To check operation of Choke-A-Matic carburetor controls, move control lever to "CHOKE" position. Carburetor choke slide or plate must be completely closed. Then, move control lever to "STOP" position. Magneto grounding switch should be making contact. With the control lever in "RUN", "FAST" or "SLOW" position, carburetor choke should be completely open. On units with remote controls, synchronize movement of remote lever to carburetor control lever by loosening screw (C–Fig. B80 or Fig. B82) and moving control wire housing (D) as required; then, tighten screw to clamp the housing securely. Refer to Fig. B83 to check remote control wire movement.

AUTOMATIC CHOKE (THERMOSTAT TYPE). A thermostat operated choke is used on some models equipped with the two-piece carburetor. To adjust choke linkage, hold choke shaft so thermostat lever is free. At room temperature, stop screw in thermostat collar should be located midway between thermostat stops. If not, loosen stop screw, adjust the collar and tighten stop screw. Loosen set screw (S–Fig.

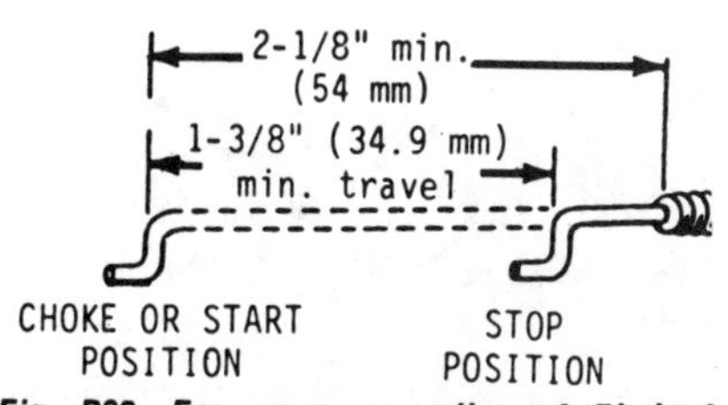

Fig. B83—For proper operation of Choke-A-Matic controls, remote control wire must extend to dimension shown and have a minimum travel of 1-3/8 inches (34.9 mm).

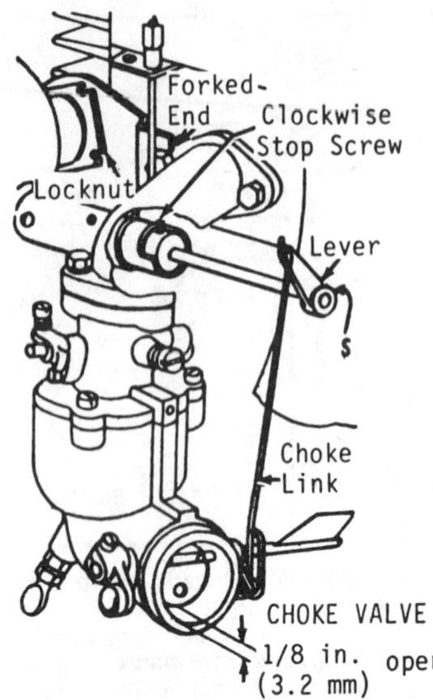

Fig. B84—Automatic choke used on some models equipped with "two-piece" Flo-Jet carburetor showing unit in "HOT" position.

B84) on thermostat lever. Then, slide lever on shaft to ensure free movement of choke unit. Turn thermostat shaft clockwise until stop screw contacts thermostat stop. While holding shaft in this position, move shaft lever until choke is open exactly 1/8 inch (3.17 mm) and tighten lever set screw. Turn thermostat shaft counterclockwise until stop screw contacts thermostat stop as shown in Fig. B85. Manually open choke valve until it stops against top of choke link opening. At this time, choke valve should be open at least 3/32 inch (2.38 mm), but not more than 5/32 inch (3.97 mm). Hold choke valve in wide open position and check position of counterweight lever. Lever should be in a horizontal position with free end towards right.

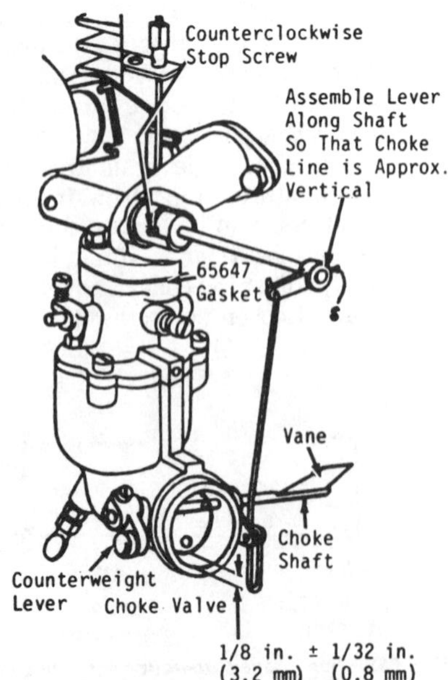

Fig. B85—Automatic choke on "two-piece" Flo-Jet carburetor in "COLD" position.

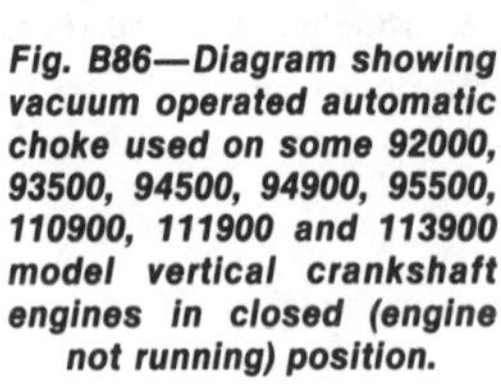

Fig. B86—Diagram showing vacuum operated automatic choke used on some 92000, 93500, 94500, 94900, 95500, 110900, 111900 and 113900 model vertical crankshaft engines in closed (engine not running) position.

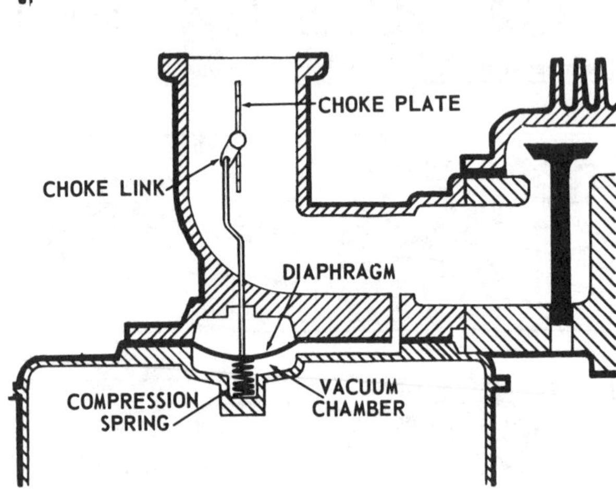

Fig. B87—Diagram showing vacuum operated automatic choke in open (engine running) position.

AUTOMATIC CHOKE (VACUUM TYPE). A spring and vacuum operated automatic choke is used on some 92000, 93500, 94500, 94900, 95500, 110900, 111900 and 113900 vertical crankshaft engines. A diaphragm under carburetor is connected to the choke shaft by a link. The compression spring works against the diaphragm, holding choke in closed position when engine is not running. See Fig. B86. As engine starts, increased vacuum works against the spring and pulls the diaphragm and choke link down, holding choke in open (running) position shown in Fig. B87.

During operation, if a sudden load is applied to engine or a lugging condition develops, a drop in intake vacuum occurs, permitting choke to close partially. This provides a richer fuel mixture to meet the condition and keeps the engine running smoothly. When the load condition has been met, increased vacuum returns choke valve to normal running (fully open) position.

FUEL TANK OUTLET. Small models with float type carburetors are equipped with a fuel tank outlet as

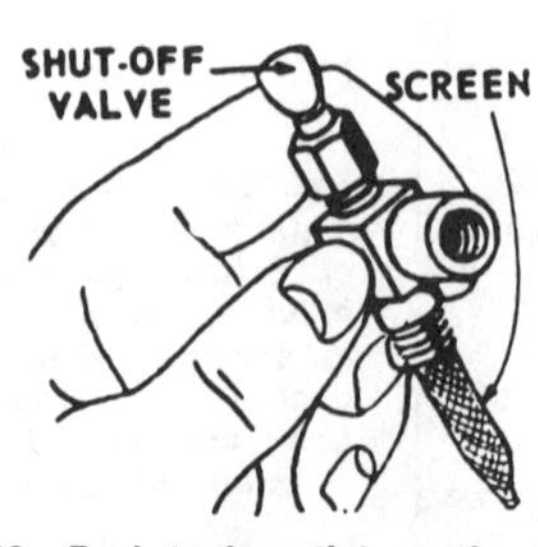

Fig. B88—Fuel tank outlet used on smaller engines with float type carburetor.

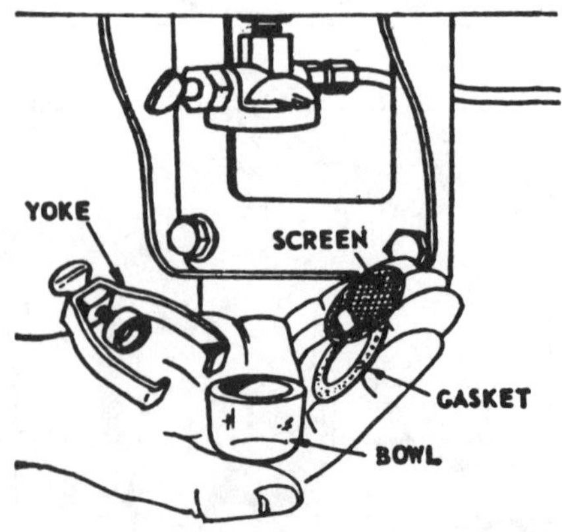

Fig. B89—Fuel tank outlet used on larger B&S engines.

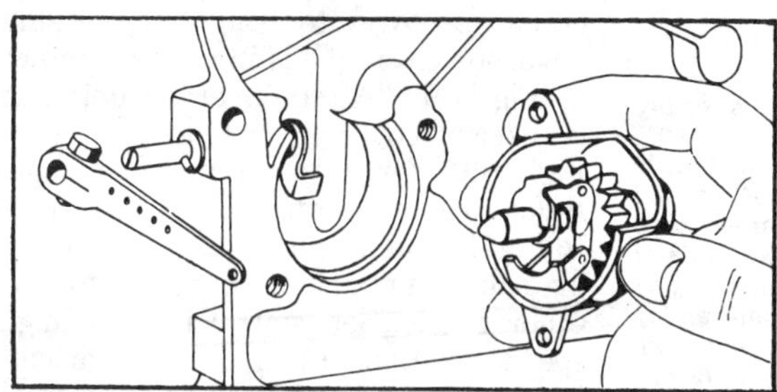

Fig. B90—Removing governor unit (except on 100000, 112200, 130000, 131000 and late 140000 models) from inside crankcase cover on horizontal crankshaft models. Refer to Fig. B91 for exploded view of governor.

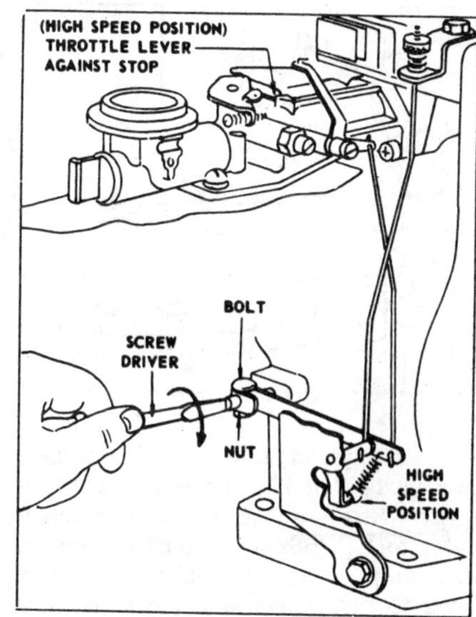

Fig. B95—Linkage adjustment on 100000, 112200, 130000, 131000 and late 140000 horizontal crankshaft mechanical governor models; refer to text for procedure.

shown in Fig. B88. On larger engines, a fuel sediment bowl is incorporated with the fuel tank outlet as shown in Fig. B89. Clean any lint and dirt from tank outlet screens with a brush. Varnish or other gasoline deposits may be removed by using a suitable solvent. Tighten packing nut or remove nut and shut-off valve, then renew packing if leakage occurs around shut-off valve stem.

FUEL PUMP. A fuel pump is available as optional equipment on some models. Refer to SERVICING BRIGGS & STRATTON ACCESSORIES section in this manual for fuel pump service information.

GOVERNOR. All models are equipped with either a mechanical (flyweight type) or an air vane (pneumatic) governor. Refer to the appropriate paragraph for model being serviced.

Mechanical Governor. Three different designs of mechanical governors are used.

On all engines except 100000, 112200, 130000, 131000 and all late 140000 models, a governor unit as shown in Fig. B90 is used. An exploded view of this governor unit is shown in Fig. B91. The governor housing is attached to inner side of crankcase cover and the governor gear is driven from the engine camshaft gear. Use Figs. B90 and B91 as a

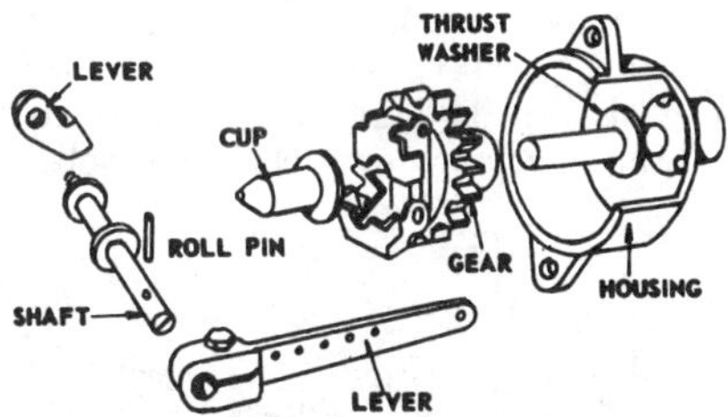

Fig. B91—Exploded view of govenor unit used on all horizontal crankshaft models (except 100000, 112200, 130000, 131000 and late 140000 models) with mechanical governor. Refer also to Figs. B90 and B92.

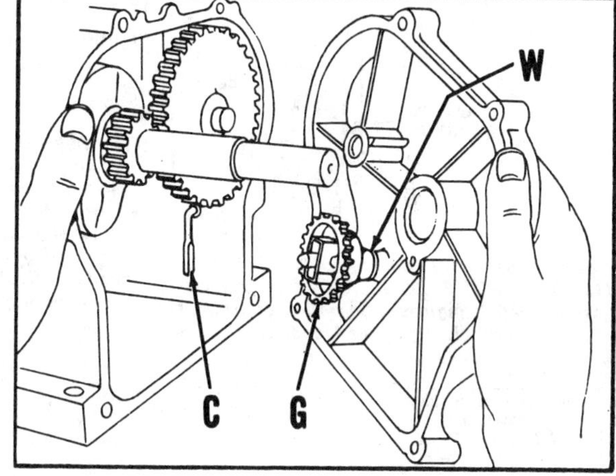

Fig. B93—Installing crankcase cover on 100000, 112200, 130000, 131000 and late 140000 models with mechanical governor. Governor crank (C) must be in position shown. A thrust washer (W) is placed between governor (G) and crankcase cover.

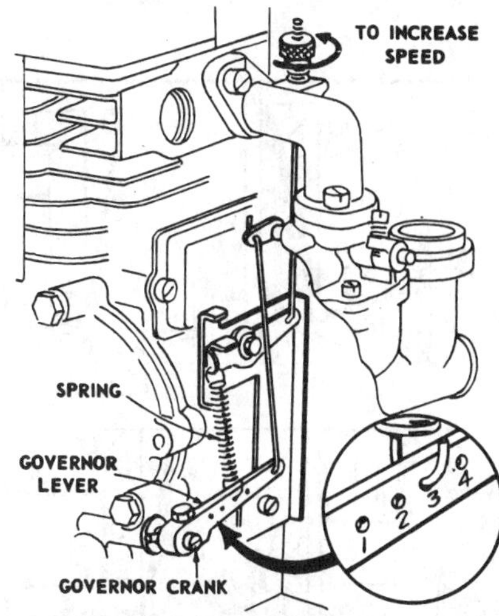

Fig. B92—View of governor linkage used on horizontal crankshaft mechanical governor models except 100000, 112200, 130000, 131000 and late 140000 models. Governor spring should be hooked in governor lever as shown in inset.

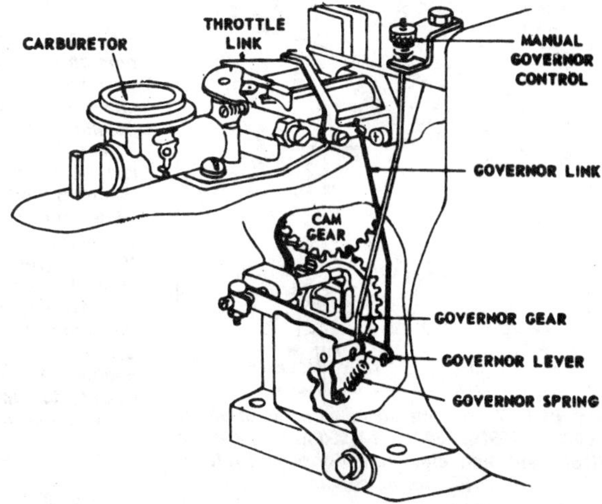

Fig. B94—Cutaway drawing of governor and linkage used on 100000, 112200, 130000, 131000 and late 140000 horizontal crankshaft models.

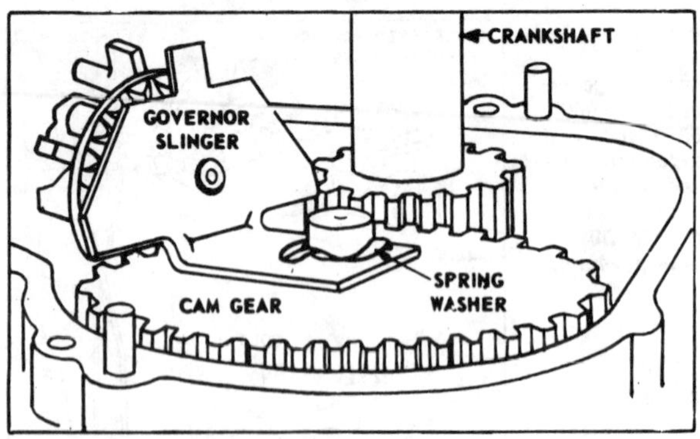

Fig. B96—View showing 95500, 100000, 113900, 130000, 131000 and 140000 vertical crankshaft model mechanical governor unit. Drawing is of lower side of engine with oil sump (engine base) removed; note spring washer location used on 100000, 130000 and 131000 models only.

disassembly and assembly guide. Renew any parts that bind or show excessive wear. After governor is assembled, refer to Fig. B92 and adjust linkage by loosening screw clamping governor lever to governor crank. Turn governor lever counterclockwise so carburetor throttle is in wide open position. Hold lever and turn governor crank as far counterclockwise as possible, then tighten screw clamping lever to crank. Governor crank can be turned with screwdriver. Check linkage to be sure it is free and that the carburetor throttle will move from idle to wide open position.

On 100000, 112200, 130000, 131000 and late 140000 horizontal crankshaft models, the governor gear and weight unit (G–Fig. B93) is supported on a pin in engine crankcase cover and the governor crank is installed in a bore in the engine crankcase. A thrust washer (W) is placed between governor gear and crankcase cover. When assembling crankcase cover to crankcase, be sure governor crank (C) is in position shown in Fig. B93. After governor unit and crankcase cover is installed, refer to Fig. B94 for installation of linkage. Before attempting to start engine, refer to Fig. B95 and adjust linkage by loosening bolt clamping governor lever to governor crank. Set control lever in high speed position. Using a screwdriver, turn governor crank as far clockwise as possible and tighten governor lever clamp bolt. Check to be sure carburetor throttle can be moved from idle to wide open position and that linkage is free.

On vertical crankshaft 95500, 100000, 113900, 130000, 131000 and 140000 with mechanical governor, the governor weight unit is integral with the lubricating oil slinger and is mounted on lower end of camshaft gear as shown in Fig. B96. With engine upside down, place governor and slinger unit on camshaft gear as shown, place spring washer (100000, 130900 and 131900 models only) on camshaft gear and install engine base on crankcase. Assemble linkage as shown in Fig. B97; then, refer to Fig. B98 and adjust linkage by loosening governor lever to governor crank clamp bolt, place control lever in high speed position, turn governor crank with screwdriver as far as possible and tighten governor lever clamping bolt.

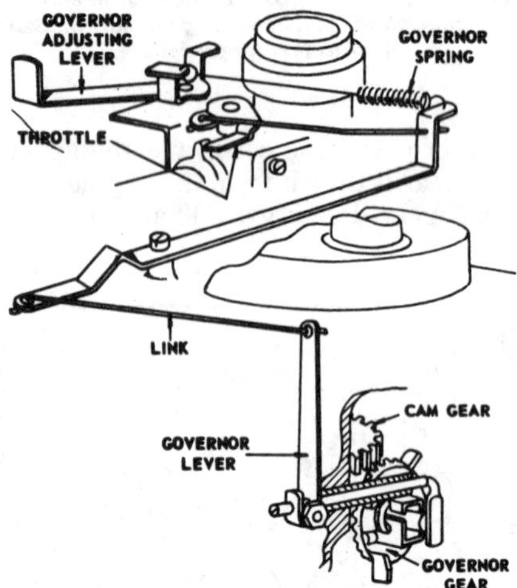

Fig. B97—Schematic drawing of 95500, 100000, 113900, 130000, 131000 and 140000 mechanical governor and linkage used on vertical crankshaft mechanical governed models.

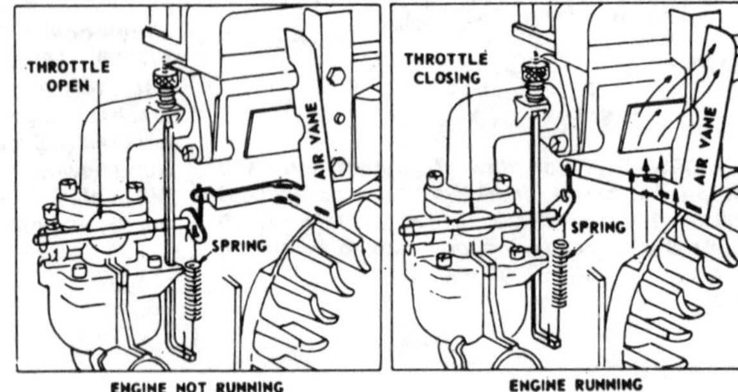

Fig. B99—Views showing operating principle of air vane (penumatic) governor. Air from flywheel fan acts against air vane to overcome tension of governor spring; speed is adjusted by changing spring tension.

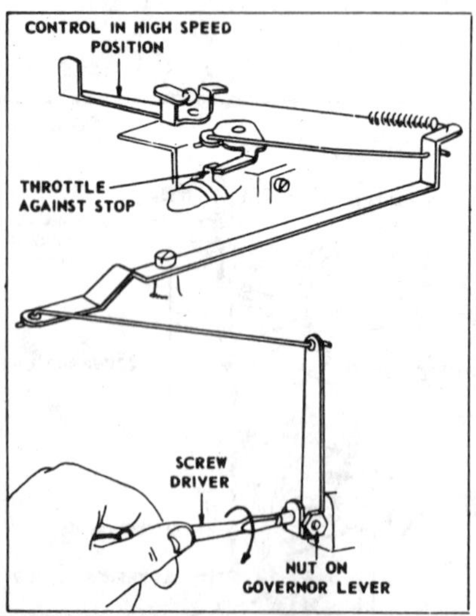

Fig. B98—View showing adjustment of 95500, 100000, 113900, 130000, 131000 and 140000 vertical crankshaft mechanical governor; refer to text for procedure.

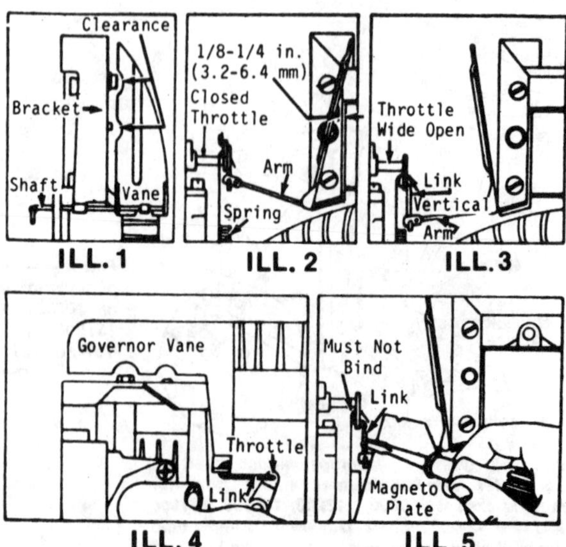

Fig. B100—Air vane governors and linkage. ILL. 1; the governor vane should be checked for clearance in all positions. ILL. 2; the vane should top 1/8 to 1/4 inch (3.175-6.35 mm) from the magneto coil. ILL. 3; with wide open throttle, the link connecting vane arm to throttle lever should be in a vertical position on vertical cylinder engines and in a horizontal position (ILL. 4) on horizontal cylinder engines. Bend link slightly (ILL. 5) to remove any binding condition in linkage.

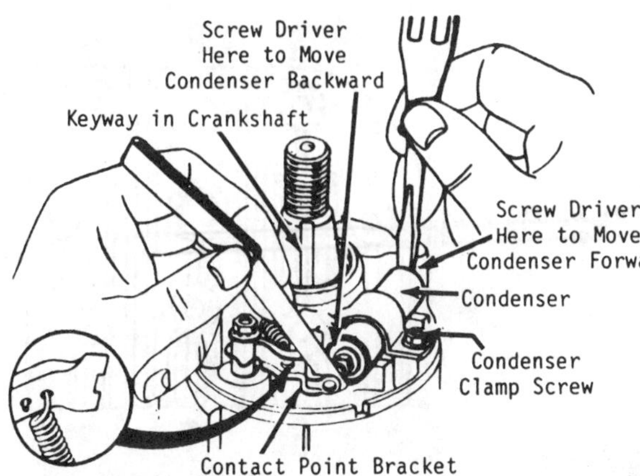

Fig. B101—View showing breaker point adjustment on models having breaker point integral with condenser. Move condenser to adjust point gap.

Air Vane (Pneumatic) Governor. Refer to Fig. B99 for views showing typical air vane governor operation.

The vane should stop 1/8 to ¼ inch (3.17-6.35 mm) from the magneto coil (Fig. B100, illustration 2) when the linkage is assembled and attached to carburetor throttle shaft. If necessary to adjust, spring the vane while holding the shaft. With wide open throttle, the link from the air vane arm to the carburetor throttle should be in a vertical position on horizontal crankshaft models and in a horizontal position on vertical crankshaft models. (Fig. B100, illustration 3 and 4.) Check linkage for binding. If binding condition exists, bend links slightly to correct. Refer to Fig. B100, illustration 5.

NOTE: Some engines are equipped with a nylon governor vane which does not require adjustment.

MAGNETO. Breaker contact gap on all models with magneto type ignition is 0.020 inch (0.51 mm). On all models except "Sonoduct" (vertical crankshaft with flywheel below engine), breaker points and condenser are accessible after removing engine flywheel and breaker cover. On "Sonoduct" models, breaker points are located below breaker cover on top side of engine.

On some models, one breaker contact point is an integral part of the ignition condenser and the breaker arm pivots on a knife edge retained by a slot in pivot post. On these models, breaker contact gap is adjusted by moving the condenser as shown in Fig. B101. On other models, breaker contact gap is adjusted by relocating position of breaker contact bracket. Refer to Fig. B102.

On all models, breaker contact arm is actuated by a plunger held in a bore in engine crankcase and rides against a cam on engine crankshaft. Plunger can be removed after removing breaker points. Renew plunger if worn to a length of 0.870 inch (22.1 mm) or less. Plunger must be installed with grooved end to the top to prevent oil seepage (Fig. B102A). Check breaker point plunger bore wear in crankcase with B&S plug gage 19055. If plug gage will enter bore ¼ inch (6.35 mm) or more, bore should be reamed and a bushing installed. Refer to Fig. B103 for method of checking bore and to Fig. B104 for steps in reaming bore and installing bushing if bore is worn. To ream bore and install bushing, it is necessary that the breaker points, armature and ignition coil and the crankshaft be removed.

On "Sonoduct" models, armature and ignition coil are inside flywheel on bottom side of engine. Armature air gap is correct if armature is installed flush with mounting boss as shown in Fig. B105. On all other models, armature and ignition coil are located outside flywheel. Armature-to-flywheel air gap should be as follows:

Models 6B, 8B, 6000–

Two Leg Armature . . . 0.006-0.010 in. (0.15-0.25 mm)

Three Leg Armature . . 0.012-0.016 in. (0.30-0.41 mm)

Models 80000, 82000, 92000, 93000, 93500, 94000, 94500, 94900, 95000, 95500, 110000, 110900–

Two Leg Armature . . . 0.006-0.010 in. (0.15-0.25 mm)

Three Leg Armature . . 0.012-0.016 in. (0.30-0.41 mm)

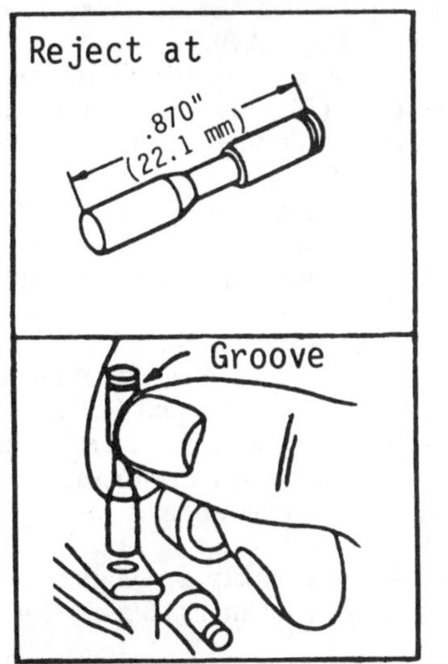

Fig. B102—Adjustment of breaker point gap on models having breaker point separate from condenser.

Fig. B102A—Insert plunger into bore with groove toward top. Refer to text.

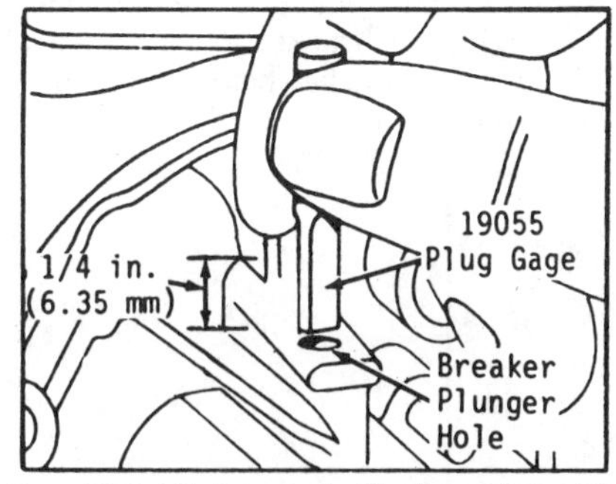

Fig. B103—If Briggs & Stratton plug gage #19055 can be inserted in breaker plunger bore a distance of ¼ inch (6.35 mm) or more, bore is worn and must be rebushed.

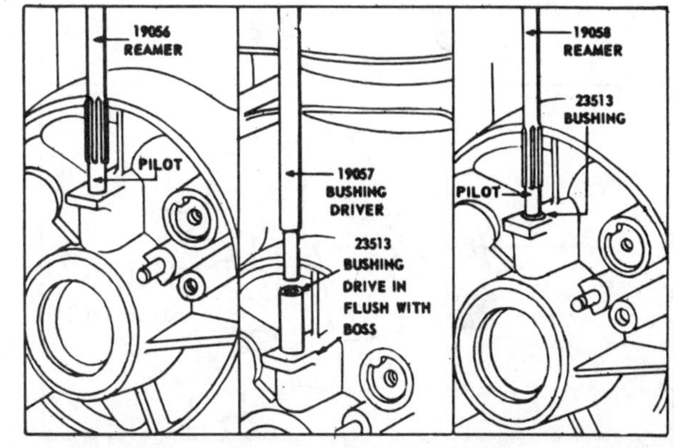

Fig. B104—Views showing reaming plunger bore to accept bushing (left view), installing bushing (center) and finish reaming bore (right) of bushing.

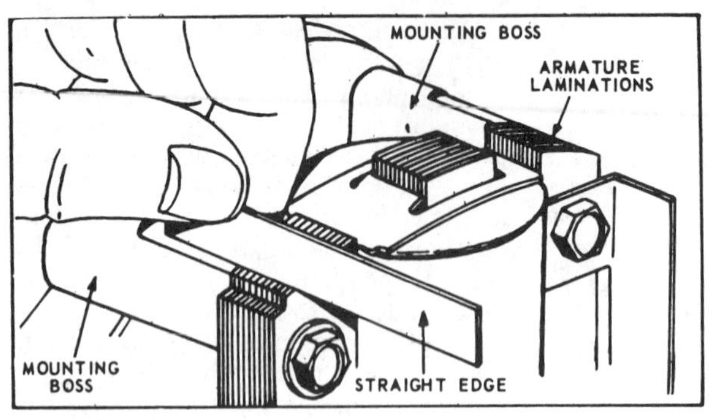

Fig. B105—On "Sonoduct" models, align armature core with mounting boss for proper magneto air gap.

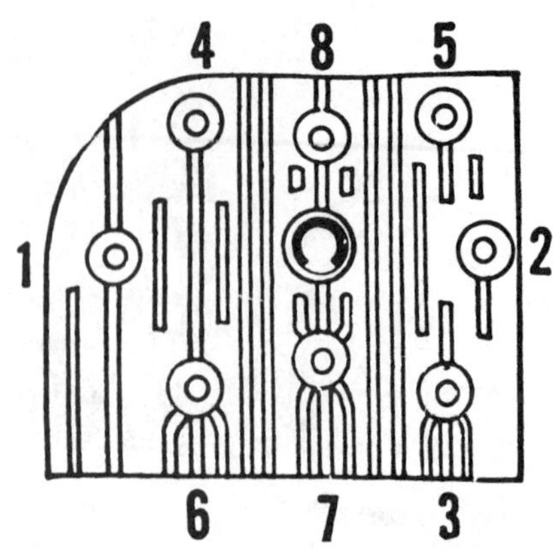

Fig. B106—Cylinder head screw tightening sequence. Long screws are used in positions 2, 3 and 7.

Models 111200, 111900, 112200, 113900 . . . . 0.010-0.016 in. (0.25-0.41 mm)
Models 100000, 130000, 131000–
  Two Leg Armature . . . 0.010-0.014 in. (0.25-0.36 mm)
  Three Leg Armature . . 0.012-0.016 in. ((0.30-0.41 mm)
Model 140000–
  Two Leg Armature . . . 0.010-0.014 in. (0.25-0.36 mm)
  Three Leg Armature . . 0.016-0.019 in. (0.41-0.48 mm)

MAGNETRON IGNITION. Magnetron ignition is a self-contained breakerless ignition system. Flywheel does not need to be removed except to check or service keyways or crankshaft key.

To check spark, remove spark plug. Connect spark plug cable to B&S tester, part 19051, and ground remaining tester lead to cylinder head. Spin engine at 350 rpm or more. If spark jumps the 0.166 inch (4.2 mm) tester gap, system is functioning properly.

To remove armature and Magnetron module, remove flywheel shroud and armature retaining screws. Use a 3/16 inch (4.76 mm) diameter pin punch to release stop switch wire from module. To remove module, unsolder wires, push module retainer away from laminations and remove module. See Fig. B105A.

Resolder wires for reinstallation and use Permatex or equivalent to hold ground wires in position.

Adjust armature air gap to 0.010-0.014 inch (0.25-0.36 mm).

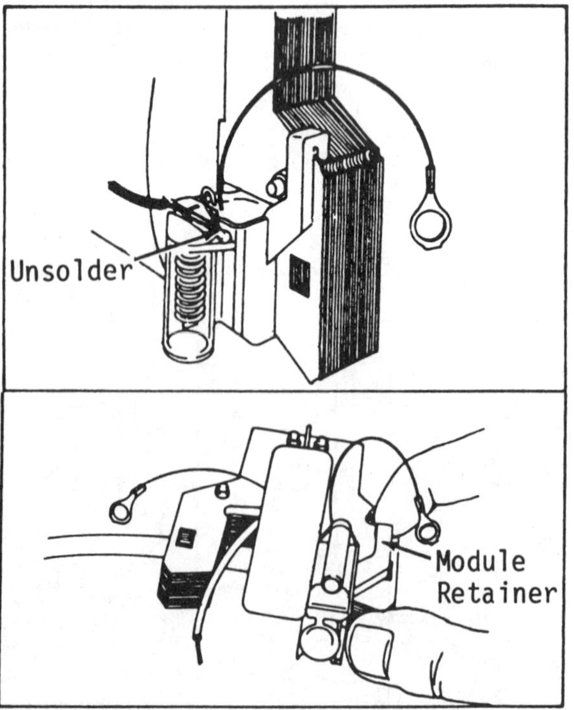

Fig. B105A—Wires must be unsoldered to remove Magnetron ignition module. Refer to text.

LUBRICATION. Vertical crankshaft engines are lubricated by an oil slinger wheel driven by the cam gear. On early 6B and 8B models, the oil slinger wheel was mounted on a bracket attached to the crankcase. Renew the bracket if pin on which gear rotates is worn to 0.490 inch (12.45 mm). Renew steel bushing in hub or gear if worn. On later model vertical crankshaft engines, the oil slinger wheel, pin and bracket are an integral unit with the bracket being retained by the lower end of the engine camshaft. On 100000, 130000 and 131000 models, a spring washer is placed on lower end of camshaft between bracket and oil sump boss. Renew the oil slinger assembly if teeth are worn on slinger gear or gear is loose on bracket. On horizontal crankshaft engines, a splash system (oil dipper on connecting rod) is used for engine lubrication.

Check oil level at five hour intervals and maintain at bottom edge of filler plug or to FULL mark on dipstick.

Recommended oil change interval for all models is every 25 hours of normal operation.

Manufacturer recommends using oil with an API service classification of SC, SD, SE, or SF. Use SAE 10W-40 oil for temperatures above 20°F (-7°C) and SAE 5W-30 oil for temperatures below 20°F (-7°C).

Crankcase capacity for all aluminum cylinder engines with displacement of 9 cubic inches (147 cc) or below, is 1¼ pint (0.6 L).

Crankcase capacity of vertical crankshaft aluminum cylinder engines with displacement of 11 cubic inches (186 cc) is 1¼ pint (0.6 L).

Crankcase capacity for all 10 and 13 cubic inch (169 and 203 cc) vertical crankshaft, aluminum cylinder engines is 1¾ pint (0.8 L).

Crankcase capacity for all 10 and 13 cubic inch (169 and 203 cc) horizontal crankshaft, aluminum cylinder engines is 1¼ pint (0.6 L).

Crankcase capacity for 14 cubic inch (231 cc) vertical crankshaft, aluminum cylinder engines is 2¼ pint (1.1 L).

Crankcase capacity for 14 cubic inch (231 cc) horizontal crankshaft, aluminum cylinder engines is 2¾ pint (1.3 L).

Crankcase capacity for engines with cast iron cylinder liner is 3 pints (1.4 L).

CRANKCASE BREATHER. The crankcase breather is built into the

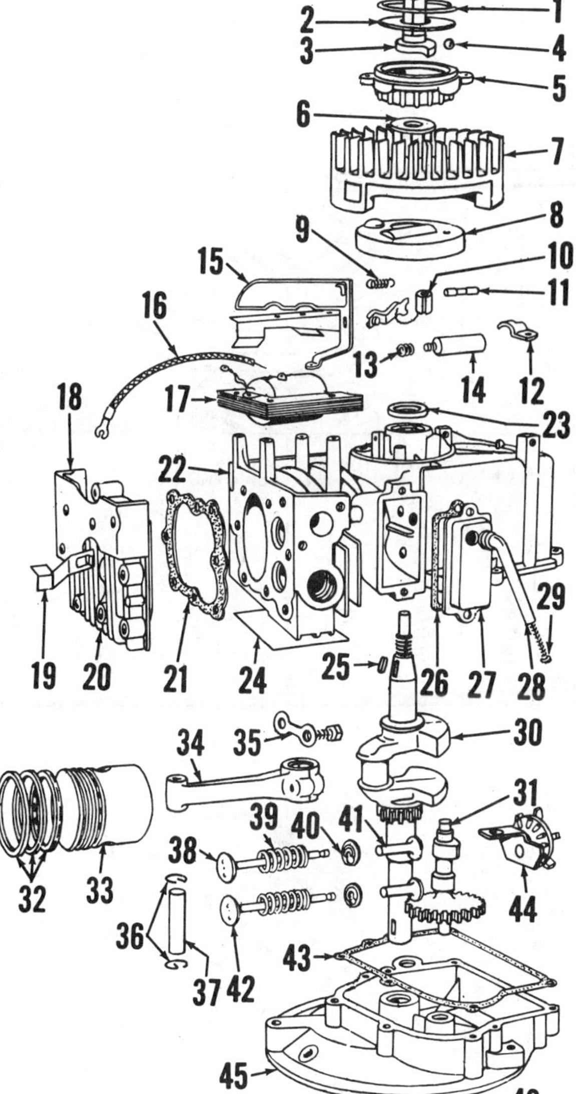

Fig. B107 – Exploded view of typical vertical crankshaft model with air vane (pneumatic) governor. To remove flywheel, remove blower housing and starter unit; then, unscrew starter clutch housing (5). Flywheel can then be pulled from crankshaft.

1. Snap ring
2. Washer
3. Ratchet
4. Steel balls
5. Starter clutch
6. Washer
7. Flywheel
8. Breaker cover
9. Breaker point spring
10. Breaker arm & pivot
11. Breaker plunger
12. Condenser clamp
13. Coil spring (primary wire retainer)
14. Condenser
15. Governor air vane & bracket assy.
16. Spark plug wire
17. Armature & coil assy.
18. Air baffle
19. Spark plug grounding switch
20. Cylinder head
21. Cylinder head gasket
22. Cylinder block
23. Crankshaft oil seal
24. Cylinder shield
25. Flywheel key
26. Gasket
27. Breather & tappet chamber cover
28. Breather tube assy.
29. Coil spring
30. Crankshaft
31. Cam gear & shaft
32. Piston rings
33. Piston
34. Connecting rod
35. Rod bolt lock
36. Piston pin retaining rings
37. Piston pin
38. Intake valve
39. Valve springs
40. Valve spring keepers
41. Tappets
42. Exhaust valve
43. Gasket
44. Oil slinger assy.
45. Oil sump (engine base)
46. Crankshaft oil seal

engine valve cover. The mounting holes are offset so the breather can only be installed one way. Rinse breather in solvent and allow to drain. A vent tube connects the breather to the carburetor air horn on certain model engines for extra protection against dusty conditions.

REPAIRS

CYLINDER HEAD. When removing cylinder head, be sure to note the position from which each of the different length screws was removed. If screws are not reinstalled in the same holes when installing the head, it will result in screws bottoming in some holes and not enough thread contact in others. Lubricate the cylinder head screws with graphite grease before installation. Do not use sealer on head gasket. When installing cylinder head, tighten all screws lightly and then retighten them in sequence shown in Fig. B106 to 165 in.-lbs. (19 N·m) on 140000 models and to 140 in.-lbs. (16 N·m) on all other models. Run the engine for 2 to 5 minutes to allow it to reach operating temperature and retighten the head screws again following the sequence and torque value specified.

NOTE: When checking compression on models with "Easy-Spin" starting, turn engine opposite the direction of normal rotation. See CAMSHAFT paragraph.

OIL SUMP REMOVAL, AUXILARY PTO MODELS. On 92000, 93500, 94900, 95500, 110900 or 113900 models with auxiliary pto, one of the oil sump (engine base) to cylinder retaining screws is installed in the recess in sump for the pto auxiliary drive gear. To remove the oil sump, refer to Fig. B108 and remove the cover plate (upper view) and then remove the shaft stop (lower left view). The gear and shaft can then be moved as shown in lower right view to allow removal of the retaining screws. Reverse procedure to reassemble.

OIL BAFFLE PLATE. 140000 MODEL ENGINES. Model 140000 engines with mechanical governor have a bafflc located in the cylinder block (crankcase). When servicing these engines, it is important that the baffle be correctly installed. The baffle must fit tightly against the valve tappet boss in the crankcase. Check for this before installing oil sump (vertical crankshaft models) or crankcase cover (horizontal crankshaft models).

CONNECTING ROD. The connecting rod and piston are removed from cylinder head end of block as an assembly. The aluminum alloy connecting rod rides directly on the induction hardened crankpin. The rod should be rejected if the crankpin hole is scored or out-of-round more than 0.0007 inch (0.018 mm) or if the piston pin hole is scored or out-of-round more than 0.0005 inch (0.013 mm). Wear limit sizes are given in the following chart. Reject the connecting rod if either the crankpin or piston pin hole is worn to, or larger than the sizes given in the following table.

REJECT SIZES FOR CONNECTING ROD

Model	Bearing Bore	Pin Bore
6B, 60000 . . .	0.876 in. (22.25 mm)	0.492 in. (12.50 mm)
8B, 80000 . . .	1.001 in. (25.43 mm)	0.492 in. (12.50 mm)
82000, 92000, 11000 . . .	1.001 in. (25.43 mm)	0.492 in. (12.50 mm)
92000, 93500 . . .	1.001 in. (25.43 mm)	0.492 in. (12.50 mm)
94500, 94900, 95500 . . .	1.001 in. (25.43 mm)	0.492 in. (12.50 mm)
100000 . .	1.001 in. (25.43 mm)	0.555 in. (14.10 mm)
110900, 111200, 111900, 112200, 113900, 130000, 131000 . .	1.001 in. (25.43 mm)	0.492 in. (12.50 mm)
140000 . .	1.095 in. (27.81 mm)	0.674 in. (17.12 mm)

NOTE: Piston pins of 0.005 inch (0.13 mm) oversize are available for service. Piston pin hole in rod can be reamed to this size if crankpin hole in rod is within specifications.

Connecting rod must be reassembled so match marks on rod and cap are aligned. Tighten connecting rod bolts to 165 in.-lbs. (19 N•m) for 140000 models and to 100 in.-lbs. (11 N•m) for all other models.

PISTON, PIN AND RINGS. Pistons for use in engines having aluminum bore ("Kool Bore") are not interchangeable with those for use in cylinders having cast-iron sleeve. Pistons may be identified as follows: Those for use in cast-iron sleeve cylinders have a plain, dull aluminum finish, have an "L" stamped on top and use an oil ring expander. Those for use in aluminum bore cylinders are chrome plated (shiny finish), do not have an identifying letter and do not use an oil ring expander.

Reject pistons showing visible signs of wear, scoring and scuffing. If, after cleaning carbon from top ring groove, a new top ring has a side clearance of 0.007 inch (0.18 mm) or more, reject the piston. Reject piston or hone piston pin hole to 0.005 inch (0.13 mm) oversize if pin hole is 0.0005 inch (0.013 mm) or more out-of-round, or is worn to a diameter of 0.554 inch (14.07 mm) or more on 100000 engines, 0.673 inch (17.09 mm) or more on 140000 engines, or 0.491 inch (12.47 mm) or more on all other models.

If the piston pin is 0.0005 inch (0.013 mm) or more out-of-round, or is worn to a diameter of 0.552 inch (14.02 mm) or smaller on 100000 engines, 0.671 inch (17.04 mm) or smaller on 140000 engines, or 0.489 inch (12.42 mm) or smaller on all other models, reject pin.

The piston ring gap for new rings should be 0.010-0.025 inch (0.25-0.64 mm) for models with aluminum cylinder bore and 0.010-0.018 inch (0.25-0.46 mm) for models with cast-iron cylinder bore. On aluminum bore engines, reject compression rings having an end gap of 0.035 inch (0.80 mm) or more and reject oil rings having an end gap of 0.045 inch (1.14 mm) or more. On cast-iron bore engines, reject compression rings having an end gap of 0.030 inch (0.75 mm) or more and reject oil rings having an end gap of 0.035 inch (0.90 mm) or more.

Pistons and rings are available in several oversizes as well as standard.

A chrome ring set is available for slightly worn standard bore cylinders. Refer to note in CYLINDER section.

CYLINDER. If cylinder bore wear is 0.003 inch (0.76 mm) or more or is 0.0025 inch (0.06 mm) or more out-of-round, cylinder must be rebored to next larger oversize.

The standard cylinder bore sizes for each model are given in the following table.

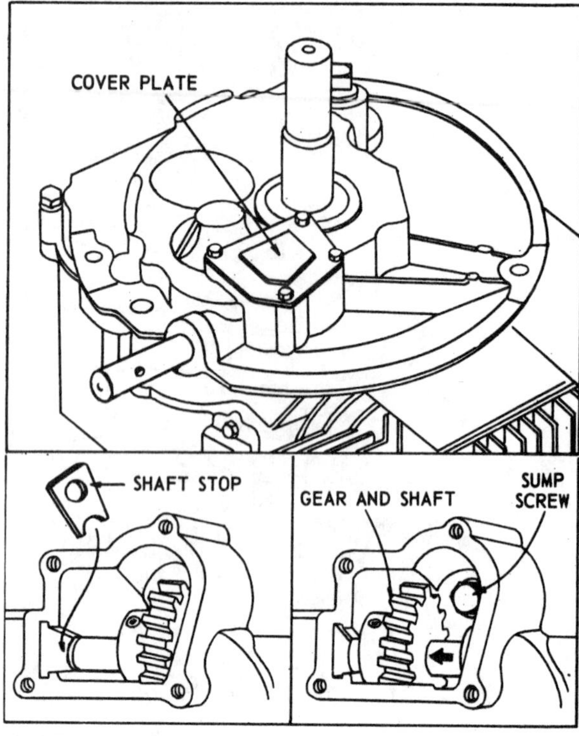

Fig. B108—To remove oil sump (engine base) on 92000, 93500, 94900, 95500, 110900 or 113900 models with auxiliary pto, remove cover plate, shaft stop and retaining screw as shown.

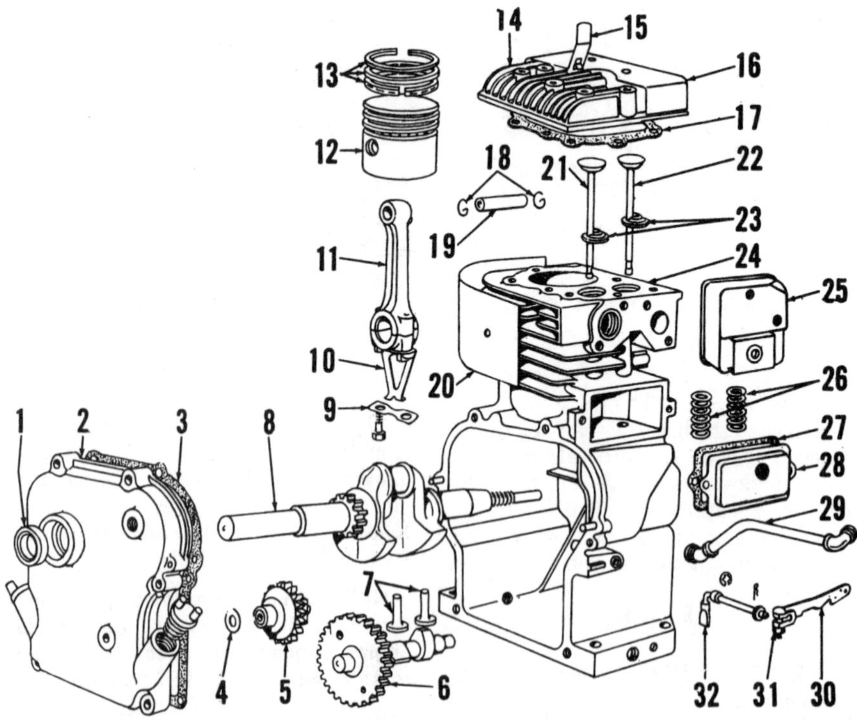

Fig. B109—Exploded view of 100000 horizontal crankshaft engine assembly. Except for 112200, 130000, and late 140000 models, other horizontal crankshaft models with mechanical governor will have governor unit as shown in Fig. B90; otherwise, construction of all other horizontal crankshaft models is similar.

1. Crankshaft oil seal
2. Crankcase cover
3. Gasket
4. Thrust washer
5. Governor assy.
6. Cam gear & shaft
7. Tappets
8. Crankshaft
9. Rod bolt lock
10. Oil dipper
11. Connecting rod
12. Piston
13. Piston rings
14. Cylinder head
15. Spark plug ground switch
16. Air baffle
17. Cylinder head gasket
18. Piston pin retaining rings
19. Piston pin
20. Air baffle
21. Exhaust valve
22. Intake valve
23. Valve spring retainers
24. Cylinder block
25. Muffler
26. Valve springs
27. Gaskets
28. Breather & tappet chamber cover
29. Breather pipe
30. Governor lever
31. Clamping bolt
32. Governor crank

STANDARD CYLINDER BORE SIZES

Model	Cylinder diameter
6B, Early 60000	2.3115-2.3125 in. (58.71-58.74 mm)
Late 60000	2.3740-2.3750 in. (60.30-60.33 mm)
8B, 80000, 81000, 82000	2.3740-2.3750 in. (60.30-60.33 mm)
92000, 93500	2.5615-2.5625 in. (65.06-65.09 mm)
94500, 94900, 95500	2.5615-2.5625 in. (65.06-65.09 mm)
100000	2.4990-2.5000 in. (63.47-63.50 mm)
110900, 111200, 111900, 112200, 113900	2.7802-2.7812 in. (70.62-70.64 mm)
130000, 131400, 131900	2.5615-2.5625 in. (65.06-65.09 mm)
140000	2.7490-2.7500 in. (69.82-69.85 mm)

A hone is recommended for resizing cylinders. Operate hone at 300-700 rpm and with an up and down movement that will produce a 45° crosshatch pattern. Clean cylinder after honing with oil or soap suds. Always check availability of oversize piston and ring sets before honing cylinder.

Approved hones are as follows: For aluminum bore, use Ammco 3956 for rough and finishing or Sunnen AN200 for rough and Sunnen AN500 for finishing. For sleeved bores, use Ammco 4324 for rough and finishing, or Sunnen AN100 for rough and Sunnen AN300 for finishing.

NOTE: A chrome piston ring set is available for slightly worn standard bore cylinders. No honing or cylinder deglazing is required for these rings. The cylinder bore can be a maximum of 0.005 inch (0.01 mm) oversize when using chrome rings.

CRANKSHAFT AND MAIN BEARINGS. Except where equipped with ball bearings, the main bearings are an integral part of the crankcase and cover or sump. The bearings are renewable by reaming out the crankcase and cover or sump bearing bores and installing service bushings. The tools for reaming the crankcase and cover or sump, and for installing the service bushings are available from Briggs & Stratton. If the bearings are scored, out-of-round 0.0007 inch (0.018 mm) or more, or are worn to or larger than the reject sizes in the following table, ream the bearings and install service bushings.

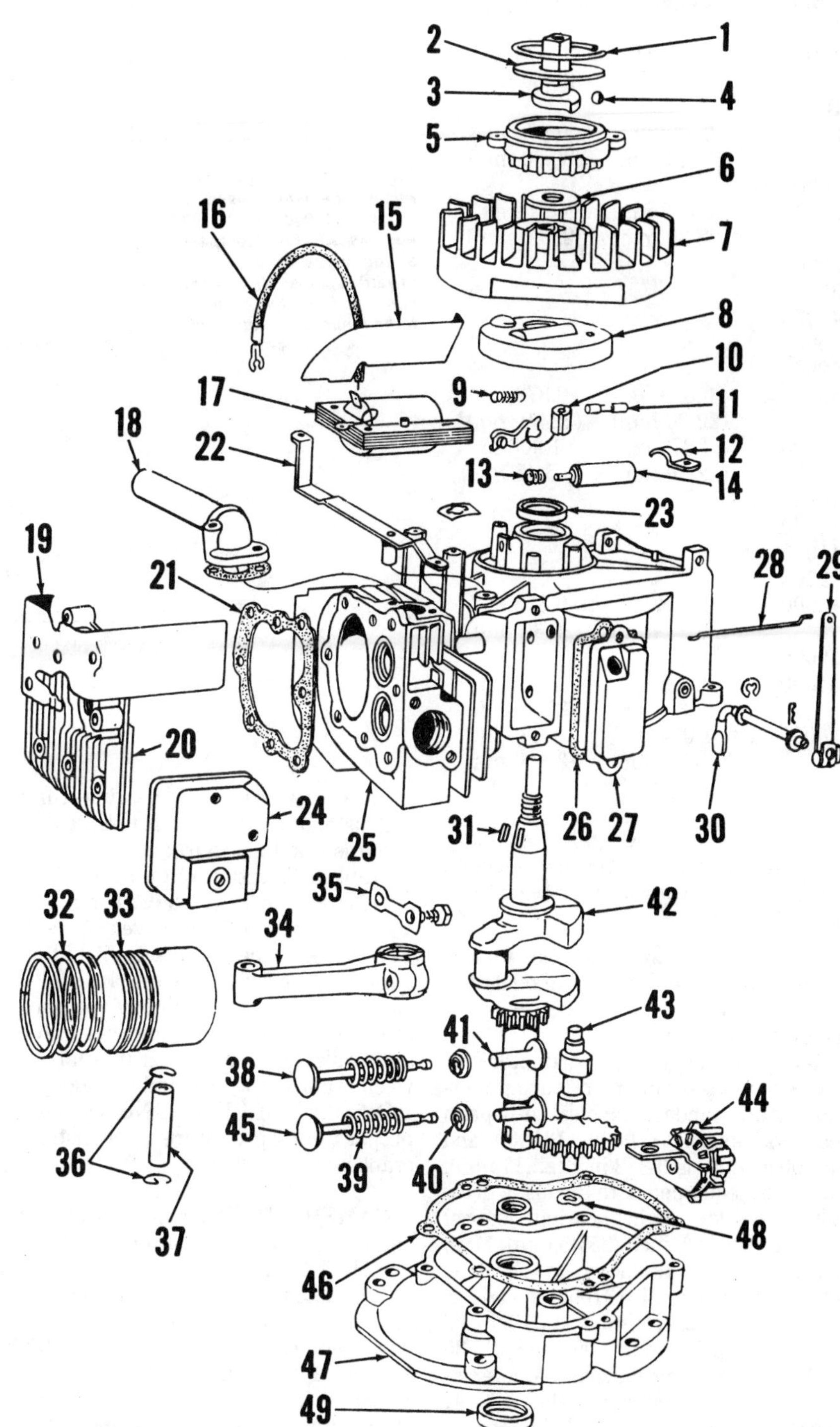

Fig. B110—Exploded view of 100000 vertical crankshaft engine with mechanical governor. Models 130000, 131000 and 140000 vertical crankshaft models with mechanical governor are similar. Refer to Fig. B107 for typical vertical crankshaft model with air vane governor.

1. Snap ring
2. Washer
3. Starter ratchet
4. Steel balls
5. Starter clutch
6. Washer
7. Flywheel
8. Breaker cover
9. Breaker arm spring
10. Breaker arm & pivot
11. Breaker plunger
12. Condenser clamp
13. Primary wire retainer spring
14. Condenser
15. Air baffle
16. Spark plug wire
17. Armature & coil assy.
18. Intake pipe
19. Air baffle
20. Cylinder head
21. Cylinder head gasket
22. Linkage lever
23. Crankshaft oil seal
24. Muffler
25. Cylinder block
26. Gasket
27. Breather & tappet chamber cover
28. Governor link
29. Governor crank
30. Governor crank
31. Flywheel key
32. Piston rings
33. Piston
34. Connecting rod
35. Rod bolt lock
36. Piston pin retaining rings
37. Piston pin
38. Intake valve
39. Valve springs
40. Valve spring retainers
41. Tappets
42. Crankshaft
43. Cam gear
44. Governor & oil slinger assy.
45. Exhaust valve
46. Gasket
47. Oil sump (engine base)
48. Thrust washer
49. Crankshaft oil seal

MAIN BEARING REJECT SIZES

Model	Magneto Bearing	PTO Bearing
6B, 60000 . . .	0.878 in. (22.30 mm)	0.878 in.* (22.30 mm)*
8B, 80000 . . .	0.878 in. (22.30 mm)	0.878 in.* (22.30 mm)*
81000, 82000 . . .	0.878 in. (22.30 mm)	0.878 in.* (22.30 mm)*
92000, 93500, 94000, 94500, 94900, 95500 . . .	0.878 in. (22.30 mm)	0.878 in.* (22.30 mm)*
100000 . .	0.878 in. (22.30 mm)	1.003 in. (25.48 mm)
110900, 111200, 111900, 112200, 113900 . .	0.878 in. (22.30 mm)	0.878 in.* (22.30 mm)*
130000, 131000 . .	0.878 in. (22.30 mm)	1.003 in. (25.48 mm)
140000 . .	1.004 in. (25.50 mm)	1.185 in. (30.10 mm)

*All models equipped with auxiliary drive unit have a main bearing rejection size for main bearing at pto side of 1.003 inch (25.48 mm).

Bushings are not available for all models; therefore, on some models the sump must be renewed if necessary to renew the bearing.

Main bearing journal diameter for flywheel and pto side main bearing journals on all standard models with plain bearings, except 130000, 131000 and 140000 models, is 0.873 inch (22.17 mm). On models equipped with auxiliary drive unit, pto side main bearing journal diameter is 0.998 inch (25.35 mm). Main bearing journal diameter at flywheel side of 130000 and 131000 models is 0.873 inch (22.17 mm) and journal diameter at pto side is 0.998 inch (25.35 mm). Main bearing journal diameter at flywheel side of 140000 model is 0.997 inch (25.32 mm) and journal diameter at pto side is 1.179 inch (29.95 mm).

Crankpin journal rejection size for 6B and 60000 models is 0.870 inch (22.10 mm), crankpin journal rejection size for 140000 model is 1.090 inch (27.69 mm) and crankpin journal rejection size for all other models if 0.996 inch (25.30 mm).

Ball bearing mains are a press fit on the crankshaft and must be removed by pressing the crankshaft out of the bearing. Reject ball bearing if worn or rough.

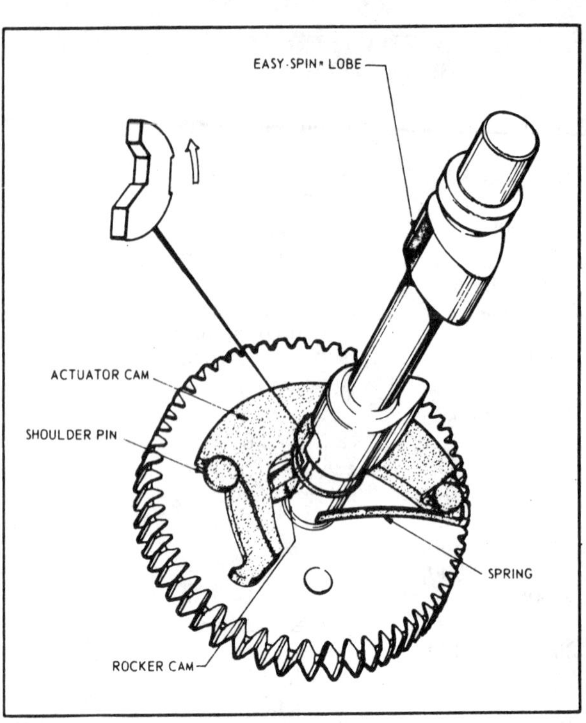

Fig. B111—Compression release camshaft used on 111200, 111900 and 112200 engines. At cranking speed, spring holds actuator cam inward against the rocker cam and rocker cam is forced above exhaust cam surface.

Expand new bearing by heating it in oil and install it on crankshaft with seal side towards crankpin journal.

Crankshaft end play on all models is 0.002-0.008 inch (0.05-0.20 mm). At least one 0.015 inch cover or sump gasket must be in place. Additional gaskets in several sizes are available to aid in end play adjustment. If end play is over 0.008 inch (0.20 mm), metal shims are available for use on crankshaft between crankshaft gear and cylinder.

Refer to VALVE TIMING section for proper timing procedure when installing crankshaft.

CAMSHAFT. The camshaft and camshaft gear are an integral casting on all models. The camshaft and gear should be inspected for wear on the journals, cam lobes and gear teeth. On all standard models except 110900, 111200, 111900, 112200 and 113900 models, if camshaft journal diameter is 0.498 inch (12.65 mm) or less, reject camshaft.

On 110900, 111200, 111900, 112200 and 113900 models, camshaft journal rejection size for journal on flywheel side is 0.436 inch (11.07 mm) and for pto side journal rejection size is 0.498 inch (12.65 mm).

On models with "Easy-Spin" starting, the intake cam lobe is designed to hold the intake valve slightly open on part of the compression stroke. Therefore, to check compression, the engine must be turned backwards.

"Easy-Spin" camshafts (cam gears) can be identified by two holes drilled in the web of the gear. Where part number of an older cam gear and an "Easy-Spin" cam gear are the same (except for an "E" following the "Easy-Spin" part number), the gears are interchangeable.

The camshaft used on 11200, 11900 and 112200 models is equipped with the "Easy-Spin" intake lobe and a mechanically operated compression release on the exhaust lobe. With engine stopped or at cranking speed, the spring holds the actuator cam weight inward against the rocker cam. See Fig. B111. The rocker cam is held slightly above the exhaust cam surface which in turn holds the exhaust valve open slightly during compression stroke. This compression release greatly reduces the power needed for cranking.

When engine starts and rpm is increased, the actuator cam weight moves outward overcoming the spring pressure. See Fig. B111A. The rocker cam is rotated below the cam surface to provide normal exhaust valve operation.

Refer to VALVE TIMING section for proper timing procedure when installing camshaft.

VALVE SYSTEM. Valve tappet clearance (cold) for 94500, 94900 and 95500 models is 0.004-0.006 inch (0.10-0.15 mm) for intake valve and 0.007-0.009 inch (0.18-0.23 mm) for exhaust valve.

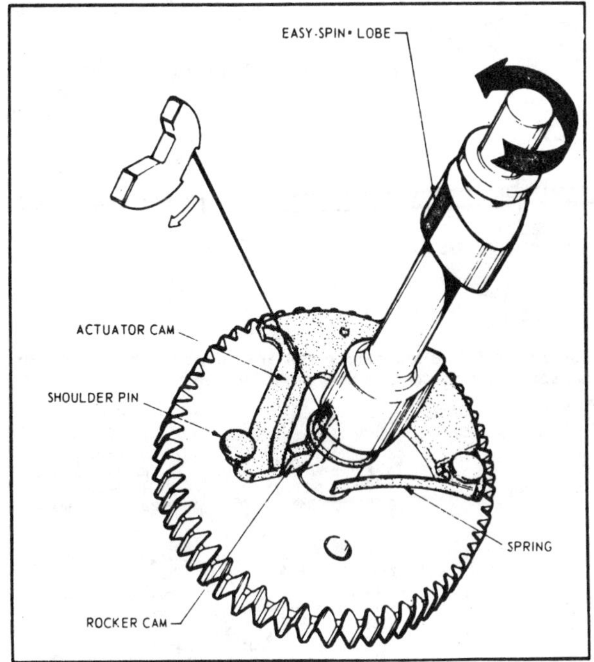

Fig. B111A – When engine starts and rpm is increased, actuator cam weight moves outward allowing rocker cam to rotate below exhaust cam surface.

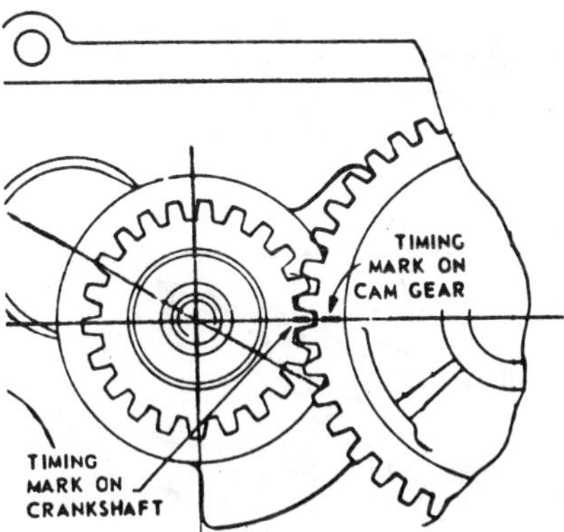

Fig. B113 – Align timing marks on cam gear and crankshaft gear on plain bearing models.

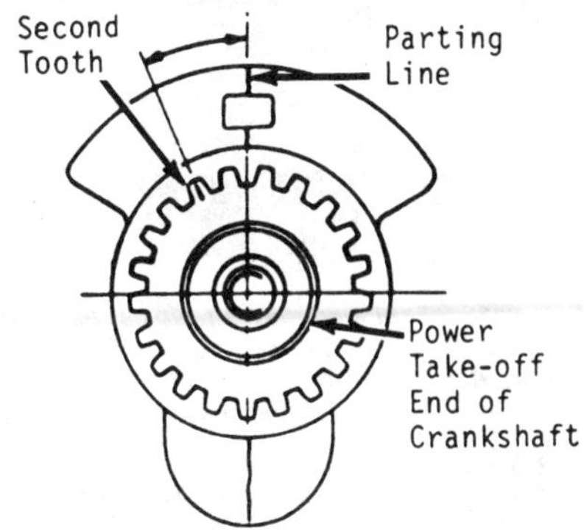

Fig. B114 – location of tooth to align with timing mark on cam gear if mark is not visible on crankshaft gear.

Valve tappet clearance (cold) for all other models is 0.005-0.007 inch (0.13-0.18 mm) for intake valve and 0.009-0.011 inch (0.23-0.28 mm) for exhaust valve.

Valve tappet clearance is adjusted on all models by carefully grinding off end of valve stem to increase clearance or by grinding valve seats deeper and/or renewing valve or lifter to decrease clearance.

Valve face and seat angle should be ground at 45°. Renew valve if margin is 1/16 inch (1.588 mm) or less. Seat width should be 3/64 to 1/16 inch (1.119-1.588 mm).

The valve guides on all engines with aluminum blocks are an integral part of the cylinder block. To renew the valve guides, they must be reamed out and a bushing installed.

On all models except 140000 model, ream guide out first with reamer (B&S part 19064) approximately 1/16 inch (1.588 mm) deeper than length of bushing (B&S part 63709). Press in bushing with driver (B&S part 19065) until it is flush with top of guide bore. Finish ream the bushing with a finish reamer (B&S part 19066).

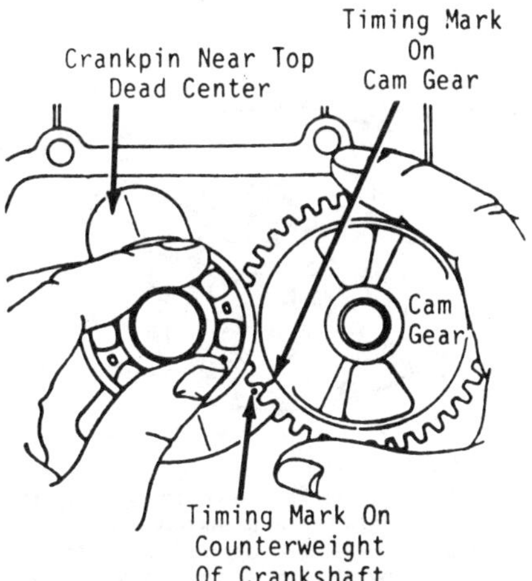

Fig. B112 – Align timing marks on cam gear with mark on crankshaft counterweight on ball bearing equipped models.

On 140000 model, ream guide out with reamer (B&S part 19183) only about 1/16 inch (1.588 mm) deeper than top end of flutes on reamer. Press in bushings with soft driver as bushing is finish reamed at the factory.

VALVE TIMING. On engines equipped with ball bearing mains, align the timing mark on the cam gear with the timing mark on the crankshaft counterweight as shown in Fig. B112. On engines with plain bearings, align the timing marks on the camshaft gear with the timing mark on the crankshaft gear as shown in Fig. B113. If the timing mark is not visible on the crankshaft gear, align the timing mark on the camshaft gear with the second tooth to the left of the crankshaft counterweight parting line as shown in Fig. B114.

SERVICING BRIGGS & STRATTON ACCESSORIES

WINDUP STARTER

STARTER OPERATION (EARLY UNITS). Refer to Figs. B115 and B116. Turning the knob (17) to "CRANK" position holds the engine flywheel stationary. Turning the crank (1) winds up the power spring within the spring and cup assembly (8). Spring tension is held by the clutch ratchet (15) and the crank ratchet pawl (2). On some starters, an additional ratchet pawl (2A – Fig. B116) allows the crank to be reversed when there is not enough room for a full turn of crank. Approximately 5½ turns of crank are required to fully wind the power spring. Turning the holding knob (17) to "START" position releases the engine flywheel allowing the power spring to turn the engine.

NOTE: Do not turn knob to "CRANK" position when engine is running.

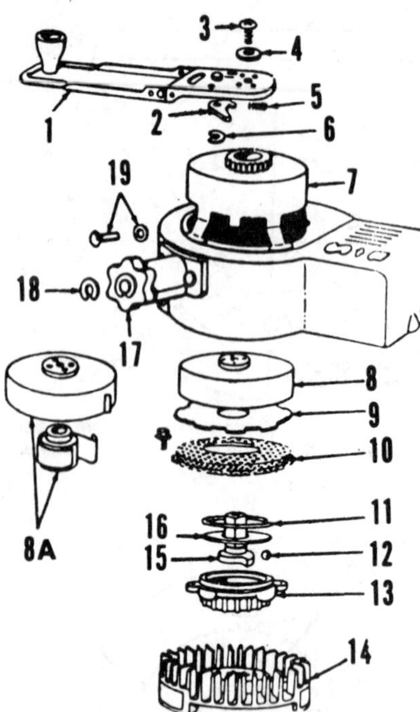

Fig. B115—Exploded view of typical early production windup starter. Refer to Fig. B116 for exploded view of starter using a ratchet type crank and to Fig. B117 for late production starter.

1. Crank
2. Pawl
3. Screw
4. Washer
5. Spring
6. Snap ring
7. Blower housing
8. Power spring & cup assy.
8A. Power spring & cup assy.
9. Retainer plate
10. Screen
11. Snap ring
12. Clutch balls
13. Clutch housing
14. Flywheel
15. Starter ratchet
16. Ratchet cover
17. Flywheel holder
18. Snap ring
19. Rivet & burr

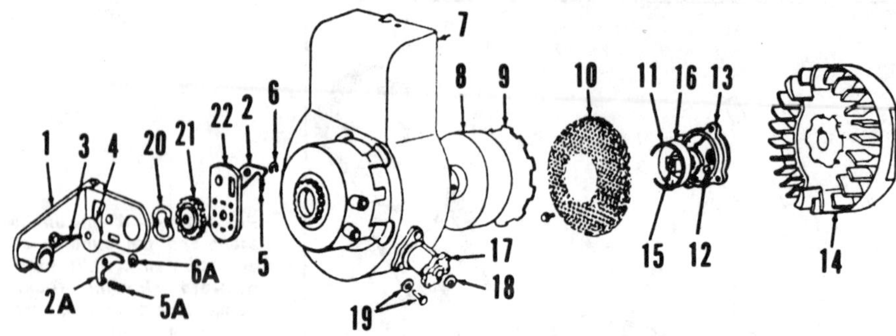

Fig. B116—Exploded view of typical Briggs & Stratton windup starter equipped with ratchet type windup crank. Starter is used where space does not permit full turn of crank.

1. Crank
2. Pawl
2A. Pawl
3. Screw
4. Washer
5. Spring
5A. Spring
6. Snap ring
6A. Snap ring
7. Blower housing
8. Power spring & cup assy.
9. Retainer plate
10. Screen
11. Snap ring
12. Clutch balls
13. Clutch housing
14. Flywheel
15. Clutch ratchet
16. Ratchet cover
17. Flywheel holder
18. Snap ring
19. Rivet & burr
20. Wave washer
21. Ratchet
22. Pawl housing

If engine does not start readily, check to be sure choke is in fully closed position and that ignition system is delivering a spark.

STARTER OPERATION (LATE UNITS). Late production windup starter assemblies have a control lever (Fig. B117) instead of a flywheel holder (17–Fig. B115 and B116). Moving the control lever to "CRANK" position locks the hub of the starter spring assembly (see lower view, Fig. B117) allowing starter spring to be wound up by turning crank. Moving the control lever to "START" position releases the spring assembly hub. The spring unwinds, engaging the starter ratchet, and cranks up engine.

CAUTION: Never work on engine or driven machinery if starter is wound up. If attached machinery such as mower blade is jammed up so starter will not turn engine, remove blower housing and starter assembly from engine to allow starter to unwind.

OVERHAUL (EARLY UNIT). Remove starter and blower housing from engine. Check condition of flywheel and flywheel holder assembly. Check action of clutch ratchet (15–Fig. B115 or B116). Remove snap ring (11) to disassemble clutch unit. Bend tangs of blower housing back to remove retainer (9). Remove screw (3) retaining crank and ratchet assembly to power spring cup (8).

Flywheel holder assembly may be renewed as a unit using new rivets and washers (19) after cutting old assembly from blower housing. Knob (17) and snap ring (18) are available separately.

On early models, power spring and cup (8A) are available separately. On later units, spring and cup (8) are available only as an assembly.

Renew parts as necessary and reassemble as follows: Place ratchet (15)

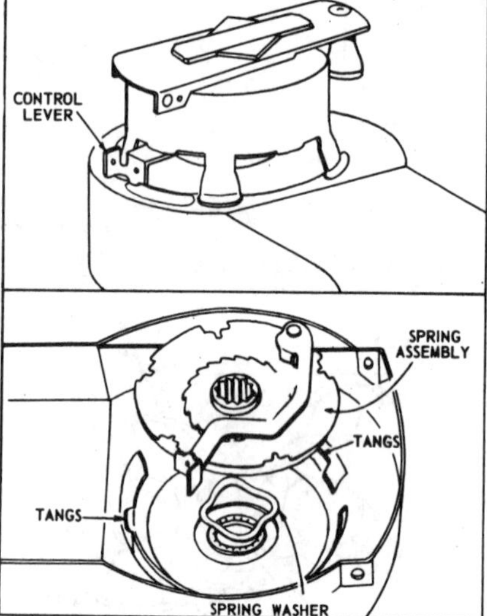

Fig. B117—On late production units, control lever (top view) locks hub of spring assembly (bottom) to allow spring to be wound up by turning crank. To remove spring assembly, first remove screw (3–Fig. B118), washer (4) and crank (1). Bend tangs of blower housing out as shown in bottom view and lift cup, spring and control lever assembly from housing. Note spring washer placed between cup and washer.

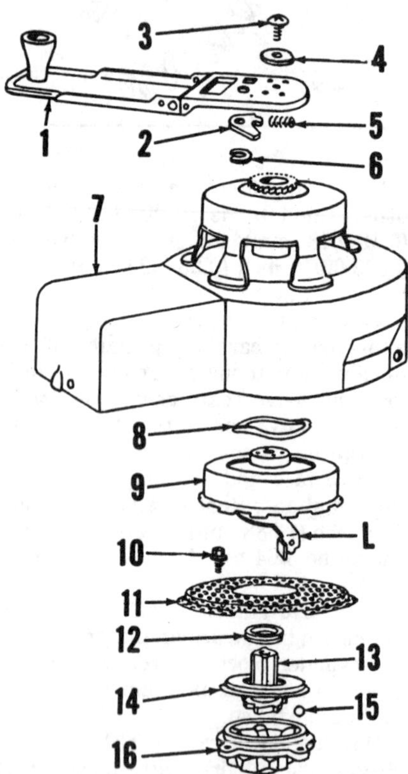

Fig. B118—Exploded view of late production windup starter assembly. Control lever (L) locks ratchet drive hub to allow starter to be wound up and releases hub to crank engine.

L. Control lever
1. Crank
2. Pawl
3. Screw
4. Washer
5. Spring
6. Snap ring
7. Blower housing
8. Spring washer
9. Cup, spring & release assy.
10. Sems screws
11. Rotating screen
12. Seal
13. Starter ratchet
14. Ratchet cover
15. Steel balls
16. Clutch housing

in cup (13), drop the balls (12) in beside ratchet and install washer (16) and snap ring (11). Reassemble blower housing and starter unit in reverse of disassembly procedure.

OVERHAUL (LATE UNIT). Refer to Fig. B118 and move control lever (L)

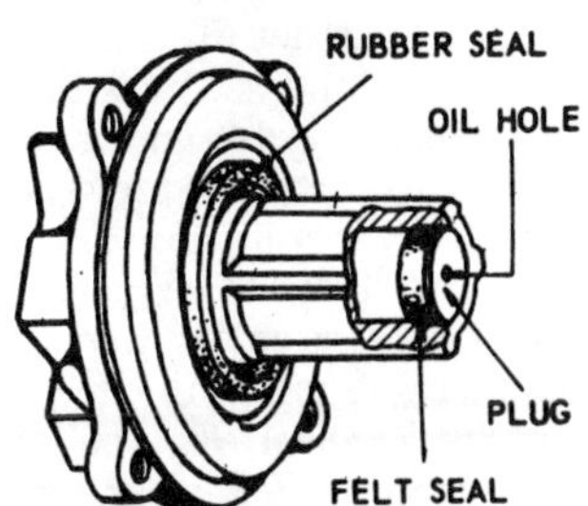

Fig. B119—Cutaway view of late type starter ratchet showing sealing felt and plug in outer end. Oil ratchet through hole in plug. A rubber seal is also used at ratchet cover.

to "START" position. If starter is wound up and will not turn engine, hold crank (1) securely and loosen screw (3). Remove blower housing and starter unit from engine and, if not already done, remove screw (3) and crank (1). Turn blower housing over and bend tangs outward as shown in bottom view of Fig. B117; then lift cup, spring and control lever assembly from the blower housing.

CAUTION: Do not attempt to disassemble the cup and spring assembly; unit is serviced as a complete assembly only.

When reassembling unit, renew the spring washer (8–Fig. B118) if damaged in any way. Grease the mating surfaces of spring cup and blower housing and stick spring washer in place with grease. Renew pawl (2) or pawl spring (5) on crank assembly (1) if necessary. Check condition of ratchet teeth on blower housing. Renew housing if teeth are broken or worn off. Reassemble by reversing disassembly procedure.

Starter ratchet cover (14) and ratchet (13) can be removed after removing the Sems screws (10) and rotating screen (11). Be careful not to lose the steel balls (15). Clutch housing (16) also is the flywheel retaining nut. To remove housing, hold flywheel and turn housing counterclockwise. A special wrench (B&S 19114) is available for removing and installing housing.

To reassemble, first be sure spring washer is in place on crankshaft with cup (hollow) side towards the flywheel. Install starter clutch housing and tighten securely. Place starter ratchet on crankshaft and drop the steel balls in place in housing. Reinstall ratchet cover with new seal (12) if required. Reinstall the rotating screen and install blower housing and starter assembly.

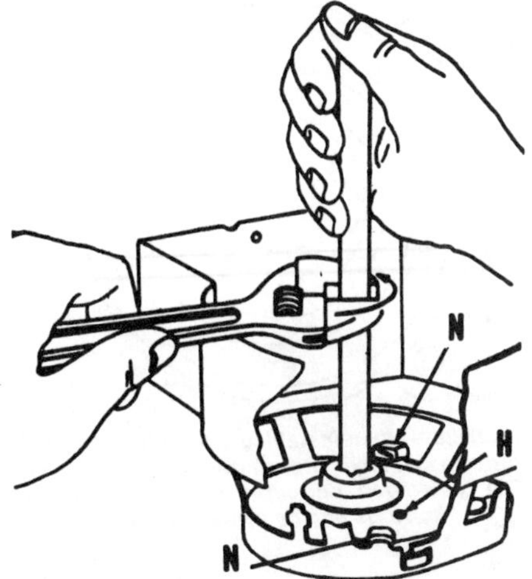

Fig. B121—Using square shaft and wrench to windup the rewind starter spring. Refer to text.

H. Hole in starter pulley
N. Nylon bumpers

REWIND STARTERS

OVERHAUL. To renew broken rewind spring, grasp free outer end of spring (S–Fig. B120) and pull broken end from starter housing. With blower housing removed, bend nylon bumpers (N) up and out of the way and remove starter pulley from housing. Untie knot in rope (R) and remove rope and inner end of broken spring from pulley. Apply a small amount of grease on inner face of pulley, thread inner end of new spring in pulley hub (on older models, install retainer in hub with split side of retainer away from spring hook) and place pulley in housing. Renew nylon bumpers if necessary and bend bumpers down to within 1/8 inch (3.175 mm) of the pulley. Insert a 3/4-inch (19 mm) square bar in pulley hub and turn pulley approximately 13½ turns in a counterclockwise

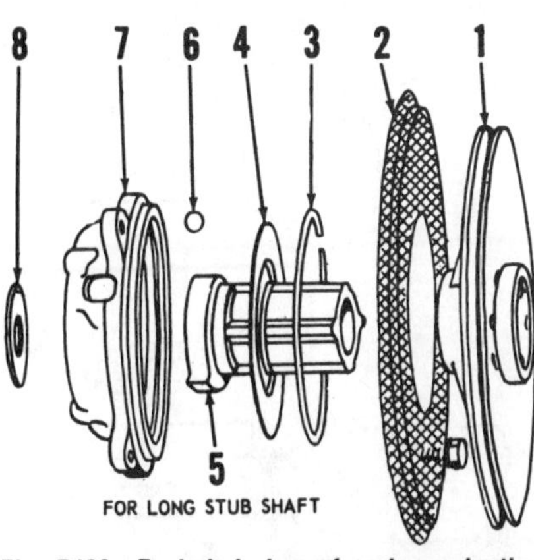

Fig. B123—Exploded view of early production starter clutch unit. Refer to Fig. B126 for view of "long stub shaft". A late type unit (Fig. B125) should be installed when renewing "long" crankshaft with "short" (late production) shaft.

1. Starter rope pulley
2. Rotating screen
3. Snap ring
4. Ratchet cover
5. Starter ratchet
6. Steel balls
7. Clutch housing (flywheel nut)
8. Spring washer

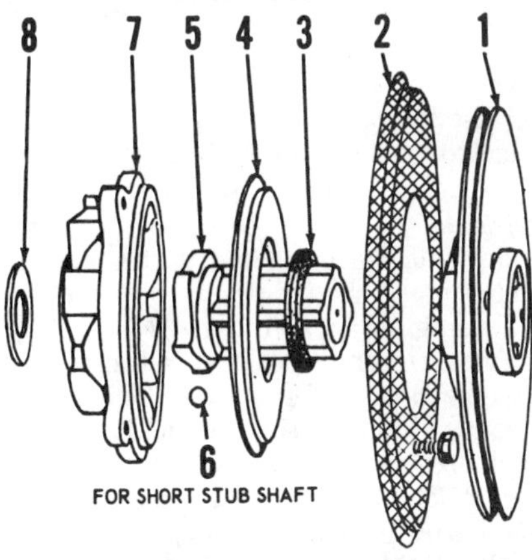

Fig. B124—View of late production sealed starter clutch unit. Late unit can be used with "short stub shaft" only. Refer to Fig. B126. Refer to Fig. B125 for cutaway view of ratchet (5).

1. Starter rope pulley
2. Rotating screen
3. Rubber seal
4. Ratchet cover
5. Starter ratchet
6. Steel balls
7. Clutch housing (flywheel nut)
8. Spring washer

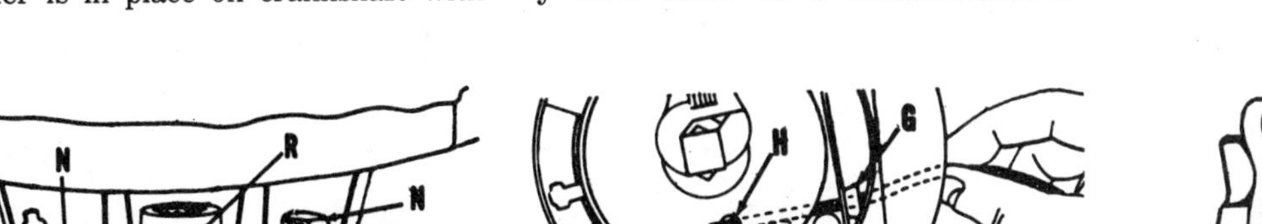

Fig. B120—View of Briggs & Stratton rewind starter assembly.

N. Nylon bumpers
OS. Old style spring
R. Starter rope
S. Rewing spring

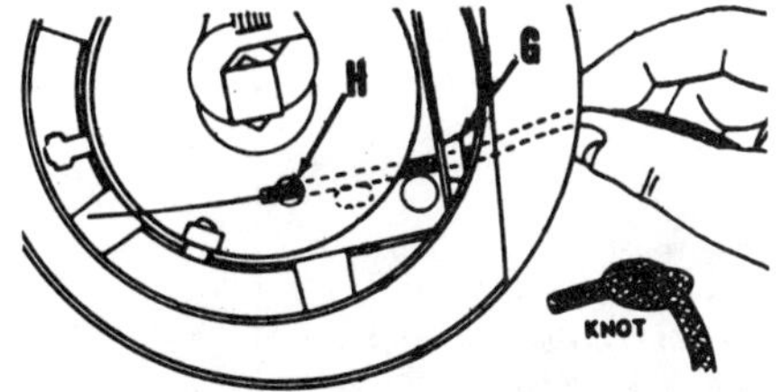

Fig. B122—Threading starter rope through guide (G) in blower housing and hole (H) in starter pulley with wire hooked in end of rope. Tie knot in end of rope as shown.

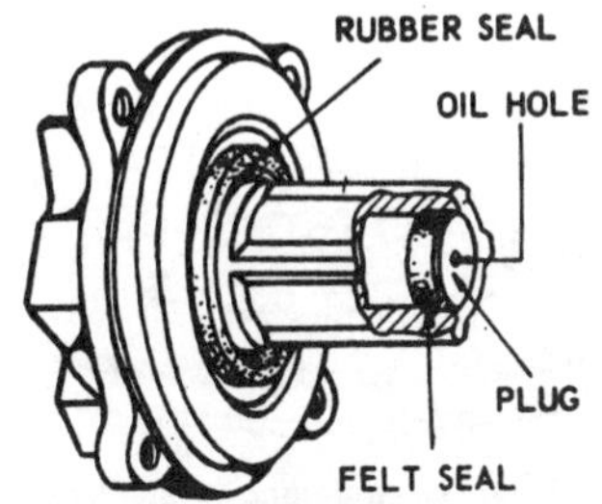

Fig. B125—Cutaway view showing felt seal and plug in end of late production starter ratchet (5–Fig. B124).

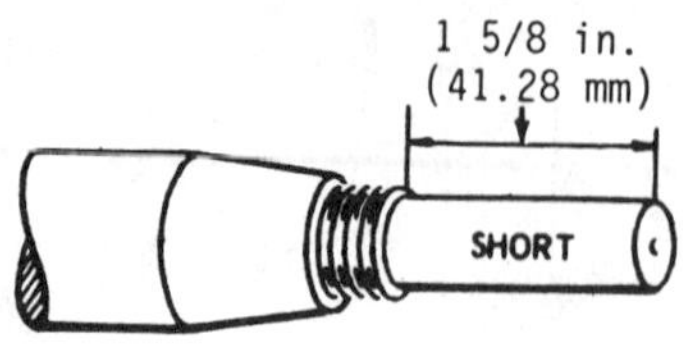

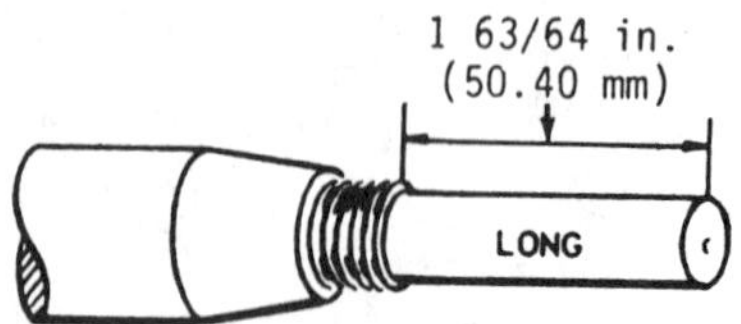

Fig. B126—Crankshaft with short stub (top view) must be used with late production starter clutch assembly. Early crankshaft (bottom view) can be modified by cutting off stub end to the dimension shown in top view and beveling end of shaft to allow installation of late type clutch unit.

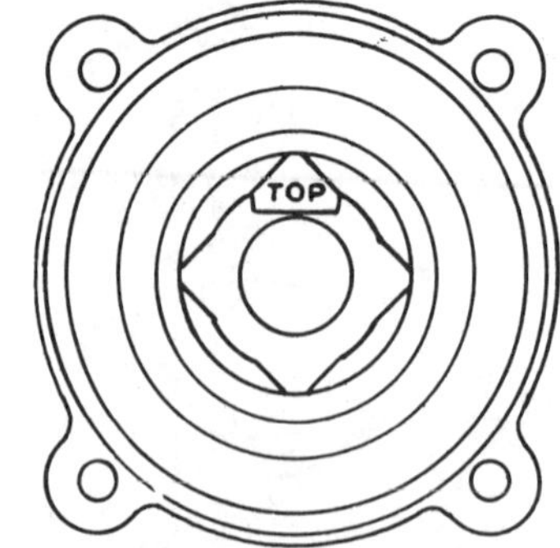

Fig. B128—When installing blower housing and starter assembly, turn starter ratchet so word "TOP" stamped on outer end of ratchet is towards engine cylinder head.

direction as shown in Fig. B121. Tie wrench to blower housing with wire to hold pulley so hole (H) in pulley is aligned with rope guide (G–Fig. B122) in housing. Hook a wire in inner end of rope and thread rope through guide and hole in pulley. Tie a knot in rope and release the pulley allowing the spring to wind the rope into the pulley groove.

To renew starter rope only, it is not generally necessary to remove starter pulley and spring. Wind up the spring and install new rope as outlined in preceding paragraph.

Two different types of starter clutches have been used. Refer to exploded view of early production unit in Fig. B123 and exploded view of late production unit in Fig. B124. The outer end of the late production ratchet (refer to cutaway view in Fig. B125) is sealed with a felt and a retaining plug and a rubber ring is used to seal ratchet to ratchet cover.

To disassemble early type starter clutch unit, refer to Fig. B123. Remove snap ring (3). Lift ratchet (5) and cover (4) from starter housing (7) and crankshaft. Be careful not to lose the steel balls (6). Starter housing (7) is also flywheel retaining nut. To remove housing, first remove screen (2) and using B&S flywheel wrench 19114, unscrew housing from crankshaft in counterclockwise direction. When reinstalling housing, be sure spring washer (8) is placed on crankshaft with cup (hollow) side towards flywheel. Install starter housing and tighten securely. Reinstall rotating screen. Place ratchet on crankshaft and into housing, then insert the steel balls. Reinstall cover and retaining snap ring.

To disassemble late starter clutch unit, refer to Fig. B124. Remove rotating screen (2) and starter ratchet cover (4). Lift ratchet (5) from housing and crankshaft and extract the steel balls (6). If necessary to remove housing (7), hold flywheel and unscrew housing in counterclockwise direction using B&S flywheel wrench 19114. When installing housing, be sure spring washer (8) is in place on crankshaft with cup (hollow) side towards flywheel. Tighten housing securely. Inspect felt seal and plug in outer end of ratchet. Renew ratchet if seal or plug if damaged as these parts are not serviced separately. Lubricate the felt with oil and place ratchet on crankshaft. Insert the steel balls and install ratchet cover, rubber seal and rotating screen.

NOTE: Crankshafts used with early and late starter clutches differ. Refer to Fig. B126. If renewing early (long) crankshaft with late (short) shaft, also install late type starter clutch unit. If renewing early starter clutch with late type unit, the crankshaft must be shortened to the dimension shown for short shaft in Fig. B126. Also, hub of starter rope pulley must be shortened to ½-inch (12.7 mm) dimension shown in Fig. B127. Bevel end of crankshaft after removing the approximate ⅜ inch (9.525 mm) from shaft.

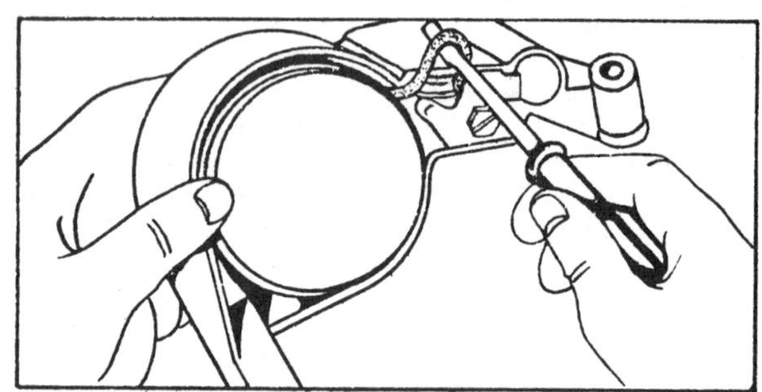

Fig. B129—Use a screwdriver to pull rope up to provide about one foot (305 mm) of slack.

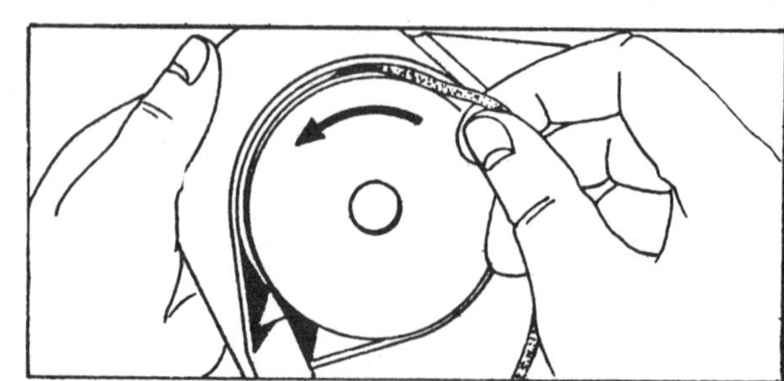

Fig. B130—Rotate pulley and rope 3 turns counterclockwise to remove spring tension from rope.

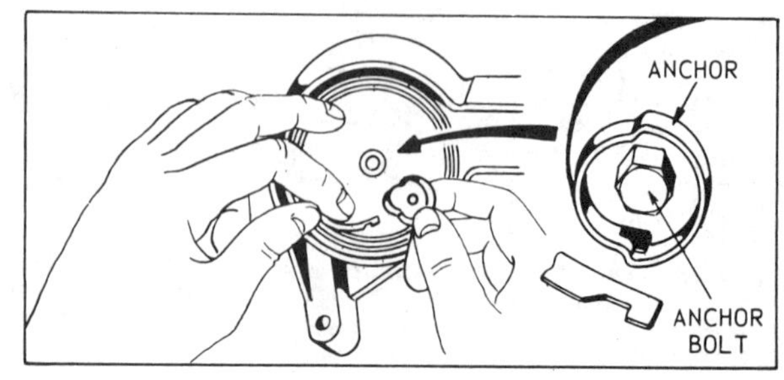

Fig. B131—Remove anchor bolt and spring anchor to release inner end of spring.

Fig. B127—When installing a late type starter clutch unit as replacement for early type, either install new starter rope pulley or cut hub of old pulley to dimension shown.

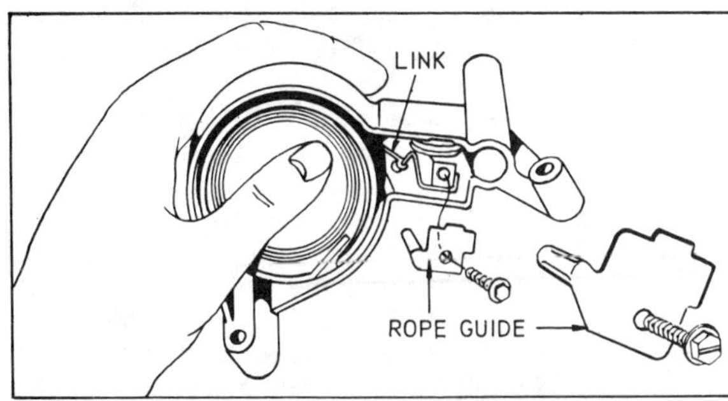

Fig. B132—Rope guide removed from Vertical Pull starter. Note position of friction link.

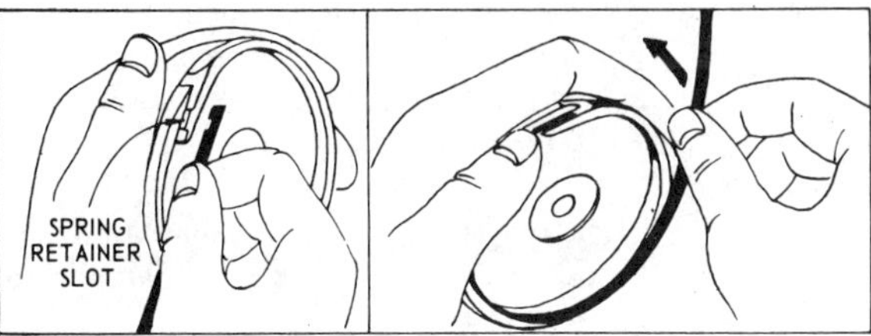

Fig. B133—Hook outer end of spring in retainer slot, then coil the spring counterclockwise in the housing.

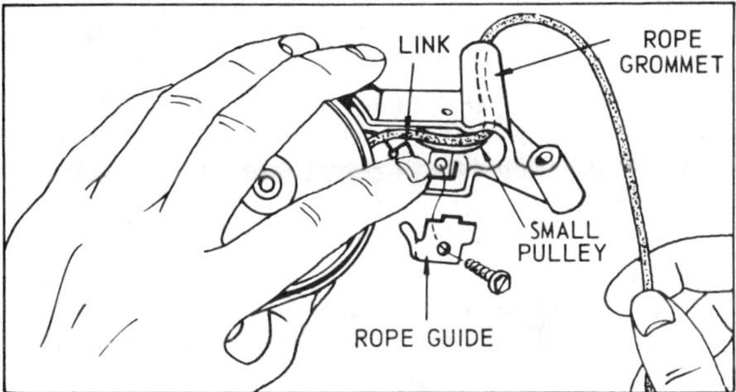

Fig. B134—When installing pulley assembly in housing, make sure friction link is positioned as shown, then install rope guide.

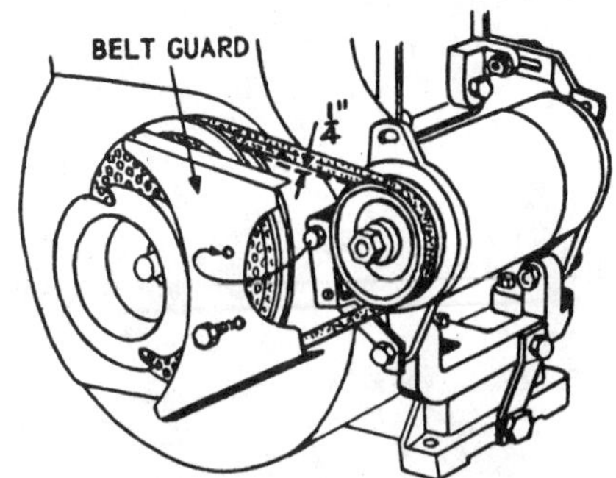

Fig. B137—View showing starter-generator belt adjustment on models so equipped. Refer to text.

When installing blower housing and starter assembly, turn starter ratchet so word "TOP" on ratchet is towards cylinder head.

VERTICAL PULL STARTER

OVERHAUL. To renew rope or spring, first remove all spring tension from rope. Using a screwdriver as shown in Fig. B129, pull rope up about one foot (305 mm). Wind rope and pulley counterclockwise 3 turns as shown in Fig. B130 to remove all tension. Carefully pry off the plastic spring cover. Refer to Fig. B131 and remove anchor bolt and spring anchor. Carefully remove spring from housing. Unbolt and remove rope guide. Note position of the friction link in Fig. B132. Remove old rope from pulley.

It is not necessary to remove the gear retainer unless pulley or gear is damaged and renewal is necessary. Clean and inspect all parts. The friction link should move the gear to both extremes of its travel. If not, renew the linkage assembly.

Install new spring by hooking end of retainer slot and winding until spring is coiled in housing. See Fig. B133. Insert new rope through the housing and into the pulley. Tie a small knot, heat seal the knot and pull it tight into the recess in pulley. Install pulley assembly in the housing with friction link in pocket of casting as shown in Fig. B134. Install rope guide. Rotate pulley counterclockwise until rope is fully wound. Hook free end of spring to anchor, install screw and tighten to 75-90 in.-lbs. (8-11 N·m). Lubricate spring with a small amount of engine oil. Snap the plastic spring cover in place. Preload spring by pulling rope up about one foot (305 mm), then winding rope and pulley 2 or 3 turns clockwise.

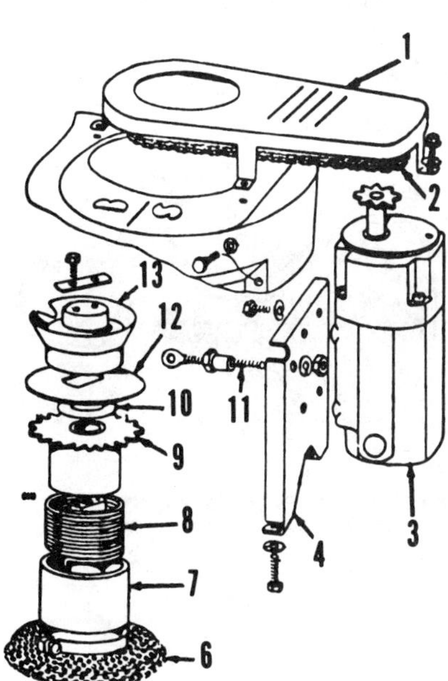
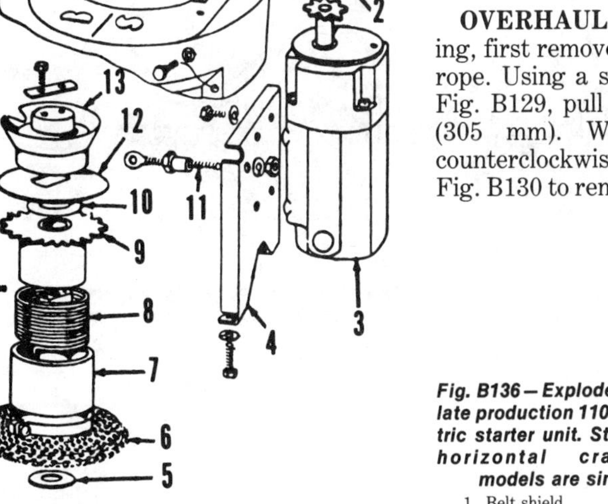

Fig. B135—Exploded view of early production 110 volt electric starter used on vertical crankshaft models. Starter used on horizontal crankshaft models was similar.

1. Chain shield
2. Drive chain
3. Electric motor
4. Mounting bracket
5. Washer
6. Rotating screen
7. Shaft adapter
8. Spring
9. Sprocket & hub
10. Thrust washer
11. Adjusting screw
12. Guard washer
13. Rope starter pulley

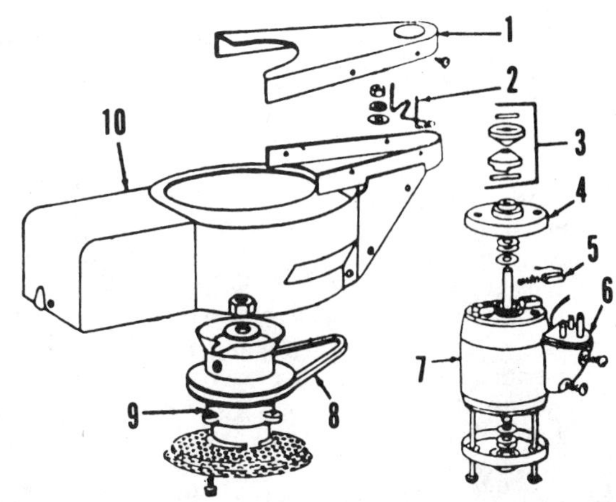

Fig. B136—Exploded view of late production 110 volt electric starter unit. Starters for horizontal crankshaft models are similar.

1. Belt shield
2. Belt guide
3. Clutch assy.
4. Cover
5. Commutator brush assy.
6. Connector plug
7. Electric motor
8. Drive belt
9. Drive pulley
10. Blower housing

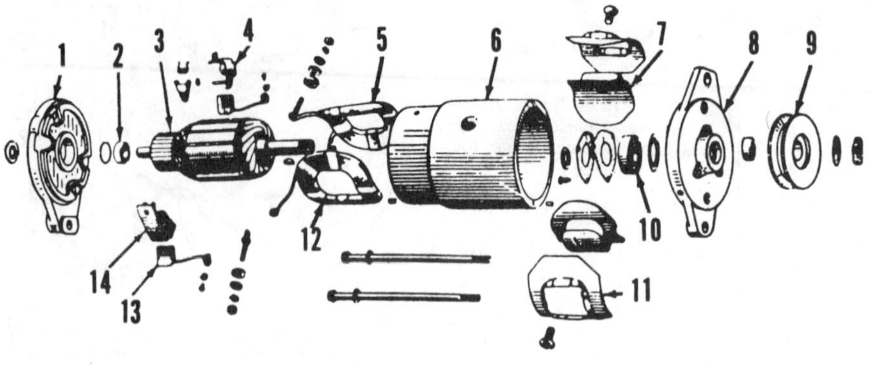

Fig. B138—Exploded view of Delco-Remy starter-generator unit used on some B&S engines.

1. Commutator end frame
2. Bearing
3. Armature
4. Ground brush holder
5. Field coil
6. Frame
7. Pole shoe
8. Drive end frame
9. Pulley
10. Bearing
11. Field coil insulator
12. Field coil
13. Brush
14. Insulated brush holder

110-VOLT ELECTRIC STARTERS (CHAIN AND BELT DRIVE)

Early production 110-volt electric starter with chain drive is shown in Fig. B135. Later type with belt drive is shown in Fig. B136. Starters for horizontal crankshaft engines are similar to units shown for vertical shaft engines.

Chain (2–Fig. B135) on early models is adjusted by changing position of nuts on adjusting stud (11) so chain deflection is approximately ¼ inch (6.35 mm) between sprockets. There should be about 1/32-inch (0.8 mm) clearance between spring (8) and sprocket (9) when unit is assembled. Clutch unit (7, 8 and 9) should disengage when engine starts.

Belt (8–Fig. B135) tension is adjusted by shifting starter motor in slotted mounting holes so clutch unit (3) on starter motor will engage belt and turn engine, but so belt will not turn starter when clutch is disengaged.

CAUTION: Always connect starter cord to starter before plugging cord into 110-volt outlet and disconnect cord from outlet before removing cord from starter connector. Do not run the electric starter for more than 60 seconds at a time.

12-VOLT STARTER-GENERATOR UNIT

The combination starter-generator functions as a cranking motor when the starting switch is closed. When engine is operating and with starting switch open, the unit operates as a generator. Generator output and circuit voltage for the battery and various operating requirements are controlled by a current-voltage regulator. On units where voltage regulator is mounted separately from generator unit, do not mount regulator with cover down as regulator will not function in this position. To adjust belt tension, apply approximately 30 pounds (14 kg) pull on generator adjusting flange and tighten mounting bolts. Belt tension is correct when a pressure of 10 pounds (5 kg) applied midway between pulleys will deflect belt ¼ inch (6.35 mm). See Fig. B137. On units equipped with two drive belts, always renew belts in pairs. A 50-ampere hour capacity battery is recommended. Starter-generator units are intended for use in temperatures above 0° F (−18° C). Refer to Fig. B138 for exploded view of starter-generator. Parts and service on the starter-generator are available at authorized Delco-Remy service centers.

12-VOLT ELECTRIC STARTER (BELT DRIVE)

Refer to Fig. B139. Adjust position of starter (10) so clutch will engage belt and turn engine when starter is operated, but so belt will not turn starter when engine is running. The 12-volt electric starter is intended for use in temperatures above 15° F (−9° C). Driven equipment should be disengaged before using starter. Parts and service on starter motor are available at Autolite service centers.

CAUTION: Do not use jumper (booster) battery as this may result in damage to starter motor.

GEAR DRIVE STARTERS

Two types of gear drive starters may be used, a 110 volt AC starter or a 12 volt DC starter. Refer to Fig. B140 for an exploded view of starter motor. A properly grounded receptacle should be used with power cord connected to 110 volt AC starter motor. A 32 ampere hour capacity battery is recommended for use with 12 volt DC starter motor.

To renew a worn or damaged flywheel ring gear, drill out retaining rivets using a 3/16-inch drill. Attach new ring gear

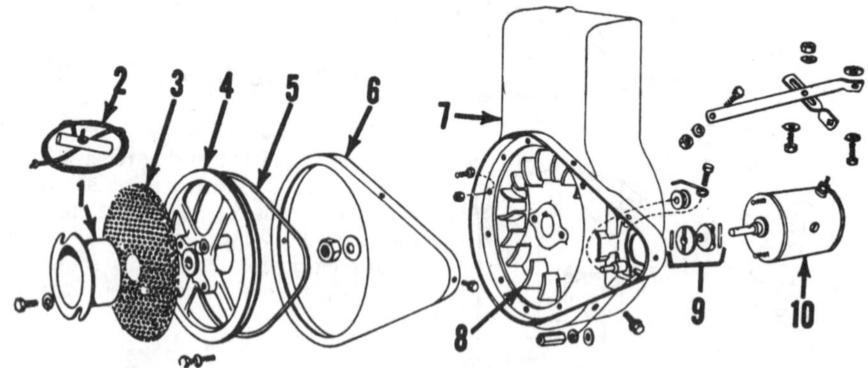

Fig. B139—Exploded view of Briggs & Stratton 12 volt starter used on some engines. Parts and/or service on starter motor are available through an authorized Autolite service center.

1. Rope pulley
2. Starter rope
3. Rotating screen
4. Starter driven pulley
5. Drive belt
6. Belt guard
7. Blower housing
8. Flywheel
9. Starter clutch
10. Starter motor.

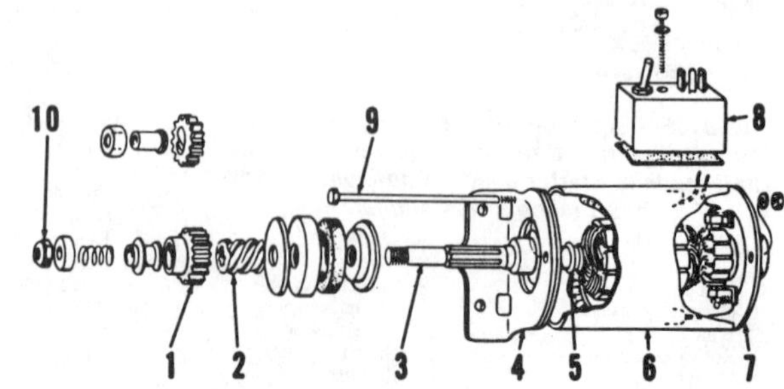

Fig. B140—View of 110 volt AC starter motor. 12 volt DC starter motor is similar. Rectifier and switch unit (8) is used on 110 volt motor only.

1. Pinion gear
2. Helix
3. Armature shaft
4. Drive cap
5. Thrust washer
6. Housing
7. End cap
8. Rectifier & switch
9. Bolt
10. Nut

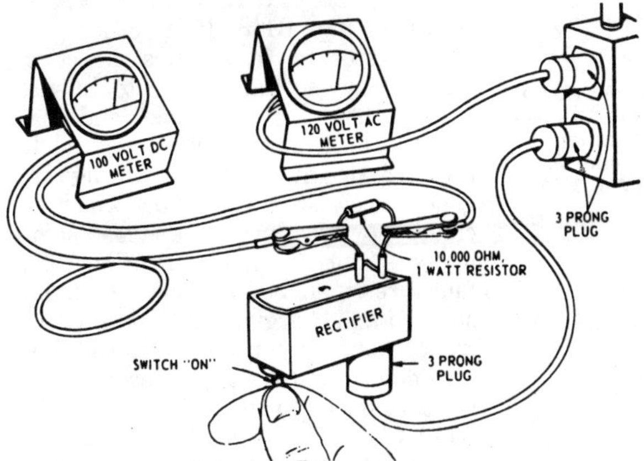

Fig. B141—Test connections for 110 volt rectifier. Refer to text for procedure.

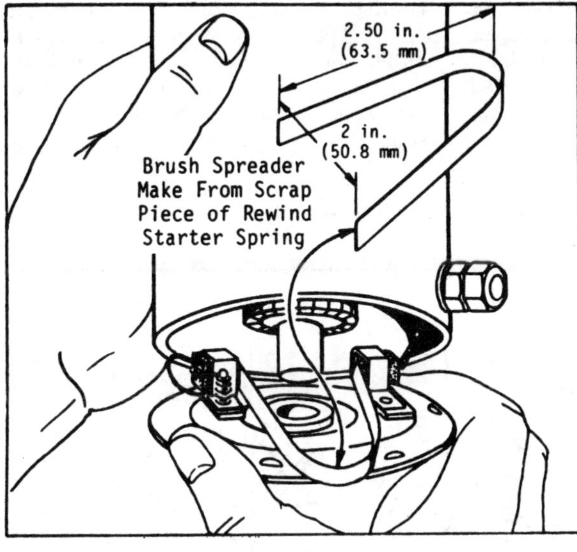

Fig. B142—Tool shown may be fabricated to hold brushes when installing motor and cap.

using screws provided with new ring gear.

To check for correct operation of starter motor, remove starter motor from engine and place motor in a vise or other holding fixture. Install a 0-5 amp ammeter in power cord to 110 volt AC starter motor. On 12 volt DC motor, connect a 6 volt battery to motor with a 0-50 amp ammeter in series with positive (+) line from battery starter motor. Connect a tachometer to drive end of starter. With starter activated on 110 volt motor, starter motor should turn at 5200 rpm minimum with a maximum current draw of 3½ amps. The 12 volt motor should turn at 5000 rpm minimum with a current draw of 25 amps maximum. If starter motor does not operate satisfactorily, check operation of rectifier or starter switch. If rectifier and starter switch are good, disassemble and inspect starter motor.

To check the rectifier used on the 110 volt AC starter motor, remove rectifier unit from starter motor. Solder a 10,000 ohm, 1 watt resistor to the DC internal terminals of the rectifier as shown in Fig. B141. Connect a 0-100 range DC voltmeter to resistor leads. Measure the voltage of the AC outlet to be used. With starter switch in "OFF" position, a zero reading should be shown on DC voltmeter. With starter switch in "ON" position, the DC voltmeter should show a reading that is 0-14 volts lower than AC line voltage measured previously. If voltage drop exceeds 14 volts, renew rectifier unit.

Disassembly of starter motor is evident after inspection of unit and referral to Fig. B140. Note position of bolts (9) during disassembly so they can be installed in their original positions during reassembly. When reassembling motor, lubricate end cap bearings with SAE 20 oil. Be sure to match the drive cap keyway to the stamped key in the housing when sliding the armature into the motor housing. Brushes may be held in their holders during installation by making a brush spreader tool from a piece of metal as shown in Fig. B142. Splined end of helix (2–Fig. B140) must be towards end of armature shaft as shown in Fig. B143. Tighten armature shaft nut to 170 in-lbs. (19 N·m).

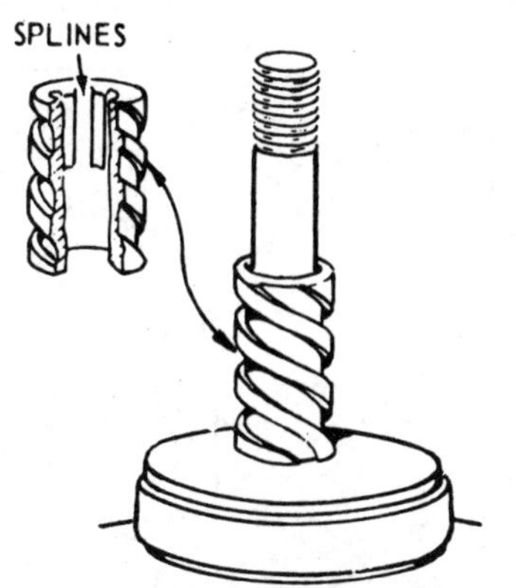

Fig. B143—Install helix on armature so splines of helix are to the outer end as shown.

FLYWHEEL ALTERNATOR

Early Type External

Refer to Fig. B144 for assembled view and to Fig. B145 for exploded view of

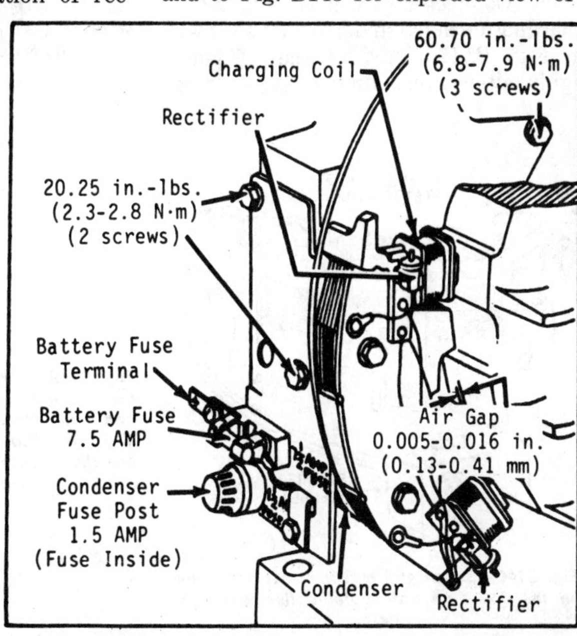

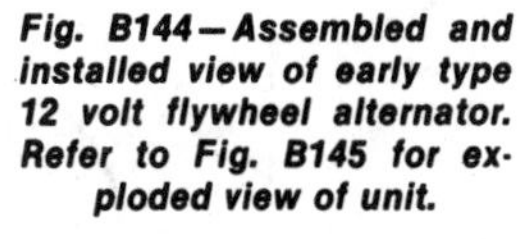

Fig. B144—Assembled and installed view of early type 12 volt flywheel alternator. Refer to Fig. B145 for exploded view of unit.

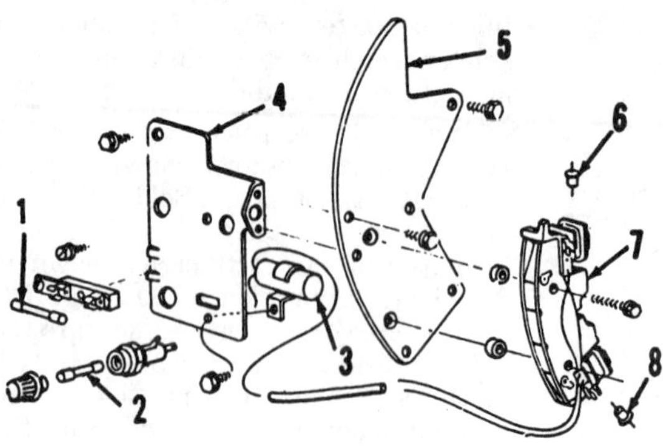

Fig. B145—Exploded view of early type 12 volt flywheel alternator designed for use with the 12 volt starter motor unit shown in Fig. B139.

1. Fuse (3AG 7½ amp)
2. Fuse (AG 1½ amp)
3. Condenser
4. Support plate
5. Alternator plate
6. Rectifier
7. Armature & coil assy.
8. Rectifier

alternator. The 12 volt flywheel alternator is designed for use with the 12 volt electric starter and a 20 to 24 ampere hour capacity battery.

Armature air gap should be 0.005 inch (0.13 mm). Rectifiers (diodes) are available separately from the armature and coil unit which is available as an assembly only. When renewing the condenser or fuses, be sure to use the correct Briggs & Stratton parts.

No cut-out is required with this alternator as the rectifiers (6) prevent reverse flow of electricity through the alternator.

1½ Amp Nonregulated

Some engines are equipped with 1½ amp nonregulated flywheel alternator. This alternator (Fig. B146) with a solid state rectifier, is designed for use with a compact battery. A 12 ampere hour battery is suggested for warm temperature operation and a 24 ampere hour battery should be used in cold service. The alternator is rated at 3600 rpm. At lower speeds, available output is reduced.

To test for alternator ouptut, disconnect charging lead from charging terminal. Connect a 12 volt lamp between charging terminal and ground as shown in Fig. B147. Start engine. If lamp lights, alternator is functioning. If lamp does not light, alternator is defective.

To check for faulty stator, disconnect charging lead from battery and rectifier. Remove rectifier box mounting screw. Turn box to expose eyelets to which red and black stator leads are soldered. Start engine and with engine operating, touch load lamp leads to eyelets as shown in Fig. B148. If lamp does not light, stator or flywheel is defective. Remove flywheel and check to be sure magnet ring is in place and has magnetism. Check wires from stator to rectifier. If flywheel and wiring are good, renew stator assembly which includes new rectifier. If the load lamp lights, stator is satisfactory. Check for faulty rectifier as follows: With engine stopped and charging lead disconnected from charging terminal, connect one ohmmeter lead to charging terminal and remaining lead to engine block. See Fig. B149. Check for continuity, then reverse ohmmeter leads and again check for continuity. Ohmmeter should show a continuity reading for one direction only. If tests show no continuity in either direction or continuity in both directions, rectifier is faulty and must be renewed. Rectifier assembly is serviced separately.

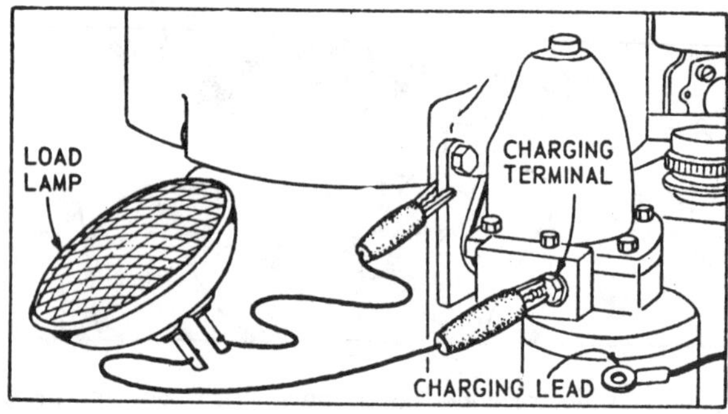

Fig. B147—Disconnect charging lead and connect load lamp to test alternatc output.

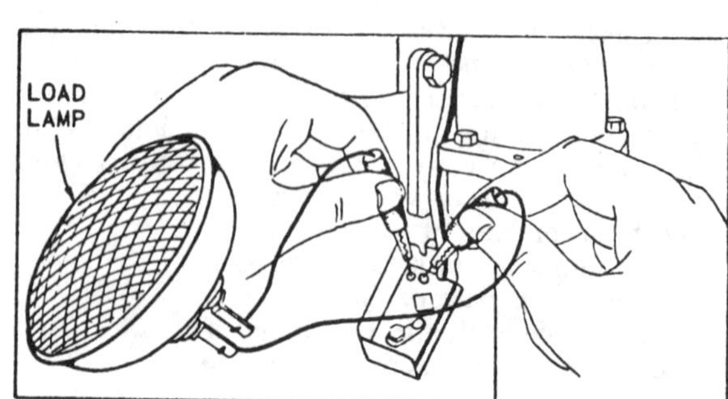

Fig. B148—Use 12 volt load lamp to test for defective stator. Refer to text.

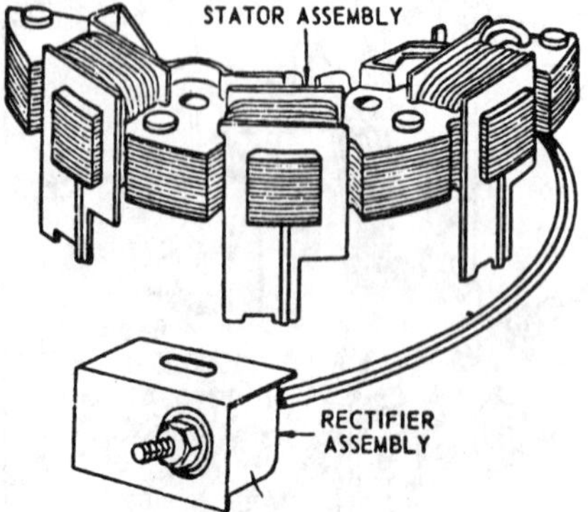

Fig. B146—Stator and rectifier assembly used on the 1½ amp nonregulated flywheel alternator.

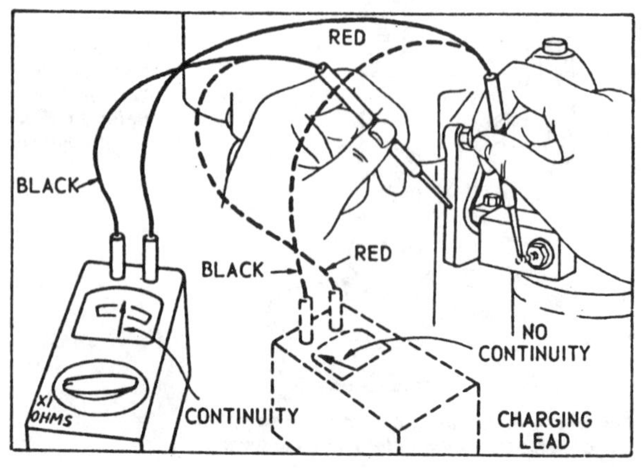

Fig. B149—When checking rectifier with an ohmmeter, and meter shows continuity both directions or neither direction, rectifier is defective.

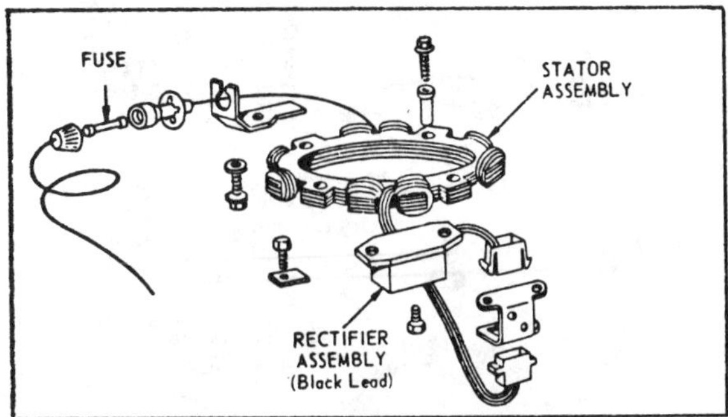

Fig. B150—Stator and rectifier assemblies used on the 4 amp nonregulated flywheel alterator. Fuse is 7½ amp AGC or 3AG.

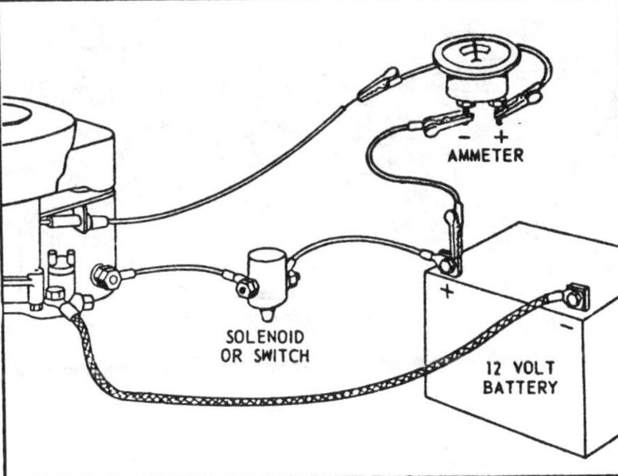

Fig. B151—install ammeter as shown for output test.

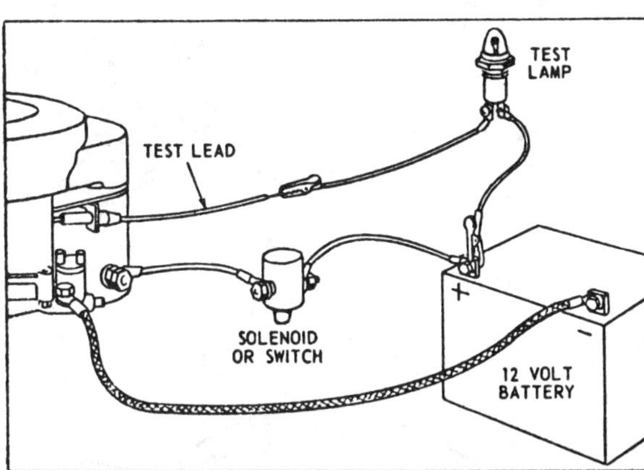

Fig. B152—Connect a test lamp as shown to test for shorted stator or defective rectifier. Refer to text.

4 Amp Nonregulated

Some engines are equipped with the 4 amp nonregulated flywheel alternator shown in Fig. B150. A solid-state rectifier and 7½ amp fuse is used with this alternator.

If battery is run down and no output from alternator is suspected, first check the 7½ amp fuse. If fuse is good, clean and tighten all connections. Disconnect charging lead and connect an ammeter as shown in Fig. B151. Start engine and check for alternator output. If ammeter shows no charge, stop engine, remove ammeter and install a test lamp as shown in Fig. B152. Test lamp should not light. If it does light, stator or rectifier is defective. Unplug rectifier plug under blower housing. If test lamp goes out, rectifier is defective. If test lamp does not go out, stator is shorted.

If shorted stator is indicated, use an ohmmeter and check continuity as follows: Touch one test lead to lead inside of fuse holder as shown in Fig. B153. Touch the remaining test lead to each of the four pins in rectifier connector. Unless the ohmmeter shows continuity at each of the four pins, stator winding is open and stator must be renewed.

If defective rectifier is indicated, unbolt and remove the flywheel blower housing with rectifier. Connect one ohmmeter test lead to blower housing and remaining lead to the single pin connector in rectifier connector. See Fig. B154. Check for continuity, then reverse leads and again test for continuity. If tests show no continuity in either direction or continuity in both directions, rectifier is faulty and must be renewed.

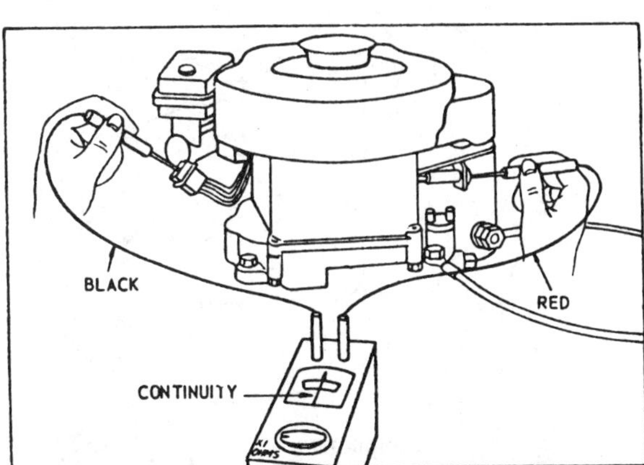

Fig. B153—Use an ohmmeter to check condition of stator. Refer to text.

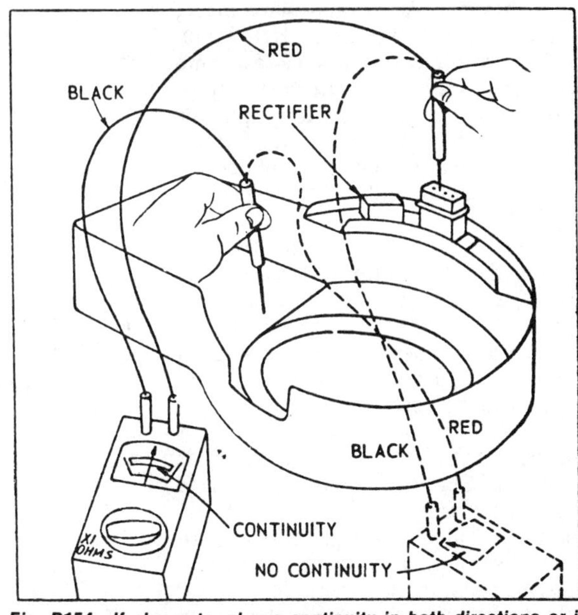

Fig. B154—If ohmmeter shows continuity in both directions or in neither direction, rectifier is defective.

7 Amp Regulated

A 7 amp regulated flywheel alterator is used with the 12 volt gear drive starter motor. The alternator is equipped with a solid state rectifier and regulator. An isolation diode is also used on most models.

If engine will not start, using electric start system, and trouble is not in starting motor, install an ammeter in circuit as shown in Fig. B156. Start engine manually. Ammeter should indicate charge. If ammeter does not show battery charging taking place, check for defective wiring and if necessary proceed with troubleshooting.

If battery charging occurs with engine running, but battery does not retain charge, then isolation diode may be defective. The isolation diode is used to prevent battery drain if alternator circuit malfunctions. After troubleshooting diode, remainder of circuit should be inspected to find reason for excessive battery drain. To check operation of diode, disconnect white lead of diode from fuse holder and connect a test lamp from the diode white lead to the negative terminal of battery. Test lamp should not light. If test lamp lights, diode is defective. Disconnect test lamp and disconnect red lead of diode. Test continuity of diode with ohmmeter by connecting leads of ohmmeter to leads of diode, then reverse lead connections. The ohmmeter should show continuity in one direction and an open circuit in the other direction. If readings are incorrect, then diode is defective and must be renewed.

To troubleshoot alternator assembly, disconnect white lead of isolation diode from fuse holder and connect a test lamp between positive (+) terminal of battery and fuse holder on engine. Engine must not be started. With connections made, test lamp should not light. If isolation diode operates correctly and test lamp does light, stator, regulator or rectifier is defective. Unplug rectifier-regulator plug under blower housing. If lamp remains lighted, stator is grounded. If lamp goes out, regulator or rectifier is shorted.

If previous test indicated stator is grounded, check stator leads for defects and repair if necessary. If shorted leads are not found, renew stator. Check stator for an open circuit by using an ohmmeter. Connect positive (+) ohmmeter lead to fuse holder as shown in Fig. B157 and remaining lead to one of the pins in the rectifier and regulator connector. Check each of the four pins in the connector. The ohmmeter should show continuity at each pin; if not, there is an open circuit in stator and stator must be renewed.

To test rectifier, unplug rectifier and regulator connector plug and remove

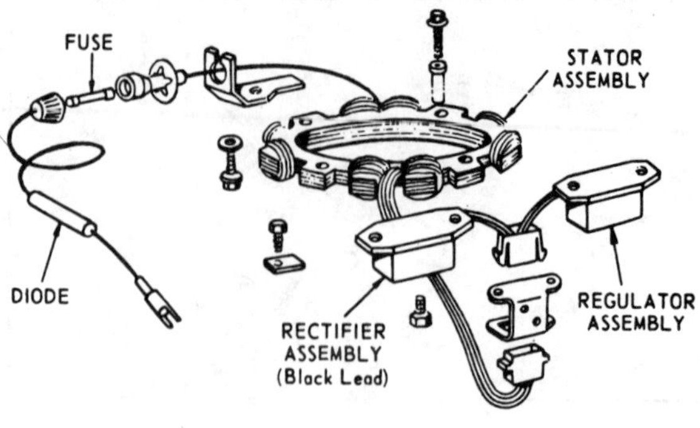

Fig. B155—Stator, rectifier and regulator assemblies used on the 7 amp regulated flywheel alternator.

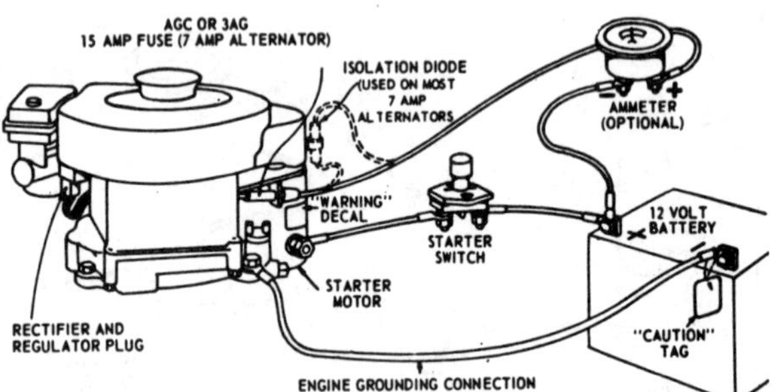

Fig. B156—Typical wiring used on engines equipped with 7 amp flywheel alternator.

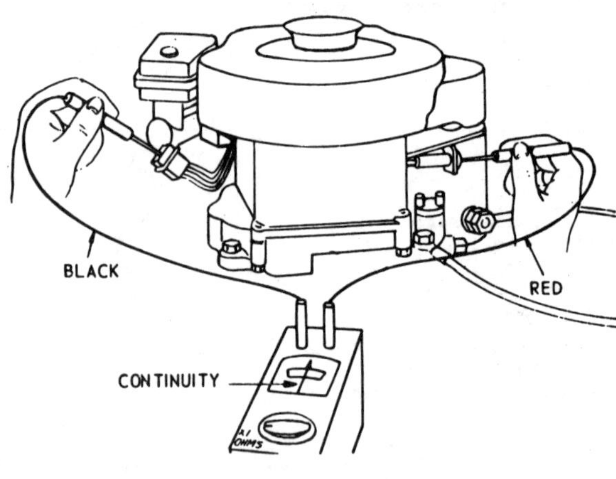

Fig. B157—Use an ohmmeter to check condition of stator. Refer to text.

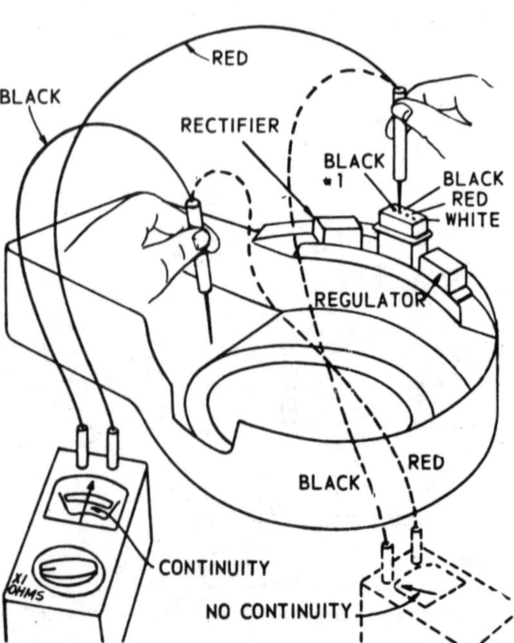

Fig. B158—Be sure good contact is made between ohmmeter test lead and metal cover when checking rectifier and regulator.

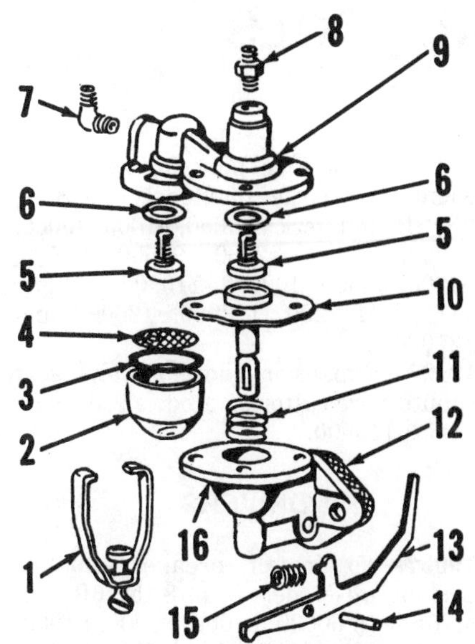

Fig. B159—Exploded view of diaphragm type fuel pump used on some Briggs & Stratton engines. Refer to Fig. B160 for assembly and disassembly views.

1. Yoke assy.
2. Filter bowl
3. Gasket
4. Filter screen
5. Pump valves
6. Gaskets
7. Elbow fitting
8. Connector
9. Fuel pump head
10. Pump diaphragm
11. Diaphragm spring
12. Gasket
13. Pump lever
14. Lever pin
15. Lever spring
16. Fuel pump body

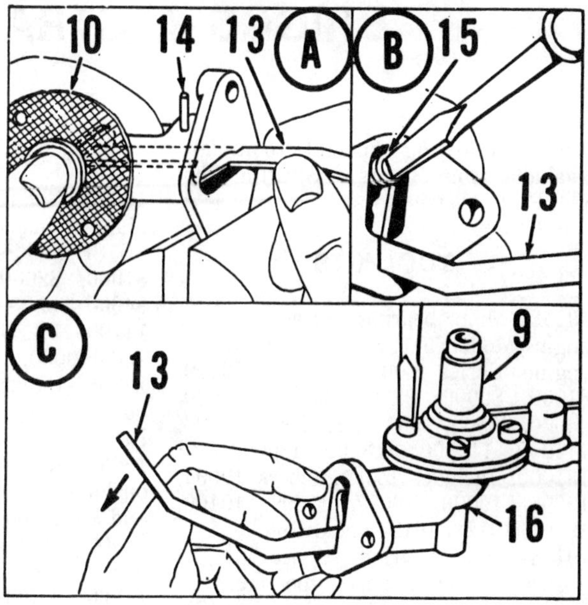

Fig. B160—Views showing disassembly and reassembly of diaphragm type fuel pump. Refer to text for procedure and to Fig. B159 for exploded view of pump and for legend.

blower housing from engine. Using an ohmmeter, check for continuity between connector pins connected to black wires and blower housing as shown in Fig. B158. Be sure good contact is made with metal of blower housing. Reverse ohmmeter leads and check continuity again. The ohmmeter should show a continuity reading for one direction only on each plug. If either pin shows a continuity reading for both directions, or if either pin shows no continuity for either direction, rectifier must be renewed.

To test regulator unit, repeat procedure used to test rectifier unit, except connect ohmmeter lead to pins connected to red wire and white wire. If ohmmeter shows continuity in either direction for red lead pin, regulator is defective and must be renewed. White lead pin should read as an open on the ohmmeter in one direction and a weak reading in the other direction. Otherwise, the regulator is defective and must be renewed.

FUEL PUMP

A diaphragm type fuel pump is available on many models as optional equipment. Refer to Fig. B159 for exploded view of pump.

To disassemble pump, refer to Figs. B159 and B160. Remove clamp (1), fuel bowl (2), gasket (3) and screen (4). Remove screws retaining upper body (9) to lower body (16). Pump valves (5) and gaskets (6) can now be removed. Drive lever retaining pin (14) out to either side of body (16). Press diaphragm (10) against spring (11) as shown in view A, Fig. B160, and remove lever (13). Diaphragm and spring (11–Fig. B159) can now be removed.

To reassemble, place diaphragm spring in lower body and place diaphragm on spring, being sure spring enters cup on bottom side of diaphragm and that slot in shaft is at right angle to pump level. Compress diaphragm against spring as in view A, Fig. B160, and insert hooked end of lever into slot in shaft. Align hole in lever with hole in lower body and drive pin into place. Insert lever spring (15) into body and push outer end of spring into place over hook on arm of lever as shown in view B. Hold lever downward as shown in view C while tightening screws holding upper body to lower body. When installing pump on engine, apply a liberal amount of grease on lever (13) at point where it contacts groove in crankshaft.

BRIGGS & STRATTON SPECIAL TOOLS

The following special tools are available from Briggs & Stratton Central Parts Distributors.

TOOL KITS

19158—Main bearing service kit for engine Models 5, 6, 8, N, 6B, 8B and for engine Series 60000, 61000, 80000, 81000, 82000, 92000, 93500, 94500, 94900, 95500, 100000, 110900, 111200, 111900, 112200, 113900, 130000 and 131000. Includes tool numbers 19094, 19095, 19096, 19097, 19099, 19100, 19101, 19123, 19124 and 19166.
19184—Main bearing service kit for Series 140000. Includes tool numbers 19096, 19168, 19169, 19170, 19171, 19172, 19173, 19174, 19175, 19178, 19179 and 19201.
291661—Dealer service tool kit. Includes tool numbers 19051, 19055, 19056, 19057, 19058, 19061, 19062, 19063, 19064, 19065, 19066, 19069, 19070, 19114, 19122, 19151, 19165, 19167, 19191 and 19203.

PLUG GAGES

19055—Check breaker plunger hole on Models 5, 6, 8, N, 6B, 8B and Series 60000, 61000, 80000, 81000, 82000, 92000, 93500, 94500, 94900, 95500, 100000, 110900, 111200, 111900, 112200, 113900, 130000, 131000 and 140000.

19117—Check main bearing bore on Models 9 and 14.

19122—Check valve guide bore on Models 5, 6, 8, N, 6B, 8B, and 60000, 61000, 80000, 81000, 82000, 92000, 93500, 94500, 94900, 95500, 100000, 110900, 111200, 111900, 112200, 113900, 130000 and 131000.

19151—Check valve guide bore on Models 9, 14 and Series 140000.

19164—Check camshaft bearings on Models 6B, 8B, and Series 60000, 61000, 80000, 81000, 82000, 92000, 93500, 94500, 94900, 95500, 100000, 110900, 111200, 111900, 112200, 113900, 130000, 131000 and 140000.

19166—Check main bearing bore on Models 5, 6, 8, N, 6B, 8B and on engine Series 60000, 61000, 80000, 81000, 82000, 92000, 93500, 94500, 94900, 95500, 100000, 110900, 111200, 111900, 112200, 113900, 130000 and 131000.

19178—Check main bearing bore on engine Series 140000.

REAMERS

19056—Ream hole to install breaker plunger bushing on Models 5, 6, 8, N, 6B, 8B, and Series 60000, 61000, 80000, 81000, 82000, 92000, 93500, 94500, 94900, 95500, 100000, 110900, 111200, 111900, 112200, 113900, 130000, 131000 and 140000.
19058—Finish ream breaker plunger bushing on same models as 19056 Reamer.
19064—Ream valve guide bore to install bushing on Models 5, 6, 8, N, 6B, 8B, and Series 60000, 61000, 80000, 81000, 82000, 92000, 93500, 94500, 94900, 95500, 100000, 110900, 111200, 111900, 112200, 113900, 130000 and 131000.
19066—Finish ream valve guide bushing on same models as listed for 19064 Reamer.
19095—Finish ream main bushings on same models as listed for 19064 Reamer.
19099—Ream counterbore for main bearings on Models 6B and 8B, and Series 60000, 61000, 80000, 81000, 82000, 92000, 93500, 94500, 94900, 95500, 100000, 110900, 111200, 111900, 112200, 113900, 131000 and 140000.
19172—Ream counterbore for main bearing on Series 100000, 130000, 131000 and 140000.
19173—Finish reamer for main bearing on Series 100000, 130000, 131000 and 140000.
19174—Ream counterbore for main bearing on Series 140000.
19175—Finish ream main bearing on Series 140000.
19183—Ream valve guide bore to install bushing on Models 9, 14 and Series 140000.

GUIDE BUSHINGS FOR VALVE GUIDE REAMERS

19191—For Models 5, 6, 8, N, 6B, 8B and Series 60000, 61000, 80000, 81000, 82000, 92000, 93500, 94500, 94900, 95500, 100000, 110900, 111200, 111900, 112200, 113900, 130000 and 131000.
19192—For Models 9, 14 and Series 140000.

PILOTS

19096—Pilot for main bearing reamer on Models 5, 6, 8, N, 6B, 8B and Series 60000, 61000, 80000, 81000, 82000, 92000, 93500, 94500, 94900, 95500, 100000, 110900, 111200, 111900, 112200, 113900, 130000, 131000, and 140000.
19126—Expansion pilot for valve seat counterbore cutter on Models 5, 6, 8, N, 6B, 8B, and Series 60000, 61000, 80000, 81000, 82000, 92000, 93500, 94500, 94900, 95500, 100000, 110900, 111200, 111900, 112200, 113900, 130000 and 131000.
19127—Expansion pilot for valve seat counterbore cutter on Models 9, 14, and Series 140000.

DRIVERS

19057—To install breaker plunger bushing on Models 5, 6, 8, N, 6B, 8B, and Series 60000, 61000, 80000, 81000, 82000, 92000, 93500, 94500, 94900, 95500, 100000, 110900, 111200, 111900, 112200, 113900, 130000, 131000 and 140000.
19065—To install valve guide bushings on Models 5, 6, 8, N, 6B, 8B and Series 60000, 61000, 80000, 81000, 82000, 92000, 93500, 94500, 94900, 95500, 100000, 110900, 111200, 111900, 112200, 113900, 130000 and 131000.
19124—To install main bearing bushings on Models 5, 6, 8, N, 6B, 8B, and Series 60000, 61000, 80000, 81000, 82000, 92000, 93500, 94500, 94900, 95500, 100000, 110900, 111200, 111900, 112200, 113900, 130000 and 131000.
19136—To install valve seat inserts on all Models and Series.
19179—Install main bearing on Series 140000.

GUIDE BUSHINGS FOR MAIN BEARING REAMERS

19094—For Models 5, 6, N, 6B, 8B and Series 60000, 61000, 80000, 81000, 82000, 92000, 93500, 94500, 94900, 95500, 100000, 110900, 111200, 111900, 112200, 113900, 130000 and 131000.
19100—For Models 6B, 8B and Series 60000, 80000, 81000, 82000, 92000, 93500, 94500, 94900, 95500, 110900, 111200 and 111900.
19101—For Models 6B, 8B, and Series 60000, 61000, 80000, 81000, 82000, 92000, 93500, 94500, 94900, 95500, 100000, 110900, 111200, 111900, 112200, 113900, 130000 and 131000.
19168—For Series 100000, 130000, 131000 and 140000.
19169—For Series 140000.
19170—For Series 100000, 130000, 131000 and 140000.
19171—For Series 140000.
19186—For Series 100000, 130000 and 131000.

COUNTERBORE CUTTERS

19131—To install intake valve seat on Model 14.
19132—To install intake valve seat insert on Models 8 and 9.
19133—To install intake valve seat on Models 5, 6 and N.

FLYWHEEL HOLDER

19167—For Models 6B, 8B and Series 60000, 61000, 80000, 81000, 82000, 92000. 93500, 94500, 94900, 95500, 110900, 111200, 111900, 112200 and 113900.

VALVE SPRING COMPRESSOR

19063—For all Models and Series.

PISTON RING COMPRESSOR

19070—For all Models and Series.

FLYWHEEL PULLERS

19069—For Models 6B, 8B and Series 60000, 61000, 80000, 81000, 82000, 92000, 93500, 94500, 94900, 95500, 110900, 111200, 111900, 112200 and 113900.
19165—For Series 140000.
19203—For Models 9 and 14.

CRANKCASE SUPPORT JACK

19123—To support crankcase when removing and installing main bearings bushings on Models 5, 6, 8, N, 6B, 8B, and Series 60000, 61000, 80000, 81000, 82000, 92000, 93500, 94500, 94900, 95500, 100000, 110900, 111200, 111900, 112200, 113900, 130000, 131000 and 140000.

IGNITION SPARK TESTER

19051—For all Models and Series.

VALVE SEAT REPAIR TOOLS

19129—Planer shank-driver for counterbore cutter on all Models and Series.
19130—T-handle for planer shank 19129 on all Models and Series.
19135—Knockout pin to remove counterbore cutter from planer shank.
19137—T-handle for expansion pilots
19138—Insert puller to remove valve seats on all Models and Series (when puller nut 19140, included with 19138, is ground to 1/32-inch thick for pulling insert on aluminum alloy models).
19182—Puller nut to adapt 19138 puller to aluminum alloy models.

STARTER WRENCH

19114—All models with rewind or wind-up starters.
19161—All models with rewind or wind-up starters.

BRIGGS AND STRATTON CENTRAL PARTS DISTRIBUTORS

(Arranged Alphabetically by States)
These franchised firms carry extensive stocks of repair parts. Contact them for name of the nearest service distributor who may have the parts you need.

BEBCO, Incorporated
2221 Second Avenue, South
Birmingham, Alabama 35233

Power Equipment Company
3141 North 35th Avenue, Unit 101
Phoenix, Arizona 85107

Pacific Power Equipment Company
1565 Adrain Road
Burlingame, California 94010

Power Equipment Company
1045 Cindy Lane
Carpinteria, California 93013

Pacific Power Equipment Company
700 W. Mississippi Avenue
Denver, Colorado 80223

Small Engine Clinic, Inc.
98019 Kam Highway
Pearl City, Hawaii 96782

Midwest Engine Warehouse
515 Romans Road
Elmhurst, Illinois 60126

Commonwealth Engine, Inc.
11421 Electron Drive
Louisville, Kentucky 40229

W. J. Connell Company
77 Green Street
Foxboro, Massachusetts 02035

Wisconsin Magneto, Inc.
8010 Ranchers Road
Minneapolis, Minnesota 55432

Diamond Engine Sales
3134 Washington
St. Louis, Missouri 63103

Original Equipment, Inc.
905 Second Avenue, North
Billings, Montana 59101

Midwest Engine Warehouse of Omaha
7706-30 "I" Plaza
Omaha, Nebraska 68127

W. J. Connel Co., Bound Brook Div.
Chimney Rock Road, Route 22
Bound Brook, New Jersey 08805

Power Equipment Company
7209 Washington Street, North East
Alburquerque, New Mexico 87109

AEA, Incorporated
700 West 28th Street
Charlotte, North Carolina 28206

Central Power Systems
2555 International Street
Columbus, Ohio 43228

Engine Warehouse, Inc.
4200 Highline Boulevard
Oklahoma City, Oklahoma 73108

Brown & Wiser, Inc.
9991 South West Avery Street
Tualatin, Oregon 97062

Three Rivers Engine Distributors
1411 Beaver Avenue
Pittsburgh, Pennsylvania 15233

Automotive Electric Corporation
3250 Millbranch Road
Memphis, Tennessee 38116

Grayson Company, Inc.
1234 Motor Street
Dallas, Texas 75207

Engine Warehouse, Inc.
7415 Empire Central Drive
Houston, Texas 77040

Frank Edwards Company
110 South 300 West
Salt Lake City, Utah 84101

RBI Corporation
101 Cedar Ridge Drive
Ashland, Virginia 23005

Air Cooled Engine Supply
South 111 Walnut Street
Spokane, Washington 99204

Wisconsin Magnetic, Inc.
4727 North Teutonial Avenue
Milwaukee, Wisconsin 53209

CANADIAN DISTRIBUTORS

Canadian Curtiss-Wright, Ltd.
2-3571 Viking Way
Richmond, British Columbia V6V 1W1

Canadian Curtiss-Wright, Ltd.
89 Paramount Road
Winnipeg, Manitoba R2X 2W6

Canadian Curtiss-Wright, Ltd.
1815 Sismet Road
Mississauga, Ontario L4W 1P9

Canadian Curtiss-Wright, Ltd.
8100-N Trans Canada Hwy.
St. Laurent, Quebec H4S 1M5

CLINTON

CLINTON ENGINES CORPORATION
Clark & Maple Streets
Maquoketa, Iowa 52060

CLINTON ENGINE IDENTIFICATION INFORMATION

To obtain the correct service replacement parts when overhauling Clinton engines, it is important that the engine be properly identified as to:

1. Model number
2. Variation number
3. Type letter

A typical nameplate from the model number series engines prior to 1961 is shown in Fig. CL1. In this example, the following information is noted from the nameplate:

1. Model number – **B-760**
2. Variation number – **AOB**
3. Type letter – **B**

In some cases, the model number may be shown as in the following example:

D-790-2124

Thus, "D-790" would be the model number of a D-700-2000 series engine in which the digits "2124" would be the variation number.

In late 1961, the identification system for Clinton engines was changed to be acceptable for use with IBM inventory record systems. A typical nameplate from a late production engine is shown in Fig. CL2. In this example, the following information is noted from the nameplate:

1. Model number – 405-0000-000
2. Variation number – 070
3. Type letter – D

In addition, the following information may be obtained from the model number on the nameplate:

First digit – identifies type of engine, i.e., 4 means four-stroke engine and 5 means two-stroke engine.

Second and third digits - completes basic identification of engine. Odd numbers will be used for vertical shaft engines and even numbers for horizontal shaft engines; i.e., 405 would indicate a four-stroke vertical shaft engine and 500 would indicate a two-stroke horizontal shaft engine.

Fourth digit – indicates type of starter as follows:

0 – Rewind starter
1 – Rope starter
2 – Impluse starter
3 – Crank starter
4 – 12 Volt electric starter
5 – 12 Volt starter-generator
6 – 110 Volt electric starter
7 – 12 Volt generator
8 – unassigned to date
9 – Short block assembly.

Fifth digit – indicates bearing type, etc., as follows:

0 – Standard bearing
1 – Aluminum or bronze sleeve bearing with flange mounting surface and pilot diameter on engine mounting face for mounting equipment concentric to crankshaft center line
2 – Ball or roller bearing
3 – Ball or roller bearing with flange mounting surface and pilot diameter on engine mounting face for mounting equipment concentric to crankshaft center line
4 – Numbers 4 through 9 are unassigned to date

Sixth digit – indicates auxiliary power takeoff and speed reducers as follows:

0 – Without auxiliary power pto or speed reducer
1 – Auxiliary power takeoff
2 – 2:1 speed reducer
3 – not assigned to date
4 – 4:1 speed reducer
5 – not assigned to date
6 – 6:1 speed reducer
7 – numbers 7 through 9 are unassigned to date

Seventh digit – if other than "0" will indicate a major design change.

Eighth, ninth & tenth digits – identifies model variations.

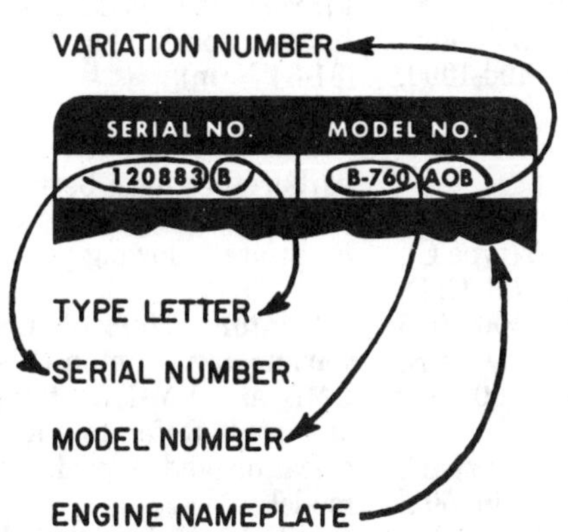

Fig. CL1—Typical nameplate from Clinton engine manufactured prior to late 1961.

Fig. CL2—Typical nameplate from Clinton engine after model identification system was changed in late 1961. First seven digits indicate basic features of engine. Engines with "Mylar" (plastic) nameplate have engine model and serial numbers stamped on the cylinder air deflector next to the nameplate.

CLINTON

Model	Bore	Stroke	Displacement
E-65 CW, E-65 CCW	2.125 in. (53.2 mm)	1.625 in. (41.3 mm)	5.8 cu. in. (92 cc)
200, A-200, AVS-200-1000, VS-200, VA-200-1000, VS-200-2000, VS-200-3000	1.875 in. (47.6 mm)	1.625 in. (41.3 mm)	4.5 cu. in. (74 cc)
VS-200-4000	2.125 in. (53.2 mm)	1.625 in. (41.3 mm)	5.8 cu. in. (92 cc)
290	1.875 in. (47.6 mm)	1.625 in. (41.3 mm)	4.5 cu. in. (74 cc)
A-400	2.125 in. (53.2 mm)	1.625 in. (41.3 mm)	5.8 cu. in. (92 cc)
A-400-1000, AVS-400, CVS-400-1000, VS-400, VS-400-1000, VS-400-2000, VS-400-3000, VS-400-4000	2.125 in. (53.2 mm)	1.625 in. (41.3 mm)	5.8 cu. in. (92 cc)
A-460*	2.125 in. (53.2 mm)	1.625 in. (41.3 mm)	5.8 cu. in. (92 cc)
490, A-490	2.125 in. (53.2 mm)	1.625 in. (41.3 mm)	5.8 cu. in. (92 cc)
500-0100-000, 501-0000-000, 501-0001-000, GK-590	2.125 in. (53.2 mm)	1.625 in. (41.3 mm)	5.8 cu. in. (92 cc)
502-0308-000, 503-0308-000	2.375 in. (60.3 mm)	1.750 in. (44.5 mm)	7.8 cu. in. (127 cc)
502-0309-000, 503-0309-000	2.500 in. (63.5 mm)	1.750 in. (44.5 mm)	8.6 cu. in. (141 cc)

*Reduction gear

ENGINE IDENTIFICATION

Clinton two-stroke engines are of aluminum alloy die-cast construction with an integral cast-in iron cylinder sleeve. Reed type inlet valves are used on all models. Refer to CLINTON ENGINE IDENTIFICATION INFORMATION section preceding this section.

MAINTENANCE

SPARK PLUG. Champion H10J or equivalent spark plug is recommended for use in E-65 CW and E-65 CCW models.

Champion H11 or H11J or equivalent spark plug is recommended for use in 200, A-200, AVS-200, AVS-200-1000, VS-200, VS-200-1000, VS-200-2000, VS-200-3000 (types A and B), 290, A-400, A-400-1000 (type A), AVS-400, AVS-400-1000, BVS-400, VS-400-1000, VS-400-2000, VS-400-3000, VS-400-4000 (types A and B), 490 and A-490-1000 (type A) models.

Champion J12J or equivalent plug is recommended for VS-200-3000 (type C), VS-200-4000, A-400-1000 (types B, C, D, E, F), CVS-400-1000, VS-400-4000 (types C, D, E, F), A-490-1000 (types B, C, D, E, F), 500-0100-000, 501-000-000 and 501-001-000 models.

Autolite A7NX or equivalent spark plug is recommended for use in 502-0308-000, 502-0309-000, 503-0308-000 and 503-0309-000 models.

On all models, set electrode gap at 0.030 inch (0.76 mm). Use graphite on threads when installing spark plug and tighten spark plug to 275-300 in.-lbs. (31-34 N·m).

CARBURETOR. Several different carburetors have been used on Clinton two-stroke engines. Refer to the appropriate following paragraphs.

Clinton (Walbro) LM Float Type Carburetor. Clinton (Walbro designed) LMB and LMG float type carburetors are used. Refer to Fig. CL5 for identification and exploded view of typical model.

Initial adjustment of idle (9) and main (22) fuel needles from a lightly seated position is 1¼ turns open for each nee-

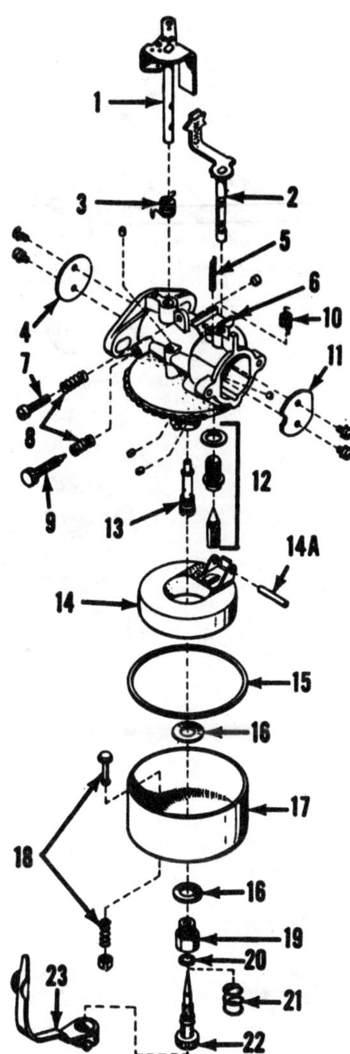

Fig. CL5—Exploded view of typical LMG series carburetor. LMV and LMB series are similar.

1. Throttle shaft
2. Choke shaft
3. Spring
4. Throttle plate
5. Spring
6. Carburetor body
7. Idle stop screw
8. Springs
9. Idle fuel needle
10. Spring
11. Choke plate
12. Inlet needle & seat
13. Main nozzle
14. Float
14A. Float pin
15. Gasket
16. Gaskets
17. Float bowl
18. Drain valve
19. Retainer
20. Seal
21. Spring
22. Main fuel needle
23. Lever (optional)

dle. Final adjustments are made with engine at operating temperature and running. Operate engine at rated speed under load and adjust main fuel mixture needle for smoothest engine operation. Operate engine at idle speed and adjust idle fuel mixture needle for smoothest

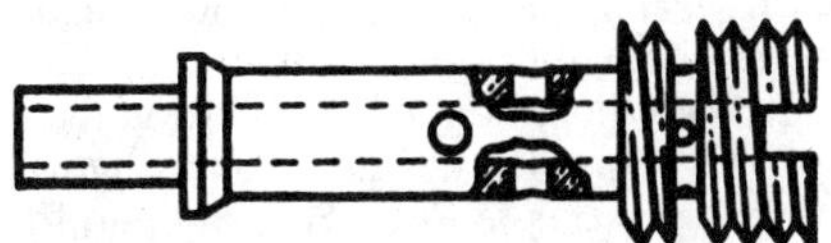

Fig. CL6—When original main fuel nozzle is removed from a LM series carburetor, it must be discarded and service type nozzle shown installed.

Fig. CL7—Exploded view of a typical Carter Model N carburetor. Design of float and float bowl may vary from that shown.

1. Idle fuel needle
2. Idle stop screw
3. Throttle shaft
4. Main fuel needle
5. Carburetor body
6. Choke shaft
7. Choke plate
8. Detent ball
9. Inlet valve seat
10. Inlet needle
11. Float pin
12. Float bowl
13. Retainer
14. Float
15. Main nozzle
16. Throttle plate

engine idle. Check acceleration from slow to fast idle and open idle needle approximately 1/8 turn (counterclockwise) if necessary for proper acceleration.

When overhauling LMB or LMG carburetors, discard main fuel nozzle (13) if removed and install a service type nozzle (see Fig. CL6). Do not install old nozzle.

To check float setting, invert carburetor throttle body and float assembly. Measure distance between top of free side of float and float bowl mating surface of carburetor. Measurement should be 5/32 inch (3.97 mm). Carefully bend float lever tang which contacts fuel inlet needle if necessary to obtain correct float level.

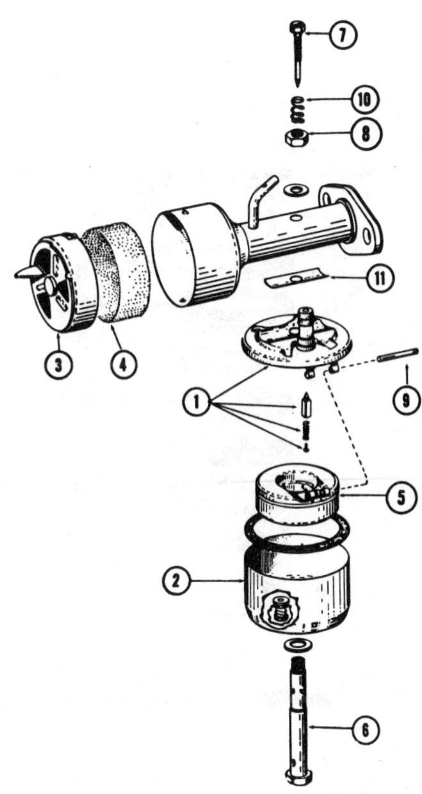

Fig. CL8—Exploded view of float type carburetor used on 501-0000-000 model series engine.

1. Inlet valve & seat
2. Float bowl
3. Choke assy.
4. Air cleaner element
5. Float
6. Main nozzle
7. Fuel needle
8. Nut
9. Float pin
10. Spring
11. Gasket

Float drop should be 3/16 inch (4.76 mm). Carefully bend tabs on float hinge to adjust float drop.

Carter Model N Float Type Carburetor. Refer to Fig. CL7 for identification and exploded view of Carter Model N float type carburetor. The following models have been used:

N-2003-S
N-2029-S
N-2087-S
N-2171-S
N-2457-S

Initial adjustment of fuel mixture needles on Models N-2003-S and N-2029-S from a lightly seated position is ¾ turn open for idle fuel needle (1) and 1 turn open for main fuel mixture needle (4).

Initial adjustment of fuel mixture needles on Models N-2087-S, N-2171-S and N-2457-S is 1½ turns open for idle mixture needle (1) and ½ turn open for main mixture needle (4).

Final adjustments on all models are made with engine at operating temperature and running. Operate engine at rated speed under load and adjust main fuel mixture needle for smoothest engine operation. Operate

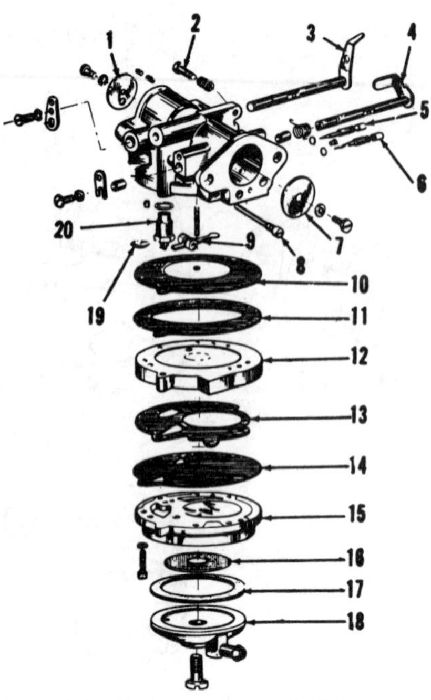

Fig. CL9—Exploded view of Tillotson HL diaphragm type carburetor.

1. Choke plate
2. Idle stop screw
3. Choke shaft
4. Throttle shaft
5. Main fuel needle
6. Idle fuel needle
7. Throttle plate
8. Inlet lever pivot
9. Inlet lever
10. Diaphragm
11. Gasket
12. Cover
13. Gasket
14. Pump diaphragm
15. Pump body
16. Screen
17. Gasket
18. Cover
19. Welch plug
20. Inlet valve & seat

engine at idle speed and adjust idle needle for smooth engine idle then open needle slightly to run a slightly over-rich mixture.

Float setting on Models N-2003-S and N-2457-S is 13/64-inch (5.16 mm) clearance between outer edge of casting and free end of flat. Float setting on Models N-2029-S, N-2087-S and N-2171-S is 11/64-inch (4.37 mm) clearance between outer edge of casting and free end of float. On all models, carefully bend lip on float hinge if necessary to obtain correct float level.

On Model N-2457-S, low speed jet and high speed fuel nozzle are permanently installed. Do not attempt to remove these parts.

Carter Model NS Float Type Carburetor. Refer to Fig. CL8 for identification and exploded view of Carter Model NS float type carburetor. This carburetor is designed for engine operation between 3000 and 3800 rpm and is not equipped with an idle fuel mixture adjusting needle.

Initial adjustment of main fuel mixture needle (7) from a lightly seated position is 1 turn open. Final adjustments are made with engine at operating temperature. Operate engine at rated speed and adjust main fuel mixture needle to obtain smoothest engine operation and acceleration.

To check float level, invert carburetor throttle body. Measure between outer edge of bowl cover and top surface of free end of float. Measurement should be 13/64 inch (5.16 mm). Fuel inlet needle, seat and spring are integral parts of bowl cover and available as an assembly only.

Fig. CL10—Exploded view of Brown CP carburetor.

1. Choke plate
2. Idle stop screw
3. Filter plug
4. Throttle shaft
5. Gasket
6. Carburetor body
7. Idle jet
8. Throttle plate
9. Inlet lever spring
10. Gasket
11. Diaphragm
12. Cover
13. Gasket
14. Diaphragm
15. Pump cover
16. Gasket
17. Inlet fitting
18. Inlet lever
19. Inlet valve
20. Idle fuel needle
21. Choke shaft
22. Main fuel needle
23. Inlet lever pivot

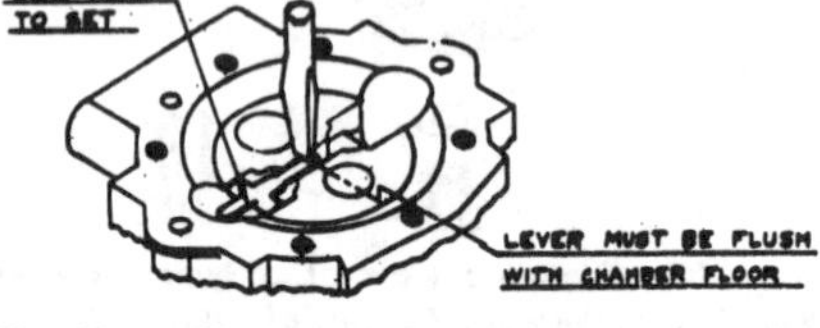

Fig. CL11—Method of checking and setting fuel inlet lever on Brown diaphragm type carburetors.

Tillotson HL Diaphragm Type Carburetor. Refer to Fig. CL9 for identification and exploded view of Tillotson HL diaphragm type carburetor. Models HL-13A and Hl-44A are used. Clockwise rotation of mixture needles on all models leans the fuel mixture.

Initial adjustment of idle (6) and main (5) fuel mixture needles from a lightly seated position is ¾ turn open for idle mixture needle and 1 to 1¼ turns open for main fuel mixture needle.

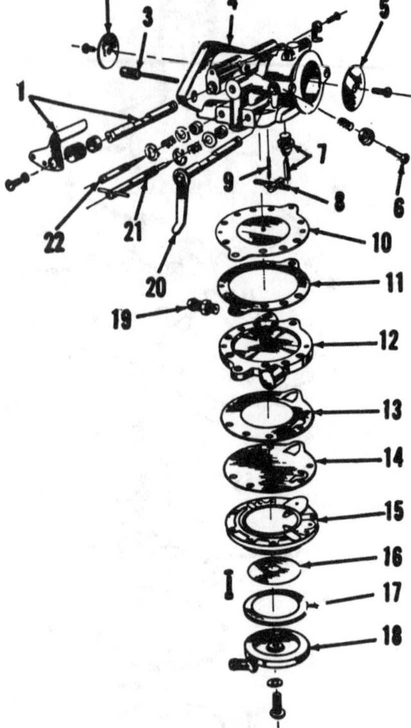

Fig. CL12—Exploded view of Brown CS carburetor.

1. Throttle shaft
2. Throttle plate
3. Inlet lever pivot
4. Carburetor body
5. Choke plate
6. Idle stop screw
7. Inlet needle valve & seat
8. Inlet lever
9. Inlet lever spring
10. Diaphragm
11. Gasket
12. Cover
13. Gasket
14. Pump diaphragm
15. Pump cover
16. Screen
17. Gasket
18. Inlet fitting
19. Connection
20. Choke shaft
21. Main fuel nozzle
22. Idle fuel needle

Final adjustments are made with engine at operating temperature and running. Operate engine at rated speed and adjust main fuel mixture needle for smoothest engine operation, then open needle slightly to provide an over-rich fuel mixture. Operate engine at idle speed and adjust idle mixture screw for smoothest engine idle. If engine fails to accelerate smoothly, open idle mixture needle slightly as necessary.

Brown CP Diaphragm Type Carburetor. Refer to Fig. CL10 for identification and exploded view of Brown Model 3-CP diaphragm type carburetor. Idle mixture needle has a slotted head and main fuel mixture needle has a "T" head. Note fuel must pass through a felt filter and screen, located under plug (3), to enter the diaphragm chamber.

Initial adjustment of idle (20) and main (22) fuel mixture needles from a lightly seated position is ½ to ¾ turn open for each needle.

Final adjustments are made with engine at operating temperature and running. Operate engine at rated speed

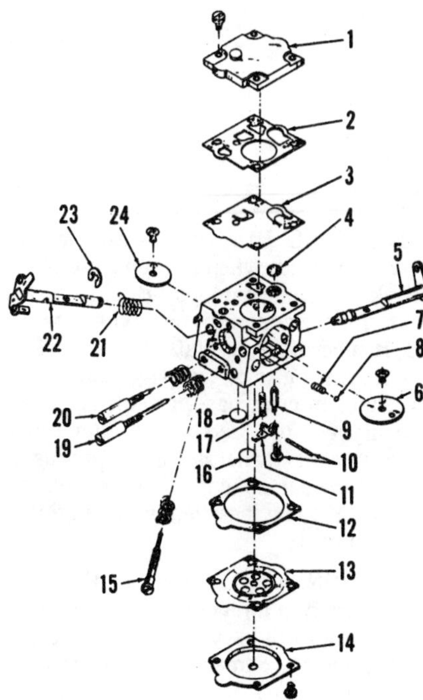

Fig. CL13—Exploded view of typical Walbro SDC, diaphragm type carburetor. Model SDC-35 used on 502-0308-000 engines has choke (5) and throttle (22) levers of slightly different configuration from those shown. Also, an additional diaphragm check valve (not shown) is fitted between gasket (2) and pump diaphragm (3).

1. Fuel pump cover
2. Gasket
3. Pump diaphragm
4. Fuel inlet screen
5. Choke shaft
6. Choke plate
7. Choke detent spring
8. Choke detent ball
9. Inlet valve needle
10. Lever pin retainer
11. Metering lever
12. Metering diaphragm gasket
13. Metering diaphragm
14. Diaphragm cover
15. Idle speed screw & spring
16. Welch plug
17. Metering lever spring
18. Welch plug
19. Main fuel needle & spring
20. Idle mixture needle & spring
21. Throttle return spring
22. Throttle shaft
23. Shaft retainer
24. Throttle plate

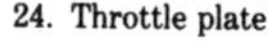

and adjust main fuel mixture needle for smoothest engine operation, then open mixture needle slightly to provide an over-rich fuel mixture. Operate engine at idle speed and adjust idle mixture needle for smooth engine idle. Check acceleration and open main fuel needle slightly if necessary to obtain smooth acceleration.

Refer to Fig. CL11 for proper diaphragm lever setting. Always renew the copper sealing gasket if inlet needle seat is removed.

Brown CS Diaphragm Type Carburetor. Refer to Fig. CL12 for identification and exploded view of Brown CS-15 diaphragm type carburetor. Idle fuel mixture needle has slotted adjustment head and main fuel adjustment needle has a "T" head.

Initial adjustment of idle (22) and main (21) fuel mixture needles from a lightly seated position is ¾ turn open for each screw.

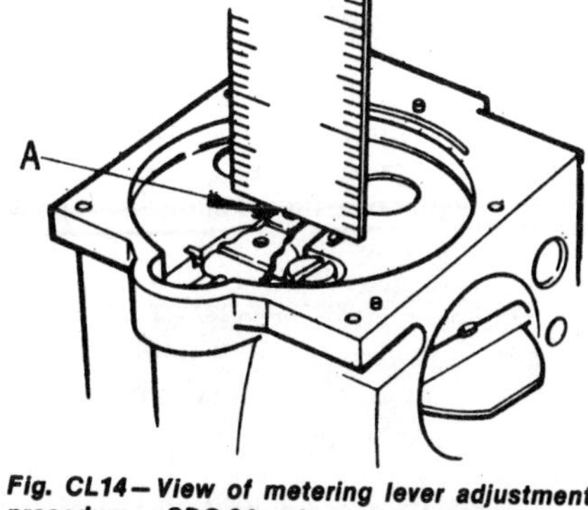

Fig. CL14—View of metering lever adjustment procedure on SDC-34 carburetors. Check engagement of inlet valve needle and diaphragm hook to lever. Ensure spring under lever is correctly seated. Metering lever should just touch straightedge (A).

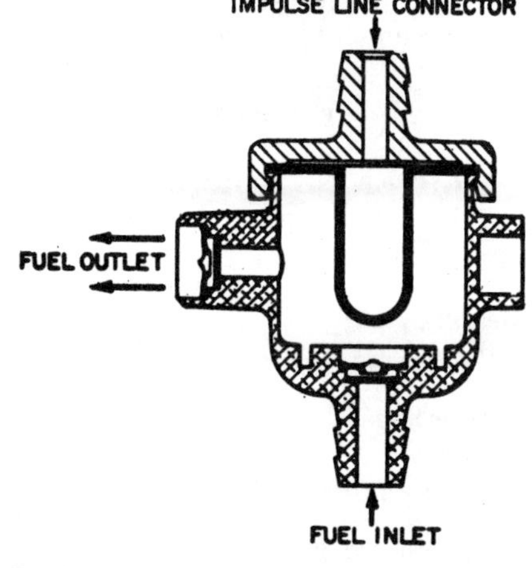

Fig. CL15—Cross section view of impulse type fuel pump used on some engines. Renew complete pump assembly if inoperative.

Final adjustments are made with engine at operating temperature and running. Operate engine at rated speed and adjust main fuel mixture needle for smoothest engine operation, then open mixture needle slightly to provide an over-rich fuel mixture. Operate engine at idle speed and adjust idle mixture needle for smooth engine idle and proper acceleration.

Refer to Fig. CL11 for proper diaphragm lever setting. Always renew the copper sealing gasket if inlet needle is removed.

Walbro Model SDC-34 Diaphragm Type Carburetor. Refer to Fig. CL13 for identification and exploded view of Walbro Model SDC-34 carburetor used on Model 502-0308-000 engine.

Initial adjustment of idle (20) and main (19) fuel mixture needles from a lightly seated position is 1 turn open for each needle.

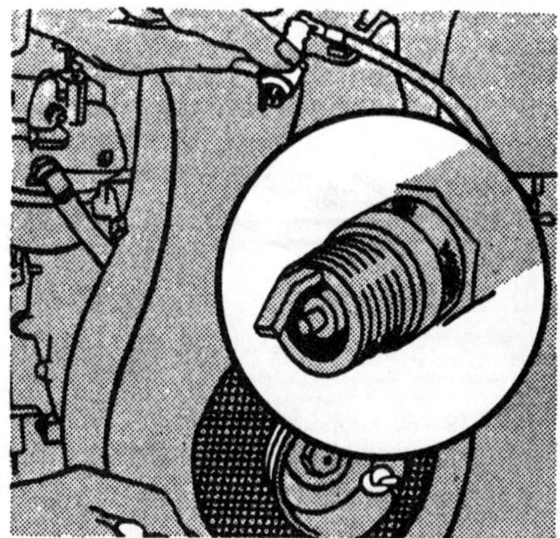

Fig. CL16—Magneto condition may be considered satisfactory if it will fire an 18 mm spark plug with electrode gap set at 0.156-0.187 inch (3.97-4.76 mm).

Final adjustments are made with engine at operating temperature and running. Operate engine at rated speed and adjust main fuel mixture needle for smoothest engine performance. Operate engine at idle speed and adjust idle fuel mixture for smooth engine idle. Note that very slight adjustments to idle and main fuel mixture needles may be necessary to obtain satisfactory engine acceleration.

Model 502-0308-000 engines rely on throttle position changes to control speed only and are not equipped with governor.

Refer to Fig. CL14 for adjustment procedure for metering lever. Be sure metering lever spring is properly seated and hook on diaphragm and inlet valve needle are correctly engaged.

FUEL PUMP. Some models are equipped with a diaphragm type fuel pump as shown in the cross-sectional view in Fig. CL15. The pump diaphragm is actuated by pressure pulsations in the engine crankcase transmitted via an impulse tube connected between the fuel pump and crankcase. The pump is

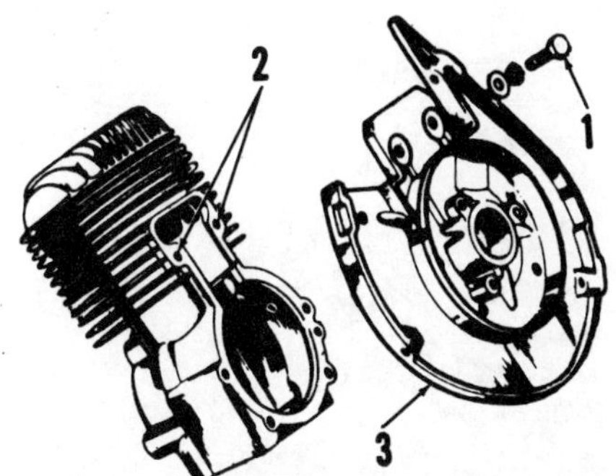

Fig. CL17—If cap screws (1) which thread into holes (2) are too long, they may bottom and damage cylinder walls.

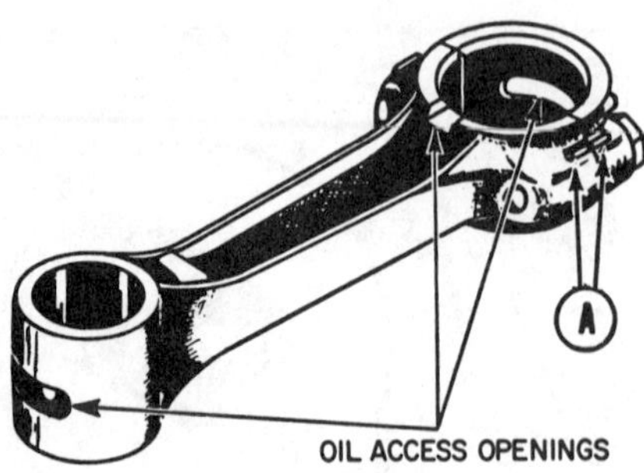

Fig. CL18—Be sure embossments (A) are aligned when assembling cap to connecting rod.

designed to lift fuel approximately 6 inches (152 mm). Service consists of renewing the complete fuel pump assembly.

GOVERNOR. Except on nongoverned karting engines, speed control is maintained by an air vane type governor. Renew bent or worn governor vane or linkage and be sure linkage does not bind in any position when moved through full range of travel.

On models with speed control lever, adjust lever stop so maximum speed does not exceed 3600 rpm. On models without speed control, renew or retension governor spring to limit maximum speed to 3600 rpm. On all models (so equipped), make sure blower housing is clean and free of dents that would cause binding of air vane.

MAGNETO AND TIMING. Ignition timing is 27° BTDC and is nonadjustable. Armature air gap should be 0.007-0.017 inch (0.18-0.43 mm). Adjust breaker point gap to 0.018-0.021 inch (0.46-0.53 mm) on all models.

Magento may be considered satisfactory if it will fire an 18 mm spark plug with gap set at 0.156-0.187 inch (3.97-4.76 mm). Refer to Fig. CL16.

LUBRICATION. Engine is lubricated by mixing a good quality two-stroke air-cooled engine oil with regular gasoline.

Oil to gasoline mixing ratio is determined by type of bearings installed in a particular two-stroke engine. Plain sleeve main and connecting rod bearings require 0.75 pint (0.36 L) of oil for each gallon of gasoline. Engines equipped with needle bearings require 0.50 pint (0.24 L) of oil for each gallon of gasoline. During break-in (first five hours of operation) a 50% increase of oil portion is advisable, especially if engine will be operated at or near maximum load or rpm.

CARBON. Power loss on two-stroke engines can often be corrected by cleaning the exhaust ports and muffler. To clean the exhaust ports, remove the muffler and turn engine until piston is below ports. Use a dull tool to scrape carbon from ports. Take care not to damage top of piston or cylinder walls.

REPAIRS

TIGHTENING TORQUES. Recommended tightening torque specifications are as follows:

Item	Torque
Bearing plate to block	75-95 in.-lbs. (8-11 N•m)
Blower housing	65-70 in.-lbs. (7-8 N•m)
Carburetor mounting	60-65 in.-lbs. (7 N•m)
Connecting rod:	
E65, GK590, 502, 503 (aluminum)	70-80 in.-lbs. (8-9 N•m)
E65, GK590 (steel)	90-100 in.-lbs. (10-11 N•m)
All others (aluminum)	35-45 in.-lbs. (4-5 N•m)
Cylinder head (502 & 503)	140-160 in.-lbs. (16-18 N•m)
Engine base	125-150 in.-lbs. (14-17 N•m)
Flywheel:	
E65 & GK590	250-300 in.-lbs. (28-34 N•m)
All others	375-400 in.-lbs. (42-45 N•m)
Muffler	40-60 in.-lbs. (4-7 N•m)
Spark plug	275-300 in.-lbs. (31-34 N•m)
Stator plate to bearing plate	50-60 in.-lbs. (5-7 N•m)

BEARING PLATE. Because of the pressure and vacuum pulsations in crankcase, bearing plate and gasket must form an airtight seal when installed. Inspect bearing plate gasket surface for cracks, nicks or warpage. Ensure oil passages in crankcase, gasket and plate are aligned and that correct number and thickness of thrust washers are used. Also make certain cap screws of the correct length are installed when engine is reassembled.

CAUTION: Cap screws (1—Fig. C17) may bottom in threaded holes (2) if incorrect screws are used. When long screws are tightened, damage to cylinder walls can result.

FLYWHEEL. On vertical shaft engines, both a lightweight aluminum flywheel and a cast iron flywheel are available. The lightweight aluminum flywheel should be used only on rotary lawnmower engines where the blade is attached directly to the engine crankshaft. The cast iron flywheel should be used on engines with belt pulley drive, etc.

Some kart engine owners will prefer using the lightweight aluminum flywheel due to the increase in accelerating performance.

CONNECTING ROD. Connecting rod and piston unit can be removed after removing the reed plate and crankshaft.

On E-65 and GK590 models (karting engines), and on 502 and 503 models, needle roller bearings are used at crankpin end of connecting rod. Renew crankshaft and/or connecting rod if bearing surfaces are scored or rough. When reassembling, place the thirteen needle rollers between connecting rod and crankpin. Using a low melting point grease, stick the remaining twelve needle rollers to connecting rod cap. Install

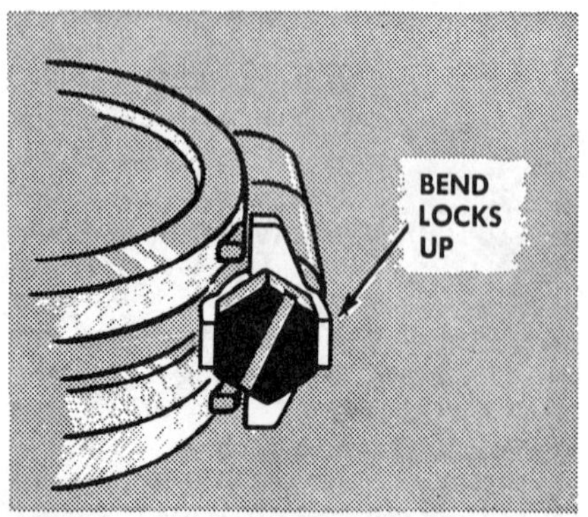

Fig. CL19—Bend connecting rod cap retaining screw locks up as shown.

Fig. CL20—View showing method of installing ring locking wires on three-ring pistons and placement of rings on two- and three-ring pistons. Some two-ring pistons do not have ring locking pins.

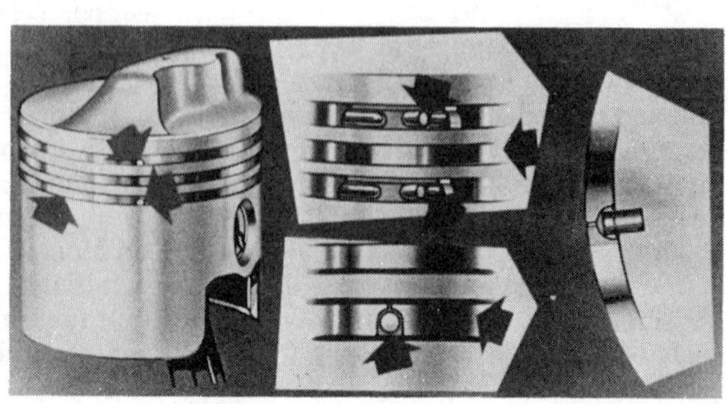

cap with the embossments (see A – Fig. CL18) on cap and rod properly aligned.

On all other models, an aluminum connecting rod with bronze bearing surfaces is used. Standard diameter of connecting rod crankpin bearing bore (plain bearing) is 0.7820-0.7827 inch (19.863-19.881 mm).

Clearance between connecting rod and crankpin journal should be 0.0026-0.0040 inch (0.07-0.10 mm). Renew connecting rod and/or crankshaft if clearance exceeds 0.0055 inch (0.14 mm).

Clearance between piston pin and connecting rod piston pin bore should be 0.0004-0.0011 inch (0.010-0.028 mm). Renew connecting rod and/or piston pin if clearance exceeds 0.002 inch (0.05 mm).

Reinstall cap to connecting rod with embossments (A – Fig. CL18) aligned and carefully tighten screws to recommended specification. Bend the locking tabs against screw heads as shown in Fig. CL19.

PISTON, PIN AND RINGS. Pistons may be equipped with either two or three compression rings. On three-ring pistons, a locking wire is fitted in a small groove behind each piston ring to prevent ring rotation on the piston. To install the three locking wires, refer to Fig. CL20 and hold piston with top up and intake side of piston to right. Install top and bottom locking wires with locking tab (end of wire ring bent outward) to right of locating hole in ring groove. Install center locking wire with locking tab to left of locating hole.

Piston and rings are available in several oversizes as well as standard size. Piston pin is available in standard size only.

When installing pistons with ring locating pins or locking wires, be sure end gaps of rings are properly located over the pins or the locking tabs and install piston using a ring compressor. Refer to Fig. CL20.

Be sure to reinstall piston and connecting rod assembly with exhaust side of

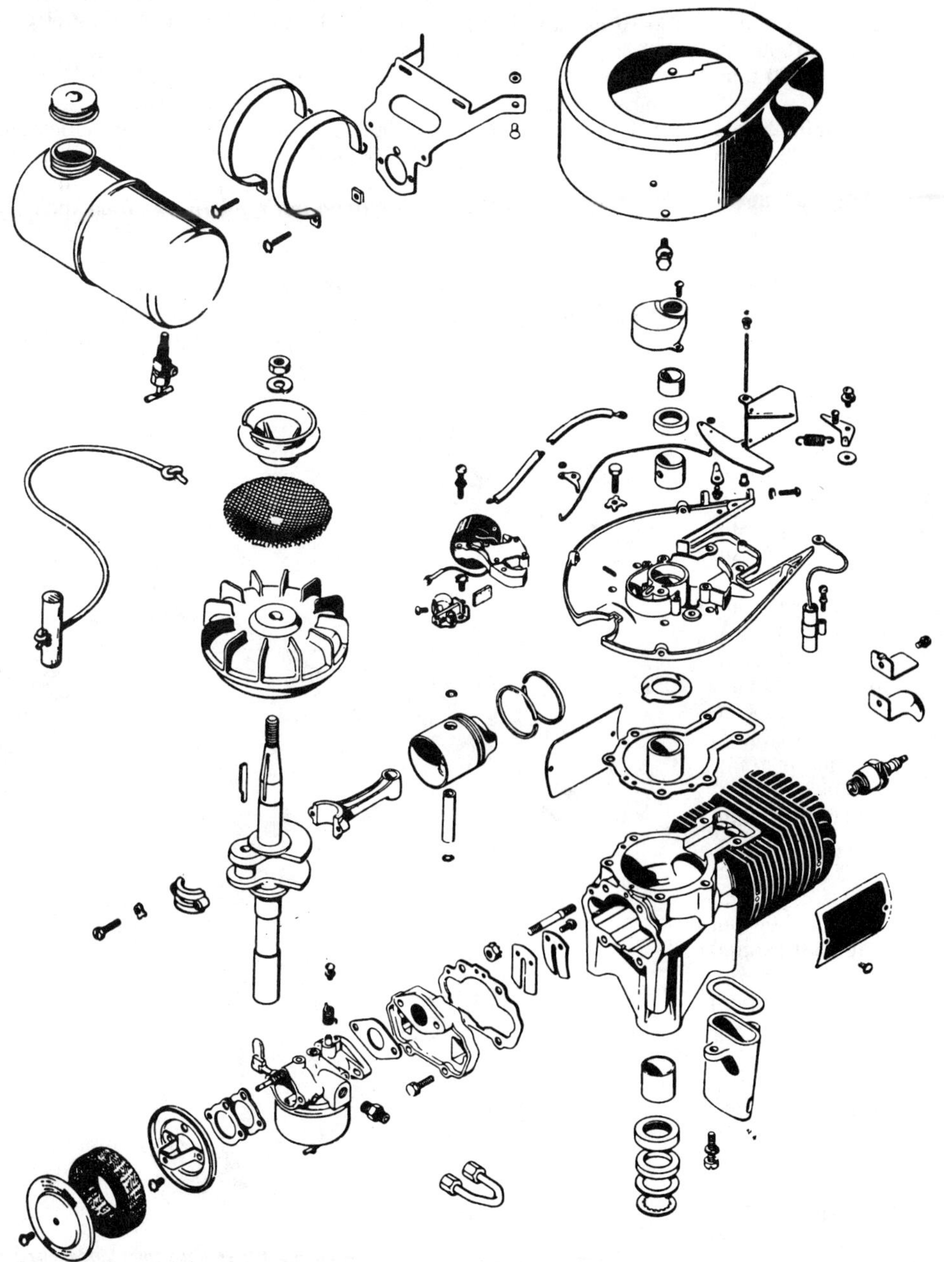

Fig. CL21—Exploded view of typical vertical shaft two-stroke engine. Model 503 differs in type of starter (see Fig. CL102—ACCESSORIES), use of an eight-petal rossette type reed valve, a detachable cylinder head and has ball and roller bearings to support crankshaft with needle bearing at crankpin.

piston (long sloping side of piston dome) towards exhaust ports in cylinder.

Check piston, pin and rings against the following specifications:

Specifications For 1.875 in. (47.6 mm) Pistons

Ring end gap:
Standard . . . . 0.005-0.013 in. (0.13-0.33 mm)
Maximum . . . . 0.020 in. (0.51 mm)
Ring side clearance:
Standard . . . . 0.0015-0.0040 in. (0.038-0.102 mm)
Maximum . . . . 0.006 in. (0.15 mm)
Piston skirt clearance:
Standard . . . . 0.0045-0.0065 in. (0.114-0.165 mm)
Maximum . . . . 0.008 in. (0.20 mm)
Piston pin diameter . . . . 0.4294-0.4296 in. (10.899-10.904 mm)
Piston pin bore diameter . . . . 0.4295-0.4298 in. (10.901-10.909 mm)

Specifications For 2.125 in. (53.2 mm) Piston

Ring end gap (two-ring piston):
Standard . . . . 0.007-0.017 in. (0.18-0.43 mm)
Maximum . . . . 0.025 in. (0.64 mm)
Ring end gap (three-ring piston):
Standard . . . . 0.010-0.015 in. (0.25-0.38 mm)
Maximum . . . . 0.017 in. (0.43 mm)
Ring side clearance (two-ring piston):
Standard . . . . 0.0015-0.0040 in. (0.038-0.102 mm)
Maximum . . . . 0.006 in. (0.15 mm)
Ring side clearance (three-ring piston):
Standard . . . . 0.002-0.004 in. (0.05-0.10 mm)
Maximum . . . . 0.0055 in. (0.140 mm)
Piston skirt clearance (two-ring piston):
Standard . . . . 0.005-0.007 in. (0.13-0.18 mm)
Maximum . . . . 0.008 in. (0.20 mm)
Piston skirt clearance (three-ring piston):
Standard . . . . 0.0045-0.0050 in. (0.114-0.127 mm)
Maximum . . . . 0.007 in. (0.18 mm)
Piston pin diameter . . . . 0.4999-0.5001 in. (12.698-12.703)
Piston pin bore diameter . . . . . . 0.5000-0.5003 in. (0.127-0.135 mm)

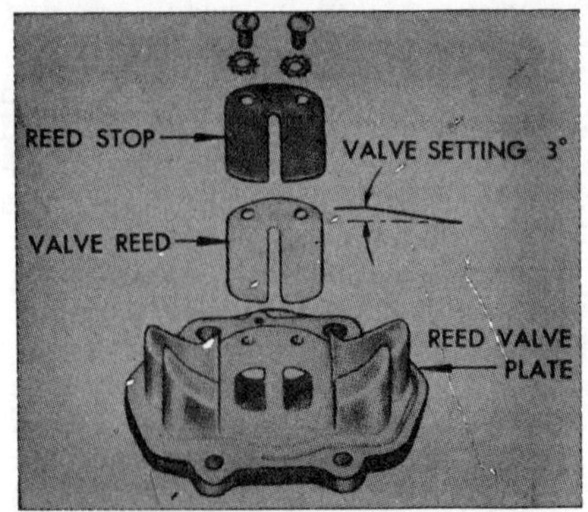

Fig. CL22—Details of two-petal reed inlet valve used on some models. Eight-petal rosette style reed valve is also used. See Fig. CL23.

Specifications For 2.375 in. (60.3 mm) & 2.500 in. (63.5 mm) Pistons

Ring end gap:
Standard . . . . 0.007-0.017 in. (0.18-0.43 mm)
Maximum . . . . 0.025 in. (0.64 mm)
Ring side clearance:
Standard . . . . 0.0015-0.0040 in. (0.038-0.102 mm)
Maximum . . . . 0.006 in. (0.15 mm)
Piston skirt clearance:
Standard . . . . 0.005-0.007 in. (0.13-0.18 mm)
Maximum . . . . 0.008 in. (0.20 mm)

CYLINDER AND CRANKCASE. The one-piece aluminum alloy cylinder and crankcase unit is integrally die-cast around a cast iron sleeve (cylinder liner). Cylinder can be rebored if worn beyond specifications. Standard cylinder bore diameter for models with 1.875 inch (47.63 mm) bore is 1.875-1.876 inches (47.63-57.65 mm); with 2.125 inch (53.98 mm) bore is 2.125-2.126 inches (53.98-54.00 mm); with 2.375 inch (60.33 mm) bore is 2.375-2.376 inches (60.33-60.35 mm) and with 2.500 inch (63.50 mm) bore is 2.500-2.501 inches (63.50-63.53 mm).

Refer to PISTON, PIN AND RINGS paragraphs for piston-to-cylinder clearance specifications.

Standard main bearing diameters (plain bushing type) should be 0.7517-0.7525 inch (19.093-19.114 mm) at flywheel end and either 0.877-0.878 inch (22.276-22.301 mm) or 1.002-1.003 inch (25.451-25.476 mm) at pto end. Standard main bearing clearance should be 0.0015-0.0030 inch (0.038-0.076 mm)

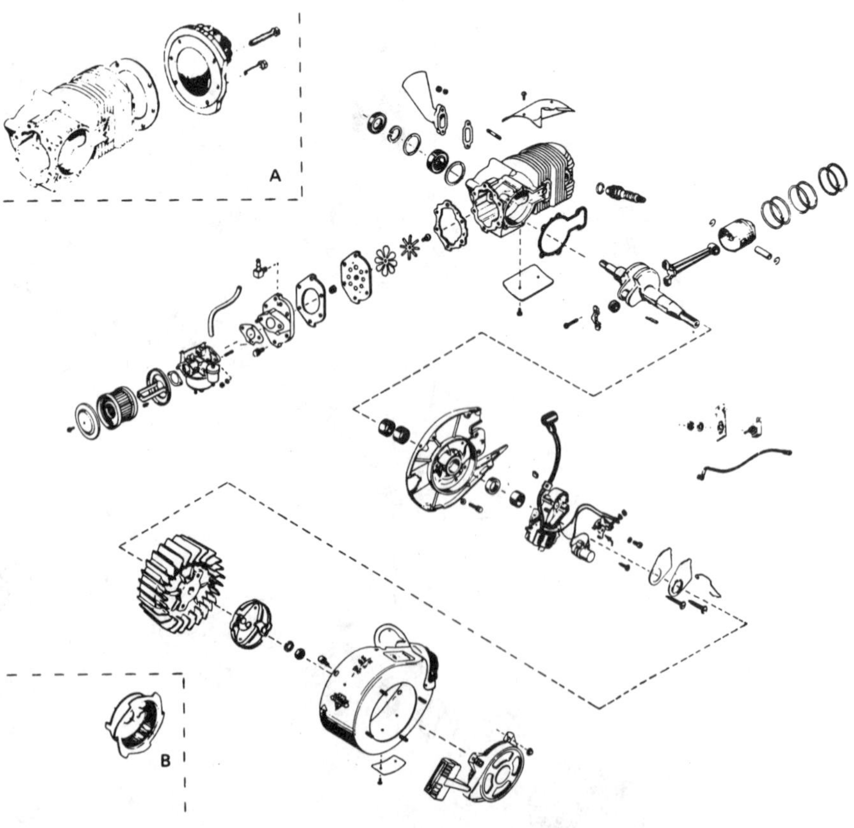

Fig. CL23—Exploded view of E-65 kart engine. Differences from Model 502 are minor: Cylinder head is detachable as in inset "A"; piston is fitted with two rings instead of three; muffler, starter cup (inset "B") and carburetor are of different design. See Fig. CL13 for carburetor used on 502-0308-000 engines. Model 502-0309-000 engines are equipped with float type carburetor.

and crankcase should be renewed if worn beyond specification or if out-of-round 0.0015 inch (0.038 mm) or more.

Main bearing bushings, bushing remover and driver, reamers and reamer alignment plate are available through Clinton parts sources for renewing bushings in crankcase and bearing plate. On vertical crankshaft engines having two bushings in the crankcase, the outer bushing must be removed towards inside of crankcase.

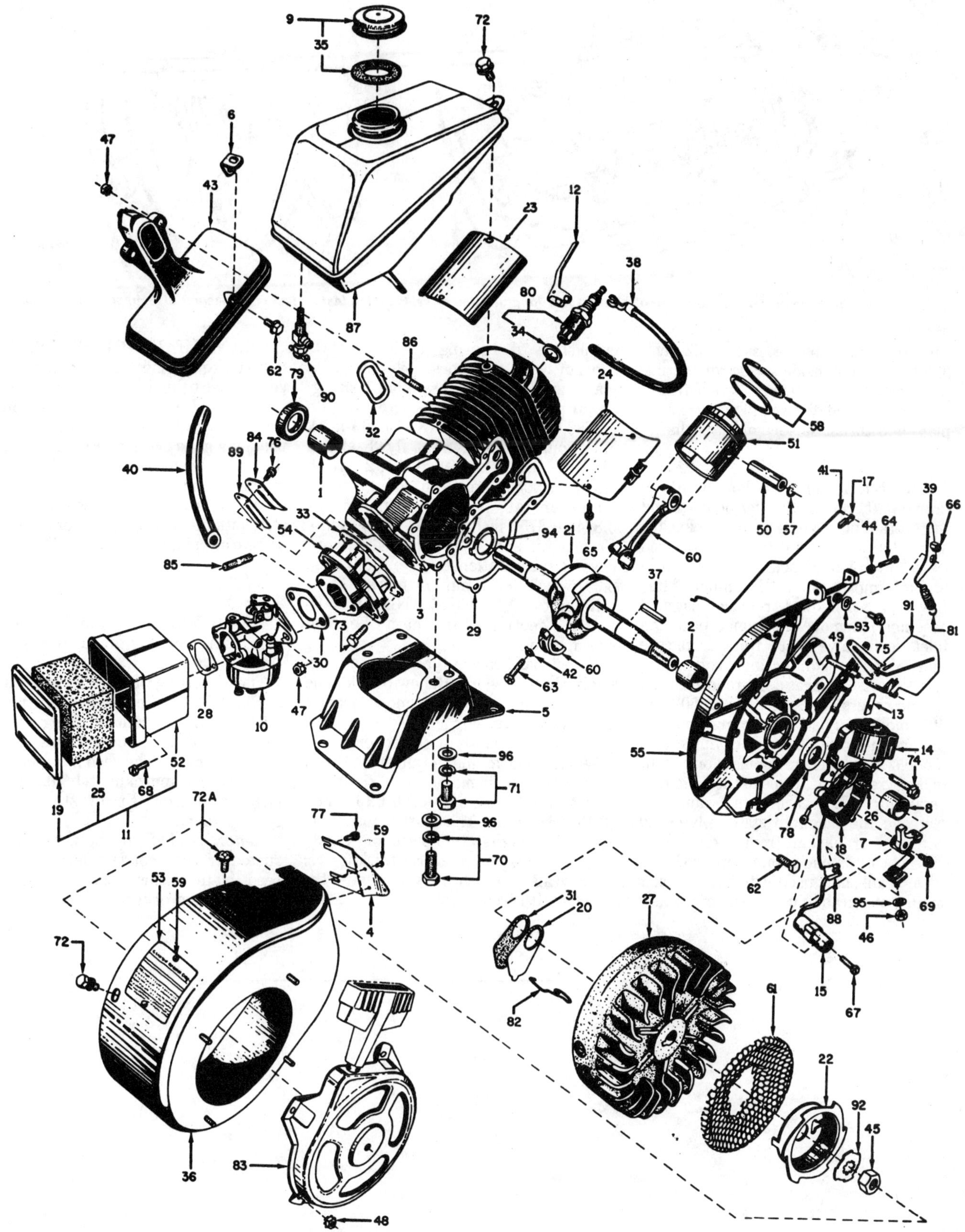

Fig. CL24—Exploded view of typical horizontal crankshaft engine.

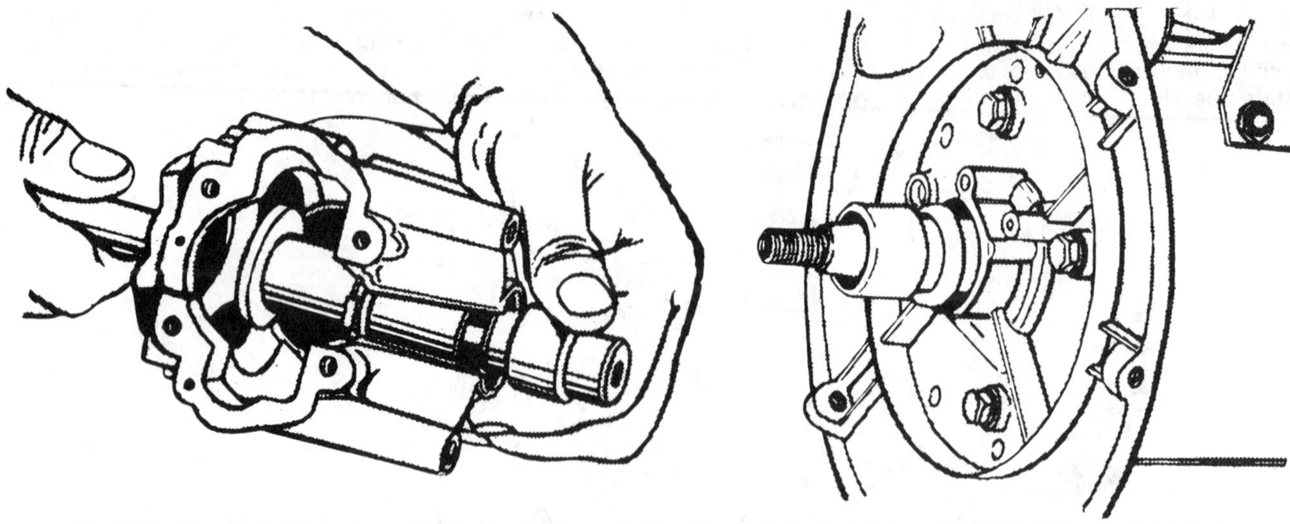

Fig. CL25—Examples of typical use of oil seal loaders for protection of seals during installation of crankshaft. Use ample lubrication.

Bushings must be recessed 1/32 inch (0.79 mm) from inside surface of bearing plate or crankcase except when thrust washer is used between crankcase or plate and thrust surface of crankshaft.

CRANKSHAFT. To remove crankshaft, remove blower housing and nut retaining flywheel to crankshaft. Thread an impact nut to within 1/8 inch (3.18 mm) of flywheel. Pull on flywheel and tap impact nut with hammer. After flywheel is removed, remove magneto assembly. Remove the engine base and inlet reed valve plate from crankcase and detach connecting rod from crankshaft. Be careful not to lose the 25 needle rollers on E-65, GK-590, 502 and 503 models. Push connecting rod and piston unit up against top of cylinder. On models with ball bearing main on output end of crankshaft, remove the snap ring retaining ball bearing in crankcase. Remove the magneto stator plate and withdraw crankshaft from engine taking care not to damage connecting rod.

On crankshaft used with needle roller main or crankpin bearings, renew crankshaft if bearing surface shows signs of wear, roughness or scoring. On mains and/or cast-in crankpin bearing, crankpin diameter should be 0.7788-0.7795 inch (19.782-19.799 mm). Main journal diameter at flywheel end should be 0.7495-0.7502 inch (19.037-19.055 mm) and main journal diameter at pto end should be 0.8745-0.8752 inch (22.212-22.230 mm). Crankshaft should be renewed if worn beyond specifications or if out-of-round 0.0015 inch (0.038 mm) or more.

Refer to CYLINDER AND CRANKCASE section for main bearing specifications.

Renew ball bearing type mains if bearing is rough or worn. Renew needle type main bearings if one or more needles show any defect or if needles can be separated the width of one roller.

When reinstalling crankshaft with bushing or needle mains, crankshaft end play should be 0.005-0.020 inch (0.13-0.51 mm). The gasket used between bearing (stator) plate and crankcase is available in several thicknesses. It may also be necessary to renew the crankshaft thrust washer.

CRANKSHAFT OIL SEALS. On all two-stroke engines, the crankshaft oil seals must be maintained in good condition to hold crankcase compression on the downward stroke of the piston. It is usually a good service practice to renew the seals whenever overhauling an engine. Apply a small amount of gasket sealer to the outer rim of the seal and install with lip towards inside of crankcase or bearing plate.

REED VALVE. Either a dual reed (Fig. CL21) or eight-petal rosette reed (Fig. CL23) is used.

The 3° setting of the dual reed shown in Fig. CL22 is the design of the reed as stamped in the manufacturing process. The 3° bend should be towards the reed seating surface. The reed stop should be adjusted to approximately 0.280 inch (7.11 mm) from tip of stop to reed seating surface.

Renew reed if any petal is cracked, rusted or does not lay flat against the reed plate. Renew the reed plate if rusted, pitted or worn.

CLINTON

Horizontal Crankshaft Engines

Model	Bore	Stroke	Displacement
300, A-300, 350	2.0 in. (50.8 mm)	1.5 in. (38.1 mm)	4.7 cu. in. (77 cc)

Vertical Crankshaft Engines

Model	Bore	Stroke	Displacement
VS-300	2.0 in. (50.8 mm)	1.5 in. (38.1 mm)	4.7 cu. in. (77 cc)

ENGINE IDENTIFICATION

All models are four-stroke, single cylinder air-cooled engines with valves located in cylinder block. Refer to CLINTON ENGINE IDENTIFICATION INFORMATION section preceding Clinton service sections of this manual.

MAINTENANCE

SPARK PLUG. All models use a 14 mm, 3/8-inch (9.53 mm) reach spark plug. Recommended plug is a Champion J8 or equivalent. Electrode gap for all models is set at 0.025-0.028 inch (0.64-0.71 mm).

When installing spark plug, apply graphite to threads and tighten to 275-300 in.-lbs. (31-34 N·m).

CARBURETOR. Clinton suction type carburetors, numbers 7100, 7120, 7080-1 and 7080-2 or Carter Model N float type carburetors are used. Refer to appropriate following paragraphs.

Clinton Suction Type Carburetor. Refer to Fig. CL30, Fig. CL31 and Fig. CL32 for identification and exploded views of Clinton suction type carburetors. Carburetor numbers 7100, 7120 and 7080 are no longer available. If renewal is necessary, carburetor number 7080-2, along with certain other parts, must be used.

Carburetor is equipped with only one fuel mixture adjustment needle (7–Figs. CL30, CL31, and CL32). Initial adjustment of needle is 3½ turns open from a lightly seated position. Final adjustments are made with engine running at operating temperature with fuel tank approximately ½ full. Adjust mixture needle for smoothest engine performance and acceleration.

Carter Float Type Carburetor. Refer to Fig. CL33 for identification and exploded view of Carter float type carburetor. Carter Model N-705S is shown; Models N-2019S and N-2147S are similar except for shape of float bowl and float.

Initial adjustment of idle (1) and main (4) fuel mixture needles from a lightly seated position is 1½ turns open for each needle. Final adjustments are made with engine at operating temperature and running. Operate engine at rated speed and adjust main fuel mixture needle for smoothest engine operation. Operate at idle speed and adjust idle fuel mixture needle for smoothest engine idle.

To check float level, invert carburetor throttle body. Measure from free side of float to float bowl mating surface of carburetor. Measurement should be 13/64 inch (5.16 mm) for Model N-705S or 11/64 inch (4.37 mm) for Models N-2019S and N-2147S. Carefully bend float lever tang which contacts fuel inlet needle if necessary to obtain correct float level.

GOVERNOR. An air vane type governor is used on all models. Recommended maximum governed speed is 3300 rpm. Do not set maximum governed speed above 3600 rpm.

Make sure governor linkage does not bind when moved through full range of travel. The blower housing should be clean and free of dents which could cause governor air vane to bind.

On models without adjustable speed control, renew governor spring (90–Fig. CL40) and link (46) to obtain correct maximum governed speed.

MAGNETO AND TIMING. Repco flywheel type magnetos are used on all

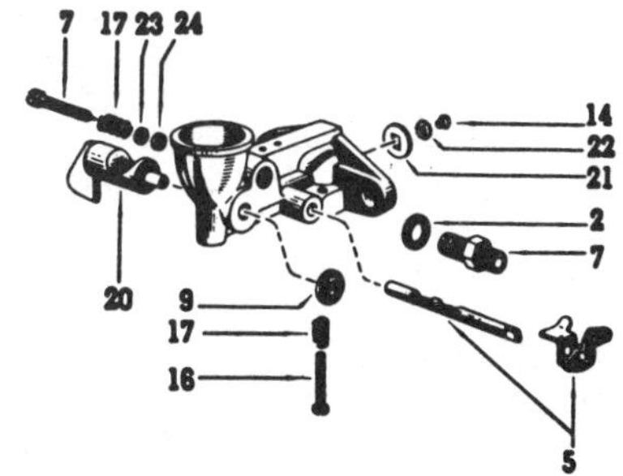

Fig. CL30—View of early production suction lift carburetor. Choke valve (20) or complete carburetor are no longer available. Only one fuel adjustment needle (7) is used.

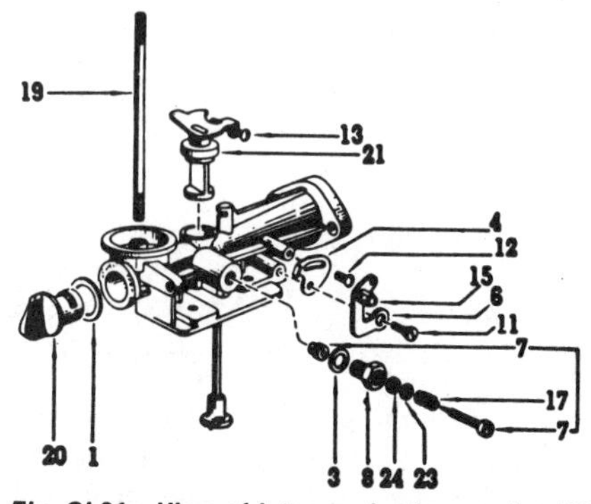

Fig. CL31—View of later production suction lift carburetor. See Fig. CL32 also.

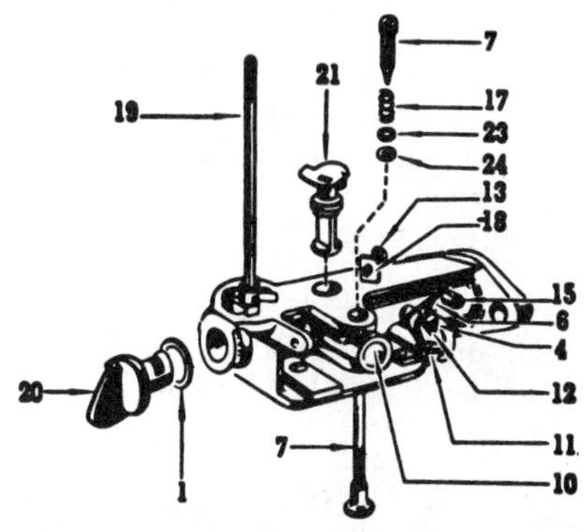

Fig. CL32—View of later production suction lift carburetor. See Fig. CL31 also.

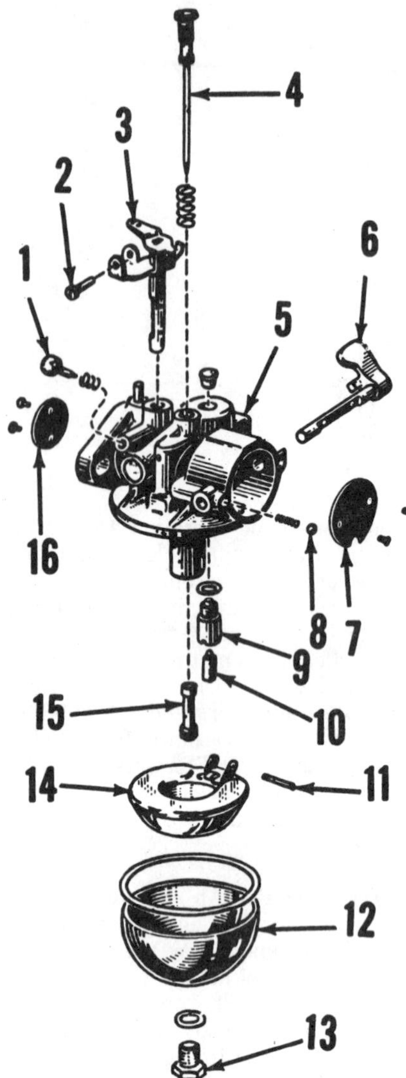

Fig. CL33—Exploded view of typical Carter Model N carburetor. Design of float and float bowl may vary from that shown.

1. Idle fuel needle
2. Idle stop screw
3. Throttle shaft
4. Main fuel needle
5. Carburetor body
6. Choke shaft
7. Choke plate
8. Detent ball
9. Inlet valve seat
10. Inlet needle
11. Float pin
12. Float bowl
13. Retainer
14. Float
15. Main nozzle
16. Throttle plate

models. Breaker points and condenser are located behind flywheel. Adjust maximum breaker point gap to 0.018-0.021 inch (0.46-0.53 mm). Magneto condition may be considered satisfactory if it will fire an 18 mm spark plug with electrode gap set to 0.156-0.187 inch (3.97-4.76 mm). See Fig. CL36.

Ignition timing is fixed and nonadjustable at 21° BTDC. Armature air gap should be 0.007-0.017 inch (0.18-0.43 mm).

LUBRICATION. Manufacturer recommends oil with an API service classification SE or SF. Use SAE 30 oil for temperatures above 32°F (0°C); SAE 10W oil for temperature range of −10° to 32°F (−23° to 0°C) and SAE 5W oil for temperatures below −10°F (−23°C).

On models equipped with reduction gearing, use SAE 30 oil in gearbox.

Fig. CL34—Suction lift carburetor adjustments. See text for procedure.

1. Choke
2. Idle speed adjustment
3. Mixture adjustment

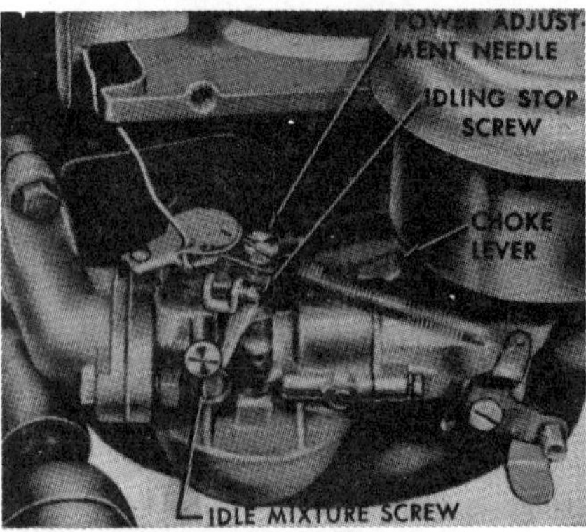

Fig. CL35—Carter Model N float type carburetor installation.

CRANKCASE BREATHER. Crankcase breather assembly (8–Fig. CL39 or 5–Fig. CL40) should be removed and cleaned if difficulty is experienced with oil loss through hole in valve cover and whenever the engine is overhauled. Be sure breather is correctly reassembled and reinstalled.

REPAIRS

TIGHTENING TORQUES. Recommended tightening torque specifications are as follows:

Base bolts:
- Model A300 . . . . 150-160 in.-lbs. (17-18 N•m)
- All others . . . . 325-375 in.-lbs. (37-42 N•m)

Bearing plate:
- Model A300 . . . . 120-150 in.-lbs. (14-17 N•m)
- All others . . . . 140-160 in.-lbs. (16-18 N•m)

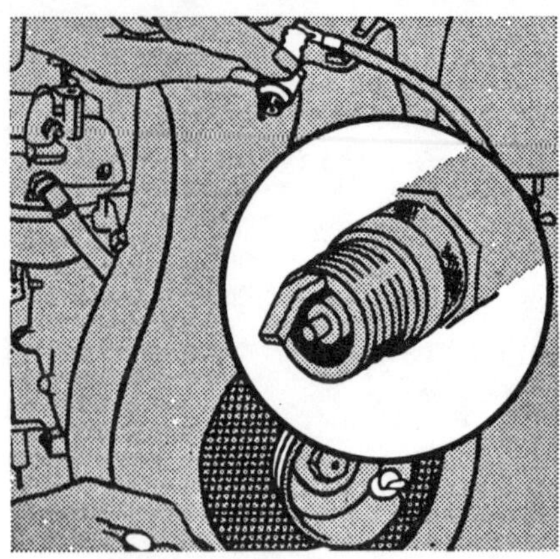

Fig. CL36—Magneto may be considered satisfactory it if will fire an 18 mm spark plug with the electrode gap set a 0.156-0.187 inch (3.97-4.76 mm).

Carburetor to manifold . . . . 35-50 in.-lbs. (4-5 N•m)

Manifold to block . . . . 60-65 in.-lbs. (6.7-7.3 N•m)

End cover or gearbox . . . . 120-150 in.-lbs. (14-17 N•m)

Flywheel:
- Model A300 . . . . 375-400 in.-lbs. (42-45 N•m)
- All others . . . . 400-450 in.-lbs. (45-52 N•m)

Pto housing:
- Models 300, 350 & VS300 . . . . 120-150 in.-lbs. (14-17 N•m)

Spark plug . . . . 275-300 in.-lbs. (31-34 N•m)

Speed reducer mount . . . 110-150 in.-lbs. (12-17 N•m)

CONNECTING ROD. Connecting rod and piston assembly can be removed from engine after cylinder head and engine base (on horizontal crankshaft models) or crankcase end cover (on vertical crankshaft models) are removed.

Recommended clearances are as follows:

Connecting rod to crankshaft:
- Standard . . . . 0.0018-0.0035 in. (0.046-0.089 mm)
- Maximum . . . . 0.0045 in. (0.114 mm)

Connecting rod to piston pin:
- Standard . . . . 0.0004-0.0011 in. (0.010-0.028 mm)
- Maximum . . . . 0.002 in. (0.05 mm)

Connecting rod side play (maximum) . . . . 0.005-0.020 in. (0.13-0.51 mm)

Connecting rod is available in standard size only. When reassembling, be sure embossments on connecting rod

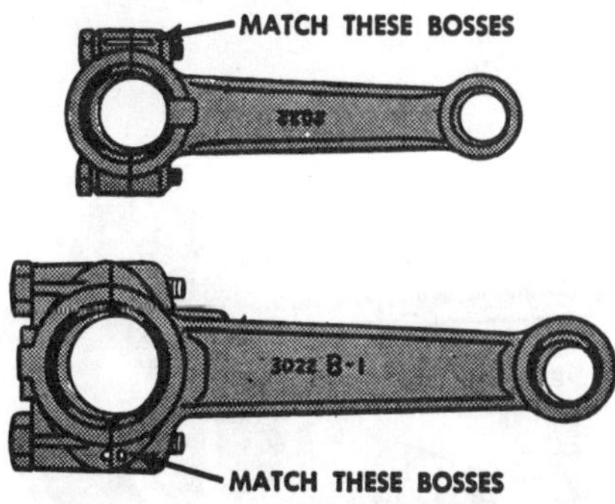

Fig. CL37—When assembling cap to connecting rod, be sure embossments on rod and cap are aligned as shown.

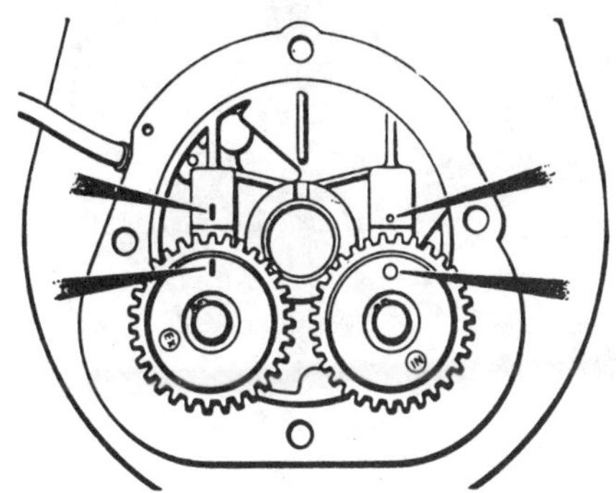

Fig. CL38—Valves are in time when piston is at top dead center and marks on cam gears are aligned as shown. Refer to text for suggested methods of reassembly.

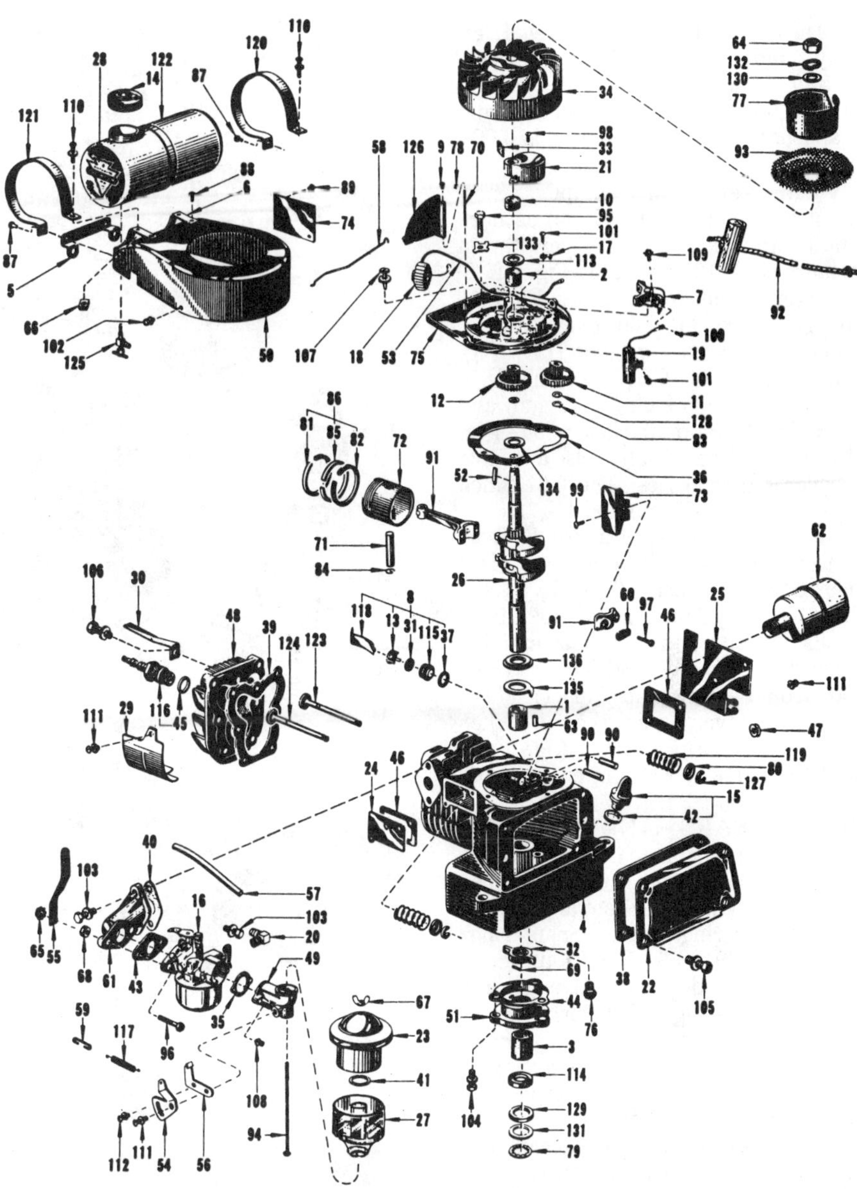

Fig. CL39—Exploded view of Model VS-300 engine. Oil pump impeller drive pin (69) must be removed as outlined in text before crankshaft can be removed from crankcase.

and cap are aligned as in Fig. CL37. Oil hole in connecting rod should face flywheel side of engine.

PISTON, PIN AND RINGS. Piston is equipped with two compression rings and one oil control ring. Recommended piston ring end gap is 0.007-0.017 inch (0.18-0.43 mm). Renew ring if end gap is 0.025 inch (0.64 mm) or more. Ring side clearance in groove should be 0.002-0.005 inch (0.05-0.13 mm). Maximum allowable ring side clearance is 0.006 inch (0.15 mm). Rings are available in several oversizes as well as standard size.

The piston pin is retained in piston with a snap ring at each end. Piston pin is available in standard size only and should be a hand push fit in piston. Piston pin diameter is 0.4999-0.5001 inch (12.698-12.703 mm) and pin bore diameter in connecting rod is 0.5005-0.5010 inch (12.713-12.725 mm).

Piston skirt clearance in cylinder bore should be 0.0045-0.0065 inch (0.114-0.165 mm). Maximum allowable skirt clearance is 0.008 inch (0.21 mm). Piston is available in several oversizes as well as standard size.

CYLINDER. Standard cylinder bore diameter is 2.000-2.001 inch (50.80-50.83 mm). If piston skirt clearance is 0.008 inch (0.21 mm) or more with new piston, or ring end gap is 0.025 inch (0.64 mm) or more with new rings, renew cylinder crankcase or recondition cylinder bore.

CRANKSHAFT. VS-300 series engines are fitted with needle roller bearings at flywheel end of crankshaft and at lower pto journal. Upper pto journal is fitted with a bushing type bearing. Main bearings on other engines are plain bushings.

Renew needle roller main bearings if any needle is pitted or has flat spot, or if needles can be separated the width of one roller.

Crankshaft main bearing journal diameters at flywheel side for all models should be 0.7483-0.7490 inch (19.007-19.025 mm). Pto side bearing journal diameter should be 0.8745-0.8752 inch (22.21-22.23 mm) for VS-300 models and 0.7483-0.7490 inch (19.007-19.025 mm) for all other models.

Clearance between main bearing journals and main bearing bores should be 0.0018-0.0035 inch (0.05-0.09 mm); maximum allowable clearance is 0.005 inch (0.12 mm). Main bearing bushings are available for service and must be reamed after installation. Bushing drivers, reamers and reamer alignment plates are available through Clinton parts sources.

Clearance between connecting rod bearing bore and crankpin journal should be 0.0018-0.0035 inch (0.05-0.09 mm). Maximum allowable clearance is

0.0045 inch (0.11 mm). Crankshaft end play should be 0.004-0.012 inch (0.10-0.30 mm). Maximum allowable end play is 0.020 inch (0.051 mm). Renew crankshaft if crankpin journal is 0.0015 inch (0.038 mm) out-of-round.

When overhauling engine, renew thrust washer (104-Fig. CL40) on horizontal crankshaft engines or thrust washers (134, 135 and 136 – Fig. CL39) on vertical crankshaft engines. For controlling crankshaft end play, the flywheel end thrust washer (134 – Fig. CL39 or 104 – Fig. CL40) is available in oversizes of 0.005 inch (0.127 mm) and 0.010 inch (0.254 mm). Gasket (36 – Fig. CL39 or 32 – Fig. CL40) is also available in several thicknesses. The 0.015 inch (0.38 mm) gasket is standard.

Prior to removing crankshaft from vertical crankshaft engines, first remove the pto housing (51 – Fig. CL39). Slide the oil pump impeller (32) from crankshaft and remove the pump impeller drive pin (69). Withdraw crankshaft from bearing plate (upper) side of crankcase. Reverse this procedure when installing crankshaft. Also see LUBRICATING SYSTEM paragraph.

CAM GEARS AND BEARING PLATE. Two cam gears are used instead of a conventional camshaft. The gears turn on axle pins that are pressed into the bearing plate. Snap rings (83 – Fig. CL39 or 63 – Fig. CL40) retain gears on the axle pins. Wave (spring) washers (128 – Fig. CL39 or 99 – Fig. CL40) are used between the snap rings and gears to eliminate all end play of cam gears on axle pins. Renew the wave washers if worn or flattened.

Renew the bearing plate and cam axle pins if pins are loose in plate. If axle pins have worked out of plate towards inside of crankcase, but are reasonably tight, press them back into the plate and drill holes through the bearing plate and axle pins so retaining pins may be installed. Note: Drill holes to size of pins available. Retaining pins are not available as a service part.

Valves are properly timed when the cam gears are aligned with marks on the bearing plate as shown in Fig. CL38 and the gears meshed with the crankshaft gear with the crankpin in TDC position. Note the "EX" and "IN" on the gears indicating the exhaust valve cam gear and intake valve cam gear. Do not reverse gears from position shown in Fig. CL38.

If engine has been completely disassembled, reinstall the bearing plate, cam gears and crankshaft as a unit. Be sure the two ridges on the thrust washer (134 – Fig. CL39 or 104 – Fig. CL40) engage tooth slots on the crankshaft gear so the washer will turn with the crankshaft.

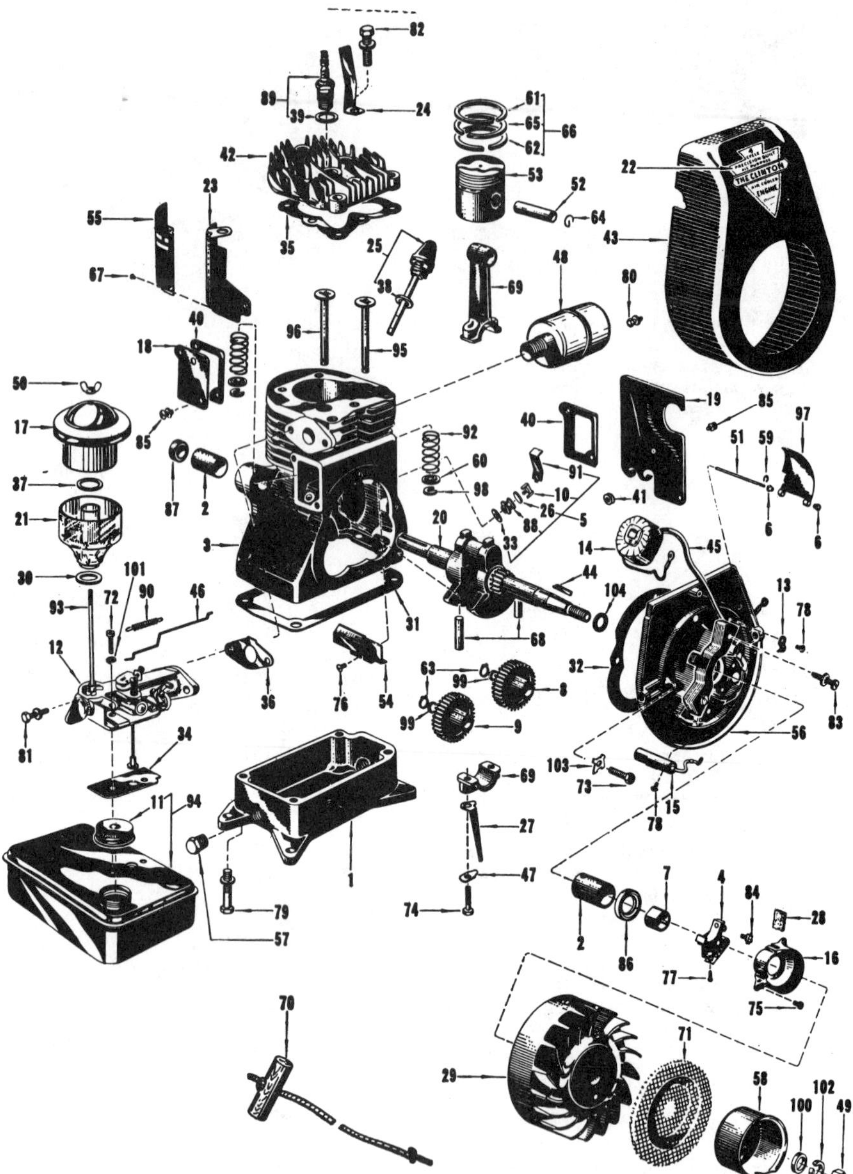

Fig. CL40—Exploded view of Model 350 engine. Models 300 and A-300 are similar.

If the bearing plate and cam gear assembly are removed without removing the crankshaft, connecting rod and piston assembly, reinstall as follows: If valves have not been removed, remove cylinder head, push the valves open and hold them in open position with end wrenches placed between the valve heads and the cylinder block. Be sure the valve tappets (90 – Fig. CL39 or 68 – CL40) are in place. Turn engine so piston is at top dead center and align cam gears with marks on bearing plate. While holding the thrust washer (134 – Fig. CL39 or 104 – Fig. CL40) in place against bearing plate, slide the assembly over the crankshaft. The cam gears and crankshaft gear can be meshed by working through the cover plate (22 – Fig. CL39) or base plate (1 – Fig. CL40) opening. Be sure the thrust washer ridges engage the crankshaft gear before tightening bearing plate to crankcase. Check valve timing after tightening bearing plate and removing the wrenches from between valve heads and cylinder block.

VALVE SYSTEM. Recommended valve tappet gap is 0.007-0.009 inch (0.18-0.23 mm) for either intake or exhaust valve. Adjust valve clearance as follows: To increase valve clearance, grind end of valve stem. To decrease valve clearance, renew valve and/or grind valve seats deeper. Valve face angle should be ground to 45°. Valve seat angle should be ground between 43½° and 44½°. Recommended seat width is 0.030-0.045 inch (0.80-1.03 mm). Maximum seat width is 0.060 inch (1.52 mm).

Recommended valve stem-to-guide clearance is 0.002-0.0045 inch (0.051-0.114 mm). Ream guides to 0.2813 inch (7.144 mm). Install valves with oversize stems if clearance is 0.006 inch (0.152 mm) or more. Install the "C" type valve keepers with sharp edge of keepers towards stem end of valves.

Valves are actuated by pin type tappets. The tappets may be removed and installed through the valve guides if valves are removed or through the crankcase openings if bearing plate and cam gears are removed.

LUBRICATION SYSTEM. Horizontal crankshaft engines are splash lubricated by an oil distributor (27 – Fig. CL40) attached to the connecting rod cap.

Vertical crankshaft engines are lubricated by an oil pump. The pump impeller (32 – Fig. CL39) is driven by a pin (69) in the engine crankshaft. Oil is sprayed up into the engine through a brass nozzle (63). Renew the nozzle if it is damaged or drilled oversize.

To remove the crankshaft on vertical shaft engines, first remove the pto housing (51), impeller (32) and impeller drive pin (69) from the bottom of the crankcase. Scribe a mark on the pto housing and the bottom of the crankcase so the housing may be reinstalled in original position. The oil return hole in the housing must be aligned with the oil return hole in the crankcase.

CLINTON

Horizontal Crankshaft Engines

Model	Bore	Stroke	Displacement
500, 650, 700-A, B-700, C-700, 350	2.000 in. (50.8 mm)	1.875 in. (47.6 mm)	5.9 cu. in. (97 cc)
D-700	2.125 in. (54.0 mm)	1.875 in. 47.6 mm)	6.7 cu. in. (109 cc)
800, A-800, 900, A-1100, B-1100, C-1100, D-1100	2.375 in. (60.3 mm)	1.875 in. (47.6 mm)	8.3 cu. in. (136 cc)
1200, A-1200, B-1200	2.469 in. (62.7 mm)	2.125 in. (54.0 mm)	10.2 cu. in. (167 cc)
494-0000-000, 494-0101-000	2.375 in. (60.3 mm)	1.875 in. (47.6 mm)	8.3 cu. in. (136 cc)
498-0000-000, 498-0301-000	2.469 in. (62.7 mm)	2.125 in. (54.0 mm)	10.2 cu. in. (167 cc)

Vertical Crankshaft Engines

Model	Bore	Stroke	Displacement
VS-700, VS750	2.000 in. (50.8 mm)	1.875 in. (47.6 mm)	5.9 cu. in. (97 cc)
VS-800, VS-900, V-1000, VS-1000	2.375 in. (60.3 mm)	1.875 in. (47.6 mm)	8.3 cu. in. (136 cc)
V-1100, VS-1100	2.375 in. (60.3 mm)	2.125 in. (54.0 mm)	9.5 cu. in. (154 cc)
V-1200, VS-1200	2.469 in. (62.7 mm)	2.125 in. (54.0 mm)	10.2 cu. in. (167 cc)
497-0000-000, 499-0000-000	2.469 in. (62.7 mm)	2.125 in. (54.0 mm)	10.2 cu. in. (167 cc)

ENGINE IDENTIFICATION

All models are four-stroke, single cylinder air-cooled engines with valves located in cylinder block.

Refer to CLINTON ENGINE IDENTIFICATION INFORMATION section preceding Clinton service sections of this manual.

MAINTENANCE

SPARK PLUG. All models use a 14 mm 3/8-inch (9.53 mm) reach spark plug. Recommended plug is a Champion J8 or equivalent. Set electrode gap to 0.025-0.028 inch (0.64-0.71 mm). When installing plug, apply graphite to threads and tighten to 275-300 in.-lbs. (31-34 N·m).

CARBURETOR. Several different carburetors, both suction lift and float

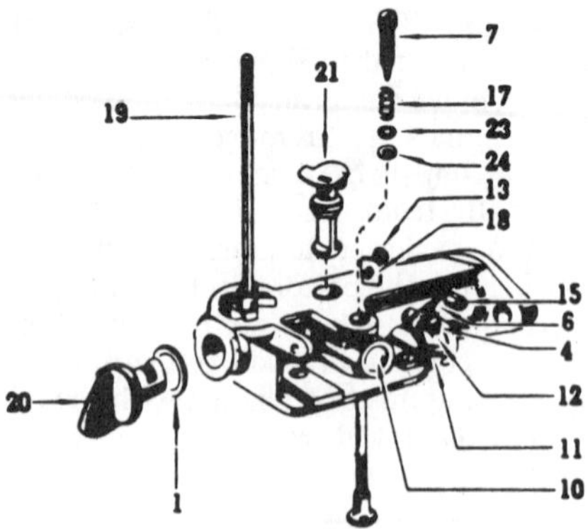

Fig. CL45—Exploded view of suction lift carburetor. Only one fuel mixture adjustment needle (7) is used.

type, have been used on this series of Clinton engines. Refer to the following paragraphs for information on each carburetor model.

The throttle shaft on some carburetors may have several holes in which the throttle link and governor backlash spring (if used) may be installed. On these carburetors, be sure to mark the holes in which spring and linkage were installed so they may be reinstalled correctly.

Clinton Suction Type Carburetors. Clinton Model 4730-2 and 4484-2 carburetors have been used. Refer to Fig. CL45 for identification and exploded view of Clinton suction type carburetors. Carburetors are equipped with only one fuel adjustment needle (7).

Initial adjustment of mixture screw from a lightly seated position is 3½ turns open. Final adjustments are made with engine at operating temperature with approximately ½ full fuel tank and running. Operate engine at rated speed and adjust fuel mixture needle for smoothest engine operation and acceleration.

Clinton Float Type Carburetor. Several different Clinton LMG and LMV series float type carburetors have been used. Identification number is located on the carburetor body as shown in Fig. CL46. Refer to Fig. CL47 for identification and exploded view of typical Clinton LM series carburetor.

Initial adjustment of idle (9) and main (22) fuel mixture needles from a lightly seated position is 1¼ turns open for each needle. Final adjustments are made with engine at operating temperature and running. Operate engine at rated speed and adjust main fuel mixture needle for smoothest engine operation. Operate engine at idle speed and adjust idle fuel mixture needle for smooth engine idle. If engine fails to accelerate smoothly, open main fuel needle slightly.

When overhauling or cleaning carburetor, do not remove the main fuel nozzle (13) unless necesary. If nozzle is removed, it must be discarded and a service type nozzle (Fig. CL48) installed.

Choke plate (11 – Fig. CL47) should be installed with the "W" or part number to the outside. Install throttle plate with the side marked "W" towards mounting flange and with the part number towards idle needle side of carburetor bore when plate is in closed position. When installing throttle plate, back idle speed adjustment screw out. Turn plate and throttle shaft to closed position and seat plate by gently tapping with small screwdriver before tightening plate retaining screws.

If either the float valve or seat is damaged, install a new matched valve and seat assembly (12) and tighten seat to 40-50 in.-lbs. (5-6 N·m).

To check float level, invert carburetor throttle body and float assembly and measure clearance between body casting and free side of float. Measurement should be 5/32 inch (3.97 mm). Adjust float level by carefully bending float lever tang which contacts fuel inlet valve if necessary to obtain correct float level. Float drop measured at free end of float should be 3/16 inch (4.76 mm). Adjust float drop by carefully bending tab that contacts the carburetor body.

Carter Float Type Carburetor. Refer to Fig. CL49 for identification and exploded view of Carter N series float type carburetor. Float and float bowl shape may vary from model to model.

Initial adjustment of idle (1) and main (4) fuel mixture needles from a lightly seated position is 1½ turns open for each mixture needle on Models N-705S, N-707S, N-707SA, N-2020S, N-2147S, N-2449S and N-2456S; 1 turn open for idle mixture needle and 2 turns open for main fuel mixture needle on Models N-2236S, N-2399S and N-2466S or 1 turn open for idle mixture needle and 1½ turns open for main fuel mixture needle for Models N-2246S and N-2459S.

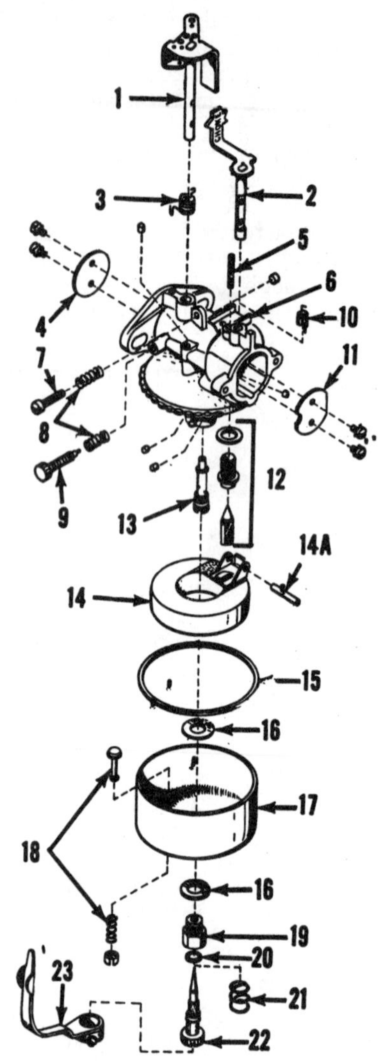

Fig. CL47—Exploded view of typical LMG series carburetor. LMV and LMB series are similar.

1. Throttle shaft
2. Choke shaft
3. Spring
4. Throttle plate
5. Spring
6. Carburetor body
7. Idle stop screw
8. Springs
9. Idle fuel needle
10. Spring
11. Choke plate
12. Inlet needle & seat
13. Main nozzle
14. Float
14A. Float
15. Gasket
16. Gaskets
17. Float bowl
18. Drain valve
19. Retainer
20. Seal
21. Spring
22. Main fuel needle
23. Lever (optional)

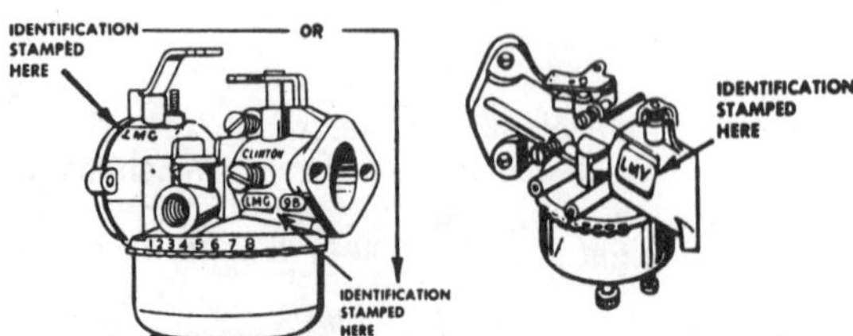

Fig. CL46—View showing location numbers on LMB, LMG and LMV carburetors. Identification number must be used when ordering service parts.

Fig. CL48—When original main fuel nozzle is removed from LM series carburetor, it must be discarded and service type nozzle shown installed.

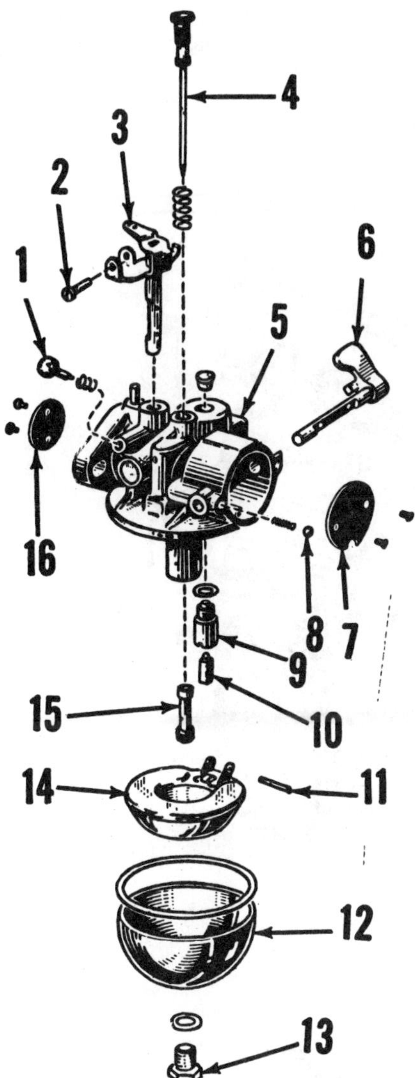

Fig. CL49—Exploded view of typical Carter Model N carburetor. Design of float and float bowl may vary from that shown.

1. Idle fuel needle
2. Idle stop screw
3. Throttle shaft
4. Main fuel needle
5. Carburetor body
6. Choke shaft
7. Choke plate
8. Detent ball
9. Inlet valve seat
10. Inlet needle
11. Float pin
12. Float bowl
13. Retainer
14. Float
15. Main nozzle
16. Throttle plate

Final adjustments on all models are made with engine at operating temperature and running. Operate engine at rated speed and adjust main fuel mixture needle for smoothest engine operation. Operate engine at idle speed and adjust idle mixture needle for smoothest engine idle.

To check float level, invert carburetor throttle body and float assembly and measure distance from free end of float to float bowl mating surface of carburetor. Measurement should be 13/64 inch (5.15 mm) for Models N-705A, N-707S, N-707SA, N-2449S, N-2456S or 11/64 inch (4.37 mm) for all other models. Carefully bend float lever tang which contacts fuel inlet needle if necessary to obtain correct float level on all models.

Tillotson Float Type Carburetor. Some early models were equipped with Tillotson ML series carburetors.

Initial adjustment of idle and main fuel mixture needles from a lightly seated position is 1 turn open for idle mixture needle and 1½ turns open for main fuel mixture needle.

Final adjustments are made with engine at operating temperature and running. Operate engine at rated speed and adjust main fuel mixture needle for smoothest engine operation. Operate engine at idle speed and adjust idle fuel mixture screw for smoothest engine idle.

To check float level, invert carburetor throttle body and float assembly. Measure distance between free end of float and bowl mating surface of carburetor. Measurement should be 1-5/64 to 1-3/32 inch (26.78-27.38 mm). Carefully bend float lever tang which contacts fuel inlet needle if necessary to obtain correct float level.

Zenith Float Type Carburetor. Models 10390, 10658 and 10665 were used on some early model engines. The 10390 model carburetor has only one fuel mixture adjustment needle. Set this fuel mixture needle ¼ turn rich (counterclockwise) from setting producing maximum high idle speed. On Models 10658 and 10665, turning the idle fuel needle clockwise will enrichen the fuel mixture while turning the main fuel needle clockwise will lean the fuel mixture. Float setting is nonadjustable on this series of Zenith carburetors.

FUEL PUMP. Some vertical shaft engines are equipped with a diaphragm type fuel pump located on the crankcase cover and operated by pressure pulsations within the crankcase. Service consists of renewing the diaphragm (79–Fig. CL53).

GOVERNOR. Either a mechanical or an air vane type governor is used. All vertical shaft engines and some horizontal shaft engines use an air vane type governor (126–Fig. CL53). Mechanical governors are of the type shown in the exploded views of engines in Figs. CL50 and CL56.

The carburetor throttle shaft, governor arm and speed control devices may have several different holes in which springs and linkage can be installed. Before removing carburetor, governor arm, springs, linkage or controls, be sure to mark location of holes in which springs and linkage were installed so they may be reassembled correctly.

Mechanical Governor. The governor weight unit (112–Fig. CL50 or 126–Fig. CL56) is driven by the cam gear and retained to the gear by a pin (59–Fig. CL50 or 25–Fig. CL56) driven into the gear. The governor collar (14–Fig. CL50 or 25–Fig. CL56) has a notch in the inner flange of the collar which fits around the weight unit retaining pin. The governor shaft (2–Fig. CL50 or 25–Fig. CL56) has a notch in the inner flange of the collar which fits around the weight unit retaining pin. The governor shaft (2–Fig. CL50 or 4–Fig. CL56) is fitted with a square cross section weight that contacts the outer flange of the governor collar. The shaft is supported in a renewable bushing (6–Fig. CL50 or 55–Fig. CL56).

It is very important that the travel of the governor and the carburetor throttle be synchronized. With the engine not running, move the governor throttle arm (1–Fig. CL50 or 3–Fig. CL56) so the governor shaft holds the governor weight unit in fully closed position. At this time, there should be 1/32 to 1/16 inch (0.79-1.59 mm) clearance between the high speed stop on the carburetor throttle arm and the carburetor casting. To obtain this adjustment, increase or decrease the loop in the carburetor to governor link (49–Fig. CL50) on older model engines or loosen the adjustment screw (89–Fig. CL56) on newer model engines and reposition the arm (3) on the adjuster (2) then tighten adjustment screw (89).

Adjust desired maximum speed (do not exceed 3600 rpm) by adjusting tension of governor spring (98–Fig. CL50 or 110–Fig. CL56). CAUTION: Do not use any spring other than correct Clinton part specified for a particular engine as spring must be balanced to governor weight unit for proper speed control.

On most models, a backlash spring (96–Fig. CL50) is used to hold any free

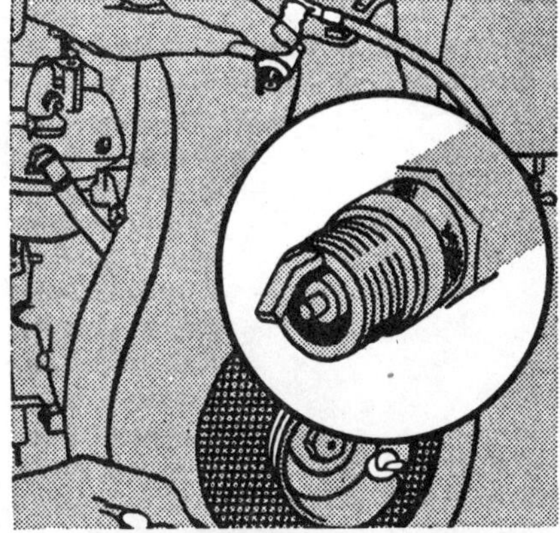

Fig. CL49A—Magneto may be considered in satisfactory condition if it will fire an 18 mm spark plug with electrode gap set at 0.156-0.187 inch (3.97-4.76 mm).

play out of governor to carburetor linkage and thereby reduce any tendency for engine to surge.

Air Vane Governor. Refer to Fig. CL53. The governor air vane (126) pivots on a renewable pin (69) that is driven into a hole in the bearing plate (72). The vane is connected to the carburetor throttle shaft by link (58).

Check to see that the vane and link are not bent from their original shape. Renew vane and/or link if damaged. To maintain proper speed control, the correct governor spring must be used. Be sure to use the correct Clinton part number when ordering the governor spring (113). Be sure vane and linkage

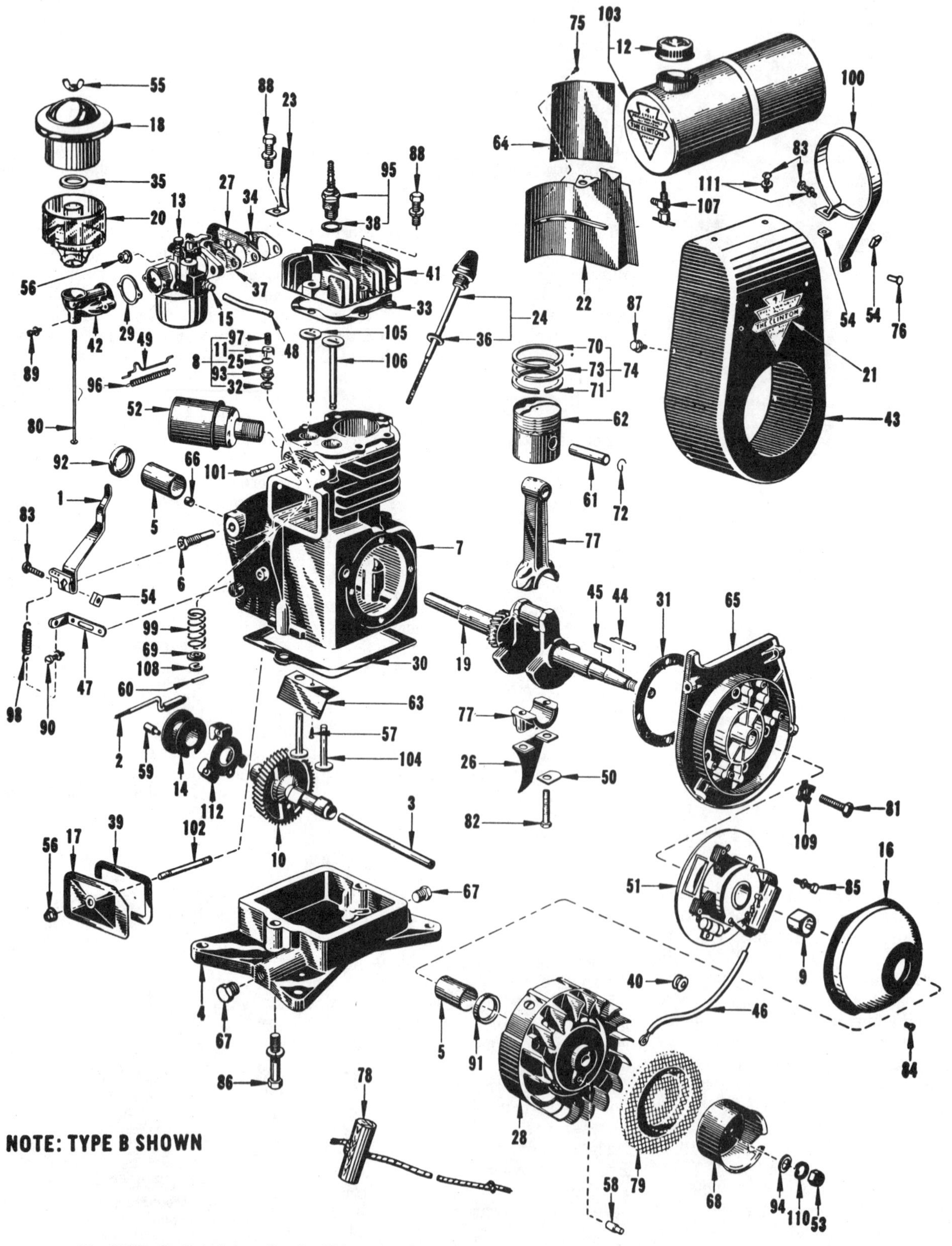

Fig. CL50—Exploded view of series 700 engine. Early models were equipped with Scintilla magneto (51) as shown.

move through their full range of travel without binding and take care when servicing engine not to bend the vane or link.

Adjust speed control (tension on governor spring) for desired maximum rpm, but do not exceed 3600 rpm.

MAGNETO AND TIMING. Some early model engines were equipped with Scintilla magnetos (51 – Fig. CL50). Condenser, breaker points and armature coil are located behind the flywheel on all models.

On Scintilla magnetos, adjust breaker contact gap to 0.015-0.018 inch (0.38-0.46 mm). Edge gap should be not less than 5°. Magneto can be rotated on early 700-A models to advance or retard timing. Timing is fixed and nonadjustable at 21° BTDC.

On all magnetos, magneto may be considered in satisfactory condition if it will fire an 18 mm spark plug with electrode gap set at 0.156-0.187 in.(3.97-4.76 mm). See Fig. CL49A.

LUBRICATION. Manufacturer recommends oil with an API service classification SE or SF. Use SAE 30 oil for temperatures above 32°F (0°C); SAE 10W oil for temperature ranges of −10° to 32°F (−23° to 0°C) or SAE 5W oil for temperatures below −10°F (−23°C).

On models equipped with reduction gearing, use SAE 30 oil in gearbox.

CRANKCASE BREATHER. Crankcase breather assembly (8 – Fig. CL50, 10 – Fig. CL53 or 14 – Fig. CL56) should be removed and cleaned if oil loss through hole in valve cover is evident or whenever the engine is overhauled. Be sure the breather is correctly reassembled and reinstalled.

REPAIRS

TIGHTENING TORQUES. Recommended torque specifications are as follows:

Adapter flange	120-150 in.-lbs. (14-18 N·m)
Base bolts	325-375 in.-lbs. (37-42 N·m)
Carburetor to manifold	35-50 in.-lbs. (4-5 N·m)
Carburetor to block	60-65 in.-lbs. (6.7-7.3 N·m)
Connecting rod	70-80 in.-lbs. (8-10 N·m)
Cylinder head	200-220 in.-lbs. (23-24 N·m)
End cover	120-150 in.-lbs. (14-18 N·m)
Flywheel	400-450 in.-lbs.* (45-52 N·m)
Spark plug	275-300 in.-lbs. (31-34 N·m)
Stator plate	50-60 in.-lbs. (5-7 N·m)

*350 in.-lbs. (40 N·m) maximum on 7/16 inch crankshaft.

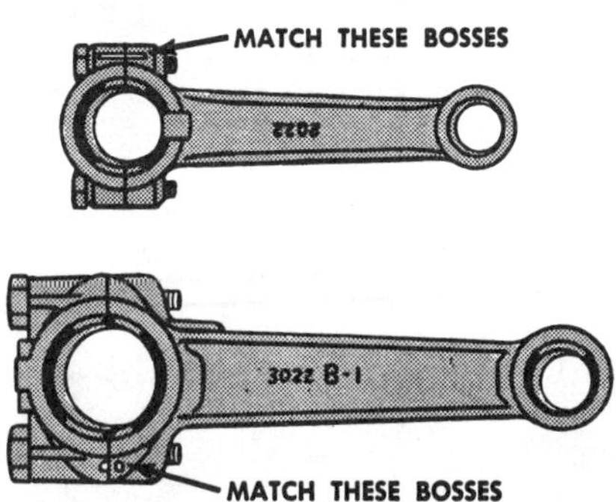

Fig. CL52—When assembling connecting rod cap to connecting rod, be sure embossments on rod and cap are aligned as shown.

CONNECTING ROD. Connecting rod and piston assembly can be removed from engine after cylinder head and engine base (on horizontal crankshaft models) or crankcase end cover (on vertical crankshaft models) are removed.

Standard connecting rod to crankshaft crankpin journal clearance is 0.0018-0.0035 inch (0.046-0.089 mm). Maximum allowable clearance is 0.0045 inch (0.114 mm). Standard connecting rod piston pin bearing bore to piston pin clearance is 0.0004-0.0011 inch (0.010-0.028 mm). Maximum allowable clearance is 0.002 inch (0.051 mm). Connecting rod side play should be 0.005-0.020 inch (0.13-0.51 mm). Connecting rod is available in standard size only.

When reassembling, be sure embossments on connecting rod and cap are aligned as shown in Fig. CL52. Oil hole or oil access slot in connecting rod should face flywheel side of engine. On some models, the connecting rod has a "clearance side" which must be towards the camshaft. Be sure connecting rod locks and oil distributor (on horizontal crankshaft models) clears the camshaft after assembly.

On horizontal crankshaft engines, refer to Fig. CL52A for view of "old" and "new" rod locks and oil distributor. "Old" type rod locks must be used with "old" type oil distributor and "new" type lock with "new" oil distributor. Also, a few oil distributors have been made incorrectly. Do not install an oil distributor if bolting flange faces in direction opposite to that shown in fig. CL52A. If installing a new connecting rod in a vertical crankshaft engine equipped with an oil pump, refer to LUBRICATION SYSTEM paragraph.

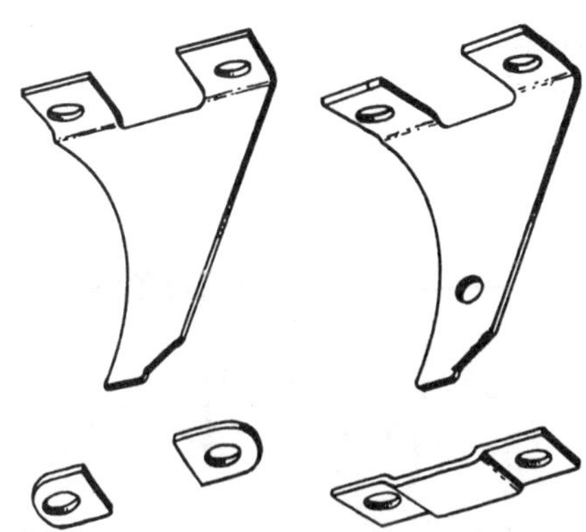

Fig. CL52A—New type oil distributor at right identified by hole drilled through it as shown. One piece rod lock shown below new type distributor should be used only with new distributor; two-piece rod lock should be used with old type distributor. Refer to text.

PISTON, PIN AND RINGS. Piston is fitted with two compression rings and one oil control ring. Recommended piston ring end gap is 0.007-0.017 inch (0.18-0.43 mm). Renew rings and/or recondition cylinder if end gap of top ring is 0.025 inch (0.64 mm) or more. Ring side clearance in groove should be 0.002-0.005 inch (0.05-0.13 mm). Maximum allowable side clearance is 0.006 inch (0.15 mm). Piston and rings are available in several oversizes as well as standard size.

The piston pin is retained in the piston with a snap ring at each end of the pin. Piston pin is available in standard size only and should be a hand push fit in piston. On models with a 2-inch (50.4 mm) bore, piston pin diameter is 0.4999-0.5001 inch (12.698-12.703 mm). Piston pin bore diameter in piston is 0.5000-0.5003 inch (12.700-12.708 mm) and piston pin bore diameter in connecting rod is 0.5005-0.5010 inch (12.713-12.725 mm).

On all other models, piston pin diameter is 0.5624-0.5626 inch (14.285-14.290 mm). Piston pin bore diameter in piston is 0.5625-0.5628 inch (14.288-14.295 mm) and piston pin bore diameter in connecting rod is 0.5630-0.5635 inch (14.300-14.313 mm).

On all models, maximum piston pin to connecting rod pin bore clearance is 0.002 inch (0.05 mm).

Piston skirt clearance should be 0.0045-0.0065 inch (0.114-0.165 mm) for models with 2-inch (50.8 mm) bore and 0.005-0.007 inch (0.13-0.18 mm) for all other models. Maximum piston skirt clearance for all models is 0.008 inch (0.20 mm). Piston is available in several oversizes as well as standard size.

CYLINDER. Standard cylinder bore diameters are 2.000-2.001 inch (50.80-

50.83 mm), 2.125-2.126 inch (53.97-53.99 mm), 2.375-2.376 inch (60.33-60.35 mm) and 2.4685-2.4695 inch (62.70-62.73 mm).

If piston skirt clearance is 0.008 inch (0.20 mm) or more with new piston, cylinder must be reconditioned or renewed.

CRANKSHAFT. Connecting rod to crankpin clearance should be 0.0018-0.0035 inch (0.05-0.09 mm). Renew connecting rod and/or crank-

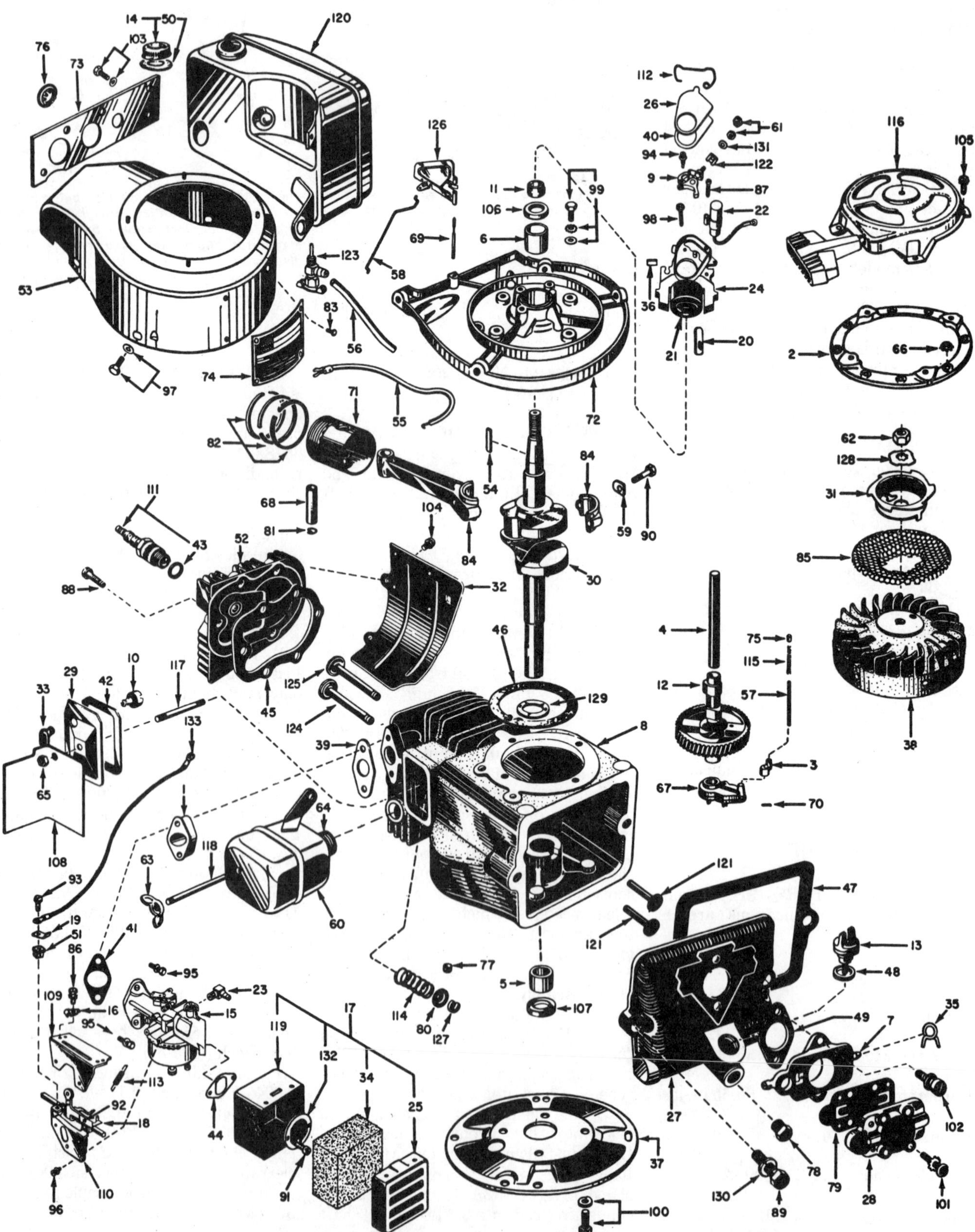

Fig. CL53—Exploded view of Model 499-0000-000 vertical shaft engine. Note the fuel pump (items 7, 28 and 79) which attaches to the crankcase cover (27). Engine is lubricated by oil pump (67) although some vertical shaft models are splash lubricated bv an oil scoop riveted to the bottom side of camshaft gear.

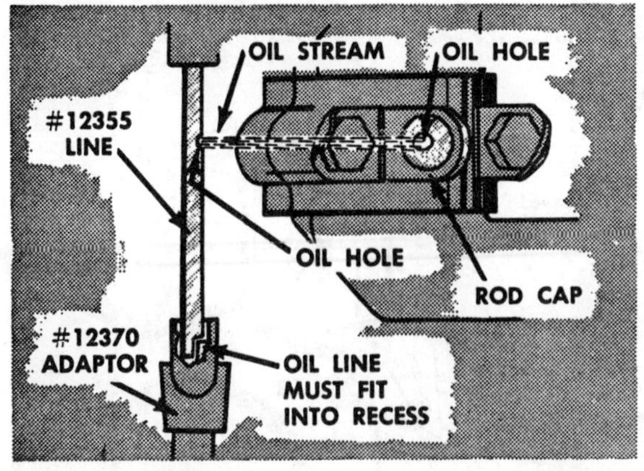

Fig. CL54—On early models, connecting rod bearing is lubricated by oil spray from hole in oil line as shown. Be sure the squared-off end of the oil line fits into the recess in oil pump adapter when reassembling engine.

shaft if clearance is 0.0045 inch (0.114 mm) or more or if journals are 0.0015 inch (0.04 mm) or more out-of-round. Standard crankpin diameter is either 0.8745-0.8752 inch (22.212-22.230 mm) or 0.9114-0.9120 inch (23.150-23.165 mm).

Main bearing clearance should be 0.0018-0.0035 inch (0.046-0.089 mm) on all plain bushing models. Renew crankshaft and/or bushings if clearance is 0.005 inch (0.13 mm) or more. Flywheel side main journal diameter is 0.8745-0.8752 inch (22.212-22.230 mm) and main journal at pto side is either 0.8745-0.8752 inch (22.212-22.230 mm) or 0.9995-1.0002 inch (25.265-25.405 mm). Main bearing bushings are available in standard size only and must be reamed after installation for proper size.

On some vertical crankshaft engines, upper crankshaft journal rides directly in aluminum bearing plate; if bearing plate wear is excessive, renew bearing plate or ream bearing out oversize to accept a bronze service bushing. Bushing driving tools, reamers and reamer alignment plates are available through Clinton parts sources.

On vertical crankshaft models with needle roller mains, renew needle roller bearing if any needle has flat spots or damage. Also renew needle bearing if needles can be separated the width of one needle. Renew crankshaft if rough, scored or shows signs of wear where needle rollers make contact. Check upper pto bushing and journal as outlined in preceding paragraph.

On models with tapered roller main bearings, renew bearing cones and cups if roller or cup is scored or rough.

On ball bearing equipped models, renew ball bearing assembly if excessive wear is noted or if bearing is rough when turned.

On 800 and 900 models, crankshaft end play should be 0.008-0.012 inch (0.20-0.31 mm). Maximum allowable end play is 0.020 inch (0.51 mm). On models with tapered roller main bearings, crankshaft end play should be 0.001-0.006 inch (0.03-0.15 mm). Maximum allowable end play is 0.008 inch (0.20 mm). On all other models with plain bushings, needle bearings or ball bearings, crankshaft end play should be 0.008-0.018 inch (0.20-0.46 mm). Maximum allowable end play is 0.025 inch (0.64 mm).

Bearing plate gaskets are available in several thicknesses to adjust end play on all models. Vertical shaft models are equipped with either one or two thrust washers between the lower thrust surface of the crankshaft and the cylinder block.

CAMSHAFT AND GEAR. The hollow camshaft and cam gear unit rotates on a cam axle that is pressed into the engine crankcase. Camshaft can be removed after removing engine crankshaft and pressing camshaft axle from crankcase. On models with mechanical governor, governor weight unit is attached to cam gear by a pin that is pressed into the gear.

Operating clearance between camshaft axle and camshaft should be 0.001-0.003 inch (0.03-0.08 mm). Renew axle and/or camshaft if clearance is 0.005 inch (0.13 mm) or more. Camshaft end play should be 0.003-0.010 inch (0.08-0.25 mm). Maximum allowable end play is 0.015 inch (0.38 mm).

VALVE SYSTEM. Recommended valve tappet gap for intake and exhaust valves on all models is 0.009-0.011 inch (0.23-0.28 mm). Adjust clearance as follows: To increase tappet clearance, grind end of valve stem. To decrease tappet clearance, renew valve and tappet and/or grind valve seat deeper.

Valve face angle is 45° and valve seat angle is 43½° to 44½°. Seat width should be 0.030-0.045 inch (0.76-1.14 mm). Narrow seat if width is 0.060 inch (1.52 mm) or more. Valve head margin for new valve is 1/32 inch (0.79 mm). Minimum valve margin is 1/64 inch (0.40 mm).

Recommended valve stem to guide clearance is 0.002-0.0045 inch (0.05-0.11 mm). If clearance is 0.006 inch (0.15 mm) or more, guides may be reamed for valve with oversize stem.

LUBRICATING SYSTEM. All horizontal crankshaft engines and some vertical crankshaft engines are splash lubricated. On horizontal crankshaft models, an oil distributor is attached to the connecting rod cap. On vertical crankshaft models, an oil scoop is riveted to the lower side of the cam gear.

A gear type oil pump, driven by a pin on the lower end of the engine crankshaft, is used on some vertical crankshaft models. Oil pump is available for service as an assembly only.

On early models equipped with an oil pump, the crankpin bearing is lubricated by oil spraying from a hole in the oil pump tube as shown in Fig. CL54. On later models, the oil pump tube does not have the spray hole and the crankpin bearing is lubricated by oil spray from the top main bearing. If renewing older type connecting rod (with oil cup on cap) using a new type connecting rod (without oil cup), also install a new oil tube adapter and oil tube without the spray hole. Refer to Fig. CL55.

When assembling engine using early type oil tube with spray hole, be sure the squared-off lower end of the oil tube fits into the recess in the oil tube adapter. Also, if renewing oil pump, adapter and/or oil tube, be sure the correct oil tube and adapter are used. Refer to Fig. CL54 and Fig. CL55.

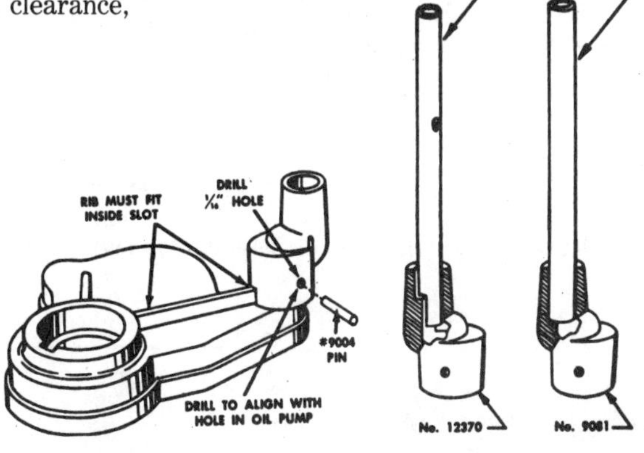

Fig. CL55—Oil pump adapters and oil lines for early and late model vertical shaft engines are shown. If late type connecting rod without oil cup is installed in early engine, it is recommended that the 12370 adapter and 12355 oil line be discarded and the 9084 line and 9081 adapter be installed as shown.

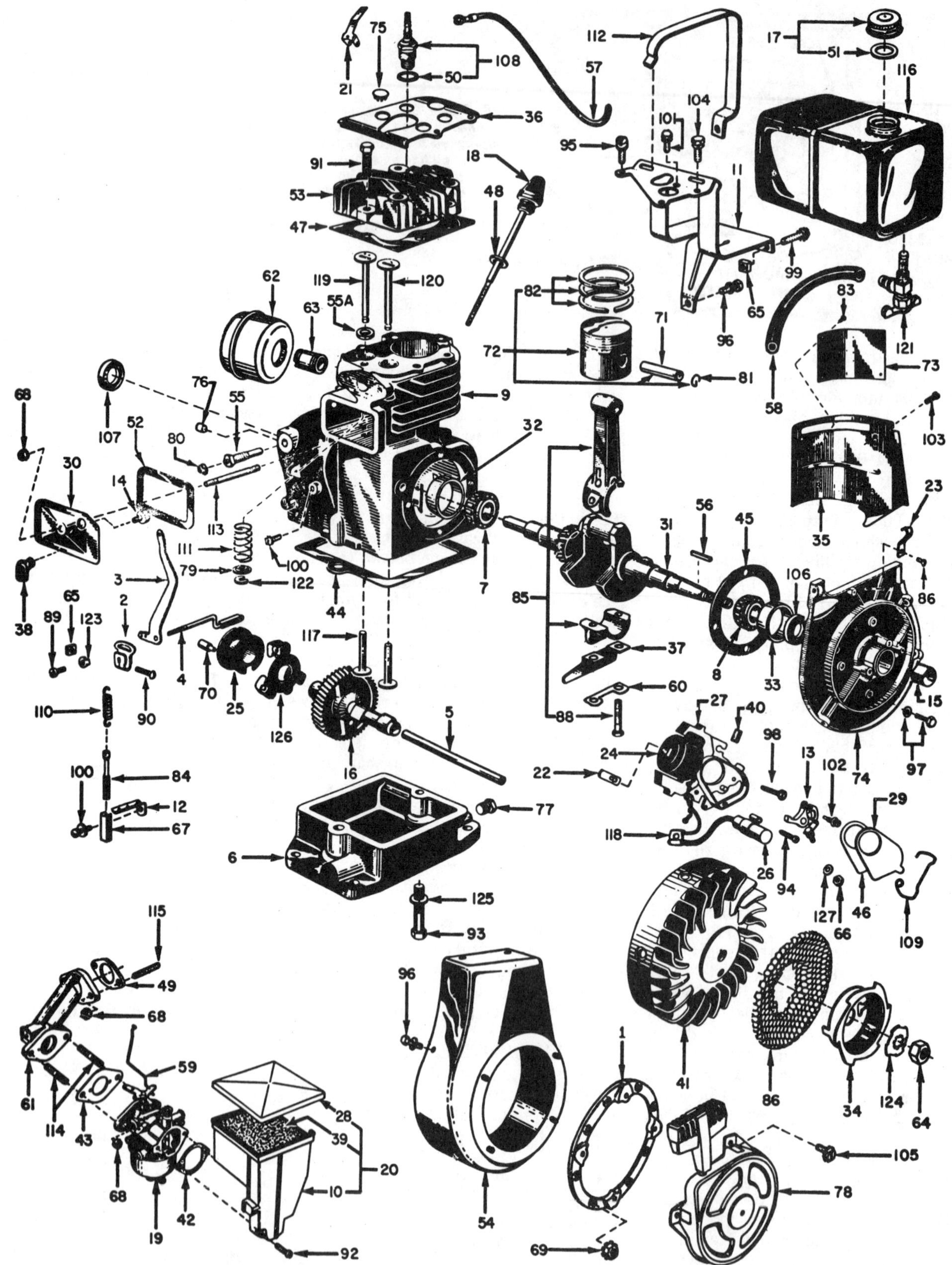

Fig. CL56—Exploded view of Model 498-0000-000 engine. Governor weight assembly (126) is retained to camshaft gear by pin (70) pressed into the gear.

CLINTON

Horizontal Crankshaft Engines

Model	Bore	Stroke	Displacement
100, 2100,A-2100, 400-0100-000, 402-0100-000	2.375 in. (60.3 mm)	1.625 in. (41.3 mm)	7.2 cu. in. (118 cc)
3100,4100, 404-0100-000, 406-0100-000, 408-0100-000, 424-0100-000, 426-0100-000, 492-0000-000	2.375 in. (60.3 mm)	1.875 in. (47.6 mm)	8.3 cu. in. (136 cc)
410-0000-000	2.500 in. (63.5 mm)	1.875 in. (47.6 mm)	9.2 cu. in. (151 cc)

Vertical Crankshaft Engines

Model	Bore	Stroke	Displacement
V-100,VS-100, VS-2100,VS300, 403-0000-000, 405-0000-000, 411-0002-000	2.375 in. (60.3 mm)	1.625 in. (41.3 mm)	7.2 cu. in. (118 cc)
AFV-3100, AV-3100, AVS-3100, FV-3100, V-3100, VS-3100, AVS-4100, VS-4100, 405-0000-000, 407-0000-000, 409-0000-000, 415-0000-000, 417-0000-000, 435-0000-000, 455-0000-000	2.375 in. (60.3 mm)	1.875 in. (47.6 mm)	8.3 cu. in. (136 mm)
419-0000-000, 429-0000-000, 431-0003-000	2.500 in. (63.5 mm)	1.875 in. (47.6 mm)	9.2 cu. in. (151 cc)

MAINTENANCE

SPARK PLUG. Recommended spark plug for all models is a Champion H10 or equivalent. Electrode gap is 0.025-0.028 inch (0.64-0.71 mm). Tighten spark plug to specified torque.

CARBURETOR. Either Clinton (Walbro) float type or Clinton suction type carburetor is used. Refer to the appropriate paragraph.

Clinton (Walbro) Float Type Carburetor. A variety of Clinton (Walbro) LMB, LMG and LMV series carburetors have been used. To identify carburetor, refer to Fig. CL61 for identification number.

Initial adjustment of idle (9) and main (22) fuel mixture needles from a lightly seated position is 1¼ turns open for each needle.

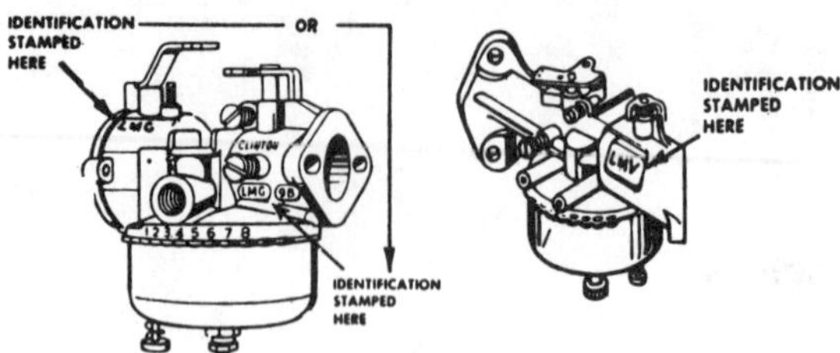

Fig. CL60—Views showing location of identification numbers on LMB, LMG and LMV series carburetors. Identification numbers must be used when ordering service parts.

Final adjustment is made with engine at operating temperature and running. Operate engine at rated speed and adjust main fuel mixture needle for smoothest engine operation. Operate engine at idle speed and adjust idle fuel mixture needle for smooth engine idle. If engine fails to accelerate smoothly it may be necessary to slightly richen main fuel needle adjustment.

When overhauling carburetor, do not remove the main fuel nozzle (13) unless necessary. If nozzle is removed, it must be discarded and a service type nozzle (Fig. CL61A) installed.

Choke plate (11 – Fig. CL61) should be installed with the "W" or part number to outside. Install throttle plate with the side marked "W" facing towards mounting flange and with the part number towards idle needle side of carburetor bore when plate is in closed position. When installing throttle plate, back idle speed adjustment screw out, turn plate and throttle shaft to closed position and seat plate by gently tapping wtih small screwdriver before tightening plate retaining screws.

If either the float valve or seat is damaged, install a new matched valve and seat assembly (12) and tighten seat to 40-50 in.-lbs. (5-6 N·m).

To check float level, invert carburetor throttle body and float assembly. Distance between free side of float and float bowl mating surface of carburetor surface should be 5/32 inch (3.97 mm). Carefully bend float lever tang which contacts fuel inlet needle if necessary to obtain correct float level.

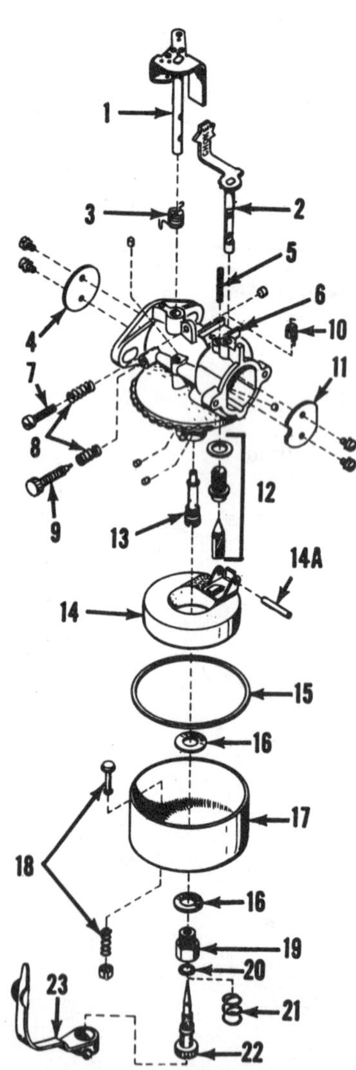

Fig. CL61 – Exploded view of typical LMG series carburetor. LMB and LMV series are similar.

1. Throttle shaft
2. Choke shaft
3. Spring
4. Throttle plate
5. Spring
6. Carburetor body
7. Idle stop screw
8. Springs
9. Idle fuel needle
10. Spring
11. Choke plate
12. Inlet needle
13. Main nozzle
14. Float
14A. Float pin
15. Gasket
16. Gaskets
17. Float bowl
18. Drain valve
19. Retainer
20. Seal
21. Spring
22. Main fuel needle
23. Lever (optional)

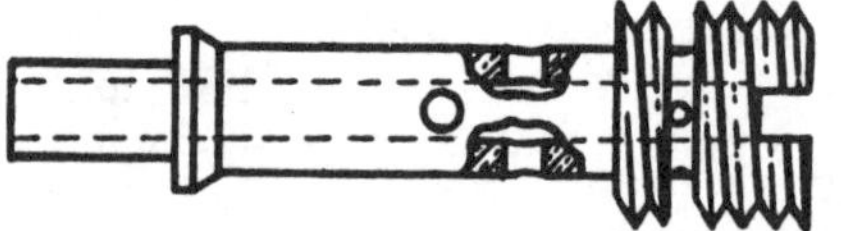

Fig. CL61A – When original main fuel nozzle is removed from LM series carburetors, it must be discarded and service type nozzle shown must be installed.

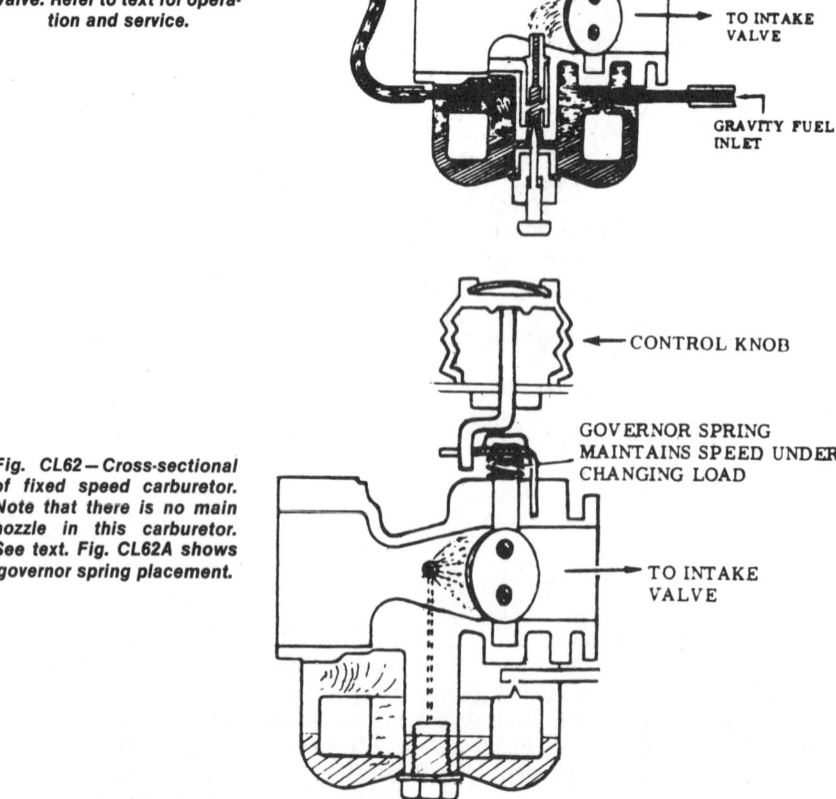

Fig. CL61B – Cross sectional view of "Touch 'N' Start" carburetor and primer bulb. Note absence of choke valve. Refer to text for operation and service.

Fig. CL62 – Cross-sectional of fixed speed carburetor. Note that there is no main nozzle in this carburetor. See text. Fig. CL62A shows governor spring placement.

Clinton "Touch 'N' Start" Carburetor. Refer to Fig. CL61B for identification and cross-sectional view of "Touch 'N' Start" carburetor. Instead of choke valve (11–Fig. CL61) carburetor is furnished with a flexible primer bulb which provides a rich charging mixture to carburetor venturi and intake manifold for easier starting. Float bowl is vented through primer tube to flexible bulb and vent is closed when operator's finger depresses primer bulb. Service procedures and specifications are the same as for Clinton (Walbro) float type carburetor.

Fig. CL62A – Top view of fixed speed carburetor showing correct installation of governor link and spring.

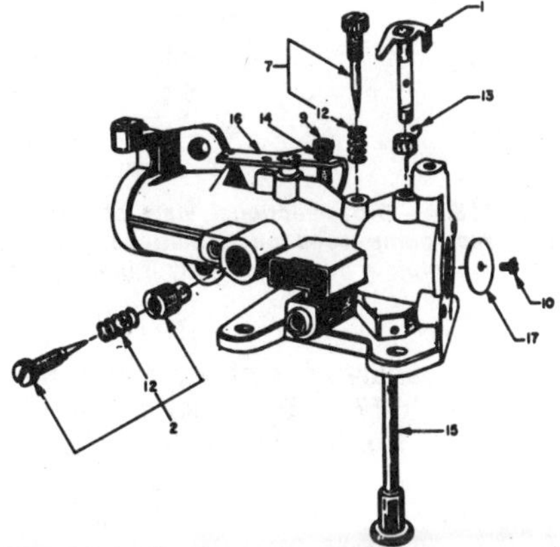

Fig. CL64—Exploded view of Clinton suction lift carburetor. On late production models, idle adjustment screw is threaded directly into carburetor body and does not have the threaded bushing shown. Choke plate (17) and throttle plate (not shown) may be attached with one screw as shown although later models have two holes for attaching screws in each plate and shaft.

1. Choke shaft
2. Idle fuel needle & bushing
7. Main fuel needle
9. Idle stop screw
10. Screw
12. Springs
13. Spring
14. Spring
16. Throttle shaft
17. Choke plate

Fixed Speed Carburetor. Refer to Fig. CL62 for identification and sectional view of LMB, LMG and LMV carburetors equipped with constant speed control. These carburetors have no main nozzle (13–Fig. CL61) and are without an idle current. Speed control knob is turned clockwise to close throttle and stop engine.

Engine speed is controlled at 3000-3400 rpm (no load) by throttle governor spring. See Fig. CL62A for spring placement.

Initial adjustment of idle and main fuel adjustment needles are the same as for Clinton (Walbro) float type carburetor with the exception that main fuel needle initial adjustment is 1¼ to 1½ turns open. Refer to CLINTON (WALBRO) CARBURETORS section for service specifications.

Carter Float Type Carburetor. Refer to Fig. CL63 for identification and ex-

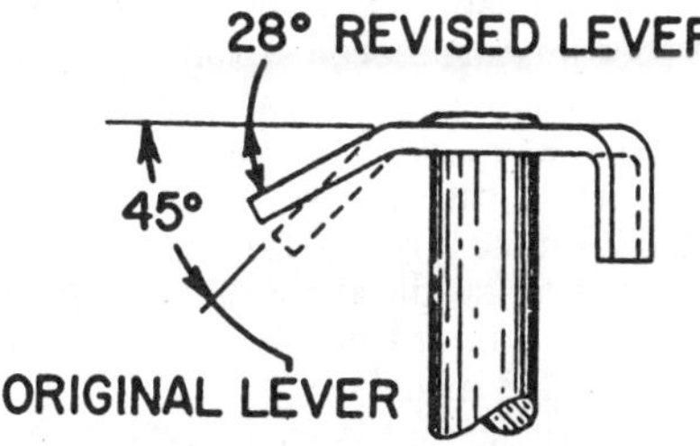

Fig. CL65—Choke lever for later production of Clinton suction carburetor shown in Fig. CL64 is modified as shown here to prevent breakage. Early models should have choke lever (1—Fig. CL64) modified as shown.

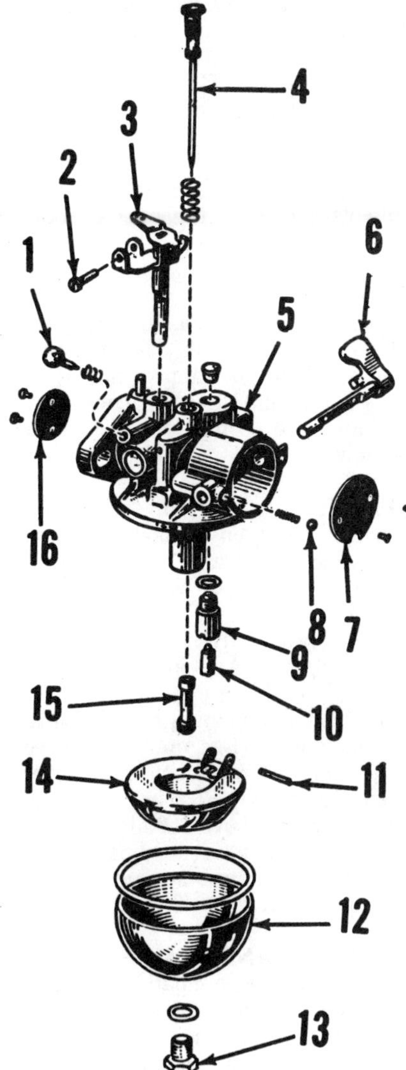

Fig. CL63 – Exploded view of typical Carter Model N carburetor. Design of float and float bowl may vary from that shown.

1. Idle fuel needle
2. Idle stop screw
3. Throttle shaft
4. Main fuel needle
5. Carburetor body
6. Choke shaft
7. Choke plate
8. Detent ball
9. Inlet valve seat
10. Inlet needle
11. Float pin
12. Float bowl
13. Retainer
14. Float
15. Main nozzle
16. Throttle plate

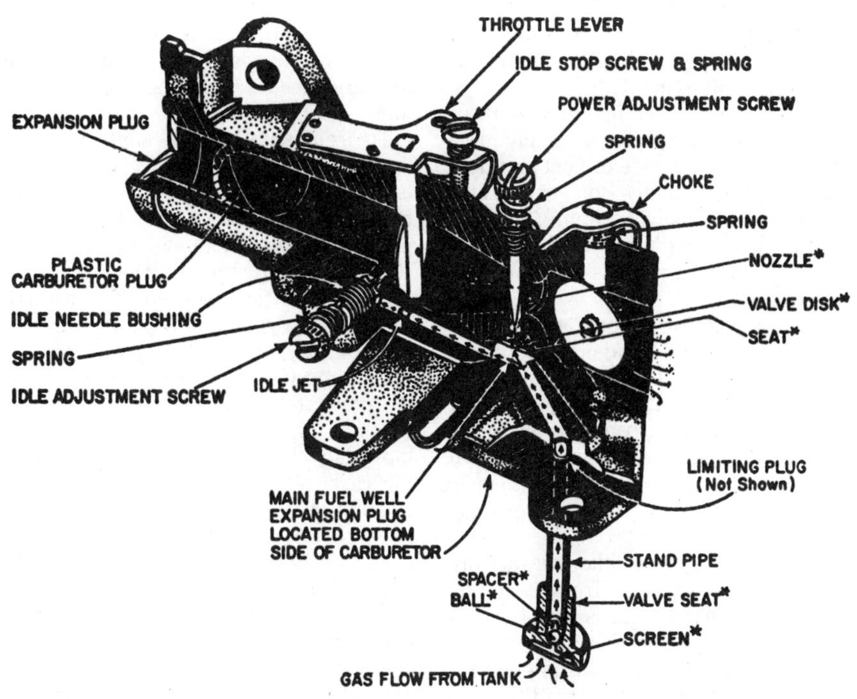

Fig. CL64A—Cut-away view of Clinton suction lift carburetor. Refer to text for disassembly procedure.

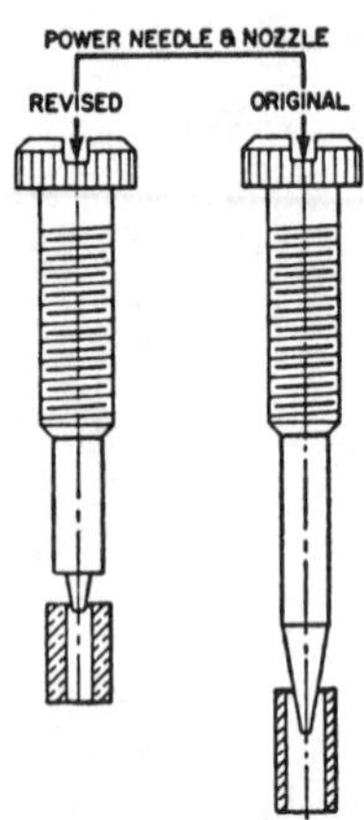

Fig. CL66—View showing early production and late production main fuel needles for Clinton suction lift carburetors. Original and revised needles are not interchangeable. Needle seats are nonrenewable; renew carburetor if seat is damaged.

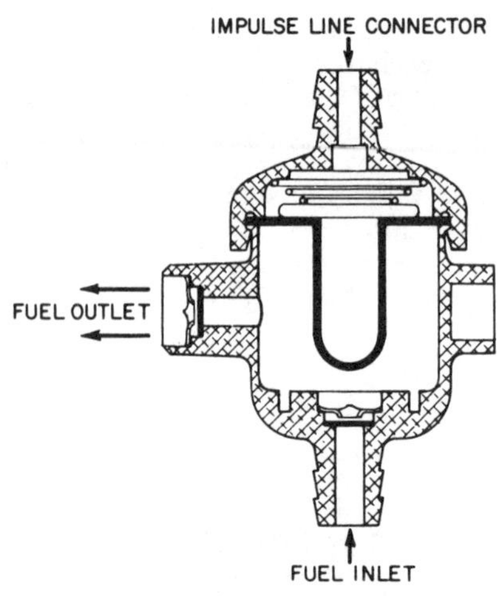

Fig. CL66A—Cross-sectional view of impulse type fuel pump used on four-stroke engines. Renew complete pump assembly if inoperative.

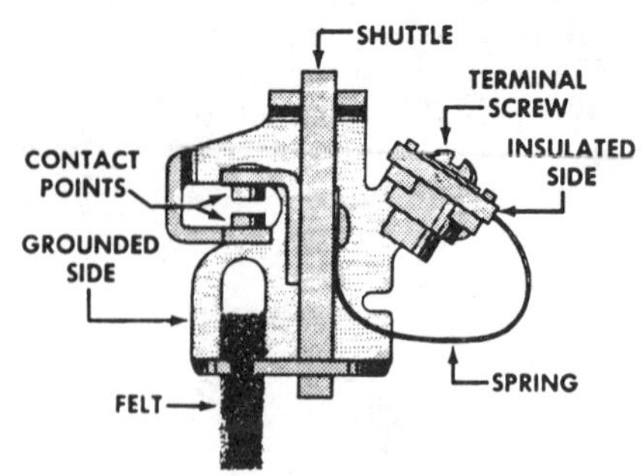

Fig. CL67—Shuttle type breaker points shown are used in early production models.

ploded view of Carter N series carburetors used on some models. Design of float and float bowl may vary.

Initial adjustment of idle (1) and main (4) fuel mixture needles from a lightly seated position is 1 turn open for idle mixture screw and 1½ turns open for main fuel mixture screw on N-2264S and N-2459S models or 1 turn open for idle fuel mixture needle and 2 turns open for main fuel mixture needles on N-2236S and N-2458S models.

Final adjustment is made with engine at operating temperature and running. Operate engine at rated speed and adjust main fuel mixture needle for smoothest engine operation. Operate engine at idle speed and adjust idle fuel mixture needle for smooth engine idle. If engine fails to accelerate smoothly it may be necessary to slightly richen main fuel needle adjustment.

To check float level, invert carburetor throttle body and float assembly. Distance between free side of float and float bowl mating surface on carburetor should be 11/64 inch (4.37 mm). Carefully bend float lever tang which contacts fuel inlet needle as necessary to obtain correct float level.

Clinton Suction Type Carburetors. Refer to Fig. CL64 for identification and exploded view of Clinton suction type carburetor used on vertical crankshaft engines. Carburetor used on horizontal crankshaft engines is similar except that expansion plug and plastic plug shown in cut-away view (Fig. CL64A) are not used.

Initial adjustment of idle fuel mixture needle (2–Fig. CL64) from a lightly seated position is 1½ turns open for mixture needle threaded directly into carburetor body or 4 to 4¼ turns open for mixture needle which thread into a bushing, and bushing is threaded into carburetor body.

Initial adjustment of main fuel mixture needle (7–Fig. CL64) adjustment needle from a lightly seated position is ¾ to 1 turn open for early style needle (refer to Fig. CL66 to identify fuel mixture needle used in early and late carburetors) or 1¼ to 1½ turns open for late style needle (Fig. CL64).

Final adjustments are made with engine at operating temperature and running with fuel tank approximately ½ full. Operate engine at rated speed (fully loaded) and adjust main fuel mixture needle for smoothest engine operation. Note that engine main fuel adjustments cannot be made without engine under load. Engine high speed operation must be adjusted for a richer fuel mixture which may not produce smooth engine operation until load is applied. Operate engine at idle speed and adjust idle fuel mixture needle for smoothest engine idle.

To remove throttle shaft on carburetors from vertical crankshaft engines, drill through the expansion plug at rear of carburetor body, insert punch in drilled hole and pry plug out. Remove plastic plug, throttle valve screws, throttle valve and the throttle shaft. When reassembling, use new expansion plug and seal plug with sealer.

Test check valve in the fuel stand pipe by alternately blowing and sucking air through pipe. Stand pipe can be removed by clamping pipe in vise and prying carburetor from pipe. Apply sealer to stem of new stand pipe and install so it projects 1.895-1.985 inches (48.13-50.32 mm) from carburetor body.

To remove idle jet (Fig. CL64A), remove expansion plug from bottom of carburetor, idle needle and if so equipped, idle needle bushing. Insert a 1/16 inch (1.59 mm) rod through fuel well and up through idle passage to push jet out into idle fuel reservoir. To install new jet, place mark on the rod exactly 1¼ inches (31.75 mm) from end and push new jet into passage until mark on rod is in exact center of fuel well.

Limiting plug (Fig. CL64A) located in fuel passageway from stand pipe should not be removed unless necessary for cleaning purposes. Use sealer on plug during installation.

FUEL PUMP. Some models are equipped with a diaphragm type fuel pump as shown in the cross-sectional view in Fig. CL66A. The pump diaphragm is actuated by pressure pulsations transmitted via an impulse tube connected between the fuel pump and intake manifold. The pump is designed to lift fuel approximately six inches (152 mm). Service consists of renewing the complete fuel pump assembly.

GOVERNOR. An air vane type governor is used on all models. The air vane, which is located in the blower housing, is linked to the throttle lever. The governor is actuated by air delivered by the flywheel fan and by governor spring. Any speed within the operating speed range of the engine can be obtained by adjusting tension on the governor spring. Use only the correct Clinton part for replacement of governor spring and do not adjust maximum governed speed above 3600 rpm.

Make sure the governor linkage does not bind when linkage is moved through full range of travel.

MAGNETO AND TIMING. Magneto coil and armature (laminations), breaker points and condenser are located under the engine flywheel. Two different types of breaker point assemblies are used. Early model engines were equipped with shuttle type breaker points as shown in Fig. CL67. Early type breaker points are enclosed in a box which is an integral part of the engine crankcase and are covered by a plate and gasket. Some shuttle type points may also be enclosed in a sealed unit within the breaker box. Magneto edge gap for models with shut-

Fig. CL67A—Magneto can be considered in satisfactory condition if it will fire an 18 mm spark plug with electrode gap set at 0.156-0.187 inch (3.96-4.75 mm).

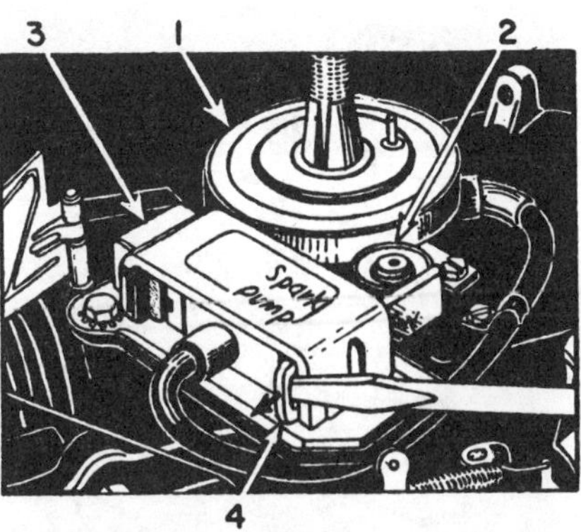

Fig. CL68—After installing spark pump assembly, remove clip (4) with screwdriver as shown. Clip (4) or a 0.10 inch (2.54 mm) spacer should be placed between spark pump lever and frame prior to removing the spark pump from engine. Refer to text.

1. Timing switch
2. Eccentric bearing
3. Spark pump
4. Spacer clip

Fig. CL70—If spark pump has been removed without installing clip or spacer (Fig. CL68), before reinstalling pump, turn load adjusting screw about ½ turn in a counterclockwise direction and hold screw in this position. Press actuating lever in direction shown and insert a 0.10 inch (2.54 mm) spacer between inner end of lever and spark pump frame.

tle type points is 0.156-0.187 inch (3.96-4.75 mm). Lubricate felt with a high melting point grease.

Later engine models are equipped with rocker type breaker points and the points are enclosed in a breaker box which is attached to the engine crankcase. Magneto edge gap for systems with rocker type breaker points is 0.094-0.250 inch (2.39-6.35 mm).

Breaker point gap for all models is 0.018-0.021 inch (0.46-0.53 mm). Ignition timing is 21° BTDC and is non-adjustable.

SPARK PUMP. On a limited number of engines, a Dyna-Spark ignition system (spark pump or "piezo-electric" ignition system) is used instead of the conventional flywheel type magneto. The system consists of a spark pump (3–Fig. CL68) and a timing switch (1).

The spark pump is actuated by an eccentric bearing (2) on the entended end of the engine camshaft and the timing switch is driven by the crankshaft. A spark is generated whenever the cam lever on the spark pump is moved.

As with the conventional magneto, the Dyna-Spark ignition system may be considered in satisfactory condition if it will fire a spark plug with electrode gap set at 0.156-0.187 inch (3.96-4.75 mm). See Fig. CL67A.

If system is inoperative, inspect wire from spark pump to timing switch and from timing switch to spark plug for shorts or breaks in wire. If inspection does not reveal open or shorted condition, the complete Dyna-Spark unit must be renewed.

To remove Dyna-Spark unit, disconnect spark plug and remove engine blower housing and flywheel. Turn engine slowly until eccentric bearing (2–Fig. CL68) on end of camshaft is at maximum lift position and install a 0.10 inch (2.54 mm) spacer between spark pump actuating lever and spark pump frame. Turn engine so the eccentric bearing is at minimum lift position and remove the spark pump and timing switch from engine. The eccentric bearing may be removed from the camshaft after removing the retaining snap ring.

To install Dyna-Spark unit, turn engine so eccentric bearing is at minimum lift position and install the spark pump and timing switch on engine. Turn engine so eccentric bearing is at maximum lift position and remove the clip (4) or spacer from spark pump. Turn the load screw (6–Fig. CL69) counterclockwise with slotted tool (7) or needle nose pliers to release tension on the spark generating element. Lever on spark pump should then follow eccentric bearing closely and smoothly as the engine is turned. Be sure pin (5) properly engages engine flywheel when reinstalling flywheel.

Fig. CL69—Turn load adjusting screw (6) counterclockwise with slotted tool (7) to remove pressure from spark generating cell; spring within spark pump will turn screw back clockwise to apply correct pressure. CAUTION: Never turn load adjusting screw in a clockwise direction. Timing switch is driven by pin (5) which engages engine flywheel.

CAUTION: Never turn load screw (6) in a clockwise direction. A coil spring within the spark pump unit will return the load screw to correct tension.

If the spark pump has been removed without a 0.10 inch (2.54 mm) spacer (or the clip as provided in a new spark pump) installed, refer to Fig. CL70 prior to installation of spark pump on engine.

LUBRICATION. Manufacturer recommends oil with an API service classification SE or SF. Use SAE 30 oil for temperatures above 32° F (0° C), SAE 10W oil for temperatures between −10° F (−23° C) and 32° F (0° C) and SAE 5W oil for temperatures below −10° F (−23° C).

On models equipped with reduction gearing, use SAE 30 oil in gearbox.

CRANKCASE BREATHER. Crankcase breather assembly located in or behind valve chamber cover should be cleaned if difficulty is experienced with oil loss through breather. Be sure the breather is correctly reassembled and reinstalled.

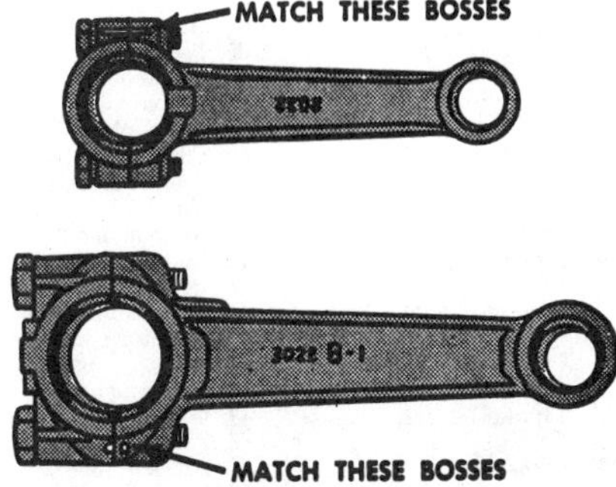

Fig. CL71—When assembling cap to connecting rod, be sure embossments on rod and cap are aligned as shown.

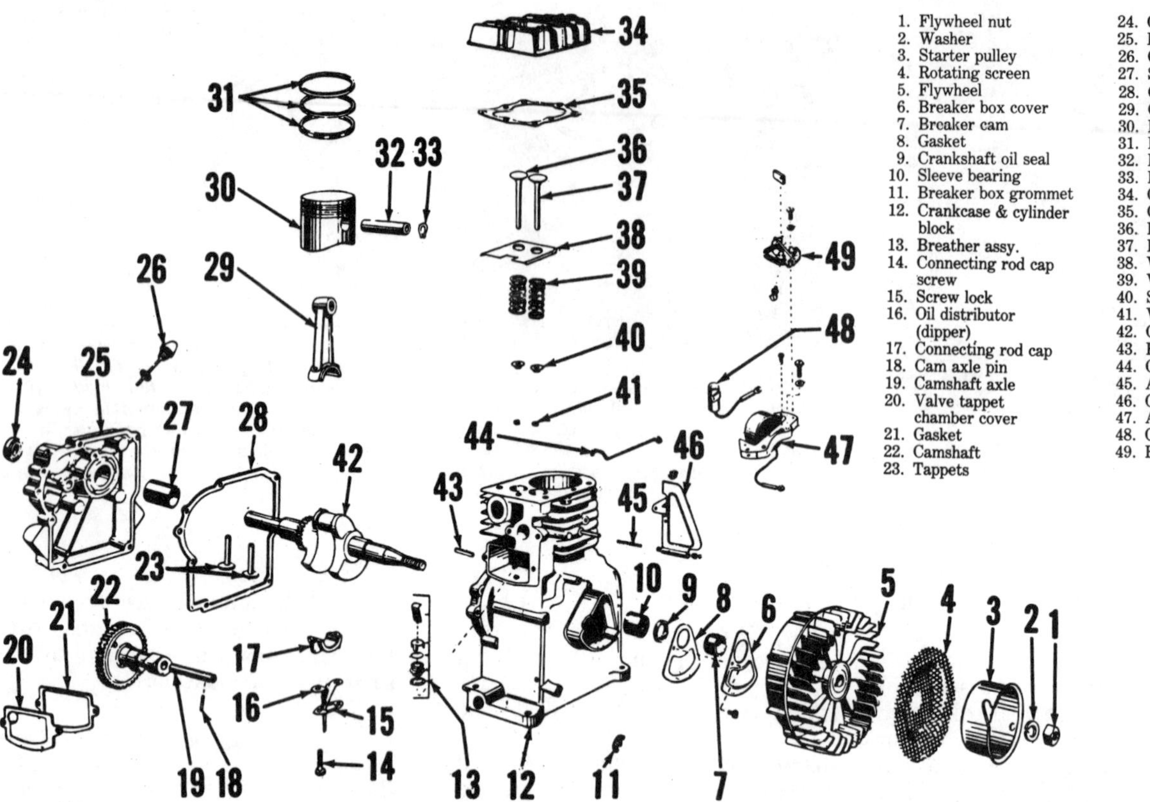

1. Flywheel nut
2. Washer
3. Starter pulley
4. Rotating screen
5. Flywheel
6. Breaker box cover
7. Breaker cam
8. Gasket
9. Crankshaft oil seal
10. Sleeve bearing
11. Breaker box grommet
12. Crankcase & cylinder block
13. Breather assy.
14. Connecting rod cap screw
15. Screw lock
16. Oil distributor (dipper)
17. Connecting rod cap
18. Cam axle pin
19. Camshaft axle
20. Valve tappet chamber cover
21. Gasket
22. Camshaft
23. Tappets
24. Crankshaft oil seal
25. Bearing plate
26. Oil dipstick
27. Sleeve bearing
28. Gasket
29. Connecting rod
30. Piston
31. Piston rings
32. Piston pin
33. Pin retainers
34. Cylinder head
35. Gasket
36. Exhaust valve
37. Intake valve
38. Valve spring retainer
39. Valve springs
40. Spring retainers
41. Valve keepers
42. Crankshaft
43. Flywheel key
44. Governor link
45. Air vane pin
46. Governor air vane
47. Armature & coil assy.
48. Condenser
49. Breaker points

Fig. CL71A—Exploded view of early production horizontal crankshaft model. Camshaft (22) rotates on cam axle (19). Breather assembly (13) is located inside valve chamber. Breaker box is integral part of crankcase and breaker points (49) are accessible after removing flywheel and cover (6). Refer to Fig. CL71B for exploded view of late production horizontal crankshaft model.

REPAIRS

TIGHTENING TORQUES. Recommended tightening torque specifications are as follows:

Base plate or side cover	75-80 in.-lbs. (8.5-9 N•m)
Blower housing	60-70 in.-lbs. (6.7-8 N•m)
Carburetor to manifold	35-50 in.-lbs. (4-6 N•m)
Carburetor (manifold) to block	60-65 in.-lbs. (6.7-7.3 N•m)

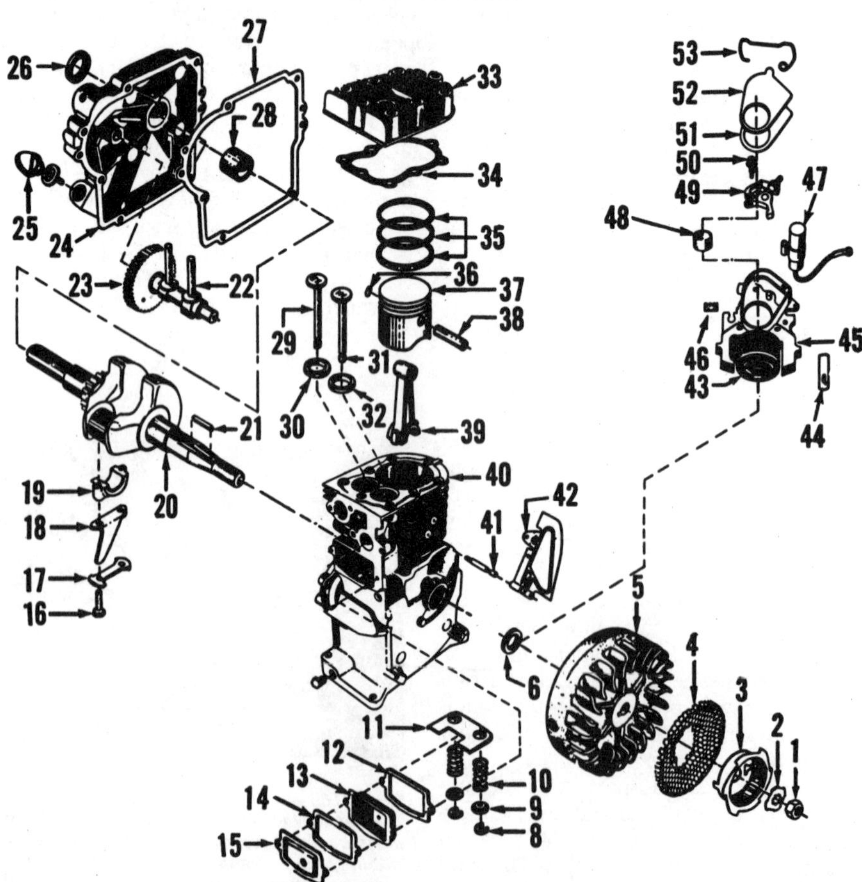

1. Flywheel nut
2. Washer
3. Starter cup
4. Rotating screen
5. Flywheel
6. Crankshaft oil seal
8. Valve keeper
9. Spring retainer
10. Valve springs
11. Spring retainer plate
12. Gasket
13. Breather assy.
14. Gasket
15. Tappet chamber cover
16. Connecting rod cap screws
17. Cap screw lock
18. Oil distributor
19. Connecting rod cap
20. Crankshaft
21. Flywheel key
22. Tappets
23. Camshaft
24. Bearing plate
25. Oil dipstick
26. Crankshaft oil seal
27. Gasket
28. Sleeve bearing
29. Exhaust valve
30. Exhaust valve seat
31. Intake valve
32. Intake valve seat
33. Cylinder head
34. Gasket
35. Piston rings
36. Piston pin retainers
37. Piston
38. Piston pin
39. Connecting rod
40. Crankcase & cylinder block
41. Air vane pin
42. Governor air vane
43. Ignition coil
44. Coil retaining clips
45. Armature & stator assy.
46. Cam wiper felt
47. Condenser
48. Breaker cam
49. Breaker points
50. Screw
51. Gasket
52. Breaker box cover
53. Retainer spring

Fig. CL71B—Exploded view of late production horizontal crankshaft model with die-cast aluminum crankcase and cylinder block assembly. Models with cast iron cylinder block are similar except for valve seat inserts (30 and 32).

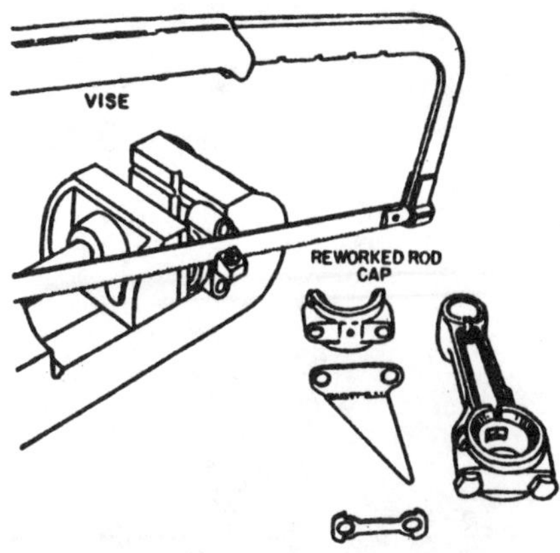

Fig. CL72—When necessary to install late type oil distributor on early type connecting rod, saw oil cup from rod cap as shown. Take care not to damage crankpin bearing surface and smooth off any burrs.

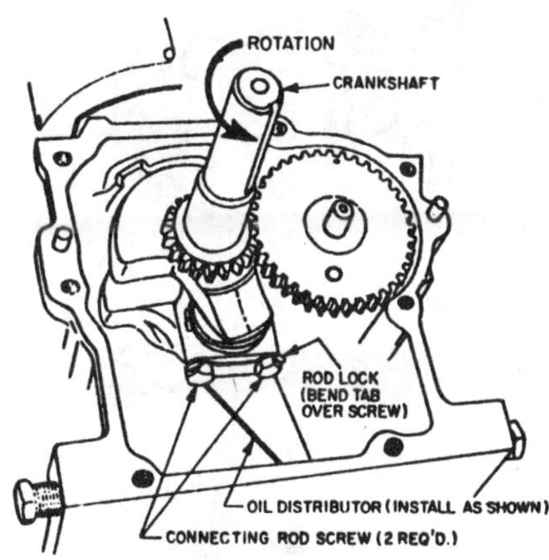

Fig. CL73—View showing correct installation of oil distributor in relation to crankshaft rotation.

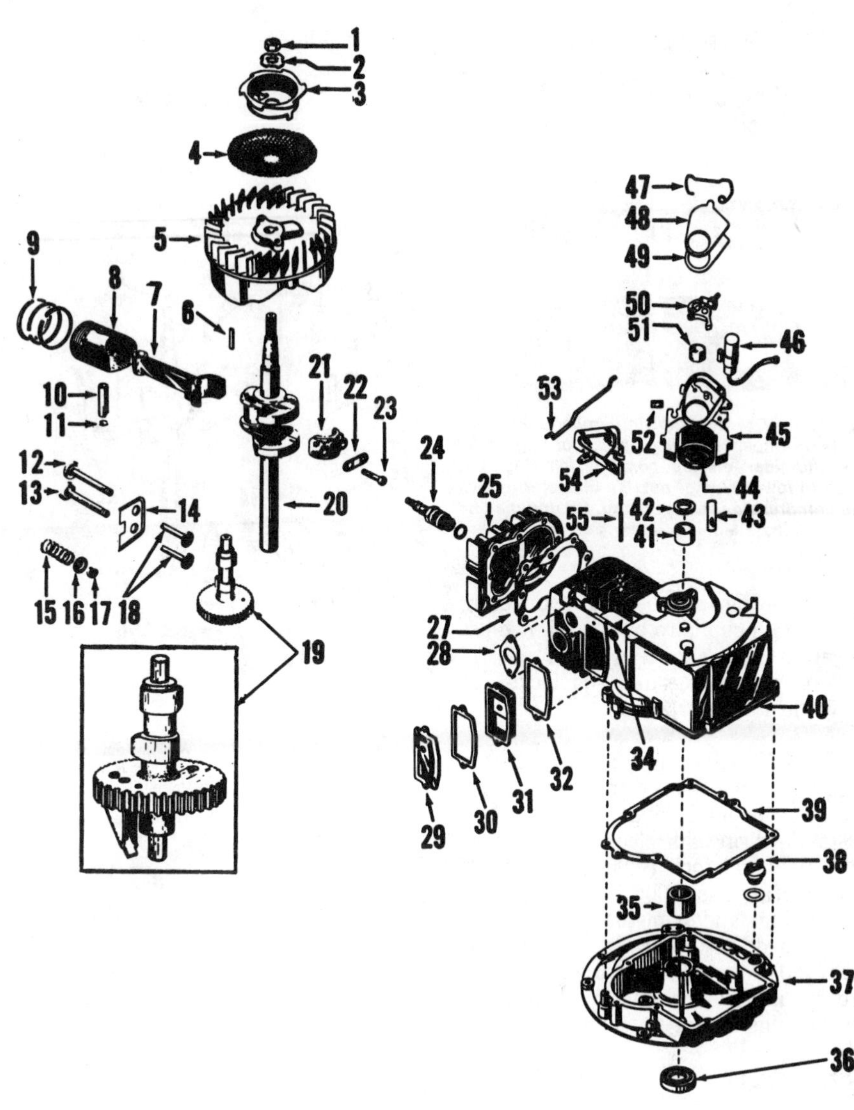

Fig. CL73A—Exploded view of vertical crankshaft model with splash lubrication system. Note oil dipper attached to lower side of camshaft gear (inset 19). Refer to Fig. CL76A for exploded view of vertical crankshaft model with oil pump.

1. Flywheel nut
2. Washer
3. Starter cup
4. Rotating screen
5. Flywheel
6. Flywheel key
7. Connecting rod
8. Piston
9. Piston rings
10. Piston
11. Pin retainers
12. Intake valve
13. Exhaust valve
14. Spring retainer plate
15. Valve springs
16. Spring retainer
17. Valve keepers
18. Valve tappets
19. Camshaft
20. Crankshaft
21. Connecting rod cap
22. Cap screw lock
23. Connecting rod cap screws
24. Spark plug
25. Cylinder head
27. Gasket
28. Gasket
29. Tappet chamber cover
30. Gasket
31. Breather assy.
32. Gasket
34. Grommet
35. Sleeve bearing
36. Crankshaft oil seal
37. Engine base
38. Oil filler plug
39. Gasket
40. Crankcase & cylinder block
41. Sleeve bearing
42. Crankshaft oil seal
43. Coil retaining clips
44. Ignition coil
45. Armature & coil assy.
46. Condenser
47. Retainer spring
48. Breaker box cover
49. Gasket
50. Breaker points
51. Breaker cam
52. Felt cam wiper
53. Governor link
54. Governor air vane
55. Air vane pin

Connecting rod	100-125 in. lbs. (11-14 N•m)
Cylinder head	225-250 in.-lbs. (25-28 N•m)
Flywheel	375-400 in.-lbs. (42-45 N•m)*
Spark plug	275-300 in.-lbs. (31-34 N•m)

*Flywheel for "Touch 'N' Start" brake is tightened to 650-700 in.-lbs. (73-79 N•m).

CONNECTING ROD. Connecting rod and piston assembly can be removed after removing cylinder head and engine base (vertical crankshaft models). On horizontal crankshaft engines having ball bearing mains, oil seal and snap ring must be removed from side cover and crankshaft before side cover can be removed from engine.

The aluminum alloy connecting rod rides directly on the crankpin. Recommended connecting rod-to-crankpin clearance is 0.0015-0.0030 inch (0.04-0.08 mm). If clearance is 0.004 inch (0.10 mm) or more, connecting rod and/or crankshaft must be renewed.

Standard connecting rod bearing bore diameter may be 0.8140-0.8145 inch (20.68-20.69 mm) or 0.8770-0.8775 inch (22.28-22.29 mm) according to model and application.

Standard connecting rod-to-piston pin clearance is 0.0004-0.0011 inch (0.01-0.03 mm) on all models. Renew piston pin and/or connecting rod if clearance is 0.002 inch (0.05 mm) or more.

Standard piston pin bore diameter in connecting rod is 0.5630-0.5635 inch (14.30-14.31 mm).

Install connecting rod with oil hole towards flywheel side of engine and with embossments on rod and cap aligned as shown in Fig. CL71.

Connecting rods on early horizontal crankshaft models had an oil cup on rod cap which was designed for use with an oil distributor which is no longer available. If necessary to renew this oil distributor, a new distributor, Clinton part 220-147, and two new rod locks,

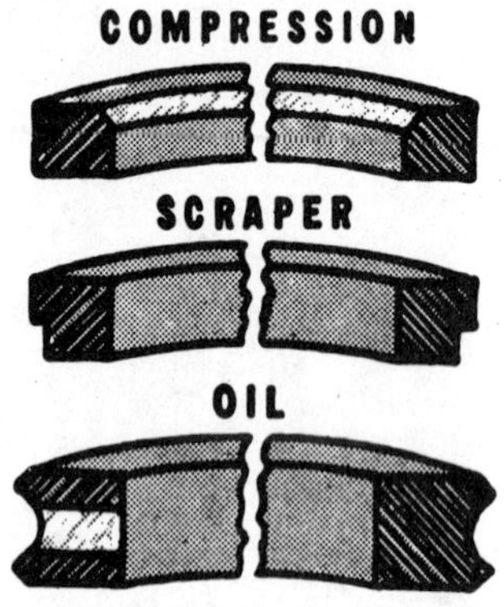

Fig. CL76—Drawing showing proper placement of piston rings. Note bevel at top of ring on inside diameter on top compression ring and notch in lower side of outside diameter of second compression (scraper) ring. Oil ring may be installed with either side up.

part 31038 must be used. If connecting rod with oil cup is to be reused, it must be reworked as shown in Fig. CL72. Install oil distributor and rod locks as shown in Fig. CL73.

PISTON, PIN AND RINGS. Piston is equipped with two compression rings and one oil control ring.

Standard ring side clearance in groove is 0.002-0.005 inch (0.05-0.12 mm). If side clearance is 0.006 inch (0.15 mm), renew rings and/or piston.

Standard piston ring end gap is 0.007-0.017 inch (0.18-0.26 mm). If ring end gap is 0.025 inch (0.64 mm) or more, renew rings and/or recondition cylinder bore.

Standard piston skirt-to-cylinder bore clearance is 0.0045-0.0065 inch (0.114-0.165 mm). If clearance is 0.008 inch (0.203 mm) or more, renew piston and/or recondition cylinder bore.

Standard pin bore diameter in piston is 0.5625-0.5628 inch (14.29-14.30 mm). Standard piston pin diameter is 0.5624-0.5626 inch (14.288-14.295 mm). Piston pin is a 0.0001 inch (0.003 mm) interference to a 0.0004 inch (0.010 mm) loose fit in piston pin bore.

Piston and rings are available in a variety of oversizes as well as standard and a special chrome ring set is available for cylinders having up to 0.010 inch (0.25 mm) taper and/or out-of-round condition.

CLYINDER AND CRANKCASE. Cylinder and crankcase are an integral unit of either an aluminum alloy die-casting with a cast-in iron cylinder liner or a shell casting of cast iron.

Standard cylinder bore diameter for 419-0003-000, 429-0003-000 and 431-0003-000 models is 2.499-2.500 inch (63.48-63.50 mm). Standard cylinder bore diameter for all other models is 2.3745-2.3755 inch (60.31-60.34 mm).

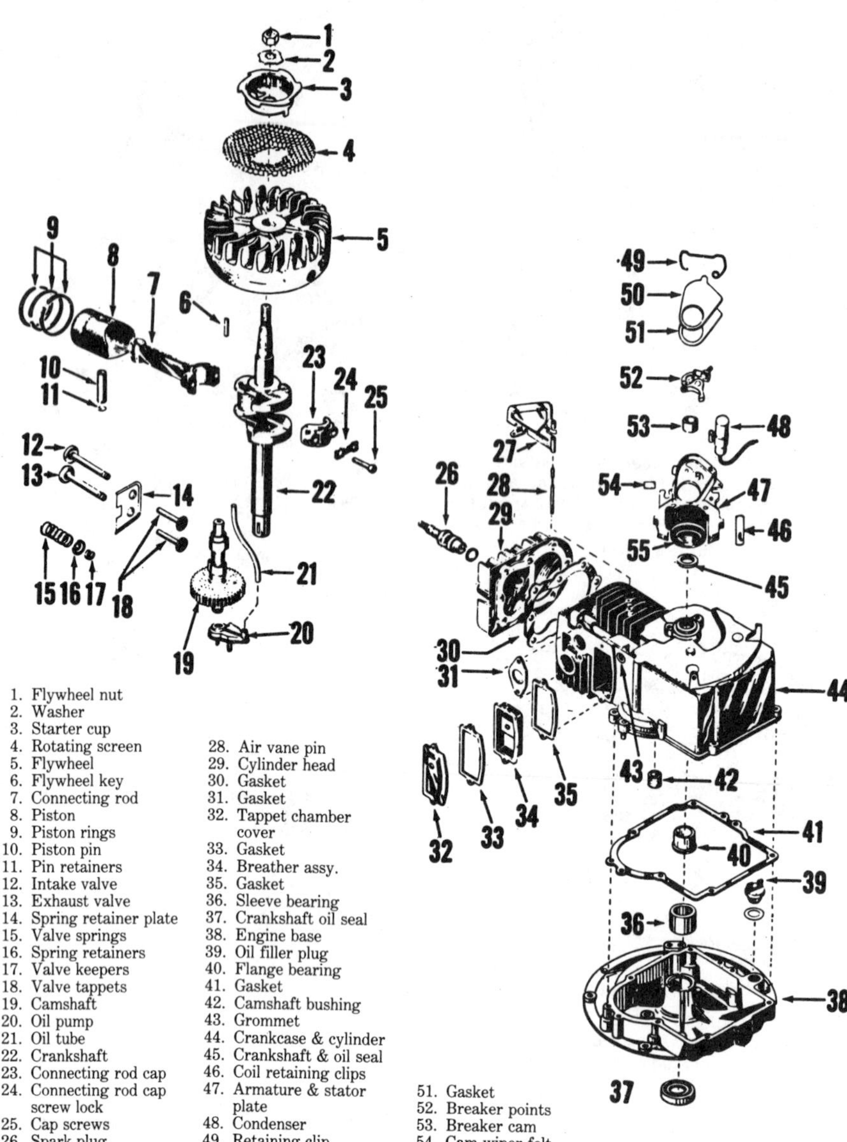

1. Flywheel nut
2. Washer
3. Starter cup
4. Rotating screen
5. Flywheel
6. Flywheel key
7. Connecting rod
8. Piston
9. Piston rings
10. Piston pin
11. Pin retainers
12. Intake valve
13. Exhaust valve
14. Spring retainer plate
15. Valve springs
16. Spring retainers
17. Valve keepers
18. Valve tappets
19. Camshaft
20. Oil pump
21. Oil tube
22. Crankshaft
23. Connecting rod cap
24. Connecting rod cap screw lock
25. Cap screws
26. Spark plug
27. Governor air vane
28. Air vane pin
29. Cylinder head
30. Gasket
31. Gasket
32. Tappet chamber cover
33. Gasket
34. Breather assy.
35. Gasket
36. Sleeve bearing
37. Crankshaft oil seal
38. Engine base
39. Oil filler plug
40. Flange bearing
41. Gasket
42. Camshaft bushing
43. Grommet
44. Crankcase & cylinder
45. Crankshaft & oil seal
46. Coil retaining clips
47. Armature & stator plate
48. Condenser
49. Retaining clip
50. Breaker box cover
51. Gasket
52. Breaker points
53. Breaker cam
54. Cam wiper felt
55. Ignition coil

Fig. CL76A—Exploded view of vertical crankshaft model with lubricating oil pump (20). Refer to Fig. CL73A for vertical crankshaft model with splash lubrication.

Refer to PISTON, PIN AND RINGS section for piston-to-cylinder specifications.

CRANKSHAFT, MAIN BEARINGS AND SEAL. Crankshaft may be supported in bushing type, integral bushing type or ball bearing type main bearings.

Standard crankpin diameter is either 0.8119-0.8125 inch (20.62-20.64 mm) or 0.8745-0.8752 inch (22.21-22.23 mm) according to model and application. If crankshaft main journal is 0.001 inch (0.03 mm) or more out-or-round, or if crankpin journal is 0.0015 inch (0.04 mm) or more out-of-round, renew crankshaft.

Standard clearance between crankshaft main bearing journals and bushing or integral type main bearings is 0.0018-0.0035 inch (0.05-0.09 mm). If clearance is 0.005 inch (0.13 mm) or

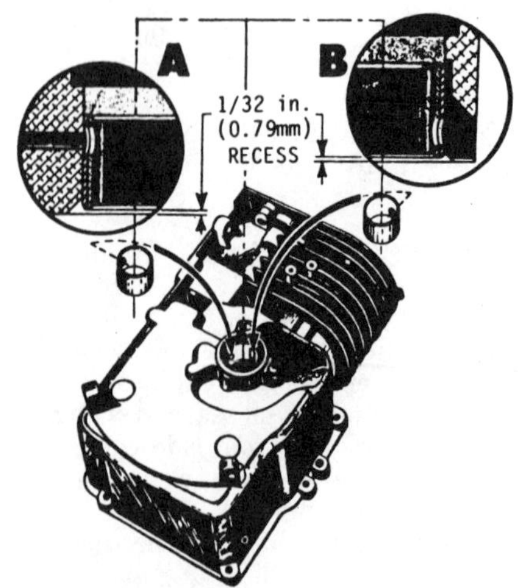

Fig. CL77—When installing crankshaft bushing in crankcase be sure oil holes are aligned as shown and that inner edge of bushing is 1/32 inch (0.79 mm) below thrust face of block.

more, renew bearings and/or crankshaft. Refer to Fig. CL77 for proper placement of bushing in crankcase bore. On models where the crankshaft rides directly in the aluminum alloy crankcase; side plate or base plate (integral type), bearing can be renewed by reaming out the bore to accept a service bushing. Contact nearest Clinton Central Parts Distributor for correct tools, reamers and parts.

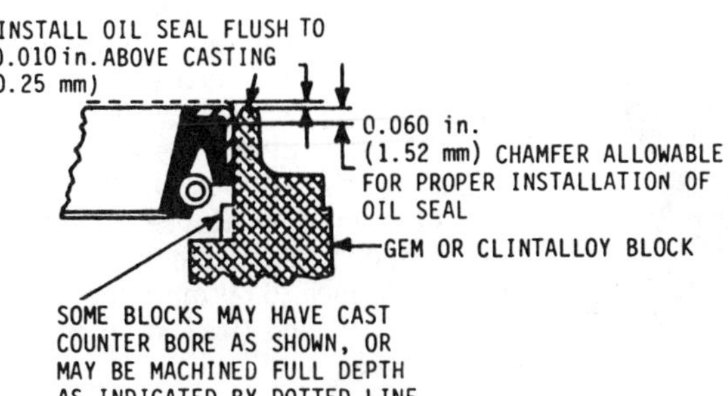

Fig. CL78—Install oil seal in block (crankcase) as shown.

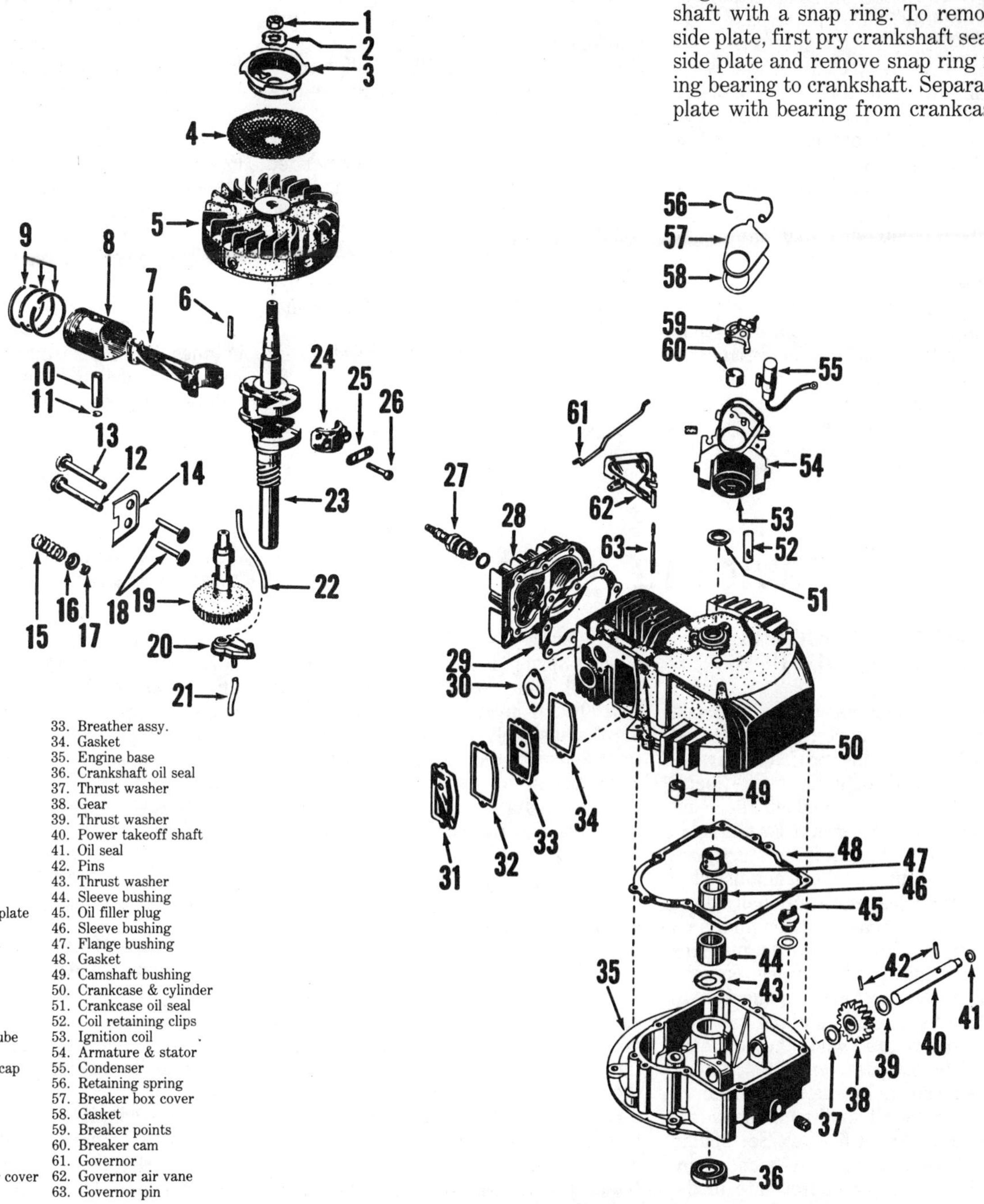

Standard crankshaft end play for models with bushing type bearings is 0.008-0.018 inch (0.20-0.46 mm). If end play is 0.025 inch (0.64 mm) or more, end play must be adjusted by varying thickness and number of shims between crankcase and side plate/base plate. Gaskets are available in a variety of thicknesses.

On horizontal crankshaft models with ball bearing type main bearings, bearing on pto (side plate) end of crankshaft is retained in the side plate with a snap ring and is also retained on the crankshaft with a snap ring. To remove the side plate, first pry crankshaft seal from side plate and remove snap ring retaining bearing to crankshaft. Separate side plate with bearing from crankcase and

1. Flywheel key
2. Washer
3. Starter cup
4. Rotating screen
5. Flywheel
6. Flywheel key
7. Connecting rod
8. Piston
9. Piston rings
10. Piston pin
11. Pin retainers
12. Exhaust valve
13. Intake valve
14. Spring retainer plate
15. Valve springs
16. Spring retainers
17. Valve keepers
18. Valve tappets
19. Camshaft
20. Oil pump assy.
21. Oil pickup tube
22. Oil distributor tube
23. Crankshaft
24. Connecting rod cap
25. Cap screw lock
26. Cap screws
27. Spark plug
28. Cylinder had
29. Gasket
30. Gasket
31. Tappet chamber cover
32. Gasket
33. Breather assy.
34. Gasket
35. Engine base
36. Crankshaft oil seal
37. Thrust washer
38. Gear
39. Thrust washer
40. Power takeoff shaft
41. Oil seal
42. Pins
43. Thrust washer
44. Sleeve bushing
45. Oil filler plug
46. Sleeve bushing
47. Flange bushing
48. Gasket
49. Camshaft bushing
50. Crankcase & cylinder
51. Crankcase oil seal
52. Coil retaining clips
53. Ignition coil
54. Armature & stator
55. Condenser
56. Retaining spring
57. Breaker box cover
58. Gasket
59. Breaker points
60. Breaker cam
61. Governor
62. Governor air vane
63. Governor pin

Fig. CL78A—Exploded view of vertical crankshaft model with auxiliary power take-off; output shaft gear (38) is driven by worm gear on crankshaft. Sleeve bearing (47) is used in upper side of engine base and bearing (44) is installed in lower side.

crankshaft. Ball bearing and retaining snap ring can then be removed from the side plate. Bearing should be renewed if rough, loose or damaged.

When installing crankshaft seal in crankcase, refer to Fig. CL78 for proper placement of seal. When installing crankshaft, make certain crankshaft and camshaft gear timing marks are aligned as shown in Fig. CL79.

CAMSHAFT. On early models, integral camshaft and gear turned on a stationary axle. Later models are equipped with a solid integral camshaft and gear which is supported at each end in bearings which are an integral part of crankcase or side plate/baseplate assemblies.

On early models, clearance between camshaft and cam axle is 0.001-0.003 inch (0.03-0.08 mm). If clearance is 0.005 inch (0.13 mm) or more, camshaft and/or axle must be renewed.

Standard axle diameter is 0.3740-0.3744 inch (9.50-9.51 mm). Axle should be a 0.0009 inch (0.02 mm) tight to 0.001 inch (0.03 mm) loose fit in crankcase and side plate/base plate.

On models where camshaft turns in bores in crankcase and base or side plate, standard clearance between camshaft journals and bearing bore is 0.001-0.003 inch (0.03-0.08 mm). If clearance is 0.006 inch (0.15 mm) or more, renew camshaft and/or crankcase, side plate or base plate.

Fig. CL79—Valves are correctly timed when marks on camshaft gear and crankshaft gear are aligned as shown.

When installing camshaft, align camshaft and crankshaft gear timing marks as shown in Fig. CL79.

VALVE SYSTEM. Recommended valve tappet gap for intake and exhaust valves on all models is 0.009-0.011 inch (0.23-0.28 mm). Adjust valve clearance by grinding off end of valve stem to increase clearance or by renewing valve and/or grinding valve seat deeper into block to reduce clearance.

Valve face angle is 45° and valve seat angle is 44°. If valve face margin is 1/64 inch (0.4 mm) or less, renew valve. On models with aluminum alloy cylinder block, seat inserts are standard. On models with cast iron cylinder block, seats are ground directly into cylinder block surface. Standard valve seat width is 0.030-0.045 inch (0.76-1.29 mm). If seat width is 0.060 inch (1.52 mm) or more, seat must be narrowed.

Standard valve stem-to-guide clearance is 0.0015-0.0045 inch (0.04-0.11 mm). If clearance is 0.006 inch (0.15 mm) or more, guide may be reamed to 0.260 inch (6.60 mm) and a valve with a 0.010 inch (0.25 mm) oversize stem installed.

LUBRICATING SYSTEM. All horizontal crankshaft models and some vertical crankshaft models are splash lubricated. An oil distributor is attached to the connecting rod cap on horizontal crankshaft models and an oil scoop is riveted to the lower side of the camshaft gear on vertical crankshaft models. Refer to CONNECTING ROD section for additional information on oil distributor on connecting rod cap.

A gear type oil pump, driven by a pin on the lower end of the engine camshaft, is used on some vertical crankshaft models. When reassembling these engines, be sure the oil tube fits into the recesses in cylinder block and oil pump before installing base plate.

SERVICING CLINTON ACCESSORIES

IMPULSE STARTERS (EARLY PRODUCTION)

Difficulty in starting engine equipped with impulse type starter may be caused by improper starting procedure or adjustment. The throttle lever must be in the full choke position and left there until engine starts. After impulse starter is fully wound, move handle to start position and push handle down against stop until starter releases. Occasionally there is a hesitation because the engine is on the compression stroke. If engine does not start readily (within 5 releases of the starter), make the following checks:

1. Remove air cleaner and be sure choke is fully closed when throttle lever is in full choke position. If not, adjust controls so choke can be fully closed.
2. Check the idle and high speed fuel mixture adjustment needles. See engine servicing section for recommended initial adjustment for appropriate model being serviced.
3. Check magneto for spark.

To disassemble unit, make certain starter spring is released, remove the four phillips head screws (3–Fig. CL100) and invert the assembly. Holding the assembly at arms length, lightly tap the legs of the starter frame against work bench to remove bottom cover, power springs and cups, plunger assembly and the large gear. Carefully separate the spring and cup assemblies from the plunger. Hold plunger and unscrew ratchet using a 3/8-inch Allen wrench. Remove the snap ring (9) and disassemble plunger unit. Renew all damaged parts or assemblies.

Prior to reassembly, coat all internal parts with light grease. Install large gear with beveled edge of teeth to bottom (open) side of starter and engage the lock pawl as shown in Fig. CL100A. Assemble the plunger unit and install it through the large gear so release button protrudes through top of starter frame. Install power spring and cup assembly with closed side of cup towards large gear and carefully work inner end of spring over the plunger. If two power springs are used, install second spring and cup assembly with closed side of cup towards first spring and cup unit. In some starters, an empty spring cup is used as a spacer. Install cup with closed side toward power spring. Install the bottom cover (4–Fig. CL100) and spring (2); then, screw ratchet into plunger bushing. It is not necessary to tighten the ratchet as normal action of the starter will do this.

IMPULSE STARTER (LATE PRODUCTION)

Troubleshooting procedure for late type impulse starter unit shown in Fig. CL101 will be similar to that for early unit.

Service is limited to renewal of handle (3), pawl (6) and/or pawl spring (9). If power spring is broken, or if drive gear that engages starter cup (2) is worn or damaged, a complete new starter assembly must be installed. Starter cup

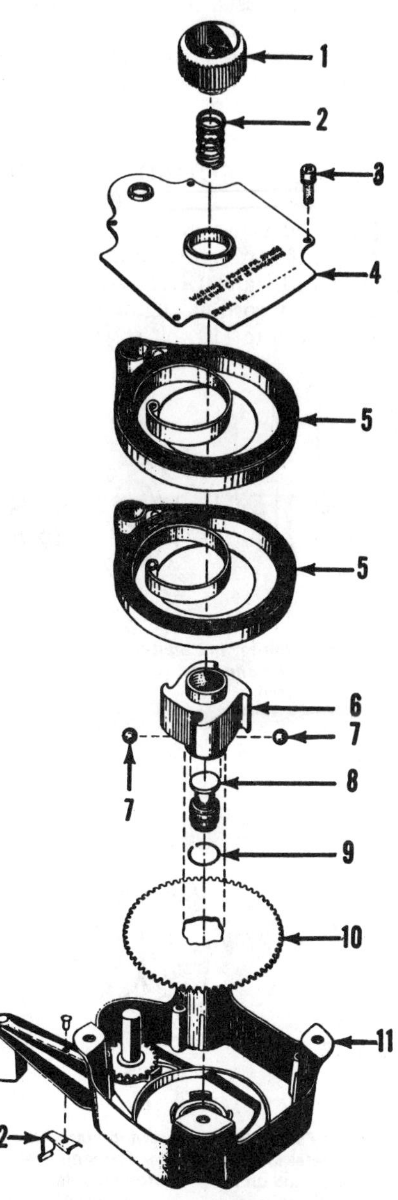

Fig. CL100 – Exploded view of early type Clinton impulse type starter. Some starters having only one power spring (5) used an empty power spring cup as a spacer.

1. Ratchet
2. Spring
3. Cap screws
4. Cover plate
5. Power spring & cup assy.
6. Plunger hub
7. Steel balls (2)
8. Plunger
9. Snap ring
10. Gear
11. Housing, crank & pinion assy.

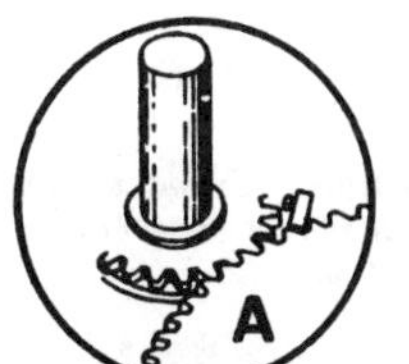

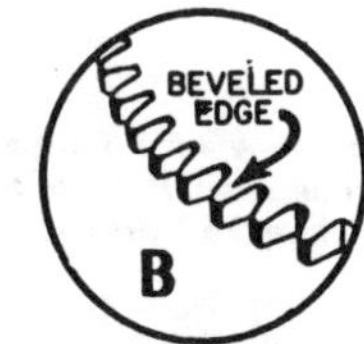

Fig. CL100A – When reinstalling large gear (10 – Fig. CL100), be sure beveled edge of gear is towards open side of housing and lock engages gear as shown.

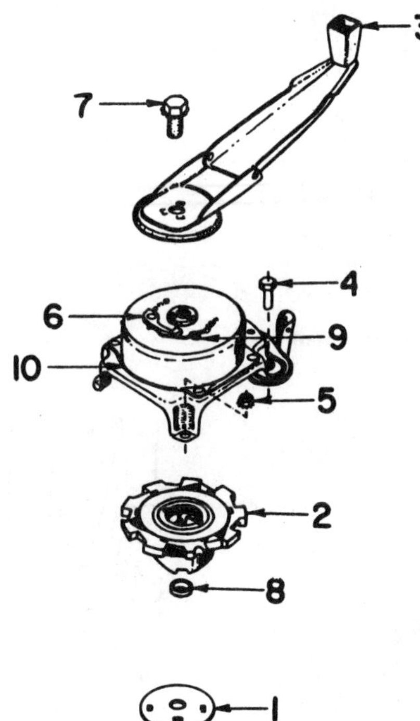

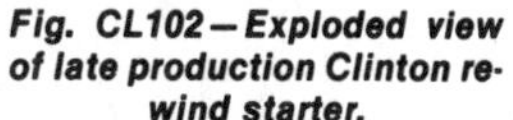

Fig. CL101 – Exploded view of late type Clinton impulse starter. Starter is serviced only in parts indicated by callouts (1) through (9).

1. Starter cup adapter
2. Starter cup
3. Handle assy.
4. Barrel nut
5. Nut
6. Starter pawl
7. Cap screw & washer
8. Starter cup spacer
9. Pawl spring
10. Housing, power spring & gear assy.

Fig. CL102 – Exploded view of late production Clinton rewind starter.

1. Starter cup
2. Snap ring
3. Actuator
4. Snap ring
5. Nut
6. Pawl & pin assy.
7. Actuator spring
8. Pulley
9. Pan head screw
10. Housing
11. Handle
12. Rope
13. Nut & washer assy.
14. Rewind spring

Fig. CL102A – Exploded view of Deluxe Model of Clinton rewind starter of the type used on chain saws and some engines. Starter will operate in either rotation by interchanging springs (5) and inverting pawls (6) and rewind spring (12).

1. Flywheel
2. Plate
3. Flywheel nut
4. Spacer
5. Spring
6. Pawl
7. Snap ring
8. Washer
9. Wave washer
10. Pulley
11. Cup
12. Rewind spring
13. Handle
14. Guide
15. Housing
16. Screen
17. Retainer

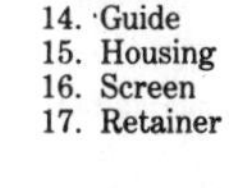

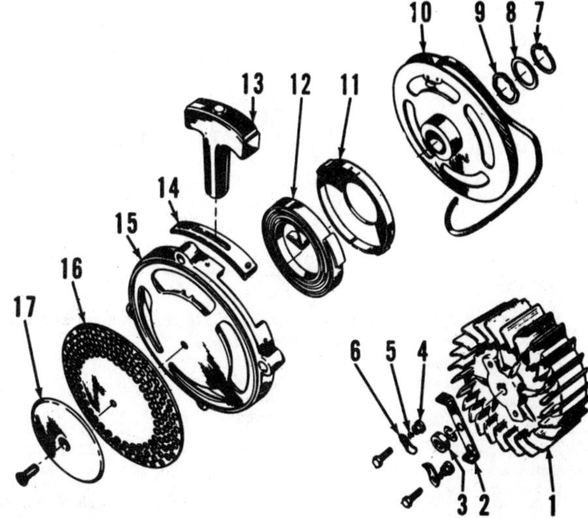

(2) is not included as part of the starter assembly.

REWIND STARTERS

Exploded view of rewind starter currently used on most Clinton engines is shown in Fig. CL102. Other starters used are shown in Figs. CL102A through CL110. Care should be taken when reassembling all starters to be sure the rewind spring is not wound too tightly. The spring should be wound tight enough to rewind the rope, but not so tight that the spring is fully wound before the rope is pulled out to full length. Coat spring and all internal parts with Lubriplate or equivalent grease when reassembling. Be sure spring, pulley and related parts are assembled for correct rotation.

12 VOLT STARTER AND LIGHTING COIL

Refer to Figs. CL113, CL114, CL115 and CL116. Service and/or parts for the 12 volt starter are available at American Bosch Service Centers.

The 12 volt lighting coils are mounted on the magneto armature core as shown in Fig. CL115 and CL116. Wiring diagram when unit is used for lighting circuit only is shown in Fig. CL115.

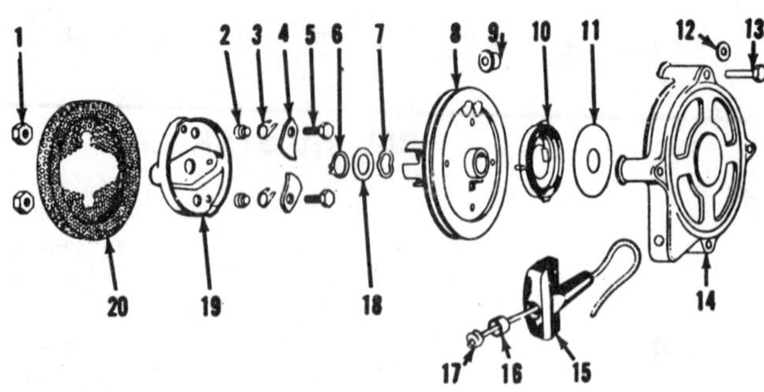

Fig. CL103—Exploded view of an earlier type Clinton rewind starter.

1. Locknut
2. Pawl spacer
3. Pawl spring
4. Pawl
5. Cap screw
6. Snap ring
7. Wave washer
8. Pulley
9. Rope bushing
10. Rewind spring & cup asy.
11. Washer
12. Lockwasher
13. Barrel nut
14. Housing
15. Handle
16. Retainer
17. Rope
18. Washer
19. Starer cup
20. Rotating screen

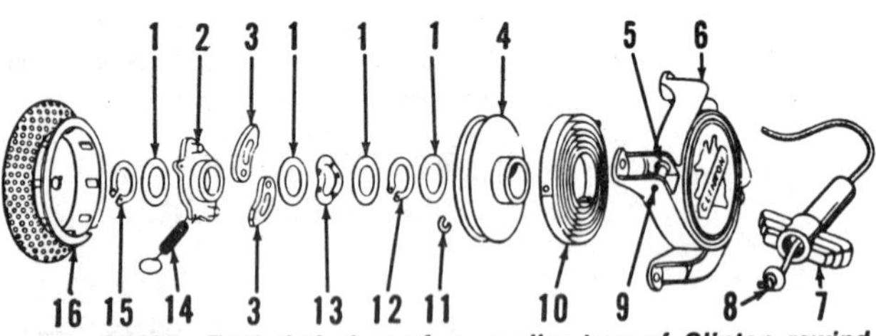

Fig. CL104—Exploded view of an earlier type of Clinton rewind starter.

1. Flat washer
2. Drive plate
3. Drive pawls
4. Pulley
5. Rope pulley
6. Housing
7. Handle
8. Rope
9. Roll pin
10. Rewind spring
11. Snap ring
12. Snap ring
13. Wave washer
14. Tension spring
15. Snap ring
16. Starter cup

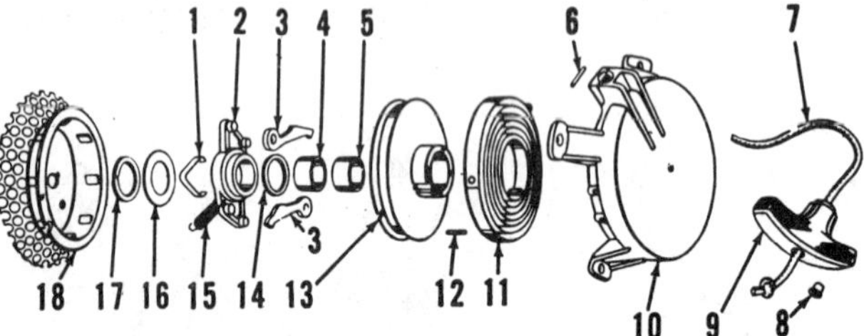

Fig. CL105—Exploded view of an earlier type Clinton rewind starter. Design of housing (10) may vary from that shown. Bushing (5) is not used in some starters.

1. Retaining spring
2. Pawl plate
3. Pawls
4. Bushing
5. Bushing
6. Roll pin
7. Rope
8. Washer
9. Handle
10. Housing
11. Rewind spring
12. Roll pin
13. Pulley
14. Spacer
15. Tension spring
16. Washer
17. Snap ring
18. Starter cup

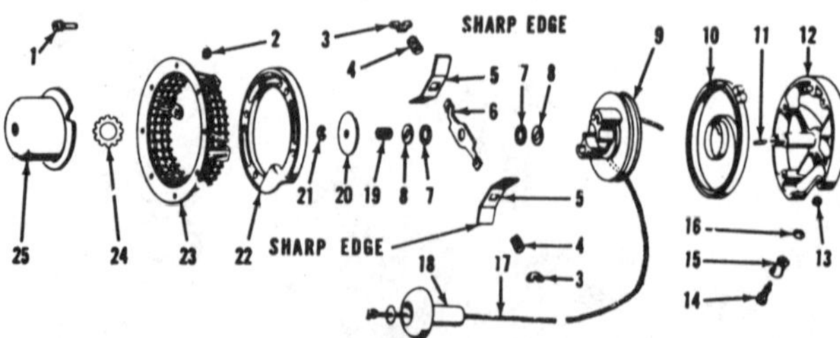

Fig. CL106—Exploded view of Bulldog starter used on some early production engines.

1. Screw & washer assy.
2. Nut & retainer assy.
3. Retainer
4. Spring
5. Friction shoe
6. Brake lever
7. Washer
8. Slotted washer
9. Pulley
10. Rewind spring
11. Centering pin
12. Housing
13. Nut
14. Shoulder screw
15. Roller
16. Washer
17. Rope
18. Handle
19. Spring
20. Washer
21. Snap ring
22. Flange
23. Base
24. Lockwasher
25. Pulley (cup)

When unit is used to provide a battery charging current, a rectifier must be installed in the circuit to convert the AC current into DC current as shown in Fig. CL116.

110 VOLT ELECTRIC STARTERS

Exploded views of the two types of 110 volt electric starters are shown in Figs. CL117 and CL118. Always connect starter cord at engine before connecting cord to 110 volt power source.

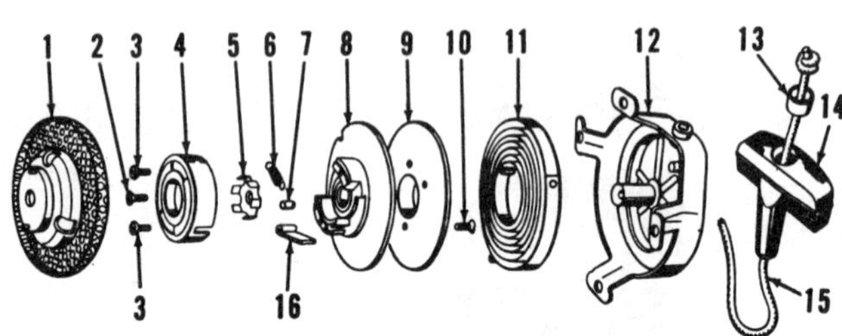

Fig. CL107—Exploded view of Eaton rewind starter used on some early production engines.

1. Starter pulley
2. Screws
3. Screws (2)
4. Retainer
5. Brake
6. Tension spring
7. Spring retainer
8. Hub
9. Plate
10. Screw
11. Rewind spring
12. Housing
13. Cup
14. Handle
15. Rope

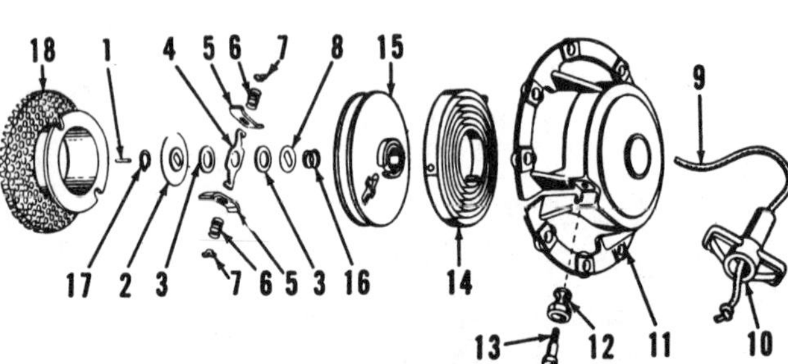

Fig. CL108—Exploded view of Fairbanks-Morse rewind starter used on early production engines. Refer to Fig. CL109 for a second type of Fairbanks-Morse starter used on other Clinton models.

1. Centering pin
2. Washer
3. Fiber washer
4. Brake lever
5. Friction shoe
6. Spring
7. Retainer
8. Washer
9. Rope
10. Handle
11. Housing
12. Roller
13. Screw
14. Rewind spring
15. Pulley
16. Spring
17. Snap ring
18. Pulley (cup)

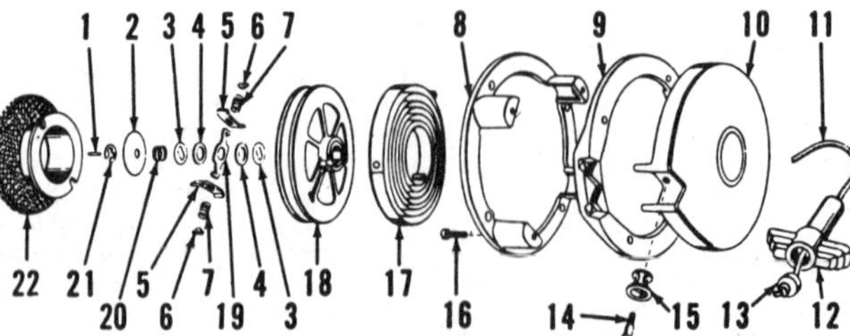

Fig. CL109—Exploded view of a Fairbanks-Morse rewind starter used on some Clinton engines. Refer to Fig. CL108 for another type of Fairbanks-Morse starter used.

1. Centering pin
2. Washer
3. Washer
4. Fiber washer
5. Friction shoe
6. Retainer
7. Spring
8. Mounting flange
9. Middle flange
10. Housing
11. Rope
12. Handle
13. Cup
14. Screw
15. Roller
16. Screws
17. Rewind spring
18. Pulley
19. Brake lever
20. Spring
21. Snap ring
22. Pulley (cup)

ELECTRA-START

Electra-Start models have a built-in battery, starter and flywheel mounted alternator. Early models used a wet cell 12 volt battey mounted on mower deck. Late models use a 12-cell, nickel-cadmium battery which attaches to crankcase, making the unit fully self-contained.

Electra-Start models are equipped with the bulb type fuel primer which pressurizes the carburetor float chamber when bulb is depressed, forcing a small amount of fuel out main nozzle.

Nominal voltage of the nickel-cadmium battery is 15 volts. Charging rate of the flywheel alternator is 0.2-0.25 DC amps. About 40-125 amp-seconds are normally required to start the engine and recovery time to full charge should be 3-10 minutes of operation. About 20 minutes running time should be allowed when battery or engine is first put into service.

Fig. CL113—View of American Bosch 12 volt starter mounted on a late production horizontal crankshaft engine. Square-shaped unit below starter motor is a rectifier to convet AC current from lighting coils on magneto armature core to DC current.

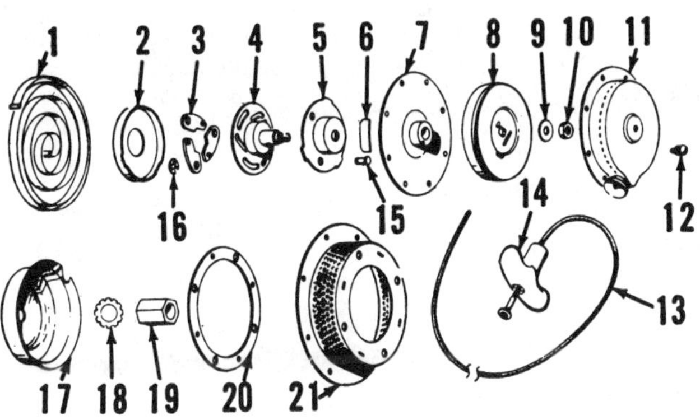

1. Rewind spring
2. Cover
3. Pawls
4. Cam plate & axle
5. Drive plate*
6. Spring*
7. Plate
8. Drum
9. Washer
10. Nut
11. Housing
12. Screws
13. Rope
14. Handle
15. Pin
16. Snap ring
17. Rope pulley
18. Lockwasher
19. Driven nut
20. Ring
21. Housing

Fig. CL110—Exploded view of typical Schnacke starter unit used on early production engines. Asterisk indicates part not used on all starters.

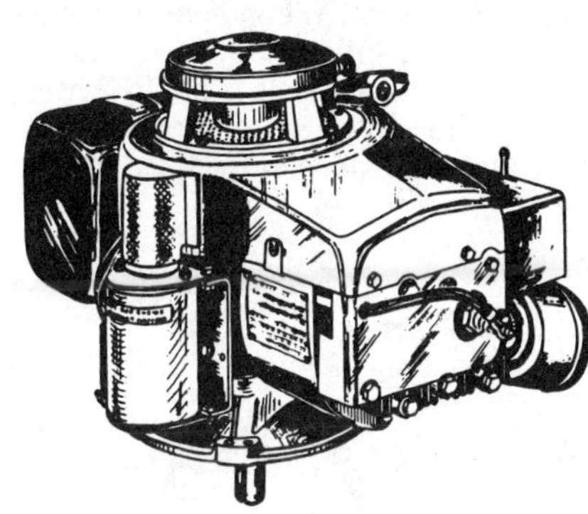

Fig. CL114—View of American Bosch 12 volt starter mounted on a late production vertical crankshaft engine.

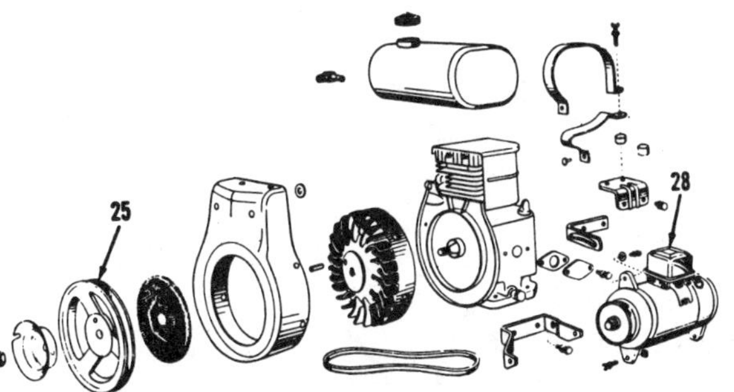

Fig. CL111—The 12 volt starter-generator (28) is driven (and drives) by a belt and pulley (25).

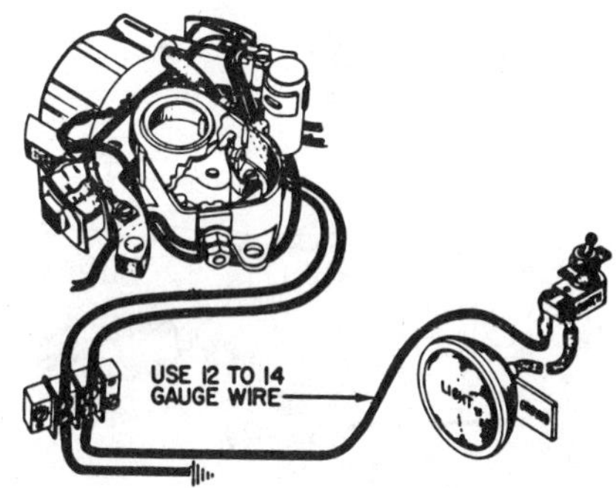

Fig. CL115—View showing wiring circuit from lighting coils when used for lighting purposes only.

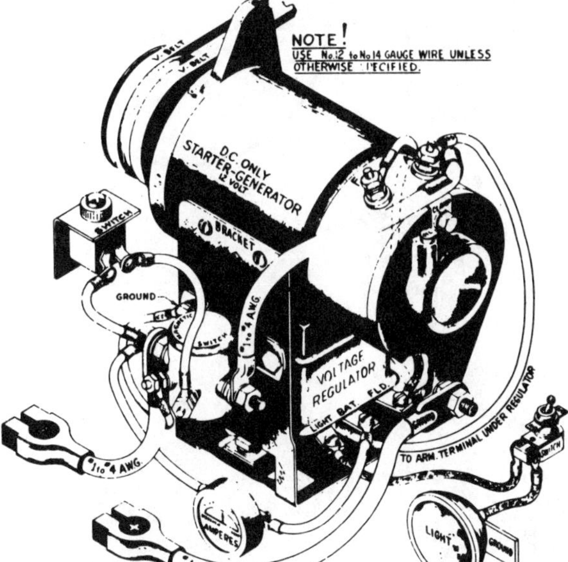

Fig. CL112—Wiring diagram for 12 volt DC starter-generator.

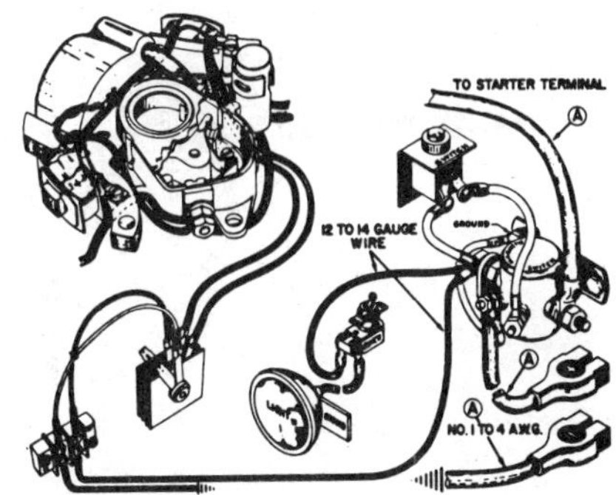

Fig. CL116—View showing wiring circuit from lighting coils. A rectifier is included in circuit to provide DC current for charging a 12 volt battery.

The alternator output (200-250 milliamperes) cannot be measured with regular shop equipment. If trouble is encountered, start the engine and operate at rated speed. Disconnect generator lead from red (positive) battery lead and check generator output using one of the following methods:

1. Connect one lead of a voltmeter or small test light to generator lead and ground the remaining lead. Voltmeter should indicate charging current or test light should glow, indicating charging current.
2. In the absence of test equipment, momentarily touch the disconnected generator lead to a suitable ground and watch for a spark.

If test equipment or spark indicated charging current, the generator should be considered satisfactory.

If a charging current is not indicated, remove the flywheel and renew the generator coil which is mounted on one leg of magneto pole shoe. The halfwave rectifier is built into magneto coil and is not renewable separately.

POLYURETHANE AIR CLEANER

Some Clinton models are equipped with a polyurethane air filter element. Element should be removed and washed in nonflammable solvent (mild solution of detergent and water) after every 10 hours of operation. Reoil element with SAE 10 motor oil and wring out excess oil. Insert the element evenly into air cleaner housing and snap cover in place. Refer to Fig. CL119.

REMOTE CONTROLS

Several different remote control options are available. Refer to Fig. CL122. Views A through G are for engines with following basic numbers: IMB basic numbers 494 and 498; old basic numbers 900, 960 990, B1260 and B1290.

View A shows standard fixed speed for hoizontal shaft engine. B, C and F show available remote control options. Typical cable and handle assembly is shown in view D. Detail of pivot pin to block is shown in E. View G shows method of securing control wire to lever on some applications.

The remote control allows the engine speed to be regulated by movement of the remove control lever some distance away from the engine. The engine speed will be maintained by the engine governor at any setting of the remote control lever within the prescribed speed range of the engine. Maximum engine speed should not exceed 3600 rpm.

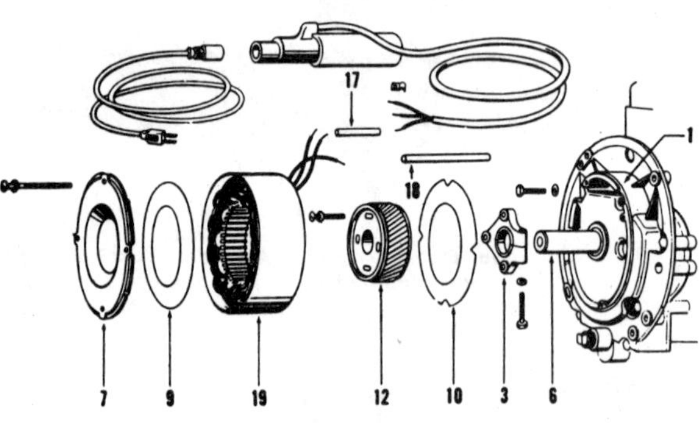

Fig. CL117—Exploded view of the 110 volt below deck electric starter, which is available on some vertical shaft engines.

1. Engine base
3. Rotor clamp
6. Crankshaft
7. Cover
9. Insulator gasket
10. Insulator gasket
12. Rotor
17. Connector sleeves (3)
18. Insulating sleeve
19. Stator

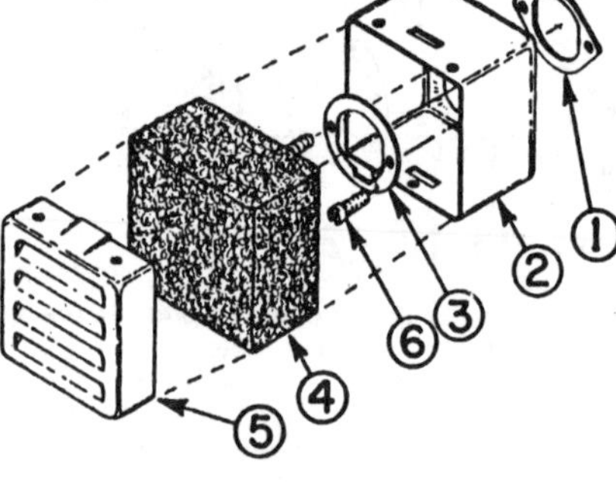

Fig. CL119—Exploded view of polyurethane air cleaner.

1. Gasket
2. Body
3. Retainer
4. Element
5. Cover
6. Screws

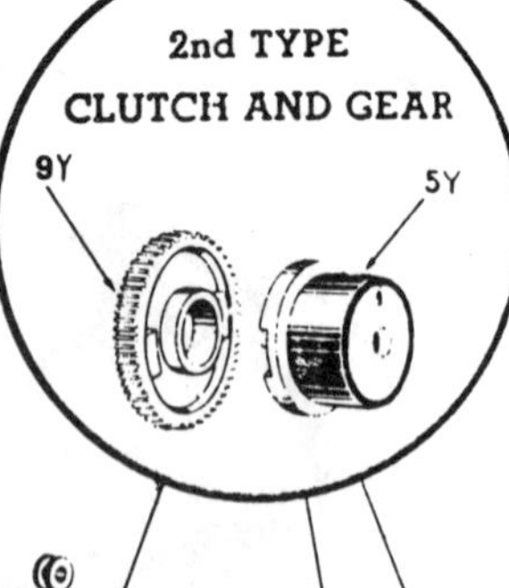

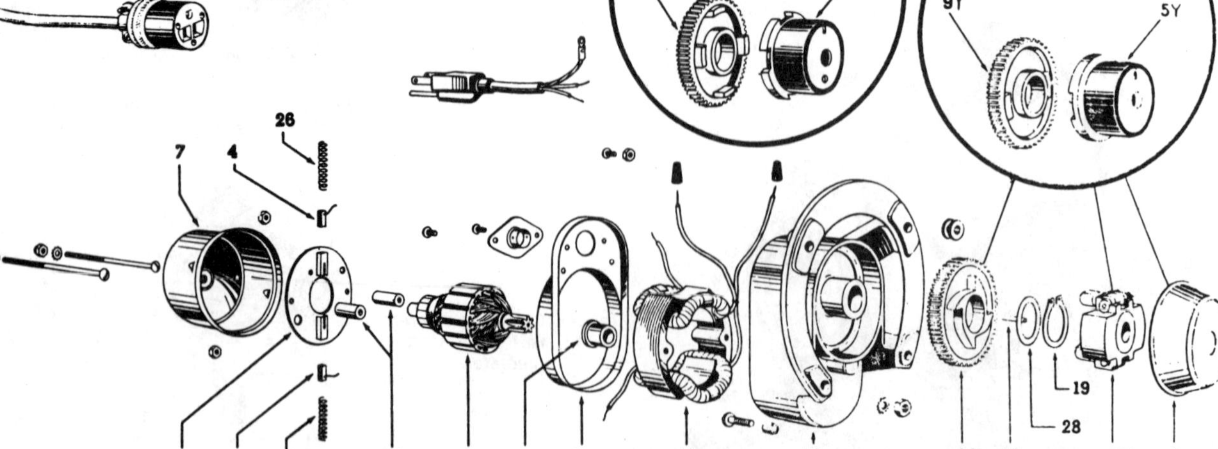

Fig. CL118—Exploded view of the 110 volt electric starter available on some models.

2. Armature
3. Bearing
4. Brushes
5X. Latest clutch assy.
5Y. Second clutch assy.
5Z. Earliest clutch assy.
7. Cover
8. Cup
9X. Latest clutch gear
9Y. Second clutch gear
9Z. Earliest clutch gear
11. Upper housing
12. Lower housing
13. Field winding
16. Centering pin
17. Brush mounting plate
24. Spacers
26. Springs
28. Fiber washer

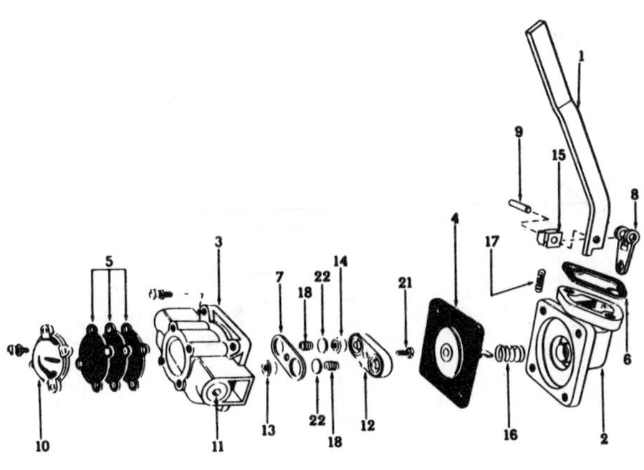

Fig. CL121—Exploded view of mechanical fuel pump used on some engines.
1. Rocker arm
2. Pump body
3. Top cover
4. Diaphragm
5. Pulsator
6. Gasket
7. Valve gasket
8. Lever link
9. Rocker arm pin
10. Cover plate
11. Pipe plug
12. Valve retainer
13. Valve seat
14. Valve seat
15. Rocker spring clip
16. Diaphragm spring
17. Rocker arm spring
18. Valve spring
21. Valve retainer
22. Valve (2)

To adjust remote control with the engine running:

1. Loosen screw in swivel nut on control lever.
2. Move control lever to high speed position (view B–Fig. CL122).
3. Move control wire through control casing and swivel nut until maximum desired engine speed is obtained. Do not exceed maximum engine speed of 3600 rpm. Lever should remain in high speed position.
4. Tighten screw in swivel nut to hold control wire.
5. Move control lever to slow speed position to be sure engine can slow down to idle speed.

PTO AND SPEED REDUCER UNITS

AUXILIARY PTO. Some vertical crankshaft models are equipped with an auxiliary pto as shown in Fig. CL123. Unit is lubricated by oil in engine crankcase. To disassemble unit, drive the pins (6, 7 and 22) partially out of shaft (11), turn shaft half-turn and pull pins. Remove shaft and gears (2 and 4). Remove output shaft (12) and gear (21) in similar manner.

Fig. CL123—Exploded view of auxiliary pto unit used on some vertical crankshaft engines. Gear ratio is either 15:1 or 30:1.
1. Flange bearing
2. Drive gear
3. Gasket
4. Worm gear
5. Thrust washer
6. Pin
7. Pin
8. Pin
9. Expansion plug
10. Needle bearing
11. Worm shaft
12. Pto shaft
13. Oil seal
14. Housing
15. Oil filler cap
16. Gasket
17. Oil drain plug
18. Cap screw
19. Expansion plug
20. Sleeve bearing
21. Pto gear
22. Pin

SPEED REDUCERS. Speed reducers (gear reduction units) with ratios of 2:1, 4:1 and 6:1 are available on a number of horizontal shaft engines.

Refer to Fig. CL124 for exploded view of typical speed reducer for engines of less than 5 horsepower (3.7 kW). Unit should be lubricated by filling with SAE 30 oil to the level plug. Oil level should be checked after every 10 hours of operation. Drain and refill the reduction unit with clean SAE 30 oil after every 100 hours of operation.

Use Fig. CL124 as disassembly and reassembly guide. Unit can be mounted on engine in any of four positions; however, be sure outer housing is installed so filler plug is up.

An exploded view of typical speed reducer used on engines of 5 horsepower (3.7 kW) and larger is shown in Fig. CL125. Units used on early engines support output end of crankshaft in sleeve bearing (1) as shown. Late units support output end of crankshaft in a tapered roller bearing.

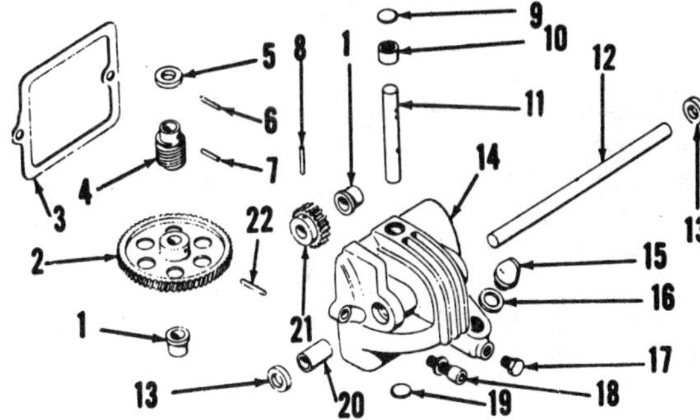

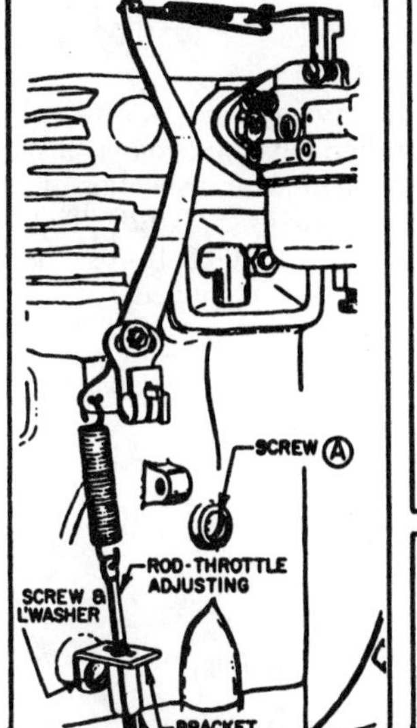

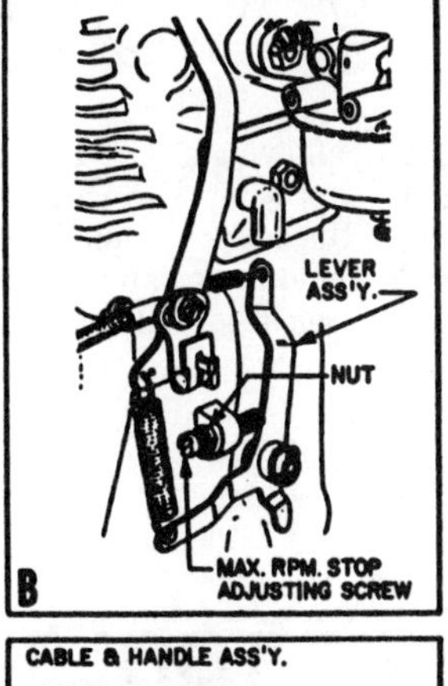

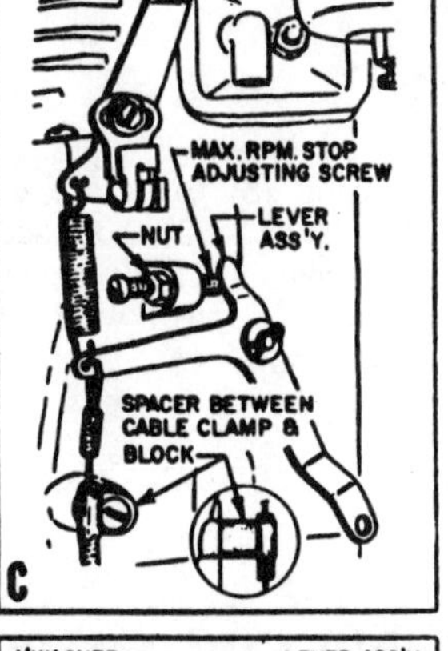

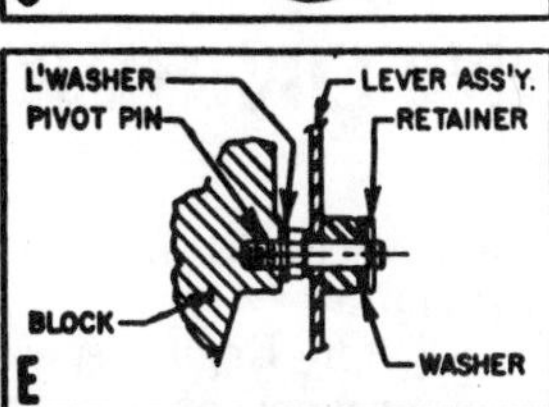

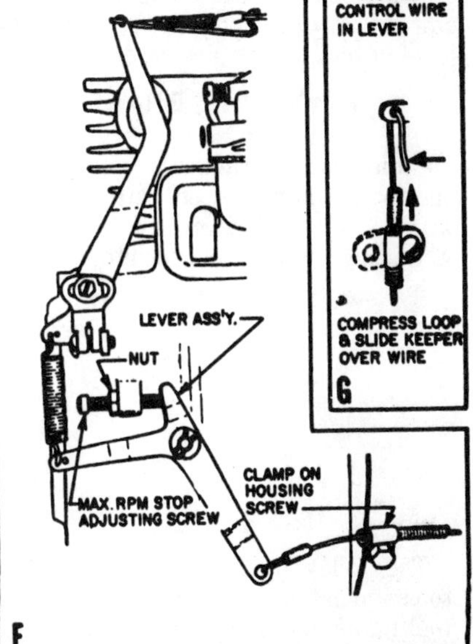

Fig. CL122—Remote control hook-ups for various Clinton engines. Refer to text for engine application and adjustment procedures.

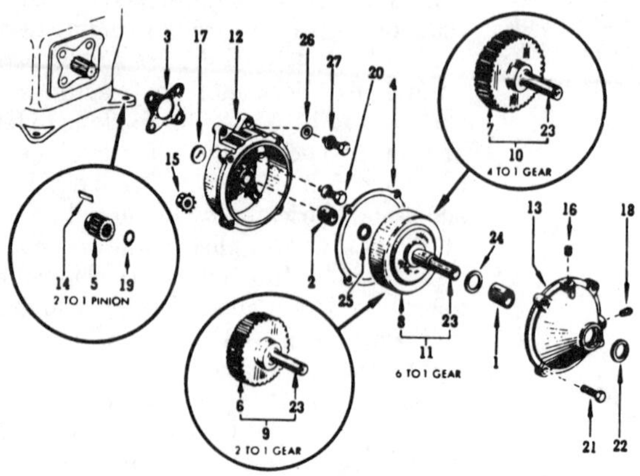

Fig. CL124—Exploded view of gear reduction unit used on engines of less than 5 horsepower (3.7 kW). Gear ratio may be 2:1, 4:1 or 6:1. Sleeve bearings must be reamed after installation.

1. Sleeve bearing
2. Sleeve bearing
3. Gasket
4. Gasket
5. 2:1 pinion
6. 2:1 gear
7. 4:1 gear
8. 6:1 gear
9. Gear & shaft
10. Gear & shaft
11. Gear & shaft
12. Inner housing
13. Outer housing
14. Key
15. Nut & washer assy.
16. Breather plug
17. Expansion plug
18. Oil level plug
19. Snap ring
20. Screw
21. Screw
22. Oil seal
23. Shaft
24. Thrust washer
25. Thrust washer
26. Flat washer
27. Lockwasher

Fig. CL125—Exploded view of gear reduction unit used on engines larger than 5 horsepower (3.7 kW). Gear ratio may be 2:1, 4:1 or 6:1. Sleeve bearing (1) must be finish reamed after installation. Units used on late production engines have tapered roller engine main bearing instead of sleeve bearing (1) shown.

1. Sleeve bearing
2. Flange bearing
3. Gasket
4. Gasket
5. 2:1 pinion
6. 2:1 gear
7. 4:1 gear
8. 6:1 gear
9. Gear & shaft
10. Gear & shaft
11. Gear & shaft
12. Inner housing
13. Outer housing
14. Key
15. Snap ring
16. Screw & washer
17. Screw & washer
18. Oil seal
19. Shaft

After unit has been drained of oil or after reassembly, initially fill unit with same oil as used in engine. Lubricating oil supply is thereafter maintained from engine crankcase through passage drilled in inner housing (12). Gaskets (4) used between inner housing and engine crankcase or bearing plate are available in several thicknesses. Use proper thickness gasket to maintain specified crankshaft end play.

As unit can be mounted on engine in any of four different positions, be sure outer housing (13) is installed with filler plug to top and oil level drain plugs down. Note that after engine has been started, oil level will usually be below level of oil level plug.

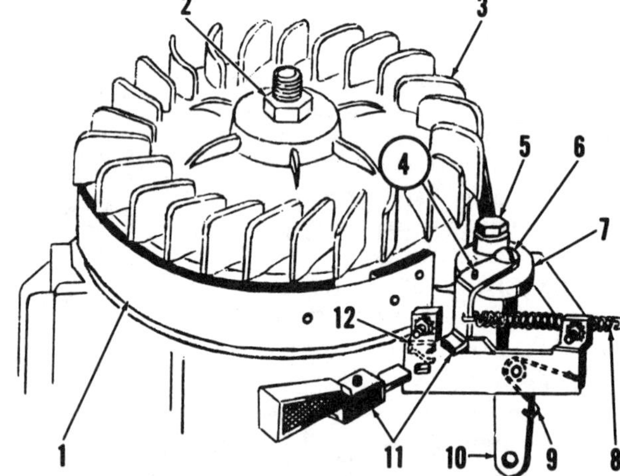

Fig. CL126—Schematic drawing of "Touch 'N' Stop" brake used on some vertical crankshaft rotary lawnmower engines.

1. Brake band
2. Flywheel nut
3. Flywheel
4. Pin
5. Pivot bolt
6. Screw
7. Cam
8. Spring
9. Spring
10. Pawl lever
11. Cocking lever
12. Flat spring

FLYWHEEL BRAKE

Some vertical crankshaft models used on rotary lawnmowers are equipped with a "Touch 'N' Stop" flywheel brake. Actuating the control assembly releases pawl lever (10–Fig. CL126) allowing spring (8) to rotate cam (7) applying brake band (1). To reset, move the cocking lever (11) to release brake band and latch the pawl with brake in released position.

Required service, such as renewal of worn or broken parts or broken springs, should be evident after inspection of unit and reference to Fig. CL126. Tighten flywheel nut (2) to 50-58 ft.-lbs. (68-79 N·m) on models so equipped.

CLINTON SERVICE TOOLS

The following list of special service tools indicates the Clinton part number, tool name and usage by engine groups. Following the tool list are the engine group listing and a cross-reference by engine model number to tool usage group.

Clinton Engine special service tools are available by part number from Clinton Central Parts Distributors listed elsewhere in this manual.

Bearing Tools

951-48 REAMER GUIDE. Three bushings included with the following inside diameters: 0.878, 1.0003 and 1.030 inch. Used to line ream bearing in blocks of engine groups II, III, IV, V, VI, VII, VIII, IX, X, XIII, XIV and XV. Used also to ream bearings in bases of engine group VIII.

951-148 GUIDE. A bushing that is used when reaming bearings in bases of engine group XV that have a 1.000-inch dia. bearing bore.

951-18 and 951-39 BEARING DRIVER. Used to remove and install bearings in two-bearing blocks, or in mounting flanges of engine group II and VIII.

951-29 REAMER. Diameter, 0.8773-0.8778 inch. Used to ream bearings in blocks of engine groups II, IV, VI, VII and VIII, and bearing plates of engine groups VII and VIII (using block for guide). Can also be used to ream bearing plates for engine group X if sleeve bearing block of a group VII or VIII engine is available for use as a guide.

951-30 REAMER. Diameter, 0.7518-0.7523 inch. Used to ream bearing in bearing plates of engine groups II, III, IV, V and VI and bearing in blocks of engine group V.

951-44 REAMER. Diameter, 1.0003-1.0008 inch. Used to enlarge bore in bases of engine group XV having a 7/8-inch bore so that a bronze bearing can be installed.

951-59 REAMER. Diameter, 1.0023-1.0028 inch. Used to ream bearing in blocks of engine groups III and X.

951-60 REAMER. Diameter, 0.8144-0.8155 inch. Used to ream bearings in blocks of engine groups XIII, XIV and XV.

951-63 REAMER. Diameter, 0.8758-0.8763 inch. used to ream bearings in bases of engine groups XIV and XV.

951-143 REAMER. Diameter, 1.1218-1.1220 inch. Used as second step to enlarge 1-inch diameter bore of base in engine group XV so that bronze bearing can be installed. Also see 951-144.

951-144 REAMER. Diameter, 1.1042-1.1062 inch. Used as first step to enlarge 1-inch diameter bore of base in engine group XV so that bronze bearing can be installed. Also see 951-143.

951-146 REAMER. Diameter, 0.9338-0.9340 inch. Used as second step to enlarge bore in aluminum blocks of engines in groups XIV and XV so that bronze bearing can be installed. Also see 951-147.

941-147 REAMER. Diameter, 0.9167-0.9187 inch. Used as first step to enlarge bore in aluminum blocks of engines in groups XIV and XV so that bronze bearing can be installed. Also see 951-146.

Oil Seal Loaders (Installation Sleeves)

For loading (installing) oil seals over shafts, the following oil seal loaders are available:

951-12 . . . . for 3/4-inch diameter shaft
951-13 . . . . for 57/64-inch diameter shaft
951-40 . . . . . . for 1 3/8-inch diameter shaft
941-49 . . . for 1-3/16 inch diameter shaft
951-55 . . . . for 7/8-inch diameter shaft
951-56 . . . . for 1 inch diameter shaft
951-57 . . . . for 3/4-inch diameter shaft
951-47 . . . . for 1/2-inch diameter shaft
951-14 . . . . for 57/64-inch diameter shaft
951-118 . . . . . for 1 1/4-inch diameter shaft
951-145 . . . . . . for 3/4-inch diameter shaft

Oil Seal Pullers and Drivers

951-50 OIL SEAL PULLER. For removing oil seal from flywheel side of engines in groups XIII, XIV and XV.

951-16 OIL SEAL DRIVER. Inside dia., 1.010 inch; outside dia., 1.250 inch; length, 4 inches.

951-17 OIL SEAL DRIVER. Inside dia., 0.885 inch; outside dia., 1.125 inch; length, 4 inches.

951-62 OIL SEAL DRIVER. Inside dia., 0.760 inch; outside dia., 1.000 inch; length 3 inches.

Valve Tools, Four-Stroke Engines

951-32 VALVE SPRING COMPRESSOR. For engines in groups V, VI, VII, VIII, IX, X, XIII, XIV and XV. Also see 951-67.

951-67 VALVE SPRING COMPRESSOR. For engines in groups V, VI, VII, VIII, IX, X, XIII, XIV and XV. Also see 951-32.

951-22 VALVE GUIDE REAMER. (0.281-inch dia.) Used to ream valve guides for installation of valves with oversize (9/32-inch diameter) stems in engine groups VI, VII, VIII, IX and X.

951-43 VALVE GUIDE REAMER. (0.261-inch dia.) Used to ream valve guides for installation of valves with 0.010-inch oversize stems in engine groups XIII, XIV and XV.

951-69 VALVE GUIDE REAMER. Used to clean standard size valve guides (0.250-inch dia.) for engines in groups V, VI, VII, VIII, IX, X, XIII, XIV and XV.

951-58-500 PILOT ASSEMBLY. A kit of valve guide pilots containing one each of the following:

951-137, 0.2490-0.2505-inch dia.
951-138, 0.2505-0.2520-inch dia.
951-139, 0.2605-0.2620-inch dia.
951-140, 0.2805-0.2820-inch dia.

Used in conjunction with 951-37 Valve Seat Cutter, 951-41 Valve Seat Counterbore Cutter and 951-61 Valve Seat Counterbore Cutter on engines in groups V, VI, VII, VIII, IX, X, XIII, XIV and XV.

951-37 VALVE SEAT CUTTER. Used to recondition valve seats on engines in groups V, VI, VII, VIII, IX, X, XIII, XIV and XV. See also 951-58-500. Set of three new cutting edges are available as Part No. 951-131.

951-136 VALVE LAPPING TOOL. For lapping valves to seats on engines in groups V, VI, VII, VIII, IX, X, XIII, XIV and XV.

951-41 CUTTER. Used to ream valve seat counterbore for installation of 0.040-inch oversize valve seat insert in engines of groups XIV and XV. Also see 951-58-500.

951-61 CUTTER. Used to cut a counterbore for installation of service valve seat insert in engines of groups V, VI, VII, VIII, IX and X. See also 951-58-500.

951-52 DRIVER. Used to drive valve seat inserts into place on engines in groups V, VI, VII, VIII, IX, X, XIII, XIV and XV.

951-53 ROLLING TOOL. Used to peen or roll metal around outside edge of valve seat insert after installing new insert in engines of groups V, VI, VII, VIII, IX, X, XIII, XIV and XV.

Piston Ring Tools

951-34 COMPRESSOR. Used for compressing piston rings on pistons of 1 3/4 to 3 1/8-inch diameter on engines in groups V, VI, VII, VIII, IX, X, XIII, XIV and XV.

951-150 SLEEVE (1 7/8-inch bore)
951-153 SLEEVE (2 1/8-inch bore).
Used for installing standard size and 0.020-inch oversize piston assembly in blocks of engines in groups I, II, III and IV.

Flywheel and Crankshaft Tools

IMPACT NUTS. Impact or knock-off nuts are available for removing flywheels as follows:

951-23, for crankshafts having 7/16-inch threads.
951-36, for crankshafts having 7/8-inch threads.
951-66, for crankshafts having 1/2-inch threads.

951-133 FLYWHEEL PULLER. For use on engines in groupe I, II, III, IV, V, VII, VIII, IX, X, XIII, XIV and XV.

941-42 FLYWHEEL HOLDER. All engines. New belt for holder available as Part No. 951-151.

951-45 HAND CRANK. Used for turning engine when adjusting breaker point gap, checking timing, etc., on engines in groups I, II, III, IV, V, VI, VII, VIII, IX, X, XIII, XIV and XV.

951-64 CRANKSHAFT RUNOUT GAGE. All engines; used to check for bent crankshafts.

Camshaft Tools

951-46 CAM AXLE DRIVER. Used to drive cam axle from block in engines of groups VII, VIII, IX and X.

ENGINE GROUPING FOR TOOL USAGE LISTING

GROUP I

E65.

GROUP II

AVS200, SVA200-1000, VS200, VS200-1000, AVS400, VS400, VS400-1000.

GROUP III

VS200-2000, VS200-3000, VS200-4000, AVS400-1000, BVS400, CVS400-1000, VS400-2000, VS400-3000, VS400-4000, 501-0000-000, 502-0000-000 and 503-0000-000.

GROUP IV

200, A200, A400, A400-1000, CK590, and 500-0000-000.

GROUP V

300, A300, 350.

GROUP VI

VS300.

GROUP VII

650, 700A, B700, C700, D700, D700-1000, D700-2000, D700-3000, 800, A800, 900, 900-1000, 900-2000, 900-3000, 900-4000, A&B1100, C1100, 494-0000-000, 494-0001-000.

GROUP VIII

VS700, VS750, VS800.

GROUP IX

D1100, 1200, 1200-1000, 1200-2000, A1200, B1290-1000, 498-0300-000, 498-0301-000.

GROUP X

VS900, V1000-1000, VS1000, V1100-1000, VS1100, VS1100-1000, V1200-1000, VS1200, 497-0000-000, 499-0000-000.

GROUP XIII

429-003-000, 431-0003-000.

GROUP XIV

100, 100-1000, 100-2000, 2100, A2100, A2100-1000, A2100-2000, 3100, 3100-1000, 3100-2000, 3100-3000, H3100-1000, 4100, 4100-1000, 4100-2000, 400-0000-000, 402-0000-000, 404-0000-000, 406-0000-000, 408-0000-000, 410-0000-000, 424-0000-000, 426-0000-000, 492-0000-000.

GROUP XV

V100-1000, VS100, VS100-1000, VS100-2000, VS100-3000, VS100-4000, VS2100, VS2100-1000, VS2100-2000, VS2100-3000, VS3000, FV3100-1000, AFV3100-1000, AV3100-1000, AVS3100-2000, AVS3100-3000, V3100-1000, V3100-2000, VS3100, VS3100-1000, VS3100-2000, VS3100-3000, AVS4100-1000, AVS4100-2000, VS4100-1000, VS4100-2000, 401-0000-000, 403-0000-000, 405-0000-000, 407-0000-000, 407-0002-000, 409-0000-000, 411-0000-000, 411-0002-000, 415-0000-000, 415-0002-000, 417-0000-000, 419-0000-000, 435-0003-000, 455-0000-000.

ENGINE TO TOOL USAGE GROUP CROSS-REFERENCE

E-65 I
100 XIV
100-1000 XIV
100-2000 XIV
V100-1000 XV
VS100 XV
VS100-1000 XV
VS100-2000 XV
VS100-3000 XV
VS100-4000 XV
200 IV
A-200 IV
AVS200 II
AVS200-1000 II
VS200 II
VS200-1000 II
VS200-2000 III
VS200-3000 III
VS200-4000 III
300 V
A300 V
VS300 VI
350 V
A-400 IV
A400-1000 IV
AVS400 II
AVS400-1000 III
BVS400 III
CVS400-1000 III
VS400 II
VS400-1000 II
VS400-2000 III
VS400-3000 III
VS400-4000 III
GK590 IV
650 VII
700A VII
B700 VII
C700 VII
D700 VII
D700-1000 VII
D700-2000 VII
D700-3000 VII
VS700 VIII
VS750 VIII
800 VII
A800 VII
VS800 VIII
900 VII
900-1000 VII
900-2000 VII
900-3000 VII
900-4000 VII
VS900 X
V1000-1000 X
VS1000 X
A&B1100 VII
C1100 VII
D1100 IX
V1100-1000 X
VS1100 X
VS1100-1000 X
1200 IX
1200-1000 IX
1200-2000, A-1200 IX
B1290-1000 IX
V1200-1000, VS1200 X
2100, A2100 XIV
A2100-1000 XIV
A2100-2000 XIV
VS2100 XV
VS2100-1000 XV
VS2100-2000 XV
VS2100-3000 XV
VS3000 XV
3100, 3100-1000 XIV
3100-2000 XIV
3100-3000 XIV
H3100-1000 XIV
FV3100-1000 XV
AFV3100-1000 XV
AV3100-1000 XV
AV3100-2000, AVS3100 XV
AVS3100-1000 XV
AVS3100-2000 XV
AVS3100-3000 XV
V3100-1000 XV
V3100-2000 XV
VS3100 XV
VS3100-1000 XV
VS3100-2000 XV
VS3100-3000 XV
4100 XIV
4100-1000 XIV
4100-2000 XIV
AVS4100-1000 XV
AVS4100-2000 XV
VS4100-1000 XV
VS4100-2000 XV
400-0000-000 XIV
401-0000-000 XV
402-0000-000 XIV
403-0000-000 XV
404-0000-000 XIV
405-0000-000 XV
406-0000-000 XIV
407-0000-000 XV
407-0002-000 XV
408-0000-000 XIV
409-0000-000 XV
410-0000-000 XIV
411-0000-000 XV
411-0002-000 XV
415-0000-000 XV
415-0002-000 XV
417-0000-000 XV
419-0000-000 XV
424-0000-000 XIV
426-0000-000 XIV
429-0003-000 XIII
431-0003-000 XIII
435-0003-000 XV
455-0000-000 XV
492-0000-000 XIV
494-0000-000 VII
497-0000-000 X
498-0300-000 IX
498-0301-000 IX
499-0000-000 X
500-0000-000 IV
501-0000-000 III
502-0000-000 III
503-0000-000 III

CLINTON CENTRAL PARTS DISTRIBUTORS

(Arranged Alphabetically by States)

These franchised firms carry extensive stocks of repair parts. Contact them for name and address of nearest distributor who many have the parts you need.

Charlie C. Jones Battery & Electric Company
Phone: (602) 272-5621
2440 West McDowell Road
P.O. Box 6654
Phoenix, Arizona 85005

Garden Equipment Company
Phone: (213) 633-8105
6600 Cherry Avenue
Long Beach, California 90805

Air Cooled Engines
Phone: (408) 295-4790
1076 North 10th
San Jose, California 95112

Radco Distributors Company
Phone: (904) 733-7956
4909 Victor Street
P.O. Box 5459
Jacksonville, Florida 32207

Power Products Distributors, Incorporated
Phone: (813) 748-0734
1523 8th Avenue
P.O. Box 7
Palmetto, Flordia 33561

Rogers Engines
Phone: (305) 689-8300
3330 West 45th Street
West Palm Beach, Florida 33407

Lanco Engines Services, Incorporated
Phone: (808) 845-2244
720 Moowoa Street
Honolulu, Hawaii 96817

Midwest Engine Warehouse
Phone: (312) 883-1200
515 Romans Road
Elmhurst, Illinois 60126

Mid-East Power Equipment Company
Phone: (606) 253-0688
185 Lisle Road
P.O. Box 40505
Lexington, Kentucky 40505

Springfield Auto Electric Service, Inc.
Phone: (413) 736-3648
117 West Gardner Street
Springfield, Massachusetts 01105

Automotive Products Company
Phone: (906) 863-2651
520 First Street
Menominee, Michigan 49858

Engine Parts Supply Company
Phone: (612) 338-7086
1220-24 Harmon Place
Minneapolis, Minnesota 55403

Johnson Big Wheel Mowers, Inc.
Phone: (601) 856-8848
Highway 51 North
P.O. Box 717
Ridgeland, Mississippi 39157

Medart Engines & Parts
Phone: (314) 343-0515
100 Larkin Williams Ind. Court
Fenton, Missouri 63026

Electrical & Magneto Service Company
Phone: (816) 421-3711
1600 Campbell
Kansas City, Missouri 64108

A & I Distributors
Phone: (406) 245-6443
2112-4th Avenue, North
Box 1999
Billings, Montana 59103

A & I Distributors
Phone: (406) 453-4314
317 Second Street, South
Box 1159
Great Falls, Montana 59403

Carl A. Anderson, Incorporated of Nebraska
Phone: (402) 339-4944
7510 "L" Street
Box 27139
Omaha, Nebraska 68127

Central Motive Power, Incorporated
Phone: (505) 345-8335
3740 Princeton Drive, North East
P.O. Box 1924
Alburquerque, New Mexico 87103

Wayne Auto Electric, Incorporated
Phone: (315) 255-1111
26 East Genesee Street
Auburn, New York 13021

H. G. S. Power House
Phone: (516) 423-1348
68-70 West Jericho Turnpike
(West of Route 116)
Huntington Station, New York 11746

E.J. Smith & Sons Company
Phone: (704) 394-3361
4250 Golf Acres Drive
P.O. Box 668887
Charlotte, North Carolina 28266

United Power Equipment
Phone: (701) 255-2450
2031 East Lee Avenue
P.O. Box 1077
Bismark, North Dakota 58502

V.E. Petersen Company
Phone: (419) 838-5911
28101 East Broadway
Walbridge (Toldeo), Ohio 43465

Victory Motors, Incorporated
Phone: (918) 863-6675
605 Cherokee
Muskogee, Oklahoma 74001

Brown & Wisner, Incorporated
Phone: (503) 692-0330
9991 South West Avery Street
Box 1109
Tualatin, Oregon 97062

Fry's Power Equipment
Phone: (814) 944-6259
4th Street & Bell Avenue
Altoona, Pennsylvania 16602

McCullough Distributing Company
Phone: (215) 288-9700
5613-19 Tulip Street
Philadelphia, Pennsylvania 19124

Locke Auto Electric Service, Inc.
Phone: (605) 336-2780
231 North Dakota Avenue
Box 1165
Souix Falls, South Dakota 57101

RCH Distributors, Incorporated
Phone: (901) 345-2200
3140 Carrier Street
Box 173
Memphis, Tennessee 38101

Automobile Electric Service
Phone: (615) 255-7515
1008 Charlotte Avenue
Nashville, Tennessee 37203

Wes Kern's Repair Service
Phone: (915) 532-2455
2423 East Missouri
El Paso, Texas 79901

McCoy Sales & Service
Phone: (817) 838-6338
4045 East Belknap Street
Fort Worth, Texas 76111

Yazoo of Texas, Incorporated
Phone: (713) 923-5979
1409 Telephone Road
Houston, Texas 77023

A-1 Engine & Mower Company
Phone: (801) 364-3653
437 East 9th, South
Salt Lake City, Utah 84111

Hunt Auto Supply
Phone: (804) 583-4553
1300 Monticello Avenue
Norfolk, Virginia 23510

R.B.I. Corporation
Phone: (804) 798-1541
101 Cedar Run Drive
P.O. Box 9318
Richmond, Virginia 23227

Fremont Electric Company
Phone: (206) 633-2323
744 North 34th Street
P.O. Box 31640
Seattle, Washington 98103

CANADIAN DISTRIBUTORS

Loveseth, Ltd.
Phone: (403) 436-8250
9570-58th Avenue
Edmonton, Alberta T6E 0B6

Western Air Cooled Engines, Ltd.
Phone: (604) 254-8121
1205 East Hastings
Vancouver, British Columbia V6A 1S5

Yetman's, Ltd.
Phone: (204) 586-8046
949 Jarvis Avenue
Winnipeg, Manitoba R2X 0A1

Longwood Equipment Company
Phone: (416) 438-3710
1940 Ellesmere Road
Unit 8
Scarborough, Ontario M1H 2V7

COX

Model	Bore	Stroke	Displacement
140	1.250 in. (31.8 mm)	1.136 in. (28.9 mm)	1.4 cu. in. (23 cc)
170	1.375 in. (34.9 mm)	1.136 in. (28.9 mm)	1.7 cu. in. (28 cc)

MAINTENANCE

SPARK PLUG. Recommended spark plug for all models is Champion CJ8, or equivalent. Electrode gap is 0.025 inch (0.635 mm) and spark plug should be tightened to 12-14 ft.-lbs. (16-19 N·m).

CARBURETOR. The diaphragm carburetor (Fig. COX 1) is equipped with both idle mixture and main adjustment needles.

Initial adjustment for main fuel adjustment needle (2) is 2 turns open from a lightly seated position.

Initial adjustment for idle mixture screw (1) is to lightly seat (close) adjustment needle.

Make final adjustment with engine at normal operating temperature and running. Place engine under load and adjust main fuel needle for leanest setting that will allow satisfactory acceleration and steady smooth operation. Idle mixture screw is normally adjusted to a lightly seated position, however, screw can be opened (turned out) slightly to lean the idle fuel mixture. Idle speed is adjusted at screw (4).

MAGNETO AND TIMING. A flywheel type magneto is used as shown in Fig. COX 2. The breaker contact points are enclosed under the flywheel. The armature, coil and condenser are mounted outside the flywheel.

To gain access to the breaker contact points, remove the recoil starter and blower housing. Loosen flywheel retaining nut one turn and, while supporting engine with flywheel, tap the nut sharply with a small hammer to unseat the tapered fit between flywheel and shaft.

NOTE: End of flywheel nut is shaft for recoil starter. Be careful not to damage nut when removing flywheel.

Remove nut, flywheel and breaker box cover. Be careful not to lose the flywheel drive key.

Ignition timing is fixed and not adjustable. Ignition breaker point gap

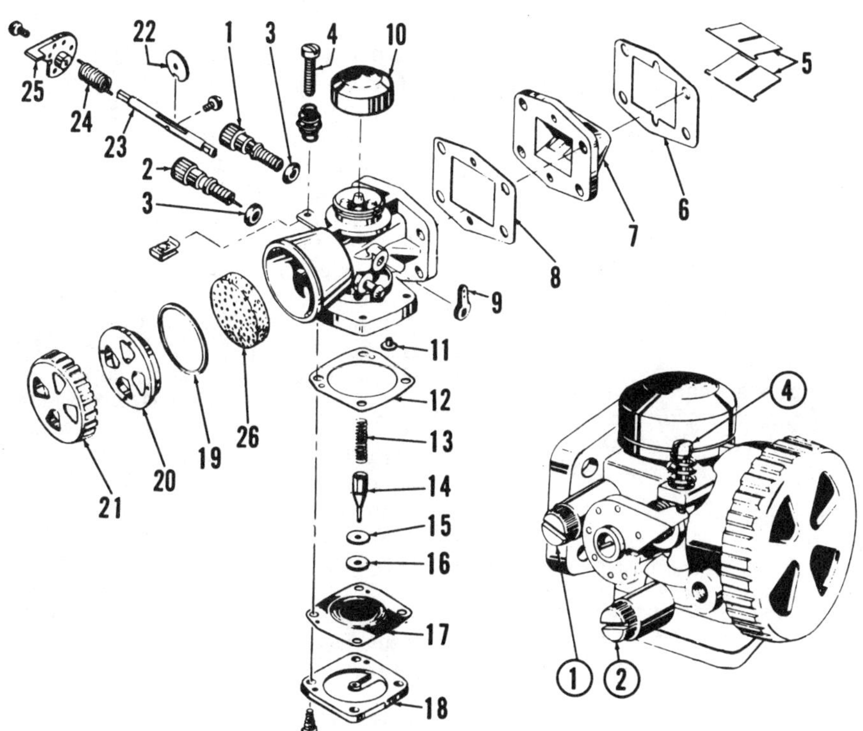

Fig. COX1—Exploded view of carburetor used. Make certain pulse chamber cap (10) and reeds (5) are in good condition and correctly installed.

1. Idle mixture screw
2. Main fuel mixture needle
3. "O" rings
4. Idle speed screw
5. Reeds
6. Gasket with cut-outs
7. Reed block
8. Gasket with holes
9. Bellcrank
10. Pulse chamber cap
11. Rivet
12. Gasket
13. Spring
14. Inlet needle
15. Needle seat
16. Seat retainer
17. Diaphragm
18. Cover
19. "O" ring
20. Choke plate
21. Choke cover
22. Throttle valve
23. Throttle shaft
24. Return spring
25. Throttle control
26. Air filter

Fig. COX 2—Exploded view of magneto assembly. End of nut (2) is shaft for starter.

1. Blower housing
2. Flywheel nut
3. Flywheel
4. Cover
5. Cam oiler felt
6. Wire to switch (9)
7. Condenser
8. Breaker points
9. Stop switch
10. Coil
11. Coil core
12. Magneto stator plate
13. High tension wire retainer

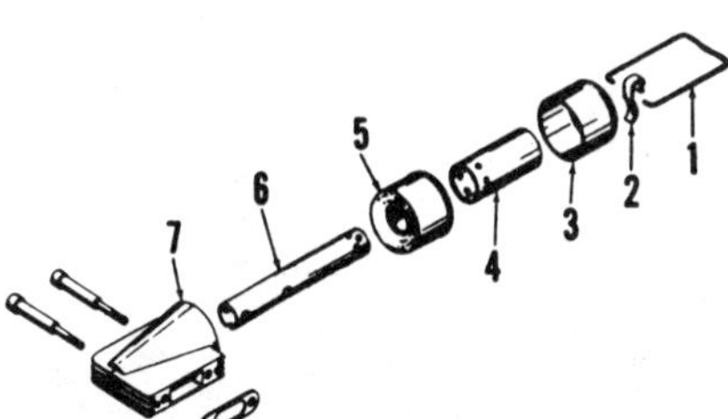

Fig. COX 4—Exploded view of muffler and exhaust collector. Refer to text when assembling.

1. Retainer
2. Clip
3. Outer cover
4. Short baffle tube
5. Inner cover
6. Long baffle tube
7. Exhaust collector

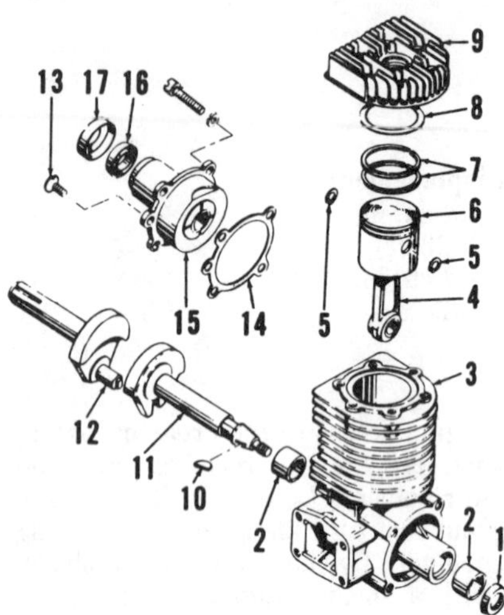

Fig. COX 5—Exploded view of the standard power head assembly. Parts (13, 14, 15 and 16) are different for unit used to drive bicycles. Refer to Fig. COX 6.

1. Seal
2. Bushings
3. Crankcase & cylinder
4. Connecting rod
5. Snap rings
6. Piston
7. Rings
8. Cylinder head gasket
9. Cylinder head
10. Flywheel drive key
11. Crankshaft (flywheel end)
12. Crankshaft (pto end)
13. Screw
14. Gasket
15. End cover & two main bearings
16. Seal
17. Seal retainer

should be 0.015 inch (0.381 mm). If necessary, the magneto stator plate can be removed after removing the four attaching screws.

LUBRICATION. The engine is lubricated by oil mixed with the fuel. Normal ratio is ½ pint (0.24 L) oil to 1 gallon (3.8 L) of gasoline (1:16 oil:fuel ratio). For new or rebuilt motors, the first gallon of fuel should have 1 pint (0.5 L) of oil mixed with 1 gallon (3.8 L) of gasoline. Use regular gasoline and SAE 30 oil with an API service classification SF. Mix oil and gasoline thoroughly in a separate container.

CARBON. Exhaust ports, muffler and exhaust collector should be cleaned after every 60 hours of operation. Use a dull nonferrous tool such as a ¼-inch wooden dowel to clean ports and crank engine several revolutions after cleaning to blow carbon out of cylinder.

Fig. COX 7—Exploded view of recoil starter. Clutch (6) drives the flywheel nut (2—Fig. COX 2) when rope is pulled.

1. Housing
2. Recoil spring
3. Snap ring
4. Pulley (2 halves)
5. Rope
6. Starter clutch
7. Thrust washers (4 used)
8. Retainers (late models)

When reassembling, make certain the two locator pins on collector slide between fins of cylinder. Slot on one end of the long baffle tube of muffler should correctly engage boss in collector. Holes in short baffle tube should be toward exhaust collector.

REPAIRS

CONNECTING ROD. The connecting rod and piston can be withdrawn after removing the cylinder head and the end cover (15–Fig. COX 5) or drive carrier (15–Fig. COX 6).

Connecting rod sintered bronze bushings are integral with connecting rod. Renew connecting rod and/or pto end of crankshaft if wear is excessive.

When assembling, coat gasket (14–Fig. COX 5 or COX 6) and connecting rod lower bearing bore with SAE 30 oil. Make certain projection on end of crankshaft (12–Fig. COX 5 or COX 6) enters slot in crankshaft end (11–Fig. COX 5). Gasket (8) should be renewed whenever cylinder head is removed.

PISTON, PIN AND RINGS. Refer to CONNECTING ROD paragraphs for removal procedure. Piston and rings are available in standard size only. When assembling, end gaps in the two rings should be 180° apart. Cylinder head gasket (8–Fig. COX 5) should be renewed whenever head is removed.

CRANKSHAFT, BEARINGS AND SEALS. A two-piece crankshaft is used as shown in Fig. COX 5. The flywheel end crankshaft rides in two renewable sintered bronze bushings (2). The pto end of crankshaft includes the crankpin on all models. On standard engines (Fig. COX 5), the sintered bronze bushings are located in the end cover (15).

On bicycle drive units, the pto end bearings (18 and 22–Fig. COX 6) are of the needle bearing type. Bearing (22) should be lubricated with Lubriplate or equivalent. On all models, use caution when sliding crankshaft through seals.

The flywheel end of the crankshaft can be removed after removing the flywheel, connecting rod and piston. Hold breaker points open and withdraw crankshaft (11–Fig. COX 5). Bearings (2) and seal (1) can be renewed at this time. When assembling, use caution to prevent damage to seal (1) and breaker points. Gaskets (8 and 14) should be renewed.

INTAKE VALVING. The reed valve assembly (5 and 7–Fig. COX 1) can be removed after removing the carburetor. Check reeds (5) carefully and renew if cracked, distorted or in any other way questionable. Make certain that gasket (6) with cutouts is installed between reed block (7) and crankcase. Gasket (8) with two small holes should be between carburetor and reed block.

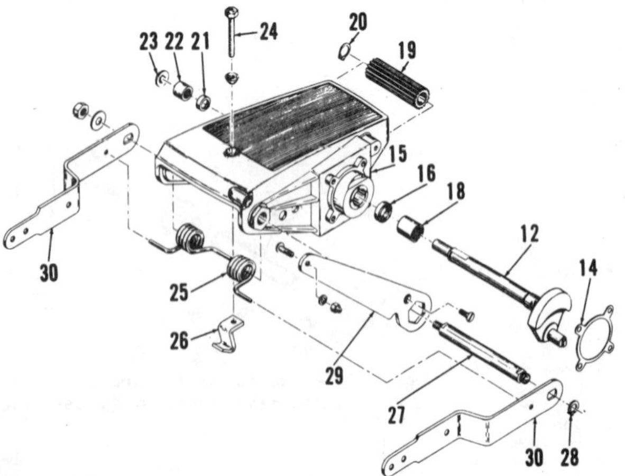

Fig. COX 6—View of parts used with bicycle drive. Drive carrier (15) is used in place of end cover (15—Fig. COX 5).

12. Crankshaft (pto end)
14. Gasket
15. Drive carrier
16. Seal
18. Needle bearing
19. Drive-roller
20. Snap ring
21. Seal
22. Needle bearing
23. Expansion plug
24. Tension adjusting screw
25. Tension spring
26. Spring hook
27. Pivot shaft
28. Snap ring
29. Lift lever
30. Support straps

SERVICING COX ACCESSORIES

RECOIL STARTER

R&R AND OVERHAUL. The complete recoil starter can be removed after removing the two retaining screws. On early models, a hooked wire should be used to hold the starter assembly (2 through 7–Fig. COX 7) in housing (1) while removing. Later models are equipped with retainers (8).

To disassemble the removed starter, apply thumb pressure to pulley to prevent pulley from turning. Pull rope free from slot in housing, release thumb pressure slightly and allow spring to slowly unwind. Carefully remove starter assembly from cover. Spring (2) should remain in the housing (1). If spring is to be renewed, remove old spring, hook outer end of new spring in housing and wind spring counterclockwise to the inside. If necessary to disassemble pulley halves (4) from clutch (6), remove snap ring (3) and separate parts.

When reassembling, install new rope (5) through pulley half (engine side) and wind rope around pulley in a counterclockwise direction. See Fig. COX 7. Carefully install assembled starter in housing, making certain that slot in clutch drive engages inner hook on recoil spring. Preload spring by rotating pulley and rope 1 to 1½ turns counterclockwise and place handle end of rope in slot in housing. Reinstall starter assembly on engine.

CRAFTSMAN

SEARS ROBUCK & CO.
Chicago, Ill. 60607

The following Craftsman engines were manufactured by the Tecumseh Products Company. The accompanying cross-reference chart will assist in identifying an engine for service requirements. Much of the service will be identical to service on similar Tecumseh engines. The Craftsman service section which follows the cross reference charts includes service which is different than repairing Tecumseh engines.

TWO-STROKE MODELS

Craftsman Model Number	Tecumseh Engine Type Number
143.17112	638-07 (AV520)
143.171112	638-07 (AV520)
143.171122	638-01A (AV520
143.171272	638-51 (AC520)
200.183112	638.07A (AV520)
200.183122	638-17A (AV520)
200.193132	638-07B (AV520)
200.193142	638-17B (AV520)
200.193152	650-44 (AV520)
200.193162	650-29A (AV520)
200.203172	670-39 (AV520)
200.203182	670-32 (AV520)
200.203192	670-18 (AV520)
200.213112	670-39A (AV520)
200.213122	670-32A (AV520)
200.243112	670-83 (AV520)
200.283012	670-100 (AV520)
200-503111	1525A (AH520)
200.583111	1499 (AH520)
200.593121	1525 (AH520)
200.613111	1553 (AH520)
200.672102	(AH520)
200.682102	(AH520)
200.692112	(AH520)
200.692122	(AH520)
200.692132	(AH520)

FOUR-STROKE ENGINES
Vertical Crankshaft

Craftsman Model Number	Basic Tecumseh Model Number
143.V20A-1-1WA	V 20 D
143.V25A-5-1-1WA	V 25 D
143.V27A-3-1-1WA	V 27 D
143.09300	LAV 25 B
143.09301	LAV 25 B
143.09302	LAV 25 B
143.10250	LAV 22 B
143.10251	LAV 22 B
143.10300	LAV 25 B
143.10301	LAV 25 B
143.11300	LAV 25 B
143.11301	LAV 25 B
143.11302	LAV 25 B
143.11350	LAV 30 B
143.11351	LAV 30 B
143.11352	LAV 30 B
143.12300	LAV 25 B
143.12301	LAV 25 B
143.12302	LAV 25 B
143.12303	LAV 25 B
143.12304	LAV 25 B
143.12350	LAV 30 B
143.12351	LAV 30 B
143.13250	LAV 22 B
143.13251	LAV 22 B
143.13300	LAV 25 B
143.13301	LAV 25 B
143.14350	LAV 30 B
143.14351	LAV 30 B
143.15300	LAV 25 B
143.15301	LAV 25 B
143.16350	LAV 30 B
143.16351	LAV 30 B
143.17350	LAV 30 B
143.17351	LAV 30 B
143.18350	LAV 30 B
143.18351	LAV 30 B
143.19400	LAV 35 B
143.19401	LAV 35 B
143.20100	V 20 D
143.20400	V 35 D
143.20401	V 35 D
143.20500	V 45 B
143.20501	V 45 B
143.20502	V 45 C
143.20503	V 45 C
143.23250	HA 22 C
143.23251	HA 22 C
143.24250	HA 22 C
143.24251	HA 22 C
143.25100	V 25 D
143.25250	HA 22 C
143.25251	HA 22 C
143.27100	V 27 D
143.27200	V 20 D
143.30250	LAV 22 B
143.30251	LAV 22 B
143.30350	LAV 30 B
143.30351	LAV 30 B
143.31600	V 55 B
143.31601	V 55 C
143.36250	HA 22 C
143.36251	HA 22 C
143.36252	H 22 D
143.36253	H 22 D
143.36254	H 22 H
143.36255	H 22 H
143.39250	LAV 22 B
143.39251	LAV 22 B
143.40250	H 22 D
143.40251	H 22 D
143.40300	LAV 25 B
143.40301	LAV 25 B
143.40350	LAV 30 B
143.40351	LAV 30 B
143.40500	V 45 B
143.40501	V 45 B
143.40502	V 45 C
143.40503	V 45 C
143.40600	V 55 B
143.41300	LAV 25 B
143.41301	LAV 25 B
143.41302	LAV 25 B
143.41350	LAV 30 B
143.41351	LAV 30 B
143.41352	LAV 30 B
143.42200	V 22 H-8
143.42201	V 22 H-8P
143.42202	V 22 H-37
143.42203	V 22 H-37P
143.42204	V 22 H-8
143.42205	V 22 H-8B
143.42500	V 22 H-8
143.42501	V 25 H-8P
143.42700	V 25 H-8
143.42701	V 25 H-8P
143.43004	V 30 H-3P
143.43200	V 30 H-8
143.43201	V 30 H-29
143.43202	V 30 H-40
143.43203	V 30 H-8P
143.43204	V 30 H-29P
143.43205	V 30 H-40P
143.43250	H 22 H
143.43251	H 22 H
143.43300	V 30 H-8
143.43301	V 30 H-8P
143.43500	V 45 B
143.43501	V 45 B
143.43700	V 32 A-28
143.43701	V 32 A-28P
143.44000	V 35 B-28
143.44400	H 35 D
143.44401	H 35 D
143.50020	V 20 G-8
143.50021	V 20 G-8P
143.50025	V 25 G-8
143.50026	V 25 G-8P
143.50030	V 30 G-14
143.50031	V 30 G-14P
143.50040	V 40 A-4
143.50045	V 40 B-4
143.50125	V 25 G-3
143.50126	V 25 G-3P
143.50130	V 30 G-8

Craftsman Model Number	Basic Tecumseh Model Number
143.50131	V 30 G-8P
143.50230	V 30 G-3
143.50231	V 30 G-3P
143.50250	HA 22 A
143.50251	HA 22 A
143.50300	H 30 D
143.50301	H 30 D
143.50400	H 35 D
143.50401	H 35 D
143.50402	H 35 H
143.50403	H 35 H
143.50700	V 25 H-8
143.52200	V 22 H-8
143.52201	V 22 H-8P
143.52701	V 25 H-8P
143.52702	V 25 H-8
143.52703	V 25 H-8P
143.53001	V 30 H-4P
143.53200	V 30 H-8
143.53201	V 30 H-8P
143.53300	V 30 H-8
143.53301	V 30 H-8P
143.54500	VX-45 A-24
143.54502	V 45 A-4
143.55250	HA 22 A
143.55251	HA 22 A
143.55300	H 30 D
143.56250	HA 22 A
143.56251	HA 22 A
143.56252	HA 22 A
143.56253	HA 22 A
143.58250	HA 22 C
143.58251	HA 22 C
143.59250	H 22 D
143.59251	H 22 D
143.60020	V 20 G-8
143.60021	V 20 G-8P
143.60025	V 25 G-14
143.60026	V 25 G-14P
143.60030	V 30 G-8
143.60031	V 30 G-8P
143.60040	VX 40 A-19
143.60125	V 25 G-8
143.60126	V 25 G-8P
143.60130	V 30 G-8
143.60131	V 30 G-8P
143.60140	VX 40 A-24
143.60225	V 25 G-8
143.60226	V 25 G-8P
143.60231	V 30 G-4P
143.60240	VX 40 A-19
143.60325	V 25 G-14
143.60326	V 25 G-14F
143.60327	V 27 G-16
143.60328	V 27 G-16P
143.60330	V 27 G-2
143.60331	V 27 G-2P
143.60340	VX 40 A-19
143.60351	LV 30 B
143.60401	LV 35 B
143.62200	V 22 H-8
143.62201	V 22 H-8P
143.62700	V 25 H-8
143.62701	V 25 H-14
143.62702	V 25 H-3
143.62703	V 25 H-8P

Craftsman Model Number	Basic Tecumseh Model Number
143.62704	V 25 H-14P
143.62705	V 25 H-3P
143.63200	V 30 H-8
143.63201	V 30 H-14
143.63202	V 30 H-3
143.63203	V 30 H-8P
143.63204	V 30 H-14P
143.63205	V 30 H-3P
143.64000	V 40 B-4
143.64500	V 45 A-4
143.65250	H 22 H
143.65251	H 22 H
143.66250	H 22 H
143.66251	H 22 H
143.66500	V 45 B
143.66501	V 45 B
143.67250	H 22 H
143.67251	H 22 H
143.71250	LAV 22 B
143.71251	LAV 22 B
143.75250	LAV 22 B
143.75251	LAV 22 B
143.76250	LAV 22 B
143.76251	LAV 22 B
143.76252	LAV 22 B
143.82000	H 20 B-15
143.82001	H 20 B-15P
143.82002	H 20 B-15
143.82003	H 20 B-15P
143.82004	H 20 B-15
143.82005	H 20 B-15P
143.82006	H 20 B-15
143.82007	H 20 G-15P
143.83000	H 30 B-15
143.83001	H 30 B-15P
143.83250	LAV 22 B
143.83251	LAV 22 B
143.83252	LAV 22 B
143.84300	LAV 25 B
143.84301	LAV 25 B
143.84302	LAV 25 B
143.86400	LAV 35 B
143.86401	LAV 36 B
143.86402	LAV 35 B
143.90020	H 20 A-15
143.90021	H 20 A-15P
143.91250	H 22 H
143.91251	H 22 H
143.93000	H 30 B-15P
143.97250	H 22 H
143.97251	H 22 H
143.101010	LAV 30 E
143.101011	LAV 30 E
143.101012	LAV 30 E
143.101020	LAV 30 E
143.101021	LAV 30 E
143.101022	LAV 30 E
143.101030	LAV 30 E
143.101031	LAV 30 E
143.101032	LAV 30 E
143.102010	LAV 30 E
143.102011	LAV 30 E
143.102012	LAV 30 E
143.102020	LAV 30 E
143.102021	LAV 30 E
143.102022	LAV 30 E

Craftsman Model Number	Basic Tecumseh Model Number
143.102030	LAV 30 E
143.102031	LAV 30 E
143.102040	LAV 30 E
143.102041	LAV 30 E
143.102041	LAV 30 E
143.102050	LAV 30 E
143.102051	LAV 30 E
143.102060	LAV 30 E
143.102061	LAV 30 E
143.102062	LAV 30 E
143.102070	LV 30 E
143.102071	LV 30 E
143.102072	LV 30 E
143.102100	LAV 30 E
143.102101	LAV 30 E
143.102102	LAV 30 E
143.102110	LAV 30 E
143.102111	LAV 30 E
143.102112	LAV 30 E
143.102120	LAV 30 E
143.102121	LAV 30 E
143.102130	LAV 30 E
143.102131	LAV 30 E
143.102132	LAV 30 E
143.102140	LAV 30 E
143.102141	LAV 30 E
143.102150	LAV 30 E
143.102151	LAV 30 E
143.102152	LAV 30 E
143.102160	LAV 25 E
143.102161	LAV 25 E
143.102162	LAV 25 E
143.102170	LAV 30 E
143.102171	LAV 30 E
143.102172	LAV 30 E
143.102190	LAV 30 E
143.102191	LAV 30 E
143.102200	LAV 30 E
143.102201	LAV 30 E
143.102202	LAV 30 E
143.102210	LV 30 E
143.102211	LV 30 E
143.102212	LV 30 E
143.102220	LAV 30 E
143.102221	LAV 30 E
143.102222	LAV 30 E
143.103010	LAV 30 E
143.103011	LAV 30 E
143.103012	LAV 30 E
143.103020	LAV 30 E
143.103021	LAV 30 E
143.103050	LAV 30 E
143.103051	LAV 30 E
143.103052	LAV 30 E
143.103060	LAV 30 E
143.103061	LAV 30 E
143.104010	LAV 35 E
143.104011	LAV 35 E
143.104020	LAV 35 E
143.104021	LAV 35 E
143.104030	LV 35 E
143.104031	LV 35 E
143.104040	LV 35 E
143.104041	LV 35 E
143.104050	LV 35 E
143.104051	LV 35 E

Craftsman Model Number	Basic Tecumseh Model Number
143.104080	LAV 35 E
143.104081	LAV 35 E
143.104090	LAV 35 E
143.104091	LAV 35 E
143.104100	LV 35 E
143.104101	LV 35 E
143.104110	LV 35 E
143.104111	LV 35 E
143.104120	LV 35 E
143.104121	LV 35 E
143.105010	V 45 D
143.105011	V 45 D
143.105020	V 45 D
143.105021	V 45 D
143.105030	V 45 D
143.105031	V 45 D
143.105040	V 55 D
143.105041	V 55 D
143.105050	V 45 D
143.105051	V 45 D
143.105060	V 45 D
143.105061	V 45 D
143.105070	V 45 D
143.105071	V 45 D
143.105090	V 45 D
143.105091	V 45 D
143.105100	V 45 D
143.105101	V 45 D
143.105110	V 55 D
143.105120	V 45 D
143.105121	V 45 D
143.106010	LAV 30 E
143.106011	LAV 30 E
143.106012	LAV 30 E
143.106020	LAV 30 E
143.106021	LAV 30 E
143.106030	LAV 30 E
143.106031	LAV 30 E
143.122011	LV 30 L
143.122012	LV 30 L
143.122021	LAV 30 L
143.122022	LAV 30 L
143.122031	LAV 30 L
143.122032	LAV 30 L
143.122041	LAV 30 L
143.122042	LAV 30 L
143.122051	LAV 30 L
143.122052	LAV 30 L
143.122061	LAV 30 L
143.122071	LAV 30 L
143.122081	LAV 30 L
143.122082	LAV 30 L
143.122091	LAV 30 L
143.122092	LAV 30 L
143.122101	LAV 25 L
143.122102	LAV 25 L
143.122201	LAV 30 L
143.122202	LAV 30 L
143.122211	LAV 30 M
143.122212	LAV 30 M
143.122221	LAV 30 M
143.122222	LAV 30 M
143.122231	LAV 30 M
143.122232	LAV 30 M
143.122242	LAV 35 M
143.122251	LV 30 M

Craftsman Model Number	Basic Tecumseh Model Number
143.122252	LV 30 M
143.122261	LAV 30 M
143.122262	LAV 30 M
143.122271	LAV 30 M
143.122272	LAV 30 M
143.122281	LAV 30 M
143.122282	LAV 30 M
143.122291	LAVT 30 C
143.122311	LV 30 N
143.122321	LAV 30 L
143.122322	LAV 30 L
143.123021	LAVR 30 L
143.123022	LAVR 30 L
143.123031	LAVR 30 L
143.123041	LAVR 30 L
143.123051	LAVR 30 L
143.123052	LAVR 30 L
143.123071	LAVR 30 L
143.123091	LAVR 30 L
143.123092	LAVR 30 L
143.124011	LV 35 L
143.124021	LAV 35 L
143.124031	LAV 35 L
143.124041	LV 35 M
143.124051	LV 35 N
143.124061	LAVT 35 C
143.124071	LV 35 N
143.125011	VT 45 A
143.125021	VT 45 C
143.125031	VT 45 A
143.125041	VT 45 C
143.125051	VT 45 C
143.125061	VT 45 C
143.126011	VT 55 A
143.126021	VT 55 B
143.126031	VT 55 B
143.126041	VT 55 C
143.126051	VT 55 C
143.126061	VT 55 C
143.131022	LAV 30
143.131032	LAV 30
143.131042	LAV 30
143.131052	LAV 30
143.131062	LAV 30
143.131072	LAV 30
143.131082	LAV 30
143.131092	LAV 30
143.131102	LAV 30
143.131112	LAV 22
143.131122	LAV 30
143.131132	LAV 30
143.131142	LAV 30
143.131152	LAV 30
143.131162	LAV 30
143.131172	LAV 30
143.131182	LAV 30
143.133012	LAV 30
143.133022	LAV 30
143.133032	LAV 30
143.133042	LV 35
143.133052	LAV 30
143.134012	LV 35
143.134022	LV 35
143.134032	LAV 35
143.134042	LV 35
143.134052	LV 35

Craftsman Model Number	Basic Tecumseh Model Number
143.135012	V 50
143.135022	V 50
143.135042	V 50
143.135052	V 50
143.135062	V 50
143.135072	V 50
143.135082	V 50
143.135092	V 50
143.136012	V 60
143.135112	V 50
143.136012	V 60
143.136052	V 60
143.136042	V 60
143.136052	V 60
143.137012	V 40
143.137032	V 40
143.141012	LAV 30
143.141022	LAV 30
143.141032	LAV 30
143.141042	LAV 30
143.141052	LAV 30
143.141062	LAV 30
143.141072	LAV 30
143.141082	LAV 30
143.141092	LAV 25
143.141102	LAV 30
143.141112	LAV 30
143.141122	LAV 30
143.141132	LAV 25
143.141142	LAV 30
143.141152	LAV 30
143.141162	LAV 30
143.141172	LAV 30
143.141182	LAV 30
143.141192	LAV 30
143.141202	LAV 30
143.141212	LAV 30
143.141222	LAV 30
143.141232	LAV 30
143.141242	LAV 30
143.141252	LAV 30
143.141262	LAV 30
143.141272	LAV 30
143.141282	LAV 30
143.141292	LAV 30
143.141302	LAV 30
143.143012	LAV 30
143.143022	LAV 30
143.143032	LAV 30
143.144012	LV 35
143.144022	LV 35
143.144032	VL 35
143.144042	VL 35
143.144052	VL 35
143.144062	LV 35
143.144072	LV 35
143.144082	LV 35
143.144092	LV 35
143.144102	LV 35
143.144112	LV 35
143.144122	LV 35
143.144132	LV 35
143.145012	V 50
143.145022	V 50
143.145032	V 50
143.145042	V 50

Craftsman Model Number	Basic Tecumseh Model Number
143.145052	V 50
143.145062	V 50
143.145072	V 50
143.146012	V 60
143.146022	V 60
143.147012	V 40
143.147022	V 40
143.147032	V 40
143.151012	LAV 30
143.151022	LAV 30
143.151032	LAV 30
143.151042	LAV 30
143.151052	LAV 30
143.151072	LAV 30
143.151082	LAV 30
143.151092	LAV 30
143.151102	LAV 30
143.151103	LAV 30
143.151104	LAV 30
143.151105	LAV 30
143.151142	LAV 30
143.151152	LAV 30
143.151162	LAV 30
143.153012	LAV 30
143.153022	LAV 30
443.153032	LAV 30
143.154012	LAV 35
143.154022	LAV 35
143.154032	LAV 35
143.154042	LAV 35
143.154052	LAV 35
143.154062	LAV 35
143.154072	LAV 35
143.154082	LAV 35
143.154092	LAV 35
143.154102	LAV 35
143.154112	LAV 35
143.154132	LAV 35
143.154142	LAV 35
143.155012	V 50
143.155022	V 50
143.155032	V 50
143.155042	V 50
143.155052	V 50
143.155062	V 50
143.156012	V 60
143.156022	V 60
143.156032	V 60
143.157012	V 40
143.157022	V 40
143.157032	V 40
143.161012	LAV 30
143.161022	LAV 30
143.161032	LAV 30
143.161042	LAV 30
143.161052	LAV 30
143.161062	LAV 30
143.161072	LAV 30
143.161082	LAV 30
143.161092	LAV 30
143.161102	LAV 30
143.161112	LAV 30
143.161132	LAV 30
143.161142	LAV 30
143.161152	LAV 30
143.161162	LAV 30

Craftsman Model Number	Basic Tecumseh Model Number
143.161172	LAV 30
143.161182	LAV 30
143.161192	LAV 30
143.161202	LAV 30
143.161212	LAV 30
143.161222	LAV 30
143.161232	LAV 30
143.161242	LAV 30
143.161262	LAV 30
143.162022	LAV 35
143.162032	LAV 35
143.162092	LAV 35
143.163012	LAV 30
143.163022	LAV 30
143.163032	LAV 30
143.163042	LAV 30
143.163052	LAV 30
143.163062	LAV 30
143.164012	LAV 35
143.164022	LAV 35
143.164032	LAV 35
143.164042	LAV 35
143.164052	LAV 35
143.164062	LAV 35
143.164072	LAV 35
143.174082	LAV 35
143.164102	LAV 35
143.164112	LAV 35
143.164122	LAV 35
143.164132	LAV 35
143.164142	LAV 35
143.164152	LAV 35
143.164162	LAV 35
143.164172	LAV 35
143.164182	LAV 35
143.164202	LAV 35
143.165012	V 50
143.165022	V 50
143.165032	V 50
143.165042	V 50
143.165052	V 50
143.166012	V 60
143.166022	V 60
143.166032	V 60
143.166042	V 60
143.166052	V 60
143.167012	V 40
143.167022	V 40
143.167032	V 40
143.167042	V 40
143.171012	LAV 30
143.171022	LAV 30
143.171032	LAV 30
143.171042	LAV 30
143.171052	LAV 30
143.171062	LAV 30
143.171072	LAV 3C
143.171082	LAV 30
143.171092	LAV 30
143.171102	LAV 30
143.171132	LAV 30
143.171142	LAV 30
143.171152	LAV 30
143.171162	LAV 30
143.171172	LAV 30
143.171202	LAV 35

Craftsman Model Number	Basic Tecumseh Model Number
143.171212	LAV 30
143.171232	LAV 30
143.171242	LAV 30
143.171252	LAV 30
143.171262	LAV 30
143.171302	LAV 30
143.171312	LAV 30
143.171322	LAV 30
143.171332	LAV 30
143.173012	LAV 30
143.173042	LAV 30
143.174012	LAV 35
143.174022	LAV 35
143.174032	LAV 35
143.174042	LAV 35
143.174052	LAV 35
143.174062	LAV 35
143.174072	LAV 35
143.174082	LAV 35
143.174092	LAV 35
143.174102	LAV 35
143.174132	LAV 35
143.174142	LAV 35
143.174152	LAV 35
143.174162	LAV 35
143.174172	LAV 35
143.174182	LAV 35
143.174192	LAV 35
143.174232	LAV 35
143.174242	LAV 35
143.174252	LAV 35
143.174272	LAV 35
143.174292	LAV 35
143.175012	V 50
143.175022	V 50
143.175032	V 50
143.175042	V 50
143.175052	V 50
143.175062	V 50
143.175072	V 50
143.176012	V 60
143.176022	V 60
143.176032	V 60
143.176042	V 60
143.176052	V 60
143.176062	V 60
143.176072	V 60
143.176082	V 60
143.176092	V 60
143.176102	V 60
143.177012	V 40
143.177022	V 40
143.177032	V 40
143.177042	V 40
143.177062	V 40
143.177072	V 40
143.181042	LAV 30
143.181052	LAV 30
143.181062	LAV 30
143.181082	LAV 30
143.181092	LAV 30
143.181102	LAV 30
143.181112	LAV 30
143.181122	LAV 30
143.181132	LAV 30
143.183042	LAV 30

Craftsman Model Number	Basic Tecumseh Model Number
143.184012	LAV 35
143.184052	LAV 35
143.184082	LAV 35
143.184092	LAV 35
143.184102	LAV 35
143.184112	LAV 35
143.184122	LAV 35
143.184132	LAV 35
143.184142	LAV 35
143.184152	LAV 35
143.184162	LAV 35
143.184172	LAV 35
143.184182	LAV 35
143.184192	LAV 35
143.184202	LAV 35
143.184212	LAV 35
143.184232	ECV 100
143.184242	ECV 100
143.184252	ECV 100
143.184262	LAV 35
143.184272	LAV 35
143.184282	LAV 35
143.184292	LAV 35
143.184302	LAV 35
143.184402	LAV 35
143.185012	V 50
143.185022	V 50
143.185032	V 50
143.185042	V 50
143.185052	V 50
143.186012	V 60
143.186052	V 60
143.186062	V 60
143.186082	V 60
143.186092	V 60
143.186102	V 60
143.186112	V 60
143.186122	V 60
143.187022	LAV 40
143.187042	LAV 40
143.187052	LAV 40
143.187062	LAV 40
143.187072	LAV 40
143.187082	LAV 40
143.187094	LAV 40
143.187102	EVC 105
143.191012	LAV 30
143.191022	LAV 30
143.191032	LAV 30
143.191042	LAV 30
143.191052	LAV 30
143.194012	LAV 35
143.194022	LAV 35
143.194032	LAV 35
143.194042	LAV 35
143.194052	LAV 35
143.194062	ECV 100
143.194072	LAV 35
143.194082	LAV 35
143.194092	LAV 35
143.194102	ECV 100
143.194112	LAV 35
143.194122	LAV 35
143.194132	LAV 35
143.194142	LAV 35
143.195012	V 50

Craftsman Model Number	Basic Tecumseh Model Number
143.195022	V 50
143.196012	V 60
143.190622	V 60
143.196032	V 60
143.196082	V 60
143.17012	LAV 40
143.197022	ECV 105
143.197032	ECV 105
143.197042	LAV 40
143.197052	LAV 40
143.197062	LAV 40
143.197072	LAV 40
143.197082	ECV 105
143.201032	LAV 30
143.201042	LAV 30
143.203012	LAV 30
143.204022	ECV 100
143.204032	LAV 35
143.204042	LAV 35
143.204052	LAV 35
143.204062	ECV 100
143.204072	LAV 35
143.204082	LAV 35
143.204092	LAV 35
143.204102	EVC 100
143.204132	ECV 100
143.204142	LAC 35
143.204162	LAV 35
143.204172	LAV 35
143.204182	LAV 35
143.204192	LAV 35
143.204202	ECV 100
143.205022	V 50
143.206012	V 60
143.206032	V 60
143.207012	LAV 40
143.207022	LAV 40
143.207032	LAV 40
143.207052	LAV 40
143.207062	ECV 105
143.207072	LAV 40
143.207082	ECV 105
143.213012	LAV 30
143.213022	LAV 30
143.213042	LAV 30
143.214012	LAV 35
143.214022	LAV 35
143.214032	LAV 35
143.214042	ECV 100
143.214052	ECV 100
143.214062	ECV 100
143.214072	ECV 100
143.214082	LAV 35
143.214092	LAV 35
143.214102	LAV 35
143.214112	LAV 35
143.214122	LAV 35
143.214132	LAV 35
143.214192	LAV 35
143.214202	LAV 35
143.214212	LAV 35
143.214222	LAV 35
143.214232	LAV 35
143.214242	LAV 35
143.214252	LAV 35
143.214262	ECV 100

Craftsman Model Number	Basic Tecumseh Model Number
143.214272	ECV 100
143.214282	ECV 100
143.214292	LAV 35
143.214302	LAV 35
143.214312	ECV 100
143.214322	ECV 100
143.214332	LAV 35
143.214342	LAV 35
143.214352	ECV 100
143.216042	V 60
143.216052	V 60
143.216062	V 60
143.216122	V 60
143.216142	V 60
143.216182	V 60
143.217012	ECV 105
143.217022	ECV 105
143.217032	ECV 105
143.217042	LAV 40
143.217052	LAV 40
143.217062	LAV 40
143.217072	LAV 40
143.217092	ECV 105
143.217102	LAV 40
143.223012	LAV 30
143.223022	LAV 30
143.223032	LAV 30
143.223042	LAV 30
143.223052	LAV 30
143.224012	LAV 35
143.224022	LAV 35
143.224032	ECV 100
143.224062	LAV 35
143.224092	LAV 35
143.224102	LAV 35
143.224112	LAV 35
143.224122	LAV 35
143.224132	LAV 35
143.224142	LAV 30
143.224162	LAV 35
143.224172	LAV 35
143.224182	LAV 35
143.224192	LAV 35
143.224202	LAV 35
143.224212	LAV 35
143.224222	LAV 35
143.224232	ECV 100
143.224242	ECV 100
143.224252	LAV 35
143.224262	LAV 35
143.224272	LAV 35
143.224282	LAV 35
143.224292	ECV 100
143.224302	ECV 100
143.224312	LAV 35
143.224322	LAV 35
143.224332	LAV 35
143.224342	LAV 35
143.224352	ECV 100
143.224362	ECV 100
143.224372	LAV 35
143.224382	LAV 35
143.224392	LAV 35
143.224402	LAV 35
143.224412	LAV 35
143.224422	LAV 35

Craftsman Model Number	Basic Tecumseh Model Number
143.225012	ECV 120
143.225022	ECV 120
143.225032	V 50
143.225042	V 50
143.225052	V 50
143.225062	ECV 120
143.225072	ECV 120
143.225082	V 50
143.225092	V 50
143.225102	V 50
143.226012	V 60
143.226032	V 60
143.266132	V 60
143.226152	V 60
143.226162	V 60
143.226182	V 60
143.226222	V 60
143.226232	V 60
143.226242	V 60
143.226262	V 60
143.262322	V 60
143.226332	V 60
143.227012	ECV 110
143.227022	ECV 110
143.227062	ECV 110
143.227072	ECV 110
143.233012	LAV 30
143.233032	LAV 30
143.233042	LAV 30
143.234022	LAV 35
143.234042	LAV 35
143.234052	LAV 35
143.234062	ECV 100
143.234072	ECV 100
143.234082	ECV 100
143.234092	ECV 100
143.234102	LAV 35
143.234112	LAV 35
143.234122	LAV 35
143.234132	LAV 35
143.234142	LAV 35
143.234162	LAV 35
143.234182	ECV 100
143.234192	LAV 35
143.234202	LAV 35
143.234212	ECV 100
143.234222	ECV 100
143.234232	ECV 100
143.234242	LAV 35
143.234252	LAV 35
143.234262	LAV 35
143.235012	ECV 120
143.235022	ECV 120
143.235032	LAV 50
143.235042	ECV 120
143.235052	ECV 120
143.235062	V 50
143.235072	LAV 50
143.236012	V 60
143.236052	V 60
143.236082	V 60
143.236102	V 60
143.236112	V 60
143.236132	V 60
143.236152	V 60
143.237012	ECV 110
143.237022	ECV 110
143.237032	ECV 105
143.237042	LAV 40
143.243012	LAV 30
143.243022	LAV 30
143.242042	LAV 30
143.243052	LAV 30
143.243062	LAV 30
143.243072	LAV 30
143.243082	LAV 30
143.244012	LAV 35
143.244022	LAV 35
143.244032	LAV 35
143.244042	ECV 100
143.244052	ECV 100
143.244062	ECV 100
143.244072	LAV 35
143.244082	LAV 35
143.244092	LAV 35
143.244102	LAV 35
143.244112	LAV 35
143.244122	ECV 100
143.244132	ECV 100
143.244142	ECV 100
143.244202	LAV 35
143.244212	ECV 100
143.244222	LAV 35
143.244232	LAV 35
143.244242	ECV 100
143.244252	ECV 100
143.244262	LAV 35
143.244272	LAV 35
143.244282	LAV 35
143.244292	ECV 100
143.244302	ECV 100
143.244312	ECV 100
143.244322	ECV 100
143.244332	ECV 100
143.245012	LAV 50
143.245042	V 50
143.245052	ECV 120
143.245062	ECV 120
143.245072	ECV 120
143.245082	ECV 120
143.245092	LAV 50
143.245102	ECV 120
143.245112	ECV 120
143.245122	ECV 120
143.245132	ECV 120
143.245142	LAV 50
143.245152	LAV 50
143.245162	ECV 120
143.245172	LAV 50
143.245182	LAV 50
143.245192	ECV 120
143.246012	V 60
143.246042	V 60
143.246352	V 60
143.246392	V 60
143.254012	LAV 35
143.254022	LAV 35
143.254032	LAV 35
143.254042	LAV 35
143.254052	LAV 35
143.254062	ECV 100
143.254072	LAV 35
143.254082	LAV 35
143.254092	LAV 35
143.254102	LAV 35
143.254112	LAV 35
143.254122	LAV 35
143.254142	ECV 100
143.254152	ECV 100
143.264162	ECV 100
143.254172	ECV 100
143.254182	ECV 100
143.254192	ECV 100
143.254212	LAV 35
143.254222	LAV 35
143.254232	ECV 100
143.254242	ECV 100
143.254252	ECV 100
143.254262	ECV 100
143.254272	ECV 100
143.254282	ECV 100
143.254292	ECV 100
143.254302	LAV 35
143.254312	LAV 35
143.254322	ECV 100
143.254332	LAV 35
143.254342	ECV 100
143.254352	ECV 100
143.254362	LAV 35
143.254372	ECV 100
143.254382	ECV 100
143.254392	LAV 35
143.254402	ECV 100
143.254412	ECV 100
143.254432	LAV 30
143.254442	ECV 100
143.254452	LAV 35
143.254462	ECV 100
143.254472	LAV 35
143.254482	LAV 35
143.254492	ECV 100
143.254502	LAV 35
143.254512	LAV 35
143.254522	LAV 35
143.254532	LAV 35
143.255012	LAV 50
143.255022	LAV 50
143.255042	LAV 50
143.255052	LAV 50
143.255062	LAV 50
143.255072	LAV 50
143.255092	LAV 50
143.255112	LAV 50
143.256022	V 60
143.256052	V 60
143.256082	V 60
143.256092	V 60
143.256122	V 60
143.257012	LAV 40
143.257022	LAV 40
143.257032	LAV 40
143.257042	LAV 40
143.257052	LAV 40
143.257062	LAV 40
143.257072	LAV 40
143.264012	LAV 35
143.264022	LAV 35
143.264032	LAV 35

Craftsman Model Number	Basic Tecumseh Model Number
143.264042	LAV 35
143.264052	ECV 100
143.264062	ECV 100
143.264072	ECV 100
143.264082	ECV 100
143.264092	LAV 35
143.264102	ECV 100
143.264232	LAV 35
143.264242	LAV 35
143.264252	LAV 35
143.264262	LAV 35
143.264272	LAV 35
143.264282	LAV 35
143.264292	LAV 35
143.264302	LAV 35
143.264312	LAV 35
143.264322	LAV 35
143.264332	LAV 35
143.265352	ECV 100
143.264362	ECV 100
143.264372	ECV 100
143.264382	LAV 35
143.264392	ECV 100
143.264402	ECV 100
143.264412	ECV 100
143.264422	LAV 35
143.264432	ECV 100
143.264452	ECV 100
143.264462	ECV 100
143.264482	ECV 100
143.264492	LAV 35
143.264052	LAV 35
143.264512	ECV 100
143.264522	LAV 35
143.264542	LAV 35
143.264562	ECV 100
143.264572	ECV 100
143.264582	ECV 100
143.264592	ECV 100
143.264602	ECV 100
143.264612	ECV 100
143.264622	ECV 100
143.264632	ECV 100
143.264642	ECV 100
143.264652	ECV 100
143.264672	ECV 100
143.264682	LAV 35
143.265012	LAV 50
143.265032	LAV 50
143.265042	LAV 50
143.265052	LAV 50
143.265062	LAV 50
143.265072	LAV 50
143.265082	LAV 50
143.265092	LAV 50
143.265112	LAV 50
143.265122	LAV 50
143.265132	LAV 50
143.265142	LAV 50
143.265152	LAV 50
143.265162	LAV 50
143.265172	LAV 50
143.265192	LAV 50
143.266032	V 60
143.266062	V 60
143.266082	V 60

Craftsman Model Number	Basic Tecumseh Model Number
143.266252	V 60
143.266372	V 60
143.266382	V 60
143.266392	V 60
143.266402	V 60
143.266412	V 60
143.266432	V 60
143.266442	V 60
143.266452	V 60
143.267012	LAV 40
143.267022	LAV 40
143.267042	LAV 40
143.274022	ECV 100
143.274032	ECV 100
143.274042	ECV 100
143.274052	ECV 100
143.274062	ECV 100
143.274072	ECV 100
143.274092	LAV 35
143.274102	LAV 35
143.274112	LAV 35
143.274122	LAV 35
143.274132	LAV 35
143.274142	ECV 100
143.274162	LAV 35
143.274172	LAV 35
143.274182	LAV 35
143.274202	ECV 100
143.274212	ECV 100
143.274222	ECV 100
143.274232	ECV 100
143.274242	ECV 100
143.274252	LAV 35
143.274262	ECV 100
143.274272	LAV 35
143.274282	LAV 35
143.274292	LAV 35
143.274302	LAV 35
143.274312	LAV 35
143.274322	LAV 35
143.274332	LAV 35
143.274342	ECV 100
143.274352	LAV 35
143.274362	ECV 100
143.274372	LAV 35
143.274392	ECV 100
143.274402	ECV 100
143.274412	ECV 100
143.274422	ECV 100
143.274432	ECV 100
143.274442	ECV 100
143.274452	ECV 100
143.274462	ECV 100
143.274472	LAV 35
143.274482	ECV 100
143.274492	LAV 35
143.274502	ECV 100
143.274512	ECV 100
143.274522	ECV 100
143.274542	ECV 100
143.274552	LAV 35
143.274562	ECV 100
143.274582	ECV 100
143.274592	LAV 35
143.274602	ECV 100
143.274612	EVC 100

Craftsman Model Number	Basic Tecumseh Model Number
143.274622	EVC 100
143.274632	EVC 100
143.274642	LAV 35
143.274652	ECV 100
143.274662	LAV 35
143.274672	ECV 100
143.274682	LAV 35
143.274692	ECV 100
143.274702	LAV 35
143.274712	ECV 100
143.274722	ECV 100
143.274732	ECV 100
143.274742	ECV 100
143.274752	ECV 100
143.274762	ECV 100
143.274772	LAV 35
143.274782	ECV 100
143.274792	LAV 35
143.275012	LAV 50
143.275022	LAV 50
143.275042	LAV 50
143.275052	LAV 50
143.275062	LAV 50
143.275072	LAV 50
143.275082	LAV 50
143.276182	V 60
143.276202	V 60
143.276252	V 60
143.276412	V 60
143.277012	LAV 40
143.277022	LAV 40
143.284012	LAV 35
143.284022	LAV 35
143.284032	LAV 35
143.284042	ECV 100
143.284052	LAV 35
143.284062	LAV 35
143.284072	ECV 100
143.284082	LAV 35
143.284092	LAV 35
143.284102	ECV 100
143.284112	LAV 35
143.284142	LAV 35
143.284152	LAV 35
143.284162	LAV 35
143.284182	LAV 35
143.284212	ECV 100
143.284222	LAV 30
143.284232	TVS 90
143.284242	LAV 35
143.284252	LAV 35
143.284262	ECV 100
143.284272	TVS 90
143.284282	LAV 35
143.284292	LAV 35
143.284302	LAV 35
143.284312	LAV 35
143.284322	LAV 35
143.284332	ECV 100
143.284342	ECV 100
143.284352	ECV 100
143.284362	ECV 100
143.284372	ECV 100
143.284382	ECV 100
143.284392	LAV 35
143.284402	LAV 35

Craftsman Model Number	Basic Tecumseh Model Number
143.284412	LAV 35
143.284422	LAV 35
143.284432	ECV 100
143.284442	LAV 30
143.284452	ECV 100
143.284462	ECV 100
143.284472	ECV 100
143.284482	LAV 35
143.284492	ECV 100
143.284502	ECV 100
143.284512	LAV 35
143.284522	LAV 35
143.284532	ECV 100
143.284542	LAV 35
143.284552	LAV 35
143.284562	LAV 35
143.284572	LAV 35
143.284582	ECV 100
143.284592	LAV 35
143.284602	ECV 100
143.284612	ECV 100
143.284622	ECV 100
143.284632	LAV 35
143.284642	ECV 100
143.284652	LAV 35
143.284672	ECV 100
143.284682	ECV 100
143.284692	ECV 100
143.284702	ECV 100
143.284712	LAV 35
143.284722	LAV 35
143.284732	LAV 35
143.284742	ECV 100
143.284752	ECV 100
143.284762	LAV 35
143.284772	ECV 100
143.284782	ECV 100
143.285012	LAV 50
143.285022	LAV 50
143.285032	LAV 50
143.285042	LAV 50
143.285052	LAV 50
143.285062	LAV 50
143.285072	LAV 50
143.285082	LAV 50
143.285092	LAV 50
143.285102	LAV 50
143.286012	V 50
143.286022	V 60
143.286342	V 60
143.287012	LAV 40
143.293012	TVS 75
143.294012	TVS 90
143.294022	TVS 90
143.294032	TVS 90
143.294042	TVS 90
143.294052	TVS 90
143.294062	TVS 90
143.294072	TVS 90
143.294092	TVS 90
143.294102	TVS 90
143.294112	TVS 90
143.294122	TVS 90
143.294132	TVS 90
143.294142	ECV 100
143.294152	ECV 100
143.294162	ECV 100
143.294172	ECV 100
143.294182	TVS 90
143.294192	TVS 90
143.294202	TVS 90
143.294212	TVS 90
143.294222	ECV 100
143.294232	ECV 100
143.294242	TVS 90
143.294252	TVS 90
143.294262	TVS 90
143.294272	TVS 90
143.294282	TVS 90
143.294292	TVS 90
143.294302	TVS 90
143.294312	TVS 90
143.294322	TVS 90
143.294332	ECV 100
143.294342	TVS 90
143.294352	ECV 100
143.294362	ECV 100
143.294372	ECV 100
143.294382	ECV 100
143.294392	ECV 100
143.294402	ECV 100
143.294412	ECV 100
143.294422	ECV 100
143.294432	TVS 90
143.294442	TVS 90
143.294452	TVS 90
143.294462	TVS 90
143.294472	ECV 100
143.294482	ECV 100
143.294492	TVS 90
143.294502	TVS 90
143.294512	TVS 90
143.294522	TVS 90
143.294532	TVS 90
143.294542	ECV 100
143.294552	TVS 105
143.294562	TVS 105
143.294572	ECV 100
143.294582	ECV 100
143.294592	ECV 100
143.294602	TVS 90
143.294612	TVS 90
143.294632	TVS 105
143.294642	TVS 105
143.294652	TVS 90
143.294662	ECV 100
143.294672	ECV 100
143.294682	ECV 100
143.294692	TVS 90
143.294702	TVS 105
143.294712	TVS 90
143.294722	ECV 100
143.294732	ECV 100
143.294742	ECV 100
143.295012	LAV 50
143.295022	LAV 50
143.295032	LAV 50
143.295042	ECV 120
143.297012	TVS 105
143.304012	TVS 90
143.304032	ECV 100
143.304042	ECV 100
143.304052	TVS 90
143.304072	ECV 100
143.304092	TVS 90
143.304102	ECV 100
143.304112	ECV 100
143.304122	ECV 100
143.305012	ECV 120
143.305022	ECV 120
143.305032	ECV 120
143.305042	LAV 50
143.305052	ECV 120
143.305062	LAV 50
143.313012	TVS 75
143.314012	ECV 100
143.314022	ECV 100
143.314032	TVS 90
143.314042	TVS 90
143.314052	TVS 90
143.314062	TVS 90
143.314072	TVS 90
143.314082	TVS 90
143.314092	TVS 90
143.314102	TVS 90
143.314112	TVS 90
143.314122	ECV 100
143.314132	ECV 100
143.314142	ECV 100
143.314152	ECV 100
143.314162	ECV 100
143.314172	ECV 100
143.314182	TVS 90
143.314192	ECV 100
143.314202	ECV 100
143.314212	ECV 100
143.314222	ECV 100
143.314232	ECV 100
143.314242	ECV 100
143.314252	ECV 100
143.314262	TVS 90
143.314272	TVS 90
143.314282	TVS 90
143.314292	TVS 90
143.314302	TVS 90
143.314312	ECV 100
143.314322	TVS 90
143.314332	TVS 90
143.314342	TVS 90
143.314372	ECV 100
143.314382	TVS 90
143.314392	ECV 100
143.314402	TVS 90
143.314412	TVS 90
143.314422	ECV 100
143.314432	LAV 35
143.314442	ECV 100
143.314452	ECV 100
143.314462	ECV 100
143.314472	ECV 100
143.314152	ECV 100
143.314522	ECV 100
143.314532	LAV 35
143.314542	TVS 90
143.314552	TVS 90
143.314562	TVS 90
143.314572	TVS 90
143.314582	ECV 100

Craftsman Model Number	Basic Tecumseh Model Number
143.314592	ECV 100
143.314612	ECV 100
143.314622	ECV 100
143.314632	ECV 100
143.314642	ECV 100
143.314652	ECV 100
143.314662	ECV 100
143.314672	ECV 100
143.314682	ECV 100
143.314692	ECV 100
143.314702	LAV 35
143.315012	ECV 120
143.315022	LAV 50
143.315032	TVS 105
143.315042	TVS 105
143.315062	LAV 50
143.315072	TVS 105
143.315082	ECV 120
143.315092	LAV 50
143.315102	LAV 50
143.315112	LAV 50
143.315122	LAV 50
143.321012	TVS 75
143.321022	TVS 75
143.324012	ECV 100
143.324022	ECV 100
143.324042	ECV 100
143.324052	TVS 90
143.324062	ECV 100
143.324072	ECV 100
143.324082	ECV 100
143.324102	ECV 100
143.324112	TVS 90
143.324132	ECV 100
143.324142	TVS 90
143.324152	TVS 90
143.324162	TVS 90
143.324172	TVS 90
143.324182	TVXL 105
143.324192	TVS 90
143.324202	ECV 100
143.324212	ECV 100
143.324222	ECV 100
143.324232	ECV 100
143.326012	TVM 195
143.331012	TVS 75
143.331022	TVS 75
143.334032	TVS 90
143.334042	ECV 100
143.334052	TVXL 105
143.334062	VS 90
143.334072	TVS 90
143.334082	ECV 100
143.334102	ECV 100
143.334112	TVS 90
143.334122	TVS 90
143.334132	ECV 100
143.334142	TVS 90
143.334152	TVS 90
143.334162	TVS 90
143.334172	ECV 100
143.334182	ECV 100
143.334192	LAV 35
143.334202	TVS 90
143.334212	ECV 100
143.334222	ECV 100

Craftsman Model Number	Basic Tecumseh Model Number
143.334232	ECV 100
143.334242	ECV 100
143.334252	ECV 100
143.334262	TVS 90
143.334272	TVS 90
143.334282	TVS 90
143.334292	TVS 90
143.334302	TVS 90
143.334312	TVS 90
143.334322	ECV 100
143.334332	TVS 90
143.334342	ECV 100
143.334352	TVS 90
143.334362	TVS 90
143.334372	TVS 90
143.334382	TVS 90
143.335012	ECV 120
143.335022	ECV 120
143.335032	LAV 50
143.335042	LAV 50
143.335052	TVS 120
143.341012	TVS 75
143.344022	TVS 90
143.344032	TVS 90
143.344042	TVS 90
143.344052	ECV 100
143.344062	ECV 100
143.344072	TVS 90
143.344082	ECV 100
143.344092	ECV 100
143.344102	TVS 90
143.344112	TVXL 105
143.344122	ECV 100
143.344132	ECV 100
143.344142	TVS 90
143.344152	ECV 100
143.344162	TVS 90
143.344172	ECV 100
143.344182	TVS 90
143.344192	TVS 90
143.344202	TVS 90
143.344212	TVS 90
143.344222	TVS 90
143.344232	ECV 100
143.344242	ECV 100
143.344262	EVC 100
143.344272	EVC 100
143.344282	ECV 100
143.344292	ECV 100
143.344302	ECV 100
143.344312	ECV 100
143.344322	ECV 100
143.344332	ECV 100
143.344342	ECV 100
143.344352	ECV 100
143.344362	ECV 100
143.344372	ECV 100
143.344382	ECV 100
143.344392	ECV 100
143.344402	TVXL 105
143.344412	TVXL 105
143.344422	TVS 90
143.344432	TVS 90
143.344442	TVS 105
143.344452	ECV 100
143.344462	TVS 105

Craftsman Model Number	Basic Tecumseh Model Number
143.344472	ECV 100
143.345012	ECV 120
143.345022	ECV 120
143.345032	TVS 120
143.345042	LAV 50
143.345052	ECV 120
143.345062	ECV 120
143.346202	TVM 125
143.351012	TVS 75
143.354012	TVS 90
143.354022	ECV 100
143.354032	ECV 100
143.354042	ECV 100
143.354052	ECV 100
143.354062	TVS 90
143.354072	ECV 100
143.354082	ECV 100
143.354092	TVS 90
143.354102	TVS 90
143.354112	TVS 100
143.354122	TVS 90
143.354132	TVXL 105
143.354142	TVS 90
143.354152	EVC 100
143.354162	TVS 90
143.354172	TVS 90
143.354182	TVS 90
143.354192	TVS 90
143.354202	TVS 90
143.354212	TVS 90
143.354222	ECV 100
143.354232	TVS 90
143.354242	ECV 100
143.354252	ECV 100
143.354262	ECV 100
143.354272	EVC 100
143.354282	LAV 35
143.354292	TVS 90
143.354302	ECV 100
143.355022	ECV 120
143.356022	TVM 125

Horizontal Crankshaft

Craftsman Model Number	Basic Tecumseh Model Number
143.501011	H22K
143.501020	H22K
143.501041	H22K
143.501051	H22K
143.501061	H22K
143.501071	H22K
143.501081	H22K
143.501090	H22K
143.501091	H22K
143.501111	H22K
143.501121	H22K
143.501131	H22K
143.501141	H22K
143.501151	H22K
143.501161	H22K
143.501170	H22K
143.501171	H22K
143.501181	H22K
143.501190	H22K
143.501191	H22K
143.501201	H22K
143.501211	H22K

Craftsman Model Number	Basic Tecumseh Model Number
143.501221	H22K
143.501231	H22K
143.501241	H22K
143.501251	H22K
143.501261	H22K
143.501270	H22K
143.501271	H22K
143.502011	H35M
143.502021	HB35K
143.502031	HB35K
143.502041	H35P
143.504011	H35K
143.505010	H55D
143.505011	H55D
143.506011	H30P
143.521011	H22R
143.521021	H22R
143.521031	H22R
143.521051	H22R
143.521061	H22R
143.521071	H22R
143.521081	HT30A
143.521091	H22R
143.521101	H22R
143.521111	H22R
143.521121	H22R
143.521131	HT30B
143.524021	HT35A
143.524031	HT35A
143.524041	H35P
143.524051	HT35A
143.524061	HT35B
143.524071	HT35B
143.524081	HT35B
143.525021	HT45C
143.526011	HT45A
143.526021	HT55B
143.526031	HT55C
143.531011	HT30B
143.531022	H25
143.531032	H22
143.531042	H22
143.531052	H30
143.531062	H22
143.531072	H30
143.531082	H30
143.531092	H22
143.531112	H22
143.531122	H30
143.531132	H30
143.531142	H22
143.531152	H30
143.531162	H22
143.531172	H30
143.531182	H30
143.534012	H35
143.534022	H35
143.534032	H35
143.534042	H35
143.534052	H35
143.534062	H35
143.534072	H35
143.535012	H50
143.535022	H50
143.535062	H50
143.536012	H60
143.536022	H60
143.536032	H60
143.536042	H60
143.536052	H60
143.536062	H60
143.537012	H40
143.541012	H30
143.541022	H22
143.541032	H22
143.541042	H30
143.541052	H30
143.541062	H30
143.541072	H22
143.541082	H22
143.541102	H22
143.541112	H30
143.541122	H30
143.541132	H30
143.541142	H25
143.541152	H30
143.541162	H22
143.541172	H25
143.541182	H25
143.541192	H30
143.541202	H30
143.541212	H22
143.541222	H30
143.541232	H22
143.541252	H22
143.541262	H22
143.541282	H30
143.541292	H25
143.541302	H30
143.544012	H35
143.544022	H35
143.544032	H35
143.544042	H35
143.544052	H35
143.545012	H50
143.545022	H50
143.545032	H50
143.545042	H50
143.546012	H60
143.546022	H60
143.546032	H60
143.546042	H60
143.546052	H60
143.546062	H60
143.546072	H60
143.546082	H60
143.546092	H60
143.546102	H60
143.546112	H60
143.546142	H60
143.546142	H60
143.547012	H40
143.547022	H40
143.547032	H40
143.551012	H30
143.551032	H30
143.551042	H22
143.551052	H30
143.551062	H30
143.551072	H30
143.551082	H25
143.551092	H25
143.551102	H30
143.551152	H30
143.551162	H30
143.551172	H22
143.551192	H30
143.554012	H35
143.554022	H35
143.554032	H35
143.554042	H35
143.554052	H35
143.554072	H35
143.554082	H35
143.555012	H50
143.555022	H50
143.555032	H50
143.555042	H50
143.555052	H50
143.556012	H60
143.556022	H60
143.556032	H60
143.556042	H60
143.556052	H60
143.556062	H60
143.556072	H60
143.556082	H60
143.556092	H60
143.556102	H60
143.556112	H60
143.556122	H60
143.556132	H60
143.556142	H60
143.556152	H60
143.556162	H60
143.556172	H60
143.556182	H60
143.556192	H60
143.556202	H60
143.556212	H60
143.556222	H60
143.556232	H60
143.556242	H60
143.556252	H60
143.556262	H60
143.556272	H60
143.556282	H60
143.557012	H40
143.557022	H40
143.557032	H40
143.557042	H40
143.557052	H40
143.557062	H40
143.557072	H40
143.557082	H40
143.561012	H30
143.561022	H25
143.561032	H30
143.561042	H30
143.561052	H30
143.561062	H30
143.561072	H25
143.561082	H25
143.561092	H30
143.561102	H30
143.561112	H30
143.561122	H22
143.561132	H30

Craftsman Model Number	Basic Tecumseh Model Number
143.561142	H22
143.561152	H22
143.561162	H25
143.561172	H30
143.561182	H30
143.561202	H30
143.561212	H30
143.561222	H30
143.564012	H35
143.564022	H35
143.564032	H35
143.564042	H35
143.564052	H35
143.564072	H35
143.564082	H35
143.564092	H35
143.564102	H35
143.564112	H35
143.565012	H50
143.565022	H50
143.566002	H60
143.556012	H60
143.566022	H60
143.566032	H60
143.566042	H60
143.566052	H60
143.566062	H60
143.566072	H60
143.566082	H60
143.566092	H60
143.566102	H60
143.566112	H60
143.566122	H60
143.566132	H60
143.566142	H60
143.566152	HH60
143.556162	H60
143.566172	H60
143.566182	H60
143.566192	H60
143.566202	H60
143.566212	H50
143.566222	H60
143.566232	H60
143.566242	H60
143.566252	H60
143.567012	H40
143.567022	H40
143.567032	H40
143.567042	H40
143.571002	H30
143.571012	H30
143.571022	H30
143.571032	H30
143.571042	H30
143.571052	H30
143.571062	H25
143.571072	H30
143.571082	H25
143.571092	H30
143.571102	H25
143.571112	H25
143.571122	H30
143.571152	H22
143.571162	H30
143.571172	H25

Craftsman Model Number	Basic Tecumseh Model Number
143.571182	H30
143.571202	H25
143.574022	H35
143.574032	H35
143.574042	H35
143.574052	H35
143.574062	H35
143.574072	H35
143.574082	H35
143.574092	H35
143.574102	H35
143.574112	H35
143.575012	H50
143.575022	H50
143.575032	H50
143.575042	H50
143.576002	H60
143.576012	HH60
143.576022	H60
143.576032	H60
143.576042	H60
143.576052	H60
143.576062	H60
143.576072	H60
143.576082	H60
143.576102	H60
143.576112	H60
143.576122	H60
143.576132	H60
143.576142	H60
143.576172	H60
143.576182	H60
143.576192	H60
143.576202	H60
143.576212	H60
143.576222	H60
143.576232	H60
143.576282	H60
143.576292	HH60
143.576302	HH60
143.576312	H60
143.576322	H60
143.576332	H60
143.577012	H40
143.577022	H40
143.577032	H40
143.577042	H40
143.577072	H40
143.577082	H40
143.581002	H30
143.581022	H25
143.581032	H30
143.581042	H25
143.581052	H30
143.581062	H30
143.581072	H30
143.581082	H30
143.581092	H30
143.581102	H25
143.584012	H35
143.584032	H35
143.584052	H35
143.584062	H35
143.584072	H35
143.584082	H35
143.584102	H35

Craftsman Model Number	Basic Tecumseh Model Number
143.584112	H35
143.584122	H35
143.584132	H35
143.584142	H35
143.585012	H50
143.585032	H50
143.585042	H50
143.586012	H60
143.586022	H60
143.586032	H60
143.586042	H60
143.586072	H60
143.586082	H60
143.586152	H60
143.586172	H60
143.586182	HH60
143.586192	H60
143.586202	HH60
143.586212	HH60
143.586222	H60
143.586232	H60
143.586242	H60
143.586262	H60
143.586272	H60
143.586282	H60
143.587012	HS40
143.587022	HS40
143.587032	HS40
143.587042	HS40
143.591012	H25
143.591022	H30
143.591032	H30
143.591042	H30
143.591052	H30
143.591062	H30
143.591072	H30
143.591082	H25
143.591092	H30
143.591102	H30
143.591112	H25
143.591122	H30
143.591132	H25
143.591142	H30
143.594012	H35
143.594022	H35
143.594032	H35
143.594042	H35
143.594052	H35
143.594060	H35
143.594072	H35
143.594082	H35
143.594092	ECH90
143.594102	H35
143.595012	H50
143.595042	H50
143.596042	H60
143.596072	H60
143.596082	H60
143.596092	H60
143.596102	HH60
143.596112	HH60
143.596122	H60
143.597012	HS40
143.597022	HS40
143.597032	HS40
143.601022	H30

Craftsman Model Number	Basic Tecumseh Model Number
143.601032	H30
143.601062	H30
143.604012	H35
143.604022	ECH90
143.604032	H35
143.604042	H35
143.604052	ECH90
143.604062	H35
143.604072	H35
143.605012	H50
143.605022	H50
143.605052	H50
143.607012	HS40
143.607022	HS40
143.607032	HS40
143.607042	HS40
143.607052	HS40
143.607062	HS40
143.611012	H30
143.611022	H30
143.611032	H30
143.611042	H30
143.611052	H30
143.611062	H25
143.611072	H30
143.611082	H30
143.611092	H30
143.611102	H25
143.611112	H30
143.614012	ECH90
143.614022	ECH90
143.614032	ECH90
143.614042	H35
143.614052	ECH90
143.614062	H35
143.614072	H35
143.614082	H35
143.614092	H35
143.614102	H35
143.614112	H35
143.614122	H35
143.614132	H35
143.614142	H35
143.614152	H35
143.614162	H35
143.615012	H50
143.615022	H50
143.615052	H50
143.615062	H50
143.615072	H50
143.615082	H50
143.615092	H50
143.616022	H60
143.616042	H60
143.616052	HH60
143.616062	H60
143.616072	H60
143.616082	H60
143.616112	H60
143.616132	HH60
143.616142	HH60
143.617012	HS40
143.617022	HS40
143.617032	HS40
143.617042	HS40
143.617062	HS40
143.617082	HS40
143.617092	HS40
143.617112	HS40
143.617132	HS40
143.617152	HS40
143.617162	HS40
143.617182	HS40
143.621012	H25
143.621022	H25
143.621032	H30
143.621042	H25
143.621052	H30
143.621062	H30
143.621082	H25
143.621092	H25
143.624012	H35
143.624022	H35
143.624032	H35
143.624042	H35
143.624092	H35
143.624102	H35
143.624112	H35
143.625012	H50
143.625022	H50
143.625032	H50
143.625042	H50
143.625052	H50
143.625072	H50
143.625082	H50
143.625092	H50
143.625102	H50
143.625112	H50
143.625122	H50
143.625132	H50
143.626022	H60
143.626042	H60
143.626132	H60
143.626162	H60
143.626182	H60
143.626202	H60
143.626222	H60
143.626232	H60
143.626242	HH60
143.626252	H60
143.626262	H60
143.626272	HH60
143.626302	H60
143.627012	HS40
143.627032	HS40
143.627042	HS40
143.631012	H30
143.631022	H25
143.631032	H25
143.631042	H25
143.631052	H25
143.631062	H25
143.631072	H25
143.631082	H25
143.631092	H30
143.634012	H35
143.634022	H35
143.634032	H35
143.635012	H50
143.635022	H50
143.635032	HS50
143.635042	H50
143.635052	H50
143.636052	H60
143.637012	HS40
143.641012	H25
143.641022	H25
143.641032	H30
143,641042	H25
143.641052	H30
143.641062	H30
143.641072	H35
143.644012	H35
143.644022	H35
143.644032	H35
143.644052	H35
143.644072	H35
143.644082	H35
143.645012	HS50
143.645022	HS50
143.645032	HS50
143.646112	HH60
143.646192	H60
143.647012	HS40
143.647022	HS40
143.647032	HS40
143.647042	HS40
143.647052	HS40
143.647062	HS40
143.651012	H30
143.651022	H30
143.651032	H30
143.651042	H30
143.651052	H30
143.651062	H30
143.651072	H30
143.654022	H35
143.654032	H35
143.654042	H35
143.654052	H35
143.654062	H35
143.654072	H35
143.654082	H35
143.654092	H35
143.654102	H35
143.654112	H35
143.654122	H35
143.654132	H35
143.654142	H35
143.654152	H35
143.654162	H35
143.654172	H35
143.654182	H35
143.654192	H35
143.654202	H35
143.654212	H35
143.654222	H35
143.654232	H35
143.654242	H35
143.654252	H35
143.654262	H35
143.654272	H35
143.654282	H35
143.654292	H35
143.564302	H35
143.654312	H35
143.654322	H35
143.655012	HS50

Craftsman Model Number	Basic Tecumseh Model Number
143.655032	HS50
143.656012	H60
143.656022	H60
143.656032	H60
143.656042	H60
143.656052	H60
143.656092	H60
143.656112	HH60
143.656182	H60
143.656202	H60
143.656252	160
143.657012	HS40
143.657022	HS40
143.657032	HS40
143.657042	HS40
143.657052	HS40
143.661012	H30
143.661022	H30
143.661032	H30
143.661042	H30
143.661062	H30
143.664032	H35
143.664042	H35
143.664052	H35
143.664062	H35
143.664072	H35
143.664082	H35
143.664092	H35
143.664102	H35
143.664112	H35
143.664122	H35
143.664132	H35
143.664142	H35
143.664152	H35
143.664162	H35
143.664172	H35
143.664182	H35
143.664192	H35
143.664202	H35
143.664212	H35
143.664222	H35
143.664232	H35
143.664242	H35
143.664252	H35
143.664262	H35
143.664272	H35
143.664282	H35
143.664292	H35
143.664302	H35
143.664312	H35
143.664332	H35
143.665012	HS50
143.665022	HS50
143.665032	HS50
143.665042	HS50
143.665052	HS50
143.665062	HS50
143.665072	HS50
143.665082	HS50
143.666102	H60
143.666112	H60
143.666122	H60
143.666132	H60
143.666142	H60
143.666172	H60
143.666182	H60
143.666192	H60
143.666202	H60
143.666242	HH60
143.666272	H60
143.666292	H60
143.666372	H60
143.667012	HS40
143.667022	HS40
143.667032	HS50
143.667042	HS40
143.667052	HS40
143.667062	HS40
143.667072	HS40
143.667082	HS40
143.675012	HS50
143.675022	HS50
143.675032	H50
143.675042	HS50
143.675052	HS50
143.675062	H50
143.676112	H60
143.676132	H60
143.676232	H60
143.676242	H60
143.677012	HS40
143.677022	HS40
143.684012	H35
143.685012	HS50
143.685022	HS50
143.685032	HS50
143.686072	H50
143.686112	H50
143,686122	H60
143.686152	H60
143.686182	H60
143.687012	HS40
143.687022	HS50
143.687032	HS50
143.687042	HS40
143.694012	H35
143.694022	H35
143.694032	H35
143.696012	H60
143.696042	H60
143.696082	HS50
143.696122	H60
143.696142	H50
143.696152	H60
143.697012	HS50
143.697022	HS50
143.697032	HS40
143.697042	HS50
143.697052	HS50
143.701012	H30
143.706012	H60
143.707012	HS50
143.707042	HS40
143.707052	HS50
143.707062	HS40
143.707072	HS50
143.707082	HS50
143.707092	HS50
143.707102	HS40
143.707112	HS50
143.707122	HS40
143.707132	HS50
143.711012	H30
143.711022	H30
143.711032	H30
143.711042	H30
143.714012	H35
143.714022	H35
143.714032	H35
143.714042	H35
143.714052	H35
143.714062	H35
143.714072	H35
143.714082	H35
143.714092	H35
143.714102	H35
143.714112	H35
143.714122	H35
143.716032	H60
143.716042	H60
143.716162	H50
143.716232	H60
143.716242	H50
143.716272	H60
143.716282	H60
143.716292	H60
143.716352	H60
143.717012	HS40
143.717022	HS40
143.717032	HS40
143.717042	HS40
143.717052	HS50
143.717062	HS50
143.717072	HS50
143.717082	HS50
143.717092	HS50
143.717102	HS40
143.717112	HS50
143.721012	H30
143.721022	H30
143.721032	H30
143.724012	H35
143.724022	H35
143.724032	H35
143.724042	H35
143.724052	HS40
143.725012	HS50
143.726092	H50
143.726302	H50
143.731012	H30
143.734012	H35
143.734022	H35
143.734032	H35
143.734042	HS40
143.735022	HS50
143.741012	H30
143.741032	H30
143.741042	H30
143.741052	H30
143.741062	H30
143.741072	H30
143.742042	H50
143.744012	H35
143.744022	H35
143.744032	H35
143.744052	H35
143.744062	H35
143.744072	H35

Craftsman Model Number	Basic Tecumseh Model Number	Craftsman Model Number	Basic Tecumseh Model Number	Craftsman Model Number	Basic Tecumseh Model Number
143.744102	HS40	143.751042	H30	143.756052	H70
143.744112	HS50	143.754012	H35	143.756072	H60
143.751012	H30	143.754022	H35	143.756092	H60
143.751022	H30	143.754062	HS50	143.756142	H50
143.751032	H30	143.756042	H70		

CRAFTSMAN

ENGINE IDENTIFICATION

Engines must be identified by the complete model number, including the serial number and type number in order to obtain correct repair parts. These numbers are located on the name plate as shown in Fig. C1A. If short block renewal is necessary, original identification plate must be transferred to replacement short block assemblies so unit can be identified when servicing at a later date.

PARTS PROCUREMENT

Parts for Craftsman engines are available from all Sears Retail Stores and Catalog Sales Offices. Be sure to give complete Craftsman model, serial and type numbers of engine when ordering parts.

SERVICE NOTES (FOUR-STROKE)

CARBURETOR. The carburetors used on many Craftsman engines are the same as used on similar Tecumseh engines. Special Craftsman fuel systems are used on some models. The tank mounted suction carburetor is shown in view (A – Fig. C1).

The Craftsman float carburetor without speed control shown in view (B) is used without a mchanical or air vane governor.

Some engines use a mechanical governor (15, 16 and 17) which controls engine speed using a throttle valve (7 and 8) located in the intake manifold (6).

Refer to the appropriate following paragraphs for servicing Craftsman fuel systems.

Fuel Tank Mounted Carburetor. Early carburetors are equipped with a Bowden wire control as shown in Fig. C2. Later fuel tank mounted carburetors have a manual control knob and a positioning spring as shown in Fig. C3. Other differences are also noted and it is important to identify the type used before servicing.

On all tank mounted carburetors, turn the control valve clockwise to the position shown in Fig. C4, then withdraw the valve. There are no check valves or check balls in the carburetor pickup tube, but the tube can be removed for more thorough cleaning. To reinstall

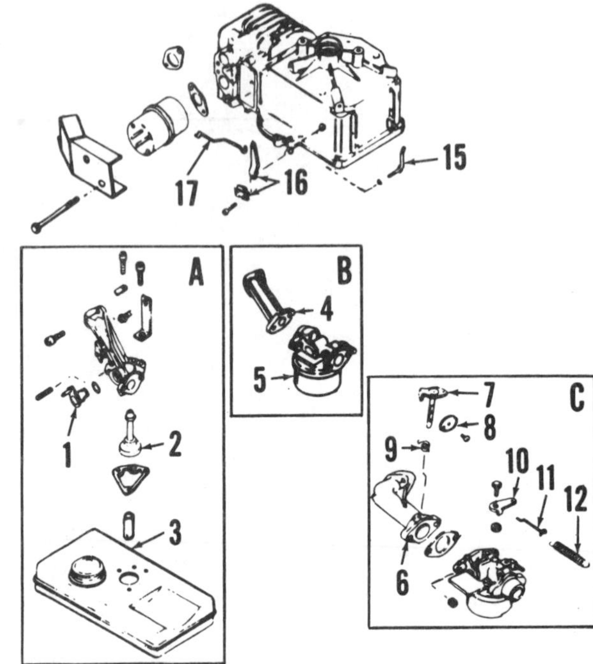

Fig. C1—Craftsman fuel systems are (A) tank mounted suction carburetor, (B) special float carburetor without governor or (C) the special float carburetor with mechanical governor linkage (15, 16 and 17) hooked up to a throttle (7 and 8) located in the intake manifold (6).

1. Control valve
2. Suction tube
3. Fuel tank
4. Intake manifold
5. Float carburetor
6. Intake manifold
7. Governor throttle shaft
8. Governor throttle plate
9. Return spring
10. Bellcrank
11. Linkage (10 to 12)
12. Governor spring
15. Internal lever
16. External lever
17. Linkage (16 to 7)

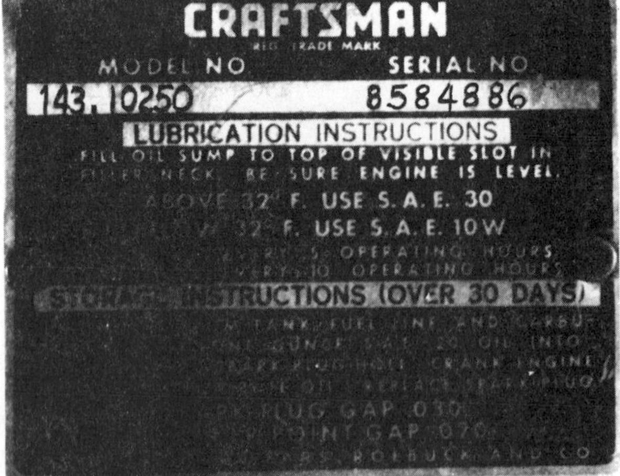

Fig. C1A—View of typical identification plate on Craftsman four-cycle engine.

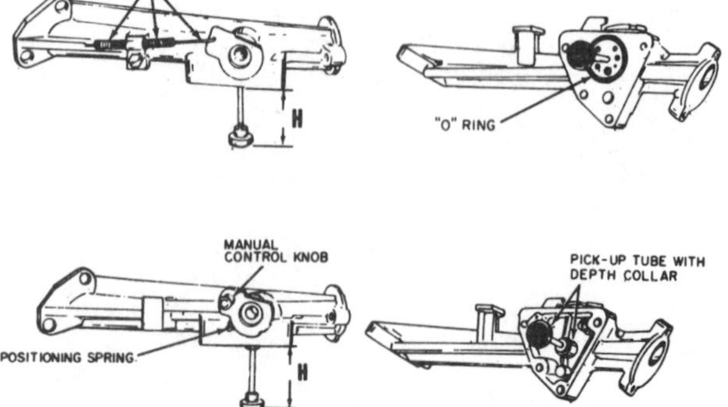

Fig. C2—View of early tank mounted suction carburetor. Notice the Bowden control cable and groove for "O" ring on lower surface. Refer to Fig. C3 for later type.

Fig. C3—View of later tank mounted suction carburetor. The unit is controlled by moving the manual control on carburetor. The lower surface is sealed with a gasket and the "O" ring used on early models is not used. The pickup tube has a collar to make sure it is installed at the correct depth.

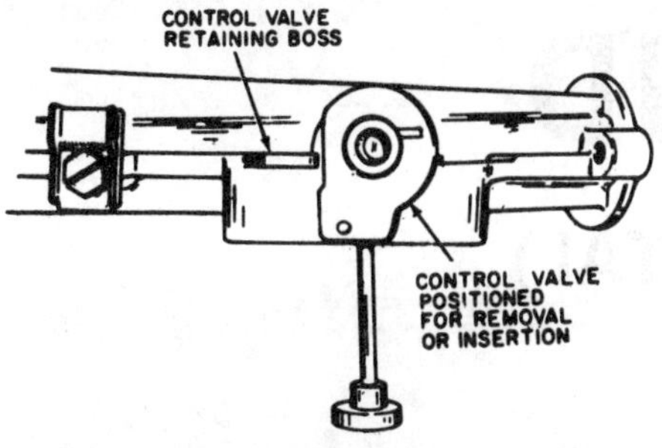

Fig. C4—The control valve must be turned clockwise to the position shown to clear the retaining rod.

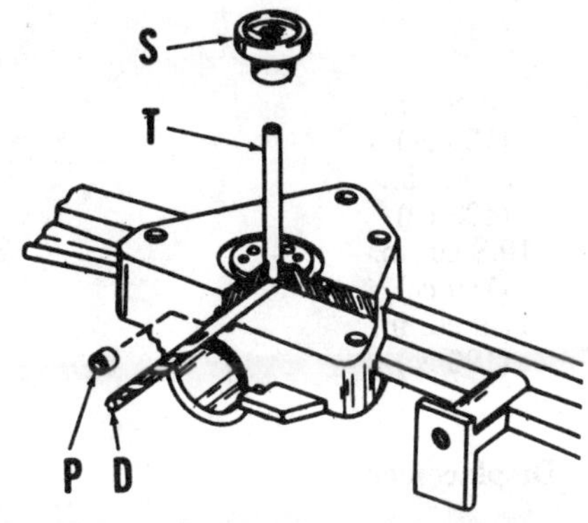

Fig. C5—A ¼-inch drill bit (D) should be used to gage the correct installed depth of tube (T). New plug (P) should be used when assembling.

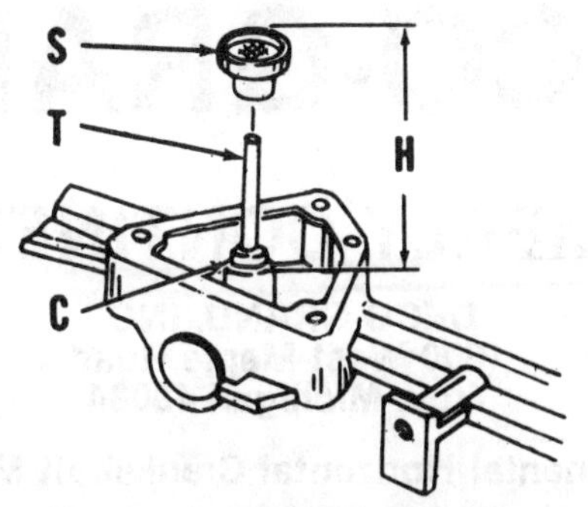

Fig. C6—Tube (T) should be pressed into bore until collar C) contacts body of late carburetor. Press strainer (S) onto tube until height (H) is correct.

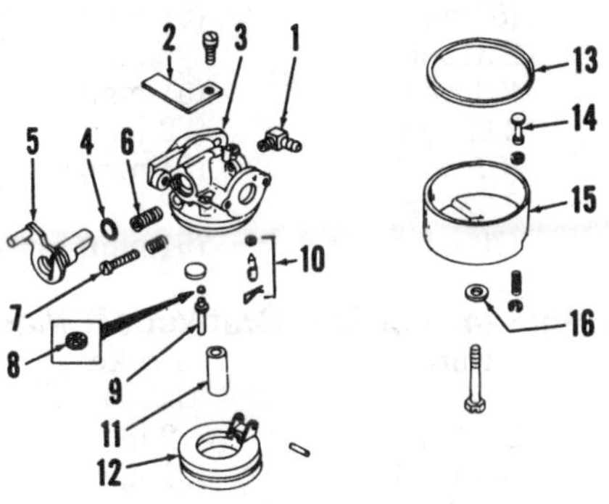

Fig. C10—Exploded view of Craftsman float type carburetor. Refer to text for service procedure.

1. Fuel inlet fitting
2. Retaining plate
3. Carburetor body
4. "O" ring
5. Control valve
6. Positioning spring
7. High speed stop
8. Screen
9. Fuel pickup tube
10. Fuel inlmt valve
11. Bowl spacer
12. Float
13. Gasket
14. Bowl drain
15. Float bowl
16. Gasket

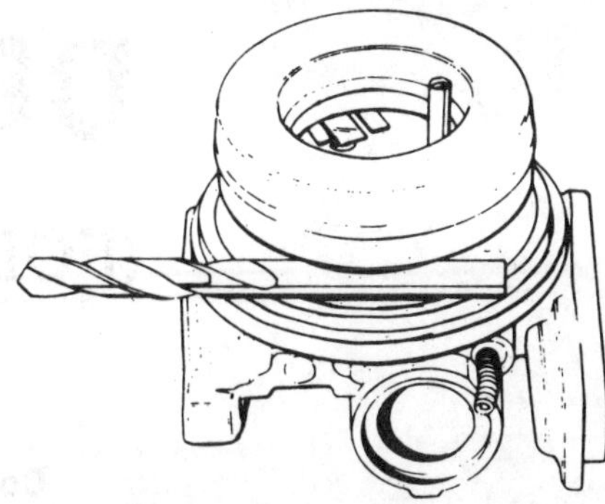

Fig. C11—Float height should be set using a number 4 drill bit positioned as shown. Drill size is 0.2090 inch (5.31 mm) diameter.

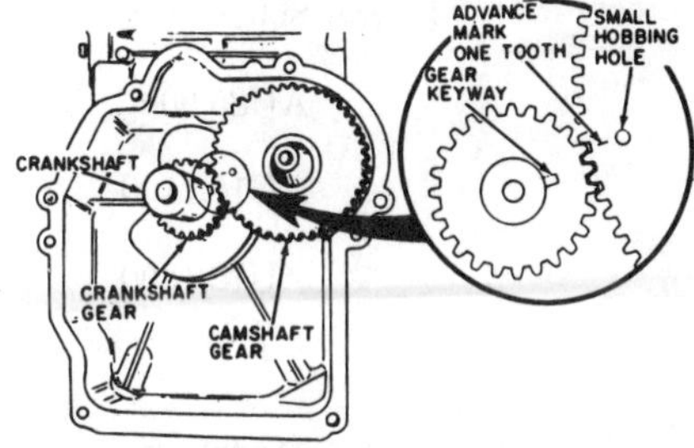

Fig. C12—Camshaft timing marks should be advanced one tooth when engine is not equipped with mechanical governor.

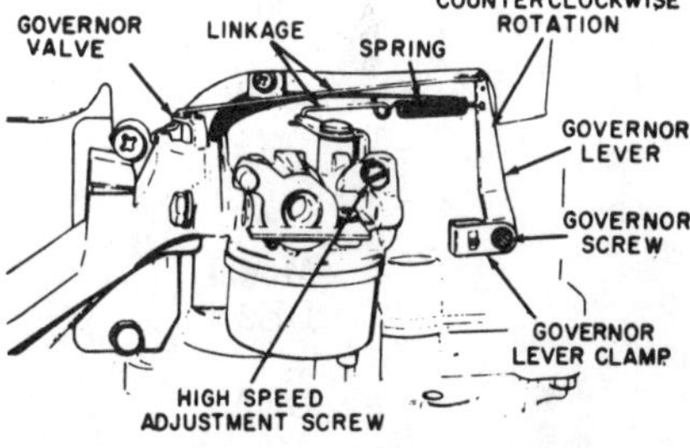

Fig. C13—View of Craftsman float type carburetor installed. The model shown uses mechanical governor controlling the governor throttle valve located in the intake manifold. This system is shown at (C—Fig. C1).

pickup tube in early models, it is necessary to remove plug (P–Fig. C5). Insert ⅛ inch drill into passage as shown and press tube into bore until seated against drill bit. Leave drill in place and press strainer (S) onto pickup tube until height (H–Fig. C2) is 1 27/64 to 1 7/16 inches (36.1-36.5 mm). Remove drill bit and install new cap plug.

On late carburetors, the tube (T–Fig. C6) has a collar near the top end. Press the tube into bore until the collar is against carburetor body casting; then press the strainer (S) onto tube until distance (H) is 1 15/32 inches (27.3 mm).

The reservoir tube in the fuel tank should have slotted end toward bottom of tank on all models.

Fuel mixture is nonadjustable, but the mixture will be changed by dirty or missing air filter. Be sure air filter is in good condition and clean.

NOTE: The camshaft timing should be advanced one tooth for engines with this carburetor. Refer to Fig. C12.

Craftsman Float Carburetor. The carburetor is serviced in manner similar to other float type carburetors. Refer to Fig. C10.

The carburetor is equipped with mixture adjustment. It is important that float height is correct. The air filter must be in good condition and clean. Refer to Fig. C11 for measuring float height with drill bit.

The fuel pickup tube (9) should be pressed into bore until collar is against carburetor body.

NOTE: Some engines equipped with this carburetor are not equipped with a variable speed governor. These models are equipped with a plain intake manifold as shown (B—Fig. C1). The camshaft timing should be advanced one tooth for engines without governor as shown in Fig. C12.

Camshaft timing marks should be aligned for models with variable speed governor. The governed high speed is adjusted by turning the adjustment screw shown in Fig. C13.

DECO-GRAND

(FORMERLY CONTINENTAL)

DECO-GRAND, INC.
1600 West Maple Road
Troy, Michigan 48084

Continental Horizontal Crankshaft Models

Model	Bore	Stroke	Displacement
AC5,AC6, AA7,AU7, 700	2.125 in. (54 mm)	2 in. (51 mm)	7.1 cu. in. (117 cc)
AA8,AU8, 800	2.250 in. (57 mm)	2 in. (51 mm)	8.0 cu. in. (130 cc)
AU85,900	2.313 in. (59 mm)	2 in. (51 mm)	8.4 cu. in. (139 cc)
AU10,1100	2.625 in. (67 mm)	2 in. (51 mm)	10.8 cu. in. (180 cc)
AU12,1200	2.750 in. (70 mm)	2 in. (51 mm)	11.9 cu. in. (196 cc)

Continental Vertical Crankshaft Models

Model	Bore	Stroke	Displacement
AV7,AD7, AW7,700	2.125 in. (54 mm)	2 in. (51 mm)	7.1 cu. in. (117 cc)
AD8,AW8, 800	2.250 in. (57 mm)	2 in. (51 mm)	8 cu. in. (130 cc)
AD85,AW85, 900	2.313 in. (59 mm)	2 in. (51 mm)	8.4 cu. in. (139 mm)
1100	2.625 in. (67 mm)	2 in. (51 mm)	10.8 cu. in. (180 cc)

Deco-Grand Models

Model	Bore	Stroke	Displacement
DE2	2.125 in. (54 mm)	2 in. (51 mm)	7.1 cu. in. (117 cc)
DEA8	2.250 in. (57 mm)	2 in. (51 mm)	8.0 cu. in. (130 cc)
DE3	2.313 in. (59 mm)	2 in. (51 mm)	8.4 cu. in. (139 cc)
DE4	2.625 in. (67 mm)	2 in. (51 mm)	10.8 cu. in. (180 cc)
DE5	2.750 in. (70 mm)	2 in. (51 mm)	11.9 cu. in. (196 cc)

ENGINE IDENTIFICATION

The engine name plate located on flywheel housing or cooling shroud shows engine model number, specification number and serial number. Always furnish engine model number, specification number and serial number when ordering parts.

Many of these engines are built to Military Specification E-11275 while others are a combination of the standard commercial engine and military version. Models produced for military applications have been treated for resistance to fungi growth and salt corrosion and are radio suppressed and equipped for cold weather operation. Make certain correct replacement parts are used.

MAINTENANCE

SPARK PLUG. Recommended spark plug for all models is a Champion XEJ8, or equivalent. Electrode gap should be set at 0.025 inch (0.64 mm).

CARBURETOR. Several different suction and float type carburetors have been used. Refer to the appropriate paragraph for model being serviced.

AC Suction Type Carburetor. Engine Models AC5 and AC6 were equipped with a suction type carburetor as shown in Fig. DG3. The carburetor is equipped with only one fuel adjustment needle (1). For initial adjustment, open the fuel needle 2½ to 3 turns. Make final adjustment with engine at operating temperature and running. Adjust fuel needle (1) to obtain smoothest engine operation under full load.

Refer to Fig. DG4 for exploded view of the late style fuel metering valve (2) and adjustment needle (7). Fiber washer (1) is not used on some carburetors. Parts shown in Fig. DG4 may be installed in early production carburetors.

Bendix-Stromberg Float Carburetor. A Bendix-Stromberg float type carburetor was used on some models. Initial adjustment of both the idle and main fuel needles is 1 turn open. Make

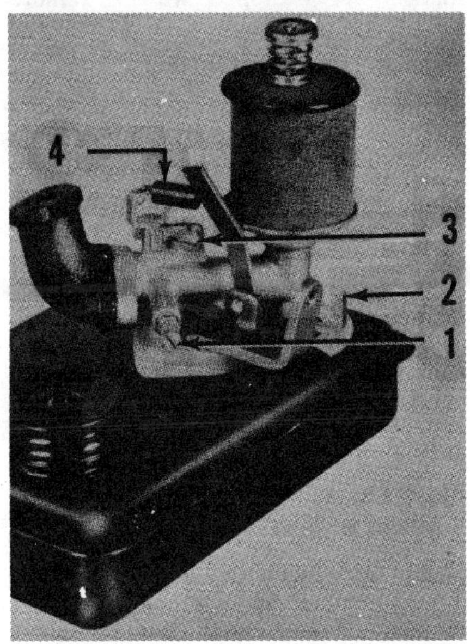

Fig. DG3—View of AC suction type carburetor used on early engines.

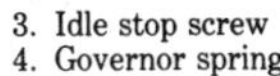

1. Fuel mixture needle
2. Choke
3. Idle stop screw
4. Governor spring

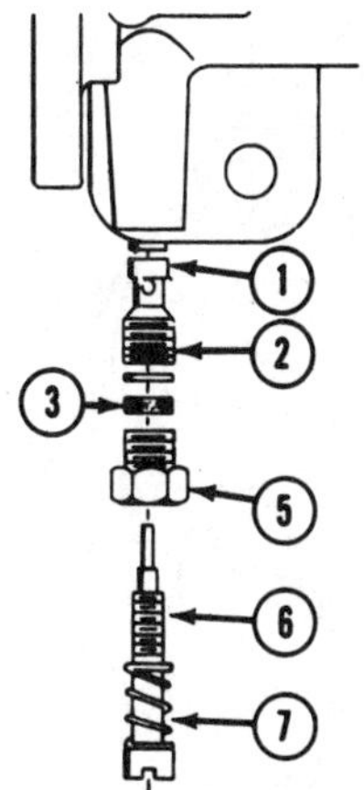

Fig. DG4—Exploded view of fuel mixture needle and related parts from AC suction type carburetor. Sealing washer (1) is not used on all models.

1. Washer
2. Metering valve
3. Washer
5. Nut
6. Needle
7. Spring

final adjustments with engine at operating temperature and running at 2500 rpm. Adjust main fuel needle for smoothest engine operation under full

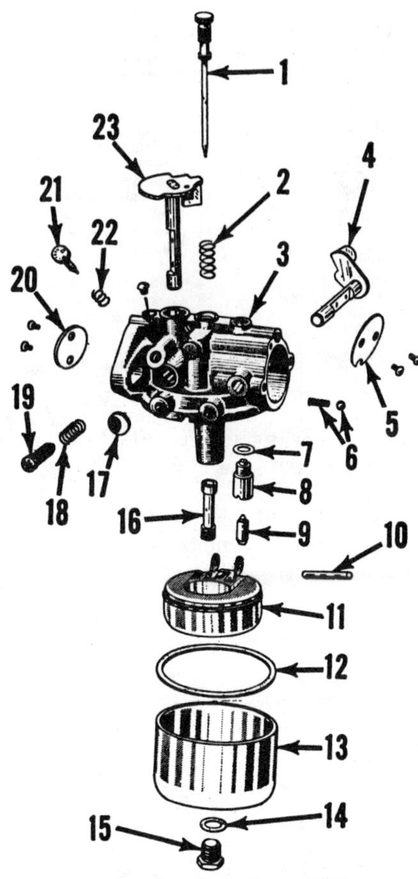

Fig. DG5—Exploded view of Model N Carter carburetor used on some engines.

1. Main fuel needle
2. Spring
3. Carburetor body
4. Choke shaft
5. Choke plate
6. Choke detent
7. Sealing washer
8. Inlet valve seat
9. Inlet valve
10. Float pin
11. Float
12. Gasket
13. Float bowl
14. Sealing washer
15. Retainer
16. Main jet
17. Plug
18. Spring
19. Idle stop screw
20. Throttle plate
21. Idle fuel needle
22. Spring
23. Throttle shaft

load. Set engine at idle speed and adjust idle fuel needle for smoothest idle performance.

Carter Model N Float Carburetor. Refer to Fig. DG5 for exploded view of typical Carter Model N float type carburetor. For initial adjustment, open idle fuel needle 1 turn and open main fuel adjustment needle 1½ turns. Make final adjustment with engine at operating temperature and running. Adjust main and idle fuel needles alternately until smoothest engine performance is obtained at both operating and idle speeds.

To check float level, invert carburetor casting and float assembly. There should be 11/64 inch (4.4 mm) clearance between free end of float and outer edge of carburetor casting. Adjust by bending tang on float lever.

Tillotson ML Float Carburetor. Series ML Tillotson float type carburetors have been used on some models. Refer to sectional view in Fig. DG6. For initial adjustment, open the idle fuel needle 1 turn and open the main fuel needle 1½ turns. Make final adjustments with engine at operating temperature and running at half throttle. Adjust main fuel needle for maximum speed and smoothest engine operation. Adjust idle fuel needle for smoothest idle performance and smooth acceleration.

Check float adjustment as shown in Fig. DG8. Dimension (D) should be 1-5/64 to 1-3/32 inch (27-28 mm). To adjust, carefully bend float tang which contacts fuel inlet needle.

Tillotson MT Float Carburetor. Refer to Fig. DG7 for exploded view of typical Tillotson Series MT float type carburetor. For initial adjustment, open idle fuel needle (14) ¾ turn and open main fuel needle (21) 1 turn. Make final adjustments with engine at operating temperature and running at half throttle. Adjust main fuel needle to obtain smoothest engine operation under full

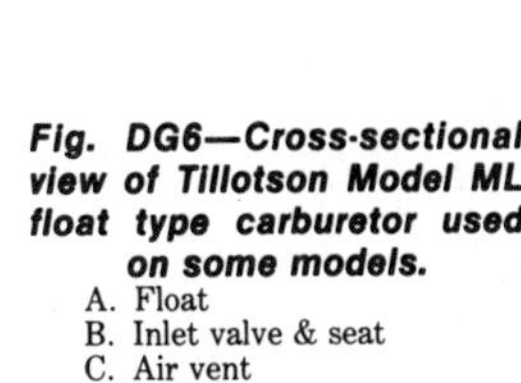

Fig. DG6—Cross-sectional view of Tillotson Model ML float type carburetor used on some models.

A. Float
B. Inlet valve & seat
C. Air vent
D. Venturi
F. Idle tube
H. Idle fuel needle
J. Idle speed screw
M. Throttle plate
N. Main fuel nozzle
O. Main fuel needle
R. Bowl drain plug
S. Choke plate
T. Gasoline level
U. Float setting

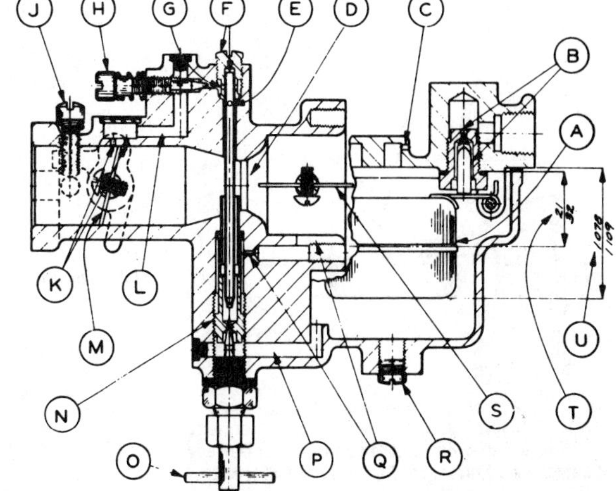

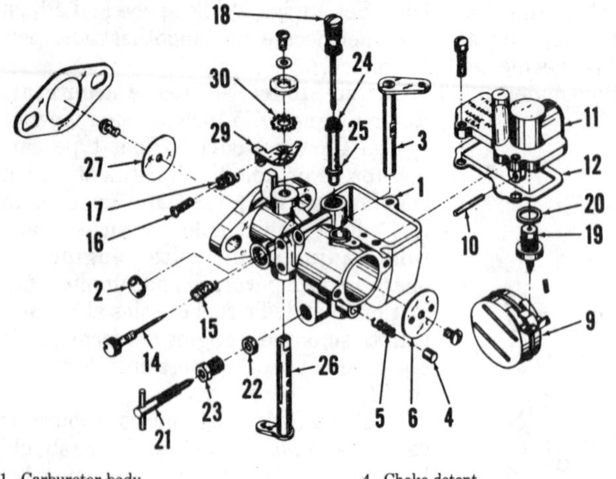

Fig. DG7—Exploded view of Tillotson Model MT float type carburetor.

1. Carburetor body
2. Plug
3. Choke shaft
4. Choke detent
5. Spring
6. Choke plate
8. Gasket
9. Float
10. Float pin
11. Float cover
12. Gasket
14. Idle fuel needle
15. Spring
16. Idle stop screw
17. Spring
18. Idle tube
19. Inlet valve assy.
20. Gasket
21. Main fuel needle
22. Packing
23. Nut
24. Main nozzle
25. Gasket
26. Throttle shaft
27. Throttle plate
29. Throttle lever
30. Lockwasher

load. Set engine at idle speed and adjust idle fuel needle to obtain smoothest engine idle and smooth acceleration.

To check float adjustment, measure float level as shown in Fig. DG8. Dimension (D) should be 1-25/64 to 1-27/64 inch (33–36 mm). To adjust, carefully bend float tang which contacts fuel inlet needle.

Zenith Float Carburetors. Refer to Fig. DG9 for exploded view of Zenith Model 210 float type carburetor. A Zenith Model 10 carburetor used prior to the Model 210 is similar except the main fuel needle is located in the float bowl in place of the plug (25).

For initial adjustment of the Zenith float carburetors, open the idle fuel needle (2) 2 turns and open the main fuel needle (7) 1 turn. Make final adjustments with engine at operating temperature and running. Adjust idle fuel needle for smoothest idle performance, then adjust main fuel needle for smoothest operation under full load. Check acceleration from idle speed to maximum governed speed and repeat final adjustment procedure if acceleration is unsatisfactory. Float lever is nonadjustable.

GOVERNOR. All models are equipped with an air vane type governor or a mechanical governor of the flyball or flyweight type. Refer to the appropriate following paragraph for model being serviced.

Air Vane (Pneumatic) Governors. Refer to Fig. DG10 for exploded view of typical late production air vane governor. The air vane governor used earlier is similar except a bellcrank is used in the linkage between the air vane and the carburetor throttle shaft arm on some models.

On models without remote controls, governed speed is adjusted by moving a slide (2) on the bracket (1) to decrease or increase tension on the governor spring (3) and thereby vary engine speed. Some models are equipped with a remote throttle control which should be adjusted for a maximum governed speed of 3600 rpm. Idle speed on all models is adjusted by the stop screw on the carburetor throttle shaft.

When servicing engines equipped with an air vane governor, be sure governor linkage is free throughout its range of travel. With the carburetor throttle in closed (idle) position, the air vane must be aligned with the edge of the opening in the blower housing back plate. Bend air vane to align it with edge of hole in back plate or to relieve binding condition. When installing blower housing, pull housing away from air vane while tightening retaining screws to provide maximum clearance for movement of the vane.

Fig. DG8—To check float level on Tillotson carburetors, invert casting and float assembly and measure dimension (D) between gasket surface of casting and farthest side of float. Refer to text for dimension "D".

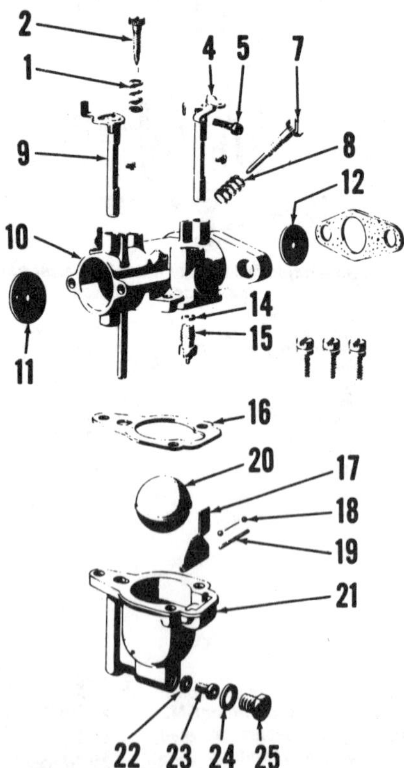

Fig. DG9—Exploded view of typical Zenith float type carburetor used on some models. On some carburetors, main fuel adjustment needle (7) is at location of plug (25) instead of in carburetor body casting.

1. Spring
2. Idle fuel needle
4. Throttle shaft
5. Idle stop screw
7. Main fuel needle
8. Spring
9. Choke shaft
10. Carburetor body
11. Choke plate
12. Throttle plate
14. Sealing washer
15. Inlet valve assy.
16. Gasket
17. Float lever
18. Lead shots
19. Float pin
20. Float
21. Float bowl
22. Sealing washer
23. Main jet
24. Gasket
25. Plug

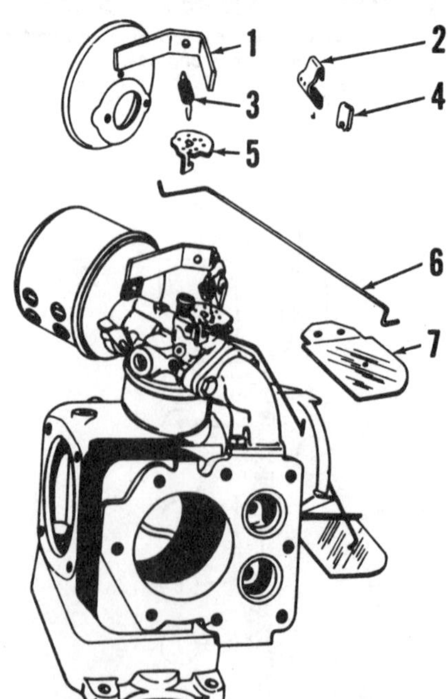

Fig. DG10—Drawing showing typical air vane governor.

1. Governor bracket
2. Slide
3. Governor spring
4. Clip
5. Throttle arm
6. Governor link
7. Air vane

Mechanical Flyball Governor. Refer to Fig. DG11 for view of the flyball governor unit. Early flyball governor units used six steel balls instead of three as shown. When servicing early units, it is recommended that three of the six steel balls be discarded and a spacer (1) be installed on the governor shaft (3). Install the spacer with protruding hub down. Press or drive the spacer down on the shaft until spacer contacts shoulder on shaft, making sure slots in the spacer are aligned with three of the grooves in the lower ball race.

Fig. DG12 shows the governor cover (4) removed. The actuating lever (3) contacts the upper flyball race (2).

On engines with carburetor located below the cylinder, linkage from governor to carburetor is as shown in Fig. DG13. To adjust engine governed speed, loosen setscrew retaining governor rod (4) in bellcrank (1) and vary the position of the bellcrank pivot on the governor rod to change engine governed speed.

On engines with carburetor located on top of the cylinder, engine governed speed can be varied by moving slide (5–Fig. DG15) on the bracket (4) to change tension on the governor spring (6). The link between the governor arm and carburetor throttle shaft can be bent, if necessary, to synchronize throttle and governor travel. Be sure there is no binding tendency throughout range of linkage travel and that the carburetor throttle can move to full open position.

Refer to Fig. DG14 for illustration of governor linkage with remote speed control used on some engines installed in David Bradley garden tractors. The

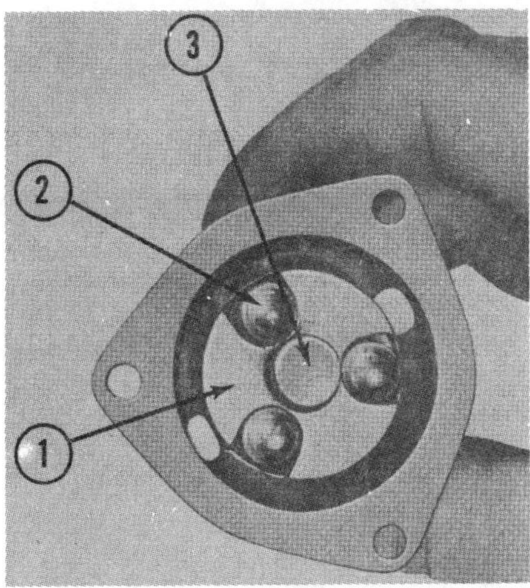

Fig. DG11—Deco-Grand mechanical governor, showing three-ball construction. The old style six-ball type should be converted to the new three-ball type.

1. Spacer
2. Flyballs
3. Governor driveshaft

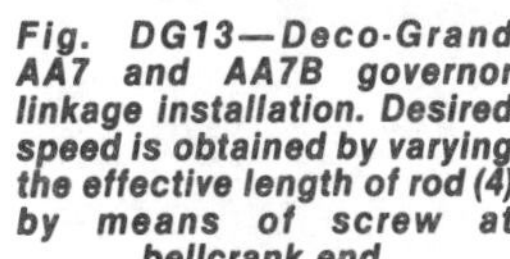

Fig. DG13—Deco-Grand AA7 and AA7B governor linkage installation. Desired speed is obtained by varying the effective length of rod (4) by means of screw at bellcrank end.

1. Bellcrank
2. Bellcrank-to-carburetor rod
3. Carburetor
4. Governor lever-to-bellcrank rod
5. Governor lever
6. Governor cover

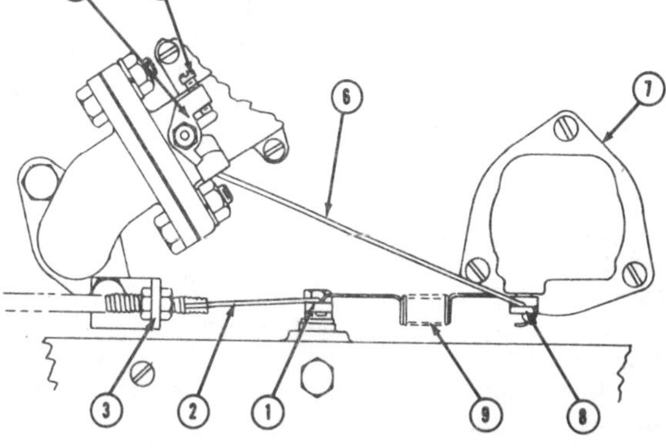

Fig. DG14—Governor linkage used on later David Bradley installations of some engines.

1. Control lever
2. Governor control
3. Control bracket
4. Throttle stop lever
5. Idle stop screw
6. Governor link rod
7. Governor flange
8. Governor lever
9. Governor spring

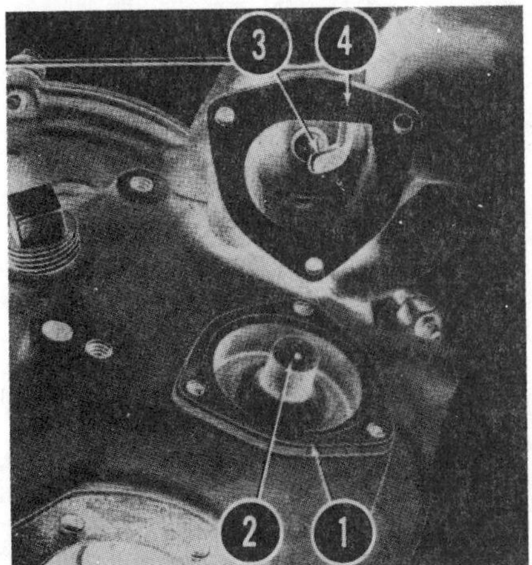

Fig. DG12—View showing mechanical governor cover assembly removed.

1. Governor housing
2. Upper ball race
3. Governor lever
4. Governor cover

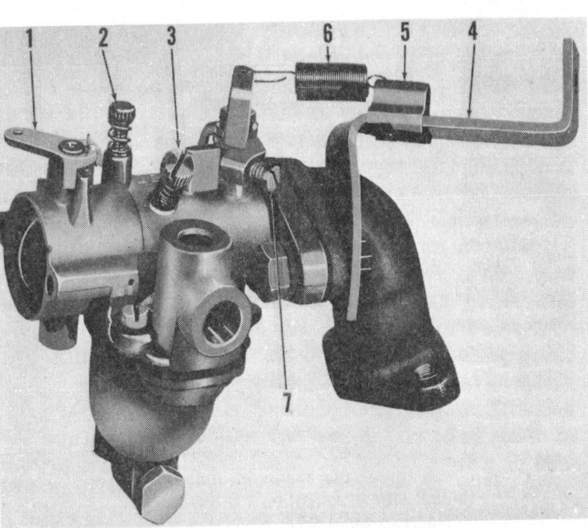

Fig. DG15—Zenith float type carburetor used on some engines.

1. Choke lever
2. Idle mixture adjustment screw
3. Main adjustment screw
4. Bracket
5. Slide
6. Governor spring
7. Idle stop screw

engine governed speed can be changed by varying the position of the control wire housing in the bracket (3). On earlier production units without remote speed control, engine governed speed can be adjusted by varying position of slide (5–Fig. DG15) on bracket (4).

Detex Flyweight Governor. The mechanical governor on late production engines is combined with the Detex ignition unit as shown in the exploded view in Fig. DG17. Refer also to Fig. DG16. Engine governed speed can be varied by remote control cable, if so equipped, or by varying the position of speed control lever (4).

NOTE: Before disconnecting governor spring (5—Fig. DG16) from governor speed control lever (4) or governor arm (3), mark the holes in lever and arm into which the spring is hooked so the spring may be reinstalled in the correct holes.

To synchronize travel of governor arm and carburetor throttle, loosen the clamp screw (2–Fig. DG16) in governor arm (3) and while holding the carburetor throttle in wide open position, turn the governor shaft (6) as far clockwise as possible and retighten the clamp screw. A screwdriver slot is provided in the outer end of the governor shaft to facilitate making this adjustment.

MAGNETO AND TIMING. Early production engines were equipped with Bendix-Scintilla, Phelon (Repco), Watts or Wico magnetos. Late production engines are equipped with Deco-Grand "Detex" ignition system. Refer to appropriate paragraph for model being serviced.

Bendix-Scintilla Magneto. On Bendix-Scintilla magnetos, the rotating magnets are a separate unit from the flywheel. To remove the rotating magnets, first remove the magneto mounting flange and the snap ring from engine crankshaft. The rotating magnet can then be pulled from the crankshaft.

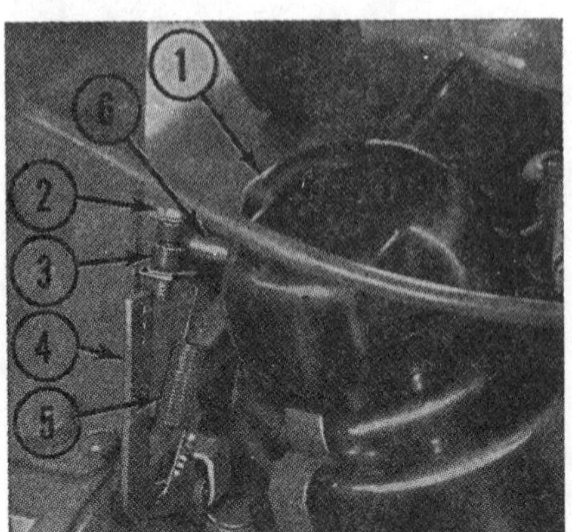

Fig. DG16—View of Detex magneto ignition system combined with flyweight type governor. Refer to Fig. DG17.

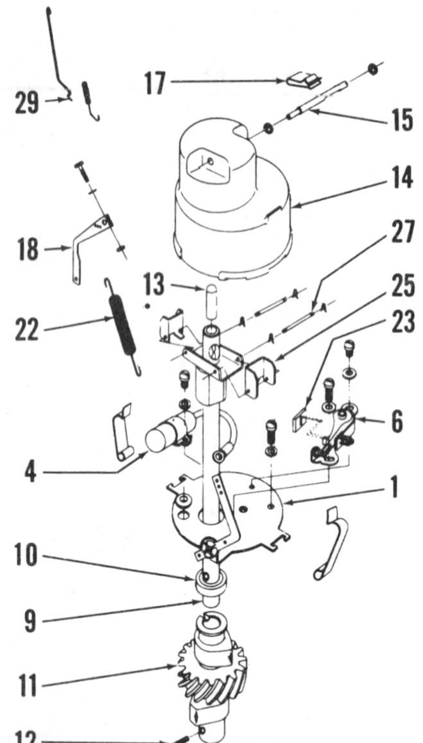

Fig. DG17—Components of Detex ignition system which in this case is combined with the flyweight type speed governor.

1. Base & lever
4. Condenser
6. Breaker points
9. Governor driveshaft & weights
10. Seal
11. Camshaft
12. Camshaft pin
13. Governor plunger
14. Housing
15. Governor shaft
17. Actuating lever
18. Governor lever
22. Governor spring
23. Cam wiper
25. Flyweight
27. Flyweight pin
29. Governor-to-carburetor rod

Breaker point gap should be adjusted to 0.015-0.018 inch (0.38-0.46 mm). Magneto edge gap should be 0.010-0.014 inch (0.25-0.36 mm).

Watts Magneto. On the Watts magneto, armature core and high tension coil are mounted outside of the flywheel and armature core is retained to back plate with two bolts and clamp washers. Breaker point gap should be adjusted to 0.015 inch (0.38 mm). Magneto edge gap should be 0.010-0.015 inch (0.25-0.38 mm).

Wico Magneto With External Armature And Coil. On Wico magneto with armature core and high tension coil mounted outside of flywheel (armature core is retained to back plate with four screws), adjust breaker point gap to 0.015 inch (0.38 mm). Armature air gap should be 0.009-0.012 inch (0.23-0.31 mm).

Wico Magneto With Internal Armature And Coil. On Wico magneto with armature core and high tension coil located inside flywheel, adjust breaker point gap to 0.020 inch (0.51 mm).

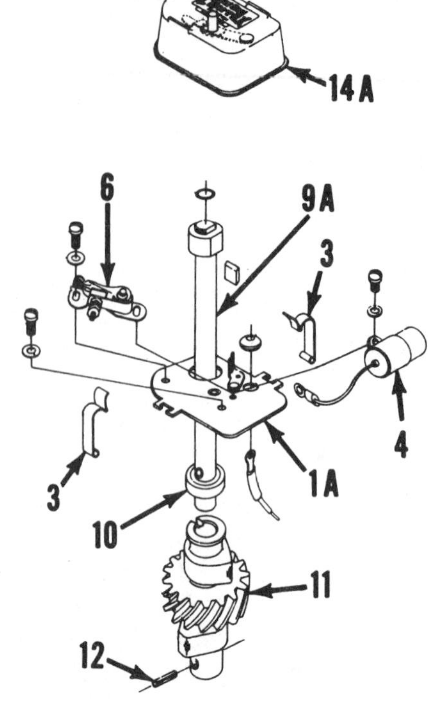

Fig. DG19—Exploded view of Detex ignition unit used with air vane type governor. Breaker cam is located on camshaft axle (9A).

1A. Breaker plate
3. Cover clips
4. Condenser
6. Breaker points
9A. Camshaft axle
10. Seal
11. Camshaft
12. Roll pin
14A. Breaker cover

Phelon (Repco) Magneto. On Phelon magneto, armature and high tension coil are located inside flywheel. Adjust breaker points to 0.018-0.020 inch (0.45-0.51 mm).

Detex Ignition System. Refer to Fig. DG17 for exploded view of Detex ignition unit combined with a flyweight mechanical governor and to Fig. DG19 for exploded view of Detex ignition unit that is used with an air vane type governor. The breaker cam shaft (9–Fig. DG17 or 9A–Fig. DG19) is also the engine camshaft axle and is driven by a pin (12) through the camshaft (11) and axle. As the breaker cam turns at one-half crankshaft speed, an ignition spark is produced every other engine revolution to ignite the air:fuel mixture for the power stroke.

Timing on engines equipped with Detex ignition is adjustable by rotating the plate to which the breaker plate is attached. Adjust timing to setting that will produce maximum power and smoothest operation for individual engine.

On engines with mechanical governor, the breaker point cover can be removed after first removing the governor spring (5–Fig. DG16) from governor arm (3) and speed control lever (4). Be sure to mark the holes in arm and lever in which the governor spring is hooked before removing spring. The ignition breaker cam is a long dwell type and a 12 volt capacitor is mounted on the backplate. Adjust breaker contact gap to 0.020 inch (0.51 mm) on all models.

In servicing the ignition coil, if the lighting generator is of the early type having six coils, one of the lighting coils is mounted behind the ignition coil. On these installations the ignition coil primary lead is assembled beneath the lighting coil windings which prohibits removal of the ignition coil. Thus in case of ignition coil failure, the entire generator plate assembly must be renewed. Later lighting generator has five coils with the ignition coil mounted on the center lower leg of the laminated plate. The ignition coil in these installations may be removed for servicing. Do not, under any circumstances, remove any of the five small coils of this assembly.

LIGHTING COILS. On some engines the flywheel has a double set of magnets and a series of additional coils on the stator plate to provide lighting current. The lighting coils are wound in series, are independent of ignition coil and are used only with Detex ignition.

Early production models have six lighting coils and later production models have five lighting coils (see Detex Ignition System paragraphs in MAGNETO AND TIMING section).

Early production lighting capacitors had a hard plastic cover and were retained in a clip bracket at each end. Latest capacitor has paper shell construction and is retained by a spring clip in the center. These capacitors are of electrolytic design and cannot be checked on the standard condenser testers. If defective capacitor is suspected, renew capacitor.

To reassemble five-coil lighting unit, locate laminated plate over the oil seal then bring the red and white lead wires under the upper coils to prevent pinching against the cylinder block. Anchor the breaker point lead wire and primary lead of ignition coil to the fibre post on stator plate. This connection must be electrically perfect as looseness can cause a loss of spark strength. Attach spark plug lead to outlet on ignition coil. Feed the four lead wires up through the large opening in the back plate. The ignition wire of this group is retained by a clip to the lower flange of the inlet manifold; route the breaker point lead through the grommet of the base plate and connect it to the breaker point post. Leads from the lighting coils, one red and one white, are brought through the grommet on the back plate and connected to the same color leads from the lighting capacitor. The long lead from the lighting capacitor is fed through the grommet and connected to the switch plate.

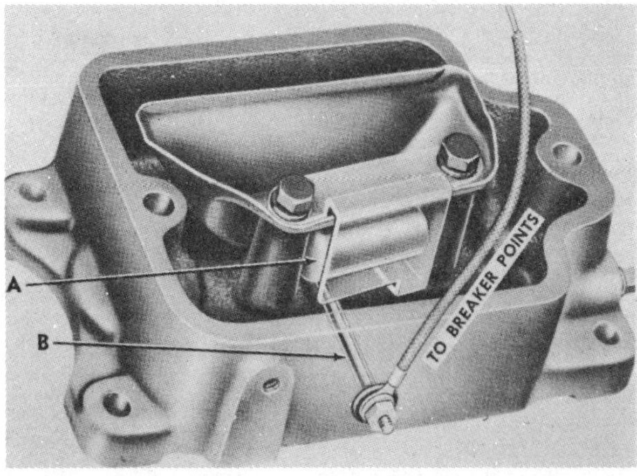

Fig. DG20—Details of Low Oil Lever Shut Off used on some later production engines. Unit is mounted in base (oil reservior) and cuts off ignition when float (A) drops into contact with stud (B). Refer to text.

LUBRICATION. Crankcase oil level should be checked after every five hours of operation in temperatures above 0°F (−18°C) and after every three hours of operation in temperatures below 0°F (−18°C). Oil level gage is located on oil fill plug and plug must be tightened into opening and then removed to check oil level.

A good quality SAE 30 oil is recommended for temperatures above 32°F (0°C), SAE 20 oil for temperatures between 32°F (0°C) and 0°F (−18°C), SAE 10W oil for temperatures between 0°F (−18°C) and −10°F (−23°C), SAE 10W oil plus 10% gasoline for temperatures between −10°F (−23°C) and −20°F (−29°C) and SAE 10W oil plus 20% gasoline for temperatures below −20°F (−29°C).

Any oil or gasoline added to crankcase in sub-zero temperatures should be added within five minutes after stopping engine.

All current production models have a splash lubrication system. Horizontal crankshaft models have an oil dipper on the connecting rod; vertical crankshaft models have a tube inserted in the lower crankshaft throw which picks up oil from a circular oil trough in the engine base.

Early vertical crankshaft models were lubricated by a gear type pump (See Fig. DG21) which delivered oil to an orifice as shown in Fig. DG22.

Early Model AA7 horizontal crankshaft engines were lubricated by a pump driven by a cam on the engine crankshaft; however, it is recommended these engines be converted to splash lubrication. Refer to SERVICING NOTES paragraphs in REPAIRS section.

LOW OIL LEVEL SHUT-OFF. This device shown in Fig. DG20 includes a float (A) which, when oil level is too low, contacts stud (B) connected to the ignition coil lead and grounds the ignition to stop engine until oil level is returned to normal.

CRANKCASE BREATHER. On some vertical crankshaft models, the crankshaft breather is located in the valve chamber cover. On all other models, the crankcase breather is located in the crankcase just below the cylinder head air baffle.

A flapper type valve is used for the crankcase breather valve on early engines. A ball check valve is used on later engines. If valve appears dirty, remove valve from engine and wash it in solvent.

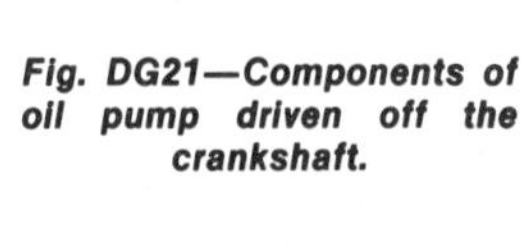

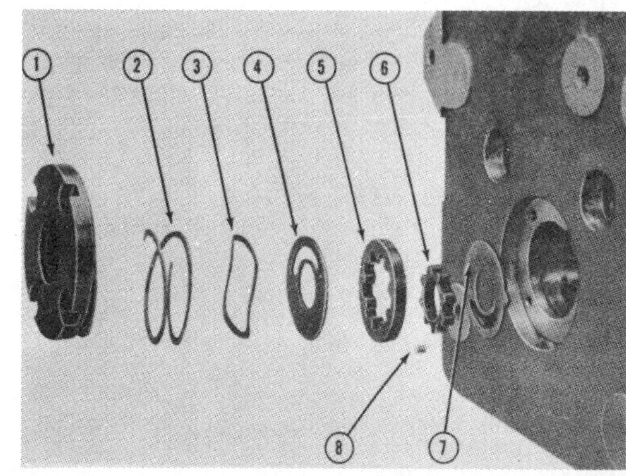

Fig. DG21—Components of oil pump driven off the crankshaft.

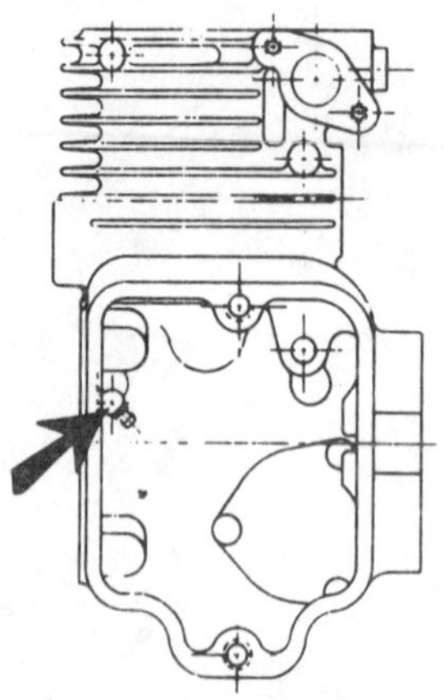

Fig. DG22—Oil orifice on models equipped with oil pump (see Fig. DG21) should clear crankshaft throw by 0.020 inch (0.25 mm) with spray directed toward the rear of engine and away from piston.

REPAIRS

TIGHTENING TORQUES. Recommended tightening torques are as follows:

Spark plug	25-30 ft.-lbs. (34-40 N·m)
Connecting rod	8-10 ft.-lbs. (11-13 N·m)
Cylinder head	21-33 ft.-lbs. (28-31 N·m)
Main bearing cap	13-15 ft.-lbs. (18-20 N·m)
Oil base	27-29 ft.-lbs. (37-39 N·m)
Flywheel nut	37-40 ft.-lbs. (50-54 N·m)

SERVICING NOTES. When servicing Deco-Grand air-cooled engines, the following precautions or procedures should be noted on certain models.

Model AA7 Oil Pump. Deco-Grand recommends discarding the cam actuated oil pump on early Model AA7 engines and installing a splash type lubricating system. To install the splash system, install new type connecting rod with oil dipper (3–Fig. DG23). Then, grind or chisel boss from center of engine base and install an oil trough similar to that shown in Fig. DG20.

Disassembly Of Model AV7 Engine. On Model AV7 vertical crankshaft engine, the crankshaft seal, oil pump gears and oil pump drive key (Fig. DG21) must be removed from the engine base before the engine base can be removed from the crankshaft and engine crankcase.

Crankshaft Removal, Early Engines. On early production engines having flywheel end main bearing bushing located in the magneto stator plate, it will be necessary to remove the stator plate before the crankshaft can be removed if the thrust collar on the crankshaft does not have a clearance flat for the camshaft gear.

Early Type Reduction Gear Drive. Early reduction gear drive units were equipped with oilite bushings. Parts are not available for these units; therefore, if necessary to renew any part of a bushing equipped gearbox, it will be necessary to install the later type needle bearing equipped reduction gear drive assembly and engine crankshaft.

VALVE SYSTEM. Valve lash (cold) should be 0.014-0.018 inch (0.36-0.46 mm) for exhaust valve and 0.012-0.016 inch (0.31-41 mm) for intake valve. Valve lash is adjusted by grinding end of valve stem to increase gap or grinding valve seats deeper to decrease gap.

Valve tappet diameter on Models with 10.8 or 11.9 cubic inch (180 or 196 cc) displacement should be 0.3110-0.3115 inch (7.899-7.912 mm) with a tappet bore of 0.3135-0.3150 inch (7.963-8.00 mm) to provide a clearance of 0.002-0.004 inch (0.05-0.10 mm). All remaining models should have a tappet diameter of 0.2820-0.2825 inch (7.163-7.176 mm) with a tappet bore of 0.2846-0.2854 inch (7.229-7.249 mm) to provide a clearance of 0.0021-0.0034 inch (0.053-0.086 mm).

Valve seat inserts are renewable and pressed into a counterbore in cylinder block. Insert should be chilled before installing. Valve seat angle should be 45°.

Valve face angle should be 45°. Intake valve stem diameter should be 0.3105-0.3115 inch (7.887-7.912 mm) and exhaust valve stem diameter should be 0.3100-0.3105 inch (7.874-7.887 mm) on models with 10.8 or 11.9 cubic inch (180-196 cc) displacement. Diameter on all remaining models should be 0.2815-0.2825 inch (7.150-7.176 mm) on intake and exhaust valve stems.

Stem-to-guide clearance on 10.8 and 11.9 cubic inch (180 and 196 mm) displacement engines should be 0.0020-0.0045 inch (0.051-0.114 mm) for intake valve and 0.003-0.005 inch (0.08-0.13 mm) for exhaust valve. Stem-to-guide clearance on all remaining models should be 0.0021-0.0039 inch (0.053-0.099 mm) for intake and exhaust valves.

CONNECTING ROD. The connecting rod and piston assembly can be removed from above after removing cylinder head and connecting rod cap. On most engines, the aluminum connecting rod rides directly on the crankpin. A heavy-duty forged aluminum connecting rod is available. Some engines may be equipped with the heavy-duty connecting rod which has renewable bronze bearing inserts.

Crankpin diameter on all models is 0.874-0.875 inch (22.20-22.23 mm). Desired running clearance between rod and crankpin is 0.0040-0.0055 inch (0.102-0.140 mm). Desired side play of connecting rod on crankpin is 0.020-0.040 inch (0.51-1.02 mm) for Models DE4, DE5, AU10, AU12, 1100 and 1200; desired connecting rod side play for all remaining models is 0.030-0.050 inch (0.76-1.27 mm). Connecting rod is available for standard and undersize crankpin journals.

If crankpin is worn more than 0.003 inch (0.08 mm), or is 0.001 inch (0.03 mm) or more out-of-round, renew or resize crankshaft.

Piston pin should have 0.0010-0.0015 inch (0.025-0.038 mm) clearance in connecting rod. Piston pin is available in a variety of oversizes as well as standard.

On horizontal crankshaft models, be sure connecting rod is installed with oil hole in rod up.

PISTON, PIN AND RINGS. On all models, the piston is equipped with two compression rings and one oil control ring. On Model DE5, AU12 and 1200 engines, an expander is used under the oil control ring.

Piston ring end gap should be 0.007-0.017 inch (0.18-0.43 mm) for all models. Piston ring side clearances are as follows:

Cylinder Diameter	Compression Ring
2.125 in. (54 mm)	0.0020-0.0035 in. (0.05-0.89 mm)
2.250 in. (57 mm)	0.0020-0.0035 in. (0.05-0.89 mm)
2.313 in. (59 mm)	0.0020-0.0040 in. (0.051-0.101 mm)
2.625 in. (67 mm)	0.0020-0.0040 in. (0.051-0.101 mm)
2.750 in. (70 mm)	0.0025-0.0045 in. (0.064-0.114 mm)

Cylinder Diameter	Oil Ring
2.125 in. (54 mm)	0.0015-0.0030 in. (0.04-0.08 mm)
2.250 in. (57 mm)	0.0015-0.0030 in. (0.04-0.08 mm)
2.313 in. (59 mm)	0.0014-0.0035 in. (0.036-0.090 mm)
2.625 in. (67 mm)	0.0015-0.0030 in. (0.04-0.08 mm)
2.750 in. (70 mm)	0.0015-0.0030 in. (0.04-0.08 mm)

Piston pin standard diameter on Models DE4, DE5, AU10, AU12, 1100 and 1200 engines is 0.4993-0.4995 inch (12.682-12.687 mm); standard piston pin diameter for all other models is 0.4190-0.4195 inch (10.643-10.655 mm). On all models, piston pins should be a zero to

0.0005 inch (0 to 0.013 mm) interference fit in the piston and a 0.0010-0.0015 inch (0.025-0.038 mm) loose fit in the connecting rod. Pin is available in standard size as well as a variety of oversizes.

Piston skirt-to-cylinder bore clearance should be 0.005-0.008 inch (0.13-0.20 mm) on Models DE4, DE5, AU10, AU12, 1100 and 1200 engines measured at bottom of piston skirt and at right angle to piston pin. On all remaining models, piston-to-cylinder clearance is correct if a ½-inch (12.7 mm) wide 0.010-inch (0.25 mm) feeler gage can be pulled out from between cylinder and piston (when placed at a right angle to piston pin) using firm, steady pull.

CYLINDER. Cylinder bore should be honed or rebored to next oversize for which piston and rings are available if taper of cylinder bore is 0.006 inch (0.15 mm) or more, if out-of-round 0.004 inch (0.10 mm) or more or if piston skirt-to-cylinder bore clearance is excessive.

CRANKSHAFT. The flywheel (magneto) end of the crankshaft is supported in either a needle bearing or a sleeve bushing. On some early models the sleeve bushing is located in the magneto stator plate. On early production engines, the flywheel end main journal diameter is 0.781-0.782 inch (19.84-19.86 mm) and the sleeve bushing inside diameter is 0.783-0.784 inch (19.89-19.91 mm) allowing a running clearance of 0.001-0.003 inch (0.03-0.08 mm). On all later models, flywheel end main journal diameter is 0.7495-0.7500 inch (19.37-19.050 mm). Bushing inside diameter, on models so equipped, is 0.753-0.754 inch (19.13-19.15 mm) allowing a running clearance of 0.0030-0.0045 inch (0.075-0.114 mm).

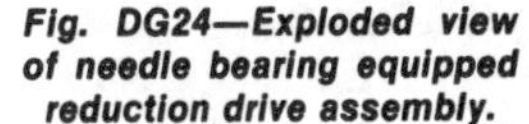

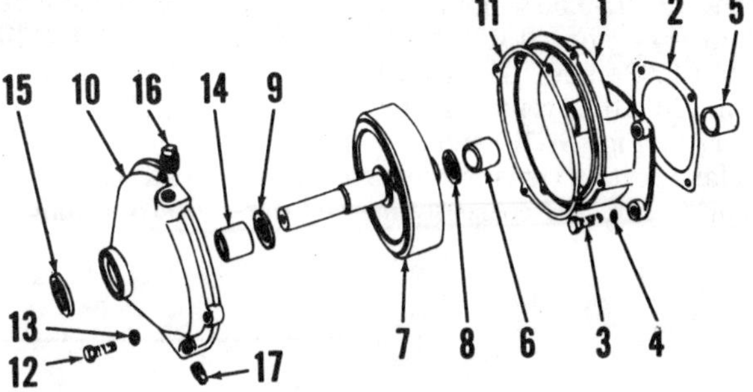

Fig. DG24—Exploded view of needle bearing equipped reduction drive assembly.

1. Adapter
2. Gasket
5. Needle bearing
6. Needle bearing
7. Drive gear
8. Thrust washer
9. Thrust washer
10. Cover
11. Gasket
14. Needle bearing
15. Oil seal
16. Oil filler plug
17. Oil level plug

On horizontal crankshaft engines, the crankshaft pto end journal rides in either a sleeve bushing or a ball bearing on direct drive models and in a needle bearing on reduction drive models. Pto end journal diameter is 0.8745-0.8750 inch (22.212-22.225 mm) for bushing equipped models. Bushing inside diameter is 0.8770-0.8774 inch (22.276-22.286 mm) allowing a clearance of 0.0020-0.0029 inch (0.025-0.074 mm). Bushing and main bearing cap are an integral unit on most models.

On vertical crankshaft engines, lower (pto) crankshaft journal rides in either a ball bearing or a sleeve bushing and needle bearing. Bushing-to-journal clearance should be the same for horizontal crankshaft models.

Ball bearing mains, on models so equipped, should be 0.0001-0.0009 inch (0.003-0.023 mm) interference fit on the crankshaft journal. On Modeld DE4, DE5, AU10, AU12, 1100 and 1200 engines, the ball bearing is retained in the main bearing cap with a snap ring.

On models so equipped, needle bearing (roller) main bearing clearance can be measured with a dial indicator and should be 0.0005-0.0029 inch (0.013-0.074 mm).

Crankshaft end play should be 0.012-0.016 inch (0.31-0.41 mm) on vertical crankshaft engines. Crankshaft end play should be 0.008-0.010 inch (0.20-0.25 mm) on all horizontal crankshaft engines except those where the ball bearing main is retained in the bearing cap with a snap ring; crankshaft end play on these models is controlled by the locked-in bearing. End play can be adjusted by varying thickness of gaskets and shims placed between the crankcase and main bearing cap or, on vertical crankshaft models, engine base. Although the end play is controlled by the locked-in ball bearing on some models, shims and/or gaskets are required between the main bearing cap and crankcase to provide clearance between the thrust surface on crankshaft and opposite main bearing.

Make certain crankshaft and camshaft timing gears are correctly meshed during installation (Fig. DG23).

CAMSHAFT. On models without "Detex" ignition, the hollow camshaft turns on an axle shaft. The camshaft is removed by driving the axle shaft out towards the open side of the crankcase. The axle shaft diameter is 0.3725-0.3735 inch (9.462-9.487 mm). The camshaft bore inside diameter is 0.3747-0.3767 inch (9.507-9.568 mm) providing a clearance of 0.0012-0.0042 inch (0.031-0.107 mm). The axle shaft has a 0.0005-0.0025 inch (0.013-0.064 mm) loose fit in the crankcase bore at open side of crankcase and a 0.0005-0.0025 inch (0.013-0.064 mm) interference fit in closed side of crankcase.

On models with "Detex" ignition system, the camshaft and cam axle shaft are pinned together and the axle turns in unbushed bores in the engine crankcase. Axle shaft diameter is 0.3730-0.3735 inch (9.474-9.487 mm); bore inside diameter is 0.3747-0.3767

Fig. DG23—Crankcase cylinder assembly with oil base removed. Note timing mark on crank gear at center face of one tooth and two timing dots on cam gear.

inch (9.507-9.568 mm) providing a clearance of 0.0012-0.0037 inch (0.031-0.094 mm). The camshaft can be removed after removing pin (12–Fig. DG17 or Fig. DG19) and axle shaft.

Make certain camshaft and crankshaft timing gears are correctly meshed during reinstallation (Fig. DG23).

REDUCTION GEAR UNIT. Horizontal crankshaft engines are available with a 6:1 ratio gear reduction unit. The standard needle bearing equipped unit is shown in Fig. DG24.

To remove the gear reduction unit from engine crankcase, the cover and drive gear and shaft assembly must first be removed to gain access to two internal mounting bolts.

Gear box should be filled to oil level plug opening with same weight oil as recommended for engine lubrication.

HOMELITE

HOMELITE DIV. OF TEXTRON
Post Office Box 7047
14401 Carowinds Blvd.
Charlotte, NC 28217

Model	Bore	Stroke	Displacement
250, 270	2½ in. (63.5 mm)	1⅝ in. (41.3 mm)	8.0 cu. in. (131 cc)

ENGINE INFORMATION

The Model 250 and 270 engines are used as a power source for centrifugal pumps, diaphragm type pumps and air blowers. The pump or blower must be at least partly disassembled to service some engine components.

MAINTENANCE

SPARK PLUG. Recommended spark plug is a Champion J6J for use on generator engine or centrifugal pump engine or a Champion UJ12 for use on diaphragm pump engine. A Champion HO8A (platinum tip) or UJ11G (gold pladium tip) spark plug may be substituted. Correct electrode gap for all models is 0.025 inch (0.635 mm).

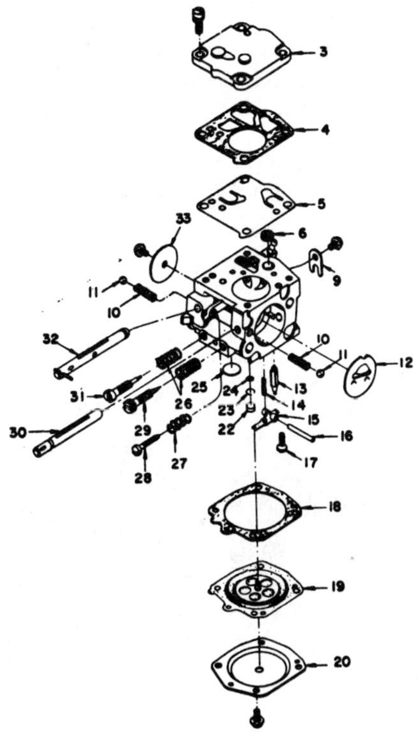

Fig. HL1 — Exploded view of Tillotson carburetor used on Models 250 and 270.

3. Fuel pump cover
4. Fuel pump gasket
5. Fuel pump diaphragm
6. Inlet screen
9. Throttle shaft clip
10. Shaft detent springs (2)
11. Shaft detent balls
12. Choke plate
13. Inlet needle
14. Inlet tension spring
15. Inlet control lever
16. Inlet pinion pin
17. Inlet pinion pin retaining screw
18. Diaphragm gasket
19. Metering diaphragm
20. Diaphragm cover
22. Welch plug
23. Retaining ring
24. Body channel screen
25. Welch plug
26. Adjusting needle springs
27. Idle speed screw spring
28. Idle speed screw
29. Main adjusting needle
31. Idle adjusting needle
32. Throttle shaft
33. Throttle plate

CARBURETOR. A Tillotson HS-45A carburetor is used on earlier engines while later models use a Tillotson Model HS-45C carburetor. Refer to Fig. HL1.

NOTE: As engine speed is controlled by governor plate on rotary inlet valve, there are no governor linkage connections to throttle shaft.

Throttle shaft has a spring-loaded detent to hold shaft in wide open position. Carburetor is accessible after removing air cleaner cover (1 – Fig. HL2).

Later model carburetor is fitted with a main nozzle check ball which allows use of larger mesh inlet screen (6 – Fig. HL1) and provides easier adjustment of fuel mixture.

The check ball assembly will be in the same bore as the small body channel welch plug (22), retaining ring (23) and screen (24) and will replace these parts.

When disassembling, slide diaphragm assembly towards adjustment needle side of carburetor body to disengage diaphragm from fuel inlet control lever. To remove welch plugs, carefully drill through plugs with a small diameter drill and pry plugs out with a pin. Caution should be taken that drill bit just goes through welch plug as deeper drilling may seriously damage carburetor. Note channel screen (24) (early model carburetor) and screen retaining ring (23) are accessible after removing welch plug (22).

Inlet control metering lever (15) should be flush with metering chamber floor of carburetor body. If not, bend diaphragm end of lever as required so lever is flush.

Normal adjustment of low speed fuel mixture needle (marked "L" on carburetor body) is ¾ turn open and main mixture adjusting needle (marked "H" on carburetor body) should be opened one full turn. On pump or blower engine, back idle speed adjusting screw out until

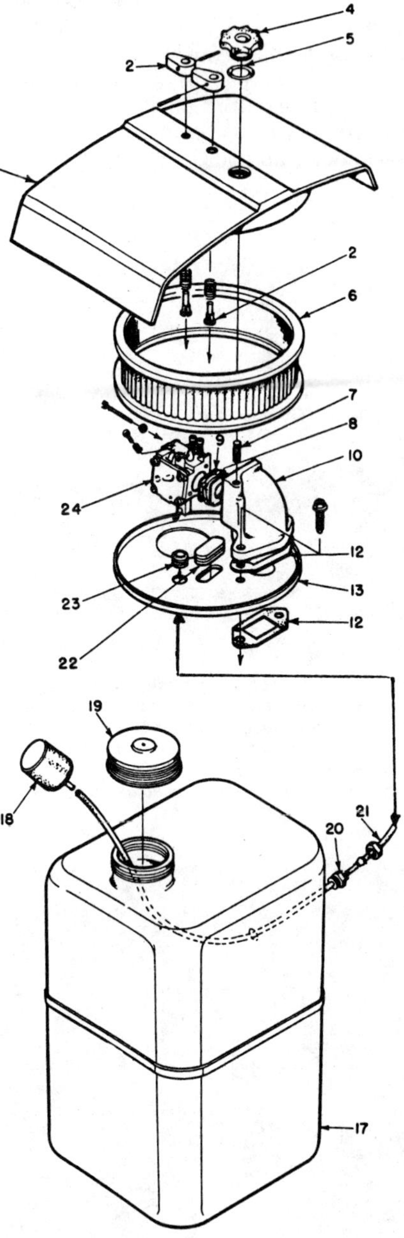

Fig. HL2 — Exploded view of air intake and fuel system. Refer to Fig. HL1 for exploded view of carburetor (24).

1. Air filter cover
2. Control shaft assy.
4. Cover retaining nut
5. Nylon washer
6. Air filter element
7. Stud
8. Gasket
9. Spacer
10. Intake manifold
12. Gearcase gasket
13. Air filter mounting plate
17. Fuel tank
18. Fuel filter
19. Fuel filler cap
20. Fitting
21. Fuel line
22. Governor adjusting hole plug
23. Grommet
24. Carburetor assy.

carburetor throttle plate will close fully, then slowly turn screw in until it just contacts pin on throttle shaft, then turn screw in an additional 1½ turns.

Start engine and allow to warm up before making final carburetor adjustments.

With carburetor throttle shaft in high speed detent position and engine running under load, adjust main (H) fuel needle for smoothest running.

Move throttle shaft to idle speed position and adjust idle fuel needle (L) for smoothest idle. Adjust idle speed stop screw to desired idle speed. If engine will not accelerate from idle speed to full throttle without hesitation, open idle fuel needle an additional 1/8 turn.

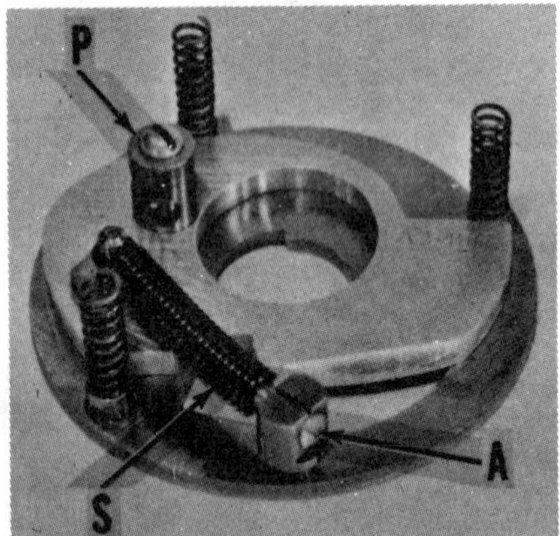

Fig. HL3 – View of rotary intake valve and governor assembly. Governor plate pivots on post (P) to close off valve opening in rotary valve plate to govern engine speed. Speed at which plate closes the opening is regulated by tension of governor spring (S) which is adjusted by turning screw (A).

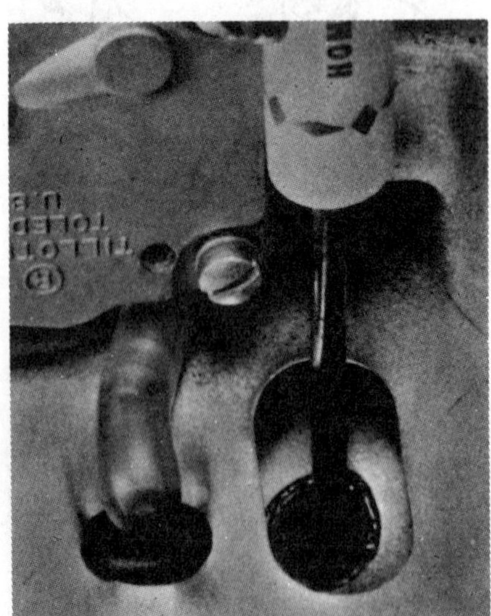

Fig. HL4 – Access to governor spring adjusting screw (A – Fig. HL3) is gained by removing air filter cover, rubber plug (22 – Fig. HL2) and the brass plug from engine crankcase. Position engine so screw can be turned with screwdriver as shown.

FUEL FILTER. The fuel filter is part of the fuel pickup inside fuel tank and can be extracted with a wire hook. Under normal operations, the filter element should be changed at one to two month intervals. If engine is run continuously, or fuel is dirty, the filter may need to be changed weekly, or more often as necessary.

AIR FILTER. The air cleaner element may be washed by immersion in a container of nonoily solvent. If engine is run continuously, clean air filter daily. After a number of cleanings, filter pores may become permanently clogged, making it necessary to renew element.

GOVERNOR. The governor is a part of the rotary inlet valve. Refer to Fig. HL3. As engine speed increases, centrifugal force pivots governor plate on pivot pin (P) against tension of spring (S). The governor plate then closes the opening in rotary valve and thus throttles the engine. Maximum governed engine speed is controlled by tension of governor spring which is adjusted by turning screw (A).

To check and adjust engine speed, bring engine to normal operating temperature and adjust carburetor for highest speed and best performance obtainable, then check engine speed with tachometer. Refer to the following table for correct governed speed:

	No Load rpm	Full Load rpm
Centrifugal Pump	3900-4000	3400-3600
Diaphragm Pump	2800-3000	2800-3000
Blower	3750-3800	3400-3600

If adjustment is necessary, stop engine and remove air filter cover and rubber plug (22 – Fig. HL2) from air filter base (13). Remove brass plug from engine crankcase through opening in filter base and turn engine so adjusting screw (A – Fig. HL3) is accessible. Then, as shown in Fig. HL4, turn adjusting screw clockwise to increase speed or counterclockwise to decrease speed. One turn of the screw will change governed speed approximately 100 rpm. Reinstall brass plug, rubber plug and air filter cover, then recheck engine speed. Readjust as necessary.

IGNITION SYSTEM AND TIMING. Breaker points, condenser and ignition coil are accessible after removing engine flywheel (magneto rotor). A hole is provided in magneto back plate and inner face of flywheel so a pin may be inserted to hold flywheel from turning. Unscrew flywheel nut and remove flywheel using Homelite puller AA-22560, or equivalent.

To adjust breaker point gap, turn engine so leading edge of breaker cam is

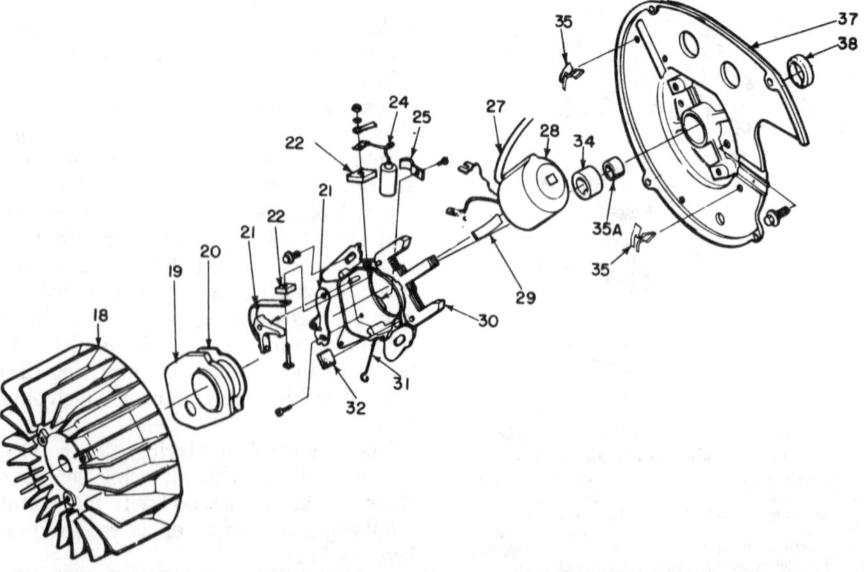

Fig. HL5 – Exploded view of Model 250 magneto assembly. Model 270 is similar. Magneto components are accessible after removing flywheel. Refer to text. A zinc flywheel (magneto rotor) should be used on diaphragm pump engine and an aluminum flywheel is used on all other Model 250 engines.

18. Flywheel
19. Breaker box cover
20. Gasket
21. Breaker points
22. Terminal connection
24. Condenser
25. Condenser clamp
27. High tension lead
28. Ignition coil
29. Coil wedge
30. Stator & armature assy.
31. Breaker box cover spring
32. Cam wiper felt
34. Breaker cam
35. Wire retaining clips
35A. Felt seal
37. Back plate
38. Felt seal

Illustrations Courtesy Homelite Div. of Textron

about 1/8 inch (3.2 mm) past breaker point cam follower, then adjust point gap to 0.020 inch (0.51 mm).

On earlier models, removal of flywheel (magneto rotor) will also require removal of the fan housing (17–Fig. HL8). Fan housing and magneto back plate are integral on later models. Removing starter and starter adapter plate will permit access to remove magneto rotor.

Service crankshafts may have two keyways for breaker cam and magneto rotor. The second keyway (painted red) is at 2 o'clock position (when considering cylinder at 12 o'clock position) for use with one-piece fan housing and back plate only when breaker points are located above crankshaft. On early engines with two-piece fan housing and back plate, breaker points are mounted below crankshaft (opposite cylinder). The breaker cam and rotor must be positioned in same keyway.

The later one-piece fan housing and back plate may be installed on earlier models by tapping the two drilled stator mounting holes and mounting magneto stator (armature and coil) in original position. When both new crankshaft and fan housing/back plate are used, remount magneto stator in new position, use red keyway for installing cam and rotor and mount starter using new adapter plate. New magneto leads, retaining clips and rivets will be required also.

CARBON. Carbon should be cleaned from exhaust ports at 100 to 200 hour intervals. Remove muffler and position piston at top dead center. Use a wooden or plastic scraper to remove carbon deposits. Avoid scratching piston or damaging edge of port.

NOTE: For easy access to exhaust port, stand engine on recoil starter end.

LUBRICATION. Engine on all models is lubricated by mixing oil with regular gasoline. If Homelite Premium SAE 40 oil is used, fuel:oil ratio should be 32:1. Fuel:oil ratio should be 16:1 if Homelite 2-Cycle SAE 30 oil or other SAE 30 oil designed for air-cooled two-stroke engines is used.

NOTE: A fuel:oil ratio of 32:1 should be used on diaphragm pump models regardless of whether Homelite Premium SAE 40 or Homelite 2-Cycle SAE 30 oil is used.

REPAIRS

CONNECTING ROD. Connecting rod and piston assembly can be removed after removing cylinder from crankcase. Refer to Fig. HL6. Be careful to remove all of the loose needle rollers when detaching rod from crankpin. Model 250 has 31 bearing rollers in rod bearing while Model 270 has 23 rollers.

PISTON, PIN AND RINGS. Piston assembly is accessible after removing cylinder assembly from crankcase. Always support piston when removing or installing piston pin. Piston is made of aluminum alloy and is fitted with three pinned piston rings.

If piston ring locating pin is worn to half the original thickness, or if there is any visible up and down play of piston pin in piston bosses, renew piston and pin assembly. Inspect piston for cracks or holes in dome and renew if any such defect is noted. Slight scoring of piston is permissible, but if rough surfaces are accompanied by deposit of aluminum on cylinder wall, renew piston.

Always use new piston pin retaining snap rings when reassembling piston to connecting rod. Fit new piston rings in grooves, aligning ring end gaps with locating pin. Be sure locating pin side of piston is away from exhaust side of engine when installing piston and connecting rod assembly.

CYLINDER. Cylinder bore is chrome plated; the plating is light gray in color and does not have the appearance of polished chrome. Renew cylinder if any part of chrome plating in bore is worn through. Worn areas are bright as the aluminum is exposed. In some instances, particles from the aluminum piston may be deposited on top of chrome plating.

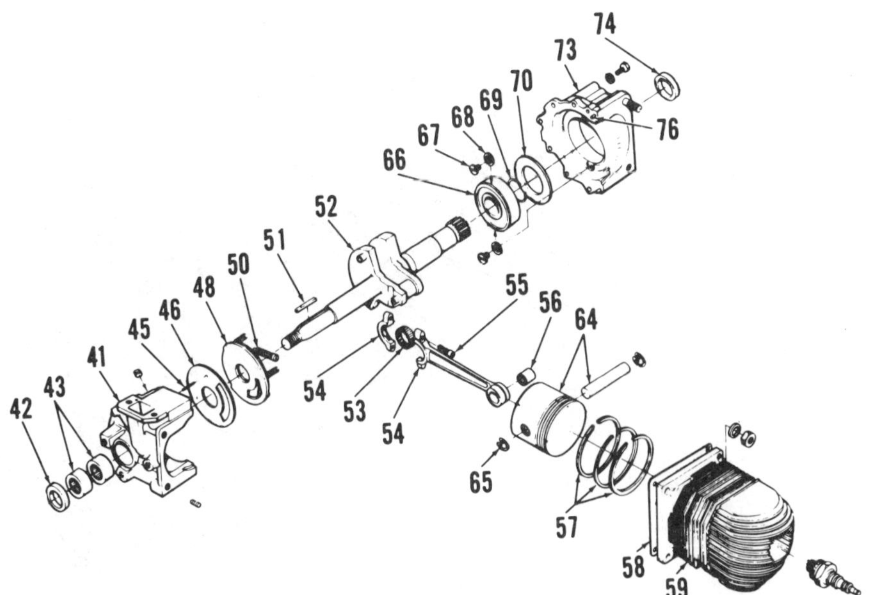

Fig. HL6–Exploded view of engine. Crankshaft end play is controlled by ball bearing (66). Three different crankshafts are used. Crankshaft shown is for diaphragm pump. Crankshaft for governor has tapered end with threaded counterbore for armature retaining bolt. Crankshaft for other applications has externally threaded end.

41. Crankcase half
42. Crankshaft seal
43. Needle roller bearings
45. Dowel pins
46. Intake valve wear plate
48. Intake valve & governor ssy.
50. Governor spring
51. Cam & flywheel key
51. Crankshaft (diaphragm pump)
53. Needle bearing rollers (31)
54. Connecting rod & cap
55. Socket head screws
56. Needle roller bearing
57. Piston rings
58. Gasket
59. Cylinder
64. Piston & piston pin
65. Snap rings
66. Roller bearing
67. Bearing retaining screws
68. Bearing retaining washers
69. Snap ring
70. Bearing gasket
73. Crankcase half
74. Crankshaft oil seal
76. Dowel pin

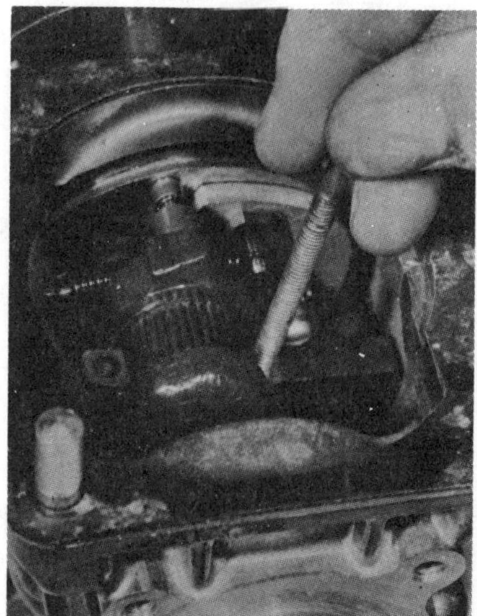

Fig. HL7—A 10-32 threaded rod or headless screw is used as tool to help in assembling connecting rod to cap. Stick the 31 loose needle rollers to crankpin with beeswax or heavy grease, then carefully position cap so that when assembled, pinned side of piston is towards intake side (upper side) of engine.

This condition is usually indicated by rough, flaky appearance and deposits can be removed by using a rubber compound buffing wheel on an electric drill.

If a screwdriver can be run over the cleaned surface without leaving marks, the cylinder is suitable for further service. If screwdriver scratches surface, renew cylinder.

Lubricate piston, rings and cylinder bore. Compress rings, then slide cylinder down over piston. Tighten cylinder retaining nuts evenly and securely.

CRANKSHAFT, BEARINGS AND SEALS. To remove crankshaft, pump must first be removed. Refer to exploded view of appropriate unit in Figs. HL9, HL10, HL11 or HL12 for required disassembly. Then refer to Figs. HL5 and HL6 and proceed as follows:

Remove flywheel, magneto assembly and magneto back plate. Remove and discard felt seals (35A and 38–Fig. HL5) from backplate (37). Remove "O" ring oil slinger (not shown) from crankshaft. Remove cylinder, piston and connecting rod assembly, then separate crankcase halves (41 and 73–Fig. HL6). Pull governor weight away from crankshaft, then carefully remove the rotary intake valve and governor assembly (48) to avoid damaging sealing surface of crankshaft. Remove two screws (67) and washers (68) retaining ball bearing (66) in crankcase half (73) and remove shaft and bearing. Tape shaft to prevent scratching sealing surface, then remove snap ring (69) and pull bearing (66) from crankshaft. Remove intake valve plate (46) from crankcase half (41), pry seal (42) out of bore and press needle roller bearing cages (43) out of crankcase half. Remove seal (74) from opposite half.

NOTE: Remove bearings from crankshaft and flywheel side crankcase half ony if renewal is indicated.

Renew crankshaft if it has damaged threads, enlarged keyways, or if runout exceeds 0.003 inch (0.08 mm). Inspect drive gear on output end of diaphragm pump engine for wear or other tooth

Fig. HL8—Exploded view of Fairbanks-Morse rewind starter. Starter cup (15) has rope notches so engine can be started with starter assembly removed. Early type fan housing (17) is shown. Later fan housing and magneto back plate (37–Fig. HL5) are integral.

2. Cover & bushing assy.
3. Starting cord
4. Starting cord grip
5. Insert
6. Rewind spring
7. Starter pulley
8. Fiber washer
9. Friction shoe assy.
10. Brake spring
11. Brake retaining washer
12. "E" ring
13. Flywheel nut
14. Lockwasher
15. Starter cup
16. Engine stop switch
17. Blower housing
18. Flywheel (magneto rotor)

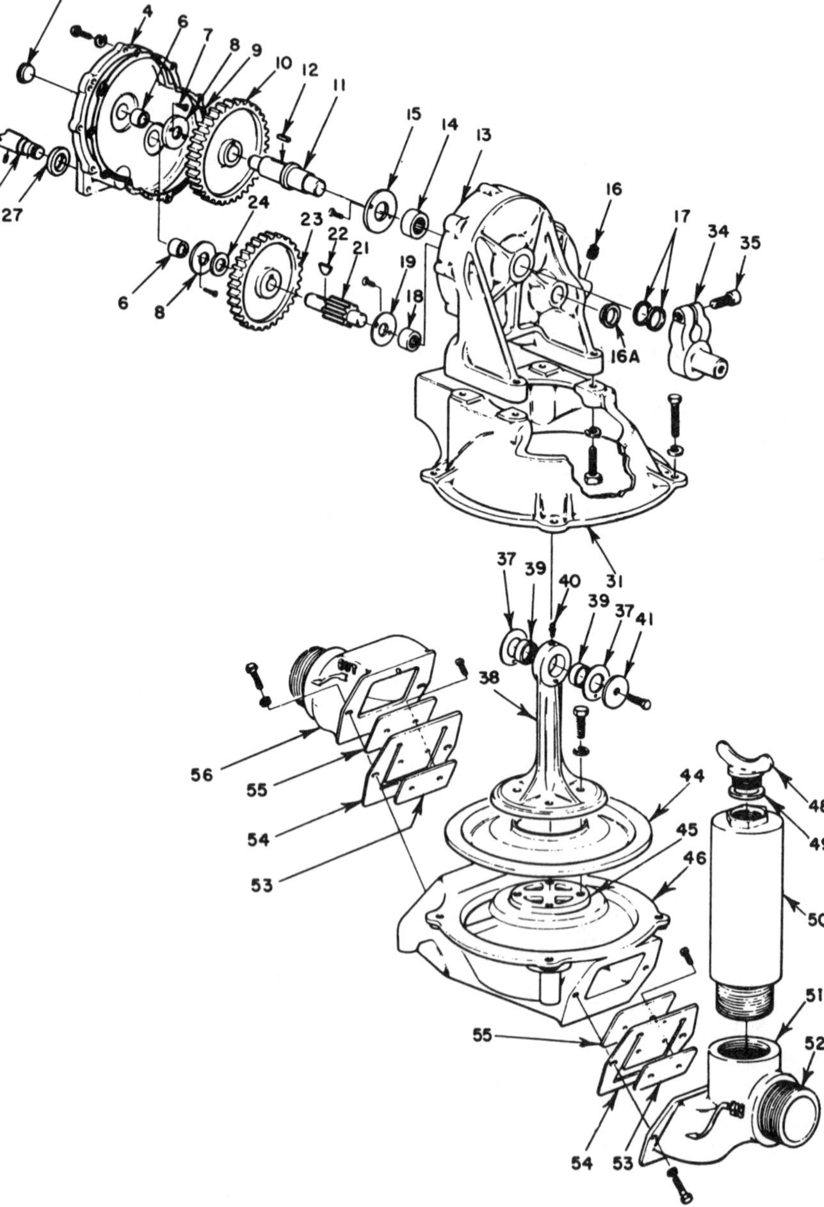

Fig. HL9—Exploded view of diaphragm pump assembly.

1. Engine crankshaft
4. Gearcase cover
5. Bearing cap
6. Needle bearing
8. Thrust washer
9. Gasket
10. Pump gear
11. Shaft & key assy.
12. Key
13. Gearcase assy.
14. Needle bearing
15. Thrust washer
16. Oil filler plug
16A. Bearing cap
17. Garlock seals
18. Needle bearing
19. Thrust washer
21. Intermediate shaft
22. Woodruff key
23. Intermediate gear
24. Spacer
31. Upper pump body
34. Crank
35. Cap screw
37. Thrust washer
38. Pump rod assy.
39. Needle bearings
40. Grease fitting
41. Washer
44. Pump diaphragm
45. Diaphragm cap
46. Pump body
48. Wing plug
49. Gasket
50. Standpipe
51. Suction fitting
52. Pipe fitting
53. Valve weight
54. Valve
55. Valve plate
56. Discharge fitting

damage. Flywheel end main journal must be free of pits, galling or heavy score marks. If journal is worn or out of round more than 0.001 inch (0.025 mm), renew crankshaft. Renew ball bearing at output end if bearing shows perceptible wear or feels rough when rotated. The caged needle roller bearings at flywheel end should be renewed if any roller end shows visible flat spot, or if rollers in either cage can be separated more than the width (diameter) of one roller.

When reassembling, soak new bearing gasket (70) in oil, then insert in crankcase half (73). Support crankshaft at throw, then press new ball bearing onto shaft and secure with snap ring (69).

NOTE: Be sure that groove in outer bearing race is towards crankshaft throw.

Lubricate seal (74), then using suitable installation tool (Homelite 24120-1 or equivalent) press seal into crankcase half with lip of seal inward. Pressing against outer race of bearing (66) only, install crankshaft and bearing assembly into crankcase half (73) and secure with the two screws (67) and washers (68).

NOTE: Use suitable seal protector (Homelite 24125-1, 24126-1 or 24127-1, or equivalent) to prevent damage to seal.

Using Homelite tool 24155-1, press outer needle bearing into crankcase half (41) with stepped end of tool, then press inner bearing into crankcase half with straight end of tool.

NOTE: Press on lettered side of bearing cage only. Install seal (42) with seal lip towards needle bearing (43) using straight end of bearing installation tool.

Fit governor and rotary intake valve assembly onto crankshaft so that thrust springs fit into proper bores of crankshaft.

NOTE: Hold governor plate away from crankshaft when installing to prevent scratching seal surface.

Lubricate all parts thoroughly and insert intake valve plate (46) in crankcase half (41) so it is properly positioned on dowel pins and intake opening. Using seal protector (Homelite 24121-1 or equivalent) assemble the crankcase half over crankshaft and governor assembly, hold assembly together against thrust spring pressure and install crankcase cap screws. Tighten cap screws to 80 in.-lbs. (9 N·m). Insert new felt seal (38 – Fig. HL5) in bore of magneto back plate, place new "O" ring oil slinger on crankshaft and install back plate. Tighten backplate retaining cap screws to 80 in.-lbs. (9 N·m). Complete assembly by reversing disassembly procedure.

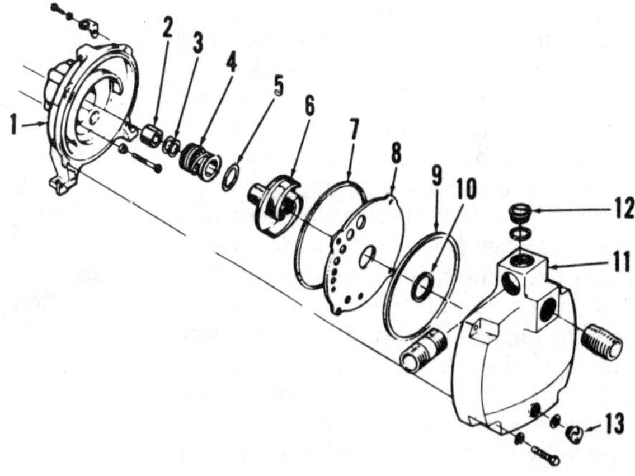

Fig. HL10 – Exploded view of centrifugal pump. End housing (11) and impeller (6) must be removed to allow removal of the four cap screws retaining impeller housing (1) to engine. Prevent flywheel from turning with pin (see magneto section) to unscrew impeller.

1. Impeller housing
2. Spacer
3. Shims
4. Seal assy.
5. Shim
6. Impeller
7. Gasket
8. Wear plate
9. Gasket
10. Gasket
11. End housing
12. Primer plug
13. Drain plug

CRANKCASE. Be sure all passages through crankcase are clean. The idle passage line which enters crankcase via the intake valve register may be restricted with carbon deposits.

If main bearing bore at output end has a worn appearance, bearing has been turning in bore. If so, crankcase half should be renewed. The mating surfaces of the two-piece crankcase must be free of all nicks and burrs as neither sealing compound nor gaskets are used at this joint.

NOTE: Fuel tank bracket mounting screws are secured in engine crankcase with Loctite.

When reinstalling bracket, clean the screw threads and threads in crankcase, then apply a drop of Loctite to each screw. Tighten screws to 120 in.-lbs. (13.5 N·m).

ROTARY INTAKE VALVE. The combination rotary intake valve and governor (Fig. HL3) should be renewed if sealing faces of valve or governor plate are worn or scored enough to produce a ridge, if spring post is loose or extended to valve seating surface or if governor pivot point has started to wear through the surface of valve. Maximum allowable clearance between governor plate and intake valve plate is 0.006 inch (0.15 mm). The governor spring and/or governor spring adjusting screw may be renewed separately from the assembly.

Slight scoring of valve face may be corrected by lapping on a lapping plate using a very fine abrasive. Lapping motion should be in the pattern of a figure eight to obtain best results. Slight scoring of the Formica wear plate is permissible. Homelite recommends soaking a new Formica plate in oil for 24 hours prior to installation.

REWIND STARTER. Refer to Fig. HL8 for exploded view of the Fairbanks-Morse starter used on all applications. Should rewind starter fail during an emergency situation, remove starter assembly and wind rope around starter cup (15) to start engine.

Refer to exploded view for proper reassembly of starter unit. Hook end of starter rope in notch of pulley and turn pulley five turns counterclockwise, then let spring wind rope into pulley for proper spring tension.

To remove starter cup, insert lock pin through hole in magneto back plate and hole in flywheel to hold flywheel from turning, then unscrew retaining nut.

DIAPHRAGM PUMP

The diaphragm pump is lubricated by filling gearcase to level of plug (16 – Fig. HL9) with SAE 90 gear lubricant and by greasing pump rod upper bearing at fitting (40) once a month with pressure gun.

When installing new diaphragm (44) or assembling upper pump body (31) to lower body (46), the diaphragm must be centered and in fully down position before tightening upper body-to-lower body bolts.

To remove pump from engine, drain gear lubricant and separate gearcase (13) from gearcase cover (4). Gear teeth are machined on end of engine crankshaft to drive intermediate gear (23). Unbolt and remove cover (4) from engine crankcase. When reassembling, use new gasket (9).

CENTRIFUGAL PUMP

Refer to Fig. HL10. To remove pump from engine, remove end housing (11) and wear plate (8) from impeller housing (1), taking care not to damage sealing gaskets. Unscrew impeller (6) from engine crankshaft in counterclockwise direction by placing wrench on hex end of impeller and striking wrench a sharp blow with hammer. Take care not to lose or damage seals or shims. Impeller housing can now be unbolted from engine.

When reassembling pump, shims (3) are available to maintain clearance between impeller and wear plate (8). When shims are added to decrease clearance, seal shims (5) of the same thickness must be installed to maintain proper tension on seal spring. Before reassembling pump, hold wear plate (without gasket) against impeller housing and turn engine by hand to be sure impeller does not rub against wear plate.

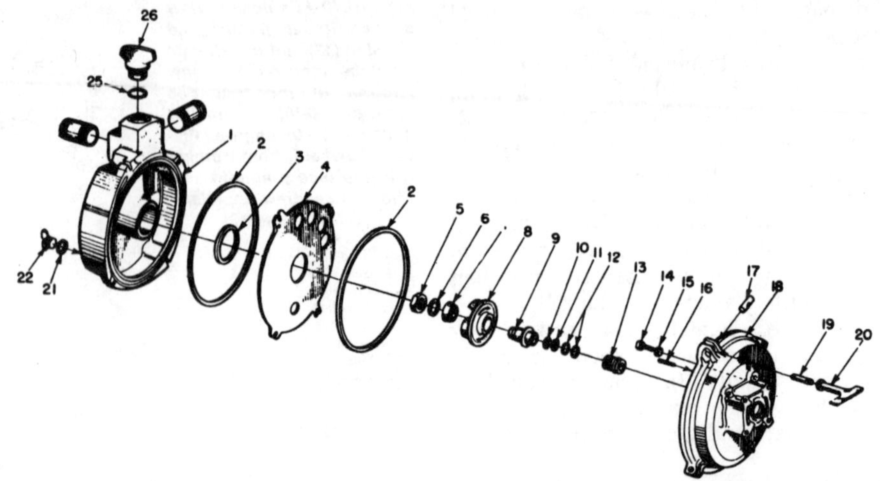

Fig. HL11—Exploded view of trash pump. End housing and impeller must be removed from impeller housing to allow housing to be unbolted from engine. Model 270 uses a spacer in place of washers (12).

1. End housing
2. Gasket
3. Gasket
4. Wear plate
5. Impeller nut
6. Washer
7. Tapered bushing
8. Impeller
9. Impeller hub
10. Shim
11. Shim
12. Spacer washers
13. Seal assy.
14. Screws
15. Sealing washers
16. Spiral pin
17. Pivot pin
18. Impeller housing
19. Studs
20. Wing nuts

TRASH PUMP

The trash pump (Fig. HL11) impeller (8) is mounted on tapered bushings so that if a solid object lodges in pump to block the impeller, shaft rotation will not halt immediately.

To remove pump from engine, remove end housing (1) and wear plate (4), then unscrew impeller retaining nut (5) and remove impeller (8), taking care not to damage or lose seal and shims (10, 11 and 12). Unbolt and remove impeller housing (18) from engine.

When reassembling pump, use shims (10 and 11) of total thickness required to maintain minimum clearance between impeller and wear plate. Shims are placed between shoulder on crankshaft and impeller hub (9). Install seal shims of same total thickness along with spacer washers (12) to maintain seal spring tension. Model 270 is equipped with a spacer in space of washers (12).

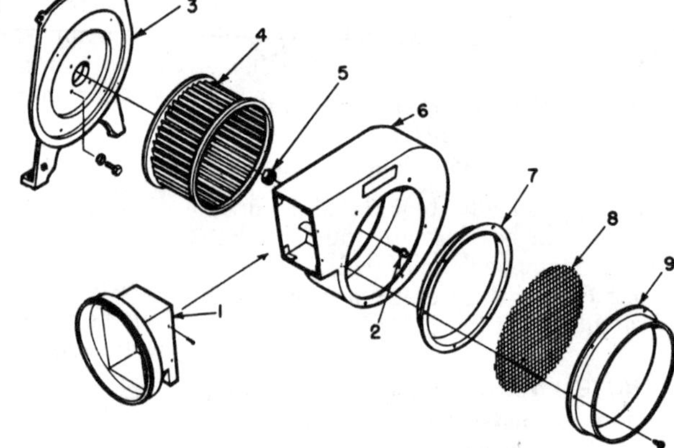

Fig. HL12—Exploded view of blower unit. Remove blower housing and rotor to allow unbolting and removal of blower plate (1) from engine.

1. Blower outlet
2. Slotted head screws
3. Mounting plate
4. Blower rotor
5. Elastic stop nut
6. Blower housing
7. Blower inlet venturi
8. Inlet screen
9. Inlet collar

HOMELITE

Model	Bore	Stroke	Displacement
10	2¾ in. (69.9 mm)	2⅛ in. (53.9 mm)	12.6 cu. in. (207 cc)
251	2¾ in. (69.9 mm)	2⅛ in. (53.9 mm)	12.6 cu. in. (207 cc)

MAINTENANCE

SPARK PLUG. Either a Champion J6J, HO8A or UJ11G spark plug may be used. The Champion HO8A platinum tip or UJ11G gold paladium tip plug will provide longer service as well as longer intervals between cleanings. Electrode gap for all plugs is 0.025 inch (0.635 mm).

CARBURETOR. Refer to Fig. HL20. The Tillotson HS carburetor can be removed from engine by removing air intake manifold (3–Fig. HL21) and air cleaner asembly as the two manifold bolts also retain carburetor and reed valve assemblies.

When disassembling carburetor, slide diaphragm assembly towards adjustment needle side of carburetor body to disengage diaphragm from fuel inlet control lever. To remove welch plug (29–Fig. HL20), carefully drill through plug with a small diameter drill and pry plug out with a pin. Caution should be taken that drill bit just goes through welch plug as deeper drilling may seriously damage carburetor. Note channel screen (6) and check valve assembly (25) located in bores of carburetor body.

Inlet control metering lever (18) should be flush with metering chamber floor of carburetor body. If not, bend diaphragm end of lever up or down as required so lever is flush.

Normal adjustment of low speed fuel mixture needle (marked "LO" on air inlet manifold or needle nearest throttle shaft) is one turn open and main mixture adjustment needle (marked "HI" on air inlet manifold or needle nearest choke shaft) should be opened ¾ turn. On pump engines, back the idle speed adjusting screw (Fig. HL22) out until carburetor throttle plate will close completely and screw is clear of pin on throttle rod. Slowly turn screw in until it just contacts pin, then turn screw in one additional turn.

Start engine and allow to warm up before making final carburetor ad-

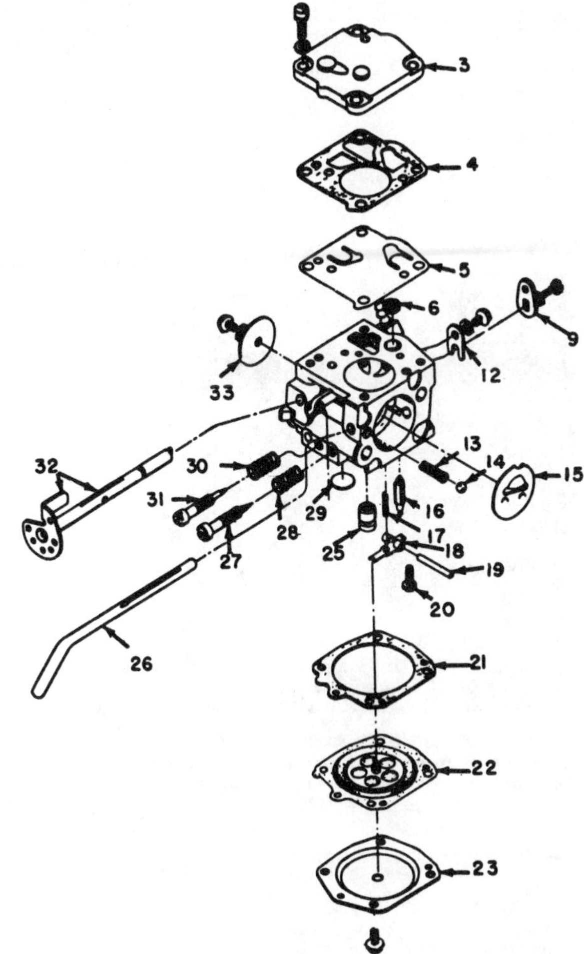

Fig. HL20—Exploded view of Tillotson Series HS carburetor used on Model 251 engine. Note that carburetor is shown inverted from actual position on engine.

3. Fuel pump cover
4. Fuel pump gasket
5. Fuel pump diaphragm
6. Inlet channel screen
9. Governor spring arm
12. Throttle shaft clip
13. Choke detent spring
14. Choke detent ball
15. Choke shutter
16. Inlet needle valve
17. Inlet lever spring
18. Inlet control lever
19. Lever pinion pin
20. Pinion pin retaining screw
21. Diaphragm gasket
22. Metering diaphragm
23. Diaphragm cover
25. Nozzle check valve
26. Choke shaft
27. Main adjusting needle
28. Main needle spring
29. Welch plug
30. Idle fuel needle spring
31. Idle adjusting needle
32. Throttle shaft
33. Throttle shutter

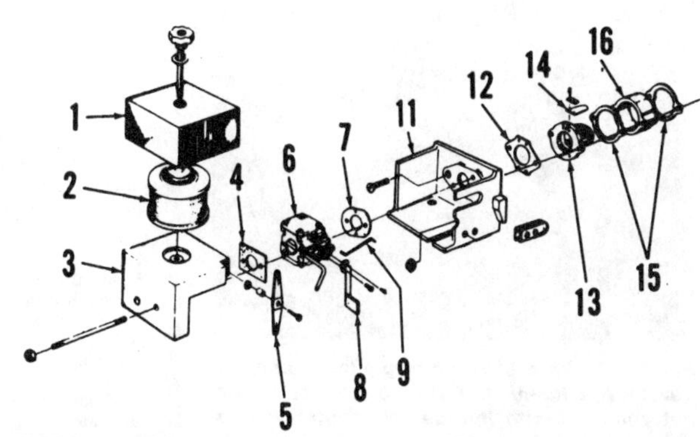

Fig. HL21—Exploded view of induction system. Components (8 and 11) are not used on models with 251 engines.

1. Filter cover
2. Air filter
3. Intake manifold
4. Gasket
5. Bellcrank
6. Carburetor
7. Gasket
8. Throttle arm
9. Link
11. Housing
12. Gasket
13. Reed valve seat
14. Reed valve
15. Gasket
16. Spacer

justments. On pump engines, pump water with throttle control rod pushed all the way in and readjust main mixture needle so engine runs best. Then, lift suction hose out of water, pull throttle rod out and adjust low idle speed for smoothest running. If necessary, readjust idle speed screw to obtain a slow idle speed of 1800-2500 rpm.

CAUTION: Do not adjust low idle speed higher than 2500 rpm. Higher idle speed will result in damage to governor.

NOTE: The main and idle speed mixture adjustments are interdependent so changing one needle setting often requires readjustment of other needle.

When reinstalling carburetor, be sure governor link is connected as shown in Fig. HL23.

GOVERNOR. Engine is equipped with a flyweight type governor mounted on engine crankshaft. Refer to Fig. HL24 for exploded view showing governor unit. External governor linkage is shown in Fig. HL23.

CAUTION: Never move governor linkage manually, or exert any pressure on lever or linkage to increase engine speed. Working governor linkage manually, even momentarily, may cause damage to governor cup and cam due to friction and burning. Also, on pump engines, do not adjust slow idle speed above 2500 rpm.

Maximum no-load speed for pump engines should be 3800-3900 rpm. If necessary to readjust governor, first remove the cover plate (43–Fig. HL24) and slightly loosen governor shaft guide (36) retaining screws. Note that screw hole in guide at carburetor side is slotted. Insert screwdriver between side of guide and shoulder machined in housing (45). Pry carburetor side of guide towards cylinder to decrease governed speed or away from cylinder to increase speed. Adjustment provided by total length of slot will change the maximum governed speed about 1000 rpm.

If engine governed speed cannot be properly adjusted, check for wear on governor shaft cam and inspect governor spring connected to carburetor throttle shaft.

To renew governor spring (32), cup (31) or back plate (25), first remove starter, magneto rotor (49) and housing (45). Snap ring (34) can then be removed from crankshaft allowing removal of spring cup and back plate.

Fig. HL22—Adjusting idle speed screw. Throttle rod friction screw is just visible at under side of air intake manifold. Idle speed must not be adjusted to above 2500 rpm.

Fig. HL23—View showing proper installation of governor bellcrank and links. Generator engine not equipped with throttle (idle control rod) is shown.

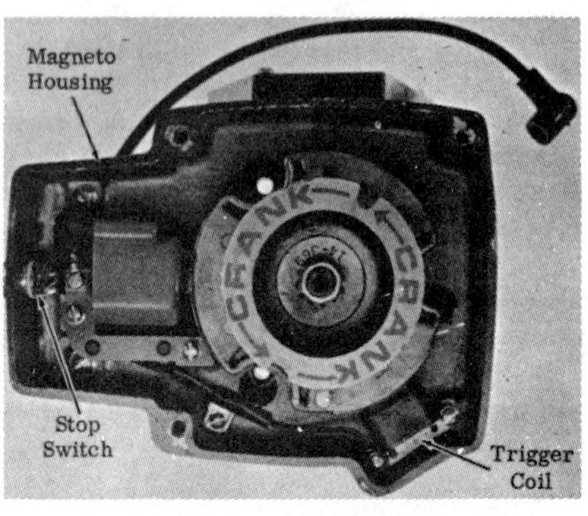

Fig. HL25—View of magneto assembly with rewind starter assembly removed. Armature-to-magneto rotor air gap and trigger coil-to-rotor air gap should be adjusted using a 0.0075 inch (0.191 mm) thick plastic shim.

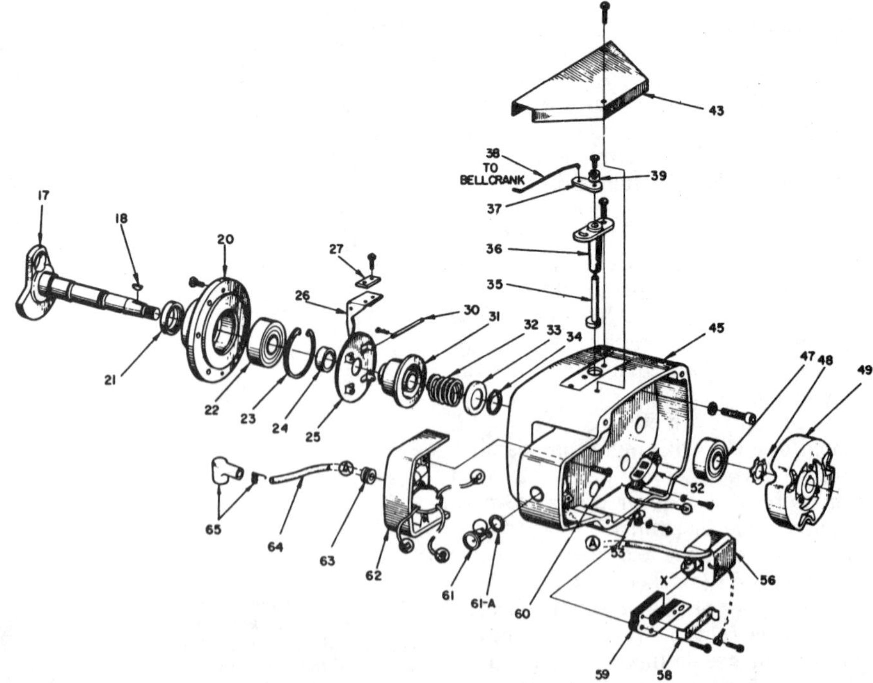

Fig. HL24—Exploded view showing magneto end of crankshaft (17), governor and magneto assemblies. The housing (45) retaining socket head screws can be removed by working through notches in outside of magneto rotor (49) allowing all parts to be removed as a unit after removing governor linkage and disconnecting spark plug wire.

17. Crankshaft, magneto end
18. Woodruff key
20. Bearing housing
21. Crankshaft seal
22. Bearing
23. Snap ring
24. Spacer
25. Governor back plate
26. Governor weight arm
27. Governor weight
30. Weight pivot pin
31. Governor cup
32. Governor spring
33. Spring retainer
34. Snap ring
35. Governor camshaft
36. Camshaft guide
37. Governor arm
38. Bellcrank link
39. Flat washer
43. Governor linkage cover
45. Magneto housing
47. Bearing
48. Loading spring
49. Magneto rotor
52. Magneto trigger coil
53. Lead clamp
56. Ignition coil
58. Coil spring clip
59. Armature core
60. Cover retaining screw
61. Stop switch
61A. Magneto cover assy. (includes condenser and solid state switchbox)
63. Grommet
64. Spark plug wire
65. Spark plug terminal

Illustrations Courtesy Homelite Div. of Textron

IGNITION. A breakerless solid state ignition system is used. Refer to Fig. HL24 for exploded view of the magneto (items 49 through 65) and to Figs. HL25 and HL26.

To check the solid state magneto, disconnect spark plug wire and remove spark plug. Insert a bolt or screw in spark plug wire terminal and while holding bolt or screw about ¼ inch (6 mm) away from engine casting, crank engine and check for spark as with conventional magneto. If no spark occurs, visually check for broken or frayed wires which would result in open circuit or short. Be sure stop switch is not permanently grounded. Inspect magneto rotor (49–Fig. HL24), trigger coil (52) and the switch box, condenser and magneto cover assembly (62) for visible damage.

To test magneto components, remove starter assembly, magneto cover and disconnect leads as shown in Fig. HL26.

To test ignition coil, refer to test instrument instructions. Readings for Graham Model 51 and Merc-O-Tronic testers are as follows:

Graham Model 51:

Maximum secondary 10,000
Maximum primary 1.7
Coil index 65
Maximum coil test 20
Maximum gap index 65

Merc-O-Tronic:

Operating amperage 1.3
Minimum primary resistance 0.6
Maximum primary resistance 0.7
Minimum secondary continuity 50
Maximum secondary continuity 60

If ignition does not meet test specifications, renew using correct part number coil. Do not substitute a coil of other specifications with the solid state ignition system. If coil test specifications are correct, use the following procedure to check switch box.

With leads and condenser disconnected as shown in Fig. HL26, connect one ohmmeter lead to one switch box test point. The ohmmeter reading should be either between 5 to 25 ohms or from one megohm to infinity. When ohmmeter test leads are reversed, the opposite reading should be observed. If these ohmmeter readings are not observed, renew magneto cover and switch box assembly. If ignition coil and switch box test specifications are correct, use the following procedure to check trigger coil.

Connect ohmmeter positive lead to junction of switch box and trigger coil leads and ohmmeter negative lead to magneto housing (see "Trigger Coil Test Points" in Fig. HL26). It is not necessary to disconnect trigger coil lead from switch box lead. The ohmmeter reading should be 22 to 24 ohms.

To check condenser, stick a pin through the condenser lead to provide a contact point, then test condenser using standard procedure to check series resistance, short and capacitance. Condenser capacitance should be 0.16-0.020 mfd.

If either the switch box or condenser tested faulty, renew the complete condenser, switch box and magneto cover assembly.

LUBRICATION. Engine on all models is lubricated by mixing oil with regular gasoline. If Homelite Premium SAE 40 chain saw oil is used, fuel:oil ratio should be 32:1. Fuel:oil ratio should be 16:1 if Homelite 2-Cycle SAE 30 oil or other SAE 30 oil designed for air-cooled two-stroke engines is used.

AIR AND FUEL FILTERS. The air cleaner element may be washed in a detergent and water solution or by immersion in a container of nonoily solvent. After a number of cleanings, the filter pores may become permanently clogged, making it necessary to renew element.

The fuel filter is a part of the fuel pickup inside fuel tank and can be extracted from tank filler opening using a wire hook. Under normal operations, the filter element should be changed at one

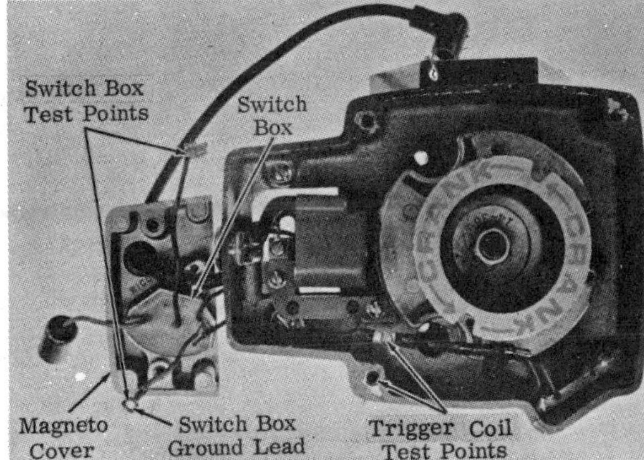

Fig. HL26—View showing magneto leads disconnected for testing purposes. Condenser must not touch any other part of the unit. Refer to text for procedure and specifications.

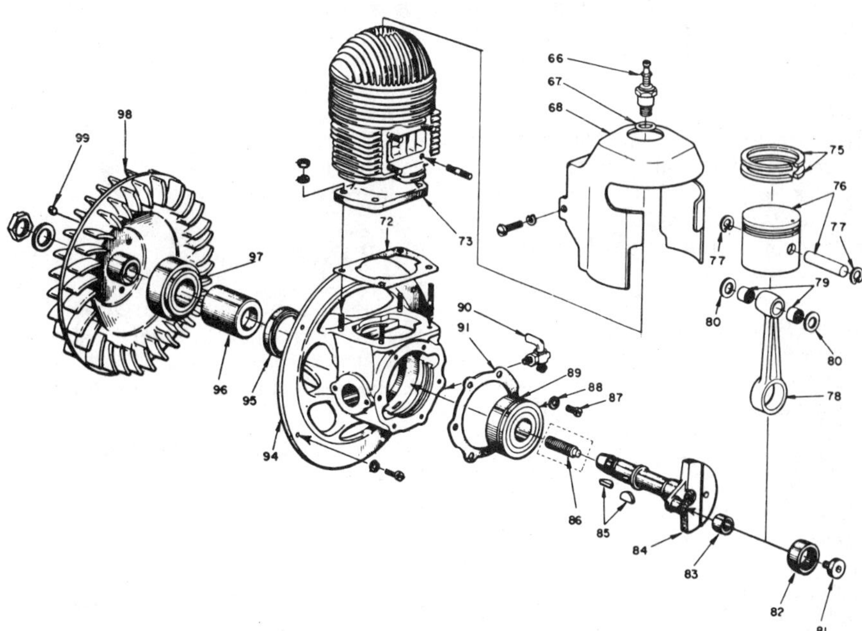

Fig. HL27—Exploded view showing Model 251 engine crankcase, cylinder, rod and piston assembly, fan, crankshaft (output end) and related parts. Value (90) may be used to drain crankcase should it become flooded with fuel. Do not attempt to start and run engine with valve open.

66. Spark plug
67. Spark plug gasket
68. Cylinder shield
72. Cylinder gasket
73. Cylinder
75. Piston rings
76. Piston & pin assy.
77. Snap rings
78. Connecting rod
79. Needle bearings
80. Thrust washers
81. Crankpin screw
82. Roller bearing
83. Inner race
84. Crankshaft (output end)
85. Woodruff keys
86. Stud bolts (251S3 pump only
87. Bearing retaining screws
88. Bearing retaining washers
89. Bearing
90. Valve
91. Crankcase gasket
94. Crankcase
95. Oil seal
96. Spacer
97. Bearing
98. Fan
99. Cork plug

to two month intervals. If engine is run continuously or fuel is dirty, filter may need to be changed weekly or at shorter intervals.

CARBON. The carbon should be cleaned from exhaust ports at 100 to 200 hour intervals. Remove muffler and position piston at top dead center. Use a wooden or plastic scraper to remove carbon deposits. Avoid scratching piston or damaging edges of port.

REPAIRS

CONNECTING ROD. Connecting rod lower end is fitted with a roller bearing (82–Fig. HL27) which rides on a renewable inner race (83). To remove piston and connecting rod assembly from crankshaft (84) crankpin, first remove cylinder and the magneto housing and crankshaft rotor end assembly. Place a block of wood between crankshaft throw and crankcase to keep crankshaft from turning, then unscrew crankpin screw (81) (counterclockwise) with 3/8 inch Allen wrench. If renewal of inner race is indicated, remove race from crankpin. Usually, race will slide off pin, however, it may be necessary to pry race from crankpin with screwdrivers.

To renew crankpin needle roller bearing in the connecting rod, press old bearing out using plug (Homelite tool 24120-1), supporting rod on sleeve (Homelite tool 24118-1). Install new bearing by supporting rod on sleeve (Homelite tool 24124-1) and pressing bearing in with same plug as used to remove old bearing. Shouldered face of sleeve (24124-1) will properly position bearing so it protrudes equally from each side of rod.

To renew piston pin bearings, support rod on sleeve (Homelite tool 24124-1) and press bearings out with plug (24131-1). New bearings are installed separately from opposite ends of bore. Support rod on sleeve (24124-1) and using straight end of plug (24131-1, end with recessed face), press new bearing in (press on lettered side of cage only) until shoulder of plug seats against rod. Turn rod over and press remaining new bearing into rod in same manner. When properly installed, recessed faces of piston pin thrust washers will clear protruding bearing races and will contact connecting rod.

When reinstalling connecting rod and bearing inner race, thoroughly lubricate all parts and tighten connecting rod cap screw to 50 ft.-lbs. (68 N•m).

NOTE: Locate connecting rod on crankpin so oil hole in upper end of rod will be towards intake side of engine.

Piston should be assembled to connecting rod so piston ring locating pin will face same (intake) side of assembly as oil hole in rod.

PISTON, PIN AND RINGS. Piston is accessible after removing cylinder from crankcase. Always support piston while removing or installing piston pin. Piston should be renewed if ring side clearance, measured with new ring installed in top groove, exceeds 0.004 inch (0.010 mm). Also, renew piston if piston skirt to cylinder bore clearance exceeds 0.007 inch (0.18 mm) when measured with new or unworn cylinder. Inspect piston ring locating pin and renew piston if pin has worn to half of its original thickness. Piston pin should be a snug push-fit to light press-fit in piston. Piston, pin and rings are available in standard size only. Homelite recommends piston rings be renewed whenever engine is disassembled for service.

When reassembling piston to connecting rod, insert new snap ring in exhaust (opposite ring locating pin) side of piston. Lubricate all parts and place piston, exhaust side down, in holding fixture.

NOTE: A used cylinder sawed in half makes a good holding fixture.

Press pin into upper (intake) side of piston, then insert connecting rod and thrust washers into piston with oil hole in rod up and recessed sides of washers next to piston pin bearings in rod. Press pin on through the assembly and secure with new snap ring. Be sure piston and rod assembly is installed on crankpin with pinned (intake) side of piston away from exhaust port side of engine.

CYLINDER. Cylinder bore is chrome plated. Inspect cylinder bore for excessive wear and damage to chrome surface of bore. A new cylinder must be installed if chrome is scored, cracked or the base metal underneath exposed.

CRANKSHAFT, BEARINGS AND SEALS. The two-piece crankshaft can be serviced as two separate parts. Refer to following paragraphs for crankshaft service.

Crankshaft, Magneto End. To service the shaft, bearings, seal or governor components, remove rewind starter as an assembly. Unscrew magneto rotor retaining nut and using suitable puller (Homelite tool AA-22560 or equivalent), remove rotor. Disconnect governor linkage and remove governor bellcrank.

Using a 3/16 inch Allen wrench, remove the six socket head screws retaining magneto housing to crankcase and remove housing and shaft assembly. Remove the two screws retaining bearing housing (20–Fig. HL24) to magneto housing (45) and separate shaft and bearing assembly from housing (45). Bearing (47) can be renewed at this time. Remove snap ring (34), retainer (33) and governor spring (32) and pry governor back plate (25), with weights, from shaft (17). Remove spacer (24), support bearing housing (20) and press shaft from housing. Remove crankshaft seal (21) and snap ring (23), then press bearing (22) from housing.

To assemble magneto end shaft and bearing assembly, install new seal (21) in housing with lip of seal towards crankcase side. Lubricate seal and insert shaft through seal and housing. Support flat inner end of shaft and press bearing (22) down over shaft and into housing until bearing inner race is seated against shoulder on shaft. Support housing and press bearing outer race into housing so retaining snap ring (23) can be installed. Place spacer (24) on shaft, then drive or press governor back plate onto shaft against spacer. Install governor cup, spring, spring retainer and snap ring. Attach bearing housing to magneto housing with the two screws, then reinstall shaft, bearing housing and magneto housing to crankcase using new gasket.

Crankshaft, Output End. First, remove magneto end crankshaft, bearing and magneto housing assembly as described in preceding paragraph. Remove cylinder and piston and connecting rod unit. Remove crankcase, output crankshaft end and blower rotor (fan) as an assembly from pump. Refer to exploded views of pump units shown in this section. Remove fan retaining nut and washer and the two crankshaft bearing retaining screws (87–Fig. HL27) and washers (88). Support magneto (open) end of crankcase, then press crankshaft (84) from fan and crankcase. Remove the three corks (99), if so equipped, from fan, insert jackscrews into the tapped holes and push bearing (97) from fan inner hub. Remove screws and reinstall corks. Corks keep threads clean but are not necessary for operation. Support outer race of bearing (89) and press shaft out of bearing. Remove spacer (96) and crankshaft seal (95) from crankcase.

To reassemble, press new seal into crankcase with lip towards inside (away from fan). Support outer hub of fan, then press new bearing (97) onto fan inner hub. Press new inner bearing (89) into crankcase with retaining groove in outer race properly positioned. Support bearing inner race with sleeve, then press crankshaft into bearing. Install bearing retaining screws and washers.

Illustrations Courtesy Homelite Div. of Textron

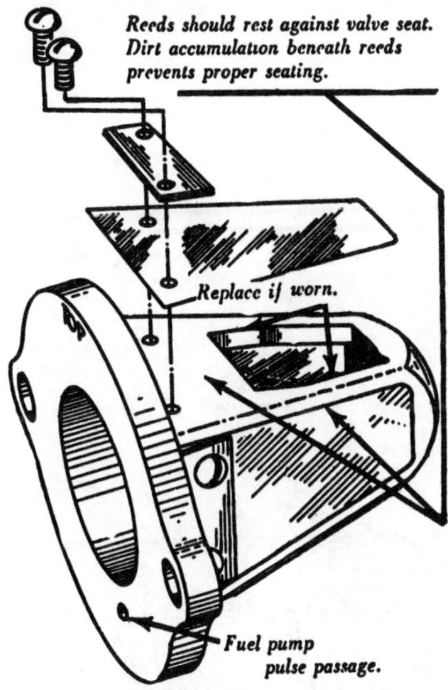

Fig. HL29—View showing reed valve assembly with one valve reed removed. Inspect seat and reeds as noted and be sure fuel pump pulse passage is open.

Place spacer (96) in position on crankshaft, then carefully press fan onto shaft making sure keys and keyways are aligned. Install fan retaining washer and nut securely. Complete reassembly by installing connecting rod and piston, cylinder and magneto end assembly.

CRANKCASE. To renew crankcase, follow procedures as outlined in previous paragraph "CRANKSHAFT, OUTPUT END"

REED INTAKE VALVE. Engine is equipped with a pyramid reed valve assembly shown in Fig. HL21. The reed valve should be inspected whenever carburetor is removed. Refer to Fig. HL29 for inspection points. When installing new reeds on pyramid seat, thoroughly clean all threads and apply Loctite to threads on screws before installing. Be sure reeds are centered before tightening screws.

REWIND STARTER. Refer to Fig. HL30 for exploded view. Should the rewind starter fail, engine can be started by removing starter and winding a rope around starter cup (1).

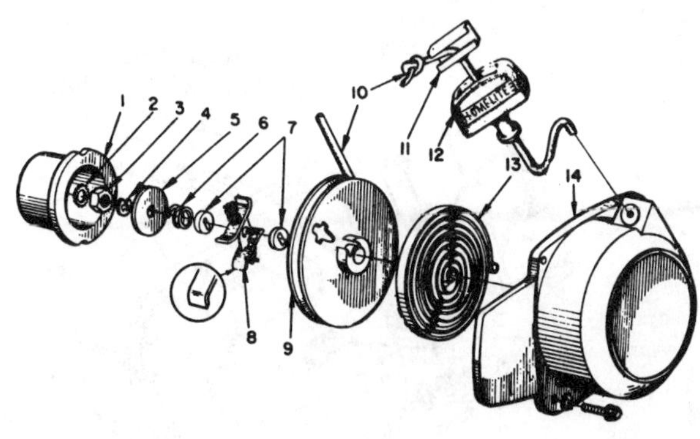

Fig. HL30—Exploded view of rewind starter used on Model 251 engine. Starter cup (1) has notches for using emergency starting rope.

1. Starter cup
2. Lockwasher
3. Crankshaft nut
4. Snap ring
5. Retaining washer
6. Brake spring
7. Brake washer
8. Friction shoe assy.
9. Starter pulley
10. Starter rope
11. Handle insert
12. Handle
13. Rewind spring
14. Starter cover

CENTRIFUGAL PUMP

Model 251S3

To remove pump from engine, remove end housing (22–Fig. HL31). Unscrew impeller clockwise by placing wrench on hex portion and striking wrench a sharp blow with hammer.

NOTE: The impeller is mounted on a stud bolt threaded into end of engine crankshaft. The impeller may unscrew off the stud or stud may be removed with impeller.

Take care not to lose the shims (14) or washer (13). After removing impeller, impeller housing can be unbolted and removed from engine fan housing (5).

Before assembling pump, lubricate seal seat and seal head with oil and make sure gaskets are in good condition. Shims (14) are used as required to maintain minimum clearance between impeller and wear plate (18). When reassembling, place wear plate against impeller housing without a gasket and turn engine by hand to be sure impeller does not rub against wear plate.

TRASH PUMP

Model 251TP3

Refer to Fig. HL32. To remove pump, first remove end housing (19) from pump body and fan housing (35). Remove impeller housing (22) if not removed with end housing. Unscrew impeller retaining cap screw (23) and remove washer (24), tapered bushing (25), impeller and seal assembly. The engine can then be unbolted and removed from housing (35).

Before reassembling pump, make sure gaskets and shaft seal are in good condition. Lubricate seal seat and seal head.

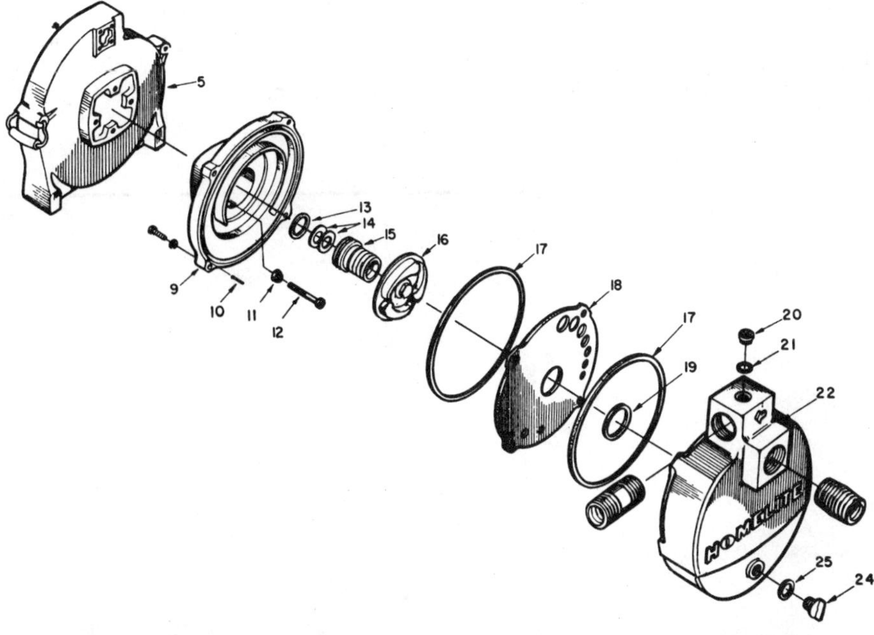

Fig. HL31—Exploded view of Model 251S pump assembly. Impeller (16) threads onto stud (86—Fig. HL27) threaded into output end of crankshaft. Use shims (14) as required to maintain 0.020-0.030 inch (0.51-0.76 mm) clearance between impeller and wear plate (18).

5. Engine fan housing
9. Impeller housing
10. Spirol pin
11. Sealing washer
12. Cap screws (to fan housing)
13. Washer
14. Shims
15. Seal assy.
16. Impeller
17. Gaskets
18. Wear plate
19. Gasket
20. Primer plug
21. Gasket
22. End housing
24. Drain plug
25. Gasket

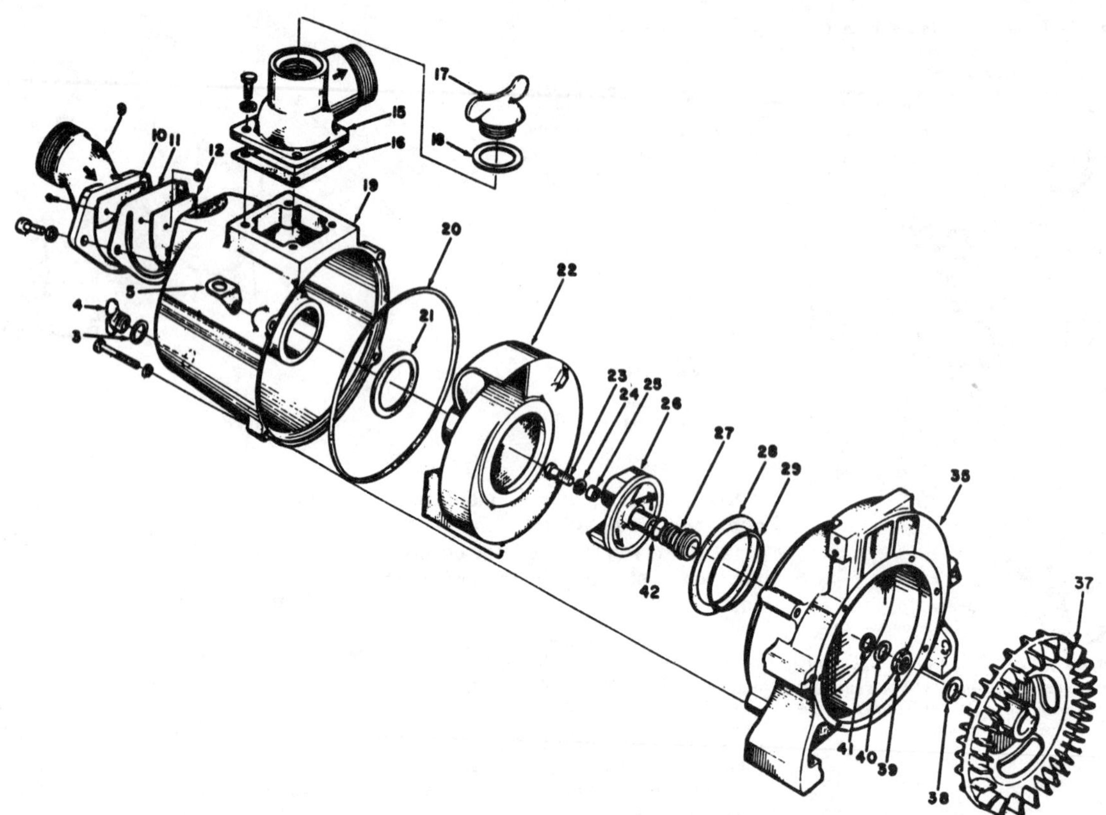

Fig. HL32—Exploded view of Model 251TP3 pump ssembly. Impeller screw (23) threads into end of engine crankshaft. Impeller is not solidly connected to engine crankshaft and can slip on tapered bushing (25) and end of crankshaft should a solid object become lodged between impeller and housing.

3. Gasket
4. Drain plug
5. Wrench holder
9. Suction fitting
10. Valve plate (small)
11. Valve
12. Valve plate (large)
15. Discharge fitting
16. Gasket
17. Primer plug
18. Gasket
19. End housing
20. Gasket
21. Gasket
22. Impeller housing
23. Impeller screw
24. Washer
25. Tapered bushing
26. Impeller
27. Seal assy.
28. Gasket
29. "O" ring
35. Pump frame & fan housing
37. Engine fan
38. Washer
39. Fan retaining nut
40. Washer
41. Shims
42. Shims

A 0.010-0.020 inch (0.25-0.51 mm) clearance between impeller and impeller housing is maintained by varying number of shims (41) between impeller hub and washer (40). Add shims until impeller just contacts impeller housing when housing is held in position, then remove one 0.015 inch (0.38 mm) thick shim. Compressed length of seal assembly (not including seat) should be 7/8 inch (22.225 mm), plus or minus 0.010 inch (0.254 mm), and is adjusted by adding or removing shims (42) between seal head and impeller. Seal length should be measured after impeller to impeller housing clearance is adjusted. After impeller clearance and seal length are correct, complete reassembly of pump.

HOMELITE

Model	Bore	Stroke	Displacement
8	2¾ in. (69.9 mm)	2⅛ in. (53.9 mm)	12.6 cu. in. (207 cc)
9	2¾ in. (69.9 mm)	2⅛ in. (53.9 mm)	12.6 cu. in. (207 cc)
20	2⅜ in. (60.3 mm)	1½ in. (38.1 mm)	6.6 cu. in. (109 cc)
23	2¼ in. (57.2 mm)	2⅛ in. (53.9 mm)	8.5 cu. in. (139 cc)
24	2⅝ in. (66.7 mm)	2⅛ in. (53.9 mm)	11.5 cu. in. (188 cc)
35	2-7/16 in. (60.9 mm)	1½ in. (38.1 mm)	7.0 cu. in. (111 cc)
36	2¼ in. (57.2 mm)	2⅛ in. (53.9 mm)	8.5 cu. in. (139 cc)

ENGINE INFORMATION

This section covers Homelite two-stroke engines used on various pumps. The unit model number (such as pump model number 8S3-1) also indicates engine model number (which would be Model 8). The pump must be at least partially disassembled to perform certain engine service operations.

CAUTION: Centrifugal type pumps should not be operated without water in pump. Water is necessary for lubrication of the pump seal while engine is running. Cap suction side of pump and fill pump with water before starting engine when pump is assembled.

MAINTENANCE

SPARK PLUG. Recommended spark plug for all models except 36S2 and 8S3-1 pumps is Champion HO8A or UJ11G. Models 36S2 and 8S3-1 pumps require a Champion UJ22 spark plug.

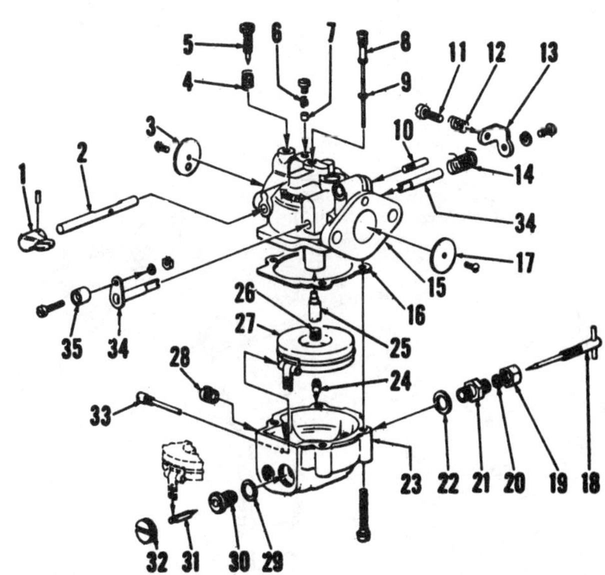

Fig. HL40—Exploded view of typical Tillotson Series MD carburetor as used on some Homelite pump models.

1. Choke lever
2. Choke shaft
3. Choke plate
4. Spring
5. Idle fuel screw
6. Spring
7. Choke friction pin
8. Bypass tube
9. Gasket
10. Throttle stop in
11. Idle speed screw
12. Spring
13. Throttle stop lever
14. Throttle return spring
15. Carburetor body
16. Gasket
17. Throttle plate
18. Main fuel needle
19. Packing nut
20. Packing
21. Packing gland
22. Gasket
23. Fuel bowl
24. Fuel bowl plug
25. Main nozzle
26. Main nozzle plug
27. Float
28. Bowl drain plug
29. Gasket
30. Inlet valve seat
31. Inlet valve needle
32. Plug
33. Float pin
34. Throttle shaft
35. Roller

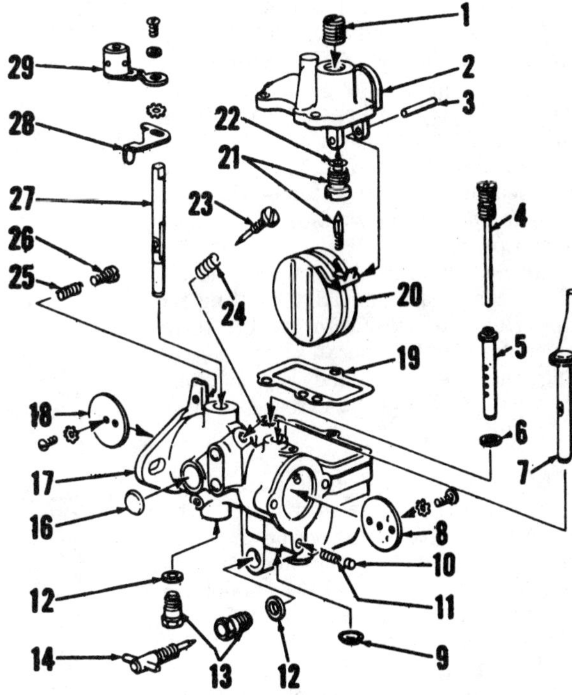

Fig. HL41—Exploded view of typical Series MT Tillotson float type carburetor used on some Homelite pump models.

1. Pipe plug
2. Float bowl cover
3. Float pin
4. Bypass tube
5. Main nozzle
6. Gasket
7. Choke shaft
8. Choke plate
9. Expansion plug
10. Choke friction pin
11. Spring
12. Gasket
13. Packing nut
14. Main fuel needle
16. Expansion plug
17. Carburetor body
18. Throttle plate
19. Gasket
20. Float
21. Inlet valve assy.
22. Gasket
23. Idle fuel needle
24. Spring
25. Spring
26. Idle stop screw
27. Throttle shaft
28. Throttle stop lever
29. Throttle lever

Champion J6 or J6J spark plugs can be substitued for HO8A or UJ11G but will be more susceptible to fouling and electrode erosion. Electrode gap for all models should be 0.025 inch (0.635 mm).

CARBURETORS. Tillotson Series MD and MT float type carburetors and Homelite carburetors have been used. Refer to appropriate exploded view in Fig. HL40 or HL41, or to the cross-sectional view in Fig. HL44.

On some models, flow from fuel tank to carburetor is by gravity with tank mounted above carburetor. With Tillotson carburetors and fuel tank mounted below carburetor, a diaphragm type fuel pump is used as shown in Fig. HL42. On models equipped with Homelite carburetor, fuel tank is pressurized.

On all carburetors, clockwise rotation of fuel mixture adjustment needles leans the mixture.

NOTE: Later models with MT carburetor are not equipped with a throttle plate or shaft.

Normal idle needle (5 – Fig. HL40) setting for Tillotson MD carburetors is 1/4 turn open. Main (high speed) needle (18) setting is 1 1/4 to 1 1/2 turns open. Float setting with fuel bowl assembly held in upside down position should be 1/32 inch (0.8 mm) from the lowest point of float at free end to rim of fuel bowl.

Normal idle needle (23 – Fig. HL41) setting for Tillotson MT carburetors is 3/4 turn open. Main (high speed) needle (14) setting is 1 turn open. Float setting, with the float bowl cover assembly (2) inverted, should be 1-13/32 inch (35.7 mm) as measured from the top of the float (20) to the gasket surface of the float bowl cover.

Normal mixture adjustment needle (32 – Figs. HL43 and HL44) setting for the Homelite carburetor is 1 1/2 turns open. Carburetor operates by pressurizing the fuel tank and loss of tank pressure will prevent carburetor from operating properly.

GOVERNOR. All models are equipped with rotary type inlet valves. The governor is a centrifugal type and is an integral part of the rotary inlet valve as indicated in Fig. HL45. On models so equipped, Formica wear plate (15 – Fig. HL50) is pinned to the engine crankcase with opening in wear plate in register with intake opening in crankcase. On all models, the inlet opening in the valve plate (IV – Fig. HL45) is controlled by the combination shutter and governor weight (GW) which when the engine is stationary, is held in the open position by spring (GS). When engine speed exceeds the designed limit, centrifugal force acting on the weight overcomes the opposing spring pressure and partially covers the manifold opening and thus throttles the engine. As the throttled engine slows down, centrifugal force diminishes and the spring again uncovers the opening to restore engine speed.

Governor Adjustment. On later production engines equipped with adjustable governor (Fig. HL45) or on earlier engines in which adjustable governor has been installed, engine governed speed may be adjusted as follows:

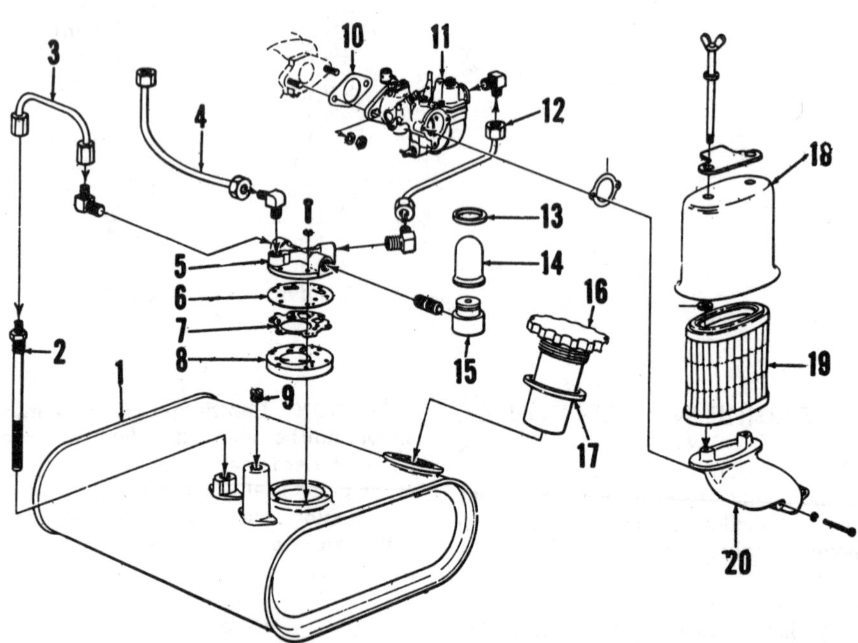

Fig. HL42—Exploded view of fuel system on models using Tillotson MT float type carburetor. Exploded view of carburetor is shown in Fig. HL41. Actuating primer bulb (14) will pump fuel into carburetor float bowl.

1. Fuel tank
2. Fuel filter
3. Pump suction line
4. Pump pulse line
5. Pump body
6. Pump diaphragm
7. Gasket
8. Diaphragm cover
9. Pipe plug
10. Gasket
11. Carburetor assy.
12. Fuel ine
13. Ring
14. Priming bulb
15. Adapter
16. Filler cap
17. Gasket
18. Air cleaner cover
19. Element
20. Adapter

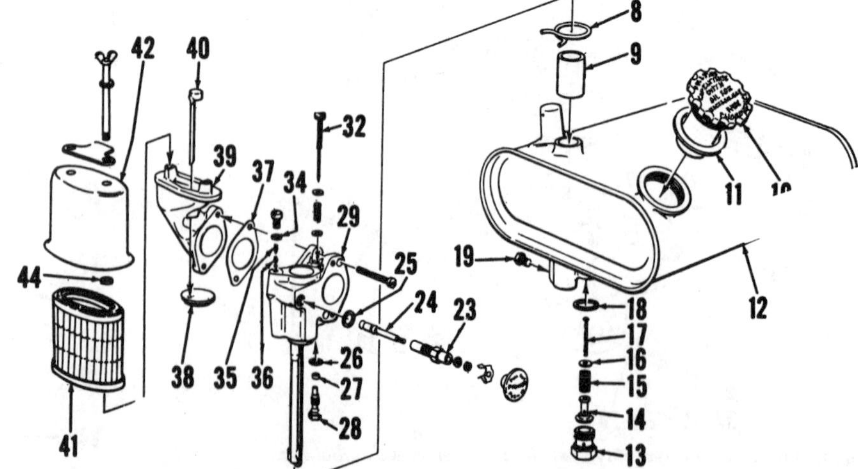

Fig. HL43—Exploded view of fuel system on models equipped with Homelite carburetor. Refer to Fig. HL44 for cross-sectional views of carburetor assembly. Fuel tank is pressurized to force fuel upward into carburetor.

8. Hose clamps
9. Adapter hose
10. Fuel cap
11. Gasket
12. Fuel tank
13. Filter plug
14. Filter adapter
15. Screen
16. Washer
17. Screw
18. Gasket
19. Drain plug
23. Plunger tube
24. Primer plunger
25. Gasket
26. Gasket
27. Sleeve
28. Jet
29. Carburetor body
32. Fuel needle
34. Gasket
35. Spring
36. Priming valve
37. Gasket
38. Carburetor cover
39. Adapter
40. Vent tube
41. Filter element
42. Air filter cover
44. Washer

Illustrations Courtesy Homelite Div. of Textron

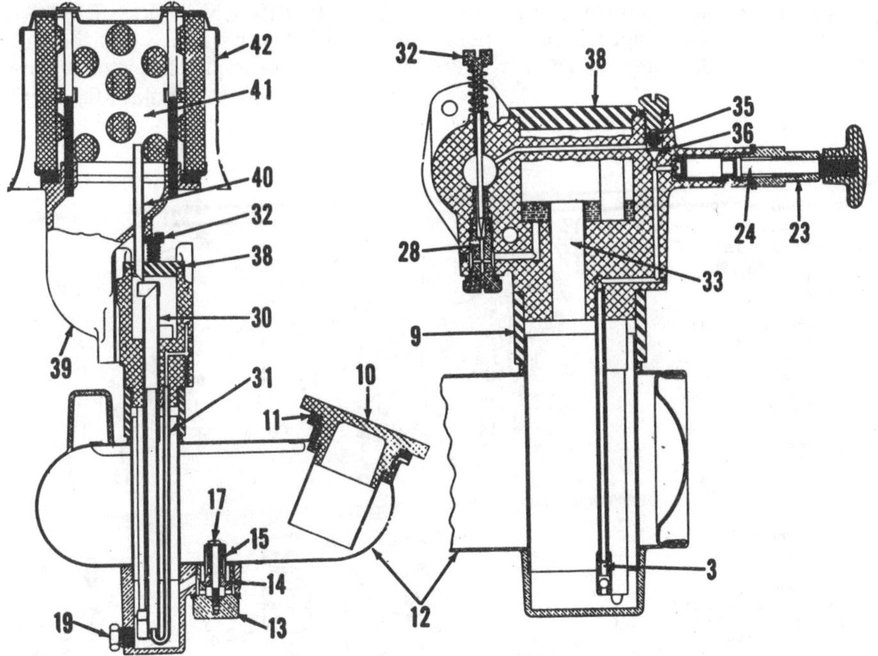

Fig. HL44 – Cross-sectional view of Homelite carburetor. Refer to Fig. HL43 for additional legend and exploded view of complete fuel system.

3. Priming tube and foot valve
30. Feed tube
31. Pressure tube
33. Overflow tube

First, bring engine to normal operating temperature and adjust carburetor for highest speed and best performance obtainable. On model 35 engine, remove drain valve (DP – Fig. HL46) on muffler side of front half of crankcase. On Models 8, 9 and 9-A, disconnect pulse line (L – Fig. HL47) (above drain valve) and unscrew the 90° elbow (E); or, on Model 20 engine, remove slotted head brass plug (P – Fig. HL48) in front half of crankcase.

Turn adjusting screw (AS – Fig. HL45) clockwise to increase no-load speed, or counterclockwise to decrease no-load speed as required. One turn of the adjusting screw will change engine no-load governed speed approximately 100 rpm.

Recommended no-load engine governed rpm for the various pump applications is as follows:

Pump Model	No-Load Rpm
20S1 ½-1A	3900-4000
20DP-3	2850-2950
35S2-1	2850-2950
8S3-1, 9S3-1	2850-2950
8S3-1P, 9S3-1P	4200-4300*
36S2	2650-2750
All other pumps	3800-3900

*3850 rpm at 60 psi pump pressure

Fig. HL47 – On engine Models 8 and 9, governor adjusting screw (see Fig. HL45) is accessible after removing pulse line (L) and elbow (E).

IGNITION AND TIMING. Breaker point gap on all models should be 0.020 inch (0.51 mm). Gap should be measured when breaker arm rub block is ⅛ inch (3.175 mm) past breaking edge of cam.

Ignition timing is fixed and nonadjustable. The cam must be installed with arrow facing out. Refer to Fig. HL49 and HL49A for typical magneto assembly.

Coil primary resistance should be 0.95 ohms and coil secondary resistance should be 5500-6000 ohms. Condenser capacity is 0.34 mfds.

Later model 9 engines are equipped with a starter plate (1 – Fig. HL49) which has countersunk screw holes and is retained by flathead screws in compliance with OSHA regulations. Models 8, 35, 36 and early Model 9 engines may be fitted with Kit A-51497 which contains the starter plate and screws needed to comply with OSHA regulations.

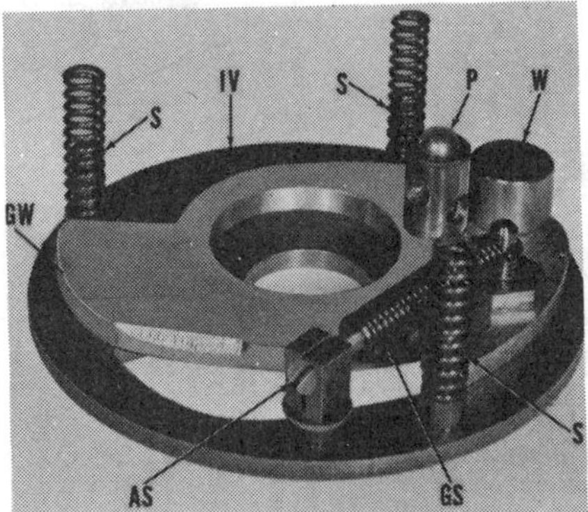

Fig. HL45 – View of late production rotary inlet valve and governor assembly. Early production units did not have adjusting screw (AS) and engine governed speed was nonadjustable.

AS. Adjusting screw
GS. Governor spring
GW. Governor weight & plate assy.
IV. Intake valve plate
P. Pivot post
S. Springs
W. Weight

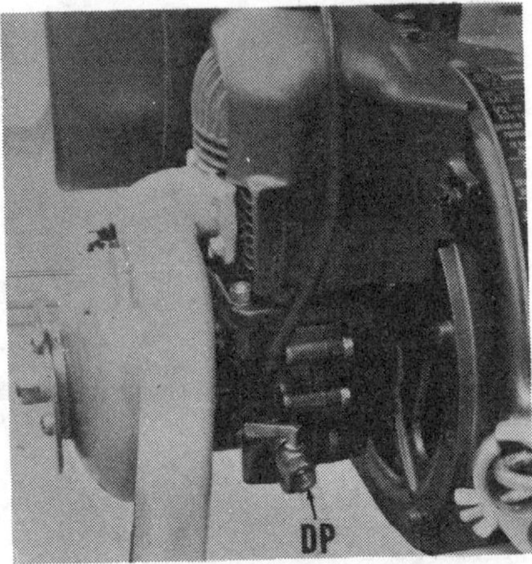

Fig. HL46 – Governor adjusting screw (see Fig. HL45) is accessible on Model 35 engines after removing drain petcock (DP).

Fig. HL48 – On engine Model 20, governor adjusting screw (see Fig. HL45) is accessible after removing slotted head plug (P).

LUBRICATION. Engine on all models is lubricated by mixing oil with regular gasoline. If Homelite Premium SAE 40 oil is used, fuel:oil ratio should be 32:1. Fuel:oil ratio should be 16:1 if Homelite 2-Cycle SAE 30 oil or other SAE 30 oil designed for air-cooled two-stroke engines is used.

NOTE: A fuel:oil ratio of 32:1 should be used on diaphragm pump models regardless of whether Homelite Premium SAE 40 or Homelite 2-Cycle SAE 30 oil is used.

CARBON. Carbon deposits should be cleaned from the exhaust ports and muffler at regular intervals. When scraping carbon be careful not to damage the finely chamfered edges of the exhaust ports.

REPAIRS

CONNECTING ROD. On Model 20 engines, the connecting rod lower bearing is a needle roller type and rod cap is detachable from rod. The rod and piston assembly can be removed from engine after removing the cylinder assembly. Be careful to avoid loss of the 31 individual needle rollers in the crankcase. Crankpin diameter is 0.6937-0.6940 inch 17.620-17.628 mm). Renew connecting rod if surfaces of rod and cap, which form the outer race for the needle rollers, are rough, scored or worn. If any one needle roller is worn or if rollers can be separated more than the width of one needle, reject the bearing. This applies also to the needle roller caged bearing assembly or assemblies mounted in the upper end of the connecting rod. Homelite recommends renewal of the crankpin (lower) connecting rod bearing at each overhaul.

Inspect crankpin and if it is scored or is out-of-round or tapered more than 0.001 inch (0.0254 mm), install a new crankshaft.

Refer to PISTON RINGS, PISTONS AND PINS section for correct reassembly of rod to piston. When reasembling rod to crankshaft, always

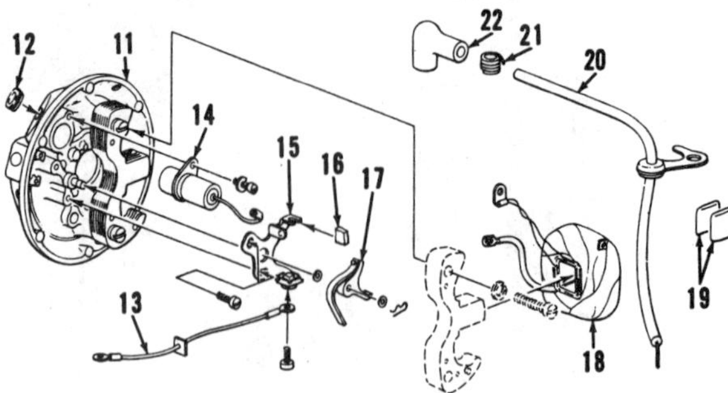

Fig. HL49A—Exploded view of typical magneto assembly (see 10—Fig. HL49).

11. Back plate
12. Grommet
13. Ground (switch) lead
14. Condenser
15. Breaker point base
16. Cam wiper felt
17. Breaker point arm
18. Ignition coil
19. Coil wedges
20. Spark plug wire
21. Terminal
22. Spark plug boot

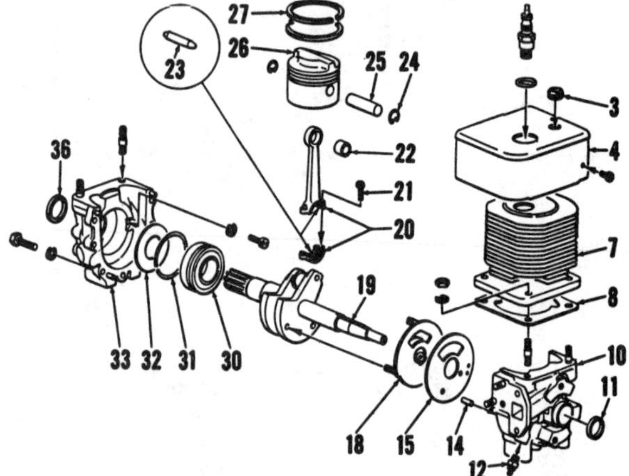

3. Grommet
4. Cylinder shield
7. Cylinder
8. Gasket
10. Crankcase front half
11. Crankshaft seal
12. Idle line connector
14. Dowel pin
15. Intake valve plate
18. Intake valve & governor assy.
19. Crankshaft
20. Connecting rod assy.
21. Allen screws
22. Bearing
23. Needle rollers (31)
24. Snap rings
25. Piston pin
26. Piston
27. Piston rings
30. Main bearing
31. Snap ring
32. Gasket
33. Rear crankcase half
36. Crankshaft seal

Fig. HL50—Exploded view of engine used on Model 20DP3. Other model 20 engines are similar in construction.

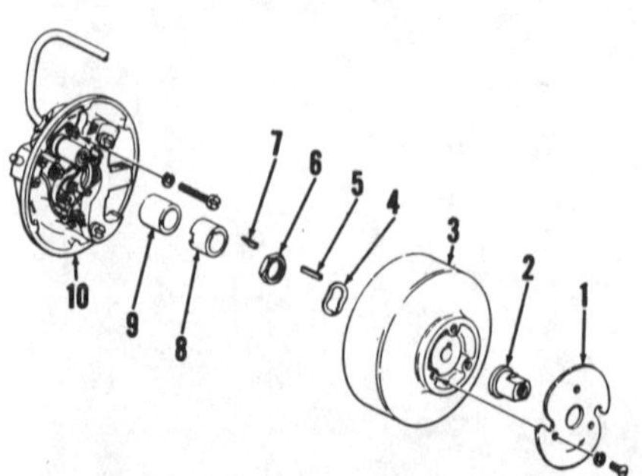

Fig. HL49—To remove magneto rotor (3), unscrew retaining nut (2) against plate (1) which acts as puller. Some models do not have magneto cam retaining nut (6). Notch in inner end of cam (8) must fit over pin in spacer (9).

1. Starter plate
2. Rotor nut
3. Magneto rotor (flywheel)
4. Wave washer
5. Rotor key
6. Breaker cam nut
7. Breaker cam key
8. Breaker cam
9. Cam spacer
10. Magneto assy.

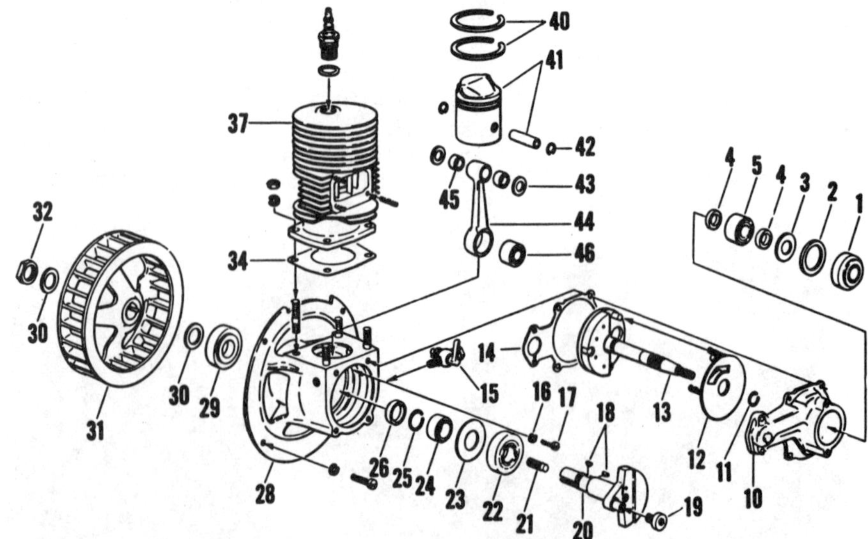

Fig. HL51—Exploded view of engine used on Model 36S2 pump. Engine used on 23S2 pump is of similar construction.

1. Bearing
2. Seal retaining ring
3. Seal
4. Spacers
5. Bearing
10. Timer bracket
11. Snap ring
12. Intake valve & governor assy.
13. Intake valve shaft
15. Drain petcock
16. Lockwasher
17. Main bearing retaining screw
18. Woodruff keys
19. Crankpin screw
20. Crankshaft
21. Pump drive stud
22. Main bearing
23. Gasket
24. Spacer
25. Snap ring
26. Spacer
28. Crankcase
29. Fan bearing
30. Washer
31. Fan
32. Fan nut
34. Gasket
37. Cylinder
40. Piston rings
41. Piston & pin
42. Snap rings
43. Washers
44. Connecting rod
45. Needle bearings
46. Crankpin bearing

Illustrations Courtesy Homelite Div. of Textron

renew the Allen type retaining cap screws and align mating marks on rod and cap.

On Models 36S2, 23S2, some 8S3-1, 8S3-1P and 8S3-1R and all 9S3-1, 9S3-1P and 9S3-1R (refer to 42–Fig. HL52), the connecting rod lower ball bearing is of the double track ball bearing type. To remove the connecting rod, it is necessary to remove the cylinder, timer bracket and the connecting rod retaining Allen head screw (43), then with the use of special pullers, remove the connecting rod and bearings. Renew bearing if it feels rough when rotated or is worn.

On Models 24S3, 24S3-1P and some 8S3-1, 8S3-1P and 8S3-1R, the lower end of the connecting rod is equipped with 25 needle rollers (52–Fig. HL52). As the lower end of the rod does not have a removable rod cap, removal of the cylinder, timer bracket and Allen head connecting rod bearing retainer is necessary before removing the connecting rod and piston assembly. Install large washer (54) with tapered side of hole toward pump and small washer (51) with tapered side toward inlet valve. Race (53) should be heated to 180°F (266°C) to aid in assembling in crankpin.

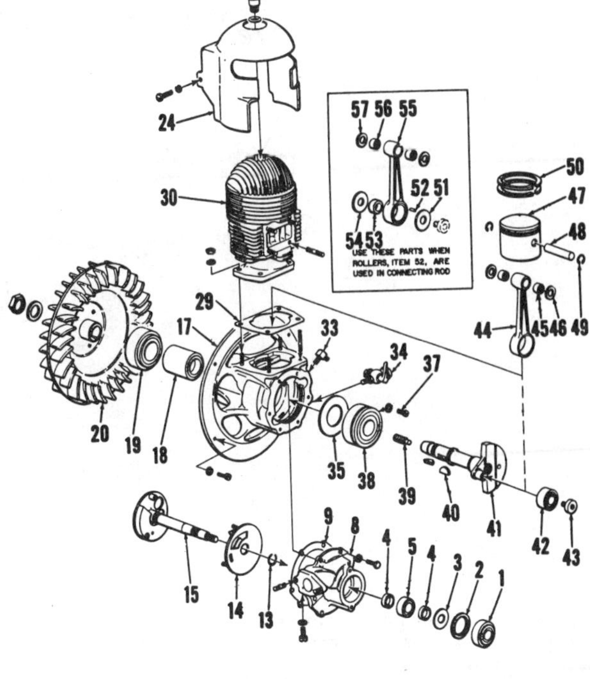

Fig. HL52—Exploded view of engine used on 8S3 series pumps. Some engines are equipped with needle roller crankpin bearings (see inset) while others have a double-row ball bearing (42) in crankpin end of connecting rod.

1. Bearing
2. Retaining ring
3. Seal
4. Seal
5. Ball bearing
8. Timer bracket
9. Gasket
13. Snap ring
14. Intake valve & governor assy.
15. Intake valve shaft
17. Crankcase
18. Spacer
19. Fan bearing
20. Fan
24. Cylinder shield
30. Cylinder
33. Pulse line connection
34. Drain petcock
35. Gasket
37. Retaining screw
38. Main bearing
39. Pump drive stud
40. Woodruff keys
41. Crankshaft
42. Ball bearing
43. Crankpin screw
44. Connecting rod
45. Needle bearings
46. Washers
47. Piston
48. Piston pin
49. Snap rings
50. Piston rings
51. Washer
52. Needle rollers (25)
53. Inner race
54. Washer
55. Connecting rod
56. Needle bearings
57. Washers

PISTON RINGS, PISTONS AND PINS. On all models, the piston assembly is accessible after removing the cylinder assembly from crankcase. Always support the piston when removing or installing the piston pin. All models are equipped with an aluminum alloy piston.

Piston ring end gap should be 0.105-0.115 inch (2.67-2.92 mm) on piston with ring locating pin and 0.008-0.020 inch (0.20-0.50 mm) on models with unpinned rings. Side clearance of piston ring in groove should be 0.0025-0.004 inch (0.0635-0.1016 mm). Clearance between piston skirt and cylinder bore, measured at right angle to piston pin, should be 0.002-0.005 inch (0.05-0.13 mm) on models with 2¼ inch (57.2 mm) bore and 0.004-0.007 inch (0.06-0.18 mm) on all other models.

On all models, reject the pin and the piston if there is any visible up and down play of pin in the piston bosses. Neither the piston nor the pin are available separately.

Install new needle roller bearing assemblies to upper end of connecting rod if any of the rollers have slight flat spots or are pitted. Do likewise if rollers can be separated more than the thickness of one roller.

Inspect piston for cracks and for holes in dome of same and reject if any are found. Slight scoring of piston walls is permissible, but if rough surfaces are accompanied by a deposit of aluminum on cylinder walls, reject the piston. Refer to CYLINDER section for methods of removing such deposits.

If piston ring locating pins in piston grooves are worn to half their normal thickness, reject the piston.

When piston and ring unit is assembled to connecting rod, the side of the piston which has the piston ring locating pins should be on the intake (away from exhaust side) side of the cylinder. Always use new piston pin retaining snap rings when reassembling piston to connecting rod. All wearing parts of the engine are supplied as replacements in standard size only.

CYLINDER. Cylinder bore is chrome plated. Inspect chrome bore for excessive wear and damage to chrome surface of bore. A new cylinder must be installed if chrome is scored, cracked or the base metal underneath exposed.

CRANKSHAFT, BEARINGS AND SEALS. If shaft has damaged threads or enlarged keyways or if run out exceeds 0.003 inch (0.08 mm), reject the shaft. Journals for roller type bearings must be free of pits, galling or heavy score marks. If they are out-of-round or worn than 0.001 inch (0.03 mm) reject the shaft.

If any needle roller shows wear or any visible flat spot, or if rollers can be separated more than the width (or diameter) of one roller, reject all of the rollers. If annular ball bearing feels rough when rotated, or has perceptible wear, reject the bearing.

Suitable pullers, pushers and mandrels (available from Homelite) should be used in removing and installing main and connecting rod bearings.

Crankshaft seals must be maintained in good condition because crankcase compression leakage through seals causes a loss of power.

Centrifugal pumps must be disassembled to remove crankshaft from crankcase. Refer to CENTRIFUGAL PUMPS section.

CRANKCASE. Be sure all passages through the crankcase are clean. This is especially true of the idle passage line (in the front half crankcase) which enters the crankcase via the intake valve register. This passage is sometimes restricted by carbon deposits which can be cleaned with a piece of wire. The same holds true of the passage for the fuel tank pressure line (on models with pressurized fuel tank) or the actuator line on pump equipped engines. If main bearing bores are worn, indicating that the bearing has been turning in the crankcase, the bearing and/or the crankcase should be rejected. On models so constructed, the mating surfaces of two-piece crankcase (Fig. HL50) must be free of all nicks and burrs as neither sealing compound nor gaskets are used at this joint. Always use new bearing seals when reassembling engine.

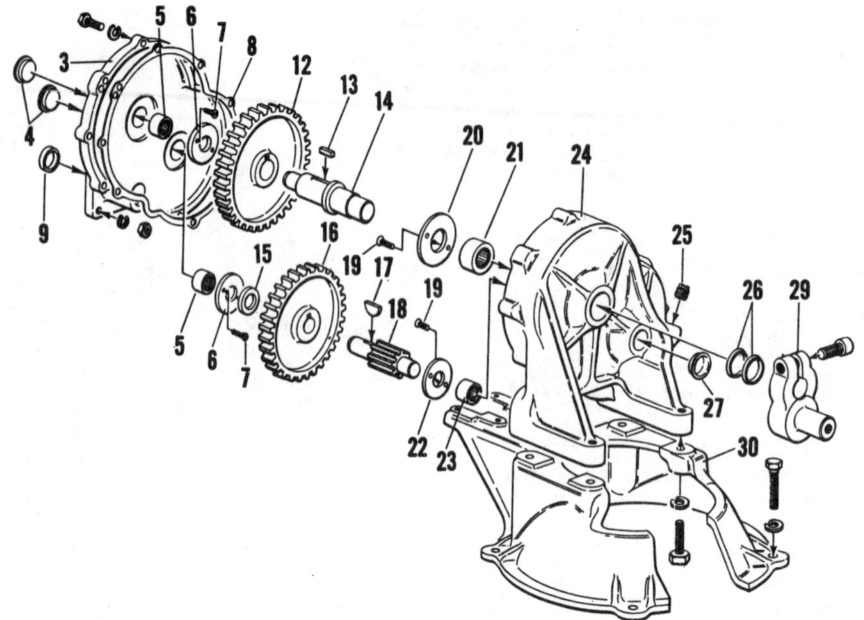

Fig. HL53 — Exploded view of drive gear case for diaphragm type pump. Refer to Fig. HL54 for exploded view of the pump assembly. Pinion machined on rear end of engine crankshaft engages the intermediate gear (16). Pinion on shaft (18) drives crankgear (12). Guard A-51135 is available to cover pump rod and crank.

3. Gear case cover
4. Bearing caps
5. Needle bearings
6. Thrust washers
7. Retainer screws
8. Gasket
9. Crankshaft seal
12. Pump crank gear
13. Key
14. Pump crankshaft
15. Spacer
16. Intermediate gear
17. Woodruff key
18. Intermediate shaft
19. Retainer screws
20. Thrust washer
21. Needle bearing
22. Thrust washer
23. Needle bearing
24. Gear case
25. Filler plug
26. Seals
27. Bearing cap
29. Pump crank
30. Pump upper body

ROTARY VALVE. The combination rotary type inlet valve and governor (Fig. HL45) should be rejected if any of the following conditions are encountered during inspection:

The sealing faces of valve are scored or worn enough to produce a ridge, if spring posts are loose or extend to valve seating surface or if the governor pivot point has started to wear through the surface of the valve.

Slight scoring of valve may be corrected by lapping on a lapping plate using very fine abrasive. Lapping motion should be in the pattern of a figure eight to obtain best results. Slight scoring of the Formica wear plate is permissible. Homelite recommends soaking a new wear plate in oil for 24 hours prior to installation.

DIAPHRAGM PUMP

The diaphragm pump is lubricated by filling gearcase to level of plug (25 – Fig. HL53) with SAE 90 gear lubricant and by greasing pump rod upper bearing once a month with pressure gun through fitting in upper end of rod.

When installing new diaphragm or assembling upper pump body to lower body, the diaphragm must be centered and in fully down position before tightening upper body to lower body bolts as shown in Fig. HL55.

To remove pump from engine, drain gear lubricant and separate gearcase

Fig. HL54 — Exploded view of diaphragm type pump assembly. Pump upper body is (30 — Fig. HL53). Refer to Fig. HL55 for pump assembly information.

1. Plug
2. Stand pipe
3. Suction fitting
4. Valve weight
5. Valve
6. Valve plate
7. Thrust washer
8. Needle bearings
9. Pump rod
10. Thrust washer
11. Washer
12. Cap screw
13. Pump diaphragm
14. Diaphragm cap
15. Valve
16. Valve plate
17. Discharge fitting
18. Valve weight
19. Pump lower body

16. Valve plate
17. Discharge fitting
18. Valve weight
19. Pump lower body

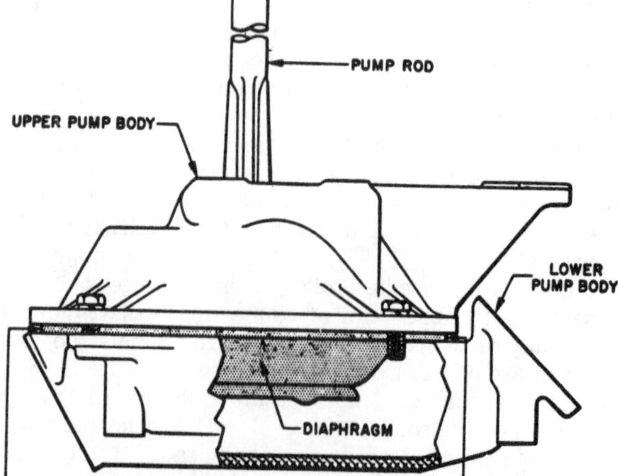

Fig. HL55 — View showing proper installation of pump diaphragm. Refer to Figs. HL53 and HL54 for exploded views of pump gear box and pump assembly.

Illustrations Courtesy Homelite Div. of Textron

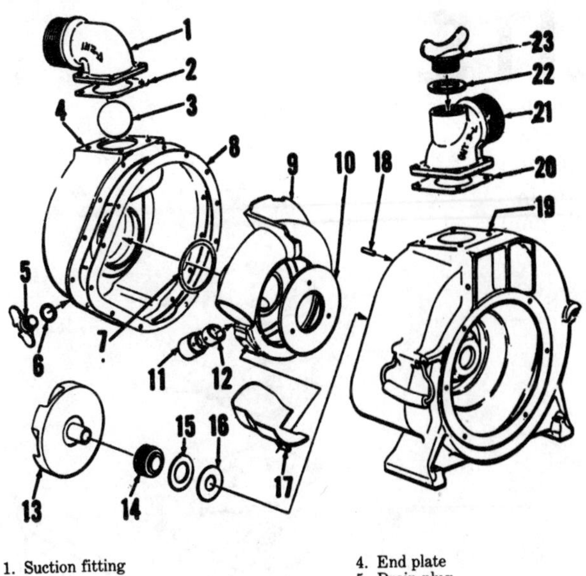

Fig. HL56 – Exploded view of early type centrifugal type pump assembly. Pump impeller (13) is mounted on engine crankshaft. Pump seals (14, 15 and 16) are the later ceramic type. Earlier pumps with rubber seal may be converted by installing new impeller (13). Refer to Figs. HL31 and HL32 in preceding section for views of later type pumps.

1. Suction fitting
2. Gasket
3. Check ball
4. End plate
5. Drain plug
6. Gasket
7. Gasket
8. Gasket
9. Impeller housing
10. Wear plate
11. Priming tube
12. Auxiliary priming tube
13. Impeller
14. Seal
15. Ceramic ring
16. Washer
17. Baffle
18. Locating pin
19. Pump body
20. Gasket
21. Discharge fitting
22. Gasket
23. Priming filler plug

(24–Fig. HL53) from gearcase cover (3). Gear teeth are machined on end of engine crankshaft to drive intermediate gear (16). Unbolt and remove cover (3) from engine crankcase. When reassembling, use new gasket (8).

CENTRIFUGAL PUMP

Refer to exploded view of early type centrifugal pump in Fig. HL56. For later type pumps, refer to Fig. HL31 and HL32 in preceding Homelite section on Model 251 engine. The pump must be disassembled and the impeller unscrewed from stud (21–Fig. HL51 or 39–Fig. HL52) in end of engine crankshaft so the crankcase can be unbolted from pump body and crankshaft be removed from crankcase. Shims are used between impeller and end of crankshaft to maintain minimum clearance between impeller and wear plate (10–Fig. HL56). Two shims are normally used. Remove shims if impeller rubs against wear plate when reassembling engine.

If necessary to renew pump seals, apply a thin coat of shellac to plain side of ceramic ring (15) and press new washer (16) against ring. Apply a thin coat of shellac on washer and press the washer and ceramic ring, washer side in, against and into recess of pump body. Slide seal (14) over hub of impeller with hy-car side of seal toward impeller. Thread impeller onto stud that is screwed into end of crankshaft.

Place the assembled impeller housing (9, 10 and 12) over impeller and in proper engagement with pin (18), press against housing and turn engine. If impeller rubs, remove impeller and discard shim or shims as required from between impeller and crankshaft so rubbing condition is eliminated. Install assembled impeller housing, then install pump end plate (4) with new gasket (8) and one or two new gaskets (7) as required so end plate holds impeller housing securely against pump body.

NOTE: End plate should rock slightly before being tightened.

TRASH PUMP

Trash pump used on the models in this section is the same as the pump shown in Fig. HL32. Refer to TRASH PUMP paragraphs in Model 251 section for overhaul of trash pump.

HOMELITE

Model	Bore	Stroke	Displacement
XL-12	1¾ in. (44 mm)	1⅜ in. (35 mm)	3.3 cu. in. (54 cc)
Super XL	1-13/16 in. (46 mm)	1⅜ in. (35 mm)	3.6 cu. in. (58 cc)

ENGINE INFORMATION

This section covers service of Homelite Model XL-12 and Super XL engines that are used in the following Homelite tools and equipment.

XL Brushcutter
XL100 Circular Saw
XL120 Circular Saw
XLS1½-1 Centrifugal Pump
XLS1½-4 Centrifugal Pump
XLS1½-4A Centrifugal Pump
DM20 Multi-Purpose Saw

MAINTENANCE

SPARK PLUG. A Champion TJ8J spark plug or equivalent is used in XL-100 Saw and XL Brushcutter application. For extended plug life, a Champion UTJ11P platinum tip spark plug may be used. A Champion CJ6 or equivalent is recommended for all other applications except the XL100A saw which should use a Champion CJ8 spark plug or equivalent. Spark plug gap for pump engines should be 0.035 inch (0.9 mm) and 0.025 inch (0.64 mm) for all other applications.

CARBURETOR. Refer to Fig. HL60 for exploded view of Tillotson HS diaphragm type carburetor with integral fuel pump used on XL12 and Super XL engines. Carburetor is accessible after removing air box cover (Fig. HL61).

NOTE: If early type cover gasket becomes damaged when air box cover is removed, install new gasket as follows: Carefully remove old gasket from air box and be sure that surface is free of all dirt, oil, etc. Apply Homelite 22788, or equivalent gasket cement to new gasket and carefully place gasket, adhesive side down, on lip around air box chamber. On later engines, gasket is bonded to filter element. Install new filter element if either gasket or filter is damaged.

When disassembling carburetor, slide the diaphragm assembly towards adjustment needle side of carburetor body to disengage diaphragm from fuel inlet control lever. To remove welch plugs, carefully drill through large plug (27–Fig. HL60) with a ⅛-inch drill or through small plug (24) with a 1/16-inch drill and pry plugs out with a pin in-

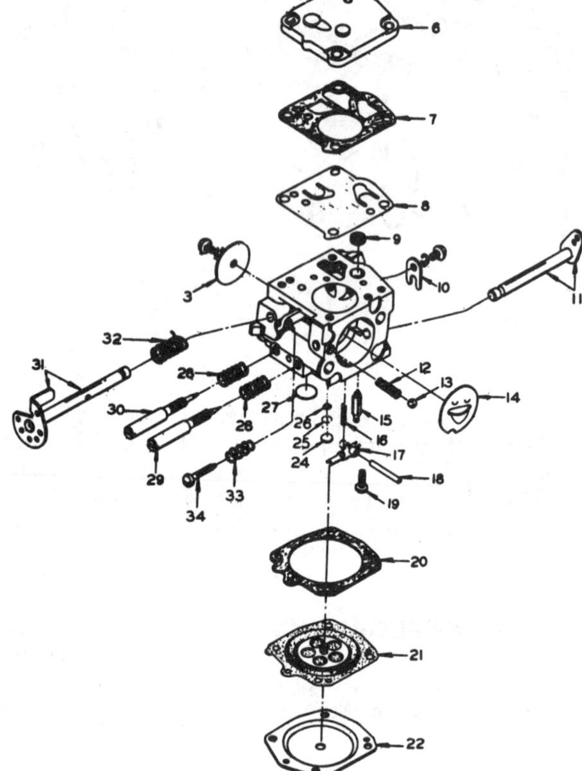

Fig. HL60—Exploded view of typical Tillotson series HS carburetor used on XL-12 and Super XL engines. On some models, idle speed adjustment screw (34) is located in air box casting and is adjustable without removing cover (8–Fig. HL61).

3. Throttle plate
6. Pump cover
7. Gasket
8. Pump diaphragm
9. Inlet screen
10. Throttle shaft clip
11. Choke shaft & lever
12. Detent spring
13. Choke detent ball
14. Choke plate
15. Inlet needle
16. Spring
17. Diaphragm lever
18. Lever pin
19. Pin retaining screw
20. Gasket
21. Diaphragm
22. Diaphragm cover
24. Welch plug
25. Retaining ring
26. Channel screen
27. Welch plug
28. Springs
29. Main fuel needle
30. Idle fuel needle
31. Throttle shaft & lever
32. Throttle spring
33. Idle speed screw spring
34. Idle speed adjustment screw

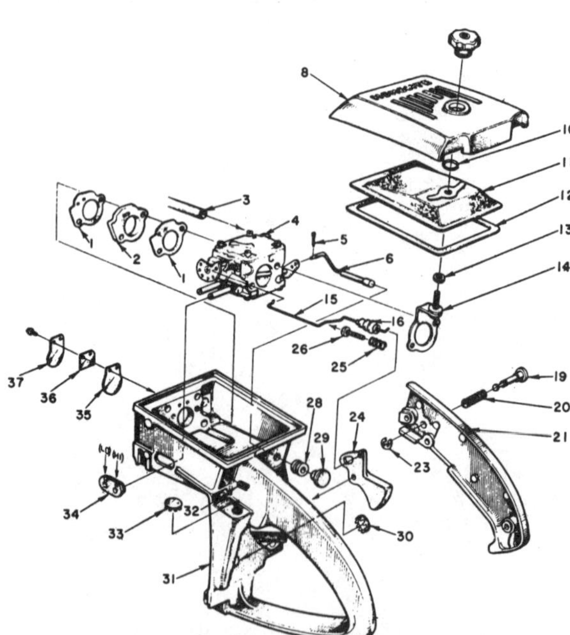

1. Gaskets
2. Spacer
3. Fuel line
4. Carburetor assy.
5. Cotter pin
6. Choke rod
8. Air filter cover
10. Retaining ring
11. Air filter element
12. Gasket
13. Gasket
14. Mounting bracket
15. Throttle rod
16. Throttle rod boot
19. Throttle lock pin
20. Spring
21. Handle cover
23. Retaining ring
24. Throttle trigger
25. Spring
26. Idle speed screw (external)
28. Grommet
29. Choke cutton
30. Plug
31. Air box housing
32. Headless screw
33. Plug
34. Adjusting needle grommet
35. Reed valve
36. Reed back-up
37. Reed stop

Fig. HL61—Exploded view of handle and air box assembly used on early XL100 saw; Model XL120 handle and air box assembly is shown in Fig. HL61A. Late Model XL100 has air vane governor similar to that shown in Fig. HL61A. Refer to Fig. HL62. Refer to Fig. HL62 for differences in brushcutter and pump air box construction.

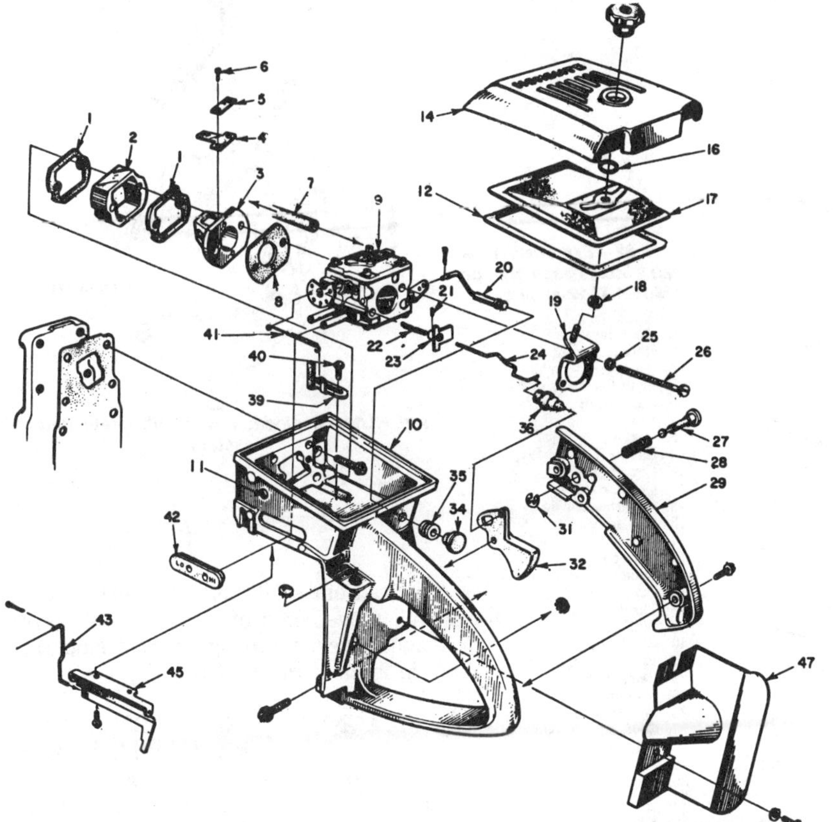

Fig. HL61A — Exploded view of Model XL120 air box and handle. Late Model XL100 is similar except a flat reed intake valve is used as shown in Fig. HL61. Note Super XL engine pyramid reed valve assembly (items 1 through 6). Tension of governor spring is adjusted by loosening screw (40) and moving speed adjusting plate (39). Refer to text.

1. Gaskets
2. Spacer
3. Reed valve seat
4. Valve reeds
5. Reed plates
6. Reed screws
7. Fuel line
8. Gasket
9. Carburetor
10. Handle & air box
11. Plug
12. Cover gasket
14. Air box cover
16. Snap ring
17. Filter element
18. Stud gasket
19. Bracket & stud
20. Choke rod
21. Set screw
22. Throttle rod spring
23. Throttle rod collar
24. Throttle rod
25. Lockwashers
26. Carburetor mounting screws
27. Throttle lock pin
28. Lock spring
29. Handle cover
31. Snap ring
32. Throttle trigger
34. Choke rod button
35. Grommet
36. Choke rod boot
39. Speed adjusting plate
40. Plate retaining screw
41. Governor spring
42. Adjusting needle grommet
43. Governor link
45. Governor air vane
47. Muffler shield

serted through the drilled hole. Caution should be taken that the drill just goes through the welch plug as deeper drilling may seriously damage the carburetor. Note channels screen (26) and screen retaining ring (25) are accessible after removing welch plug (24).

Inlet control lever (17) should be flush with metering chamber floor of carburetor body. If not, bend diaphragm end lever up and down as required so lever is flush.

On brushcutter, turn idle speed stop screw in until it just contacts throttle lever tab, then turn screw in 3/4 turn further. Turn idle and main fuel adjustment needles in gently until they just contact seats, then back each needle out 5/8 turn. With engine warm and running, adjust idle fuel needle so engine runs smoothly, then adjust idle stop screw so engine runs at 2600 rpm, or just below clutch engagement speed. Check engine acceleration and open idle fuel needle slightly if engine will not accelerate properly. Adjust main fuel needle under load so engine will neither slow down or smoke excessively.

On pump engine, set idle stop screw so it does not interfere with full travel of throttle stop lever. Open idle fuel adjustment needle one turn and leave needle at this setting. All adjustment should be made with main needle. Open main fuel adjustment needle one turn for initial adjustment, be sure pump is filled with water and start engine. Pump water until engine is at full operating temperature, then turn main fuel needle in slowly until engine begins to lose speed under load. Correct final adjustment is 1/8 turn open from this point.

On circular saw, turn idle speed stop screw in until it just contacts throttle stop lever plus 3/4 turn additional. Open both the idle fuel needle and the main fuel needle one turn each.

On multi-purpose saw open the idle fuel needle one turn from a lightly seated position. With engine warm and running, adjust idle speed stop screw so engine runs at just below clutch engagement speed, or about 2600 rpm. Check engine acceleration and open idle fuel needle slightly if enigne does not accelerate properly on initial adjustment. High speed mixture is not adjustable. A fuel governor is installed in the carburetor which enrichens the mixture at high speed to prevent overspeeding.

GOVERNOR. The early XL100 Circular Saw engine and Model DM20 Multi-Purpose Saw are not governed. Refer to appropriate following paragraph for information on other units which are equipped with governors.

Circular Saw Governor. Refer to exploded view of air box and handle assembly in Fig. HL61A. To adjust governed speed, loosen screw (40) and move slotted speed adjusting plate (39) to obtain desired maximum speed, then tighten screw. Maximum governed no-load speed should be 5000 rpm.

Brushcutter Governor. The engine used in brushcutter application is equipped with an air-vane type governor. Refer to Fig. HL62. Air vane (27) is connected to throttle shaft through link (26) and is balanced by tension of spring (12). When throttle trigger is fully depressed, collar (16) on remote control cable moves away from throttle shaft lever allowing governor to control engine speed.

To adjust governor using vibrating reed or electronic tachometer, proceed as follows: With engine warm and running and throttle trigger released, adjust position of collar on remote control cable so engine slow idle speed is 2500 rpm, or just below clutch engagement speed. Then when throttle trigger is fully depressed, collar should move away from throttle shaft lever and engine no-load speed should be 6300 rpm. To adjust maximum governed no-load speed, loosen screw (14 – Fig. HL62) and move speed adjusting plate (13) as required to obtain no-load speed of 6300 rpm. When adjusting maximum no-load speed, be sure throttle trigger is fully depressed and that collar (16) clears carburetor throttle shaft lever.

If carburetor or linkage has been removed, be sure governor link is reconnected at hole "A" in carburetor throttle shaft lever as indicated in Fig. HL62. Governor spring (12) is connected to third hole away from hole "A" (two open holes between link and spring). Be sure governor linkage moves smoothly throughout range of travel.

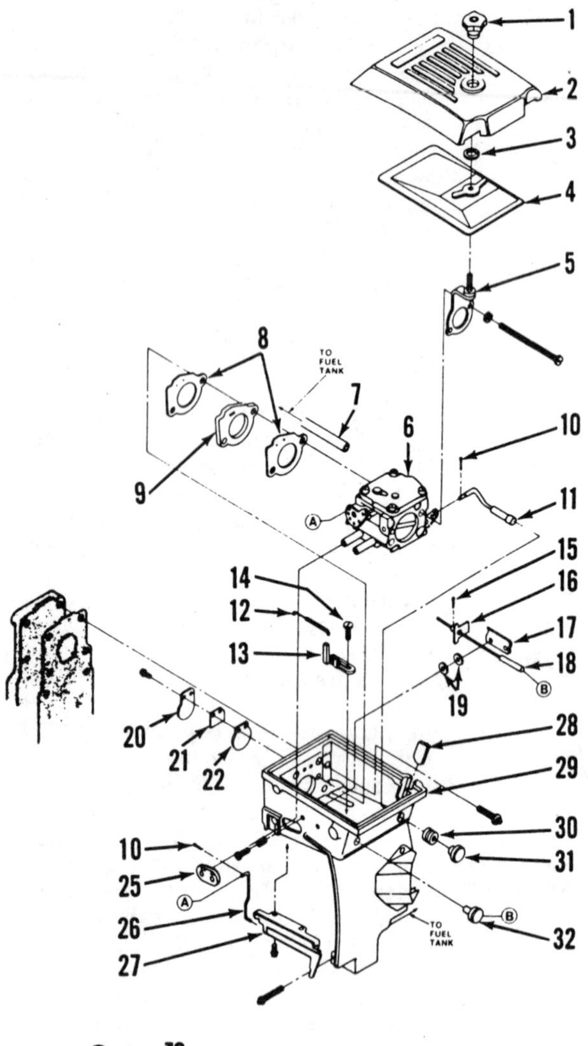

Fig. HL62—Exploded view of air box assembly on Model XL12 engine.

1. Nut
2. Air filter cover
3. Retaining ring
4. Air filter element
5. Mounting bracket
6. Carburetor
7. Fuel line
8. Gasket
9. Spacer
10. Cotter pin
11. Choke rod
12. Governor spring
13. Adjusting plate
14. Screw
15. Cotter pin
16. Collar
17. Clamp
18. Throttle cable
19. Washers
20. Reed stop
21. Reed backup
22. Reed valve
25. Grommet
26. Governor link
27. Governor air vane
28. Felt plug
29. Air box
30. Grommet
31. Choke button
32. Throttle button

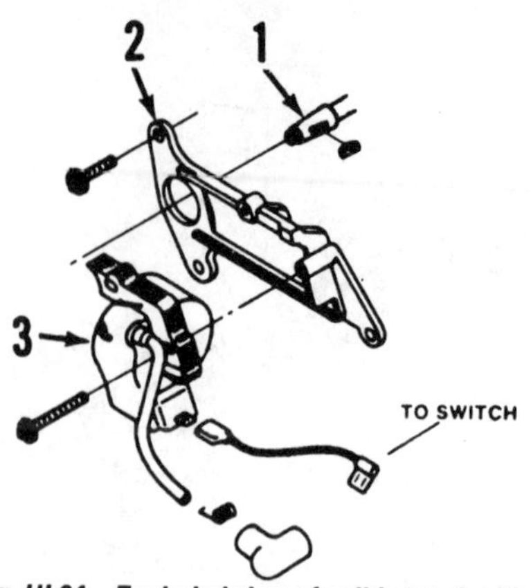

Fig. HL64—Exploded view of solid-state ignition system.

1. Crankshaft
2. Stator plate
3. Ignition module

Fig. HL63—Exploded view of the flywheel type magneto. Connect condenser and low tension leads as indicated by letters "B" and "C".

24. Flywheel
25. Condenser
26. Condenser clamp
28. Breaker box cover
29. Terminal block
30. Breaker point set
31. Washer
32. Pivot post clip
34. Gasket
37. Plate & armature assy.
38. Gasket
40. Felt seal
41. Coil retaining clip
42. Ignition coil assy.
43. High tension wire
44. Connector spring
45. Spark plug both

Pump Governor. On pump applications, engine is equipped with air-vane type governor as shown in Fig. HL62.

With engine running under no load and pump housing filled with water, engine speed should be 6400 to 6600 rpm. If not, loosen screw (14) and move speed adjusting plate (13) as necessary so engine governed speed is 6500 rpm.

If carburetor has been removed or linkage has been disconnected, be sure governor link (26) is reconnected in hole "A" in carburetor throttle shaft lever. Be sure idle speed screw on carburetor is backed out so it will not interfere with full movements of throttle shaft lever and governor linkage moves smoothly throughout range of travel.

MAGNETO AND TIMING. Early engines are equipped with a conventional breaker point, flywheel magneto (Fig. HL63). Breaker point gap should be 0.015 inch (0.38 mm). Armature air gap is not adjustable. Condenser capacitance should test 0.16-0.20 mfd.

CAUTION: Be careful when installing breaker points not to bend tension spring any more than necessary. If spring is bent excessively, spring tension may be reduced causing improper breaker point operation.

Later pump engines and all saw engines are equipped with a solid-state ignition (Fig. HL64). Ignition service is accomplished by replacing ignition components until faulty part is located. Air gap between ignition module (3) and flywheel is adjustable and should be 0.015 inch (0.38 mm). Loosen ignition module mounting screws and adjust module position to set air gap.

LUBRICATION. The engine on all models is lubricated by mixing oil with regular gasoline. Fuel:oil ratio should be 32:1 when Homelite SAE 40 Premium Motor Oil is mixed with fuel. Fuel:oil ratio should be 16:1 if Homelite SAE 30 2-Cycle oil or another good grade of oil designed for air-cooled two-stroke engines is used.

Gearcase of circular saw must be filled to proper level with SAE 90 gear lubricant (Homelite part 55291-C) as follows: Remove saw blade and working through opening in retractable blade guard (49–Fig. HL69), remove drain plug (1). Remove filler plug (9) and with base plate of saw level, pour oil into filler

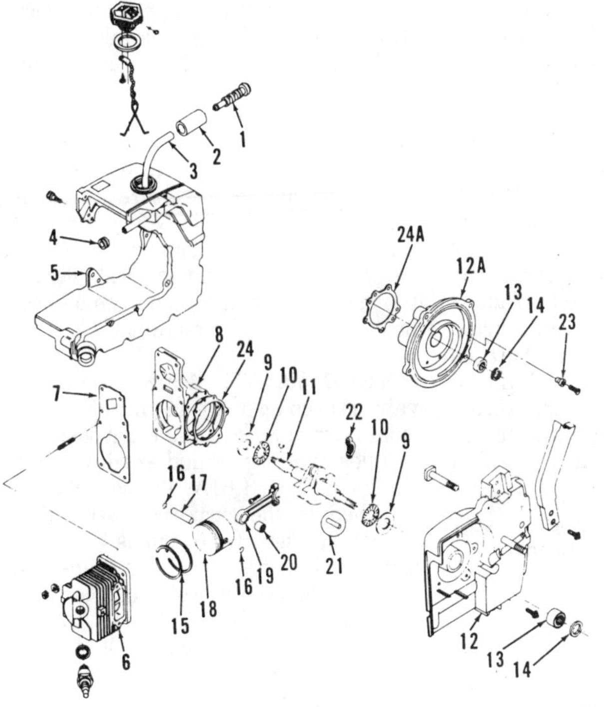

Fig. HL65—Exploded view of engine. Drive case (12) is used on saw engines while pump housing (12A) is used on pump engines.

1. Fuel pickup
2. Fuel filter
3. Fuel line
4. Grommet
5. Fuel tank
6. Cylinder
7. Gasket
8. Crankcase
9. Thrust washer
10. Thrust bearing
11. Crankshaft
12. Drive case
12A. Pump housing
13. Needle bearing
14. Seal
15. Piston rings
16. Retaining ring
17. Piston pin
18. Piston
19. Connecting rod
20. Needle bearing
21. Crankpin rollers (31)
22. Rod cap
23. Sealing washer
24. Gasket
24A. Gasket

plug opening until it starts to run out drain plug. Reinstall plugs and blade. Oil in gearcase should be changed after each 100 hours of operation.

To lubricate brushcutter, proceed as follows: Each 10 hours of operation, squirt a few drops of SAE 30 oil into hole marked "OIL" on underside of upper head casting. Remove drive shaft each 50 hours of operation, clean the shaft and lubricate full length of shaft with a good grade of wheel bearing grease. Lower gear head should be repacked with a good grade of wheel bearing grease after every 50 hours of operation.

NOTE: Do not wash lower head and bearing housing in solvent as this will wash lubricant from the sealed ball bearings, causing premature bearing failure.

Pump is lubricated by water in pump.

CAUTION: Do not start pump engine without filling pump with water.

CARBON. Muffler, manifold and cylinder exhaust ports should be cleaned periodically to prevent loss of power through carbon buildup. Remove muffler and scrape free of carbon. A bent wire can be inserted through hole in housing pump and generator mufflers to clean outer shell. With muffler or manifold removed, turn engine so piston is at top dead center and carefully remove carbon from exhaust ports with a wooden scraper. Be careful not to damaged chamfered edges of exhaust ports or to scratch piston. Do not run engine with muffler removed.

REPAIRS

CONNECTING ROD. Connecting rod and piston assembly can be removed after removing cylinder from crankcase. Refer to Fig. HL65. Be careful to remove all needle rollers when detaching rod from crankpin. Early models have 28 loose needle rollers while later models have 31 needle rollers.

Renew connecting rod if bent or twisted, or if crankpin bearing surface is scored, burned or excessively worn. The caged needle roller piston pin bearing can be renewed by pressing old bearing out and pressing new bearing in with Homelite tool 23756. Press on lettered end of bearing cage only.

The crankpin needle rollers should be renewed as a set whenever engine is disassembled for service.

On early models with 28 needle rollers, stick 14 needle rollers in the rod and remaining 14 rollers in rod cap with light grease or beeswax.

On later models with 31 rollers, stick 16 rollers in rod and 15 rollers in rod cap.

PISTON, PIN AND RINGS. The piston is fitted with two pinned compression rings. Renew piston if scored, cracked or excessively worn, or if ring side clearance in top ring groove exceeds 0.0035 inch (0.089 mm).

Recommended piston ring end gap is 0.070-0.080 inch (1.78-2.03 mm) and maximum allowable ring end gap is 0.085 inch (2.16 mm). Desired ring side clearance in groove is 0.002-0.003 inch (0.05-0.08 mm).

Piston, pin and rings are available in standard size only. Piston and pin are available as a matched set, and are not available separately.

Piston pin has one open and one closed end and may be retained in piston with snap rings or a spirol pin. A wire retaining ring is used on exhaust side of piston on some models and should not be removed.

To remove piston pin on all models, remove the snap ring at intake side of piston. On piston with spirol pin at exhaust side, drive pin from piston and rod with slotted driver (Homelite tool A-23949). On all other models, insert a 3/16-inch (4.76 mm) pin through snap ring at exhaust side and drive piston pin out.

When reassembling, be sure closed end of piston pin is to exhaust side of piston (away from piston ring locating pin). Install Truarc snap ring with sharp edge out.

The cylinder bore is chrome plated. Renew the cylinder if chrome plating is worn away exposing the softer base metal.

CRANKCASE, BEARING HOUSING AND SEALS. CAUTION: Do not lose crankcase screws. New screws of same length must be installed in place of old screws. Refer to parts book if correct screw length is unknown.

The crankshaft is supported in two caged needle roller bearings and crankshaft end play is controlled by a roller bearing and hardened steel thrust washer at each end of the shaft. Refer to Fig. HL65.

The needle roller main bearings and crankshaft seals in crankcase and drive case of pump housing can be renewed using Homelite tool 23757 and 23758. Press bearings and seals from

Fig. HL66—When installing flat reed, reed backup and reed stop, be sure reed is centered between two points indicated by black arrows.

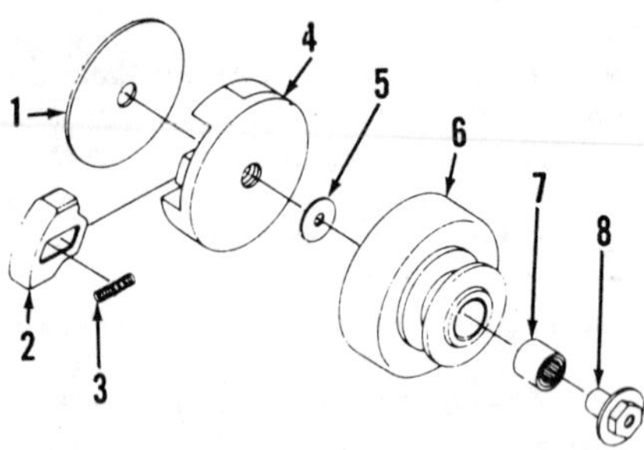

Fig. HL67—Exploded view of clutch used on saw engines.

1. Cover
2. Clutch shoe
3. Spring
4. Plate
5. Thrust washer
6. Clutch drum
7. Needle bearing
8. Nut

crankcase or bearing housing with large stepped end of tool 23757, pressing towards outside of either case.

To install new needle bearings, use the shouldered short end of tool 23757 and press bearings into bores from inner side of either case. Press on lettered end of bearing cage only.

To install new seals, first lubricate the seal and place seal on long end of tool 23758 so lip of seal will be towards needle bearing as it is pressed into place.

To install crankshaft, lubricate thrust bearings (10) and place on shaft as shown. Place a hardened steel thrust washer to the outside of each thrust bearing. Insert crankshaft into crankcase being careful not to damage seal in crankcase. Place a seal protector sleeve (Homelite tool 23759) on crankshaft and gasket on shoulder of drive case or pump housing. Lubricate seal protector sleeve, seal and needle bearing and mate drive case or pump housing to crankshaft and crankcase. Use NEW retaining screws. Clean the screw threads and apply Loctite to threads before installing screws. Be sure the screws are correct length; screw length is critical. Tighten the screws alternately and remove seal protector sleeve from crankshaft.

REED VALVE. A flat reed intake valve is used on pump engines. Early circular saw and late multi-purpose saw engines use a pyramid reed valve as shown in Fig. HL61A. The reed valve is attached to the carburetor air box as shown in Fig. HL66 and is accessible after removing air box from crankcase.

Check the reed seating surface on air box to be sure it is free of nicks, chips or burrs. Renew valve reed if rusted, pitted or cracked, or if it does not seat flatly against its seat.

The reed stop is curved so measurement of reed lift distance is not practical. However, be sure reed is centered over opening in air box and reed stop is aligned with reed.

NOTE: If air box has been removed to service reed valve, inspect gasket between air box and crankcase. If gasket is damaged and cylinder is not being removed for other purposes, it is suggested the exposed part of the old gasket be carefully removed and the new gasket be cut to fit between the air box and crankcase.

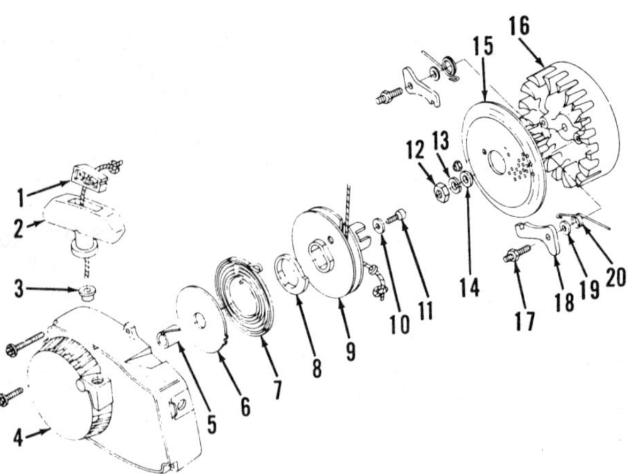

Fig. HL68—Exploded view of rewind starter.

1. Rope retainer
2. Handle
3. Bushing
4. Starter housing
5. Bushing
6. Washer
7. Rewind spring
8. Spring lock
9. Rope pulley
10. Washer
11. Screw
12. Nut
13. Lockwasher
14. Washer
15. Screw
16. Flywheel
17. Stud
18. Pawl
19. Washer
20. Spring

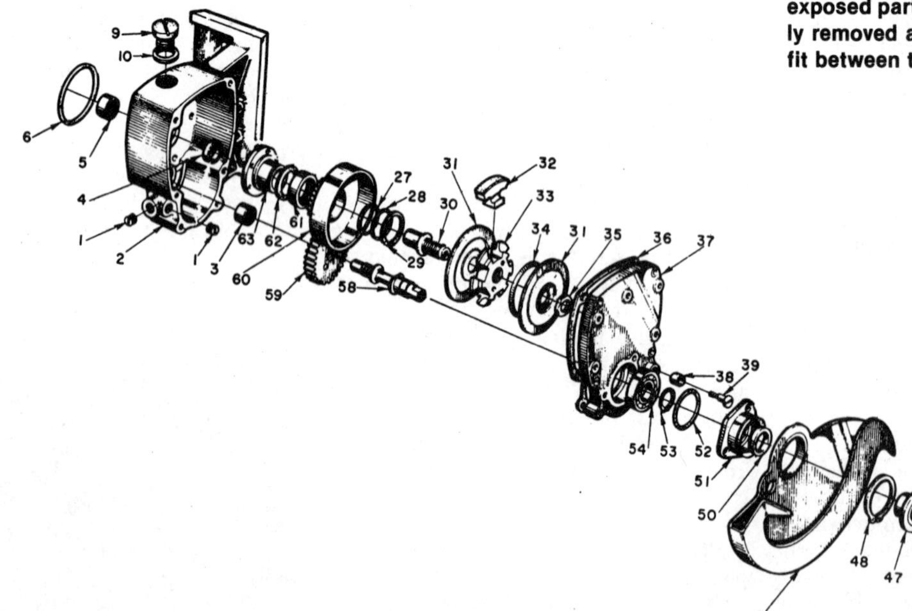

Fig. HL69—Exploded view of circular saw drive case assembly. Drive case (2) carries crankshaft needle roller bearing (5) and crankshaft seal (4). "O" ring (6) seats drive case to engine crankcase.

1. Drain plugs
2. Drive case
3. Needle bearing
4. Crankshaft seal
5. Needle (main) bearing
6. "O" ring
9. Filler plug
10. Gasket
27. Phenolic washer
28. Steel washer
29. Snap ring
30. Extension shaft (clutch adapter)
31. Clutch covers
32. Clutch shoes (3)
33. Clutch spider
34. Clutch springs (2)
35. Nut
36. Cover gasket
37. Drive case cover
38. Guard bumper
39. Bumper screw
44. Blade bolt
45. Blade outer washer
46. Blade
47. Blade inner washer
48. Snap ring
49. Rotating guard
50. Seal
51. Seal & bearing retainer
52. "O" ring
53. Snap ring
54. Ball bearing
58. Blade shaft
59. Driven gear
60. Clutch drum & driven gear
61. Needle bearing
62. Thrust washer
63. Inner race

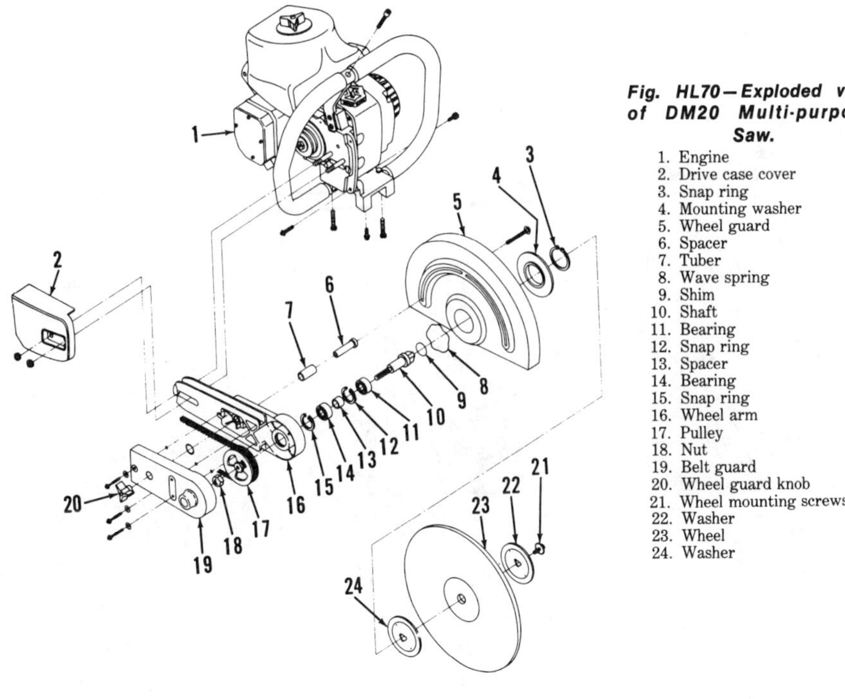

Fig. HL70—Exploded view of DM20 Multi-purpose Saw.

1. Engine
2. Drive case cover
3. Snap ring
4. Mounting washer
5. Wheel guard
6. Spacer
7. Tuber
8. Wave spring
9. Shim
10. Shaft
11. Bearing
12. Snap ring
13. Spacer
14. Bearing
15. Snap ring
16. Wheel arm
17. Pulley
18. Nut
19. Belt guard
20. Wheel guard knob
21. Wheel mounting screws
22. Washer
23. Wheel
24. Washer

CLUTCH. Super XL engines used on Model DM20 saws are equipped with the centrigual clutch shown in Fig. HL67. Clutch bearing (7) should be cleaned, inspected and repacked with grease after every 100 hours of saw operation. Use a good quality lithium base grease.

To remove the clutch assembly, first remove the blade and arm asembly. The clutch retaining nut rod bearing inner race (8–Fig. HL67) unscrews clockwise (left-hand threads). Remove inner race with impact wrench, or if impact wrench is not available, use a ¾ inch socket wrench and strike wrench handle a sharp blow to loosen threads.

Remove clutch drum, pulley and bearing assembly and remove thrust washer (5). Unscrew clutch plate (4) using a spanner wrench (Homelite tool A-23934 or equivalent) in clockwise direction. Remove clutch cover (1).

Inspect clutch drum and pulley for excessive wear or scoring. Inspect all needle bearing rollers for scoring, excessive wear of flat spots, and renew bearing if such defect is noted. Bearing is excessively worn if rollers can be separated more than the width of one roller.

REWIND STARTER. To disassemble starter, refer to exploded view in Fig. HL68. Pull starter rope out fully, hold pulley (9) and place rope in notch of pulley. Let pulley rewind slowly. Hold pulley while removing screw (11) and washer (10). Turn pulley counterclockwise until disengaged from spring, then carefully lift pulley off starter post. Turn open side of housing down and rap housing sharply against top of work bench to remove spring.

CAUTION: Be careful not to dislodge spring when removing pulley as spring could cause injury if it should uncoil rapidly.

Install new spring with loop in outer end over pin in blower housing and be sure spring is coiled in direction shown in Fig. HL68. Install pulley (9), turning pulley clockwise until it engages spring and secure with washer and screw. Insert new rope through handle and hole in blower housing. Knot both ends of the rope and harden the knots with cement. Turn pulley clockwise eight turns and slide knot in rope into slot and keyhole in pulley. Let starter pulley rewind slowly.

Starter pawl spring outer ends are hooked behind air vanes on flywheel in line with starter pawls when pawls are resting against flywheel nut. Pull

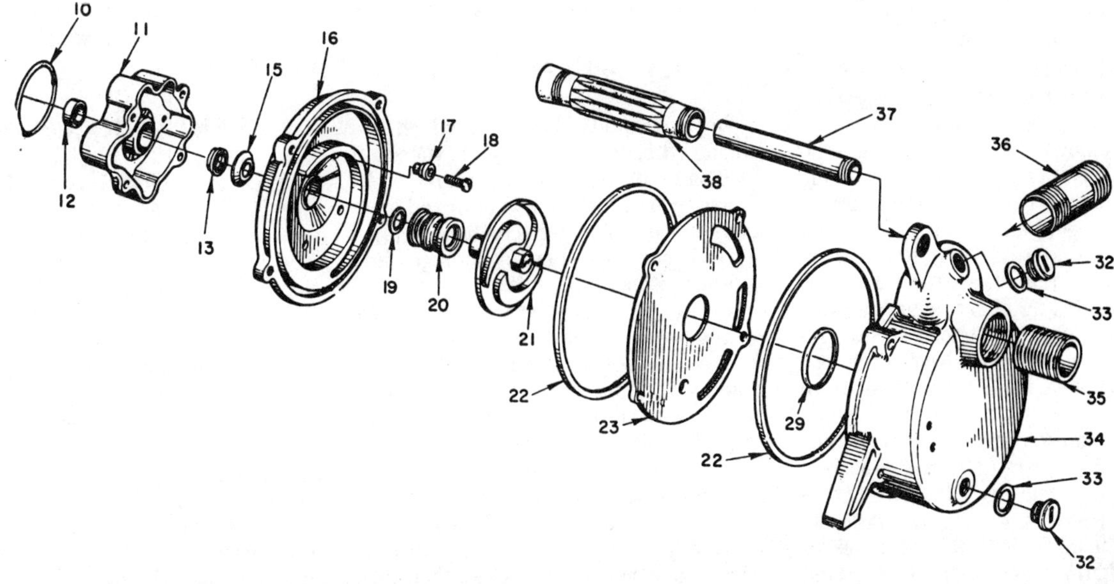

Fig. HL71—Exploded view of centrifugal pump assembly. Pump rotor (21) must be unscrewed from engine crankshaft and back plate (16) must be removed to allow removal of bearing housing (11) and engine crankshaft. Bearing housing retains crankshaft needle roller bearing (12) and crankshaft seal.

10. "O" ring
11. Bearing housing
12. Needle bearing
13. Crankshaft seal
15. Water slinger
16. Back plate
17. Sealing washers (4)
18. Screws (4)
19. Shims (as required)
20. Seal assy.
21. Pump rotor (impeller)
22. Gaskets
23. Wear plate
29. Gasket
32. Drain & filler caps
33. Gaskets
34. Pump end plate
35. Inlet fitting
36. Discharge fitting
37. Carrying handle
38. Rubber grip

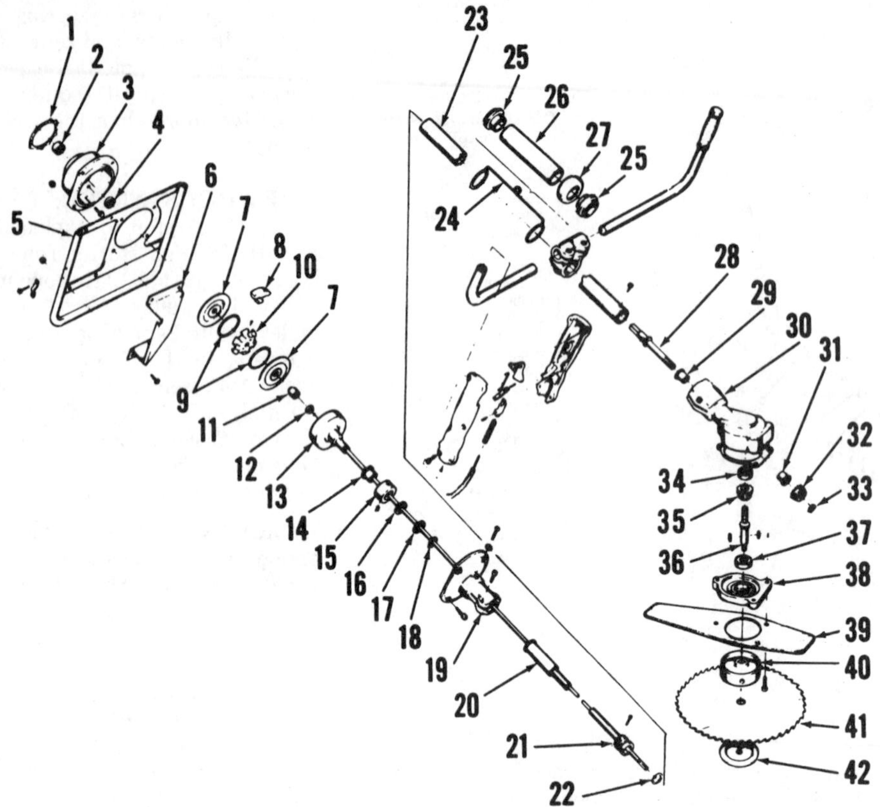

1. Gasket
2. Needle roller bearing
3. Bearing housing
4. Seal
5. Frame assy.
6. Exhaust deflector
7. Clutch cover
8. Clutch shoe
9. Clutch spring
10. Clutch hub
11. Spacer
12. Nut
13. Driveshaft & clutch drum
14. Flanged bearing
15. Collar
16. Thrust washer
17. Spacer
18. Snap ring
19. Upper head
20. Shaft casing
21. Split heading
22. "O" ring
23. Tube
24. Hanger
25. Collar
26. Hanger tube
27. Bumper
28. Gear shaft
29. Bearing
30. Lower head
31. Bearing
32. Bevel gear
33. Snap ring
34. Ball bearing
35. Bevel gear
36. Blade shaft
37. Ball bearing
38. Bearing retainer
39. Glade guard
40. Spindle head
41. Blade
42. Nut

Fig. HL72—Exploded view of early type brushcutter assembly. Refer to Fig. HL73 for later brushcutter.

starter rope slowly when installing blower housing so starter cup will engage pawls.

CIRCULAR SAW DRIVE CASE

Refer to exploded view of circular saw drive case assembly in Fig. HL69. To gain access to engine crankshaft, the unit must be disassembled as follows:

Drain all lubricating oil from gearbox. Hold hex blade washer (45), unscrew cap screw (44) and remove washer and saw blade. Remove blower housing and starter assembly. Remove snap ring (48), bumper screw (39) and bumper (38). Disconnect guard return spring and remove rotating guard (49). Remove upper saw guard (not shown) and the bearing retainer (51). Remove snap ring (53) from shaft (58) and remove screws retaining cover (37) to drive case (2). Remove ball bearing (54) and cover from shaft.

While holding flywheel, remove nut (35), clutch cover (31) and unscrew clutch rotor using Homelite tool A-23696. Remove extension shaft (30) from crankshaft with a socket head screw (Allen) wrench. Remove snap ring (29) and pull clutch drum (60) from flanged bushing (63). Withdraw driven gear (59) and shaft (58); clamp flat outer end of shaft in smooth-jawed vise, if necessary to remove gear from shaft, and unscrew gear.

The drive case can be removed from engine crankcase. Need and procedure for further disassembly should be obvious from inspection of parts. Reassemble unit by reversing disassembly procedure. Refer to LUBRICATION section for refilling unit with oil.

MULTI-PURPOSE SAW

The saw should be equipped with a cutting wheel rated for spindle speeds of 6000 rpm or higher. Maximum wheel diameter is 12 inches (305 mm) and maximum wheel thickness is 0.160 inch (4.06 mm). Be sure wheel is not damaged and is otherwise safe for use.

The saw arm may be installed so the cutting wheel is inboard or outboard of the saw arm. After installing saw arm, install wheel guard so it is properly positioned, then adjust belt tension as outlined in the following paragraph.

To adjust belt tension, loosen nuts securing saw arm then slide arm forward to remove belt slack. Turn tension adjusting screw until it contacts drive case cover or crankcase, then turn screw an additional 3 full turns. Tighten saw arm mounting nuts. Belt will stretch after an hour or two of operation and belt tension should be readjusted.

SAW ARMS

To remove saw arm, back off belt tension screw, remove saw arm mounting nuts and separate saw arm from engine. Refer to Fig. HL70 and disassemble as required.

When installing saw arm on engine, position belt on arm then loop belt around clutch pulley while mating arm with engine.

CENTRIFUGAL PUMP

Refer to exploded view of pump in Fig. HL71. Remove pump end plate (34) and wear plate (23). While holding flywheel, unscrew the pump impeller (21). Remove seal (20), back plate (16), shims (19) and water slinger (15). Bearing housing (11), which carries crankshaft needle roller bearing (12) and seal (13) can then be removed.

When reassembling unit, use all new seals and gaskets. Shims (19) are available in thicknesses of 0.010 and 0.015 inch. Install shims to provide minimum clearance between impeller (21) and wear plate (23) without causing impeller to rub against the plate.

Refill pump assembly with water before attempting to start engine. Water in pump is necessary to lubricate pump seal (20).

1. Gasket
2. Needle roller bearing
3. Bearing housing
4. Seal
5. Clutch cover
6. Clutch assy.
7. Clutch shoe
8. Spring
9. Clutch hub
10. Clutch drum
11. Snap ring
12. Bearing
13. Frame
14. Snap ring
15. Drive shaft
16. Shaft casing
17. Split bushing
18. "O" ring
19. Tube
20. Collar
21. Hanger tube
22. Hanger
23. Handle clamp
24. Bumper
25. Gear shaft
26. Bearing
27. Lower head
28. Bushing
29. Bevel gear
30. Snap ring
31. Ball bearing
32. Bevel gear
33. Blade shaft
34. Ball bearing
35. Bearing retainer
36. Blade guard
37. Spindle head
38. Blade
39. Nut

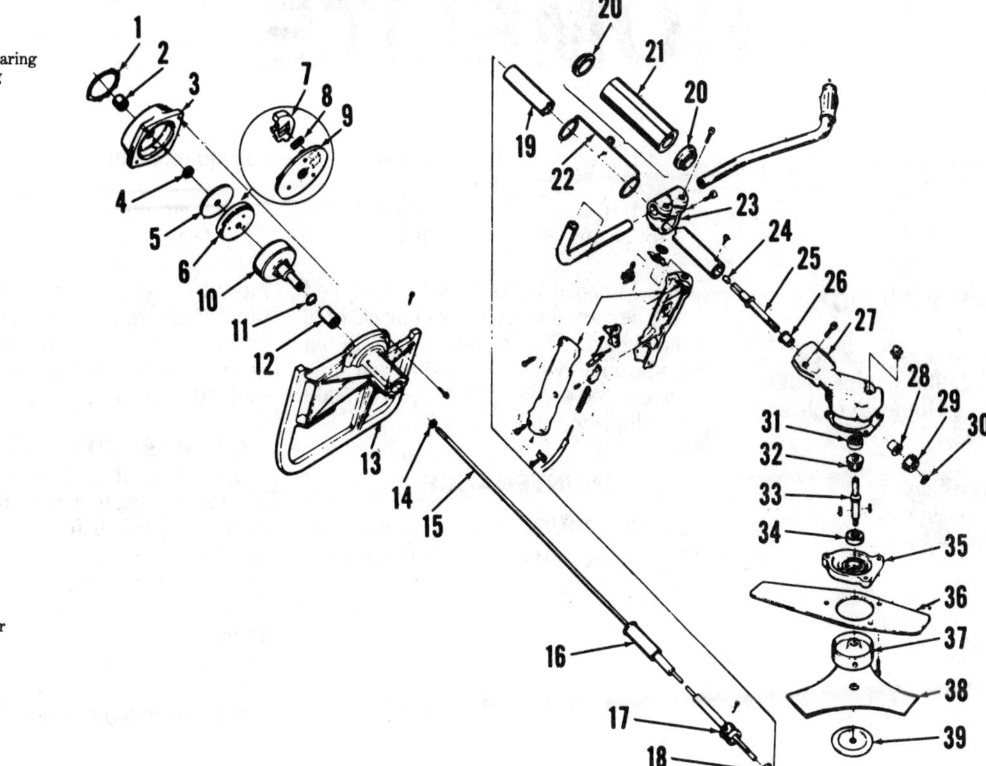

Fig. HL73—Exploded view of late model brushcutter assembly used on XL series.

BRUSHCUTTER

To gain access to crankshaft bearing housing (3–Fig. HL72 or HL73), or engine crankshaft, proceed as follows:

On early models, unbolt upper head (19–Fig. HL72) and remove brushcutter unit and frame (5). While holding engine flywheel, remove nut (12), spacer (11) and outer clutch cover. Unscrew clutch hub (10) from engine crankshaft. Remove inner clutch cover. Bearing housing can now be removed from engine crankcase.

NOTE: Some early models have a flexible drive shaft instead of the rigid drive shaft shown in Fig. HL72.

To remove nut (42) from blade shaft, insert a ¼-inch (6.35 mm) steel rod in hole in spindle head (40) to hold shaft from turning.

On later models, unscrew cap screws securing frame (13–Fig. HL73) to bearing housing (3) and separate brushcutting unit from engine. Remove snap ring (11) and clutch drum (10). Rotate clutch hub in counterclockwise direction to remove clutch assembly. Bearing housing can now be removed from engine as previously outlined. To remove nut (39) from blade shaft, insert a ¼-inch (6.35 mm) steel rod in hole in spindle head (27) to prevent shaft from turning.

HOMELITE

Model	Bore	Stroke	Displacement
XL-925	2-1/16 in. (52.4 mm)	1½ in. (38.1 mm)	5.0 cu. in. (82 cc)

ENGINE INFORMATION

This section covers service of Homelite Model XL-925 engine that is used in the following Homelite tools and equipment.

XL88 Multi-Purpose Saw
XL98 Multi-Purpose Saw
XL98A Multi-Purpose Saw
XLS2-1 Pump
XLS2-1A Pump
XLS2-1B Pump

Multi-Purpose Saws may be equipped with either an abrasive wheel or carbide tipped wheel. Caution should be taken that abrasive wheels installed on this unit are rated for spindle speeds of 5000 rpm or higher.

MAINTENANCE

SPARK PLUG. Models equipped with a solid-state, one-piece ignition module use a Champion DJ6Y spark plug. Early models use a Champion CJ6 or Champion UJ11G for heavy duty operation. Set electrode gap to 0.025 inch (0.6 mm) on all models.

CARBURETOR. All models are equipped with a Tillotson Model HS diaphragm carburetor. Refer to Figs. HL75 and HL76 for exploded views of carburetor.

Carburetor used on Model XL98A (Fig. HL76) has a fixed main jet (19) and a governor valve (6) which is designed to maintain a goverened speed of about 5000 rpm. Neither main jet nor governor valve is adjustable. Carburetors on

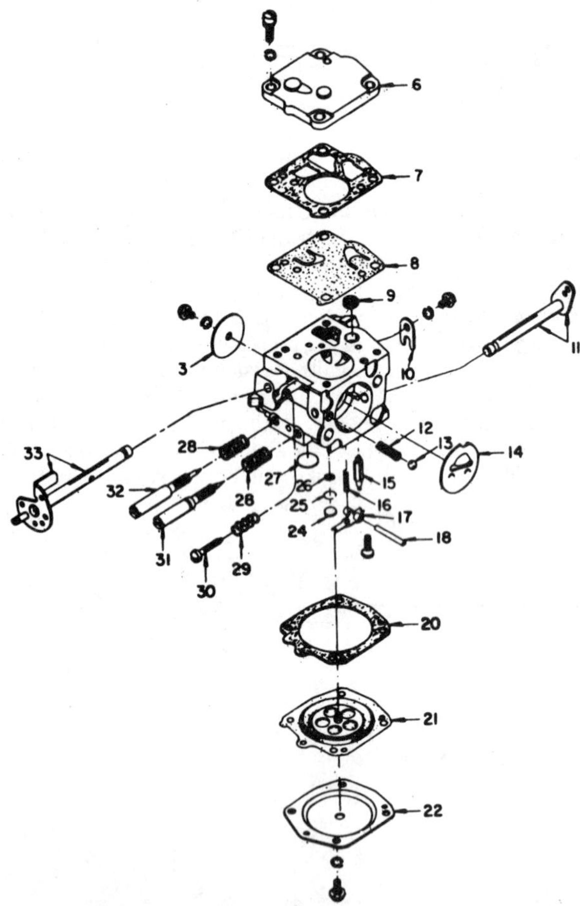

Fig. HL75—Exploded view of Tillotson series HS carburetor used on all models except XL98A. Model XLS2-1A or XLS1-1B do not have mixture needles (31 and 32).

3. Throttle plate
6. Diaphragm cover
7. Gasket
8. Fuel pump diaphragm
9. Filter screen
10. Throttle shaft clip
11. Choke shaft
12. Detent spring
13. Choke detent ball
14. Choke plate
15. Inlet valve needle
16. Inlet valve spring
17. Inlet control lever
18. Lever hinge pin
19. Pin retainer screw
20. Diaphragm gasket
21. Metering diaphragm
22. Diaphragm cover
24. Welch plug (3/16 in.)
25. Retainer ring
26. Channel screen
27. Welch plug (11/32 in.)
28. Mixture needle springs (2)
29. Idle speed screw spring
30. Idle speed screw
31. Main fuel needle
32. Idle fuel needle
33. Throttle shaft

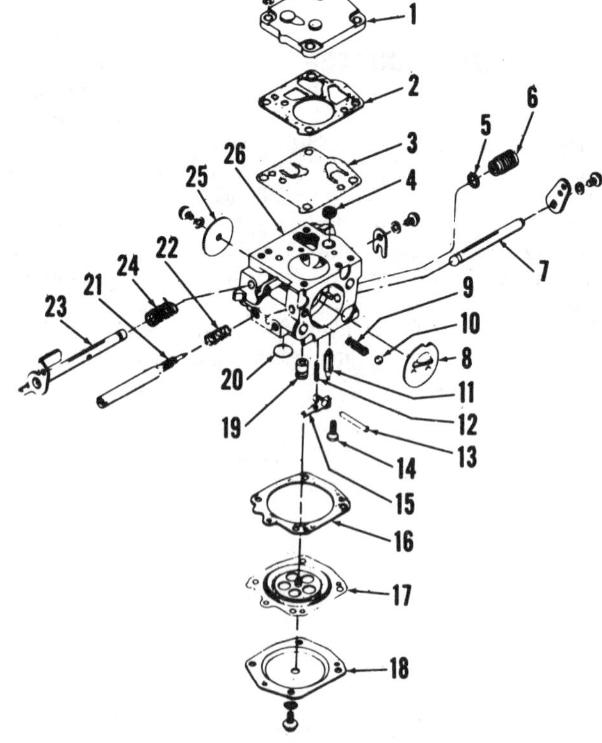

Fig. HL76—Exploded view of Tillotson Model HS carburetor used on Model XL98A. Note governor valve (6). There is no high speed mixture needle on this carburetor.

1. Diaphragm cover
2. Gasket
3. Fuel pump diaphragm
4. Filter screen
5. Gasket
6. Governor valve
7. Choke shaft
8. Choke plate
9. Spring
10. Choke detent ball
11. Inlet valve needle
12. Inlet lever spring
13. Lever pin
14. Screw
15. Inlet control lever
16. Diaphragm gasket
17. Metering diaphragm
18. Diaphragm cover
19. Main jet
20. Welch plug
21. Idle mixture screw
22. Spring
23. Throttle shaft
24. Spring
25. Throttle plate
26. Body

Models XLS2-1A and XLS2-1B do not have adjustable idle or high speed mixture screws.

Initial carburetor adjustment on saw models is one turn open for idle mixture and high speed mixture screw, if so equipped. Adjust idle mixture screw and idle speed screw to obtain smooth idle with engine warm and running at 2400-2600 rpm which should be just below clutch engagement speed. Adjust high speed mixture screw, on models so equipped, to obtain optimum engine performance at cutting speed with saw under normal load. Do not adjust mixture screws too lean as engine damage may result.

To adjust carburetor on XLS2-1 pump models, turn idle and high speed mixture screws until they are 1½ turns open. Start engine and allow to run until warm. Adjust high speed mixture screw for highest pumping speed obtainable then turn high speed mixture screw ⅛ turn counterclockwise. Governor is designed to limit maximum no-load speed to 6400 rpm. Adjust idle mixture screw by pulling throttle button all the way out and turn idle mixture screw to obtain highest and smoothest idle speed. Turn idle speed screw to obtain idle speed of approximately 3000 rpm. Adjustment of one mixture screw will require checking adjustment of remaining mixture screw as operation of mixture needles is related.

Carburetor may be disassembled after inspecting unit and referring to exploded view in Fig. HL76. Clean filter screen (4). Welch plugs may be removed by drilling plug with a suitable size drill bit and prying out. Care must be taken not to drill into carburetor body.

Inspect inlet lever spring (12) and renew if stretched or damaged. Inspect diaphragms for tears, cracks or other damage. Renew idle and high speed ad-

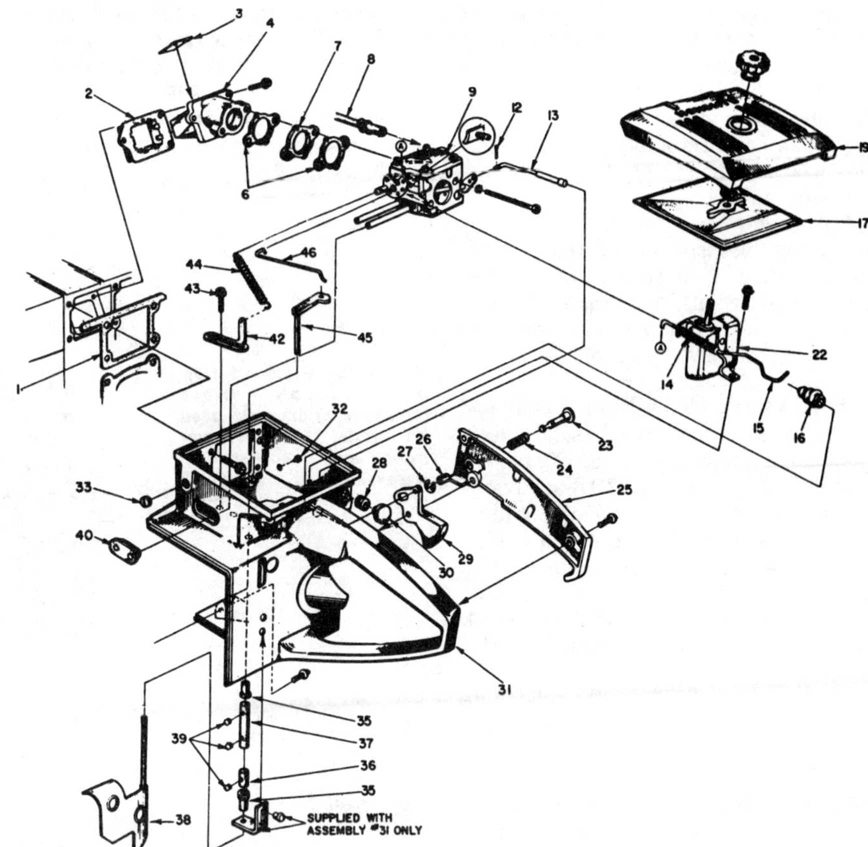

Fig. HL78—Exploded view of Model XL88 air box and handle assembly. Models XL98, XLS2-1, XLS2-1A and XLS2-1B are similar. Governed speed is adjusted by loosening screw (43) and moving slotted speed adjusting plate (42). Reed retainer (2) should be installed in engine intake port, then install reed seat (4) with reeds (3).

1. Gasket
2. Reed retainer
3. Valve reeds
4. Reed seat
6. Gaskets
7. Spacer
8. Fuel line
9. Carburetor
12. Cotter pin
13. Choke rod
14. Return spring
15. Throttle rod
16. Boot
17. Filter element
19. Air filter cover
22. Air deflector
23. Throttle lock pin
24. Throttle lock spring
25. Handle cover
26. Nylon bushing
27. Snap ring
28. Grommet
29. Throttle trigger
30. Choke button
31. Handle & air box
32. Plug
33. Plug
35. Bushing
36. Collar
37. Coupling
38. Air vane & shaft
39. Set screws
40. Grommet
42. Speed adjusting plate
43. Adjusting plate screw
44. Governor spring
45. Governor arm & shaft
46. Carburetor link

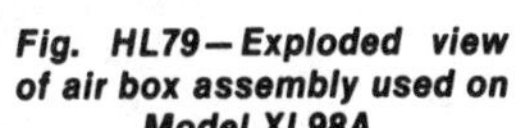
Fig. HL79—Exploded view of air box assembly used on Model XL98A.

1. Gasket
2. Reed retainer
3. Reed petal
4. Reed seat
5. Gasket
6. Intake manifold
7. Gasket
8. Carburetor
9. Choke button
10. Throttle rod
11. Throttle trigger
12. Pin
13. Snap ring
14. Nylon bushing
15. Lock spring
16. Throttle lock
17. Air cleaner base
18. Gasket
19. Filter element
20. Filter mount
21. Filter cover

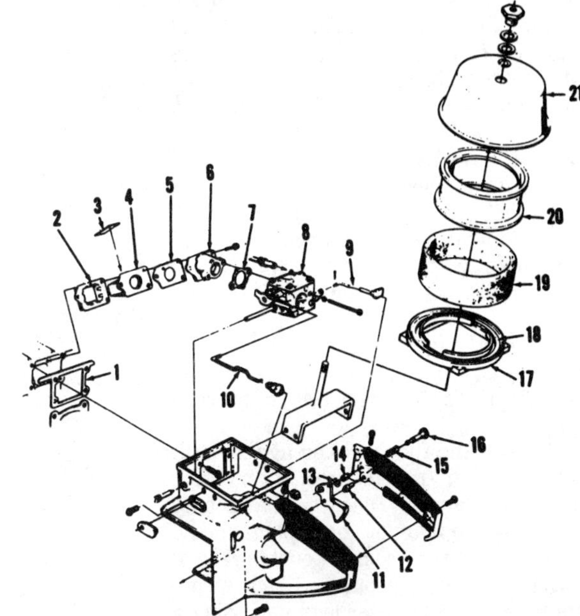

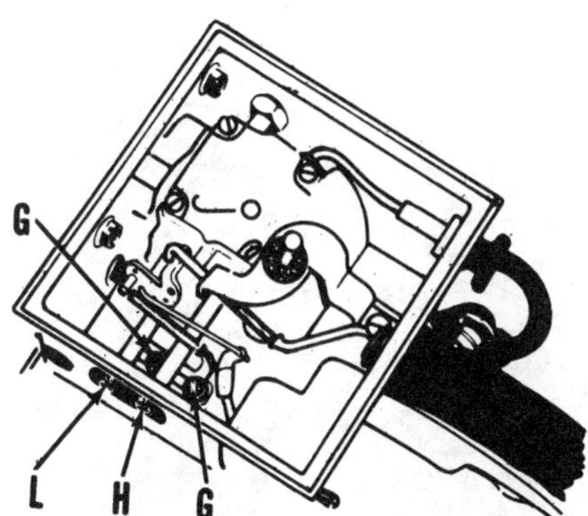

Fig. HL77—Drawing showing locations of idle (L) and high speed (H) mixture screws and governor plate screws (G) on Model XL88. Models XL98, XLS2-1, XLS2-1A and XLS2-1B are similar.

justing needles if needle points are grooved or broken. Carburetor body must be renewed if needle seats are damaged. Fuel inlet needle has a rubber tip and seats directly on a machined orifice in carburetor body. Inlet needle or carburetor body should be renewed if worn excessively.

Carburetor may be reassembled by reversing disassembly procedure. Adjust position of inlet control lever so lever is flush with diaphragm chamber floor. Bend lever adjacent to spring to obtain correct lever position.

MAGNETO. Model XL88 is equipped with Wico breakerless solid state magneto as shown in Fig. HL80, Models XL98, XL98A and XLS2-1B are equipped with the capacitor discharge ignition system shown in Fig. HL82 and Models XLS2-1 and XLS2-1A are equipped with a conventional flywheel magneto ignition system as shown in Fig. HL84. Refer to appropriate paragraph for model being serviced.

Model XL88. To check the solid state magneto, disconnect spark plug wire, turn ignition switch on and crank engine while holding terminal about 1/4 inch (6 mm) away from ground and check for spark as with conventional magneto. If no spark occurs, refer to the following inspection and test procedure:

Visually inspect rotor (flywheel) for damage. Check for broken or frayed wires.

To test the ignition coil, disconnect the wires at insulating sleeve (39–Fig. HL80) and test coil according to tester procedure. Specifications for testing with either Graham Model 51 or Merc-O-Tronic tester are as follows:

Graham Model 51:

Maximum secondary	10,000
Maximum primary	1.7
Coil index	65
Maximum coil test	20
Maximum gap index	65

Merc-O-Tronic:

Operating amperage	1.3
Minimum primary resistance	0.6
Maximum primary resistance	0.7
Minimum secondary continuity	50
Maximum secondary continuity	60

Renew the ignition coil if found faulty and again check for spark. If no specifications were as specified, proceed as follows.

Remove the flywheel and again check for broken or frayed wires. If no defect is noted, remove the screw attaching condenser to magneto back plate and be sure condenser is not touching back plate or other ground. Push a pin through the condenser lead and using a condenser tester, check for short, series resistance and capacitance. Condenser capacitance should be 0.16-0.20 mfd. If condenser is faulty, renew the switch box and condenser assembly (26). If condenser tested to specifications, proceed as follows:

Disconnect coil primary lead at insulating sleeve (39) and disconnect switch box ground lead from back plate (Fig. HL81). Remove the screw attaching condenser to back plate and be sure condenser is insulated from any ground. Be sure the switch box ground

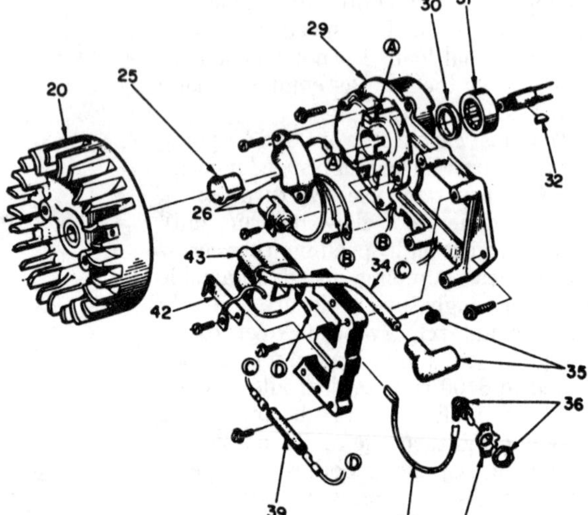

Fig. HL80—Exploded view of the solid state (breakerless) magneto used on Model XL88. Trigger switch and condenser (26) are available as an assembly only. The trigger coil is molded into back plate (29).

20. Flywheel (rotor)
25. Dust cap
26. Trigger switch & condenser
29. Back plate & trigger coil
30. Crankshaft seal
31. Roller bearing
32. Woodruff key
34. Spark plug wire
35. Spark plug terminal
36. Magneto grounding switch
37. "ON-OFF" plate
38. Ground lead
39. Insulating sleeve
42. Coil retaining clip
43. Ignition coil

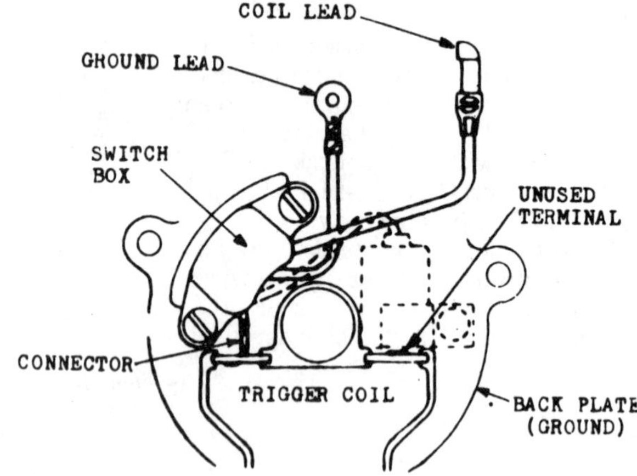

Fig. HL81—Drawing showing points for ohmmeter test lead connections for checking solid state magneto trigger coil, condenser and switch box. Refer to text for procedure and specifications.

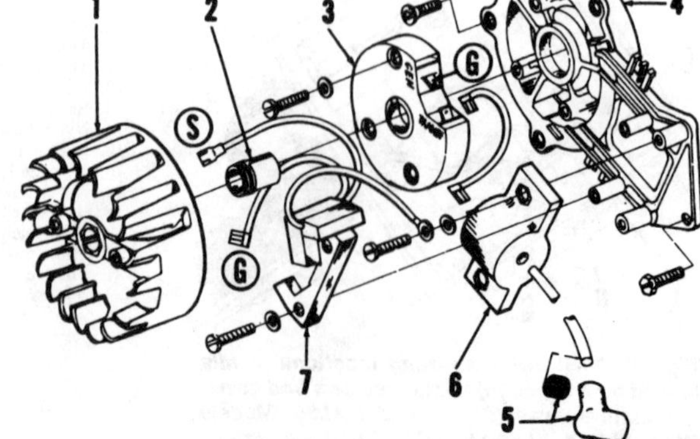

Fig. HL82—Exploded view of Phelon capacitor discharge type magneto used on Models XL98, XL98A and XLS2-1B.

G. Connector to "GEN" terminal
S. Connector to "ON-OFF"
1. Magneto rotor (flywheel)
2. Dust cap
3. Ignition module
4. Back plate
5. High tension wire & terminal
6. Transformer coil
7. Generator coil & armature

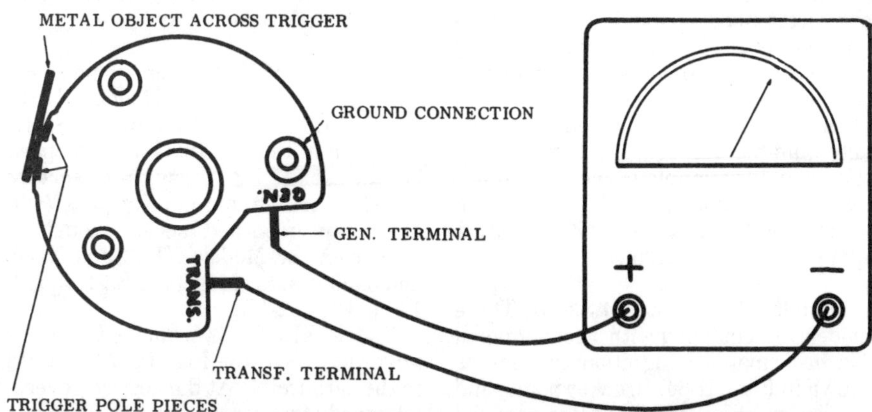

Fig. HL83—Drawing showing volt-ohmmeter connections to ignition module (3—Fig. HL82) for checking module. It should be noted that this is not a conclusive test and module should be renewed in event of spark failure when other magneto components test to specifications.

lead and ignition coil lead are not touching anything and connect leads of an ohmmeter to the two leads. Meter should read either between 1 megohm and infinity or between 5 and 25 ohms. With ohmmeter leads reversed, reading should be opposite that of preceding test. That is, if first reading obtained was 5-25 ohms, second reading should be 1 megohm to infinity. If ohmmeter readings are not as specified, renew the switch box tested to specifications, test trigger coil as follows:

Connect one ohmmeter lead to connector between switch box and trigger coil and remaining ohmmeter lead to back plate (ground). Reading should be either between 0 to 85 ohms or between 85 to 150 ohms. Reverse the leads; second reading on ohmmeter should be opposite first reading. That is, if first reading was in specified range of 0-85 ohms, second reading should be within range of 85-150 ohms. Then, connect the ohmmeter leads to unused terminal of trigger coil and to magneto back plate. Ohmmeter reading should then be 20-26 ohms. If trigger coil does not test within specifications, renew the magneto back plate and trigger coil assembly.

When reassembling magneto, check back plate and remove any sharp edges especially where wires may contact the back plate. Be sure all leads are in place as shown in Fig. HL81. Be sure the back plate is clean and check all screws for tightness. If there is any doubt about the strength of the rotor (flywheel) magnets, install a new flywheel; be sure to remove "keeper" plates from new flywheel before installing it.

Models XL98, XL98A, XLS2-1B. These models are equipped with the capacitor discharge ignition system shown in Fig. HL82. Refer to CAPACITOR DISCHARGE IGNITION SYSTEM section of this manual for explanation of ignition system operation.

The capacitor discharge magneto is operating satisfactorily if spark will jump a 3/8-inch (9.5 mm) air gap when engine is turned at cranking speed. If magnet fails to produce spark, service consists of locating and renewing inoperative unit; no maintenance is necessary.

To check magneto with volt-ohmmeter, remove starter housing and disconnect wire from ignition switch. Check to be sure there is no continuity through switch when in "ON" position to be sure a grounded switch is not cause of trouble and inspect wiring to be sure it is not shorted.

CAUTION: Be sure storage capacitor is discharged before touching connections. Switch ignition switch to "OFF" position and ground switch lead (S—Fig. HL82).

Resistance through secondary (high tension) winding of transformer coil should be 2400-2900 ohms and resistance through primary winding should be 0.2-0.4 ohms. Connect ohmmeter leads between high tension (spark plug) wire and ground and then between input terminal and ground. If transformer coil does not test within specifications, renew coil and recheck for spark at cranking speed. If magneto still does not produce spark, check generator as follows: Remove rotor (flywheel) and disconnect lead from generator to generator (G) terminal on module (3) and switch lead (S) at ignition switch. Connect negative lead of ohmmeter to ground wire from generator and the positive lead of ohmmeter to generator (G) wire. The ohmmeter should register showing continuity through generator. Reverse leads from ohmmeter; ohmmeter should then show no continuity (infinite resistance) through generator. Renew generator if continuity is noted with ohmmeter leads connected in both directions. A further check can be made using voltmeter if continuity checked correctly. Remove spark plug and reinstall rotor leaving

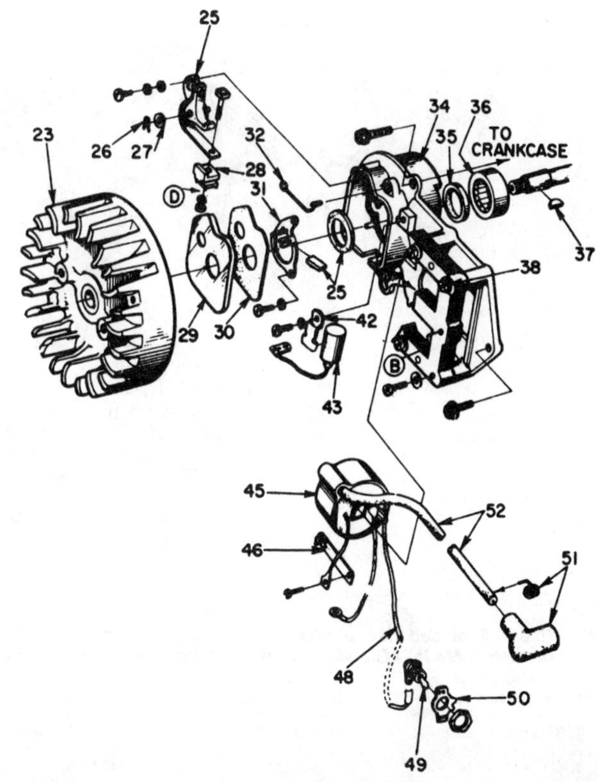

Fig. HL84—Exploded view of conventional flywheel type magneto used on Models XLS2-1 and XLS2-1A. Coil clip retaining screw location is shown by letter "B". Condenser lead and ignition coil primary lead are attached to terminal block (28) at "D".

23. Rotor (flywheel)
25. Breaker point set
26. Clip
27. Washer
28. Terminal block
29. Breaker box cover
30. Gasket
31. Felt retainer
32. Cover spring clip
34. Back plate
35. Crankshaft seal
36. Roller bearing
37. Rotor key
38. Coil core (armature)
42. Clamp
43. Condenser
45. Ignition coil
46. Coil retaining clip
48. Ground lead
49. Ignition switch
50. "ON-OFF" plate
51. Spark plug terminal
52. Spark plug wire

wire (G) from generator disconnected. Connect positive (red) lead from voltmeter to wire (G) from generator and negative (black) lead of voltmeter to magneto back plate. Wires must be routed so starter can be reinstalled. A firm pull on starter rope should spin engine at about 500 rpm and voltmeter should show minimum reading of 4 volts. If both generator and transformer coil tested to specifications, a faulty ignition module (3) should be suspected.

A partial check of ignition module can be made using an ohmmeter. With ohmmeter set to R x 1000 scale, connect positive (red) lead of ohmmeter to module terminal marked "GEN" and negative ohmmeter lead to module ground connection (Fig. HL83). An instant deflection of ohmmeter needle should be noted. If no deflection of needle is noted with ohmmeter leads connected in either direction, module is faulty and should be renewed. If needle deflection is observed, select R x 1 (direct reading) scale of ohmmeter and connect positive (red) lead to module terminal marked "GEN" and place negative (black) lead against terminal marked "TRANS". Place a screwdriver across the two trigger poles (Fig. HL83). The ohmmeter needle should deflect and remain deflected until the ohmmeter lead is released from the module terminal. If the desired result are obtained with ohmmeter checks, the module is probably suitable for service, however, as this is not a complete check, if other magneto components and wiring check satisfactory, renew module if no ignition spark can yet be obtained.

Models XLS2-1 and XLS2-1A. These models are equipped with a conventional flywheel magneto ignition system as shown in Fig. HL84. Breaker points and condenser are accessible after removal of starter housing, flywheel and breaker box cover. Adjust breaker point gap to 0.015 inch (0.38 mm).

Ignition timing is fixed at 30° BTDC. After reinstalling flywheel, check armature air gap which should be 0.005-0.007 inch (0.13-0.18 mm). To adjust air gap, turn flywheel so magnets are below legs of armature core and place plastic shim (Homelite part 23987) between armature and magnets. Loosen then tighten armature retaining screws and remove shim.

Condenser capacity should test 0.18-0.22 mfd.

GOVERNOR. An air vane type governor is used on Models XL88, XL98, XLS2-1, XLS2-1A and XLS2-1B as shown in Fig. HL78. Governed speed is adjusted by loosening screws (43–Fig. HL78) and moving plate (42). Maximum governed speed should be 5000 rpm for Models XL88 and XL98 and 6400 rpm for Models SLS2-1, XLS2-1A and XLS2-1B.

Model XL98A is equipped with a governor valve (6–Fig. HL76) located in the carburetor. At the desired governed speed, the governor valve will open and allow additional fuel into the engine. This excessively rich fuel mixture will prevent engine overspeeding. Governed speed should be 5000 rpm and is not adjustable. If valve (6) does not function properly, it must be renewed as a unit.

LUBRICATION. Engine on all models is lubricated by mixing oil with regular gasoline. If Homelite Premium SAE 40 chain saw oil is used, fuel:oil ratio should be 32:1. Fuel:oil ratio should be 16:1 if Homelite 2-Cycle SAE 30 oil or

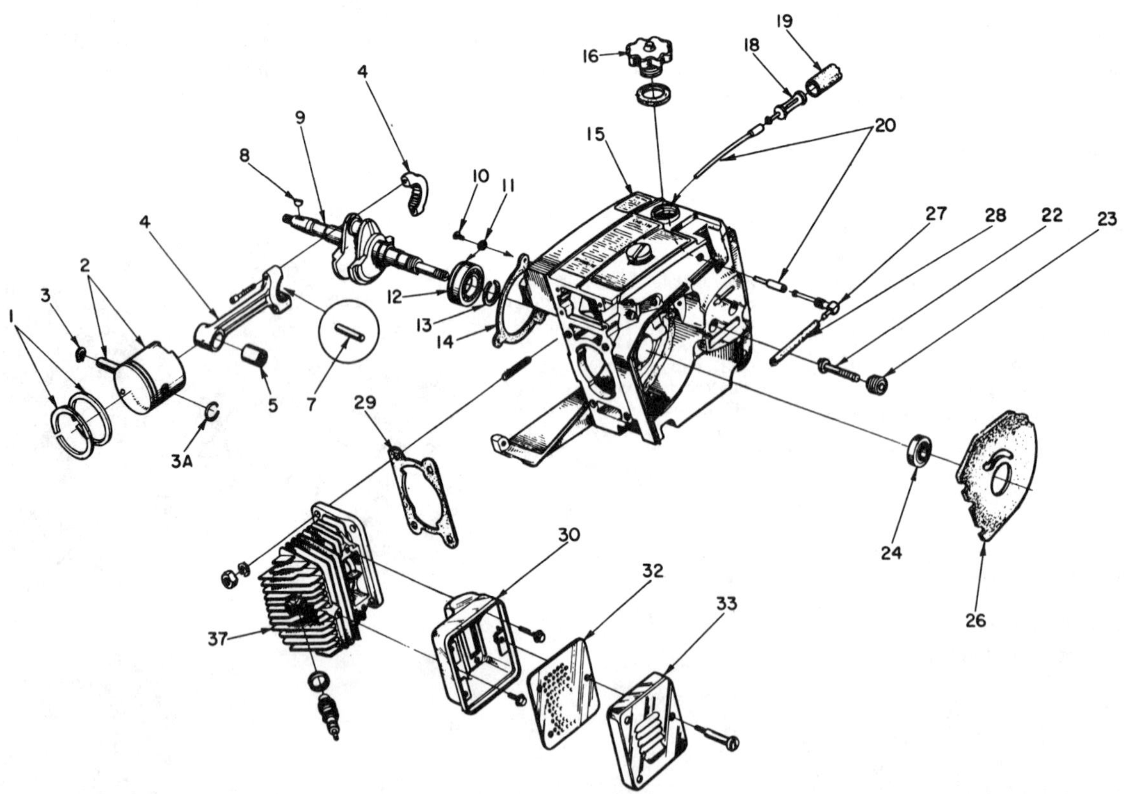

Fig. HL85—Exploded view of Model XL88 engine assembly; other models are similar. Magneto back plate (29—Fig. HL80) carries magneto end of crankshaft. Model XL88 saw arm stud retainers (23) must be unscrewed to remove the captive studs (22). Do not attempt to unscrew the studs.

1. Piston rings
2. Piston & pin
3. Truarc snap ring
3A. Wire section ring
4. Connecting rod
5. Needle bearing
7. Needle rollers (28)
8. Woodruff key
9. Crankshaft
10. Retaining screw
11. Bearing retainer washers
12. Ball bearing
13. Snap ring
14. Gasket
15. Crankcase
16. Fuel filler cap
18. Filter assy.
19. Filter element
20. Fuel line
22. Studs
23. Threaded inserts
24. Crankshaft seal
26. Dust shield
27. Elbow fitting
28. Fuel line
29. Gasket
30. Muffler body
32. Baffle
33. Cap
37. Cylinder

other SAE 30 oil designed for air-cooled two-stroke engines is used.

CARBON. Muffler and cylinder exhaust ports should be cleaned periodically to prevent loss of power due to carbon buildup. Remove muffler cover and baffle plate and scrape muffler free of carbon. With muffler cover removed, turn engine so piston is at top dead center and carefully remove carbon from exhaust ports with wooden scraper. Be careful not to damage the edges of exhaust ports or to scratch piston. Do not attempt to run engine with muffler baffle plate or cover removed.

REPAIRS

CONNECTING ROD. Connecting rod and piston assembly can be removed after removing cylinder from crankcase. Be careful to remove all of the 28 loose needle rollers when detaching rod from crankpin.

Renew connecting rod if bent or twisted, or if crankpin bearing surface is scored, burned or excessively worn. The caged needle roller piston pin bearing can be renewed by pressing old bearing out and pressing new bearing in with Homelite tool 23955. Press on lettered side of bearing cage only.

It is recommended the crankpin needle rollers be renewed as a set whenever engine is disassembled for service. Stick 14 needle rollers in rod and the remaining 14 needle rollers in rod cap with light grease or beeswax. Assemble rod to cap with match marks aligned and with open end of piston pin towards flywheel side of engine. Make certain fractured surfaces of connecting rod are aligned as rod cap retaining screws are being tightened.

PISTON, PIN AND RINGS. The piston is fitted with two pinned compression rings. Renew piston if scored, cracked or excessively worn, or if ring side clearance in top ring groove exceeds 0.0035 inch (0.089 mm).

Fig. HL86—Installing shoes and springs on clutch spider plate.

Recommended piston ring end gap is 0.070-0.080 inch (1.78-2.03 mm). Maximum allowable ring end gap is 0.085 inch (2.16 mm). Desired ring side clearance in groove is 0.002-0.003 inch (0.05-0.08 mm).

Piston, pin and rings are available in standard size only. Piston and pin are available as a matched set and are not available separately.

Piston pin on Models XL88 and XL98 is retained in piston by a wire type snap ring on exhaust side and by a Truarc snap ring on opposite end. Disassemble piston and rod by removing the Truarc snap ring and pushing pin out with a 3/16-inch (4.8 mm) diameter pin. Piston pin on Models XL98A, XLS2-1, XLS2-1A and XLS2-1B is retained by Truarc snap rings at each end.

When reassembling piston to connecting rod, be sure closed end of pin is towards exhaust port. Install piston pin retaining Truarc snap rings with sharp side out. Rotate snap ring to be sure it is secure in retaining groove, then turn gap toward closed end of piston.

CRANKSHAFT. Flywheel end of crankshaft is supported in a roller bearing in magneto back plate and drive end is supported in a ball bearing located in crankcase. End play is controlled by the ball bearing.

Renew the crankshaft if the flywheel end main bearing or crankpin bearing surface or sealing surfaces are scored, burned or excessively worn. Renew the ball bearing if excessively loose or rough. Also, reject crankshaft if flywheel keyway is beat out or if threads are badly damaged.

CYLINDER. The cylinder bore is chrome plated. Renew cylinder if chrome plating is worn away exposing the softer base metal.

To remove cylinder, first remove the blower (fan) housing, carburetor and air box (handle) assemblies and remove the screw retaining magneto back plate to flywheel side of cylinder. The cylinder can then be unbolted from crankcase and removed from the piston.

CRANKCASE, MAGNETO BACK PLATE AND SEALS. To remove the magneto back plate, first remove the blower (fan) housing and flywheel. Loosen the cylinder retaining stud nuts on flywheel side of engine to reduce clamping effect on back plate boss, then unbolt and remove the back plate assembly from crankcase.

To remove crankshaft from crankcase, first remove the cylinder, connecting rod and piston assembly and the magneto back plate as previously

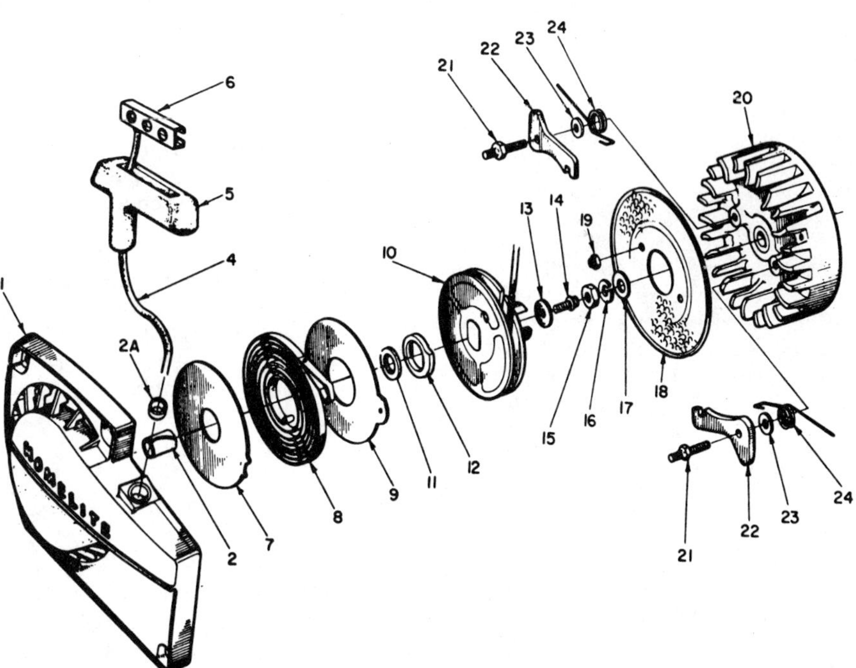

Fig. HL87—Exploded view of typical rewind starter assembly and engine flywheel.

1. Starter housing
2. Bushing
2A. Rope bushing
4. Starter rope
5. Hand grip
6. Insert
7. Inner spring shield
8. Rewind spring
9. Outer spring shield
10. Pulley & cup assy.
11. Bushing
12. Spring lock
13. Washer
14. Cap screw
15. Flywheel nut
16. Lockwasher
17. Flat washer
18. Rotating screen
19. Self-locking nuts
20. Flywheel (magneto rotor)
21. Shoulder studs
22. Starter pawls
23. Washers
24. Pawl springs

outlined. Remove drive clutch assembly and dust shield (26 – Fig. HL85) on saws and pump assembly (Fig. HL89) on Models XLS2-1, XLS2-1A and XLS2-1B. Then, remove the two ball bearing retaining screws (10 – Fig. HL85) from inside of crankcase and remove crankshaft and ball bearing assembly from crankcase. Remove snap ring (13) and press crankshaft from bearing if necessary.

REED VALVES. The pyramid type reed valve seat is made of "Delrin" plastic and reeds are located by pins molded on seat. The reeds are held in place by a molded retainer that also serves as a gasket between reed seat and crankcase. Reeds are 0.004 inch (0.10 mm) thick.

When installing reed valve assembly, it is important that reed retainer be installed in crankcase first, then install reed seat with reeds in place. Oil can be used to stick reeds to seat. Also, special type shoulder retaining screws must be used.

CLUTCH. All models except Models XLS2-1, XLS2-1A and XLS2-1B are equipped with a centrifugal clutch. To remove the clutch assembly, first remove the blade and arm assembly (see Fig. HL88 or HL89). The clutch bearing inner race (27) unscrews counterclockwise (left-hand thread). Remove inner race with impact wrench, or if impact wrench is not available, use a ¾-inch socket wrench and strike wrench handle a sharp blow to loosen threads.

Remove clutch drum, pulley and bearing assembly and remove thrust washer (30). Unscrew clutch spider plate (31) using a spanner wrench (Homelite tool A-23934 or equivalent) in counterclockwise direction. Remove clutch cover (34.)

Inspect clutch drum and pulley for excessive wear or scoring. Inspect all needle bearing rollers for scoring, excessive wear or flat spots and renew bearing if such defect is noted. Bearing is excessively worn if rollers can be separated more than the width of one roller.

Pry clutch shoes from spider plate with screwdriver. To install new shoes and/or springs, refer to Fig. HL86. Reinstall clutch by reversing removal procedure. Lubricate needle roller bearing in clutch drum with a small amount of high temperature multi-purpose grease. The bearing should be cleaned and repacked at 45 to 50 hour intervals.

REWIND STARTER. Refer to Fig. HL87 for exploded view of rewind starter. To disassemble starter after removing housing and starter assembly from engine, pull rope out a short distance, hold rope nd pry retainer (6) from hand grip. Untie knot in end of rope, then allow pulley to rewind slowly. Remove hex head screw (14) and remove rope pulley.

CAUTION: Be careful not to dislodge spring (8) while removing pulley as the rapidly uncoiling spring could cause injury.

Check all starter parts for wear or other damage and renew as necessary. Rope bushing (2A) in housing should be renewed if rope notch is worn in bushing. When reassembling starter, lubricate starter post lightly and install spring without lubrication. Refer to exploded view in Fig. HL87 for reassembly guide. Rotate pulley 2 to 4 turns with rope in notch to place tension on spring.

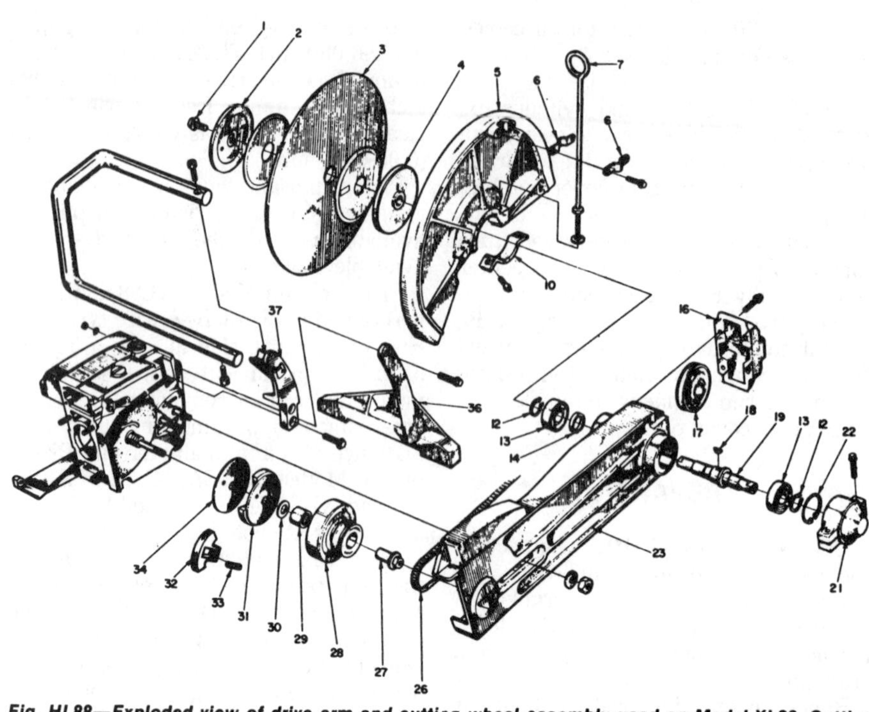

Fig. HL88—Exploded view of drive arm and cutting wheel assembly used on Model XL88. Cutting wheel or saw blade can be mounted on either side of arm. Shaft guard (21) is used to cover opposite end of shaft. Belt can be renewed after removing arm from engine and the cap (16) from end of arm.

1. Cap screw
2. Outer blade washer
3. Cutting wheel
4. Inner washer
5. Guard
6. Clamps
7. Eye or hook bolt
10. Guard clamp
12. Snap ring
13. Ball bearing
14. Spacer
16. Arm cap
17. Driven pulley
18. Woodruff key
19. Blade shaft
21. Shaft guard
22. Snap ring
23. Arm
26. Drive belt
27. Clutch bearing race
28. Clutch drum & pulley
29. Needle bearing
30. Thrust washer
31. Spider plate
32. Clutch shoes
33. Clutch springs
34. Clutch cover
36. Saw support
37. Handle bar bracket

BLADE SHAFT, BEARINGS AND PULLEY

Refer to exploded view of unit in Fig. HL88 or HL89. To renew the blade drive mechanism, proceed as follows:

On Model XL88, unbolt and remove the arm assembly from engine. Remove shaft guard (21 – Fig. HL88), blade or cutting wheel (3) and blade guard (5).

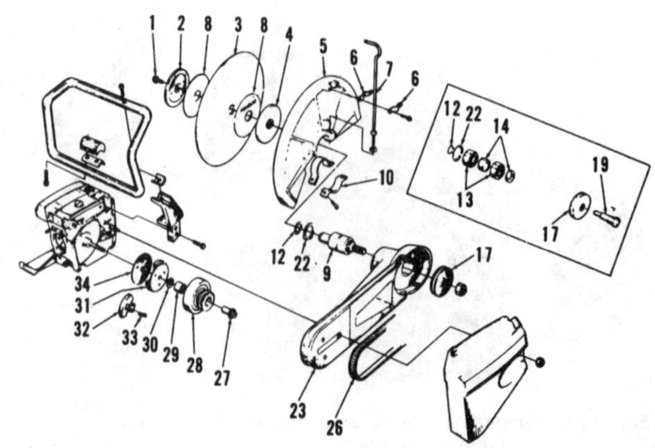

Fig. HL89—Exploded view of clutch and cutting wheel assembly used on Model XL98A. Blade shaft assembly on Model XL98 is shown in inset. Refer to Fig. HL88 for parts identification except for: 8. Spacer; 9. Bearings & shaft.

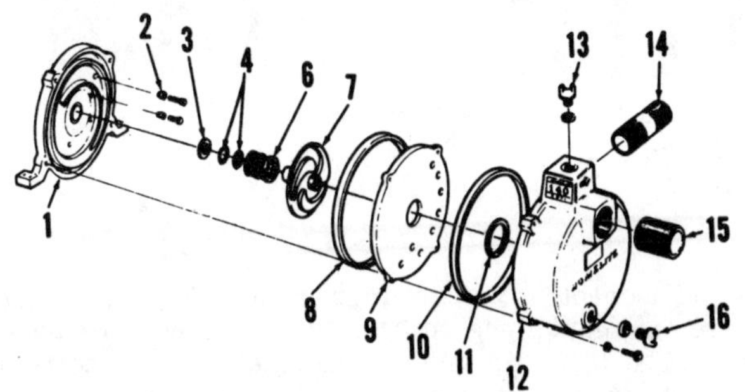

Fig. HL90—Exploded view of centrifugal pump used on Models XLS2-1, XLS2-1A and XLS2-1B.

1. Impeller housing
2. Seal washer
3. Washer
4. Shims
6. Seal
7. Impeller
8. Gasket
9. Wear plate
10. Gasket
11. Gasket
12. End housing
13. Fill plug
14. Discharge fitting
15. Inlet fitting
16. Drain plug

Remove arm cap (16) and drive belt (26). Remove the large internal snap ring (22) from arm and external snap ring (12) from shaft at opposite side of arm. Support the arm and press shaft and outer bearing out towards outside of arm. Remove snap ring retaining outer bearing to shaft, then press shaft from bearing.

To disassemble blade drive on Models XL98 and XL98A, remove blade arm from engine and detach drive cover and components (1 through 8 – Fig. HL89) as well as clamp (10). Remove driven pulley (17) and drive belt (26). Note that pulley on Model XL98 is retained by a nut while blade shaft (19) holds pulley (17) in position and shaft must be removed with pulley. Remove snap rings and press bearings out of blade arm.

To reassemble, reverse disassembly procedure. Place belt over drive pulley on engine, then mount arm to engine crankcase. A wrench with tapered handle (Homelite tool A-24085) can be inserted between arm and crankcase to help tension the drive belt. Be sure belt is tight enough to drive blade or wheel under full cutting load without slippage, then tighten arm retaining nuts.

CENTRIFUGAL PUMP

Refer to Fig. HL90 for an exploded view of centrifugal pump used on Models XLS2-1, XLS2-1A and XLS1-1B. Disassembly of pump is evident after referring to Fig. HL90 and inspection of unit. Pump impeller (7) is threaded on engine crankshaft with left-hand threads.

Use new gaskets and seals to reassemble pump. Assemble components (1 through 7) and place a straightedge across face of impeller housing (1). Measure distance between impeller (7) and straightedge. Install shims (4) to obtain desired clearance of 0.015-0.025 inch (0.38-0.64 mm). Shims are available in thicknesses of 0.010 and 0.015 inch.

Refill pump with water before attempting to start engine. Water is necessary in pump to lubricate pump seal (6).

HOMELITE

Model	Bore	Stroke	Displacement
EZ-10, Chipper	1.4375 in. (36.5 mm)	1.3 in. (33 mm)	2.1 cu. in. (34.6 cc)

MAINTENANCE

SPARK PLUG. A Champion DJ6J spark plug with tapered seat is used; no gasket is required. Adjust electrode gap to 0.025 inch (0.64 mm).

CARBURETOR. A Walbro Model HDC diaphragm type carburetor is used on all models. Refer to Fig. HL91 for exploded view of carburetor.

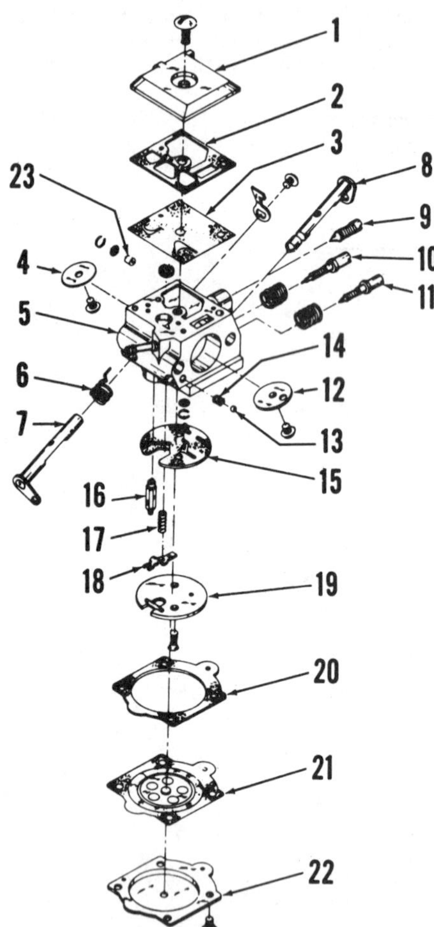

Fig. HL91 — Exploded view of Walbro Model HDC carburetor.

1. Pump cover
2. Gasket
3. Fuel pump diaphragm & valves
4. Throttle plate
5. Body
6. Return spring
7. Throttle shaft
8. Choke shaft
9. Idle speed screw
10. Idle mixture screw
11. High speed mixture screw
12. Choke plate
13. Choke friction ball
14. Spring
15. Gasket
16. Fuel inlet valve
17. Spring
18. Diaphragm lever
19. Circuit plate
20. Gasket
21. Metering diaphragm
22. Cover
23. Limiting jet

For initial carburetor adjustment, back idle speed adjusting screw out until throttle valve will completely close, then turn screw back in until it contacts idle stop plus ½ turn additional. Turn both fuel adjusting needles in until lightly seated, then back main fuel needle (located to left and marked "HI" on grommet when viewing adjustment needle side of throttle handle) out about ¾ turn. Start engine, readjust idle speed and fuel needles so engine idles at just below clutch engagement speed. With engine running at full throttle under load, readjust main fuel needle to obtain optimum performance at high speed. Do not adjust main fuel needle too lean as engine may be damaged.

MAGNETO. Refer to exploded view of magneto in Fig. HL94. Breaker points and condenser are accessible after removing starter assembly, magneto rotor (flywheel) and breaker box cover.

Condenser capacity should test approximately 0.2 mfd. Adjust breaker points to 0.015 inch (0.38 mm). After reinstalling magneto rotor (flywheel), check magneto armature core to rotor air gap. Air gap should be 0.008-0.012 inch (0.20-0.30 mm) and can be adjusted using plastic shim stock (Homelite part 24306).

CARBON. Carbon deposits should be removed from muffler and exhaust ports at regular intervals. When scraping carbon, be careful not to damage chamfered edges of exhaust ports or scratch piston. A wooden scraper should be used. Turn engine so piston is at top dead center so carbon will not fall into cylinder. Do not attempt to run engine with muffler removed.

LUBRICATION. Engine on all models is lubricated by mixing oil with regular gasoline. If Homelite Premium SAE 40 chain saw oil is used, fuel:oil ratio should be 32:1. Fuel:oil ratio should be 16:1 if Homelite 2-Cycle SAE 30 oil or other SAE 30 oil designed for air-cooled two-stroke engines is used.

The clutch needle roller bearing should be cleaned and lubricated after every 100 hours of operation. A high temperature multi-purpose grease should be used.

REPAIRS

TIGHTENING TORQUES. Recommended tightening torques are as follow:

6/32 compression clamp . . . . . 20 in.-lbs. (2.3 N·m)
6/32 compression release post nut . . . . . 20 in.-lbs. (2.3 N·m)
6/32 breaker box . . . . . 20 in.-lbs. (2.3 N·m)
6/32 breaker point adjustable arm . . . . . 20 in.-lbs. (2.3 N·m)

Fig. HL92 — Exploded view of handle and related assemblies.

1. Air filter
2. Carburetor
3. Throttle rod
4. Oil line
5. Spacer
6. Gasket
7. Reed valve seat
8. Reed retainer
9. Reed petals
10. Spring post
11. Spring
12. Choke rod
13. Throttle stop
14. Spring
15. Trigger
16. Bushing
17. Spring
18. Throttle latch
19. Air filter bracket

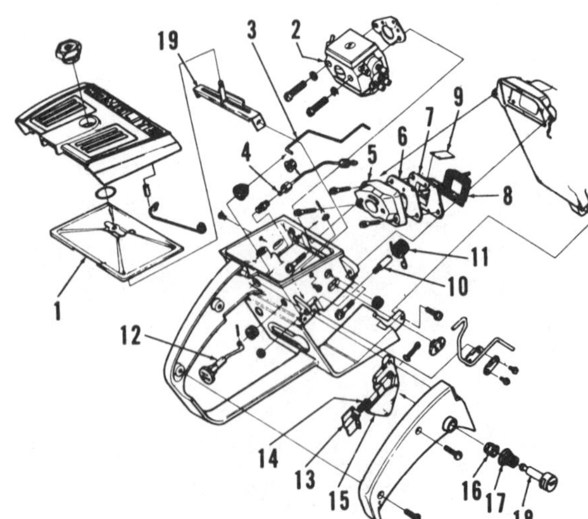

6/32 condenser ....20 in.-lbs. (2.3 N•m)
8/32 air filter bracket........25 in.-lbs (2.8 N•m)
8/32 connecting rod ....55 in.-lbs. (6.2 N•m)
8/32 throttle handle cover ....35 in.-lbs. (4 N•m)
8/32 rewind spring cover.....35 in.-lbs. (4 N•m)
8/32 intake manifold (reed spacer) ....20 in.-lbs. (2.3 N•m)
8/32 coil assembly ....20 in.-lbs. (2.3 N•m)
8/32 fuel tank ....35 in.-lbs. (4 N•m)
10/32 main bearing retainer screw............50 in.-lbs. (5.7 N•m)
10/32 muffler body..........50 in.-lbs. (5.7 N•m)
10/32 starter housing........50 in.-lbs. (5.7 N•m)
10/32 carburetor ....20 in.-lbs. (2.3 N•m)
10/32 starter pawl studs .....50 in.-lbs. (5.7 N•m)
10/32 handle bar............50 in.-lbs. (5.7 N•m)
12/24 throttle handle ....80 in.-lbs. (9.7 N•m)
12/24 fuel tank to crankcase ..75 in.-lbs. (8.5 N•m)
12/24 drive case ....75 in.-lbs. (8.5 N•m)
1/4-28 cylinder nuts ....100 in.-lbs. (11.3 N•m)
5/16-24 rotor (flywheel nut)..150 in.-lbs. (17 N•m)
14 mm spark plug...........120 in.-lbs. (13.6 N•m)
Clutch hub ....180 in.-lbs. (20.3 N•m)

SPECIAL SERVICE TOOLS. Special service tools which may be required are listed as follows:
24299– Anvil, crankshaft installation
24300– Sleeve, crankshaft bearing
24294– Plug, needle bearing
24292– Plug, seal removal
24298– Plug, bearing and seal
24320– #3 Pozidriv screwdriver bit
A24290–Bracket, rotor remover
A24060–Wrench, clutch spanner
A24309–Jackscrew, crankshaft and-bearing
23136-1–Body for A24309
24295– Bearing collar for A24309
24291– Sleeve, drive case seal
24297– Sleeve, crankcase seal

CYLINDER. The cylinder can be unbolted and removed from crankcase after removing starter housing and throttle handle. Be careful not to let piston strike crankcase as cylinder is removed.

The cylinder bore is chrome plated and cylinder should be renewed if the chrome plating has worn through exposing the softer base metal. Also inspect for cracks and damage to compression release valve bore.

PISTON, PIN AND RINGS. Model EZ-10 and Chipper piston has one Head land type ring. Piston ring should be renewed if ring end gap exceeds 0.016 inch (0.4 mm); desired ring end gap is 0.006-0.016 inch (0.15-0.40 mm). The base side of the ring has a cut-out at the end gap to fit the ring locating pin in piston ring groove.

Piston pin is retained by snap rings at both ends of piston pin. Open end of snap ring should be towards closed end of piston.

Assemble piston to connecting rod so piston ring locating pin is towards intake side of cylinder (away from exhaust port).

CONNECTING ROD. Connecting rod and piston assembly can be detached from crankshaft after removing cylinder; refer to Fig. HL95. Be careful to remove all of the 28 loose needle bearing rollers.

Renew connecting rod if bent, twisted or if crankpin bearing surface shows visible wear or is scored. The needle roller bearing for piston pin should be renewed if any roller shows flat spot or if worn so any two rollers can be separated the width equal to thickness of one roller and if rod is otherwise serviceable. Press on lettered side of bearing cage only when removing and installing bearing.

The crankpin needle rollers should be renewed at each overhaul. To install connecting rod, refer to Fig. HL96.

Fig. HL93—Before tightening screws retaining air filter bracket (19—Fig. HL92) in throttle handle, place air filter element on bracket stud and align filter with edges of air box.

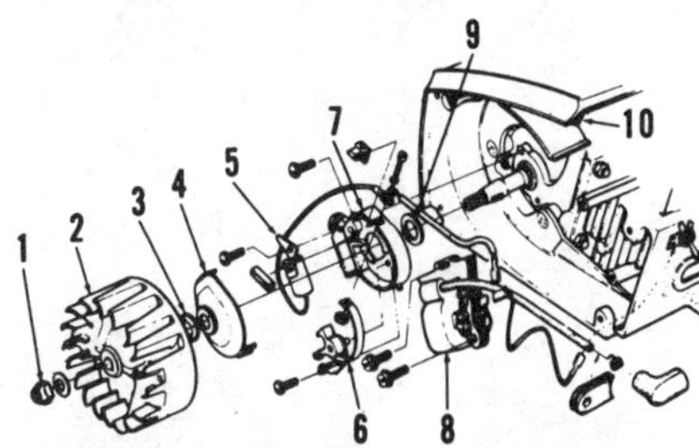

Fig. HL94—Exploded view of ignition assembly. Felt seal (3) is cemented to breaker box cover (4).

1. Nut
2. Flywheel
3. Felt seal
4. Box cover
5. Condenser
6. Breaker points
7. Breaker box
8. Ignition coil
9. Felt seal
10. Fuel tank

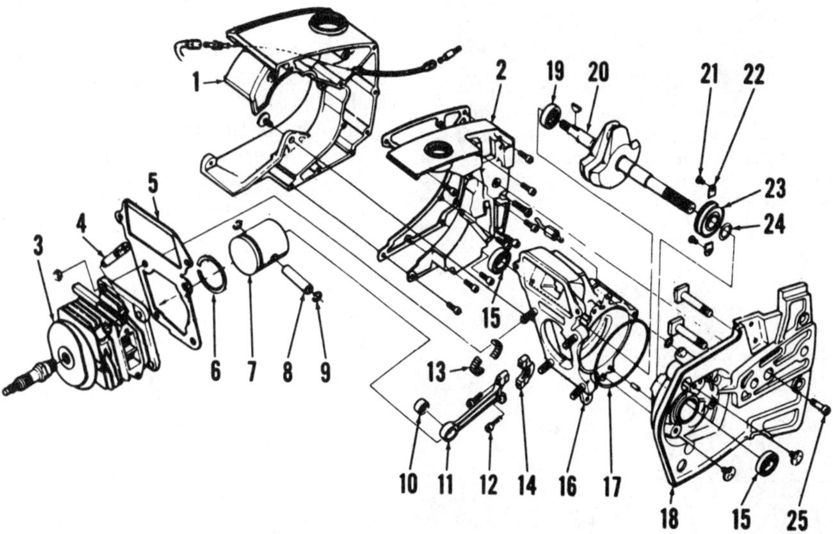

Fig. HL95—Exploded view of typical engine assembly.

1. Fuel tank
2. Oil tank
3. Cylinder
4. Compression release valve
5. Gasket
6. Piston rings
7. Piston
8. Piston pin
9. Pin retainer
10. Needle bearing
11. Connecting rod
12. Cap screw
13. Bearing rollers (28)
14. Connecting rod cap
15. Seal
16. Crankcase
17. "O" ring
18. Drive case
19. Roller bearing
20. Crankshaft
21. Screw
22. Bearing retainer
23. Ball bearing
24. Snap ring
25. Screw

Stick 14 rollers in cap with grease. Support rod cap in crankcase, then place rod over crankpin and to cap with match marks aligned and install new retaining cap screws.

CRANKSHAFT, BEARINGS AND SEALS. Crankshaft is supported by a roller bearing (19 – Fig. HL95) mounted in crankcase bore and by a ball bearing (23) mounted in drive case (18).

To remove crankshaft, first remove blade arm, clutch assembly, starter housing, magneto rotor, throttle handle, cylinder, piston and connecting rod assembly and the fuel/oil tank assembly. Remove retaining screws and separate drive case and crankshaft from crankcase. Use "Pozidriv" screwdriver bit only when removing drive case to fuel tank cover screw (25). Remove the two main bearing retaining screws (21) and special washers (22), then push crankshaft and ball bearing (23) from drive case. Remove snap ring (24) and press crankshaft from ball bearing.

When reassembling, be sure groove in outer race of ball bearing is towards crankpin and that retaining snap ring is seated in groove on crankshaft. Install new seals (15) with lip of seal inward. Using protector sleeve to prevent damage to seal, press the crankshaft and ball bearing into drive case and install new retaining screws and washers. Assemble crankcase to crankshaft and drive case using new "O" ring (17) and protector sleeve to prevent damage to crankcase seal. Be sure bar studs are in place before installing fuel tank.

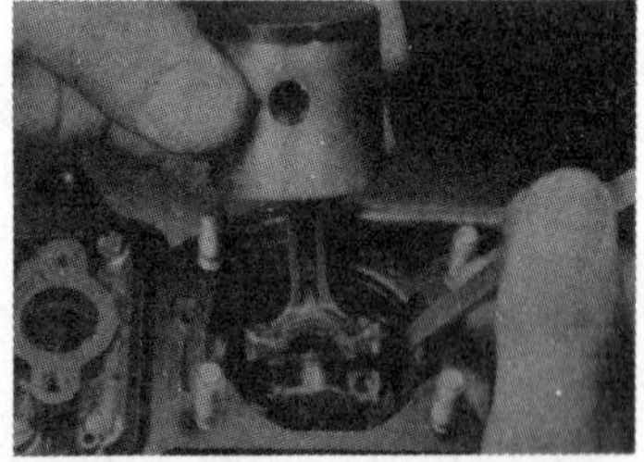

Fig. HL96 – Installing piston and connecting rod assembly using locally made tool to hold rod cap in position. Tool can be made from flat strip of metal. Using grease, stick 14 rollers in cap and 14 rollers in rod. Make sure match marks on rod and cap are aligned.

Fig. HL97 – Roller type main bearing used at flywheel end of crankshaft is marked on one side, "PRESS OTHER SIDE". Be sure to observe this precaution when installing bearing in crankcase.

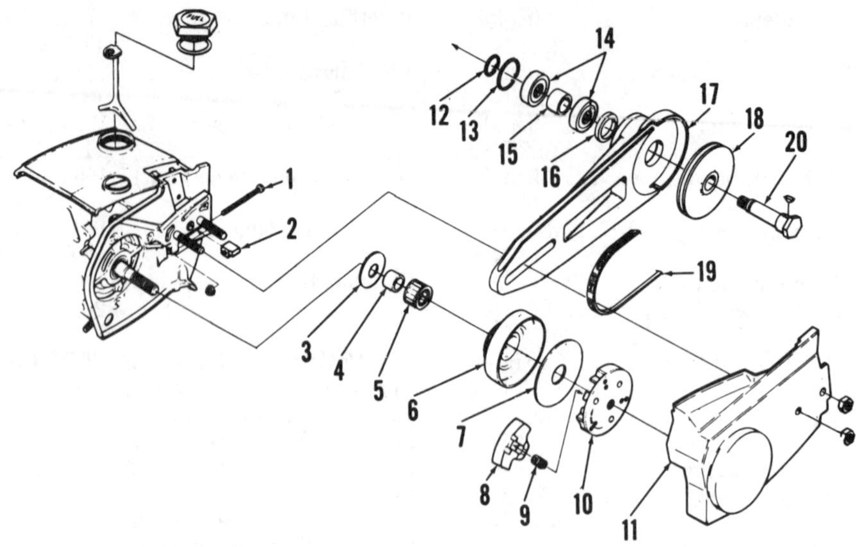

Fig. HL98 – Exploded view of clutch and wheel arm assemblies.

1. Belt tension screw
2. Arm locating block
3. Thrust washer
4. Inner race
5. Needle bearing
6. Clutch drum
7. Thrust washer
8. Clutch shoe
9. Spring
10. Clutch hub
11. Side cover
12. Snap ring
13. Snap ring
14. Ball bearing
15. Spacer
16. Spacer
17. Wheel arm
18. Driven pulley
19. Belt
20. Shaft

COMPRESSION RELEASE. When throttle lock is pushed in, a lever connected to throttle lock lifts away from compression release valve (4 – Fig. HL95). When engine is cranked, compression forces valve open and compression is partly relieved through port in cylinder. Squeezing throttle trigger after engine is running releases throttle lock, allowing spring (11 – Fig. HL92) to snap lever against release valve, closing the valve.

Service of compression release valve usually consists of cleaning valve seat and port in cylinder as carbon may gradually fill the port.

When overhauling engine, cylinder should be inspected for any damage to compression release port.

PYRAMID REED VALVE. A "Delrin" plastic pyramid type reed intake valve seat and four reeds are used. Reeds are retained on pins projecting from the reed seat by a moulded retainer. Inspect reed seat, retainer and reeds for any distortion, excessive wear or other damage.

To reinstall, use a drop of oil to stick each reed to the plastic seat, then push reed retainer down over the seat and reeds. Install the assembly in crankcase; never install retainer, then attempt to install reed seat and reeds.

CLUTCH. Refer to Fig. HL98 for exploded view of the shoe type clutch. The

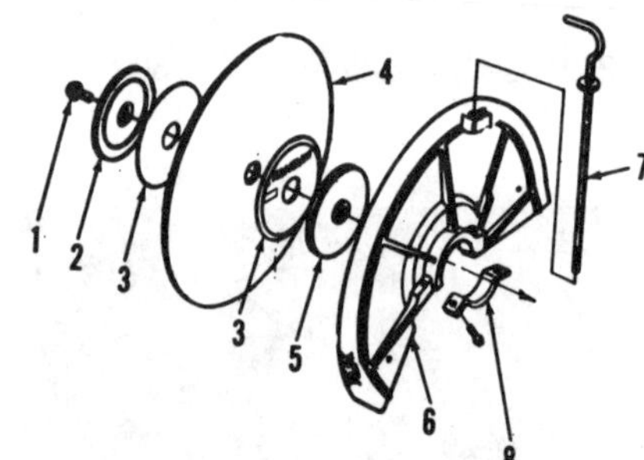

Fig. HL100 – Exploded view of Model EZ-10 wheel assembly.

1. Cap screw
2. Outer wheel washer
3. Spacer
4. Wheel
5. Inner wheel washer
6. Guard
7. Hook bolt
8. Clamp

Fig. HL99 – Although clutch shown above is not used on Model EZ-10 or Chipper, easy method of clutch spring and shoe installation is shown.

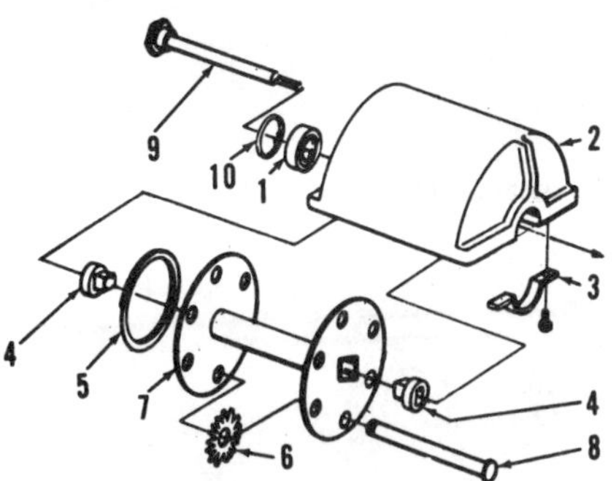

Fig. HL101—Exploded view of cutting assembly on Chipper.

1. Bearing
2. Guard
3. Clamp
4. Insert
5. Retainer
6. Cutters
7. Hub
8. Cutter pin
9. Shaft
10. Snap ring

clutch hub (10) is threaded to crankshaft. Turn clutch hub in clockwise direction to remove from crankshaft.

If clutch slips with engine running at high speed under load, check clutch shoes for excessive wear. If clutch will not release (cutting unit is driven at normal idle speed), check for broken or weak clutch springs.

Refer to Fig. HL99 for easy method of installing clutch shoes and springs on clutch hub.

Fig. HL103—View showing proper installation of pawl springs.

CUTTING WHEEL AND CHIPPER. Model EZ-10 is equipped with an abrasive wheel (4–Fig. HL100) for cutting while Chipper models are equipped with the chipping device shown in Fig. HL101. The abrasive wheel or chipper is attached to the drive assembly shown in Fig. HL98.

To disassemble drive mechanism, remove wheel or chipper assembly from arm (17–Fig. HL98). Remove side cover (11) and belt (19). Separate arm assembly from engine. Remove snap rings (12 and 13), shaft (20) and driven pulley (18). Inspect components (14, 15 and 16) and remove if necessary. Bearings (14) are a press fit in arm.

Belt tension is adjusted by turning adjusting screw (1) with side cover (11) retaining nuts loosed to allow arm (17) to move. Belt tension should be adjusted so belt is tight enough to prevent slippage under load.

REWIND STARTER. Exploded view of rewind starter is shown in Fig. HL102. Starter can be removed as a complete unit by removing housing retaining screws. To disassemble starter, hold cover (7) while removing retaining screws, then allow cover to turn slowly until spring tension is released. Remainder of disassembly is evident from inspection of unit and with reference to exploded view.

Refer to Fig. HL103 to correctly install starter dogs on flywheel. Rewind spring is wound in clockwise direction in cover (7–Fig. HL102). When installing a new starter rope, knot rope ends as shown in Fig. HL104, pull the knots tight and coat with Duxseal (Homelite part 24352). Before installing cover (7–Fig. HL102) retaining screws, turn cover to pull rope handle against starter housing, then continue turning cover three turns to properly tension the rewind spring.

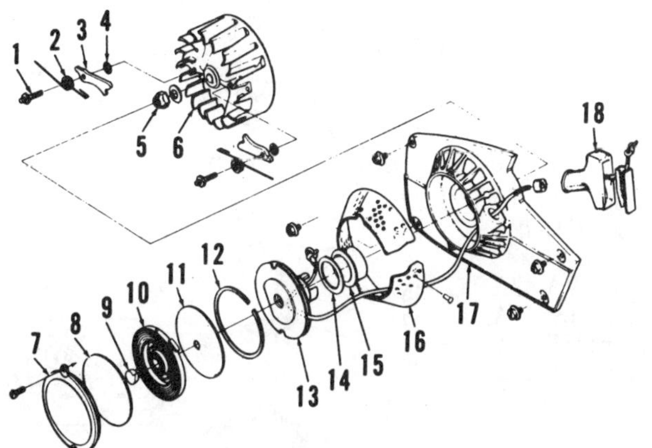

Fig. HL102—Exploded view of recoil starter.

1. Stud
2. Spring
3. Pawl
4. Washer
5. Nut
6. Flywheel
7. Cover
8. Spring shield
9. Spring lock
10. Rewind spring
11. Spring shield
12. Snap ring
13. Rope pulley
14. Washer
15. Bushing
16. Screen
17. Starter housing
18. Rope handle

Fig.HL104—When installing new starter rope, knot ends as shown. Seal rope holes in pulley and coat knots with Duxseal (Homelite part 24382).

HOMELITE

Model	Bore	Stroke	Displacement
HB-280, P100-1, ST-80, ST-100, ST-120	1-5/16 in. (33.3 mm)	1-3/16 in. (30.2 mm)	1.6 cu. in. (26.2 cc)
HB-480, HB-680, ST-200, ST-210	1-7/16 in. (36.5 mm)	1-3/16 in. (30.2 mm)	1.9 cu. in. (31.2 cc)

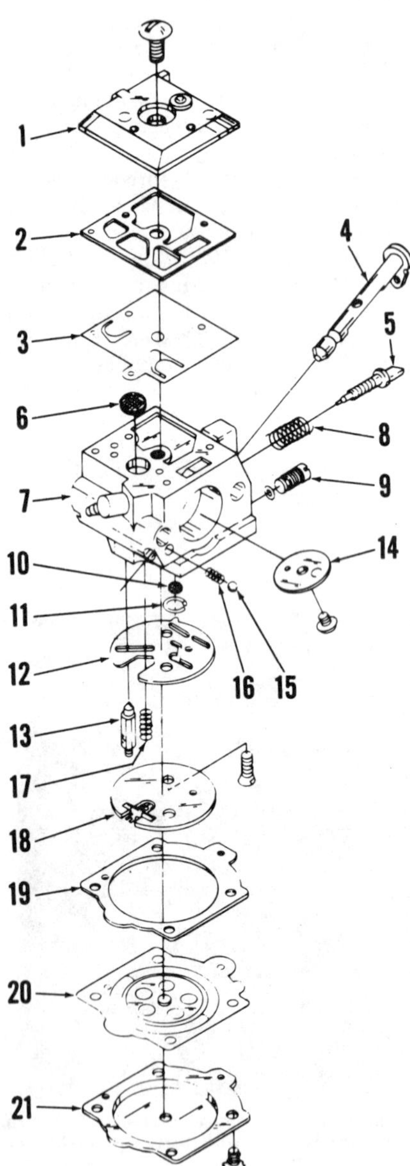

Fig. HL105—Exploded view of Walbro Model HDC carburetor.

1. Pump cover
2. Gasket
3. Fuel pump diaphragm
4. Throttle shaft
5. Idle mixture screw
6. Screen
7. Body
8. Spring
9. High speed mixture screw
10. Check valve screen
11. Retainer
12. Gasket
13. Fuel inlet valve
14. Throttle plate
15. Choke friction ball
16. Spring
17. Spring
18. Circuit plate
19. Gasket
20. Metering diaphragm
21. Cover

ENGINE INFORMATION

These engines are used on Model P100-1 water pumps, Model ST-80, ST-100, ST-200 and ST-210 string trimmers and brushcutters and Model HB-280, HB-480 and HB-680 blowers and sprayers. Early Models HB-480 and HB-680 are equipped with 1.6 cubic inch (26.2 cc) displacement engine.

MAINTENANCE

SPARK PLUG. Recommended spark plug is a Champion DJ7J. Spark plug electrode gap should be 0.025 inch (0.64 mm).

CARBURETOR. Refer to the following table for carburetor applications:

Model	Carburetor
HB-280, HB-480, HB-680	Walbro WA-95
P100-1	Walbro HDC 68
ST-80	Walbro WA-83
ST-100	Tillotson HU-42A Walbro WA-43
ST-120	Walbro WA-43A Zama C1S-H2A
ST-200	Walbro HDC-59 Walbro HDC-69
ST-210	Walbro HDC-69

Refer to Figs. HL105 through HL108 for exploded view of carburetors.

Adjust the carburetor on Models ST-80, ST-100, ST-120, ST-200 and ST-210 using the following procedure. Unit must be shut off when making adjustments and throttle cable must move freely. Initial setting of idle mixture screw is 1¼ turns open. Adjust idle mixture screw so trimmer idles at highest speed and accelerates smoothly. Adjust idle speed screw so engine idles at 2800-3200 rpm on Models ST-80, ST-100 and ST-120, or below clutch engagement speed on Models ST-200 and ST-210. The idle speed screw on Models ST-80, ST-100 and ST-120 is located in trimmer handle. High speed mixture is not adjustable.

There is no idle mixture or idle speed adjustment on P100-1 water pump as pump engine runs at wide-open throttle during operation. To adjust high speed mixture screw, run pump while pumping water so engine is loaded. Turn high speed mixture screw in until engine speed drops, then back screw out approximately ¼ turn or until maximum engine speed is obtained. Backing screw out too far will cause engine speed to drop and adjustment must be repeated.

Before adjusting carburetor on HB-280, HB-480 and HB-680 blowers,

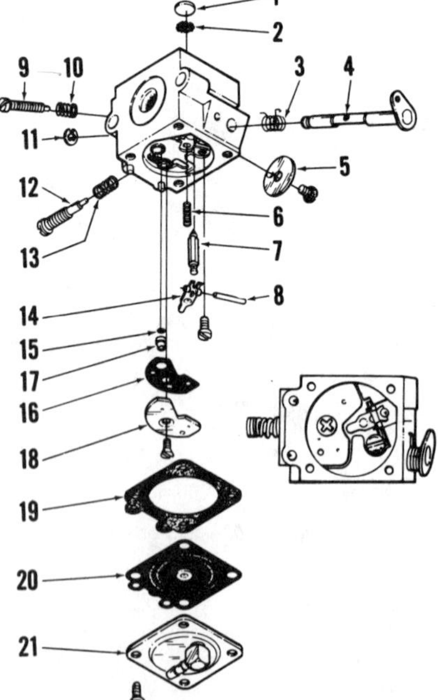

Fig. HL106—Exploded view of typical Walbro WA carburetor.

1. Welch plug
2. Screen
3. Spring
4. Throttle shaft
5. Throttle plate
6. Spring
7. Fuel inlet valve
8. Lever pin
9. High speed mixture screw
10. Spring
11. Clip
12. Idle mixture screw
13. Spring
14. Diaphram lever
15. Screen
16. Gasket
17. Check valve
18. Circuit plate
19. Gasket
20. Metering diaphragm
21. Fuel inlet cover

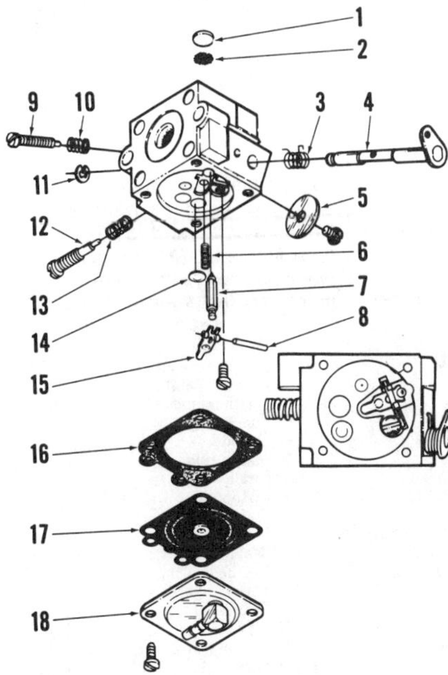

Fig. HL107—Exploded view of typical Tillotson Model HU carburetors.

1. Welch plug
2. Screen
3. Spring
4. Throttle shaft
5. Throttle plate
6. Spring
7. Fuel inlet valve
8. Lever pin
9. High speed mixture screw
10. Spring
11. Clip
12. Idle adjust screw
13. Spring
14. Welch plug
15. Inlet control lever
16. Gasket
17. Metering diaphragm
18. Fuel inlet cover

first be sure throttle cable is adjusted properly. With throttle trigger fully depressed, the carburetor throttle plate should be completely open; if not, adjust throttle cable. Initial adjustment of idle and high speed mixture screws is one turn open. Adjust idle mixture screw so engine will accelerate smoothly. Adjust idle speed screw so idle speed is not less than 3350 rpm or more than 3500 rpm. Adjust high speed mixture screw to obtain maximum engine speed which should be between 6800 rpm and 7400 rpm. To disassemble and overhaul carburetor, refer to the appropriate following paragraph for model being serviced.

Walbro Models HDC and WA. Carburetor may be disassembled after inspection of unit and referral to exploded views in Figs. HL105 or HL106. Care should be taken not to lose ball and spring which will be released when choke shaft is withdrawn.

Clean and inspect all components. Inspect diaphragms for defects which may affect operation. Examine fuel inlet needle and seat. Inlet needle is renewable, but carburetor body must be renewed if needle seat is excessively worn or damaged. Sharp objects should not be used to clean orifices or passages as fuel flow may be altered. Compressed air should not be used to clean main nozzle as check valve may be damaged. A check valve repair kit is available to renew a damaged valve. Fuel mixture needles must be renewed if grooved or broken. Inspect mixture needle seats in carburetor body and renew body if seats are damaged or excessively worn. Screens should be clean.

To reassemble carburetor, reverse disassembly procedure.

On Model HDC, fuel metering lever should be flush with a straightedge laid across carburetor body. On Model WA, lever should be flush with bosses on chamber floor. Be sure lever spring correctly contacts locating dimple on lever before measuring lever height. Bend lever to obtain correct lever height.

Tillotson Model HU. Carburetor may be disassembled after inspecting unit and referral to exploded view in Fig. HL107. Clean filter screen. Welch plugs may be removed by drilling plug with a suitable size drill bit and prying out. Care must be taken not to drill into carburetor body.

Inspect inlet lever spring and renew if stretched or damaged. Inspect diaphragms for tears, cracks or other damage. Renew idle and high speed adjusting needles if needle points are grooved or broken. Carburetor body must be renewed if needle seats are damaged. Fuel inlet needle has a rubber tip and seats directly on a machined orifice in carburetor body. Inlet needle or carburetor body should be renewed if worn excessively.

Carburetor may be reassembled by reversing disassembly procedure. Adjust position of inlet control lever so lever is flush with diaphragm chamber floor. Bend lever adjacent to spring to obtain correct lever position.

MAGNETO AND TIMING. Early Model ST-100 is equipped with a conventional breaker point, flywheel magneto. Breaker point gap should be 0.015 inch (0.38 mm). Ignition timing is not adjustable, however, an incorrect breaker point gap setting will affect ignition timing.

A solid-state ignition is used on all models except early ST-100. The ignition module is attached to the side of the engine cylinder. Ignition service is accomplished by renewing ignition components until faulty part is located. Air gap between ignition module and flywheel is adjustable and should be 0.015 inch (0.38 mm). Loosen ignition module mounting screws and adjust module position to set air gap.

Refer to Fig. HL112 and install flywheel key in crankshaft as shown.

LUBRICATION. The engine is lubricated by mixing oil with regular gasoline. A good quality oil designed for two-stroke engines mixed at a 16:1 fuel:oil ratio is recommended.

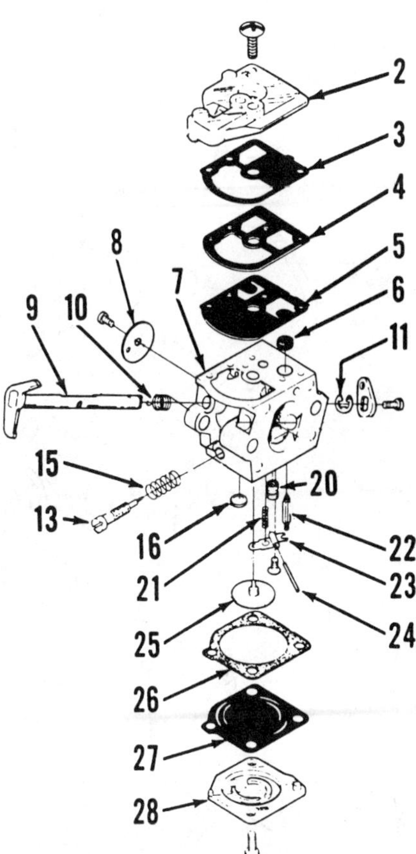

Fig. HL108—Exploded view of typical Zama carburetor used on ST-120 trimmer.

2. Fuel pump cover
3. Gasket
4. Plate
5. Fuel pump diaphragm
6. Screen
7. Body
8. Throttle plate
9. Throttle shaft
10. Spring
11. "E" ring
13. Idle mixture screw
15. Spring
16. Plug
20. Check valve
21. Spring
22. Fuel inlet valve
23. Metering lever
24. Pin
25. Metering disc
26. Gasket
27. Metering diaphragm
28. Cover

MUFFLER. Outer screen of muffler should be cleaned of debris every week or as required. Carbon should be removed from muffler and engine ports to prevent excessive carbon buildup and power loss. Do not allow loose carbon to enter cylinder and be careful not to damage exhaust port or piston.

REPAIRS

TIGHTENING TORQUES. Recommended tightening torques are as follows:

Flywheel . . . . 100 in.-lbs. (11.3 N·m)
Clutch hub . . . . 100 in.-lbs. (11.3 N·m)
Spark plug . . . . 150 in.-lbs. (17 N·m)
Crankcase screws (socket head) . . . . 35 in.-lbs. (4 N·m)
Starter pulley screw . . . . . . 40-50 in.-lbs. (4.5-5.7 N·m)

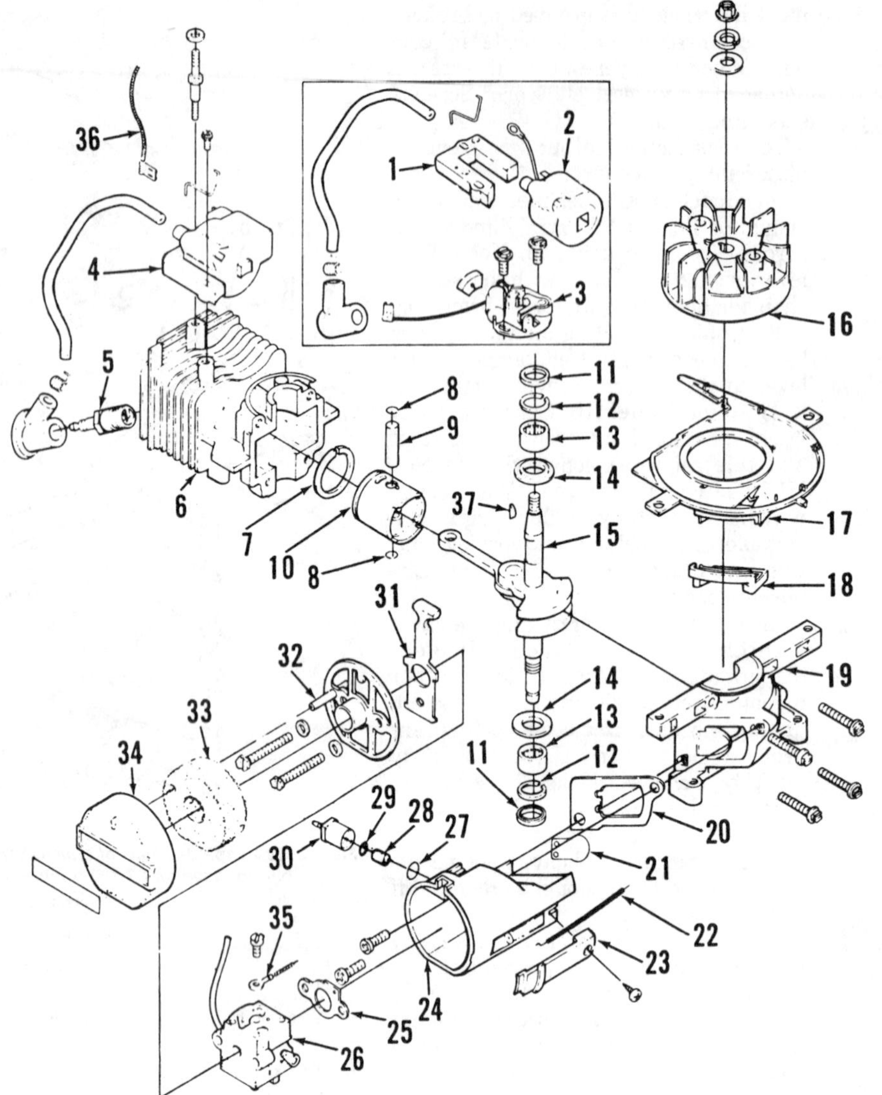

Fig. HL109—Exploded view of engine used on Models ST-80, ST-100 and ST-120. Early Model ST-100 trimmers are equipped with breaker point ignition shown in inset.

1. Coil core
2. Ignition coil
3. Breaker point assy.
4. Ignition module
5. Spark plug
6. Cylinder
7. Piston ring
8. Circlip
9. Piston pin
10. Piston
11. Seal
12. Seal spacer
13. Needle bearing
14. Thrust washer
15. Crankshaft
16. Flywheel
17. Air cover
18. Seal
19. Crankcase
20. Gasket
21. Reed valve petal
22. Throttle cable
23. Cable clamp
24. Carburetor housing
25. Gasket
26. Carburetor
27. "O" ring
28. Filter
29. Gasket
30. Fuel inlet
31. Choke
32. Filter support
33. Air filter
34. Cover
35. Ground wire (early ST-100)
36. Ground wire (solid state ignition)
37. Key

CYLINDER, PISTON, PIN AND RINGS. Cylinder may be removed after unscrewing four screws in bottom of crankcase (19–Fig. HL109, HL110 or HL111). Be careful when removing cylinder as crankshaft assembly will be loose in crankcase. Care should be taken not to damage mating surfaces of cylinder and crankcase.

Inspect crankshaft bearings and renew if scored or worn. Thrust washers (14) should be installed with shoulder to outside. On P100-1 water pump models, note tangs on thrust washers (14) which must index in crankcase (19). Crankshaft seals are installed with a seal lip to inside. Cylinder and crankcase mating surfaces should be flat and free of nicks and scratches. Mating surfaces should be cleaned then coated with silicone sealer before assembly.

Bearings, seals and thrust washers must be positioned correctly on crankshaft before final assembly.

To install crankshaft with piston assembly installed on rod, insert piston

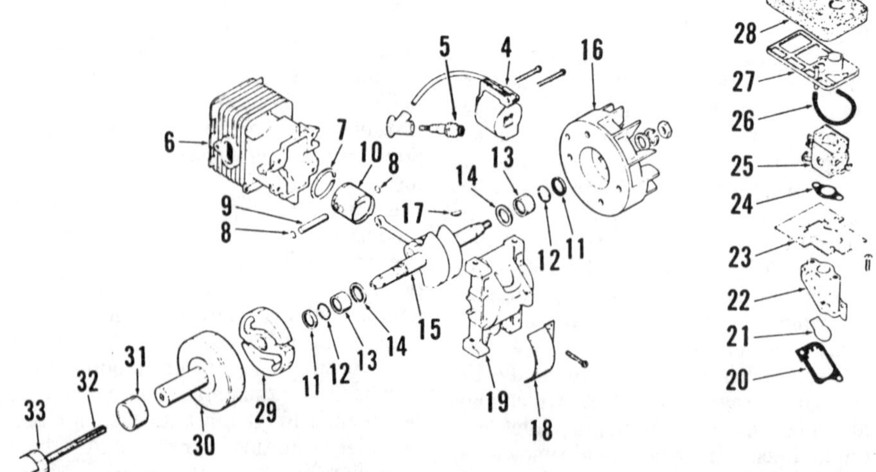

Fig. HL110—Exploded view of engine used on Models ST-200 and ST-210. Model P100-1 water pump is similar but clutch components are not used. Thrust washers (14) on Model P100-1 are equipped with tangs which must index in crankcase (19).

4. Ignition module
5. Spark plug
6. Cylinder
7. Piston ring
8. Circlip
9. Piston pin
10. Piston
11. Seal
12. Seal spacer
13. Needle bearing
14. Thrust washer
15. Crankshaft
16. Flywheel
17. Key
18. Shroud
19. Crankcase
20. Gasket
21. Reed valve petal
22. Carburetor spacer
23. Air baffle
24. Gasket
25. Carburetor
26. Tubing
27. Filter support
28. Air filter
29. Clutch hub
30. Clutch drum
31. Bushing
32. Driveshaft
33. Drive tube

in cylinder being sure piston ring is aligned on locating pin. Install thrust washers (14), bearings (13), seal spacers (12) and seals (11) on crankshaft. Place 0.0125 (0.32 mm) thick shims shown in Fig. HL113 between thrust washers and bearings as shown in Fig. HL114. Gently push seals toward crankshaft counterweights until assemblies are snug. Remove shims and complete assembly being careful not to disturb position of thrust washers, bearings and seals.

Before final tightening of crankcase screws, lightly tap both ends of crankshaft to obtain proper crankshaft end play, then tighten crankcase screws. On P100-1 water pump models, be sure tangs on thrust washers index in recesses of crankcase.

REED VALVE. All models are equipped with a reed valve induction system. Renew reed petal (21–Fig. HL109, HL110 or HL111) if cracked, bent or otherwise damaged. Do not attempt to straighten a bent reed petal. Seating surface for reed petal should be flat, clean and smooth.

CLUTCH. Models ST-200 and ST-210 are equipped with centrifugal clutch (29–Fig. HL110) which is accessible after removing engine housing. Clutch hub (29) has left-hand threads. Inspect bushing (31) and renew if excessively worn. Install clutch hub (29) while noting "OUTSIDE" marked on side of hub.

REWIND STARTER

To service the rewind starter on all models except Model HB-280, HB-480 and HB-680 blower/sprayers, remove starter housing (10–Fig. HL115 or HL116). Pull starter rope and hold rope pulley with notch in pulley adjacent to rope outlet. Pull rope back through outlet so it engages notch in pulley and allow pulley to completely unwind. Unscrew pulley retaining screw (5) and remove rope pulley being careful not to dislodge rewind spring in housing. Care must be taken if rewind spring is removed to prevent injury if spring is allowed to uncoil uncontrolled.

Rewind spring is wound in clockwise direction in starter housing. Rope is wound on rope pulley in clockwise direction as viewed with pulley in housing. To place tension on rewind spring, pass rope through rope outlet in housing and install rope handle. Pull rope out and hold rope pulley so notch on pulley is adjacent to rope outlet. Pull rope back through outlet between notch in pulley and housing. Turn rope pulley clockwise

POINT IGNITION

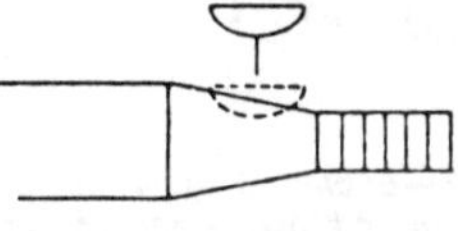

SOLID STATE IGNITION

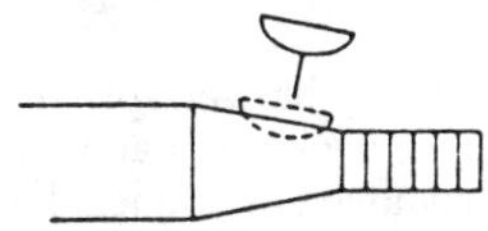

Fig. HL112—Flywheel key must be installed in crankshaft as shown above according to ignition type.

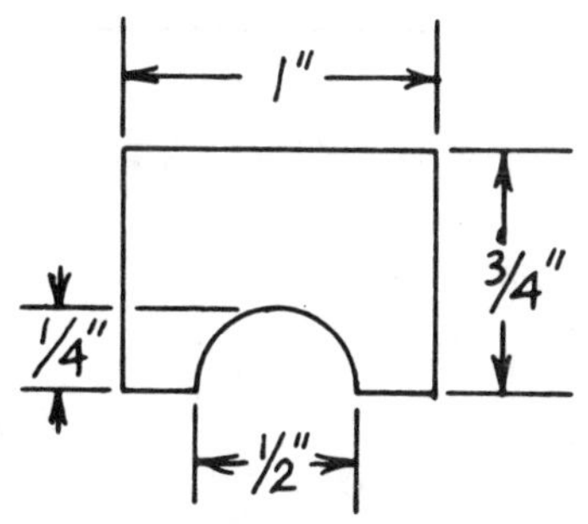

Fig. HL113—Shims used in crankshaft assembly may be made by cutting 0.0125-inch (0.32 mm) thick plastic, metal or other suitable material in the outline shown above. Refer to Fig. HL114 and text.

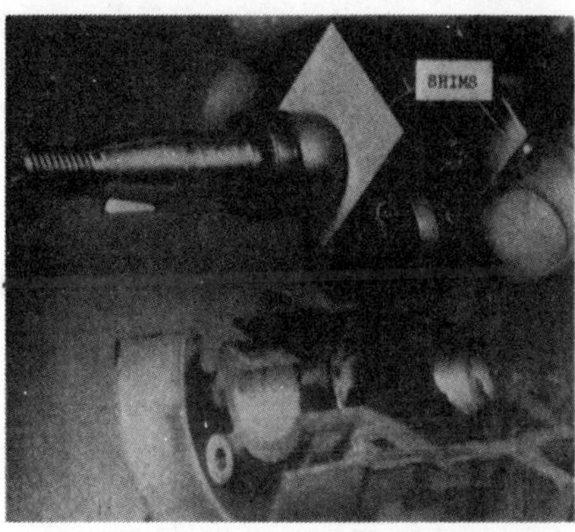

Fig. HL114—View showing placement of shims (Fig. HL113) between thrust washers and bearings for correct crankshaft assembly. Refer to text.

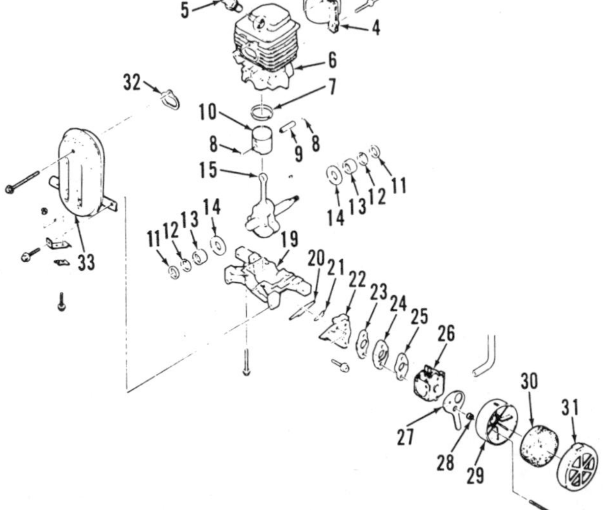

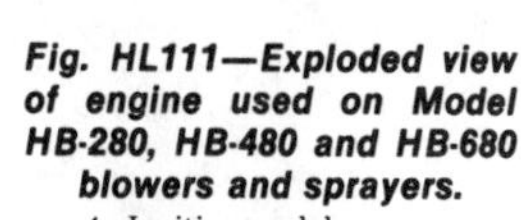

Fig. HL111—Exploded view of engine used on Model HB-280, HB-480 and HB-680 blowers and sprayers.

4. Ignition module
5. Spark plug
6. Cylinder
7. Piston ring
8. Circlip
9. Piston pin
10. Piston
11. Seal
12. Seal spacer
13. Needle bearing
14. Thrust washer
15. Crankshaft assy.
19. Crankcase
20. Gasket
21. Reed valve petal
22. Carburetor spacer
23. Gasket
24. Intake manifold
25. Spacer
26. Carburetor
27. Choke lever
28. Spacer
29. Air filter housing
30. Filter element
31. Cover
32. Gasket
33. Muffler

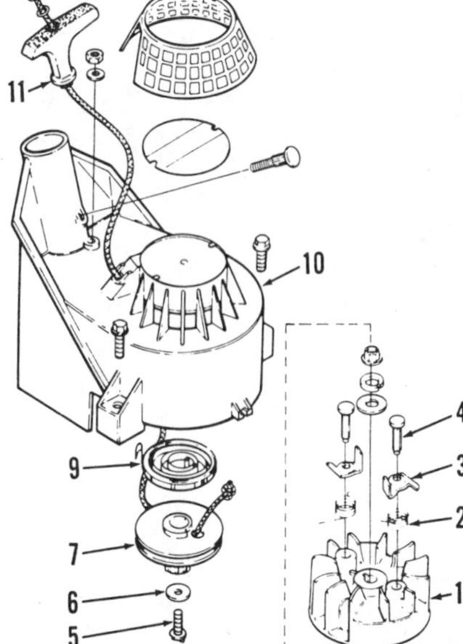

Fig. HL115—Exploded view of rewind starter used on Models ST-80, ST-100 and ST-120.

1. Flywheel
2. Spring
3. Pawl
4. Pawl pin
5. Screw
6. Washer
7. Rope pulley
9. Rewind spring
10. Housing
11. Rope handle

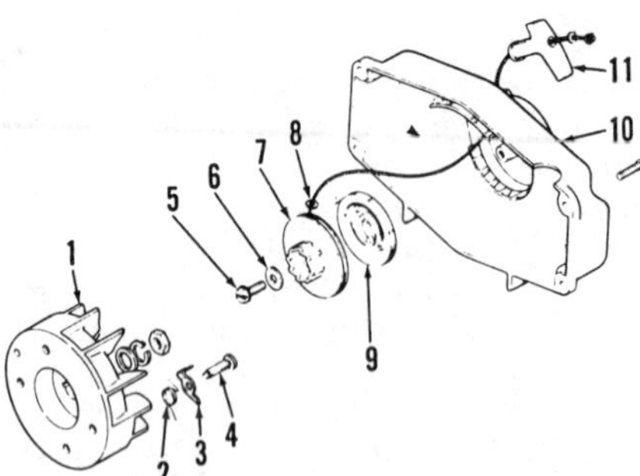

Fig. HL116—Exploded view of rewind starter used on Models ST-200 and ST210.

1. Flywheel
2. Spring
3. Pawl
4. Pawl pin
5. Screw
6. Washer
7. Rope pulley
8. Nylon washer
9. Rewind spring
10. Housing
11. Rope handle

to place tension on spring. Release pulley and check starter action. Do not place more tension on rewind spring than is necessary to draw rope handle up against housing.

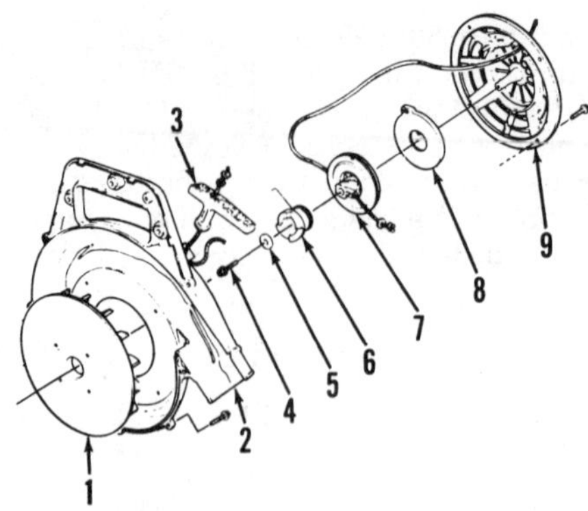

Fig. HL117—Exploded view of recoil starter used on blowers and sprayers.

1. Flywheel
2. Housing
3. Handle
4. Cap screw
5. Washer
6. Ratchet
7. Pulley
8. Spring retainer
9. Starter housing

To service blower/sprayer rewind starters, the back rest pad and back rest must be removed on Models HB-480 and HB-680 for access to starter. On all models, unscrew mounting screws and remove starter. Remove rope handle and allow rope to wind into starter. Unscrew retaining screw (4–Fig. HL117) and remove ratchet (6) and pulley while being careful not to dislodge rewind spring in housing.

When assembling starter, wind rope around rope pulley in a clockwise direction as viewed with pulley in housing. To place tension on rewind spring, pass rope through rope outlet in housing and install rope handle. Pull rope out and hold rope pulley so notch on pulley is adjacent to rope outlet. Pull rope back through outlet between notch in pulley and housing. Turn rope pulley clockwise to place tension on spring. Release pulley and check starter action. Do not place more tension on rewind spring than is necessary to draw rope handle up against housing. Install ratchet (6) with hooked end of ratchet lever wire up and between posts of starter housing. Install screw (4) and washer (5) then install starter housing.

HOMELITE

Model	Bore	Stroke	Displacement
DM50	1.875 in. (47.6 mm)	1.625 in. (41.3 mm)	4.5 cu. in. (74 cc)

ENGINE INFORMATION

This engine is used as the power unit on the Model DM50 Multi-Purpose Saw.

MAINTENANCE

SPARK PLUG. Recommended spark plug is Champion DJ6J, or equivalent. Spark plug electrode gap should be 0.025 inch (0.64 mm). A Champion CJ4 may be used for heavy duty operation in hot temperatures.

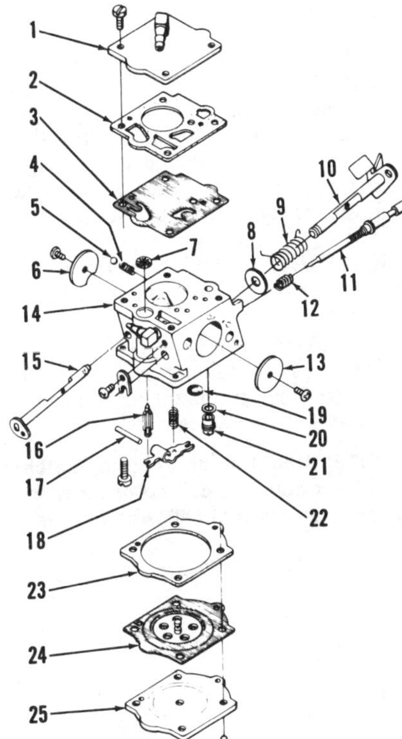

Fig. HL120—Exploded view of Walbro SDC carburetor used on DM50 Multi-Purpose Saw.

1. Fuel inlet
2. Gasket
3. Fuel pump diaphram
4. Spring
5. Friction ball
6. Choke plate
7. Screen
8. Spacer
9. Spring
10. Throttle shaft
11. Idle adjust screw
12. Spring
13. Throttle plate
14. Body
15. Choke shaft
16. Fuel inlet valve
17. Lever pin
18. Diaphram lever
19. Welch plug
20. Gasket
21. Check valve
22. Spring
23. Gasket
24. Metering diaphgram
25. Cover

CARBURETOR. Model DM50 is equipped with a Walbro SDC diaphragm carburetor. Refer to exploded view of carburetor in Fig. HL120.

Initial adjustment of idle speed mixture screw is one turn open. High speed mixture is not adjustable. Adjust idle speed screw (Fig. HL121) so engine idles at 2400-2600 rpm. Adjust idle mixture screw so engine will accelerate smoothly without stalling. If necessary, readjust idle speed screw to obtain engine idle speed of 2400-2600 rpm.

Starting speed is adjusted by turning slotted head adjustment screw in fast idle latch. See Fig. HL122. Turning screw clockwise raises starting speed while turning screw counterclockwise lowers starting speed. Adjust starting speed by latching trigger in start position, start engine and turn screw until desired engine speed is obtained.

Carburetor may be disassembled after inspection of unit and referral to exploded view in Fig. HL120. Care should be taken not to lose ball and spring which will be released when choke shaft is withdrawn.

Clean and inspect all components. Inspect diaphragms for defects which may affect operation. Examine fuel inlet needle and seat. Inlet needle is renewable but carburetor body must be renewed if needle seat is excessively worn or damaged. Sharp objects should not be used to clean orifices or passages as fuel flow may be altered. Compressed air should not be used to clean main nozzle as check valve may be damaged. A check valve repair kit is available to renew damaged valve. Fuel mixture needle must be renewed if grooved or broken. Inspect mixture needle seats in carburetor body and renew body if seats are damaged or excessively worn. Screws should be clean.

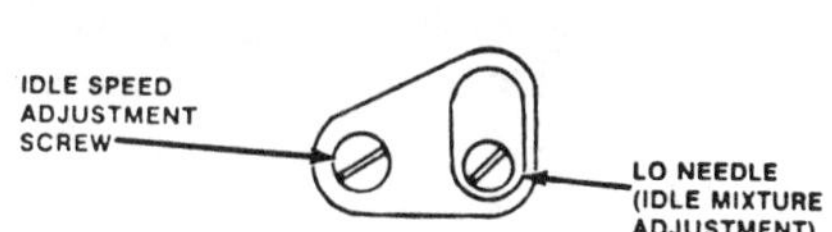

Fig. HL121—View showing location of idle mixture screw and idle speed screw.

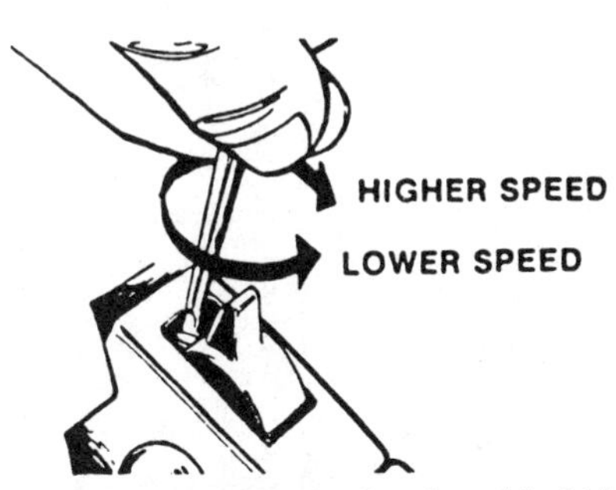

Fig. HL122—View showing location of fast idle screw.

To reassemble carburetor, reverse disassembly procedure. Diaphragm lever (18) should be flush with bosses on chamber floor. Be sure lever spring (22) correctly contacts locating dimple on lever before measuring lever height. Bend lever to obtain correct lever height.

MAGNETO AND TIMING. A solid state ignition is used. The ignition module is mounted adjacent to the flywheel while the high tension coil covers the spark plug and is mounted on the cylinder shield. The high tension coil must be removed for access to spark plug.

The ignition system is serviced by renewing the spark plug, ignition module, high tension coil or wires with new components. The ignition system can be checked by taping or glueing a neon lamp (#NE-2) to the high tension coil as shown in Fig. HL123. The leads

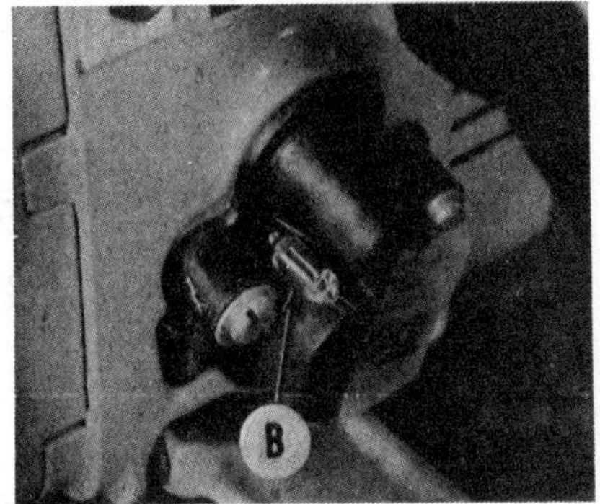

Fig. HL123—Neon bulb (B) may be used to determine if ignition system is operating correctly. Refer to text.

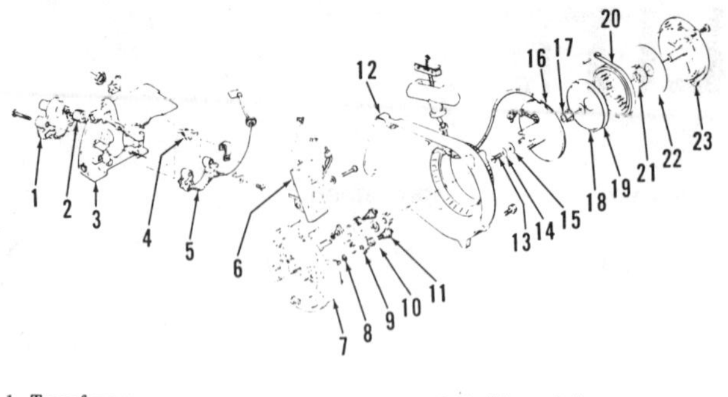

Fig. HL124—Exploded view of ignition system and rewind starter.

1. Transformer
2. Grommet
3. Shield
4. Ignition switch
5. Transformer receptacle
6. Ignition module
7. Flywheel
8. Lockwasher
9. Starter pawl
10. Spring
11. Stud
12. Starter housing
13. Screw
14. Washer
15. Washer
16. Rope pulley
17. Bushing
18. Snap ring
19. Outer spring shield
20. Rewind spring
21. Spring lock
22. Inner spring shield
23. Cover

of the lamp are not attached to the system. When the rewind starter handle is pulled briskly, the lamp should flash with each revolution of the flywheel. It is helpful to compare the lamp flash on a saw with a good ignition system when analyzing the ignition system on a faulty saw. If the lamp flash is dim, irregular or nonexistent, check wires and connections, then renew ignition components by starting with the spark plug until the problem is corrected.

High tension coil and leads may be checked by disconnecting wires at ignition module which lead from ignition module to coil receptacle and connecting an ohmmeter to end of wires. There should be continuity between wire ends. If continuity does not exist, disassemble rear of saw until access is possible to the two coil receptacle leads and disconnect leads. Check continuity of each wire and terminal.

To check ignition switch and lead, connect one probe of ohmmeter to switch terminal and ground remaining probe to ignition module core. Check continuity of ignition switch and lead with switch in "RUN" and "STOP" positions. If continuity exists when switch is in "RUN" position, switch or lead is shorted and must be renewed or repaired. Continuity should exist with switch in "STOP" position. If continuity is not present in "STOP" position, check connection of switch lead and renew lead and switch as necessary.

Air gap between ignition module and flywheel is adjustable. Adjust air gap by loosening module retaining screws and place 0.020 inch (0.51 mm) shim stock between flywheel and module. Load crankshaft bearings during adjustment by applying pressure to flywheel in direction of ignition module.

LUBRICATION. The engine is lubricated by mixing a good quality oil designd for air-cooled two-stroke engines at a 16:1 fuel:oil ratio with regular or non-leaded gasoline which has an 85 or higher octane rating.

REPAIRS

CYLINDER, PISTON, PIN AND RINGS. The cylinder may be separated from crankcase after removing nuts securing cylinder to crankcase. Care should be used when separating cylinder and crankcase as crankshaft may be dislodged from crankcase. Inspect cylinder bore and discard cylinder if excessively worn or damaged. Piston, rings and cylinders are available in standard size only. Refer to CONNECTING ROD, CRANKSHAFT AND CRANKCASE section when installing cylinder.

The piston is equipped with two piston rings. The piston pin rides in nonrenewable needle bearings in piston. Piston and bearings are available only as a unit assembly.

CONNECTING ROD, CRANKSHAFT AND CRANKCASE. Refer to preceding section and remove cylinder. Separate crankshaft assembly from crankcase and disassemble as required. Inspect components and renew any which are damaged.

Connecting rod (11–Fig. HL125) rides on twelve caged bearing rollers (12). The crankshaft is supported by roller bearings (16) which are installed so lettered end is towards snap rings (15).

Tighten connecting rod screws to 65-75 in.-lbs. (7.3-8.5 N·m). When assembling crankcase and cylinder, use a suitable sealant on mating surfaces. Be sure components are properly assembled and snap rings (15) engage grooves in cylinder and crankcase. Tighten retaining screws to 60-70 in.-lbs. (7-8 N·m).

CLUTCH. Model DM50 is equipped with clutch shown in Fig. HL126. Clutch hub has left-hand threads. Clutch shoes should be renewed only as a set. Inspect bearing and lubricate with a good quality lithium base multi-purpose grease.

RECOIL STARTER. Refer to Fig. HL124 for en exploded view of starter assembly.

To disassemble starter, hold cover (23) and unscrew retaining screws. Allow cover to turn until spring tension is relieved and remove cover.

NOTE: If outer hook of spring catches on starter housing, pull cover away from housing until cover is allowed to turn.

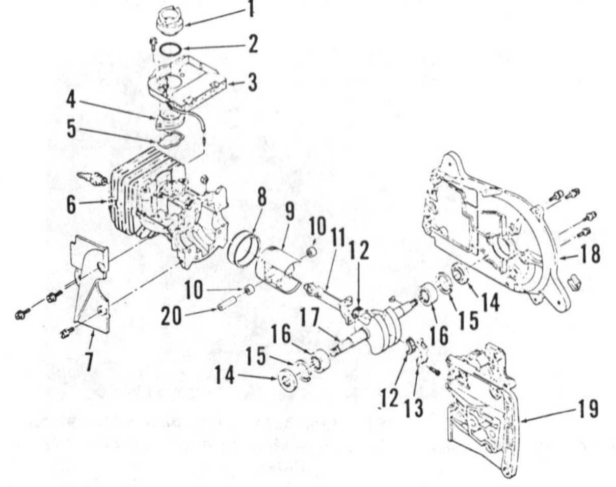

Fig. HL125—Exploded view of engine.

1. Connector
2. Carter spring
3. Air deflector & seal
4. Intake manifold
5. Gasket
6. Cylinder
7. Shield
8. Piston rings
9. Piston
10. Piston pin needle bearings
11. Connecting rod
12. Bearing
13. Rod cap
14. Seal
15. Snap ring
16. Bearing
17. Crankshaft
18. Backplate
19. Crankcase
20. Piston pin

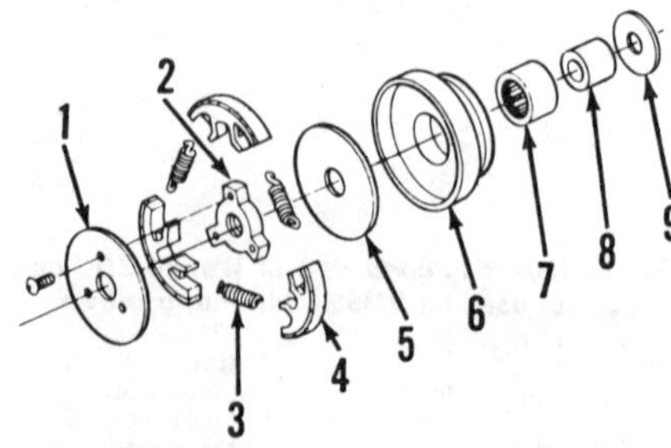

Fig. HL126—Exploded view of clutch used on DM50 Multi-Purpose Saw.

1. Cover plate
2. Hub
3. Spring
4. Shoe
5. Washer
6. Clutch drum
7. Bearing
8. Inner race
9. Thrust washer

Unscrew screw (13) to separate rope pulley (16) from cover. Remove snap ring (18) for access to rewind spring. If starter pawl assemblies must be removed, unscrew housing screws and remove starter housing (12). Threaded inserts are available if stud holes are damaged in flywheel.

Clean and inspect components. Lubricate sides of rewind spring with a small amount of a good quality litium base, multi-purpose grease. Do not oil spring. Install inner spring shield (22), rewind spring (20) and spring lock (21) in cover with spring wound as shown in Fig. HL127. Install outer spring shield (19–Fig. HL124) and snap ring (18). Install inner washer (15). Insert bushing (17) in rope pulley (16) being sure knobs on bushings align with notches in pulley. Slide pulley onto post in cover and check to be sure splines on pulley engage splines in spring lock. Install and tighten cap screws (13) to 45 in.-lbs. (5 N•m). Wind rope around pulley in clockwise direction as viewed from screw end of pulley. Set cover in housing. Pull rope handle and then allow rope to rewind so starter pawls will be forced open and pulley hub can slide between them into place. Turn cover clockwise 2 or 3 turns to preload rewind spring, snap plastic screen into place and install cover screws. Check starter operation.

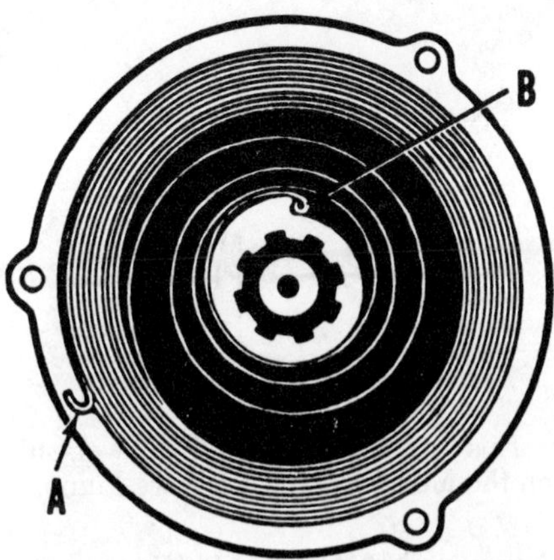

Fig. HL127–View of rewind spring installation in starter cover. Hook outer loop (A) of spring in notch as shown. Inner loop (B) of spring must be curved inward to engage notch of spring lock.

HOMELITE

Model	Bore	Stroke	Displacement
FP-150	2.250 in. (57.2 mm)	1.720 in. (43.7 cc)	6.8 cu.in. (112 cc)

ENGINE INFORMATION

This engine is used as the power unit on the Model FP-150 Pressure Pump.

MAINTENANCE

SPARK PLUG. Recommended spark plug is Champion CJ3, or equivalent. Spark plug electrode gap should be 0.025 inch (0.64 mm).

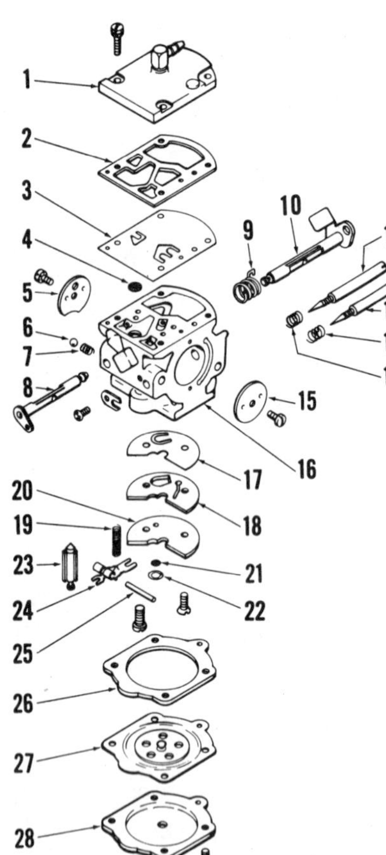

Fig. HL128—Exploded view of Walbro WB-12 carburetor used on Model FP-150 pump engine.

1. Fuel inlet
2. Gasket
3. Pump diaphragm
4. Screen
5. Choke plate
6. Friction ball
7. Spring
8. Choke shaft
9. Spring
10. Throttle shaft
11. High speed mixture screw
12. Idle mixture screw
13. Spring
14. Spring
15. Throttle plate
16. Body
17. Gasket
18. Circuit plate
19. Spring
20. Cover plate
21. Screen
22. Retainer
23. Fuel inlet valve
24. Inlet control lever
25. Lever pin
26. Gasket
27. Metering diaphragm
28. Cover

CARBURETOR. Model FP-150 pump is equipped with a Walbro WB-12 diaphragm carburetor. Refer to Fig. HL128 for exploded view of carburetor.

Before attempting carburetor adjustments, be sure pump is primed. Initial adjustment of idle and high speed mixture screws is one turn open. Adjust idle speed screw (S–Fig. HL129) so engine idles at 2600-3400 rpm. Adjust idle mixture screw (I) to obtain maximum idle speed, then readjust idle speed screw so engine idles as 2600-3400 rpm.

To adjust high speed mixture screw, close discharge valve until engine speed is 7000-75000 rpm or pump discharge pressure is 150-160 psi (1034-1103 kPa). Adjust high speed mixture screw so best lean power is obtained.

Carburetor may be disassembled after inspection of unit and referral to exploded view in Fig. HL128. Care should be taken not to lose ball and spring which will be released when choke shaft is withdrawn.

Clean and inspect all components. Inspect diaphragms for defects which may affect operation. Examine fuel inlet needle and seat. Inlet needle is renewable, but carburetor body must be renewed if needle seat is excessively worn or damaged. Sharp objects should not be used to clean orifices or passages as fuel flow may be altered. Compressed air should not be used to clean main nozzle as check valve may be damaged. A check valve repair kit is available to renew a damaged valve. Fuel mixture needles must be renewed if grooved or broken. Inspect mixture needle seats in carburetor body and renew body if seats are damaged or excessively worn. Screens should be clean.

Fig. HL129—View showing location of idle mixture screw (I), high speed mixture screw (H) and idle speed screw (S).

To reassemble carburetor, reverse disassembly procedure. Fuel metering lever should be flush with bosses on chamber floor. Be sure lever spring (19) correctly contacts locating dimple on lever (24) before measuring lever height. Bend lever to obtain correct lever height.

MAGNETO AND TIMING. A solid state ignition is used on Model FP-150 engine. The ignition module is mounted adjacent to the flywheel while the high tension coil covers the spark plug and is mounted on the cylinder shield. The high tension coil must be removed for access to spark plug.

The ignition system is serviced by renewing the spark plug, ignition module, high tension coil or wires with new components. The ignition system can be checked by taping or glueing a neon lamp (#NE-2) to the high tension coil as shown in Fig. HL130. The leads of the lamp are not attached to the system. When the rewind starter handle is pulled briskly, the lamp should flash with each revolution of the flywheel. It is helpful to compare the lamp flash on a pump with a good ignition system when analyzing the ignition system on a faulty pump. If the lamp flash is dim, irregular or nonexistent, check wire connections,

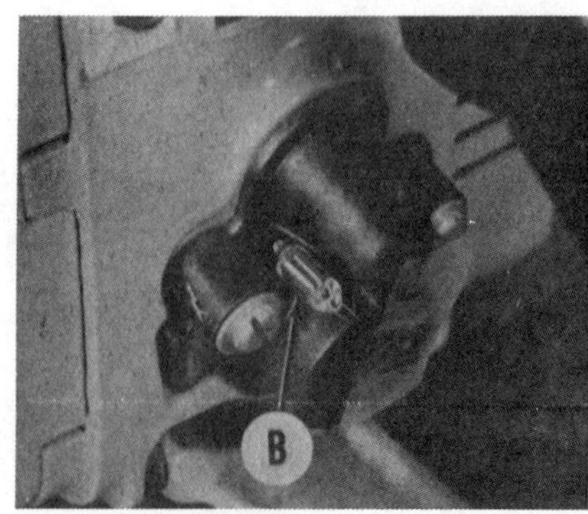

Fig. HL130—A neon bulb may be used to determine if ignition system is operating correctly. Refer to text.

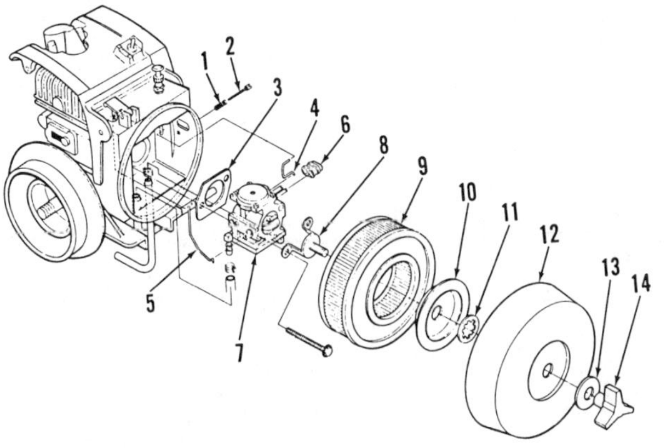

Fig. HL131—Exploded view of intake system.
1. Spring
2. Idle speed screw
3. Spacer
4. Throttle link
5. Choke link
6. Grommet
7. Carburetor
8. Stud
9. Air filter
10. Retainer
11. Lockwasher
12. Cover
13. Washer
14. Knob

then renew ignition components by starting with the spark plug until the problem is corrected.

High tension coil and leads may be checked by disconnecting wires at ignition module which lead from ignition module to coil receptacle and connecting an ohmmeter to end of wires. There should be continuity between wire ends. If continuity does not exist, disassemble rear engine until access is possible to the two coil receptacle leads and disconnect leads. Check continuity of each wire and terminal.

To check ignition switch and lead, connect one probe of ohmmeter to switch terminal and ground remaining probe to ignition module core. Check continuity of ignition switch and lead switch in "RUN" and "STOP" positions. If continuity exists when switch is in "RUN" position, switch or lead is shorted and must be renewed or repaired. Continuity should exist with switch in "STOP" position. If continuity is not present in "STOP" position, check connection of switch lead and renew lead and switch as necessary.

Air gap between ignition module and flywheel is adjustable. Adjust air gap by loosening module retaining screws and place 0.015 inch (0.64 mm) shim stock between flywheel and module. Load crankshaft bearings during adjustment by applying pressure to flywheel in direction of ignition module.

MUFFLER. Muffler should be disassembled and periodically cleaned. Carbon should be removed from muffler and exhaust port to prevent excessive carbon buildup and power loss. Do not allow loose carbon to enter cylinder.

LUBRICATION. The engine is lubricated by mixing a good quality oil designed for air-cooled two-stroke engines at a 16:1 fuel:oil ratio with regular or nonleaded gasoline which has an 85 or higher octane rating.

REPAIRS

TIGHTENING TORQUES. Recommended tightening torques are as follows:

Spark plug	150 in.-lbs. (17 N•m)
Flywheel nut	250-300 in.-lbs. (28-34 N•m)
Cylinder screws	80 in.-lbs. (9 N•m)
Connecting rod screws	70-80 in.-lbs. (8-9 N•m)
Starter housing screws	35 in.-lbs. (4 N•m)
Ignition module screws	35 in.-lbs. (4 N•m)
Transformer coil screws	27 in.-lbs. (3 N•m)
Air box to tank	45 in.-lbs. (5 N•m)
Carburetor mounting screws	45 in.-lbs. (5 N•m)
Starter pawl studs	80-90 in.-lbs. (9-10 N•m)

CYLINDER, PISTON, PIN AND RINGS. Cylinder has a chrome bore which should be inspected for wear or damage. Piston and rings are available in standard sizes only. Piston pin is pressed in rod and rides in two needle roller bearings in piston. Piston and bearings are available as a unit assembly only.

Note that one piston pin boss is marked with an arrow and "EXH". Install piston with side indicated by arrow towards exhaust port.

CONNECTING ROD. Connecting rod is fractured type secured by two socket head screws. Connecting rod rides on a split-caged needle bearing at big end. Marks at big end of rod must be aligned and cap and rod properly mated during assembly. Needle bearings may be held around crankpin with a suitable grease to aid in assembly.

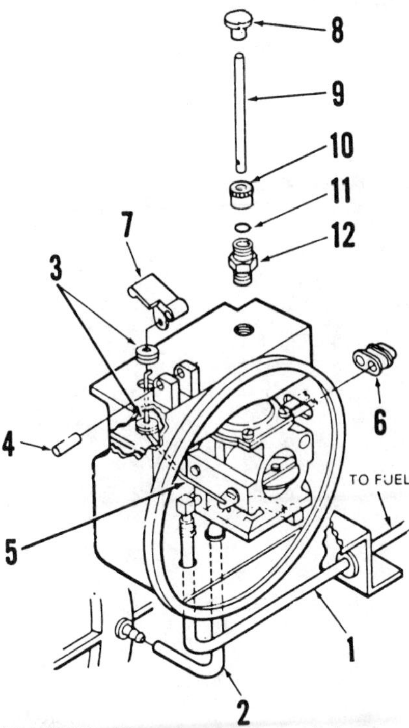

Fig. HL132—View of carburetor and chamber.
1. Fuel line
2. Crankcase pulse line
3. Grommets
4. Pin
5. Choke link
6. Grommet
7. Choke lever
8. Throttle knob
9. Rod
10. Nut
11. "O" ring
12. Bushing

CRANKSHAFT, CRANKCASE AND SEALS. Crankshaft is supported by a roller bearing (13–Fig. HL134) in crankcase (9) and by a ball bearing (23) in back plate (27). Bearing (23) and crankshaft will remain in back plate (27) when separating crankcase and back plate. Remove bearing retainers (24), heat back plate and remove crankshaft with bearing (23). Remove snap ring (25) and pull bearing off crankshaft. Roller bearing and seals may be pressed out of crankcase and back plate. Inspect bearings, seals and "O" ring (26) for damage or excessive wear. Be sure "O" ring is properly seated during assembly of crankcase. Install bearings so unstamped side is towards inside of crankcase and back plate.

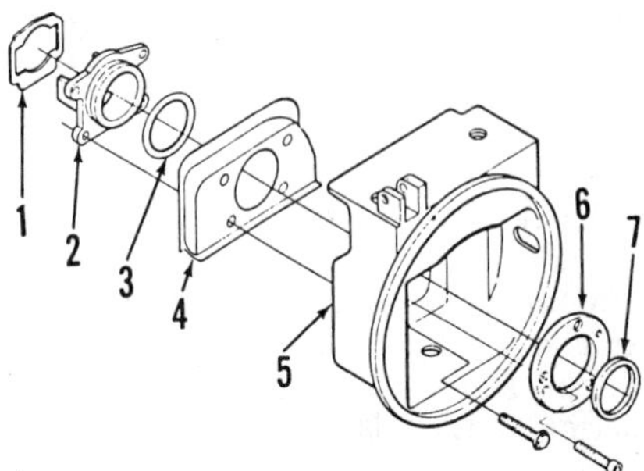

Fig. HL133—Exploded view of carburetor chamber assembly.
1. Gasket
2. Intake adapter
3. "O" ring
4. Air deflector
5. Carburetor chamber
6. Plate
7. Square ring

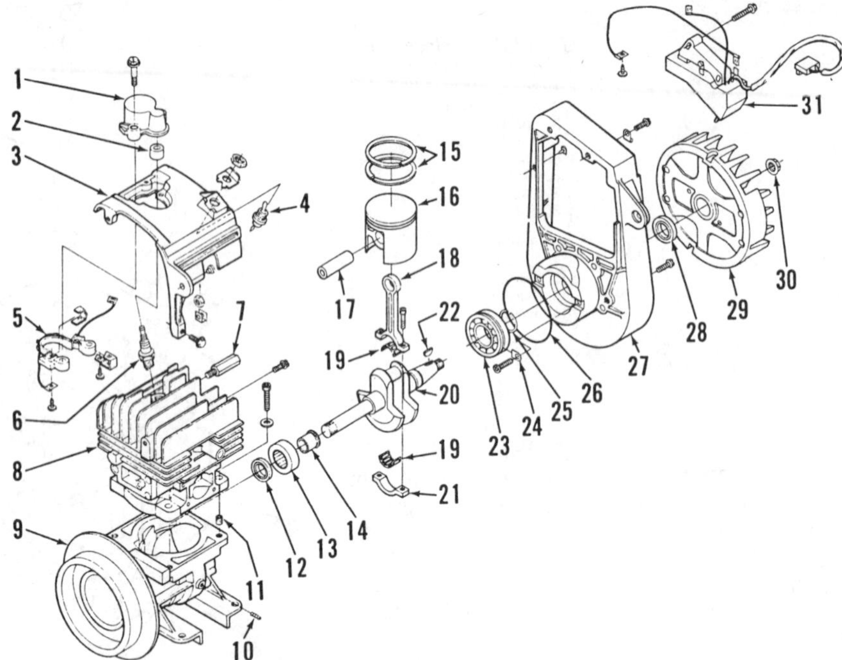

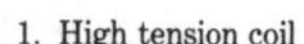

1. High tension coil
2. Grommet
3. Shield
4. Ignition switch
5. Coil receptacle
6. Spark plug
7. Stud
8. Cylinder
9. Crankcase
10. Nipple
11. Helicoil insert
12. Seal
13. Roller bearing
14. Sleeve
15. Piston rings
16. Piston
17. Piston pin
18. Connecting rod
19. Split-cage bearing
20. Crankshaft
21. Rod cap
22. Key
23. Ball bearing
24. Bearing retainer
25. Snap ring
26. "O" ring
27. Back plate
28. Seal
29. Flywheel
30. Nut
31. Ignition module

Fig. HL134—Exploded view of engine used on Model FP-150 pump.

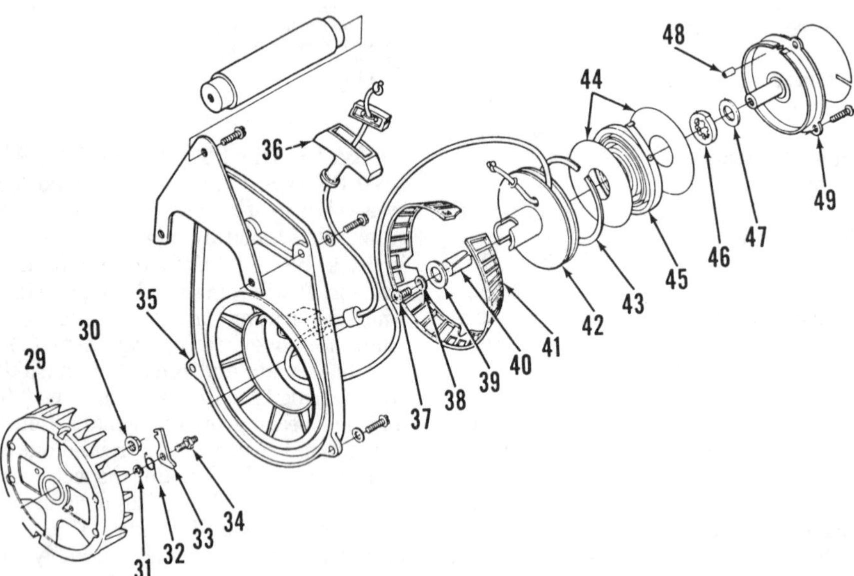

Fig. HL135—Exploded view of rewind starter.

29. Flywheel
30. Nut
31. Lockwasher
32. Spring
33. Pawl
34. Stud
35. Starter housing
36. Rope handle
37. Screw
38. Lockwasher
39. Washer
40. Bushing
41. Screen
42. Rope pulley
43. Snap ring
44. Spring shields
45. Rewind spring
46. Spring lock
47. Washer
48. Pin
49. Cover

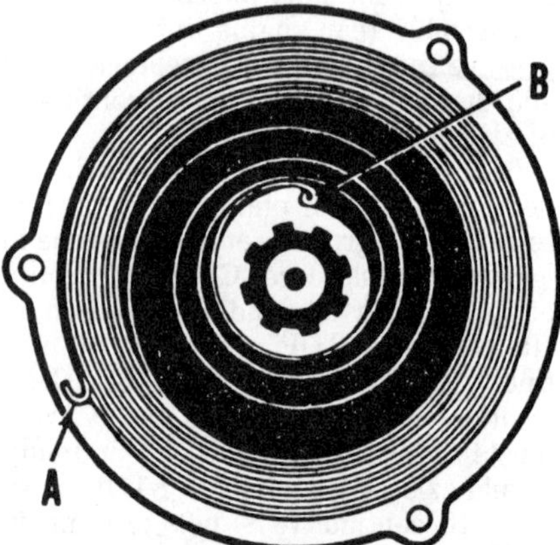

Fig. HL136—View of rewind spring installation in starter cover. Hook outer loop (A) of spring in notch as shown. Inner loop (B) of spring must be curved inward to engage notch of spring lock.

REWIND STARTER. Refer to Fig. HL135 for an exploded view of rewind starter assembly. To disassemble starter, hold cover (49) and unscrew retaining screws. Allow cover to turn until spring tension is relieved and remove cover.

NOTE: If outer hook of spring catches on starter housing, pull cover away from housing until cover is allowed to turn.

Unscrew screw (37) to separate rope pulley (42) from cover. Remove snap ring (43) for access to rewind spring. If starter pawl assemblies must be removed, unscrew housing screws and remove starter housing (35). Threaded inserts are available if stud holes are damaged in flywheel.

Clean and inspect components. Lubricate sides of rewind spring with a small amount of lithium base, multi-purpose grease. Do not oil spring. Install washer (47), inner spring shield (44), rewind spring (45) and spring lock (46) in cover with spring wound as shown in Fig. HL136. Install outer spring shield (44–Fig. HL135) and snap ring (43). Insert bushing (40) in rope pulley (42). Slide pulley onto post in cover and check to be sure splines on pulley engage splines in spring lock. Install and tighten screw (37) to 45 in.-lbs. (5 N•m). Wind rope around pulley in clockwise direction as viewed from screw end of pulley. Set cover in housing. Pull rope handle and then allow rope to rewind so starter pawls will be forced open and pulley hub can slide between them into place. Turn cover clockwise 3 or 4 turns to preload rewind spring; never turn cover more than 5 turns against spring tension. Snap plastic screen into place and install cover screws. Check starter operation.

HOMELITE CAPACITOR DISCHARGE IGNITION SYSTEM

OPERATING PRINCIPLES

The Homelite capacitor discharge ignition system used in the Multi-Purpose Saw Models XL98 and XL98A generates alternating current which is rectified into direct current. The current is stored as electrical energy in a capacitor (condenser) and is discharged on timing signal into the transformer (coil) that steps up the voltage to fire the spark plug. Instead of using breaker points as in a conventional magneto, ignition timing is done by magnetically triggered solid state switch components. Refer to Fig. HL150 for schematic diagram of capacitor discharge ignition system. Ignition system components are as follows:

1. **SPARK PLUG**—A conventional Champion spark plug with a 0.025 inch (0.64 mm) electrode gap.

2. **FLYWHEEL (ROTOR)**—The CD flywheel is slightly different from breaker ignition flywheels in that it has two pole pieces to trigger the solid state components. One pole piece triggers the switch for normal starting and operating ignition timing and the second pole piece is a safety device to prevent the engine from running backwards. The flywheel magnet passes the generator coil and core to generate electrical current.

3. **GENERATOR**—The generator is an alternator type similar to that used in a battery charging circuit. The generator coil module is a permanently sealed unit mounted on a core. The module generates electrical current to charge the capacitor.

4. **CAPACITOR**—The capacitor stores electrical energy which is discharged into the transformer on signal from the switch module.

5. **TRANSFORMER**—The transformer increases the voltage discharged from the capacitor to a voltage high enough to fire the spark plug. The transformer is mounted on the generator core and can be renewed separately.

6. **TIMING SWITCH MODULE**—The timing switch module, which is mounted on the backplate under the flywheel, consists of two magnetic devices which will trigger the silicon controlled rectifier (SCR) switch contained in the module. One magnetic device will trigger the switch for retarded timing at cranking speed and the second will trigger the switch for advanced timing when the engine is running.

The advance triggering device is located 16° ahead of the retard device. At cranking speed, the advance triggering device will not generate enough electricity to trigger the SCR switch, but the retard device is stronger and will trigger the SCR switch at cranking speed, thus allowing the electrical energy stored in the capacitor to be discharged into the transformer and fire the spark plug.

When the engine is running, the increased speed at which the pole piece in the flywheel passes the advanced triggering device generates enough electrical energy to trigger the SCR switch, thus the capacitor is discharged into the transformer 16° sooner than at cranking speed. When the pole piece passes the retard device, a triggering current is also created, but the capacitor has already been discharged and no ignition spark will occur.

Should the engine be turned backwards far enough to charge the capacitor, a second "safety" pole piece in the flywheel will trigger the SCR switch and the spark plug will be fired when the engine exhaust port is open. Thus, when the engine is turned backwards, a "poof" may be heard from the exhaust but a power stroke will not be created.

TESTING THE CD IGNITION SYSTEM

A number of tests to indicate condition of the ignition system components can be made using a volt-ohmmeter. To

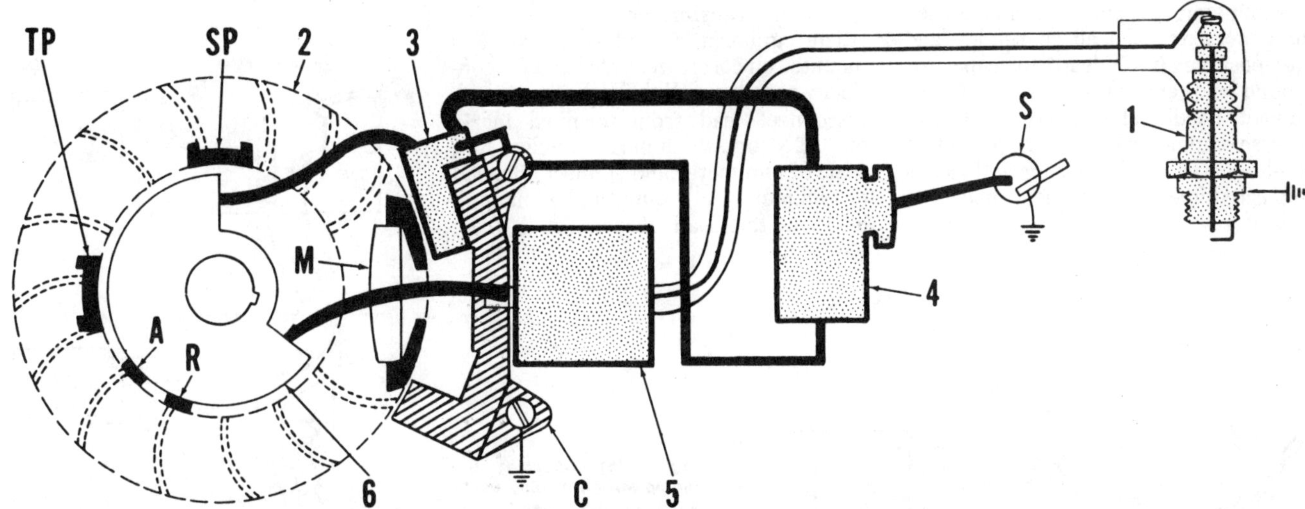

Fig. HL150—Schematic diagram of CD ignition system. Electrical energy created by magnet (M) in flywheel (2) passing generator coil (3) and core (C) is stored in the capacitor (4) until SCR switch in timing module (6) is turned on by timing pole piece (TP) passing magnetic triggering device (A or R) in module. The capacitor will then discharge the stored electrical energy into the transformer (5) which builds up the voltage to fire the spark plug (1). SCR switch is actuated by retard magnetic triggering device (R) at cranking speeds and by advance triggering device (A) when engine is running. Safety pole piece (SP) will cause spark plug to be fired on exhaust stroke if engine is turned backwards. "ON-OFF" switch (S) is used to stop engine by grounding the capacitor.

make the volt-ohmmeter tests, refer to Figs. HL151 through HL155.

Turn the ignition switch to "ON" position, disconnect lead terminal from spark plug and insert a screw into the terminal. Hold terminal insulating boot to position the screw head 1/4 inch (6.4 mm) from engine ground and observe for spark while pulling the starter rope. If a spark is observed, the magneto is performing satisfactorily. If no spark is observed, discharge capacitor by switching ignition to "OFF" position or by touching the switch lead to ground if disconnected from switch. Remove fan housing and flywheel and thoroughly inspect to see that all wires are properly connected and there are no broken or loose connections. If all wires are secure and in place, make the following tests:

NOTE: Except where rotor is required to be in place during the generator coil test, components may be tested on or off the unit.

Select Rx1 scale of ohmmeter and connect one lead of ohmmeter to timing switch module marked "TRANS" and remaining lead to terminal marked "GEN" as shown in Fig. HL151. Strike the pole pieces with a screwdriver as shown. The ohmmeter needle should show a deflection and remain deflected until the leads are disconnected. If no deflection is noted, reverse the leads and again strike pole pieces with screwdriver. If no needle deflection is noted with ohmmeter leads connected in either manner, renew the switch module.

To check capacitor, select Rx1000 scale of ohmmeter and disconnect ignition switch lead or turn switch to "ON" position. Connect negative (black) lead of ohmmeter to capacitor terminal used for generator coil lead connection and the positive (red) lead to capacitor ground lead terminal. An instant deflection of needle should occur; if not, reverse ohmmeter leads. If no deflection of ohmmeter needle occurs with leads connected in either direction, renew the capacitor.

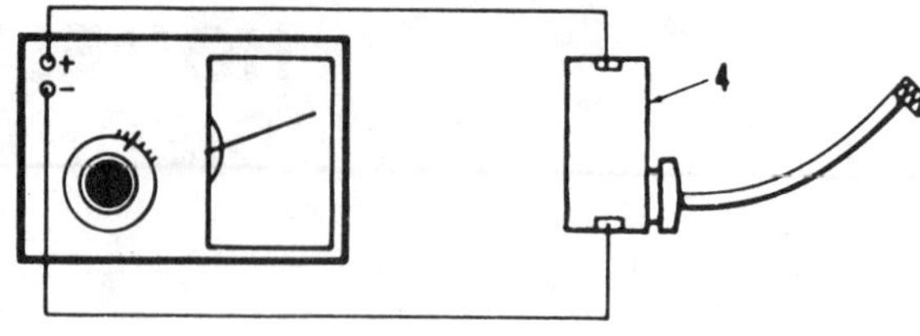

Fig. HL152—View showing ohmmeter connections for checking capacitor; refer to text for procedure.

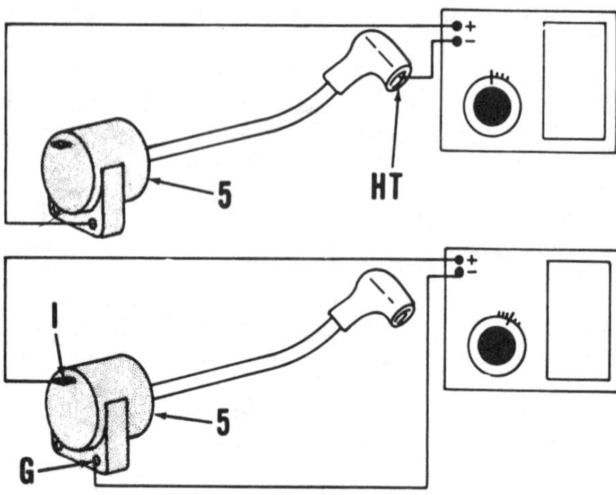

Fig. HL153—High tension coil of transformer should have between 2400 and 2900 ohms resistance; make ohmmeter connections as shown in top view. Resistance of input coil should be between 0.2 and 0.24 ohms with leads connected as shown in bottom view.

Again using Rx1000 scale of ohmmeter, test transformer by connecting either lead of ohmmeter to transformer high voltage lead and remaining ohmmeter lead to transformer ground; the resulting reading should be between 2400 and 2900 ohms. If proper reading is obtained, disconnect leads, select Rx1 scale of ohmmeter and connect one ohmmeter lead to transformer input terminal and remaining lead to transformer ground; reading should be between 0.2 and 0.24 ohms. If either reading is not within desired range, renew the transformer.

The generator coil (square coil mounted on core) can be checked for continuity as follows: With flywheel removed, disconnect lead from terminal marked "GEN" on switch module. Select Rx1 scale of ohmmeter and connect one lead of ohmmeter to ground and remaining lead to the lead disconnected from switch module; then reverse leads. The ohmmeter should show continuity (by deflection of needle) with the leads connected in one direction, but not in the opposite. If the continuity is not observed in either direction, or if the needle deflects showing continuity in both directions, renew the generator coil. The generator coil can be tested for output by using the voltmeter as follows: Refer to Fig. HL155. Remove spark plug, disconnect lead from ignition switch and bring lead out through switch hole in throttle handle. Disconnect ground lead

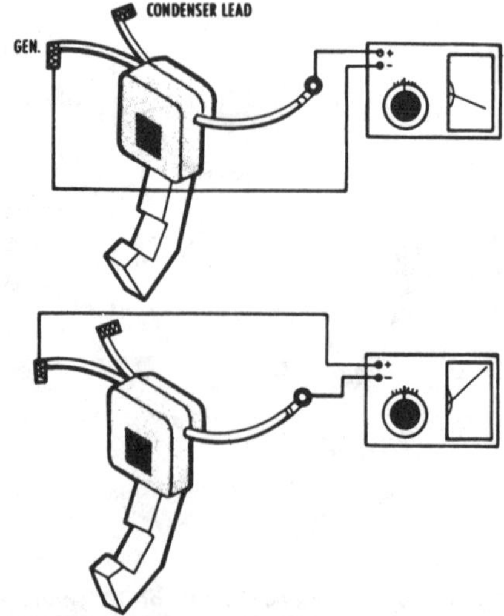

Fig. HL154—Generator coil module should show continuity in one direction only; reversing the leads should cause opposite reading to be observed.

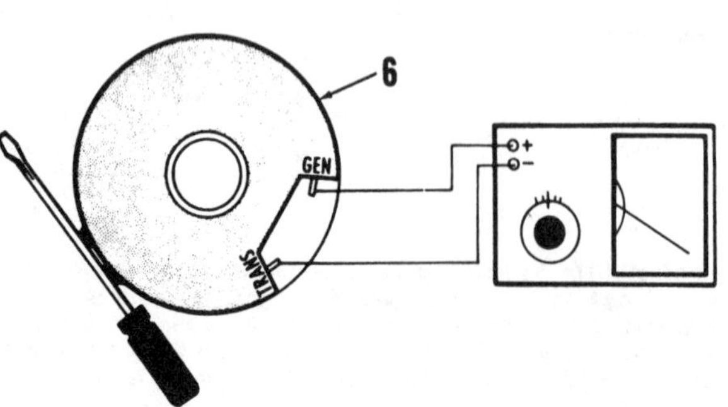

Fig. HL151—Checking the timing switch module using an ohmmeter; refer to text for procedure.

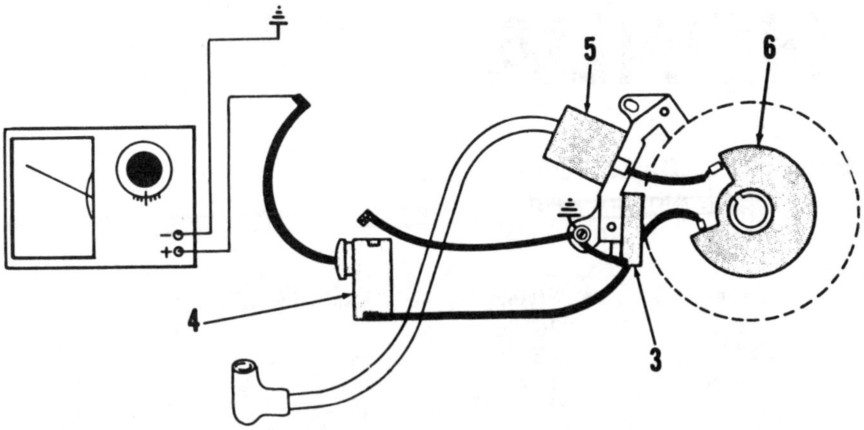

Fig. HL155 – Testing output of generator with voltmeter; refer to text for procedure. Minimum output should be 4 volts.

from capacitor. Select lowest "DC" scale on voltmeter. Connect positive (red) lead of voltmeter to switch wire and the negative (black) lead to engine ground. Spin engine by pulling firmly on starter rope. A minimum of 4 volts should be observed on voltmeter.

It is possible for some capacitor discharge ignition system components to be faulty, but undetected using the volt-ohmmeter tests. If after testing, a faulty component is not located, renew the components one at a time until the trouble is located. The components should be renewed in the following order:

1. Capacitor
2. Generator coil and core
3. Transformer
4. Timing switch module

HONDA

AMERICAN HONDA MOTOR CO., INC.
100 W. Alondra Blvd.
Gardena, California 90247

Model	Bore	Stroke	Displacement
G100	46 mm (1.84 in.)	46 mm (1.84 in.)	76 cc (4.6 cu. in.)
G150	64 mm (2.5 in.)	45 mm (1.8 in.)	144 cc (8.8 cu. in.)
GV150	64 mm (2.5 in.)	45 mm (1.8 in.)	144 cc (8.8 cu. in.)
G200	67 mm (2.6 in.)	56 mm (2.2 in.)	197 cc (12.0 cu. in.)
GV200	67 mm (2.6 in.)	56 mm (2.2 in.)	197 cc (12.0 cu. in.)

ENGINE IDENTIFICATION

Honda G series engines are four-stroke, air-cooled, single-cylinder engines. Valves are located in cylinder block and crankcase casting. Model G100 is rated at 1.5kW (2 hp) at 3600 rpm, Models G150 and GV150 are rated at 2.6kW (3.5 hp) at 3600 rpm and Models G200 and GV200 are rated at 3.7 kW (5 hp) at 3600 rpm.

The "G" prefix indicates horizontal crankshaft model and "GV" prefix indicates vertical crankshaft model.

Engine model number decal is located on cooling shroud just above or beside recoil starter. Engine serial number for Models G100, G150 (after serial number 1181478), G200 (after serial number 1286400) and all GV200 engines is located on crankcase or crankcase cover near oil filler and dipstick opening. Engine serial number for all Model GV150 engines is located on oil pan just below cylinder head. Engine serial number for all other models is located on lower left edge (facing pto side) of crankcase. See Fig. HN1.

Always furnish engine model and serial number when ordering parts or service information.

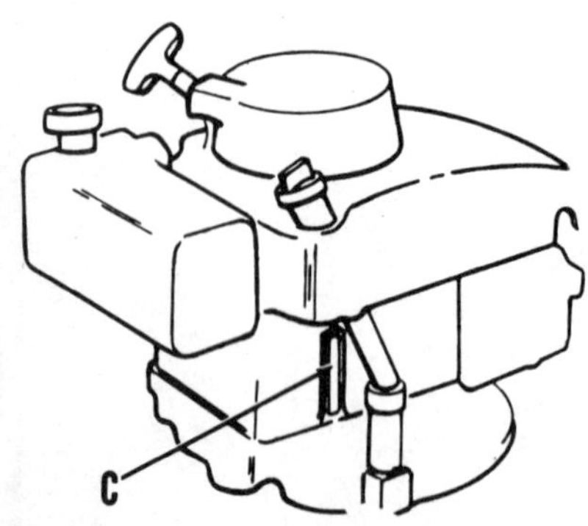

Fig. HN1—View showing serial number locations for all models. (A)—All Model G100, Models G150 (after serial number 1181478) and G200 (after serial number 1286400). (B)—Models G150 (prior to serial number 1181479) and G200 (prior to serial number 1286401). (C)—All model GV200.

MAINTENANCE

SPARK PLUG. Recommended spark plug is as follows:

Model	Standard	Resistor
G100	NGK BM4A	NGK BMR4A
G150*	NGK B4HS	NGK BR4HS
GV150*	NGK BM6A	NGK BMR6A
G200*	NGK B4HS	NGK BR4HS
G150**	NGK BP4HS	NGK BPR4HS
GV150**	NGK BPM6A	NGK BPMR6A
G200*	NGK BP4HS	NGK BPR4HS
GV200*	NGK BM6A	NGK BMR6A
GV200**	NGK BPM6A	NGK BPMR6A

*Breaker point ignition system
**CDI ignition system

Spark plug should be removed and cleaned after every 100 hours of operation. Set electrode gap at 0.6-0.7 mm (0.024-0.028 in.) for models with breaker point ignition system or 0.9-1.0 mm (0.035-0.039 in.) for models with CDI ignition system.

NOTE: Caution should be exercised if abrasive type spark plug cleaner is used. Inadequate cleaning procedure may allow the abrasive cleaner to be deposited in engine cylinder causing rapid wear and part failure.

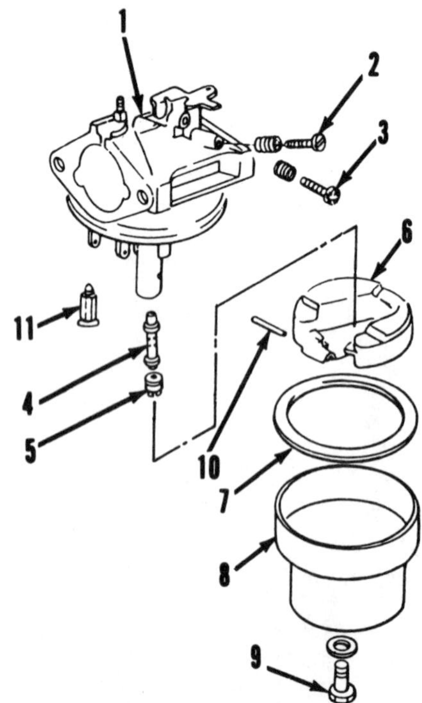

Fig. HN2—Exploded view of carburetor used on Models G150, G200, GV150 and GV200. Carburetor used on Model G100 is similar.

1. Carburetor throttle body
2. Idle mixture screw
3. Throttle stop screw
4. Nozzle
5. Main jet
6. Float
7. Gasket
8. Float bowl
9. Bolt
10. Float pin
11. Fuel inlet needle

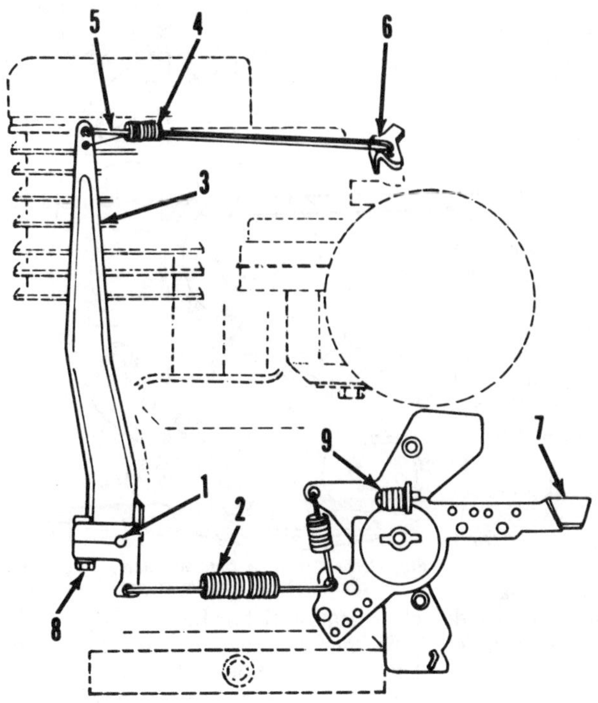

Fig. HN4—View of external governor linkage on Models G100, G150 and G200. Models GV150 and GV200 are similar.

1. Governor shaft
2. Tension spring
3. Governor lever
4. Spring
5. Carburetor-to-governor lever rod
6. Throttle pivot
7. Throttle lever
8. Clamp bolt
9. Maximum speed screw

CARBURETOR. All models are equipped with a float type side draft carburetor. Carburetor is equipped with idle fuel mixture screw. High speed fuel mixture is controlled by a fixed jet.

Engine idle speed is 1400 rpm for Models G100, G150 and G200 and 1700 rpm for Models GV150 and GV200. Idle speed is adjusted by turning throttle stop screw (3–Fig. HN2). Initial adjustment of idle fuel mixture screw (2) from a lightly seated position is 1³/₈ turns open on Models G100 and G150, 3¼ turns open on Model G200 and 1¾ turns open on Models GV150 and GV200. On all models, final adjustment is made with engine at operating temperature and running. Adjust idle mixture screw to attain smoothest engine operation. Recheck engine idle speed and adjust if necessary.

Main jet (5) controls fuel mixture for high speed operation. Standard main jet size is #55 for Model G100, #65 for Models G150 and GV150, #72 for Model G200 and #75 for Model GV200.

Float level should be 6.7-9.7 mm (0.26-0.38 in.) on Models G100 and GV200 and 8.2 mm (0.32 in.) on Models G100, G150 and G200. To measure float level, invert carburetor throttle body and float assembly. Measure distance from top of float to float bowl mating surface. If dimension is not as specified, renew float.

FUEL FILTER. A fuel filter screen is located in sediment bowl below fuel shut-off valve. To remove sediment bowl, shut off fuel, unscrew threaded ring and remove ring, sediment bowl and gasket. Make certain gasket is in place before reassembly. To clean screen, fuel shut-off valve must be disconnected from fuel line and unscrewed from fuel tank.

AIR FILTER. Engines may be equipped with one of four different types of air cleaner (filter); single element type, dual element type, foam (semidry) type or oil bath type. On all models, air filter should be removed and serviced after every 20 hours of operation. Refer to appropriate paragraph for model being serviced.

Dry Type Air Filter. To remove element, loosen the two wing nuts and remove air cleaner cover. Remove element and separate foam element from paper element. Direct low pressure air from inside filter elements toward the outside to remove all loose dirt and foreign material. Reinstall elements.

Dual Element Type Air Filter. Remove wing nut, cover and elements. Separate foam outer element from paper element. Wash foam element in warm soapy water and thoroughly rinse. Allow element to air dry. Dip dry foam element in clean engine oil and gently squeeze our excess oil.

Direct low pressure air from inside paper element toward the outside to remove all loose dirt and foreign material. Reassemble elements and reinstall.

Foam (Semidry) Type Air Filter. Remove air cleaner cover and element. Clean element in nonflammable solvent and squeeze dry. Soak element in new engine oil and gently squeeze our excess oil. Reinstall element and cover.

Oil Bath Type Air Filter. Remove air cleaner assembly and separate cover, element and housing. Clean element in nonflammable solvent and air dry. Drain old oil and thoroughly clean housing. Fill housing to oil level mark with new engine oil and reassemble air cleaner.

GOVERNOR. The internal centrifugal flyweight governor assembly is located inside crankcase and is either gear or chain driven.

To adjust governor, first stop engine and make certain all linkage is in good condition and tension spring (2–Fig. HN4) is not stretched or damaged. Spring (4) must pull governor lever (3) toward throttle pivot (6).

On all models except Models GV150 and GV200, loosen clamp bolt (8) and

Fig. HN5—"F" mark on flywheel should align with timing (index) mark on crankcase when points just begin to open. Timing tool number 07974-8830001 is available to adjust timing with flywheel removed.

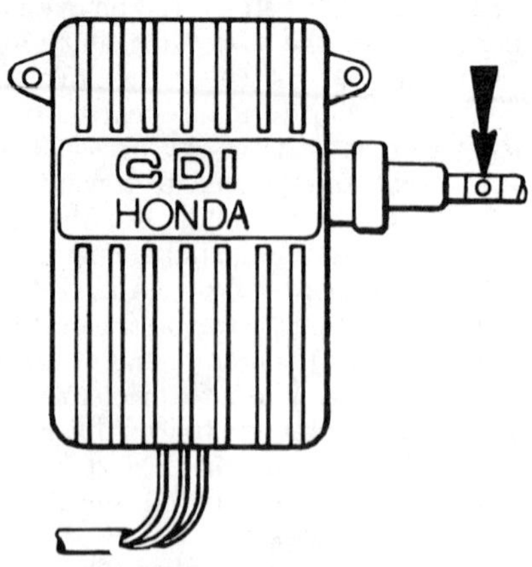

Fig. HN6—Exciter coil must be timed according to timing number on CDI unit.

turn governor shaft clockwise as far as it will go and move governor lever as far to the right as it will go. Tighten retaining bolt. Start and run engine until it reaches operating temperature. Adjust stop screw to obtain 4000 rpm for Models G150 and G200 or 4200 rpm for Model G100.

On Models GV150 and GV200, loosen clamp bolt, pull governor shaft all the way to the right and rotate governor shaft clockwise as far as possible. Tighten clamp bolt. With engine stopped, governor is assembled properly when governor lever is spring-loaded and moves freely. Maximum engine speed is 3600 rpm.

IGNITION SYSTEM. Engines may be equipped with a breaker point ignition system or a capacitor discharge ignition (CDI) system. Refer to appropriate paragraph for model being serviced.

Breaker Point Ignition System. Breaker points and ignition coil are located underneath the flywheel on all models. Breaker points should be checked after every 300 hours of operation. Initial breaker point gap should be 0.3-0.4 mm (0.012-0.016 in.) and can be varied to obtain 20° BTDC timing setting.

NOTE: Timing tool number 07974-8830001 is available from Honda Motor Company to allow timing adjustment with flywheel removed.

To check ignition timing, connect positive ohmmeter lead to engine stop switch wire and connect remaining lead to engine ground. Rotate flywheel until ohmmeter needle deflects. "F" mark on flywheel should align with index mark on crankcase (Fig. HN5). Remove flywheel and vary point gap to obtain correct timing setting. If timing tool is used, ignition timing can be checked with flywheel removed.

To check Model G100 ignition coil, connect positive ohmmeter lead to black wire and remaining lead to coil laminations. Ohmmeter should register 0.49 ohms. Disconnect positive lead and reconnect lead to spark plug wire. Ohmmeter should register 4 ohms.

To check Models G150, GV150, G200 and GV200 ignition coil, connect positive ohmmeter lead to spark plug wire and remaining lead to coil laminations. Ohmmeter should register 6.6 ohms.

CDI Ignition System. The capacitor discharge ignition (CDI) system consists of the flywheel magnets, exciter coil (located under flywheel) and CDI unit. CDI system does not require regular maintenance.

To test exciter coil on all models, disconnect the blue and black exciter coil leads on Model GV200 or the red and black exciter coil leads on all other models. Connect an ohmmeter lead to each exciter coil lead. Ohmmeter should register continuity, if not, renew coil. To renew exciter coil, remove engine cooling shrouds and flywheel. Remove exciter coil.

NOTE: Exciter coil must be installed and timed according to timing number on CDI unit. Refer to Fig. HN6.

Determine CDI unit timing number (Fig. HN6) and position coil so its inner edge aligns with scale position indicated by the CDI unit timing number (Fig. HN7). Distance between scale ridges indicates 2°. Tighten exciter coil mounting bolts and recheck alignment. Reinstall flywheel and cooling shrouds.

If timing, ignition stop switch and exciter coil check satisfactory, but an ignition system problem is still suspected, renew CDI unit. Make certain ground terminal on CDI unit is making an adequate connection.

VALVE ADJUSTMENT. Valves and seats should be refaced and stem clearance adjusted after every 300 hours of operation. Refer to REPAIRS section for service procedures and specifications.

CYLINDER HEAD AND COMBUSTION CHAMBER. Cylinder head, combustion chamber and piston should be cleaned and carbon and other deposits removed after every 300 hours of operation. Refer to REPAIRS section for service procedure.

LUBRICATION. Engine oil should be checked prior to each operating interval. Oil level should be maintained between reference marks on dipstick with dipstick just touching first threads. Do not screw dipstick in to check oil level.

Manufacturer recommends SAE 10W-40 oil with an API service

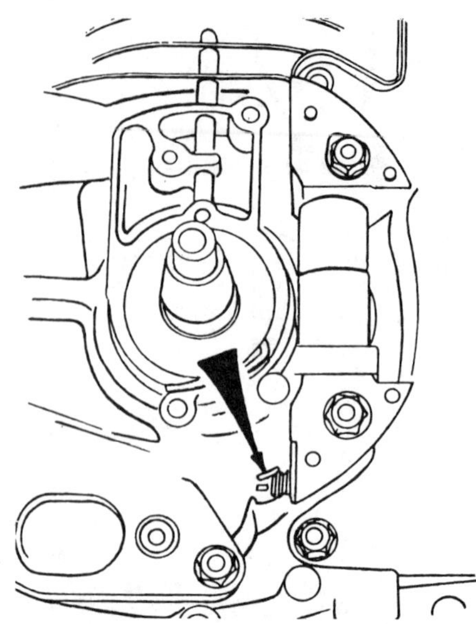

Fig. HN7—Distance between scale ridges indicate 2° ignition timing positions.

classification SE or SF.

Oil should be changed after the first 20 hours of operation and after every 100 hours of operation thereafter. Crankcase capacity is 0.45 L (0.95 pt.) for Model G100 and 0.7 L (1.48 pt.) for all other models.

GENERAL MAINTENANCE. Check and tighten all loose bolts, nuts and clamps prior to each operating interval. Check for fuel and oil leakage and repair if necessary.

Clean dust, dirt, grease and any foreign material from cylinder head and cylinder block cooling fins after every 100 hours of operation. Inspect fins for damage and repair if necessary.

REPAIRS

TIGHTENING TORQUES. Recommended tightening torque specifications are as follows:

Flywheel nut:
G100 .... 4.8 N·m (3.5 ft.-lbs.)
All others .... 73 N·m (54 ft.-lbs.)
Crankcase cover .... 8-12 N·m (6-8 ft.-lbs.)
Cylinder head bolts:
G100 .... 10 N·m (7 ft.-lbs.)
All others .... 24-26 N·m (18-19 ft.-lbs.)
Connecting rod bolts:
G100 .... 3 N·m (2.2 ft.-lbs.)
All others .... 9-11 N·m (6-9 ft.-lbs.)
Oil pump cover
(GV150 & GV200) .... 10 N·m (7 ft.-lbs.)

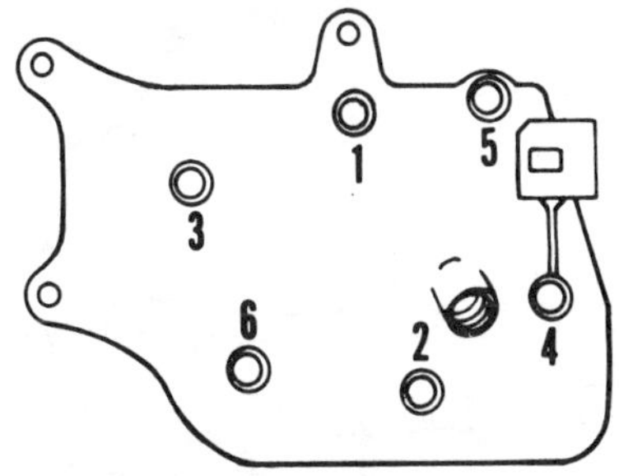

Fig. HN8—Tighten head bolts on Model G100 to specified torque following sequence shown.

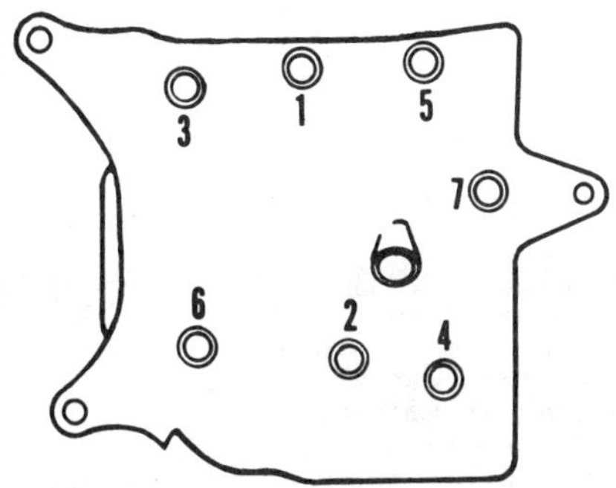

Fig. HN9—Tighten head bolts on all models except Model G100 to specified torque following sequence shown.

CYLINDER HEAD. To remove cylinder head, first remove cooling shrouds. Clean engine to prevent entrance of foreign material. Remove spark plug. Loosen cylinder head bolts in 1/4 turn increments following the sequence shown in Fig. HN8 for Model G100 or Fig. HN9 for all other models until all bolts are loose enough to remove by hand. Remove cylinder head. Clean carbon and deposits from cylinder head.

Reinstall cylinder head and new gasket. Tighten head bolts to specified torque using the correct sequence shown (Fig. HN8 or HN9) according to model being serviced.

CONNECTING ROD. Connecting rod rides directly on crankshaft crankpin journal on all models. Piston and connecting rod are accessible after cylinder head removal and crankcase cover or oil pan is separated from crankcase. Remove the two connecting rod bolts, lock plate and connecting rod cap. Push piston and connecting rod assembly out through the top of cylinder block. Remove snap rings and piston pin to separate piston from connecting rod.

Standard diameter for piston pin bore in connecting rod small end is 10.006-10.017 mm (0.3939-0.3944 in.) for Model G100 or 15.005-15.020 mm (0.5907-0.5913 in.) for all other models. If dimension exceeds 10.050 mm (0.3957 in.) for Model G100 or 15.070 mm (0.5933 in.) for all other models, renew connecting rod.

Standard clearance between connecting rod bearing surface and crankpin journal is 0.016-0.033 mm (0.0006-0.0013 in.) for Model G100 and 0.040-0.066 mm (0.0016-0.0026 in.) for all other models. If clearance exceeds 0.1 mm (0.0004 in.) for Model G100 or 0.120 mm (0.0047 in.) for all other models, renew connecting rod and/or recondition crankshaft.

Standard connecting rod side play on crankpin journal is 0.2-0.9 mm (0.008-0.035 in.) for Model G100 or 0.10-0.80 mm (0.004-0.031 in.) for all other models. If side play exceeds 1.1 mm (0.043 in.) for Model G100 or 1.20 mm (0.047 in.) for all other models, renew connecting rod.

To install piston on connecting rod, refer to appropriate paragraph for model being serviced.

Model G100. Install piston on connecting rod so number 896 stamped on top of piston is toward long side of connecting rod (Fig. HN10). With match marks on connecting rod and cap aligned, long side of rod is installed toward valve side of engine.

Models G150, GV150 And G200. Piston may be installed on connecting rod either way. Install connecting rod and piston assembly in engine so marked side of piston top is toward valve side of engine. Align match marks on connecting rod and rod cap.

Model GV200. Install piston on connecting rod with mark on top of piston toward ribbed side of connecting rod. Install connecting rod so ribbed side is towards oil pan. Align connecting rod and rod cap match marks.

All models. Install oil dipper if equipped. Tighten connecting rod bolts to specified torque and lock bolts with lock plate if equipped.

PISTON, PIN AND RINGS. Piston and connecting rod are removed as an assembly. Refer to CONNECTING ROD section for removal and installation procedure.

After separating piston and connecting rod, carefully remove rings. Clean carbon and deposits from piston surface and ring lands.

CAUTION: Extreme care should be exercised when cleaning ring lands. Do not damage squared edges or widen ring grooves. If ring lands are damaged, piston must be renewed.

Measure piston diameter at piston thrust surfaces, 90° from piston pin. Standard piston diameters and service limits are as follows:

Model	Standard Diameter	Service Limit
G100	45.98-46.00 mm (1.810-1.811 in.)	45.92 mm (1.808 in.)
G150, GV150	63.98-64.00 mm (25.18-2.520 in.)	63.88 mm (2.515 in.)
G200, GV200	66.98-67.00 mm (2.637-2.638 in.)	66.88 mm (2.633 in.)

If piston diameter is less than service limit, renew piston.

Before installing rings, install piston in cylinder bore and use a suitable feeler gage to measure clearance between piston and cylinder bore. Standard clearance and service limits are as follows:

Model	Standard Clearance	Service Limit
G100	0.030 mm (0.001 in.)	0.130 mm (0.005 in.)
G150, GV150	0.060 mm (0.0024 in.)	0.285 mm (0.0112 in.)
G200, GV200	0.040 mm (0.0016 in.)	0.045 mm (0.0018 in.)

If clearance exceeds service limit dimension, renew piston and/or recondition cylinder bore.

Standard piston bore diameter in piston is 10.000-10.006 mm (0.3937-0.3939 in.) for Model G100 or 15.000-15.006 mm (0.5906-0.5908 in.) for all other models. Service limit for piston pin bore diameter is 10.046 mm (0.4096 in.) for Model G100 or 15.046 mm (0.5924 in.) for all other models. If diameter exceeds service limit, renew piston.

Standard piston pin outside diameter is 9.994-10.000 mm (0.3935-0.3937 in.) for Model G100 or 14.994-15.000 mm (0.5903-0.5906 in.) for all other models. Service limit for piston pin outside diameter is 9.950 mm (0.3917 in.) for Model G100 or 14.954 mm (0.5887 in.) for all other models. If diameter is less than service limit, renew piston pin.

Standard piston ring to piston groove side clearance for Model G100 is 0.025-0.055 mm (0.0009-0.0022 in.) for top ring and 0.010-0.040 mm (0.0004-0.0016 in.) for all remaining rings. Standard piston ring to piston groove side clearance for all other models is 0.01-0.05 mm (0.0004-0.0020 in.). If ring side clearance exceeds 0.10 mm (0.0039 in.) on Model G100 or 0.15 mm (0.0059 in.) on all other models, renew rings and/or piston.

On all models, if piston ring end gap exceeds 1.0 mm (0.039 in.) with ring

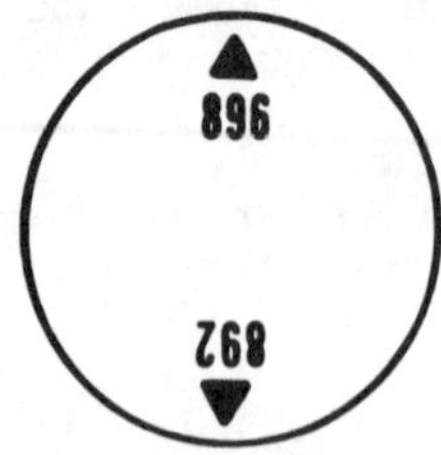

Fig. HN10—Note numbers on piston crown and install piston of Model G100 so "896" on piston is towards long side of rod.

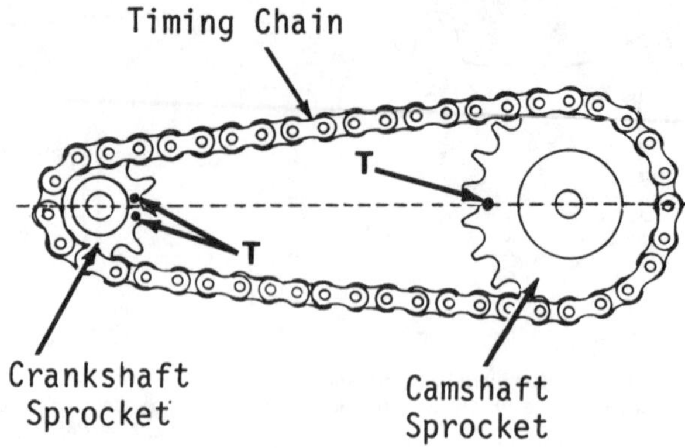

Fig. HN12—On models with extended camshaft for pto drive shaft, engines are equipped with timing sprockets and timing chain. Timing sprockets must be aligned with timing dots "T" as shown before installing timing chain and sprocket assembly on crankshaft and camshaft.

squarely installed in cylinder bore, renew ring and/or recondition cylinder bore.

Install piston rings with marked side towards top of piston. Stagger ring end gaps equally around circumference of piston.

CYLINDER AND CRANKCASE. Cylinder and crankcase are an integral casting. Standard cylinder bore diameters are as follows:

Model	Standard Diameter	Service Limit
G100	46.00-46.01 mm (1.811-1.812 in.)	46.05 mm (1.813 in.)
G150, GV150	64.00-64.02 mm (2.519-2.520 in.)	64.17 mm (2.526 in.)
G200, GV200	67.00-67.02 mm (2.638-2.639 in.)	67.17 mm (2.644 in.)

If cylinder bore diameter at any point in cylinder bore exceeds the service limit, recondition cylinder bore.

CRANKSHAFT, MAIN BEARINGS AND SEALS. Crankshaft is supported by ball bearing type main bearings at each end. To remove crankshaft, remove all cooling shrouds, flywheel, cylinder head and crankcase cover or oil pan. Remove piston and connecting rod assembly. Carefully remove crankshaft and camshaft. Remove main bearings and crankshaft oil seals if necessary.

Standard crankpin diameter is 17.973-17.984 mm (0.7076-0.7080 in.) for Model G100 or 25.967-25.980 mm (1.0223-1.0228 in.) for all other models. If crankpin diameter is less than 17.940 mm (0.7063 in.) for Model G100 or 25.197 mm (1.0204 in.) for all other models, renew or recondition crankshaft.

Main bearings are a light press fit on crankshaft and in bearing bores of crankcase and crankcase cover. It may be necessary to slightly heat crankcase or crankcase cover to reinstall bearings.

Inspect main bearings for roughness and looseness. Also check bearings for a loose fit on crankshaft journals or in crankcase and crankcase cover. Renew bearings if any of the previously described conditions are evident.

If crankshaft oil seals have been removed, use suitable seal driver to install new seals. Seals should be pressed in evenly until 4.5 mm (0.18 in.) below flush for seal in crankcase cover or oil pan and until 2.00 mm (0.08 in.) below flush for seal in crankcase.

Make certain crankshaft gear (sprocket) and camshaft gear (sprocket) timing marks are aligned (Fig. HN11 or Fig. HN12) during crankshaft installation.

CAMSHAFT, BEARINGS AND SEAL. Camshaft is supported at each end by bearings which are an integral part of crankcase or crankcase cover casting. An extended camshaft pto which utilizes a ball bearing for crankcase bearing, is available on some models. Refer to CRANKSHAFT, MAIN BEARINGS AND SEALS section for camshaft removal procedure.

Standard camshaft lobe height is 18.1-18.5 mm (0.71-0.73 in.) for intake and exhaust lobes on Model G100 or 33.4-33.6 mm (1.31-1.32 in.) for intake lobe, and 33.7-33.9 mm (1.33-1.34 in.) for exhaust lobe on all other models. If intake or exhaust lobe on Model G100 is less than 17.940 mm (0.7063 in.), renew camshaft. On all other models, if intake lobe is less than 33.25 mm (1.309 in.) or exhaust lobe is less than 33.55 mm (1.321 in.) renew camshaft.

Standard clearance between camshaft bearing journal and integral type bearings is 0.013-0.043 mm (0.0005-0.0017 in.). If clearance exceeds 0.1 mm (0.004 in.), renew camshaft and/or crankcase and crankcase cover.

On models with extended camshaft, ball bearing should be a light press fit on camshaft journal and in crankcase cover. Camshaft seal should be pressed into crankcase cover 2.0 mm (0.08 in.).

Make certain camshaft gear (sprocket) and crankshaft gear sprocket timing marks are aligned during installation.

GOVERNOR. The internal centrifugal flyweight governor is gear driven off of the camshaft gear on Model

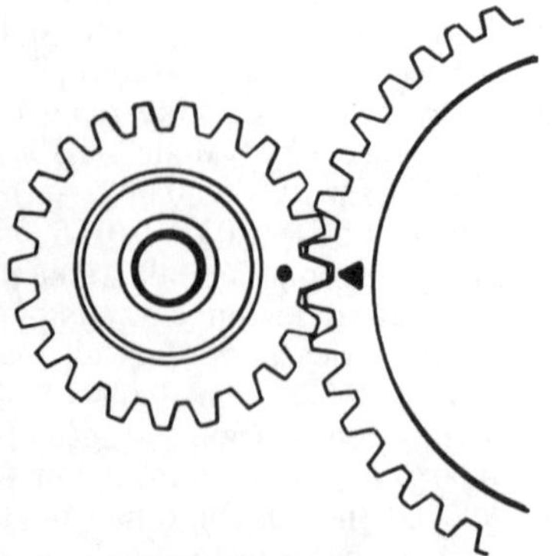

Fig. HN11—When installing crankshaft or camshaft, make certain timing marks on gears are aligned as shown.

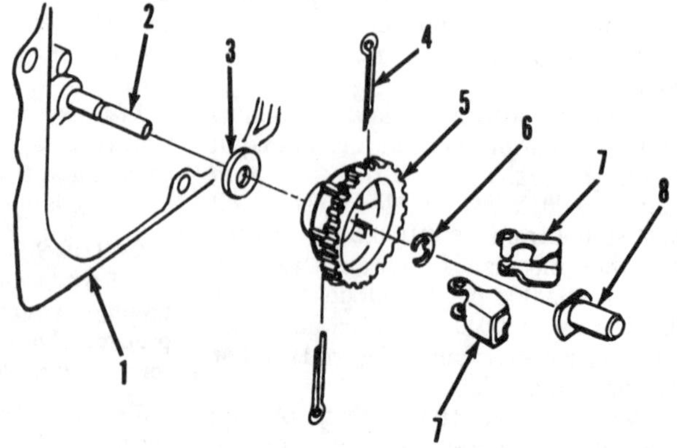

Fig. HN13—Exploded view of governor assembly used on Model G100. On all other models, governor weight assembly is mounted on camshaft gear or sprocket.

1. Crankcase cover
2. Governor stud
3. Thrust washer
4. Pin
5. Gear
6. "E" clip
7. Weights
8. Sleeve

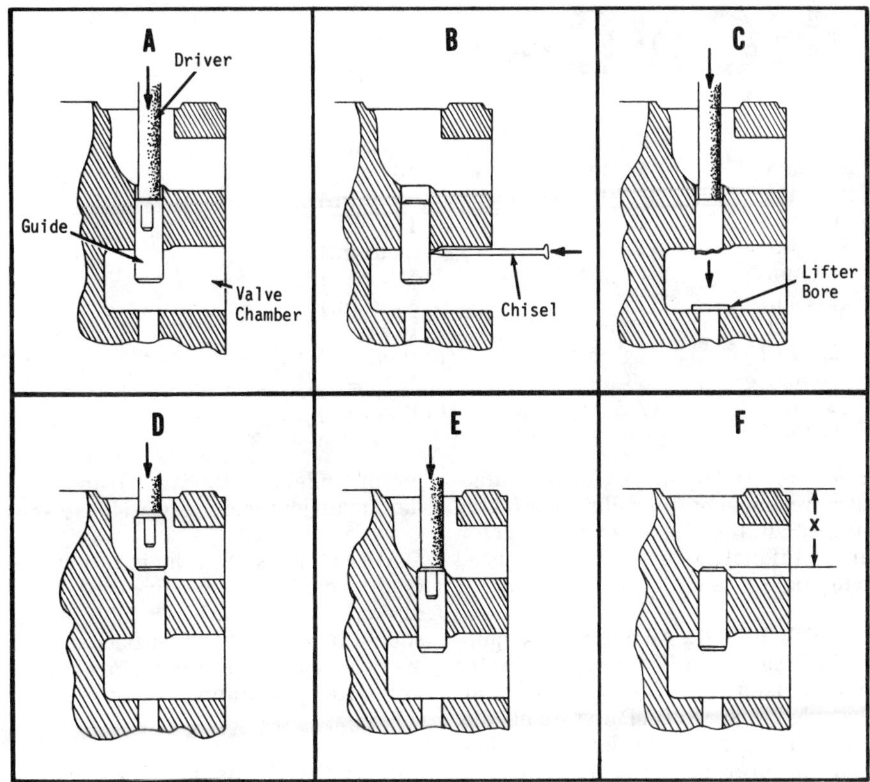

Fig. HN14—View showing valve guide removal and installation sequence on all models except Model G100. Refer to text.

G100 and is located on camshaft gear (sprocket) for all other models. Refer to GOVERNOR paragraphs in MAINTENANCE section for external governor adjustments.

To remove governor assembly, remove external linkage, metal cooling shrouds and crankcase cover or oil pan. On G100 models, remove "E" clip (Fig. HN13) and slide governor assembly off of shaft. On all other models, governor assembly is mounted on camshaft gear or sprocket.

When reassembling, make certain governor sliding sleeve and internal governor linkage is correctly positioned.

OIL PUMP. Models GV150 and GV200 are equipped with an internal oil pump located inside crankcase. All other models are splash lubricated.

Oil pump rotors may be removed and checked without disassembling the engine. Remove oil pump cover and withdraw outer and inner rotors. Remove "O" ring. Make certain oil passages are clear and pump body is thoroughly clean. Inner-to-outer rotor clearance should be 0.15 mm (0.006 in.). If clearance is 0.20 mm (0.008 in.) or more, renew rotors. Clearance between oil pump body and outer rotor diameter should be 0.15 mm (0.006 in.). If clearance is 0.26 mm (0.010 in.) or more, renew rotors and/or oil pan. Outside diameter of outer rotor should be 23.15-23.28 mm (0.911-0.917 in.). If diameter is 23.23 mm (0.915 in.) or less, renew rotors.

Reverse disassembly procedure for reassembly and tighten oil pump cover bolts to specified torque.

VALVE SYSTEM. Clearance between valve stem and valve tappet (cold) should be 0.04-0.10 mm (0.002-0.004 in.) for intake and exhaust valves on Model G100; 0.04-0.12 mm (0.002-0.005 in.) for intake and exhaust valves on Models G150 (prior to serial number 1100543) and G200 (prior to serial number 1168820); 0.05-0.11 mm (0.002-0.004 in.) for intake valve and 0.09-0.15 mm (0.004-0.006 in.) for exhaust valve on Models G150 (serial number 1100543 to 1380694) and G200 (serial number 1168820 to 1556007); 0.08-0.16 mm (0.003-0.006 in.) for intake valve and 0.16-0.24 mm (0.006-0.009 in.) for exhaust valve on Models G150 (after serial number 1380694) and G200 (after serial number 1556007); or 0.05-0.11 mm (0.002-0.004 in.) for intake valve and 0.09-0.015 mm (0.004-0.006 in.) for exhaust valve on all GV200 models.

On all models, valve clearance is adjusted as follows: To increase valve clearance, grind off end of stem. To reduce valve clearance, renew valve and/or grind valve seat deeper.

Valve face and seat angles are 45° for all models. Standard valve seat width is 0.42-0.78 mm (0.016-0.031 in.) for Model G100 while standard valve seat width is 0.7 mm (0.028 in.) for all other models. If valve seat width exceeds 1.0 mm (0.039 in.) for Model G100 or 2.0 mm (0.08 in.) for all other models, seats must be narrowed.

Standard valve stem diameters on Model G100 are 5.480-5.490 mm (0.2157-0.2161 in.) for intake valve stem and 5.435-5.445 mm (0.2140-0.2144 in.) for exhaust valve stem. If intake valve stem diameter is less than 5.450 mm (0.2146 in.) or exhaust valve stem diameter is less than 5.400 mm (0.2126 in.), renew valve.

Standard valve stem diameters on all models except Model G100 are 6.955-6.970 mm (0.2738-0.2744 in.) for intake valve stem and 6.910-6.925 mm (0.2720-0.2726 in.) for exhaust valve stem. If intake valve stem diameter is less than 6.805 mm (0.2679 in.) or exhaust valve stem diameter is less than 6.760 mm (0.2661 in.), renew valve.

Standard valve guide inside diameter for both intake and exhaust valve guides is 5.500-5.512 mm (0.2165-0.2170 in.) for Model G100 and 7.00-7.015 mm (0.2756-0.2762 in.) for all other models. If inside diameter of guide exceeds 5.560 mm (0.2189 in.) for Model G100 or 7.080 mm (0.2787 in.) for all other models, guides must be renewed.

To remove and install Model G100 valve guides, use Honda valve guide tool 07969-8960000 to pull guide out of guide bore and press new guide in. New guide is pressed in to a depth of 18 mm (0.7 in.) measured from end of guide to cylinder head surface as shown in section (F–Fig. HN14) at (X). Finish ream guide after installation with reamer 07984-2000000.

To remove and install valve guides on all models except Model G100, use the following procedure and refer to the sequence of illustrations in Fig. HN14. Use driver 07942-8230000 and drive valve guide down into valve chamber slightly (A). Use a suitable chold chisel and sever guide adjacent to guide bore (B). Cover tappet opening to prevent fragments from entering crankcase. Drive remaining piece of guide into valve chamber (C) and remove from chamber. Place new guide on driver and start guide into guide bore (D). Alternate between driving guide into bore and measuring guide depth below cylinder head surface (E). Guide is driven in to a depth of 27.5 mm (1.08 in.) measured from end of guide to cylinder head surface as shown in section (F) at (X). Finish ream guide after installation with reamer 07984-5900000.

HONDA

Model	Bore	Stroke	Displacement
GX110	57 mm (2.2 in.)	42 mm (1.7 in.)	107 cc (6.6 cu. in.)
GX140	64 mm (2.5 in.)	45 mm (1.8 in.)	144 cc (8.8 cu. in.)
GXV120	60 mm (2.4 in.)	42 mm (1.7 in.)	118 cc (7.2 cu. in.)
GXV160	68 mm (2.7 in.)	45 mm (1.8 in.)	163 cc (10.0 cu. in.)
GX240	73 mm (2.9 in.)	58 mm (2.3 in.)	242 cc (14.8 cu. in.)

ENGINE INFORMATION

All models are four-stroke, overhead valve, single-cylinder air-cooled engines. Models GXV120 and GXV160 are vertical crankshaft engines and all other models are horizontal crankshaft engines with cylinder inclined 25 degrees.

Model GX110 is rated at 2.6 kW (3.5 hp) at 3600 rpm, Model GX140 is rated at 3.8 kW (5.0 hp) at 3600 rpm, Model GXV120 is rated at 2.9 kW (4.0 hp) at 3600 rpm, model GXV160 is rated at 4.1 kW (5.5 hp) at 3600 rpm and Model GX240 is rated at 5.9 kW (8.0 hp) at 3600 rpm.

Engine model number is cast into side of crankcase (Fig. HN30) and engine serial number is stamped into crankcase (Fig. HN31). Always furnish engine model and serial number when ordering parts.

MAINTENANCE

SPARK PLUG. Spark plug should be removed, cleaned and inspected at 100 hour intervals.

Recommended spark plug for Model GXV160 is a Champion N11YC or equivalent. If resistor plug is required, use a Champion RN11YC or equivalent. Recommended spark plug for all other models is a Champion N9YC or equivalent. Recommended spark plug electrode gap is 0.7-0.8 mm (0.028-0.031 in.) for all models.

Fig. HN30—Engine model number (MN) is cast into side of engine crankcase.

When installing spark plug, manufacturer recommends installing spark plug fingertight, then for a new plug, tighten an additional ½ turn and for a used plug, tighten an additional ¼ turn.

CARBURETOR. All models are equipped with a Keihin float type carburetor with a fixed main fuel jet and an adjustable low speed fuel mixture needle.

Initial adjustment of low speed fuel mixture screw (LS—Fig. HN32) from a lightly seated position is 3 turns open for Models GX110 and GXV120, 1-5/8 turns open for Model GX140, 2 turns open for Model GXV160 and 2-1/2 turns open for Models GX240.

For final adjustment, engine must be at operating temperature and running. Operate engine at idle speed (1400 rpm) and adjust low speed mixture screw (LS) to obtain a smooth idle and satisfactory acceleration. Adjust idle speed to 1400 rpm by turning throttle stop screw (TS).

To check float level, remove float bowl and invert carburetor throttle body and float assembly. Measure from top edge of float to float bowl mating edge of carburetor throttle body. Measurement should be 12.2-15.2 mm (0.48-0.60 in.). Renew float if float height is incorrect.

Standard main jet for Models GX110 and GXV120 is a #65, standard main jet for Model GX140 is a #68, standard main jet for Model GXV160 is a #70 and standard main jet for Model GX240 is a #88.

Fig. HN31—Engine serial number (SN) is stamped on raised portion of crankcase.

AIR CLEANER. Engine may be equipped with either a dry type, semi-dry type or oil bath air filter which should be cleaned and inspected after every 50 hours of operation. Refer to appropriate paragraph for type being serviced.

Dry Type Air Cleaner. Remove foam and paper air filter elements from air filter housing. Foam element should be washed in a mild detergent and water solution, rinsed in clean water and allowed to air dry. Soak foam element in clean engine oil. Squeeze out excess oil.

Paper element may be cleaned by directing low pressure compressed air stream from inside filter toward the outside. Reinstall elements.

Semi-Dry Type Air Cleaner. Remove element and clean in solvent. Wring out excess solvent and allow element to air dry. Dip element in clean engine oil and wring out excess oil. Reinstall element.

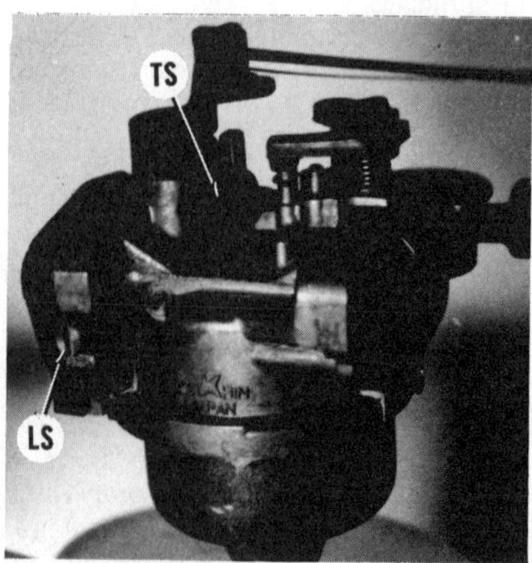

Fig. HN32—View of Keihin float type carburetor used on all models showing location of low speed mixture screw (LS) and throttle stop screw (TS).

Oil Bath Type Air Cleaner. Remove elements and clean in suitable solvent. Foam element should be washed in mild detergent and water solution and allowed to air dry. Discard old oil and clean oil reservoir with solvent. Refill oil reservoir with 60 mL (1.3 pt.) clean engine oil. Soak foam element in clean engine oil and wring out excess oil. Reassemble air cleaner.

GOVERNOR. The mechanical flyweight type governor is located inside engine crankcase. To adjust external linkage, stop engine and make certain all linkage is in good condition and tension spring (5—Fig. HN33) is not stretched or damaged. Spring (2) must pull governor lever (3) and throttle pivot toward each other. Loosen clamp bolt (7) and move governor lever (3) so throttle is completely open. Hold governor lever in this position and rotate governor shaft (6) in the same direction until it stops. Tighten clamp bolt.

Start engine and operate at an idle until operating temperature has been reached. Attach a tachometer to engine and move throttle so engine is operating at maximum speed (3800-4000 rpm). Adjust throttle stop screw (8) so throttle movement is limited to correct maximum engine rpm.

IGNITION SYSTEM. Breakerless ignition system requires no regular maintenance. Ignition coil unit is mounted outside the flywheel and air gap between flywheel and coil should be 0.2-0.6 mm (0.008-0.024 in.).

To check ignition coil primary side, connect one ohmmeter lead to primary (block) coil lead and touch iron coil laminations with remaining lead. Ohmmeter should register 0.7-0.9 ohm.

To check ignition coil secondary side, connect one ohmmeter lead to the spark plug lead wire and remaining lead to the iron coil laminations. Ohmmeter should register 6.3-7.7 ohms. If ohmmeter readings are not as specified, renew ignition coil.

VALVE ADJUSTMENT. Valve stem clearance should be checked and adjusted at 300 hour intervals.

To adjust valve stem clearance, refer to Fig. HN34 and remove rocker arm cover. Rotate engine so piston is at top dead center (TDC) on compression stroke. Insert a feeler gage between rocker arm (3) and end of valve stem. As necessary, loosen rocker arm jam nut (1) and turn adjusting nut (2) to obtain 0.10 mm (0.004 in.) clearance on intake valve and 0.15 mm (0.006 in.) clearance on exhaust valve on Model GVX120, 0.10 (0.004 in.) clearance on intake valve and 0.05 mm (0.002 in.) clearance on exhaust valve on Model GX240 or 0.15 mm (0.006 in.) clearance on intake valve and 0.20 mm (0.008 in.) clearance on exhaust valve for all other models. Tighten jam nut and recheck clearance. Install rocker arm cover.

CYLINDER HEAD AND COMBUSTION CHAMBER. It is recommended that cylinder head be removed and carbon and lead deposits be cleaned from head and combustion chamber and valves and seats be refaced at 300 hour intervals. Refer to CYLINDER HEAD paragraphs in REPAIRS section for service procedure.

LUBRICATION. Engine oil level should be checked prior to each operating interval. Maintain oil level at top of reference marks (Fig. HN35) when checked with cap not screwed in, but just touching first threads.

Oil should be changed after the first 20 hours of engine operation and at 100 hour intervals thereafter.

Manufacturer recommends oil with an API service classification SE or SF. Use SAE 10W-30 or 10W-40 oil.

Crankcase capacity is 0.6 L (0.63 qt.) for all models except Model GX240. Crankcase capacity for Model GX240 is 1.1 L (1.16 qt.).

GENERAL MAINTENANCE. Check and tighten all loose bolts, nuts or clamps daily. Check for fuel or oil leakage and repair as necessary.

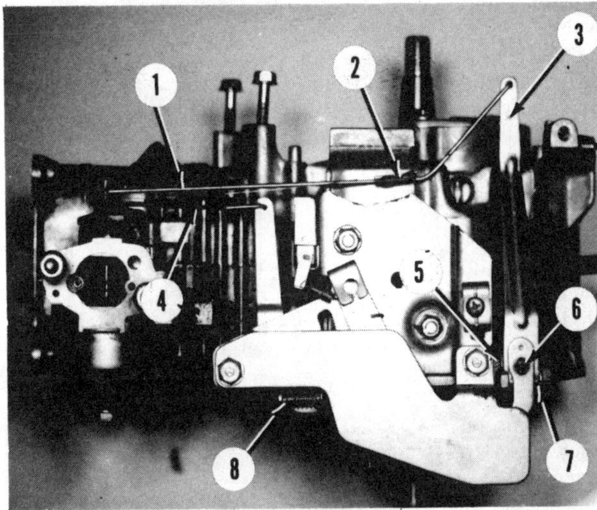

Fig. HN33—View of governor linkage.
1. Governor-to-carburetor rod
2. Spring
3. Governor lever
4. Choke rod
5. Tension spring (behind plate & lever)
6. Governor shaft
7. Clamp bolt
8. Throttle stop screw

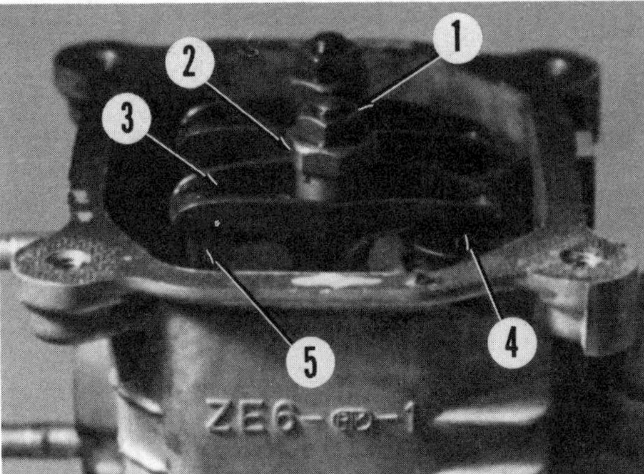

Fig. HN34—View of rocker arm and related parts.
1. Jam nut
2. Adjustment nut
3. Rocker arm
4. Valve stem clearance
5. Push rod

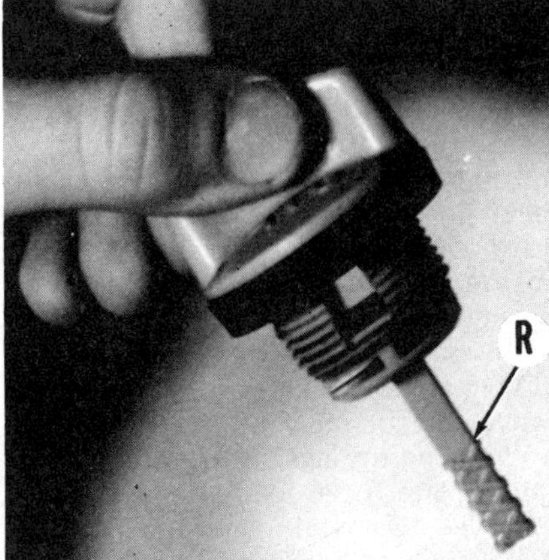

Fig. HN35—Do not screw in oil plug and gage when checking oil level. Maintain oil level at top edge of reference marks (R) on dipstick.

Clean dust, dirt, grease or any foreign material from cylinder head and cylinder block cooling fins at 100 hour intervals. Inspect fins for damage and repair as necessary.

REPAIRS

TIGHTENING TORQUES. Recommended tightening torque specifications are as follows:

Rocker arm cover..... 8-12 N·m (6-9 ft.-lbs.)
Cylinder head:
Models GX110, GX140, GXV120, GXV160............ 22-26 N·m (16-19 ft.-lbs.)
Model GX240....... 32-38 N·m (23-27 ft.-lbs.)
Oil pan.............. 10-14 N·m (7-10 ft.-lbs.)
Crankcase cover...... 22-26 N·m (16-19 ft.-lbs.)
Connecting rod bolts:
Models GX110, GX140, GXV120, GXV160.... 10-14 N·m (7-10 ft.-lbs.)
Model GX240 (serial number 1000001-1020232 .... 22-26 N·m (16-19 ft.-lbs.)
Model GX240 (after serial number 1020232 .... 12-16 N·m (9-11 ft.-lbs.)
Oil drain plug........ 15.20 N·m (11-15 ft.-lbs.)
Rocker arm jam nut... 8-12 N·m (6-9 ft.-lbs.)
Flywheel nut:
Models GX110, GX140, GXV120, GXV160............ 70-80 N·m (51-58 ft.-lbs.)
Model GX240....... 110-120 N·m (80-86 ft.-lbs.)

CYLINDER HEAD. To remove cylinder head, remove cooling shroud, disconnect and remove carburetor linkage and carburetor. Remove muffler. Remove rocker arm cover and the four head bolts. Remove cylinder head. Use care not to lose push rods.

Remove rocker arms, compress valve springs and remove valve retainers. Note exhaust valve on Models GX110, GX140 and GX240 are equipped with a valve rotator on valve stem. Remove valves and springs. Remove push rod guide plate if necessary.

Valve face and seat angle is 45 degrees. Standard valve seat width is 0.8 mm (0.032 in.). Narrow seat if seat width is 2.0 mm (0.079 in.) or more.

Standard valve spring free length for all models except Model GX240 is 34.0 mm (1.339 in.). Renew valve spring if free length is 32.5 mm (1.280 in.) or less. Standard valve spring free length for Model GX240 is 39.0 mm (1.54 in.). Renew valve spring if free length is 37.5 mm (1.48 in.) or less.

Standard valve guide inside diameter for all models except Model GX240 is 5.50-5.51 mm (0.2165-0.2170 in.). Renew guide if inside diameter is 5.562 mm (0.219 in.) or more. Standard valve guide inside diameter for Model GX240 is 6.60 mm (0.260 in.). Renew guide if inside diameter is 6.66 mm (0.262 in.) or more.

Valve stem-to-guide clearance for all models except Model GX240 should be 0.02-0.04 mm (0.001-0.002 in.) for intake valve and 0.06-0.09 mm (0.002-0.003 in.) for exhaust valve. Renew valve and/or guide if clearance is 0.10 mm (0.004 in.) or more for intake valve or 0.12 mm (0.005 in.) or more for exhaust valve.

Valve stem-to-guide clearance for Model GX240 should be 0.010-0.037 mm (0.0004-0.0015 in.) for intake valve and 0.050-0.077 mm (0.002-0.003 in.) for exhaust valve. Renew valve and/or guide if clearance is 0.10 mm (0.004 in.) or more for intake valve or 0.12 mm (0.005 in.) or more for exhaust valve.

To renew valve guide on all models except Model GX240, heat entire cylinder head to 150° F (300° C) and use valve guide driver 07942-8920000 to remove and install guides. Drive guides into cylinder head until top of guide is 23.0 mm (0.905 in.) below cylinder head surface (cylinder head-to-cylinder bore mating surface) for Models GX110 and GXV120 or 25.5 mm (1.004 in.) below cylinder head surface for Models GX140 and GXV160. New valve guides must be reamed after installation using reamer 07984-4600000 or 07984-2000000.

To renew valve guide on Model GX240, heat entire cylinder head to 150° C (300° F) and use valve guide driver 07942-6570100 to remove and install guides. Drive guides into cylinder head until top of intake valve guide is 9.0 mm (0.35 in.) from top of valve guide bore and exhaust valve guide is 7.0 mm (0.28 in.) from top of valve guide bore. New valve guides must be reamed after installation using reamer 07984-ZE2000A or 07984-ZE2000B.

When installing cylinder head on all models, tighten head bolts to specified torque following sequence shown in Fig. HN36. Adjust valves as outlined under VALVE ADJUSTMENT paragraphs in MAINTENANCE section.

CONNECTING ROD. The aluminum alloy connecting rod rides directly on crankpin journal on all models. Connecting rod cap for all models except Models GXV120 and GXV160 is equipped with an oil dipper (Fig. HN38). Models GXV120 and GXV160 have no dipper on connecting rod cap (Fig. HN37).

To remove connecting rod, remove cylinder head and crankcase cover or oil pan. Remove connecting rod cap screws and cap. Push connecting rod and piston assembly out of cylinder. Crankshaft and camshaft may also be removed if required. Remove piston pin retaining rings and separate piston from connecting rod.

Fig. HN36—Tighten cylinder head bolts in sequence shown.

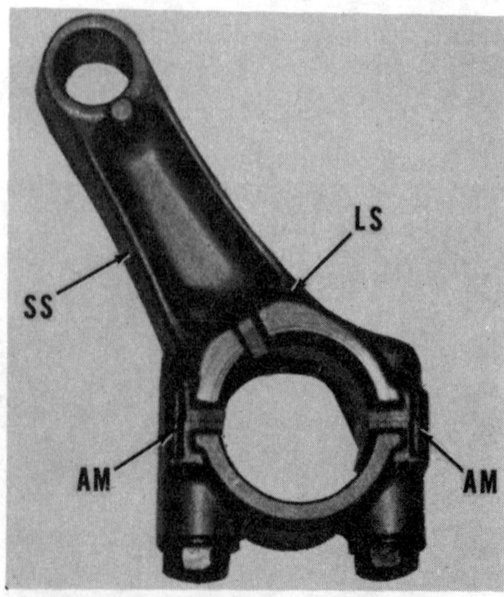

Fig. HN37—View of Models GXV120 and GXV160 connecting rod. Models GX110, GX140 and GX240 are equipped with an oil dipper as shown in Fig. HN38.

Standard piston pin bore diameter in connecting rod is 13.005-13.020 mm (0.512-0.513 in.) for Models GX110, GXV120 and GXV160 and 18.005-18.020 mm (0.7089-0.7094 in.) for Models GX140 and GX240. Renew connecting rod if diameter is 13.07 mm (0.5 in.) or more for Models GX110, GXV120 and GXV160 or 18.07 mm (0.711 in.) or more for Models GX140 and GX240.

Standard piston pin-to-connecting rod pin bore clearance is 0.005-0.026 mm (0.0002-0.001 in.) for all models. Renew connecting rod and/or pin if clearance is 0.08 mm (0.0031 in.) or more.

Standard connecting rod bearing bore-to-crankpin clearance 0.040-0.063 mm (0.0015-0.0025 in.) for all models. Renew connecting rod and/or crankshaft if clearance is 0.12 mm (0.0047 in.) or more.

Connecting rod side play on crankpin should be 0.1-0.7 mm (0.004-0.028 in.) for all models. Renew connecting rod if side play is 1.1 mm (0.043 in.) or more.

When reassembling piston on connecting rod, long side (LS—Fig. HN37) of connecting rod and arrowhead on piston top (Fig. HN39) must be on the same side.

When reinstalling piston and connecting rod assembly in cylinder, arrowhead on piston top must be on push rod side of engine. Align connecting rod cap and connecting rod match marks (AM—Fig. HN37), install connecting rod bolts and tighten to specified torque.

Fig. HN38—Exploded view of rod and piston assembly used on Models GX110, GX140 and GX240. Note location of oil dipper on connecting rod cap.

1. Retaining rings
2. Piston pin
3. Piston
4. Connecting rod
5. Rod cap & dipper
6. Bolts

PISTON, PIN AND RINGS. Piston and connecting rod are removed as an assembly. Refer to previous CONNECTING ROD paragraphs for removal and installation procedure.

Standard piston diameter measured at lower edge of skirt and 90 degrees from piston pin bore is 56.965-56.985 mm (2.242-2.243 in.) for Model GX110, 63.965-63.985 mm (2.518-2.519 in.) for Model GX140, 59.965-59.985 mm (2.360-2.361 in.) for Model GXV120, 67.965-67.985 mm (2.675-2.677 in.) for Model GXV160 and 72.965-72.985 mm (2.872-2.873 in.) for Model GX240. Renew piston if diameter is 56.55 mm (2.226 in.) or less for Model GX110, 63.55 mm (2.502 in.) or less for Model GX140, 59.55 mm (2.344 in.) or less for Model GXV120, 67.845 mm (2.671 in.) for Model GXV160 or 72.620 mm (2.859 in.) for Model GX240. Standard piston-to-cylinder clearance is 0.015-0.050 mm (0.0006-0.0020 in.) for all models. Renew piston and/or cylinder if clearance is 0.12 mm (0.0047 in.) or more.

Standard piston pin bore diameter is 13.002-13.008 mm (0.5118-0.5120 in.) for Models GX110, GXV120 and GXV160 and 18.002-18.048 mm (0.709-0.711 in.) or more for Models GX140 and GX240. Standard piston pin diameter is 12.994-13.000 mm (0.5115-0.5118 in.) for Models GX110, GXV120 and GXV160 and 17.994-18.000 mm (0.7084-0.7087 in.) for Models GX140 and GX240. If pin diameter is 12.954 mm (0.510 in.) or less for Models GX110, GXV120 and GXV160, or 17.954 mm (0.707 in.) or less for Models GX140 and GX240, renew piston pin. Standard clearance between piston pin and pin bore in piston is 0.002-0.014 mm (0.0001-0.0006 in.) for all models. If clearance is 0.08 mm (0.0031 in.) or more, renew piston and/or pin.

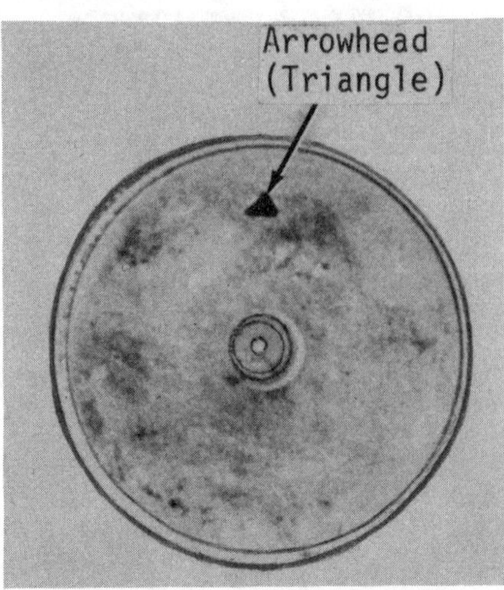

Fig. HN39—Arrowhead (triangle) on top of piston must be on push rod side of engine after installation. Refer to text.

Ring side clearance should be 0.015-0.045 mm (0.0006-0.0018 in.) for all models. Ring end gap for compression rings on all models should be 0.2-0.4 mm (0.008-0.016 in.). Ring end gap for oil control ring on all models should be 0.15-0.35 mm (0.006-0.014 in.). If ring end gap for any ring is 1.0 mm (0.039 in.) or more, renew ring and/or cylinder.

Install marked piston rings with marked side toward top of piston and stagger ring end gaps equally around circumference of piston.

CYLINDER AND CRANKCASE. Cylinder and crankcase are an integral casting. Standard cylinder bore diameter is 57.000-57.015 mm (2.244-2.245 in.) for Model GX110, 64.000-64.015 mm (2.519-2.520 in.) for Model GX140, 60.000-60.015 mm (2.262-2.363 in.) for Model GXV120, 68.000-68.015 mm (2.667-2.668 in.) for Model GXV160 and 73.000-73.015 mm (2.874-2.875 in.) for Model GX240.

If cylinder diameter is 57.165 mm (2.251 in.) or more for Model GX110, 64.165 mm (2.526 in.) or more for Model GX140, 60.165 mm (2.370 in.) or more for Model GXV120, 68.165 mm (2.680 in.) or more for Model GXV160 or 73.170 mm (2.881 in.) or more for Model GX240, renew cylinder.

CRANKSHAFT, MAIN BEARINGS AND SEALS. Crankshaft for Models GX110, GX140 and GX240 is supported at each end in ball bearing type main bearings. Crankshaft for Models GXV120 and GXV160 is supported at flywheel end in a ball bearing type main bearing and at pto end in a bushing type main bearing which is an integral part of the oil pan casting. To remove crankshaft, refer to previous CONNECTING ROD paragraphs.

Standard crankpin journal diameter is 26.0 mm (1.024 in.) for Models GX110, GXV120 and GXV160, 30.0 mm (1.181 in.) for Model GX140 and 32.985 mm (1.2986 in.) for Model GX240. If crankpin diameter is 25.92 mm (1.020 in.) or less for Models GX110, GXV120 and GXV160, 29.92 mm (1.178 in.) or less for Model GX140 or 32.92 mm (1.296 in.) for Model GX240, renew crankshaft.

Ball bearing type main bearings are a press fit on crankshaft journals and in bearing bores of crankcase and cover. Renew bearings if loose, rough or loose fit on crankshaft or in bearing bores.

Bushing type bearing in oil pan of Models GXV120 and GXV160 is an integral part of oil pan. Renew oil pan if bearing is worn, scored or damaged.

Seals should be pressed into seal bores until outer edge of seal is flush with seal bore.

Fig. HN41—Align crankshaft gear and camshaft gear timing marks (F) during installation.

When installing crankshaft, make certain crankshaft gear and camshaft gear timing marks are aligned as shown in Fig. HN41.

CAMSHAFT. Camshaft and camshaft gear are an integral casting equipped with a compression release mechanism (Fig. HN42). To remove camshaft, refer to previous CONNECTING ROD paragraphs.

Standard camshaft bearing journal diameter for Models GX110, GX140, GXV120 and GXV160 is 13.984 mm (0.551 in.). Renew camshaft if journal diameter is 13.916 mm (0.548 in.) or less. Standard camshaft bearing journal diameter for Model GX240 is 15.984 mm (0.6293 in.). Renew camshaft if journal diameter is 15.92 mm (0.627 in.) or less.

Standard camshaft lobe height for all models except Model GX240 is 27.7 mm (1.091 in.) for intake lobe and 27.75 mm (1.093 in.) for exhaust lobe (Fig. HN44). If intake lobe measures 27.45 mm (1.081 in.) or less or exhaust lobe measures 27.50 mm (1.083 in.) or less, renew camshaft. Standard camshaft lobe height for Model GX240 is 31.2 mm (1.23 in.) for intake lobe and 31.1 mm (1.22 in.) for exhaust lobe. If intake lobe measures 30.95 mm (1.291 in.) or less or exhaust lobe measures 30.85 mm (1.215 in.) or less, renew camshaft.

Inspect compression release mechanism for damage. Spring must pull weight tightly against camshaft so decompressor lobe holds exhaust valve slightly open. Weight overcomes spring tension at 1000 rpm and moves decompressor lobe away from cam lobe to release exhaust valve.

When installing camshaft, make certain camshaft and crankshaft gear timing marks are aligned as shown in Fig. HN41.

GOVERNOR. Centrifugal flyweight type governor controls engine rpm via external linkage. Governor is located in oil pan on Models GXV120 and GXV160, on flywheel side of crankcase on Models GX110 and GX140 and on crankcase cover on Model GX240. Refer to GOVERNOR paragraphs in MAINTENANCE section for adjustment procedure.

To remove governor assembly on Models GXV120 and GXV160, remove oil pan. Remove governor assembly retaining bolt (1—Fig. HN45) and remove governor gear and weight assembly. Governor sleeve, thrust washer, retaining clip and gear may be removed from shaft.

To remove governor assembly on Models GX110, GX140 and GX240, refer to previous CONNECTING ROD paragraphs and on Models GX110 and GX140, remove crankshaft and camshaft. Remove governor sleeve and washer. Remove retaining clip from governor gear shaft, then remove gear and weight assembly and remaining thrust washer.

Reinstall governor assemblies by reversing removal procedure. Adjust external linkage as outlined under GOVERNOR in MAINTENANCE section.

Compression Release

Auxiliary Drive Gear

Fig. HN42—Camshaft and gear are an integral casting equipped with a compression release mechanism. Camshaft shown is for Model GXV120 with auxilliary driveshaft.

Fig. HN43—Compression release mechanism spring (1) and weight (2) installed on camshaft gear.

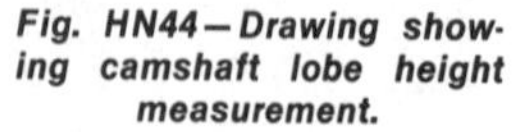

Fig. HN44—Drawing showing camshaft lobe height measurement.

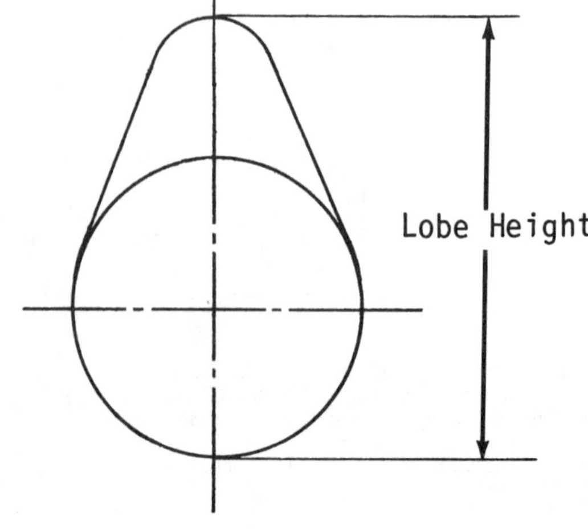

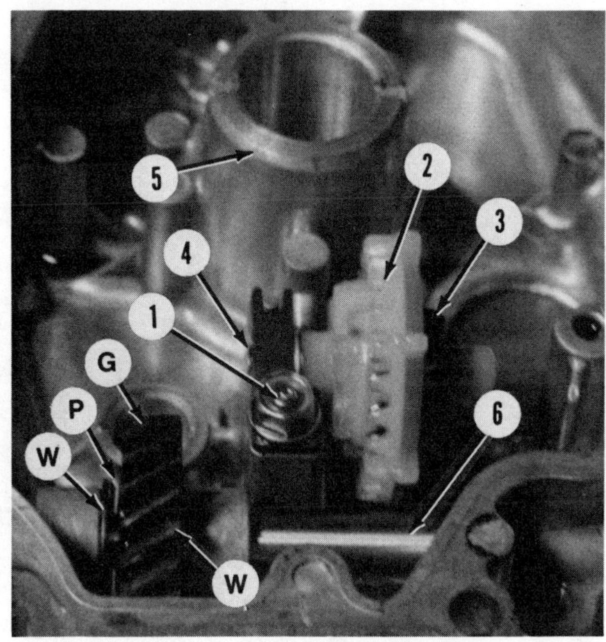

Fig. HN45—Governor assembly on Models GXV120 and GXV160 is mounted in oil pan. Refer to Fig. HN46 also. Governor assembly for all other models is similar except assembly is located in crankcase.

W. Thrust washers
G. Auxillary drive gear
P. Retaining hair pin
1. Bolt
2. Governor gear
3. Weight
4. Governor gear shaft
5. Oil pan
6. Auxillary drive shaft

Fig. HN46—Auxiliary drive shaft is mounted in oil pan and driven by a gear which is an integral part of camshaft. See Fig. HN42 and HN43.

1. Bolt
2. Governor gear
6. Governor gear shaft
7. Auxiliary drive shaft
8. Auxiliary drive gear
9. Retaining hair pin
10. Thrust washer

AUXILIARY DRIVE (MODEL GXV120). Auxiliary drive shaft (Fig. HN45 and HN46) is mounted in oil pan and is driven by a gear which is an integral part of the camshaft.

Drive shaft is retained in oil pan by retaining pin (9—Fig. HN46). When reassembling, carefully slide drive shaft through oil pan seal, thrust washer, gear and remaining thrust washer. Insert retaining hair pin in hole in drive shaft so it is located between gear (8) and outer thrust washer (10).

HONDA

Model	Bore	Stroke	Displacement
G20	58 mm (2.28 in.)	50 mm (1.97 in.)	132 cc (8.1 cu. in.)
G30	66 mm (2.59 in.)	50 mm (1.97 in.)	170 cc (10.4 cu. in.)
G35	64 mm (2.52 in.)	45 mm (1.77 in.)	144 cc (8.8 cu. in.)
GV35	64 mm (2.52 in.)	45 mm (1.77 in.)	144 cc (8.8 cu. in.)

ENGINE INFORMATION

Engines covered in this section are four-stroke, single-cylinder, air-cooled engines with intake and exhaust valves located in cylinder block.

All models except Model GV35 have a horizontal crankshaft and a splash lubrication system. Model GV35 has a vertical crankshaft and is lubricated by a trochoid type oil pump.

MAINTENANCE

SPARK PLUG. Spark plug should be removed, cleaned and adjusted after every 100 hours of operation. Recommended spark plug is a NKG B6HS, or equivalent. Electrode gap should be 0.6-0.7 mm (0.024-0.028 in.).

CARBURETOR. All models are equipped with a Keihin float type carburetor equipped with an idle fuel mixture screw (11–Fig. HN50). High speed fuel mixture is controlled by a fixed jet (30) in carburetor.

For initial adjustment of idle fuel mixture screw, lightly seat mixture screw then open screw 1³/₈ turns for Model G35 or ⁷/₈ turn for all other models.

Make final adjustments with engine at operating temperature and running. Adjust throttle stop screw (16) so engine idles at 1400 rpm. Adjust idle mixture screw to obtain smoothest engine idle. Check idle speed and adjust if necessary.

Standard main jet size is #70 for Models G35 and GV35, #75 for Model G20 or #85 for Model G30.

To check float level, remove float bowl and invert carburetor throttle body and float assembly. Correct float level with float lever just touching the spring loaded fuel inlet needle, is 17.5 mm (0.68 in.) for Models G20 and G30 or 11.5 mm (0.45 in.) for Models G35 and GV35 when measured from top edge of float to float bowl mating surface of carburetor. Bend float stop tab on float lever to obtain correct level.

AIR CLEANER. Air cleaner element should be removed and cleaned after every 50 hours of operation. Remove element and clean in solvent. Squeeze out solvent and allow element to air dry. Soak element in a 10:1 gasoline:engine oil solution and squeeze out excess solution. Reinstall element.

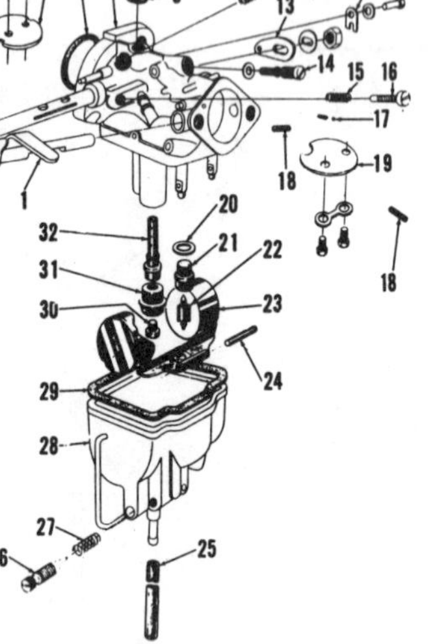

Fig. HN50—Exploded view of Keihin float type carburetor used on Models G20 and G30. Carburetor used on Models G35 and GV35 is similar except a square type float is used.

1. Choke shaft
2. Throttle stop lever
3. Throttle stop spring
4. Throttle shaft
5. Throttle plate
6. "O" ring
7. Carburetor body
8. Plug
9. Gasket
10. Spring
11. Idle mixture screw
12. Clip
13. Throttle shaft lever
14. Idle jet
15. Spring
16. Idle stop screw
17. Choke friction ball & spring
18. Air jets
19. Choke plate
20. Gasket
21. Inlet valve seat
22. Inlet valve
23. Float
24. Float pin
25. Overflow tube
26. Drain valve
27. Spring
28. Float bowl
29. Gasket
30. Main jet
31. Main jet holder
32. Main nozzle

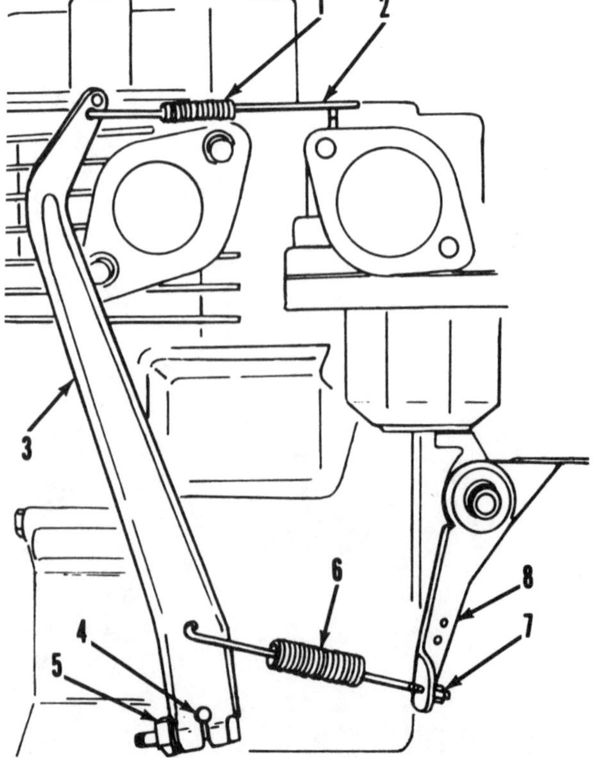

Fig. HN51—View of external governor linkage similar to all models.

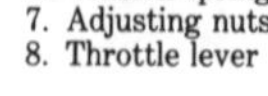

1. Spring
2. Governor lever-to-carburetor rod
3. Governor lever
4. Governor shaft
5. Clamp bolt
6. Tension spring
7. Adjusting nuts
8. Throttle lever

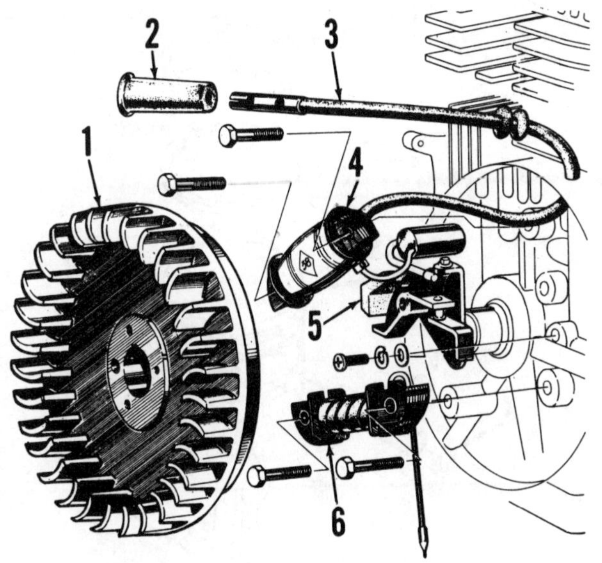

Fig. HN52—View of flywheel magneto type ignition system used on Models G20 and G30. Models G35 and GV35 are similar.
1. Flywheel
2. Spark plug boot
3. Spark plug wire
4. Ignition coil
5. Breaker point and condenser assembly
6. Lighting coil (optional)

GOVERNOR. All models are equipped with a mechanical flyweight type governor. Governor flyweight assembly is attached to camshaft gear.

To adjust external linkage, stop engine and make certain all linkage is in good condition and tension spring (6–Fig. HN51) is not stretched or damaged. Spring (1) must pull governor lever (3) and throttle pivot toward each other. Loosen clamp bolt (5) and push governor lever (3) toward carburetor as far as possible. Hold lever in this position and rotate governor shaft (4) clockwise until it stops. Tighten clamp bolt.

To adjust throttle for maximum speed, connect tachometer to engine and start engine. Move throttle lever (8) until engine is at full throttle or 3600 rpm, whichever comes first. Adjust nuts (7) so maximum engine speed which throttle lever movement allows is 3600 rpm.

IGNITION SYSTEM. A breaker point ignition system is standard on all models. Breaker points, condenser and ignition coil are located under engine flywheel. Breaker point gap is 0.3-0.4 mm (0.012-0.016 in.).

Correct engine timing is 20° BTDC. To set timing, rotate engine so both the intake and the exhaust valves are closed and the "F" mark on flywheel is aligned with mating joint between oil pan and upper crankcase on Models G20 and G30, or "F" mark is aligned with timing index mark on crankcase of Models G35 and GV35. Remove flywheel carefully so engine does not rotate. Adjust breaker point plate so breaker points just begin to open.

NOTE: Special tool 07974-8780000 is available to set timing while flywheel is removed.

VALVE ADJUSTMENT. Valve stem clearance should be checked and adjusted after every 300 hours of operation. Refer to VALVE SYSTEM paragraphs in REPAIRS section for service procedure.

CYLINDER HEAD AND COMBUSTION CHAMBER. It is recommended that cylinder head be removed and carbon and other deposits cleaned from cylinder head, combustion chamber and valves after every 300 hours of operation. Refer to REPAIRS section for service procedure.

LUBRICATION. Engine oil level should be checked prior to each

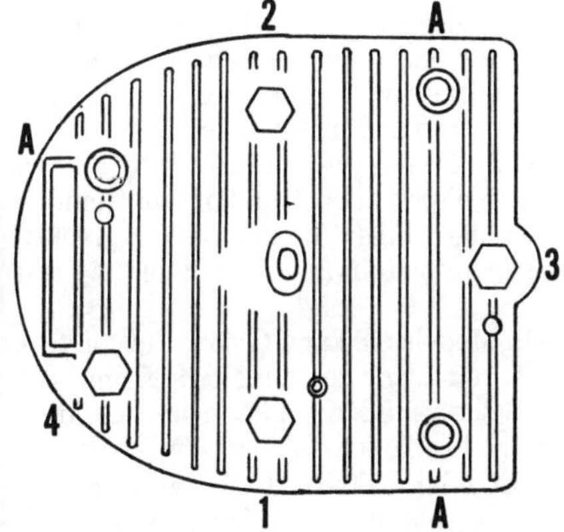

Fig. HN53—On Models G20 and G30, tighten the four short cylinder head bolts to specified torque in sequence shown. Install fuel tank and tighten the long bolts (A) to specified torque.

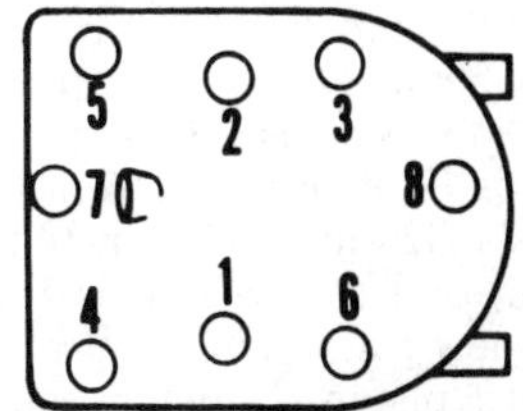

Fig. HN54—On Models G35 and GV35, tighten head bolts to specified torque in sequence shown.

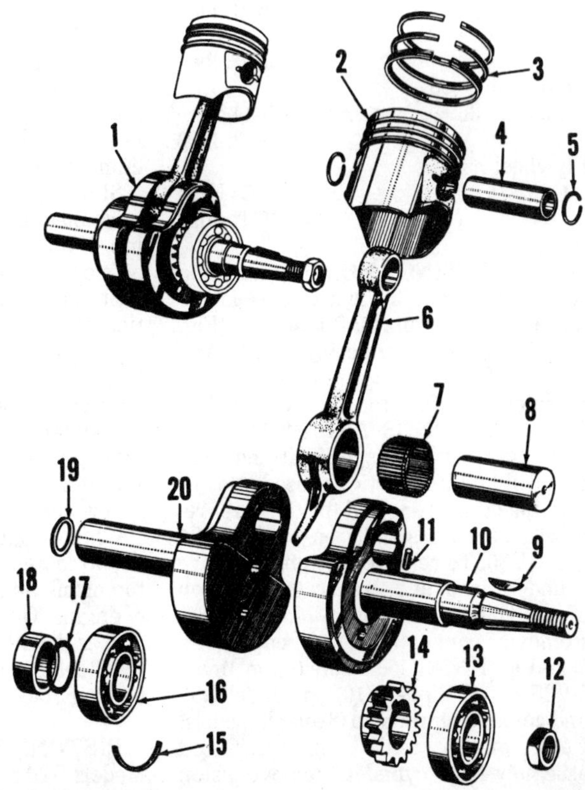

Fig. HN55—Exploded view of crankshaft, connecting rod and piston assembly used on Models G20 and G30. Crankshaft and connecting rod are serviced as an assembly only, even though breakdown shows individual parts. Refer to text.
1. Crankshaft, connecting rod & piston assy.
2. Piston
3. Piston rings
4. Piston pin
5. Pin retaining rings
6. Connecting rod
7. Crankpin roller bearings
8. Crankpin
9. Flywheel key
10. Right crankshaft
11. Timing gear pin
12. Flywheel nut
13. Ball bearing
14. Timing gear
15. Bearing set ring
16. Ball bearing
17. "O" ring (water pump engine)
18. Crankshaft collar
19. Washer
20. Left crankshaft

operating interval. Maintain oil level at top of reference marks on gage. Oil level is checked with cap unscrewed and just touching first threads.

Oil should be changed after the first 20 hours of engine operation and after every 100 hours of operaton thereafter.

Manufacturer recommends SAE 10W-30 or 10W-40 oil with an API service classification SE or SF.

Crankcase capacity of Model GV35 is 0.7 L (1.5 pt.). Crankcase capacity for all other models is 0.6 L (1.27 pt.).

On engines equipped with speed reduction unit, use same weight oil as used in engine crankcase. Reduction case capacity is 118 mL (1/4 pt.).

GENERAL MAINTENANCE. Check and tighten all loose bolts, nuts or clamps after each day's use. Check for fuel or oil leakage and repair if necessary.

Clean dust, dirt, grease or any foreign material from cylinder head and cylinder block cooling fins after every 100 hours of operation. Inspect fins for damage and repair if necessary.

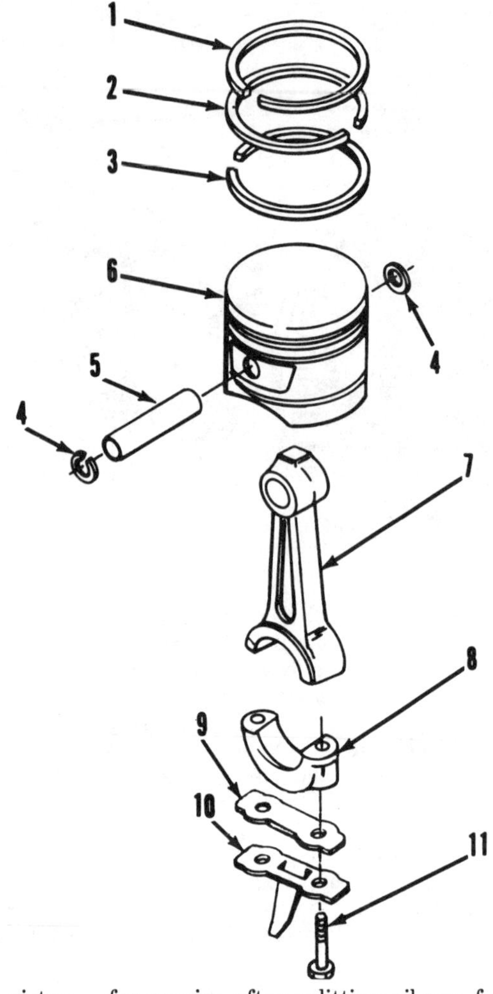

Fig. HN56—Exploded view of piston and connecting rod assembly used on Models G35 and GV35.

1. Compression ring
2. Compression ring
3. Oil control ring
4. Piston pin retaining rings
5. Piston pin
6. Piston
7. Connecting rod
8. Connecting rod cap
9. Lock plate (GV35)
10. Dipper/lock plate (G35)
11. Bolts

REPAIRS

TIGHTENING TORQUES. Recommended tightening torque specifications are as follows:

Cylinder head:
- G20, G30–
  - Short bolts .... 38-40 N·m (28-30 ft.-lbs.)
  - Long bolts .... 22-24 N·m (16-18 ft.-lbs.)
- G35, GV35 .... 19-27 N·m (14-20 ft.-lbs.)

Connecting rod
- G35, GV35 .... 10-11 N·m (7-8 ft.-lbs.)

Flywheel nut
- G35, GV35 .... 60-64 N·m (43-47 ft.-lbs.)

CONNECTING ROD. Connecting rod on Models G20 and G30 is an integral part of crankshaft assembly (Fig. HN55) and connecting rod and crankshaft must be renewed as an assembly. Refer to CRANKSHAFT, MAIN BEARINGS AND SEALS section for service specifications and removal and installation procedures.

Connecting rod and piston are removed as an assembly on Models G35 and GV35. To remove assembly, remove cylinder head and crankcase cover on Model G35 or oil pan on Model GV35. Remove connecting rod cap bolts (11 – Fig. HN56), lock plate (9) on Model GV35 or oil dipper (10) on Model G35 and connecting rod cap (8) on all models. Push piston and connecting rod assembly from cylinder. Remove piston pin retaining rings (4) and push piston pin (5) from piston (6) to separate piston from connecting rod (7).

On Models G35 and GV35, standard connecting rod bearing bore to crankpin clearance is 0.027 mm (0.001 in.). If clearance is 0.1 mm (0.004 in.) or more, renew connecting rod and/or crankshaft.

Standard diameter of piston pin bore in connecting rod is 15.005 mm (0.5907 in.). If diameter is 15.05 mm (0.5925 in.) or more, renew connecting rod.

Standard clearance between connecting rod piston pin bore and piston pin is 0.005 mm (0.0002 in.). If clearance is 0.05 mm (0.002 in.) or more, renew connecting rod and/or piston pin.

Standard connecting rod side play on crankpin journal is 0.1 mm (0.004 in.). If side play is 1.0 mm (0.04 in.) or more, renew connecting rod.

When installing connecting rod and piston assembly, stamped triangle on top of piston must be toward valve side of engine. Oil dipper plate used on Model G35 is installed so dipper is closer to flywheel side of crankcase. Tighten connecting rod bolts to specified torque.

PISTON, PIN AND RINGS. On Models G20 and G30, piston is accessible for service after splitting oil pan from cylinder and upper crankcase assembly and withdrawing the crankshaft, connecting rod and piston assembly (1 – Fig. HN55). Piston pin is retained by a snap ring in piston at each end of pin. Remove retaining rings (5) and piston pin (4). Separate piston (2) from connecting rod (6).

On Models G35 and GV35, the connecting rod can be unbolted from the crankshaft and the piston and connecting rod removed as an assembly. Refer to CONNECTING ROD section for correct piston and connecting rod removal and installation procedures.

Piston pin should be a hand push fit in piston and connecting rod. Piston pin to piston clearance should be 0-0.012 mm (0-0.0005 in.) on all models.

Standard piston pin diameter is 17.994-18.000 mm (0.7084-0.7087 in.) on Models G20 and G30 and 15.00 mm (0.591 in.) for Models G35 and GV35. If diameter is 17.95 mm (0.7067 in.) or less for Models G20 and G30, or 14.97 mm (0.589 in.) or less for Models G35 and GV35, renew piston pin.

Standard piston diameter, measured at lower edge of piston skirt and 90° from piston pin bore, is 57.96 mm (2.282 in.) for Model G20, 65.96 mm (2.597 in.)

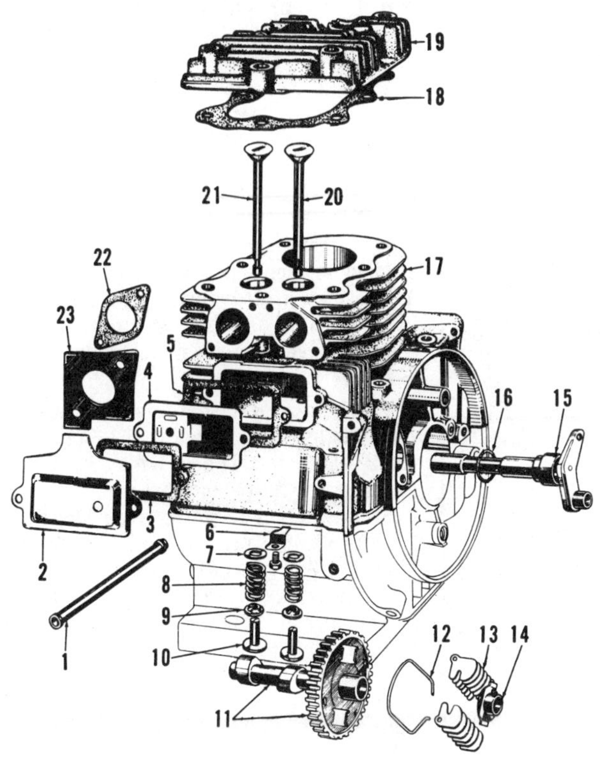

Fig. HN58—Exploded view of cylinder and upper crankcase assembly for Models G20 and G30. Models G35 and GV35 do not have a split crankcase.

1. Breather tube
2. Tappet cover
3. Gasket
4. Breather assy.
5. Gasket
6. Oil separator
7. Valve spring seats
8. Valve springs
9. Spring retainers
10. Valve lifters
11. Camshaft
12. Governor weight center
13. Governor weights
14. Governor sleeve
15. Camshaft holder
16. "O" ring
17. Cylinder & upper crankcase
18. Cylinder head gasket
19. Cylinder head
20. Exhaust valve
21. Intake valve
22. Gasket
23. Carburetor insulator

for Model G30 and 63.96 mm (2.518 in.) for Models G35 and GV35. If piston diameter is 57.7 mm (2.272 in.) or less for Model G20, 65.7 mm (2.587 in.) or less for Model G30 or 63.7 mm (2.508 in.) or less for Models G35 and GV35, renew piston.

Standard piston-to-cylinder wall clearance measured at lower edge of piston skirt and 90° from piston pin bore is 0.07-0.10 mm (0.0028-0.0039 in.) for Models G20 and G30 or 0.04 mm (0.002 in.) for Models G35 and GV35. If clearance is 0.25 mm (0.0098 in.) or more for Models G20 and G30 or 0.12 mm (0.005 in.) or more for Models G35 and GV35, renew piston and/or cylinder.

Standard top ring side clearance in ring groove is 0.02-0.06 mm (0.0008-0.0024 in.) for Models G20 and G30 and 0.03 mm (0.001 in.) for Models G35 and GV35. If ring side clearance is 0.15 mm (0.006 in.) or more for Models G20 and G30 or 0.1 mm (0.004 in.) or more for Models G35 and GV35 on any ring, renew piston and/or rings.

Standard ring end gap is 0.2-0.4 mm (0.008-0.016 in.) for Models G20 and G30 or 0.3 mm (0.012 in.) for Models G35 and GV35. If ring end gap is 1.5 mm (0.059 in.) or more for Models G20 and G30 or 0.6 mm (0.024 in.) or more for Models G35 and GV35, renew rings and/or cylinder.

Piston rings with a mark on one side are installed with marked side up. Stagger piston ring end gaps equally around circumference of piston. When installing piston and connecting rod assembly, triangle stamped on piston top must be towards valve side of engine.

CYLINDER AND CRANKCASE. Cylinder and crankcase are an integral casting on all models. Standard cylinder bore diameter is 58.00-58.01 mm (2.2835-2.2839 in.) for Model G20, 66.00-66.01 mm (2.5984-2.5988 in.) for Model G30 and 64.00-64.01 mm (2.5197-2.5201 in.) for Models G35 and GV35. If cylinder bore diameter is 58.1 mm (2.298 in.) or more for Model G20, 66.1 mm (2.603 in.) or more for Model G30 or 64.1 mm (2.524 in.) or more for Model G35 and GV35, renew cylinder. Oversize piston and ring sets are available for Models G20 and G30 if cylinder bore is worn beyond service limit.

OIL PUMP. MODEL GV35. Model GV35 is equipped with an oil pump in oil pan. To remove pump, remove oil pan. Remove bolt and washer. Lift pump assembly from housing in oil pan.

Standard clearance between pump body and outer rotor is 0.15 mm (0.006 in.). If clearance is 0.3 mm (0.012 in.) or more, renew pump. Standard clearance between inner rotor tip and outer rotor is 0.15 mm (0.006 in.). If clearance is 0.20 mm (0.008 in.) or more, renew pump. Standard inside diameter of oil pump body is 16.0 mm (0.630 in.). If inside diameter is 16.07 mm (0.633 in.) or more, renew oil pump.

When assembling oil pan to crankcase, make certain oil pump drive shaft enters drive slot in camshaft end. If slot and shaft do not engage, there will be a 3-5 mm (0.12-0.20 in.) gap between oil pan and crankcase. If engaged, oil pan will meet crankcase. Do not force oil pan and crankcase together.

CRANKSHAFT, MAIN BEARINGS AND SEALS. Crankshaft on all models is supported at each end in ball bearing type main bearings.

MODELS G20 AND G30. On Models G20 and G30, crankshaft and connecting rod are integral assemblies and must be removed together. Remove all cooling shrouds and split crankcase (Fig. HN58). Remove crankshaft, connecting rod and piston as an assembly. Separate piston and connecting rod by removing piston pin retaining rings and piston.

Connecting rod bearing clearance should be 0.004-0.036 mm (0.00016-0.0014 in.). If clearance is 0.1 mm (0.0039 in.) or more, renew crankshaft and connecting rod assembly. Side play of connecting rod on crankpin should be 0.1-0.35 mm (0.0039-0.0138 in.). If side play is 0.85 mm (0.0335 in.) or more, renew crankshaft and connecting rod assembly. Crankpin bearing wear and connecting rod side play can be checked by measuring connecting rod small end "shake" or side movement. If shake is 3.0 mm (0.118 in.) or more, renew crankshaft and connecting rod assembly.

When reinstalling crankshaft in Models G20 and G30, be sure piston ring end gaps are placed 120° apart and lubricate the cylinder bore, piston, rings and bearings. Align the crankshaft and camshaft gear timing marks during installation and be sure bearing on left crankshaft is fitted on the bearing set ring (15–Fig. HN55) in upper crankcase. Place oil seals on crankshaft before installing the assembly in upper crankcase. Coat upper and lower crankcase mating surfaces with gasket sealer.

MODELS G35 AND GV35. On Models G35 and GV35, the connecting rod can be unbolted from the crankshaft and the connecting rod and piston removed as an assembly. Remove connecting rod and piston assembly. Separate crankcase cover on Model G35 or oil pan on Model GV35 from crankcase. Remove crankshaft and camshaft.

Standard crankpin journal diameter is 25.980 mm (1.0228 in.). If diameter is 25.7 mm (1.0118 in.) or less, renew crankshaft.

Ball bearing type main bearings are a press fit on crankshaft and in bearing bores of crankcase and crankcase cover/oil pan. Renew bearings if they are loose, rough or damaged.

Crankshaft oil seals should be installed in crankcase and crankcase cover/oil pan before crankshaft installation to avoid damaging seals. Press seals in until

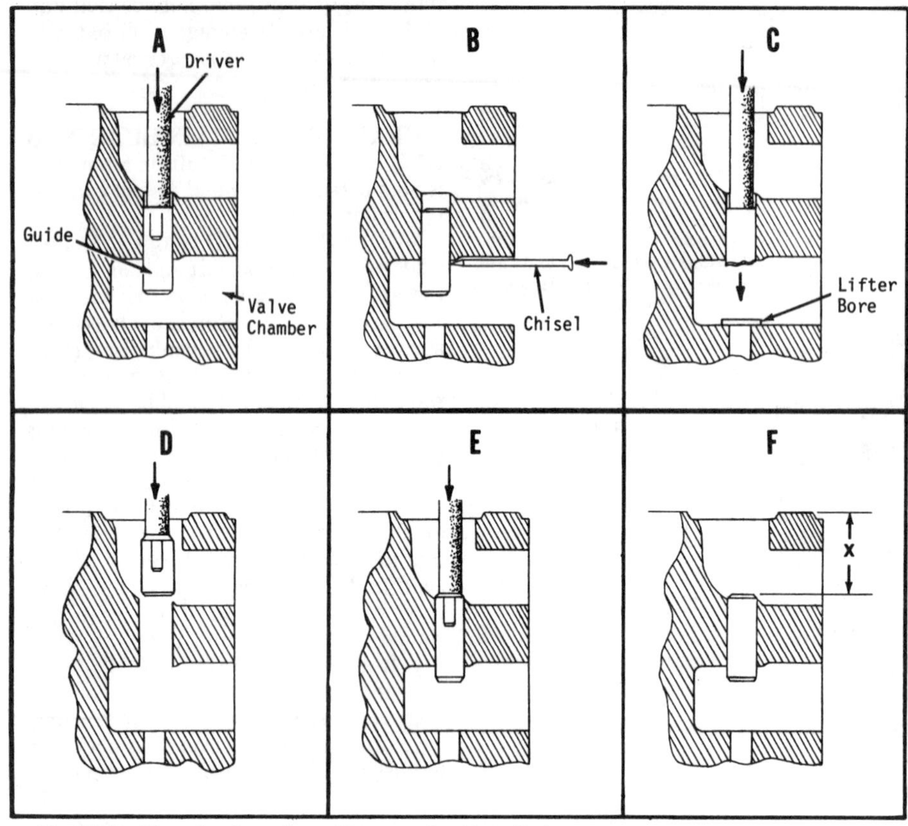

Fig. HN59—View showing valve guide removal and installation sequence on all models. Refer to text.

outer surface of seal is flush to slightly below flush with seal bore.

When installing crankshaft, use care not to damage oil seals and make certain crankshaft and camshaft gear timing marks are aligned.

CAMSHAFT AND GOVERNOR. Governor flyweight assembly (12, 13 and 14–Fig. HN58) is attached to camshaft gear on all models. Governor weights may be removed from camshaft gear on all models except those equipped with extended camshaft for camshaft pto drive.

To remove camshaft on Models G20 and G30, remove crankshaft assembly and camshaft holder (15–Fig. HN58). Withdraw camshaft (11) and governor weight assembly from upper crankcase (17). Remove tappets (10). Separate governor weight center (12), weights (13) and sleeve (14) if necessary.

To remove camshaft on Models G35 and GV35, remove all necessary cooling shrouds. Remove crankcase cover on Model G35 or oil pan on Model GV35. Remove camshaft and separate governor assembly from camshaft if necessary.

On Models G20 and G30, standard clearance between camshaft holder and camshaft is 0.016-0.061 mm (0.0006-0.0024 in.). If clearance is 0.1 mm (0.0039 in.) or more, renew camshaft and/or holder. Standard camshaft holder diameter is 10.966-10.984 mm (0.4317-0.4324 in.) at inner end and 11.00-11.027 mm (0.4331-0.4342 in.) at outer end. If diameter is 10.92 mm (0.4299 in.) at inner end or 10.954 mm (0.4313 in.) or less at outer end, renew camshaft holder. Standard camshaft lobe height is 29.32-29.40 mm (1.1543-1.1575 in.). If lobe height is 29.0 mm (1.1417 in.) or less, renew camshaft. Standard camshaft end play is 0.1-0.5 mm (0.0039-0.0197 in.).

On Models G35 and GV35, standard clearance between camshaft bearing and camshaft is 0.02 mm (0.001 in.). If clearance is 0.1 mm (0.004 in.) or more, renew camshaft. Standard camshaft journal diameter for Model G35 is 17.03 mm (0.670 in.). If journal diameter is 16.5 mm (0.650 in.) or less, renew camshaft. Standard camshaft journal diameter for Model GV35 is 15.98 mm (0.630 in.) at oil pump body end and 17.03 mm (0.670 in.) at opposite end. If journal diameter is 15.92 mm (0.628 in.) or less at oil pump body end or 16.5 mm (0.650 in.) or less at opposite end, renew camshaft. Standard camshaft lobe height for Models G35 and GV35 is 28.1 mm (1.106 in.). If lobe height is 27.6 mm (1.087 in.) or less, renew camshaft.

When reinstalling camshaft on all models, make certain crankshaft and camshaft gear timing marks are aligned. On Model GV35, refer to OIL PUMP section and make certain slot in camshaft end is aligned with oil pump drive shaft as oil pan and crankcase are joined.

VALVE SYSTEM. Valve stem clearance (cold) for all models is 0.05-0.10 mm (0.002-0.004 in.). Valve stem clearance is adjusted as follows: To increase stem clearance, grind off end of valve stem. To reduce stem clearance, renew valve and/or grind valve seat deeper.

Valve face and seat angles are 45°. Standard valve seat width is 0.7-1.0 mm (0.028-0.039 in.). If seat width is 1.5 mm (0.06 in.) or more, narrow seat.

Standard valve spring free length is 27 mm (1.063 in.) for Models G20 and G30 or 27.9 mm (1.098 in.) for Models G35 and GV35. If spring length is 25 mm (0.985 in.) or less for Models G20 and G30 or 26.0 mm (1.023 in.) or less for Models G35 and GV35, renew spring.

On all models, standard valve stem diameter is 6.965-6.980 mm (0.2742-0.2748 in.) for intake valve and 6.940-6.973 mm (0.2732-0.2745 in.) for exhaust valve. If stem diameter is 6.92 mm (0.2724 in.) or less for intake valve or 6.90 mm (0.2716 in.) or less for exhaust valve, renew valve. On all models, standard valve guide inside diameter is 7.0 mm (0.2756 in.). If diameter is 7.065 mm (0.2781 in.) or more, renew or recondition guide.

Valve guides are renewable on Models G35 and GV35. To remove and install valve guides on G35 and GV35 models, use the following procedure and refer to the sequence of illustrations in Fig. HN59. Use driver 07942-8230000 and drive old guide down into valve chamber slightly (A). Use a suitable cold chisel and sever guide adjacent to guide bore (B). Cover tappet opening to prevent fragments from entering crankcase. Drive remaining piece of guide into valve chamber (C) and remove from chamber. Place new guide on driver and start guide into guide bore (D). Alternate between driving guide into bore and measuring guide depth below cylinder head surface (E). Guides are driven in to a depth of 28 mm (1.102 in.) for intake valve guide and 25 mm (0.984 in.) for exhaust valve guide measured from end of guide to cylinder head surface as shown in section (F) at (X). Finish ream guide after installation with reamer 07984-5900000.

REWIND STARTER. Refer to Fig. HN60 for exploded view of the rewind starter assembly used on some models. Starter can be disassembled as follows: Pull rope out slightly and untie knot at handle end while holding the rope pulley

(6) from turning. Remove handle from rope and allow spring to unwind slowly. Remove cap screw (1), friction plate (2), spring (3) and friction disc (5). Remove ratchets (4) from pulley and remove pulley and spring from housing. Reassemble by reversing disassembly procedure and check action of starter before reinstalling.

REDUCTION DRIVE. Refer to exploded view of optional reduction drive unit in Fig. HN61. To disassemble unit, remove cover (7) and pull drive sprocket (11), driven sprocket (13) and chain (14) as a unit. Remove reduction case (15) from engine. Reassemble by reversing disassembly procedure. Refer to LUBRICATION section for proper oil to refill unit.

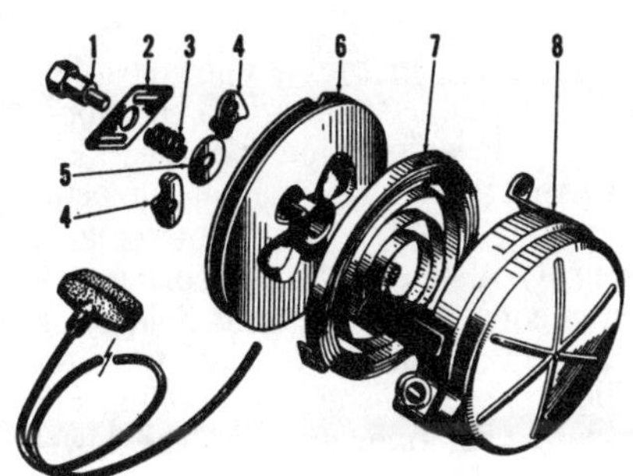

Fig. HN60—Exploded view of rewind starter assembly. Starter cup (attached to engine crankshaft) is not shown.

1. Special bolt
2. Friction plate
3. Friction spring
4. Ratchets
5. Washer
6. Rope pulley
7. Rewind spring
8. Housing

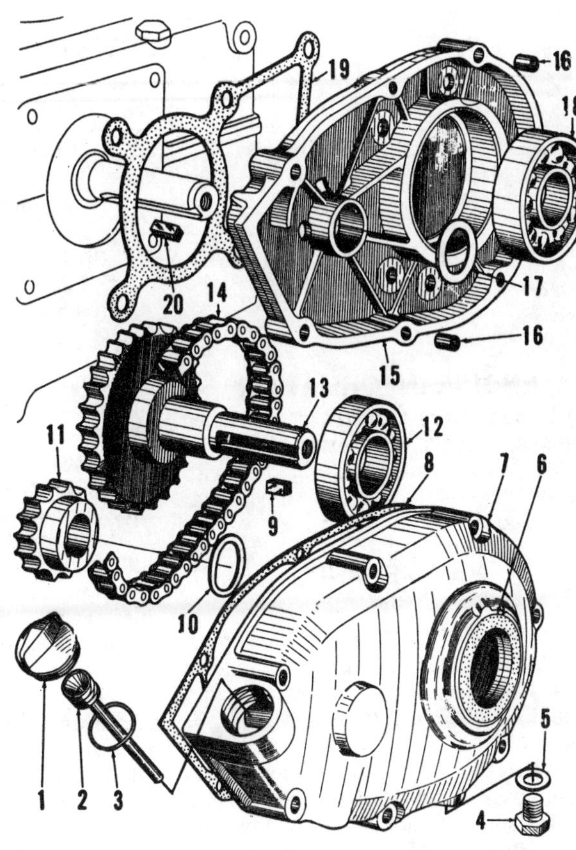

Fig. HN61—Exploded view of optional reduction unit.

1. Oil filler cap
2. Oil dipstick
3. "O" ring
4. Drain plug
5. Gasket
6. Oil seal
7. Cover
8. Gasket
9. Key
10. Thrust washer
11. Drive sprocket
12. Ball bearing
13. Driven sprocket
14. Drive chain
15. Reduction case
16. Hollow dowels
18. Ball bearing
19. Gasket
20. Key

JACOBSEN

JACOBSEN DIVISION OF TEXTRON INC.
1721 Packard Ave.
Racine, Wisconsin 53403

Model	Bore	Stroke	Displacement
J-100	2.00 in. (50.8 mm)	1.50 in. (38.1 mm)	4.7 cu. in. (77 cc)
J-150	2.25 in. (57.2 mm)	1.75 in. (44.5 mm)	7.0 cu. in. (114 cc)
J-200	2.00 in. (50.8 mm)	2.00 in. (50.8 mm)	6.3 cu. in. (103 cc)

ENGINE INFORMATION

All models are two-stroke, air-cooled engines. Models with letter "V" suffix have vertical crankshafts. Models with letter "H" suffix have horizontal crankshafts.

MAINTENANCE

SPARK PLUG. Recommended spark plug for normal engine use for all models is a Champion J11J. Electrode gap should be 0.035 inch (0.89 mm).

CARBURETOR. Float feed type carburetors are used. Model J-100 is equipped with either Tillotson MT-52A or MT-77A carburetor, Model J-150 is equipped with a Tillotson MT-4A carburetor and Model J-200 is equipped with either a Tillotson MT-26A or MT-36A carburetor.

Clockwise rotation of the knurled head idle needle and of the T-head main jet needle leans the mixture.

Initial mixture needle adjustments, from a lightly seated position, are ¾ turn open for the idle mixture needle and 1¼ turn open for the main fuel mixture needle.

Make final adjustments with engine at operating temperature and running. Operate engine at ¼ to ½ throttle and adjust the main fuel mixture needle until engine speed drops, then turn main fuel mixture needle ⅛ turn counterclockwise to richen mixture. Lower engine rpm to idle speed and adjust idle mixture needle to obtain smoothest idle while maintaining steady acceleration. As each adjustment affects the other, adjustment procedure may have to be repeated.

Carburetor inlet needle is spring loaded. Float setting is 3/32 inch (2.38 mm) from bowl cover flange to nearest edge of float when bowl cover assembly is in inverted position.

GOVERNOR. Speed control on all models is maintained by a pneumatic (air vane) type governor. The air vane which is linked directly to the carburetor upper throttle lever is actuated by air from the flywheel fan. Make certain linkage does not bind in any position when moved through full range of travel.

To adjust Models J-100 and J-150 governors, stop engine and place throttle control knob in the closed position. Upper throttle shaft lever (2–Fig. JAC4) should be in the ½ open position. If lever is not in the ½ open position, bend the balance spring (1), at upper end, until lever position is correct. The lower throttle shaft lever is connected by linkage to the governor balance spring (1). A small antisurge spring (5) is used in conjunction with this linkage arrangement. Correct installation of the spring is accomplished by hooking one end into the inner hole in the lower

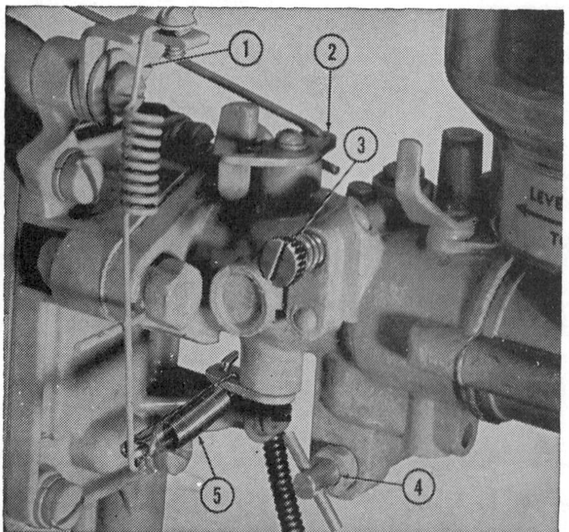

Fig. JAC4—Jacobsen Models J-100 and J-150 carburetor and governor linkage.
1. Balance spring
2. Upper throttle shaft lever
3. Idle needle
4. Main jet needle
5. Antisurge spring

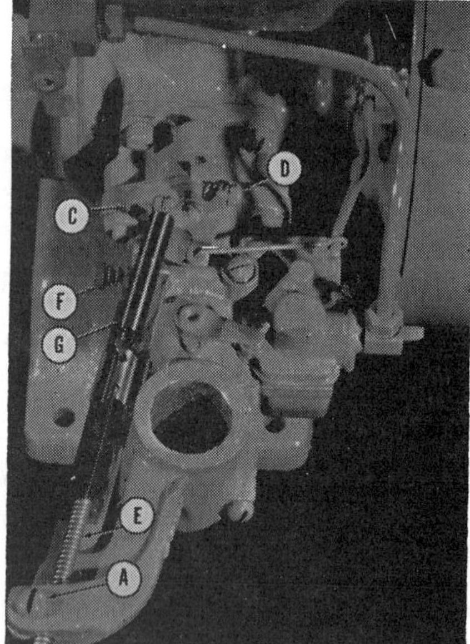

Fig. JAC5—Jacobsen Model J-200 carburetor and governor linkage.
A. Screw
C. Throttle lever
D. Throttle stop screw
E. Cable housing
F. Idle needle
G. Governor spring

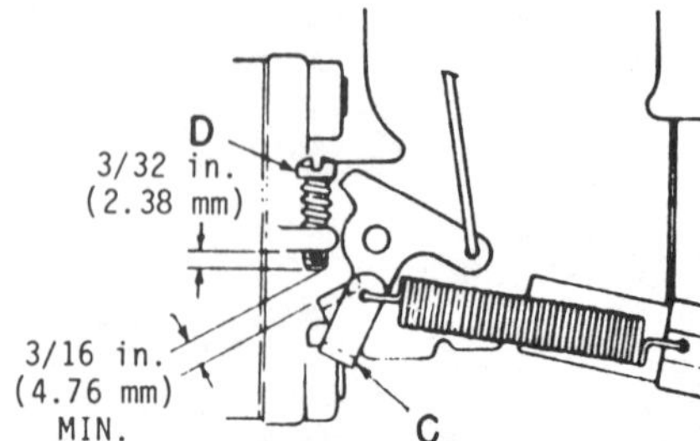

Fig. JAC6—Jacobsen Model J-200 governor linkage details. See Fig. JAC5 for legend.

throttle shaft lever from the top side. Hook the remaining end into loop of governor balance spring so it is underneath the throttle lever link as shown. To adjust Model J-200 governor, stop engine and place the throttle control in the closed position and adjust the throttle stop screw (D–Fig. JAC5) so 3/32 inch (2.38 mm) of the screw protrudes beyond the boss as shown. With all parts so positioned, the governor spring should pull the throttle lever (C) open a minimum of 3/16 inch (4.76 mm) from stop screw (D) as shown in Fig. JAC6. If adjustment is not as specified, loosen screw (A–Fig. JAC5) and move cable housing (E) in or out as required.

MAGNETO AND TIMING. Flywheel type magnetos are used. Early production J-100 engines use Wico FW-1753 magneto while later production models use Wico FW-2124 magneto. Special application engines designated J-100A were equipped with Jack & Heintz H23-370 and H28-118 magneto. Latest Model J-100 engines use Wico FW-2489 magneto. Wico FW-1754 magneto is used on Model J-150 engine and Wico FW-2259 magneto is used on the Model J-200 engine.

On all models, breaker point gap is 0.020 inch (0.51 mm). Recommended timing is 30° BTDC on Model J-100 and J-150 engines and 29° BTDC on Model J-200 engine.

Ignition timing is satisfactory if breaker contacts just begin to open when piston is 0.125 inch (3.18 mm) BTDC. Piston position can be determined by inserting a suitable dial indicator through spark plug hole. Reposition magneto stator plate to obtain desired timing.

LUBRICATION. Mix regular gasoline and SAE 30 nondetergent oil at a 16:1 ratio to provide internal engine lubrication. Use SAE 10 oil in the gear reduction unit.

CYLINDER HEAD AND CARBON. It is recommended the muffler, exhaust manifold and cylinder head be removed and carbon cleaned after each 25 hours of operation. Crank engine until piston is in bottom dead center position and thoroughly clean carbon from the exhaust and intake ports.

CAUTION: Make certain all carbon particles are removed from cylinder bore before reassembly.

REPAIRS

CONNECTING ROD. Connecting rod and piston assembly are removed from above all engines. Crankshaft crankpin diameter is 0.623 inch (15.82 mm) for Model J-100, 0.7475 inch (18.987 mm) for Model J-150 and 0.750 inch (19.05 mm) for Model J-200.

On J-100 and J-150 models, the forged bronze connecting rod rides directly on the crankshaft crankpin. Recommended diametral rod-to-crankpin clearance is 0.0015-0.0025 inch (0.038-0.064 mm) on J-100H and J-150 models and 0.0025-0.0035 inch (0.064-0.089 mm) on J-100V models. Obtain the recommended clearance by filing rod and cap and reaming bearing bore to proper size. On all J-100 and J-150 models, recommended connecting rod side play is 0.005-0.015 inch (0.12-0.38 mm). When reassembling, make certain index marks on rod and rod cap are aligned.

On J-200 models, the jeweled connecting rod is equipped with 28 needle bearings between connecting rod and crankpin. The steel connecting rod and rod cap are a single casting which has been fractured between rod and rod cap mating surface. Make certain connecting rod cap is correctly seated to connecting rod during assembly.

PISTON, PIN AND RINGS. The cam and taper ground piston is equipped with three 1/8 inch (3.18 mm) wide compression rings. Piston ring end gap should be 0.005-0.010 inch (0.13-0.25 mm) and ring side clearance in ring groove should be 0.004-0.006 inch (0.10-0.15 mm). Renew rings if end gap is 0.020 inch (0.51 mm) or more, or if piston side clearance is 0.010 inch (0.25 mm) or more. Piston and rings are available in several oversizes as well as standard size.

Piston-to-cylinder wall clearance, checked 90° from piston pin at bottom of skirt, should be 0.004-0.005 inch (0.10-0.13 mm) for J-100H models; 0.005-0.006 inch (0.13-0.15 mm) for J-100V models; 0.0020-0.0035 inch (0.05-0.09 mm) for J-150 models or 0.003-0.004 inch (0.08-0.10 mm) for J-200 models.

The piston pin should be a 0.0002-0.0004 inch (0.005-0.010 mm) tight fit in piston and a 0.0005-0.0015 inch (0.013-0.038 mm) loose fit in connecting rod. Piston is available in several oversizes as well as standard size.

On all models, the long taper on head of piston should face exhaust side of engine after installation.

CRANKSHAFT, MAIN BEARINGS AND SEALS. The crankshaft is supported by ball bearings which are a press fit on the shaft. To avoid bending the crankshaft during bearing installation, support crankshaft between the throws and heat the bearings in oil.

On direct drive engines, install crankcase seals so lips face toward inside of engine. On engines equipped with speed reduction unit, the crankcase oil seal on the speed reducer side should be installed with lip facing the reducer.

CARBURETOR REED VALVE. The carburetor is mounted on an adapter plate which, on the engine side, carries a small spring steel leaf or reed that acts as an inlet valve. Blow back through carburetor can be caused by foreign matter holding reed open or by a damaged or improperly installed reed. Reed has a bend in it and must be installed so as to force reed firmly against mounting plate.

REDUCTION UNIT. Reduction gear cover and gear must be removed before housing can be removed from engine. Crankcase oil seal on the speed reducer side should be installed with the lip facing the reducer. On the power takeoff side of the reduction unit, the oil seal lip faces the reduction unit.

JACOBSEN

Model	Bore	Stroke	Displacement
J-125	2.00 in. (50.8 mm)	1.50 in. (38.1 mm)	4.7 cu. in. (77 cc)
J-175	2.125 in. (54.0 mm)	1.75 in. (44.5 mm)	6.2 cu. in. (102 cc)
J-225	2.25 in. (57.2 mm)	2.00 in. (50.8 mm)	8.0 cu. in. (131 cc)
J-321	2.125 in. (54.0 mm)	1.75 in. (44.5 mm)	6.2 cu. in. (102 cc)
J-501	2.125 in. (54.0 mm)	1.75 in. (44.5 mm)	6.2 cu. in. (102 cc)

ENGINE INFORMATION

All models are two-stroke, air-cooled engines. Models with letter "V" suffix have vertical crankshafts. Models with letter "H" suffix have horizontal crankshafts.

MAINTENANCE

SPARK PLUG. Refer to the following chart for recommended Champion spark plug.

J-125, rotary mower J12J
J-125, reel mower UJ12
J-175, rotary mower J8J
J-225, rotary mower J8J
J-321 & J-501, rotary mower ... J-17LM
J-321 & J-501, reel mower* UJ12
*For short spark plug, use TJ8.

Set electrode gap to 0.030 inch (0.76 mm) for all models and applications.

CARBURETOR. Tillotson MT58A carburetor is used on Model J-125, Tillotson MT59A carburetor is used on Model J-175, Tillotson MT54A carburetor is used on Model J-225 and either a Walbro LMB or LMG carburetor is used on Models J-321 and J-501. Refer to appropriate paragraph for model being serviced.

Tillotson Carburetors. Initial adjustment of fuel mixture screws from a lightly seated position, is 1 turn open for idle mixture screw (23–Fig. JAC8) and 1¼ turns open for main fuel mixture screw (26).

Make final adjustments with engine at operating temperature and running. Operate engine at ½ throttle and turn main fuel adjustment screw in until engine loses speed. Turn main fuel screw counterclockwise until maximum rpm is obtained. Operate engine just above idle speed. Turn idle mixture screw in until engine loses speed and misses, then back screw out until engine is idling smoothly. Set idle speed as listed in chart (Fig. JAC7) for model and application by adjusting idle speed screw (18).

Carburetor inlet needle is spring loaded. Float setting is 1/16 to 3/32 inch

Engine Application	Engine RPM Idle Speed	Top Speed
9" Edge-R-Trim (32A9, 32B9 & 32C9)	1500-2000	3000-3300
9" Edge-R-Trim (50012)	1500-1800	Up to 3600
10" Trimo (3110-8610, 86A & 86B)	1500-2000	3400-3600
10" Trimo (50035)		3500-3700
18" Pacer (52C18, 42D18 & 42E-18)	1300-1600	3500 max.
18" Pacer (11814)	1500-1800	Up to 3600
18" Turbo-Cut (3418, 34B18, 34C18 & 34D18)	1300-1800	3400-3500
18" Turbo-Cut (7518, 75A18 & 75B18)	1500-2000	3400-3500
18" Turbo-Vac (31817)	1500-1800	3200-3400
18" Turbo-Vac (31819)*	1500-1800	3200-3400
18" Turbo-Vac (31819)	2400-2600	3200-3400
18" 4-Blade Rotary (31809)	2400-2600	3200-3400
18" Turbo Cone (117-18)	1500-1600	3200-3400
20" Scepter (8020 & 80A20)	1500-1800	3000-3200
20" Commercial Rotary (35025)	1500-1600	3200-3400
20" Commercial Rotary (32028)	2400-2600	3200-3400
20" Commercial Rotary (32028)*	1500-1600	3200-3400
20" Robust (32031)	1500-1700	3200-3400
20" Snow Jet (9620 & 96A20)	1500-1800	3600-3800
20" Snow Jet (52002, 52003)	1700-1900	3600-3800
21" 4-Blade Rotary (32114)	2400-2600	3200-3400
21" 4-Blade Rotary S.P. (42114, 42118, 42119)	2400-2600	3200-3400
21" Turbo-Cut (3921, 39B21 & 39C21)	1500-1800	3200-3400
21" Turbo Cone (119-21)	1500-1600	3200-3400
21" Turbo-Cut (3521, 35C21, 35D21, 35E21 & 35F21)	1500-1800	3500 max.
21" Turbo Cone (121-21)	1500-1600	3200-3400
21" Lawn Queen (2C21, 2D21 & 2E21)	1300-1600	3500 max.
21" Lawn Queen (12113)	1500-1800	Up to 3600
21" Manor (28F21 & 28G21)	1300-1600	3500 max.
21" Manor (22114, 32121-7B1)	1500-1800	Up to 3600
22" Putting Green (9A22 & 9B22)	1500-1800	3400 max.
22" Greensmower (62203, 62208)	1500-1800	Up to 3800
22" Scepter (8022 & 80A22)	1500-1800	3000-3200
24" Estate (8A24 & 8B24)	1300-1600	3800 max.
24" Rotary S. P. (40A24)	1900-2100	3000 max.
26" Estate (8A26, 8B26, 8C26 & 8D26)	1300-1600	3800 max.
26" Estate R.R. (22601, 22605-7B1)	1500-1800	Up to 3800
26" Estate F.R. (22611, 22615-7B1)	1500-1800	Up to 3800
26" Lawn King (12A26 & 12B26)	1300-1600	3200 max.
26" Lawn King (12601)	1500-1800	Up to 3800

*Speed control on handle.

*Speed control on handle.

Fig. JAC7—Chart showing engine idle and top speed for various engine applications. Refer to text.

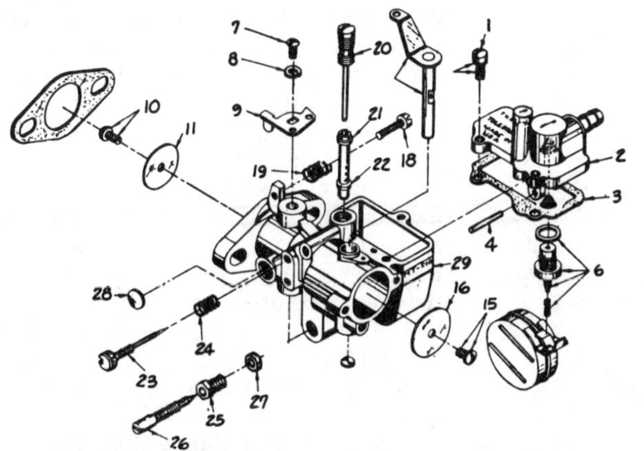

Fig. JAC8—Exploded view of typical Tillotson carburetor.

1. Screw & lockwasher
2. Bowl cover
3. Gasket
4. Lever pin
6. Inlet needle assy.
9. Throttle stop lever
10. Shutter screw
11. Throttle shutter
15. Shutter screw
16. Choke shutter
18. Idle speed screw
19. Spring
20. Idle tube
21. Main nozzle
22. Gasket
23. Idle mixture screw
24. Spring
25. Packing nut
26. Main adjusting screw
27. Packing
28. Welch plug
29. Carburetor body

(1.59-2.38 mm) from bowl cover flange to top of float with bowl cover assembly inverted and float resting lightly on inlet needle. Refer to Fig. JAC9. Float level should be 5/16 inch (7.94 mm) if a plastic float is used. Bend tab on float if necessary to obtain correct setting.

Walbro Carburetor. Refer to Figs. JAC10, JAC11 and JAC12 for exploded view of typical Walbro carburetors used on J-321 and J-501 models.

Walbro carburetor shown in Fig. JAC12 does not have adjustable idle or main fuel orifices. Note some carburetors shown in Fig. JAC10 or JAC11 are not equipped with an idle mixture adjusting screw. Refer to the following paragraphs for carburetor adjustments on carburetors equipped with idle adjusting screw and/or main fuel mixture screw.

For initial adjustment, open the idle and/or main fuel adjustment screw 1 to 1¼ turns from a lightly seated position. Make final adjustments with engine at operating temperature and running at "FAST" throttle position. Slowly turn main fuel mixture screw clockwise until engine begins to lose speed, then turn the main screw counterclockwise ⅛ to ¼ turn. Move throttle to "IDLE" position and slowly turn idle fuel screw on carburetors so equipped, clockwise until engine begins to lose speed, then turn idle fuel screw counterclockwise ⅛ to ¼ turn. Engine will have a slight "stutter" or intermittent exhaust sound at both idle and high speed when fuel mixture adjustments are correct.

Idle speed is controlled by proper adjustment of the governor rather than by adjustment of the idle speed stop screw on carburetor throttle. This is to prevent engine stalling when traction and/or reel clutch is engaged with engine at idle speed. Refer to GOVERNOR section.

To check float setting on Walbro carburetor, invert the body casting and float level should be 5/32 inch (3.97mm)

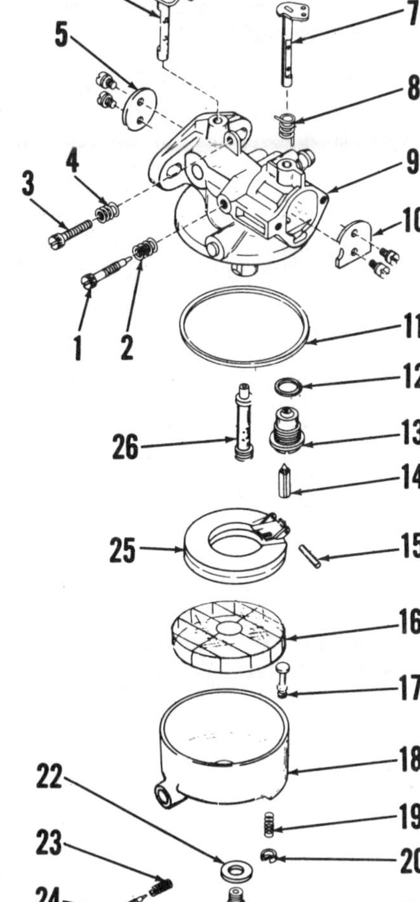

Fig. JAC10—Exploded view of Walbro series LMB float carburetor used on some J-321 and J-501 models. Note the fuel bowl screen (16) used on this typed carburetor.

1. Idle fuel needle
2. Spring
3. Idle speed screw
4. Spring
5. Throttle plate
6. Throttle shaft
7. Choke shaft
8. Choke return spring
9. Carburetor body
10. Choke plate
11. Gasket
12. Gasket
13. Inlet valve seat
14. Inlet valve
15. Float pin
16. Fuel bowl screen
17. Drain valve
18. Fuel bowl
19. Spring
20. Retainer
21. Bowl retainer
22. Gasket
23. Spring
24. Main fuel nozzle
25. Float
26. Main nozzle

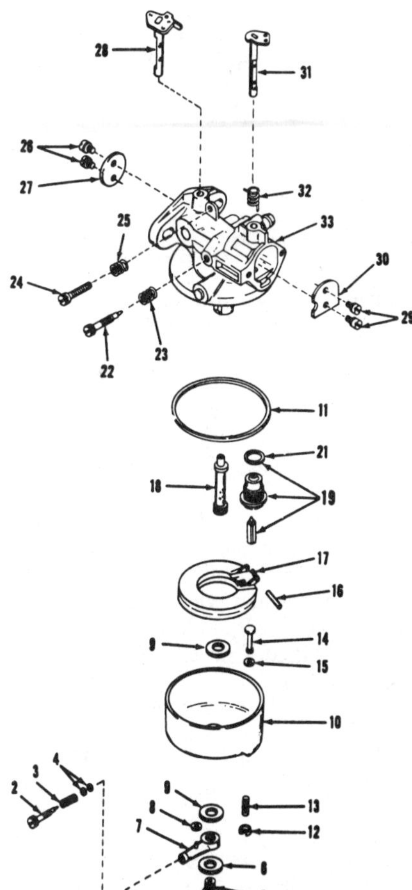

Fig. JAC11—Exploded view of typical Walbro series LMG carburetor used on some Model J-321 and J-501 engines.

2. Main fuel nozzle
3. Spring
4. Seals
5. Bowl retainer
6. Washer
7. Adapter
8. Seal
9. Washer
10. Float bowl
11. Gasket
12. Retainer
13. Spring
14. Drain valve
15. Gasket
16. Float pin
17. Float
18. Main nozzle
19. Inlet valve
21. Gasket
22. Idle fuel needle
23. Spring
24. Idle stop screw
25. Spring
27. Throttle plate
28. Throttle shaft
30. Choke plate
31. Choke shaft
32. Choke spring
33. Carburetor body

Fig. JAC9—Distance (D) between bowl cover flange and nearest edge of float with bowl cover assembly inverted should be 1/16 to 3/32 inch (1.59-2.38 mm).

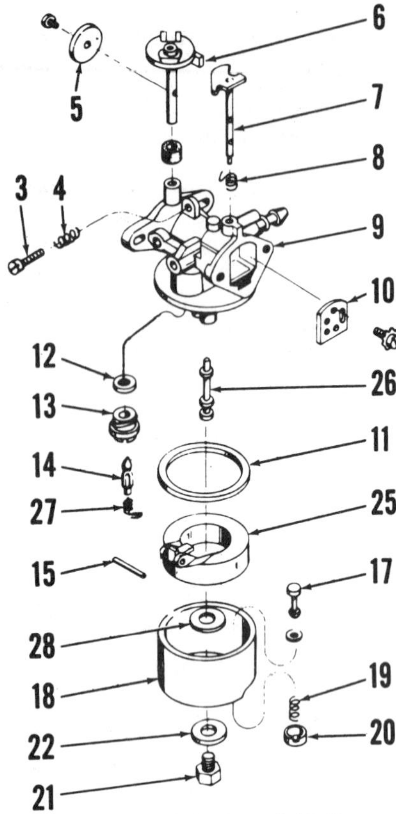

Fig. JAC12—Exploded view of Walbro carburetor with fixed idle and main fuel jets. Refer to Fig. JAC10 for parts identification except for retainer (27) and gasket (28).

measured between carburetor body casting and free side of float. On carburetor with fixed main fuel orifice, float level should be 1/16-3/32 inch (1.59-2.38 mm). Adjust clearance by bending tab on float which contacts fuel inlet needle on all models.

Check float free travel. Float should have 3/16 inch (4.76 mm) of free movement. Adjust free travel by bending tab on float which contacts float stop.

Fixed main fuel orifice on carburetors so equipped, may be cleaned with a #63 drill bit. Be careful not to damage orifice during cleaning.

Fig. JAC13—View of governor adjustment screw on Model J-321 or J-501 engine.

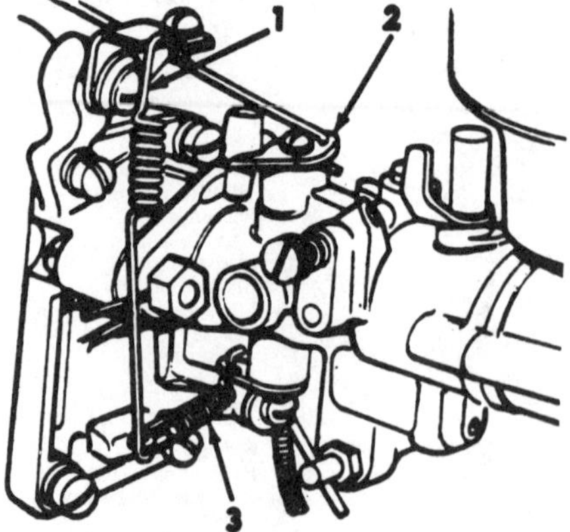

Fig. JAC14—Drawing showing correct installation of Tillotson carburetor controls. Refer to text.

GOVERNOR. Speed control on all models is maintained by a pneumatic (air vane) type governor. The air vane which is linked directly to the carburetor upper throttle lever is actuated by air from the flywheel fan. Make certain the linkage does not bind in any position when moved through the full range of travel.

To adjust governor linkage, refer to chart (Fig. JAC7) for recommended idle and top speed for various engine applications and refer to appropriate paragraphs for model being serviced.

Models J-321 And J-501 With Fixed Speed. Governor adjustment on models with fixed governed speed is accomplished by turning governor adjustment screw shown in Fig. JAC13. Turning adjustment screw provides adjustment range of approximately 400 rpm. Refer to recommended governor speeds.

Governor Adjustments With Tillotson Carburetor. With engine stopped, close the throttle control lever. The upper throttle shaft lever (2–Fig. JAC14) should be in the ½ open position. If not in this position, bend the balance spring (1) at upper end until the desired position is obtained. The lower throttle shaft lever is connected to the governor balance spring (1) by a small antisurge spring (3). Correct spring (3) installation is accomplished by hooking one end into the inner hole of the lower throttle shaft lever from the top side. Hook the other end into the governor balance spring loop so it is underneath the throttle lever link.

Governor Adjustment With Remote Control Walbro Carburetor. Set the throttle control to "FAST" position. The hook in the choke lever (link) should just touch the first loop in choke spring, or clear the first loop by 1/16 inch (1.59 mm) as shown in Fig. JAC15. If choke is partially closed, or clearance between hook in link and loop in spring exceeds 1/16 inch (1.59 mm), loosen the engine control cable clamp. Slide cable back or forward until choke control lever (link) just touches the loop in choke spring, then tighten the control cable clamp. Test setting by moving control to "STOP" position. The carburetor control lever should touch the contact point of "STOP" switch. Move the control to "CHOKE (START)" position. Carburetor choke should be completely closed.

MAGNETO AND TIMING. Model J-321 may be equipped with a breakerless ignition system. All other models are equipped with a flywheel magneto ignition system with breaker points. Refer to appropriate section for model being serviced.

Breakerless Ignition System. A Blaser ignition system is used on some engines and is identified by use of a

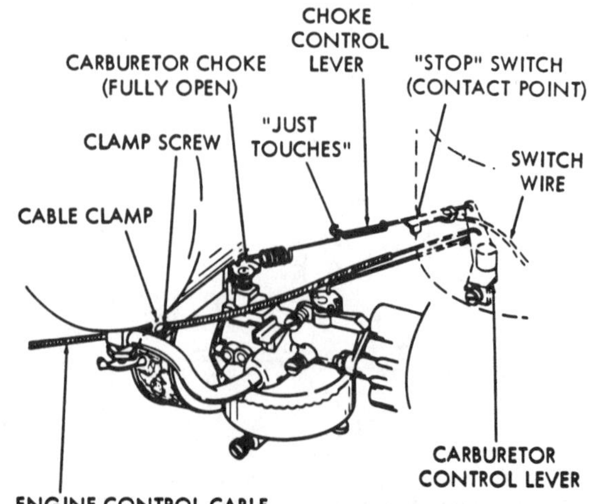

Fig. JAC15—Drawing showing correct installation of Walbro carburetor controls. Refer to text.

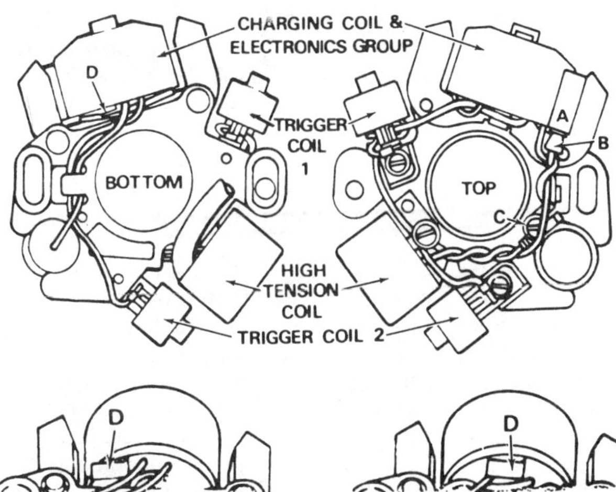

Fig. JAC16—Top and bottom views of CD ignition system used on some J-321 engines.

Fig. JAC17—Location of terminal (D) will identify early and late production CD ignition systems. Refer also to Fig. JAC16 for location of terminal (D).

single trigger coil instead of two trigger coils used in CD ignition system shown in Fig. JAC16. Individual components and complete assembly are not available for Blaser unit. Blaser ignition must be replaced with either a standard breaker type ignition system or with other CD ignition system if renewal is necessary. Flywheel must be renewed when installing different ignition system.

Some J-321 engines are equipped with the CD ignition system shown in Fig. JAC16. Components are available for this ignition. If service is required, the following troubleshooting procedure may be used. Be sure ignition is defective by using Wico test plug #14281 and rotating engine with starter. If test plug does not fire, ignition should be tested.

Remove ignition system from engine. To test high tension coil, connect positive lead of an ohmmeter to coil terminal (B – JAC16) and connect negative lead to ground screw (C). Ohmmeter should read 0.1-0.5 ohms. Disconnect ohmmeter and connect positive lead to high tension lead of coil and connect negative lead to ground screw (C). Ohmmeter should read 900-1100 ohms. Renew high tension coil if coil fails either of the two tests. Disconnect leads to each trigger coil. Connect ohmmeter leads to the leads of a trigger coil and read resistance of each coil. Ohmmeter should indicate 15-25 ohms resistance. Renew trigger coil if incorrect reading was obtained. Check connections and leads of trigger coils for continuity.

Before checking charging coil and electronic unit, note difference in early and late production units as shown in Fig. JAC17. Remove plug-in terminal (B – Fig. JAC16) from coil tower and disconnect "ENGINE STOP" wire from terminal (D). Connect positive lead of ohmmeter to terminal (A) and negative lead to terminal (D). Ohmmeter should read 5-25 ohms on early units and 1 megohm to infinity on late units. Reverse leads of ohmmeter. Ohmmeter should read infinity on early units and 1 megohm to infinity on late units. Disconnect ohmmeter leads and connect positive meter lead to terminal (D) and negative meter lead to ground screw (C). Ohmmeter should read infinity on early production units and 560-760 ohms on late units. Reverse meter lead connections. Ohmmeter should read 1500-2000 ohms on early units. On late units, ohmmeter reading should be slightly less than reading (560-760) obtained previously. If any of these tests are failed, charging coil windings are defective and charging coil and electronic unit must be renewed. Electronic circuit of unit cannot be tested except by substitution with a new charging coil and electronic unit.

Ignition timing on models with CD or Blaser ignition system is fixed and cannot be adjusted.

Breaker Point Ignition. Some J-321 models and all other models are equipped with a flywheel magneto ignition system using ignition breaker points. Magneto components are accessible after removing flywheel.

Breaker point gap for all models is 0.020 inch (0.51 mm). Recommended ignition timing is 30° BTDC on J-125 models, 28° BTDC on J-175 models and 27° BTDC on J-225 models. On Model

Fig. JAC18—View showing use of special Wico Test Plug number S14281 to check magneto output. Refer to text.

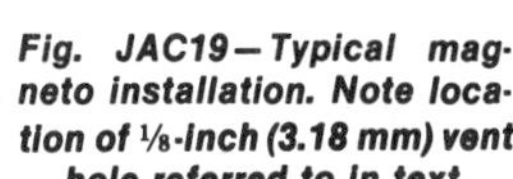
Fig. JAC19—Typical magneto installation. Note location of 1/8-inch (3.18 mm) vent hole referred to in text.

J-321 engine used on Snow Jet, rotate stator to full counterclockwise (advanced) position. On J-501 models and all other J-321 models with breaker point ignitions, rotate stator plate to full clockwise (retarded) position.

On J-125, J-175 and J-225 models, ignition timing is satisfactory if breaker points just begin to open when piston is 0.125 inch (3.18 mm) BTDC. Piston position can be determined by inserting a suitable dial indicator through spark plug hole. Reposition magneto stator plate if necessary to obtain desired timing.

Magneto output at starting speeds can be checked without removing magneto from engine using a special Wico Test Plug S14281. Refer to Fig. JAC18. Note engine spark plug remains in place when using test plug.

On early models, the manufacturer recommends drilling an 1/8-inch (3.18 mm) vent hole in the lower left-hand corner of magneto breaker box to aid in ventilating the contact points (refer to Fig. JAC19).

AIR CLEANER. If the engine is operated under dry or dusty conditions, the manufacturer recommends servicing the air cleaner after every 25 hours of operation. Refer to the appropriate paragraphs for type being serviced.

Foil Type Cleaner. Wash and rinse thoroughly in a suitable nonflammable solvent. Shake off solvent and immerse in clean SAE 30 oil. Allow excess oil to drain off before reinstallation.

Paper Filter. Brush or wipe outside of cleaner. Tap gently to loosen dirt from inside of filter. Do not oil or wash filter. Filter can also be cleaned by gently blowing compressed air from the inside. If, after extended use, filter is too dirty to clean properly, renew the filter.

Oil Bath Air Cleaner. Remove cover from oil reservoir and discard old oil. Wash cleaner in a suitable nonflammable solvent. Dry cleaner thoroughly. Refill to level indicated by arrow with clean SAE 30 motor oil.

Foam Type Filter. Remove filter and wash and rinse thoroughly in a nonflammable solvent. Soak in clean SAE 30 engine oil and carefully compress element to remove excess oil.

LUBRICATION. A good quality two-stroke, air-cooled engine oil should be mixed with regular grade gasoline at a ratio of 50:1 for Model J-501; 30:1 for Model J-321 and 16:1 for all other models.

Use SAE 10 oil in the reduction gear box on models so equipped. Gearbox is fitted with an oil level plug.

CLEANING CARBON. Power loss can often be corrected by cleaning carbon from exhaust ports and muffler. To clean, remove spark plug and muffler. Turn engine so piston is below bottom of exhaust ports and remove carbon from ports with a dull knife or similar tool. Clean out the muffler openings, then replace muffler and spark plug.

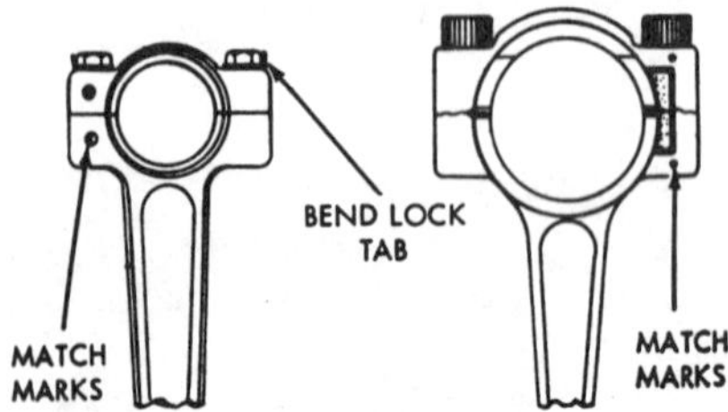

Fig. JAC20—Be sure match marks are aligned when reassembling rod cap to connecting rod. Bend lock tab against screw heads on models so equipped.

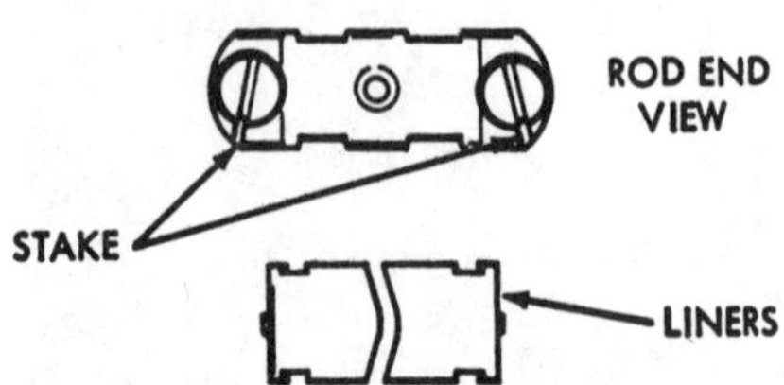

Fig. JAC21—On Model J-321 and J-501 engines with aluminum connecting rod, be sure bearing liners fit together as shown. Stake cap retaining screws as shown.

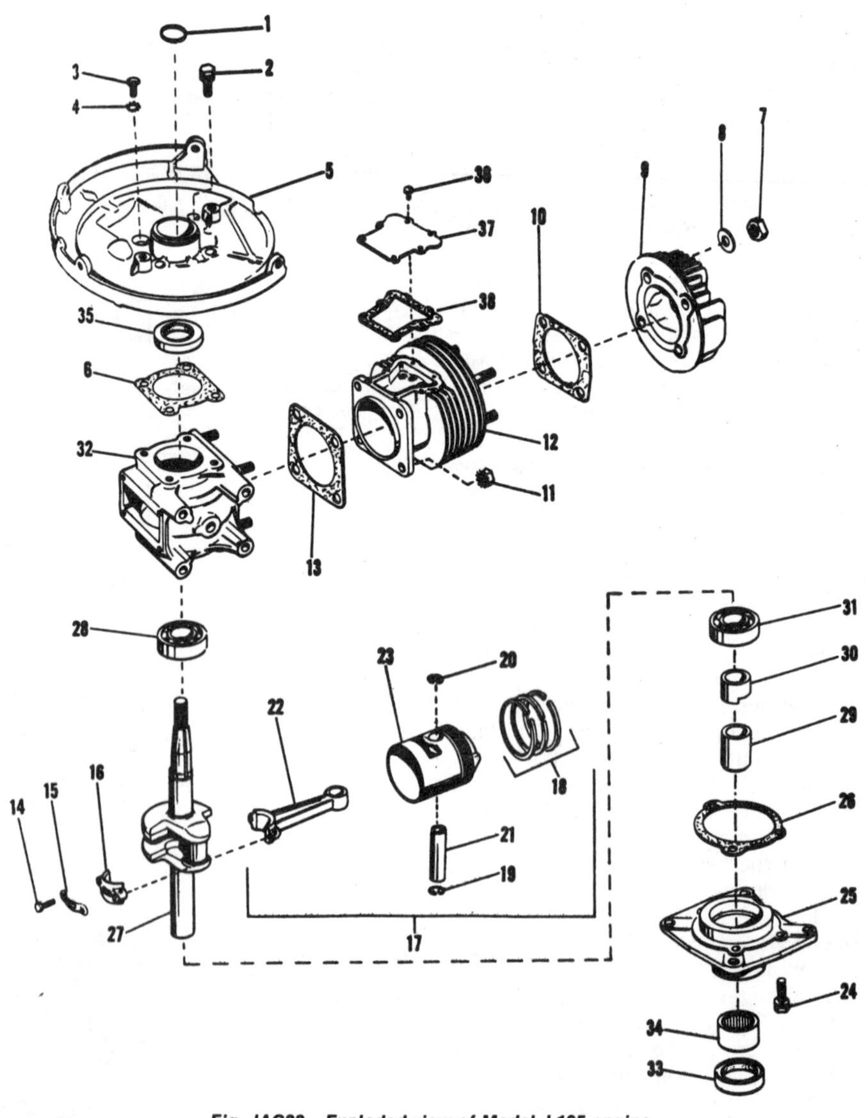

Fig. JAC22—Exploded view of Model J-125 engine.

1. Snap ring
5. Back plate
6. Gasket
9. Cylinder head
10. Gasket
12. Cylinder
13. Gasket
15. Lock tab
16. Rod cap
18. Piston rings
19. Snap ring
20. Snap ring
21. Piston pin
22. Connecting rod
23. Piston
25. Crankcase head
26. Gasket
27. Crankshaft
28. Ball bearing
29. Bearing race
30. Spacer
31. Ball bearing
32. Crankcase
33. Crankshaft seal
34. Needle bearing
35. Crankshaft seal
37. Transfer cover
38. Gasket

REPAIRS

TIGHTENING TORQUES. Recommended tightening torques are as follows:

Back plate screws . . . . 60-80 in.-lbs. (7-9 N•m)
Carburetor adapter screws:
  J-321 & J-501 with needle bearing connecting rod . . 60-80 in.-lbs. (7-9 N•m)
  All other models . . . . 30-50 in.-lbs. (3-6 N•m)
Connecting rod screws:
  J-321 & J-501 with needle bearing connecting rod . . . . 45-50 in.-lbs. (5-6 N•m)
  All other models . . . . 65-75 in.-lbs. (7-9 N•m)
Crankcase bearing plate bolts . . . . 80-100 in.-lbs. (9-11 N•m)
Cylinder to crankcase nuts (J-125) . . . . 150-180 in.-lbs. (17-20 N•m)
Cylinder head bolts or nuts . . . . 150-180 in.-lbs. (17-20 N•m)
Engine mounting bolts . . . . 120-150 in.-lbs. (13-17 N•m)
Muffler mounting bolts:
  Rotary mower engine . . . 60-80 in.-lbs. (7-9 N•m)
  Reel mower engine . . . . 80-100 in.-lbs. (9-11 N•m)
Fan housing screws . . . . . . 60-80 in.-lbs. (7-9 N•m)
Flywheel nut . . . . 300-360 in.-lbs. (34-41 N•m)
Gearbox cover:
  1/4 inch screws . . . . 60-80 in.-lbs. (7-9 N•m)
  5/16 inch screws . . . . 80-100 in.-lbs. (9-11 N•m)
Spark plug . . . . 180-200 in.-lbs. (20-23 N•m)
Stator plate screws . . . . 60-80 in.-lbs. (7-9 N•m)

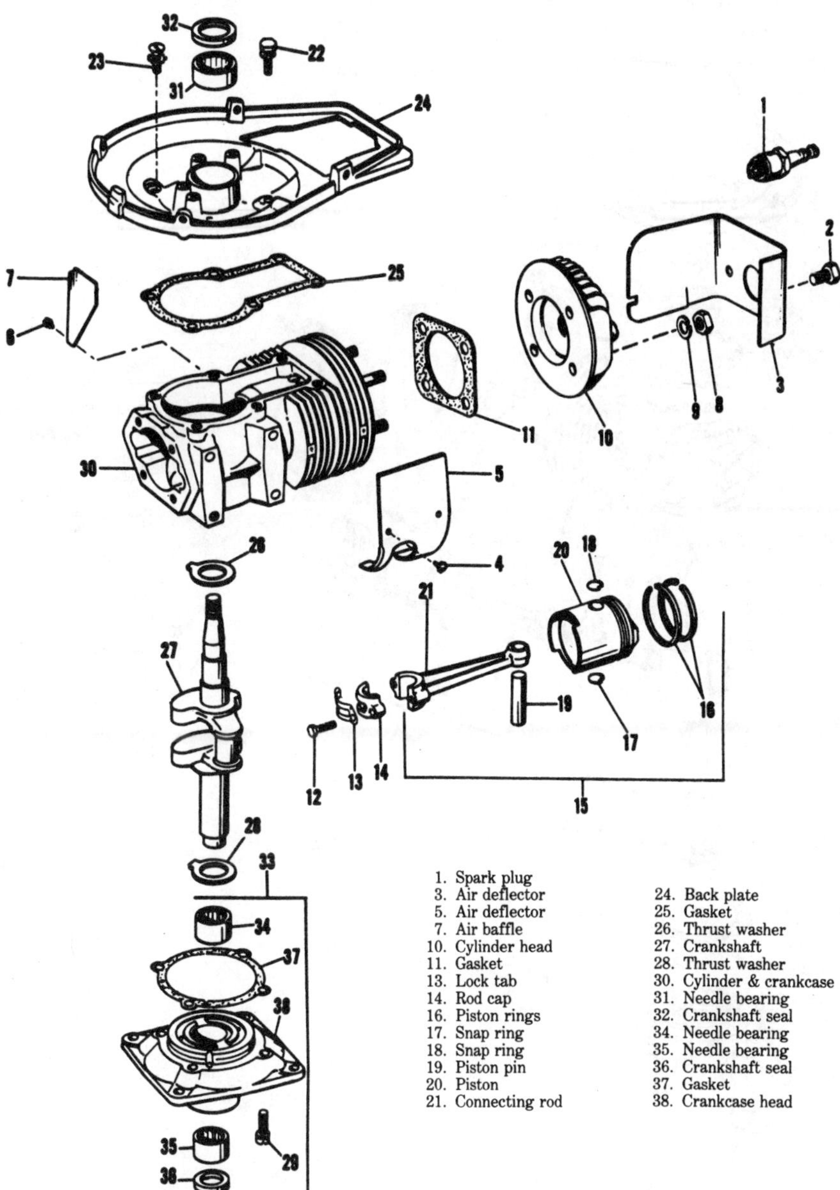

Fig. JAC23—Exploded view of Model J-175 engine.

PISTON, PIN AND RINGS. To remove piston, remove gas tank, air cleaner and interfering shrouds preventing access to cylinder head or carburetor. Disconnect governor link to carburetor and remove carburetor and reed valve plate. Remove cylinder head. Unscrew connecting rod cap screws and remove connecting rod and piston. Be careful not to lose loose bearing rollers on models so equipped. Refer to CONNECTING ROD section to service connecting rod.

The cam ground aluminum piston is fitted with either two or three compression rings depending upon engine model. Renew piston if badly worn or scored, or if side clearance of new ring in top ring groove is 0.010 inch (0.25 mm) or more. Install top ring with chamfer down and second ring with chamfer up.

Piston skirt diameters (measured at right angle to piston pin) are listed in the following table.

J-125 . . . . 1.9965-1.9970 in. (50.711-50.724 mm)
J-175 . . . . 2.1235-2.1240 in. (53.937-53.950 mm)
J-225 . . . . 2.2475-2.2480 in. (57.087-57.099 mm)
J-321 (early) . . . . 2.1225-2.2480 in. (53.912-53.929 mm)
J-321 (late) . . . . 2.1220-2.1227 in. (53.899-53.917 mm)
J-501 . . . . 2.1220-2.1227 in. (53.899-53.917 mm)

Piston skirt-to-cylinder wall clearances (new, measured at right angle to piston pin) are shown in the following table.

J-125 . . . . 0.004-0.005 in. (0.10-0.30 mm)
J-175 . . . . 0.002-0.003 in. (0.05-0.08 mm)
J-225 . . . . 0.003-0.004 in. (0.08-0.10 mm)
J-321 (early) . . . . 0.0028-0.0040 in. (0.071-0.101 mm)
J-321 (late) . . . . 0.0033-0.0045 in. (0.084-0.114 mm)
J-501 . . . . 0.0033-0.0045 in. (0.084-0.114 mm)

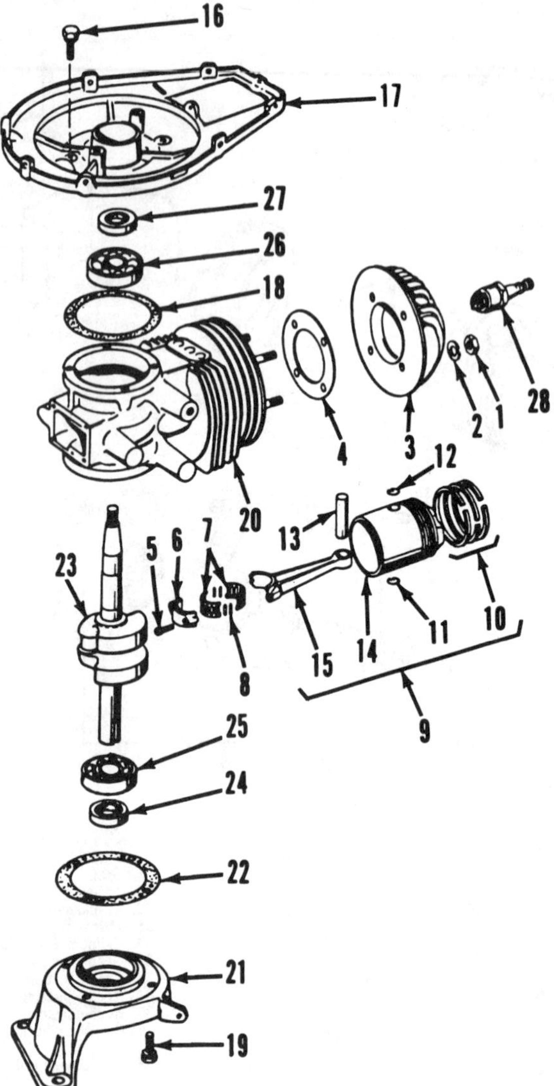

Fig. JAC24—Exploded view of Model J-225 engine.

3. Cylinder head
4. Gasket
6. Connecting rod cap
7. Bearing cage
8. Needle rollers
10. Piston rings
11. Snap ring
12. Snap ring
13. Piston pin
14. Piston
15. Connecting rod
17. Back plate
18. Gasket
21. Crankcase head
22. Gasket
23. Crankshaft
24. Crankshaft seal
25. Ball bearing
26. Ball bearing
27. Crankshaft seal

Piston ring end gap specifications are shown in the following table.

J-125	0.005-0.010 in. (0.13-0.25 mm)
J-175 (un-pinned ring)	0.005-0.013 in. (0.13-0.33 mm)
J-175 (pinned ring)	0.052-0.060 in. (1.32-1.52 mm)
J-225, J-321, J-501	0.005-0.013 in. (0.13-0.33 mm)

Piston pin diameter for Model J-321 and J-501 engines is 0.4999-0.5001 inch (12.697-12.726 mm). Piston pin diameter for all other models is 0.49975-0.50025 in. (12.693-12.705 mm).

On Model J-225 engines (needle bearing in connecting rod) renew piston pin if scoring or excessive wear evident.

Piston pin-to-connecting rod clearances for all models without needle bearings are listed in the following table.

J-125	0.0005-0.0015 in. (0.013-0.038 mm)
J-175	0.00055-0.00175 in. (0.0140-0.0445 mm)
J-321, J-501	0.0007-0.0016 in. (0.018-0.041 mm)

On Models J-321 and J-501, piston pin should be 0.0003 inch (0.008 mm) tight to 0.0002 inch (0.005 mm) loose fit in pin bore of piston. On all other models, piston pin should be 0.00025 inch (0.006 mm) tight to 0.00055 inch (0.013 mm) loose fit in pin bore of piston.

Piston pins are available in several oversizes as well as standard size. If oversize pin is used, refer to the previous paragraphs for the appropriate pin to piston and rod bore specifications and ream the bores accordingly.

CONNECTING ROD. Connecting rod and piston unit can be removed from above after removing cylinder head and carburetor adapter.

Several types of connecting rods have been used according to engine model and application. Refer to appropriate paragraphs for model being serviced.

MODEL J-225. Model J-225 engine is equipped with a connecting rod which has needle roller crankpin and piston pin bearings. The crankpin needle rollers are separated by a split type cage and use crankpin and connecting rod surfaces as bearing races. The caged piston pin needle bearing is not renewable except by renewing connecting rod assembly.

Renew connecting rod is crankpin bearing surface is worn or scored and if any of the piston pin needle rollers have a flat spot or if two needles can be separated the width of one needle. Connecting rod crankpin bearing bore diameter is 0.9399-0.9403 inch (23.874-23.884 mm). Connecting rod side play on crankpin is 0.008-0.018 inch (0.20-0.46 mm).

MODELS J-321 AND J-501. Models J-321 and J-501 engines may be equipped with either a plain bearing bronze connecting rod (see All Other Models paragraphs) or an aluminum connecting rod with needle roller crankpin bearing and plain piston pin bearing. Model J-501 is equipped with an aluminum connecting rod.

The aluminum connecting rod has renewable steel bearing inserts for needle roller outer race. Twenty eight loose needle bearing rollers are used. Bearing guides (10–Fig. JAC25) are used to center bearing rollers on crankpin. Bearing rollers and guides may be held on crankpin with grease before installation of connecting rod.

Piston pin-to-connecting rod bearing bore clearance should be 0.0007-0.0016 inch (0.018-0.041 mm) on all models. On bronze connecting rod, crankpin-to-connecting rod clearance is 0.0035-0.0045 inch (0.09-0.11 mm) and connecting rod side play on crankpin should be 0.004-0.017 inch (0.10-0.43 mm).

On aluminum connecting rods, rod should have a crankpin bore diameter of 0.9819-0.9824 inch (24.940-24.953 mm). Connecting rod side play on crankpin should be 0.005-0.008 inch (0.13-0.46 mm).

ALL OTHER MODELS. A bronze connecting rod with plain crankpin and piston pin bearings is used in all models except J-225, J-501 and some J-321 engines.

Piston pin-to-connecting rod clearance should be 0.0005-0.0015 inch (0.013-0.038 mm) on Models J-125H and

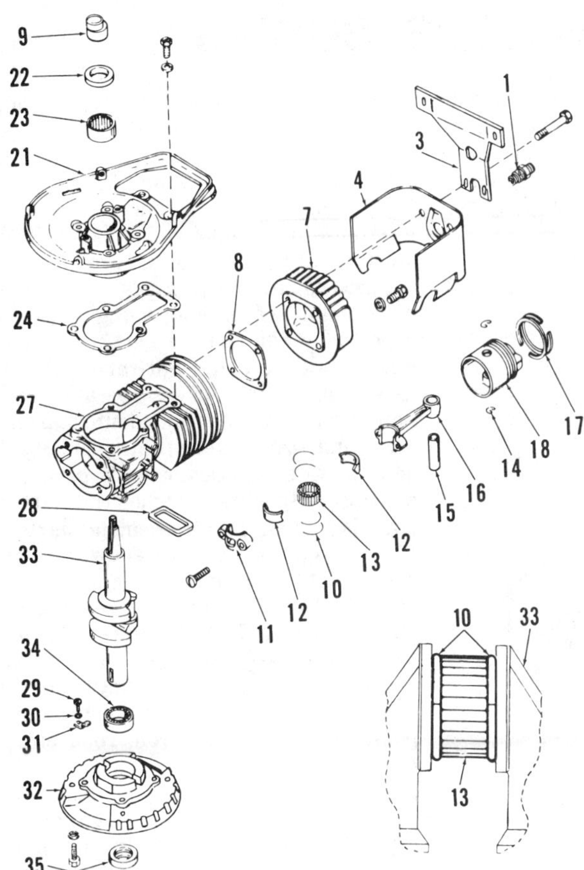

Fig. JAC25—Exploded view of vertical crankshaft Model J-321 and J-501 engine. Horizontal crankshaft models are similar. Bearing (34) on J-501 is retained by a snap ring and not screw retainer (29, 30 and 31). Note location of bearing rollers (13) and guides (10) in inset. Design of crankcase head (32) will vary with engine application.

1. Spark plug
3. Bracket
4. Air deflector
7. Cylinder head
8. Gasket
9. Breaker cam
10. Bearing roller guides
11. Rod cap
13. Bearing rollers (28)
14. Snap rings
15. Piston pin
16. Connecting rod
17. Piston rings
18. Piston
21. Back plate
22. Crankshaft seal
23. Needle bearing
24. Gasket
27. Cylinder & crankcase
28. Gasket
31. Bearing retainer
32. Crankcase head
33. Crankshaft
34. Ball bearing
35. Crankshaft seal

J-125V and 0.00055-0.00175 inch (0.0140-0.0445 mm) on Model J-175.

Crankpin-to-connecting rod bearing clearance should be 0.0015-0.0030 inch (0.038-0.076 mm) for Model J-125H, 0.0025-0.0040 inch (0.064-0.102 mm) for Model J-125V and 0.0035-0.0045 inch (0.090-0.114 mm) for Model J-175.

Connecting rod side play on crankpin for Models J-125H, J-125V and J-175 should be 0.004-0.017 inch (0.10-0.43 mm).

On all models, make certain match marks on connecting rod and cap are aligned as shown in Fig. JAC20 during installation. Bend lock tabs, if equipped, against cap screw heads. On Model J-321 and J-501 with aluminum connecting rod, steel bearing inserts must be installed correctly and screw heads staked as shown in Fig. JAC21.

CYLINDER. Cylinder bore should be resized if scored, out-of-round more than 0.0015 inch (0.038 mm) or worn more than 0.002 inch (0.05 mm). Standard cylinder bore diameter is 2.0010-2.0015 inch (50.825-50.838 mm) for Model J-125 engines, 2.1260-2.1265 inch (54.000-54.013 mm) for Model J-175 engines, 2.2510-2.2515 inch (57.175-57.188 mm) for Model J-225 engines and 2.1260-2.1265 inch (54.000-54.013 mm) for Models J-321 and J-501 engines.

Several oversize pistons as well as standard size are available. If reboring is not necessary, the manufacturer recommends deglazing the bore to aid in seating new rings.

CRANKSHAFT AND SEALS. On all models except Models J-175, J-321 and J-501, crankshaft is supported at each end by ball type main bearings which are pressed onto crankshaft. On Models J-175, crankshaft is supported by needle type main bearings and on Models J-321 and J-501, crankshaft is supported by a needle type main bearing at one end and a ball type main bearing at the opposite end. Model J-125 and J-175 engines have a third bearing (needle bearing) supporting outer end of crankshaft.

On all models, use an approved bearing puller to remove bearings. When installing new bearings, be sure crankshaft is supported between the throws and the bearing is heated in oil to prevent bending of the crankshaft.

On Model J-321 engines, it is necessary to remove the two bearing retainers (31–Fig. JAC25) before the crankcase head (32) can be removed from the crankshaft. Remove snap ring which retains bearing on Model J-501.

Install crankcase head (32) on crankcase (27) of Model J-321 or J-501 so flat spot on inner portion of crankcase head will be towards cylinder.

On Model J-175 engines, maintain crankshaft end play of 0.004-0.018 inch (0.10-0.46 mm) by renewing thrust washers (26 and 28–Fig. JAC23); or gaskets (25 and 37) are available with less thickness than standard gaskets to reduce crankshaft end play.

Standard crankpin diameters are shown in the following table.

Model J-125H . . . . 0.6240-0.6245 in.
(15.850-15.862 mm)
Model J-125V . . . . 0.6230-0.6235 in.
(15.824-15-837 mm)
Model J-175 . . . . 0.7485-0.7490 in.
(19.012-19.025 mm)
Model J-225 . . . . 0.7500-0.7503 in.
(19.050-19.058 mm)
Model J-321 . . . . 0.7485-0.7490 in.
(19.012-19.025 mm)
Model J-501 . . . . 0.7496-0.7501 in.
(19.040-19.053 mm)

On all models with direct drive, install crankcase seals so lip faces toward inside of engine. On models with speed reducer, the crankcase oil seal on the speed reducer side should be installed with lip facing the reducer.

REED VALVE. On all models the carburetor is mounted on an adapter plate. The plate carries a small spring steel leaf or reed, on the engine side, which acts as an inlet valve. The reed has a bend in it and must be installed so reed is forced firmly against the mounting plate. Blow-back through the carburetor may be caused by foreign matter holding reed valve open or by a damaged or improperly installed reed.

SERVICING JACOBSEN ACCESSORIES

REWIND STARTER

All models may be equipped with one of the rewind starters shown in Fig. JAC26, Fig. JAC27 or Fig. JAC28. Refer to the appropriate following paragraph and exploded view when servicing rewind starter.

To disassemble dog type starter shown in Fig. JAC26, remove starter from engine and pull starter rope out of starter approximately 12 inches (305 mm). Prevent pulley (3) from rewinding and pull slack rope back through rope outlet. Hold rope away from pulley and allow rope pulley to unwind. Unscrew retaining screw (11) and remove dog assembly and rope pulley being careful not to disturb rewind spring (2) in cover. Note direction of spring winding and carefully remove spring from cover.

To reassemble starter, reverse disassembly procedure. Rewind spring must be preloaded by installing rope through rope outlet of housing and hole in rope pulley, tie a knot at each end of rope and turn rope pulley approximately four turns in direction that places load on rewind spring. Hold pulley and pull remainder of rope through rope pulley. Attach handle to rope end and allow rope to rewind into starter. If spring is properly preloaded, rope will fully rewind.

Fig. JAC26—Exploded view of dog type rewind starter.

1. Cover
2. Rewind spring
3. Rope pulley
4. Dog
5. Washer
6. Brake
7. Spring
8. Retainer
9. Washer
10. Washer
11. Screw

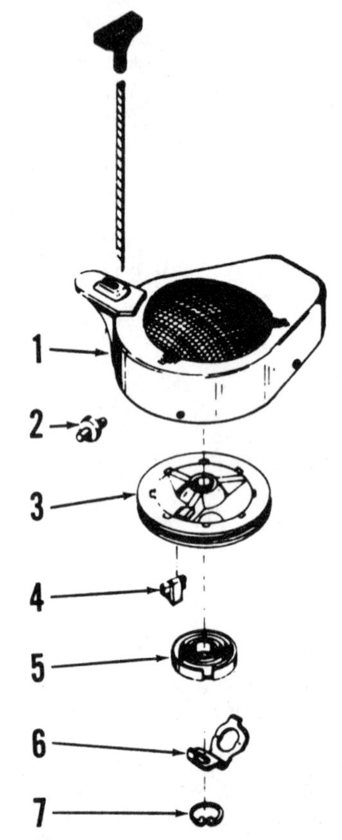

Fig. JAC28—Exploded view of rewind starter used on some later models.

1. Starter housing
2. Rope guide
3. Rope pulley
4. Pawl
5. Rewind spring & Cup
6. Pawl plate
7. Snap ring

To disassemble starter shown in Fig. JAC27, remove rope handle and allow rope to wind into starter. Remove roll pin (1) and washers (2 and 3). Remove rope pulley and remainder of starter components. When reassembling starter, turn rope pulley approximately 4 turns counterclockwise (viewed from open side) before connecting rope to rope pulley. Clip at rope end should have open side facing outward. After assembling starter, check operation and note if rope is fully rewound into starter.

Rewind starter shown in Fig. JAC28 is a pawl type starter with pawl (4) engaging a flywheel fin when the starter rope is pulled. To disassemble starter, remove rope handle and allow rope to rewind into starter. Remove snap ring (7) and remove starter components from starter housing. Inspect components for wear and damage and install in reverse order of disassembly. Spring cup (5) should be installed so spring attachment point on cup is 180° from rope entry hole in starter housing (1). Turn rope pulley approximately 3½ turns against spring tension before passing end of rope through hole in starter housing. Note that four embossed projections on starter housing adjacent to rope pulley are friction devices and should not be lubricated.

GEAR REDUCTION UNIT

Before the housing can be removed from the engine, the reduction gear cover and gear must be removed. Crankcase oil seal on the speed reducer side should be installed with the lip facing the reducer. On the power takeoff side of the reducer unit, the oil seal lip faces the reduction unit. Refer to Fig. JAC29.

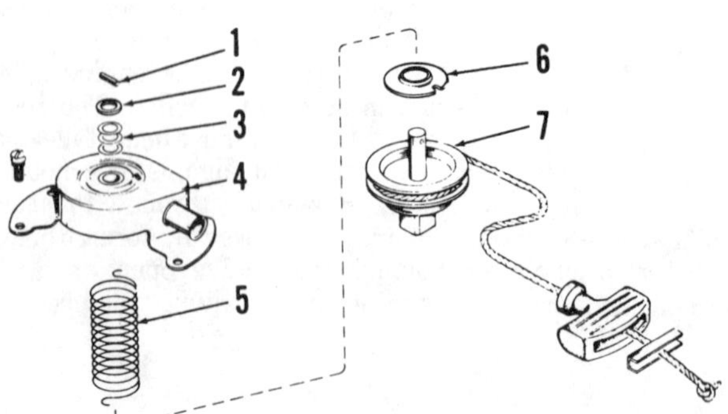

Fig. JAC27—Exploded view of rewind starter used on some engines. Some models use a snap ring in place of pin (1).

1. Pin
2. Washer
3. Washer
4. Cover
5. Spring
6. Washer
7. Rope pulley

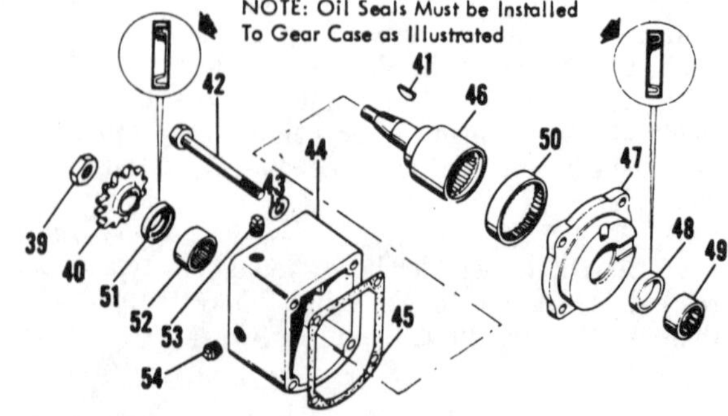

Fig. JAC29—Exploded view of reduction gear assembly used on horizontal crankshaft engines.

40. Drive sprocket
41. Woodruff key
44. Housing
45. Gasket
46. Drive gear
47. Crankcase head
48. Crankshaft seal
49. Needle bearing
50. Needle bearing
51. Oil seal
52. Needle bearing
53. Oil filler plug
54. Oil level plug

JACOBSEN CENTRAL PARTS DISTRIBUTORS

(Arranged Alphabetically by States)

These franchised firms carry extensive stocks of repair parts. Contact them for name of dealer in their area who will have replacement parts.

Tieco, Inc.
Phone: (205) 328-6464
913 North 21st Street
Birmingham, Alabama 35203

Tieco, Inc.
Phone: (205) 834-3705
3231 Thomason Avenue
Montgomery, Alabama 36108

Totem Equipment & Supply Inc.
Phone: (907) 272-9573
2636 Commercial Drive
Anchorage, Alaska 99501-3096

Lawn & Garden Supply Co.
Phone: (602) 278-8585
2222 N. 27th Avenue
Phoenix, Arizona 85009

Lawn & Garden Supply Co.
Phone: (602) 294-3177
5455 S. Nogales Highway
Tucson, Arizona 85706

Capital Turf & Equipment Company
Phone: (501) 847-3057
P.O. Box 300
Alexander, Arkansas 72002

H.V. Carver Co., Inc.
Phone: (209) 224-7626
P.O. Box 9185
Hughes Station
Fresno, California 93791

H. V. Carter Co., Inc.
Phone: (415) 536-9300
P.O. Box 12006
Oakland, California 94604

H. V. Carter Co., Inc.
Phone: (916) 927-3824
2309 Lexington Street
Sacramento, California 95815

B. Hayman Company, Inc.
Phone: (213) 946-6411
9525 Sorensen Avenue
Santa Fe Springs, California 90670

Boyd Distributing Co., Inc.
Phone: (303) 629-7701
378 Osage Street
Denver, Colorado 80223

The Magovern Company, Inc.
Phone: (203) 348-8211
P.O. Box 4820
911 Hope Street
Stamford, Connecticut 06907

The Magovern Company, Inc.
Phone: (203) 623-2508
P.O. Box 270
27 Lawnacre Road
Windsor Locks, Connecticut 06096

Debra Equipment Co.
Phone: (813) 332-4663
2857 Hanson Street
Fort Myers, Florida 33901

Debra Turf & Industrial Equipment Co.
Phone: (305) 987-1400
5921 N. Oak Street
Hollywood, Florida 33021

Tresca Industries, Inc.
Phone: (904) 268-6741
10639 Phillips Highway
Jacksonville, Florida 32224

Tresca Industries, Inc.
Phone: (305) 291-1717
3132 John Young Parkway
Orlando, Florida 32804

Tieco Gulf Coast, Inc.
Phone: (904) 434-5475
3840 Hopkins Street
Pensacola, Florida 32505

Debra Turf & Industrial Equipment Co.
Phone: (813) 621-3077
6025 U.S. Highway 301
Tampa, Florida 33610

Lawn & Turf, Inc.
Phone: (404) 483-4743
P.O. Box 480
Conyers, Georgia 30207

B. Hayman Company, Ltd.
Phone: (808) 671-2811
94-062 Leokane Street
Waipahu, Hawaii 96797

Western Power Equipment, Inc.
Phone: (208) 375-0260
8190 Overland Road
Boise, Idaho 83707

Illinois Lawn Equipment, Inc.
Phone: (312) 349-8484
16450 S. 104th Avenue
Orland Park, Illinois 60462

Illinois Lawn Equipment, Inc.
Phone: (312) 459-7008
291 E. Messner
Wheeling, Illinois 60090

Indiana Turf Equipment Corporation
Phone: (317) 291-2280
6810 Guion Road
Indianapolis, Indiana 46268

Big Bear Turf Equipment Co.
Phone: (319) 285-4440
900 East Franklin Street
Eldridge, Iowa 52748

Robinson's Lawn & Golf, Inc.
Phone: (316) 942-2224
4611 W. Harry
Wichita, Kansas 67209

Tieco Louisville, Inc.
Phone: (502) 499-9300
1829 Laser Lane
Louisville, Kentucky 40299

Southern Specialty Sales Co., Inc.
Phone: (504) 486-6101
P.O. Box 19965
617 N. Broad Avenue
New Orleans, Louisiana 70179

G. L. Cornell Company
Phone: (301) 948-2000
P.O. Box 729
16031 Industrial Drive
Gaithersburg, Maryland 20877-0729

Sawtelle Brothers, Inc.
Phone: (617) 599-4856
P.O. Box 267
565 Humphrey Street
Swampscott, Massachusetts 01907

W. F. Miller Company
Phone: (313) 647-7700
P.O. Box 625
1593 S. Woodward
Birmingham, Michigan 48011

Miller West, Inc.
Phone: (616) 241-4481
274 Mart Street S.W.
Grand Rapids, Michigan 49508

R. L. Gould & Company
Phone: (612) 484-8411
3711 N. Lexington Parkway
St. Paul, Minnesota 55112

Robinson's Lawn & Golf, Inc.
Phone: (816) 765-3333
11918 Grandview Road
Grandview, Missouri 64080

Outdoor Equipment Company
Phone: (314) 569-3232
160 Weldon Parkway
Maryland Heights, Missouri 63043

Turf-Aid Distributing Co.
Phone: (406) 656-8008
2015 S. 56th St. W.
Billings, Montana 59102

Rio Grande Turf Supply, Inc.
Phone: (505) 243-7797
P.O. Box 27670
Albuquerque, New Mexico 87125

Big Bear Equipment, Inc.
Phone: (402) 331-0200
10405 J Street
Omaha, Nebraska 68127

Wilfred Macdonald, Inc.
Phone: (201) 471-0244
340 Main Avenue
Clifton, New Jersey 07014

J.E.P. Sales, Inc.
Phone: (609) 585-2300
P.O. Box 11126
211 Yardville-Hamilton Square Road
Yardville, New Jersey 08620

S.V. Moffett Co., Inc.
Phone: (518) 783-0668
10 Green Mountain Drive
Cohoes, New York 12047

Malvese Mowers & Equipment, Inc.
Phone: (516) 681-7600
Box 295
530 Old Country Road
Hicksville, L.I., New York 11802

Malvese Mowers & Equipment, Inc.
Phone: (516) 369-1147
232 E. Old Country Road (Rt. 58)
Riverhead, L.I., New York 11901

S.V. Moffett Co., Inc.
Phone: (716) 334-0100
Thruway Park Drive
West Henrietta, New York 14586

Porter Brothers, Inc.
Phone: (704) 482-3424
P.O. Box 520
1005 E. Dixon Blvd.
Shelby, North Carolina 28150

Ohio Turf Equipment Co.
Phone: (513) 772-7441
11472 Gondola Street
Cincinnati, Ohio 45241

Krigger & Company, Inc.
Phone: (614) 445-7101
852 Marion Road
Columbus, Ohio 43027

Krigger & Company, Inc.
Phone: (216) 467-2176
380 E. Highland Road
Macedonia, Ohio 44056

Toledo Turf Equipment Co.
Phone: (419) 473-2503
4329 W. Alexis Road
Toledo, Ohio 43623

Paul Blakeney Company
Phone: (918) 366-9446
14716 S. Grant Street
Bixby, Oklahoma 74008

Paul Blakeney Company
Phone: (405) 528-7577
330 Northeast 38th Street
Oklahoma City, Oklahoma 73105

Farwest Turf Equipment Company
Phone: (503) 224-6100
2305 N.W., 30th Avenue
Portland, Oregon 97210-2011

Lawn & Golf Supply Co., Inc.
Phone: (215) 933-5801
647 Nutt Road
Phoenixville, Pennsylvania 19460

Krigger & Company, Inc.
Phone: (412) 931-2176
3025 Babcock Blvd.
Pittsburgh, Pennsylvania 19460

Dakota Turf Supply, Inc.
Phone: (605) 336-1873
3910 Southwestern Avenue
Sioux Falls, South Dakota 57105

Bob Ladd, Inc.
Phone: (901) 324-8801
P.O. Box 12269
Memphis, Tennessee 38112

Tieco Tennessee, Inc.
Phone: (615) 244-9871
118 Park South Court
Nashville, Tennessee 37210

Colonial Motor Company
Phone: (214) 556-1191
P.O. Box 59225
2030 Walnut Hill Lane
Dallas, Texas 75229

Watson Distributing Co., Inc.
Phone: (713) 771-5771
P.O. Box 36211
6335 Southwest Freeway
Houston, Texas 77236-6211

Watson Distributing Co., Inc.
Phone: (512) 654-7065
P.O. Box 47505
5511 Brewster Drive
San Antonio, Texas 78265-7505

Stan Bonham Company, Inc.
Phone: (801) 262-2574
P.O. Box 7246
4410 S. Main Street
Salt Lake City, Utah 84107

Northwest Mowers, Inc.
Phone: (206) 542-7484
926 N. 165th Street
Seattle, Washington 98133

Northwest Mowers, Inc.
Phone: (509) 467-6604
North 7718 Market Street
Spokane, Washington 99207

Horst Distributors, Inc.
Phone: (414) 849-2341
444 North Madison Street
Chilton, Wisconsin 53014

Wisconsin Turf Equipment Corporation
Phone: (608) 752-8766
Box 708
1917 W. Court Street
Janesville, Wisconsin 53545

Wisconsin Turf Equipment Corporation
Phone: (414) 544-6421
21520 W. Greenfield Avenue
New Berlin, Wisconsin 53151

CANADIAN DISTRIBUTORS

Interprovincial Turf, Ltd.
Phone: (403) 227-3300
P.O. Box 2010
4404 42nd Avenue
Innisfail, Alberta T0M 1A0

Fallis Turf Equipment, Ltd.
Phone: (604) 277-1314
11951 Forge Place
Richmond, British Columbia V7A 4V9

Consolidated Turf Equipment, Ltd.
Phone: (204) 633-7276
972 Powell Avenue
Winnipeg, Manitoba R3H 0H6

The Halifax Seed Company, Ltd.
Phone: (902) 454-0938
Box 8026, Station A
Halifax, Nova Scotia B3K 5L8

Ontario Turf Equipment Co., Ltd.
Phone: (519) 452-3540
50 Charterhouse Crescent
London, Ontario N5W 5V5

O.J. Company, Ltd.
Division of Otto Jangle Co., Ltd.
Phone: (514) 861-5379
294 Rang St. Paul
Sherrington, Quebec J0L 2N0

Brandt Industries, Ltd.
Phone: (306) 525-1314
705 Toronto Street
Regina, Saskatchewan S4R 8G1

KAWASAKI

KAWASAKI MOTORS CORP. USA
P.O. Box 504
Shakopee, Minnesota 55379

Model	Bore	Stroke	Displacement
KT43	54 mm (2.13 in.)	48 mm (1.89 in.)	110 cc (6.7 cu. in.)

ENGINE INFORMATION

KT43 models are two-stroke, air-cooled engines. Cylinder head and cylinder bore are bolted to an aluminum die-cast crankcase. The three-piece crankshaft is supported at each end by ball bearing type main bearings.

Model KT43 is rated at 3.2 kW (4.3 hp) at 5000 rpm.

Engine model identification decal is located on cooling shroud and serial number is stamped in crankcase (Fig. KW1). Always furnish engine model and serial number when ordering parts or service material.

MAINTENANCE

SPARK PLUG. Recommended spark plug is NGK BM7, or equivalent.

Spark plug should be removed and cleaned and electrode gap set at 0.6-0.7 mm (0.024-0.027 in.) after every 25 hours of operation. Renew spark plug if electrode is severely burnt or damaged. Tighten spark plug to 27 N·m (20 ft.-lbs.).

NOTE: Caution should be exercised if abrasive type spark plug cleaner is used. Inadequate cleaning procedure may allow the abrasive cleaner to be deposited in engine cylinder accelerating wear and part failures.

CARBURETOR. Model KT43 is equipped with a Mikuni float type carburetor equipped with idle mixture adjustment needle. Main fuel mixture is controlled by a fixed jet.

Initial adjustment of idle mixture screw (2 – Fig. KW2) is 1 turn open from a lightly seated position. Make final adjustment with engine at operating temperature and running. Set engine idle at 1400 rpm by adjusting idle stop screw (1). Adjust idle mixture screw (2) to obtain the smoothest idle and acceleration.

Standard main jet size is #82.5.

To check float level, remove carburetor and separate float bowl from carburetor throttle body. Invert throttle body and float assembly (Fig. KW3). Float should contact float valve lever when bottom edge of float is just level with edge of throttle body. Float lever is spring-loaded and float should just touch lever, but not compress fuel inlet valve spring.

Fig. KW1—View showing location of engine serial number.

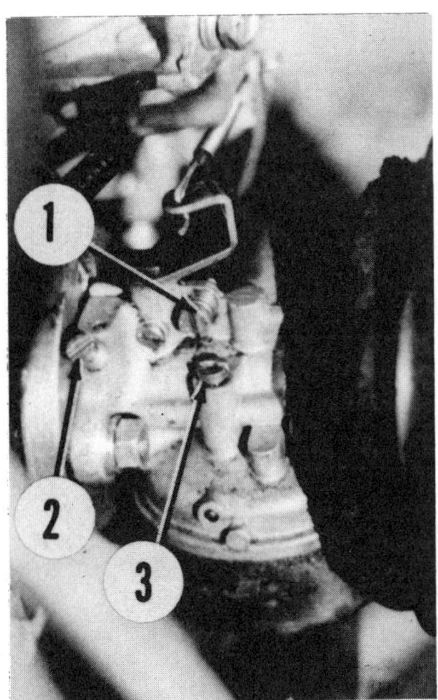

Fig. KW2—Model KT43 engines are equipped with Mikuni float type carburetor.

1. Throttle stop screw
2. Idle mixture screw
3. Pilot air jet

GOVERNOR. Model KT43 is equipped with a flyweight type governor. To adjust external linkage, stop engine. Make certain all linkage is in good condition and tension spring (3 – Fig. KW4) is not stretched. Spring (2) around governor-to-carburetor rod (1) must pull governor lever (4) and throttle arm toward each other. Loosen clamp bolt (5) and push governor lever until throttle is in full open position. Hold lever in this position and use a screwdriver in governor shaft slot to rotate governor shaft (6) clockwise until it stops (range is very slight). Tighten governor clamp bolt (5).

To set minimum speed, engine should be at operating temperature and running, no-load. Close throttle valve by hand until it hits throttle stop screw (1 – Fig. KW2). Adjust throttle stop screw so engine is running at 1400 rpm.

To set maximum speed, engine should be at operating temperature and running, no-load. Attach a tachometer to engine. Slowly move throttle lever increasing engine speed while observing tachometer.

CAUTION: At no time, even during adjustment, should engine be allowed to run above 5000 rpm. Stop throttle movement when engine reaches 4950 rpm and continue as outlined.

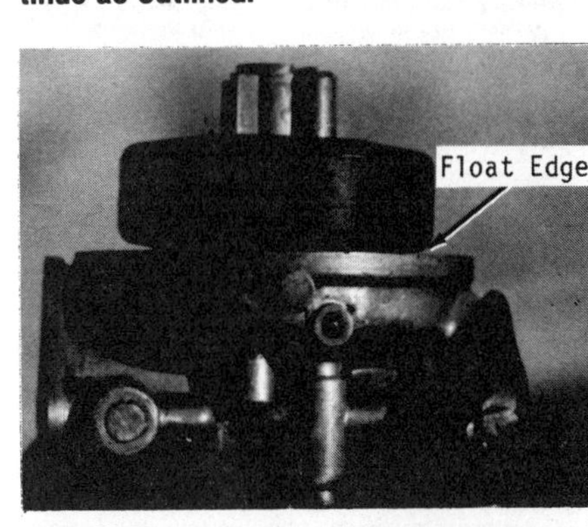

Fig. KW3—Carburetor float must be level with lower edge of carburetor throttle body without compressing fuel inlet needle spring.

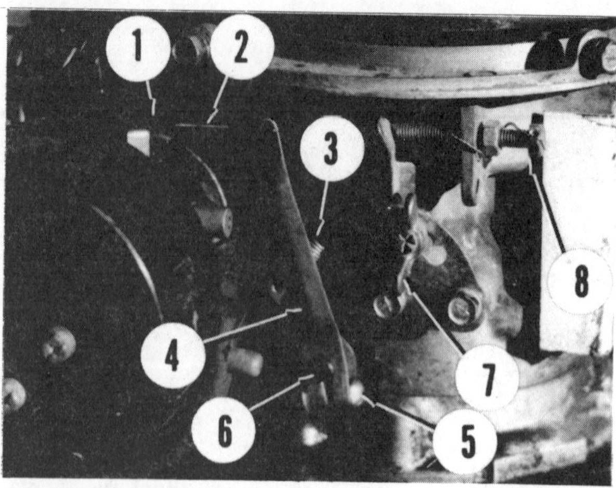

Fig. KW4—View of external governor linkage.
1. Carburetor-to-governor rod
2. Spring
3. Tension spring
4. Governor lever
5. Clamp bolt
6. Governor shaft
7. Throttle & lock
8. Maximum speed limit bolt

Fig. KW8—Loosen or tighten cylinder head bolts in sequence shown. Refer to text for torque specifications.

Adjust maximum speed limit screw (8–Fig. KW4) to obtain 4850-5000 rpm and lock it securely in place with jam nut.

IGNITION SYSTEM. A breaker point ignition system is standard. Ignition coil, breaker points and condenser are located behind flywheel (Fig. KW5). Flywheel must be removed to perform service or adjustment. Breaker point gap should be 0.3-0.5 mm (0.012-0.020 in.).

To check timing, first remove flywheel, then gently place flywheel on crankshaft. Rotate engine so "P" mark on flywheel is aligned with timing mark on crankcase (Fig. KW7). Very carefully remove flywheel without rotating engine. Breaker points should just begin to open. Loosen breaker plate screw and move points if necessary. This sets ignition timing at 25° BTDC. When the "T" mark on flywheel is aligned with timing mark, piston is at top dead center.

CYLINDER HEAD AND COMBUSTION CHAMBER. Cylinder head, combustion chamber and piston should be cleaned and carbon and other deposits removed after every 300 hours of operation. Refer to REPAIR section of this manual for service procedure.

LUBRICATION. Manufacturer recommends mixing a good quality two-stroke, air-cooled engine oil with regular grade gasoline at a 25:1 ratio. Always mix fuel in a separate container and add only mixed fuel to engine fuel tank.

GENERAL MAINTENANCE. Check and tighten all loose bolts, nuts or clamps prior to each day of operation. Check for fuel or oil leakage and repair if necessary.

Clean dust, dirt, grease or any foreign material from cylinder head and cylinder block cooling fins after every 100 hours of operation. Inspect fins for damage and repair if necessary.

REPAIRS

TIGHTENING TORQUES. Recommended tightening torques are as follows:

Spark plug .... 27 N·m (20 ft.-lbs.)
Flywheel nut .... 39-44 N·m (30-33 ft.-lbs.)
Cylinder head .... 20-22 N·m (15-16 ft.-lbs.)
Crankcase .... 7-9 N·m (5-6 ft.-lbs.)

CYLINDER HEAD. To remove cylinder head, disconnect spark plug. Remove rewind starter and cooling shroud assembly. Unbolt cylinder head retaining bolts in sequence shown in Fig. KW8.

Reinstall by reversing removal procedure and tighten head retaining bolts to specified torque in sequence shown in Fig. KW8.

Fig. KW5—View of ignition coil (1), breaker points (2) and crankshaft (4). Condenser is not shown but is attached to wire leads (3).

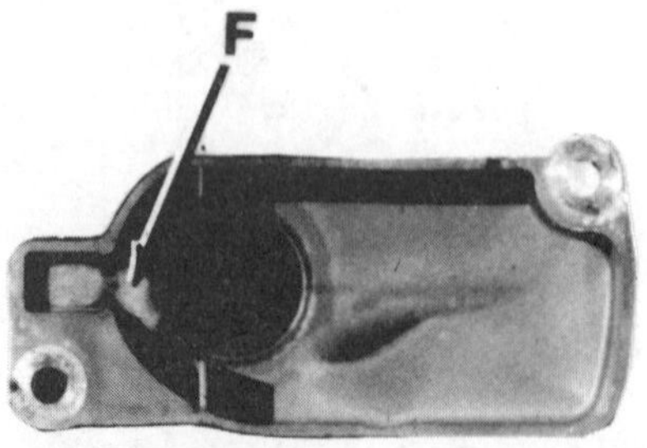

Fig. KW6—Lubricating felt (F) in point cover should receive a drop of oil whenever points are renewed.

Fig. KW7—When "P" mark on flywheel is aligned with timing mark on crankcase cover, breaker points should just begin to open.

CONNECTING ROD. Connecting rod is forged steel with needle bearings in piston pin bore and crankpin bearing bore. To remove connecting rod, remove cylinder head, cylinder and flywheel. Remove piston pin retainers and piston pin. Separate piston from connecting rod. Use care not to lose needle bearings in connecting rod piston pin bore. Loosen upper crankcase cover bolts in sequence shown in fig. KW9. Loosen lower crankcase cover bolts in a criss-cross pattern and separate crankcase from crankcase covers. Remove connecting rod and crankshaft assembly.

CAUTION: Do not separate crankshaft from connecting rod. Crankshaft and connecting rod are available only as an assembly. Kawasaki does not recommend crankshaft disassembly.

If connecting rod side play at crankpin bearing is 0.5 mm (0.02 in.) or more, or if radial play at piston pin bearing or crankpin bearing is 0.05 mm (0.002 in.) or more, renew crankshaft and connecting rod assembly.

Use light grease to retain needle bearings in their bore during reassembly. Tighten crankcase cover bolts evenly, in graduated steps, to specified torque in sequence shown in Fig. KW9 for upper crankcase bolts. Use a criss-cross pattern when tightening lower crankcase cover bolts.

PISTON, PIN AND RINGS. Piston may be removed after removing cylinder head and cylinder. Refer to CONNECTING ROD paragraphs.

Check clearance between piston and cylinder. If clearance is 0.2 mm (0.008 in.) or more, renew piston and/or cylinder. Maximum side clearance for new ring in ring groove is 0.15 mm (0.006 in.). Maximum ring end gap is 1.0 mm (0.04 in.). Maximum clearance between piston pin and piston is 0.03 mm

Fig. KW9—Upper crankcase bolts must be tightened evenly, in graduated steps, in sequence shown. Refer to text for torque specifications.

(0.001 in.). Renew piston if specifications are not as specified.

When installing piston rings, marked side of ring is towards top of piston. Align ring end gaps with pins in piston ring grooves.

Piston should be installed on connecting rod with the arrow mark on piston top toward the exhaust port side of engine. Lubricate and install piston pin and retaining rings. Use a suitable ring compressor to avoid damaging rings and very carefully install cylinder over piston and ring assembly.

CYLINDER. The cylinder is a low pressure aluminum alloy with iron sleeve. To remove cylinder, remove cylinder head and carefully work cylinder up off of piston.

Standard cylinder bore diameter is 54 mm (2.126 in.) with a service limit of 0.1 mm (0.004 in.). If cylinder bore wear exceeds service limit, renew cylinder.

Refer to PISTON, PIN AND RINGS section for piston-to-cylinder clearance and reassembly procedure.

CRANKSHAFT, MAIN BEARINGS AND SEALS. Refer to CONNECTING ROD section for crankshaft removal. Note that Kawasaki does not recommend separating connecting rod from crankshaft and that crankshaft and connecting rod are available as an assembly only. Ball bearing type main bearings are a slight press fit in crankcase covers. It may be necessary to slightly heat crankcase covers for removal and installation. Maximum axial play for main bearings is 0.15 mm (0.006 in.). Renew main bearings if loose, rough or a loose fit on crankshaft.

GOVERNOR. Refer to MAINTENANCE section for adjustment procedure of external governor linkage.

To remove governor, remove crankcase cover, loosening bolts in sequence shown in Fig. KW9. Separate cover from crankcase. Slide governor holder off the crankshaft and remove the pin which will allow governor sleeve and thrust washer removal.

Check governor tip contact surface and governor weight contact surface of the thrust washer for wear and roughness.

Remove governor lever from external end of governor shaft and governor tip from internal end of governor shaft. Pull governor shaft from crankcase bore. Check governor sleeve contact surface of governor tip for wear. Check governor shaft for wear where it is supported by crankcase bushing bore.

When reassembling governor, lubricate all parts with engine oil and make certain pin is correctly positioned in notch of governor sleeve. Tighten lower crankcase cover bolts equally in sequence shown in Fig. KW9 to specified torque.

KAWASAKI

Model	Bore	Stroke	Displacement
FA76D	52 mm (2.05 in.)	36 mm (1.42 in.)	76 cc (4.8 cu. in.)
FA130D	62 mm (2.44 in.)	43 mm (1.69 in.)	129 cc (7.9 cu. in.)
FA210D	72 mm (2.83 in.)	51 mm (2.01 in.)	207 cc (12.7 cu. in.)

ENGINE IDENTIFICATION

Kawasaki FA series engines are four-stroke, single-cylinder, air-cooled horizontal crankshaft engines. Model FA76D develops 1.3 kW (1.7 hp) at 4000 rpm, Model FA130D develops 2.3 kW (3.1 hp) at 4000 rpm and Model FA210D developes 3.9 kW (5.2 hp) at 4000 rpm.

Engine model number is located on the cooling shroud just above the rewind starter. Engine model number and serial number are both stamped on crankcase cover (Fig. KW20). Always furnish engine model and serial numbers when ordering parts or service information.

MAINTENANCE

SPARK PLUG. Recommended spark plug for all models is a NGK BM-6A, or equivalent.

Spark plug should be removed and cleaned and electrode gap set at 0.6-0.7 mm (0.024-0.027 in.) after every 100 hours of operation. Renew spark plug if electrode is severely burnt or damaged. Tighten spark plug to 27 N·m (20 ft.-lbs.).

NOTE: Caution should be exercised if abrasive type spark plug cleaner is used. Inadequate cleaning procedure may allow the abrasive cleaner to be deposited in engine cylinder accelerating wear and part failures.

CARBURETOR. Engines may be equipped with either a pulse type (Fig. KW21) or a float type (Fig. KW22) carburetor. Engine idle speed on all models should first be adjusted to 1500 rpm using idle stop screw (IS–Fig. KW21 or KW22) on carburetor, then, adjust idle limit screw (IL–Fig. KW25) on throttle plate assembly so engine idles at 1600 rpm. Adjust maximum speed limit screw (ML) so maximum engine speed is 4000 rpm.

CAUTION: Maximum engine speed must not exceed 4000 rpm even during adjustment procedure. Running engine at speeds in excess of 4000 rpm may result in engine damage and possible injury to operator.

For initial carburetor adjustment of pulse type carburetor, lightly seat pilot air screw (P–Fig. KW21) and then open pilot screw 1¼ turns on FA130D and FA210D models, or 1 turn on FA76D models. Make final adjustments with engine at operating temperature and running. Back out idle limit screw (IL–Fig. KW25) so it does not limit idle speed, then, adjust idle stop screw (IS–Fig. KW21) on carburetor so engines runs at lowest speed possible. Adjust pilot air screw (P) to obtain highest engine idle speed. Readjust idle stop screw (IS) and idle limit screw (IL–Fig. KW25) for correct speeds.

For initial carburetor adjustment of float type carburetor, lightly seat pilot

Fig. KW22—View of float type carburetor used on some models. Fuel mixture needle (FM) has a slotted head on Models FA76D and FA210D as shown. Model FA130D has a larger wheel type knob.

P. Pilot air screw
IS. Throttle stop screw
FM. Main fuel mixture screw

Fig. KW20—View showing location of engine model and serial number stamped on engine crankcase cover.

Fig. KW21—View of pulse type carburetor used on some models showing location of throttle stop screw (IS) and pilot air screw (P). Refer to text for adjustment procedure.

air screw (P – Fig. KW22) and fuel mixture needle (FM). Open pilot air screw (P) 1½ turns on FA130D and FA210D models and 1¼ turns on FA76D models. Open fuel mixture screw (FM) 1½ turns on all models. Make final adjustments with engine at operating temperature and running. Back out idle limit screw (IL – Fig. KW25) so it does not limit idle speed, then, adjust idle stop screw (IS – Fig. KW22) on carburetor so engine runs at lowest speed possible. Adjust pilot air screw (P) to obtain highest engine idle speed. Readjust idle stop screw and idle limit screw for correct speeds. Operate engine under a load at maximum engine speed (4000 rpm) and readjust fuel mixture needle (FM) to obtain smoothest engine operation.

To check float level, invert carburetor throttle body and float assembly. Float surface should be parallel to carburetor throttle body. If float is equipped with metal tang which contacts inlet needle,

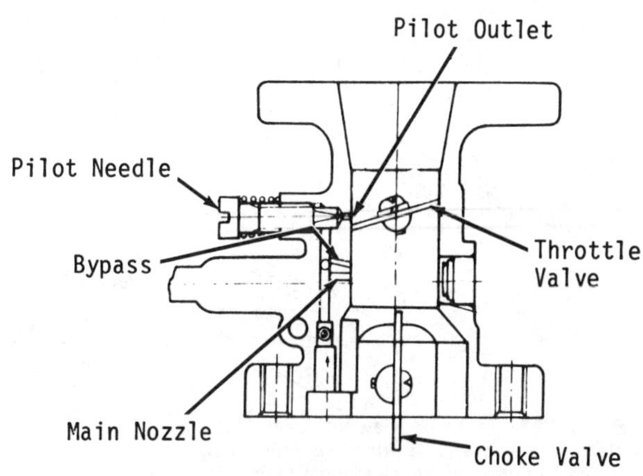

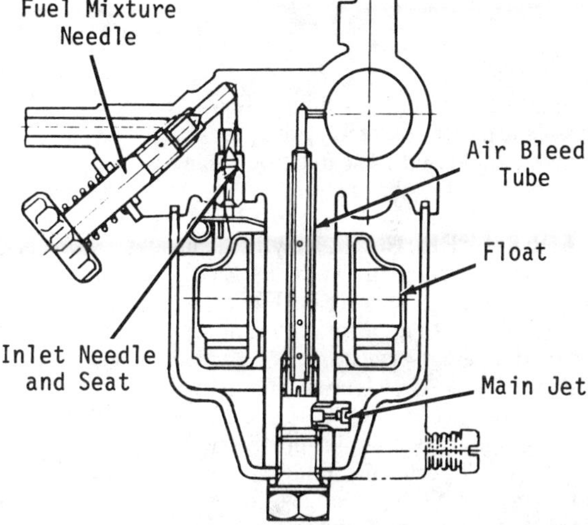

Fig. KW24—Cross-sectional view of float type carburetor used on Model FA130D. Carburetor on Models FA76D and FA210D is similar.

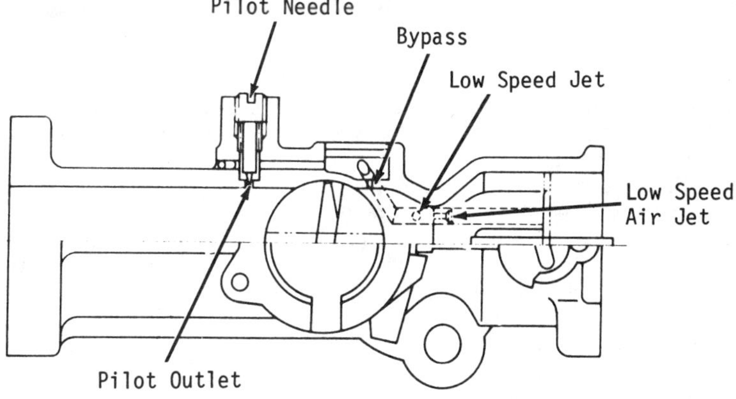

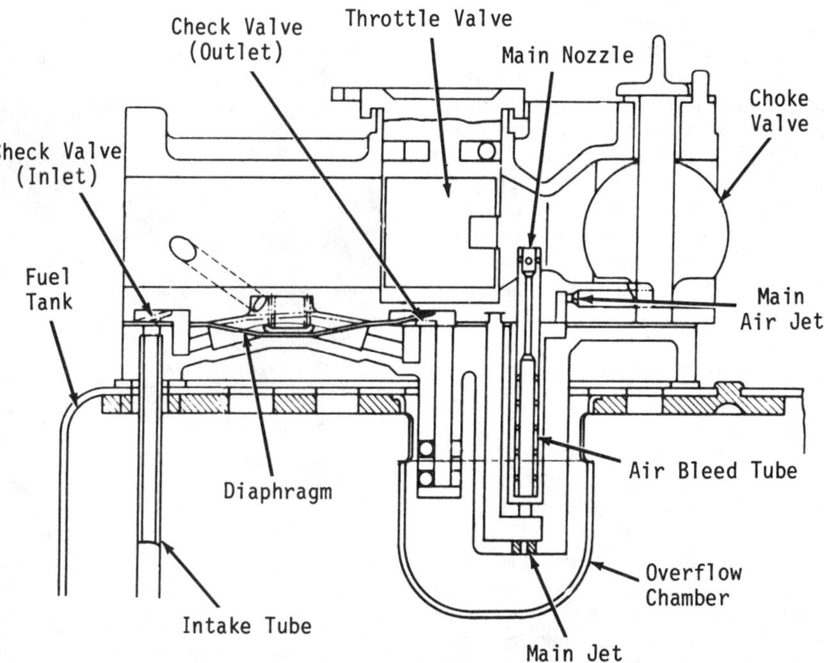

Fig. KW23—Cross-sectional view of pulse type carburetor.

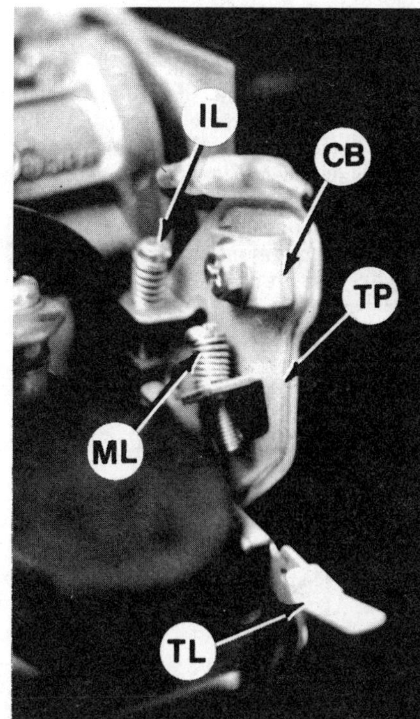

Fig. KW25—View of throttle plate assembly (TP) showing idle limit screw (IL) and maximum speed limit screw (ML). Throttle lever (TL) and remote throttle cable clamp (CB) are also shown.

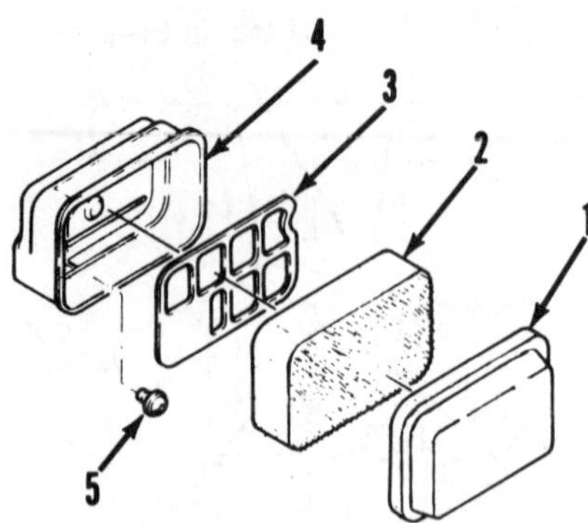

Fig. KW26—Exploded view of foam type air cleaner. Element (2) is installed with filament (hair like protrusions) away from carburetor.

1. Cover
2. Element
3. Plate
4. Element case
5. Screw

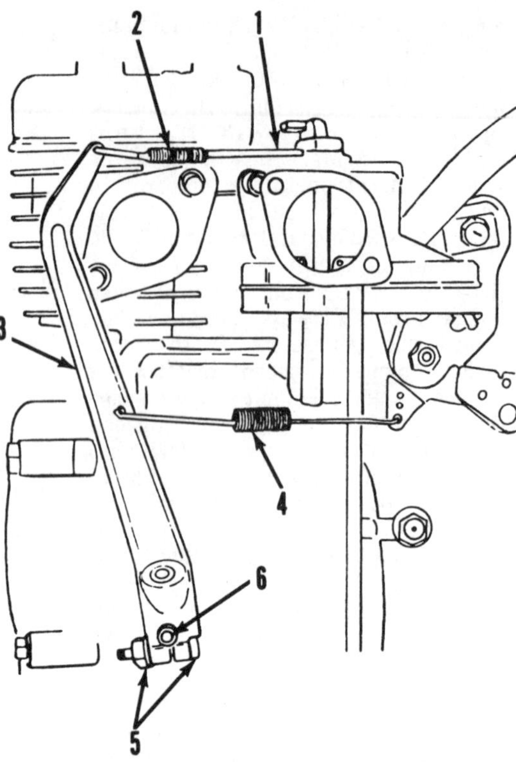

Fig. KW28—View of external governor linkage. Refer to text for adjustment procedure.

1. Governor-to-carburetor rod
2. Spring
3. Governor lever
4. Tension spring
5. Clamp bolt & nut
6. Governor shaft

carefully bend tang to adjust float. If float is a plastic assembly, float level is nonadjustable and float must be renewed if float level is incorrect.

AIR FILTER. Engine may be equipped with either a foam type (Fig. KW26) or a paper (dry) type (Fig. KW27) air cleaner. Refer to appropriate paragraph for model being serviced.

Foam Type Air Cleaner. Foam type air cleaner (Fig. KW26) should be removed and cleaned after every 25 hours of operation under normal operating conditions. Remove foam element and clean in a suitable nonflammable solvent. Carefully squeeze out solvent and allow element to air dry. Soak element in clean engine oil and squeeze out the excess. Reinstall element making certain filament (hair like protrusions) side is toward cover (Fig. KW26).

Paper (Dry) Type Air Cleaner. Paper type air cleaner (Fig. KW27) should be removed and cleaned after every 100 hours of operation under normal operating conditions. Remove foam and paper element, separate foam precleaner from paper element, wash element and foam precleaner in water and detergent and rinse in clean water. Allow to air dry. Soak foam precleaner in clean engine oil and carefully squeeze out excess oil. Slide precleaner over paper element and reinstall elements in air cleaner case.

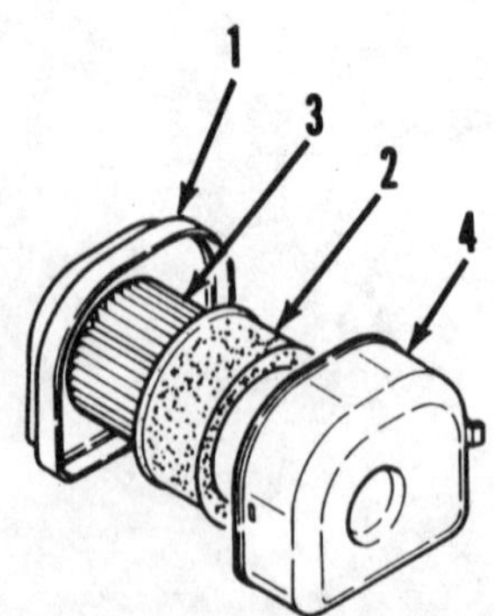

Fig. KW27—Exploded view of paper (dry) type air cleaner.

1. Cover
2. Precleaner (foam)
3. Paper element (filter)
4. Element case

GOVERNOR. A gear driven flyweight governor assembly is located inside engine crankcase on crankcase cover. To adjust external linkage, place engine throttle control in idle position. Make certain all linkage is in good condition and tension spring (4–Fig. KW28) is not stretched. Spring (2–Fig. KW29) around governor-to-carburetor rod must pull governor lever (3) and throttle pivot (4) toward each other. Loosen clamp bolt (5–Fig. KW28) and rotate governor shaft (6) clockwise as far as possible. Hold shaft in this position and pull governor lever (3) as far right as possible. Tighten clamp bolt (5). Install tension spring (4) in correct hole for speed desired.

IGNITION SYSTEM. Ignition coil (1–Fig. KW30) is located just above flywheel and breaker point set and condenser are located behind flywheel. Ignition timing should be set at 23° BTDC in the following manner. Remove flywheel and then reinstall it loosely on crankshaft. Rotate flywheel until edge of coil laminations are aligned with edge of notch in flywheel as shown in Fig. KW31. Remove flywheel making certain crankshaft does not turn. Slowly move breaker point plate assembly until points just begin to open. Tighten breaker point plate screw. Turn crankshaft clockwise until point gap is at widest open position and adjust gap to 0.3-0.5 mm (0.012-0.020 in.). Place a small amount of point cam lube on lubricating felt (4–Fig. FW30) and reinstall flywheel. Coil edge gap should be 0.05 mm (0.020 inch).

Fig. KW29—Spring (2) around governor-to-carburetor rod (1) must pull governor lever (3) toward throttle pivot (4).

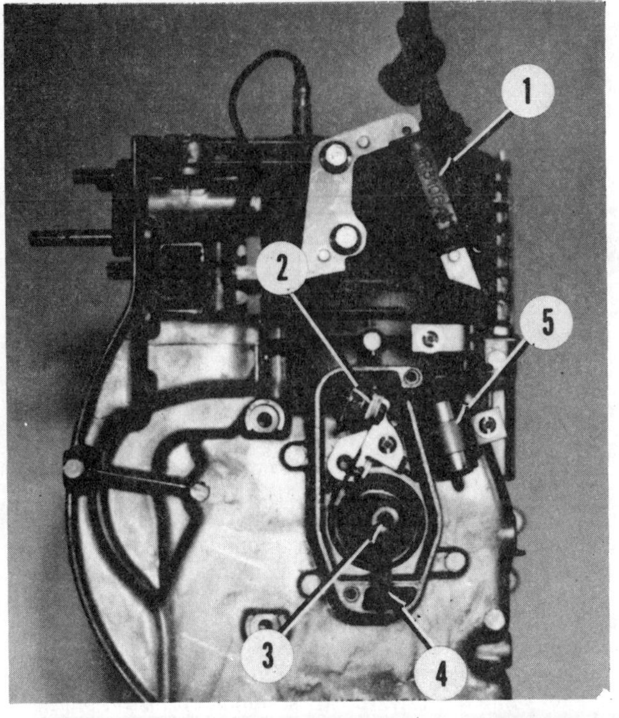

Fig. KW30—View showing location of breaker point ignition system components.
1. Ignition coil
2. Breaker points
3. Crankshaft
4. Lubricating felt
5. Condenser

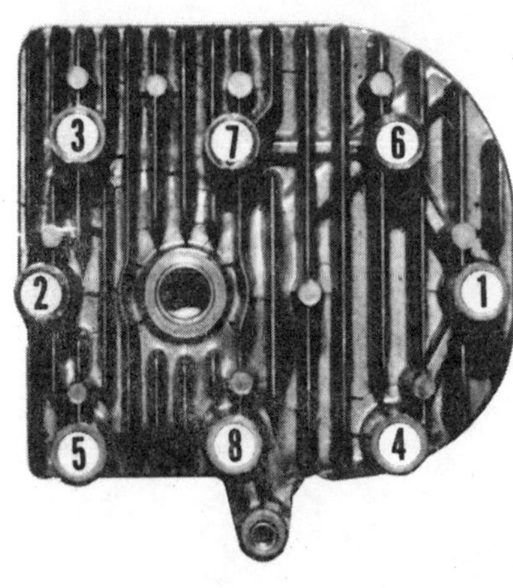

Fig. KW33—On all models, loosen or tighten cylinder head bolts using the sequence shown.

VALVE ADJUSTMENT. Valves and seats should be refaced and stem clearance adjusted after every 300 hours of operation. Refer to REPAIRS section for service procedure and specifications.

CYLINDER HEAD AND COMBUSTION CHAMBER. Cylinder head, combustion chamber and piston should be cleaned and carbon and other deposits removed after every 300 hours of operation. Refer to REPAIRS section for service procedure.

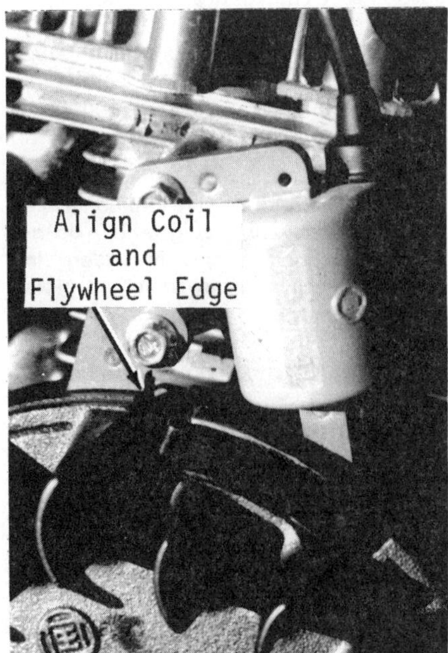

Fig. KW31—Align edge of flywheel notch with coil core laminations as shown to adjust timing. Refer to text.

LUBRICATION. All models are splash lubricated. Engine oil level should be checked prior to each operating interval. Oil level should be maintained between reference marks on gage with oil fill plug just touching first threads. Do not screw oil fill plug in to check oil level (Fig. KW32).

Manufacturer recommends oil with an API service classification SC, SD or SE. Use SAE 20 oil in winter and SAE 30 oil in summer.

Oil should be changed after the first 20 hours of operation and every 100 hours of operation thereafter. Crankcase capacity is 320 cc (0.68 pt.) on FA76D models, 500 cc (1.06 pt.) on FA130D models and 600 cc (1.27 pt.) on FA210D models.

GENERAL MAINTENANCE. Check and tighten all loose bolts, nuts or clamps prior to each day of operation. Check for fuel or oil leakage and repair if necessary.

Clean dirt, dust, grease or any foreign materal from cylinder head and cylinder block cooling fins after every 100 hours of operation. Inspect fins for damage and repair if necessary.

REPAIRS

TIGHTENING TORQUES. Recommended tightening torques are as follows:

Spark plug . . . . 16 N·m (12 ft.-lbs.)
Head bolts:
FA76D . . . . 7 N·m (5 ft.-lbs.)
All other models . . . . 19 N·m (14 ft.-lbs.)
Connecting rod bolts:
FA76D . . . . 7 N·m) (5 ft.-lbs.)
All other models . . . . 11 N·m (8 ft.-lbs.)
Crankcase cover bolts . . . . 5 N·m (4 ft.-lbs.)
Flywheel nut:
FA76D . . . . 34 N·m (25 ft.-lbs.)
All other models . . . . 60 N·m (44 ft.-lbs.)
Drain plug . . . . 13 N·m (10 ft.-lbs.)

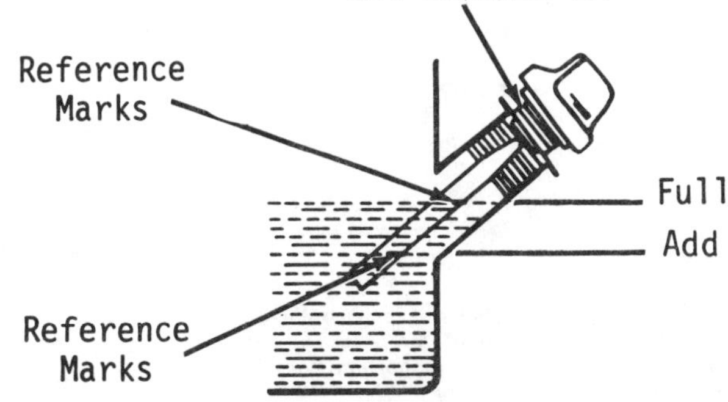

Fig. KW32—View showing oil fill plug and oil gage. Plug is not screwed in, but is just touching threads when checking oil level.

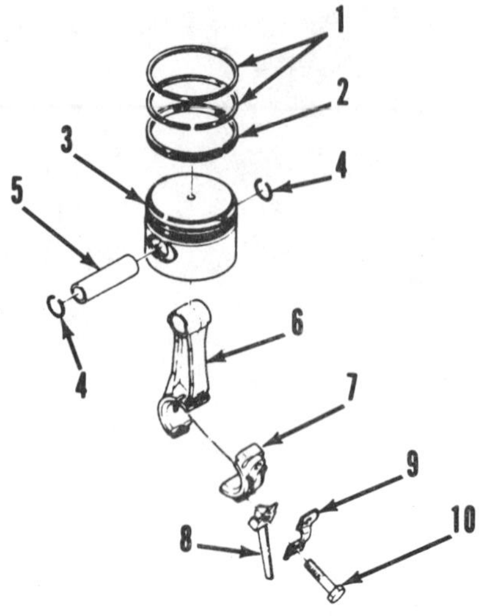

Fig. KW34—Exploded view of piston and connecting rod assembly.

1. Compression rings
2. Oil control ring
3. Piston
4. Retaining rings
5. Pin
6. Connecting rod
7. Connecting rod cap
8. Oil dipper
9. Lock plate
10. Bolts (2)

CYLINDER HEAD. To remove cylinder head, first remove cylinder head shroud. Clean engine to prevent entrance of foreign material. Loosen cylinder bolts in 1/4 turn increments in sequence shown in Fig. KW33 until all bolts are loose enough to remove by hand.

Remove spark plug and clean carbon and other deposits from cylinder head. Place cylinder head on a flat surface and check entire sealing surface for warpage. If warpage exceeds 0.25 mm (0.010 in.), cylinder head must be renewed. Slight warpage may be repaired by lapping cylinder head. In a figure eight pattern, lap head against 200 grit and then 400 grit emery paper on a flat surface.

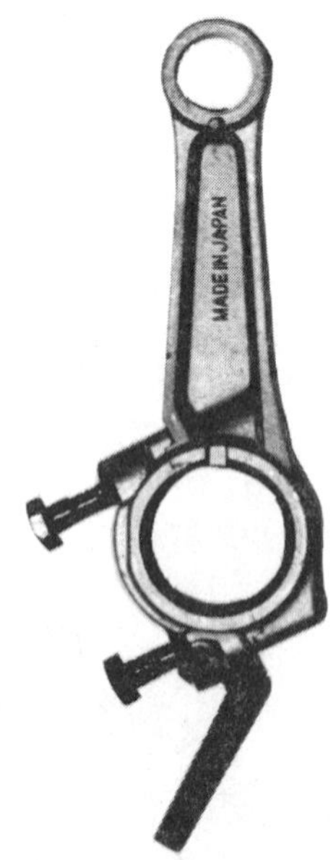

Fig. KW35—Connecting rod is assembled on piston so the side marked "MADE IN JAPAN" is toward "M" stamped on piston pin boss (Fig. KW36).

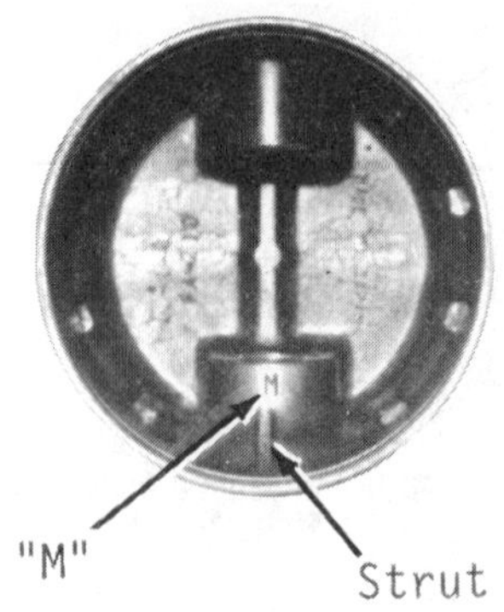

Fig. KW36—Piston pin boss is stamped with an "M". Note "M" is on boss with the strengthening strut.

Reinstall cylinder head and tighten bolts evenly to specified torque in sequence shown in Fig. KW33.

CONNECTING ROD. Connecting rod rides directly on crankshaft journal on all models. Piston and connecting rod are removed as an assembly after cylinder head and crankcase cover have been removed. Bend lock plate up to release connecting rod bolts and remove

Fig. KW37—On some models, piston top has a mark which must be toward flywheel side of engine after installation.

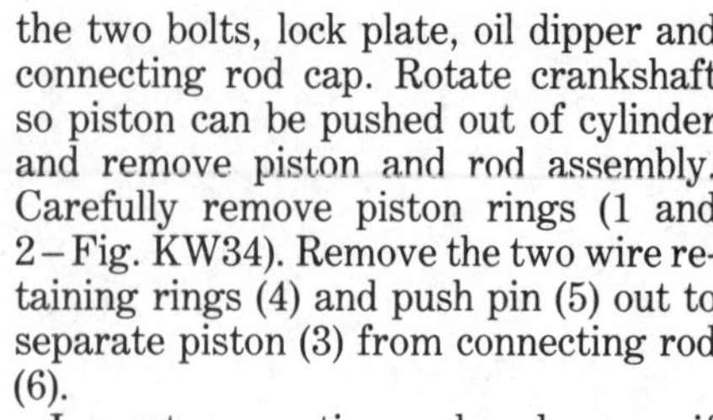
the two bolts, lock plate, oil dipper and connecting rod cap. Rotate crankshaft so piston can be pushed out of cylinder and remove piston and rod assembly. Carefully remove piston rings (1 and 2–Fig. KW34). Remove the two wire retaining rings (4) and push pin (5) out to separate piston (3) from connecting rod (6).

Inspect connecting rod and renew if piston pin bore or crankpin bearing bore are scored or damaged. If clearance between piston pin and connecting rod piston pin bore exceeds 0.05 mm (0.002 in.), renew connecting rod.

If clearance between crankpin journal and connecting rod bearing bore exceeds 0.07 mm (0.003 in.) on FA76D models or 0.10 mm (0.004 in.) on all other models, renew connecting rod and/or crankshaft. Correct connecting rod side play on crankpin should be 0.70 mm (0.028 in.).

Install piston on connecting rod so the side of connecting rod which is marked "MADE IN JAPAN" (Fig. KW35) is toward "M" stamped on piston pin boss (Fig. KW36). Secure piston pin with wire retaining rings. Install piston and connecting rod assembly so side of connecting rod marked "MADE IN JAPAN" is toward flywheel side of engine. Lubricate crankpin journal and install connecting rod cap and oil dipper as shown in Fig. KW38. Install new lock plate and bolts. Tighten bolts to specified torque and lock by bending lock plate.

PISTON, PIN AND RINGS. Piston and connecting rod are removed as an assembly. Refer to CONNECTING ROD section for removal and installation procedure.

After carefully removing piston rings and separating piston and connecting rod, carefully clean carbon and other deposits from piston surface and ring lands.

CAUTION: Extreme care should be exercised when cleaning ring lands. Do not damage squared edges or widen ring grooves. If ring lands are damaged, piston must be renewed.

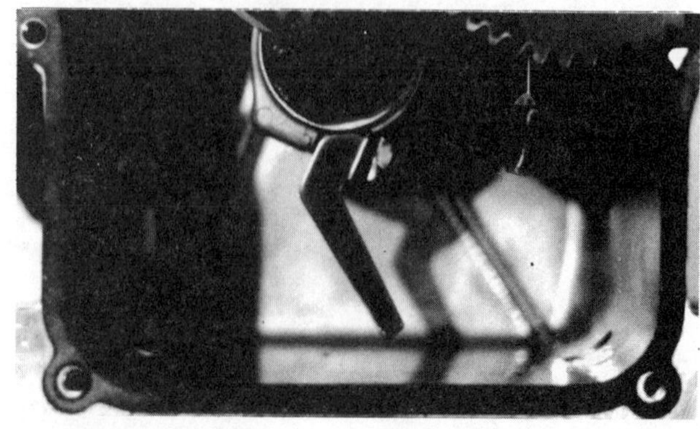

Fig. KW38—Install connecting rod and piston assembly with oil dipper in position shown.

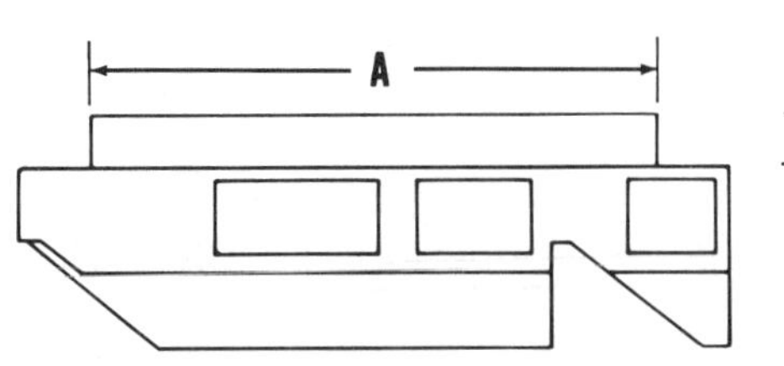

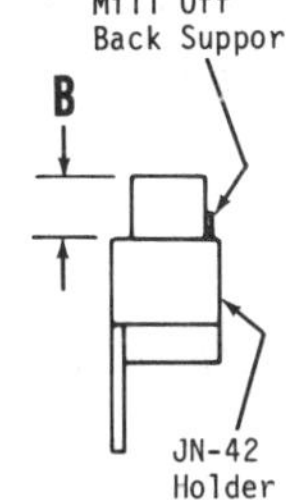

Fig. KW39—Felt strips must be altered for each model. Dimension A is 51.003 mm (2.80 in.) for FA76 models or 76.251 mm (3.20 in.) for FA130 models. Modification of felt strip length (A) for FA210 models is not required. Dimension B is 0.205 inch (0.18 mm) for FA76 and FA130 models or 0.305 inch (0.889 mm) for FA210 models.

Before installing rings on piston, install piston in cylinder and use a suitable feeler gage to measure clearance between piston and cylinder bore. If clearance exceeds 0.20 mm (0.008 in.) on FA76D models or 0.25 mm (0.010 in.) on all other models, renew piston and/or renew cylinder bore.

If ring side clearance between new ring and piston groove exceeds 0.15 mm (0.006 in.), renew piston. If ring end gap on new rings exceeds 0.10 mm (0.004 in.) when installed, renew cylinder. If clearance between piston pin and piston pin bore exceeds 0.05 mm (0.002 in.), renew piston.

Refer to CONNECTING ROD section to properly assemble piston to rod. Piston is marked "M" on piston pin boss (Fig. KW36) and on top of piston (Fig. KW37). Mark on top of piston must be toward flywheel side of engine after piston installation.

CYLINDER AND CRANKCASE. Cylinder and crankcase are an integral casting. Standard cylinder bore diameter is 52 mm (2.05 in.) on FA76D models, 62 mm (2.44 in.) on FA130D models and 72 mm (2.83 in.) on FA210D models. If cylinder bore diameter exceeds these dimensions by 0.07 mm (0.003 in.), renew cylinder.

Because of the high silicone content of the die-cast aluminum block, special honing procedures, stones and conditioning compounds are required to hone FA series engine cylinders. Procedures outlined and material prescribed are recommended to produce a cylinder bore finish comparable to the factory finish.

The following procedures outlined are designed to remove 0.258 mm (0.0102 in.) total material from cylinder bore diameter upon completion. Use 38.1 mm (1½ in.) strokes throughout the honing process. Honing time should be varied to remove only the amount of material specified to obtain desired bore size.

The cylinder bore is silicone impregnated. Refacing a cylinder bore removes the thin silicone contact surface and exposes both aluminum and silicone particles. If the aluminum particles are not removed, damage to the cylinder and piston may result. To erode the aluminum particles so only a silicone surface is present in bore, the last honing procedure must be performed exactly as outlined.

Model FA76. Use a Sunnen JN-95 hone with a spindle speed of approximately 150 rpm. Equip hone with four T20-J85-75xxx stones, without guides, and hone cylinder until 0.229 mm (0.009 in.) of material is removed. Hone should remove the specified amount of material in approximately 50 seconds.

Equip hone with four T20-C05-75xxx stones, without guides, and hone cylinder until 0.025 mm (0.001 in.) of material is removed. Hone should remove specified amount of material in approximately 20 seconds.

Alter T20-F05-75xxx felt set as shown in Fig. KW39. Saturate altered felt set with MB-30 honing oil. Mix AN-30 conditioning compound thoroughly and coat felts and entire cylinder wall surface with compound. Equip hone with altered felt set and insert hone into cylinder. Tighten wing nut finger tight. Hone cylinder, periodically tightening feed pinion, until 0.004 mm (0.0002 in.) of material is removed. Hone should remove specified amount of material in approximately 1½ minutes.

CAUTION: Do not use additional honing oil as it will wash away the AN-30 conditioning compound.

Model FA130. Use a Sunnen JN-95 hone with a spindle speed of approximately 150 rpm. Equip hone with four V24-J85-85xxx stones and hone cylinder until 0.229 mm (0.009 in.) of material is removed. Hone should remove specified amount of material in 1½ minutes.

Equip hone with four V24-J85-85xxx stones and hone cylinder until 0.025 mm (0.001 in.) of material is removed. Hone should remove specified amount of material in 20 seconds.

Alter V24-J85-85xxx felt set as shown in Fig. KW39. Saturate altered felt set with MB-30 honing oil. Mix AN-30 conditioning compounds thoroughly and coat felts and entire cylinder wall surface with compound. Equip hone with altered felt set and insert hone into cylinder. Tighten wing nut finger tight. Hone cylinder, periodically tightening feed pinion, until 0.004 mm (0.0002 in.) of material is removed. Hone should remove specified amount of material in approximately 1½ minutes.

CAUTION: Do not use additional honing oil as it will wash away the AN-30 conditioning compound.

Model FA210. Use Sunnen AN-112 hone with a spindle speed of approximately 150 rpm. Equip hone with GG25-J65 stones and hone cylinder until 0.152 mm (0.006 in.) of material has been removed. Hone should remove specified amount of material in approximately 45 seconds.

Fig. KW40—Loosen or tighten crankcase cover bolts in sequence shown. Refer to text for torque specifications.

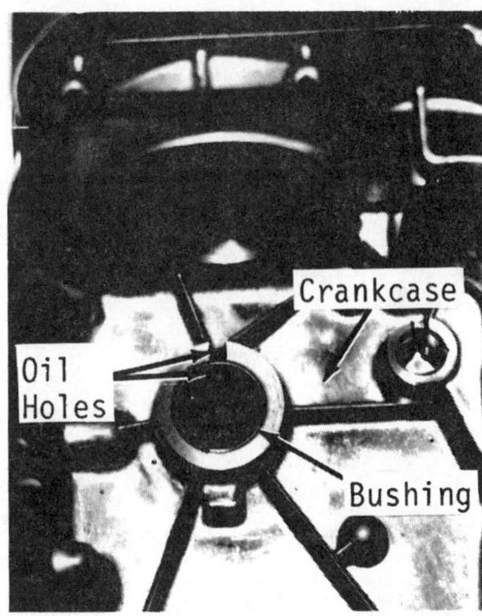

Fig. KW41—View of crankcase bushing. Oil holes must be aligned during installation. Refer to text.

Equip hone with GG25-J85 stones and hone cylinder until 0.076 mm (0.003 in.) of material is removed. Hone should remove specified amount of material in approximately 35 seconds.

Equip hone with GG25-C05 stones and hone cylinder until 0.025 mm (0.001 in.) material is removed. Hone should remove specified amount of material in approximately 20 seconds.

Alter GG25-F05 felt set as shown in Fig. KW39. Saturate altered felt set with MB-30 honing oil. Mix AN-30 conditioning compound thoroughly and coat felts and entire cylinder wall surface with compound. Equip hone with altered felt set and insert hone into cylinder. Tighten wing nut finger tight. Hone cylinder, periodically tightening feed pinion, until 0.005 mm (0.0002 in.) of material is removed. Hone should remove specified amount of material in approximately 1½ minutes.

CAUTION: Do not use additional honing oil as it will wash away the AN-30 conditioning compound.

All Models. Thoroughly clean cylinder and block after honing procedure. Lightly oil and assemble engine.

If crankcase cover has been removed, refer to CRANKSHAFT, MAIN BEARINGS AND SEALS section for bearing, bushing and seal installation. Tighten crankcase cover bolts to specified torque in sequence shown in Fig. KW40.

CRANKSHAFT, MAIN BEARINGS AND SEALS. Crankshaft on FA76D models is supported at each end in bushing type main bearings which are pressed into crankcase and crankcase cover. Crankshaft on all other models is supported in a ball bearing at crankcase cover (pto) end and in a bushing type main bearing which is pressed into crankcase at flywheel end.

To remove crankshaft, remove all metal shrouds, flywheel, cylinder head and crankcase cover. Remove connecting rod and piston assembly and remove crankshaft and camshaft. Remove valve tappets.

Minimum crankpin diameter is 19.95 mm (0.785 in.) for FA76D models, 23.92 mm (0.942 in.) for FA130D models and 26.92 mm (1.060 in.) for FA210D models. If dimensions are not as specified, renew crankshaft.

Maximum clearance between bushing type main bearing and crankshaft journal is 0.13 mm (0.005 in.). When renewing bushing type main bearing, make sure oil hole in bushing is aligned with oil hole in crankcase or cover. Press bushing in 1 mm (0.039 in.) below inside edge of bushing bore. See Fig. KW41.

Ball bearing type main bearing is a light press fit on crankshaft and maximum bearing play is 0.30 mm (0.001 in.). It may be necessary to slightly heat crankcase cover to remove or install ball bearing.

Crankcase cover oil seals should be pressed into seal bores until seal is flush with outer edge of seal bore on FA76D models or 4 mm (0.158 in.) below outer edge of seal bore on all other models. Crankcase crankshaft seal should be pressed into seal bore until flush with outer edge of seal bore on all models.

To reinstall crankshaft, reverse removal procedure making sure crankshaft and camshaft gear timing marks are aligned during installation. Refer to Fig. KW42.

Fig. KW42—Align crankshaft and camshaft gear timng marks as shown during assembly.

CAMSHAFT AND BEARINGS. Camshaft is supported at each end in bushing type bearings which are an integral part of crankcase and crankcase cover. To remove camshaft, refer to CRANKSHAFT, MAIN BEARINGS AND SEALS section.

Maximum clearance between camshaft and bushing type bearings is 0.10 mm (0.004 in.). If clearance exceeds specifications, renew camshaft and/or crankcase cover or crankcase.

Make certain camshaft lobes are smooth and free of scoring, scratches and other damage. Make certain camshaft and crankshaft gear timing marks are aligned during installation. Refer to Fig. KW42.

GOVERNOR. The internal centrifugal flyweight governor is gear driven off of the camshaft gear. Refer to GOVERNOR paragraph in MAINTENANCE section for external governor adjustments.

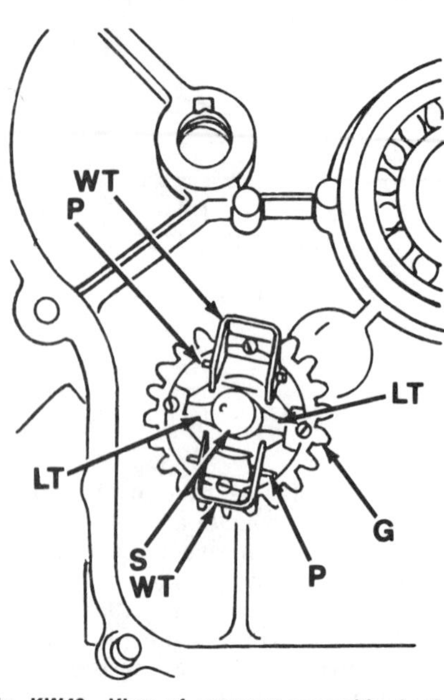

Fig. KW43—View of governor assembly. Lock tabs (LT) must be pressed in to remove sleeve and governor assembly. Renew sleeve prior to reassembly to ensure lock tabs will retain sleeve and governor assembly.

G. Gear
P. Pins
S. Sleeve
LT. Lock tabs
WT. Flyweights

To remove governor assembly, remove external linkage, metal cooling shrouds and crankcase cover. Governor gear and flyweights are attached to crankcase cover. Remove governor sleeve (S–Fig. KW43), gear (G) and flyweight assembly only if renewal is necessary. Use a screwdriver to press lock tabs (LT) in to remove sleeve. Flyweights are pinned to gear. Remove thrust washer located between gear and crankcase cover.

Reverse removal procedure for reinstallation. Adjust external linkage as outlined in MAINTENANCE section.

VALVE SYSTEM. Clearance between valve stem and valve tappet (cold) on FA76D models should be 0.1-0.2 mm (0.004-0.008 in.) for intake valve and 0.1-0.3 mm (0.004-0.012 in.) for exhaust valve. Clearance between valve stem and valve tappet (cold) on all other models should be 0.12-0.18 mm (0.005-0.007 in.) for intake valve and 0.10-0.034 mm (0.004-0.013 in.) for exhaust valve. If clearance is not as specified, valves must be removed and end of stems ground off to increase clearance or seats ground deeper to reduce clearance.

Valve seats are ground directly into crankcase casting and valve face and seat angles are 45°. Correct valve seat width for FA76D models is 0.5-1.1 mm (0.020-0.043 in.) and correct valve seat width for all other models is 1.0-1.6 mm (0.039-0.063 in.).

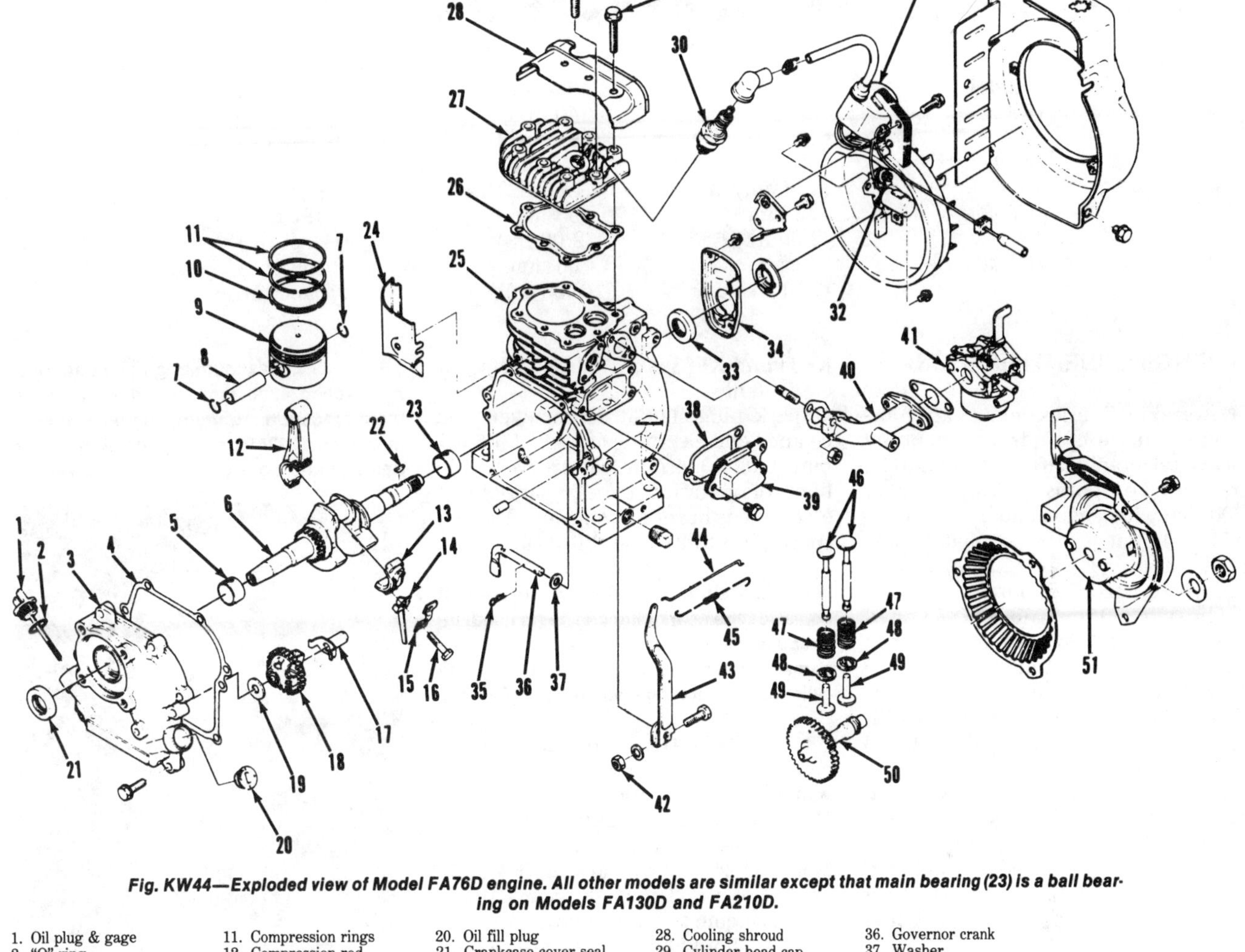

Fig. KW44—Exploded view of Model FA76D engine. All other models are similar except that main bearing (23) is a ball bearing on Models FA130D and FA210D.

1. Oil plug & gage
2. "O" ring
3. Crankcase cover
4. Gasket
5. Main bearing bushing
6. Crankshaft
7. Retaining ring
8. Piston pin
9. Piston
10. Oil control ring
11. Compression rings
12. Compression rod
13. Connecting rod cap
14. Oil plate
15. Lock plate
16. Connecting rod bolts
17. Governor sleeve
18. Governor gear & weight assy.
19. Thrust washer
20. Oil fill plug
21. Crankcase cover seal
22. Crankshaft key
23. Main bearing bushing (FA76D)
24. Cooling shroud
25. Cylinder block (crankcase)
26. Gasket
27. Cylinder head
28. Cooling shroud
29. Cylinder head cap screws
30. Spark plug
31. Ignition coil
32. Points & condenser
33. Seal
34. Point & condenser cover
35. Retaining clip
36. Governor crank
37. Washer
38. Gasket
39. Valve chamber cover
40. Manifold
41. Carburetor
42. Clamp bolt nut
43. Governor lever
44. Governor-to-carburetor rod
45. Spring
46. Valves
47. Valve springs
48. Valve spring retainers
49. Valve lifters
50. Camshaft & gear assy.
51. Rewind starter assy.

KAWASAKI

Model	Bore	Stroke	Displacement
KF24	56 mm (2.20 in.)	40 mm (1.58 in.)	98 cc (6 cu. in.)
KF34	60 mm (2.36 in.)	47 mm (1.85 in.)	132 cc (8.1 cu. in.)
KF53	66 mm (2.60 in.)	53 mm (2.09 in.)	181 cc (11 cu. in.)
KF68	74 mm (2.91 in.)	60 mm (2.36 in.)	258 cc (15.7 cu. in.)

ENGINE IDENTIFICATION

Kawasaki KF series engines are four-stroke, single-cylinder, air-cooled horizontal crankshaft engines. Model KF24 rated output is 1.2 kW (1.6 hp) at 3600 rpm with a maximum output of 1.7 kW (2.3 hp) at 4200 rpm. Model KF34 rated output is 1.7 kW (2.3 hp) at 3600 rpm with a maximum output of 2.5 kW (3.4 hp) at 4200 rpm. Model KF53 rated output is 2.7 kW (3.6 hp) at 3600 rpm with a maximum output of 3.7 kW (5.0 hp) at 4200 rpm. Model KF68 rated output is 7.0 kW (5.2 hp) at 3600 rpm with a maximum output of 9.4 kW (7.0 hp) at 4200 rpm.

Engine model number is located on the cooling shroud just above the rewind starter. Model number and serial number are both stamped on crankcase cover (Fig. KW50). Always furnish engine model and serial numbers when ordering parts.

MAINTENANCE

SPARK PLUG. Recommended spark plug for Models KF24 and KF53 is a Champion L86C, or equivalent and recommended spark plug for Models KF34 and KF68 is a Champion J8C, or equivalent.

Spark plug should be removed and cleaned and electrode gap set at 0.6-0.7 mm (0.024-0.027 in.) after every 100 hours of operation. Renew spark plug if electrode is burnt or damaged. Tighten spark plug to 27 N.m (20 ft.-lbs.).

NOTE: Caution should be exercised if abrasive type spark plug cleaner is used. Inadequate cleaning procedure may allow the abrasive cleaner to be deposited in engine cylinder accelerating wear and part failures.

CARBURETOR. All models are equipped with a float type carburetor which has an idle fuel mixture needle. Main fuel mixture is controlled by a fixed main jet. Main jet size is a #65 on Model KF24, #72.5 on Model KF34, #85 on Model KF53 and #87.5 on Model KF68. Engine idle and maximum speed settings should be checked after every 50 hours of operation. Adjust engine idle speed to 1300 rpm for Model KF24 and 1200 rpm for all other models at idle stop screw (IS—Fig. KW51). On Model KF53, readjust idle speed to 1300 rpm with idle limit screw (IL—Fig. KW52) on throttle plate after initial idle speed adjustment as previously described. On all models, adjust maximum speed limit screw (ML) so maximum engine speed is 4200 rpm.

CAUTION: Maximum engine speed must not exceed 4200 rpm even during adjustment procedure. Running engine at speeds in excess of 4200 rpm may result in engine damage and possible injury to operator.

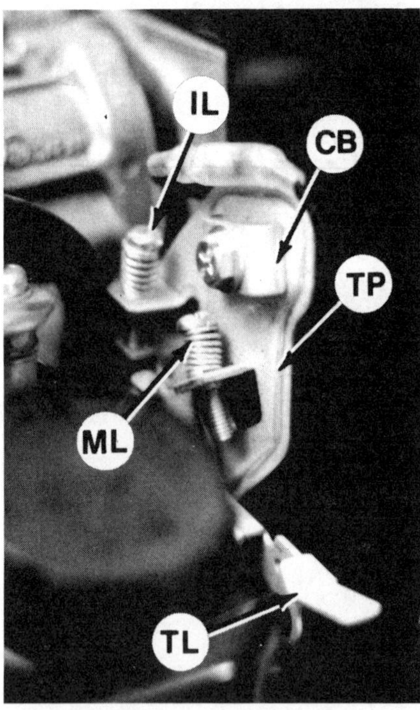

Fig. KW52—View of throttle plate assembly (TP) showing idle limit screw (IL) and maximum speed limit screw (ML). Throttle lever (TL) and remote throttle cable clamp (CB) are also shown.

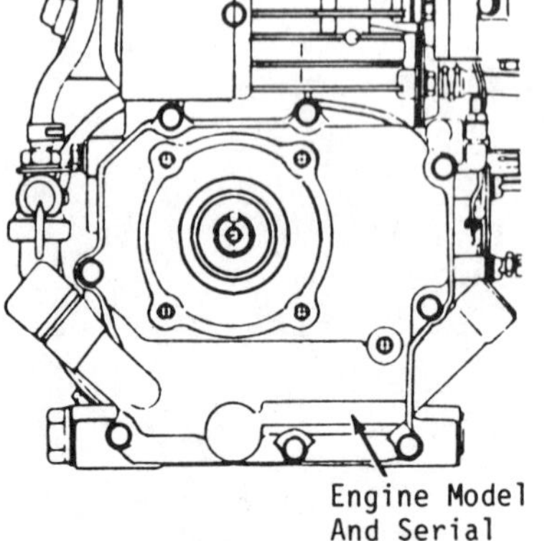

Fig. KW50—Engine model and serial number are located on crankcase cover.

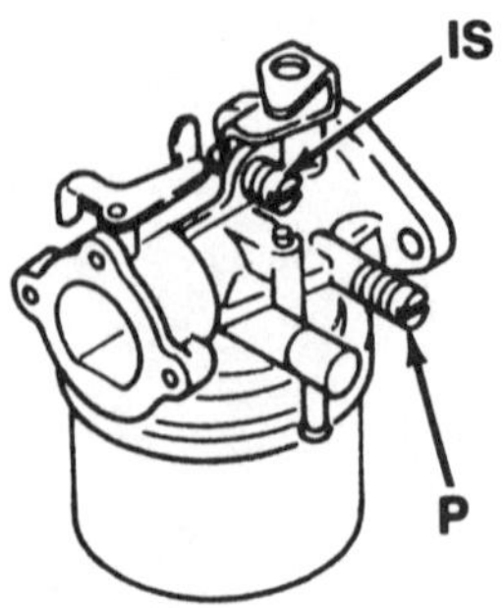

Fig. KW51—Model KF24 float type carburetor. Throttle stop screw (IS) and pilot air screw (P) adjustment procedure is outlined in text. Models KF34, KF53 and KF68 are similar.

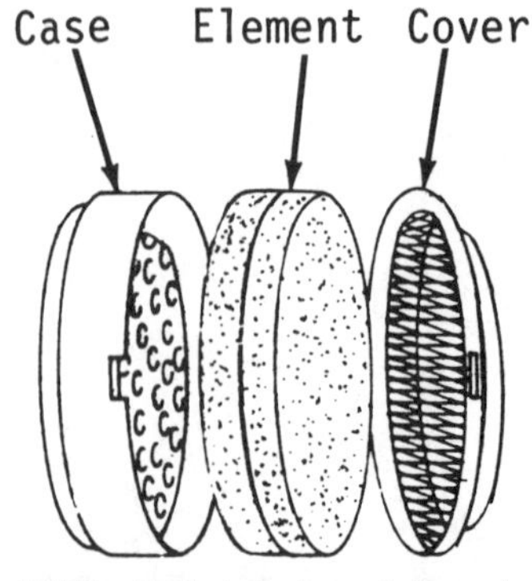

Fig. KW53—Exploded view of foam type air cleaner. Refer to text for service procedure and interval.

For initial carburetor adjustment, lightly seat pilot air screw (P—Fig. KW51) and open pilot screw 1 turn on Models KF24 and KF68, 7/8 turn on Model KF34 and 3/4 turn on Model KF53. Make final adjustment with engine at operating temperature and running. Back out idle limit screw on throttle plate so it does not limit idle speed. Adjust idle stop screw (IS—Fig. KW51) so engine runs at lowest speed possible. Adjust pilot screw to obtain highest engine speed possible. Readjust idle stop screw and idle limit screw for correct speeds.

To check float level, invert carburetor throttle body and float assembly. Float surface should be parallel to carburetor throttle body. If float is equipped with metal tang which contacts inlet needle, carefully bend tang to adjust float. If float is a plastic assembly, float level is nonadjustable and float must be renewed if float level is incorrect.

AIR FILTER. Engines are equipped with a foam type air cleaner (Fig. KW53). Foam type air cleaner should be removed and cleaned after every 50 hours of operation under normal operating conditions. Remove foam element and clean in a nonflammable solvent. Carefully squeeze out solvent and allow element to air dry. Soak element in clean engine oil and squeeze out the excess. Reinstall element making certain filament (hair-like protrusions) side is toward cover (Fig. KW53).

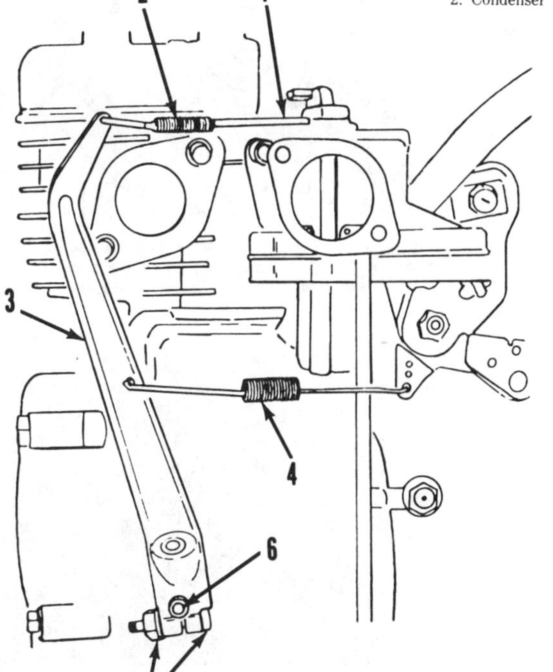

Fig. KW54—View of external governor linkage. Refer to text for adjustment procedure.
1. Governor-to-carburetor rod
2. Spring
3. Governor lever
4. Tension spring
5. Clamp bolt
6. Governor shaft

GOVERNOR. The mechanical flyweight type governor is located internally in engine with flyweight assembly mounted on camshaft gear. To adjust external governor linkage, stop engine. Make certain all linkage is in good condition and tension spring (4—Fig. KW54) is not stretched. Spring (2) must pull governor lever (3) and throttle arm toward each other. Loosen clamp bolt (5) and push governor lever to fully open throttle. Rotate governor shaft (6) counterclockwise as far as possible while holding governor lever so throttle is in full speed position. Tighten clamp bolt (5).

IGNITION SYSTEM. Ignition coil on Models KF24, KF34 and KF68 is located behind flywheel (Fig. KW55) while ignition coil on Model KF53 is located externally adjacent to flywheel (Fig. KW56). Breaker-point set and condenser are located behind flywheel on all models. Ignition timing is 23 degrees BTDC and nonadjustable. Breaker-point gap should be set at 0.3-0.5 mm (0.012-0.020 in.). Ignition coil edge gap for Model KF53 should be 0.5 mm (0.010 in.) at each side of coil.

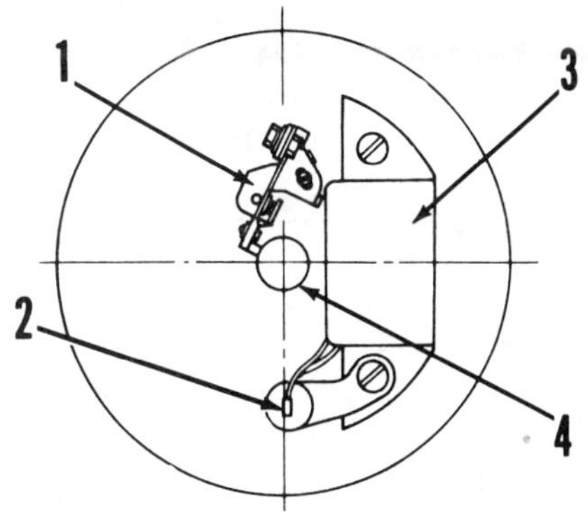

Fig. KW55—Ignition coil on Models KF24, KF34 and KF68 is located under flywheel.
1. Breaker points
2. Condenser
3. Coil
4. Crankshaft

VALVE ADJUSTMENT. Valves and seals should be refaced and stem clearance adjusted after every 200 hours of operation. Refer to REPAIRS section for service procedure.

CYLINDER HEAD AND COMBUSTION CHAMBER. Cylinder head, combustion chamber and piston should be cleaned and carbon and other deposits removed after every 300 hours of operation. Refer to REPAIRS section for service procedure.

LUBRICATION. All models are splash lubricated and engine oil should be checked prior to each operating interval. Oil level should be maintained at top edge of reference marks on gage with oil fill plug just touching first threads. Do not screw oil fill plug and gage in to check oil level (Fig. KW57).

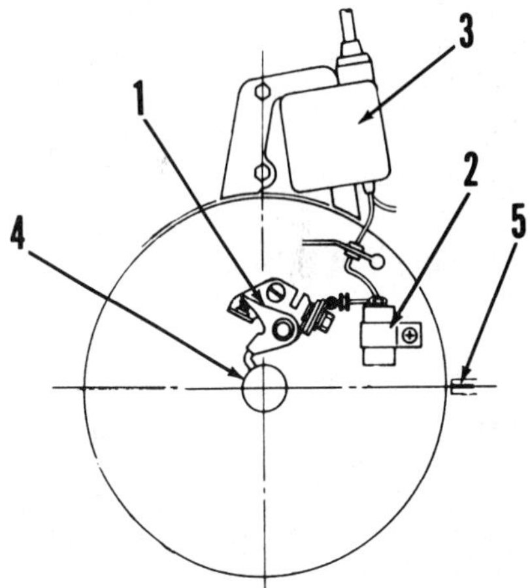

Fig. KW56—Ignition coil on Model KF53 is located just above flywheel.
1. Breaker points
2. Condenser
3. Coil
4. Crankshaft
5. Timing pointer

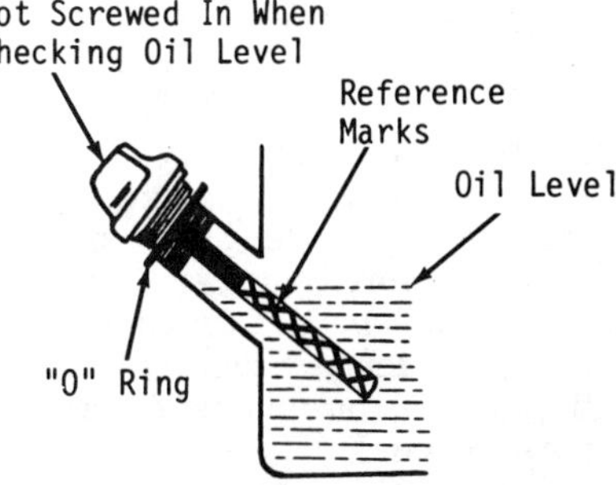

Fig. KW57—View of oil fill plug and gage. Do not screw plug in to check oil level and make certain "O" ring is in place on plug after checking oil level.

Manufacturer recommends oil with an API service classification SC, SD or SE. Use SAE 20 oil in winter and SAE 30 oil in summer.

Oil should be changed after the first 20 hours of operation and every 50 hours of operation thereafter. Crankcase capacity is 450 cc (0.48 qt.) on Model KF24, 500 cc (0.53 qt.) on Model KF34, 700 cc (0.74 qt.) on Model KF53 and 900 cc (0.95 qt.) on Model KF68.

GENERAL MAINTENANCE. Check and tighten all loose bolts, nuts or clamps prior to each day of operation. Check for fuel or oil leakage and repair if necessary.

Clean dust, dirt, grease or any foreign material from cylinder head and cylinder block cooling fins after every 100 hours of operation. Inspect fins for damage and repair if necessary.

REPAIRS

TIGHTENING TORQUES. Recommended tightening torques are as follows:

Spark plug.................27 N·m (20 ft.-lbs.)
Head bolts:
KF53.................23-24 N·m (17-18 ft.-lbs.)
KF68.............23.5-24.5 N·m (17-18 ft.-lbs.)
All other models.......19-20 N·m (14-15 ft.-lbs.)
Connecting rod bolts:
KF24..................8-10 N·m (6-7 ft.-lbs.)
KF34................15-16 N·m (11-12 ft.-lbs.)
KF53.................16-18 N·m (12-13 ft.-lbs.)
KF68..............18.6-19.6 N·m (14-15 ft.-lbs.)
Flywheel nut:
KF24.................39-43 N·m (29-32 ft.-lbs.)
KF68.............58.8-63.7 N·m (43-47 ft.-lbs.)
All other models.......57-62 N·m (42-46 ft.-lbs.)
Crankcase cover bolts........5 N·m (4 ft.-lbs.)

CYLINDER HEAD. To remove cylinder head, first remove cylinder head shroud. Clean engine to prevent entrance of foreign material. Loosen cylinder bolts in ¼ turn increments following sequence shown in Fig. KW58 until all bolts are loose enough to remove by hand.

Remove spark plug and clean carbon and other deposits from cylinder head. Place cylinder head on a flat surface and check entire sealing surface for warpage. If warpage exceeds 0.25 mm (0.010 in.), cylinder head must be renewed. Slight warpage may be repaired by lapping cylinder head. In a figure eight pattern, lap head against 200 grit and then 400 grit emery paper on a flat surface.

Reinstall cylinder head and tighten bolts evenly to specified torque following sequence shown in Fig. KW58.

CONNECTING ROD. Connecting rod rides directly on crankshaft journal on all models. Piston and connecting rod are removed as an assembly after cylinder head and crankcase cover have been removed. Remove the two connecting rod bolts (9—Fig. KW59), lockplate (8) and connecting rod cap (7). Rotate crankshaft so piston can be pushed out of cylinder and remove piston and rod assembly. Carefully remove rings (1 and 2), piston pin retaining rings (4), piston pin (5) and separate piston (3) from connecting rod (6).

Inspect connecting rod and renew if piston pin bore or crankpin bearing bore are scored or damaged.

If clearance between piston pin and connecting rod piston pin bore exceeds 0.05 mm (0.0020 in.), renew connecting rod. If clearance between crankpin journal and connecting rod bearing bore exceeds 0.07 mm (0.0028 in.), renew connecting rod and/or crankshaft. Correct connecting rod side play on crankpin should be 0.5 mm (0.002 in.) for Models KF24, KF34 and KF53 and 0.7 mm (0.003 in.) for Model KF68.

Piston crown is marked with a "D", "G", "L" or "R." Pistons marked with a "D" or "R" are for engines with clockwise crankshaft rotation as viewed from flywheel side. Pistons marked with a "G" or "L" are for engines with counterclockwise crankshaft rotation as viewed from flywheel side. On all models, install piston on connecting rod with letter mark on piston crown on the same side as the Japanese characters stamped on connecting rod.

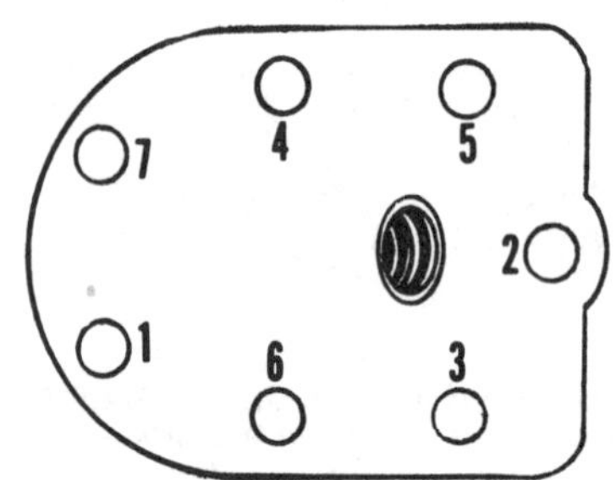

Fig. KW58—Tighten or loosen head bolts in sequence shown. Refer to text for correct torque specification.

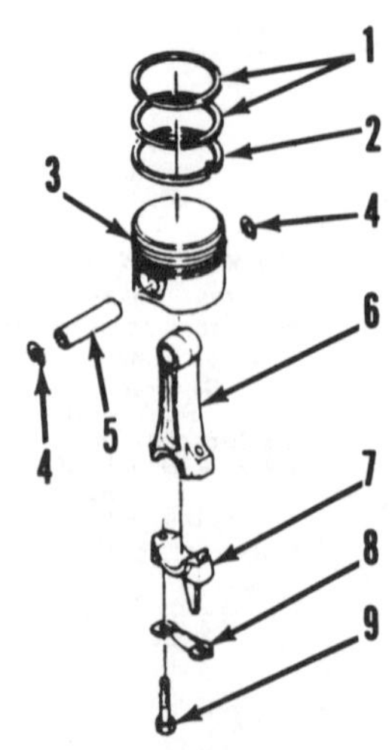

Fig. KW59—Exploded view of piston and rod assembly.

1. Compression rings
2. Oil control ring
3. Piston
4. Retaining rings
5. Piston pin
6. Connecting rod
7. Connecting rod cap
8. Lock plate
9. Bolts

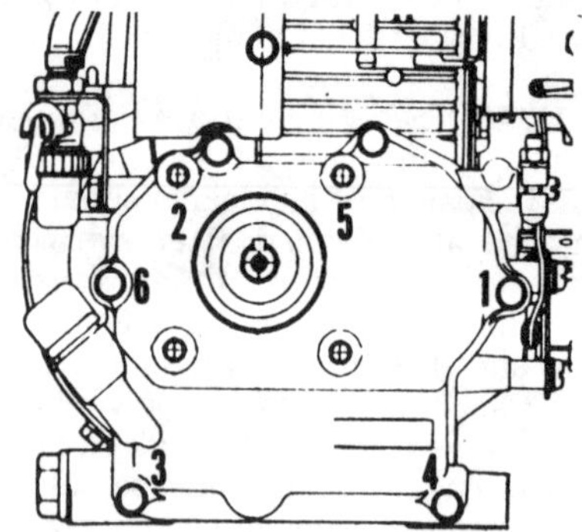

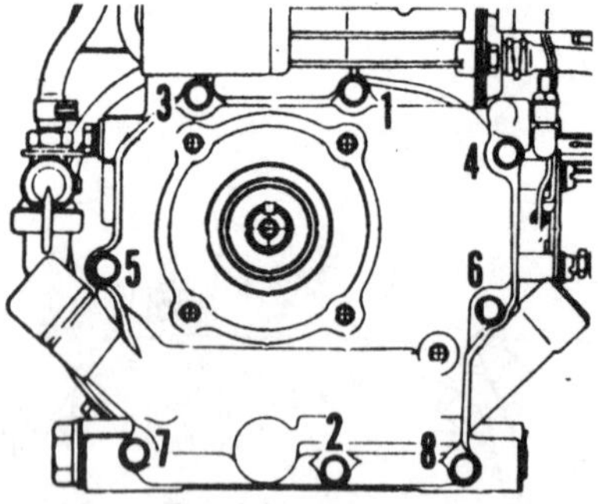

Fig. KW60—Tighten crankcase cover bolts to specified torque following sequence shown in relation to number of retaining bolts.

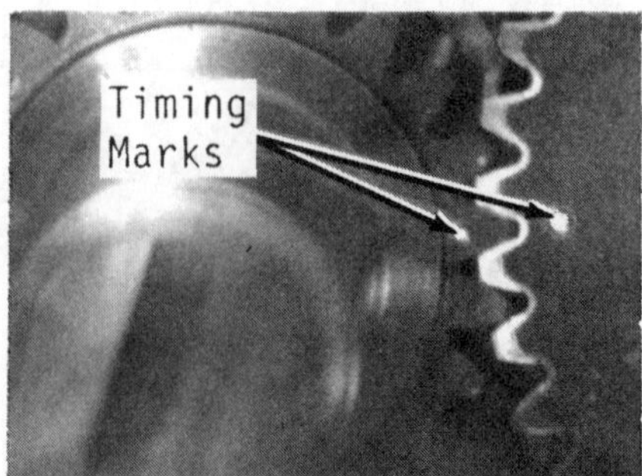

Fig. KW61—Timing marks on crankshaft and camshaft gear must be aligned during installation.

Install piston and connecting rod assembly into cylinder so Japanese characters on connecting rod (Fig. KW64) are toward flywheel side of engine. Align match marks on connecting rod and connecting rod cap and install cap. Install oil dipper so dipper is on open side of crankcase. Tighten connecting rod bolts to specified torque.

PISTON, PIN AND RINGS. Piston and connecting rod are removed as an assembly. Refer to CONNECTING ROD section for removal and installation procedure.

After carefully removing piston rings and separating piston and connecting rod, carefully clean carbon and other deposits from piston surface and ring lands.

CAUTION: Extreme care should be exercised when cleaning ring lands. Do not damage squared edges or widen ring grooves. If ring lands are damaged, piston must be renewed.

Before installing rings on piston, install piston into cylinder and use a suitable feeler gage to measure clearance between piston and cylinder bore. If clearance exceeds 0.25 mm (0.010 in.), renew piston and/or cylinder.

If ring side clearance between new ring and piston groove exceeds 0.15 mm (0.006 in.), renew piston. If ring end gap on new rings exceeds 1.0 mm (0.04 in.) when installed, renew cylinder bore.

If clearance between piston pin and piston pin bore exceeds 0.05 mm (0.002 in.), renew piston.

Refer to CONNECTING ROD section to properly assemble piston on connecting rod.

CYLINDER AND CRANKCASE. Cylinder and crankcase are an integral casting. Standard cylinder bore diameter is 56 mm (2.20 in.) for Model KF24, 60 mm (2.36 in.) for Model KF34, 66 mm (2.60 in.) for Model KF53 and 74 mm (2.91 in.) for Model KF68. If cylinder bore diameter exceeds these dimensions by 0.15 mm (0.006 in.), renew cylinder.

If crankcase cover has been removed, refer to following CRANKSHAFT, MAIN BEARINGS AND SEALS for bearing and seal installation. Tighten crankcase cover bolts to specified torque following sequence shown in Fig. KW60.

CRANKSHAFT, MAIN BEARINGS AND SEALS. Crankshaft is supported at each end in a ball type main bearing. To remove crankshaft, remove all metal shrouds, flywheel, cylinder head and crankcase cover. Remove connecting rod and piston assembly and remove crankshaft and camshaft. Remove valve tappets.

Minimum crankpin diameter is 22.5 mm (0.886 in.) for Model KF24, 24.5 mm (0.965 in.) for Model KF34, 26 mm (1.024 in.) for Model KF53 and 28 mm (1.10 in.) for Model KF68. If dimensions are not as specified, renew crankshaft.

Ball bearing type main bearings are a light press fit on crankshaft and maximum bearing play is 0.30 mm (0.012 in.). It may be necessary to slightly heat crankcase cover to remove or install ball bearings. Crankcase and crankcase cover oil seals should be pressed into seal bores until flush or slightly below flush.

To reinstall crankshaft, reverse

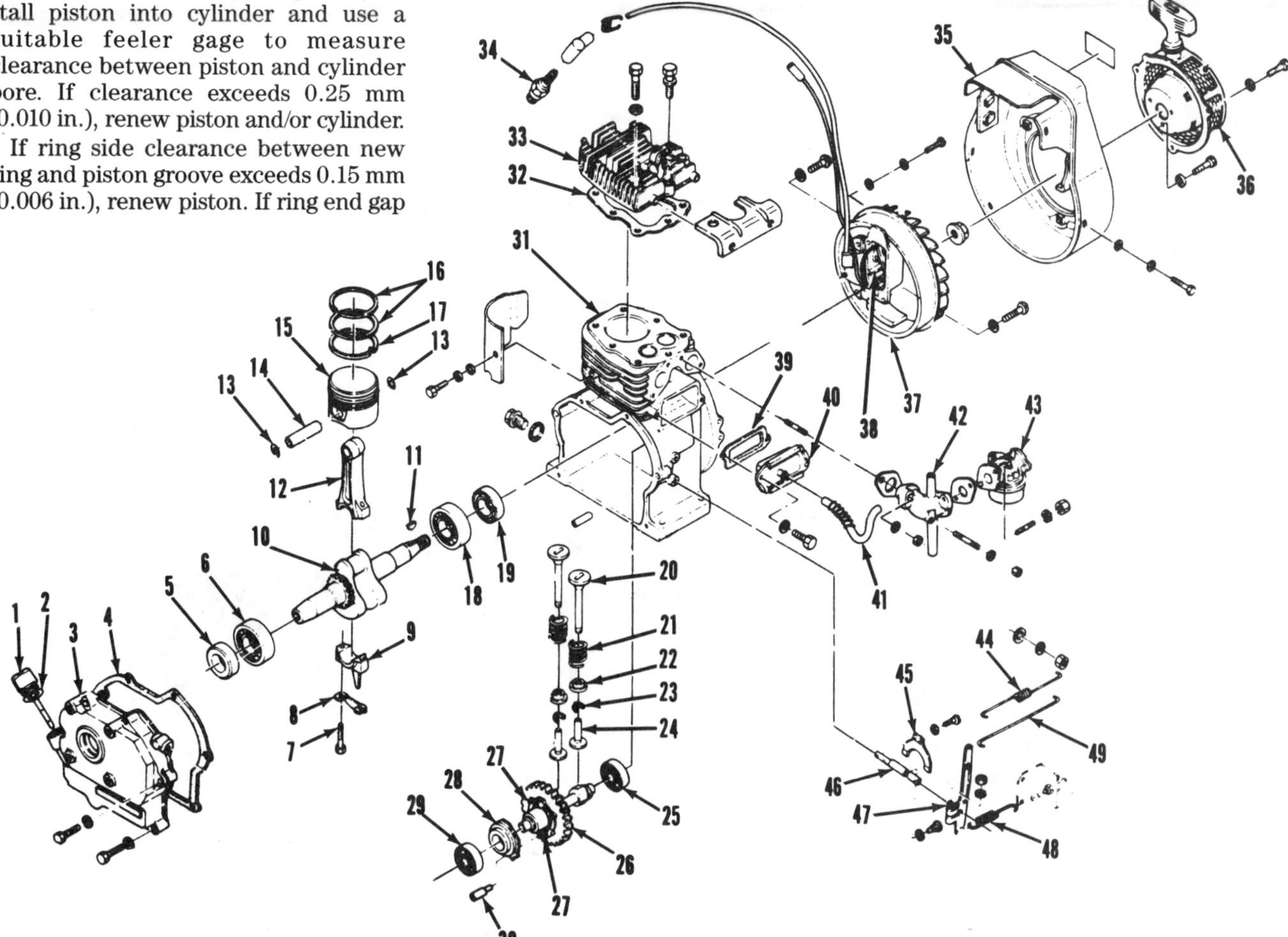

Fig. KW62—Exploded view of Model KF24 engine.

1. Oil plug & gage
2. "O" ring
3. Crankcase cover
4. Gasket
5. Seal
6. Main bearing
7. Connecting rod bolt
8. Lock plate
9. Connecting rod cap
10. Crankshaft
11. Key
12. Connecting rod
13. Retaining rings
14. Piston pin
15. Piston
16. Compression rings
17. Oil control ring
18. Main bearing
19. Seal
20. Valves
21. Valve springs
22. Valve spring retainers
23. Valve keepers
24. Valve tappets
25. Cam bearing
26. Camshaft & gear assy.
27. Governor weights
28. Governor sleeve
29. Cam bearing
30. Governor sleeve pin
31. Cylinder block (crankcase)
32. Gasket
33. Cylinder head
34. Spark plug
35. Cooling shroud
36. Rewind starter
37. Flywheel
38. Points, condenser & ignition coil
39. Gasket
40. Valve chamber cover
41. Breather tube
42. Manifold
43. Carburetor
44. Spring
45. Governor yoke
46. Governor yoke shaft
47. Governor lever
48. Tension spring
49. Governor-to-carburetor rod

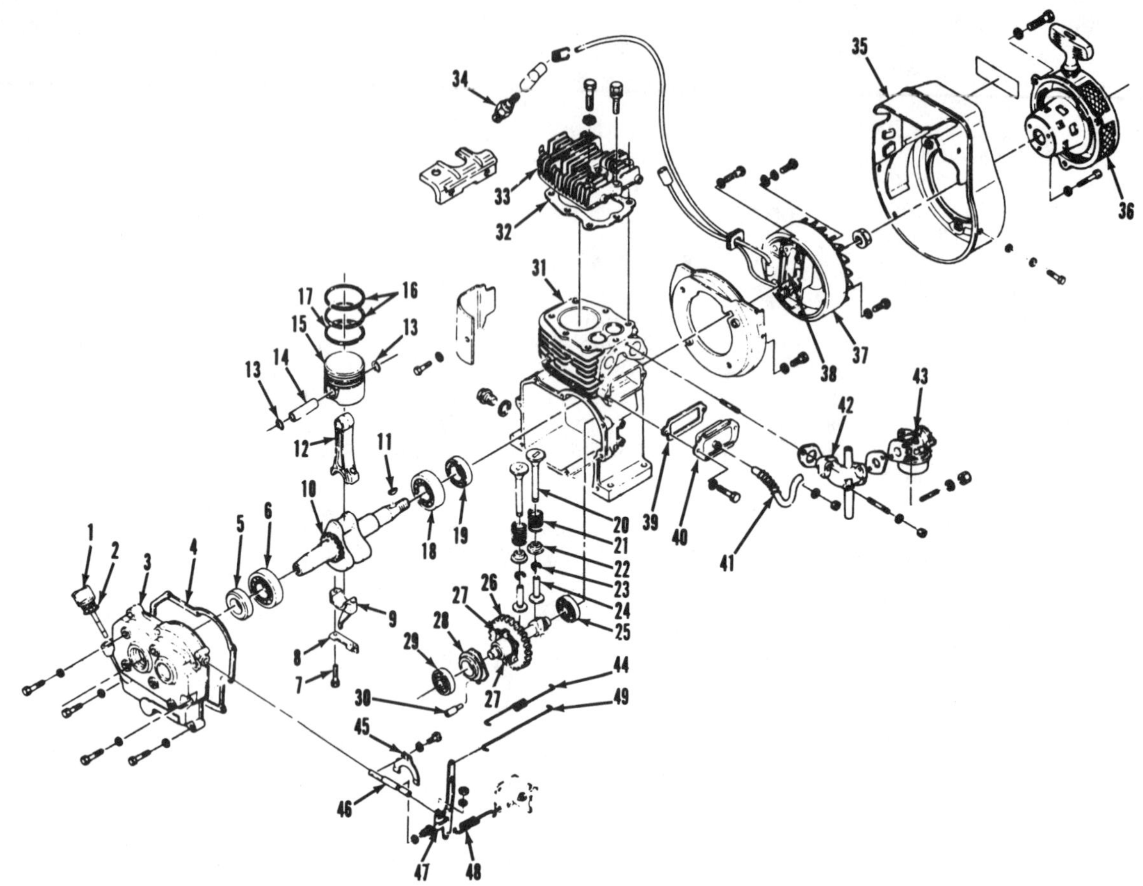

Fig. KW63—Exploded view of Model KF34 and KF68. Model KF53 is similar except ignition coil is located outside flywheel and valve retainers (22) are also the valve keepers.

1. Oil plug & gage
2. "O" ring
3. Crankcase cover
4. Gasket
5. Seal
6. Main bearing
7. Connecting rod bolt
8. Lock plate
9. Connecting rod cap
10. Crankshaft
11. Key
12. Connecting rod
13. Retaining rings
14. Piston pin
15. Piston
16. Compression rings
17. Oil control ring
18. Main bearing
19. Seal
20. Valves
21. Valve springs
22. Valve spring retainers
23. Valve keepers
24. Valve tappets
25. Cam bearing
26. Camshaft & gear assy.
27. Governor weights
28. Governor sleeve
29. Cam bearing
30. Governor sleeve pin
31. Cylinder block (crankcase)
32. Gasket
33. Cylinder head
34. Spark plug
35. Cooling shroud
36. Rewind starter
37. Flywheel
38. Points, condenser & ignition coil
39. Gasket
40. Valve chamber cover
41. Breather tube
42. Manifold
43. Carburetor
44. Spring
45. Governor yoke
46. Governor yoke shaft
47. Governor lever
48. Tension spring
49. Governor-to-carburetor rod

removal procedure making sure crankshaft and camshaft gear timing marks are aligned during installation. Refer to Fig. KW61.

CAMSHAFT AND BEARINGS. Camshaft is supported at each end in ball bearing type cam bearings which are a slight press fit in crankcase and crankcase cover. To remove camshaft, refer to previous CRANKSHAFT, MAIN BEARINGS AND SEALS. It may be necessary to slightly heat crankcase or crankcase cover to remove or install bearings.

Make certain camshaft lobes are smooth and free of scores, scratches and damage. Make certain camshaft and crankshaft gear timing marks are aligned during installation. Refer to Fig. KW61.

GOVERNOR. The mechanical flyweight governor is located internally in engine with flyweight assembly mounted on camshaft gear. Refer to GOVERNOR paragraph in MAINTENANCE section for external governor adjustments.

Refer to previous CRANKSHAFT, MAIN BEARINGS AND SEALS for camshaft and governor removal procedure.

VALVE SYSTEM. Clearance between valve stem and valve tappet (cold) on all models should be 0.12 mm (0.005 in.). If clearance is not as specified, valves must be removed and end of stems ground off to increase clearance or seats ground deeper to reduce clearance.

Valve face and seat angles are 45 degrees. Correct valve seat width for all models is 1.0 mm (0.04 in.).

Valve spring free length should be 30.9 mm (1.22 in.) for Model KF24, 33 mm (1.30 in.) for Models KF34 and KF53 and 37.8 mm (1.49 in.) for Model KF68. If valve spring free length is 30 mm (1.18 in.) or less for Model KF24, 32 mm (1.26 in.) or less for Model KF34 and KF53 or 36.8 mm (1.45 in.) or less for Model KF68, renew valve spring.

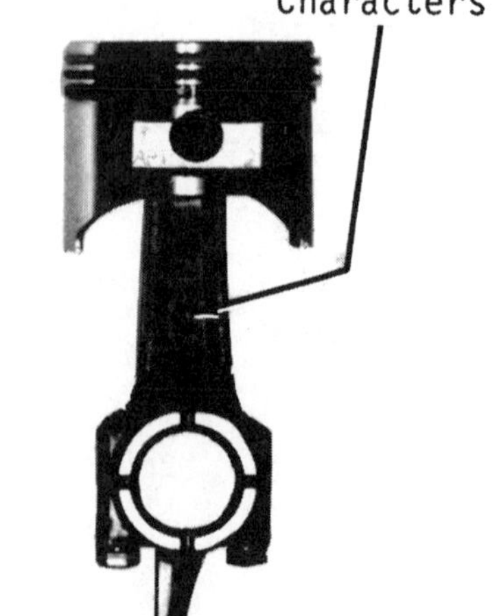

Fig. KW64—Side of connecting rod which has Japanese character markings must be toward flywheel side of engine after installation.

KAWASAKI

Model	Bore	Stroke	Displacement
FG150	64 mm (2.51 in.)	47 mm (1.85 in.)	151 cc (9.2 cu. in.)
FG200	71mm (2.79 in.)	51mm (2.01 in.)	201 cc (12.3 cu. in.)

ENGINE IDENTIFICATION

Kawasaki FG150 and FG200 models are four-stroke, single-cylinder side valve gasoline engines. Model FG150 rated output is 2.1 kW (2.8 hp) at 3600 rpm with a maximum output of 2.7 kW (3.6 hp) at 4000 rpm. Model FG200 rated output is 2.8 kW (3.8 hp) at 3600 rpm with a maximum output of 3.7 kW (5.0 hp) at 4000 rpm.

Engine model number is located on the shroud adjacent to rewind starter and engine model and serial number are located on crankcase cover (Fig. KW70). Always furnish engine model and serial numbers when ordering parts or service information.

Fig. KW70—View showing location of engine model and serial numbers.

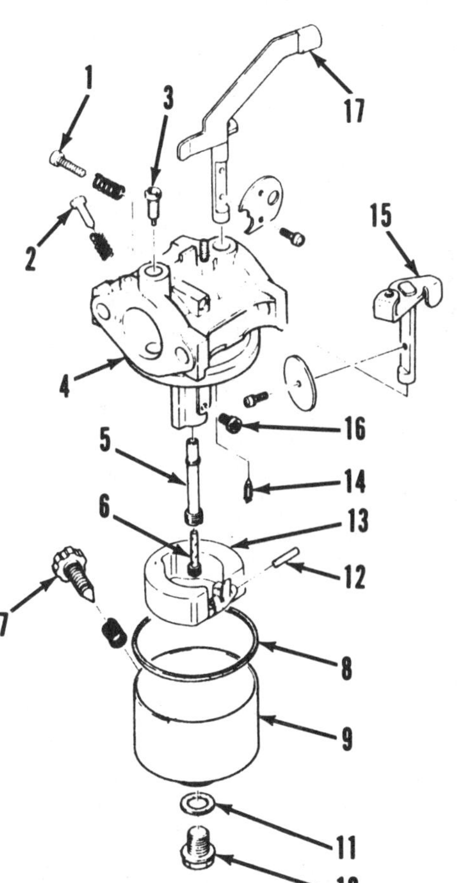

Fig. KW71—Exploded view of carburetor used on Model FG150 engine.

1. Idle speed screw
2. Pilot screw
3. Pilot jet
4. Carburetor body
5. Nozzle
6. Bleed tube
7. Drain screw
8. Gasket
9. Float bowl
10. Plug
11. Gasket
12. Float pin
13. Float
14. Fuel inlet needle
15. Throttle shaft
16. Main jet
17. Choke shaft

MAINTENANCE

SPARK PLUG. Recommended spark plug for all models is a NGK BP4HS, or equivalent.

Spark plug should be removed, cleaned and electrode gap set at 0.6-0.7 mm (0.024-0.027 in.) after every 100 hours of operation. Renew spark plug if electrode is severely burnt or damaged.

NOTE: Caution should be exercised if abrasive type spark plug cleaner is used. Inadequate cleaning procedure may allow the abrasive cleaner to be deposited in engine cylinder accelerating wear and part failures.

CARBURETOR. All models are equipped with a float type side draft carburetor. Carburetor adjustment should be checked after every 50 hours of operation or whenever poor or erratic performance is noted.

Refer to Fig. KW71 for exploded view of carburetor used on Model FG150 and to Fig. KW72 for exploded view of carburetor used on Model FG200.

Engine idle speed, no-load, is 1300 rpm. Adjust idle speed by turning throttle stop screw (1–Fig. KW73) clockwise to increase idle speed or counterclockwise to decrease idle speed. Initial adjustment of pilot screw (2) from a lightly seated position is 7/8 turn open for Model FG150 or 1 turn open for Model FG200.

On all models, final adjustment is made with engine at operating temperature and running. Adjust pilot screw to obtain smoothest idle and acceleration. Main fuel mixture is controlled by a fixed jet.

Standard pilot jet size is #37.5 for Model FG150 and #45 for Model FG200. Standard main jet size is #72.5 for Model FG150 and #87.5 for Model FG200.

Float should be parallel to carburetor float bowl mating surface when float tab just does touch valve needle. Carefully bend float tab to adjust.

FUEL FILTER. A fuel filter screen is located in sediment bowl below fuel shut-off valve (Fig. KW74). Sediment bowl and screen should be cleaned after every 50 hours of operation. To remove sediment bowl, shut off fuel valve. Unscrew sediment bowl and clean bowl and

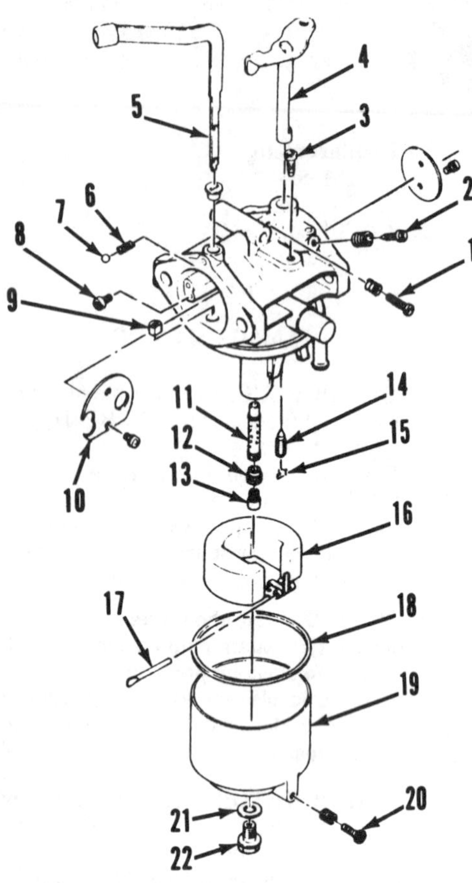

Fig. KW72—Exploded view of carburetor used on Model FG200 engine.

1. Idle speed screw
2. Pilot screw
3. Pilot air jet
4. Throttle shaft
5. Choke shaft
6. Spring
7. Ball
8. Air jet
9. Collar
10. Choke valve
11. Nozzle
12. Main jet holder
13. Main jet
14. Fuel inlet needle
15. Clip
16. Float
17. Float pin
18. Gasket
19. Float bowl
20. Drain screw
21. Gasket
22. Bolt

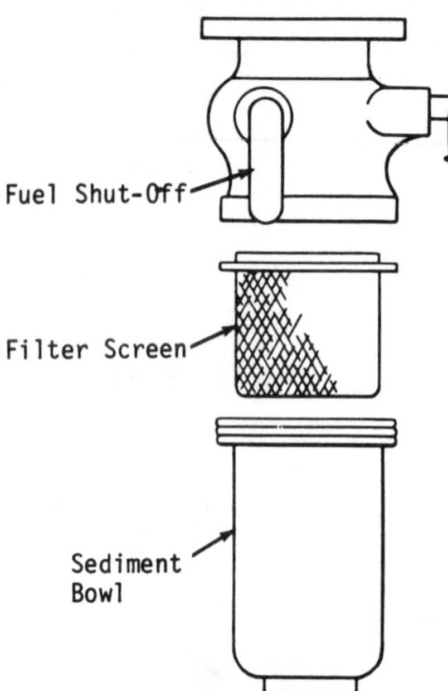

Fig. KW74—Exploded view of fuel filter assembly used on most models.

filter screen. Make certain bowl gasket is in position before bowl installation.

AIR FILTER. The air filter element should be removed and cleaned after the first 10 hours of operation and every 50 hours of operation thereafter.

To remove filter elements (2 and 3–Fig. KW75), unsnap air filter cover and pull out elements. Clean elements in nonflammable solvent, squeeze out excess solvent and allow to air dry. Soak elements in SAE 30 motor oil and squeeze out excess oil. Reinstall by reversing removal procedure.

GOVERNOR. A gear driven flyweight governor assembly is located inside engine crankcase. To adjust external linkage, place engine throttle control in idle position. Make certain all linkage is in good condition and that tension spring (4—Fig. KW76) is not stretched. Spring (2) around governor-to-carburetor rod must pull governor lever (3) and throttle lever toward each other. Loosen clamp bolt (5) and rotate governor shaft (6) clockwise as far as possible. Hold shaft in this position and pull governor lever (3) as far right as possible. Tighten clamp bolt (5). Install tension spring (4) in correct hole for speed desired.

IGNITION SYSTEM. All models are equipped with transistor ignition system and regular maintenance is not required. Ignition timing is nonadjustable. Ignition coil is located outside flywheel. Ignition coil edge air gap should be 0.3 mm (0.01 in.).

To test ignition coil, remove cooling shrouds and ignition coil. Connect one ohmmeter lead to coil core and remaining lead to high tension (spark plug) terminal. Ohmmeter reading should be 12.0-15.0 ohms. Remove the lead connected to coil core and reconnect lead to primary terminal. Ohmmeter reading should be 0.58-0.72 ohms. If readings are not as specified, renew coil.

To test control unit, disconnect all electrical leads and connect positive

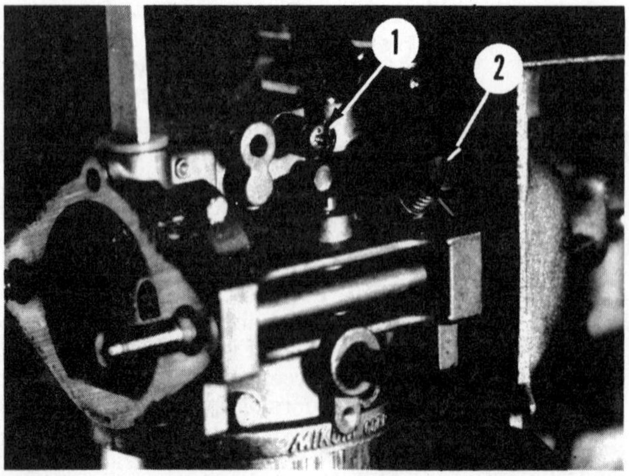

Fig. KW73—View of carburetor used on Model FG200 showing location of throttle stop screw (1) and pilot screw (2). Carburetor for Model FG150 is similar.

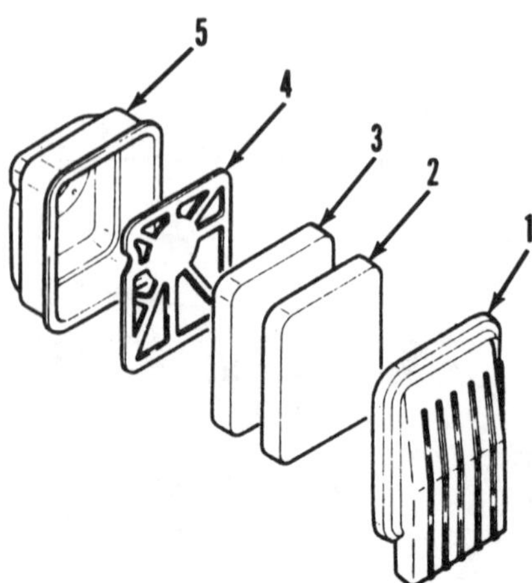

Fig. KW75—Exploded view of air cleaner assembly. Refer to text for service procedure and intervals.

1. Cover
2. Element
3. Element
4. Plate
5. Housing (case)

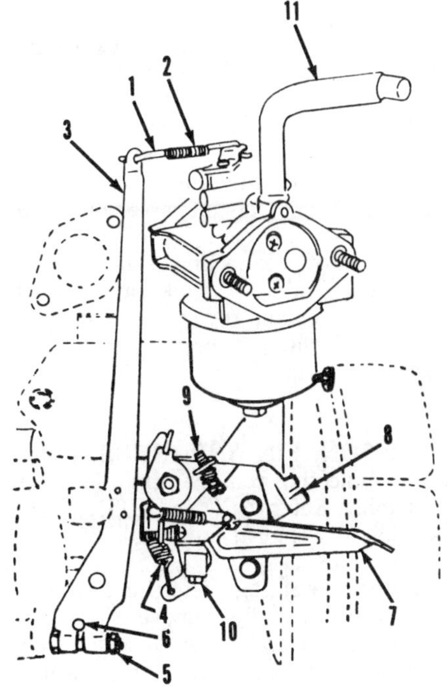

Fig. KW76—View of external governor linkage. Refer to text for adjustment procedure.

1. Governor-to-carburetor rod
2. Spring
3. Governor lever
4. Tension spring
5. Clamp bolt
6. Governor shaft
7. Throttle lever
8. Throttle plate
9. Maximum speed limit screw
10. Idle speed limit screw
11. Throttle pivot

ohmmeter lead to terminal and negative lead to case. Ohmmeter reading should be 10-40 ohms. Reverse lead positions. Ohmmeter reading should be 400-3000 ohms.

If ohmmeter readings are not as specified, renew control unit.

VALVE ADJUSTMENT. Valves and seats should be refaced and stem clearance adjusted after every 300 hours of operation. Refer to REPAIRS section for service procedure and specifications.

CYLINDER HEAD AND COMBUSTION CHAMBER. Cylinder head, combustion chamber and piston should be cleaned and carbon and other deposits removed after every 300 hours of operation. Refer to REPAIRS section for service procedure.

LUBRICATION. Engine oil level should be checked prior to each operating interval. Oil should be maintained between reference marks on dipstick with dipstick just touching first threads. Do not screw dipstick in to check oil level (Fig. KW77).

Manufacturer recommends oil with an API service classification SC. Use SAE 10W-30 oil when temperature is below 35° C (95° F) and SAE 40 oil when temperature is above 35° C (95° F).

Oil should be changed after the first 20 hours of operation and every 100 hours

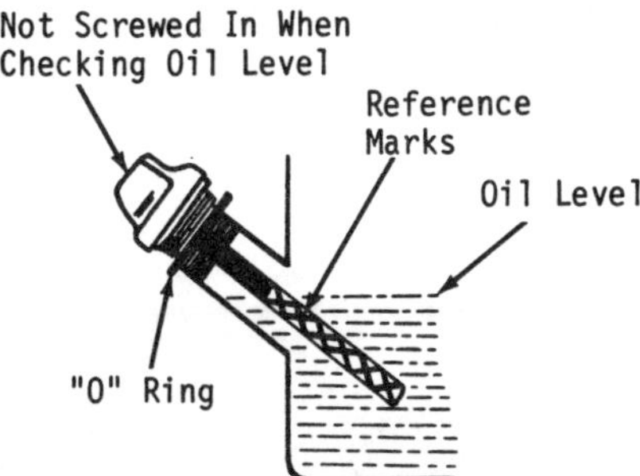

Fig. KW77—Oil plug and gage should not be screwed into crankcase when checking oil level. Refer to text.

of operation thereafter. Crankcase capacity is 500 cc (1.05 pt.) for Model FG150 and 700 cc (1.48 pt.) for Model FG200.

GENERAL MAINTENANCE. Check and tighten all loose bolts, nuts or clamps prior to each day of operation. Check for fuel or oil leakage and repair if necessary.

Clean dust, dirt, grease or any foreign material from cylinder head and cylinder block cooling fins after every 100 hours of operation. Inspect fins for damage and repair if necessary.

REPAIRS

TIGHTENING TORQUES. Recommended tightening torques are as follows:

Spark plug	27 N·m (20 ft.-lbs.)
Head bolts	23-24 N·m) (17-18 ft.-lbs.)
Connecting rod bolts	19-20 N·m (14-15 ft.-lbs.)
Crankcase cover bolts	9-11 N·m (6-8 ft.-lbs.)
Flywheel nut	59-64 N·m (44-47 ft.-lbs.)

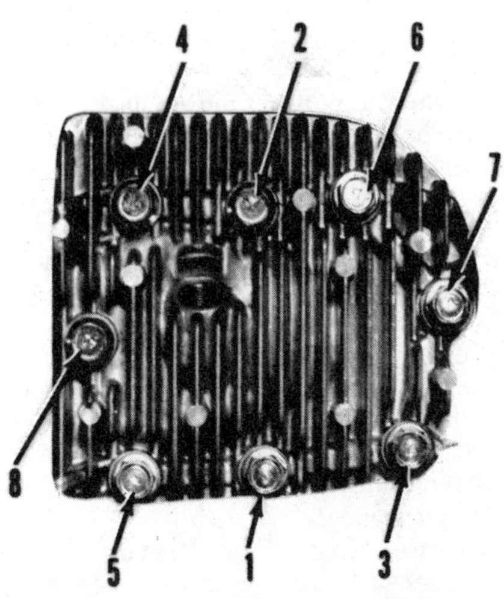

Fig. KW78—Tighten and loosen cylinder head bolts to specified torque in sequence shown.

CYLINDER HEAD. To remove cylinder head, first remove cylinder head shroud. Clean engine to prevent entrance of foreign material. Loosen cylinder bolts in ¼ turn increments in sequence shown in Fig. KW78 until all bolts are loose enough to remove by hand.

Remove spark plug and clean carbon and other deposits from cylinder head. Place cylinder head on a flat surface and check entire sealing surface for warpage. If warpage exceeds 0.25 mm (0.010 in.), cylinder head must be renewed. Slight warpage may be repaired by lapping cylinder head. In a figure eight pattern, lap head against 200 grit and then 400 grit emery paper on a flat surface.

Reinstall cylinder head and tighten bolts evenly to specified torque in sequence shown in Fig. KW78.

CONNECTING ROD. Connecting rod rides directly on crankshaft journal on all models. Piston and connecting rod are removed as an assembly after removing cylinder head and splitting crankcase. Remove the two connecting rod bolts (9 – Fig. KW79) and connecting rod cap (8). Push piston and connecting rod assembly about through the top of cylinder block. Remove snap rings (4) and push pin (5) out of piston to separate piston (6) from connecting rod (7).

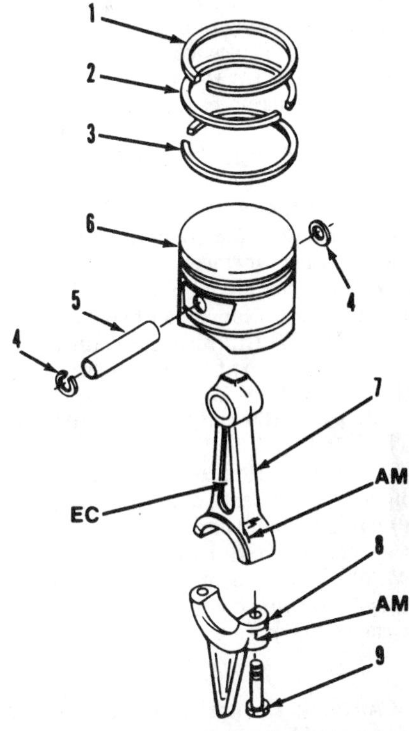

Fig. KW79—Exploded view of piston and connecting rod assembly.

1. Compression ring
2. Compression ring
3. Oil control ring
4. Retaining ring
5. Piston pin
6. Piston
7. Connecting rod
8. Connecting rod cap
9. Connecting rod bolts
EC. Engine code numbers
AM. Alignment marks

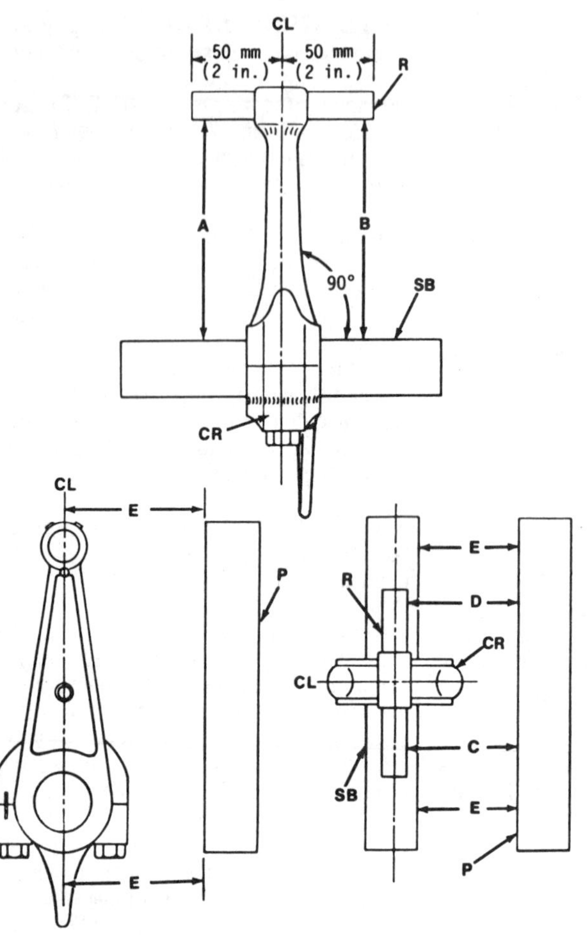

Fig. KW80—Refer to text for procedure used to measure connecting rod to assure it is suitable for service.

Before installing any connecting rod, check the following dimensions to assure it is suitable for service.

Small End Inside Diameter. Maximum inside diameter for connecting rod small end is 13.043 mm (0.5135 in.) for Model FG150 or 15.027 mm (0.5916 in.) for Model FG200.

Big End Inside Diameter. Maximum inside diameter for connecting rod big end bearing surface is 24.553 mm (0.9667 in.) for Model FG150 or 27.050 mm (1.0650 in.) for Model FG200.

Big End Width. Measure connecting rod big end width at flat thrust surfaces. Minimum width is 23.30 mm (0.9173 in.) for Model FG150 and 24.30 mm (0.9566 in.) for Model FG200.

Connecting Rod Bend. To measure connecting rod bend, obtain a 100 mm (4 in.) straight round bar which is a snug fit in connecting rod piston pin bore (Fig. KW80). Mount connecting rod on a straight bar which is the same diameter as the crankshaft connecting rod journal.

CAUTION: Do not distort connecting rod big end bearing bore by over tightening connecting rod cap on a bar which is to large in diameter.

Center line of rod should be at a 90° angle to bar on which connecting rod is mounted (Fig. KW80). Center the 100 mm (4 in.) bar in connecting rod piston pin bore and measure dimensions (A) and (B). Dimensions must not vary by more than 0.12 mm (0.0047 in.).

Connecting Rod Twist. Mount connecting rod and center a 100 mm (4 in.) straight round bar in connecting rod piston pin bore as outlined in Connecting Rod Bend paragraphs. Position a straight, flat plate as shown in (P–Fig. KW80). When dimension (E) at all locations is equal, measure dimensions (C) and (D). Dimensions (C) and (D) must not vary by more than 0.12 mm (0.0047 in.).

Connecting Rod Side Play. With crankshaft removed from crankcase, mount connecting rod on crankpin journal. Use suitable feeler gage to measure clearance between flat connecting rod big end thrust surface and crankshaft. Clearance must not exceed 0.7 mm (0.0018 in.).

Connecting Rod/Crankpin Clearance. Clearance between connecting rod bearing surface and crankshaft journal must not exceed 0.10 mm (0.0040 in.). Use micrometers or Plastigage to determine clearance.

Connecting Rod/Piston Pin Clearance. Clearance between connecting rod small end bushing bore and piston pin must not exceed 0.05 mm (0.002 in.).

If dimensions from any check are not as specified, connecting rod must be renewed.

Piston must be reinstalled on connecting rod so "R" on top of piston is on the unmarked side of the connecting rod. Install rod and piston assembly with engine code numbers (EC–Fig. KW79) on side of connecting rod opposite flywheel end of crankshaft. Match marks on rod and rod cap (AM) must be aligned. Tighten connecting rod bolts to specified torque.

PISTON, PIN AND RINGS. Piston and connecting rod are removed as an assembly. Refer to CONNECTING ROD section for removal and installation procedure.

After separating piston and connecting rod, carefully remove rings and clean carbon and other deposits from piston surface and ring lands.

CAUTION: Extreme care should be exercised when cleaning ring lands. Do not damage squared edges or widen ring grooves. If ring lands are damaged, piston must be renewed.

Measure piston diameter 6 mm (0.24 in.) above, and 90° from, piston pin center lines. If diameter is less than 63.810 mm (2.2526 in.) for Model FG150 or less than 70.810 mm (2.7878 in.) for Model FG200, renew piston.

Before installing rings, install piston in cylinder bore and use a suitable feeler gage to measure clearance between piston and cylinder bore. If clearance exceeds 0.25 mm (0.010 in.) renew piston and/or recondition cylinder bore.

Piston pin outside diameter should be 12.986 mm (0.511259 in.) for Model FG150 or 14.977 mm (0.58964 in.) for Model FG200.

If piston pin bore inside diameter exceeds 13.043 mm (0.5135 in.) for Model FG150 or 15.027 mm (0.5916 in.) for Model FG200, renew piston. If top ring groove exceeds 2.120 mm (0.0835 in.), second ring groove exceeds 2.115 mm (0.0832 in.) or bottom ring groove exceeds 3.06 mm (0.1205 in.), renew piston. If piston ring to piston groove side clearance exceeds 0.17 mm (0.0067 in.), renew piston and/or rings.

Piston ring end gap must not exceed 0.80 mm (0.031 in.) for Model FG150 or 1 mm (0.004 in.) for Model FG200 when rings are squarely installed in cylinder bore. If specification is not as specified, renew rings and/or recondition cylinder bore.

Install piston rings which are marked with marked side toward top of piston. Stagger ring end gaps equally around circumference of piston.

Fig. KW81—Crankshaft and camshaft gear timing marks "T" must be aligned during crankshaft or camshaft installation.

CYLINDER AND CRANKCASE. Cylinder and crankcase are an integral casting. Standard cylinder bore diameter is 63.98-64.00 mm (2.5189-2.5196 in.) for Model FG150 or 70.98-71.00 mm (2.7945-2.7953 in.) for Model FG200. Maximum bore out-of-round is 0.045 mm (0.0018 in.) for Model FG150 or 0.50 mm (0.0019 in.) for Model FG200.

When installing crankcase cover, tighten bolts evenly in sequence shown in Fig. KW82.

Several oversize piston and ring sets are available, as well as standard size.

CRANKSHAFT, MAIN BEARINGS AND SEALS. Crankshaft is supported by ball bearings at each end. To remove crankshaft, remove all metal shrouds, flywheel, fan housing and cylinder head. Remove crankcase cover, disconnect connecting rod and remove piston and connecting rod assembly. Turn crankcase upside down so tappets fall away from camshaft journals and remove crankshaft and camshaft. It may be necessary to heat crankcase or crankcase cover slightly to remove crankshaft bearings. Remove crankshaft seals using a suitable puller.

Minimum crankshaft main bearing journal diameter for Model FG150 is 19.963 mm (0.7859 in.). Minimum crankpin diameter is 24.447 mm (0.9625 in.).

Minimum crankshaft main bearing journal diameter for Model FG200 is 24.962 mm (0.9827 in.). Minimum crankpin diameter is 26.95 mm (1.061 in.).

Maximum crankshaft runout when measured at center of crankshaft main bearing journals is 0.045 mm (0.0017 in.) for Model FG150 or 0.50 mm (0.00196 in.) for Model FG200.

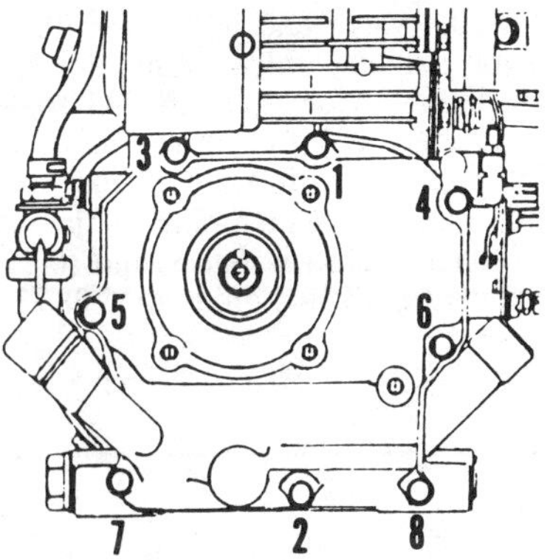

Fig. KW82—Tighten crankcase cover bolts to specified torque in sequence shown.

Ball bearing main bearings must be a light press fit on crankshaft and in bearing bores of crankcase and crankcase cover. Bearings should spin smoothly and have no rough spots. It may be necessary to heat crankcase or crankcase cover slightly to remove or install bearings.

Oil seal in crankcase cover should be pressed in until seal is 5 mm (0.2 in.) below flush with cover seal bore. Oil seal in crankcase should be pressed in until seal is flush to 1 mm (0.004 in.) below flush.

Make certain crankshaft gear and camshaft gear timing marks are aligned during crankshaft installation (Fig. KW81).

CAMSHAFT AND BEARINGS. Camshaft is supported at crankcase end in a bearing which is an integral part of crankcase casting and is supported at crankcase cover end in a ball bearing. Refer to CRANKSHAFT, BEARINGS AND SEALS section for camshaft removal.

On Model FG150, if camshaft lobe height is less than 27.3 mm (1.07 in.) for intake lobe or 27.1 mm (1.06 in.) for exhaust lobe, renew camshaft. Camshaft bearing surface minimum diameter is 14.940 mm (0.58818 in.) for pto side and 14.942 mm (0.58825 in.) for flywheel side. If dimensions are not as specified, renew camshaft.

On Model FG200, if camshaft lobe height is less than 31.7 mm (1.25 in.) for either lobe, renew camshaft. Camshaft bearing surface minimum diameter is 14.966 mm (0.7858 in.) for pto side and 16.942 mm (0.6670 in.) for flywheel side. If dimensions are not as specified, renew camshaft.

Camshaft ball bearing must be a light press fit on camshaft and in crankcase cover. It may be necessary to slightly heat crankcase cover to remove or install ball bearing. Bearing should spin smoothly with no rough spots or looseness.

Camshaft bearing bore maximum diameter in crankcase is 15.043 mm (0.5922 in.) for Model FG150 or 17.043 mm (0.6710 in.) for Model FG200.

Make certain camshaft and crankshaft gear timing marks are aligned during installation (Fig. KW81).

GOVERNOR. The internal centrifugal flyweight governor is gear driven off of the camshaft gear. Refer to GOVERNOR paragraph in MAINTENANCE section for external governor adjustments.

To remove governor assembly, remove external linkage, metal cooling shrouds, flywheel and crankcase cover. Remove governor gear cover mounting screws and remove cover. Lift gear assembly from governor shaft.

To reinstall governor assembly, reverse removal procedure.

LOW OIL SENSOR. Float type low oil sensor is located inside the crankcase (Fig. KW83). To remove or check float gap, it is necessary to remove crankcase cover.

To check oil level sensor switch, remove bearing plate and float cover. Disconnect electrical leads and connect switch leads to ohmmeter leads. Slide float to top of shaft. Ohmmeter should indicate infinite resistance. Slide float slowly down the shaft until ohmmeter just deflects. Gap between top of float and lower nut (Fig. KW83) should be 9.5-15.5 mm (0.37-0.61 in.); if not, renew switch.

VALVE SYSTEM. Clearance between valve stem and valve tappet (cold) should be 0.15 mm (0.006 in.) for intake valve and 0.22 mm (0.009 in.) for exhaust valve. If clearance is not as specified, valves must be removed and end of stems ground off to increase

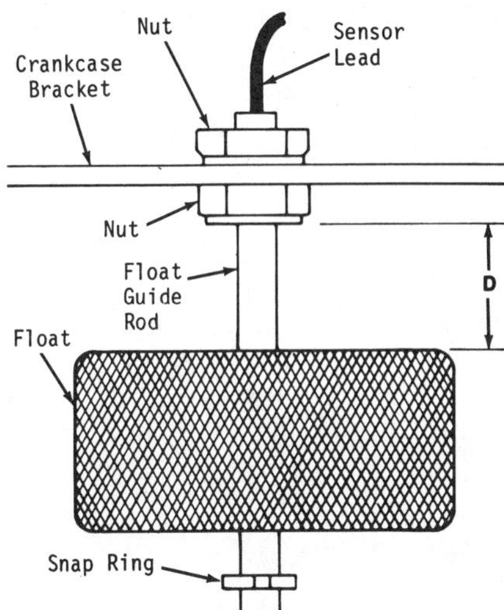

Fig. KW83—Low oil sensor float and switch assembly is mounted on crankcase cover. Float gap "D" must be 9.5-15.5 mm (0.37-0.61 in.). Refer to text.

clearance or seats ground deeper to reduce clearance.

Valve face and seat angle is 45° and maximum valve seat width is 1.0-1.6 mm (0.039-0.063 in.). Minimum valve margin is 0.6mm (0.02 in.).

Minimum valve stem diameter for Model FG150 is 5.948 mm (0.2342 in.) for intake valve and 5.935 mm (0.2336 in.) for exhaust valve. Minimum valve stem diameter for Model FG200 is 6.942 mm (0.2733 in.) for intake valve and 6.952 mm (0.2737 in.) for exhaust valve.

Maximum valve guide inside diameter for Model FG150 is 6.078 mm (0.2393 in.) for intake valve guide or 6.085 mm (0.2395 in.) for exhaust valve guide. Maximum valve guide inside diameter for Model FG200 is 7.072 mm (0.2784 in.) for intake valve guide or 7.075 mm (0.2785 in.) for exhaust valve guide. Worn valve guides may be removed and replaced using a special valve guide tool available from Kawasaki.

KIORITZ-ECHO

ECHO, INC
400 Oakwood Rd.
Lake Zurich, Illinois 60047

Model	Bore	Stroke	Displacement
Kioritz	26.0 mm (1.02 in.)	26.0 mm (1.02 in.)	13.8 cc (0.84 cu. in.)
Kioritz	28.0 mm (1.10 in.)	26.0 mm (1.02 in.)	16.0 cc (0.98 cu. in.)
Kioritz	32.2 mm (1.27 in.)	26.0 mm (1.02 in.)	21.2 cc (1.29 cu. in.)
Kioritz	28.0 mm (1.10 in.)	37.0 mm (1.46 in.)	30.1 cc (1.84 cu. in.)
Kioritz	40.0 mm (1.56 in.)	32.0 mm (1.26 in.)	40.2 cc (2.45 cu. in.)

ENGINE INFORMATION

Kioritz two-stroke, air-cooled, gasoline engines are used by several equipment manufacturers.

MAINTENANCE

SPARK PLUG. Recommended spark plug is a Champion CJ8, or equivalent. Specified electrode gap for all models is 0.6-0.7 mm (0.024-0.28 in.).

CARBURETOR. Kioritz engines may be equipped with a Walbro diaphragm type carburetor with a built in fuel pump (Fig. KZ50) or a Zama diaphragm type carburetor (Fig. KZ52). Refer to appropriate paragraph for model being serviced.

Walbro Diaphragm Type Carburetor. Initial adjustment of fuel mixture needle from a lightly seated position is 1-1/8 turn open for low speed mixture needle (18—Fig. KZ50) and 1-1/4 turn open for high speed mixture needle.

Final adjustments are made with trimmer line at recommended length. Engine should be at operating temperature and running. Turn idle speed screw (31) to obtain 2500-3000 rpm, or just below clutch engagement speed. Adjust low speed mixture needle (18) to obtain consistent idling and smooth acceleration. Readjust idle speed screw (31) as required. Open throttle fully and adjust high speed mixture needle (17) to obtain highest engine rpm, then turn high speed mixture needle counterclockwise 1/8 turn.

To disassemble carburetor, remove the four screws retaining cover (1) to carburetor body. Remove cover (1), diaphragm (2) and gasket (3). Remove screw (4), pin (5), fuel inlet lever (6), spring (8) and fuel inlet needle (7). Remove screw (9) and remove circuit plate (10), check valve (11) and gasket (12). Remove screw (29), cover (28), gasket (27) and diaphragm (26). Remove inlet screen (24). Remove high and low speed mixture needles and springs. Remove throttle plate (13) and shaft (20) as required.

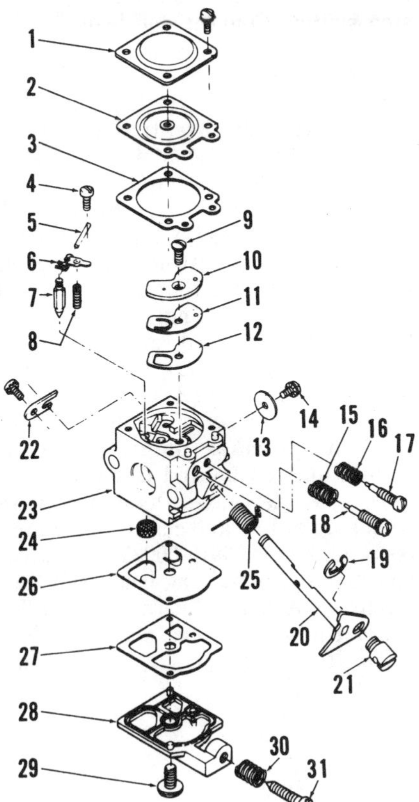

Fig. KZ50—Exploded view of Walbro carburetor used on some models.

1. Cover
2. Metering diaphragm
3. Gasket
4. Metering lever screw
5. Metering lever pin
6. Metering lever
7. Fuel inlet valve
8. Metering lever spring
9. Circuit plate screw
10. Circuit plate
11. Check valve
12. Gasket
13. Throttle valve
14. Shutter screw
15. Spring
16. Spring
17. High speed mixture needle
18. Idle mixture needle
19. "E" clip
20. Throttle shaft
21. Swivel
22. Throttle shaft clip
23. Body
24. Inlet screen
25. Return spring
26. Fuel pump diaphragm
27. Gasket
28. Cover
29. Cover screw
30. Spring
31. Idle speed screw

Carefully inspect all parts. Diaphragms should be flexible and free of cracks or tears. When reassembling, fuel inlet lever (6) should be flush with carburetor body (Fig. KZ51). Carefully bend fuel inlet lever to obtain correct setting.

Zama Diaphragm Type Carburetor. Initial adjustment of low speed (16—Fig. KZ52) and high speed (17) mixture needles is one turn open from a lightly seated position.

Final adjustments are made with trimmer line at recommended length. Engine should be at operating temperature and running. Turn idle speed screw (27) to obtain 2500-3000 rpm, or just below clutch engagement speed. Adjust low speed mixture needle (16) to obtain consistent idling and smooth acceleration. Readjust idle speed screw (27) as required. Open throttle fully and adjust high speed mixture needle (17) to obtain highest engine rpm, then turn high speed mixture needle counterclockwise 1/8 turn.

To disassemble carburetor, remove screw (26), pump cover (25), gasket (24) and diaphragm (23). Remove fuel inlet screen (22). Remove the two screws (1)

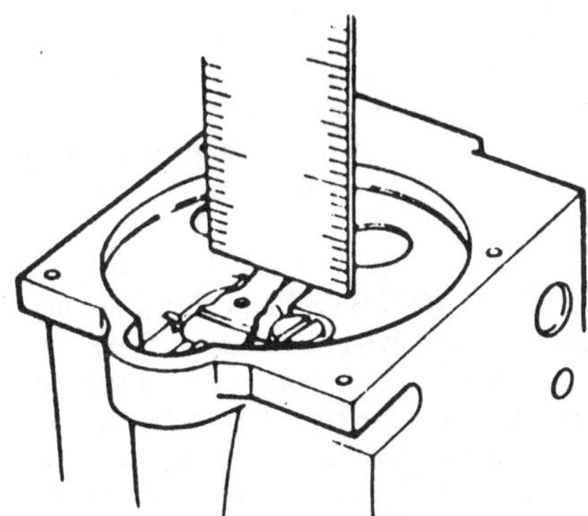

Fig. KZ51—Fuel inlet lever should be flush with carburetor body.

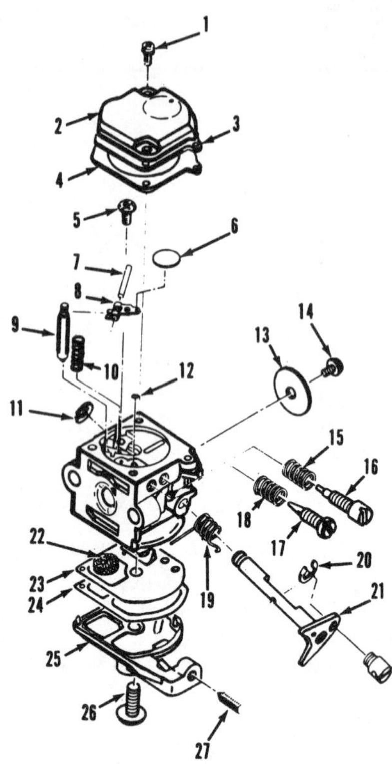

Fig. KZ52—Exploded view of Zama diaphragm carburetor used on some models.

1. Screw
2. Cover
3. Diaphragm
4. Gasket
5. Screw
6. Fuel metering lever disc
7. Pin
8. Fuel inlet lever
9. Fuel inlet needle
10. Spring
11. "E" ring
12. Disc
13. Throttle plate
14. Screw
15. Spring
16. Low speed mixture needle
17. High speed mixture needle
18. Spring
19. Spring
20. "E" ring
21. Throttle shaft
22. Screen
23. Diaphragm
24. Gasket
25. Cover
26. Screw
27. Idle speed screw

and remove cover (2), diaphragm (3) and gasket (4). Remove screw (5), pin (7), metering disc (6), fuel inlet lever (8), fuel inlet needle (9) and spring (10). Remove fuel mixture needles and springs. Remove "E" clip (11), throttle plate (13) and throttle shaft (21) as required.

Inspect all parts for wear or damage. Diaphragms should be flexible with no cracks or wrinkles. Fuel inlet lever (8) with metering disc (6) removed, should be flush with carburetor fuel chamber floor (Fig. KZ51).

IGNITION SYSTEM. Engines may be equipped with a magneto type ignition with breaker points and condenser or a CDI (capacitor-discharge) ignition system which requires no regular maintenance and has no breaker points. Ignition module (CDI) is located outside of flywheel. Refer to appropriate paragraph for model being serviced.

Breaker Point Ignition System. Breaker points may be located behind recoil starter or behind flywheel. Note location of breaker points and refer to following paragraphs for service.

To inspect or service magneto ignition with **breaker points and condenser located behind recoil starter and cover,** first remove fan cover. Refer to Fig. KZ53 and check pole gap between flywheel and ignition coil laminations. Gap should be 0.35-0.40 mm (0.014-0.016 in.). To adjust pole gap, loosen two retaining screws in coil (threads of retaining screws should be treated with Loctite or equivalent), place correct gage between flywheel and coil laminations. Pull coil to flywheel by hand as shown, then tighten screws and remove feeler gage.

Breaker points are accessible after removal of starter case, nut and pawl carrier. Nut and pawl carrier both have left-hand threads.

To adjust point gap and ignition timing, disconnect stop switch wire from stop switch. Connect one lead from a timing tester (Kioritz 990510-00031) or equivalent to stop switch wire. Ground remaining timing tester lead to engine body. Position flywheel timing mark as shown in Fig. KZ54 and set breaker point gap to 0.3-0.4 mm (0.012-0.014 in.). Rotate flywheel and check timing with timing tester. Readjust as necessary. Be accurate when setting breaker point gap. A gap greater than 0.4 mm (0.014 in.) will advance timing; a gap less than 0.3 mm (0.012 in.) will retard timing.

With point gap and flywheel correctly set, timing will be 23 degrees BTDC.

To inspect or service magneto ignition with **breaker points and condenser located behind flywheel,** first remove starter case. Disconnect wire from stop switch and connect to timing tester (Kioritz 990510-00031) or equivalent. Ground remaining lead from timing tester to engine body. Position flywheel timing mark as shown in Fig. KZ55. If timing is not correct, loosen retaining screw on breaker points through flywheel slot openings. Squeeze screwdriver in crankcase groove to adjust point gap. Breaker point gap should be 0.3-0.4 mm (0.012-0.016 in.) with flywheel at TDC.

If breaker points require renewal, flywheel must be removed. When installing breaker points note that stop lead (black) from condenser must pass under coil. White wire from coil must pass between condenser bracket and oiler felt. Push wire down to avoid contact with flywheel.

Ignition coil air gap between coil and flywheel should be 0.4 mm (0.016 in.). Retaining screws should be treated with Loctite or equivalent.

Be accurate when setting breaker point gap. A gap greater than 0.4 mm (0.016 in.) will advance timing; a gap less than 0.3 mm (0.012 in.) will retard timing.

With point gap and flywheel correctly set, timing will be 30 degrees BTDC. Tighten flywheel nut.

Primary and secondary ignition coil resistance may be checked with an ohmmeter without removing coil.

To check primary coil resistance, connect ohmmeter to disconnected ignition coil primary lead and ground remaining lead to engine. Primary coil resistance should register 0.5-0.8 ohms.

To check secondary coil resistance connect one lead of ohmmeter to spark plug cable and remaining lead to ground on engine. Secondary coil resistance should register 5-10 ohms.

CD Ignition System. CD ignition system requires no regular maintenance and has no moving parts except for magnets cast in flywheel.

Ignition timing is fixed at 30 degrees BTDC and clearance between CDI module laminations and flywheel should be 0.3 mm (0.012 in.). To check ignition timing, refer to Fig. KZ56. If timing is not correct, CDI module must be renewed.

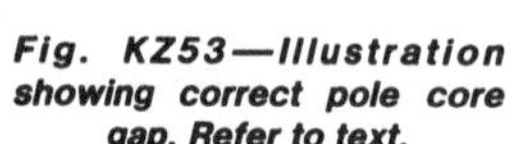
Fig. KZ53—Illustration showing correct pole core gap. Refer to text.

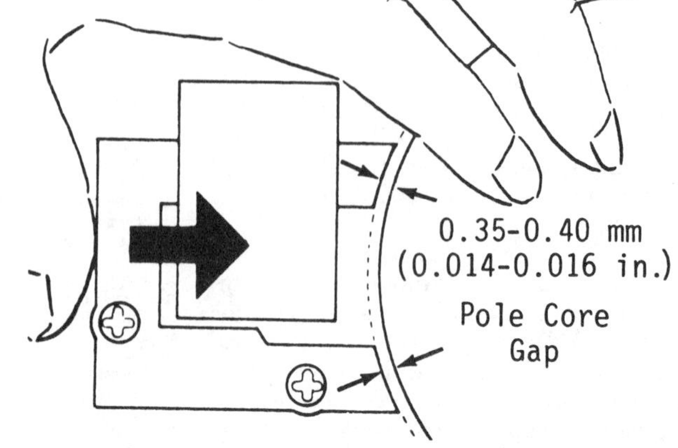

Ignition coil may be checked by connecting one ohmmeter lead to spark plug cable and remaining lead to coil primary lead; ohmmeter should register 0.0-0.2 ohms for all models. Connect ohmmeter lead to coil primary lead (red) wire and remaining ohmmeter lead to either the coil exciter lead (grounded wire), if so equipped, or to coil laminations. Ohmmeter should register 0.5-1.5k ohms for models with exciter lead or 1.5-3.0k ohms for remaining models. Connect one ohmmeter lead to exciter lead, of models so equipped, and remaining ohmmeter lead to coil laminations; ohmmeter should register 120-200 ohms.

CARBON. Muffler and cylinder exhaust ports should be cleaned periodically to prevent loss of power due to carbon build-up. Remove muffler cover and baffle plate and scrape muffler free of carbon. With muffler cover removed, position engine so piston is at top dead center and carefully remove carbon from exhaust ports with wooden scraper. Be careful not to damage the edges of exhaust ports or to scratch piston. Do not attempt to run engine without muffler baffle plate or cover.

LUBRICATION. Engine lubrication is obtained by mixing a good quality two-stroke, air-cooled engine oil with gasoline. Manufacturer recommends mixing regular grade gasoline (unleaded is an acceptable substitute) with a good quality two-stroke air-cooled engine oil at a 25:1 ratio. Do not use fuel containing alcohol.

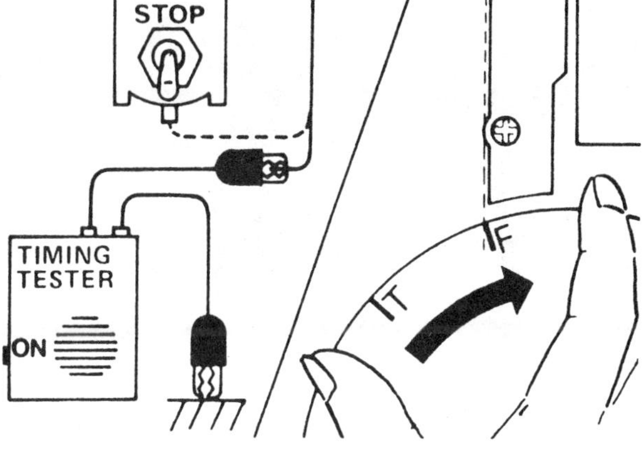

Fig. KZ54—Illustration showing proper timing mark setting for engines with breaker points located behind recoil starter and cover. Refer to text.

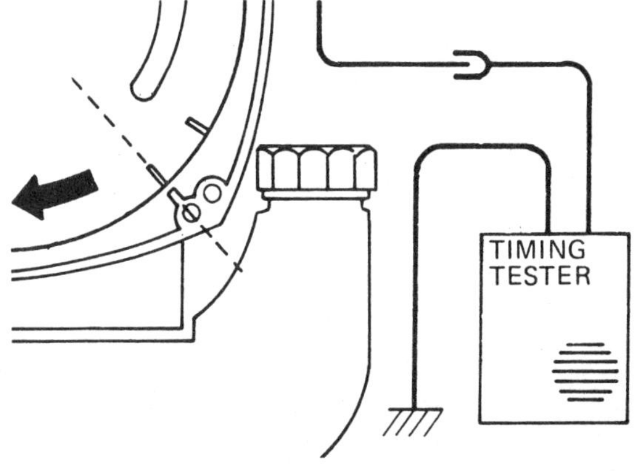

Fig. KZ55—Illustration showing proper timing mark setting for engines with breaker points located behind flywheel. Refer to text.

REPAIRS

COMPRESSION PRESSURE. For optimum performance, minimum compression pressure should be 621 kPa (90 psi). Compression test should be performed with engine cold and throttle and choke plates at wide open positions.

TIGHTENING TORQUES. Recommended tightening torque specifications are as follows:

Spark plug:
All models . . . . 14-15 N·m (168-180 in.-lbs.)

Cylinder cover:
All models so equipped 1.5-1.9 N·m (13-17 in.-lbs.)

Cylinder:
13.8, 16.0 & 21.2 cc engine . . . .5.7-6.0 N·m (50-55 in.-lbs.)
All other engines . . . .7.0-8.0 N·m (65-75 in.-lbs.)

Crankcase:
All models . . . . .4.5-5.7 N·m (40-50 in.-lbs.)

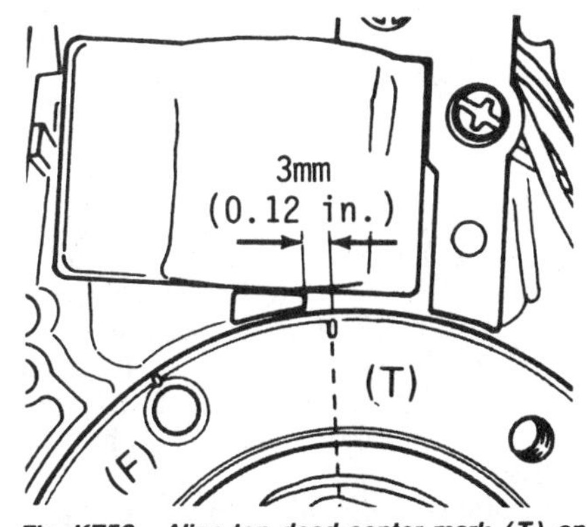

Fig. KZ56—Align top dead center mark (T) on flywheel with laminations as shown to check timing on models with CD ignition. Refer to text.

Fig. KZ57—Exploded view of engine similar to the 13.8, 16.0 and 21.2 cc (0.84, 0.98 and 1.29 cu. in.) displacement engines.

1. Clutch drum
2. Hub
3. Ignition module
4. Clutch springs
5. Clutch shoes
6. Plate
7. Flywheel
8. Seal
9. Key
10. Crankcase
11. Gasket
12. Bearing
13. Crankshaft & connecting rod assy.
14. Bearing
15. Crankcase cover
16. Seal
17. Ratchet assy.
18. Nut
19. Thrust washer
20. Bearing
21. Thrust washer
22. Retainer
23. Piston pin
24. Piston
25. Ring
26. Gasket
27. Cylinder

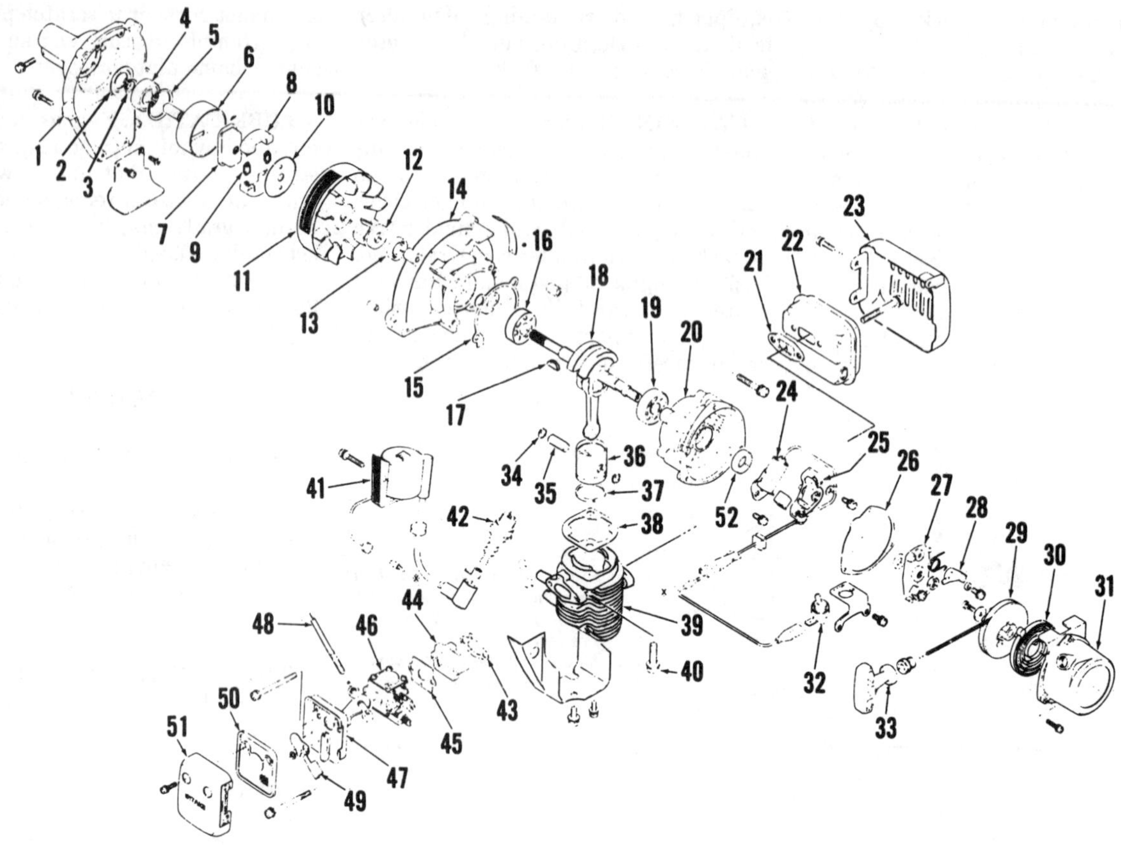

Fig. KZ58 — Exploded view of engine similar to 30.1 and 40.2 cc (1.84 and 2.45 cu. in.) displacement engines.

1. Fan cover
2. Stopper
3. Snap ring
4. Ball bearing
5. Snap ring
6. Clutch drum
7. Clutch hub
8. Clutch shoe
9. Clutch spring
10. Side plate
11. Flywheel/fan
12. Spacer
13. Seal
14. Crankcase
15. Crankcase packing
16. Ball bearing
17. Woodruff key
18. Crankshaft/connecting rod assy.
19. Ball bearing
20. Crankcase half
21. Gasket
22. Muffler
23. Cover
24. Condenser
25. Breaker points
26. Gasket
27. Pawl carrier
28. Pawl
29. Starter hub
30. Spring
31. Starter housing
32. Stop switch
33. Handle
34. Snap ring
35. Piston pin
36. Piston
37. Piston ring
38. Gasket
39. Cylinder
40. Screw
41. Coil assy.
42. Spark plug
43. Gasket
44. Insulator
45. Gasket
46. Carburetor
47. Case
48. Throttle cable
49. Choke shutter
50. Air filter
51. Cover
52. Seal

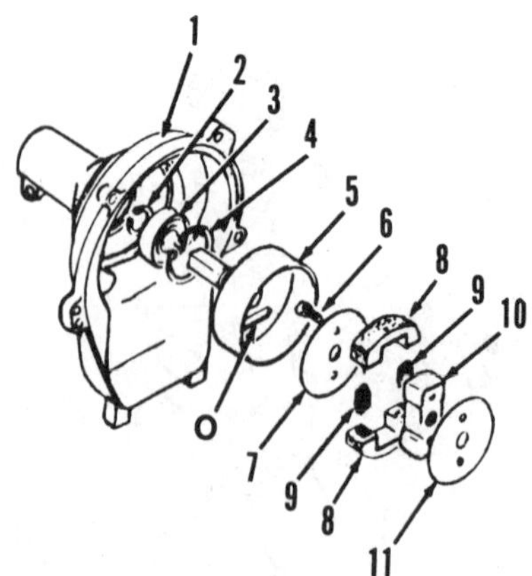

Fig. KZ59—Exploded view of clutch assembly.

0. Opening
1. Clutch housing
2. Snap ring
3. Bearing
4. Snap ring
5. Clutch drum
6. Screw
7. Plate
8. Clutch shoes
9. Clutch springs
10. Hub
11. Plate

CRANKSHAFT AND CONNECTING ROD. The crankshaft and connecting rod are serviced as an assembly only and crankshaft and connecting rod are not available separately.

To remove crankshaft and connecting rod assembly remove fan housing and clutch assembly. Remove carburetor, muffler and fuel tank. Remove starter assembly, flywheel and ignition module. Carefully remove cylinder (27—Fig. KZ57). Remove retaining rings (22) and use a suitable piston pin puller to remove pin (23). Note thrust washers (19 and 21) are installed on some engines and will fall when piston pin and piston are removed. Separate crankcase halves and remove crankshaft and connecting rod assembly. Carefully press ball bearings (12 and 14) off crankshaft as required. If bearings remain in crankcase, it may be necessary to heat crankcase halves slightly to remove bearings.

Inspect all parts for wear or damage. Standard connecting rod side clearance on crankpin journal is 0.55-0.60 mm (0.022-0.026 in.). If side clearance is 0.7 mm (0.028 in.) or more, crankshaft and connecting rod assembly must be renewed. Crankshaft runout should not exceed 0.05 mm (0.002 in.). It may be necessary to heat crankcase halves slightly to install ball bearings (12 and 14).

PISTON, PIN AND RINGS. Rings are pinned to prevent ring rotation on all models. Piston may be equipped with one or two compression rings according to model and application.

Standard ring end gap is 0.3 mm (0.012 in.). If ring end gap is 0.5 mm (0.02 in.) or more, renew rings and/or cylinder.

Standard ring side clearance in piston groove is 0.06 mm (0.0024 in.). If ring side clearance is 0.10 mm (0.004 in.)

or more, renew ring and/or piston.

When installing piston, rings must be correctly positioned over pins in ring grooves. Install piston on connecting rod with arrow on top of piston towards exhaust side of engine. Make certain piston pin retaining rings are correctly seated in piston pin bore of piston. Use care when installing piston into cylinder that cylinder does not rotate.

CYLINDER. The cylinder bore has a chrome plated surface and must be renewed if plating is worn through or scored excessively. Worn spots will appear dull and may be easily scratched while chrome plating will be bright and much harder.

CLUTCH. Refer to Fig. KZ59 for an exploded view of centrifugal clutch assembly used on 13.8, 16.0 and 21.2 cc (0.84, 0.98 and 1.29 cu. in.) engines. Snap ring (4) is removed from snap ring bore through opening (O) in clutch drum. Drive clutch drum (5) out and remove snap ring (2). Remove ball bearing (3). Check shoes for wear or oil, check clutch springs for fatigue and renew parts as necessary. If removal of clutch hub is necessary, it is screwed on crankshaft with left-hand threads.

To remove clutch from all other engines, remove fan cover. Remove screw retaining clutch drum to crankshaft. Remove ball bearing from clutch drum. Check shoes for wear or oil, check clutch springs for fatigue and renew parts as necessary.

SERVICING KIORITZ-ECHO ACCESSORIES

REWIND STARTER

To dissemble rewind starter, remove starter case from engine. Pull starter rope out all the way and tie a temporary knot at the middle. Pull knot out of handle then untie or cut rope. Remove handle, untie temporary knot and allow rope to wind into starter.

NOTE: Do not release rope suddenly or damage to spring may occur.

Remove rope pulley and spring and check for damage and renew if necessary. Note direction of spring wrap during disassembly. Rewind spring from outside as shown in Fig. KZ65.

After assembly, grease both sides of spring, then install spring with inner end loop of spring approximately 2.03 mm (0.08 in.) from starter case center post and spring wrapping in correct direction as shown in Fig. KZ66. Wind rope in correct direction onto pulley. Grease center post, install pulley and starter case and secure in position with washer and screw. Use Loctite or equivalent on screw. Rotate pulley three turns in appropriate direction to tension spring. Hold pulley and pass rope through guide and handle and knot rope.

Removal and assembly of starter pawls is obvious.

Use Loctite or equivalent on screws securing pawls.

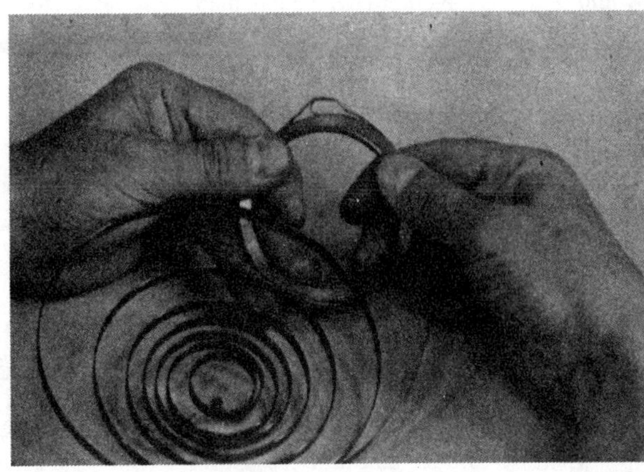

Fig. KZ65—View showing starter rewind spring being wound clockwise. Note direction of spring wrap during disassembly. Spring on some models is wound counterclockwise.

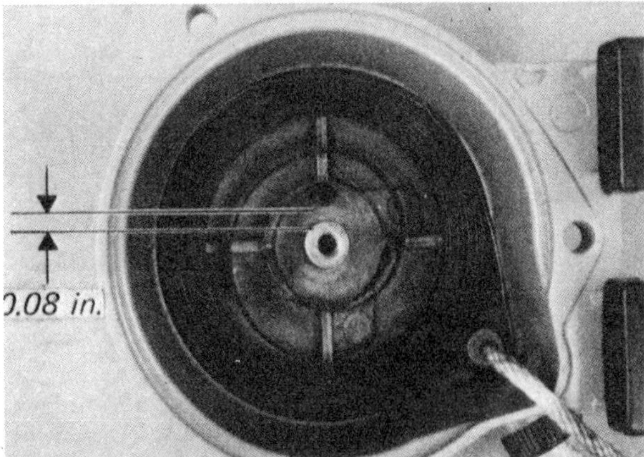

Fig. KZ66—View showing rewind spring correctly installed on models with clockwise spring wrap. Starter on all other models is similar except spring is installed counterclockwise.

KOHLER

KOHLER COMPANY
444 Highland Drive
Kohler, Wisconsin 53044

Model	Bore	Stroke	Displacement
K90, K91	2⅜ in. (60.3 mm)	2 in. (50.8 mm)	8.9 cu. in. (145 cc)

ENGINE IDENTIFICATION

Kohler engine identification and serial number decals are located on engine shrouding (Fig. KO1). The model number designates displacement (in cubic inches), digit after displacement on late models indicates number of cylinders and the letter suffix indicates the specific version. No suffix indicates a rope start model. Therefore, K91 would indicate a Kohler (K), 9 cubic inch (9), single-cylinder (1) engine with rope start (no suffix). Suffix interpretation is as follows:

C	Clutch model
A	Oil pan type
EP	Generator set
P	Pump model
R	Gear reduction
S	Electric start
T	Retractable start
G	Housed with fuel tank
H	Housed without fuel tank

Always furnish engine model, specification and serial numbers when ordering parts.

MAINTENANCE

SPARK PLUG. Recommended spark plug for all models is Champion J8, or equivalent.

KOHLERengine
HP
Spec. no.
Model no.
Refer to owners manual for operation and maintenance instructions.
KOHLER COMPANY
KOHLER WISCONSIN USA

Fig. KO1—View showing nameplate and serial number plate on Kohler engine. Refer to text.

After every 100 hours of operation, the spark plug should be removed and cleaned and the electrode gap should be checked. Electrode gap should be 0.025 inch (0.7 mm).

NOTE: Manufacturer does not recommend the use of abrasive grit type spark plug cleaners. Improper cleaning allows entrance of grit into cylinder causing premature cylinder wear and damage.

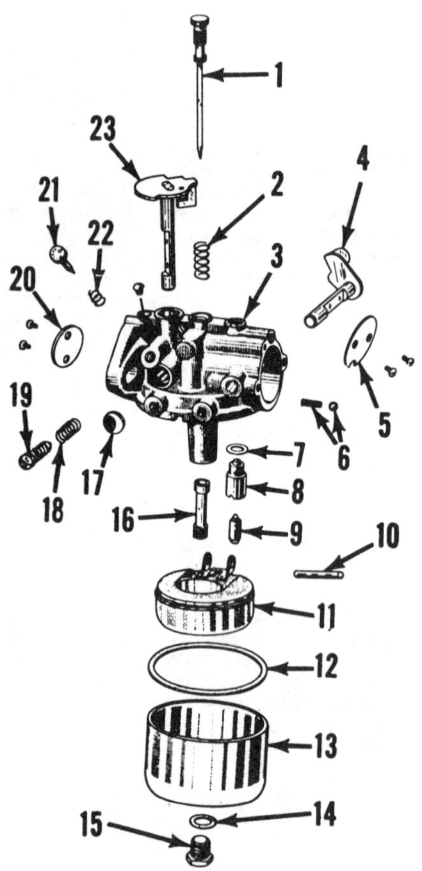

Fig. KO2—Exploded view of typical Carter Model N carburetor used on early production K90 and K91 engines.

1. Main fuel needle
2. Spring
3. Carburetor body
4. Choke shaft
5. Choke plate
6. Choke detent
7. Sealing washer
8. Inlet valve seat
9. Inlet valve
10. Float pin
11. Float
12. Gasket
13. Float bowl
14. Sealing washer
15. Retainer
16. Main jet
17. Plug
18. Spring
19. Idle stop screw
20. Throttle plate
21. Idle fuel needle
22. Spring
23. Throttle shaft

Renew spark plug if electrode is severely burnt or if other damage is apparent. Tighten spark plug to 18-22 ft.-lbs. (24-30 N·m).

CARBURETOR. Refer to Fig. KO2 for an exploded view of Carter Model N carburetor used on early production models. Late production K91 engines are equipped with the Kohler carburetor shown in Fig. KO3. Carburetor adjust-

Fig. KO3—Exploded view of Kohler carburetor used on late production engines.

1. Main fuel needle
2. Spring
3. Carburetor body assy.
4. Spring
5. Idle speed stop screw
6. Spring
7. Idle fuel needle
8. Sealing washer
9. Inlet valve seat
10. Inlet valve
11. Float pin
12. Float
13. Gasket
14. Float bowl
15. Sealing washer
16. Bowl retainer

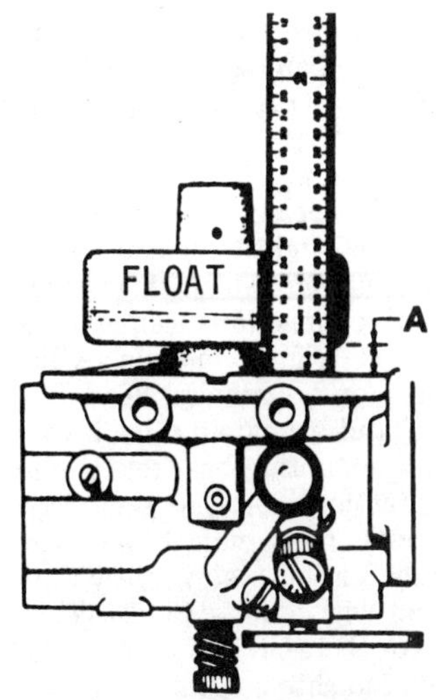

Fig. KO3A—Float height should be 11/64 inch (4.36 mm) measured at (A). Bend tang on float arm to adjust.

ment should be checked whenever poor or erratic engine operation is apparent.

Engine idle speed is 1200 rpm and is adjusted by turning idle speed screw (19–Fig. KO2 or 5–Fig. KO3) clockwise to increase idle speed or counterclockwise to decrease idle speed.

Initial adjustment of idle mixture screw is 1½ turns out from a lightly seated position. Initial setting of main fuel mixture screw is 2 turns out from a lightly seated position. Make final adjustments with engine at normal operating temperature and running. Place engine under load and adjust main fuel mixture screw for leanest setting that will allow satisfactory acceleration and steady governor operation. Set engine at idle speed, no-load, and adjust idle mixture screw to obtain smoothest idle operation.

As each adjustment affects the other, adjustment procedure may have to be repeated.

To check float level, invert carburetor throttle body and float assembly. There should be 11/64 inch (4.36 mm) clearance between free side of float and machined surface of body casting (Fig. KO3A). Carefully bend float lever tang that contacts inlet valve as necessary to provide correct clearance.

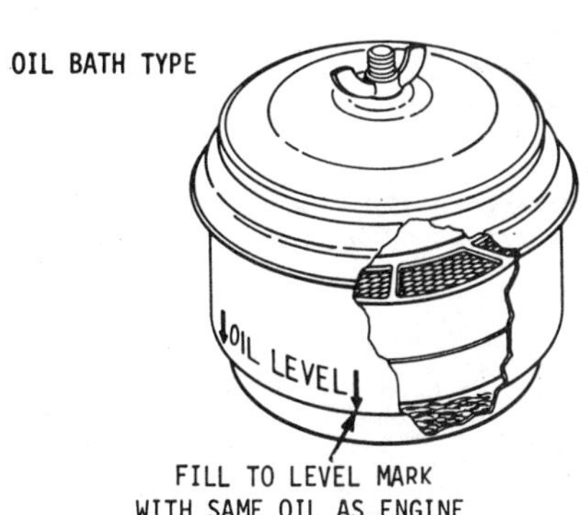

Fig. KO4—View of oil bath type air cleaner. Refer to text for service intervals and procedure.

FUEL FILTER. A fuel filter screen is located in sediment bowl below fuel shut off valve. Fuel filter screen and sediment bowl should be removed and cleaned after every 100 hours of operation. To remove, loosen bail nut, swing bail out of the way and remove sediment bowl, filter screen and gasket. Make certain sediment bowl gasket is in place before reinstalling sediment bowl.

AIR FILTER. Engines may be equipped with either an oil bath air filter (Fig. KO4) or a dry element filter (Fig. KO5). Refer to appropriate paragraph for model being serviced.

Oil Bath Type. Oil bath air filter should be serviced at 25 hour intervals of normal operation. To service, remove cover, lift element from bowl and drain oil. Thoroughly clean bowl and cover and rinse element in clean solvent. Allow element to dry, then lightly coat element and fill bowl to correct level with the same grade and type oil used in crankcase (see LUBRICATION section).

NOTE: Do not use air pressure to force dry element as element material may be damaged.

Make certain gasket is in place on air horn and reinstall air filter assembly.

Dry Type Filter. Element should be removed and cleaned after every 50 hours of operation under normal operating conditions. Tap element lightly on a flat surface to dislodge loose dirt and foreign material.

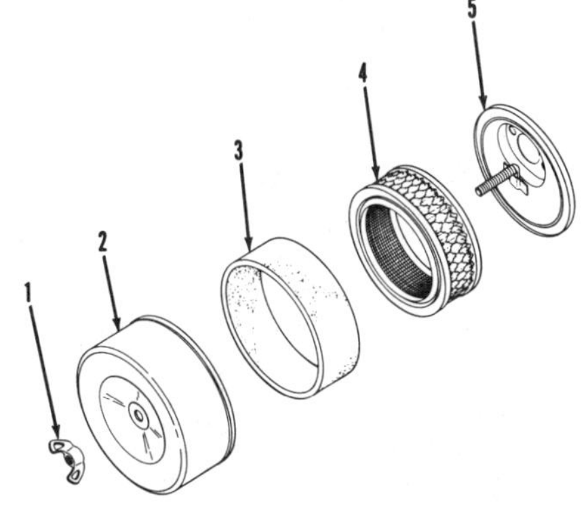

Fig. KO5—Exploded view of dry type air cleaner (filter). Foam precleaner (3) is optional. Refer to text for service interval and procedure.

1. Wing nut
2. Cover
3. Precleaner
4. Filter
5. Adapter

Fig. KO6—Points of adjustment of governor on Series K90 and K91 engines.

NOTE: Do not wash element or use air pressure to blow dirt out as element material may be damaged.

Renew element after every 150 hours of operation or whenever element is damaged or excessive dirt cannot be correctly removed.

Some dry type filters may be equipped with a foam type precleaner which fits over dry element and extends dry element service intervals. Precleaner should be cleaned and re-oiled after every 25 hours of operation. To clean, remove precleaner and wash in a mild detergent and water solution. Rinse thoroughly and squeeze away excess water. Allow to air dry, then soak precleaner in the same type and grade oil as used in engine crankcase (see LUBRICATION section). Squeeze out excess oil and reinstall precleaner.

CAUTION: Do not wring element or use air pressure to dry as element material may be stretched or damaged.

CRANKCASE BREATHER. The crankcase breather is attached to cylinder block on carburetor side of engine on early production model and to valve cover plate on late production model. Crankcase breather is designed to maintain a slight vacuum in crankcase to eliminate oil leakage at seals.

On late production model, crankcase breather should be disassembled, reed clearance checked and filter element cleaned after every 500 hours of operation. Reed clearance should be 1/64 to 1/32 inch (0.40-0.80 mm) between reed valve and its seat.

GOVERNOR. A centrifugal flyball type governor mounted within the crankcase and driven by the camshaft gear is used on all models. To adjust governed speed, first sychronize linkage by loosening clamp bolt nut (N–Fig. KO6), turn shaft (H) counterclockwise

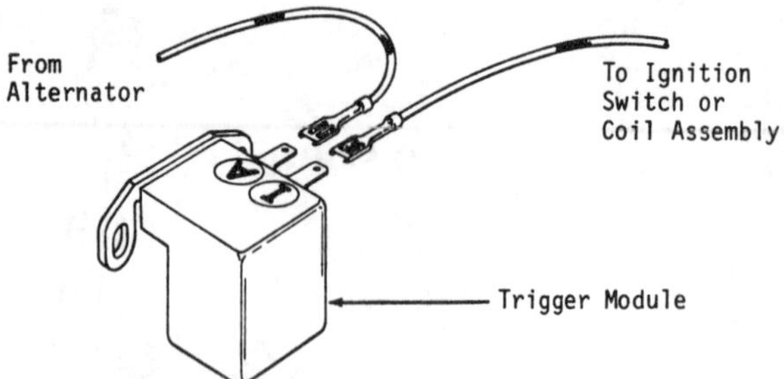

Fig. KO7—View of trigger module used with CD ignition system. Terminal "A" must be connected to alternator and terminal "I" must be connected to ignition switch.

until internal resistance is felt. Pull arm (L) completely to the left (away from carburetor) and tighten the clamp bolt nut. To increase or decrease maximum engine speed, vary the tension of governor spring (B). On engine with remote throttle control, spring (B) tension is varied by moving bracket (F) up or down. On engines without remote throttle control, rotate disc (D) after loosening bushing (C) to vary spring (B) tension.

IGNITION SYSTEM. Engines may be equipped with a magneto ignition system, capacitor-discharge (CDI) or battery ignition system. Refer to appropriate paragraph for model being serviced.

Magneto and Battery Ignition System. Breaker point cover should be removed and condenser and breaker points checked and renewed or adjusted after every 500 hours of operation. Breaker point gap is adjusted to 0.020 inch (0.51 mm) on all models.

Late production engine has a timing port in left side of bearing plate and a timing light is used to precisely set timing. There are two marks on flywheel, a "T" for top dead center and either "S", "SP" or "20" mark for the spark point. With engine running at 1200 to 1800 rpm, light should flash when spark point mark ("S", "SP" or "20") is centered in timing port. Vary point gap to align marks.

CD Ignition System. Capacitor-discharge (CDI) ignition system does not have breaker points and timing is not adjustable. A trigger module (Fig. KO7) is used instead of breaker points and must be installed correctly to prevent damage to internal components. Terminal marked "A" must be connected to the alternator and terminal "I" must be connected to the ignition switch. DO NOT reverse these leads.

If a faulty trigger module is suspected, remove module from engine. To test trigger module diodes, connect one lead of an ohmmeter to the "I" terminal and connect remaining lead to the "A" terminal. Observe ohmmeter reading. Reverse leads. Observe ohmmeter reading. Ohmmeter should indicate continuity with leads in one position only. If ohmmeter indicates continuity when connected both ways, or an open circuit for both connections, renew module.

To test trigger module SCR switch, connect one ohmmeter lead to "I" terminal and remaining lead to trigger module mounting bracket. If ohmmeter indicates continuity, reverse the leads as ohmmeter must indicate an open circuit initially for this test. Lightly tap module magnet with a metal object. This should activate the SCR switch and ohmmeter should indicate continuity. If ohmmeter indicates continuity when initially attached to module in both directions or if SCR switch will not activate as previous-

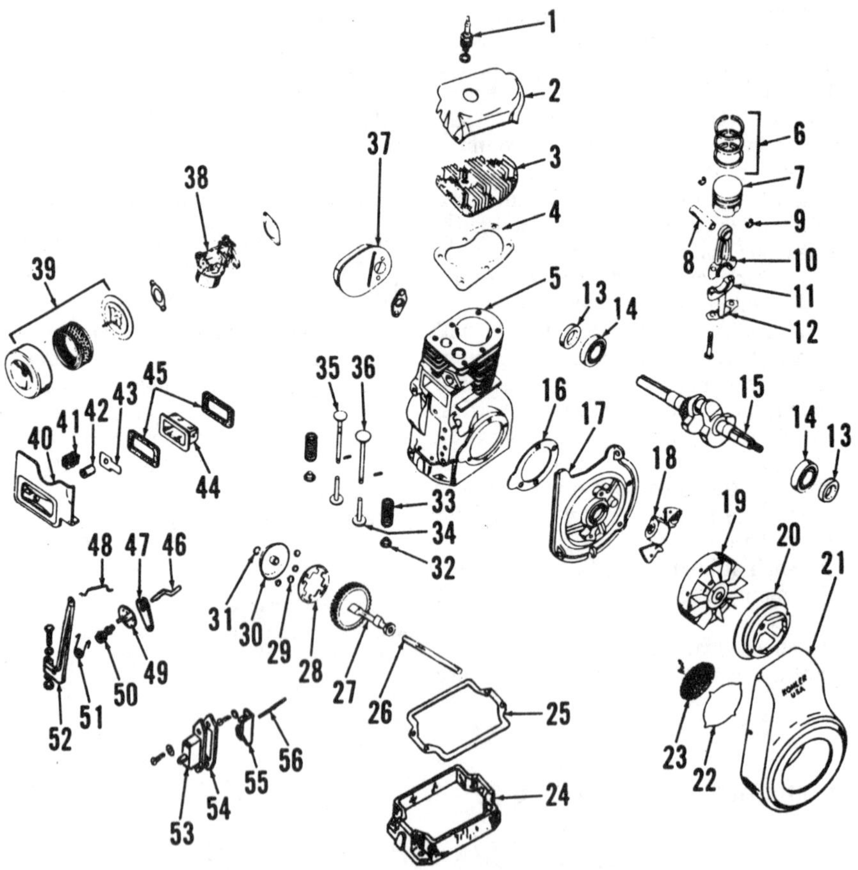

Fig. KO9—Exploded view of K90 and K91 engine. Breaker points (55) are actuated by cam on right end of camshaft (27) through push rod (56).

1. Spark plug
2. Air baffle
3. Cylinder head
4. Head gasket
5. Cylinder block
6. Piston rings
7. Piston
8. Piston pin
9. Retaining rings
10. Connecting rod
11. Rod cap
12. Rod bolt lock
13. Oil seal
14. Ball bearing
15. Crankshaft
16. Gasket
17. Bearing plate
18. Magneto
19. Flywheel
20. Pulley
21. Shroud
22. Screen retainer
23. Screen
24. Oil pan
25. Gasket
26. Camshaft pin
27. Camshaft
28. Flyball retainer
29. Steel balls
30. Thrust cone
31. Snap ring
32. Spring retainer
33. Valve spring
34. Valve tappets
35. Exhaust valve
36. Intake valve
37. Muffler
38. Carburetor
39. Air cleaner assy.
40. Valve cover
41. Filter
42. Breather seal
43. Breather reed
44. Breather plate
45. Gaskets
46. Governor shaft
47. Bracket
48. Link
49. Speed disc
50. Bushing
51. Governor spring
52. Governor lever
53. Breaker cover
54. Gasket
55. Breaker points
56. Push rod

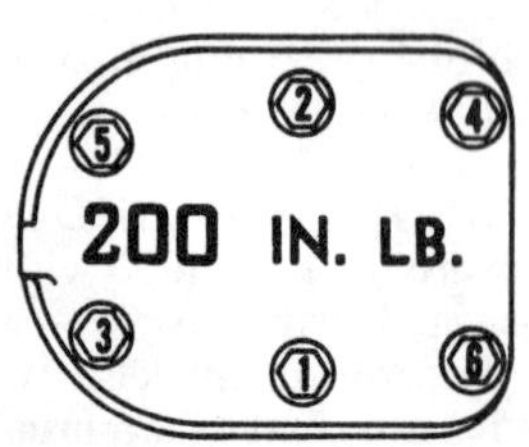

Fig. KO8—Tighten cylinder head cap screws to 200 in.-lbs. (22.6 N·m) in sequence shown.

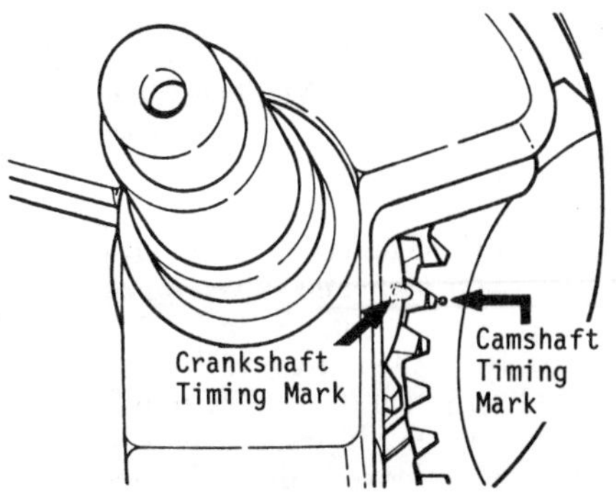

Fig. KO10—Make certain timing marks are aligned during crankshaft and camshaft installation.

ly described, the trigger module may be considered defective and should be renewed.

When installing the trigger module, set air gap between trigger assembly and flywheel projection at 0.005-0.010 inch (0.13-0.25 mm).

VALVE ADJUSTMENT. Valve stem clearance should be checked after every 500 hours of operation. Clearance for intake valve stem whould be 0.005-0.009 inch (0.13-0.23 mm) and clearance for exhaust valve stem should be 0.011-0.015 inch (0.28-0.38 mm). If clearance is not as specified, refer to REPAIR section for valve service procedure.

CYLINDER HEAD AND COMBUSTION CHAMBER. Cylinder head should be removed and carbon and lead deposits cleaned after every 500 hours of operation. Refer to REPAIR section for cylinder head removal procedure.

LUBRICATION. Engine oil should be checked daily and oil level maintained between the "F" and "L" mark on dipstick. Insert threaded plug type dipstick only until threads touch opening; do not screw in to check oil level. On models with extended oil fill tube, push dipstick all the way down in tube to obtain reading.

Manufacturer recommends oil having API service classification SC, SD, SE or SF. Use SAE 5W-30 oil for temperatures below 32°F (0°C) and use SAE 30W oil for temperatures above 32°F (0°C).

Oil should be changed after the first 5 hours of operation and at 25 hour intervals thereafter. Crankcase oil capacity is 0.5 quart (0.47 L).

GENERAL MAINTENANCE. Check and tighten all loose bolts, nuts or clamps daily. Check for fuel and oil leakage and repair if necessary. Clean cooling fins and external surfaces at 50 hour intervals.

REPAIR

TIGHTENING TORQUES. Recommended tightening torque specifications are as follows:

Spark plug . . . . 18-22 ft.-lbs. (24-30 N·m)
Flywheel nut . . . . 40-50 ft.-lbs. (54-68 N·m)
Cylinder head . . . . 200 in.-lbs.* (22.6 N·m)*
Connecting rod . . . . 140 in.-lbs.* (15.8 N·m)*

*With threads lightly lubricated.

CYLINDER HEAD. To remove cylinder head, first remove all necessary metal shrouds. Clean engine to prevent entrance of foreign material and remove cylinder head retaining bolts.

Always use a new head gasket when installing cylinder head. Tighten cylinder head bolts evenly and in graduated steps using the sequence shown in Fig. KO8 until specified torque is obtained.

CONNECTING ROD. The aluminum alloy connecting rod rides directly on the crankpin journal. Connecting rod and piston are removed as an assembly after cylinder head and oil pan (engine base) have been removed. Remove the two connecting rod bolts and connecting rod cap and push piston and rod assembly out the top of block. Remove snap rings (9–Fig. KO9) and push pin (8) out of piston (7). Separate connecting rod and piston.

Before installing connecting rod, check the following dimensions to ensure rod is suitable for service.

Small End Inside Diameter. Connecting rod piston pin bore is 0.5630-0.5633 inch (14.300-14.308 mm) and piston pin-to-connecting rod pin bore clearance should be 0.0007-0.0008 inch (0.018-0.020 mm). Renew connecting rod if dimensions are not as specified.

Big End Inside Diameter. Connecting rod big end (bearing) diameter for standard connecting rod is 0.9384-0.9387 inch (23.835-23.843 mm) and connecting rod-to-crankpin journal clearance should be 0.0010-0.0025 inch (0.025-0.063 mm). If dimensions are not as specified or if bearing clearance exceeds 0.0035 inch (0.090 mm), renew connecting rod and/or recondition crankpin journal.

Connecting Rod Side Clearance. With crankshaft removed from crankcase, mount connecting rod on crankpin journal.

Use a suitable feeler gage to measure clearance between crankshaft and flat thrust surface on connecting rod big end. Side play should be 0.005-0.016 inch (0.13-0.41 mm). If clearance is not as specified, renew connecting rod.

Piston may be installed on connecting rod either way (always use new retaining rings), however, connecting rod cap and connecting rod match marks must align and be towards flywheel side of engine after installation. Tighten connecting rod bolts to specified torque.

PISTON, PIN AND RINGS. The aluminum alloy piston is fitted with two compression rings and one oil control ring. Refer to CONNECTING ROD section for removal and installation procedure.

After separating piston and connecting rod, carefully remove rings and clean carbon and lead deposits from piston surface and ring lands.

CAUTION: Extreme care should be exercised when cleaning ring lands. Do not damage squared edges or widen ring grooves. If ring lands or grooves are damaged, piston must be renewed.

Measure piston diameter just below oil control ring groove 90° from piston pin center. Standard piston diameter is 2.369-2.371 inches (60.173-60.223 mm) and minimum piston diameter is 2.366 inches (60.096 mm). Renew piston if dimensions are not as specified.

Install piston in cylinder bore before installing rings and use a suitable feeler gage to measure clearance between piston thrust surface (90° from piston pin) and cylinder bore. Clearance should be 0.0035-0.0060 inch (0.089-0.152 mm). If dimension is not as specified, renew piston and/or recondition cylinder bore.

Piston pin fit in piston pin bore should be 0.0002 inch (0.005 mm) interference to 0.0002 inch (0.005 mm) loose. Piston pin is available in oversize if dimension is not as specified.

If piston ring-to-piston groove side clearance exceeds 0.006 inch (0.152 mm) for compression rings or 0.0015-0.0035 inch (3.81-8.89 mm) for oil control ring, renew piston.

Piston ring end gap for compression rings, measured with ring squarely installed in cylinder bore, is 0.007-0.017 inch (0.18-0.43 mm) for new bore and 0.007-0.027 inch (0.18-0.68 mm) for used bore. If dimensions are not as specified, renew piston and/or recondition cylinder bore.

Install piston rings, which are marked, with marked side toward top of piston. If compression ring has a groove or bevel on outside surface, install ring with groove or bevel down. If groove or bevel is on inside surface of compression ring, install ring with groove or bevel up. Stagger ring end gaps equally around circumference of piston before installation.

CYLINDER AND CRANKCASE. Cylinder and crankcase are integral castings. Standard cylinder bore diameter is 2.3745-2.3755 inches (60.31-60.33 mm). If cylinder bore exceeds 2.378 inches (60.40 mm), cylinder taper exceeds 0.003 inch (0.076 mm) or if cylinder is out-of-round more than 0.005 inch (0.13 mm), recondition cylinder bore to nearest oversize for which piston and rings are available.

CRANKSHAFT, MAIN BEARINGS AND SEALS. The crankshaft is supported at each end by a ball bearing type main bearing. Renew bearings (14–Fig. KO9) if excessively rough or loose. Crankshaft end play should be 0.0038-0.0228 inch (0.096-0.579 mm) and is adjusted by varying number and thickness of shim gaskets (16).

Standard crankpin journal diameter is 0.9355-0.9360 inch (23.762-23.774 mm) and minimum diameter is 0.9350 inch (23.75 mm). Maximum crankpin journal out-of-round is 0.0005 inch (0.013 mm) and maximum journal taper is 0.001 inch (0.025 mm). If crankpin dimensions are not as specified, renew or recondition crankshaft.

Main bearings should be a light press fit on crankshaft journals and in crankcase and bearing plate bores. If not, renew bearings and/or crankshaft or crankcase and bearing plate.

Front and rear crankshaft oil seals should be pressed into seal bores so outside edge of seal is 1/32 inch (0.80 mm) below seal bore surface.

Make certain crankshaft gear and camshaft gear timing marks are aligned (Fig. KO10) as crankshaft and camshaft are reinstalled.

CAMSHAFT AND BEARINGS. The hollow camshaft and integral cam gear (27–Fig. KO9) rotate on pin (26). Camshaft can be removed after removing bearing plate (17) and crankshaft, then drive pin (26) out towards bearing plate side of crankcase. When reinstalling camshaft, make certain camshaft gear and crankshaft gear timing marks are aligned (Fig. KO10).

Camshaft pin (26) is a press fit in closed (crankcase) side and a slip fit with 0.0005-0.0012 inch (0.013-0.030 mm) clearance in bearing plate side.

Camshaft-to-camshaft pin clearance should be 0.001-0.0025 inch (0.025-0.063 mm). Camshaft end play should be 0.005-0.020 inch (0.13-0.50 mm) and is controlled by varying number and thickness of shim washers between camshaft and bearing plate side of crankcase.

Governor flyball retainer (28), flyballs (29), thrust cone (30) and snap ring (31) are attached to, and rotate with, the camshaft assembly.

VALVE SYSTEM. Clearance between valve stem and valve tappet (cold) should be 0.005-0.009 inch (0.13-0.23 mm) for the intake valve and 0.011-0.015 inch (0.28-0.38 mm) for the exhaust valve. If clearance is not as specified, remove valves. To increase clearance, shorten valve length by grinding end of stem. To reduce clearance, grind valve seat.

The exhaust valve seats on a renewable seat insert and the intake valve seat is machined directly into cylinder block surface. Valve face and seat angle is 45 degrees and seat width should be 0.037-0.045 inch (0.94-1.14 mm) for intake and exhaust valve.

Minimum valve stem diameter is 0.2478 inch (6.29 mm) for intake valve and 0.2458 inch (6.24 mm) for exhaust valve.

Valve stem clearance in guide should be 0.0005-0.0020 inch (0.013-0.050 mm) for intake valve and 0.0020-0.0035 inch (0.050-0.089 mm) for exhaust valve. Excessive valve stem-to-guide clearance is corrected by reaming guides and installing valves with 0.005 inch (0.13 mm) oversize stems.

SERVICING KOHLER ACCESSORIES

REWIND STARTERS

Fairbanks-Morse or Eaton rewind starters are used on some Kohler engines. When servicing the starters, refer to appropriate paragraph.

Fairbanks-Morse Starter. To disassemble the starter, remove retainer ring, retainer washer, brake spring, friction washer, friction shoe assembly and second friction washer as shown in Fig. KO30. Hold the rope handle in one hand and the cover in the other and allow rotor to rotate and unwind the rewind spring preload. Lift rotor from cover, shaft and recoil spring noting direction of recoil spring installation for reassembly. Remove rewind spring from cover and unwind rope from rotor.

When reassembling the unit, lubricate rewind spring, cover shaft and its bore in rotor with grease. Install the rope on rotor and the rotor to the shaft and engage the rewind spring inner end hook. Preload the rewind spring four turns and install middle flange and mounting flange. Check friction shoe sharp ends and renew if necessary. Install friction washers, friction shoe assembly, brake spring and retainer ring. Make certain friction shoe assembly is installed properly for correct starter rotation. If properly installed, sharp ends of friction shoe plates will extend when rope is pulled.

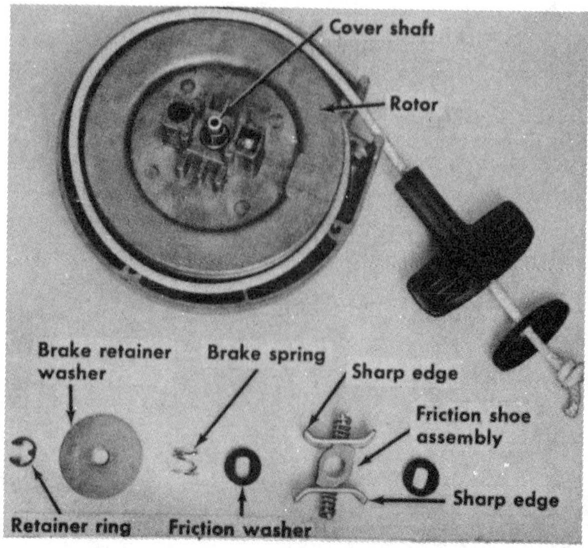

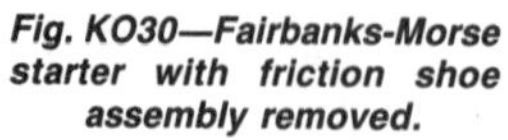
Fig. KO30—Fairbanks-Morse starter with friction shoe assembly removed.

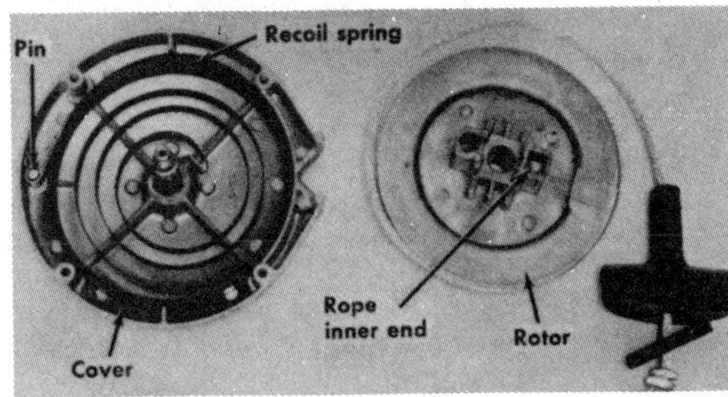

Fig. KO31—View showing rewind spring and rope installed for counterclockwise starter operation.

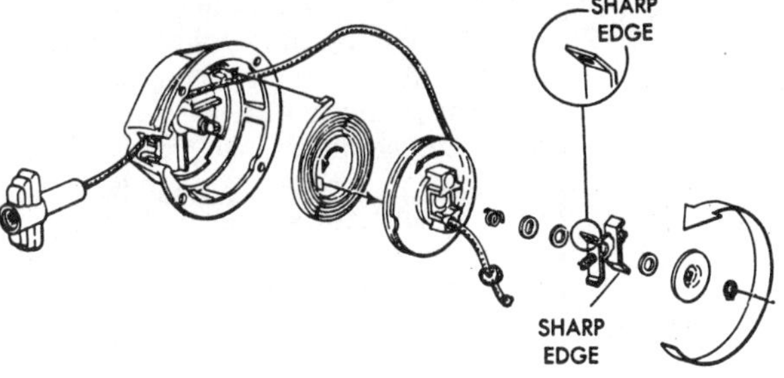

Fig. KO32—For clockwise starter operation, reverse friction shoe assembly and wind rope and rewind spring in opposite direction.

Starter operation can be reversed by winding rope and rewind spring in opposite direction and turning the friction shoe assembly upside down. See Fig. KO31 for counterclockwise assembly and Fig. KO32 for clockwise assembly.

Eaton Starter. To disassemble the starter, first release tension of rewind spring by pulling starter rope until notch in pulley is aligned with rope hole in cover. Use thumb pressure to prevent pulley from rotating. Engage rope in notch of pulley and slowly release thumb pressure to allow spring to unwind until all tension is released.

When removing rope pulley, use extreme care to keep starter spring confined in housing. Check starter spring for breaks, cracks and distortion. If starter spring is to be renewed, carefully remove it from housing, noting the direction of rotation of spring before removing. Exploded view of clockwise starter is shown in Fig. KO33.

Check the pawl, brake, spring, retainer and hub for wear and renew if necessary. If starter rope is worn or frayed, remove from pulley, noting the direction it is wrapped on pulley. Renew rope and install pulley in housing, aligning notch in pulley hub with hook in end of spring. Use a wire bent to form a hook to aid in positioning spring on hub.

After securing the pulley assembly in housing, align notch in pulley with rope bushing in housing. Engage rope in notch and rotate pulley at least two full turns in same direction it is pulled to properly preload starter spring. Pull rope to fully extended position. Release handle and if spring is properly preloaded, the rope will fully rewind.

Before installing starter on engine, check teeth in starter driven hub

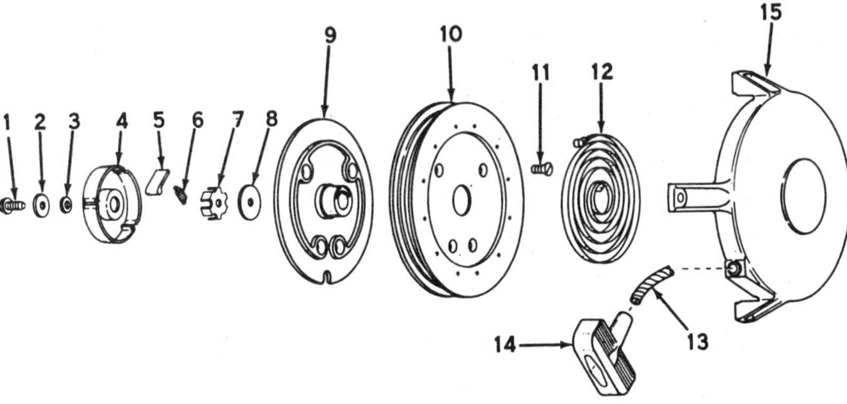

Fig. KO33—Exploded view of Eaton rewind starter.

1. Retainer screw
2. Brake washer
3. Spacer
4. Retainer
5. Pawl
6. Spring
7. Brake
8. Thrust washer
9. Pulley hub
10. Pulley
11. Screw
12. Recoil spring
13. Rope
14. Handle
15. Starter housing

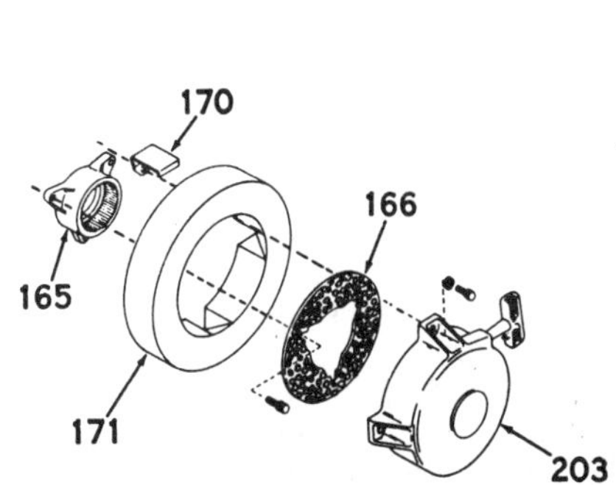

Fig. KO34—View showing rewind starter and starter hub.

165. Starter hub
166. Screen
170. Bracket
171. Air director
203. Rewind starter assy.

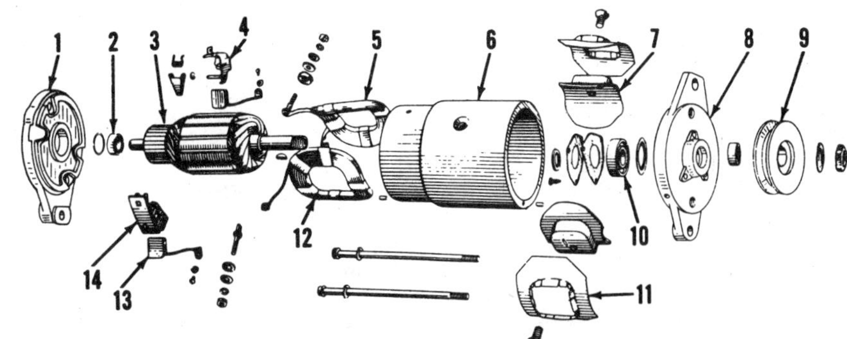

Fig. KO35—Exploded view of typical Delco-Remy starter-generator assembly.

1. Commutator end frame
2. Bearing
3. Armature
4. Ground brush holder
5. Field coil
6. Frame
7. Pole shoe
8. Drive end frame
9. Pulley
10. Bearing
11. Field coil insulator
12. Field coil
13. Brush
14. Insulated brush holder

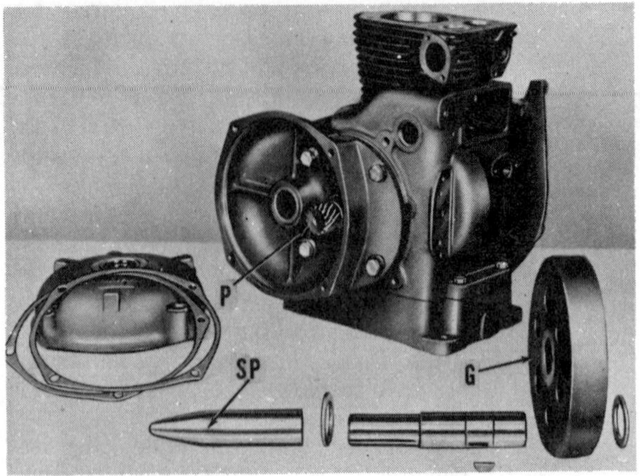

Fig. KO36—Elements of reduction gear unit used on some K90 and K91 engines. Pinion (P) is on crankshaft; internally toothed gear is (G). Seal protecting tool (SP) is used when installing cover to housing.

(165—Fig. KO34) for wear and renew hub if necessary.

12-VOLT STARTER-GENERATOR

The combination 12-volt starter-generator manufactured by Delco-Remy is used on some Kohler engines. The starter-generator functions as a cranking motor when the starting switch is closed. When the engine is operating and with starting switch open, the unit acts as a generator. Generator output and circuit voltage for the battery and various operating requirements are controlled by a current-voltage regulator.

Kohler recommends that starter-generator belt tension be adjusted until about 10 lbs. (44.5 N) of pressure applied midway between pulleys will deflect belt ¼ inch (6 mm). Refer to Fig. KO35 for exploded view of starter-generator assembly. Parts are available from Kohler as well as authorized Delco-Remy service stations.

REDUCTION DRIVE

The reduction drive unit consists of a driven gear (G—Fig. KO36) which is pressed on the power takeoff shaft. Drive gear (P) is an integral part of engine crankshaft. The pto shaft is supported by two bushings; one in the housing and the other in the cover. Oil seals are used at both ends of pto shaft. To disassemble the unit, first remove plug and drain the unit. Unbolt and remove the cover, then remove driven gear (G). Remove cap screws securing housing to engine block and lift off housing. Clean and inspect all parts and renew any showing excessive wear or damage.

When reassembling, use tape or seal protector to prevent damage to oil seals. Use copper washers on the two internal cap screws and lockwashers on external cap screws. Adjust gear reduction (pto) shaft end clearance to 0.001-0.006 inch (0.02-0.15 mm) by varying total thickness of cover gaskets. Fill housing to oil level hole with same grade of oil as used in engine.

KOHLER CENTRAL PARTS DISTRIBUTORS

(Arranged Alphabetically by States)

These franchised firms carry extensive stocks of repair parts. Contact them for name of dealer in their area who will have replacement parts.

Auto Electric & Carburetor Company
Phone: (205) 323-7113
2625 4th Avenue South
Birmingham, Alabama 35233

Charlie C. Jones, Incorporated
Phone: (602) 272-5621
2440 West McDowell Road
Phoenix, Arizona 85009

H.G. Makelim Company
Phone: (714) 978-7515
1520 South Harris Ct.
Anaheim, California 92806

H.G. Makelim Company
Phone: (415) 873-4753
219 Shaw Road
South San Francisco, California 94083-2827

Spitzer Industrial Products Company
Phone: (303) 287-3414
6601 North Washington Street
Thornton, Colorado 80229

Spencer Engine, Incorporated
Phone: (904) 262-1661
5200 Sunbeam Road
Jacksonville, Florida 32217

Small Engine Clinic
Phone: (808) 488-0711
98019 Kam Highway
Pearl City, Hawaii 96782

Medart Engines of Kansas
Phone: (913) 888-8828
15500 West 109th Street
Lenexa, Kansas 66219

The Grayson Company of Louisiana
Phone: (318) 222-3871
215 Spring Street
Shreveport, Louisiana 71162

W. J. Connell Company
Phone: (617) 543-3600
65 Green Street
Foxboro, Massachussetts 02035

Central Power Distributors, Incorporated
Phone: (612) 633-5179
2976 North Cleveland Avenue
St. Paul, Minnesota 55113

Medart Engines of St. Louis
Phone: (314) 343-0505
100 Larkin Williams Industrial Crt.
Fenton, Missouri 63026

Original Equipment, Incorporated
Phone (406) 245-3081
905 Second Avenue, North
Billings, Montana 59103

Power Distributors, Incorporated
Phone: (201) 225-4922
71 Northfield Avenue
Edison, New Jersey 08837

Gardner, Incorporated
Phone: (614) 488-7951
1150 Chesapeake Avenue
Columbus, Ohio 43212

MICO, Incorporated
Phone: (918) 627-1448
7450 East 46th Place
Tulsa, Oklahoma 74115

E.C. Distributing Co.
Phone: (503) 224-3623
2122 N.W. Upshur
Portland, Oregon 97210

Pitt Auto Electric Company
Phone: (412) 766-9112
2900 Stayton Street
Pittsburgh, Pennsylvania 15212

Medart Engines of Memphis
Phone: (901) 795-4365
4365 Old Lamar
Memphis, Tennessee 38118

Tri-State Equipment Company
Phone: (915) 532-6931
410 South Cotton Street
El Paso, Texas 79901

Waukesha-Pearch Industries, Inc.
Phone: (713) 723-1050
12320 South Main Street
Houston, Texas 77235

Diesel Electric Service & Supply Co.
Phone: (801) 972-1836
652 West 1700 South
Salt Lake City, Utah 84104

Chesapeake Engine Distributors
Division of RBI Corporation
Phone: (804) 798-1596
103 Sycamore Drive
Ashland, Virginia 23005

Kohler Company
Engine Division
Phone: (414) 457-4441
Kohler, Wisconsin 53044

CANADIAN DISTRIBUTORS

Lotus Equipment Sales, Ltd.
Phone: (403) 295-1040
908 53rd Avenue, N.E. Bay H
Calgary, Alberta T2E 6N9

Yetman's Ltd.
Phone: (204) 586-8046
949 Jarvis Avenue
Winnipeg, Manitoba R2X 0A1

W.N. White Company, Ltd.
Phone: (902) 443-5000
2-213-215 Bedford Highway
Halifax, Nova Scotia B3M 2J9

CPT Canada Power Technology, Ltd.
Phone: (416) 624-6200
5466 Timberlea Blvd.
Mississauga, Ontario L4W 2T7

CPT Canada Power Technology, Ltd.
Phone: (514) 731-3559
226 Migneron Street
Ville Saint-Laurent, P.Q. H4T 1Y7

McCULLOCH

McCULLOCH CORP.
900 Lake Havasu Avenue
Lake Havasu City, Arizona 86403

Model	Bore	Stroke	Displacement
MC-2, MC-7, MC-8, MC-9, MC-30, MC-40, MC-45	2.165 in. (55 mm)	1.635 in. (41.5 mm)	6.1 cu. in. (99 cc)
MC-5	2.125 in. (54 mm)	1.375 in. (35 mm)	4.9 cu. in. (80 cc)
MC-6, MC-10	2.125 in. (54 mm)	1.500 in. (38.1 mm)	5.3 cu. in. (87 cc)
MC-20	2.125 in. (54 mm)	1.635 in. (41.5 mm)	5.8 cu. in. (95 cc)
MC-70	2.217 in. (56.3 mm)	1.835 in. (46.6 mm)	7.1 cu. in. (116 cc)
MC-75	2.250 in. (57.2 mm)	1.835 in. (46.6 mm)	7.3 cu. in. (120 cc)

MAINTENANCE

SPARK PLUG. Recommended spark plug for average conditions is a Champion J6J. For light racing, use Champion J4J and for lengthy racing, use Champion J2J. Set electrode gap to 0.025 inch (0.63 mm) on all models.

CARBURETOR. Models MC-5 and MC-6 use Tillotson Series HL carburetor as shown in Fig. MC1. All other models use McCulloch carburetors as shown in Fig. MC2 or Fig. MC3. All carburetors are diaphragm type with integral fuel pump.

Initial carburetor adjustment for Models MC-40 and MC-70 is 2 turns open on both idle and main fuel mixture needles on both carburetors. On remaining models, initial adjustment is 1¼ turns open on idle fuel mixture needle and 1½ turns open on main fuel mixture needle. Make final adjustments with engine at operating temperature and running. Place kart on blocks so rear wheels can turn freely. Adjust idle speed regulating screw so engine idles at 1500-1700 rpm. Turn idle fuel needle slowly in clockwise direction until engine idles smoothly. If engine starts to accelerate while turning needle in clockwise direction, turn the needle back counterclockwise until engine slows down. Take kart off blocks and test for acceleration. If engine misfires during acceleration richen fuel mixture by turn-

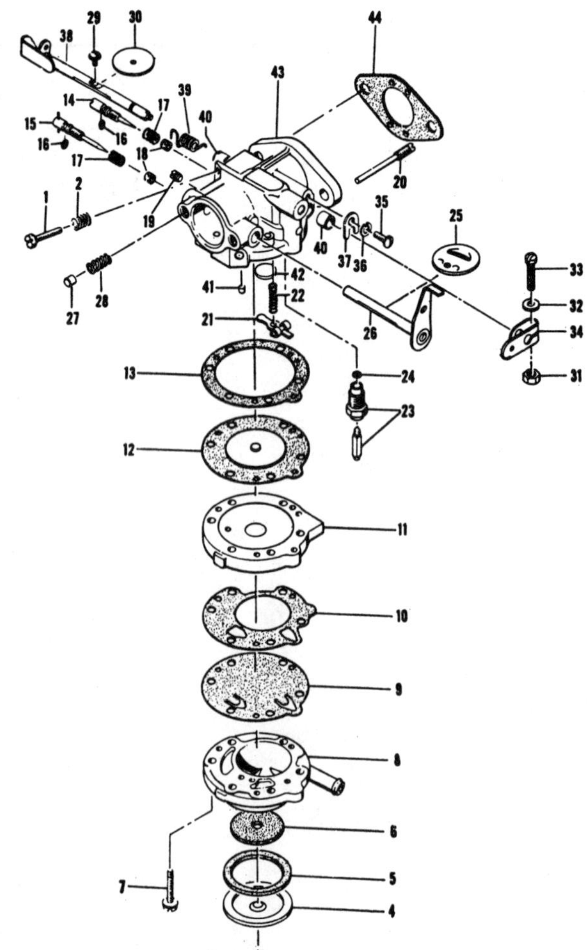

Fig. MC1—Exploded view of typical Tillotson carburetor used on Models MC-5 and MC-6.

1. Idle speed screw
2. Spring
3. Screw
4. Strainer cover
5. Gasket
6. Strainer screen
7. Screw
8. Fuel pump body
9. Pump diaphragm
10. Gasket
11. Diaphragm cover
12. Diaphragm
13. Gasket
14. Idle fuel needle
15. Main fuel needle
16. Washer
17. Spring
18. Packing
19. Plug
20. Inlet lever pin
21. Inlet lever
22. Spring
23. Inlet needle & seat
24. Gasket
25. Choke plate
26. Choke shaft & lever
27. Friction pin
28. Spring
29. Screw
30. Throttle plate
31. Nut
32. Washer
33. Screw
34. Throttle lever
35. Screw
36. Washer
37. Clip
38. Throttle shaft & lever
39. Spring
40. Bushing
41. Plug
42. Plug
43. Carburetor body
44. Gasket

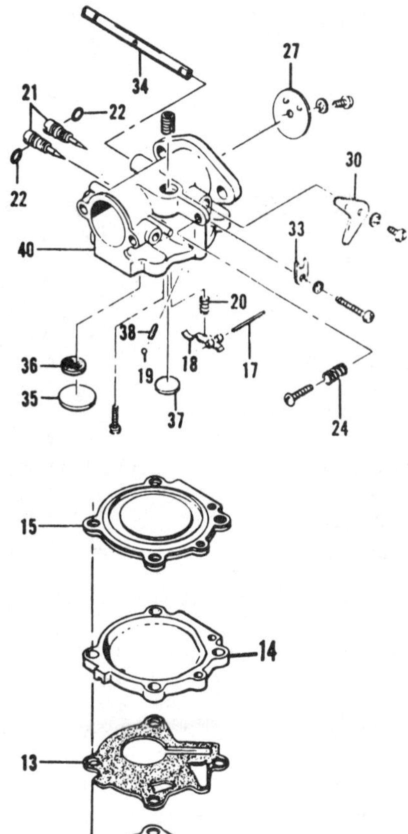

Fig. MC2—Exploded view of McCulloch 50070C diaphragm carburetor and diaphragm type fuel pump assembly used on MC-10 kart engine.

11. Fuel pump body
12. Pump diaphragm
13. Pump gasket
14. Diaphragm plate
15. Carburetor diaphragm
17. Inlet lever pin
18. Inlet control lever
19. Ball
20. Inlet lever spring
21. Metering needles
22. "O" rings
27. Throttle plate
30. Throttle shaft arm
33. Clip
34. Throttle shaft
35. Expansion plug
36. Filter
37. Expansion plug
38. Ball check seat
40. Carburetor body

ing idle fuel needle counterclockwise a small fraction of a turn at a time until engine accelerates smoothly and rapidly. If engine appears sluggish on acceleration, a leaner fuel mixture may be required. Turn idle needle a small fraction of a turn at a time in clockwise direction until acceleration of engine is at maximum. As main fuel needle has not had final adjustment at this time, acceleration may not be at maximum peak.

Test engine at high speed under load. If engine four-strokes (fires every other stroke), lean the fuel mixture by turning the main fuel needle in small increments in clockwise direction until engine fires

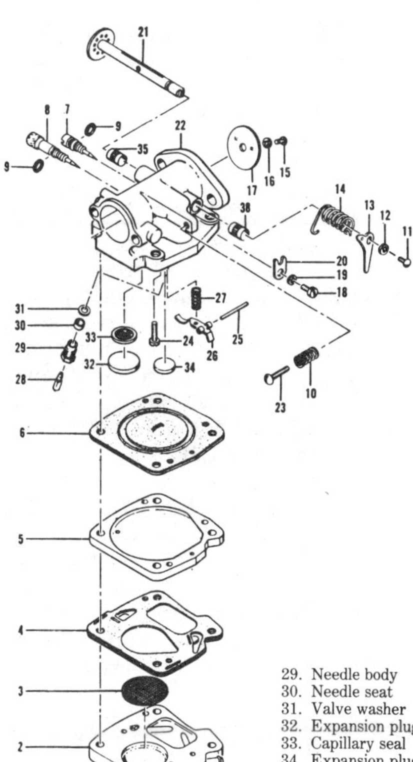

Fig. MC3—Exploded view of McCulloch carburetor used on MC-7, MC-8, MC-20, MC-30, MC-40 and MC-70 models. Carburetors used on Models MC-2, MC-45 and MC-75 are similar.

1. Screw
2. Fuel pump body
3. Fuel pump filter
4. Pump diaphragm
5. Diaphragm plate
6. Diaphragm
7. Idle fuel needle
8. Main fuel needle
9. "O" ring
10. Spring
11. Screw
12. Washer
13. Throttle arm
14. Spring
15. Screw
16. Washer
17. Throttle plate
18. Screw
19. Washer
20. Clip
21. Throttle shaft
22. Carburetor body
23. Idle speed screw
24. Screw
25. Inlet lever pin
26. Inlet lever
27. Spring
28. Needle valve
29. Needle body
30. Needle seat
31. Valve washer
32. Expansion plug
33. Capillary seal
34. Expansion plug
35. Bushing
38. Bushing

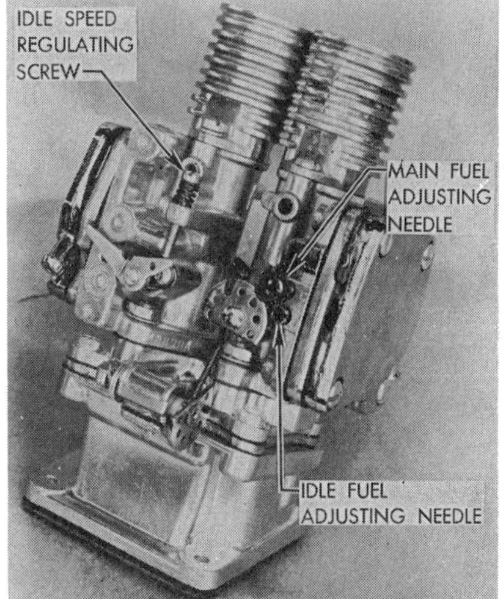

Fig. MC4—View of dual carburetor installation on Models MC-40 and MC-70 showing carburetor adjustment points. Main fuel adjusting needle and idle fuel adjusting needle are on rear side of left carburetor.

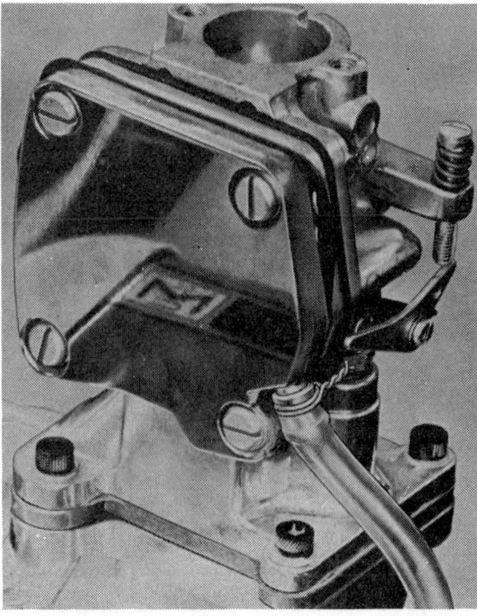

Fig. MC5—Always secure fuel tube with wire at each end as shown.

Fig. MC6—On Tillotson carburetor, be sure diaphragm lever is flush with diaphragm chamber surface as shown.

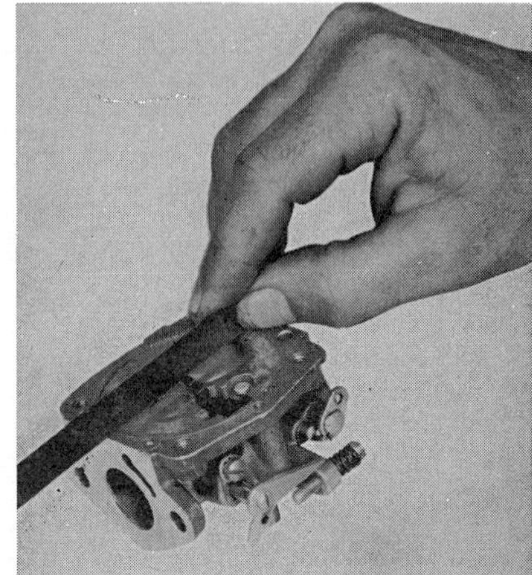

Fig. MC7—On McCulloch carburetors, be sure diaphragm lever is flush with machined gasket surface of carburetor body as shown.

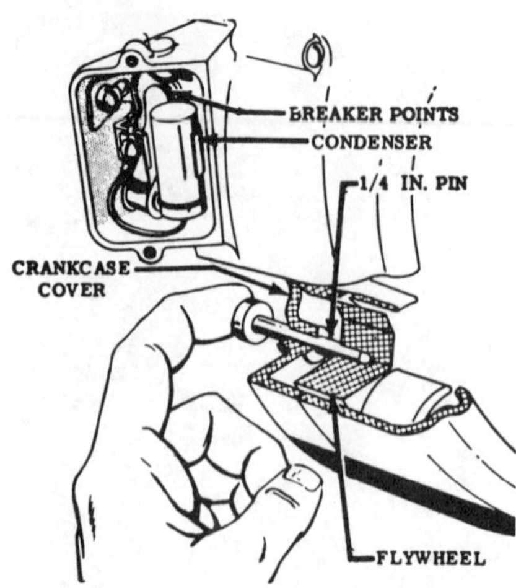

Fig. MC8 – On MC-10 engines, crankshaft can be locked in firing position by inserting a ¼-inch (6.35 mm) pin into bored hole in flywheel.

on every stroke at peak rpm. Never attempt to adjust the main fuel mixture needle unless engine is under load.

After adjusting the main fuel needle, it may be necessary to readjust the idle fuel needle as previously outlined.

MAGNETO AND TIMING. The magnet is cast into the flywheel on all models. The breaker points and condenser on the Model MC-10 are mounted in a breaker box outside the engine crankcase. The flywheel must be removed on all other models to gain access to the breaker points and condenser. Breaker point gap on all models is 0.018 inch (0.46 mm).

To adjust breaker point gap on Model MC-10, insert a ¼-inch (6.35 mm) pin through the hole in the crankcase cover (Fig. MC8) and turn engine until pin goes into timing hole in flywheel. Disconnect primary lead from breaker point terminal. Connect timing light in series between breaker point terminal and engine ground as shown in Fig. MC9. Adjust breaker points so light is on, then slowly adjust breaker points so light just goes out. Tighten breaker point mounting screws.

To adjust breaker points on all other models, remove flywheel and install degree wheel on engine crankshaft. Turn engine to 25° BTDC. Hook up timing light and adjust breaker points as previously outlined. If degree wheel is not available, set breaker point gap to 0.018 inch (0.46 mm) using a feeler gage.

Clearance between coil laminations and flywheel is 0.007-0.012 inch (0.18-0.31 mm) on all models. Desired clearance of 0.010 inch (0.25 mm) is set as shown in Fig. MC10.

LUBRICATION. For Model MC-2, mix 1 pint (0.47 L) of SAE 40 two-stroke engine oil with each gallon of regular gasoline. For all other models, mix one part SAE 40 two-stroke engine oil to 10 parts of 100 octane white marine gasoline or regular grade gasoline.

CARBON. Lack of power may indicate exhaust ports need cleaning. Clean ports with wooden scraper.

REPAIRS

TIGHTENING TORQUES. Recommended tightening torques are as follows:

Breaker point screws	30-35 in.-lbs. (3-4 N·m)
Carburetor to adapter	90-100 in.-lbs. (10-11.3 N·m)
Adapter to manifold	60-65 in.-lbs. (6.8-7.3 N·m)
Manifold to cylinder	60-65 in.-lbs. (6.8-7.3 N·m)
Coil and lamination screws	55-60 in.-lbs. (6.2-6.8 N·m)
Condenser screw	30-35 in.-lbs. (3-4 N·m)
Connecting rod – MC-70 & MC-75	90-95 in.-lbs. (10-10.7 N·m)
All other models	65-70 in.-lbs. (7.3-8 N·m)
Crankcase bottom screws	95-100 in.-lbs. (10.7-11.3 N·m)
Cylinder head screws	55-60 in.-lbs. (6.2-6.8 N·m)
Exhaust stack	55-60 in.-lbs. (6.2-6.8 N·m)
Fan housing	55-60 in.-lbs. (6.2-6.8 N·m)
Flywheel nut	300-360 in.-lbs. (34-41 N·m)
Reed valve clamp	30-35 in.-lbs. (3-4 N·m)
Spark plug	216-264 in.-lbs. (24-30 N·m)
Clutch or sprocket nut	260-300 in.-lbs. (29-34 N·m)

CONNECTING ROD. On Models MC-2 and MC-70, the cylinder head is nonremovable and the piston and connecting rod assembly must be removed from the bottom. Remove engine bottom cover, detach rod from crankshaft and remove crankshaft and engine cover assembly. Withdraw rod and piston through bottom opening. On all other models, rod and piston assembly are removed from above.

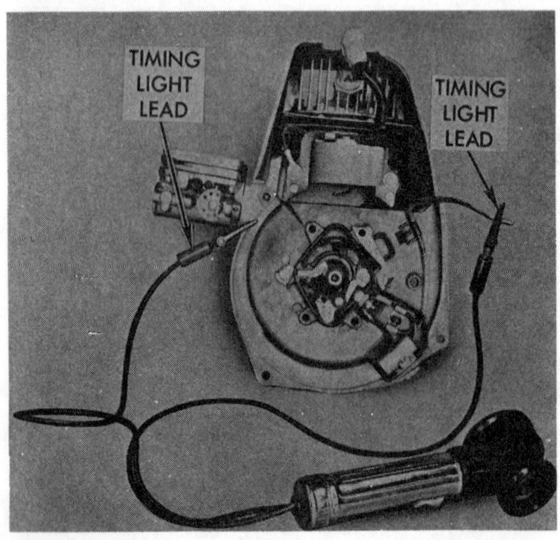

Fig. MC9 – Illustrating proper use of static timing light in adjusting breaker point gap on McCulloch kart engines.

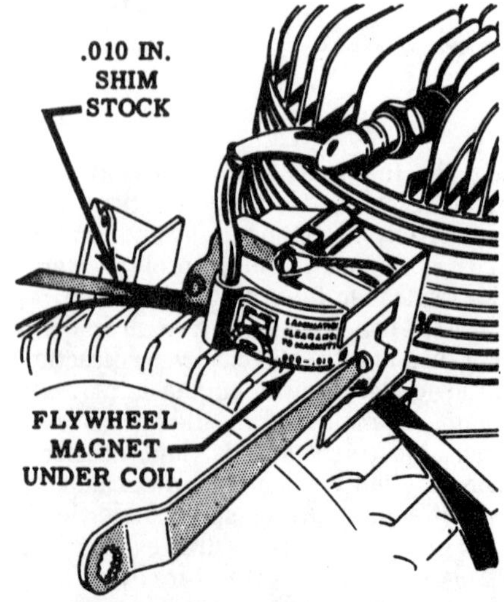

Fig. MC10 – Recommended magneto air gap is obtained by shifting coil on mounting screws.

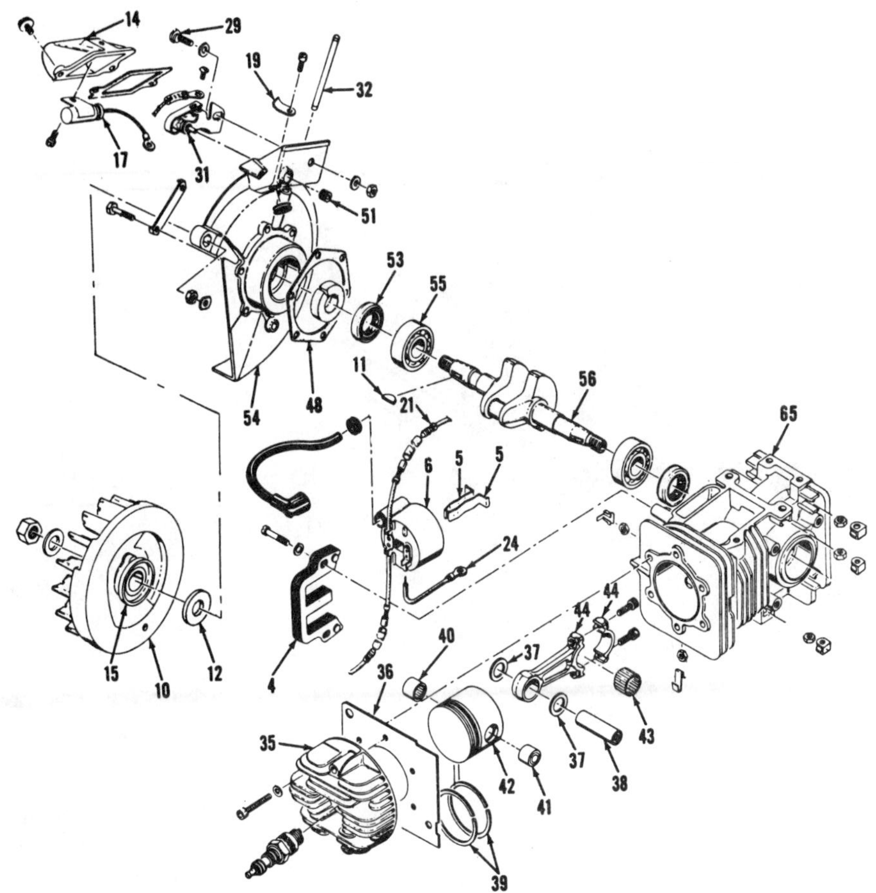

Fig. MC11—Exploded view of MC-10 engine. Note external breaker box on crankcase cover (54) and push rod (32) that actuates breaker points.

4. Coil laminations
5. Coil retainers
6. Coil
10. Flywheel
11. Woodruff key
12. Felt dirt shield
14. Breaker box cover
17. Condenser
24. Ground wire
29. Eccentric screw
31. Breaker assy.
32. Push rod
35. Cylinder head
36. Gasket
37. Thrust washers
38. Piston pin
39. Rings
40. Open needle bearing
41. Closed end needle bearing
42. Piston
43. Needle rollers (24)
44. Connecting rod
48. Gasket
51. Bushing
53. Crankcase seal
54. Crankcase cover
55. Ball bearing (2)
56. Crankshaft
65. Crankcase and cylinder

The crankpin bearing on Model MC-2 is a two-piece floating bushing. On all other models, 24 uncaged needle roller bearing are used. Renew rod, bushings, or crankshaft on Model MC-2 if scored or worn excessively. On all other models, crankpin nominal diameter is 0.6298 inch (15.997 mm). Clearance between end of rollers and side of crankpin journal (needle end clearance) should be 0.008-0.018 inch (0.20-0.46 mm). Accumulative clearance between rollers should be 0.008-0.010 inch (0.20-0.025 mm). Side clearance between rod and crankshaft should be 0.100-0.110 inch (2.54-2.79 mm). Renew rod and/or crankshaft if scored or if any wear spots are visible.

Install crankpin needle rollers by sticking 12 rollers in the rod and 12 in the rod cap with grease. Align "pips" (Fig. MC12) on rod and cap when installing on crankpin. Parting faces of connecting rod and cap are fractured to provide the dowel effect on the meshing of the consequent uneven surface. It is advisable to wiggle the rod cap back and forth while tightening to make sure the surfaces of the fractured joint are in perfect mesh. When properly meshed, no "catch points" will be felt when fingernail is rubbed along parting line of rod and cap (Fig. MC13).

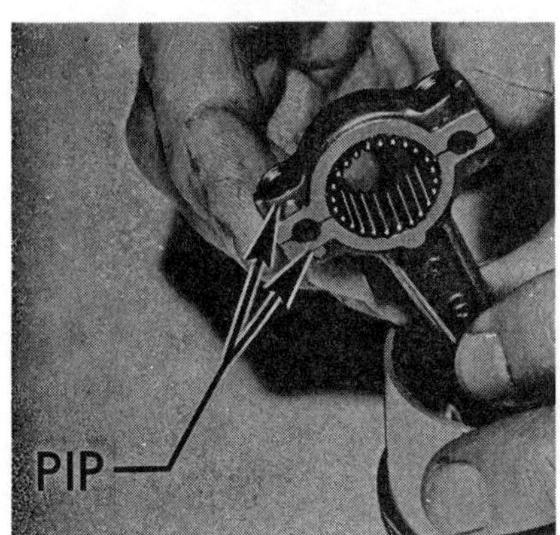

Fig. MC12—Install cap on connecting rod with "pips" on rod and cap aligned.

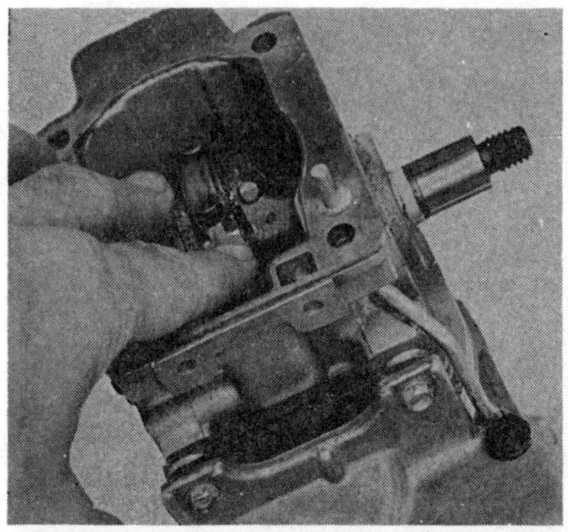

Fig. MC13—When properly assembled, parting line on rod and cap is practically invisible and no "catch points" can be felt with fingernail.

PISTON, PIN AND RINGS. Piston in early MC-5 and MC-10 model engines was fitted with two thick compression

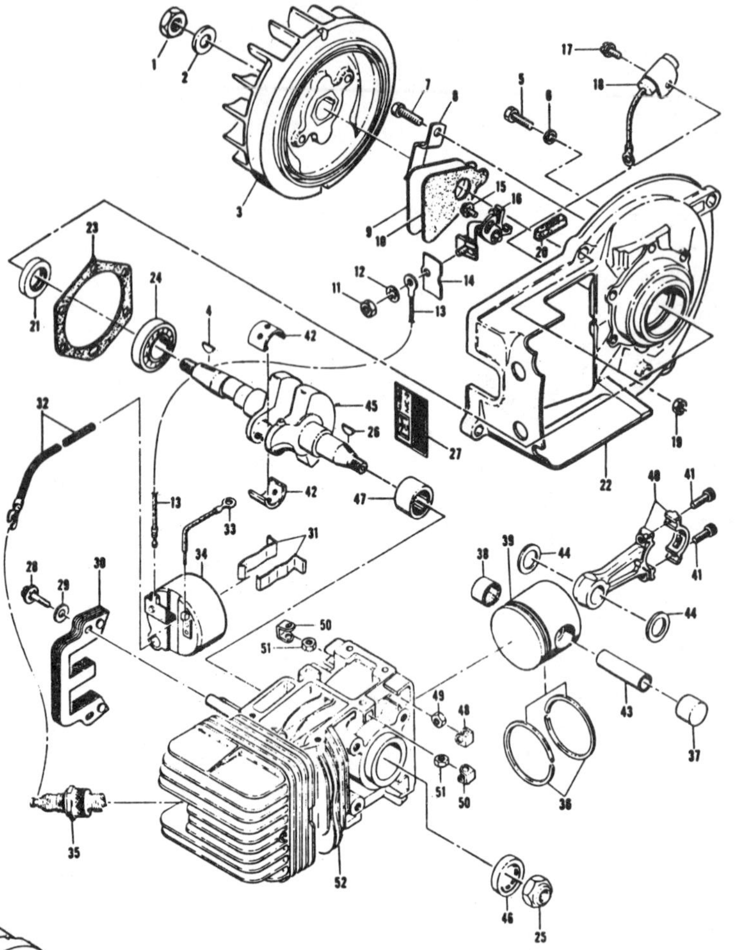

Fig. MC14—Exploded view of Model MC-2 bushing engine.

3. Flywheel
4. Woodruff key
8. Retainer
9. Breaker box cover
10. Gasket
16. Breaker assy.
18. Condenser
20. Felt wiper
21. Oil seal
22. Crankcase cover
23. Gasket
24. Ball bearing
26. Woodruff key
30. Coil lamination
31. Coil retainers
34. Coil assy.
35. Spark plug
36. Piston ring set
37. Closed end bearing
38. Open end bearing
39. Piston
40. Connecting rod
42. Floating rod bushing
43. Piston pin
44. Thrust washers
45. Crankshaft
46. Oil seal
47. Needle bearing
52. Crankcase

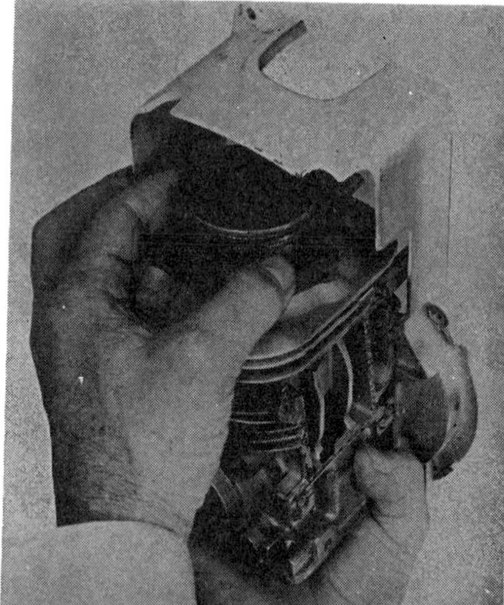

Fig. MC15—Removing connecting rod and piston assembly from models with removable cylinder head. It is not necessary to remove crankcase cover or crankshaft to remove piston and rod on models with removable cylinder head.

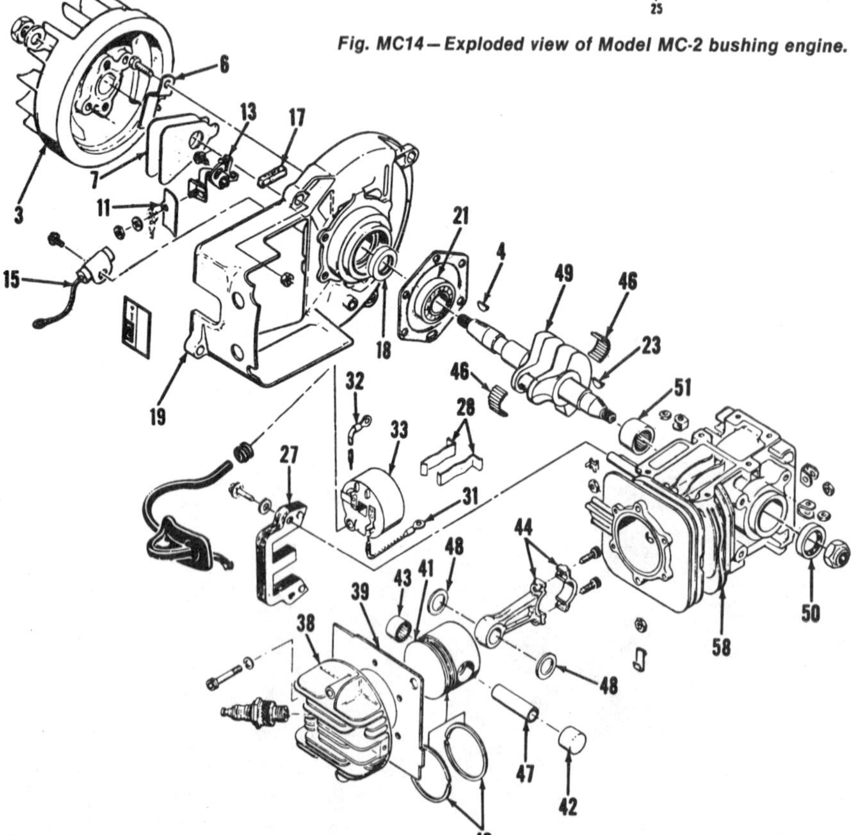

Fig. MC16—Exploded view of MC-6 engine; MC-5, MC-7 and MC-8 are similar except that MC-7 and MC-8 use an "O" ring in place of crankcase cover gasket shown adjacent to ball bearing (21).

3. Flywheel
4. Woodruff key
6. Breaker box cover retainer
7. Breaker box cover
11. Insulator
13. Breaker points
15. Condenser
17. Felt wiper
18. Crankcase seal
19. Crankcase cover
21. Ball bearing
23. Woodruff key
27. Coil laminations
28. Coil retainers
31. Ground wire
32. Primary wire
33. Coil
38. Cylinder head
39. Gasket
40. Rings
41. Piston
42. Closed-end needle bearing
43. Open needle bearing
44. Connecting rod
46. Needle rollers (24)
47. Piston pin
48. Thrust washers
49. Crankshaft
50. Crankcase seal
51. Needle bearing
58. Crankcase & cylinder

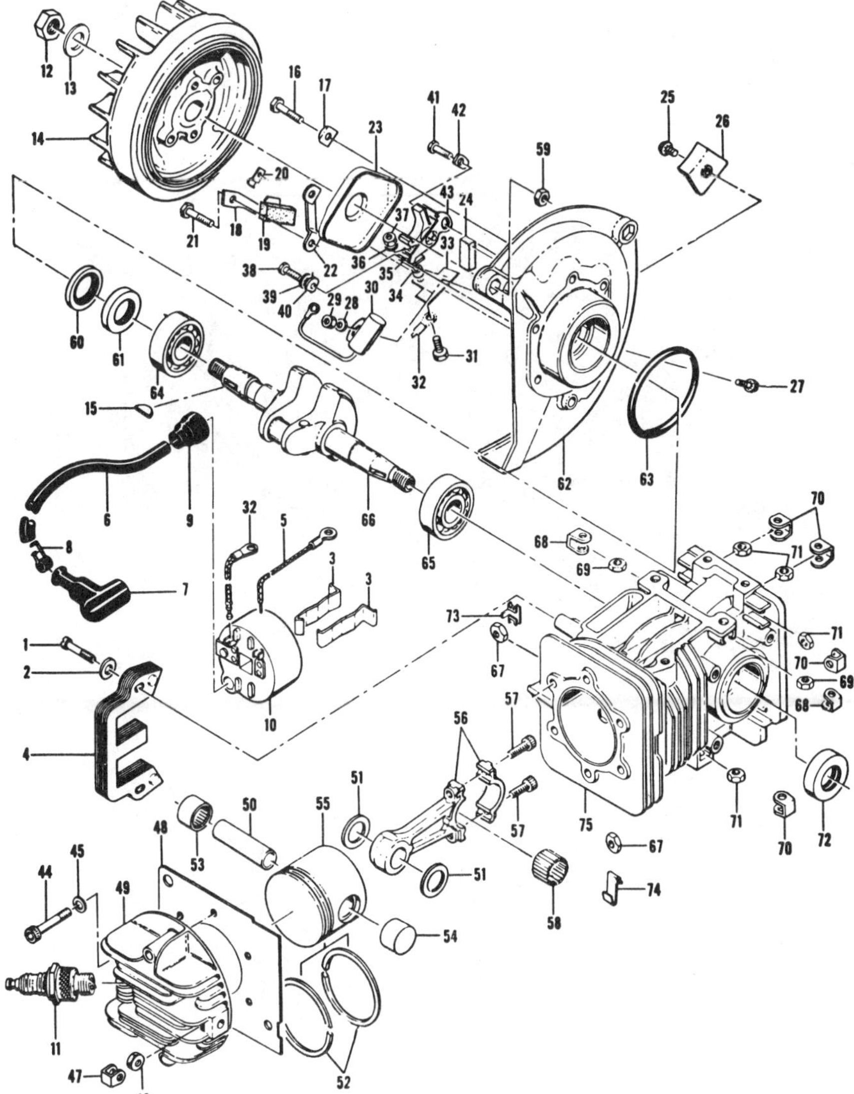

Fig. MC17—Exploded view showing construction of Models MC-20, MC-30, MC-40 engines. Model MC-45 is similar except for nine-port crankcase. Model MC75 is also similar except for nine-port crankcase and gasket is used instead of "O" ring (63).

4. Coil laminations
5. Grounding wire
6. Spark plug wire
10. Coil assy.
11. Spark plug
14. Flywheel
23. Breaker box cover
24. Felt cam wiper
30. Condenser
32. Primary coil wire
43. Breaker point assy.
49. Cylinder
50. Piston pin
51. Thrust washers
52. Piston rings
53. Needle bearing
54. Needle bearing
55. Piston
56. Connecting rod
58. Needle rollers
60. Oil seal
62. Crankcase cover
64. Ball bearing
65. Ball bearing
66. Crankshaft
75. Crankcase

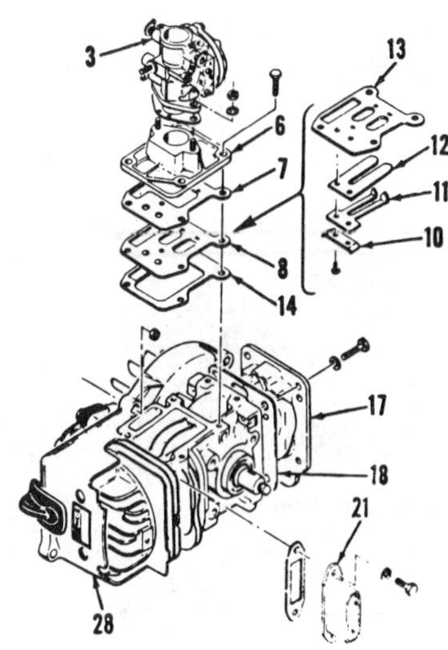

Fig. MC18—View of MC-6 reed assembly.

3. Carburetor
6. Inlet manifold
7. Gasket
8. Reed valve assy.
10. Lockplate
11. Reed guard
12. Valve reed
13. Reed plate
14. Gasket
17. Crankcase bottom
18. Gasket
21. Exhaust header

rings. Ring end gap should be 0.007-0.010 inch (0.18-0.25 mm). If piston skirt-to-cylinder wall clearance exceeds 0.005 inch (0.13 mm) or ring end gap exceeds 0.010 inch (0.25 mm) with new piston ring, hone or rebore cylinder to next oversize or renew cylinder. If ring side clearance in top ring groove exceeds 0.004 inch (0.10 mm), renew the piston. Piston rings should always be renewed whenever engine is disassembled for service. Piston and rings are available in oversizes. Install the chrome plated ring in top ring groove and install cast iron ring in second ring groove.

On late production MC-5 and MC-10 model engines and all other models, piston is fitted with two thin chrome compression rings. Ring end gap should be 0.051-0.091 inch (1.27-2.31 mm) on Models MC-40, MC-45, MC-70 and MC-75 which have ring retaining pins in the ring grooves. Ring end gap on all other thin-ring pistons is 0.004-0.050 inch (0.10-1.27 mm). If piston skirt-to-cylinder wall clearance exceeds 0.007 inch (0.18 mm) or new rings cannot be fitted within end gap tolerance, hone or rebore cylinder to next oversize for which piston and rings are available or renew cylinder. If ring side clearance in top ring groove exceeds 0.004 inch (0.10 mm) with new piston ring, renew piston. Piston rings should always be renewed

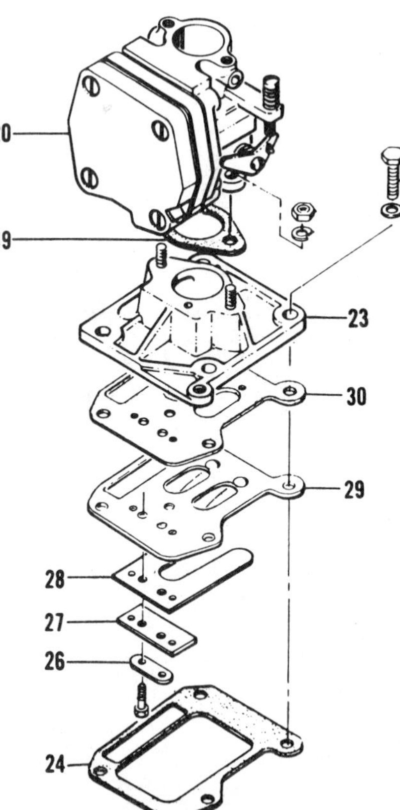

Fig. MC19—Carburetor mounting and inlet reed valve system on Models MC-2 and MC-20. Some other models are similar except reed guards are used.

19. Gasket
20. Carburetor
23. Intake manifold
24. Gasket
26. Lockplate
27. Reed clamp
28. Reed valve
29. Reed plate

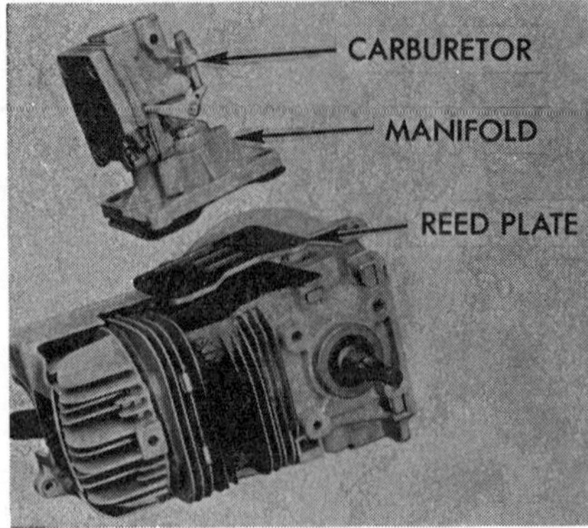

Fig. MC20—View of carburetor and manifold removed from Model MC-8 engine showing reed plate. Refer to Fig. MC19 for exploded view of reed plate assembly.

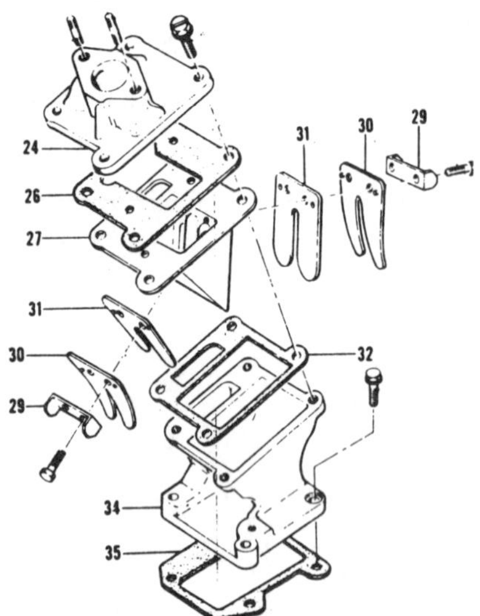

Fig. MC21—Reed inlet valve installation used on Model MC-30. Two sets of reed valves are used to increase intake capacity. Installations on dual carburetor MC-40 and MC-70 models are similar. Refer to Fig. MC22.

24. Carburetor adapter
26. Gasket
27. Reed valve block
29. Lockplate
30. Reed valve guard
31. Reed valve
32. Gasket
34. Manifold body
35. Gasket

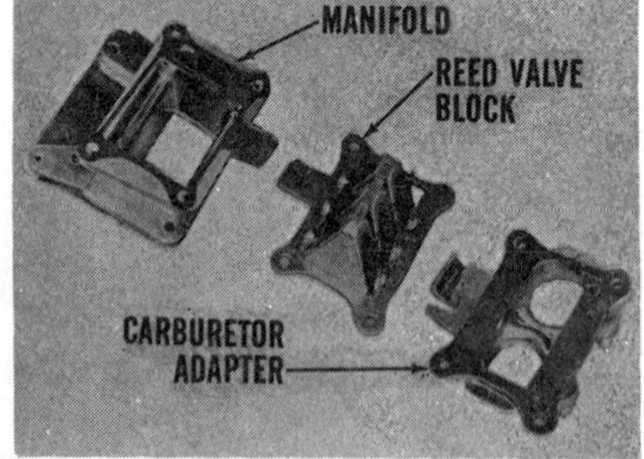

Fig. MC22—View of manifold, reed valve block with six reeds and carburetor adapter used with dual carburetor MC-40 and MC-70 models. Refer to Fig. MC21 for exploded view of reed valve block for single carburetor MC-30 engine which is of similar construction.

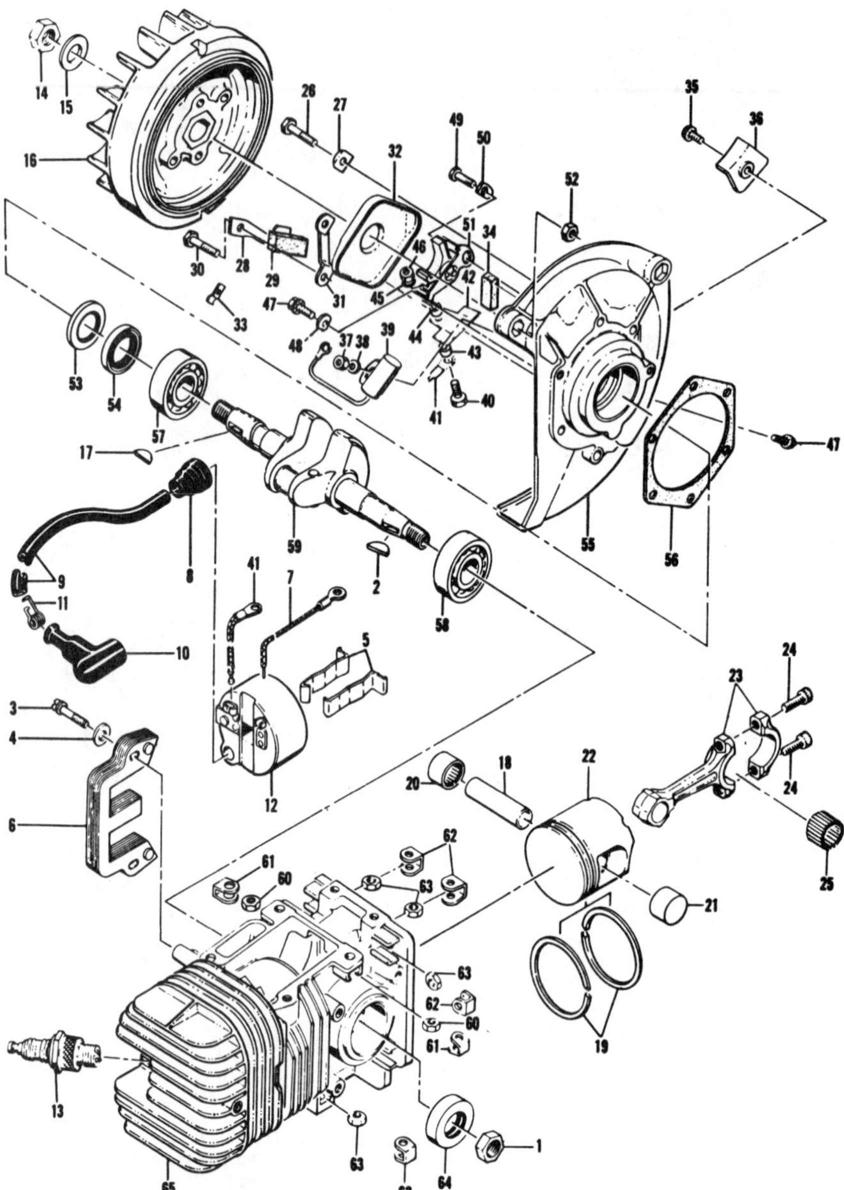

Fig. MC23—Exploded view of Model MC-70 engine.

whenever engine is disassembled for service. Piston and rings are available in a variety of oversizes as well as standard size.

Piston pin is a press fit in connecting rod of all models. Model MC-5 piston has nonrenewable oilite bushings in piston pin bore; all other models use one open needle bearing and one closed end needle bearing in piston. Piston must be supported in special support block available from McCulloch Corporation when pressing pin in or out of piston and rod. The closed end needle bearing used in the piston of all models except MC-5 must be installed in the side of the piston towards the exhaust port in cylinder wall.

CRANKSHAFT. The crankshaft is supported by a ball bearing at the flywheel

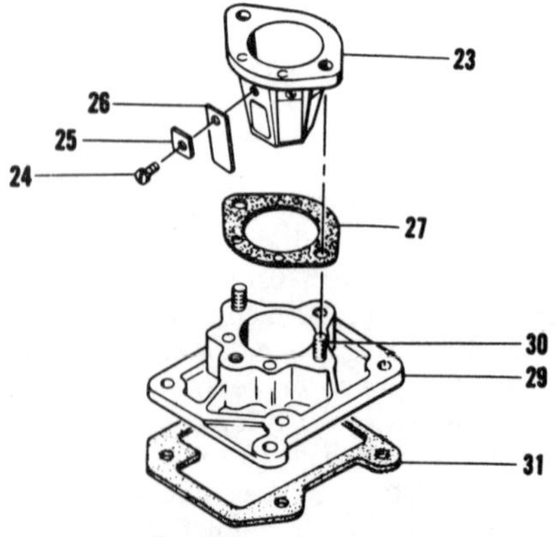

Fig. MC24—Exploded view of pyramid type reed inlet valve assembly and carburetor adapter used on Models MC-45 and MC-75. Six steel reeds (26) are used.

23. Valve block
24. Screws
25. Reed clamp plate
26. Reed valves
27. Gasket
29. Adapter
30. Stud bolts
31. Gasket

end and a caged needle roller bearing at the pto end on Models MC-2, MC-5, MC-6, MC-7 and MC-8. On Models MC-10, MC-20, MC-40 and MC-70, the crankshaft is supported by ball bearings at each end. The ball type main bearings are a press fit on the crankshaft and both the ball and needle type main bearings are a press fit in the crankcase or cover. Inspect pto end journal where needle bearing is used and crankpin journal for scoring or wear spots and renew crankshaft if these conditions are noted.

Crankcase and crankcase cover must be heated to 200° F (93° C) in an oven when installing ball or needle type bearings to prevent damage to bearing bore.

VALVE SYSTEM. A combination of reed and third port (piston porting) valve system is used. The smooth side of the valve reed should face the valve plate.

McCULLOCH

Model	Bore	Stroke	Displacement
MC-49, MC-49C, MC-49E	2.125 in. (54 mm)	1.375 in. (35 mm)	4.9 cu. in. (80 cc)
MC-90, MC-91A, MC-91B, MC-91C	2.165 in. (55 mm)	1.635 in. (41.5 mm)	6.1 cu. in. (99 cc)
MC-100	2.250 in. (57.2 mm)	1.835 in. (46.6 mm)	7.3 cu. in. (120 cc)
MC-101, MC-101A, MC-101C	2.280 in. (57.9 mm)	1.835 in. (46.6 mm)	7.5 cu. in. (123 cc)

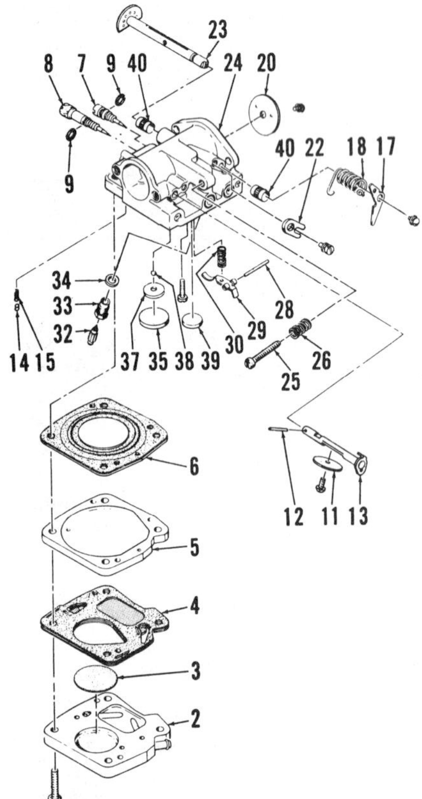

Fig. MC25—Exploded view of McCulloch diaphragm type carburetor with integral fuel pump used on Models MC-49, MC-49C, MC-90 and MC-100. On some models, a screen (capillary seal) is used instead of check valve ball (38) and seat (37). Some carburetors do not have choke.

2. Fuel pump body
3. Fuel pump filter
4. Fuel pump diaphragm
5. Carburetor diaphragm
6. Carburetor diaphragm
7. Idle fuel needle
8. Main fuel needle
9. "O" ring
11. Choke plate
12. Groove pin
13. Choke shaft
14. Choke friction pin
15. Choke friction spring
17. Throttle shaft arm
18. Throttle return spring
20. Throttle plate
22. Throttle shaft clip
23. Throttle shaft
24. Carburetor body
25. Idle speed screw
26. Idle speed screw spring
28. Inlet lever pin
29. Inlet control lever
30. Inlet lever spring
32. Inlet needle valve
33. Valve body
34. Washer
35. Expansion plug
37. Valve seat
38. Check valve ball
39. Expansion plug
40. Throttle shaft bushings

MAINTENANCE

SPARK PLUG. For normal operating conditions, Models MC-49, MC-49C and MC-49E are equipped with Champion J8J spark plug, Models MC-91B and MC-101A are equipped with Champion L78 spark plug and all other models use Champion J4J spark plug. Electrode gap for all models should be 0.025 inch (0.64 mm). Spark plug heat range may have to be changed for certain operating conditions.

CARBURETOR. McCulloch diaphragm or pressure pulse type carburetors are used on these models. Refer to Figs. MC25, MC26, MC27 and MC28 for exploded view of carburetors.

For initial carburetor adjustment on Models MC-49, MC-49C, MC-90 and MC-100, turn fuel adjusting needles clockwise until they are lightly seated. Open idle mixture needle (7–Fig. MC25) 1¼ turns and main fuel needle (8) 1½ turns. With engine at operating temperature and running, adjust idle speed screw (25) until engine idles at a speed of 1500-1700 rpm. Check engine for acceleration and open main fuel needle a slight amount if engine misfires, or turn needle in a slight amount if engine smokes and is sluggish on acceleration.

On Model MC-49E, lightly seat idle mixture needle (19–Fig. MC26), then open the needle 1¼ turns. Main fuel jet is nonadjustable. Standard jet (17) has an orifice diameter of 0.031 inch (0.79 mm). Optional jets with orifice diameters of 0.028 inch (0.71 mm) (lean), 0.029 inch (0.73 mm) and 0.033 inch (0.84 mm) (rich) are available. With engine at operating temperature and running, adjust idle speed screw (23) to obtain an engine idle speed of 1500-1700 rpm. Place a load on engine and check engine for acceleration. If engine falters or misfires on acceleration, open idle mixture needle a small amount. If engine runs rough and smokes heavily on acceleration, turn needle in (clockwise) slightly.

On Models MC-91, MC-91A, MC-91C, MC-101 and MC-101C, turn both fuel ad-

Fig. MC26—Exploded view of McCulloch diaphragm type carburetor used on Model MC-49E.

1. Cover
2. Gasket
3. Check valve diaphragm
4. Fuel pump diaphragm
5. Fuel screen
6. Choke plate
7. Carburetor body
8. Choke shaft
9. Throttle shaft clip
10. Fuel inlet needle
11. Lever pin
12. Inlet lever
13. Gasket
14. Carburetor diaphragm
15. Cover
16. Lever spring
17. Main fuel orifice
18. Throttle plate
19. Idle mixture needle
20. Throttle shaft
21. Throttle plate
22. Throttle return spring
23. Idle speed screw

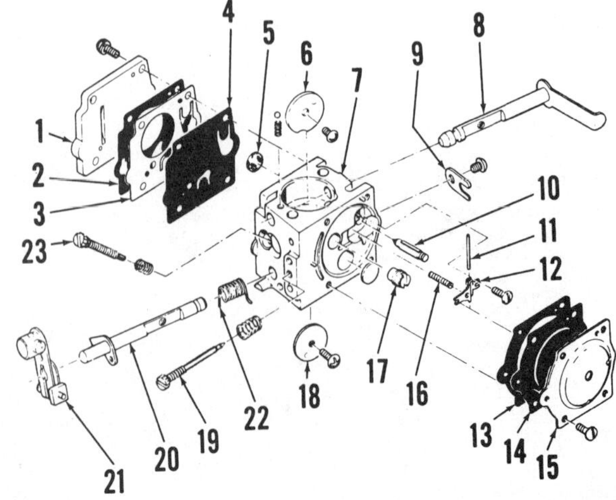

justing needles until they are lightly seated, then open both fuel needles (16 and 20 – Fig. MC27) 1 turn. With engine warm and running, adjust idle speed screw (8) to obtain an engine idle speed of 1500-1700 rpm. Check engine or acceleration and if engine falters or misfires, open main fuel needle slightly. If engine smokes and is sluggish on acceleration, turn main fuel needle in slightly.

On Models MC-91B and MC-101A, turn idle mixture needle (23 – Fig. MC28) and main fuel needle (24) clockwise until they are lightly seated. Turn idle mixture needle counterclockwise 1¼ turns, then turn main fuel needle counterclockwise 1½ to 2 turns. With engine at operating temperature and running, adjust idle speed screw (14) until engine idle speed is 1500-1700 rpm. Place a load on engine and check engine acceleration. If engine runs rough or smokes heavily on acceleration, close main fuel needle a slight amount. If engine falters or misfires on acceleration, open main fuel needle a slight amount.

CAUTION: On all models, avoid carburetor settings that will run the engine too lean. If setting is too lean, overheating and lack of lubrication will seriously damage the engine.

MAGNETO AND TIMING. Proper ignition timing is obtained by adjusting breaker point gap. If degree wheel and static timing light are available, adjust breaker points so they start to open at 25° BTDC on Models MC-49, MC-49C and MC-90, 26° BTDC on Models MC-49E, MC-91, MC-91A, MC-91B and MC-91C or 22° BTDC on Models MC-100, MC-101, MC-101A and MC-101C. An alternate method of setting timing is to adjust breaker point opening to 0.018 inch (0.46 mm) for 25° BTDC, 0.019 inch (0.48 mm) for 26° BTDC or 0.015 inch (0.38 mm) for 22° BTDC.

Refer to Fig. MC30 for magneto used on Models MC-49, MC-49C, MC-49E, MC-90, MC-91, MC-91A, MC-91B and MC-91C. Model MC-100, MC-101, MC-101A and MC-101C use magneto shown in Fig. MC31. Adjust armature air gap to 0.010 inch (0.25 mm) on all models.

LUBRICATION. Mix one part SAE 40 two-stroke engine oil with 20 parts of regular grade automotive gasoline or 100 octane white marine gasoline.

CARBON. Lack of power may indicate exhaust ports need cleaning. Clean ports with a wooden scraper.

Fig. MC28—Exploded view of McCulloch pressure pulse type carburetor used on Models MC-91B and MC-101A.

2. Fuel pump body
3. Fuel filter
4. Fuel pump
5. Diaphragm plate
6. Carburetor diaphragm
7. Circuit plate
8. Check valve diaphragm
9. Gaskets
10. Inlet lever
11. Lever pin
12. Fuel inlet needle
13. Lever spring
14. Idle speed screw
15. Spring
16. Throttle shaft clip
17. Throttle shaft arm
18. Spring
19. Bushing
20. Carburetor body
21. Throttle plate
22. Throttle shaft
23. Idle mixture needle
24. Main fuel needle
25. Spring
26. "O" ring washer
27. "O" ring
28. One-way valve

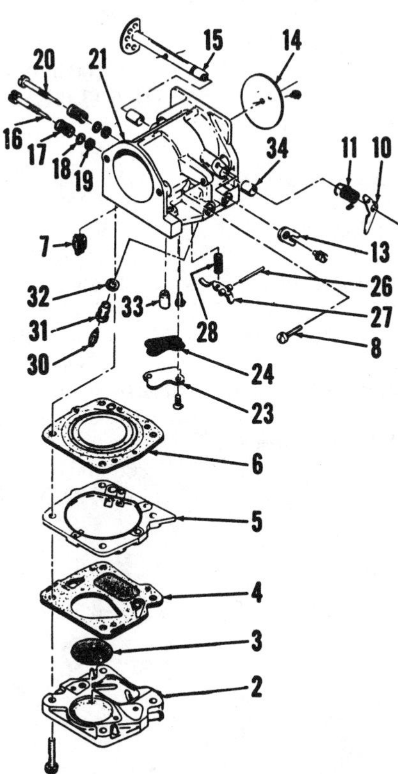

Fig. MC27—Exploded view of McCulloch pressure pulse type carburetor used on Models MC-91, MC-91A, MC-91C, MC-101 and MC-101C.

2. Fuel pump body
3. Fuel pump filter
4. Fuel pump diaphragm
5. Diaphragm plate
6. Carburetor diaphragm
7. One-way "duck-bill" valve
8. Idle speed screw
10. Throttle shaft arm
11. Throttle shaft spring
13. Throttle shaft clip
14. Throttle plate
15. Throttle shaft
16. Main fuel needle
17. Needle spring
18. "O" ring washer
19. "O" ring
20. Idle fuel needle
21. Carburetor body
23. Idle cluster cover
24. Gasket
26. Control lever pin
27. Inlet lever
28. Inlet lever spring
30. Inlet valve needle
31. Inlet valve body
32. Inlet valve gasket
33. Check valve
34. Throttle shaft bushing

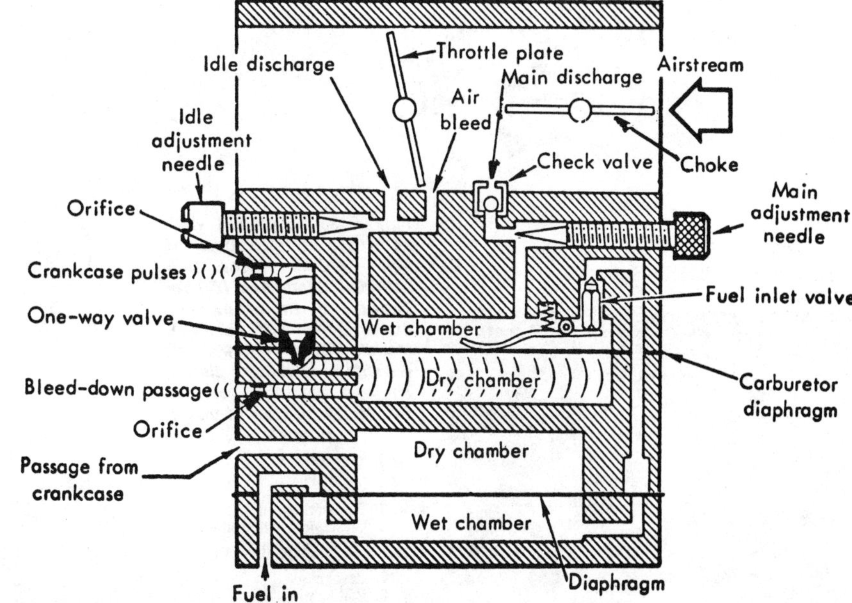

Fig. MC29—Cross-sectional diagram of McCulloch pressure pulse type carburetor. Note straight bore through carburetor; venturi is not required as fuel is ejected into carburetor bore under pressure from fuel pump. Pulse pressure from engine crankcase works against fuel pressure and force of inlet lever spring to regulate inlet valve opening, thus metering fuel ejected into carburetor bore to meet engine requirements.

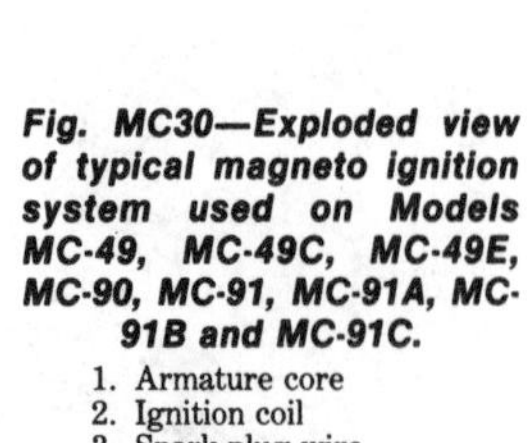

Fig. MC30—Exploded view of typical magneto ignition system used on Models MC-49, MC-49C, MC-49E, MC-90, MC-91, MC-91A, MC-91B and MC-91C.

1. Armature core
2. Ignition coil
3. Spark plug wire
4. Coil retainer clips
5. Breaker box felt
6. Primary wire
7. Insulator
11. Breaker points
12. Cam wiper felt
14. Crankcase cover

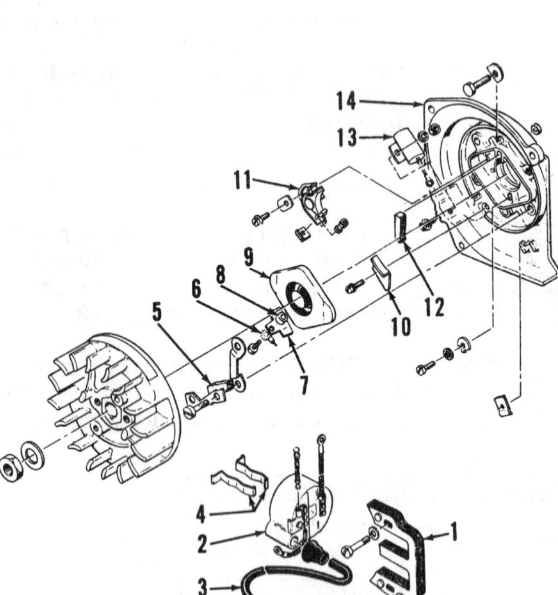

Fig. MC31—Exploded view of magneto ignition system used on Models MC-100, MC-101, MC-101A and MC-101C.

1. Armature core
2. Ignition coil
3. Spark plug wire
4. Coil retaining clips
5. Breaker box felt
6. Primary wire
7. Insulator
8. Insulator
9. Breaker box cover
10. Wire retaining clip
11. Breaker points
12. Cam wiper felt
13. Condenser
14. Crankcase cover

REPAIRS

TIGHTENING TORQUES. Recommended tightening torques are as follows:

Breaker point screws	30-35 in.-lbs. (3-4 N·m)
Carburetor to adapter	90-100 in.-lbs. (10-11 N·m)
Adapter to manifold	60-65 in.-lbs. (6.8-7.3 N·m)
Manifold to cylinder	60-65 in.-lbs. (6.8-7.3 N·m)
Coil and lamination screws	55-60 in.-lbs. (6.2-6.8 N·m)
Condenser screw	30-35 in.-lbs. (3-4 N·m)
Connecting rod– MC-100, MC-101, MC-101A, MC-101C	90-95 in.-lbs. (10.0-10.5 N·m)
MC-91, MC-91A MC-91B, MC-91C	105-110 in.-lbs. (12-13 N·m)
All other models	65-70 in.-lbs. (7.3-8 N·m)
Crankcase bottom screws	95-100 in.-lbs. (10.5-11.0 N·m)
Cylinder head screws	55-60 in.-lbs. (6.2-6.8 N·m)
Exhaust stack	55-60 in.-lbs. (6.2-6.8 N·m)
Fan housing	55-60 in.-lbs. (6.2-6.8 N·m)
Flywheel nut	300-360 in.-lbs. (34-41 N·m)
Reed valve clamp	30-35 in.-lbs. (3-4 N·m)
Spark plug	216-264 in.-lbs. (24-30 N·m)
Clutch or sprocket nut	260-300 in.-lbs. (22-34 N·m)

Fig. MC32—When installing connecting rod cap, "pips" on rod and cap must be aligned. If "pips" are not aligned as shown at right, rod and bearing failure will result.

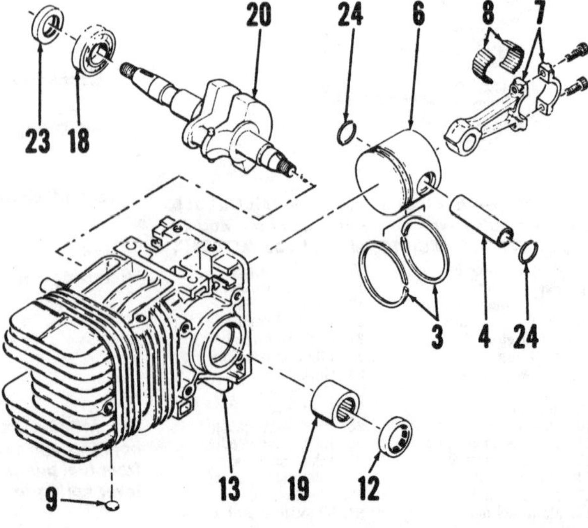

Fig. MC33—Exploded view showing construction of Model MC-49 engine. Ball bearing (18) and seal (23) are supported in crankcase cover (14—Fig. MC30). Models MC-49C and MC-49E are similar except for piston, pin and bearings; refer to Fig. MC34 for view of piston assembly.

3. Piston rings
4. Piston pin
6. Piston
7. Connecting rod
8. Crankpin needle rollers
9. Expansion plug
12. Crankshaft seal
13. Crankcase
18. Ball bearing
19. Needle roller bearing
20. Crankshaft
23. Crankshaft seal
24. Snap rings

CONNECTING ROD. On Models MC-49, MC-49C and MC-49E, the cylinder head is not removable and the piston and connecting rod assembly must be removed from bottom of crankcase after removing bottom cover, crankcase cover and the crankshaft. On all other models, piston and connecting rod assembly can be removed after removing bottom cover and cylinder head.

On all models, 24 uncaged needle rollers are used for crankpin bearing. Renew the needle rollers as a complete set if any roller has a flat spot, or if it has burn or score marks. Renew crankshaft and/or connecting rod if bearing surfaces are scored or rough. When reassembling, use light grease to stick 12 needle rollers in connecting rod and 12 needle rollers in cap. Fit connecting rod to crankpin, then install cap with "pips" aligned as shown in Fig. MC32.

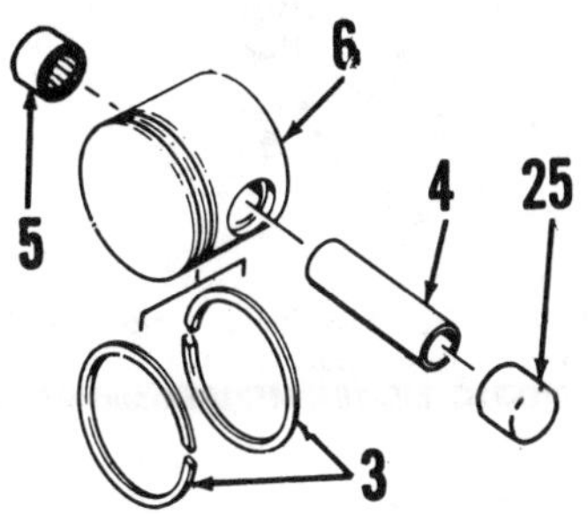

Fig. MC34—View showing type of piston, pin and piston pin bearings used on Models MC-49C, MC-90 and MC-100. Refer to Fig. MC33 for view showing construction, except for piston assembly, typical of models MC-49C and MC-49E and to Fig. MC37 for view of assembly typical of Models MC-90 and MC-100 except for piston.

3. Piston rings
4. Piston pin
5. Open end needle bearing
6. Piston
25. Closed end needle bearing

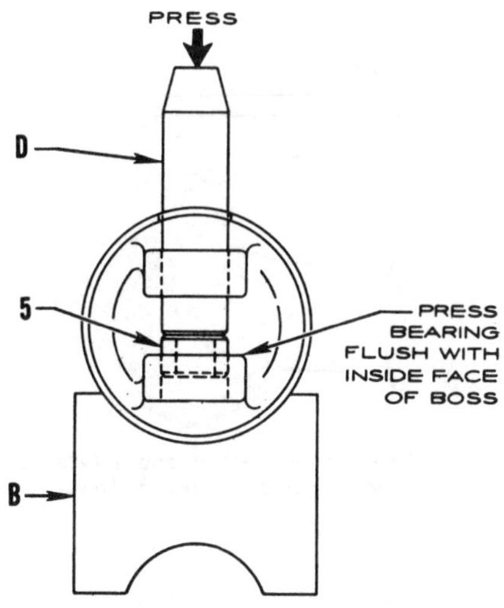

Fig. MC36—Using driver (D) and support block (B) to install piston pin needle bearings. After bearings are installed, a similar driver can be used to install piston pin.

PISTON, PIN AND RINGS. Piston is fitted with two thin steel rings. Ring end gap should be 0.004-0.050 inch (0.10-1.27 mm) on pistons that do not have ring locating pin, or 0.051-0.091 inch (1.30-2.31 mm) on pistons with pinned rings. If ring side clearance in groove exceeds 0.004 inch (0.10 mm) with new ring, the piston should be renewed. Also, renew piston if scored or if cylinder to piston skirt clearance exceeds 0.007 inch (0.18 mm). Piston and rings are available in standard size for all models as well as a variety of oversizes.

Piston pin is retained by snap rings at each end of pin bore in piston on Model MC-49 (Fig. MC33), and by being a press fit in connecting rod on all other models. On Model MC-49, piston pin rides in nonrenewable bushings in piston; pistons on all other models are fitted with caged needle roller bearings as

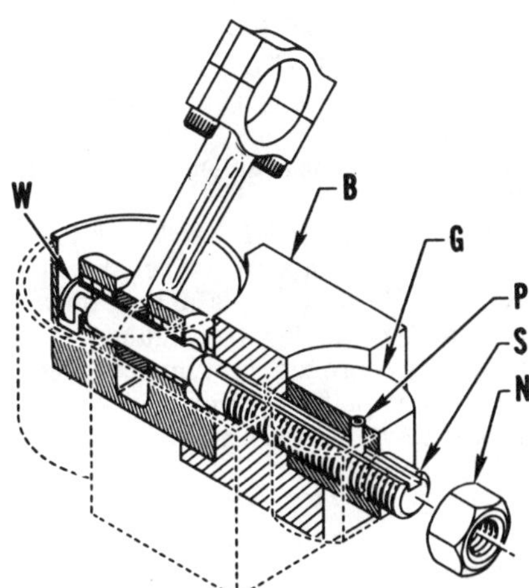

Fig. MC35—Cross-sectional view showing use of special McCulloch piston pin and needle bearing removal tool. A small washer (W) is inserted between piston skirt and pin boss to pull piston pin. After removing pin, a larger washer is used with tool to remove blind piston pin needle bearing. Refer to Fig. MC36 for pin and bearing installation.

B. Support block
G. Swivel guide
N. Nut
S. Puller shaft
W. Horseshoe washer

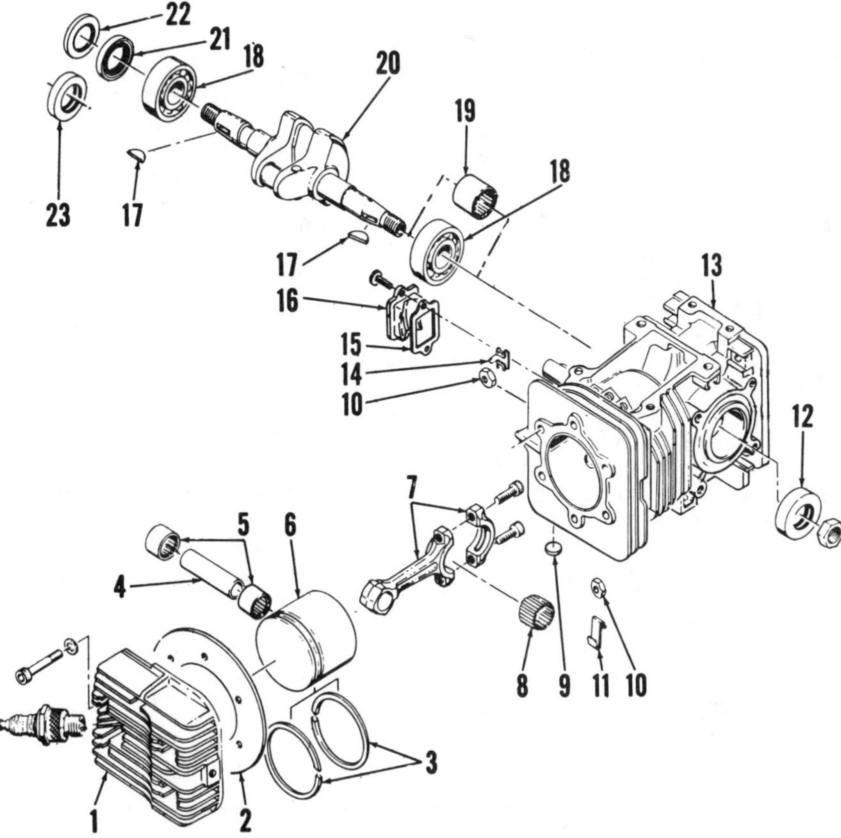

Fig. MC37—Exploded view showing typical construction of Models MC-91, MC-91A, MC-91B, MC-91C, MC-101, MC-101A and MC-101C. Models MC-90 and MC-100 are of similar construction except piston assembly shown in Fig. MC34 is used and Models MC-90 and MC-100 do not have port cover (16). Needle bearing (19) is used at output end of crankshaft on all models except MC-100, MC-101, MC-101A and MC-101C. Latest type crankshaft seals (12 and 23) are installed with sealing lip outward instead to towards crankpin.

1. Cylinder head
2. Head gasket
3. Piston rings
4. Piston pin
5. Open end needle bearings
6. Piston (exhaust side blind)
7. Connecting rod
8. Crankpin needle rollers
9. Expansion plug
10. Nuts
11. Nut retainer
12. Crankshaft seal
13. Crankcase
14. Nut retainer
15. Gasket
16. Port cover plate
17. Woodruff keys
18. Ball bearing
19. Needle roller bearing
20. Crankshaft
21. Inner crankshaft seal
22. Outer crankshaft seal
23. Single crankshaft seal

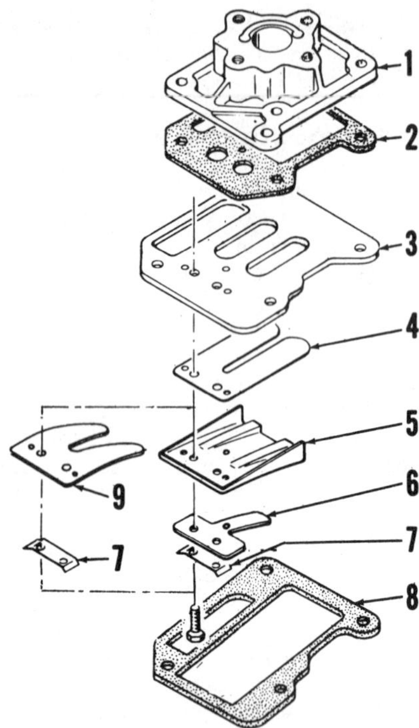

Fig. MC38—Exploded view of carburetor adapter and reed valve plate assembly used on Models MC-49, MC-49C and MC-49E. Model MC-49C and MC-49E use reed guard (9); Model MC-49 is equipped with fuel deflector (5) and reed clamp (6).

1. Carburetor adapter
2. Gasket
3. Reed plate
4. Valve reeds
5. Fuel deflector
6. Reed clamp
7. Lockplate
8. Gasket
9. Reed guard

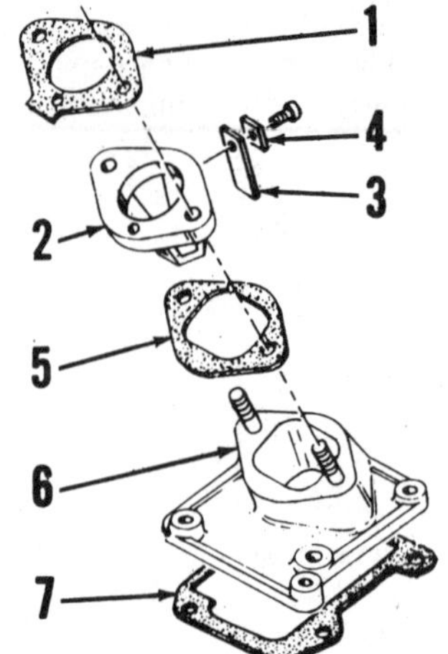

Fig. MC39—Models MC-90 and MC-100 are equipped with pyramid reed intake valve.

1. Gasket
2. Pyramid reed seat
3. Valve reeds
4. Reed clamp
5. Gasket
6. Carburetor adapter
7. Gasket

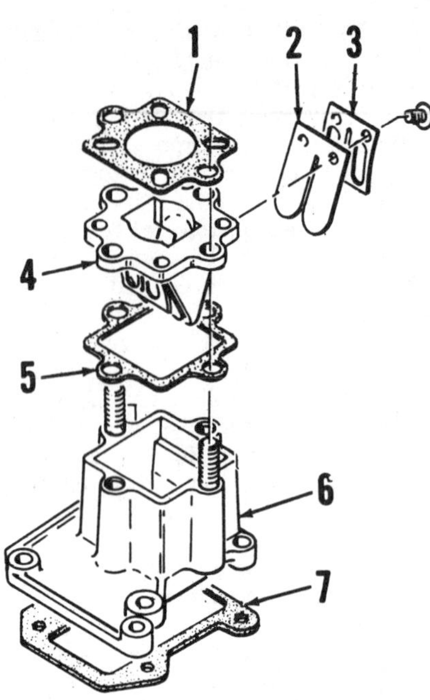

Fig. MC40—Exploded view of V-block reed valve assembly used on Models MC-91, MC-91A, MC-91B, MC-91C, MC-101, MC-101A and MC-101C.

1. Gasket
2. Valve reeds
3. Reed clamp plate
4. Valve block
5. Gasket
6. Adapter manifold
7. Gasket

shown in Fig. MC34. To remove or install piston pin and needle roller bearings in pistons used in Models MC-91, MC-91A, MC-91B, MC-91C, MC-101, MC-101A and MC-101C, refer to Figs. MC35 and MC36 for views showing use of tool kit for piston pin and bearing removal and installation.

On pistons with one closed end bearing, press pin out towards closed end bearing, forcing the bearing out with pin. Then, press open end bearing from piston. Install new bearings in piston so inner ends of bearing cages are flush with piston pin boss. Insert piston pin through open bearing and press through connecting rod until end of pin is flush with outer end of open end needle bearing cage. Always support piston in fixture when pressing pin in or out. Heating the connecting rod to approximately 180°F (82°C) will permit easier piston pin installation.

CRANKCASE. Where cylinder is worn so piston skirt to cylinder clearance is 0.007 inch (0.18 mm) or more with new piston or if ring end gap is excessive with new piston rings, the cylinder must be rebored to nearest oversize for which piston and rings are available or new crankcase and cylinder assembly must be installed.

CRANKSHAFT. On Models MC-100, MC-101, MC-101A and MC-101C, crankshaft is supported in ball bearings at each end. On all other models, flywheel end of crankshaft is supported in a ball bearing and output end is supported in a needle roller bearing. The ball bearings are a press fit on crankshaft and both the needle and ball bearings are a press fit in crankcase and crankcase cover. Inspect output end main journal for scoring or wear spots on models with needle bearing.

Crankcase and crankcase cover must be heated to 180°F (82°C) prior to installing crankshaft and ball bearing assembly, or installing needle bearing in crankcase on models so equipped.

REED VALVES. Flat valve reeds as shown in Fig. MC38 are used on Models MC-49, MC-49C and MC-49E. Models MC-90 and MC-100 are equipped with pyramid reeds as shown in Fig. MC39. Models MC-91, MC-91A, MC-91B, MC-91C, MC-101, MC-101A and MC-101C, which use McCulloch pressure pulse type carburetor, are equipped with V-type reed block as shown in Fig. MC40.

Correct seating of reed valves is essential to maintain crankcase pressure. Be sure reeds and seats are clean and reeds lay flat against seat. Renew any broken or chipped reeds. Renew reed plate, pyramid seat or reed block if worn in seat area.

SERVICING McCULLOCH ACCESSORIES

RECOIL STARTER

To disassemble the recoil starter, first unbolt and remove starter assembly from engine. Refer to Fig. MC41, hold handle (4) and remove cover (5). Allow rope pulley to rotate slowly until tension is removed from recoil spring. Remove ratchet (11), then carefully withdraw rope pulley from starter base (10). Recoil spring (8) should remain in starter base. Renew rope or recoil spring if necessary and reassemble starter. Preload recoil spring about two turns. Inspect the two spring loaded pawls which are pinned to the flywheel and renew any parts that show damage and excessive wear. Reinstall starter assembly.

1. Plate
2. Starter rope
3. Handle cup
4. Starter handle
5. Cover
6. Rope pulley
7. Rivet
8. Recoil spring
9. Bushings
10. Starter base
11. Ratchet

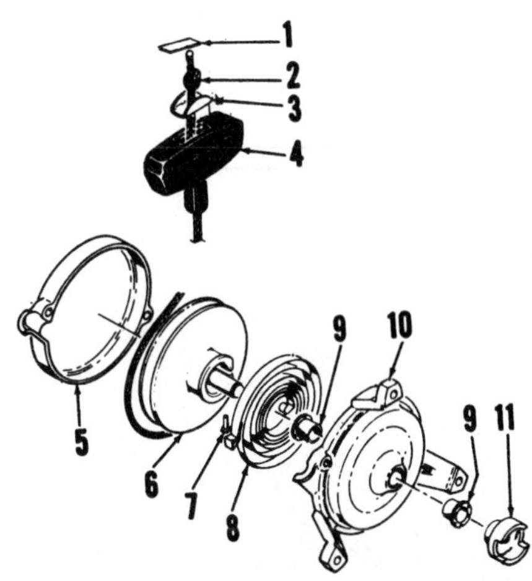

Fig. MC41—Exploded view of recoil starter assembly used on some McCulloch engines. A kit is available for installation of similar recoil starter on other models.

OLYMPYK

OLYMPYK SAWS
4575 N. Chatsworth Street
St. Paul, Minnesota 55126

Model	Bore	Stroke	Displacement
250 TTA	46 mm (1.81 in.)	32 mm (1.26 in.)	54 cc (3.3 cu. in.)
272 TTA	49 mm (1.93 in.)	35 mm (1.38 in.)	67 cc (4.1 cu. in.)

ENGINE INFORMATION

Both models are two-stroke, air-cooled, single-cylinder gasoline engines. Engines are equipped with a chrome plated cylinder. Crankshaft and connecting rod on both models is serviced only as an assembly.

MAINTENANCE

SPARK PLUG. Recommended spark plug for both models is a Bosch WS7F or equivalent. Spark plug electrode gap should be 0.51 mm (0.020 in.).

CARBURETOR. Both models are equipped with a Tillotson HS diaphragm type carburetor with a vibration type governor located in carburetor body near choke shaft.

Initial adjustment of idle and high speed mixture screws should be approximately one turn open from a lightly seated position. Final adjustment should be made with engine running and at operating temperature. Adjust low speed mixture screw so engine will accelerate cleanly without hesitation. To correctly adjust high speed mixture screw, remove carburetor mounting bolts and tilt carburetor to one side to gain access to governor valve (16—Fig. OL1). Remove governor valve and install a solid plug. Reinstall carburetor. With engine running and at operating temperature, attach an accurate tachometer and adjust high speed mixture screw (14) to obtain a maximum no-load speed of 11,000 rpm. Remove solid plug and reinstall governor valve (16).

NOTE: DO NOT use cut-off saws with governor valve blanked off as excessive blade speed will result. When adjusting maximum no-load engine speed with governor valve blanked off, do not allow maximum no-load engine speed to exceed 11,500 rpm.

Fig. OL1—Exploded view of Tillotson HS series carburetor used on all models. Carburetor is equipped with a vibration type governor valve (16) located in carburetor body.

1. Pump cover
2. Gasket
3. Fuel pump diaphragm & valves
4. Screen
5. Carburetor body
6. Throttle plate
7. Bushing
8. Spring
9. Throttle shaft
10. Spring
11. Throttle stop screw
12. Low speed mixture screw
13. Spring
14. High speed mixture screw
15. "E" clip
16. Governor valve
17. "O" ring
18. Choke shaft
19. Choke plate
20. Ball
21. Spring
22. Welch plug
23. Screen
24. Retainer
25. Welch plug
26. Spring
27. Fuel inlet valve
28. Lever
29. Pin
30. Retaining screw
31. Gasket
32. Metering diaphragm
33. Cover

With governor valve installed maximum no-load engine speed should not exceed 7500-8100 rpm at wide-open throttle.

Refer to Fig. OL1 and disassemble carburetor. Clean filter screen (4). Welch plugs (22 and 25) can be removed by drilling plug with a suitable size drill bit, then use a small punch to pry plugs out. Care must be exercised not to drill into carburetor body.

Inspect inlet lever spring (26) and renew if stretched or damaged. Inspect diaphragms for tears, cracks or other damage. Renew idle and high speed adjusting screws (12 and 14) if they are grooved or damaged. Fuel inlet needle has a rubber tip and seats directly on a machined orifice in carburetor body. Carburetor body must be renewed if seat is damaged.

When assembling carburetor, adjust position of inlet control lever (28) so lever is flush with diaphragm chamber floor as shown in Fig. OL2. Carefully bend lever adjacent to spring to obtain correct lever position.

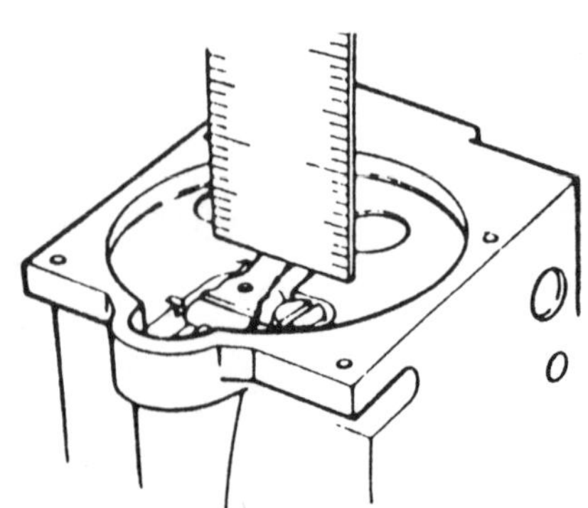

Fig. OL2—Fuel inlet valve lever should be parallel with chamber floor for correct height.

GOVERNOR. Both models are equipped with a vibration type governor valve (16—Fig. OL1) located in carburetor body. Maximum no-load engine speed is limited to 7500-8100 rpm at wide-open throttle. Valve consists of a spring loaded check ball which is vibrated off its seat when engine speed reaches 7500-8100 rpm. Fuel is then routed directly to carburetor venturi and engine is flooded down. As speed decreases, check ball seats and engine clears and resumes normal operation. This allows a richer fuel mixture under loaded conditions while providing maximum blade speed limit under no or light load conditions.

IGNITION SYSTEM. Both models are equipped with a breakerless capacitor discharge ignition system. Olympyk module and coil tester 60.00893 can be used to test ignition module. If tester 60.00893 is not available, trouble-shoot ignition module by process of elimination. Air gap between ignition module and flywheel magnets should be 0.35 mm (0.014 in.). Loosen ignition module mounting screws and move module to adjust air gap. Use Loctite 242 on threads of module mounting screws. Ignition timing is not adjustable. If flywheel requires removal, use Olympyk tool 60.00301 or a suitable equivalent puller to remove flywheel. Do not strike flywheel with hammer. It may be necessary to remove one starter pawl, if pawl hinders usage of flywheel puller.

LUBRICATION. The engine is lubricated by mixing oil with the fuel. The recommended oil is BIA certified two-stroke, air-cooled engine oil. Fuel:oil mixture should be a 40:1 ratio when using recommended oil. If BIA certified two-stroke, air-cooled engine oil is not available, use a good quality oil designed for use in air-cooled, two-stroke engines mixed at a 20:1 ratio. Use a separate container when mixing fuel and oil.

CARBON. Carbon deposits should be cleaned from muffler and exhaust port at regular intervals. Position piston at top dead center to prevent loosened carbon from entering cylinder. Use a wooden scraper to remove carbon to prevent damage to piston or cylinder.

GENERAL MAINTENANCE. Check and tighten all loose bolts, nuts or clamps prior to each day of operation. Check for fuel leakage and repair if necessary.

Clean dust, dirt, grease or any foreign material from cylinder head and cylinder block cooling fins as needed. Inspect cooling fins for damage and repair if necessary.

Due to the fine texture and abrasive nature of the dust created when cutting concrete and masonry products with cut-off saws, check condition of air filter and all filter connections frequently to make certain dust and foreign material do not enter engine through the air filter system.

REPAIRS

TIGHTENING TORQUES. Recommended tightening torques are as follows:

Crankcase screws .... 10 N·m (88 in.-lbs.)
Cylinder screws .... 10 N·m (88 in.-lbs.)
Clutch nut .... 19.6 N·m (168 in.-lbs.)

CYLINDER, PISTON, PIN AND RINGS. Cylinder bore is chrome plated and should be renewed if cylinder bore is cracked, scored or excessively worn. Piston is equipped with two piston rings. Locating pins are present in ring grooves to prevent ring rotation. Make certain ring end gaps are properly positioned around locating pins when installing cylinder. Piston pin (5—Fig. OL3) rides in needle bearing (7) and is held in position with two wire retainers (6). Thrust washers (8) are used on both models. Manufacturer recommends using Olympyk tool 60.00366 with the proper size drivers to remove and install piston pin. Use a suitable support tool between piston skirt and top of crankcase when removing pin. Remove pin by pressing from flywheel side of engine toward clutch side. Install piston with arrow on piston crown pointing toward exhaust port. Press pin into piston from flywheel side. Use a drift pin or suitable tool to align piston with needle bearing (7) and thrust washer (8).

Piston and cylinder are graded during production to obtain desired piston-to-cylinder clearance. Piston and cylinder grade marks are "A," "B," "C" and "D" with "A" being the largest size. Piston grade mark is stamped on piston crown while cylinder grade mark is stamped on cylinder head. Piston and cylinder marks should match although one size smaller piston can be used if matched piston is not available. For instance, a piston marked "B" can be installed in a cylinder mark "A."

CRANKCASE, CRANKSHAFT AND CONNECTING ROD. Crankcase must be split on both models to remove crankshaft. Remove right side crankcase seal (16—Fig. OL3) and cam (15) prior to

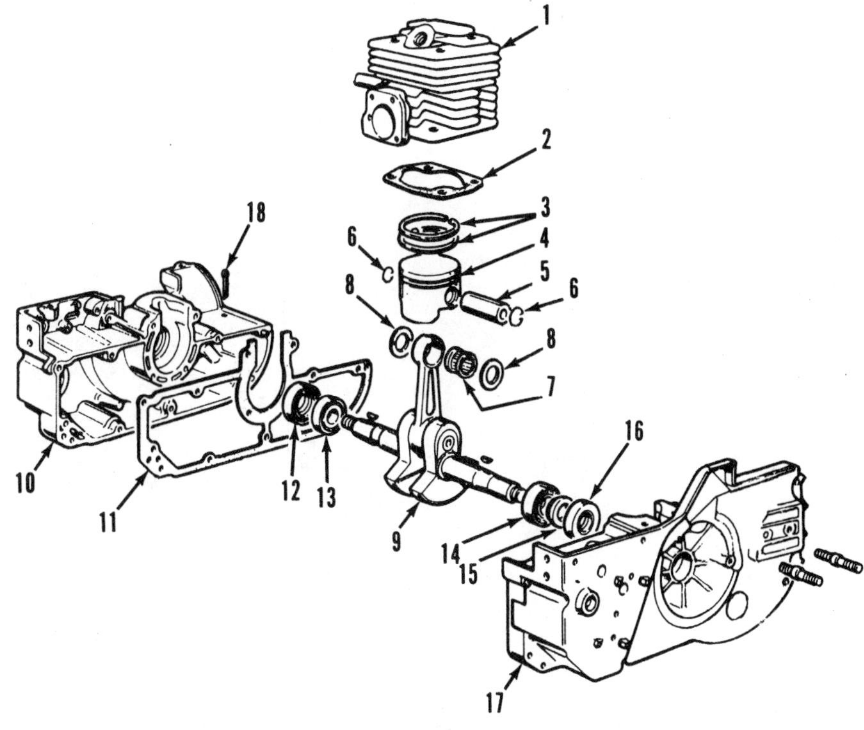

Fig OL3—Exploded view of engine assembly similar to both models.

1. Cylinder
2. Gasket
3. Piston rings
4. Piston
5. Piston pin
6. Retainer
7. Needle bearing
8. Thrust washer
9. Crankshaft & connecting rod assy.
10. Left crankcase half
11. Gasket
12. Seal
13. Main bearing
14. Main bearing
15. Cam
16. Seal
17. Right crankcase half
18. Oil tank vent pin

separating crankcase halves. Use Olympyk tool 11283 or a suitable equivalent to separate crankcase halves. Use caution not to damage crankcase mating surfaces. Once crankcase is split, use a soft-faced mallet to tap out crankshaft.

Crankshaft and connecting rod are available as a unit assembly only. Check rotation of connecting rod around crankpin and renew crankshaft assembly if roughness, excessive play or other damage is noted. Crankshaft is supported at both ends with ball type main bearings (13 and 14—Fig. OL3). Use the proper size drivers to remove and install main bearings into crankcase halves. If main bearings remain on crankshaft during disassembly, use a suitable jaw-type puller to remove bearings from crankshaft. Install main bearings into crankcase halves prior to reassembly.

Tighten crankcase screws in small increments following a crisscross pattern until torque specified in TIGHTENING TORQUES is obtained.

CLUTCH. Both models are equipped with the three-shoe centrifugal clutch shown in Fig. OL4. Clutch hub (3) is a press fit on crankshaft with a Woodruff key. Nut (9) has left-hand threads. Hub has three threaded holes to accommodate Olympyk tool 60.00301 or a suitable bolt-type puller.

Clutch shoes (2) should be renewed in complete sets to prevent an unbalanced condition. Springs (1) should also be renewed as a set. Needle bearing (6) should be lubricated with a suitable high temperature grease. Tighten clutch nut (9) to torque specified in TIGHTENING TORQUES during reassembly.

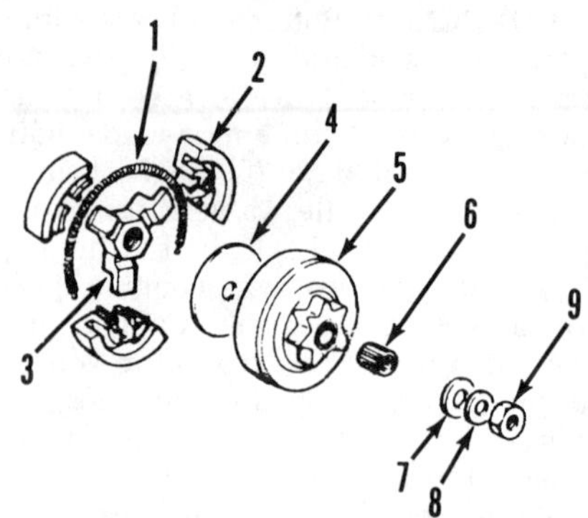

Fig. OL4—Exploded view of three-shoe centrifugal clutch assembly used on both models. Nut (9) has left-hand threads.

1. Spring
2. Shoe
3. Hub
4. Washer
5. Drum
6. Needle bearing
7. Washer
8. Washer
9. Nut

ONAN

A DIVISION OF ONAN CORPORATION
1400 73rd Avenue N.E.
Minneapolis, Minnesota 55432

Model	Bore	Stroke	Displacement
AJ	2¾ in. (69.9 mm)	2½ in. (63.5 mm)	14.9 cu. in. (243.7 cc)
AK	2½ in. (63.5 mm)	2½ in. (63.5 mm)	12.2 cu. in. (201.1 cc)

MAINTENANCE

SPARK PLUG. Recommended spark plug for all models is a Champion H8, or equivalent. Electrode gap is 0.025 inch (0.64 mm) for gasoline fuel or 0.018 inch (0.46 mm) for LP or natural gas fuel.

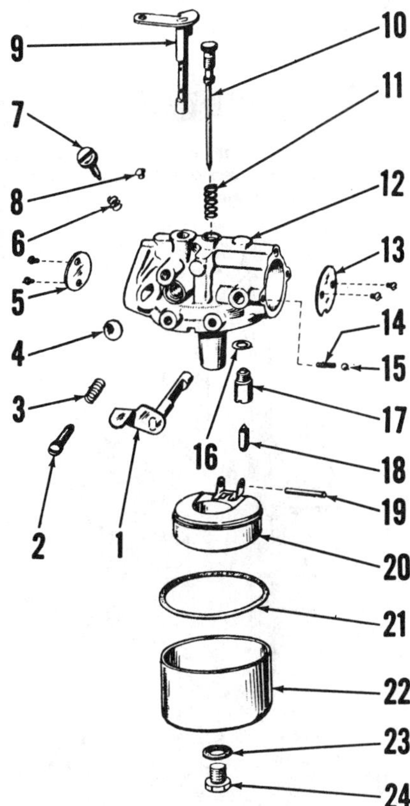

Fig. O1—Exploded view of typical Carter Model N carburetor used on gasoline fuel engines. Refer to Fig. O3 for checking float level.

1. Choke shaft
2. Throttle stop screw
3. Spring
4. Welch plug
5. Throttle plate
6. Spring
7. Idle fuel needle
8. Idle passage plug
9. Throttle shaft
10. Main fuel needle
11. Spring
12. Carburetor body
13. Choke plate
14. Spring
15. Friction ball, choke shaft
16. Gasket
17. Inlet valve seat
18. Inlet valve
19. Float pin
20. Float
21. Gasket
22. Float bowl
23. Gasket
24. Screw

CARBURETOR (GASOLINE). Refer to Figs. O1 and O2 for exploded views of typical Carter Model N carburetor and Zenith carburetor used on Model AJ and AK Onan engines. Clockwise rotation of main fuel needle and idle fuel needle leans the fuel mixture.

For initial adjustment, open main fuel needle approximately 2½ turns on Carter carburetor or 1⅛ turns on Zenith carburetor and open idle fuel needle about one turn on either carburetor. Make final adjustments with engine running and at operating temperature.

With engine operating at full rated load, turn main fuel needle in slowly until engine begins to lose speed, then turn needle out until engine will carry full load. With engine operating at no load, turn idle fuel needle in slowly until engine loses speed, then turn needle out to point of smoothest engine operation.

To help prevent governor "hunting" during load changes, operate engine at rated speed, no load, and turn throttle stop screw in until it just touches throttle lever, then turn screw out one turn.

Refer to Fig. O3 and adjust float so dimension (A) is 11/64 inch (4.37 mm).

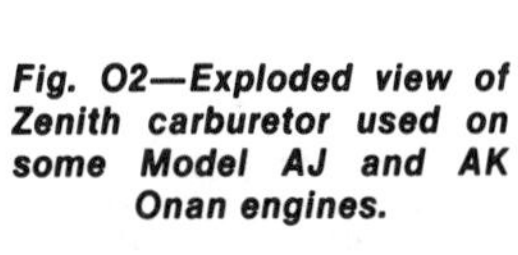

Fig. O2—Exploded view of Zenith carburetor used on some Model AJ and AK Onan engines.

Some gasoline fuel models are equipped with an automatic electrically operated choke. Refer to exploded view

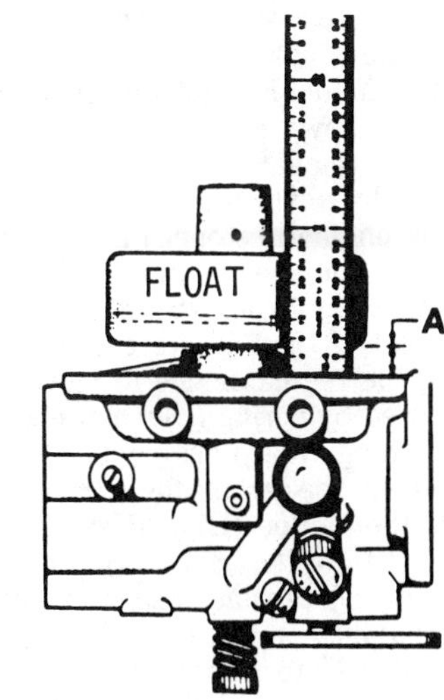

Fig. O3—Float level can be checked by measuring distance (A) between free end of float and carburetor body when body and float assembly are held in inverted position. Carter carburetor is shown; measurement on Zenith carburetor is the same.

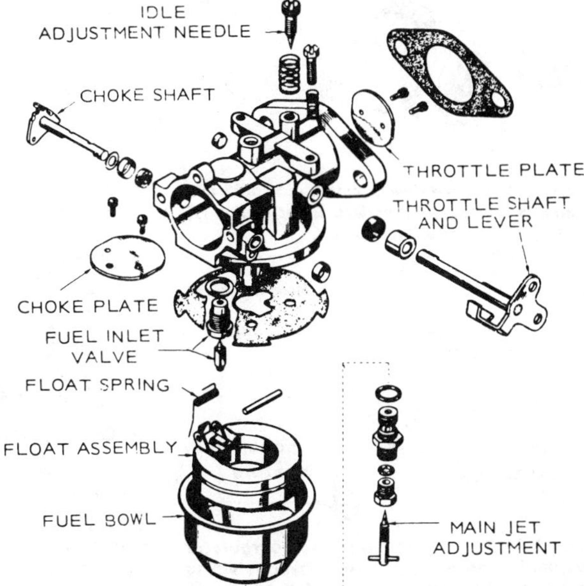

of the choke assembly in Fig. O4. When cold, the bimetal element (2) turns the carburetor choke to closed position. After engine is started, heat from electric heating coil in cover (1) causes bimetal element to move choke to open position.

At a temperature of 70°F (68°C), the choke should be approximately 1/8 inch (3.18 mm) from the fully closed position. Extreme temperature variation may require adjustment of the choke. Refer to Fig. O5.

A fuel lift pump (Fig. O6) is installed on some models. All parts except upper and lower bodies (1 and 14) are serviced separately. Actuating primer lever (10) will pump fuel into carburetor.

CARBURETOR (LP OR NATURAL GAS). On some models equipped to burn LP or natural gas, an Ensign single diaphragm Model F regulator and a carburetor as shown in Fig. O7 are used. The Ensign regulator shown in Fig. O8 automatically shuts off the fuel supply when the engine is stopped and is provided with idle fuel mixture adjusting needle (I). Gas from the supply line enters inlet (A) but is prevented from entering the regulator due to valve (C) being closed by spring (B) when engine is not running. When engine is started, vacuum from the carburetor is transmitted via the outlet (E) and regulator passage (F) to the right side of the diaphragm (G) causing diaphragm and lever (H) to move to the right thus opening the valve (C) to control flow of gas to carburetor. At slow idling speeds the movement of the diaphragm (and the quality of the idling mixture) is controlled by the idle mixture adjusting needle (I). An idle tube from the carburetor is connected to the regulator at (J). A primer (P) instead of a choke is provided to facilitate starting.

Regulator may be disassembled for overhaul. Carefully inspect all parts and renew any that show wear. The diaphragm is bulged and cemented between gaskets at the factory. DO NOT attempt to flatten it. When reassembling valve (C) and lever (H), do not tighten the pivot block screw (L) more than enough to hold the block in position. If screw is tightened too much, it will bind the lever.

Adjust carburetor in normal manner keeping in mind idle mixture is controlled by needle (I) mounted on the regulator while power mixture is controlled by the knurled adjusting needle on the carburetor.

Choke position is determined by counterweight on choke shaft. On some models, a choke adjusting screw is provided; refer to Fig. O9. The weighted choke should just close, but be free to open with air stream through carburetor when engine is running. Turn adjusting screw in to reduce choking.

On some models equipped to burn LP or natural gas, a Garretson regulator (Fig. O10) and Walbro carburetor (Fig. O11) are used. The regulator is a "demand" type and performs no metering function in regard to gas flow. It is only open or closed. The Garretson regulator (Fig. O10) should be adjusted so it will close off gas flow to carburetor when there is no demand. This prevents gas leaks when engine is not running and provides maximum regulator sensitivity.

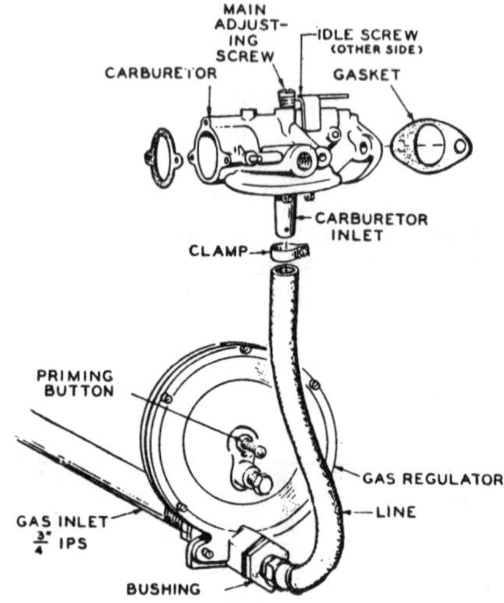

Fig. O7—View of Carter carburetor and Ensign pressure regulator used on some LP or natural gas fueled engines.

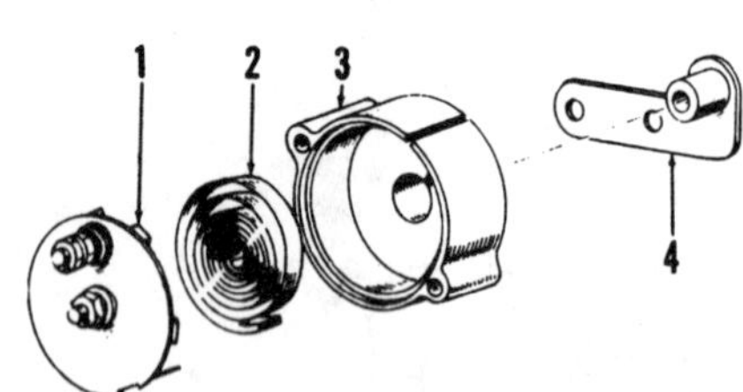

Fig. O4—Exploded view of automatic choke unit available on gasoline models. Refer to Fig. O5 for adjusting unit.

1. Cover & heating element assy.
2. Bimetal element
3. Housing
4. Mounting bracket

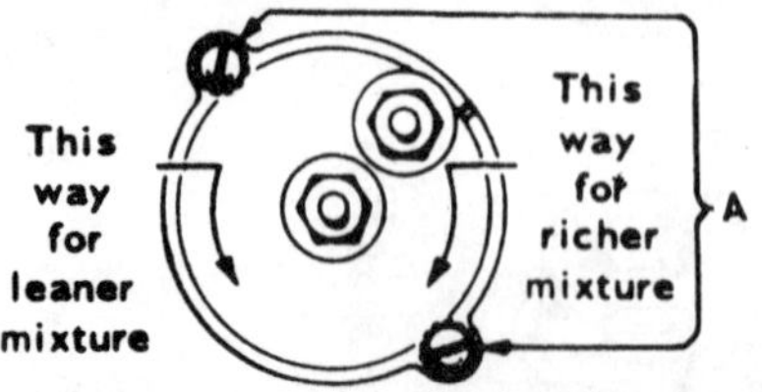

Fig. O5—To adjust gasoline carburetor automatic choke unit, loosen cover retaining screws and turn cover as required. Refer to text.

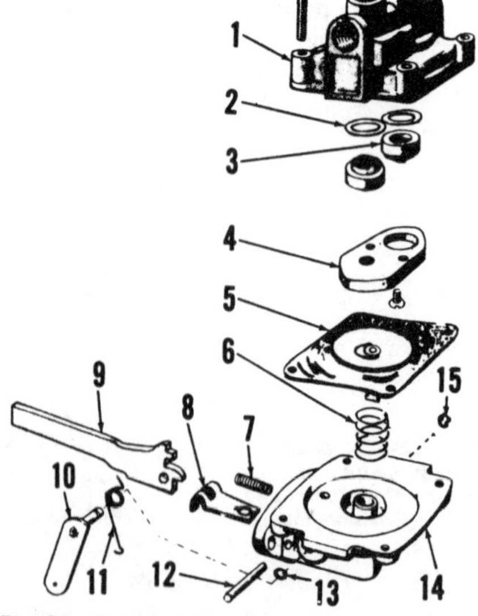

Fig. O6—Exploded view of fuel pump assembly used on some models.

1. Upper body
2. Gaskets
3. Valves
4. Valve retainer
5. Diaphragm
6. Spring
7. Spring
8. Link
9. Arm
10. Primer lever
11. Spring
12. Pin
13. "O" ring
14. Lower body
15. Snap ring

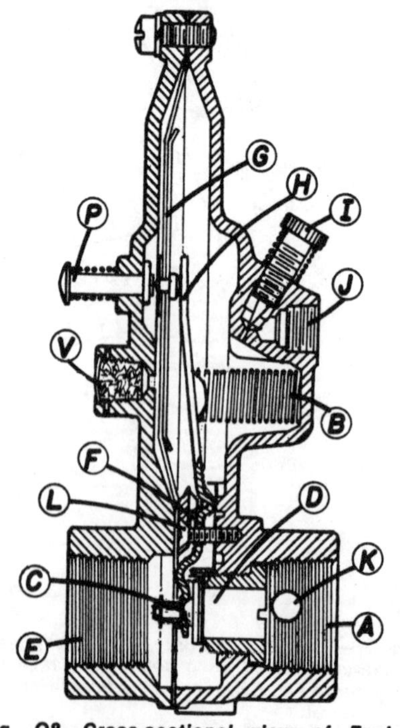

Fig. O8—Cross-sectional view of Ensign pressure regulator (see Fig. O7) used on LP or natural gas fueled engines. Idle fuel mixture adjusting screw is (I).

A. Inlet
B. Spring
C. Valve
D. Valve seat
E. Outlet
F. Passage
G. Diaphragm
H. Lever
I. Idle mixture adjustment
J. Idle connection
L. Pivot screw
P. Primer
V. Vent

The Garretson regulator may be disassembled for overhaul. Inspect all parts for damage and excessive wear and renew parts as necessary.

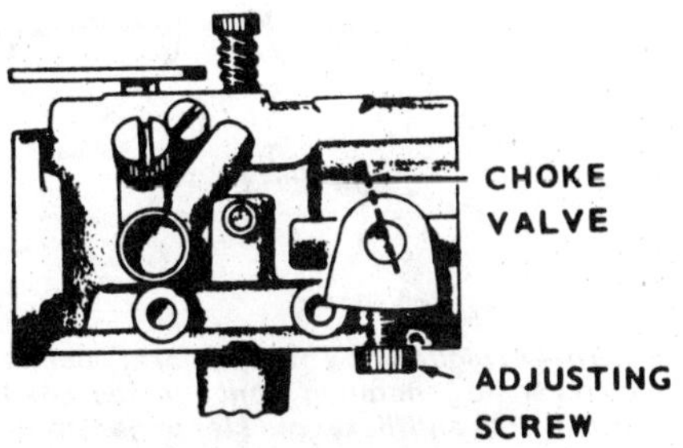

Fig. O9—View showing choke adjusting screw location on some Carter LP or natural gas carburetors.

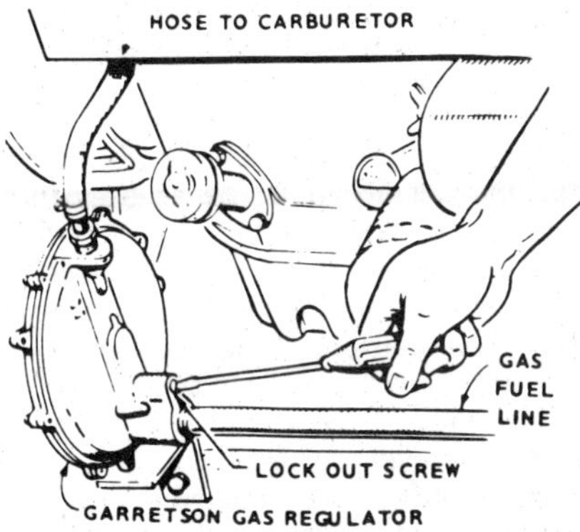

Fig. O10—View showing location of lockout screw on Garretson gas regulator used on some engines.

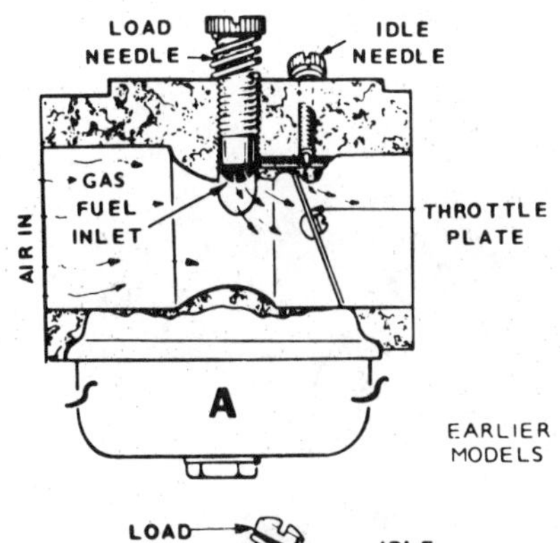

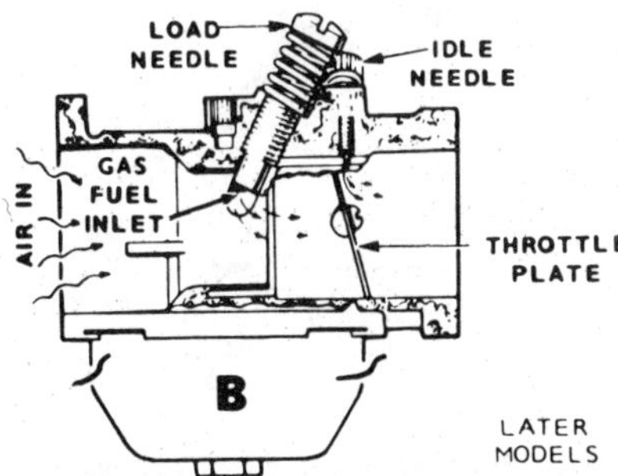

Fig. O11—Cross-sectional view of Walbro carburetor used on some models. View "A" is used on early models and view "B" is used on late models.

A Walbro carburetor shown in Fig. O11 is used with the Garretson regulator. Note adjustment points. Adjustment of vapor fuel carburetors is essentially the same as for gasoline fueled models. If carburetor is entirely out of adjustment to such extent that engine will not start or run, turn idle adjustment and main adjustment needles inward until lightly seated. Then, open idle needle between 1 and 2 turns and crank engine while opening main adjustment needle a little at a time until engine starts. Final adjustment is made to provide a smooth running engine at both idle speed and rated rpm.

GOVERNOR. Engines which drive electric plants are governed at 1500, 1600, 1800, 2400, 2600, 3000 or 3600 rpm as indicated on generating plant nameplate. Industrial engines are usually governed at 2400 rpm.

To adjust governor linkage on models with automatic idle control (Fig. O13), move control toggle switch to "off" position. With engine stopped, the tension of the governor spring should hold the throttle arm in the wide open position and the throttle lever on the carburetor throttle shaft should just clear the carburetor body by not more than 1/32 inch (0.8 mm). This setting can be obtained by adjusting the ball joint on the governor control linkage shown at upper left in Fig. O12.

On industrial engines, start engine and adjust no load speed 50-100 rpm higher than desired full load speed by turning the speed adjusting nut. Apply full load and if speed drop is too great correct by adjusting sensitivity screw to move end of governor spring closer to governor shaft. If engine tends to hunt, adjustment is too close. Any change in sensitivity adjustment will require a speed readjustment. For governor repairs refer to CAMSHAFT AND GOVERNOR section.

MAGNETO AND TIMING. Refer to Fig. O14 for view of flywheel type magneto used on some engines. Breaker points are located under flywheel. Recommended breaker point gap of 0.022 inch (0.56 mm) can be obtained by loosening the point assembly retaining screw and shifting the point set. Timing on engines operated at 1800 rpm or less should be 19° BTDC; engines operated above 1800 rpm should be timed at 25° BTDC. Magneto back plate has elongated mounting screw holes to permit timing adjustment. Breaker points should just start to open when index mark on flywheel is aligned with the correct degree mark on gear cover. Air gap between coil pole shoes and flywheel should be 0.010-0.015 inch (0.25-0.38 mm).

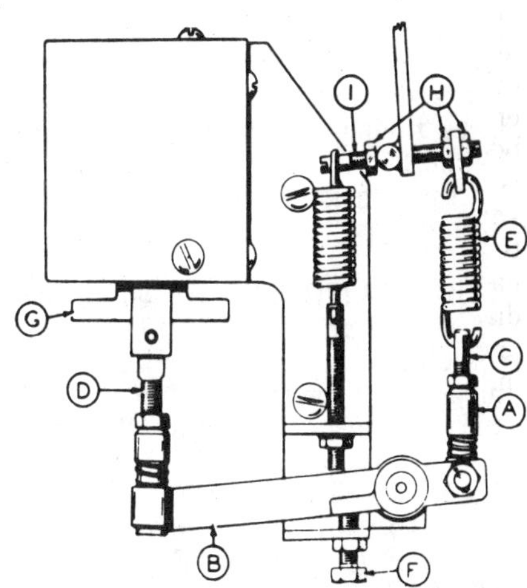

Fig. O13—View of automatic idle control used on some generating plate engines.

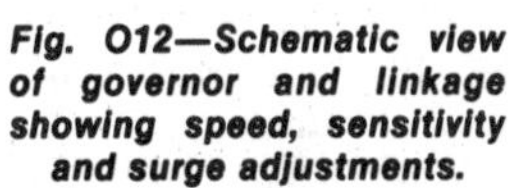

Fig. O12—Schematic view of governor and linkage showing speed, sensitivity and surge adjustments.

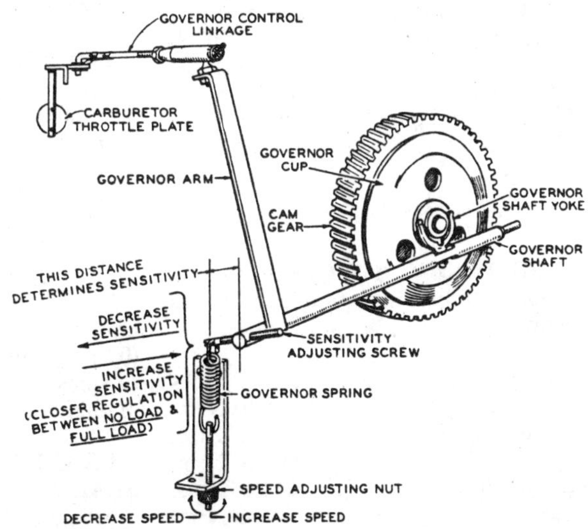

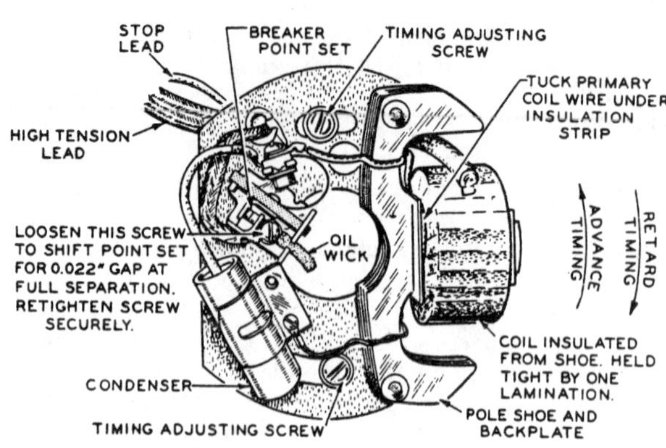

Fig. O14—View of magneto ignition system. Timing is adjustable by rotating magneto assembly on slotted mounting holes.

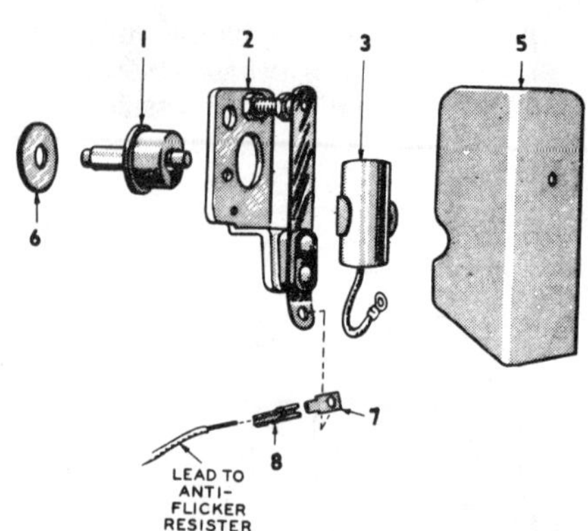

Fig. O15—Exploded view of "antiflicker" points used on some generating plant engines. Lead is attached to antiflicker resister in generator and has no connection with engine ignition system. Adjust antiflicker point (2) gap to 0.020 inch (0.51 mm).

1. Plunger assy.
2. Point set
3. Condenser
5. Cover
6. Gasket
7. Terminal
8. Terminal

BATTERY IGNITION AND TIMING. Some engines are equipped with a battery ignition system. The ignition breaker points and condenser are located on side of engine. Timing is adjusted by varying the breaker point gap. The point gap can be varied from 0.016 to 0.024 inch (0.41 to 0.61 mm) to obtain a spark advance of 19°-25° BTDC. Recommended timing is 19° BTDC on engines that run at 1800 rpm or less and 25° BTDC on engines which run above 1800 rpm. Decreasing the point gap retards timing and increasing gap advances timing. A reference mark and 19° and 25° marks on flywheel can be seen through an opening in the blower housing.

LUBRICATION. Recommended oil is API classification SE or SF. During break-in or for operation in temperatures below 32°F (0°C), CC rated oils may be used. With air temperatures below 0°F (−18°C) use SAE 5W-30 oil. From 0°F (−18°C) to 32°F (0°C), use 5W-30 or 10W-30 oil and from 32°F (0°C) to 90°F (32°C) use SAE 30 oil.

Crankcase capacity is 3 pints (0.16 L). Pressure lubrication is optional. Pressure lubricated engines utilize a gear type oil pump (21–Fig. O17), an intake tube (20) and a nonadjustable oil pressure relief valve (18 and 19). If pump is to be removed it must be unscrewed from the oil intake tube (20). If the oil pump fails, renew complete pump assembly.

REPAIRS

TIGHTENING TORQUES. Recommended tightening torques are as follows:

Spark plug . . . . 25-30 ft.-lbs (34-40 N·m)
Connecting rod . . . . 10-12 ft.-lbs. (13-16 N·m)
Cylinder head . . . . 24-26 ft.-lbs. (33-35 N·m)
Gear cover . . . . 15-20 ft.-lbs. (20-27 N·m)
Oil base . . . . 25-30 ft.-lbs. (34-40 N·m)
Oil pump mounting screws . . . . 7-9 ft.-lbs. (10-12 N·m)

PISTON, PIN AND RINGS. The aluminum piston is fitted with two compression rings and one oil control ring. Tapered compression rings should be installed with the word "TOP" or other identifying mark up. Recommended piston ring end gap for all rings is 0.006-0.018 inch (0.15-0.46 mm) on Model AK and 0.006-0.024 inch (0.15-0.61 mm) on Model AJ.

Desired piston skirt to cylinder bore clearance is 0.003-0.005 inch (0.08-0.13 mm) for Model AK engines and 0.006-0.008 inch (0.15-0.20 mm) for Model AJ engines.

Standard cylinder bore diameter is 2.502-2.503 inches (63.55-63.57 mm) for AK engine and 2.754-2.755 inches (69.95-69.98 mm) for AJ engine. Oversize piston and ring sets are available.

The floating type piston pin is retained by snap rings. Pin should be hand push fit in piston and a thumb push fit in connecting rod at 72°F (22°C). Piston pin is available in standard size as well as 0.002 inch (0.05 mm) oversize.

CONNECTING ROD. Rod and piston are removed from above as an assembly. The aluminum rod rides directly on the crankshaft crankpin. Crankpin diameter is 1.3745-1.3750 inches (33.912-34.925 mm). Recommended bearing clearance is 0.0015-0.0025 inch (0.038-0.064 mm). Oversize rod assembly is available for reground crankpin journals.

Connecting rod side play on the crankpin should be 0.012-0.035 inch (0.31-0.89 mm). Note dipper is installed to splash oil towards the camshaft side of engine on splash lubricated models.

CRANKSHAFT, BEARINGS AND SEALS. The crankshaft rides in two renewable sleeve type bearings. Some models require flange type bearings (bushings) which must be pressed into bore from inside of block or bearing plate. Bearings used in early production engines required line boring or reaming after installation; however, current service parts are precision type bearings which require no reaming and are used to renew the earlier type bearings.

Crankshaft main bearing journal diameter is 1.6860-1.6865 inches (42.830-42.843 mm) and desired bearing running clearance is 0.003-0.004 inch (0.08-0.10 mm). Renew bearings if clearance is excessive. If crankshaft main journals are worn, journals may be

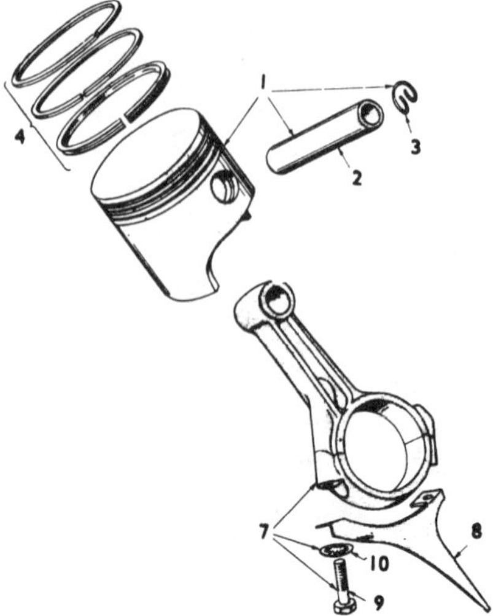

Fig. O16—Exploded view of piston and connecting rod assembly. Note position of oil dipper (8).

1. Piston & pin assy.
2. Piston pin
3. Retaining rings
4. Piston rings
7. Connecting rod assy.
8. Oil dipper
9. Rod cap screws
10. Lockwashers

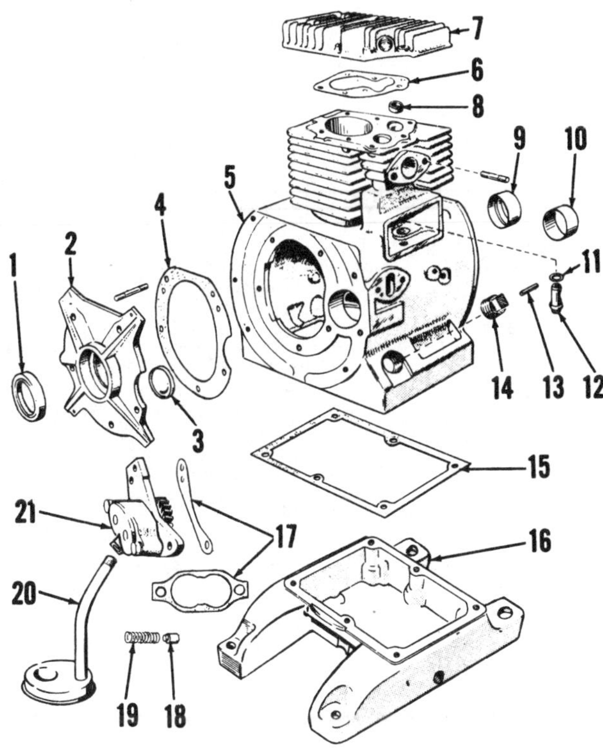

Fig. O17—Exploded view of crankcase assembly.

1. Oil seal
2. Bearing plate
3. Plug
4. Gaskets
5. Cylinder block
6. Gasket
7. Cylinder head
8. Exhaust valve seat insert
9. Crankshaft bearings
10. Camshaft bearings
11. Gasket
12. Valve guides
13. Dowel pin (timing gear cover)
14. Oil filler plug
15. Gasket
16. Oil pan
17. Gaskets
18. Oil pressure relief valve
19. Relief valve spring
20. Oil intake tube & screen
21. Oil pump assy.

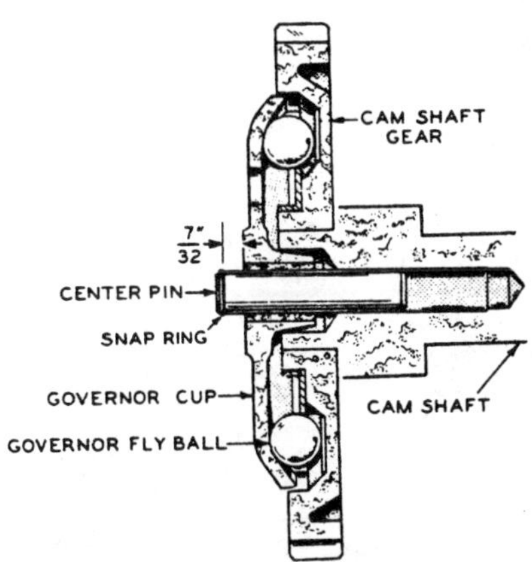

Fig. O19—When governor cup is pushed flush against camshaft gear, there should be 7/32 inch (5.56 mm) clearance between snap ring and cup.

ground undersize as bearings are available.

Desired crankshaft end play is 0.008-0.012 inch (0.20-0.31 mm) for industrial engines and 0.010-0.015 inch (0.25-0.38 mm) for generating plant engines. Obtain desired end play by varying thickness of gaskets (4–Fig. O17) between bearing plate and crankcase.

When installing new crankshaft main bearings, be sure oil hole in bearing sleeve is aligned with oil supply hole in bearing bore. On splash lubricated engines, oil hole will be upward. On pressure lubricated engines, oil hole will be opposite from the camshaft. Bearing plate and crankcase should be heated to 200°F (93°C) in an oven or in hot water before pressing bearings into place.

Renewal of front oil seal requires removal of the timing gear cover. Rear oil seal removal requires removal of bearing plate. Open side (lip) of oil seals must be installed to inside of engine. Rear seal should be flush with face of boss. Use shim stock or pilot sleeve to avoid damage to seals when installing timing gear cover or bearing plate.

CAMSHAFT AND GOVERNOR. Cam gear is press fit on camshaft and should be removed from engine as a single unit with camshaft. Assembly can be removed after first removing cylinder head, gear cover, valves, tappets (fuel pump if used) and the crankshaft gear lock ring and washer. Early camshaft bearings were babbitt-lead lined bushings which can be renewed using latest precision type bearings that do not need to be align bored or reamed after installation. Recommended running clearance is 0.0015-0.0030 inch (0.038-0.076 mm). Install bushings with oil groove at top. Front bushing should be installed flush with cylinder block; rear bushing flush with bottom of counter bore for expansion plug. Shaft should have minimum end play of 0.003 inch (0.08 mm) (measured at front bearing) which is controlled by a thrust washer behind the cam gear.

Governor weight unit is mounted on front face of cam gear as shown in Fig. O19. Make sure the distance from outer face of cup sleeve (bushing), or cup itself if bushing is flush, to inner face of snap ring is 7/32 inch (5.56 mm) as shown when cup is held against flyballs. If less than specified amount, grind end of sleeve as required being sure to remove all burrs from sleeve bore after grinding. If dimension is more than specified amount, press pin (15–Fig. O18) further into camshaft (13). Engines designed for 3600 rpm have 5 flyballs; others have 10.

When installing the gear cover, make sure pin (3–Fig. O20) in cover engages the hole in governor cup which has a brass bushing. If no brass bushing is installed, pin may be engaged in any of the three holes in cup (7–Fig. O18).

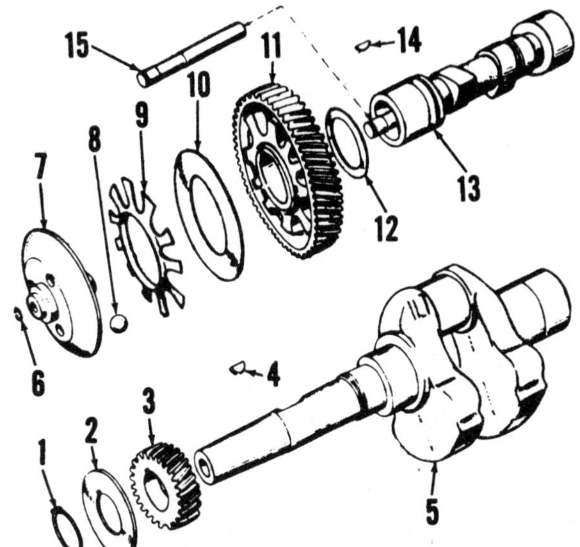

Fig. O18—Exploded view of crankshaft, camshaft and governor units. Pin (15) is pressed into camshaft (13); refer to Fig. O19. Items (9) and (10) are not used on all models.

1. Snap ring
2. Washer
3. Crankshaft gear
4. Woodruff key
5. Crankshaft
6. Snap ring
7. Governor cup
8. Steel balls
9. Spacer
10. Plate
11. Camshaft gear
12. Thrust washer
13. Camshaft
14. Woodruff key
15. Pin

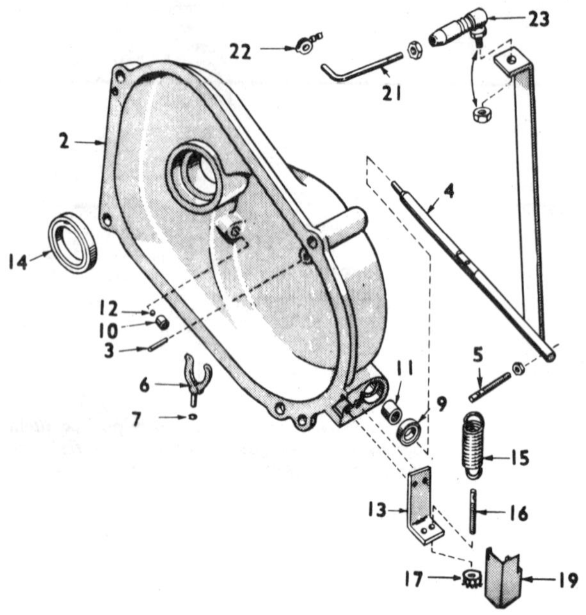

Fig. O20—Exploded view of timing gear cover assembly. Pin (3) engages a hole in governor cup. Refer to text.

2. Cover
3. Roll pin
4. Governor arm & shaft
5. Sensitivity adjustment stud
6. Yoke
7. Snap ring (not on all models)
9. Oil seal
10. Bearing
11. Bearing
12. Thrust ball
13. Spring bracket
14. Crankshaft oil seal
15. Governor spring
16. Spring adjusting stud
17. Adjusting nut
19. Spring cover
21. Governor link
22. Clip
23. Ball joint

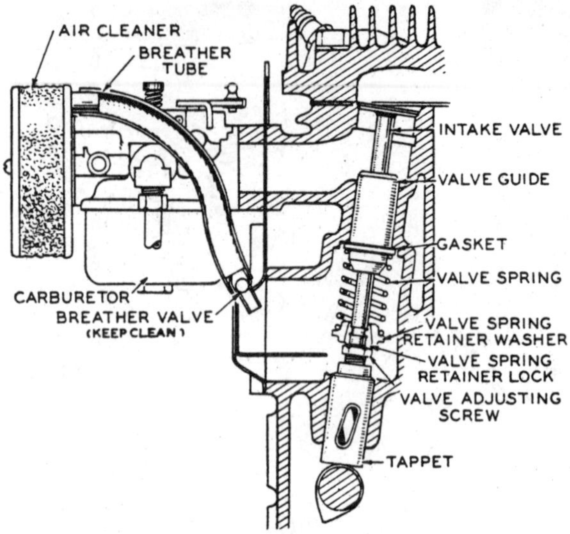

AIR CLEANER
BREATHER TUBE
INTAKE VALVE
VALVE GUIDE
GASKET
VALVE SPRING
CARBURETOR
BREATHER VALVE (KEEP CLEAN)
VALVE SPRING RETAINER WASHER
VALVE SPRING RETAINER LOCK
VALVE ADJUSTING SCREW
TAPPET

Fig. O21—Cross-sectional view of engine valve system and breather.

TIMING GEARS. Timing gears should always be renewed in pairs. To remove cam gear, first remove camshaft and gear as a unit as outlined in the preceding paragraphs then press gear off shaft. Crankshaft gear can be removed using two No. 10-32 screws threaded into holes in gear to push gear from shaft. The "O" marks on gears must be in register for correct valve timing.

VALVE SYSTEM. Valve tappet clearance (cold) for both intake and exhaust valves is 0.010-0.012 inch (0.25-0.31 mm). Obtain recommended clearance by turning the self-locking adjusting screws as needed (Fig. O21).

Valve face angle is 44° and valve seat angle is 45°. Valve seat width is 1/32 to 3/64 inch (0.79 to 1.19 mm). Renewal of valve seat inserts requires the use of special euipment and should not be attempted unless proper equipment is available.

Valve stem clearance in guides is 0.0010-0.0025 inch (0.025-0.064 mm) for intake valves; 0.0025-0.0040 inch (0.064-0.101 mm) for exhaust valves. Install valve guides so shoulder on guide is flush against gasket at valve guide openings in cylinder block casting.

SERVICING ONAN ACCESSORIES

RECOIL STARTER

Refer to Fig. O22 for exploded view of friction shoe type recoil starter. To disassemble starter, hold rope pulley (17) securely with thumb and remove four screws securing ring (3) and flange (5) to cover (20). Remove ring and flange and release thumb pressure until spring (18) is unwound. Remove snap ring (6), washer (7), spring (8), slotted washer (9) and fiber washer (10). Lift out friction shoe assembly (11, 12, 13 and 14), then remove remaining washers. Withdraw rope pulley (17) from cover. Remove rewind spring from cover if necessary and note direction of windings.

When reassembling, lubricate rewind spring cover shaft and center bore in rope pulley with a light coat of grease.

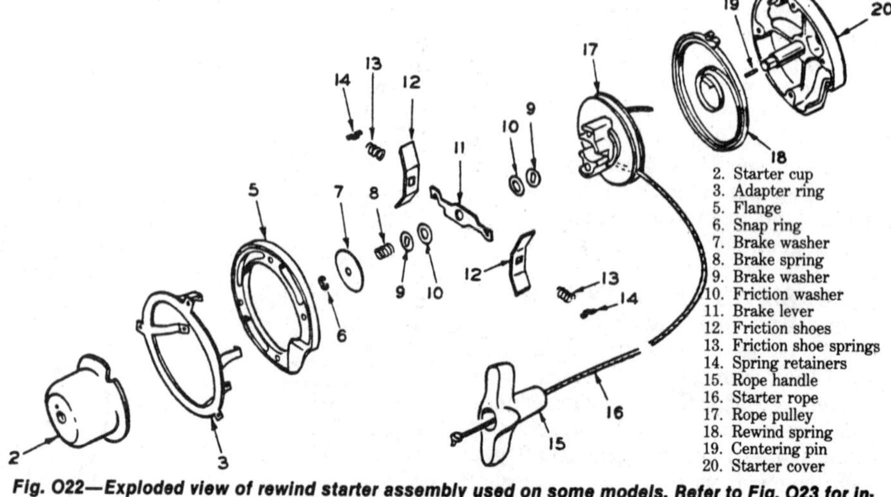

Fig. O22—Exploded view of rewind starter assembly used on some models. Refer to Fig. O23 for installation of friction shoes (12).

Install rewind spring so windings are in same direction as removed spring. Install rope on pulley and place pulley on cover shaft. Make certain inner and outer ends of spring are correctly hooked on cover and rotor. Preload the rewind spring by rotating the rope pulley two full turns. Hold pulley in preload position and install flange (5) and ring (3). Check sharp end of friction shoes (12) and sharpen or renew as necessary. Install washers (9 and 10), friction shoe assembly, spring (8), washer (7) and snap ring (6). Make certain friction shoe assembly is installed properly for correct starter rotation. Refer to Fig. O23. If properly installed, sharp ends of friction shoes will extend when rope is pulled.

Remove brass centering pin (19–Fig. O22) from cover shaft, straighten pin if necessary, then reinsert pin 1/3 of its length into cover shaft. When installing starter on engine, centering pin (19) will align starter with hole in starter cup retaining cap screw.

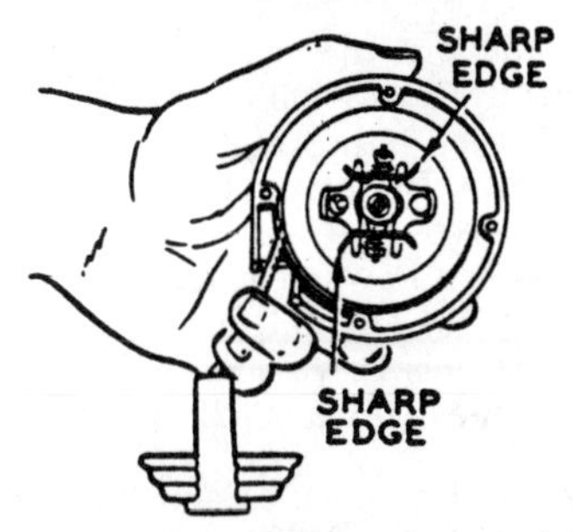

Fig. O23—Install friction shoe and lever assembly with sharp edges of friction shoes pointing in direction shown.

ONAN CENTRAL PARTS DISTRIBUTORS

(Arranged Alphabetically by States)

These franchised firms carry extensive stock of repair parts. Contact them for name and address of nearest dealer who may have the parts you need.

Atchison Equipment Company, Inc.
Phone: (205) 591-2328
4724 First Avenue North
P.O. Box 2971
Birmingham, Alabama 35212

Delhomme Industries
Phone: (205) 473-6626
3422 Georgia Pacific Avenue
Mobile, Alabama 36607

Fremont Electric Company
Phone: (907) 277-9558
140 East Dowling Road
Anchorage, Alaska 99502

Harrison Industries of Arizona Inc.
Phone: (602) 243-7222
3502 East Broadway Road
Phoenix, Arizona 85040

Mecelec of Arkansas
Phone: (501) 490-1801
1701 Dixon Road
P.O. Box 9297
Little Rock, Arkansas 72219

Equipment Service Company
Phone: (714) 562-2804
1954 Friendship Drive
El Cajon, California 92020

Equipment Service Company
Phone: (213) 426-0311
3431 Cherry Avenue
P.O. Box 1307
Long Beach, California 90801

Cal-West Electric, Inc.
Phone: (415) 873-7710
1341 San Mateo Avenue
P.O. Box 2364
San Francisco (South), California 94080

Cal-West Electric, Inc.
Phone: (916) 372-5522
3939 West Capitol Avenue
West Sacramento, California 95691

C.W. Silver Company, Inc.
Phone: (303) 399-7440
3945 East 50th Avenue
Denver, Colorado 80216

GLT Industries, Inc.
Phone: (203) 528-9944
29 Mascolo Road
P.O. Box 307
South Windsor, Connecticut 06074

Advanced Power Systems
Phone: (904) 355-4563
710 Haines Street
P.O. Box 38039
Jacksonville, Florida 32206

R.B. Grove, Inc.
Phone: (305) 854-5420
261 South West 6th Street
Miami, Florida 33130

Tampa Armature Works, Inc.
Phone: (305) 843-8250
3400 Bartlett Boulevard
Orlando, Florida 32805

Tampa Armature Works, Inc.
Phone: (813) 621-5661
440 South 78th Street
P.O. Box 3381
Tampa, Florida 33601

Blalock Machinery & Equipment Company
Phone: (912) 436-1507
700 South Westover Road
Albany, Georgia 31702

Blalock Machinery & Equipment Company
Phone: (404) 766-2632
5112 Blalock Industrial Boulevard
College Park, Georgia 30349

Atlas Electric Company, Inc.
Phone: (808) 524-5866
1151 Mapunapuna Street
Honolulu, Hawaii 96819

Power Systems, Division of E. C. Distributing
Phone: (208) 342-6541
4499 Market Street
Boise, Idaho 83705

Power Systems, Division of E. C. Distributing
Phone: (208) 234-2442
1060 South Main
Pocatello, Idaho 83204

Stannard Power Equipment Company
Phone: (312) 597-5500
4901 West 128th Place
Alsip, Illinois 60658

Service Automotive Warehouse, Inc.
Phone: (309) 794-0400
111 4th Avenue
Rock Island, Illinois 61201

Meco-Indiana, Inc.
Phone: (219) 262-4611
23900 County Road 6
Elkhart, Indiana 46514

Evansville Auto Parts, Inc.
Phone: (812) 425-8264
5 East Riverside Drive
Evansville, Indiana 47713

Meco-Indiana, Inc.
Phone: (317) 873-5005
5005 West 106th Street
Zionsville, Indiana 46077

Midwestern Power Products Company
Phone: (515) 278-5521
10100 Dennis Drive
Des Moines, Iowa 50322

Anderson Equipment Company, Inc.
Phone (712) 255-8033
300 South Virginia Street
Sioux City, Iowa 51102

Mecelec of Kansas
Phone: (316) 522-4767
4631 Palisade Street
Wichita, Kansas 67217

Southern Power Systems
Phone: (502) 459-5060
2025 Old Shepherdsville Road
Louisville, Kentucky 40218

Delhomme Industries, Inc.
Phone: (318) 234-9837
337 Mecca Drive
Lafayette, Louisiana 70505

Delhomme Industries, Inc.
Phone: (318) 439-9700
2506 Elaine Street
Lake Charles, Louisiana 70601

Delhomme Industries, Inc.
Phone: (318) 365-5476
Northside Road
P.O. Box 266
New Iberia, Louisiana 70560

Menge Pump and Machinery Company, Inc.
Phone: (504) 888-8830
2740 North Arnoult
P.O. Box 8210
New-Orleans-Metairie, Louisiana 70011

Menge Pump & Machinery Company, Inc.
Phone: (318) 222-5781
1510 Grimmet Drive
P.O. Box 7323
Shreveport, Louisiana 71107

The Leen Company
Phone: (207) 989-7363
54 Wilson Street
Brewer, Maine 04412

The Leen Company
Phone: (207) 774-6266
366 West Commercial Street
Portland, Maine 04101

Curtis Engine & Equipment Company
Phone: (301) 633-5161
6120 Holabird Avenue
Baltimore, Maryland 21224

New England Engine Corporation
Phone: (617) 948-7331
RR 1
Rowley, Massachusetts 01962

Standby Power, Inc.
Phone: (616) 949-7990
2745 South East 29th Street
Grand Rapids, Michigan 49508

Standby Power, Inc.
Phone: (313) 348-6400
4700 12 Mile Road
Novi, Michigan 48050

Flaherty Equipment Corporation
Phone: (612) 338-8796
2525 East Franklin Avenue
Minneapolis, Minnesota 55406

Interstate Detroit Diesel Allison, Inc.
Phone: (612) 854-5511
2501 East 80th Street
Minneapolis, Minnesota 55420

Northstar Detriot Diesel Allison, Inc.
Phone: (218) 749-4484
1921 West 16th Avenue
Virginia, Minnesota 55792

Menge Pump & Power Systems Division
Phone: (601) 969-9333
1327 South Gallatin
Jackson, Mississippi 39202

Comet Industries, Inc.
Phone: (816) 245-9400
4800 Deramus Avenue
Kansas City, Missouri 64120

National Industrial Supply Company
Phone: (314) 621-0350
1100 Martin Luther King Drive
St. Louis, Missouri 62201

Anderson Equipment Company, Inc.
Phone: (402) 558-1200
5532 Center Street
Omaha, Nebraska 68106

Anderson Industrial Engines
Phone: (402) 558-8700
2123 South 56th Street
Omaha, Nebraska 68106

Equipment Service Company of Nevada
Phone: (702) 382-3852
1916 Highland Avenue
Las Vegas, Nevada 89102

R. C. Equipment Company, Inc.
Phone: (609) 742-0200
522 South Broadway
Gloucester City, New Jersey 08030

GLT Industries, Inc.
Phone: (201) 767-9751
411 Clinton Avenue
Northvale, New Jersey 07647

C. W. Silver Company, Inc.
Phone: (505) 881-2454
4812 North East Jefferson
Albuquerque, New Mexico 87109

Power Plant Equipment Corporation
Phone: (518) 783-1991
6 Northway Lane
Latham, New York 12110

Ronco Communications & Electronics
Phone: (716) 424-3890
230 Metro Park
Rochester, New York 14623

Power Plant Equipment Corporation
Phone: (315) 475-7251
929 South Salina Street
Syracuse, New York 13202

Ronco Communications & Electronics, Inc.
Phone: (716) 873-0760
595 Sheridan Drive
Tonawanda-Buffalo, New York 14150

Owsley & Sons
Phone: (919) 668-2454
Interstate 40
P.O. Box 8627
Greensboro, North Carolina 27410

Interstate Detroit Diesel Allison, Inc.
Phone: (701) 258-2303
3801 Miriam Gateway
Bismarck, North Dakota 58501

Interstate Detroit Diesel Allison, Inc.
Phone: (701) 282-6556
3902 North 12th Avenue
Fargo, North Dakota 58102

Southern Ohio Power Systems, Inc.
Phone: (513) 821-6305
8148 Vine Street
Cincinnati, Ohio 45216

McDonald Equipment Company
Phone: (216) 951-8222
37200 Vine Street-Willoughby
Cleveland, Ohio 44094

Tuller Corporation
Phone: (614) 224-8246
947 West Goodale Boulevard
Columbus, Ohio 43212

G & R Equipment Company
Phone: (405) 685-5534
3826 Newcastle Road
Oklahoma City, Oklahoma 73119

Mechanical & Electrical Equipment Company
Phone: (918) 582-7777
712 Wheeling
P.O. Box 50323
Whitter Station
Tulsa, Oklahoma 74150

EC Distributing Division Electrical Construction
Phone: (503) 224-3623
2122 North West Thurman Street
Portland, Oregon 97208

A. F. Shane Company
Phone: (412) 781-8000
654 Alpha Drive RIDC Industrial Park
Pittsburg, Pennsylvania 15238

Winter Engine Generator Service
Phone: (717) 848-3777
1600 Pennsylvania Avenue
York, Pennsylvania 17404

Owsley & Sons, Inc.
Phone: (803) 548-3636
1-77 & SC Exit 72
Gold Hill Road
Fort Mill, South Carolina 29715

Hobbs Equipment Company
Phone: (615) 894-8400
6203 Provence Street
Chattanooga, Tennessee 37421

Hobbs Equipment Company
Phone: (615) 966-7550
10625 Lexington Drive
Knoxville, Tennessee 37922

Maritime & Industrial, Inc.
Phone: (901) 775-1204
P.O. Box 9397
292 East Mallory
Memphis, Tennessee 38109

Hobbs Equipment
Phone: (615) 244-4933
1327 Foster Avenue
Nashville, Tennessee 37211

Lightbourn Equipment Company
Phone: (214) 233-5151
P.O. Box 401870
13649 Beta Road
Dallas, Texas 75240

Harris Equipment Company
Phone: (713) 879-2600
1100 West Airport Boulevard
Houston, Texas 77001

Power Support Systems
Phone: (915) 332-1429
P.O. Box 4417
3208 Kermit Highway
Odessa, Texas 79760

Lightbourn Equipment Company
Phone: (512) 333-7542
4260 Dividend Drive
San Antonio, Texas 78219

C. W. Silver Company, Inc.
Phone: (801) 355-5373
550 West 7th South
Salt Lake City, Utah 84101

T & L Electric Company
Phone: (802) 295-3114
Sykes Avenue
White River Junction, Vermont 05001

Curtis Engine & Equipment
Phone: (804) 627-9470
1114 Ballentine Boulevard
Norfolk, Virginia 23504

Owsley & Sons
Phone: (804) 275-2603
10300 Jefferson
Davis Highway
P.O. Box 34508
Richmond, Virginia 23234

Fremont Electric Company
Phone: (206) 633-2323
P.O. Box 31640
744 North 34th Street
Seattle, Washington 98103

Lay & Nord
Phone: (509) 453-5591
P.O. Box 472
511 South 3rd Street
Yakima, Washington 98901

Call Detroit Diesel Allison, Inc.
Phone: (304) 744-1511
P.O. Box 8245
Charleston Ordinance Center
Charleston, West Virginia 25303

Inland Diesel, Inc.
Phone: (414) 781-7100
13015 Custer Avenue
Butler, Wisconsin 53007

Morley-Murphy
Phone: (414) 499-3171
Box 3640
700 Morley Road
Green Bay, Wisconsin 54303

Clymar, Inc.
Phone: (414) 781-0700
N55 W13787 Oak Lane
Menomenee Falls
Milwaukee, Wisconsin 53051

Power Service Company
Phone: (307) 237-3773
P.O. Box 2880
5201 West Yellowstone Highway
Casper, Wyoming 82606

CANADIAN DISTRIBUTORS

Simson-Maxwell
6447-2nd Street S.E.
Calgary, Alberta T2H 1J5

Simson-Maxwell
Box 4446
10375 59th Avenue
South Edmonton, Alberta T6H 1E7

Simson-Maxwell
1380 W. 6th Avenue
Vancouver V6H 1A7, B.C.

Simson-Maxwell
729 4th Avenue
Prince George, B.C.

Kipp Kelly, Ltd.
68 Higgins Avenue
Winnipeg, Manitoba R3B 0A6

Sansom Equipment, Ltd.
Woodstock Rd., P.O. Box 1263
Fredericton, New Brunswick

Sansom Equipment, Ltd.
P.O. Box 152
80 Glenwood Drive
Truro, Nova Scotia

Algonma Truck & Tractor Sales
815 Great Northern Road
Sault Ste. Marie, Ontario

Total Power, Ltd.
330 Nantucket Boulevard
Scarborough, Ontario

L.O.W.E. Power Systems, Ltd.
P.O. Box 81
Kenora, Ontario

Hawker Siddley Diesels & Electric
355 Wyecroft Road
Oakville, Ontario L6K 2H2

J.A. Faguy & Sons, Ltd.
2544 Sheffield Road
Ottawa, Ontario K1B 3V7

Lake of the Woods Electric, Ltd.
1177 Roland Street
Postal Station F
Thunder Bay, Ontario P7B 5M5

J.A. Faguy & Sons, Ltd.
750 Montee de Liesse
Montreal, Quebec H4T 1P3

Pryme Power & Diesels, Ltd.
5-2949 3rd Avenue North
Phone: (306) 665-8044
Saskatoon, Saskatchewan S7K 1L2

POWER BEE

U.S. MARINE POWER CORPORATION
105 Marina Drive
Hartford, Wisconsin 53027

Model	Bore	Stroke	Displacement
61001,'002, '003, '005, '006, '008, '010, '012, '014	2-3/16 in. (55.6 mm)	1⅝ in. (41.3 mm)	6.1 cu. in. (100 cc)
82001, '002, '003, '004, '005, '006, '007, '008, '009, '010, '011, '012 '014, '015, '017, '019, '021, '023, '026, '027, '029, '030, '032	2-17/32 in. (64.3 mm)	1⅝ in. (41.3 mm)	8.2 cu. in. (134 cc)

ENGINE INFORMATION

The engines listed above are known as the "L" series engines. The cylinder head is removable and the piston and connecting rod assembly can be removed without completely disassembling the engine. Cylinder and crankcase are integral castings.

MAINTENANCE

SPARK PLUG. Recommended spark plug for Model 82012 is a Champion L85, or equivalent. Recommended spark plug for Model 82003 is Champion RL4J, or equivalent. Recommended spark plug for Models 61005, 61014, 82009, 82011, 82015, 82017, 82019, 82021 and 82023 is Champion L7J, or equivalent. Recommended spark plug for Models 82007, 82026, 82027, 82029, 82030 and 82032 is Champion RL7J, or equivalent. Recommended spark plug for all other "L" series engines is a Champion L4J, or equivalent. Electrode gap for all models is 0.030 inch (0.76 mm).

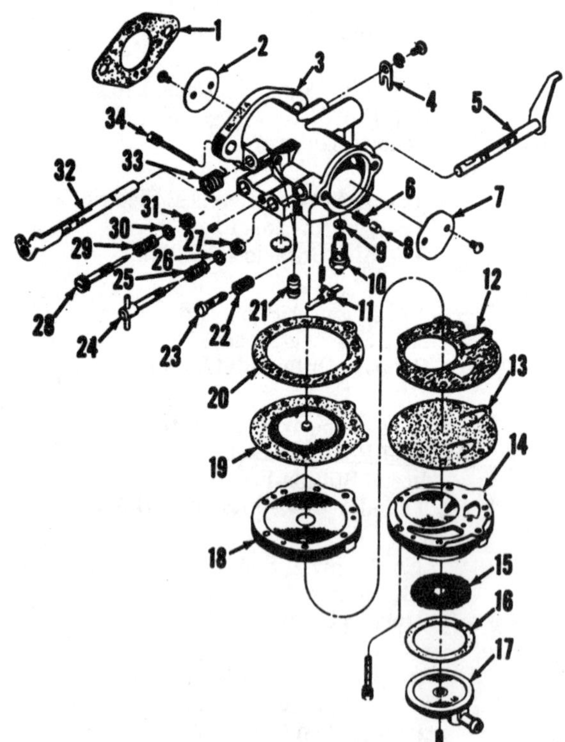

Fig. C1—Exploded view of typical HL series Tillotson carburetor used on "L" series engines.

1. Gasket
2. Throttle disc
3. Carburetor body
4. Throttle shaft retainer
5. Choke shaft
6. Spring
7. Choke disc
8. Choke detent
9. Gasket
10. Inlet valve
11. Valve lever
12. Gasket
13. Pump diaphragm
14. Cover
15. Screen
16. Gasket
17. Fuel inlet
18. Diaphragm cover
19. Fuel diaphragm
20. Gasket
21. Check valve
22. Spring
23. Idle stop screw
24. Main fuel needle
25. Spring
26. Washer
27. Seal
28. Idle fuel needle
29. Spring
30. Washer
31. Seal
32. Throttle shaft
33. Throttle spring
34. Pin for valve lever

CARBURETOR. All models are equipped with a Tillotson Series HL carburetor. Refer to Fig. C1 for exploded view of typical carburetor.

Clockwise rotation of both the idle and high speed fuel needles leans the mixture. For initial adjustment, open both needles approximately 1⅛ turns. Make final adjustment with engine running and at operating temperature. Adjust idle needle for smoothest operation at idle speed, then check acceleration from idle speed to approximately 6000 rpm. Readjust idle fuel needle if engine does not accelerate properly; then, adjust high speed needle with engine under load and at operating speed. Turn the high speed needle clockwise until engine just stops four-stroking (starts to fire on every power stroke).

NOTE: Do not attempt to operate engine with a too-lean fuel mixture; improper lubrication and engine seizure may result.

MAGNETO AND TIMING. A Wico flywheel type magneto is used on all models. Refer to Fig. C2.

To inspect or service magneto, unbolt

and remove fan housing and recoil starter assembly as a unit. On engines with model number ending in zero or even number (such as Model 82002), turn flywheel retaining nut clockwise to remove; on engines with model number ending in an uneven number, turn nut counterclockwise to remove. After removing flywheel retaining nut, remove flywheel using special knock-off nut.

Breaker contact points are accessible after removing flywheel and dust cover from stator plate. Adjust breaker point gap to 0.015 inch (0.38 mm) on all models. Condenser capacity is 0.16-0.020 mfd. Refer to the following chart for timing specifications and note MR = Midrange, FR = Full Retard and FA = Full Advance.

Model Number	Stator Position	Degrees BTDC	Measured BTDC
61001	MR	26°	0.102 in. (2.6 mm)
61002	MR	26°	0.102 in. (2.6 mm)
61003	FR	22°	0.070 in. (1.8 mm)
61005	MR	26°	0.102 in. (2.6 mm)
61006	FR	22°	0.070 in. (1.8 mm)
61008	FR	22°	0.070 in. (1.8 mm)
61010	FR	22°	0.070 in. (1.8 mm)
61012	MR	26°	0.102 in. (2.6 mm)
61014	MR	26°	0.102 in. (2.6 mm)
81001	MR	26°	0.102 in. (2.6 mm)
82002	MR	26°	0.102 in. (2.6 mm)
82003	MR-FA	28°	0.118 in. (3.0 mm)
82004	FR	22°	0.070 in. (1.8 mm)
82005	FR	22°	0.070 in. (1.8 mm)
82006	FR	22°	0.070 in. (1.8 mm)
82007	MR	26°	0.102 in. (2.6 mm)
82008	MR	26°	0.102 in. (2.6 mm)
82009	FR	22°	0.070 in. (1.8 mm)
82010	FR-MR	24°	0.090 in. (2.3 mm)
82011	MR	26°	0.102 in. (2.6 mm)
82012	FR	22°	0.070 in. (1.8 mm)
82014	FR	22°	0.070 in. (1.8 mm)
82015	MR	26°	0.102 in. (2.6 mm)
82017	MR	26°	0.102 in. (2.6 mm)
82019	MR	26°	0.102 in. (2.6 mm)
82021	MR	26°	0.102 in. (2.6 mm)
82023	MR	26°	0.102 in. (2.6 mm)
82026	MR	26°	0.102 in. (2.6 mm)
82027	MR	26°	0.102 in. (2.6 mm)
82029	MR	26°	0.102 in. (2.6 mm)
82030	MR	26°	0.102 in. (2.6 mm)
82032	MR	26°	0.102 in. (2.6 mm)

Fig. C2—View of magneto stator plate assembly removed from engine.

1. Breaker cover spring
2. Armature core
3. Ignition coil
4. Breaker point gap adjusting screw
5. Breaker cam
6. Condenser

Midrange setting of timing is when stator mounting screws are tightened in center of slotted stator mounting holes.

Full retarded is when magneto stator is moved all the way to end of slots in direction of flywheel rotation.

Full advance is when magneto stator is moved all the way to end of slots in direction opposite flywheel rotation.

The specifications FR-MR or MR-FA indicates a stator setting midway between the two positions.

LUBRICATION. Engine is lubricated by mixing oil with fuel. Thoroughly mix 1/3 pint (0.16 L) of SAE 40 two-stroke motor oil with each gallon of regular gasoline.

REPAIRS

TIGHTENING TORQUES. Recommended tightening torques are as follow:

Connecting rod . . . . 80-90 in.-lbs. (9-10 N·m)

Cylinder head:
- All 610 models . . . . 90-100 in.-lbs. (10-11 N·m)
- All 820 models except 82012 . . . . 90-100 in.-lbs. (10-11 N·m)
- 82012 model . . . . 120-130 in.-lbs. (13.5-14.7 N·m)

Flywheel . . . . 420 in.-lbs. (48 N·m)

Spark plug . . . . 120-180 in.-lbs. (14-20 N·m)

Standard screws:
- No. 10-24 thread . . . . 30 in.-lbs. (3 N·m)
- No. 10-32 thread . . . . 35 in.-lbs. (4 N·m)
- No. 12-24 thread . . . . 45 in.-lbs. (5 N·m)
- 1/4 x 20 thread . . . . 70 in.-lbs. (8 N·m)
- 5/16 x 18 thread . . . . 160 in.-lbs. (18 N·m)
- 3/8 x 16 thread . . . . 270 in.-lbs. (31 N·m)

CYLINDER HEAD. Cylinder head is removable on all models and is retained to cylinder with socket head cap screws and plain washers. The cylinder head gasket is available in various thicknesses and care should be taken to select the correct thickness gasket for the engine being serviced. Refer to the following table.

Engine Model	Gasket Thickness
61001, '002 & '003	0.032 in.
61005 & '006	0.062 in.
61008 & '010	0.062 in.
61012	0.020 in.
61014	0.062 in.
82001 & '002	0.032 in.
82003	0.094 in.
82004, '005 & '006	0.062 in.
82007	0.094 in.
82008, '009, '010 & '011	0.125 in.
82012	0.062 in.
82014	0.094 in.
82015	0.125 in.
82017	0.125 in.
82019	0.125 in.
82021	0.125 in.
82023	0.125 in.
82026	0.094 in.
82027	0.094 in.
82029	0.125 in.
82030	0.094 in.
82032	0.125 in.

When reinstalling cylinder head, be sure gasket, cylinder head and cylinder

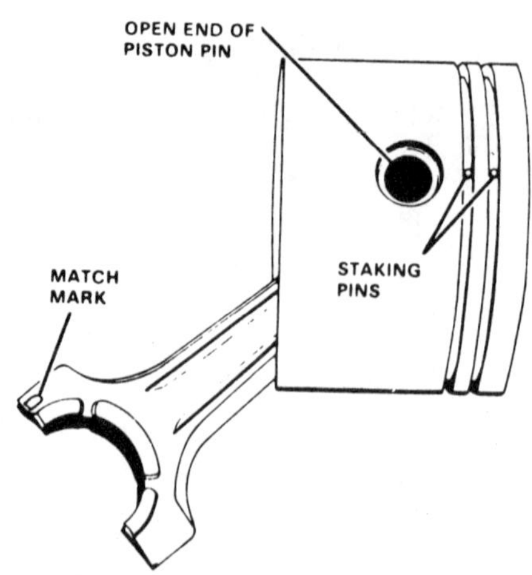

Fig. C3—Open end of piston pin, piston ring staking pins and match mark on rod must face towards support plate.

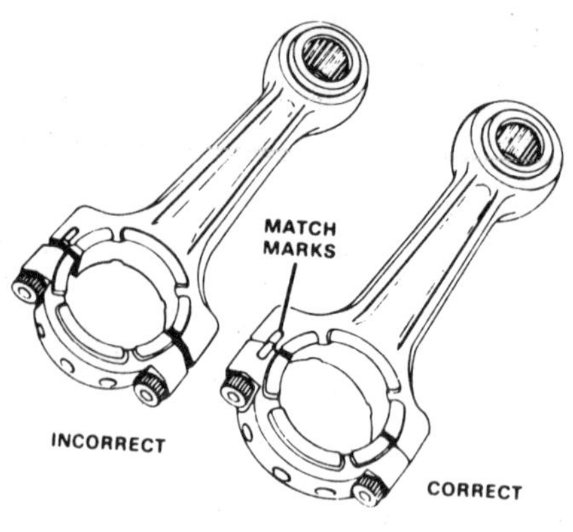

Fig. C4—Match marks on connecting rod and cap.

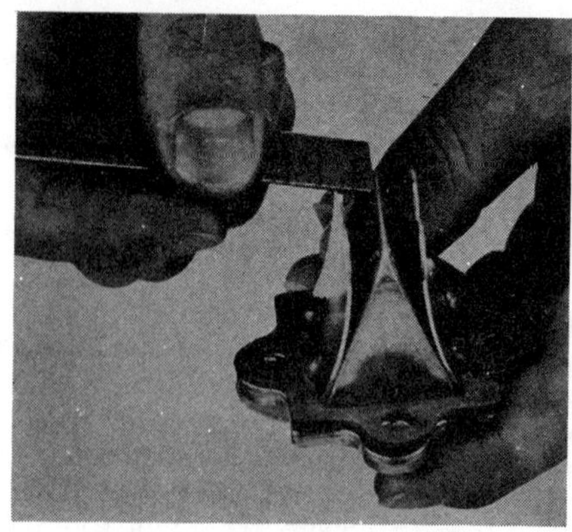

Fig. C5—Reed stop spacing should be ¼ inch (6.35 mm) as shown. Refer to text.

surfaces are clean and smooth. Install the socket head cap screws with plain washers and tighten screws evenly to specified torque.

CONNECTING ROD. Piston and connecting rod assembly can be removed after removing cylinder head, crankcase cover and rod cap. Be careful not to lose any of the 28 loose needle rollers when removing connecting rod cap.

The caged needle roller bearing in pin end of connecting rod is not renewable. Renew the connecting rod assembly if any roller in pin and bearing is rough or flat, if pin end rollers can be separated the width of one roller or if crankpin bearing surface is rough or shows signs of wear.

Match marks on connecting rod and cap and piston ring anchor pins (Fig. C3) should be toward tapered (flywheel) end of crankshaft. When assembling connecting rod to crankshaft, coat crankpin with a light grease and place 14 of the rollers between rod an crankpin; then, stick remaining 14 rollers to crankpin and carefully reinstall rod cap with match marks on rod and cap aligned. See Fig. C4. The mating surfaces of rod and cap are fractured and if cap is correctly installed, the parting line will be almost invisible and cannot be felt with fingernail. Tighten connecting rod cap screws to specified torque.

PISTON, PIN AND RINGS. Piston is equipped with two pinned compression rings. Piston and rings are available in standard size only. Renew piston if scored, ring side clearance is excessive or pin is loose in piston bosses.

The floating type piston pin is retained by a snap ring at each end of pin in piston bosses. Install flat snap ring with square-cut edge away from pin. The piston should be heated slightly prior to removing or installing piston pin to prevent pin boss damage. Install closed end of piston pin towards side of piston with piston ring anchor pin. Piston pin is available in standard size only.

Be sure piston ring end gaps are aligned with anchor pin on ring grooves. Rings are available in standard size only. Install beveled edge of rings towards top of piston. Rings can be compressed with fingers when installing the piston and connecting rod assembly.

CRANKCASE, CRANKSHAFT, BEARINGS AND SEALS. The crankcase and cylinder are an integral unit of die-cast aluminum with either a cast-in iron cylinder liner or chrome plated cylinder bore. Renew crankcase and cylinder if chrome plating has worn through or if cast iron cylinder bore is worn or out-of-round 0.002 inch (0.51 mm) or more.

Crankshaft is supported in two ball bearings. Bearings should be a light press fit on crankshaft and in crankcase and support plate. However, a slip fit in crankcase or support plate is permissible if there is no bearing end play. Crankshaft must be supported on flat surface inside crankpin throw when pressing bearings onto shaft.

As with all two-stroke engines, sealing of crankcase is very important. Renew crankshaft seals using a driver that will contact outer edge of seal only. Install seals in crankcase and bearing support plate with lip of seal to inside. Do not scrape gasket material from crankcase or support plate as this may damage the sealing surfaces. Soak old gasket material loose with solvent.

REED INLET VALVE. Reeds must seat lightly against reed plate along full length of reed and reed must entirely cover hole in reed plate. Renew reeds if rusted, cracked, broken or warped. Renew the complete reed valve unit if reed plate seats are rough, pitted or worn. Dimension measured from reed plate to reed stop as shown in Fig. C5 should be ¼ inch (6.35 mm).

POWER BEE

Model	Bore	Stroke	Displacement
2700, 2704	1¾ in. (44.5 mm)	1-9/16 in. (39.6 mm)	3.8 cu. in. (62.3 cc)
2706	1¾ in. (44.5 mm)	1⅝ in. (41.3 mm)	3.9 cu. in. (63.9 cc)
2723	2 in. (50.8 mm)	1⅝ in. (41.3 mm)	5.1 cu. in. (83.7 cc)
2725, 2726, 2727, 2728, 2729, 2730, 2731, 2732, 2733, 2734, 2735, 2736, 2738, 2740, 2742, 2744	2¼ in. (57.2 mm)	1⅝ in. (41.3 mm)	6.5 cu. in. (106 cc)
2752	1¾ in. (44.5 mm)	1-9/16 in. (39.6 mm)	3.8 cu. in. (62.3 cc)
2756, 27562, 27563, 27564	1¾ in. (44.5 mm)	1⅝ in. (41.3 mm)	3.9 cu. in. (63.9 cc)
2760, 2761, 27612, 2762, 2763, 2764	2¼ in. (57.2 mm)	1¾ in. (44.5 mm)	7.0 cu. in. (114.4 cc)
2770, 2771, 27712, 2772, 27722, 2774 27742, 2775	2 in. (50.8 mm)	1⅝ in. (41.3 mm)	5.1 cu. in. (83.7 cc)
2777	2 in. (50.8 mm)	1-9/16 in. (39.6 mm)	4.9 cu. in. (80.5 cc)
2778, 2779, 27792, 2781, 27812, 2782, 27822, 27823, 27824, 27825, 2783, 2784, 2785, 27852, 27853, 27854, 2786,2787, 2788, 27882, 27883	2 in. (50.8 mm)	1⅝ in. (41.3 mm)	5.1 cu. in. (83.7 cc)
2790	2¼ in. (57.2 mm)	1¾ in. (44.5 mm)	7.0 cu. in. (114.4 cc)
51001, 51005, 51007, 51009, 51011	2 in. (50.8 mm)	1⅝ in. (41.3 mm)	5.1 cu. in. (83.7 cc)
58003, 58004, 58007, 58008, 58009, 58010, 58011, 58012, 58013, 58015, 58019, 58021, 58023, 58024, 58025, 58029, 58031, 58033, 58035, 58037, 58039, 58041, 58045, 58047	2-1/16 in. (52.4 mm)	1¾ in. (44.5 mm)	5.8 cu. in. (95 cc)

Model (Cont.)	Bore (Cont.)	Stroke (Cont.)	Displacement (Cont.)
70003, 70007, 70008, 70009, 70010, 70012, 70013, 70015, 70017, 70019, 70021, 70023, 70025, 70027, 70029, 70031, 70033, 70035, 70037, 70039, 70041	2¼ in. (57.2 mm)	1¾ in. (44.5 mm)	7.0 cu. in. (114.4 cc)

ENGINE INFORMATION

These two-stroke engines, employing reed type inlet valves, are available with plain type, combination antifriction and plain type or all antifriction bearings for the crankshaft. One casting forms the cylinder, cylinder head and crankcase.

MAINTENANCE

SPARK PLUG. Refer to TABLE 1 for recommended spark plug and electrode gap.

TABLE 1

Engine Model	Spark Plug	Gap
2700,2704, 2706,2722, 2723,2752, 2756,27562, 27563,27564, 2770,2771, 27712,2772, 27722,2774, 27742,2775, 2777,2778, 2779,27792, 2780,2781, 2782,2783, 2784,2785, 2786,2788,	J12J	0.040 in. (1.0 mm)
2725,2726, 2727,2728, 2729,2730, 2731,2732, 2733,2734, 2735,2736, 2738,2740, 2742,2744, 27812,27822, 27823,27824, 27825,27852, 27853,27854, 27882,27883, 51001,58024, 58025,70029, 70031	H12J	0.030 in. (0.76 mm)
2760, 2761, 27612, 2762, 2763, 2764, 2787, 2790, 58003, 58004, 58007, 58008, 58009, 58010, 58011, 70003, 70007, 70008, 70009, 70010	H8J	0.030 in. (0.76 mm)
51005, 51007, 51009, 51011, 58029, 58031, 58033, 58035, 58039, 58045, 58047, 70177, 70021, 70023, 70025, 70027, 70035, 70037, 70039, 70041	UJ11G	0.040 in. (1.0 mm)
58012, 58013, 70012, 70013	H4,H03	0.030 in. (0.76 mm)
58015	H10J	0.030 in. (0.76 mm)
58019, 58021, 58023, 70019	H12	0.030 in. (0.76 mm)
70015	H08	0.030 in. (0.76 mm)
58041,70033	RH10	0.030 in. (0.76 mm)

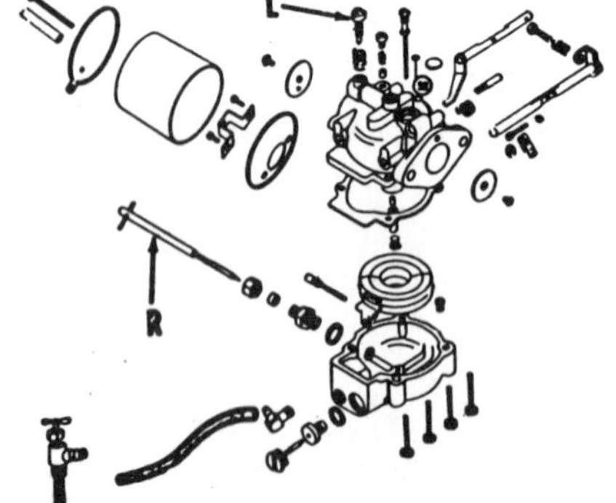

Fig. C6—Components of Tillotson type MD carburetor. Clockwise rotation of idle needle (L) will enrich the mixture.

CARBURETOR. Engines may be equipped with Tillotson MD or MT float carburetor of Tillotson H, HL, or HP diaphragm type carburetor. On MD carburetor, clockwise rotation of the idle needle (L – Fig. C6) will enrich the idle mixture. On MT, H, HL and HP carburetors, clockwise rotation of the idle needle (L – Fig. C7 or C8) will lean the mixture. On all carburetors, clockwise

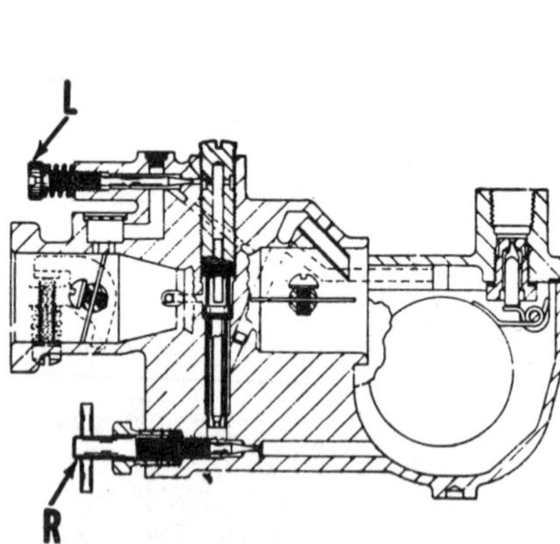

Fig. C7—Cross-sectional view of Tillotson type MT carburetor. Clockwise rotation of idle needle (L) leans the idle mixture.

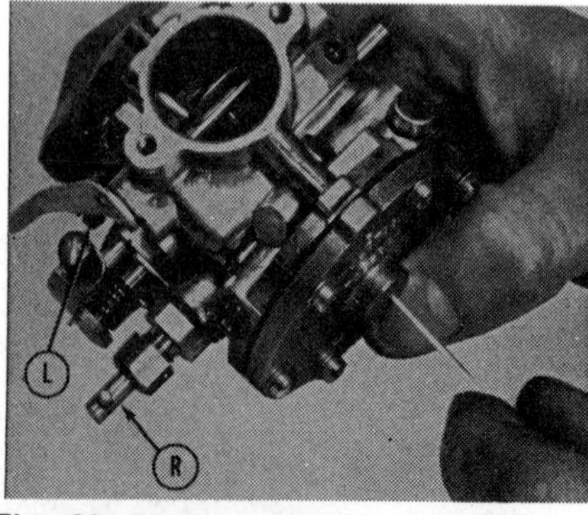

Fig. C8—Flushing diaphragm chamber of diaphram type carburetor. Clockwise rotation of idle needle (L) leans the idle mixture.

rotation of the main adjustment needle (R–Fig. C6, C7 or C8) will lean the mixture. Normal settings for both idle mixture and main adjustments is ¾ to 1 turn open for all carburetors.

Correct float setting on MT type is 3/32 inch (2.4 mm) from flange on bowl cover to nearest face of float when inlet needle is lightly seated. On MD type, farthest face of float (at free end of same) should extend approximately 1/64 to 3/64 inch (0.4-1.2 mm) beyond body casting when float needle is lightly seated as shown in Fig. C9. On diaphragm type carburetors, a certain amount of fuel leakage may occur through discharge nozzle in a normal carburetor for a short period of time after engine is stopped. Fuel inlet needle and seat should be renewed if carburetor shows a tendency to run rich or if leakage is excessive.

GOVERNOR. Three versions of air vane type governors have been used as shown in Figs. C10, C11 and C12. On early engines, the governor spring is contained inside the governor knob (8–Fig. C10). To adjust the governed speed on these installations, loosen set screw (9) and rotate knob (8) to obtain desired operating speed. Clockwise rotation of knob (8) increases the speed.

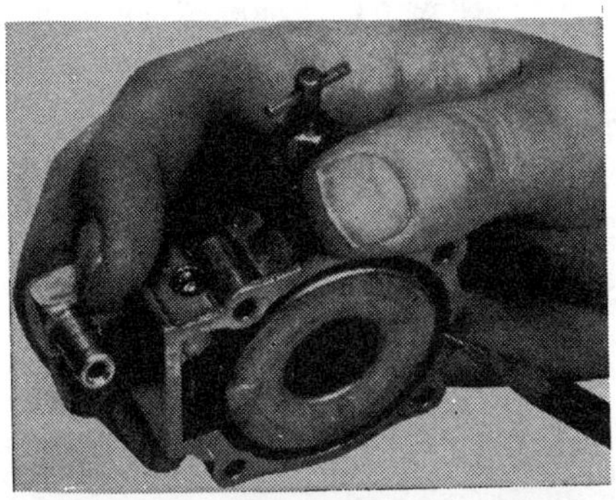

Fig. C9—Correct float setting for type MD Tillotson carburetor is when float extends 1/64 to 3/64 inch (0.4-0.12 mm) beyond body casting as shown.

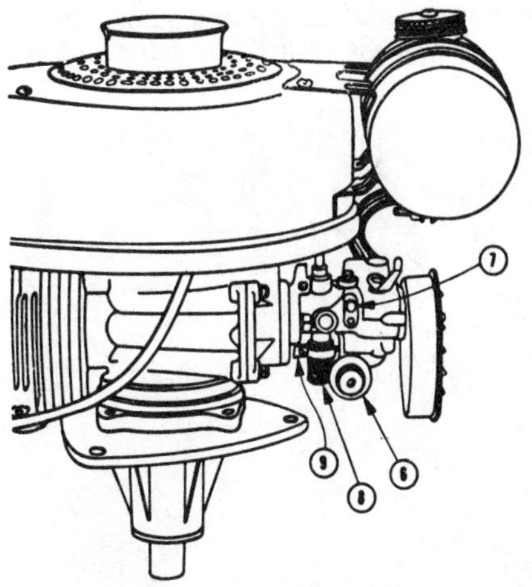

Fig. C10—Governed speed on early production engines is adjusted by knob (8).

On later type engines, the air vane (2–Fig. C11) is connected to the carburetor throttle shaft arm (5) and to the governor spring (4) by a link (3). To ad-

TABLE II

Engine Model	Stator Position	Measurement BTDC	Rotation (Drive End)
2700, 2704, 2706, 2752, 2756, 27562, 27563, 27564, 2777	FA	0.125 in. (3.18 mm)	CCW
2722, 2723	MR	0.094 in. (2.38 mm)	CCW
2725, 2727, 2729, 2731, 2733, 2725, 58011, 58019, 58023, 58025, 58029, 58031, 58033, 58035, 58039, 58041, 58043, 58045, 58047, 70017, 70019, 70021, 70023, 70025, 70027, 70029, 70031, 70033, 70035, 70037	FR	0.156 in. (3.97 mm)	CCW
2726, 2728, 2730, 2732, 2734, 2736, 2760, 2762, 2764, 2790, 58004, 58008, 58010, 58012, 70008, 70010, 70012	MR	0.156 in. (3.97 mm)	CW
2738, 2740, 2742, 2744, 51007, 51009 58024	FR	0.156 in. (3.97 mm)	CW
2761, 27612, 2763, 51001, 51005, 58003, 58007, 58009, 58013, 58015, 70003, 70007, 70009, 70013, 70015, 70039, 70041	MR	0.156 in. (3.97 mm)	CCW
2770, 2771, 27712, 2774, 27742, 2775, 2782, 27822, 27823, 27824, 27825, 2783	FA	0.250 in. (6.35 mm)	CCW
2272	MR	0.219 in. (5.56 mm)	CW
27722, 2779, 27792, 2780, 2781, 27812, 2787, 2788, 27882, 27883	MR	0.219 in. (5.56 mm)	CCW
2778, 2785, 27852, 27853, 27854, 2786	FA	0.219 in. (5.56 mm)	CW
2784	FA	0.219 in. (5.56 mm)	CCW
58021, 58037	FA	0.156 in. (3.97 mm)	CCW

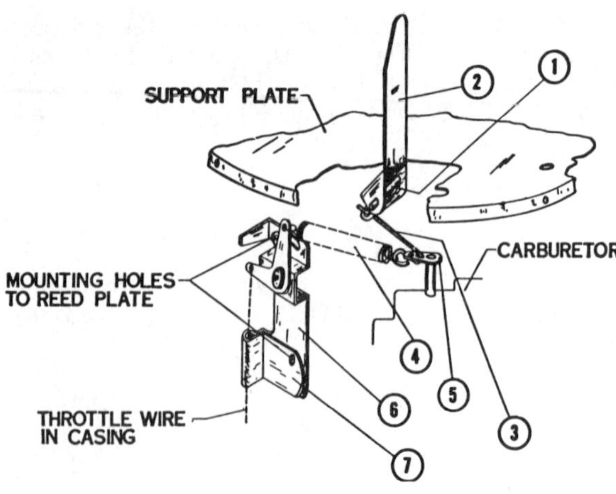

Fig. C11—Governed speed on later production engines is adjusted by moving bracket (6) which varies tension on spring (4).

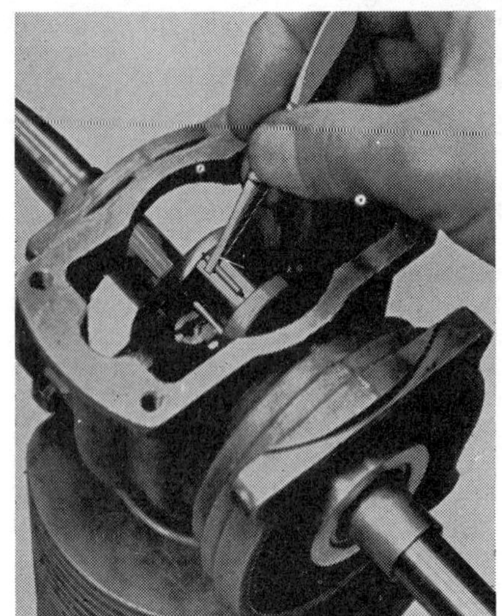

Fig. C15—Inserting needle rollers to upper half of connecting rod.

just governed speed, loosen the two screws which fasten the bracket (6) to the reed plate and move bracket toward or away from carburetor. Moving the bracket away from carburetor increases the speed.

NOTE: In order to remove governor link (3) from throttle arm, first remove the air vane hinge pin (1).

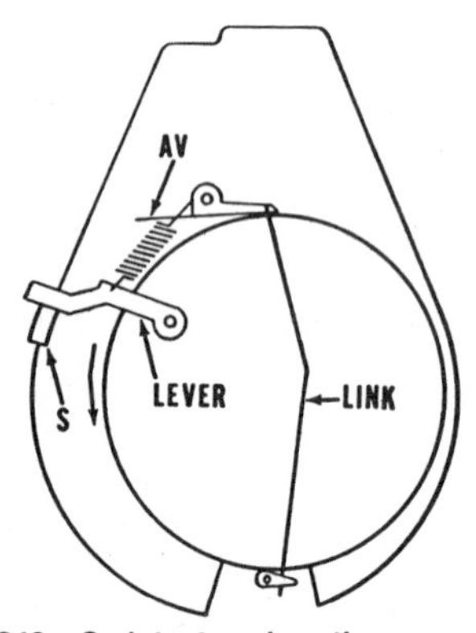

Fig. C12—On latest engines the governed speed is increased by moving the stop (S) in direction indicated by arrow. Air vane is (AV).

Fig. C13—Magneto is correctly timed when stator plate is in the proper location given in TABLE II.

On latest engines, the governor mechanism is located inside the fan housing as shown in Fig. C12. To adjust governed speed, move the control lever stop (S) as required. Stop is locked in position by prongs which bite into support plate. If stop is moved in direction shown by arrow in illustration, the speed is increased.

MAGNETO AND TIMING. Breaker contacts are accessible after removing the flywheel. Recommended gap is 0.020 inch (0.51 mm). Position of magneto stator plate controls ignition timing. Refer to TABLE II for stator position, ignition timing measurement and direction of drive end rotation. Note FA = full advance, MR = midrange and FR = full retard. Position of stator plate at full advance or full retard depends upon direction of crankshaft rotation.

LUBRICATION. Engine is lubricated by mixing oil with the fuel. Use 1/3 pint (0.16 L) of SAE 30 or 40 two-stroke engine oil to each gallon of regular grade gasoline.

REPAIRS

CONNECTING ROD. Piston and connecting rod unit is removed from open (crankcase) end of cylinder after first removing the crankshaft. If connecting rod has an oil hole in the piston pin end, assemble the rod to the piston with the oil hole toward the inlet port of the cylinder. Piston pin should be assembled to piston with closed end of pin toward the inlet port. Piston and rod unit should be installed with long tapered side of upper end of piston (2–Fig. C14) toward the exhaust port (1) as shown. Nominal crankpin diameter on all models is 0.750 inch (19.05 mm).

Steel rods with needle bearings and aluminum rods with plain bearings have each been used. Refer to appropriate paragraph for type being serviced.

Steel Rods With Needle Bearings. To insert the needle rollers, locate the rod about 1/16 inch (1.59 mm) away from crankpin. On models with bearing cages and 16 needles, install cage on crankpin and coat cage with grease. Install bearings into upper half of rod, install other half of rod in a similar manner. On models with 28 individual uncag-

Fig. C14—Tapered surface (2) of piston head should be toward the exhaust port (1).

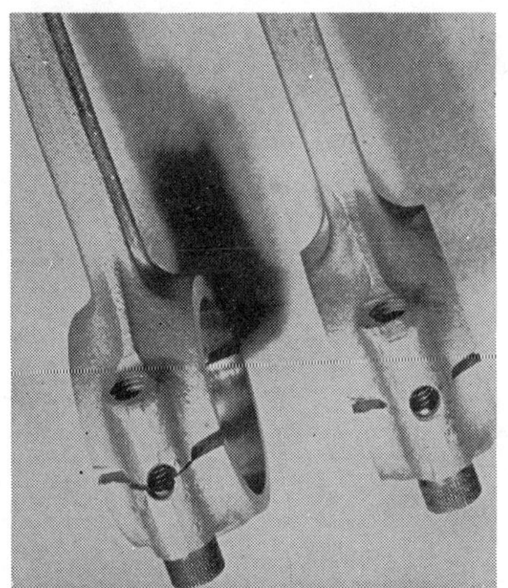

Fig. C16—Correct and incorrect connecting rod cap joints. Rod split line is established by deliberate fracturing. Refer to text.

ed needle rollers, place a light coat of grease on crankpin, then install the 28 individual uncaged needle rollers as shown in Fig. C15.

On all steel rods with either type of needle bearing, mating surfaces of rod and cap are fractured and if cap is correctly installed, the parting line will be almost invisible and cannot be felt with fingernail (Fig. C16). Tighten connecting rod cap screws to 80-90 in.-lbs. (9-10 N•m).

Side clearance of rod on crankshaft should be 3/32 inch (2.38 mm) and is controlled by length of needle rollers. End float of upper end of rod in piston bosses should not exceed 0.015 inch (0.38 mm).

Any steel connecting rod which is rusted, pitted or shows evidence of overheating should be renewed.

Aluminum Rods With Plain Bearings. Aluminum alloy connecting rods are equipped with cast-in bronze inserts which are nonrenewable. Recommended diametral clearance of lower bearing is 0.0015-0.0025 inch (0.038-0.064 mm). Install new rod assembly and/or crankshaft when clearance exceeds 0.0035 inch (0.089 mm). Desired side clearance on crankpin is 0.010-0.020 inch (0.25-0.51 mm). Renew the worn parts when side clearance exceeds 0.040 inch (1.02 mm). Piston pin should be a thumb push fit in the nonrenewable bushing at top end of rod and a palm push fit in the piston bosses. Connecting rod cap screws should be tightened to 80-90 in.-lbs. (9-10 N•m).

PISTON, PIN, RINGS AND CYLINDER. Pistons are equipped with two compression rings which are pinned to limit their rotation in the grooves.

Recommended end gap is 0.003-0.008 inch (0.08-0.20 mm) with a wear limit of 0.013 inch (0.33 mm). Recommended side clearance is 0.0035-0.005 inch (0.089-0.127 mm). Reject the piston if a new ring has more than 0.007 inch (0.18 mm) side clearance in groove. Beveled edge of rings should be installed toward top end of piston. All gaps should be lined up at the ring anchor pin.

Floating type piston pins are retained by lock rings. Two types of lock rings are used. Wire type rings are used in pistons with rounded lock grooves. In pistons with square bottom grooves, install flat lock ring with square cut side of ring away from pin. Pin should be installed with its closed end toward inlet port in cylinder. On some engines, thrust washers are installed at the ends of the piston pin. Pin should have a slightly tighter fit in the piston than in the connecting rod. Oversize pins are not available.

Desired clearance of aluminum piston in engines with steel connecting rods and needle roller bearings is 0.003-0.004 inch (0.08-0.10 mm) with a wear limit of 0.006 inch (0.15 mm). In engines with aluminum rods and plain bearings, desired piston clearance is 0.0025-0.004 inch (0.064-0.101 mm) with wear limit of 0.005 inch (0.13 mm). Cylinder should be renewed if out-of-round or taper exceeds 0.002 inch (0.05 mm). Oversize pistons and rings are not available.

CRANKSHAFT AND SEALS. Desired running clearance of plain type main bearings is 0.001-0.002 inch (0.03-0.05 mm) and wear limit for flywheel is 0.003 inch (0.08 mm) and for pto end is 0.004 inch (0.10 mm). Crankshaft end play for plain type main

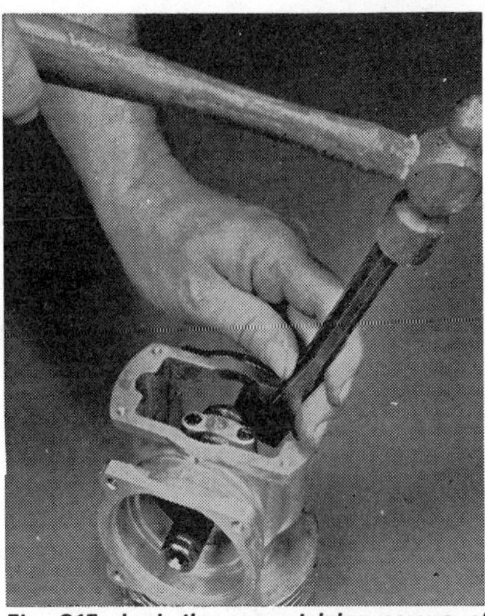

Fig. C17—Lock the cap retaining screws of aluminum rods by staking.

bearings is 0.004-0.010 inch (0.10-0.25 mm) with a wear limit of 0.020 inch (0.51 mm).

On engines with plain type main bearings, nominal diameter of flywheel (upper) main journal is 0.750 inch (19.05 mm) and pto end main journal nominal diameter is 0.875 inch (22.23 mm).

On engines with two antifriction main bearings, nominal diameter of flywheel (upper) main journal is 0.750 inch (19.05 mm) and pto end main journal nominal diameter is 0.781 inch (19.84 mm).

On late engines with ball bearing on output end and needle bearing on flywheel end, nominal crankshaft end journal diameters are 0.787 inch (19.99 mm).

All crankcase seals must be maintained in good condition because leakage through the seals releases compression and vacuum and causes loss of power.

REED VALVE UNIT. This removable unit is located between the carburetor and the cylinder block. Reeds should seat lightly against reed plate throughout their entire length with the least possible initial tension. Check seating by blowing and drawing air through ports with light vacuum/air source.

POWER BEE

Model	Bore	Stroke	Displacement
50000, '002, '004, '006, '008, '010, '014, '052, '054	2 in. (50.8 mm)	1-19/32 in. (40.5 mm)	5.0 cu. in. (82 cc)

ENGINE INFORMATION

The two-stroke engines are known as "L" series engines. All models have a detachable cylinder and a two-piece crankcase. Cylinder has an integral cylinder head and cylinder bore is chrome plated. Crankshaft is supported by two ball bearing mains and connecting rod has a nonrenewable needle bearing at piston pin end and renewable loose roller bearings at crankpin end. All engines except Models 50052 and 50054 are loop-scavenged, utilizing reed valves and flat top pistons. Models 50052 and 50054 are cross-scavenged, utilizing a 3rd port and deflector top piston. All models use a diaphragm type carburetor which allows engine to be operated in any position.

MAINTENANCE

SPARK PLUG. Recommended spark plug for all models except 50000 model is Champion CJ8, or equivalent and recommended spark plug for 50000 model is Champion H8J, or equivalent. Electrode gap for all models is 0.030 inch (0.76 mm). Tighten spark plug to 15 ft.-lbs. (20 N·m).

CARBURETOR. Engines are equipped with a Tillotson Series HS diaphragm type carburetor. Refer to Fig. C19 for exploded view of carburetor.

Clockwise rotation of both idle and high speed fuel needles leans the mixture. For initial adjustment, open both needles approximately 1¼ turns. Make final adjustment with engine running and at operating temperature.

Adjust idle needle for smoothest operation at idle speed, then accelerate engine to about 4000 rpm. If engine tends to accelerate slowly and bog down (too rich), turn idle needle clockwise (lean) as necessary. If engine tends to stall (too lean), turn idle needle counterclockwise (richer) as necessry. Average needle setting is 1-⅛ turns open. Adjust engine high idle speed as follows: Be sure idle adjustment is satisfactory, then run engine at 4000 rpm and turn high speed needle clockwise to a point where engine stops four-stroking (engine fires on every power stroke).

CAUTION: Do not go any leaner with the high speed needle adjustment as it could result in improper lubrication and engine seizure could result.

If engine idle rpm is too high, turn carburetor stop screw counterclockwise until desired rpm is obtained.

MAGNETO AND TIMING. A Wico flywheel type magneto is used on all models. Refer to Fig. C20.

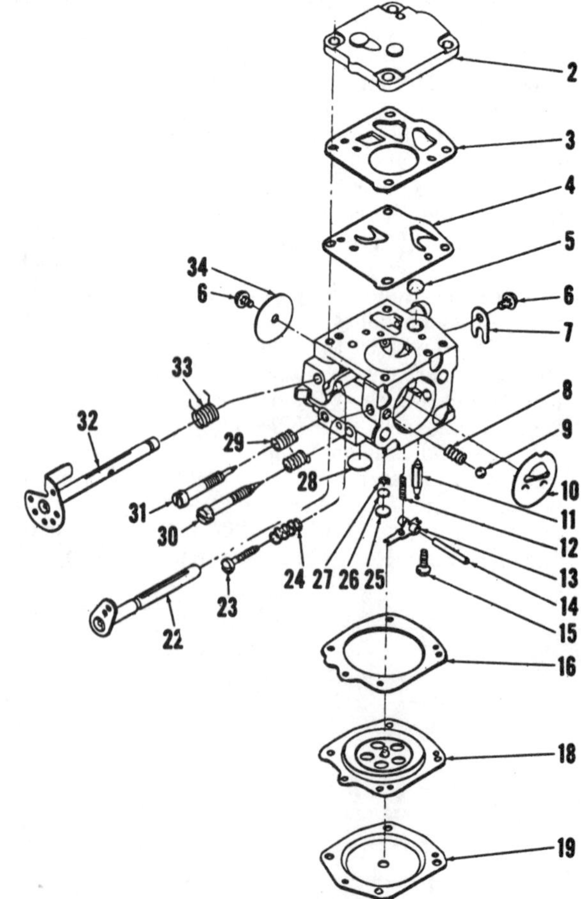

Fig. C19—Exploded view of Tillotson Series HS carburetor used on all Series 500 engines.

2. Fuel pump cover
3. Gasket
4. Fuel pump diaphragm
5. Inlet screen
7. Throttle shaft clip
8. Friction spring
9. Friction ball
10. Choke shutter
11. Inlet needle
12. Tension spring
13. Inlet control lever
14. Pin
16. Gasket
18. Diaphragm
19. Diaphragm cover
22. Choke shaft & lever
23. Idle speed screw
24. Spring
25. Welch plug (small)
26. Retaining ring
27. Screen
28. Welch plug (large)
29. Spring
30. Main adjusting screw
31. Idle adjusting screw
32. Throttle shaft & lever
33. Spring
34. Throttle shutter

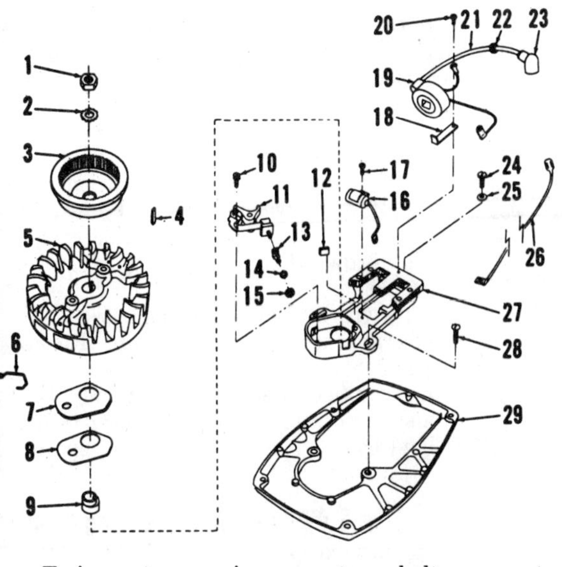

Fig. C20—Exploded view of flywheel magneto used on all Series 500 engines.

1. Nut
2. Lockwasher
3. Starter cup
4. Flywheel key
5. Flywheel
6. Cover clamp
7. Cover
8. Gasket
9. Breaker point cam
11. Breaker point set
12. Breaker cam felt
16. Condenser
18. Coil clip
19. Coil
21. High tension lead
22. Grommet
23. Boot
26. Ground lead
27. Stator & core
29. Support plate

To inspect or service magneto, unbolt and remove fan housing and recoil starter assembly as a unit. Remove flywheel retaining nut and starter cup, then remove flywheel using special knock-off nut.

Breaker contact points are accessible after removing flywheel and the dust cover from stator plate. Breaker point gap is 0.015 inch (0.38 mm). When adjusting points, be sure breaker point cam follower is positioned at index mark on breaker cam.

NOTE: Be accurate when setting the breaker gap as gap affects ignition timing.

If stator plate has been loosened or removed, breaker contact point gap must be reset. When installing stator plate on all models except Model 50014, it must be installed in the midrange position (mounting screws in center of slotted stator mounting holes). With stator installed in midrange position and breaker point gap correctly set, ignition timing will be 32° BTDC.

When installing stator plate on Model 50014, it must be installed in the full retard position. With stator installed in full retard position and breaker point gap correctly set, ignition timing will be 28° BTDC.

LUBRICATION. Engine is lubricated by mixing oil with the fuel. Thoroughly mix 1/3 pint (0.16 L) of SAE 30 or 40 two-stroke engine oil with each gallon of regular gasoline for Model 50014 or 1/2 pint (0.24 L) of SAE 30 or 40 two-stroke engine oil with each gallon of regular gasoline for all other models. Do not use premium grade gasoline.

REPAIRS

CONNECTING ROD. To remove the aluminum alloy connecting rod, remove starter, flywheel, magneto assembly, magneto support plate and cylinder. Separate crankcase halves, remove crankshaft, then separate rod cap from connecting rod using care not to lose any of the loose needle rollers. Remove piston pin retaining rings and bump out piston pin.

The caged needle roller bearing in pin end of connecting rod is not renewable. Renew the connecting rod assembly if any roller in pin end is rough or flat, if pin end rollers can be separated the width of one roller, or if crankpin bearing surface is rough or shows signs of wear.

When installing connecting rod to crankshaft, coat crankpin with grease, then install the 25 rollers. Install connecting rod assembly with match marks toward flywheel end of crankshaft. Tighten connecting rod cap screws to 70-80 in.-lbs. (8-9 N·m).

The mating surface of rod and cap are fractured and if cap is correctly installed, the parting line will be almost invisible and cannot be felt with fingernail.

If piston assembly is installed on connecting rod at this time, install deflector top pistons so longer taper side is toward exhaust side. Pistons with flat top are installed with piston ring anchor pin toward flywheel side.

PISTON, PIN AND RINGS. Pistons are fitted with two compression rings which are pinned to limit rotation in their grooves. Rings are available in standard sizes only. Recommended ring end gap is 0.009-0.012 inch (0.23-0.31 mm). Recommended side clearance is 0.0025-0.0050 inch (0.064-0.127 mm). Beveled edge of rings should be installed toward top of pistons. All ring gaps must be aligned at the ring anchor pin.

The 0.437 inch (11.09 mm) diameter floating type piston pin is retained by lock rings located at each end. Piston pins are available in standard size only. It is recommended piston be slightly heated prior to removing or installing piston pin to prevent damage to piston pin bosses.

CRANKCASE, CRANKSHAFT, BEARINGS AND SEALS. The aluminum die-cast cylinder with integral cylinder head is bolted to a two-piece magnesium die-cast crankcase. Cylinder

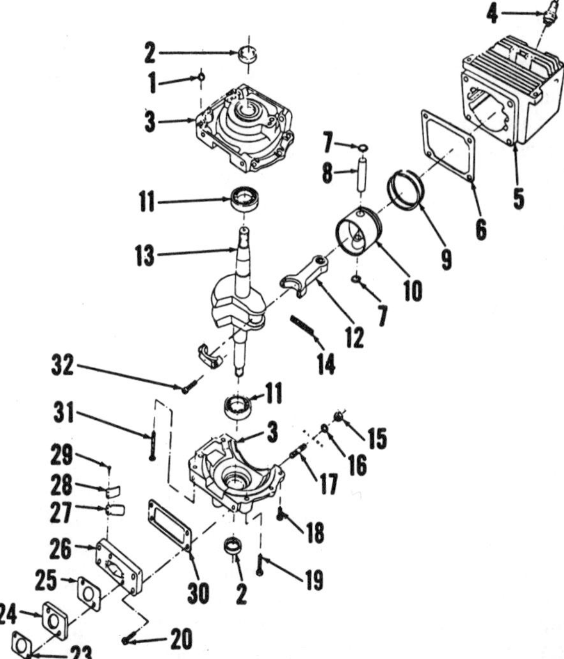

Fig. C21—Exploded view of the loop scavenged Series 500 engine. Carburetor and reed valve assembly may be located either on bottom or side of crankcase. This loop scavenged engine uses a flat top piston.

2. Crankcase seals
3. Crankcase halves
4. Spark plug
5. Cylinder
6. Gasket
7. Retaining rings
8. Piston pin
9. Piston rings
10. Piston
11. Ball bearing
12. Connecting rod
13. Crankshaft
14. Crankpin roller set
17. Cylinder stud
23. Carburetor gasket
24. Carburetor adapter
25. Adapter gasket
26. Reed plate
27. Reed valve
28. Reed stop
30. Gasket
32. Connecting rod screw

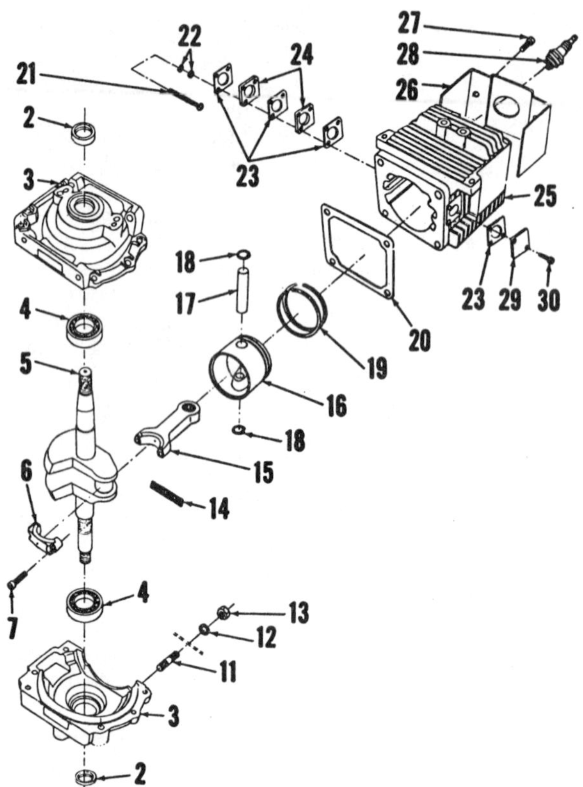

Fig. C22—Exploded view of the cross scavenged (3rd port) Series 500 engine. Carburetor is mounted to side of cylinder. This cross scavenged engine uses a deflector top piston.

2. Crankcase seals
3. Crankcase halves
4. Ball bearing
5. Crankshaft
6. Connecting rod cap
7. Connecting rod screw
11. Cylinder stud
14. Crankpin roller set
15. Connecting rod
16. Piston
17. Piston pin
18. Retaining rings
19. Piston rings
20. Gasket
23. Carburetor gasket
24. Carburetor adapter
25. Cylinder
26. Cylinder cover
28. Spark plug
29. 3rd port cover

bore is chrome plated and cylinder should be renewed if chrome plating is worn through or if cylinder bore is excessively worn or out-of-round.

Crankshaft is supported by two ball type main bearings. Bearings should be a light press fit on crankshaft and in crankcase halves. Crankshaft must be supported on flat surface inside crankpin throw when pressing bearings on crankshaft.

As with all two-stroke engines, sealing of crankcase is very important. Renew crankshaft seals using a driver which will contact outer edge of seal only. Use a seal protector, or tape all threads and keyways and install seals with lips toward inside.

REED VALVE UNIT. All 500 series engines except Models 50052 and 50054 are equipped with a reed valve assembly located between carburetor and crankcase. Reeds must seat lightly against reed plate along full length of reed and reed must completely cover hole in reed plate. Check reed seating by blowing and drawing air through ports with a light vacuum/air source. Dimension from reed plate to reed stop should be 1/4 inch (6.35 mm).

SERVICING POWER BEE ACCESSORIES

REWIND STARTERS

Three types of rewind starters are used on Chrysler two-stroke engines. Type A shown in Fig. C23 is the friction shoe type. Type B shown in Fig. C24 is the single dog type. Type C shown in Fig. C25 is the three dog type.

OVERHAUL TYPE A. To disassemble the starter, refer to Fig. C23, remove rope handle and allow pulley (3) to rotate slowly until spring (2) is unwound. Remove retaining ring (12), washer (11), spring (10), washers (8) and friction shoe assembly. Remove pulley and rope from cover and spring. Remove rewind spring from cover and unwind rope from pulley.

When reassembling, lubricate rewind spring, cover shaft and center bore in pulley with a light coat of grease. Install rewind spring so windings are in same direction as removed spring. Install rope on pulley, then place pulley on cover shaft. Make certain inner and outer ends of spring are properly hooked on cover and rotor. Preload the rewind spring by rotating the pulley 2½ to 3 turns.

Check sharp end of friction shoes (7) and sharpen or renew as necessary. Install washers (8), friction shoe assembly, spring (10), washer (11) and retaining ring (12). Make sure the friction shoe assembly is installed properly for correct starter rotation. If properly installed, sharp ends of friction shoes will extend when rope is pulled.

OVERHAUL TYPE B. To disassemble starter, refer to Fig. C24, remove rope handle and allow pulley (4) to rotate slowly until all tension is removed from spring (2). Remove screw and dog retainer (8), dog (7) and spring (6). Remove screw (11), cam (10) and spring (9). Carefully lift pulley about ½ inch (13 mm) from starter cover and unhook inner spring loop from pulley. Remove pulley from cover, then remove spring (2) and keeper (3) from cover.

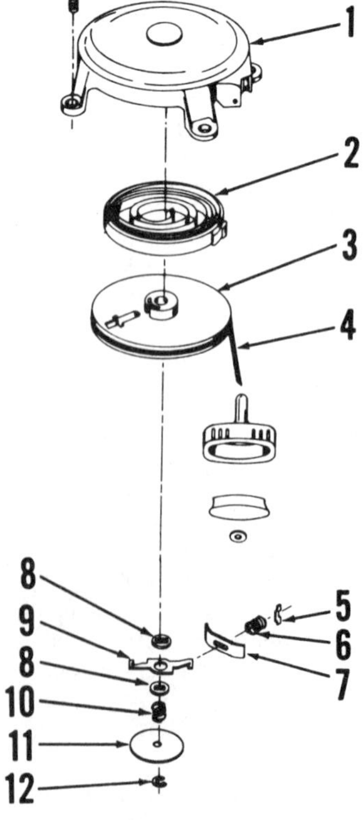

Fig. C23—Exploded view of Type A rewind starter used on some models.

1. Cover
2. Rewind spring
3. Pulley
4. Rope
5. Retainer (2)
6. Spring (2)
7. Friction shoe (2)
8. Washers
9. Actuating lever
10. Spring
11. Washer
12. Retaining ring

When reassembling, place spring with keeper in cover. Apply a few drops of oil to spring and a light coat of grease to cover shaft. Install rope on pulley, then place pulley on cover shaft. Make certain spring ends are properly hooked and preload rewind spring by rotating pulley 1½ to 2 turns. Install spring (9), cam (10) and screw (11), then install spring (6), dog (7) and retainer (8) with its screw. Check operation of starter by pulling rope. Dog should extend when rope is pulled and retract when released.

OVERHAUL TYPE C. To disassemble the starter, refer to Fig. C25, remove rope handle and allow pulley (4) to rotate slowly to unwind spring (3). Remove screw (10), dog retainer (9), spring (8), dogs (7) and springs (6). Carefully lift pulley about ½ inch (13 mm) from cover (1) and detach inside spring loop from pulley. Remove pulley from cover, then remove spring (3) and keeper (2) from cover.

When reassembling, place spring with keeper in cover. Apply a few drops of oil to the spring and a light coat of grease to cover shaft. Install rope on pulley, then place pulley on cover shaft. Make certain spring ends are properly hooked and preload rewind spring by rotating pulley 1½ to 2 turns. Install springs (6), dogs (7), spring (8) and dog retainer (9) and secure with screw (10). Check starter operation by pulling rope. Dogs should extend when rope is pulled and retract when rope is released.

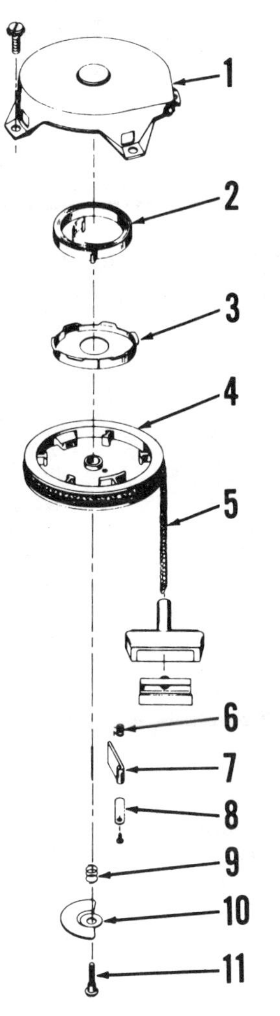

Fig. C24—Exploded view of Type B rewind starter used on some models.

1. Cover
2. Rewind spring
3. Keeper
4. Pulley
5. Rope
6. Spring
7. Dog
8. Retainer
9. Spring
10. Cam
11. Screw

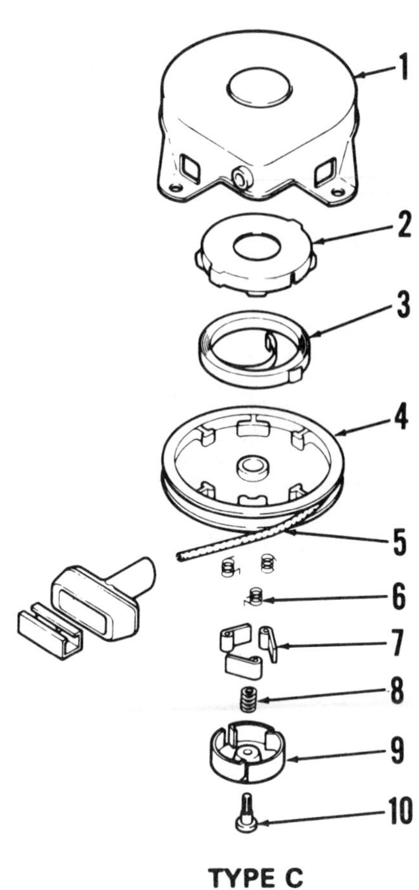

Fig. C25—Exploded view to Type C rewind starter used on some models.

SACHS

SACHS MOTOR CORP., LTD.
9615 Cote De Liesse
Dorval, Quebec, Canada H9P 1A3

FICHTEL & SACHS AG
852 Schweinfurt Postfach 52
West Germany

Model	Bore	Stroke	Displacement
ST30	38 mm (1.496 in.)	29 mm (1.14 in.)	33 cc (2 cu. in.)
ST76	48 mm (1.890 in.)	42 mm (1.65 in.)	75 cc (4.6 cu. in.)
ST96	52 mm (2.047 in.)	44 mm (1.732 in.)	90 cc (5.7 cu. in.)
ST102	52 mm (2.047 in.)	48 mm (1.890 in.)	102 cc (6.2 cu. in.)
ST126	56 mm (2.205 in.)	54 mm (2.126 in.)	132 cc (7.0 cu. in.)
ST151	60 mm (2.362 in.)	54 mm (2.126 in.)	151 cc (9.3 cu. in.)
ST204	65 mm (2.560 in.)	61 mm (2.402 in.)	201 cc (12.4 cu. in.)
ST251	72 mm (2.835 in.)	61 mm (2.402 in.)	246 cc (15.2 cu. in.)
SB102	52 mm (2.047 in.)	48 mm (1.890 in.)	102 cc (6.2 cu. in.)
SB126	56 mm (2.205 in.)	54 mm (2.126 in.)	132 cc (7.0 cu. in.)
SB151	60 mm (2.362 in.)	54 mm (2.126 in.)	151 cc (9.3 cu. in.)

MAINTENANCE

SPARK PLUG. Recommended spark plug is a Champion L88A, or equivalent. Electrode gap is 0.5 mm (0.020 in.).

CARBURETOR. Several different carburetors are used on these engines. Refer to appropriate paragraph for model being serviced.

Tillotson Diaphragm Carburetor. Idle mixture is adjusted at needle (13–Fig. SA1). High speed mixture is adjusted at main fuel needle (26). Initial setting is one turn open for both needles and clockwise rotation will lean the mixture. Make final mixture adjustment with engine warm and operating with the normal amount of load. Adjust the main fuel needle (26) for smoothest operation at governed speed, then adjust idle needle (13) for smoothest operation at idle (slow) speed.

After considerable service, it may be necessary to overhaul the carburetor and renew worn parts to restore operating efficiency. Before disassembling carburetor, thoroughly clean all external surfaces. Disassemble carburetor, clean all parts and inspect for damage and excessive wear. Renew parts if necessary and reassemble.

Bing Diaphragm Carburetor. The diaphragm carburetor shown in Fig. SA2 is typical of the carburetor used on some ST30, ST76 and ST96 engines. The setting of the carburetor is determined by selection of main jet (13) size

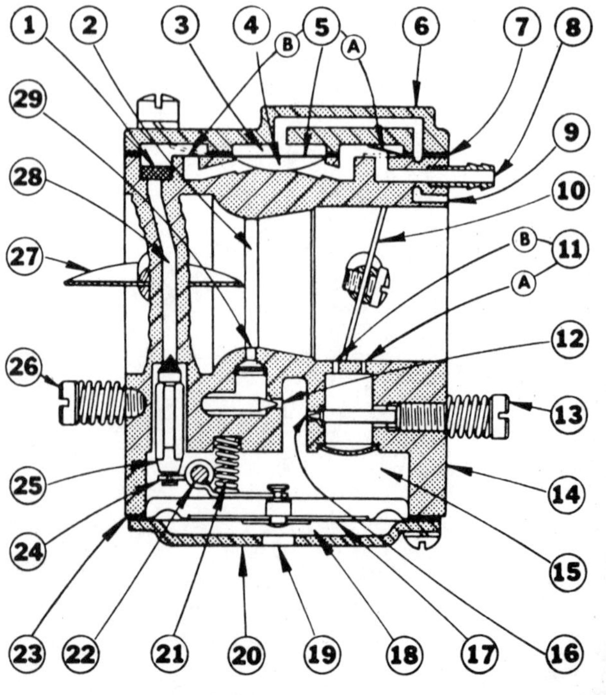

Fig. SA1—Cross-sectional view of Model HS Tillotson diaphragm carburetor.
1. Inlet screen
2. Venturi throat
3. Pulsation chamber
4. Fuel chamber
5. Pump diaphragm
5A. Inlet valve
5B. Outlet valve
6. Fuel pump body
7. Gasket
8. Inlet nipple
9. Impulse duct
10. Throttle plate
11A. Pilot outlet passage
11B. Bypass passage
12. Main jet orifice
13. Idle mixture needle
14. Carburetor body
15. Control chamber
16. Idle jet
17. Diaphragm
18. Air chamber
19. Air vent
20. Diaphragm cover
21. Spring
22. Pivot pin
23. Gasket
24. Inlet control lever
25. Inlet needle valve
26. Main fuel needle
27. Choke plate
28. Fuel duct
29. Main fuel outlet

for the operating speed of the engine. Carburetor adjustment is accomplished only by installing larger or smaller size jet.

Disassembly of the carburetor is obvious after examination of the unit and reference to Fig. SA2. Clean and inspect all parts and renew any showing damage and excessive wear.

Bing Butterfly Valve Carburetor. The Bing butterfly valve carburetor (Fig. SA3) is used on some ST76 and ST96 engines. The fuel setting of the carburetor is determined by choice of idle (4) and main (13) jet sizes. Idle mixture is also controlled by adjustment of the pilot air adjustment needle (5). Disassemble carburetor, clean all parts and inspect for damage and excessive wear. Renew parts if necessary and reassemble using Fig. SA3 as a guide.

Bing Slide Valve Carburetor. The Bing slide valve carburetor (Fig. SA4) is used on some ST30, ST76 and ST96 engines. Adjustment of this carburetor, in addition to those obtained by changing needle jet (11) and main jet (13), can be made by varying the position of the jet needle (8). Jet needle is equipped with four grooves as shown in inset. If needle is moved to a higher position, a richer air:fuel mixture is formed. The jet needle setting will affect the fuel mixture only in the lower and medium speed range of the engine and will have no effect at wide open throttle. Some carburetors may be equipped with a hand control lever instead of a control cable.

Bing Diaphragm Carburetor. The diaphragm butterfly valve carburetor shown in Fig. SA5 is of the type used on some Model ST102 and SB102 engines. The fuel setting of the carburetor is determined by the selection of idle (1) and main (3) jet sizes. Idle mixture is also controlled by adjustment of the pilot jet adjustment needle (4).

Disassembly of the carburetor is obvious after examination of the unit and reference to Fig. SA5. Clean and inspect all parts and renew any showing damage and excessive wear.

Bing Float Carburetor. The float type butterfly valve carburetor shown in Fig. SA6 is used on some Model ST102 engines. The fuel setting of the carburetor is determined by choice of idle (20) and main (18) jet sizes. Idle mixture is also controlled by adjustment of the pilot air adjustment needle (1).

Disassemble carburetor, clean all parts and inspect for excessive wear or other damage. Renew parts if necessary and reassemble using Fig. SA6 as a guide.

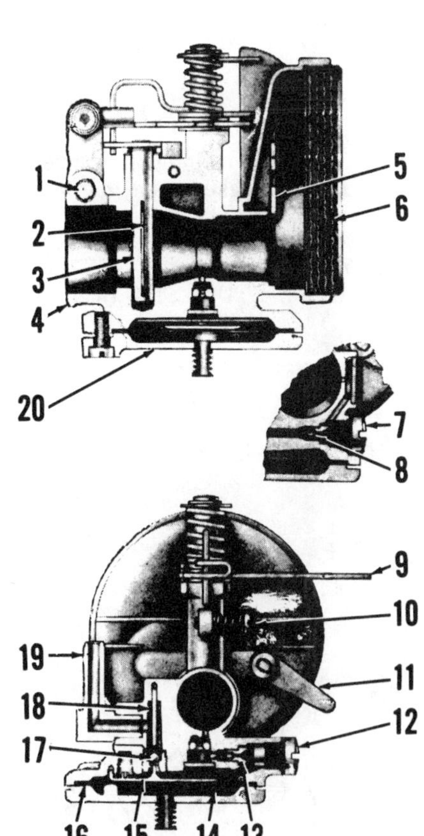

Fig. SA2—Partial cutaway views of typical Bing diaphragm carburetor used on some Model ST30, ST76 and ST96 engines.

1. Clamp screw
2. Throttle valve
3. Throttle shaft
4. Carburetor body
5. Choke
6. Filter
7. Plug
8. Idle jet
9. Actuating lever
10. Stop screw
11. Choke lever
12. Plug
13. Main jet
14. Diaphragm
15. Diaphragm plate
16. Seal ring
17. Ball check valve
18. Valve spring
19. Fuel inlet
20. Cover

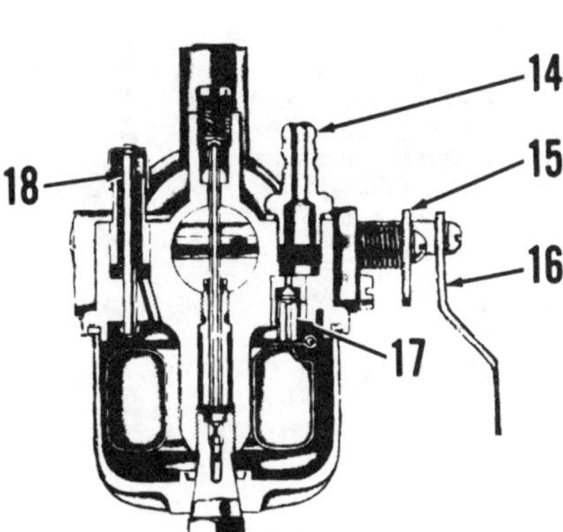

Fig. SA3—Cross-sectional views of the Bing butterfly valve carburetor used on some ST76 and ST96 engines.

1. Carburetor body
2. Butterfly (throttle) valve
3. Throttle shaft
4. Idle jet
5. Air adjustment needle
6. Spring
7. Choke shaft
8. Choke plate
9. Mixing tube
10. Seal ring
11. Float
12. Float bowl
13. Main jet
14. Fuel inlet nipple
15. Throttle lever
16. Choke lever
17. Inlet needle valve
18. Primer

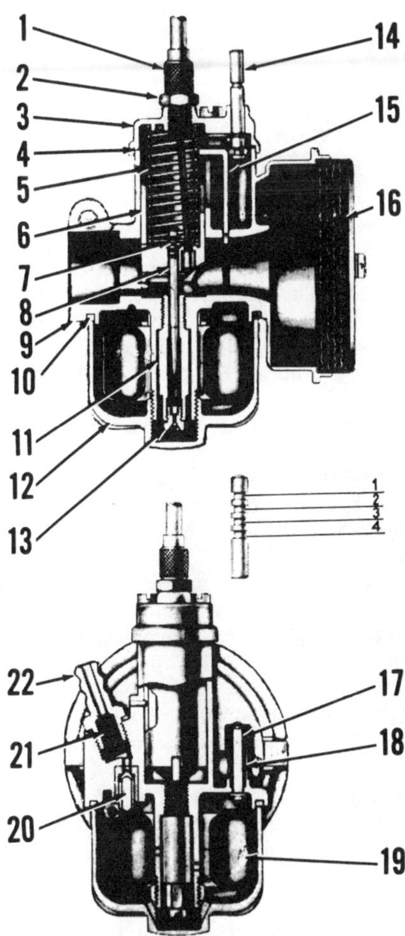

Fig. SA4—Cross-sectional view of a typical Bing slide valve carburetor used on Models ST30, ST76 and ST96.

1. Adjusting screw
2. Locknut
3. Cover
4. Gasket
5. Spring
6. Slide (throttle) valve
7. Retaining plate
8. Metering needle
9. Carburetor body
10. Seal body
11. Needle jet
12. Float bowl
13. Main jet
14. Thrust pin
15. Choke shutter
16. Filter
17. Primer
18. Spring
19. Float
20. Inlet needle valve
21. Screen
22. Fuel inlet nipple

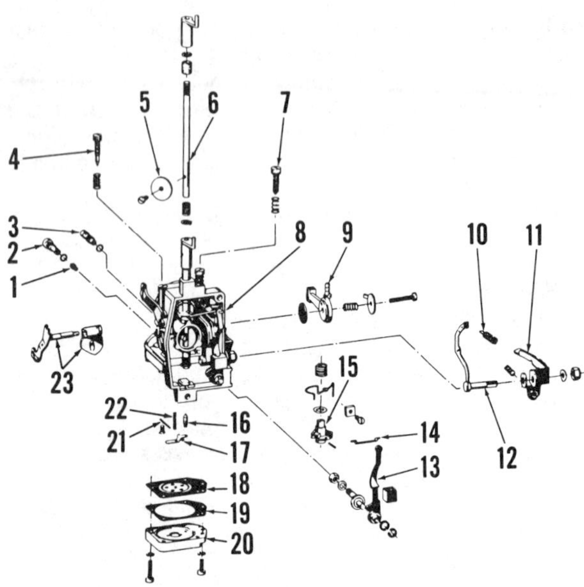

Fig. SA5—Exploded view of typical Bing diaphragm butterfly valve carburetor used on some Models ST102 and SB102 engines.

1. Idle jet
2. Plug
3. Main jet
4. Air adjusting screw
5. Throttle butterfly
6. Throttle shaft
7. Idle speed screw
8. Carburetor body
9. Fuel shut-off valve
10. Governor spring
11. Control lever
12. Governor lever
13. Pivot arm
14. Link
15. Actuating arm
16. Inlet needle valve
17. Inlet lever
18. Diaphragm
19. Gasket
20. Cover
21. Pin
22. Spring
23. Starter lever assy.

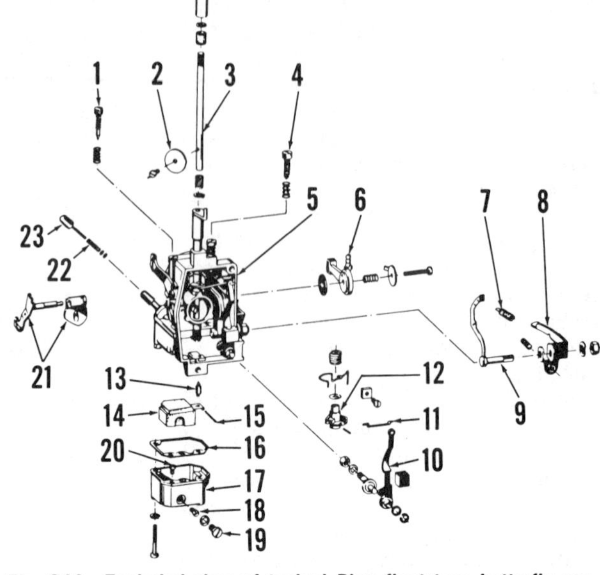

Fig. SA6—Exploded view of typical Bing float type butterfly carburetor used on some Model ST102 engines.

1. Air adjusting screw
2. Throttle butterfly
3. Throttle shaft
4. Idle speed screw
5. Carburetor body
6. Fuel shut-off valve
7. Governor spring
8. Control lever
9. Governor lever
10. Pivot arm
11. Link
12. Actuating arm
13. Inlet needle valve
14. Float
15. Float pin
16. Gasket
17. Float bowl
18. Main jet
19. Plug
20. Idle jet
21. Starting lever assy.
22. Spring
23. Primer

Bing Butterfly Valve Float Carburetor. This carburetor shown in Fig. SA7 is used on some ST126, ST151, ST204, ST251, SB126 and SB151 model engines. The fuel setting of the carburetor is determined by the selection of idle (4) and main (14) jet sizes. Idle mixture is also controlled by adjustment of the pilot air adjustment needle (3).

Disassembly of the carburetor is obvious after examination of the unit and reference to Fig. SA7. Clean and inspect all parts and renew any showing excessive wear or other damage.

Bing Slide Valve Float Carburetor. The Bing slide valve float type carburetor (Fig. SA8) is used on some ST126, ST151, ST204 and ST251 model engines. Adjustment of fuel setting on the carburetor is determined by the choice of idle (13) and main (15) jet sizes. Idle mixture is also controlled by adjustment of the pilot air adjustment needle (24). Main fuel adjustment can also be changed by varying the position of the jet needle (9). Jet needle is equipped with several notches for retainer clip (7). Raising jet needle position will richen air:fuel mixture. The jet needle setting will affect the fuel mixture only in the lower and medium speed range of the engine and will have no effect at wide open throttle.

Disassembly of the carburetor is obvious after examination of the unit and reference to Fig. SA8. Clean and inspect all parts and renew any showing excessive wear or other damage.

MAGNETO AND TIMING. Breaker point magneto ignition is used on Models ST30, ST76, ST96, ST102 and SB102. Bosch, Ducati and Motoplat magnetos are used and breaker point gap on all models is 0.4 mm (0.02 in.). Ignition timing on these models as is follows: Standard ST76 and ST96 models, 2.5-3.0 mm (0.10-0.12 in.) BTDC; ST76 (high per-

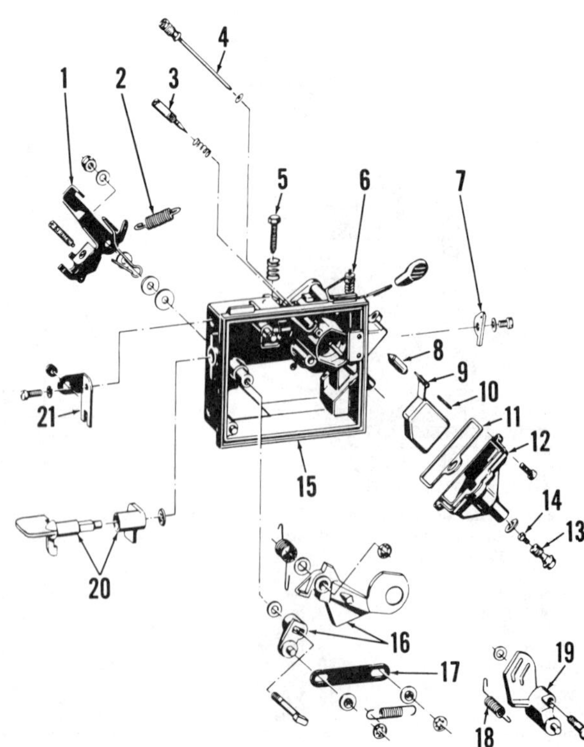

Fig. SA7—Exploded view of Bing butterfly valve float type carburetor used on some ST126, ST151, ST204, ST251, SB126 and SB151 engines.

1. Control lever
2. Tension spring
3. Air adjusting screw
4. Idle jet
5. Idle speed screw
6. Primer
7. Throttle arm
8. Inlet needle valve
9. Float
10. Float pin
11. Gasket
12. Float bowl
13. Jet holder
14. Main jet
15. Carburetor body
16. Choke assy.
17. Link
18. Governor spring
19. Pivot arm
20. Choke lever assy.
21. Bracket

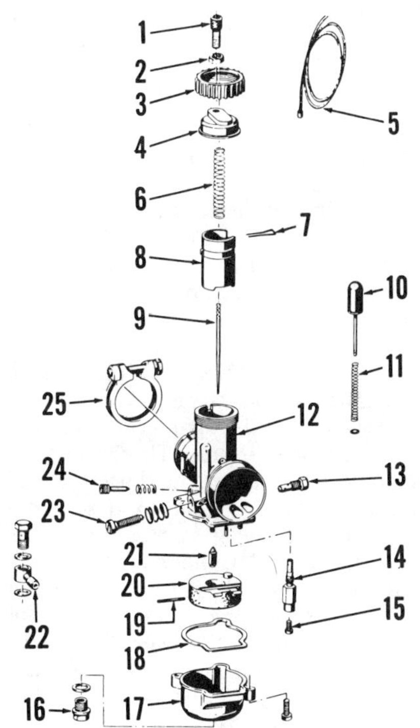

Fig. SA8—Exploded view of Bing slide valve float type carburetor used on some Models ST126, ST151, ST204 and ST251.

1. Adjuster screw
2. Jam nut
3. Cover retainer
4. Cover
5. Control cable
6. Spring
7. Jet needle retainer
8. Slide (throttle) valve
9. Jet needle
10. Primer
11. Spring
12. Body
13. Idle jet
14. Needle jet
15. Main jet
16. Plug
17. Float bowl
18. Gasket
19. Float pin
20. Float
21. Inlet needle valve
22. Inlet fuel connector
23. Idle speed screw
24. Air adjustment needle
25. Clamping ring

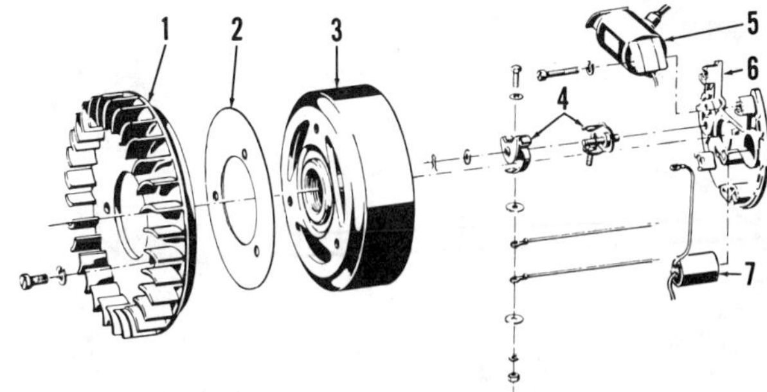

Fig. SA9—Exploded view of the Bosch magneto ignition system used on some Models ST30, ST76 and ST96 engines.

1. Fan
2. Gasket
3. Flywheel
4. Ignition point set
5. Coil armature assy.
6. Base plate
7. Condenser

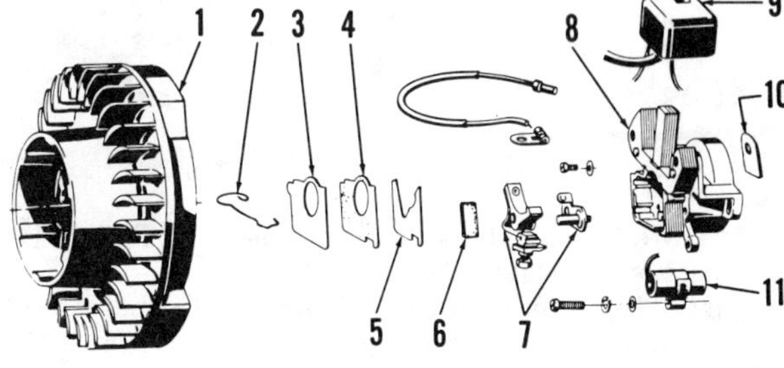

Fig. SA10—Exploded view of the Ducati magneto ignition system used on some ST76 and ST96 engines.

1. Flywheel
2. Clip
3. Cover
4. Gasket
5. Gasket
6. Lubricating felt
7. Ignition point set
8. Armature base
9. Coil assy.
10. Plate

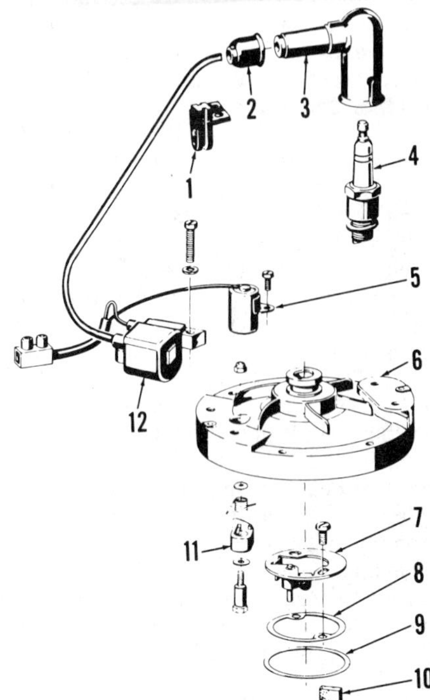

Fig. SA11—Exploded view of typical Motoplat magneto ignition system used on ST102 and SB102 engines.

1. Grommet
2. Rain cap
3. Plug connector
4. Spark plug
5. Condenser
6. Flywheel
7. Ignition breaker point assy.
8. Gasket
9. Spring washer
10. Lubricating felt
11. Starter pawl
12. Coil assy.

formance with cylinder marked H), 3.8-4.0 mm (0.15-0.16 in.) BTDC; ST76 (slow running with cylinder marked L), 2.0 mm (0.08 in.) BTDC; ST30, 3.2-3.6 mm (0.13-0.14 in.) BTDC; ST102 and SB102, 2.5-3.5 mm (0.10-0.14 in.) BTDC.

Breakerless electronic ignition is used on Models ST126, ST151, ST204, ST251, SB126 and SB151. Ignition timing on these models is as follows: ST126, ST151, SB126 and SB151, 2.2-3.0 mm (0.08-0.12 in.) BTDC; ST204 and ST251, 2.4-3.4 mm (0.09-0.13 in.) BTDC.

On all models, timing is adjusted by moving the coil-armature assembly (breaker point type) or ignition unit (breakerless type). Moving the coil-armature or ignition unit in direction opposite to flywheel rotation will advance timing.

AIR CLEANER. If the engine is operated under dusty conditions, the manufacturer recommends that the air cleaner be serviced daily. Three types of filters have been used. Refer to appropriate paragraph for type being serviced.

Oil Bath Air Cleaner. Remove cover from air cleaner and discard old oil. Wash cleaner in gasoline or other solvent, then dry cleaner thoroughly. Refill to level line or arrow mark with SAE 30 motor oil.

Foam Type Filter. Remove filter, then wash and rinse thoroughly in gasoline or similar solvent. Reoil with clean SAE30 motor oil. Compress filter lightly to distribute oil evenly throughout filter element.

Coco Fiber Filter. Remove filter, then wash and rinse thoroughly in gasoline or similar solvent. Shake off gasoline or solvent and immerse filter in clean SAE 30 motor oil for approximately five mintues. Allow excess oil to drain from filter, then reinstall.

LUBRICATION. Fuel and lubricating oil should be thoroughly mixed in a separate container. Fuel:oil ratio for Model ST30 is 25:1 or 50:1 as specified on decal on fuel tank. Fuel:oil ratio for Models SB126 and SB151 is 50:1. Fuel:oil ratio for all other models is 25:1

Manufacturer recommends a good quality two-stroke oil for two-stroke engines be mixed at appropriate ratio with regular grade gasoline.

CLEANING CARBON. After approximately 200 hours of operation or when engine performance drops or when engine tends to four-stroke although the carburetor is properly adjusted, carbon deposits should be cleaned from combustion chamber, exhaust port and muffler.

On all models, remove the muffler. On Model ST102, do not attempt to remove carbon from muffler as it is filled with basalt wool. Renew this muffler if required. On all other models, disassemble muffler, if possible, then heat muffler until red-hot. Scrape or knock off remaining carbon deposits.

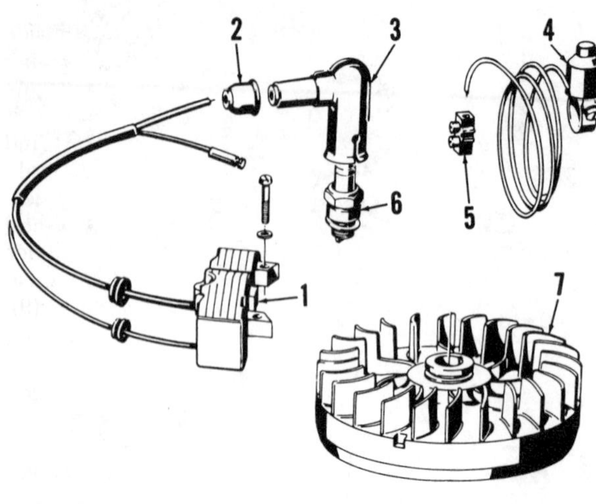

Fig. SA12—Exploded view of breakerless electronic ignition system used on ST126, ST151, ST204, ST251, SB126 and SB151 engines.
1. Ignition unit
2. Rain cap
3. Plug connector
4. Shut-off switch
5. Connector
6. Spark plug
7. Flywheel

CAUTION: Manufacturer recommends the exhaust system should not be modified. Any modification will harm engine performance, increase fuel consumption and alter exhaust noise.

On Model ST30, remove the muffler, then move piston to bottom of cylinder. Clean exhaust port from outside. Blow out any carbon that might have dropped down on the piston.

On Models ST76, ST102 and SB102, remove the muffler, then unbolt and remove the cylinder. Scrape the carbon from the combustion chamber and exhaust port. Carefully remove only the thicker carbon deposits from piston crown. Do not attempt to polish top of piston. Use new cylinder gasket when reassembling.

On Models ST96, ST126, ST151, ST204, ST151 and SB151, remove the muffler, then unbolt and remove the cylinder head. Scrape carbon from cylinder head. Move piston to bottom of cylinder and carefully remove carbon from cylinder ports. Move piston to top of cylinder and remove only the thick carbon deposits from piston. Do not attempt to polish the piston crown. Reinstall cylinder head and muffler.

REPAIRS

TIGHTENING TORQUES. Recommended tightening torques are as follows:

Stator plate (ST30) . . . . 2-3 N·m (3-4 ft.-lbs.)
Ignition coil . . . . 5-6 N·m (7-8 ft.-lbs.)
Cylinder to crankcase:
- ST76, ST96 . . . . 5-7 N·m (7-10 ft.-lbs.)
- ST102, SB120 . . . . 9-10 N·m (12-13 ft.-lbs.)
- ST126, ST151, SB126, SB151 . . . . 18-20 N·m (24-27 ft.-lbs.)
- ST204, ST251 . . . . 20-22 N·m (27-30 ft.-lbs.)

Cylinder head:
- ST96 . . . . 11-13 N·m (15-18 ft.-lbs.)
- ST126, ST151, ST204, ST251, SB126, ST251 . . . . 20-25 N·m (27-34 ft.-lbs.)

Crankcase halves:
- ST30 . . . . 5-7 N·m (7-10 ft.-lbs.)
- ST76, ST96, ST102, ST126, ST151, SB102, SB126, SB151, ST204, ST251 . . . . 8-10 N·m (11-13 ft.-lbs.)

Flywheel:
- ST30 . . . . 23-25 N·m (31-35 ft.-lbs.)
- ST76, ST96 . . . . 38-40 N·m (52-54 ft.-lbs.)
- ST102, SB102 . . . . 40-45 N·m (54-61 ft.-lbs.)
- ST126, ST151, ST204, ST251, SB126, SB151 . . . . 70-80 N·m (95-108 ft.-lbs.)

Recoil starter . . . . 2-3 N·m (3-4 ft.-lbs.)

CYLINDER HEAD. Cylinder head is removable from Models ST96, ST126, ST151, ST204, ST251, SB126 and SB151. When reinstalling cylinder head, make certain the mating surfaces of head and cylinder are clean and smooth. Refer to preceding paragraph for tightening torques.

PISTON, PINS, RINGS AND CYLINDER. On Model ST30 engines, the cylinder and upper crankcase half (7–SA13) is one unit. To disassemble, remove the fuel tank, carburetor, muffler and reduction gearbox if so equipped. Remove recoil starter, end cover (26), flywheel (24), magneto and stator plate assembly (23), plate (22) and ring (21). Invert engine, then unbolt and remove crankcase lower half (20). Carefully withdraw crankshaft and piston assembly, taking care not to lose the semi-circular retainers (6). Remove retaining rings (12), push out piston pin (13) and remove piston (11). Remove oil seals (15), then using a suitable puller, remove ball bearings (16). Identify and lay aside shims (17) and spacer (18) for each end of shaft.

Clean and inspect all parts and renew any showing excessive wear or other damage. Cylinder can be rebored to nearest oversize for which piston and rings are available if excessively worn.

Pin bushing (14) must be pressed into connecting rod and reamed after installation. Oiled piston pin should just slide through the reamed bushing under its own weight.

Piston must be installed so arrow on piston crown will point toward exhaust port after assembly.

If oil seals (15) are renewed, fill the

Fig. SA13—Exploded view of Model ST30 engine.
1. Air vane
2. Governor lever
3. Carburetor adapter pipe
4. Gaskets
5. Reed valve
6. Semicircular bearing retainer
7. Cylinder & upper crankcase half
8. Spark plug
9. Stud (4)
10. Piston rings
11. Piston
12. Retaining ring (2)
13. Piston pin
14. Bushing
15. Oil seals
16. Ball bearings
17. Shims
18. Spacer
19. Crankshaft & connecting rod assy.
20. Lower crankcase half
21. Ring
22. Centering plate
23. Magneto assy.
24. Flywheel
25. Flywheel cover
26. End cover

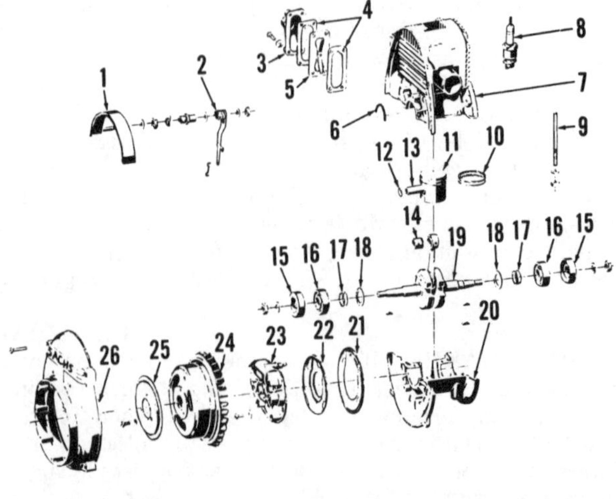

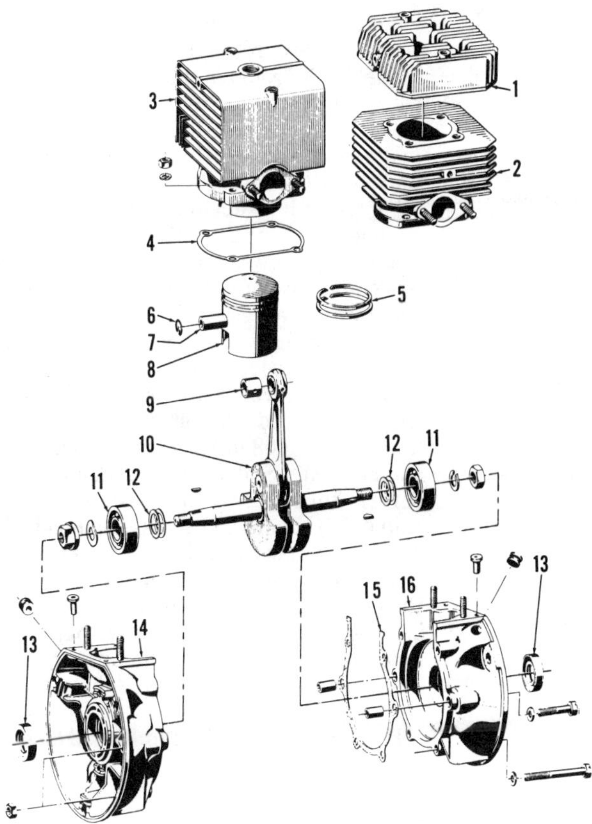

Fig. SA14—Exploded view of Model ST76 and ST96 engines. Cylinder (2) with removable cylinder head (1) is used on ST96 engines.

1. Cylinder head (ST96)
2. Cylinder (ST96)
3. Cylinder (ST76)
4. Gasket
5. Piston rings
6. Retaining ring (2)
7. Piston pin
8. Piston
9. Bushing
10. Crankshaft & connecting rod assy.
11. Ball bearings
12. Shims
13. Oil seals
14. Crankcase half (flywheel end)
15. Gasket
16. Crankcase half (pto end)

grooves of seals and lightly coat the seal lips with high melting point grease. Install seals in original positions so lube holes in crankcase will not be blocked. Coat crankcase mating surfaces with sealing compound and install crankcase lower half.

On Models ST76 and ST96, refer to Fig. SA14 and remove fuel tank, air cleaner, carburetor and muffler. On Model ST96, remove cylinder cover, then unbolt and remove the cylinder head (1) and cylinder (2). On Model ST76, unbolt and remove cylinder (3). On both models, remove retaining rings (6), push out piston pin (7) and remove piston (8).

Clean and inspect all other parts and renew any showing excessive wear or other damage. Cylinder can be rebored to nearest oversize for which piston and rings are available if excessively worn.

Pin bushing (9) must be pressed into connecting rod and reamed after installation. Oiled piston pin should just slide through the reamed bushing under its own weight.

Piston must be installed so arrow on piston crown will point toward exhaust port after assembly. Install new cylinder gasket (4) with the graphited side towards the crankcase. Tighten cylinder retaining nuts to specified torque. On Model ST96, install cylinder head, tighten retaining bolts to specified torque and continue reassembly.

On Models ST102 and SB102, refer to Figs. SA15 and SA16 and remove fuel tank, air cleaner and muffler. Unbolt and carefully remove the cylinder. Remove the piston pin retaining rings, push out piston pin and remove piston. Slide needle bearing from connecting rod.

Clean and inspect all parts and renew any showing excessive wear or other damage. The cylinder can be rebored to nearest oversize for which piston and rings are available if excessively worn.

Install piston so arrow on piston crown will point towards exhaust port after assembly.

Coat crankcase mating surface with sealing compound and install new cylinder gasket with graphited side towards cylinder. Tighten retaining screws to specified torque and continue reassembly.

On Models ST126, ST151, ST204 and ST251, unbolt and remove the fuel tank, air cleaner, carburetor and muffler. Refer to Fig. SA17, then unbolt and remove cylinder head (14) and cylinder (13). Remove retaining rings (10), push out piston pin (8) and remove piston (9). Slide needle bearing (7) from connecting rod.

Clean and inspect all parts and renew any showing excessive wear or other damage.

Cylinder can be rebored to nearest oversize for which piston and rings are available if excessively worn.

Install piston so arrow on piston crown will point towards exhaust port after assembly. Install new cylinder gasket (12) with graphited side towards crankcase and install cylinder. Tighten cylinder retaining nuts to specified torque. Install cylinder head and tighten retaining nuts to specified torque and continue reassembly.

On Models SB126 and SB151, unbolt and remove the fuel tank, air cleaner, carburetor and muffler. Refer to Fig. SA18 and unbolt and remove cylinder head (1) and cylinder (2). Remove retaining rings (5), push out piston pin (4) and remove piston (6). Slide needle bearing (8) from connecting rod.

Clean and inspect all parts and renew any showing excessive wear or other damage.

Cylinder can be rebored to nearest oversize for which piston and rings are available if excessively worn.

Piston must be installed so arrow on piston crown will point towards exhaust port after assembly. Install new cylinder gasket (3) with graphited side towards crankcase and install cylinder. Tighten cylinder retaining nuts to specified torque. Install cylinder head and tighten retaining nuts to specified torque and continue reassembly.

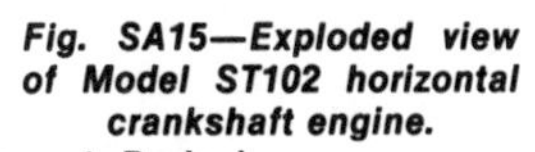

Fig. SA15—Exploded view of Model ST102 horizontal crankshaft engine.

1. Reed valve
2. "O" ring
3. Gasket
4. Bracket
5. Crankcase half
6. Oil seals
7. Ball bearings
8. Shims
9. Spacer
10. Crankshaft & connecting rod assy.
11. Needle bearing
12. Retaining ring (2)
13. Piston pin
14. Piston
15. Piston rings
16. Crankcase half
17. Gasket
18. Cylinder
19. Spark plug

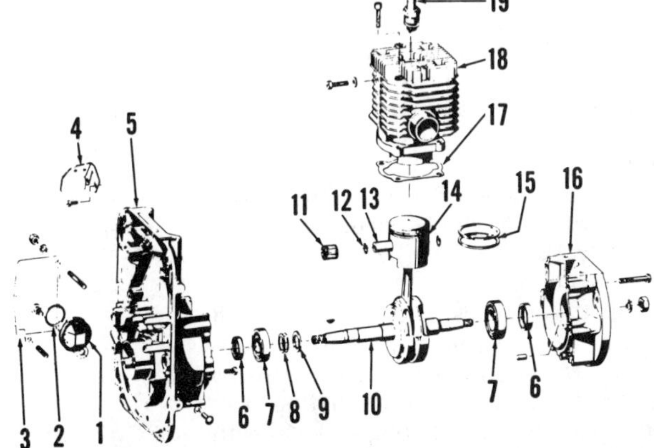

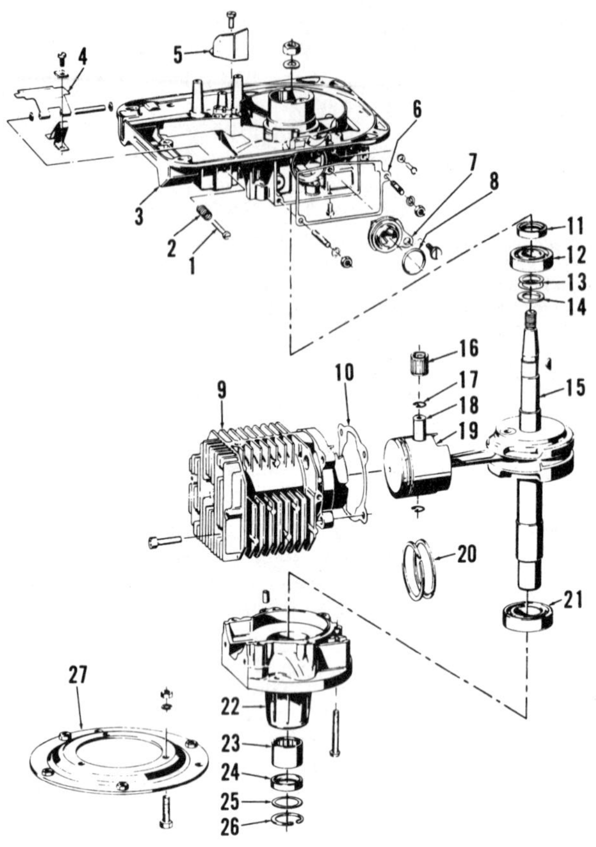

Fig. SA16—Exploded view of Model SB102 vertical crankshaft engine.

1. Adjuster screw
2. Spring
3. Crankcase half (flywheel end)
4. Air vane
5. Bracket
6. Gasket
7. Reed valve
8. "O" ring
9. Cylinder
10. Gasket
11. Oil seal
12. Ball bearing
13. Shims
14. Spacer
15. Crankshaft & connecting rod assy.
16. Needle bearing
17. Retaining rings
18. Piston pin
19. Piston
20. Piston rings
21. Ball bearing
22. Crankcase half (pto end)
23. Needle bearing
24. Oil seal
25. Washer
26. Retaining ring
27. Adapter flange

CONNECTING ROD, CRANKSHAFT AND CRANKCASE. The crankshaft and connecting rod are available only as an assembly on all models. Refer to the appropriate paragraph for model being serviced.

Model ST30. On Model ST30, the cylinder and upper crankcase half (7–Fig. SA13) are one unit. Refer to PISTON, PIN, RINGS AND CYLINDER section, disassemble engine and remove cylinder. Remove piston pin (13) and piston (11). Remove oil seals (15), then using a suitable puller, remove ball bearings (16). Identify and lay aside shims (17) and spacer (18) for each end of crankshaft.

Clean and inspect all parts and renew if necessary.

Crankshaft end play should be 0.1-0.2 mm (0.004-0.008 in.) and is controlled by shims (17). Shims are available in a variety of thicknesses. One spacer (18) should be installed against each crankshaft web and equal thickness of shims (17) should be used at each side of shaft. Use a spacer block between the crankshaft webs when installing the ball bearings. Heat the bearings to approximately 120°C (250°F) and press them into position.

Fill the grooves of new oil seal and lightly coat seal lips with light grease. Install the seals in original positions so lube holes in crankcase will not be locked. Coat crankcase mating surfaces with sealing compound and reassemble engine. Tighten crankcase nuts diagonally to specified torque.

Models ST76 and ST96. On Models ST76 and ST96, to disassemble the engine, remove the fuel tank, air cleaner, carburetor and muffler. On Model ST96, remove the cylinder cover. On both models, remove the recoil starter, flywheel, fan housing and magneto assembly. Refer to Fig. SA14 and to PISTON, PIN, RINGS AND CYLINDER section and remove cylinder and piston. Remove the six crankcase screws, heat crankcase to approximately 150°C (300°F), then separate crankcase halves (14 and 16) from crankshaft. Remove oil seals (13) from crankcase halves. Using a suitable puller, remove ball bearings (11). Crankshaft end play should be 0.1-0.2 mm (0.004-0.008 in.) and is controlled by shims (12). Shims are available in several thicknesses. Equal thickness of shims (12) should be used at each end of crankshaft between ball bearing and crank web. Use a spacer block between the crankshaft webs when installing ball bearings. Heat the bearings to approximately 150°C (300° F) and press them into position on crankshaft. Allow bearings to cool.

Heat crankcase halves to 150° C (300° F). Apply a coat of suitable sealing compound to mating surfaces of both crankcase halves and install new gasket (15). Install crankcase halves on crankshaft. Tighten crankcase screws to specified torque. Fill grooves of new oil seals (13) and lightly coat seal lips with high melting point grease. Press seals (lip first) into crankcase halves until flush. Refer to PISTON, PIN, RINGS

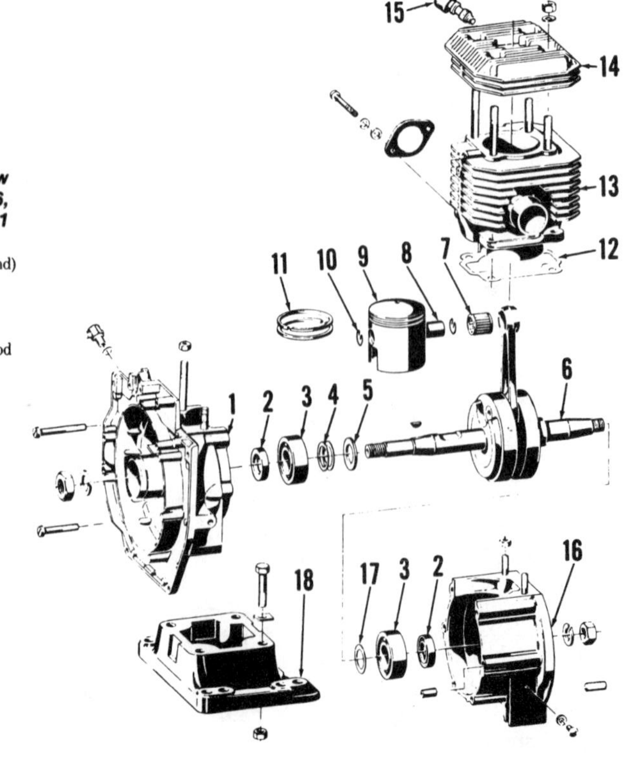

Fig. SA17—Exploded view of typical Model ST126, ST151, ST204 and ST251 engines.

1. Crankcase half (flywheel end)
2. Oil seals
3. Ball bearings
4. Shims
5. Spacer
6. Crankshaft & connecting rod assy.
7. Needle bearing
8. Piston pin
9. Piston
10. Retaining rings
11. Piston rings
12. Gasket
13. Cylinder
14. Cylinder head
15. Compression release
16. Crankcase half (pto end)
17. Spacer
18. Engine base

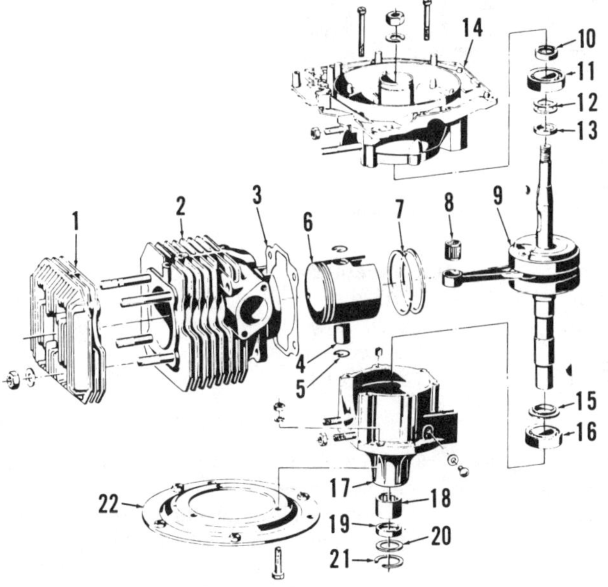

Fig. SA18—Exploded view of Model SB126 or SB151 vertical crankshaft engine.

1. Cylinder head
2. Cylinder
3. Gasket
4. Piston pin
5. Retaining rings
6. Piston
7. Piston rings
8. Needle bearing
9. Crankshaft & connecting rod assy.
10. Oil seal
11. Ball bearing
12. Shims
13. Spacer
14. Crankcase half (flywheel end)
15. Spacer
16. Ball bearing
17. Crankcase half (pto end)
18. Needle bearing
19. Oil seal
20. Washer
21. Retaining ring
22. Adapter flange

AND CYLINDER section and install piston and cylinder and continue reassembly.

Models ST102 And SB102. On Models ST102 and SB102, remove fuel tank, air cleaner, muffler, carburetor and reed valve. Unbolt and remove recoil starter, fan, flywheel and magneto ignition system. Refer to Figs. SA15 and SA16 and PISTON, PIN, RINGS AND CYLINDER section and remove cylinder and piston. Remove the six crankcase screws, heat crankcase to approximately 150° C (300° F), then separate the crankcase halves from crankshaft. On Model ST102, remove oil seals (6–Fig. SA15) from crankcase halves (5 and 16). On Model SB102, remove retaining ring (26–Fig. SA16) and washer (25), then press needle bearing (23) and oil seal (24) out toward outside of case. On both models, use a suitable puller and remove ball bearings from crankshaft. Remove shims and spacer from crankshaft.

Clean and inspect all parts and renew any showing excessive wear or other damage.

Crankshaft end play should be 0.2-0.3 mm (0.008-0.012 in.) and is controlled by shims (8–Fig. SA15) or (13–Fig. SA16). Shims are available in a variety of thicknesses. Install one 1.5 mm spacer against crankshaft web (flywheel end) and install required shims against the spacer. Heat ball bearings to 150° C (300° F) and press them into position on crankshaft. Use a spacer block between the crankshaft webs when installing ball bearings. On Model SB102, install new needle bearing (23–Fig. SA16) and oil seal (24) to their original installation depth, then install washer (25) and retaining ring (26). Install oil seal (11) in crankcase half (3). On Model ST102, install oil seals (6–Fig. SA15) in crankcase halves (5 and 16).

On both models, fill grooves of oil seals and lightly coat the seal lips with high melting point grease.

Heat crankcase halves to 150° C (300° F), apply a coat of suitable sealing compound to mating surfaces of crankcase halves and install on the preassembled crankshaft. Tighten the crankcase screws to specified torque. Refer to PISTON, PIN, RINGS AND CYLINDER section to install piston and cylinder. Continue reassembly.

Models ST126, ST151, ST204 And ST251. To disassemble, remove fuel tank, air cleaner, carburetor and muffler. Unbolt and remove electronic ignition coil, flywheel and recoil starter assembly. Refer to Fig. SA17 and to PISTON, PIN, RINGS AND CYLINDER section to remove piston and cylinder. Remove the six crankcase screws, heat crankcase to approximately 150° C (300° F), then separate the crankcase halves (1 and 16) from crankshaft. Remove oil seals (2) from crankcase halves. Using a suitable puller, remove ball bearings (3). Remove shims (4) and spacer rings (5 and 17).

Clean and inspect all parts for damage and excessive wear and renew as necessary.

Crankshaft end play should be 0.15-0.25 mm (0.006-0.010 in.) and is controlled by shims (4) installed between spacer (5) and ball bearing (3) at flywheel end. Shims are available in several thicknesses. Use only one spacer (5) and one spacer (17). Heat ball bearings (3) to 150° C (300° F) and press them into position on crankshaft. Use a spacer block between the crankshaft webs when installing ball bearings.

Fill grooves of new oil seal and lightly coat the seal lips with high melting point grease. Press seals in from outside of crankcase halves with sealing lip pointing inward, until seals are flush with outer surface.

Heat crankcase halves to 150° C (300° F), apply a coat of suitable sealing compound to mating surfaces of crankcase halves and install on the preassembled crankshaft. Tighten crankcase screws to specified torque. Refer to PISTON, PIN, RINGS AND CYLINDER section and install piston and cylinder. Continue reassembly.

Models SB126 And SB151. To disassemble, remove fuel tank, air cleaner, carburetor and muffler. Unbolt and remove the electronic ignition coil, flywheel and recoil starter assembly. Refer to Fig. SA18 and to PISTON, PIN, RINGS AND CYLINDER section and remove cylinder and piston. Remove the six screws, heat crankcase approximately 120° C (250° F), then separate the crankcase halves (14 and 17) from crankshaft. Remove retaining ring (21) and washer (20), then press oil seal (19) and needle bearing (18) out toward outside of case. Remove oil seal (10) from crankcase half (14). Using a suitable puller, remove ball bearings (11 and 16) from crankshaft. Remove shims (12) and spacers (13 and 15).

Clean and inspect all parts for damage and excessive wear and renew as necessary.

Crankshaft end play should be 0.15-0.25 mm (0.006-0.010 in.) and is controlled by shims (12) installed between spacer (13) and ball bearing (11) at flywheel end. Shims are available in a variety of thicknesses. Use only one 2.6 mm thick spacer (13) and one 5.9 mm thick spacer (15).

Heat ball bearings (11 and 16) to 120° C (250° F) and press them into position on crankshaft. Use a spacer block between the crankshaft webs when installing ball bearings. Degrease the seat for needle bearing (18) in crankcase half (17) and carefully coat with Loctite Retaining Compound #609. Press needle bearing into crankcase half from inside the case until bearing is 11.0 mm (0.43 in.) from outer surface.

Fill the grooves of new oil seals (10 and 19) and lightly coat seal lips with high melting point grease. Press seals into position and install washer (20) and retaining ring (21).

Heat crankcase halves to approximately 120° C (250° F), apply a coat of sealing compound to mating surfaces of crankcase halves and install on preassembled crankshaft. Tighten the crankcase screws to specified torque. Refer to PISTON, PIN, RINGS AND CYLINDER section and install piston and cylinder. Continue reassembly.

SACHS-DOLMAR

SACHS-DOLMAR
P.O. Box 78526
Shreveport, Louisiana 71137

Model	Bore	Stroke	Displacement
309	47 mm (1.850 in.)	40 mm (1.575 in.)	70 cc (4.2 cu. in.)
343	55 mm (2.165 in.)	40 mm (1.575 in.)	95 cc (5.8 cu. in.)

ENGINE INFORMATION

The engines used on Models 309 and 343 cut-off saws are two-stroke air-cooled. Crankshaft on both models is a pressed together assembly which should be disassembled only by a shop with the tools and experience necessary to assemble and realign crankshaft. Carburetor on all models is equipped with a vibration type governor valve to limit maximum engine speed, thus prevent blade speed from exceeding manufacturer's recommended specification.

MAINTENANCE

SPARK PLUG. Recommended spark plug for both models is a Bosch WSR6F. Spark plug electrode gap should be 0.51 mm (0.020 in.).

CARBURETOR. Both models are equipped with a Tillotson HS diaphragm type carburetor equipped with a vibration type governor located in carburetor body near choke shaft.

Initial adjustment of idle and high speed mixture screws should be approximately one turn open from a lightly seated position. Final adjustment should be made with engine running and at operating temperature. Adjust low speed mixture screw so engine will accelerate cleanly without hesitation. To correctly adjust high speed mixture screw, remove carburetor mounting bolts and tilt carburetor to one side to gain access to governor valve (16—Fig. D2). Remove governor valve and install a solid plug. Reinstall carburetor. With engine running and at operating temperature, attach an accurate tachometer and adjust high speed mixture screw (14) to obtain a maximum no-load engine speed of 11,000 rpm. Remove solid plug and reinstall governor valve (16).

NOTE: DO NOT use cut-off saws with governor valve blanked off as excessive blade speed will result. When adjusting maximum no-load engine speed with governor valve blanked off, do not allow maximum no-load engine speed to exceed 11,500 rpm.

With governor valve installed, maximum no-load engine speed should not exceed 7500-8100 rpm at wide-open throttle.

Refer to Fig. D2 and disassemble carburetor. Clean filter screen (4). Welch plugs (22 and 25) can be removed by drilling plug with a suitable size drill bit, then use a small punch to pry plugs out. Care must be exercised not to drill into carburetor body.

Inspect inlet lever spring (26) and renew if stretched or damaged. Inspect diaphragms for tears, cracks or other damage. Renew idle and high speed adjusting screws (12 and 14) if they are grooved or damaged. Fuel inlet needle has a rubber tip and seats directly on a machined orifice in carburetor body. Carburetor body must be renewed if seat is damaged.

When assembling carburetor, adjust position of inlet control lever (28) so lever is flush with diaphragm chamber floor as shown in Fig. D3. Carefully

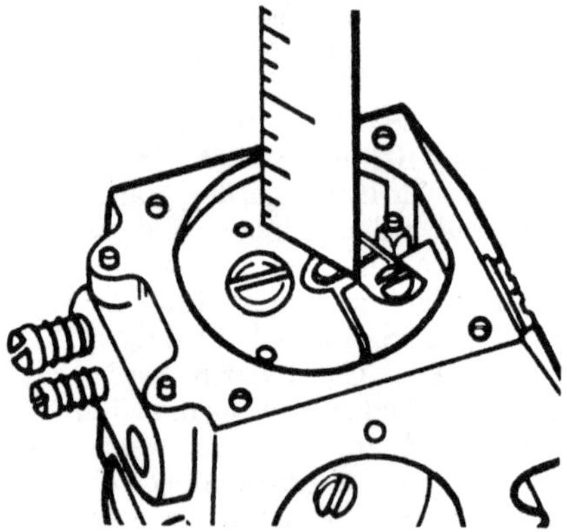

Fig. D3—Fuel inlet valve lever should be parallel with chamber floor for correct height.

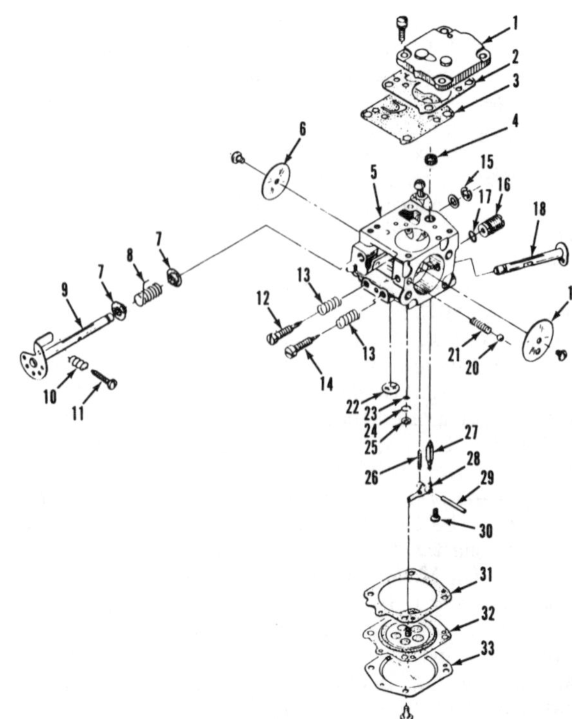

Fig. D2—Exploded view of Tillotson HS series carburetor used on both models. Carburetor is equipped with a vibration type governor valve (16) located in carburetor body.

1. Pump cover
2. Gasket
3. Fuel pump diaphragm & valves
4. Screen
5. Carburetor body
6. Throttle plate
7. Bushing
8. Spring
9. Throttle shaft
10. Spring
11. Throttle stop screw
12. Low speed mixture screw
13. Spring
14. High speed mixture screw
15. "E" clip
16. Governor valve
17. "O" ring
18. Choke shaft
19. Choke plate
20. Ball
21. Spring
22. Welch plug
23. Screen
24. Retainer
25. Welch plug
26. Spring
27. Fuel inlet valve
28. Lever
29. Pin
30. Retaining screw
31. Gasket
32. Metering diaphragm
33. Cover

bend lever adjacent to spring to obtain correct lever position.

GOVERNOR. Both models are equipped with a vibration type governor valve (16—Fig. D2) located in carburetor body. Maximum no-load engine speed is limited to 7500-8100 rpm at wide-open throttle. Valve consists of a spring loaded check ball which is vibrated off its seat when engine speed reaches 7500-8100 rpm. Fuel is then routed directly to carburetor venturi and engine is flooded down. As speed decreases, check ball seats and engine clears and resumes normal operation. This allows a richer fuel mixture under loaded conditions while providing maximum blade speed limit under no or light load conditions. Maximum blade speed should be 5100 rpm for Model 309 and 4360 rpm for Model 343.

IGNITION SYSTEM. Both models are equipped with a breakerless ignition system. Ignition spark should occur at 3.1-3.5 mm (0.122-0.138 in.) BTDC on Model 309 and at 2.1-2.5 mm (0.083-0.098 in.) BTDC on Model 343.

To check timing, position engine at recommended ignition point using a dial indicator assembly protruding into spark plug hole. Make reference marks on flywheel and starter housing. Check timing with a power timing light and engine operating at 8000 rpm. If reference marks are not aligned with engine operating at 8000 rpm, rotate stator plate to adjust.

LUBRICATION. Both models are lubricated by mixing oil with regular grade gasoline. Recommended oil is Sachs-Dolmar two-stroke engine oil. Fuel and oil mixture should be a 40:1 ratio when using recommended oil. During break-in period (first ten hours of operation), mix fuel and oil at a 30:1 ratio. If Sachs-Dolmar two-stroke engine oil is not available, a good quality oil designed for use in air-cooled, two-stroke engines may be used when mixed at a 25:1 ratio after break-in period or 20:1 ratio during break-in. Use a separate container when mixing oil and fuel.

GENERAL MAINTENANCE. Check and tighten all loose bolts, nuts or clamps prior to each day of operation. Check for fuel leakage and repair if necessary.

Clean dust, dirt, grease or any foreign material from cylinder head and cylinder block cooling fins as needed. Inspect cooling fins for damage and repair if necessary.

Due to the fine texture and abrasive nature of the dust created when cutting concrete and masonry products with cut-off saws, check condition of air filter and all filter connections frequently to make certain dust and foreign material do not enter engine through the air filter system.

REPAIRS

TIGHTENING TORQUES. Recommended tightening torques are as follows:

Spark plug . . . . 14-15 N·m (124-133 in.-lbs.)
Crankcase screws . . . . 7 N·m (62 in.-lbs.)
Cylinder screws . . . . 10 N·m (88 in.-lbs.)
Clutch nut . . . . 26 N·m (228 in.-lbs.)

PISTON, PIN, RINGS AND CYLINDER. The cylinder can be removed after removal of airbox cover, air cleaner, carburetor, cylinder shroud and muffler. The cylinder is chrome plated and should be inspected for excessive wear or damage to chrome plating.

Piston pin is retained by wire retainers and rides in a roller bearing in the small end of the connecting rod. The piston is equipped with two piston rings which are retained in position by locating pins in the ring grooves. The piston must be installed with the arrow, on the piston crown, pointing toward the exhaust port. Some pistons have a letter "A" stamped near the arrow which must not be confused with the letter stamped on piston crown to indicate piston size.

Cylinder and piston are marked "A," "B" or "C" during production to obtain desired piston-to-cylinder clearance. Cylinder and piston must have same letter marking to obtain proper clearance. Cylinders are stamped on top of cylinder or on cylinder frame. Pistons are stamped on piston crown but letter indicating piston size should not be confused with letter "A" which is stamped on some piston crowns to indicate which side of piston must be installed adjacent to exhaust port. Tighten cylinder screws to torque specified in TIGHTENING TORQUES.

CRANKSHAFT AND CONNECTING ROD. The crankshaft is supported at each end in a roller bearing type main bearing. On Model 309, main bearing (9—Fig. D4) and bearing race (8) are separate parts. On Model 343, main bearing and race (8—Fig. D5) are an assembly. On both models, connecting rod and crankshaft (10—Figs. D4 and D5) are a pressed together assembly which should be disassembled only by a shop with the necessary tools and experience required to assemble and realign crankshaft. Connecting rod, bearing and crankpin are available as a unit assembly.

Piston is retained on connecting rod by retaining rings (5—Figs. D4 and D5). Connecting rod is equipped with a roller type bearing (7) in the small end.

To install main bearings on crankshaft, heat bearings in oil and install on crankshaft main bearing journals. To install main bearing races in crankcase of Model 309, heat crankcase halves and install races. To install crankcase halves on crankshaft and main bearing assembly of Model 343, heat crankcase halves and assemble on crankshaft and main bearings.

On both models, tighten crankcase screws to torque specified in TIGHTENING TORQUES.

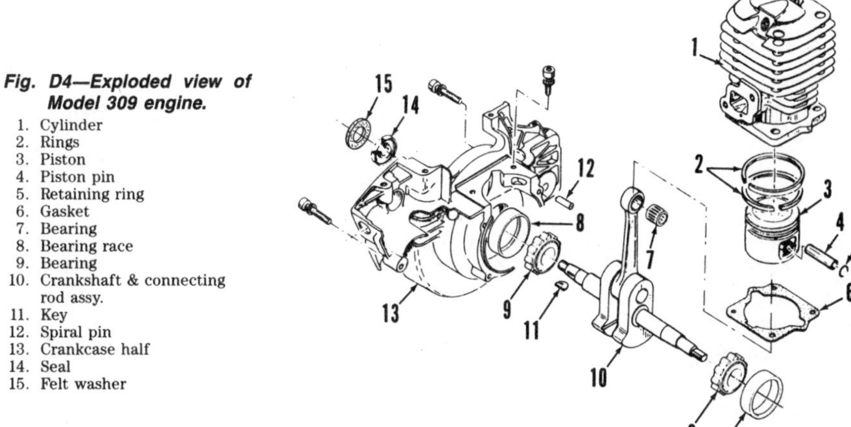

Fig. D4—Exploded view of Model 309 engine.

1. Cylinder
2. Rings
3. Piston
4. Piston pin
5. Retaining ring
6. Gasket
7. Bearing
8. Bearing race
9. Bearing
10. Crankshaft & connecting rod assy.
11. Key
12. Spiral pin
13. Crankcase half
14. Seal
15. Felt washer

CRANKCASE AND SEALS. Crankcase on both models must be split to remove the crankshaft and connecting rod assembly. On Model 309, crankcase should be heated to remove the main bearing races. On both models, it may be necessary to heat crankcase halves to install main bearing races or install crankcase halves on crankshaft and main bearing assembly.

Crankshaft seal (14—Figs. D4 and D5) at flywheel side should be pressed into seal bore until seated. Install felt washer (15). Crankshaft seal at pto side should be pressed into seal bore until seated.

When joining crankcase halves, install a new gasket and tighten crankcase retaining screws to torque specified in TIGHTENING TORQUES.

CLUTCH. A three-shoe clutch is used on both models. Refer to Fig. D6 for an exploded view of the clutch used on both models. The clutch on both models is retained by nut (9). Nut (9) has left-hand threads. A puller should be used to remove clutch hub from crankshaft. Tighten clutch nut (9) to torque specified in TIGHTENING TORQUES during reassembly.

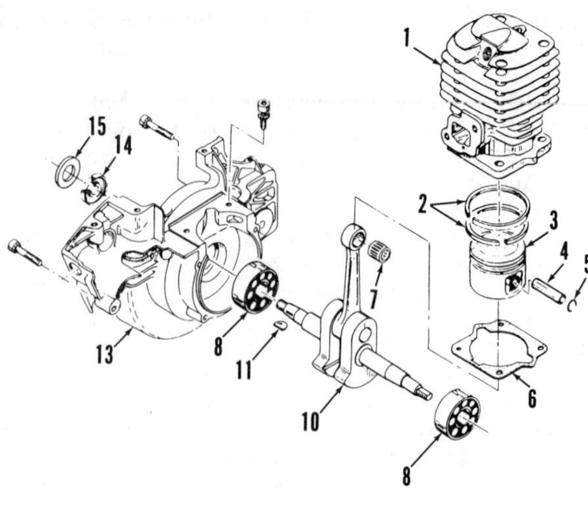

Fig. D5—Exploded view of Model 343 engine.

1. Cylinder
2. Rings
3. Piston
4. Piston pin
5. Retaining ring
6. Gasket
7. Bearing
8. Bearing & race assy.
10. Crankshaft & connecting rod assy.
11. Key
13. Crankcase half
14. Seal
15. Felt washer

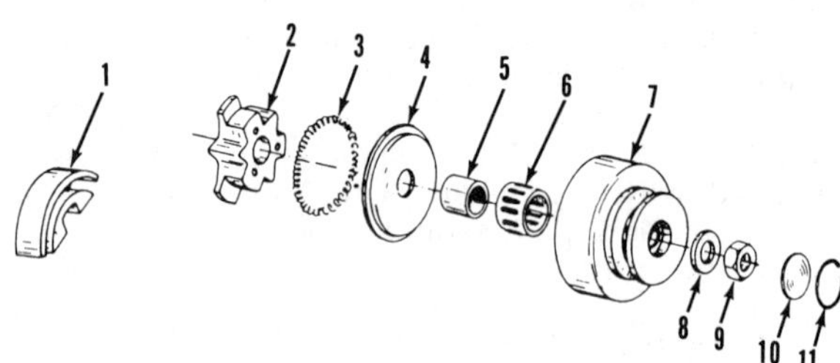

Fig. D6—Exploded view of three-shoe centrifugal clutch assembly used on both models. Nut (9) has left-hand threads.

1. Shoe
2. Clutch hub
3. Spring
4. Guide disc
5. Inner ring
6. Bearing
7. Clutch drum
8. Washer
9. Nut
10. Cover
11. Retaining ring

SHINDAIWA

SHINDAIWA, INC.
P.O. Box 1090
Tualatin, Oregon 97062

Model	Bore	Stroke	Displacement
EC350	35 mm (1.38 in.)	33 mm (1.30 in.)	35.5 cc (2.2 cu. in.)

ENGINE INFORMATION

Model EC350 engine is a two-stroke, air-cooled engine. The cylinder is an integral part of one crankcase half. Connecting rod and crankshaft are serviced as an assembly only.

MAINTENANCE

SPARK PLUG. Recommended spark plug is a Champion CJ8Y or equivalent. Specified electrode gap is 0.6 mm (0.024 in.).

CARBURETOR. A Walbro WA diaphragm type carburetor is used. Initial adjustment for both low speed and high speed mixture screws is 1-1/4 turns open from a lightly seated position. Make final adjustment with engine running and at operating temperature. Adjust idle speed screw so engine idles just below clutch engagement speed (approximately 2800 rpm.). Adjust low speed mixture screw so engine will accelerate cleanly without hesitation. Adjust high speed mixture screw to obtain optimum performance under cutting load. During first 10 hours of operation, the manufacturer recommends rotating mixture screws an additional 1/8 turn toward rich setting.

To disassemble carburetor, remove pump cover retaining screw (1—Fig. SH1), pump cover (2), diaphragm (5) and gasket (4). Remove inlet screen (6). Remove the four metering cover screws (27), metering cover (26), diaphragm (25) and gasket (24). Remove screw (20), pin (19), fuel lever (18), spring (17) and fuel inlet needle (16). Remove screw (23), circuit plate (22) and gasket (21). Remove high and low speed mixture screws and springs. Remove throttle plate (8) and throttle shaft (9) as required.

Clean parts thoroughly and inspect all diaphragms for wrinkles, cracks or tears. Diaphragms should be flexible and soft. Fuel inlet lever (18) should be flush with carburetor body (Fig. SH2).

IGNITION SYSTEM. Model EC350 is equipped with a breakerless electronic ignition system. Ignition timing is not adjustable. Air gap between ignition coil (8—Fig. SH3) legs and flywheel should be 0.30-0.35 mm (0.012-0.014 in.). Tighten flywheel retaining nut to specified torque.

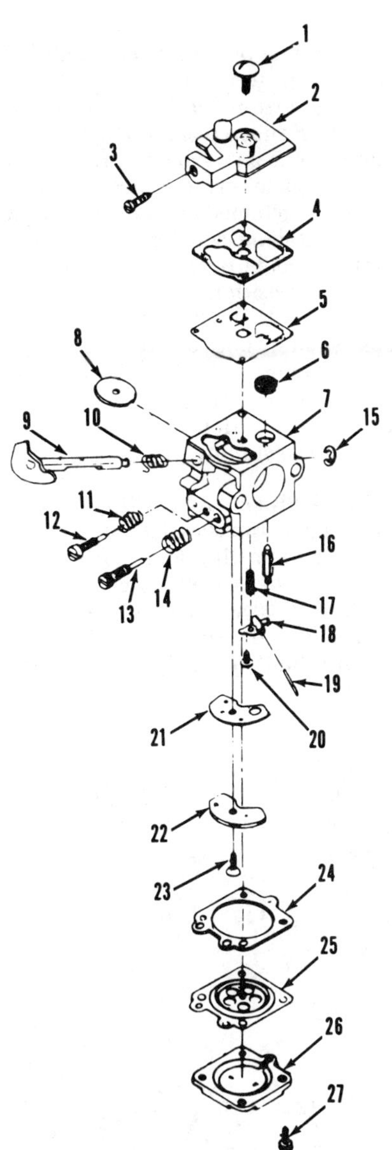

Fig. SH1—Exploded view of Walbro WA diaphragm type carburetor.

1. Screw
2. Pump cover
3. Idle speed screw
4. Gasket
5. Pump diaphragm
6. Screen
7. Body
8. Throttle plate
9. Throttle shaft
10. Spring
11. Spring
12. Low speed mixture screw
13. High speed mixture screw
14. Spring
15. "E" clip
16. Fuel inlet needle
17. Spring
18. Fuel inlet lever
19. Pin
20. Screw
21. Gasket
22. Circuit plate
23. Screw
24. Gasket
25. Diaphragm
26. Cover
27. Screw

LUBRICATION. Engine is lubricated by mixing engine oil with premium grade unleaded gasoline. Recommended oil is Shindaiwa Premium 2-Cycle Oil mixed at a ratio of 40:1. If Shindaiwa Premium 2-Cycle Oil is not available, a good quality oil designed for air-cooled, two-stroke engines may be used when mixed at a 25:1 ratio. Use a separate container when mixing the oil and gas.

CARBON. Carbon and other combustion deposits should be cleaned from muffler and exhaust ports at regular intervals. When scraping carbon, be careful not to damage the chamfered edges of the exhaust ports.

GENERAL MAINTENANCE. Check and tighten all loose bolts, nuts or clamps prior to each day of operation. Check for fuel leakage and repair if necessary.

Due to the fine texture and abrasive nature of the dust created when cutting concrete and masonry products with cut-off saws, check condition of air filter and all filter connections frequently to make certain dust and foreign material do not enter engine through the air filter system.

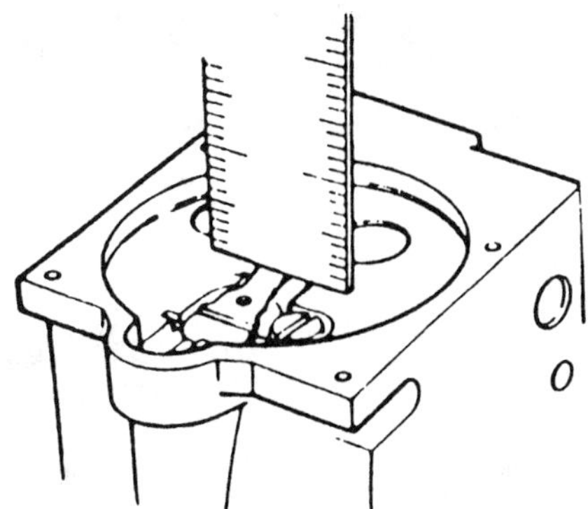

Fig. SH2—Fuel inlet lever should be flush with carburetor body.

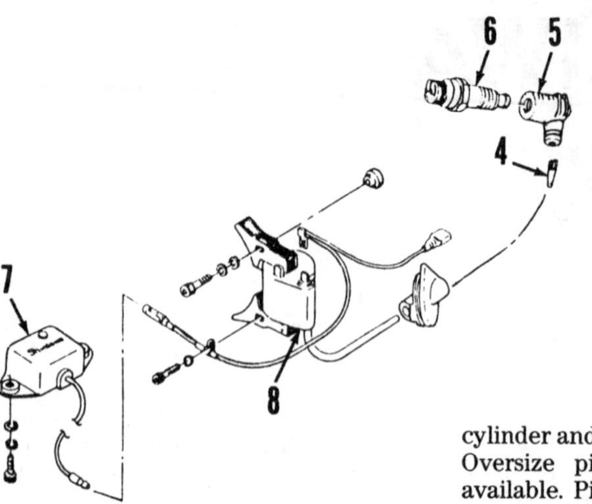

Fig. SH3—Exploded view of breakerless electronic ignition components.

4. Terminal
5. Spark plug boot
6. Spark plug
7. Module assy.
8. Ignition coil

REPAIRS

TIGHTENING TORQUES. Recommended tightening torques are as follows:

Flywheel nut . . . . 11.7-13.8 N·m (104-122 in.-lbs.)
Crankcase screws . . . . . . 6.9-7.9 N·m (61-70 in.-lbs.)

CYLINDER, PISTON, PIN AND RINGS. Cylinder (20—Fig. SH4) bore is chrome plated and should be renewed if cracking, flaking or other damage to cylinder is noted. Recommended piston-to-cylinder clearance is 0.025-0.061 mm (0.0010-0.0024 in.) with a maximum limit of 0.18 mm (0.0071 in.). Lower crankcase half (1) is renewable with cylinder and upper crankcase half (20). Oversize piston and rings are not available. Piston (14) is equipped with two piston rings (16). Locating pins are located in ring grooves to prevent ring rotation. Make certain ring end gaps are properly positioned around locating pins before installing cylinder. Piston pin (15) rides in needle bearing (12) and is retained in position with two wire retainer clips (13). Once removed, wire clips should not be reused. Install piston on connecting rod so arrow on piston crown points toward exhaust port. Note location of thrust washers (11) when installing piston.

CRANKSHAFT AND CONNECTING ROD. Upper crankcase half and cylinder are integral casting and crankshaft is loose in crankcase when cylinder and crankcase assembly is removed. Crankshaft and connecting rod (10—Fig. SH4) are a unit assembly and supported at both ends with roller type main bearings (6). Rotate connecting rod around crankpin and renew assembly if roughness, excessive play or other damage is noted. Maximum crankshaft runout measured on main bearing journals and supported between lathe centers is 0.068 mm (0.0027 in.).

New seals (5) should be installed on crankshaft when reassembling upper and lower crankcase halves. Oil holes in roller bearings (6) must face toward cylinder. With crankshaft assembly installed in lower crankcase half (1), crankshaft-to-side plate clearance should be 0.07-0.24 mm (0.003-0.009 in.). Renew side plate (7) or thrust washer (8) as necessary to obtain recommended clearance. Use a suitable silicone gasket compound on mating surface of upper and lower crankcase halves. Be sure mating surfaces are not damaged during assembly. Apply a thread locking compound on threads of crankcase screws and tighten to specified torque.

CLUTCH. Model EC350 is equipped with a four-shoe centrifugal clutch. Clutch retaining nut has left-hand threads. Clutch spring should hold shoes tight in clutch hub. Inspect shoes and drum for wear or damage and renew as necessary.

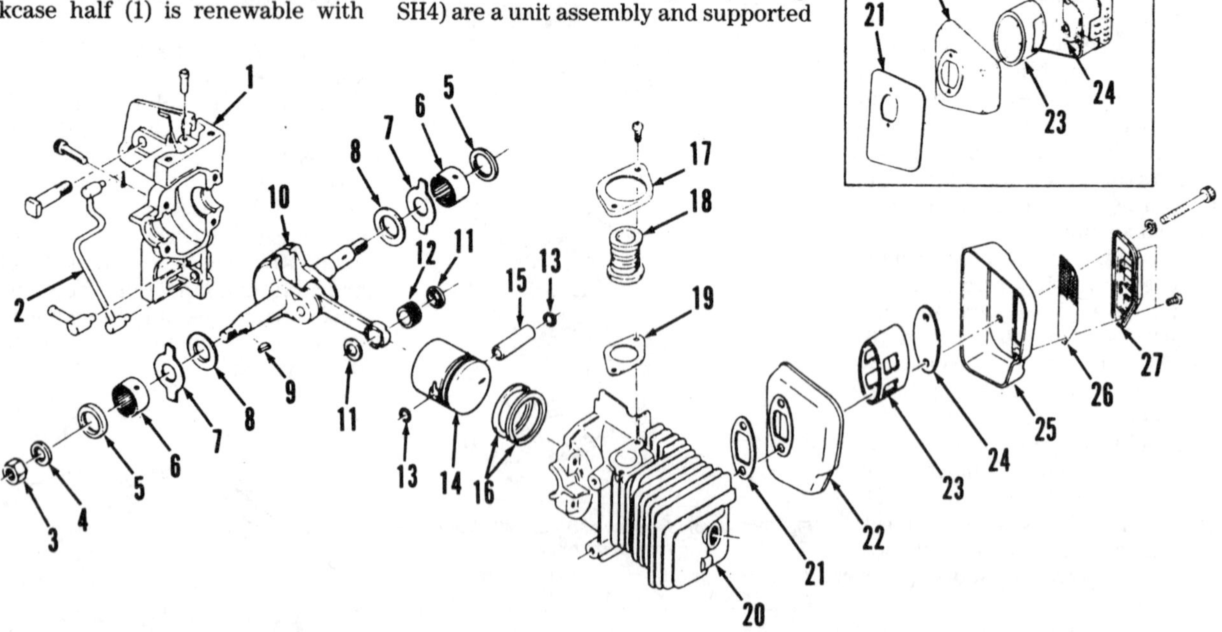

Fig. SH4—Exploded view of engine and muffler assemblies.

1. Lower crankcase half
2. Oil tube
3. Flywheel nut
4. Lockwasher
5. Oil seal
6. Roller bearing
7. Side plate
8. Thrust washer
9. Key
10. Crankshaft & connecting rod assy.
11. Thrust washer
12. Needle bearing
13. Retainer clip
14. Piston
15. Piston pin
16. Piston rings
17. Plate
18. Boot
19. Gasket
20. Cylinder & upper crankcase half
21. Gasket
22. Muffler half
23. Baffle
24. Plate
25. Muffler half
26. Spark arrestor
27. Cover

SOLO

SOLO INC.
5100 Chestnut Avenue
Copeland Industrial Park
Newport News, Virginia 23605

Model	Bore	Stroke	Displacement
206	2.56 in. (65 mm)	2.13 in. (54 mm)	10.9 cu. in. (179 cc)
209	2.80 in. (71 mm)	2.13 in. (54 mm)	13.4 cu. in. (214 cc)

MAINTENANCE

SPARK PLUG. Recommended spark plug is Bosch W175T1, Champion L85, AC 42F, Autolite AE3 or equivalent. Electrode gap is 0.018-0.020 inch (0.46-0.51 mm).

CARBURETOR. Tillotson Model HR19A carburetor is normally used; however, some motors may be equipped with Model HL287A Tillotson carburetor.

For initial adjustment, open idle mixture needle (24 – Fig. S1) 1 turn and main fuel mixture needle (23) 1½ turns from a lightly seated position.

Final adjustments should be made with engine at operating temperature and running. Adjust idle speed stop screw (4) to obtain engine idle speed of 1800-2200 rpm. Readjust idle mixture needle to obtain smooth and even idle speed operation. Adjust main fuel mixture needle so engine will accelerate without hesitation and will run smoothly without smoking.

To diassemble the carburetor, remove idle mixture needle (24) and main fuel needle (23). Remove cover (12), gasket (11) and screen (10), then unbolt and remove fuel pump body (9), valve diaphragm (8), pulse diaphragm (7) and gasket (22). Remove diaphragm cover (21), metering diaphragm (20) and gasket (19). Remove the control lever retaining screw, control lever pin (18), control lever (16) and spring (15).

NOTE: Care must be used while removing parts due to spring pressure on inlet control lever. The spring must be handled carefully to prevent stretching or compressing. Any alteration to the spring will cause improper carburetor operation. If in doubt as to its condition, renew spring.

Remove inlet needle and seat assembly (14). If necessary to remove nozzle check valve (17) or economizer check ball (13), drill through welch plugs with a ⅛ inch drill. Allow drill to just break through welch plugs. If drill travels too deep in cavity, the casting may be ruined. Pry welch plugs out of seats, using a small punch.

Throttle and choke shafts (3 and 5) can be removed for inspection if there is evidence of wear on these parts. Mark throttle and choke plates (1 and 6) so they can be reassembled in their original positions.

Clean and inspect all parts and renew any showing excessive wear or other damage. When reassembling, tighten

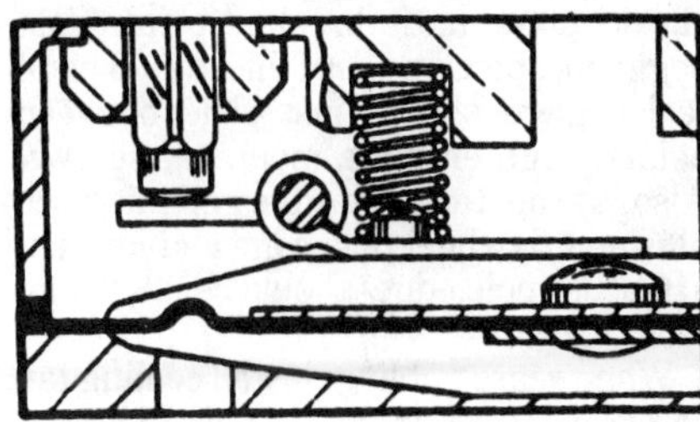

Fig. S2 – Diaphragm end of fuel inlet lever must be flush with metering chamber floor as shown. Adjust by bending control lever.

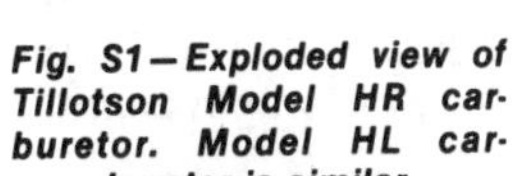

Fig. S1 – Exploded view of Tillotson Model HR carburetor. Model HL carburetor is similar.

1. Throttle plate
2. Carburetor body
3. Throttle shaft
4. Idle speed stop screw
5. Choke shaft
6. Choke plate
7. Pump diaphragm
8. Pump valves
9. Fuel pump body
10. Fuel strainer screen
11. Gasket
12. Cover
13. Economizer check ball
14. Inlet needle & seat
15. Inlet tension spring
16. Inlet control lever
17. Nozzle check valve
18. Control lever pin
19. Gasket
20. Diaphragm
21. Diaphragm cover
22. Gasket
23. Main fuel needle
24. Idle mixture needle

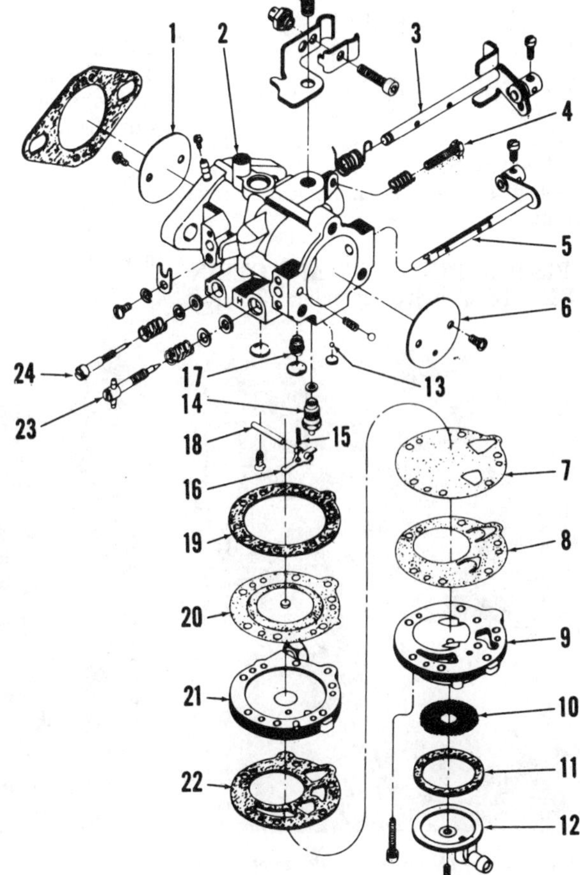

fuel inlet needle cage to 25-30 in.-lbs. (2.8-3.4 N•m). Adjust inlet control lever (16) so diaphragm end of lever is flush with metering chamber floor as shown in Fig. S2. Use Fig. S1 as a guide. Assemble gaskets, diaphragms and castings in the correct order.

MAGNETO AND TIMING. A Bosch flywheel type magneto is used on all models. Breaker point gap can be checked or adjusted after removing the recoil starter, cooling fan, felt ring, inner fan cover and flywheel. Breaker point gap should be 0.014-0.018 inch (0.36-0.46 mm). To adjust ignition timing, position piston 0.118 inch (3 mm) BTDC, loosen stator plate mounting screws and rotate stator plate until breaker points just begin to open. A 0.001 inch (0.03 mm) feeler gage should just slide between points. Tighten plate mounting screws. Also, at this time the edge gap (distance between flywheel pole edge and ignition armature pole edge) should be 0.3-0.4 in. (7.6-10.2 mm).

When reassembling, install cooling fan so arrow (A–Fig. S3) is toward top of cylinder when piston is at TDC.

LUBRICATION. The engine is lubricated by mixing a good quality SAE 30 two-stroke oil with regular grade gasoline at a 25:1 ratio.

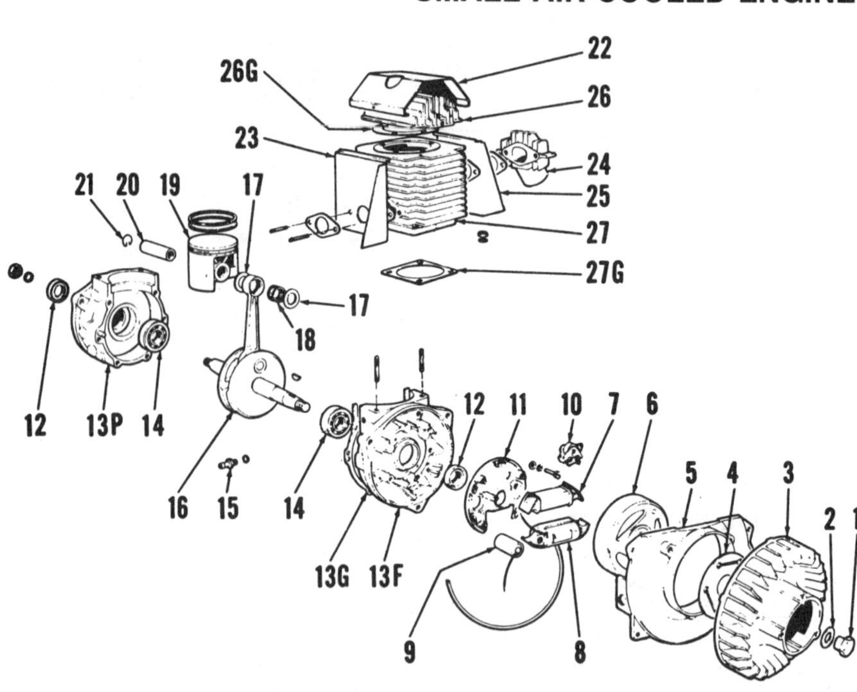

Fig. S4—Exploded view typical of Models 206 and 209 engines. Seals (12) and main bearings (14) are identical at both ends of crankshaft.

1. Flywheel nut
2. Washer
3. Cooling fan
4. Felt ring
5. Fan cover
6. Flywheel
7. Lighting coil
8. Ignition coil
9. Condenser
10. Breaker points
11. Stator plate
12. Crankcase seals
13F. Crankcase half (flywheel side)
13G. Gasket
13P. Crankcase half (output side)
14. Main bearing
15. Crankcase pulse fitting
16. Crankshaft assy.
17. Side washers
18. Piston pin bearing
19. Piston
20. Piston pin
21. Retaining ring (2)
22. Top air shroud
23. Air shroud (inlet side)
24. Exhaust manifold
25. Air shroud (exhaust side)
26. Cylinder head
26G. Gasket
27. Cylinder
27G. Gasket

REPAIRS

CYLINDER HEAD. Cylinder head (26–Fig. S4), is removable from all models. When reinstalling cylinder head, always use new head gasket (26G) and tighten head retaining cap screws evenly.

PISTON, PIN, RINGS AND CYLINDER. To remove cylinder (27–Fig. S4), unbolt and remove the carburetor, exhaust manifold and air shrouds. Remove the four cylinder retaining nuts and washers, then pull cylinder straight out from crankcase until free from piston. Inspect cylinder for scoring, cracks, damage and excessive wear.

Remove retaining rings (21), piston pin (20) and withdraw piston (19). Inspect piston pin (20), piston pin bearing (18) and side washers (17) for excessive wear. Check piston for scoring, damage and excessive wear.

Check ring end gap as shown in Fig. S5. Ring end gap should be 0.012-0.015 inch (0.30-0.38 mm). Remove all carbon from ring grooves in piston, install new rings and measure ring side clearance as shown in Fig. S6. Recommended side clearance is 0.002-0.003 inch (0.05-0.08 mm). Place piston (without rings) in cylinder close to TDC position. Using a feeler gage, measure piston side clearance in cylinder, which should be a minimum of 0.006 inch (0.15 mm). Piston and rings are available in standard size only. Grooves in pistons are equipped with pins and rings must be correctly installed to engage the pins. Install piston on rod so arrow on top of piston is pointed toward exhaust side of

Fig. S3—Arrow (A) on fan should point toward top of cylinder when piston is at top dead center (TDC).

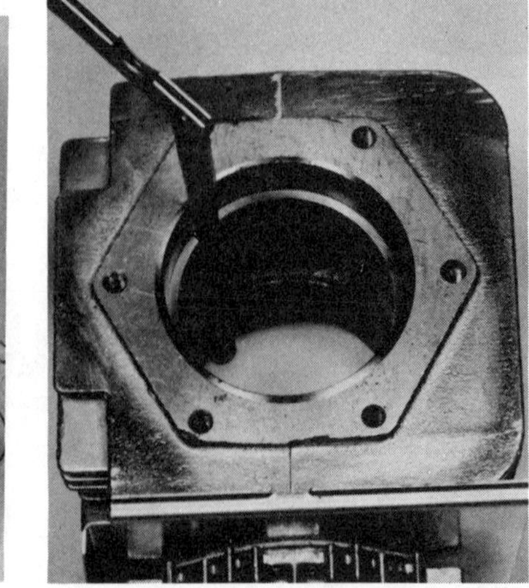

Fig. S5—Piston ring end gap should be 0.012-0.015 inch (0.30-0.38 mm) measured just above ports as shown.

Fig. S6—Use a feeler gage to measure piston ring side clearance in grooves. Recommended clearance is 0.002-0.003 inch (0.05-0.08 mm).

cylinder. A drop of sealant should be applied to crankcase joints before installing new gasket (27G – Fig. S4) and cylinder.

CONNECTING ROD, CRANKSHAFT AND CRANKCASE. To remove the connecting rod, it is necessary to separate the crankcase halves. After removing the cylinder, piston and magneto assemblies, remove cap screws securing case halves together, heat drive side of crankcase (13P – Fig. S4) around bearing area to approximately 350° F (177° C) and carefully remove the crankcase half. Heat and remove flywheel side of crankcase (13F). Crankshaft and connecting rod are available only as an assembled unit and disassembly is not recommended. Use a suitable puller to remove main bearings (14) from crankshaft using extreme care not to distort the crankshaft. Main bearings are a tight fit on crankshaft journals.

Bearings should be heated to approximately 300° F (149° C) before installing. Allow bearings to cool before installing crankshaft assembly into heated crankcase halves. Always renew gaskets and seals (12). Lips of seals should be toward inside and outside edge of seal should be flush with crankcase flange.

SERVICING SOLO ACCESSORIES

RECOIL STARTER

To disassemble the recoil starter, refer to Fig. S7 and remove pin (1), spring washer (2), thrust washer (3), snap rings (4) and lift off pawls (5) and thrust washer (6). Remove anchor (9) and handle (10). Allow rope to rewind into housing. Remove snap ring (7) and washer (8), then lift pulley (12) out. Remove spring (13) and washer (14).

Clean and inspect all parts and renew any showing excessive wear or other damage. When reassembling, install washer (14), spring (13) and pulley (12) in housing (16). Turn pulley counterclockwise (as viewed from engine side) approximately 5 turns and insert rope through pulley and exit hole in housing. Rope (11) should be approximately 43 inches (1092 mm) long. Press knot (at pulley end of rope) into recess in pulley until knot is flush with face of pulley. Install handle (10), anchor (9), washer (8) and snap ring (7). Assemble thrust washer (6), pawls (5) and snap rings (4). Install thrust washer (3), spring washer (2) and pin (1). Check operation before installing starter on engine.

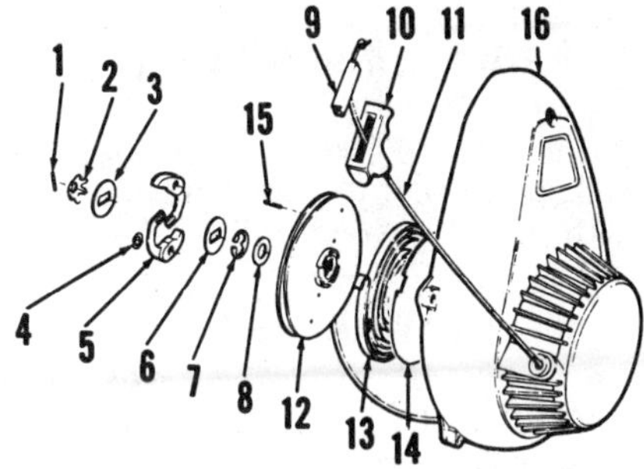

Fig. S7 – Exploded view of recoil starter used on Solo Models 206 and 209 engines. Washers (3 and 6) are identical.

1. Pin
2. Spring washer
3. Thrust washer
4. Snap rings (2)
5. Pawls
6. Thrust washer
7. Snap ring
8. Washer
9. Anchor
10. Handle
11. Rope
12. Pulley
13. Recoil spring
14. Washer
15. Pin
16. Starter housing & fan cover

STIHL

STIHL INC.
536 Viking Drive
Virginia Beach, Virginia 23452

Model	Bore	Stroke	Displacement
O20*	38.0 mm (1.50 in.)	28.0 mm (1.10 in.)	32.0 cc (1.96 cu. in.)
O41	44.0 mm (1.73 in.)	40.0 mm (1.57 in.)	61.0 cc (3.72 cu. in.)

* Some O20 engines have a displacement of 35.0 cc (2.13 cu. in.).

ENGINE INFORMATION

Stihl O20 and O41 series engines are used on Stihl cut-off saws and one man earth, ice and wood drills.

MAINTENANCE

SPARK PLUG. Recommended spark plug for all models is a Bosch WSR6F or equivalent. Specified electrode gap is 0.5 mm (0.020 in.).

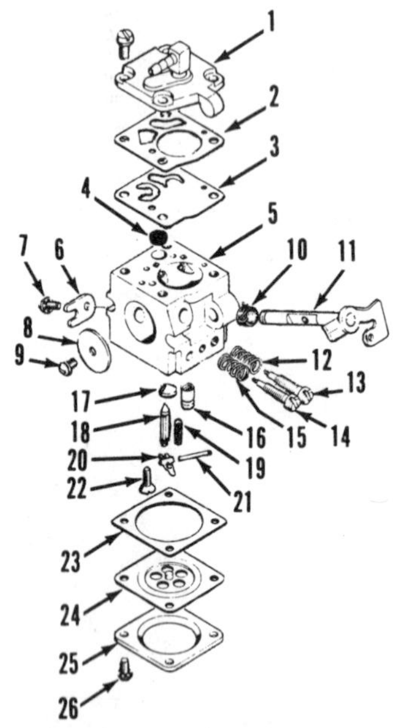

Fig. SL60—Exploded view of diaphragm type carburetor used on all models.

1. Pump cover
2. Gasket
3. Diaphragm
4. Screen
5. Body
6. Clip
7. Screw
8. Throttle plate
9. Screw
10. Spring
11. Throttle shaft
12. Spring
13. High speed mixture needle
14. Low speed mixture needle
15. Spring
16. Check valve
17. Welch plug
18. Fuel inlet needle
19. Spring
20. Fuel inlet lever
21. Pin
22. Screw
23. Gasket
24. Diaphragm
25. Cover
26. Screw

CARBURETOR. A diaphragm type carburetor is used on all models (Fig. SL60). Initial adjustment of low and high speed mixture needles for O20 engine is 1-1/4 turns open from a lightly seated position. Initial adjustment of O41 engine low and high speed mixture needles from a lightly seated position is 1-1/4 turns open for low speed mixture needle and 7/8 turn open for the high speed mixture needle.

To disassemble carburetor, refer to Fig. SL60. Clean filter screen (4). Welch plugs (18 and 24—Fig. SL61) may be removed by drilling plug with a suitable size drill bit, then pry out as shown in Fig. SL62. Care must be taken not to drill into carburetor body.

Inspect inlet lever spring (19—Fig. SL60) and renew if stretched or damaged. Inspect diaphragms for tears, cracks or other damage. Renew low and high speed mixture needles if needle points are grooved or broken. Carburetor body must be renewed if needle seats are damaged. Fuel inlet needle has a rubber tip which seats directly on a machined orifice in carburetor body. Inlet needle or carburetor body should be renewed if worn excessively.

Adjust position of inlet control lever so lever is flush with diaphragm chamber floor as shown in Fig. SL63. Bend lever adjacent to spring to obtain correct position.

IGNITION SYSTEM. Engine may be equipped with a conventional magneto type ignition system, a Bosch capacitor discharge ignition system or a Sems transistorized ignition system. Refer to the appropriate paragraph for model being serviced.

Magneto Ignition System. Ignition breaker point gap should be set at 3.5-4.0 mm (0.014-0.016 in.). Ignition timing is adjusted by loosening stator mounting screws and rotating stator plate. Ignition timing should occur when piston is 2.0-2.3 mm (0.08-0.087 in.) BTDC for O20 engine and 2.4-2.6 mm (0.095-0.102 in.) BTDC for O41 engine.

Ignition coil air gap should be 4-6 mm (0.16-0.24 in.) for O20 engine and 0.2-0.3 mm (0.008-0.012 in.) for O41 engine.

Ignition coil primary and secondary windings may be checked using an ohmmeter. Primary winding resistance should register 1.5-1.9 ohms for O20 engine, 1.9-2.5 ohms for O41 engine with Bosch date code 523 and 1.2-1.7 ohms for O41 engine with Bosch date code 524.

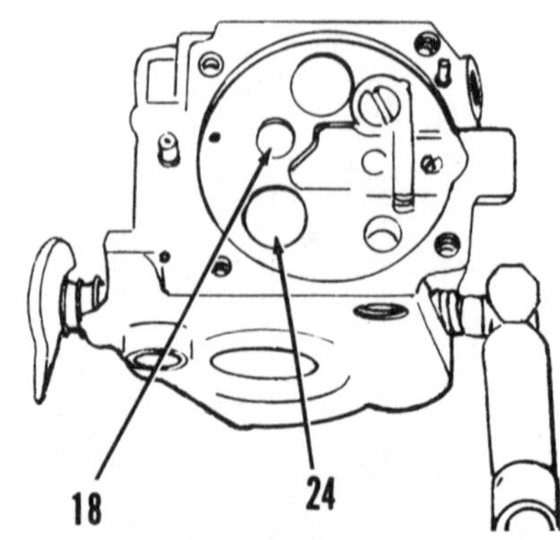

Fig. SL61—View showing location of Welch plugs (18 and 24).

Fig. SL62—A punch can be used to remove Welch plugs after drilling a hole in plug. Refer to text.

Bosch Breakerless Ignition System. Ignition module air gap should be 0.2-0.3 mm (0.008-0.012 in.). Ignition should occur when piston is 2.5 mm (0.098 in.) BTDC.

Sems Breakerless Ignition System. Ignition module air gap should be 0.2-0.3 mm (0.008-0.012 in.). Ignition should occur when piston is 2.5 mm (0.098 in.) BTDC.

Ignition module primary and secondary windings may be checked using an ohmmeter. Primary winding resistance should be 0.4-0.5 ohms. Secondary winding resistance should be 2.7-3.3k ohms.

LUBRICATION. The engine is lubricated by mixing oil with the fuel. Recommended fuel:oil ratio is 40:1 when using STIHL two-stroke engine oil. If STIHL two-stroke engine oil is not available, a good quality oil designed for two-stroke, air-cooled engines may be used when mixed at a 25:1 ratio. Use a separate container when mixing the oil and gasoline.

REPAIRS

TIGHTENING TORQUES. Recommended tightening torque specifications are as follows:

Clutch nut	39 N·m (29 ft.-lbs.)
Clutch hub	34 N·m (25 ft.-lbs.)
Clutch hub carrier (O20)	28 N·m (21 ft.-lbs.)
Flywheel nut:	
O20	24 N·m (18 ft.-lbs.)
O41	35 N·m (26 ft.-lbs.)

CRANKSHAFT AND SEALS. All models are equipped with a split type crankcase from which cylinder may be removed separately (Fig. SL64 and SL65). It may be necessary to heat crankcase halves slightly to remove main bearings if they remain in crankcase during disassembly.

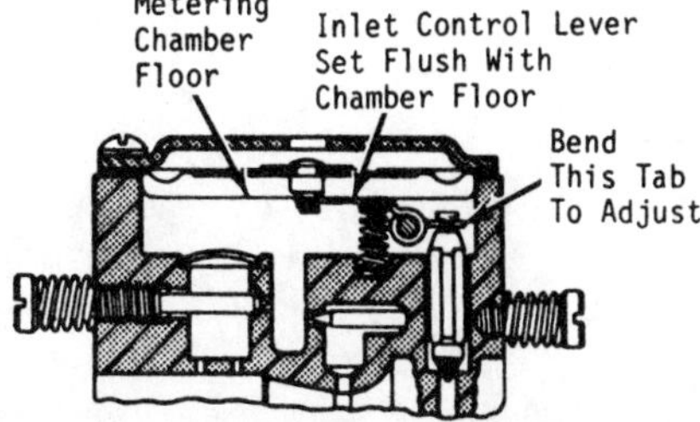

Fig. SL63—Diaphragm lever should be flush with fuel chamber floor as shown.

CONNECTING ROD. The connecting rod and crankshaft are considered an assembly and individual parts are not available separately. Do not remove connecting rod from crankshaft.

Connecting rod big end rides on a roller bearing and should be inspected for excessive wear or damage. If rod, bearing or crankshaft is damaged, complete crankshaft and connecting rod assembly must be renewed.

CYLINDER, PISTON, PIN AND RINGS. The aluminum alloy piston is equipped with two piston rings. The floating piston pin is retained in the piston with a snap ring at each end. The pin bore of the piston is unbushed; the connecting rod has a caged needle roller piston pin bearing.

Cylinder bore and cylinder head are cast as one-piece. The cylinder assembly is available only with a fitted piston. Pistons and cylinders are grouped into different size ranges. Each group is marked with letters "A" to "E." Letter "A" denotes smallest size with "E" being largest. The code letter is stamped on the top of the piston and on the top of the cylinder. The code letter of the piston and the cylinder must be the same for proper fit of a new piston

Fig. SL64—Exploded view of O20 engine.

1. Seal
2. Crankcase half
3. Bearing
4. Crankshaft & rod assy.
5. Snap ring
6. Bearing
7. Gasket
8. Crankcase half
10. Cylinder
11. Gasket
12. Piston rings
13. Piston
14. Piston pin
15. Pin retainer
16. Roller bearing

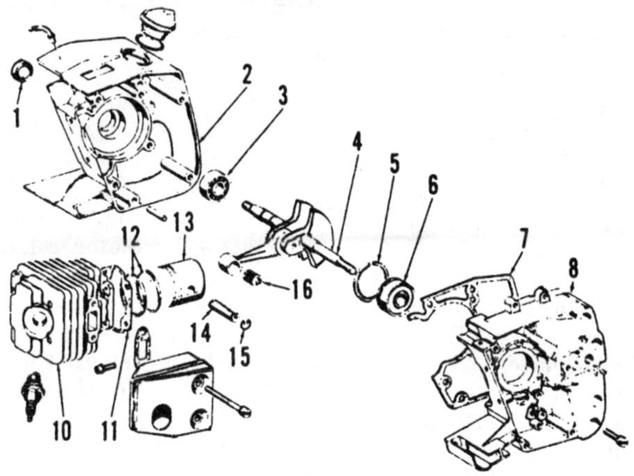

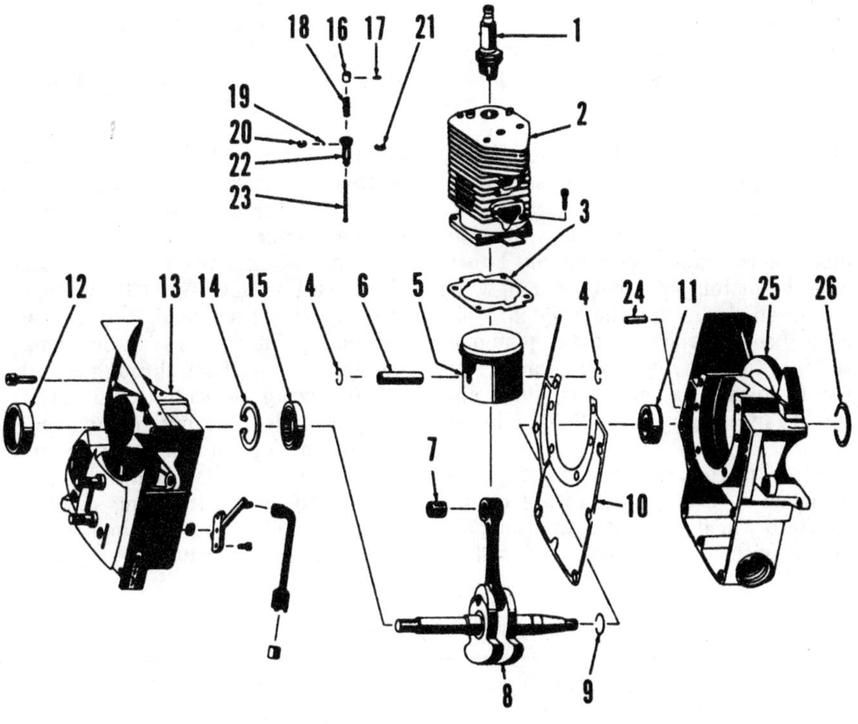

Fig. SL65—Exploded view of O41 engine.

1. Spark plug
2. Cylinder
3. Head gasket
4. Snap ring
5. Piston
6. Piston pin
7. Bearing
8. Crankshaft & rod assy.
9. Snap ring
10. Gasket
11. Bearing
12. Seal
13. Crankcase half
14. Snap ring
15. Bearing
16. Knob
17. Pin
18. Spring
19. Pin
20. Spring
21. Tab washer
22. Housing
23. Plunger
24. Pin
25. Crankcase half
26. Snap ring

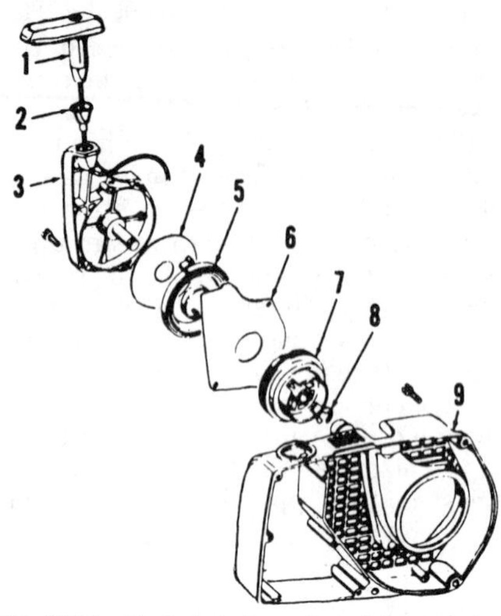

Fig. SL66—Exploded view of pawl type starter used on 020 engine.

1. Rope handle
2. Bushing
3. Housing
4. Washer
5. Rewind spring
6. Cover
7. Rope pulley
8. "E" ring
9. Fan housing

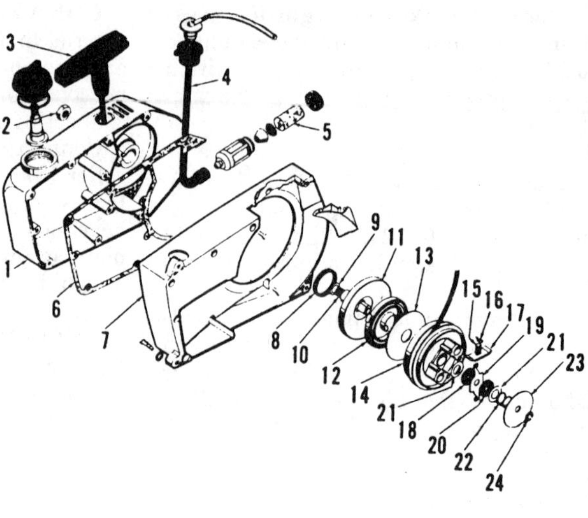

Fig. SL67—Exploded view of typical rewind starter used on 041 engine.

1. Fuel tank
2. Nut
3. Rope handle
4. Fuel pick-up
5. Filter
6. Gasket
7. Fan cover
8. Felt ring
9. Spring washer
10. Pulley shaft
11. Cover
12. Rewind spring
13. Washer
14. Rope pulley
15. Spring
16. Spring retainer
17. Friction shoe
18. Slotted washer
19. Brake lever
20. Slotted washer
21. Washer
22. Spring
23. Washer
24. "E" ring

in a new cylinder. However, new pistons are available for installation in used cylinders. Used cylinders with code letters "A" or "B" may use piston with code letter "A". Used cylinder with code letters "A", "B" or "C" may use piston with code letter "B". Used cylinder with code letters "B", "C" or "D" may use piston with code letter "C". Used cylinder with code letters "C", "D" or "E" may use piston with code letter "D". Used cylinder with code letters "D" or "E" may use piston with code letter "E".

Cylinder bore on all models is chrome plated. Cylinder should be renewed if chrome plating is flaking, scored or worn away.

To reinstall piston on connecting rod, install one snap ring in piston. Lubricate the piston pin needle bearing with motor oil, then slide bearing into pin bore of connecting rod. Install piston on rod so that arrow on piston crown points toward exhaust port. Push piston in far enough to install second snap ring.

After piston and rod assembly is attached to crankshaft, rotate crankshaft to top dead center and support piston with a wood block that will fit between piston skirt and crankcase when cylinder gasket is in place. A notch should be cut in the wood block so that it will fit around the connecting rod. Lubricate piston and rings with motor oil, then compress rings with compressor that can be removed after cylinder is pushed down over piston. On some models it may be necessary to remove the cylinder and install an additional gasket between the cylinder and crankcase if piston strikes top of cylinder.

REWIND STARTER. Friction shoe type and pawl type starters have been used. Refer to Fig. SL66 for an exploded view of starter assembly used on 020 engine and to Fig. SL67 for an exploded view of starter assembly used on 041 engine. Refer to Fig. SL68 for proper method of assembly of friction shoe plates to starter brake lever on 041 engine starter.

To place tension on rope of starter shown in Fig. SL66 for 020 engine, pull starter rope, then hold rope pulley to prevent spring from rewinding rope on pulley. Pull rope back through rope outlet and wrap two additional turns of rope on pulley without moving pulley. Release pulley and allow rope to rewind. Rope handle should be pulled against housing, but spring should not be completely wound when rope is pulled to greatest length.

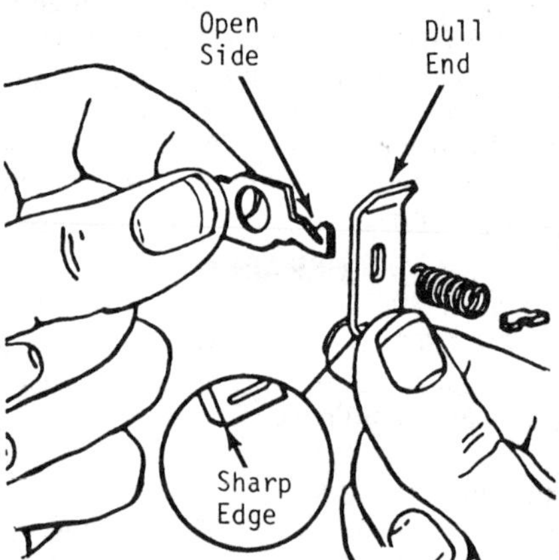

Fig. SL68—Illustration showing proper method of assembly of friction shoe plates to starter brake lever.

To place tension on 041 engine starter rope, pull rope out of handle until notch in pulley is adjacent to rope outlet, then hold pulley to prevent rope from rewinding. Pull rope back through outlet and out of notch in pulley. Turn rope pulley two turns clockwise and release rope back through notch. Check starter operation. Rope handle should be held against housing by spring tension, but spring should not be completely wound when rope is pulled to greatest length.

STIHL

Model	Bore	Stroke	Displacement
O8S	47.0 mm (1.85 in.)	32.0 (1.26 in.)	56.0 cc (3.39 cu. in.)

ENGINE INFORMATION

Stihl O8S series engines are used on Stihl cut-off saws and one man earth, ice and wood drills.

MAINTENANCE

SPARK PLUG. Recommended spark plug is a Bosch W175T7 or equivalent. Specified electrode gap is 0.5 mm (0.020 in.).

CARBURETOR. Model O8S engine is equipped with a Tillotson diaphragm type carburetor. Initial adjustment of idle and high speed mixture needles is one turn open from a lightly seated position.

Final adjustments are made with engine at operating temperature and running. Adjust low speed mixture needle and idle speed screw until engine idles just below clutch engagement speed. Adjust high speed mixture needle to obtain optimum performance under load. Do not adjust high speed mixture needle too lean as engine may be damaged.

To disassemble carburetor, refer to Fig. SL70. Remove and clean filter screen (4). Welch plugs (18 and 24—Fig. SL71) can be removed by drilling plug with a suitable size drill bit, then pry out as shown in Fig. SL72. Care must be taken not to drill into carburetor body.

Inspect inlet lever spring (19—Fig. SL70) and renew if stretched or damaged. Inspect diaphragms for tears, cracks or other damage. Renew idle and high speed mixture needles if needle points are grooved or broken. Carburetor body must be renewed if needle seats are damaged. Fuel inlet needle has a rubber tip which seats directly on a machined orifice in carburetor body. Inlet needle or carburetor body should be renewed if worn excessively.

Inlet control lever should be adjusted so control lever is flush with fuel chamber floor as shown in Fig. SL73. Carefully bend lever adjacent to spring to obtain correct position.

Fig. SL70—Exploded view of diaphragm type carburetor.

1. Pump cover
2. Gasket
3. Diaphragm
4. Screen
5. Body
6. Clip
7. Screw
8. Throttle plate
9. Screw
10. Spring
11. Throttle shaft
12. Spring
13. High speed mixture needle
14. Low speed mixture needle
15. Spring
16. Check valve
17. Welch plug
18. Fuel inlet needle
19. Spring
20. Fuel inlet lever
21. Pin
22. Screw
23. Gasket
24. Diaphragm
25. Cover
26. Screw

GOVERNOR. Model O8S is equipped with an air vane type governor. The governor linkage is attached to the carburetor choke shaft lever. Maximum speed is controlled by the air vane governor closing the choke plate.

Governed speed is adjusted by changing the tension of the governor spring. The adjusting plate is mounted on the engine behind the starter housing as shown in Fig. SL74. After maximum governed speed is adjusted at factory, position of spring is secured by a lead seal. If necessary to readjust governor, new position of governor spring should be sealed or wired securely. Maximum no-load governed speed is 8000 rpm.

IGNITION SYSTEM. Model OS8 engine is equipped with a conventional magneto type ignition system.

Coil air gap should be 0.5 mm (0.020 in.). Ignition breaker point gap should be 3.5-4.0 mm (0.014-0.016 in.). Ignition timing is adjusted by loosening stator mounting screws and rotating stator plate. Ignition should occur when piston is 1.9-2.1 mm (0.075-0.083 in.) BTDC.

Coil primary and secondary windings may be checked using an ohmmeter. Primary winding resistance should be 1.9-2.5 ohms for Bosch coil with date code 523 or 1.2-1.7 ohms for Bosch coil with date code 524. Secondary winding resistance should be 5.0-6.7k ohms for

Fig. SL71—View showing location of Welch plugs (18 and 24).

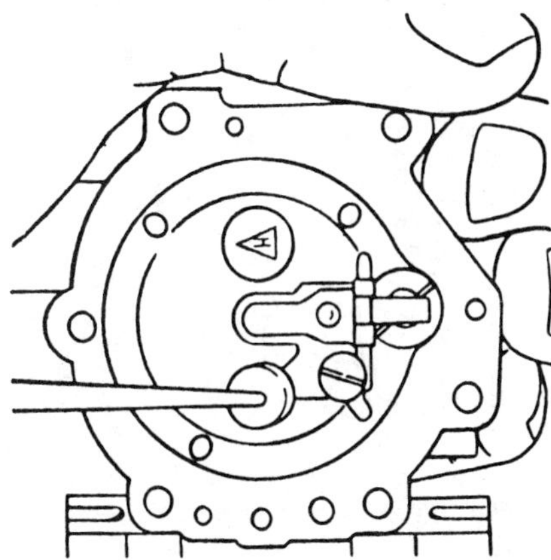

Fig. SL72—A punch can be used to remove Welch plugs after drilling a hole in plug. Refer to text.

Bosch coil with date code 523 or 5.0-6.7k ohms for Bosch coil with date code 524.

LUBRICATION. The engine is lubricated by mixing oil with the fuel. Regular leaded gasoline is recommended. Recommended oil is STIHL two-stroke engine oil mixed at a 40:1 ratio. If STIHL two-stroke engine oil is not available, a good quality oil designed for use in air-cooled, two-stroke engines may be used when mixed at a 25:1 ratio. Use a separate container to mix oil and gasoline.

REPAIRS

TIGHTENING TORQUES. Recommended tightening torque specifications are as follows:

Clutch nut . . . . 34 N·m (25 ft.-lbs.)
Flywheel nut . . . . 29 N·m (22 ft.-lbs.)
Spark plug . . . . 24 N·m (18 ft.-lbs.)
Cylinder screws . . . . 10 N·m (7 ft.-lbs.)

CRANKSHAFT AND SEALS. All models are equipped with a split type crankcase from which cylinder may be removed separately (Fig. SL75). Crankshaft is supported at both ends in ball type main bearings and crankshaft end play is controlled by shims (11) installed between the bearing (10) and shoulders on the crankshaft on early models, or between bearing and crankcase on later models. Shims are available in a variety of thicknesses. Correct crankshaft end play is 0.2-0.3 mm (0.008-0.012 in.). It may be necessary to heat crankcase halves slightly to remove main bearings if they remain in crankcase during disassembly.

CONNECTING ROD. The connecting rod and crankshaft are considered an assembly and individual parts are not available separately. Do not remove connecting rod from crankshaft.

Connecting rod big end rides on a roller bearing and should be inspected for excessive wear or damage. If rod, bearing or crankshaft is damaged, complete crankshaft and connecting rod assembly must be renewed.

CYLINDER, PISTON, PIN AND RINGS. The aluminum alloy piston is equipped with two piston rings. The floating piston pin is retained in the piston with a snap ring at each end. The pin bore of the piston is unbushed; the connecting rod has a caged needle roller piston pin bearing.

Cylinder bore and cylinder head are cast as one-piece. The cylinder is available only with a fitted piston. Pistons and cylinders are grouped into different size ranges. Each group is marked with letters "A" to "E". Letter "A" denotes smallest size with "E" being largest. The code letter is stamped on the top of the piston and on the top of the cylinder on all models. The code letter of the piston and the cylinder must be the same for proper fit of a new piston in a new cylinder. However, new pistons are available for installation in used cylinders. Used cylinders with code letters "A" or "B" may use piston with code letter "A". Used cylinder with code letters "A", "B" or "C" may use piston with code letter "B". Used cylinder with code letters "B", "C" or "D" may use piston with code letter "C". Used cylinder with code letters "C", "D" or "E" may use piston with code letter "D". Used cylinder with code letters "D" or "E" may use piston with code letter "E".

Cylinder bore on all models is chrome plated. Cylinder should be renewed if chrome plating is flaking, scored or worn away.

To reinstall piston on connecting rod, install one snap ring in piston. Lubricate the piston pin needle bearing with motor oil, then slide bearing into pin bore of connecting rod. Install piston on rod so that arrow on piston crown points toward exhaust port. Push piston in far enough to install second snap ring.

After piston and rod assembly is attached to crankshaft, rotate crankshaft to top dead center and support piston with a wood block that will fit between piston

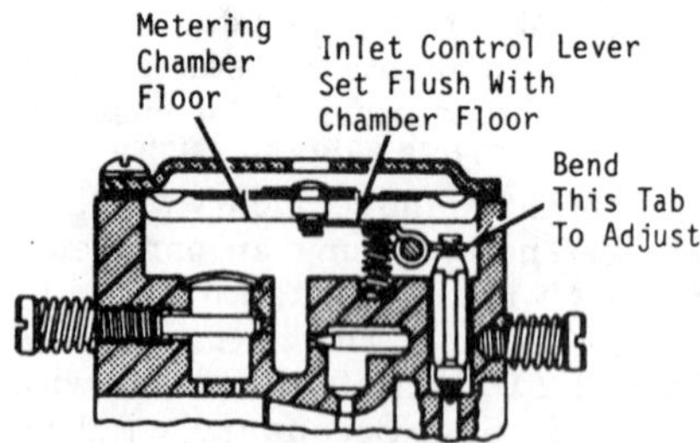

Fig. SL73—Diaphragm lever should be flush with fuel chamber floor as shown.

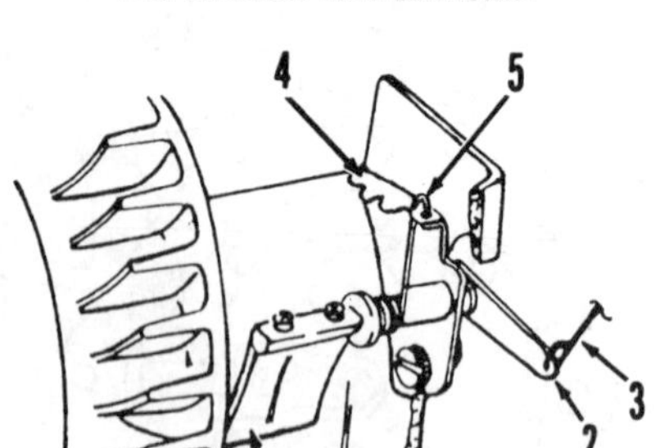
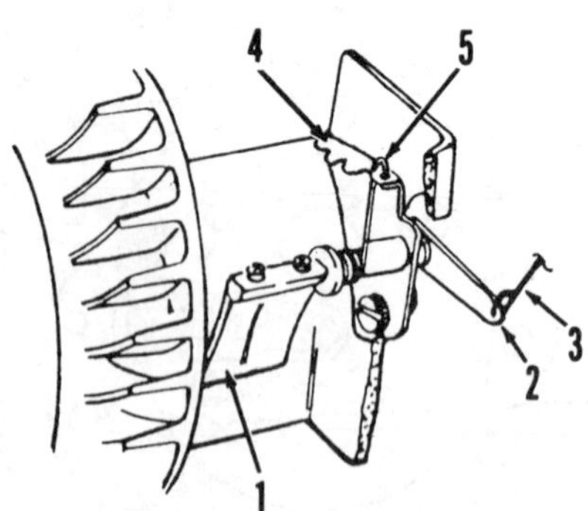

Fig. SL74—View of air vane type governor.

1. Air vane
2. Lever
3. Throttle link
4. Notched plate
5. Governor spring

Fig. SL75—Exploded view of O8S engine.

1. Spark plug
2. Cylinder
3. Head gasket
4. Snap ring
5. Piston pin
6. Piston
7. Crankcase half
8. Seal
9. Snap ring
10. Ball bearing
11. Shim
12. Gasket
13. Dowel pin
14. Crankcase half
15. Washer
16. Bearing
17. Crankshaft & rod assy.

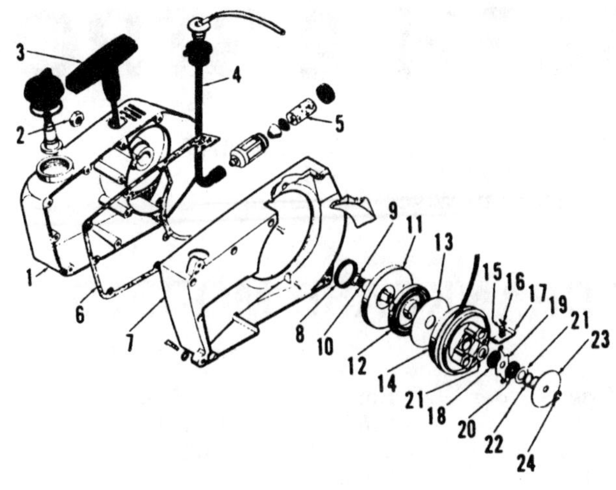

Fig. SL76—Exploded view of pawl type starter.

1. Fuel tank
2. Nut
3. Rope handle
4. Fuel pick-up
5. Filter
6. Gasket
7. Fan cover
8. Felt ring
9. Spring washer
10. Pulley shaft
11. Cover
12. Rewind spring
13. Washer
14. Rope pulley
15. Spring
16. Spring retainer
17. Friction shoe
18. Slotted washer
19. Brake lever
20. Slotted washer
21. Washer
22. Spring
23. Washer
24. "E" ring

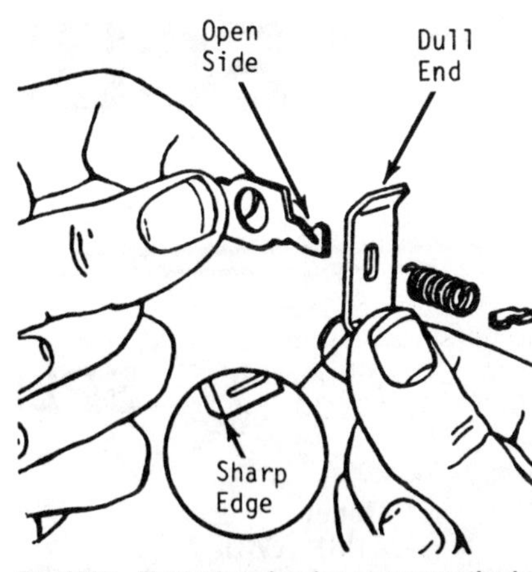

Fig. SL77—Illustration showing proper method of assembly of friction shoe plates to starter brake lever.

skirt and crankcase when cylinder gasket is in place. A notch should be cut in the wood block so that it will fit around the connecting rod. Lubricate piston and rings with motor oil, then compress rings with compressor that can be removed after cylinder is pushed down over piston. On some models it may be necessary to remove the cylinder and install an additional gasket between the cylinder and crankcase if piston strikes top of cylinder.

REWIND STARTER. Model O8S engine is equipped with a pawl type rewind starter assembly similar to Fig. SL76.

To place tension on engine starter rope, pull rope out of handle until notch in pulley is adjacent to rope outlet, then hold pulley to prevent rope from rewinding. Pull rope back through outlet and out of notch in pulley. Turn rope pulley two turns clockwise and release rope back through notch. Check starter operation. Rope handle should be held against housing by spring tension, but spring should not be completely wound when rope is pulled to greatest length.

Refer to Fig. SL77 for correct assembly of starter shoes on brake lever. Make certain assembly is installed as shown in exploded views and leading edges of friction shoes are sharp or shoes may not properly engage drum.

TECUMSEH 2-STROKE

TECUMSEH PRODUCTS COMPANY
900 North Street
Grafton, Wisconsin 53024

SPLIT CRANKCASE MODELS

Model	Cyls.	Bore	Stroke	Displacement
AH31, AV31	1	1.625 in. (41.28 mm)	1.500 in. (38.10 mm)	3.1 cu. in. (51 cc)
AH36, AV36	1	1.750 in. (43.55 mm)	1.500 in. (38.10 mm)	3.6 cu. in. (57 cc)
AH47, AV47	1	2.000 in. (50.80 mm)	1.500 in. (38.10 mm)	4.7 cu. in. (77 cc)
AH51, AV51	1	2.000 in. (50.80 mm)	1.625 in. (41.28 mm)	5.1 cu. in. (84 cc)
AH58, AV58	1	2.090 in. (53.09 mm)	1.680 in. (42.67 mm)	5.8 cu. in. (95 cc)
AH61, AV61	1	2.090 in. (53.09 mm)	1.769 in. (44.93 mm)	6.1 cu. in. (100 cc)
AH80, AV80	1	2.250 in. (57.15 mm)	2.000 in. (50.80 mm)	8.0 cu. in. (130 cc)
AH81, AV81	1	2.500 in. (63.50 mm)	1.625 in. (41.28 mm)	7.98 cu. in. (130 cc)
AH82	1	2.500 in. (63.50 mm)	1.680 in. (42.67 mm)	8.3 cu. in. (135 cc)
BH60, BV60	2	1.625 in. (41.28 mm)	1.438 in. (35.51 mm)	6.0 cu. in. (95 cc)
BH69, BV69	2	1.750 in. (44.45 mm)	1.438 in. (35.51 mm)	6.9 cu. in. (110 cc)

ENGINE IDENTIFICATION

All models are two-stroke air-cooled engines with a split crankcase design.

Prefixes indicate the following information for engine identification: A-Single cylinder, B-Two cylinder, V-Vertical crankshaft, H-Horizontal crankshaft.

Engine model number designation represents ten times the displacement. For example, 31 means 3.1 cubic inch displacement. Engine prefix and model number are located as shown in Fig. TP1-1.

NOTE: Model designation may not always indicate the exact displacement of the engine, for instance, Model AH81 has 7.98 cubic inches displacement.

Engine type number provides more specific information needed to correctly identify replacement parts. Type number is located as shown in Fig. TP1-1.

Early engines listed the type number as a suffix of the serial number. For example, if the number found on the engine is 365341 P 238, the serial number is 365341, P indicates the engine is equipped with a Phelon magneto and 238 is the engine type number. Another example is 1234567 1210. The serial number is 1234567 and the type number is 1210.

Always furnish engine model, type and serial numbers when ordering parts.

Refer to the following type numbers and engine model number cross-reference chart to identify engine for service procedure.

TYPE NUMBER	ENGINE MODEL
46 through 68	AH-31
69 through 73	AH-36
75 through 78	AH-36
79	AH-31
80 through 83	AH-36
84	AH-31
85	AH-36
86 through 88	AH-31
89 through 93	AH-36
94 through 98	AH-47
99	AH-36
209 through 244	AV-31
245	AV-36

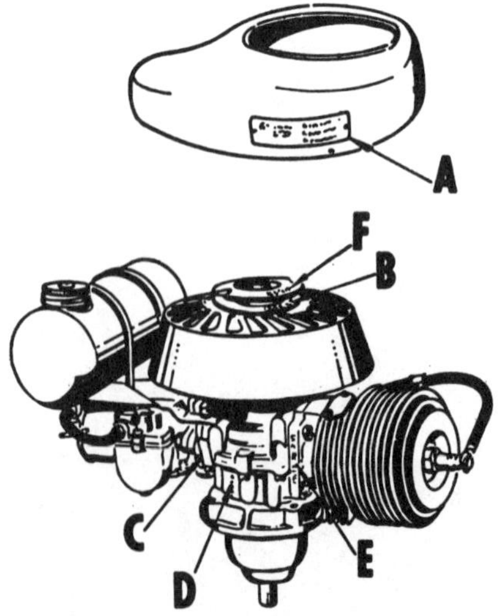

Fig. TP1-1—The engine serial number and type number may be found at one of the locations shown above.

A. Nameplate on air shroud
B. Model & type number plate
C. Metal tab on crankcase
D. Stamped on crankcase
E. Stamped on cylinder flange
F. Stamped on starter pulley

TYPE NUMBER	ENGINE MODEL
246 through 248	AV-31
249 through 251	AV-36
252	AV-31
253 through 255	AV-36
256 through 261	AV-31
262 through 265A	AV-36
266 through 267	AV-31
268 through 272	AV-36
273 through 275	AV-31
276	AV-36
277 through 279	AV-31
280 through 281	AV-36
282	AV-31
283 through 284	AV-36
285 through 286	AV-31
287	AV-36
288 through 291B	AV-31
292 through 294	AV-36
295 through 296	AV-31
297	AV-47
298 through 299	AV-31
321 through 321A	BH-60
330 through 330B	BH-69
331 through 332	BH-60
333	BH-69
334	BH-60
335 through 337	BH-69
351	BV-60
353 through 355A	BH-60
356	BV-69
357	BV-60
359 through 363A	BV-69
364	BH-69
365 through 366	BH-60
367 through 368A	BV-69
369 through 370	BH-60
371 through 375	BH-69
401 through 402	AH-80
403	AV-80
404 through 406B	AH-80
407 through 407C	AV-80
408 through 410B	AH-80
411 through 411B	AV-80
412 through 423	AH-80
424 through 427	AV-80
501 through 509A	AV-47
601 through 601-01	AV-47
602 through 602-02B	AV-47
603 through 603-23	AV-47
604 through 604-25	AV-47
605 through 605-18	AV-47
606 through 606-13A	AV-47
608 through 608C	AV-47
610 through 610-20	AV-47
611	AV-47
614 through 614-06A	AV-47
615 through 616-40	AV-47
617 through 617-06	AV-47
618 through 618-18	AV-47
619 through 619-03	AV-47
621 through 621-15	AV-47
622 through 622-07	AV-47
623 through 623-36	AV-47
624 through 630-09	AV-51
632-02	AV-61
632-02A	AV-58
632-03	AV-61

TYPE NUMBER	ENGINE MODEL
632-04 & 05A	AV-58
633 through 633-78	AV-51
634 through 634-09D	AV-51
635	AV-61
635A & B	AV-58
635-01 through 635-03A	AV-61
635-03B	AV-58
635-04 through 635-04A	AV-61
635-04B	AV-58
635-05 through 635-06	AV-61
635-06A through 635-14	AV-58
636 through 636-11	AV-51
637 through 637-17	AV-51
701 through 701-24	AH-47
1001	AH-36
1002 through 1002D	AH-47
1003	AH-36
1004 through 1004A	AH-47
1005 through 1005A	AH-36
1006	AH-47
1007 through 1008B	AH-36
1009 through 1020	AH-47
1021 through 1022C	AH-36
1023 through 1026A	AH-47
1027 through 1027B	AH-36
1028 through 1031C	AH-47
1032	AH-36
1033 through 1034G	AH-47
1035 through 1039C	AH-36
1040 through 1044E	AH-47
1045 through 1045F	AH-36
1046 through 1054C	AH-47
1055 through 1056B	AH-47
1057 through 1058B	AH-47
1059	AH-36
1060 through 1060C	AH-47
1061 through 1062	AH-36
1063 through 1066A	AH-47
1067 through 1067C	AH-36
1068 through 1069	AH-47
1070 through 1070B	AH-36
1071 through 1099	AH-47
1101 through 1185	AH-47
1185A through 1185C	AH-51
1186	AH-47
1186A through 1186C	AH-51
1187 through 1192	AH-47
1192A through 1196B	AH-51
1197 through 1198A	AH-47
1199 through 1199A	AH-51
1206 through 1208C	AH-51
1210 through 1215C	AH-58
1216 through 1216A	AH-47
1217 through 1220A	AH-51
1221	AH-47
1222 through 1223	AH-51
1224 through 1225A	AH-47
1226 through 1227B	AH-58
1228 through 1229A	AH-51
1230 through 1232	AH-47
1233 through 1237B	AH-58
1238 through 1238C	AH-47
1239 through 1243	AH-51
1244 through 1245A	AH-58
1246 through 1247	AH-51
1248	AH-47
1249 through 1251A	AH-51

TYPE NUMBER	ENGINE MODEL
1252 through 1254A	AH-58
1255	AH-51
1256 through 1262B	AH-58
1263	AH-51
1264 through 1265E	AH-58
1266 through 1267	AH-47
1269 through 1270D	AH-58
1271 through 1271B	AH-47
1272 through 1275	AH-58
1276	AH-47
1277 through 1279D	AH-58
1280	AH-51
1283 through 1284D	AH-58
1286 through 1286A	AH-47
1287	AH-58
1288	AH-51
1289 through 1289A	AH-47
1290 through 1293A	AH-58
1294 through 1295	AH-51
1296 through 1298A	AH-58
1299	AH-61
1300 through 1303	AH-58
1304 through 1307	AH-51
1308 through 1316A	AH-58
1317 through 1317B	AH-51
1318 through 1320B	AH-58
1321 through 1322	AH-51
1323	AH-58
1324	AH-51
1325	AH-47
1326 through 1326F	AH-58
1327 through 1327B	AH-47
1328 through 1328C	AH-58
1329	AH-61
1330	AH-47
1331	AH-51
1332	AV-51
1333 through 1334A	AH-47
1335 through 1342	AH-61
1343 through 1344A	AH-47
1345	AH-51
1346 through 1347	AH-61
1348 through 1350C	AH-47
1351 through 1351B	AH-58
1352 through 1352B	AH-47
1353	AH-58
1354 through 1355A	AH-47
1356 through 1356B	AH-58
1357 through 1358	AH-47
1359 through 1362B	AH-58
1363 through 1369A	AH-47
1370 through 1371B	AH-58
1372 through 1375A	AH-47
1376 through 1376A	AH-51
1377 through 1378	AH-58
1379 through 1379A	AH-47
1380 through 1380B	AH-58
1381 through 1382	AH-47
1383 through 1383B	AH-58
1384 through 1386	AH-47
1387 through 1388	AH-58
1389 through 1390B	AH-47
1391	AH-58
1392	AH-51
1393 through 1395B	AH-47
1396	AH-58
1397 through 1397B	AH-47

TYPE NUMBER	ENGINE MODEL
1403 through 1404A	AH-58
1405 through 1406A	AH-47
1407 through 1408	AH-58
1409	AH-51
1410 through 1416A	AH-47
1417 through 1417A	AH-58
1418 through 1418A	AH-47
1419 through 1419A	AH-58
1420 through 1420A	AH-47
1421 through 1421A	AH-58
1422 through 1423	AH-47
1424	AH-58
1426 through 1429A	AH-47
1431	AH-47
1433	AH-58
1434 through 1436A	AH-47
1437	AH-58
1439	AH-47
1441	AH-47
1443	AH-47
1445	AH-47
1447	AH-47
1451	AH-47
1453	AH-58
1455 through 1456	AH-58
1461	AH-58
1463	AH-58
1467 through 1468	AH-58
1477	AH-47
1478	AH-51
1481	AH-51
1487	AH-51
S-1801 through 1824	AV-47
2001	AV-31
2003	AV-31
2004 through 2006	AV-36
2007 through 2007B	AV-31
2008 through 2008B	AV-36
2009	AV-31
2010 through 2011	AV-36
2012	AV-47
2013 through 2014	AV-31
2015 through 2018	AV-36
2020	AV-47
2021	AV-36
2022 through 2022B	AV-47
2023 through 2026	AV-36
2027	AV-31
2028 through 2029	AV-36
2030 through 2031	AV-31
2032 through 2033A	AV-36
2034 through 2035C	AV-47
2036 through 2037	AV-31
2038	AV-47
2039 through 2044	AV-36
2045 through 2046B	AV-47
2047 through 2048	AV-36
2049 through 2049A	AV-47
2050 through 2050A	AV-36
2051 through 2052A	AV-36
2053 through 2055	AV-47
2056	AV-36
2057	AV-47
2058 through 2058B	AV-36
2059 through 2063	AV-47
2064 through 2064A	AV-36
2065	AV-47
2066	AV-36
2067 through 2071B	AV-47
2200 through 2201A	AV-36
2202 through 2204	AV-47
2205 through 2205A	AV-36
2206 through 2206A	AV-47
2207	AV-36
2208	AV-47
2768	AV-36
40001 through 40028	AH-81
40029 through 40032C	AH-82
40033	AH-81
40034 through 50045A	AH-82
40046	AH-81
40047 through 40052	AH-82
40053 through 40054A	AH-81
40056 through 40060B	AH-82
40061 through 40062B	AH-81
40063 through 40064	AH-82
40065 through 40066	AH-81
40067 through 40068	AH-82
40069 through 40069A	AH-81
40070	AV-81
40070A through 40075	AH-81
710101 through 710116	AV-47
710124 through 710149	AV-47
710150 through 710152	AH-47
710154	AV-80
710155	AH-80
710156	BH-69
710157	AH-47
710201 through 710227	AV-47
710228	AV-80
710229	AH-47
710230 and 710231	AH-81
710232 and 710233	AH-47
710234	AH-81
710235	AV-51
710244	AH-81
710251 and 710252	AH-47
710257 and 710258	AH-58
710296	AV-51
710298 and 710299	AH-58
710302	AV-51
710303 through 710306	AH-47
710307	AV-51
710309	AV-51
710312	AV-51
710313	AH-58
710314 and 710316	AH-47
710317 and 710318A	AH-58
710319 through 710323	AH-47

*Six digit type numbers beginning with 710 (such as 710101) are service replacement short block assemblies.

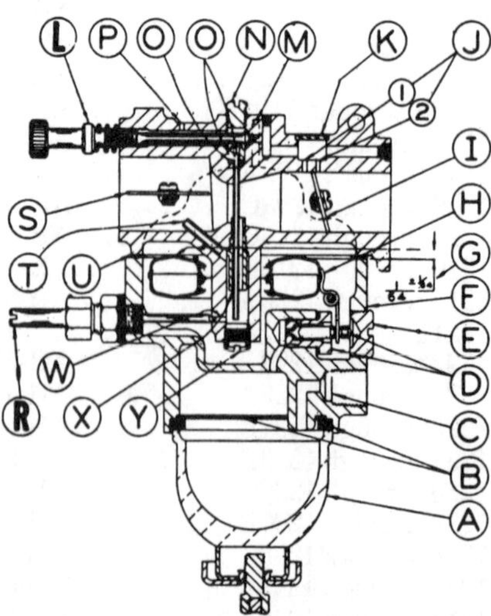

Fig. TP1-2—Section through Tillotson MD series type "B" carburetor utilizing the long idle adjusting needle (L) which leans the mixture when turned clockwise.

MAINTENANCE

SPARK PLUG. Recommended Champion spark plug numbers are shown in the following chart:

AH36 (type 70)	CJ8
AH47, AH58, AH81 (chain saw)	J8J
Go-Kart engines, short races–	
AH51, AH82	J4J
AH61 (bushing engine)	J6J
AH61 (std. & super)	L4J
AH58 with 3/8 inch (9.53 mm) reach	J4J
AH58 (type 1370 & 1371)	L7
AH58 with 1/2 inch (12.70 mm) reach	L4J
Go-Kart engines, long races–	
AH51, AH82	J2J
AH61 (bushing engine)	J4J
AH61 (std. & super)	L57R
AH58 with 3/8 inch (9.53 mm) reach	J2J
AH58 (type 1370 & 1371)	L4J
AH58 with 1/2 inch (12.7 mm) reach	L57R
All other models	J11J

Set electrode gap on L57R spark plugs at 0.020 inch (0.51 mm). Electrode gap should be 0.030 inch (0.76 mm) for all other spark plugs.

Apply a small amount of graphite grease on spark plug threads before installing. Tighten the spark plug to 15-20 ft.-lbs. (20-27 N·m).

If plug has been cleaned, be sure all sand blast grit or other abrasive materials have been removed from plug prior to installation in engine.

CARBURETOR. Several different makes of both float type and diaphragm

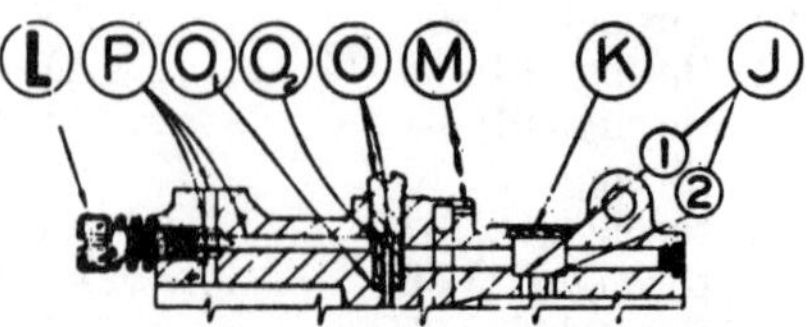

Fig. TP1-3—Some Tillotson Series MD carburetors are fitted with the short air bleed type idle mixture adjusting needle (L) which enriches the mixture when turned clockwise.

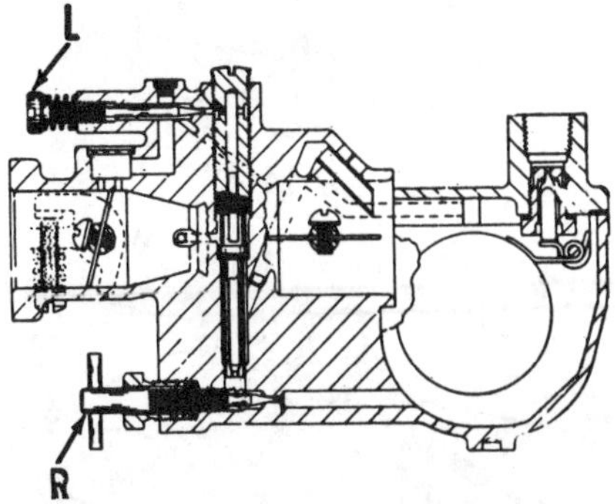

Fig. TP1-4—Section through Tillotson Series MT carburetor showing idle mixture adjusting needle (L) and main needle (R) both of which lean the mixture when turned clockwise.

type carburetors have been used. Refer to the appropriate paragraphs for servicing and adjustment procedure.

Tillotson Float Carburetors. Early Tillotson MD and MT carburetors having the long idle adjustment needle (L–Fig. TP1-2) require clockwise rotation of needle to lean mixture. When equipped with short idle fuel mixture needle (L–Fig. TP1-3) some models require counterclockwise rotation of needle to lean mixture.

On all MD and MT carburetors, clockwise rotation of the main fuel mixture needle (R) leans the fuel mixture. Refer also to Fig. TP1-4.

Some engines use Tillotson MT carburetors with only a high speed fuel circuit. Except that these units do not have an idle adjustment needle nor idle transfer and discharge ports, they are similar to other MT models.

Initial adjustment of both idle and main fuel mixture needles on MD and MT carburetors is 1 turn open from a lightly seated position.

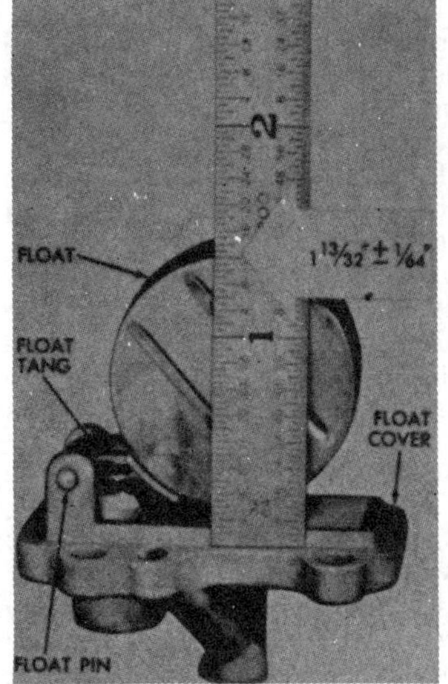

Fig. TP1-5—Checking float level on Tillotson MT carburetor.

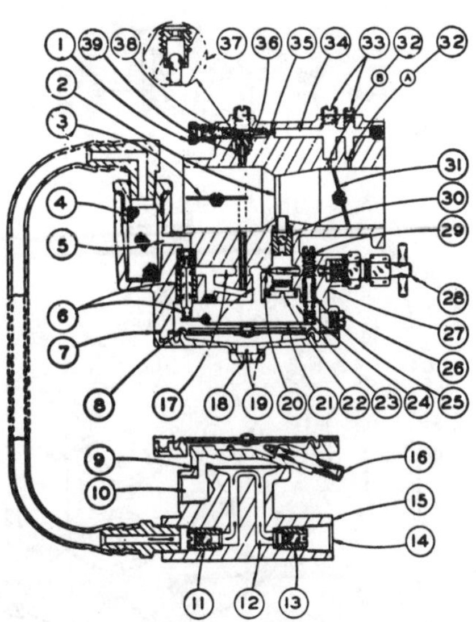

Fig. TP1-6—Cross-sectional view of Tillotson Series H and HP diaphragm type carburetor. Carburetor may not be equipped with fuel pump.

1. Idle needle
2. Venturi
3. Choke plate
4. Fuel inlet
5. Inlet channel
6. Inlet valve
7. Gasket
8. Diaphragm
9. Fuel pump diaphragm
10. Impulse channel
11. Outlet valve
12. Inlet channel
13. Inlet valve
14. Fuel inlet
15. Pump body
16. Flushing plunger
17. Supply channel
18. Vent hole
19. Vent orifices
20. Main fuel orifice
21. Diaphragm lever
22. Plug
23. Diaphragm chamber
24. Tension spring
25. Screw
26. Drain screw
27. Body
28. Main needle
29. Washers
30. Main nozzle
31. Throttle plate
32A. Primary idle port
32B. Secondary idle port
33. Plugs
34. Bypass
35. Idle orifice
36. Idle tube
37. Alternate check valve
38. Idle tube orifice
39. Gasket

Final adjustments are made with engine at operating temperature and running. Adjust main fuel needle by holding throttle halfway open and turning needle in or out to obtain smoothest engine operation; then turn needle to a slightly richer position. Back off idle speed screw and hold throttle in position to obtain slowest engine speed without stalling. Adjust idle needle for smoothest engine operation while holding throttle in this slow idle position. Adjust idle speed screw so engine idles at 1800 rpm. (On MT61 carburetor, no idle fuel adjustment needle is used. Adjust the one needle on this carburetor as described for main fuel needle.)

When disassembling Tillotson MD carburetors, remove main adjusting needle (R–Fig. TP1-2) before attempting to separate carburetor halves to prevent damage to needle.

For adjustment of float level on Tillotson MD carburetor, refer to Fig. TP1-2. Invert bowl (F) and float (H) assembly. With float tang resting against inlet valve, float level (G) measured from face of float opposite inlet valve should be 1/64 inch (0.40 mm) above face (U) of

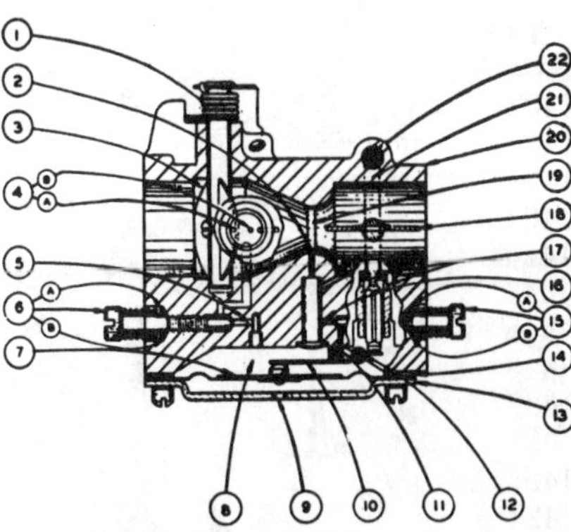

Fig. TP1-7—Cross-sectional view of Tillotson Series HC and HN diaphragm type carburetor.

1. Throttle shaft spring
2. Main fuel port
3. Throttle plate
4A. Primary idle port
4B. Secondary idle port
5. Idle fuel channel & orifice
6. Idle needle
6A. Washer
6B. Packing
7. Metering-diaphragm
8. Fuel chamber
9. Air vent hole
10. Diaphragm lever
11. Spring
12. Fulcrum
13. Cover plate
14. Gasket
15. Main needle
15A. Washer
15B. Packing
16. Main fuel channel & orifice
17. Inlet valve
18. Choke shutter
19. Venturi
20. Metering body
21. Inlet
22. Inlet screen

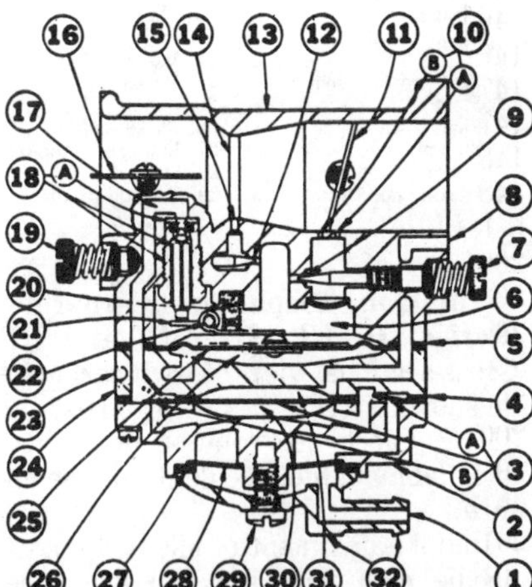

Fig. TP1-8—Cross-sectional view of Tillotson Series HL diaphragm carburetor.

1. Fuel inlet
2. Pump body
3. Pump diaphragm
3A. Pump inlet valve
3B. Pump outlet valve
4. Gasket
5. Gasket
6. Metering chamber
7. Idle needle
8. Impulse channel
9. Idle fuel orifice
10A. Primary idle port
10B. Secondary idle port
11. Throttle plate
12. Main fuel orifice
13. Body
14. Venturi
15. Main fuel port
16. Choke plate
17. Inlet channel
18. Inlet valve
19. Main needle
20. Spring
21. Diaphragm lever
22. Fulcrum pin
23. Vent hole
24. Cover
25. Diaphragm
26. Atmospheric chamber
27. Gasket
28. Screen
29. Screw
30. Fuel chamber
31. Pulse chamber
32. Strainer cover

bowl. If float level is not within specification, remove float and carefully bend tang to correct float level.

Check float level of Tillotson MT carburetors as shown in Fig. TP1-5. If float level is not within limits shown, remove float and bend tang to correct float level.

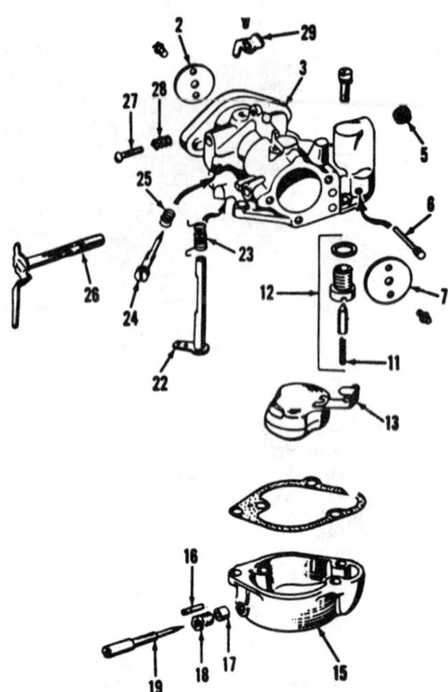

Fig. TP1-9—Exploded view of Marvel-Schebler AH float type carburetor.

2. Throttle plate
3. Body
5. Fuel strainer
6. Float shaft
7. Choke shutter
11. Spring
12. Inlet needle & seat
13. Float
15. Float bowl
16. Stop pin
17. Packing
18. Packing nut
19. Main adjustment needle
22. Throttle shaft
23. Spring
24. Low speed mixture needle
25. Spring
26. Choke shaft
27. Low speed stop screw
28. Spring
29. Throttle stop lever

Fig. TP1-10—Some Tecumseh carburetors use low speed stop screw as shown at (6), others as shown at (61).

H. Main mixture needle
L. Idle mixture needle
1. Throttle shaft
2. Spring
3. Throttle plate
5. Spring
6. Low speed stop screw
7. Choke shaft
8. Choke positioning spring
9. Choke plate
11. Springs
12. Washers
13. "O" rings
14. Cup plug
15. Welch plug
16. Cup plug
17. Welch plug
18. Fuel valve spring
20. Inlet needle & seat
21. Gasket
22. Diaphragm
23. Cover

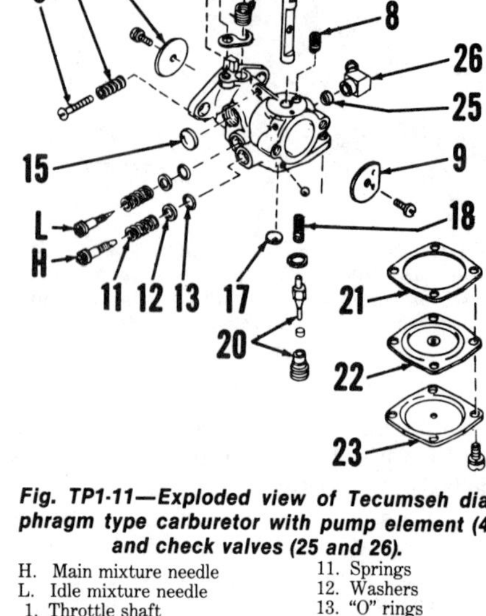

Fig. TP1-11—Exploded view of Tecumseh diaphragm type carburetor with pump element (4) and check valves (25 and 26).

H. Main mixture needle
L. Idle mixture needle
1. Throttle shaft
2. Spring
3. Throttle shutter
4. Pumping element
5. Spring
6. Low speed stop screw
7. Choke shaft
8. Choke positioning spring
9. Choke plate
11. Springs
12. Washers
13. "O" rings
15. Welch plug
17. Welch plug
18. Fuel valve spring
20. Fuel inlet valve
21. Gasket
22. Diaphragm
23. Cover
25. Outlet check valve
26. Inlet check valve

Tillotson Diaphragm Carburetors (Series, H, HC, HL, HN and HP). Cross-sectional view of each carburetor is shown in Figs. TP1-6, TP1-7 and TP1-8. Clockwise rotation of both the idle and main fuel needles leans the mixture.

Initial adjustment of idle fuel mixture needle and main fuel mixture needle from a lightly seated position is 3/4 turn open for idle mixture needle and 1 1/4 turns open for main fuel mixture needle.

Final adjustments are made with engine at operating temperature and running. Operate engine at rated speed and adjust main fuel mixture needle for smoothest engine operation. Back idle speed regulating screw off and hold throttle so engine runs at slowest speed possible without stalling and adjust idle needle for smoothest engine idle operation. Adjust idle speed regulating screw so engine idles at 1800 rpm.

Marvel-Schebler Float Carburetor. Refer to Fig. TP1-9 for identification and exploded view of Marvel-Schebler carburetor. Clockwise rotation of the idle and main fuel mixture needles leans the mixture.

Initial adjustment of idle fuel mixture needle and main fuel mixture needle from a lightly seated position is 1 turn open for idle fuel mixture needle (24) and 1 1/4 turns open for main fuel mixture needle (19).

Final adjustments are made with engine at operating temperature and running. Operate engine at rated speed and adjust main fuel mixture needle for smoothest engine operation. Set engine at lowest idle speed at which engine will run without stalling and adjust idle mixture screw for smoothest engine idle. Adjust low speed stop screw (27) so engine idles at 1800 rpm.

To check float level, invert carburetor throttle body and float assembly and measure from free end of float to float bowl mating surface of carburetor. Measurement should be 3/32 inch (2.38 mm). Carefully bend float lever tang which contacts fuel inlet needle as necessary to obtain correct float level.

Tecumseh Diaphragm Carburetor. Refer to Fig. TP1-10 and TP1-11 for identification and exploded view of Tecumseh diaphragm type carburetors. Clockwise rotation of idle and main fuel mixture needles leans the fuel mixture.

Initial adjustment of idle (L) and main (H) fuel mixture needles with knurled heads from a lightly seated position is 1 turn open for each mixture needle.

Initial adjustment of idle (L) and main (H) fuel mixture needles with slotted heads from a lightly seated position is 5/8 turn open for each mixture needle.

Final adjustments are made on all models with engine at operating temperature and running. Operate engine at rated speed and adjust main fuel mixture needle for smoothest engine operation. Operate engine at lowest idle speed at which engine will run without stalling and adjust idle fuel mixture needle for smoothest engine idle. Adjust engine idle to 1800 rpm at low speed throttle stop screw (6).

The fuel inlet fitting is pressed into carburetor body and fuel strainer behind inlet fitting can be cleaned by reverse flushing with compressed air after inlet needle and seat are removed.

The inlet needle seat fitting is metal with a neoprene seat, so fitting (and enclosed seat) should be removed before carburetor is cleaned with a commercial solvent.

Diaphragm (22) should be installed with head of rivet toward inlet needle regardless of position or size of metal discs. Make certain correct diaphragm is installed.

On carburetor Models 0234-252, 265, 266, 269, 270, 271, 282, 297, 303, 327, 333, 334, 344, 348, 349 and 356 the gasket (21) should be installed between cover (23) and diaphragm (22). Other models should be assembled as shown in Fig. TP1-10 or TP1-11 with gasket between carburetor body and diaphragm.

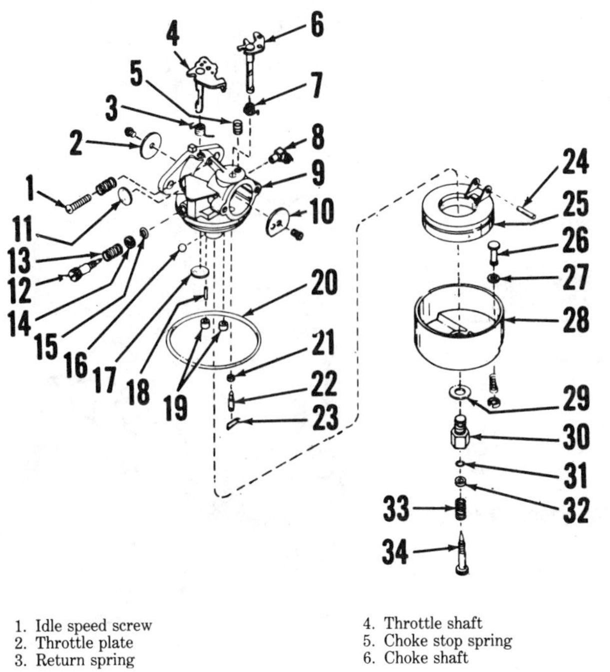

Fig. TP1-13—Exploded view of Tecumseh float type carburetor

1. Idle speed screw
2. Throttle plate
3. Return spring
4. Throttle shaft
5. Choke stop spring
6. Choke shaft
7. Return spring
8. Fuel inlet fitting
9. Carburetor body
10. Choke plate
11. Welch plug
12. Idle mixture needle
13. Spring
14. Washer
15. "O" ring
16. Ball plug
17. Welch plug
18. Pin
19. Cup plugs
20. Bowl gasket
21. Inlet needle seat
22. Inlet needle
23. Clip
24. Float shaft
25. Float
26. Drain stem
27. Gasket
28. Bowl
29. Gasket
30. Bowl retainer
31. "O" ring
32. Washer
33. Spring
34. Main fuel needle

On all models, the throttle plate (3) should be installed with short line stamped on plate to top of carburetor and facing out. The choke plate should be installed with flat toward fuel inlet side of carburetor as shown.

On some models, a fuel pumping element (4–Fig. TP1-11) is used. The element expands and contracts due to changes in crankcase pressure. The pump inlet check valve is located in fuel inlet fitting (26) and outlet check valve (25) is pressed into carburetor body behind the fuel inlet fitting.

To renew the fuel pump valves, clamp inlet fitting in vise and twist carburetor body from fitting. Using a 9/64 inch drill, carefully drill into outlet valve (25) to a depth of 1/8 inch (3.18 mm). Take care not to drill into carburetor body. Thread an 8-32 tap into the outlet valve. Using a proper size nut and flat washer, convert the tap into a puller to remove the outlet valve from carburetor body. Press new outlet valve into carburetor body until face of valve is flush with surrounding base of fuel inlet chamber. Press new inlet fitting about 1/3 of the way into carburetor body, coat the exposed 2/3 of the fitting shoulder with Loctite, then press fitting fully into carburetor body.

Tecumseh Float Type Carburetor. Refer to Fig. TP1-13 for identification and exploded view of Tecumseh float type carburetor. Clockwise rotation of the idle and main fuel mixture needles leans the fuel mixtures.

Initial adjustment of idle (12) and main (34) fuel mixture needles from a lightly seated position is 1 turn open for each mixture needle.

Final adjustments are made with engine at operating temperature and running. Operate engine at rated speed and adjust main fuel mixture needle for smoothest engine operation. Operate engine at lowest idle speed at which engine will run without stalling and adjust idle mixture screw for smoothest engine idle. Adjust idle speed to 1800 rpm at idle speed stop screw (1).

When overhauling, check mixture needles for excessive wear and other damage.

The fuel inlet needle (22) seats against a Viton rubber seat (21) which is pressed into the carburetor body. Remove the rubber seat before cleaning carburetor in a commercial cleaning solvent. The seat should be installed grooved side first.

Install throttle plate (2) with the two stamped lines facing out and at 12 and 3 o'clock positions. Install choke plate (10) with flat side towards bottom of carburetor. Fuel inlet fitting (8) is pressed into the body. When installing the fuel inlet fitting, start the fitting into the bore, then apply a light coat of Loctite to the shank and press the fitting into position.

To check float level, invert carburetor throttle body and float assembly and measure from free end of float to carburetor body and float bowl mating surface. Measurement should be 7/32 inch

Fig. TP1-14—Normal setting of governor is when distance (A) is 3/32 inch (2.38 mm).

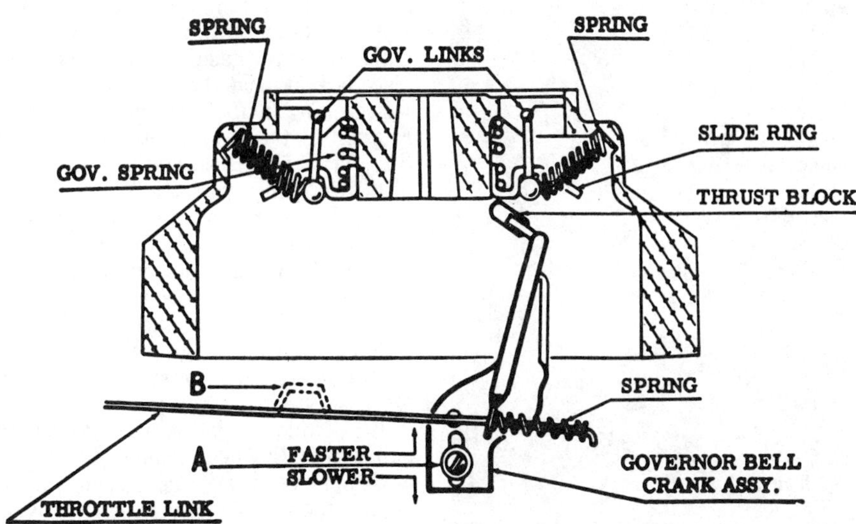

Fig. TP1-15—On some engines the centrifugal governor weight element is mounted in the flywheel. Note speed changes are obtained by raising or lowering the bellcrank assembly as indicated at (A). Bending the throttle link as shown at (B) will increase the engine speed.

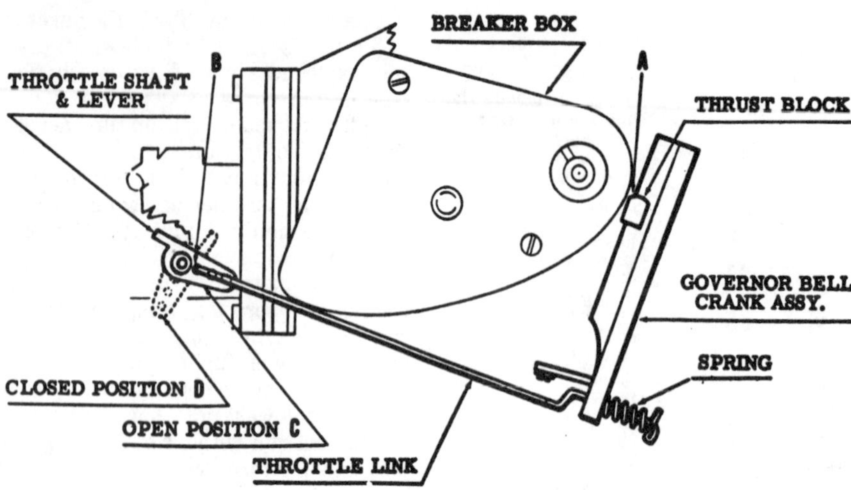

Fig. TP1-16—The gap (A) between the thrust block and breaker box should be within limits 1/32 to 1/16 inch (0.79-1.59 mm) when carburetor throttle is wide open.

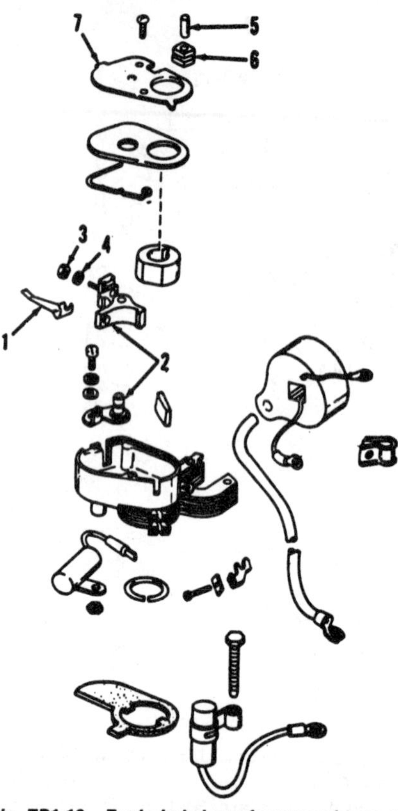

Fig. TP1-18—Exploded view of a second type of ignition cutout switch used with flywheel centrifugal governor unit to limit engine top speed.

(5.56 mm). Carefully bend float tang which contacts fuel inlet needle as necessary to obtain correct float level.

GOVERNOR. Pneumatic (air vane) and centrifugal type governors are used. Some early centrifugal governors were mounted on the pto end of crankshaft. On some later installations, the governor weight elements are in the flywheel. Refer to Figs. TP1-14, TP1-15 and TP1-16.

Refer to Figs. TP1-11 and TP1-20 for hookup and adjustment of air vane type governors.

To change governed engine rpm on engines having a fixed speed centrifugal governor installation like that shown in Fig. TP1-14, first remove the governor housing located on crankcase at drive end of engine. To increase speed, loosen the set screw (GS) which retains the fixed ring of weight unit to crankshaft and move the ring toward the engine. Moving the ring 1/16 inch (1.59 mm) changes the speed approximately 250 rpm. Recommended setting is to have 3/32 inch (2.38 mm) gap at (A) when weight unit is held in compressed positions as shown. The slide ring (R) must move freely on the shaft and bellcrank and linkage to carburetor must be free to allow proper operation of the carburetor throttle.

When governor is mounted in the flywheel, movement of the bellcrank upward as shown at (A) in Fig. TP1-15 increases the speed. A speed increase can also be obtained by bending the throttle link as shown by the dotted lines. Remove the link before bending it.

When carburetor throttle is wide open (Fig. TP1-16) the thrust block must clear the breaker point box by not less than 1/32 inch (0.79 mm) and not more than 1/16 inch (1.59 mm) as indicated at (A). If throttle lever has more than one hole, place link in inside hole as shown.

Beginning in 1959, the governor slide ring of metal with a plastic follower (Fig. TP1-15) was discontinued in favor of a nylon ring with a follower arm of metal. Parts are not interchangeable in service, hence nylon slide rings must be used with metal followers and metal rings with plastic followers.

Occasionally an engine may be encountered with a governor that is operating erratically or sticking in either the low or high speed position. Likely cause of this trouble is a rough center hub in the flywheel around which the governor ring moves up and down. To correct the trouble, dismantle the governor and using emery cloth, smooth out the center hub of the flywheel so as to allow the governor ring (plastic or metal) to move up and down freely.

Fig. TP1-17—View of ignition cutout switch used with centrifugal governor unit mounted under flywheel. Governor operates plunger (P) to short out ignition to limit top speed.

Two different types of ignition cutout switches are used in conjunction with a flywheel mounted centrifugal governor to limit engine top speed by shorting out the ignition.

In Fig. TP1-17, the governor ring contacts the plunger (P) which operates the ignition cut-out switch. The factory setting on this type unit is 4400-5000 rpm. If adjustment is necessary, loosen screw (A) and slide bracket (B) closer to flywheel to increase engine speed, or away from the flywheel to decrease speed.

In Fig. TP1-18, the governor ring contacts the plunger (5) which rides against the cutoff spring (1). The plunger is retained in a nylon block (6) which is held in a notch in the breaker box cover (7). If plunger (5) is removed, reinstall with rounded end towards governor ring. This type speed limiting device is preset at 4500-4700 rpm.

Both types of ignition cutout switches are operated by a centrifugal governor similar to that shown in Fig. TP1-15. Do not adjust either type switch for engine speeds above factory setting.

On some models the flyweight (centrifugal) governor may be the type shown in Fig. TP1-19. To adjust the models, remove the flywheel, magnets and cover (4). Hold bellcrank (7) against slide (2), loosen screw that attaches

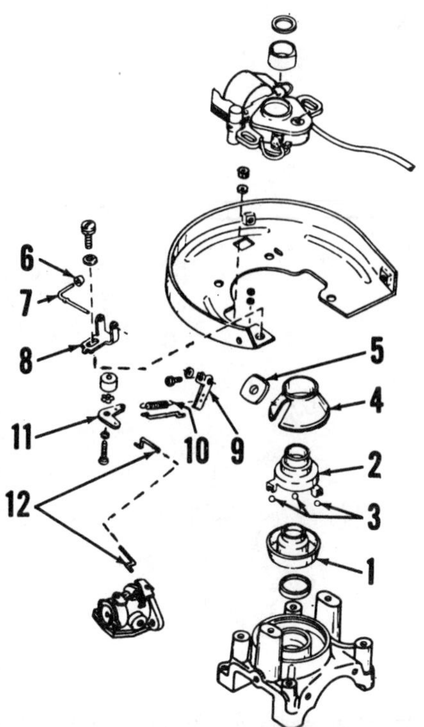

Fig. TP1-19—Exploded view of the centrifugal governor used on some models.

1. Governor cup
2. Slide ring
3. Governor balls
4. Cover
5. Cover seal
6. Follower
7. Governor bellcrank
8. Bracket
9. Lever
10. Governor spring
11. Bellcrank
12. Link

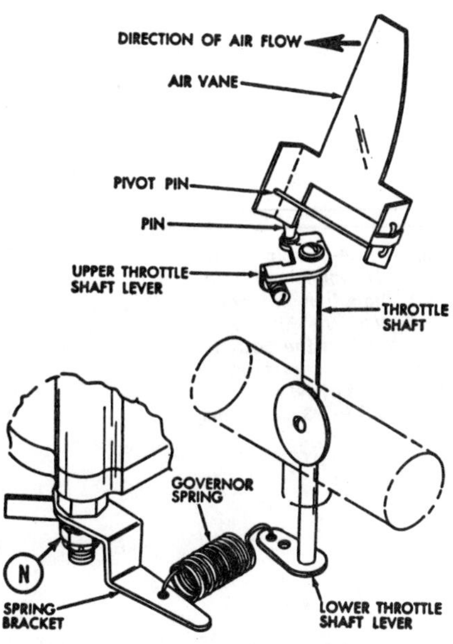
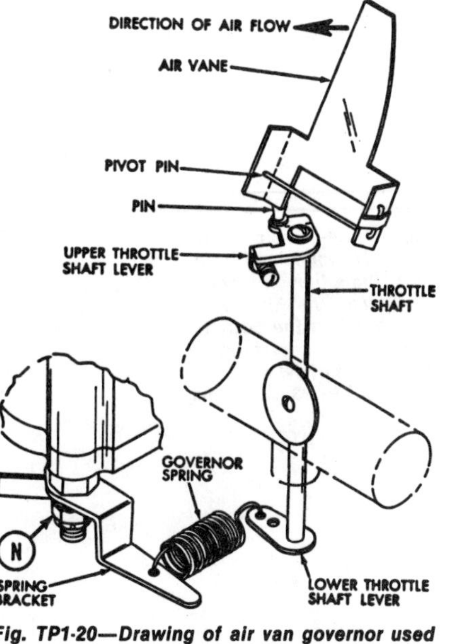

Fig. TP1-20—Drawing of air van governor used on some models. Loosen nut (N) and move the spring bracket to vary governed speed.

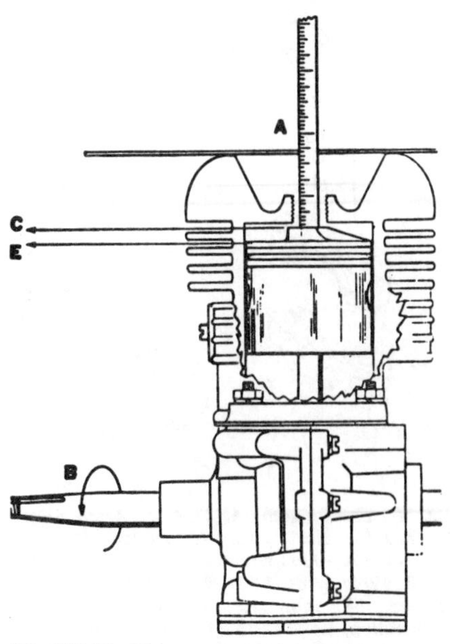

Fig. TP1-21—Recommended ignition timing can be set by measuring piston travel before top dead center (BTDC) as shown.

lever (9) to bellcrank and hold governor lever (9) up to open the carburetor throttle valve, then tighten the lever retaining screw. To check operation, push lever (9) to throttle closed position, then release. The carburetor throttle must return to the full open position. If throttle does not return to full open position, check for binding of the linkage.

The air vane governor shown in Fig. TP1-20 is adjusted by moving the spring bracket after loosening nut (N). On some engine models, the governor air vane is attached to the carburetor throttle shaft (1-Fig. TP1-11).

MAGNETO AND TIMING. Breaker point gap at maximum opening should be set before adjusting the ignition timing. Refer to the following specifications for point gap and ignition timing, piston position before top dead center (BTDC) at which point breaker points should just begin to open.

AH31, AV31:
Point gap . . . . 0.018-0.020 in. (0.46-0.51 mm)
BTDC . . . . 0.156 in. (3.96 mm)

AH36, AV36:
Point gap . . . . 0.018-0.020 in. (0.46-0.51 mm)
BTDC (aluminum rod) . . . . . . 0.156 in. (3.96 mm)
BTDC (needle bearing rod) . . 0.250 in. (6.35 mm)

AV47:
Point gap . . . . 0.015-0.021 in. (0.38-0.53 mm)
BTDC . . . . 0.156 in. (3.96 mm)

AV47 (aluminum rod):
Point gap . . . . 0.015-0.019 in. (0.38-0.48 mm)
BTDC (chain saw) . . . . 0.175 in. (4.45 mm)
BTDC (others) . . . . 0.156 in. (3.96 mm)

AH47 (needle bearing rod):
Point gap . . . . 0.015-0.019 in. (0.38-0.48 mm)
BTDC (super & slim line) . . . . 0.175 in. (4.45 mm)
BTDC (other models) . . . . 0.250 in. (6.35 mm)

AH51, AV51:
Point gap . . . . 0.017-0.023 in. (0.43-0.58 mm)
BTDC . . . . 0.175 in. (4.45 mm)

AH58, AV58:
Point gap . . . . 0.015-0.019 in. (0.38-0.48 mm)
BTDC . . . . 0.095 in. (2.41 mm)

AH61, AV61:
Point gap . . . . 0.015-0.019 in. (0.38-0.48 mm)
BTDC . . . . 0.100 in. (2.54 mm)

AH80, AV80:
Point gap . . . . 0.018-0.020 in. (0.46-0.51 mm)
BTDC (AH80) . . . . 0.125 in. (3.18 mm)
BTDC (AV80) . . . . 0.095 in. (2.41 mm)

AH81, AV81, AH82:
Point gap . . . . 0.015-0.021 in. (0.38-0.53 mm)
BTDC . . . . 0.175 in. (4.45 mm)

Ignition timing should be set by moving magneto stator plate so points just begin to open when piston is correct distance BTDC. Piston position can be measured through spark plug hole with narrow ruled scale as shown in Fig. TP1-21. If timing is not as specified, rotate the stator plate either way after loosening the stator plate mounting screw (or screws).

On late model Tecumseh magneto, a condenser may or may not be used in connection with the breaker points. On models without a condenser, capacitance is built into the magneto coil and a condenser is not required. The magneto coil furnished as repair parts for all Tecumseh magnetos may or may not require the use of a separate condenser. Specific instructions are included with the replacement coil and should be followed carefully.

LUBRICATION. Engines are lubricated by mixing a good quality two-stroke engine oil with regular grade gasoline. For engines operating below 3600 rpm, mix 0.5 pint (0.24 L) of oil with each gallon of gasoline. For engines operating above 3600 rpm, mix 0.75 pint (0.36 L) of oil with each gallon of gasoline.

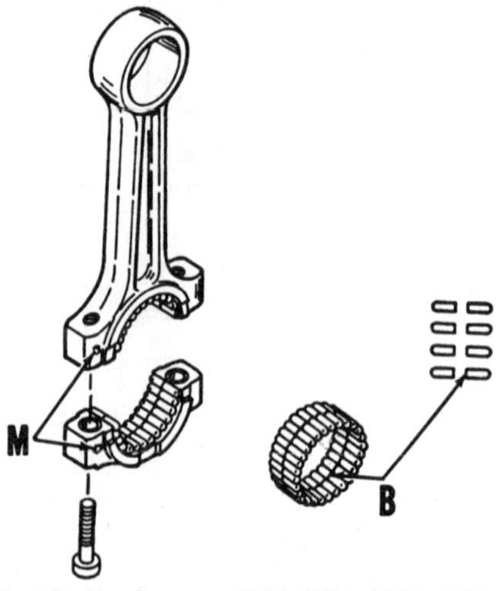

Fig. TP1-22—On some AH31, AV31, AH36, AV36, AH47, AV47, AH80 and AV80 models, 60 short rollers are used in two rows of 30 each. On some AH58, AV58, AH61, AV61, AH81, AV81 and AH82 models, 58 short rollers are used. When assembling, make certain flat ends of rollers are butted together as shown at "B". Match marks (M) on connecting rod and cap must be aligned.

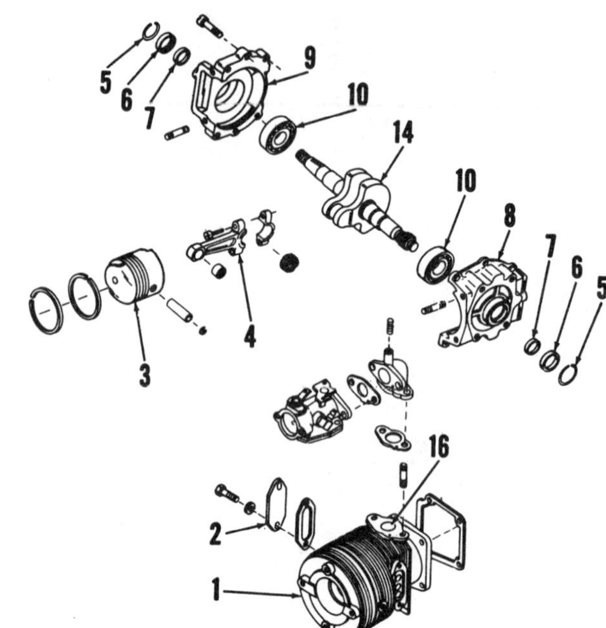

Fig. TP1-24—Exploded view of AH47 engine with third port induction instead of reed valve. The third port (16) is opened as the bottom of the piston skirt moves toward top of cylinder. Refer to Fig. TP1-23 for legend.

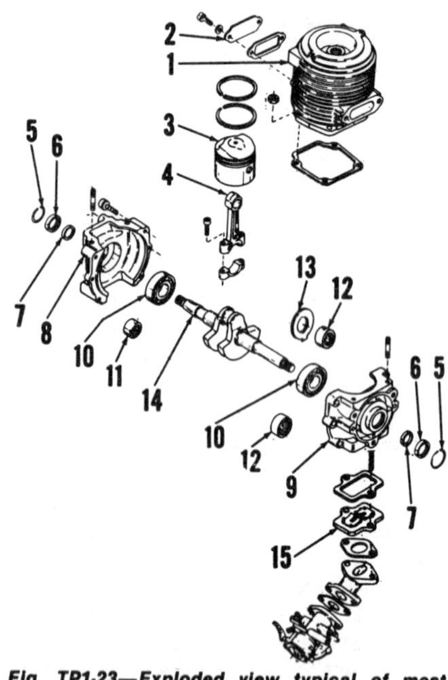

Fig. TP1-23—Exploded view typical of most models with less than 5.1 cubic inch (84 cc) displacement. Crankshaft main bearings (10, 11 & 12) may be needle roller, ball or bushing type. Thrust washer (13) is used at output end of crankshaft on some models with needle roller or bushing type main bearing. Models AH80 and AV80 are similar.

1. Cylinder
2. Transfer port cover
3. Piston
4. Connecting rod
5. Snap ring
6. Seal retainer
7. Crankshaft seal
8. Crankcase (magneto half)
9. Crankcase (pto half)
10. Ball bearings
11. Needle roller bearing
12. Needle roller bearing
13. Thrust washer
14. Crankshaft
15. Reed valve

CARBON. Muffler and exhaust ports should be cleaned after every 50 hours of operation if engine is operated continuously at full load. If operated at light or medium load, the cleaning interval can be extended to 100 hours.

REPAIRS

CONNECTING ROD. Piston and connecting rod are removed as a unit with the crankshaft after removing the cylinders and separating the halves of the crankcase.

Bronze and aluminum connecting rods ride directly on the crankshaft crankpins. Forged steel rods have needle roller bearings between connecting rod bearing bore and crankpin.

Recommended plain connecting rod bearing running clearance on crankpin is obtained when the connecting rod (without piston) is bolted to the crankpin. It should not fall of its own weight when extended horizontally. Excessive running clearance of rod lower bearings is preferably corrected by renewal of the affected parts. In an emergency, the cap may be lapped slightly but in no case should more than 0.002 inch (0.05 mm) be removed from cap. No undersize or oversize engine parts are available.

On AH31, AV31, AH36, AV36, AH47, AH80 and AV80 models with needle roller crankpin bearings, 28 long rollers or 56 short rollers are used. The short rollers are latest type and can be used when renewing the long rollers.

When installing short rollers, two rows are used and flat ends of rollers should be toward middle of crankpin. The crankpin needle rollers should be renewed as a set if any roller is damaged. If rollers are damaged, be sure to check condition of crankpin and connecting rod carefully and renew if damaged.

New rollers are serviced in a strip and can be installed by wrapping the strip around crankpin. After new needle rollers and connecting rod cap are installed, force lacquer thinner into needles to remove the beeswax, then lubricate rollers with SAE 30W oil.

On all models, make certain match marks on connecting rod and cap are aligned. On vertical shaft models, make certain lubrication hole in side of connecting rod is toward top.

On all models with deflector type piston, the long sloping side of piston head should be towards the exhaust port.

Tighten the connecting rod cap retaining screws to 40-50 in.-lbs. (5-6 N·m) on bronze and aluminum connecting rods and to 70-80 in.-lbs. (8-9 N·m) on steel connecting rod.

PISTON, PIN AND RINGS. Pistons are equipped with either 3/32 inch (2.38 mm) thick or 1/16 inch (1.59 mm) thick piston rings. On all models, the top and second rings are interchangeable. Rings should be installed on piston with beveled inner edge towards top of piston.

Recommended ring end gap is 0.005-0.010 inch (0.13-0.25 mm) for Models AH31, AV31, AH36, BH60, BV60, BH69, BV69, and 0.006-0.011 inch (0.15-0.28 mm) for AH47, AV47, AH51, AV51, AH58, AV58, AH81, AV81 and AH82.

Piston ring side clearance in groove should be 0.0015-0.003 inch (0.04-0.08 mm) for all models with 3/32 inch (2.38 mm) rings. On models with 1/16 inch (1.59 mm) rings, side clearance should be 0.003-0.005 inch (0.08-0.13 mm) in top groove and 0.002-0.004 inch (0.05-0.10 mm) in second groove.

On models with pin in ring groove, make certain ends of ring correctly engage pin when installing cylinders.

On all models the piston pin is retained by snap rings located in the piston pin bores. On all engines which are equipped with needle bearing in upper end of rod, the piston pin should be a press fit in heated piston. On engines without needle bearing in connecting rod piston pin bore, the pin should be a palm push fit in the piston and a thumb push fit in the connecting rod pin bore.

Pistons are made of aluminum alloy. Renew the piston and/or cylinder if skirt-to-cylinder clearance exceeds 0.003 inch (0.08 mm).

On all models except Models AH58, AV58, AH61, AV61, AH81, AV81 and AH82, piston should be assembled to crankshaft with the sloping or exhaust side of the piston head opposite the tapered end of the crankshaft.

On Models AH58, AV58, AH61, AH81 AV81 and AH82 with pinned rings, ends of rings should be away from exhaust port.

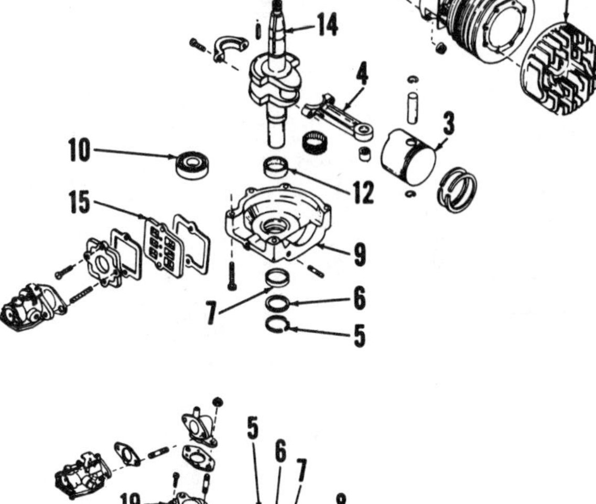

Fig. TP1-25—Exploded view typical of most 5.8 and 6.1 cubic inch (95 and 100 cc) models. Crankshaft main bearings (10, 11 and 12) may be needle roller or ball type. Seal (17) is pressed in on some models instead of retainer, snap ring and seal (5, 6 and 7).

1. Cylinder
3. Piston
4. Connecting rod
5. Snap ring
6. Seal retainer
7. Crankshaft seal
8. Crankcase (magneto half)
9. Crankcase (pto half)
10. Ball bearings
11. Needle roller bearing
12. Needle roller bearing
14. Crankshaft
15. Reed valve
17. Seal
18. Cylinder head

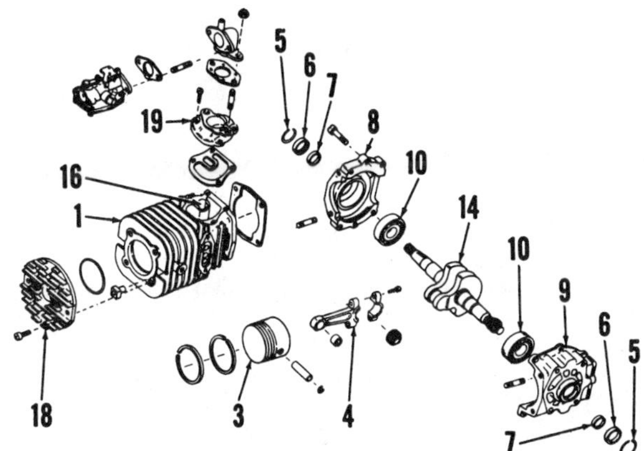

Fig. TP1-26—Exploded view of AH58 engine with third port induction instead of reed valve. The third port (16) is opened as bottom of piston skirt moves toward top of cylinder. Gasket between cover (19) and cylinder must seal around the third port as well as the crankcase opening. Refer to Fig. TP1-25 for legend.

CRANKSHAFT AND CRANKCASE. Crankshaft is induction hardened and rides in two main bearings. Some engines have ball bearings, some have needle type and on others one main may be roller type and the other a plain bushing.

To remove crankshaft it is necessary to separate the crankcase halves and bump shaft with soft mallet.

Crankshaft journal diameters are as follows:

Main Bearing Journal Diameters

AH31, AV31, AH36,
AV36, AH47, AV47,
AH51, AV51, AH58,
AV58, AH61, AV61,
BH60, BV60,
BH69, BV69:
Ball type . . . . 0.6689-0.6695 in. (16.990-17.005 mm)
Needle roller type:
Magneto end . . . . . 0.7495-0.7505 in. ((19.037-19.663 mm)
Pto end . . . . 0.9995-1.0000 in. (25.387-25.400 mm)
Bushing type . . . . . . 0.9995-1.0000 in. (25.387-25.400 mm)
AH80, AV80:
Ball type . . . . 0.9839-0.9842 in. (24.991-24.999 mm)
AH81, AV81, AH82:
Magneto end (needle bearing) . . 0.7498-0.7501 in. (19.045-19.053 mm)
Pto end (ball bearing) . . . . . 0.7871-0.7875 in. (19.992-20.003 mm)

Crankpin Journal Diameter

AH31, AV31, AH36,
AV36, AH47, AV47,
AH51, AV51, BH60,
BV60, BH69, BV69:
Aluminum or bronze rod . . . . 0.6860-0.6865 in. (17.424-17.437 mm)
Needle bearing rod . . 0.5615-0.5618 in. (14.262-14.270 mm)
AH58, AV58:
Aluminum rod . . . . . 0.8115-0.8120 in. (20.612-20.625 mm)
Needle bearing rod . . 0.7499-0.7502 in. (19.048-19.055 mm)
AH61, AV61:
Aluminum rod . . . . . 0.8740-0.8745 in. (22.200-22.212 mm)
Needle bearing rod . . 0.7499-0.7502 in. (19.048-19.055 mm)
AH80, AV80:
Needle bearing rod . . 0.8096-0.8099 in. (20.564-20.572 mm)
AH81, AV81, AH82:
Needle bearing rod . . 0.7499-0.7502 in. (19.048-19.055 mm)

Outer races of ball and needle roller main bearings should be a tight fit in crankcase. If either bearing is loose in its seat, the bearings and crankcase assembly should be renewed.

If new ball type main bearings are to be installed, heat the crankcase when removing the old bearings and installing new ones.

CAUTION: On some engines, crankcase is made of cast magnesium; therefore, when heating crankcase do not use an open flame.

Case should be heated to such temperature as will permit new cold bearings to drop into position freely.

Ball and needle roller type main bearings should be installed with printed face of the bearing race toward the center of the engine.

Main bearing oil seals must be maintained in good condition in a two-stroke engine because leakage through the seals releases compression and causes loss of power. It is important, therefore, to exercise extreme care when renewing seals to prevent their being damaged during installation. If a sleeve is not available, use tape to cover any splines, keyways, shoulders or threads over which the seal must pass during installation.

On all models, except outboard motors, both seals should be installed

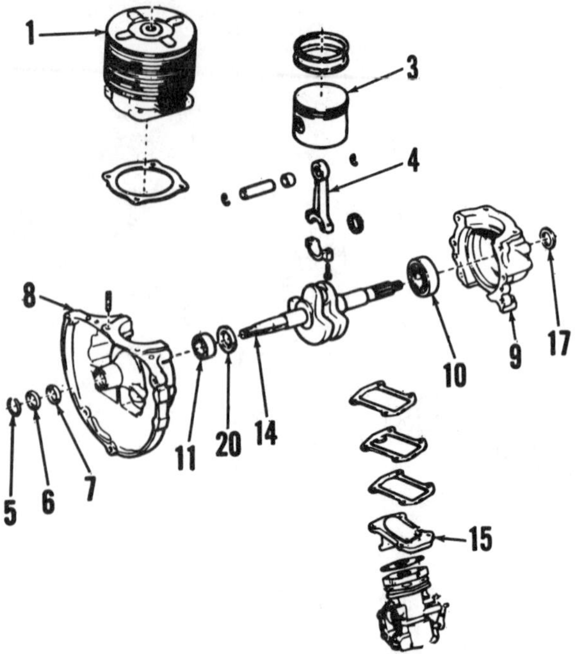

Fig. TP1-27—Exploded view typical of AH81, AV81 and AH82 models. Thrust washer (20) is used only on early models.

1. Cylinder
3. Piston
4. Connecting rod
5. Snap ring
6. Seal retainer
7. Crankshaft seal
8. Crankcase (magneto half)
9. Crankcase (pto half)
10. Ball bearing
11. Needle roller bearing
14. Crankshaft
15. Reed valve
17. Seal
20. Thrust washer

with the channel groove of seal towards the inside (center of engine). On engines used on outboard motors, the crankshaft lower (pto end) seal should be installed with lip towards outside of crankcase.

If a crankshaft thrust washer (13–Fig. TP1-23) is used on Models AH47, AV47, AH51 or AV51, it should be installed on the output (pto) end of crankshaft with grooved side toward crankshaft.

Before assembling the crankcase halves, carefully clean and inspect the mating surfaces of halves. Mating surfaces can be polished to remove slight imperfections, but the surface MUST NOT BE LOWERED. Crankcase halves are matched assembly and must not be interchanged with a half from another engine. Install one half of crankcase on the crankshaft, then coat the mating surface of other half with gasket sealer and position over other end of crankshaft. Before tightening the screws or stud nuts that attach halves together, make certain joint that seals to the cylinder is perfectly flat.

REED VALVES. Reed type inlet valves are used on most of the engines in this section. Valves may have one reed petal (1–Fig. TP1-28), four reed petals (4), six reed petals (6) or twelve reed petals (12). The twelve-petal reed valve (12–Fig. TP1-28 and TP1-29) is used on both single and dual carburetor models.

Reed petals should not stand out more than 0.010 inch (0.25 mm) from the reed plate and should not be bent, distorted or cracked. The reed plate must be smooth and flat. Renew petals or complete valve assembly if valve does not seal completely.

OPTIONAL EQUIPMENT. Some optional (high speed) equipment is available from Tecumseh.

Cylinder and Piston Kit. Cylinder and piston kit is available for AH81 and AH82 engines. The exhaust port is one large opening and rings are pinned to prevent end of ring from catching in exhaust port. Part number of complete kit including cylinder to crankcase gasket, piston, piston pin, rings and cylinder 730103.

Pistons. Pistons with narrow pinned rings are available for Models AH58, AH61, AH81 and AH82 engines. Part number of complete kit including piston, piston pin, rings and cylinder to crankcase gasket is 730114 for Models AH58 and AH81 and 730107 for Models AH81 and AH82.

Lightened Rod. A lightened connecting rod is available for Models AH58, AH61, AH81 and AH82. Holes are drilled through the I-beam section of rod. Part number for kit which includes rod, cylinder to crankcase gasket, piston pin needle bearing and crankpin needle bearings is 730110.

Twelve-Petal Reed Valve. Special 12-petal reed valve assembly is available for AH81 and AH82 engines. Kit for single carburetor installation, including reed valve assembly, manifold, fuel line, gaskets and carburetor fuel pump body etc., to convert Tillotson diaphragm carburetor from inside pulsation chamber to outside hose connection, is part number 730106.

Kit for dual carburetor installation is shown exploded in Fig. TP1-29. Kit part number 730105 includes one complete Tillotson carburetor, necessary parts to

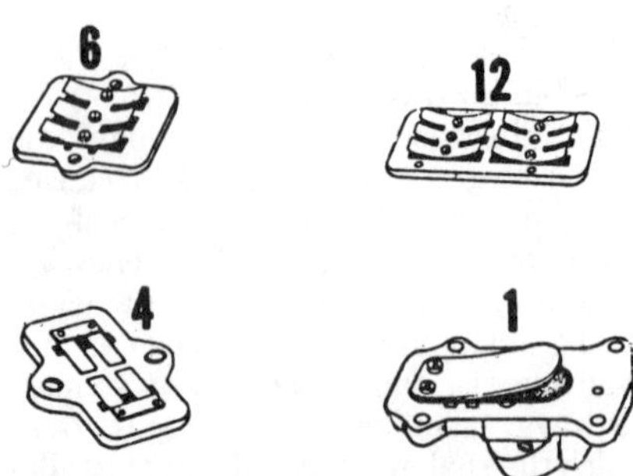

Fig. TP1-28—Reed valve may have one petal (1), four petals (4), six petals (6) or twelve petals (12). Reed plate is not available separately.

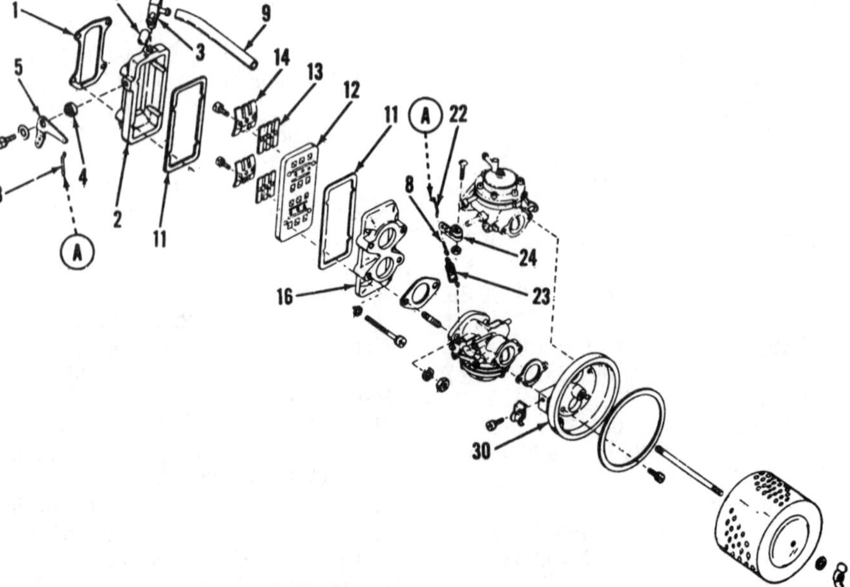

Fig. TP1-29—Exploded view of the twelve petal reed valve assembly with dual carburetors which is available for AH81 and AH82 models.

convert present Tillotson carburetor inside pulsation chamber to outside hose connection, 12-petal reed plate, manifold, air cleaner, linkage and gaskets. Engine starts on one carburetor, and second carburetor cuts in at approximately 4500 rpm.

Stroked Crankshaft. Crankshaft kits including gaskets and spacers are available to increase displacement of AH58 engines to 6.1 cubic inches (100 cc). Kit part 730122 is used for counterclockwise rotation engines and 730123 is used for engines turning clockwise.

TECUMSEH 2-STROKE

UNIT BLOCK MODELS

Model	Bore	Stroke	Displacement
AH440	2.09 in. (50.8 mm)	1.25 in. (31.8 mm)	4.30 cu. in. (65 cc)
AH480	2.09 in. (50.8 mm)	1.41 in. (35.8 mm)	4.85 cu. in. (73 cc)
AH490	2.09 in. (50.8 mm)	1.41 in. (35.8 mm)	4.85 cu. in. (73 cc)
AH520, AV520	2.09 in. (50.8 mm)	1.50 in. (38.1 mm)	5.16 cu. in. (77 cc)
AV600, TVS600	2.09 in. (50.8 mm)	1.75 in. (44.5 mm)	6.00 cu. in. (90 cc)
AH750, AV750	2.38 in. (60.3 mm)	1.68 in. (42.7 mm)	7.50 cu. in. (122 cc)
AH817, AV817	2.44 in. (62.0 mm)	1.75 in. (44.5 mm)	8.17 cu. in. (134 cc)

ENGINE IDENTIFICATION

All models are two-stroke, single cylinder air-cooled engines with a unit block type construction.

Prefixes before type and serial number indicate the following information.

T-Tecumseh
A-Aluminum
V, VS-Vertical crankshaft
H-Horizontal crankshaft

Engine prefix, type number and serial numbers are located as shown in Fig. TP2-1. Figs. TP2-1A and TP2-1B show number interpretations.

Always furnish engine model, type and serial numbers when ordering parts.

MAINTENANCE

SPARK PLUG. Recommended spark plug for all engines installed on lawnmowers and snowblowers is a Champion J17LM or equivalent. Recommended spark plug for all other applications is a Champion CJ8 or J8J for Models AH440, AH480, AH290, AH520, AV520, AV600 and TVS600; UJ12Y for Model AH750; UJ10Y for Model AV750; J13Y for Models AH817 and AV817.

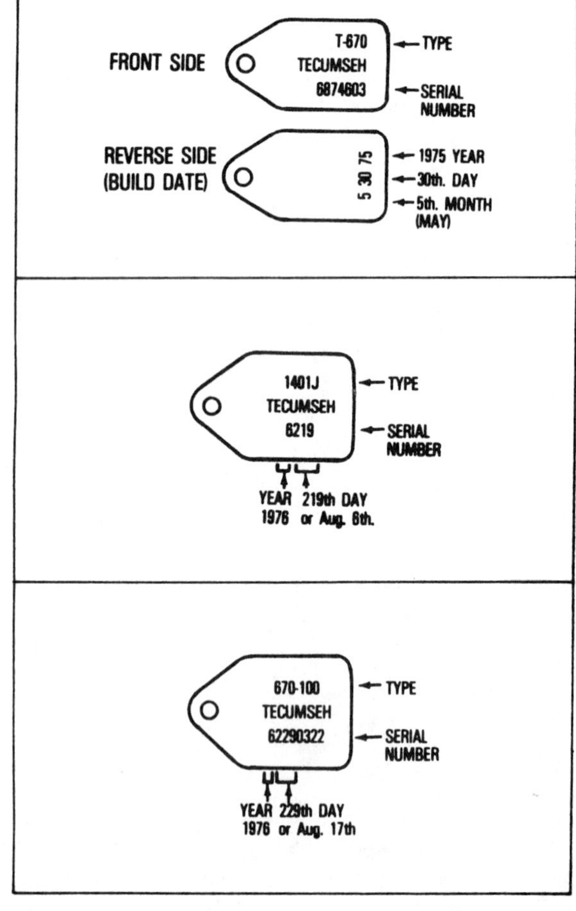

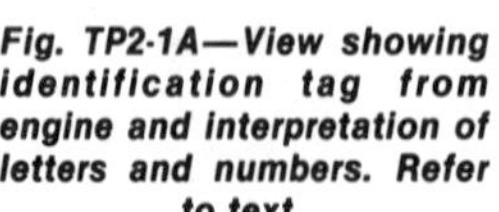
Fig. TP2-1A—View showing identification tag from engine and interpretation of letters and numbers. Refer to text.

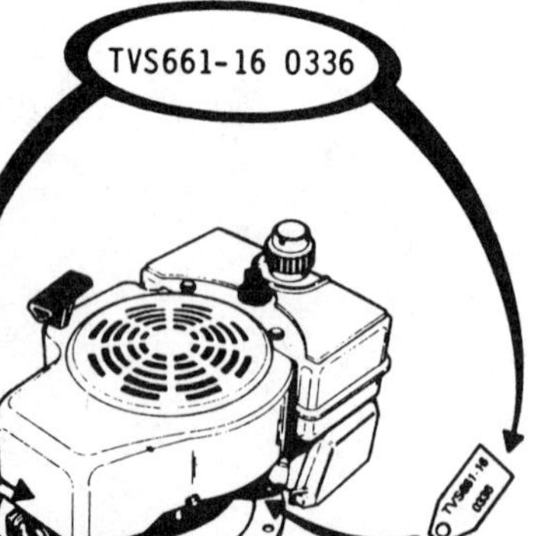

Fig. TP2-1—View showing location of engine model, type and serial number. Refer also to Figs. TP2-1A and TP2-1B.

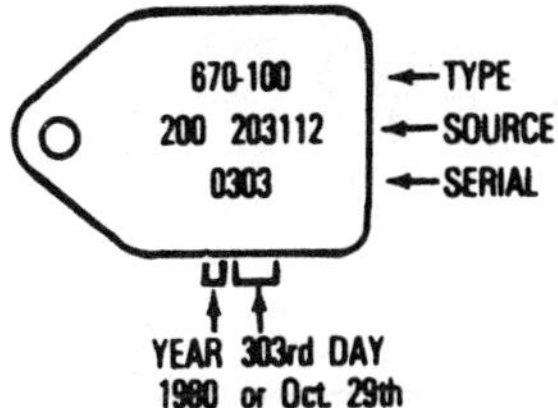

Fig. TP2-1B—View of identification tag from replacement short block and interpretation of letters and numbers.

Electrode gap should be 0.030 inch (0.76 mm) for Model TVS600 and Model AH520 with type numbers 1600 through 1617. Electrode gap for all other models is 0.035 inch (0.89 mm).

CARBURETOR. Tillotson HS, Tecumseh diaphragm and Tecumseh float type carburetors are used. Refer to the appropriate following paragraphs for information on specific carburetors.

Tillotson HS Type Carburetor. Refer to Fig. TP2-1C for identification and exploded view of Tillotson HS type carburetor.

Initial adjustment of idle and main fuel mixture screws from a lightly seated position is 1 turn open for idle mixture screw (6) and 1¼ turns open for main fuel mixture screw (5).

Final adjustments are made with engine at operating temperature and running. Operate engine at idle speed and adjust idle mixture screw to obtain smoothest engine idle. Operate engine at rated speed, under a load and adjust main fuel mixture screw for smooth engine operation. If engine fails to accelerate smoothly, it may be necessary to open idle mixture screw slightly past previous setting until engine accelerates properly.

Tecumseh Diaphragm Type Carburetor. Refer to Fig. TP2-2 for identification and exploded view of Tecumseh diaphragm type carburetor. Carburetor identification numbers are also stamped on carburetor as shown in Fig. TP2-2A.

Initial adjustment of idle mixture and main fuel mixture screws from a lightly seated position for models with solid adjustment (L and H) is 1 turn open for each mixture screw. There is no initial adjustment for idle and main fuel mixture screws which have a hole in inner end as hole size determines initial idle fuel mixture. When renewing adjustment screws, make certain hole diameter of replacement screw matches hole diameter of new screw.

Final adjustments are made with engine at operating temperature and running. Operate engine at idle speed and adjust idle mixture screw to obtain smoothest engine idle. Operate engine at rated speed, under load and adjust main fuel mixture screw for smooth engine operation. If engine cannot be operated under load, adjust main fuel mixture needle so engine runs smoothly at rated rpm and accelerates properly when throttle is opened quickly.

When overhauling, observe the following: The fuel inlet fitting is pressed into bore of body of some models. On these models, the fuel strainer behind inlet fitting can be cleaned by reverse flushing with compressed air after inlet needle and seat are removed. The inlet needle seat fitting is metal with a neoprene seat, so fitting (and enclosed seat) should be removed before carburetor is cleaned with a commercial solvent. The throttle plate (3) should be installed with short line stamped on plate to top of carburetor and facing out. The choke plate (9) should be installed with flat toward fuel inlet side of carburetor as shown.

When installing diaphragm (18), head of rivet should be against fuel inlet valve (16) regardless of size or placement of washers around rivet. On carburetor Models 0234-252, 265, 266, 269, 270, 271, 282, 297, 303, 322, 327, 333, 334, 344, 345, 348, 349, 350, 351, 352, 356, 368, 371, 375, 378, 379, 380, 404, 405 and 441 and all models with "F" embossed on carburetor body, gasket (17–Fig. TP2-2) must be installed between diaphragm (18) and cover (19). All other models are assembled with gasket, diaphragm and cover positioned as shown in Fig. TP2-2.

On models equipped with a fuel pump, pumping element is a rubber boot (20) which expands and contracts due to changes in crankcase pressure. The pump inlet check valve is located in the fuel inlet fitting (11). The pump outlet check valve (10) is pressed into the car-

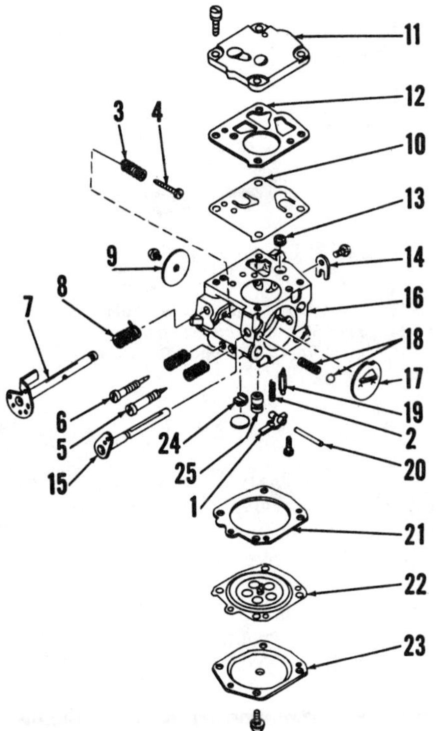

Fig. TP2-1C—Exploded view of Tillotson HS diaphragm type carburetor used on some models.

1. Inlet control lever
2. Spring
3. Spring
4. Idle speed stop screw
5. High speed mixture screw
6. Idle mixture needle
7. Throttle shaft
8. Spring
9. Throttle plate
10. Fuel pump diaphragm
11. Pump cover
12. Gasket
13. Fuel screen
14. Throttle shaft
15. Choke shaft
16. Carburetor body
17. Choke plate
18. Choke detent
19. Inlet needle
20. Lever pin
21. Gasket
22. Diaphragm
23. Cover
24. Channel reducer
25. Main nozzle & check ball

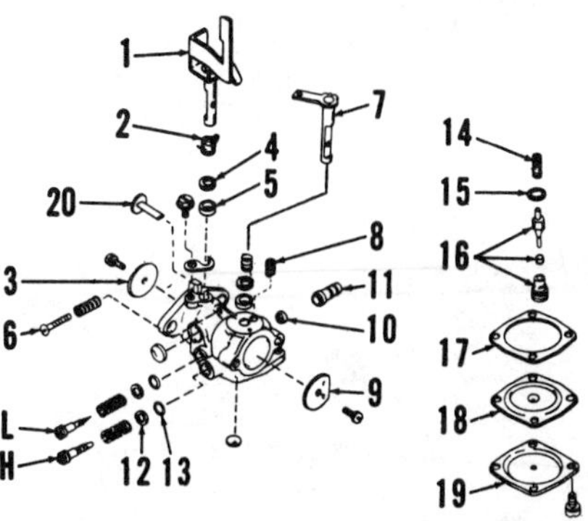

Fig. TP2-2—Exploded view of typical Tecumseh diaphragm type carburetor. Some models do not use fuel pump (20) and check valve (10).

H. High speed mixture screw
L. Idle mixture screw
1. Throttle shaft
2. Spring
3. Throttle plate
4. Felt washer
5. Flat washer
6. Idle stop screw
7. Choke shaft
8. Choke retainer
9. Choke plate
10. Outlet check valve
11. Inlet fitting
12. Washer
13. "O" ring
14. Spring
15. Gasket
16. Inlet valve
17. Gasket
18. Diaphragm
19. Cover
20. Pump element

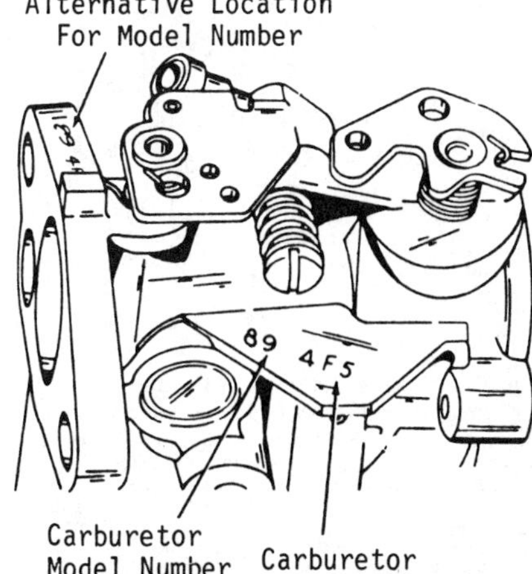

Fig. TP2-2A—View showing location of carburetor identification number on Tecumseh diaphragm type carburetor.

Fig. TP2-2B—The fuel pumping element should be installed at 45° angle as shown for Tecumseh diaphragm type carburetor (left) and float type carburetor (right).

buretor body behind the fuel inlet fitting. Engines equipped with this carburetor will operate in any position and the pump will deliver fuel to the carburetor when the fuel supply is below the carburetor.

NOTE: The fuel pumping element should be installed at 45° angle as shown in Fig. TP2-2B. Incorrect installation may interfere with pumping action.

Two types of fuel pump valves are used. Flap type of valve may be located behind plate attached to side of carburetor body. Renew flap type valves after detaching the plate from side of carburetor. Fuel pump valves are pushed into carburetor body of some models (Fig. TP2-2). Clamp the inlet fitting (11) in a vise and twist carburetor body from fitting. Using a 9/64 inch drill, carefully drill into outlet valve (10) to a depth of 1/8 inch (3.18 mm). Take care not to drill into carburetor body. Thread an 8-32 tap into the outlet valve. Using a proper size nut and flat washer, convert the tap into a puller to remove the outlet valve from carburetor body. Press new outlet valve into carburetor body until face of valve is flush with surrounding base of fuel inlet chamber. Press new inlet fitting about 1/3 of the way into carburetor body, coat the exposed 2/3 of the fitting shoulder with Loctite, then press fitting fully into carburetor body.

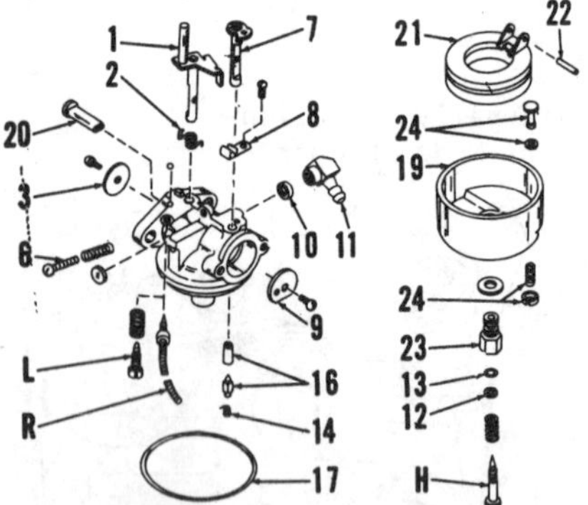

Fig. TP2-3—Exploded view of typical Tecumseh float type carburetor used on some models.

- H. High speed mixture screw
- L. Idle mixture screw
- 1. Throttle shaft
- 2. Spring
- 3. Throttle plate
- 6. Idle stop screw
- 7. Choke shaft
- 8. Choke retainer
- 9. Choke plate
- 10. Outlet check valve
- 11. Inlet fitting
- 12. Brass washer
- 13. "O" ring
- 14. Spring
- 16. Fuel inlet needle & seat
- 17. Seal
- 19. Fuel bowl
- 20. Pumping element
- 21. Float
- 22. Pivot pin
- 23. Bowl retaining nut
- 24. Bowl drain

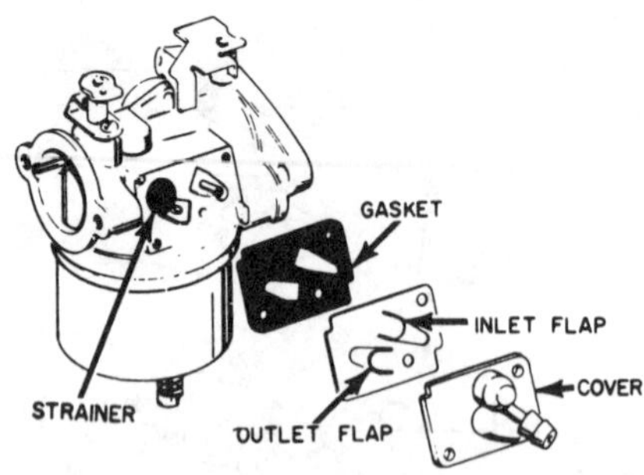

Fig. TP2-4—View of late type fuel valves. Earlier type is pressed into carburetor body as shown in Figs. TP2-2 and TP2-3.

Tecumseh Float Type Carburetor. Refer to Fig. TP2-3 for identification and exploded view of Tecumseh float type carburetor.

Initial adjustment of idle and main fuel mixture screws from a lightly seated position is 1 turn open for each screw.

Final adjustments are made with engine at operating temperature and running. Operate engine at rated speed and adjust main fuel mixture screw (H) for smoothest engine operation. Operate engine at idle speed and adjust idle mixture screw for smoothest engine idle. Set idle speed at idle stop screw (6). If engine fails to accelerate smoothly, slight adjustment of main fuel mixture screw may be necessary.

When overhauling, check adjustment screws for excessive wear or other damage. The inlet valve fuel needle seats against a Viton rubber seat (16) which is pressed into carburetor body. Remove the rubber seat before cleaning carburetor in commercial cleaning solvent. The seat is removed using a 10-32 or 10-24 tap and must be renewed after removing. Install new seat using a punch that will fit into bore of seat and is large enough to catch the shoulder inside the seat. Drive the fuel inlet needle seat into bore until the Viton rubber seat is against bottom of bore. Install throttle plate (3) with the two stamped lines facing out and at 12 and 3 o'clock positions. Install choke plate (9) with flat side toward bottom of carburetor.

Float setting should be 7/32 inch (5.56 mm), measured with body and float assembly in inverted position, between free end of float and rim on carburetor body.

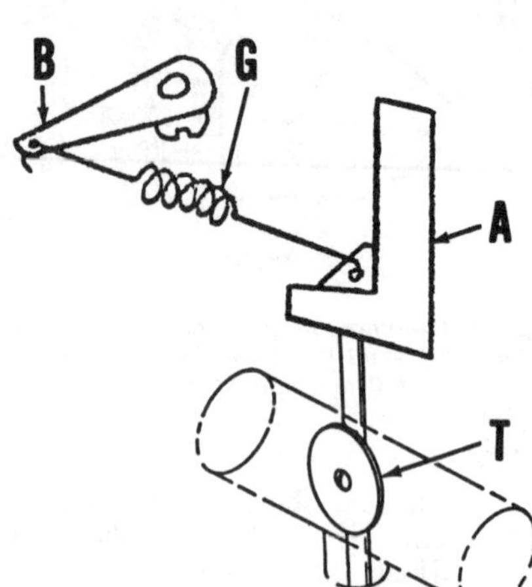

Fig. TP2-5—On some models, the governor air vane (A) is attached to the carburetor throttle shaft and tension of governor spring (G) holds throttle (T) open.

On some models, the fuel inlet fitting (11) is pressed into body. When installing and fitting, start fitting into bore, then apply a light coat of Loctite to the shank and press the fitting into position.

When installing float bowl (19), make certain that correct "O" ring (17) is used. Some "O" rings are round section, others are square.

Fuel hole and the annular groove in retaining nut (23) must be clean. The flat stepped section of fuel bowl (19) should be below the fuel inlet fitting (11). Tighten retaining nut (23) to 50-60 in.-lbs. (6-7 N·m). The high speed mixture screw (H) must not be installed when tightening nut (23).

Some models are equipped with a fuel pump. The fuel pumping element (20) is a rubber boot which expands and contracts due to changes in crankcase pressure. The pumping element should be at 45° as shown in Fig. TP2-2B. Incorrect installation may interfere with pumping action.

On early models, fuel is drawn in through a check valve in the fuel inlet fitting (11). The outlet check valve (10) is pressed into bore behind the inlet fitting. The inlet fitting (11) is removed and installed in normal manner. To renew the outlet check valve (10) drill into the outlet check valve with a 9/64 inch drill to a depth of 1/8 inch (3.78 mm). Do not drill into carburetor body. Thread an 8-32 tap into the outlet valve and pull valve from the carburetor body. Press new outlet valve into carburetor body until face of valve is flush with base of the fuel chamber.

TVS Carburetor. Refer to Fig. TP2-6 for identification and exploded view of TVS carburetor. No idle or main fuel adjustment screws are provided on one

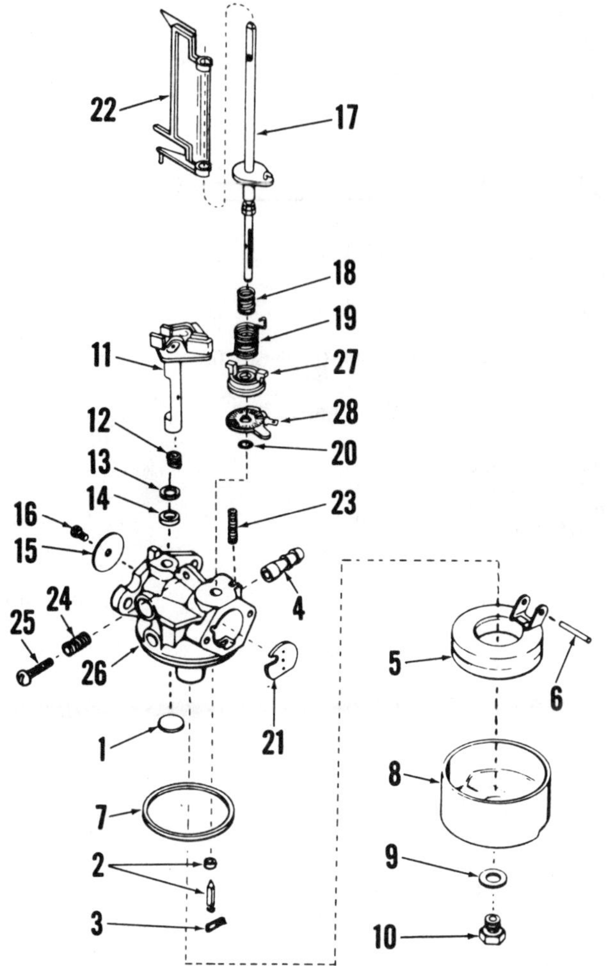

Fig. TP2-6—Exploded view of typical TVS type carburetor.

1. Welch plug
2. Fuel inlet valve & seat
3. Clip
4. Fuel inlet fitting
5. Float
6. Float shaft
7. Bowl gasket
8. Float bowl
9. Gasket
10. Bowl retainer
11. Throttle shaft
12. Return spring
13. Washer
14. Seal
15. Throttle plate
16. Screw
17. Choke/governor shaft
18. Spring
19. Governor spring
20. Washer
21. Choke plate
22. Air vane
23. Spring
24. Spring
25. Idle speed stop screw
26. Body
27. Sleeve
28. Serrated disc

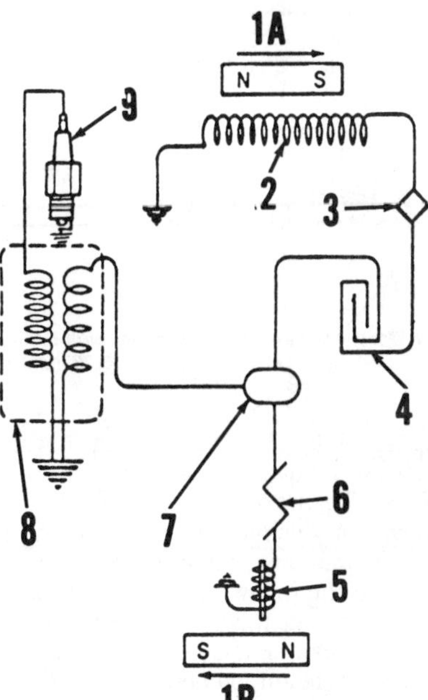

Fig. TP2-7—Diagram of Tecumseh solid state ignition system. Items (3, 4, 5, 6 and 7) are encased as shown at (10–Fig. TP2-8). Refer to text for description of operation.

model and only an idle fuel adjustment screw is provided on others.

Initial adjustment of idle mixture screw (Fig. TP2-6A as equipped) is 1 turn open from a lightly seated position. Idle fuel mixture on models without idle fuel mixture screw and main fuel mixture on all models is controlled by a fixed jet within carburetor. With engine at operating temperature and running, adjust idle mixture (as equipped) screw to obtain smooth engine idle.

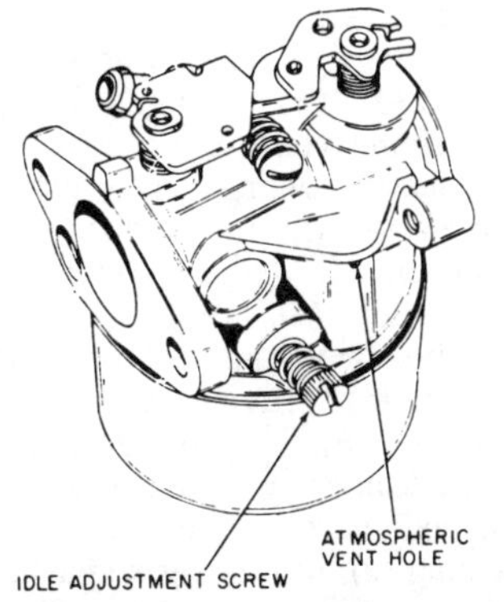

Fig. TP2-6A—View showing location of vent hole on TVS type fixed main jet carburetor.

To check float level, invert carburetor and measure from free end of float to float bowl mating surface of carburetor. Measurement should be 7/32 inch (5.56 mm). Carefully bend float lever tang which contacts fuel inlet needle as necessary to obtain correct float level.

GOVERNOR. An air vane type governor located on the carburetor throttle shaft is used on some models. On models with variable speed governor, the high speed (maximum) stop screw is located on the speed control lever. On models with fixed engine speed, rpm is adjusted by moving the governor spring bracket (B–Fig. TP2-5). To increase engine speed, bracket must be moved to increase governor spring (G) tension holding throttle (T) open.

On some TVS series engines a highly sensitive air vane (22–Fig. TP2-6) is attached to the choke shaft, independent of the choke shutter, but attached to the throttle arm. A spring on the choke shaft can be adjusted to vary the tension, causing the engine to run faster or slower as desired. One increment movement of sleeve (27) will vary engine speed approximately 100 rpm.

IGNITION SYSTEM. Engines may be equipped with either a magneto type (breaker point type) or a solid state ignition system. Refer to the appropriate paragraph for model being serviced.

Breaker Point Ignition System. Breaker point gap at maximum opening should be set before adjusting the ignition timing. On some models, ignition timing is not adjustable.

Some models may be equipped with external coil and magnet laminations. Air gap between external ignition coil laminations and flywheel magnet should be 0.015 inch (0.38 mm).

Ignition points and condenser are

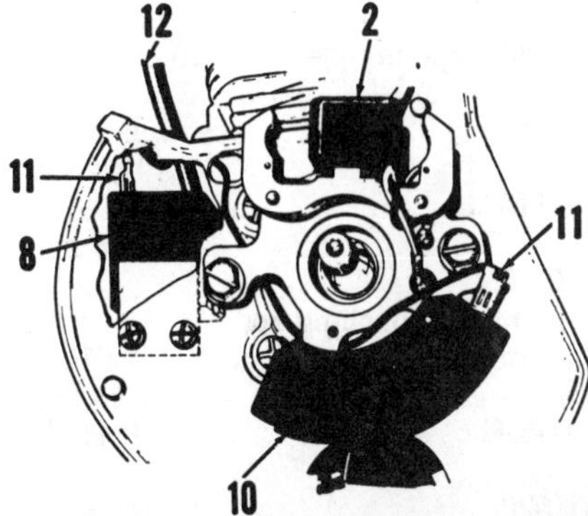

Fig. TP2-8—The solid state ignition charging coil, triggering system and mounting plate (2 and 10) are available only as an assembly and cannot be serviced.

2. Charging coil
8. Pulse transformer
10. Trigger system
11. Low tension lead
12. High tension lead

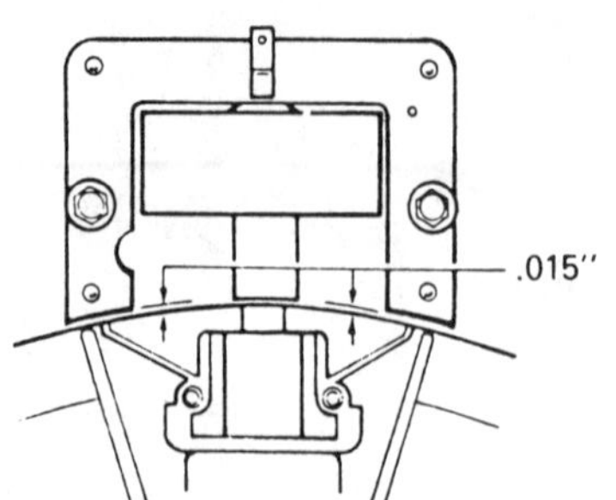

Fig. TP2-9—Set air gap (G) between coil laminations and flywheel at 0.015 inch (0.38 mm) as shown. Refer to text.

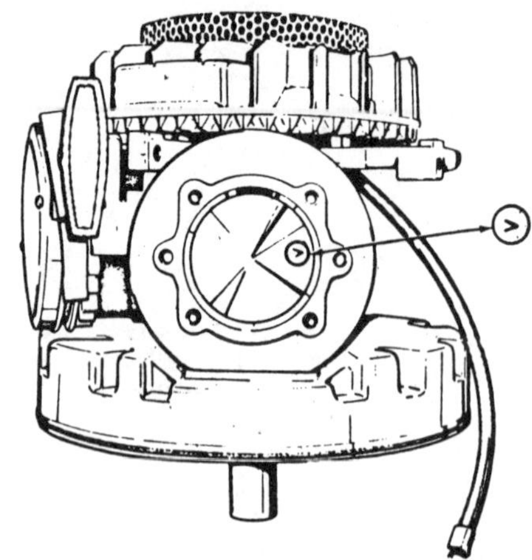

Fig. TP2-10—The "V" or "1111" mark stamped on top of piston must be toward side shown. Lubrication hole in side of connecting rod must be toward top on all vertical shaft models.

located under the flywheel and flywheel must be removed for service.

Use Tecumseh gage 670259 or equivalent thickness plastic sheeting to set air gap as shown in Fig. TP2-9. Refer to the following specifications for point gap and flywheel position before top dead center (BTDC) when breaker points just begin to open to correctly set ignition timing.

AH440:
- Point gap . . . . . 0.017 in. (0.43 mm)
- BTDC . . . . . 0.122 in. (3.10 mm)

AH480, AH490:
- Point gap . . . . . 0.017 in. (0.43 mm)
- BTDC (aluminum bushing rod) . . . . . 0.100 in. (2.54 mm)
- BTDC (needle bearing rod) . . . . . 0.135 in. (3.43 mm)

AH520 (type number 1500 to 1549):
- Point gap . . . . . 0.020 in. (0.51 mm)
- BTDC . . . . . 0.062 in. (1.575 mm)

AH520 (type number 1500 to 1577 & 1581 to 1582A):
- Point gap . . . . . 0.017 in. (0.43 mm)
- BTDC . . . . . 0.110 in. (2.79 mm)

AH520 (type number 1583 to 1589):
- Point gap . . . . . 0.017 in. (0.43 mm)
- BTDC . . . . . 0.062 in. (1.58 mm)

AH520 (aluminum bushing rod):
- Point gap . . . . . 0.017 in. (0.43 mm)
- BTDC . . . . . 0.110 in. (2.79 mm)

AH520 (needle bearing rod):
- Point gap . . . . . 0.017 in. (0.43 mm)
- BTDC . . . . . 0.185 in. (4.70 mm)

AH520 (type number 1601 to 1617):
- Point gap . . . . . 0.020 in. (0.51 mm)
- BTDC . . . . . *

AV520 (type number 638 to 650):
- Point gap . . . . . 0.018 in. (0.48 mm)
- BTDC . . . . . 0.100 in. (2.54 mm)

AV520 (type number 642-02E, F, 642-07C, 642-08, 642-13 to 14C & 642-15 to 23):
- Point gap . . . . . 0.020 in. (0.51 mm)
- BTDC . . . . . 0.085 in. (2.16 mm)

AV520 (type number 642 to 642-14C):
- Point gap . . . . . 0.018 in. (0.46 mm)
- BTDC . . . . . 0.110 in. (2.79 mm)

AV520 (type number 670 to 670-110):
- Point gap . . . . . 0.020 in. (0.51 mm)
- BTDC . . . . . 0.070 in. (1.78 mm)

AV600 (type number 641):
- Point gap . . . . . 0.018 in. (0.46 mm)
- BTDC . . . . . 0.100 in. (2.54 mm)

AV600 (type number 643-needle bearing rod):
- Point gap . . . . . 0.020 in. (0.51 mm)
- BTDC . . . . . 0.085 in. (2.16 mm)

AV600 (type number 643-aluminum bushing rod):
- Point gap . . . . . 0.018 in. (0.46 mm)
- BTDC . . . . . 0.090 in. (2.29 mm)

AV600 (type 660-2 to 660-38):
- Point gap . . . . . 0.020 in. (0.51 mm)
- BTDC . . . . . 0.070 in. (1.78 mm)

TVS600 (type 661-02 to 661-29):
- Point gap . . . . . 0.020 in. (0.51 mm)
- BTDC . . . . . *

AH750:
- Point gap . . . . . 0.017 in. (0.43 mm)
- BTDC . . . . . 0.100 in. (2.54 mm)

AV750:
- Point gap . . . . . 0.018 in. (0.46 mm)
- BTDC . . . . . 0.100 in. (2.54 mm)

AH817:
- Point gap . . . . . 0.018 in. (0.46 mm)
- BTDC . . . . . 0.113 in. (2.87 mm)

AV817:
- Point gap . . . . . 0.020 in. (0.51 mm)
- BTDC . . . . . 0.118 in. (2.80 mm)

*External coil, timing is nonadjustable. Set coil air gap to specified dimension.

Solid State Ignition. The Tecumseh solid state ignition system does not use ignition points. The only moving part of the system is the rotating flywheel with the charging magnets. As the flywheel magnet passes position (1A–Fig. TP2-7), a low voltage AC current is induced into input coil (2). The current passes through rectifier (3) converting this current to DC current. It then travels to the capacitor (4) where it is stored. The flywheel rotates approximately 180° to position (1B). As it passes trigger coil (5), it induces a very small electric charge into coil. This charge passes through resistor (6) and turns on the SCR (silicon controlled rectifier) switch (7). With SCR switch closed, the low voltage current stored in capacitor (4) travels to the pulse transformer (8). The voltage is stepped up instantaneously and the current is discharged across the electrodes of spark plug (9), producing a spark.

If system fails to produce a spark at spark plug, first check the high tension

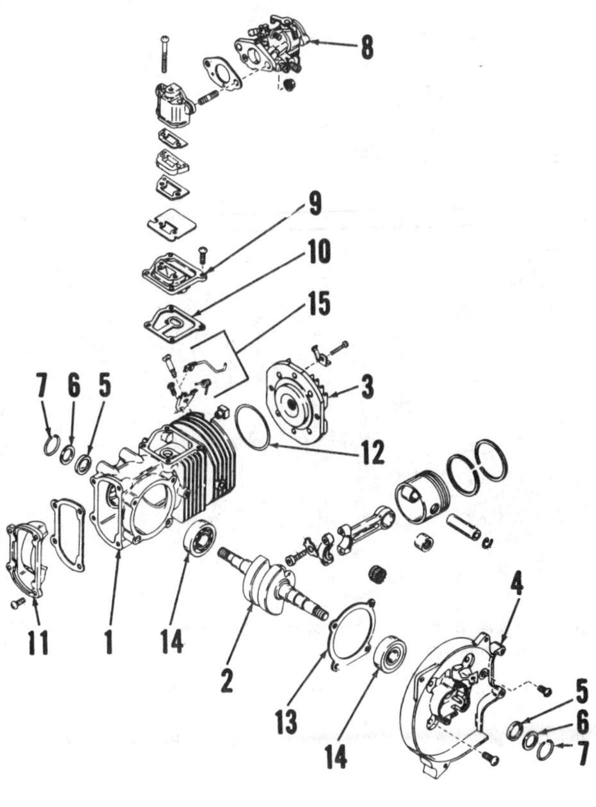

Fig. TP2-12—Exploded view typical of AH440 and AH490 models.

1. Cylinder & crankcase
2. Crankshaft
3. Cylinder head
4. Magneto end plate
5. Seal
6. Retainer
7. Snap ring
8. Carburetor
9. Third port cover
10. Gasket
11. End cover
12. Seal ring
13. Gasket
14. Main bearings (ball)
15. Compression release

lead (12–Fig. TP2-8). If condition of the high tension lead is questionable, renew the pulse transformer and high tension lead assembly. Check the low tension lead (11) and renew if insulation is faulty. The ignition charging coil, electronic triggering system and mounting plate are available only as an assembly. If necessary to renew this assembly, place the unit in position on the engine. Start the retaining screws, turn the mounting plate counterclockwise as far as possible, then tighten retaining screws to 5-7 ft.-lbs. (7-10 N•m).

LUBRICATION. Engines are lubricated by mixing a good quality two-stroke air-cooled engine oil with regular grade gasoline. For engines operating below 3600 rpm, mix 0.5 pint (0.24 L) of oil with each gallon of gasoline. For engines operating above 3600 rpm, mix 0.75 pint (0.36 L) of oil with each gallon of gasoline.

CARBON. Muffler and exhaust ports should be cleaned after every 50 hours of operation if engine is operated continuously at full load. If operated at light or medium load, the cleaning interval can be extended to 100 hours.

REPAIRS

DISASSEMBLY. Unbolt and remove the cylinder shroud, cylinder head and crankcase covers. Remove connecting rod cap and push the rod and piston unit out through top of cylinder. It may be necessary to remove the ridge from top of cylinder bore before removing piston and connecting rod assembly.

To remove the crankshaft, remove starter housing, flywheel and the bolts securing shroud base (bearing housing) to crankcase. Strike drive end of crankshaft with a leather mallet to dislodge crankshaft and bearing housing.

CONNECTING ROD. A steel connecting rod equipped with needle roller bearings at the crankpin and at the piston pin ends is used on some models.

An aluminum connecting rod which rides directly on crankpin journal is used on some models. Piston pin rides directly in connecting rod bore.

On other models with an aluminum connecting rod, a steel insert (liner) is used on inside of connecting rod and needle bearing rollers are used at the crankpin end.

On models with aluminum bushing type connecting rod, clearance on crankpin should be 0.0011-0.0020 inch (0.028-0.051 mm). Crankpin journal diameter should be 0.6857-0.6865 inch (17.417-17.437 mm). Piston pin diameter is 0.3750-0.3751 inch (9.525-9.528 mm). Only standard size parts are available.

On models with aluminum connecting rod with steel liners and bearing needles at crankpin, standard journal diameter is 0.8442-0.8450 inch (21.443-21.463 mm). There are 74 bearing rollers and ends of liners must correctly engage when match marks on connecting rod and cap are aligned. Piston pin diameter is 0.4997-0.4999 inch (12.692-12.698 mm) and piston pin rides in a cartridge type needle bearing that is pressed into connecting rod piston pin bore.

All models with steel connecting rod are provided with loose needle rollers at crankpin end of connecting rod and a

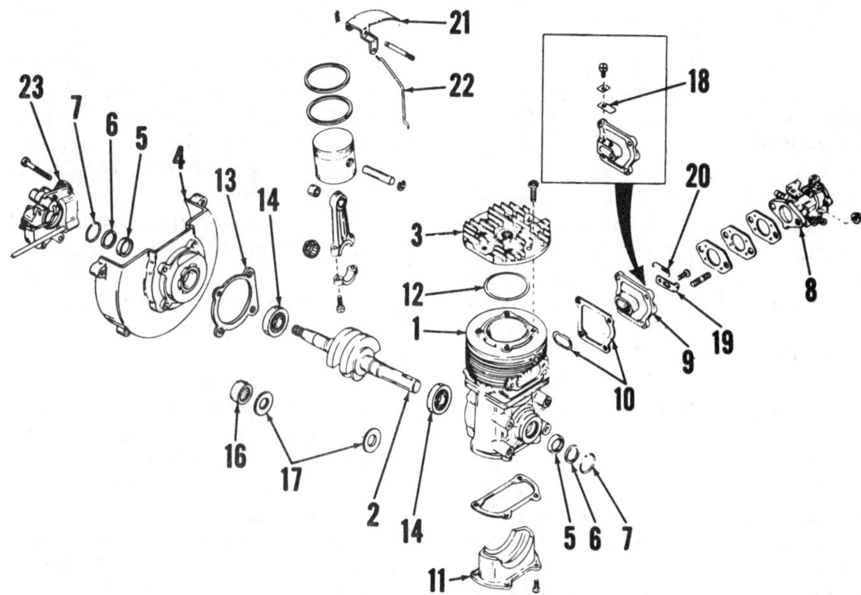

Fig. TP2-13—Exploded view of AH520 engine. Supplementary reed valve is shown at (18) and governor parts at (19 through 22).

1. Cylinder & crankcase
2. Crankshaft
3. Cylinder head
4. Magneto end plate
5. Seal
6. Retainer
7. Snap ring
8. Carburetor
9. Third port cover
10. Gaskets
11. End cover
12. Seal ring
13. Gasket
14. Main bearings (ball)
16. Main bearing (roller)
17. Thrust washers
18. Reed valve
19. Spring bracket
20. Governor spring
21. Air vane
22. Carburetor link
23. Magneto

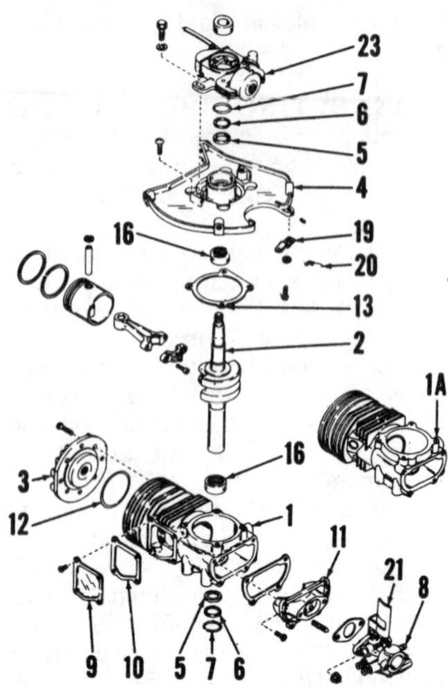

Fig. TP2-14—Exploded view typical of AV520 and AV600 engines. The cylinder and crankcase for some AV520 and all AV600 models is shown at (1A).

1. Cylinder & crankcase
2. Crankshaft
3. Cylinder head
4. Magneto end plate
5. Seal
6. Retainer
7. Snap ring
8. Carburetor
9. Cover
10. Gasket
11. Reed valve
12. Seal ring
13. Gasket
16. Main bearings (roller)
19. Spring bracket
20. Governor spring
21. Air vane (on carburetor)
23. Magneto

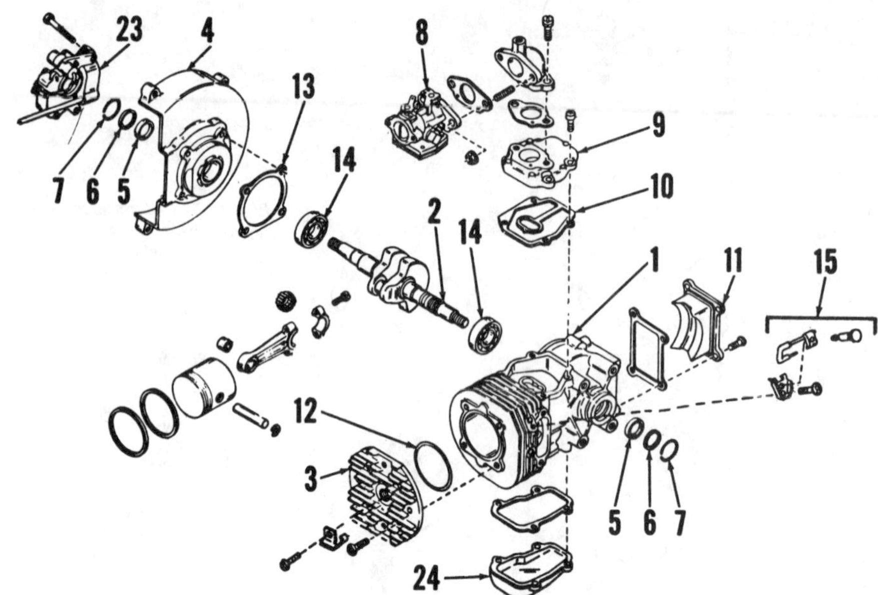

Fig. TP2-15—Exploded view of AH750 engine. Compression release is shown at (15). Model AH817 engines are similar.

1. Cylinder & crankcase
2. Crankshaft
3. Cylinder head
4. Magneto end plate
5. Seal
6. Retainer
7. Snap ring
8. Carburetor
9. Third port cover
10. Gasket
11. End cover
12. Seal ring
13. Gasket
14. Main bearings (ball)
15. Compression release
23. Magneto
24. Transfer port cover

cartridge needle bearing at piston pin bore.

Standard crankpin journal diameter is 0.5611-0.5618 inch (12.252-12.270 mm) for Model AH440 and 0.5614-0.5621 inch (12.260-12.277 mm) for Models AH480 and AH490. Crankpin bearing for Models AH440, AH480 and AH490 may use one row of 30 needle rollers or 60 short needle rollers placed in two rows.

On early Model AH520, the crankpin journal diameter is 0.5611-0.5628 inch (12.252-12.270 mm) and 56 short (half length) needle rollers are placed in two rows. Later Model AH520 has a standard crankpin journal diameter of 0.6922-0.6927 inch (17.582-17.595 mm) and uses 31 needle roller bearings in the 0.8524 inch (21.650 mm) connecting rod bearing bore.

Two different types of crankshafts, bearing rollers and connecting rods are used on Models AH750 and AV750. Early models are equipped with 0.6240-0.6243 inch (15.850-15.857 mm) diameter crankpin journal and use 32 needle rollers in the 0.7566-0.7569 inch (19.218-19.225 mm) diameter connecting rod bearing bore. Later Model AV750 is equipped with 0.6259-0.6266 inch (15.898-15.916 mm) diameter crankpin journal and uses 33 needle rollers in the 0.7588-0.7592 inch (19.274-19.284 mm) diameter connecting rod bearing bore.

Models AH817 and AV817 have standard crankpin journal diameters of 0.6259-0.6266 inch (15.898-15.916 mm) and use 66 short (half length) needle rollers placed in the two rows in the 0.7588-0.7592 inch (19.274-19.284 mm) diameter connecting rod bearing bore.

Model TVS600 has a standard crankpin journal diameter of 0.8113-0.8118 inch (20.607-20.620 mm) and uses 30 roller bearings in the 1.0443-1.0448 inch (26.525-26.538 mm) diameter connecting rod bearing bore.

On models with short (half length) needle rollers, bearing needles are placed in two rows around crankpin with flat ends together toward center of crankpin.

On all models equipped with needle bearing at crankpin, rollers should be renewed only as a set. Renew bearing set if any roller is damaged. If rollers are damaged, check condition of crankpin and connecting rod carefully and renew if bearing races are damaged. New rollers are serviced in a strip and can be installed by wrapping the strip around crankpin. After new needle rollers and connecting rod cap are installed, force lacquer thinner into needles to remove the beeswax, then lubricate bearing with SAE 30 oil.

On all models, make certain match marks on connecting rod and cap are aligned.

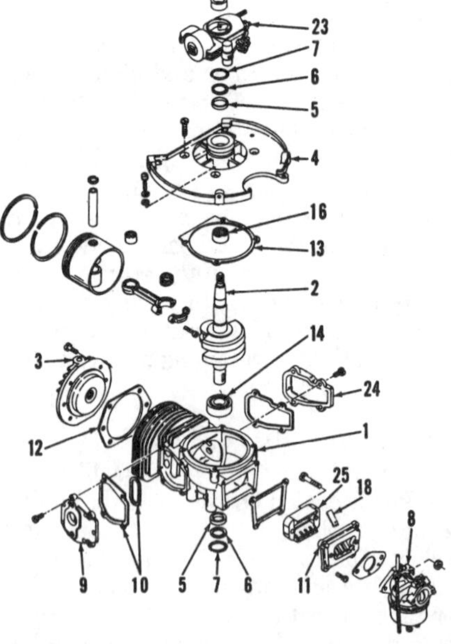

Fig. TP2-16—Exploded view of typical AV750 engine. Refer to text when assembling reed valve (11, 18 and 25). Model AV817 engines are similar.

1. Cylinder & crankcase
2. Crankshaft
3. Cylinder head
4. Magneto end plate
5. Seal
6. Retainer
7. Snap ring
8. Carburetor
9. Cover
10. Gasket
11. Reed plate
12. Head gasket
13. Gasket
14. Main bearing (ball)
16. Main bearing (roller)
18. Reed valve petal (8)
23. Magneto
24. Transfer port cover
25. Reed petal stop

On vertical shaft models, make certain lubrication hole in side of connecting rod is toward top. Some AV520 and AV600 models are stamped with "V" or "1111" mark on piston as shown in Fig. TP2-10. On models so equipped, make certain mark stamped on top of piston is toward the right side as shown.

On models with aluminum connecting rod, tighten connecting rod cap retaining screws to 40-50 in.-lbs. (5-6 N·m) and lock with the tab washer.

On all models with a steel connecting rod, tighten the connecting rod cap retaining screws (self locking) to 70-80 in.-lbs. (8-9 N·m).

PISTON, PINS AND RINGS. Refer to CONNECTING ROD section for piston removal procedure. Piston and cylinder bore specifications are as follows:

Models AH440, AH480, AH490, AH520, AV520, AV600 and TVS600

Cylinder bore diameter . . 2.093-2.094 in. (53.162-53.188 mm)

Piston-to-cylinder clearance:

AV520 (type number 642-02E & F, 642-07C, 642-08, 642-13 to 14C, 642-15 to 23) . . . . 0.005-0.012 in. (0.13-0.30 mm)

All other 440, 480, 490, 520, 600 models . . 0.005-0.007 in. (0.13-0.18 mm)

Ring end gap:

AV520 (type number 638 to 638-100 & 650) . . 0.006-0.014 in. (0.15-0.41 mm)

AV520 (type number 642-02E & F, 642-07C, 642-08, 642-13 to 14C, 642-15 to 23, 642-24 to 33, 661, 670-01 to 670-110) . . . . 0.006-0.016 in. (0.15-0.41 mm)

AV600 (type number 641 to 641-14) . . . . 0.006-0.014 in. (0.15-0.36 mm)

AV600 (type number 643-03B & C, 643-05B, 643-14A, B & C, 643-15A to 31) . . . 0.006-0.016 in. (0.15-0.41 mm)

TVS600 (type number 661-02 to 661-29) . . . . . 0.006-0.016 in. (0.15-0.41 mm)

AH480 (type number 1471 to 1471B, 1484 to 1484D, 1489 to 1490B, 1498, 1500, 1509, 1511, 1515 to 1516C, 1517, 1527, 1529A & B, 1531, 1535B, 1542, 1554, 1554A, 1573) . . . . 0.006-0.011 in. (0.15-0.28 mm)

AH520 (type number 1401 to 1401J, 1448 to 1450F, 1466 to 1466A, 1482 to 1482C, 1483, 1499, 1506 to 1507A, 1525A, 1534A, 1551, 1553, 1555, 1556, 1574 to 1577, 1581 to 1582A, 1583 to 1599, 1600 to 1620) . . . . 0.006-0.016 in. (0.15-0.41 mm)

All other 440, 480, 490, 520 & 600 models . . . . 0.007-0.017 in. (0.18-0.43 mm)

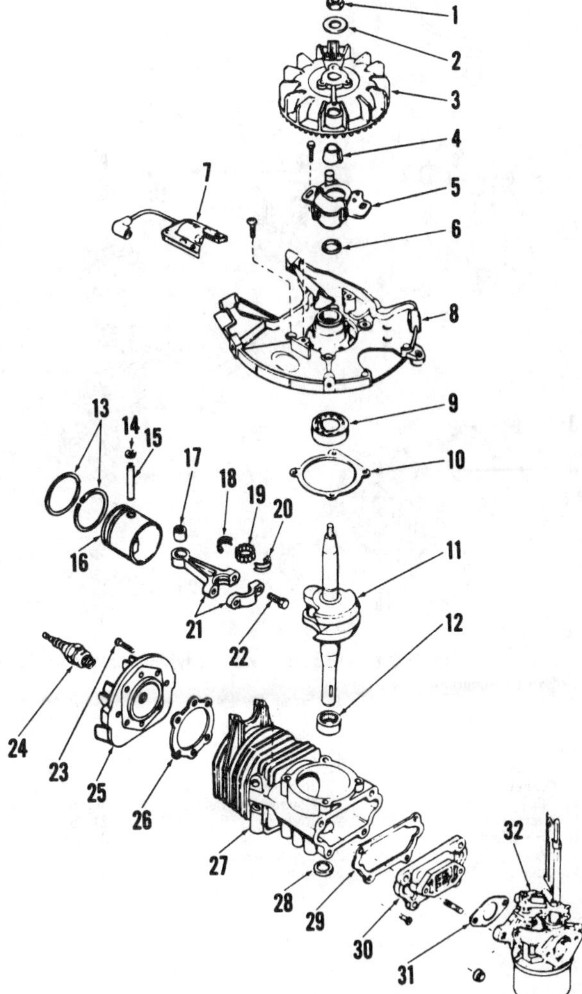

Fig. TP2-17 — Exploded view of typical Model TVS600 type 661 engine.

1. Flywheel nut
2. Washer
3. Flywheel
4. Flywheel sleeve
5. Magneto
6. Seal
7. Coil
8. End plate
9. Bearing (ball)
10. Gasket
11. Crankshaft
12. Bearing (roller)
13. Piston rings
14. Clip
15. Piston pin
16. Piston
17. Bearing
18. Liner
19. Roller bearings
20. Liner
21. Connecting rod
22. Rod bolt
23. Head bolt
24. Spark plug
25. Cylinder head
26. Head gasket
27. Cylinder
28. Seal
29. Gasket
30. Reed valve
31. Gasket
32. Carburetor

Models AH750 and AV750

Cylinder bore diameter . . 2.375-2.376 in. (60.389-60.391 mm)

Piston-to-cylinder clearance:

AH750 . . . . 0.005-0.006 in. (0.13-0.15 mm)

AV750 . . . . 0.0055-0.0075 in. (0.140-0.191 mm)

Ring end gap . . . . 0.005-0.013 in. (0.13-0.33 mm)

Models AH817 and AV817

Cylinder bore diameter . . 2.437-2.438 in. (61.900-61.925 mm)

Piston-to-cylinder clearance . . . . 0.0063-0.0083 in. (0.160-0.211 mm)

Ring end gap . . . . 0.007-0.017 in. (0.18-0.43 mm)

Piston, rings and piston pin are available in standard size only. The piston pin should be a press fit in heated piston on models with needle bearing in connecting rod pin bore. On models without needle bearing, the piston pin should be a palm push fit in piston and thumb push fit in rod.

Refer to CONNECTING ROD section for standard piston pin diameters.

When assembling piston in connecting rod, on vertical crankshaft models, lubrication hole in side of connecting rod must be toward top of engine. On Models AV520 and AV600 equipped with offset piston, make certain that "V" mark or "1111" mark stamped on top of piston is toward the right side as shown in Fig. TP2-10.

Use the old cylinder head sealing ring and a ring compressor to compress piston rings, when sliding piston into cylinder.

NOTE: Make certain rings do not catch in recess at top of cylinder.

Always renew the cylinder head metal sealing ring. The cylinder head retaining screws should be tightened to 90-100 in.-lbs. (10-11 N·m). Refer to CONNECTING ROD section for installation of the connecting rod.

CRANKSHAFT AND CRANKCASE. The crankshaft can be removed

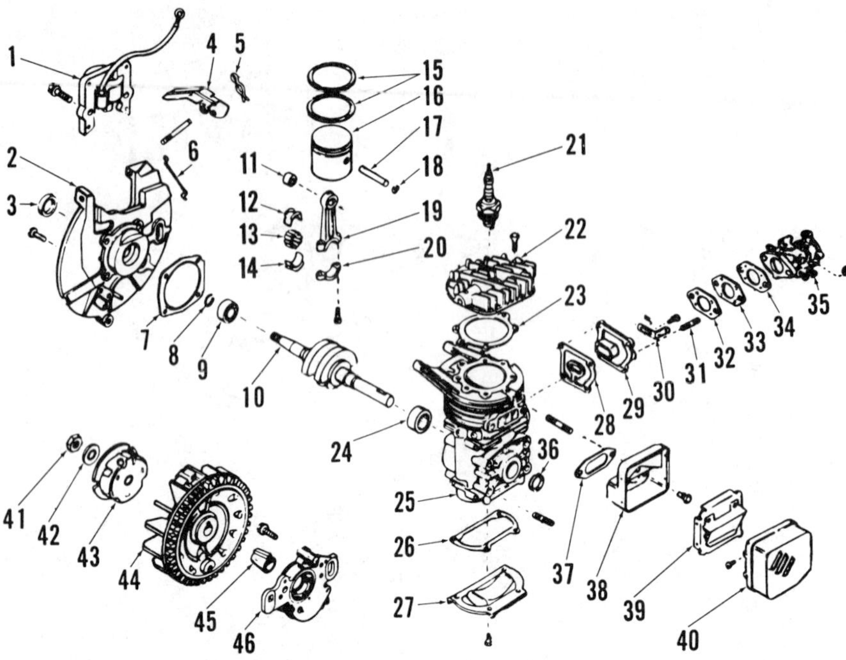

Fig. TP2-18—Exploded view of typical Model AH520 type 1600 engine.

1. External ignition
2. End plate
3. Seal
4. Air vane
5. Clip
6. Governor link
7. Gasket
8. Clip
9. Bearing
10. Crankshaft
11. Bearing
12. Liner
13. Roller needles
14. Liner
15. Piston rings
16. Piston
17. Piston pin
18. Retainer
19. Connectijng rod
20. Rod cap
21. Spark plug
22. Cylinder head
23. Head gasket
24. Bearing
25. Crankcase
26. Gasket
27. Crankcase cover
28. Gasket
29. Reed plate
30. Bracket
31. Stud
32. Gasket
33. Gasket
34. Gasket
35. Carburetor
36. Seal
37. Gasket
38. Muffler base
39. Baffle
40. Cover
41. Nut
42. Washer
43. Starter cup
44. Flywheel
45. Flywheel sleeve
46. Magneto

after the piston, connecting rod, flywheel and the magneto end bearing plate are removed.

Crankshaft main bearings may be either ball type or cartridge needle roller. If ball type main bearings are used, it should be necessary to bump the crankshaft out of the bearing inner races. Ball and roller bearing outer races should be a tight fit in bearing bores. If new ball bearings are to be installed, heat the crankcase when removing or installing bearings. Do not use open flame.

On all models, bearings should be installed with printed face on race toward center of engine.

If the crankshaft is equipped with thrust washers at ends, make certain they are installed when assembling.

Crankshaft end play should be ZERO for all AH440, AH480, AH490, AH520, AH750, AH817, AV750, AV817 and TVS600 models. Crankshaft end play should be 0.003-0.016 inch (0.08-0.41 mm) for Models AV520 and AV600 equipped with two needle roller main bearings.

It is important to exercise extreme care when renewing crankshaft seals to prevent their being damaged during installation. If a protector sleeve is not available, use tape to cover any splines, keyways, shoulders or threads over which the seal must pass during installation. Seals should be installed with channel groove of seal toward inside (center) of engine on all models except for use on outboard motors. If engine is used on an outboard motor, lip of lower seal should be toward outside (bottom) and the top (magneto end) seal should be toward inside (center) of engine.

REED VALVES. Some engines are equipped with a reed type inlet valve located in the lower crankcase cover. Reed petals should not stand out more than 0.010 inch (0.25 mm) from the reed plate and must not be bent, distorted or cracked. The reed plate must be smooth and flat. Renew petals (AV750 only) or complete valve assembly if valve does not seal completely.

Reed petals are available separately only for AV750 models. Petals must be installed with rounded edge against sealing surface of reed plate and the sharp edge must be toward reed stop. When installing the reed stop, apply Loctite to threads of retaining screws and tighten the two screws to 50-60 in.-lbs. (6-7 N·m).

TECUMSEH 2-STROKE

Model	Bore	Stroke	Displacement
TC200	1.4375 in. (36.51 mm)	2.250 in. (57.15 mm)	2.0 cu. in. (32.8 cc)

ENGINE INFORMATION

Engine type and model numbers are stamped into blower housing base as indicated in Fig. TP3-1. Always furnish engine model and type number when ordering parts.

MAINTENANCE

SPARK PLUG. Recommended spark plug is a Champion RCJ-8Y, or equivalent. Specified electrode gap is 0.030 inch (0.76 mm).

AIR CLEANER. Air cleaner element should be removed and cleaned at eight hour intervals of use. Polyurethane element may be washed in a mild detergent and water solution and squeezed until all dirt is removed. Rinse thoroughly. Wrap in clean dry cloth and squeeze until completely dry. Apply engine oil to element and squeeze out excess. Clean air cleaner body and cover and dry thoroughly.

CARBURETOR. Tecumseh TC200 engines are equipped with a diaphragm type carburetor with a single idle mixture needle. Initial adjustment of idle mixture needle is one turn open from a lightly seated position.

Final carburetor adjustment is made with engine at operating temperature and running. Operate engine at idle speed and turn idle mixture needle slowly clockwise until engine falters. Note this position and turn idle mixture needle counterclockwise until engine begins to run unevenly. Note this position and turn adjustment screw until it is halfway between first (lean) and last (rich) positions.

To disassemble carburetor, refer to Fig. TP3-2. Remove idle speed stop screw (9) and spring. Remove pump cover (10), gasket (11) and diaphragm (12). Remove cover (1), diaphragm (2) and gasket (3). Carefully remove pin (15), metering lever (4), inlet needle valve (5) and spring (6). Remove screws retaining throttle plate to throttle shaft (8). Remove screws retaining choke plate to choke shaft (14). Remove "E" clip from throttle shaft and choke shaft and remove shafts. Remove all non-metallic parts, idle mixture needle, fuel inlet screen and fuel inlet (13). Remove all Welch plugs.

Clean and inspect all parts. Do not allow parts to soak in cleaning solvent longer than 30 minutes.

To reassemble, install fuel inlet needle, metering lever spring and pin. Metering lever hooks onto the inlet needle and rests on the metering spring. Entire assembly is held in place by metering lever pin screw. Tip of metering lever must be 0.060-0.070 inch (1.52-1.78 mm) from the face of carburetor body (Fig. TP3-3).

Install diaphragm gasket so tabs of gaskets align with the bosses on the carburetor body. After gasket is in place, install the diaphragm again aligning tabs to bosses. The head of the rivet in the diaphragm must be toward the carburetor body. Check the atmospheric vent hole in the diaphragm cover to make certain it is clean. Install cover on carburetor.

Install pump diaphragm with the corner holes aligning with the same holes in the carburetor body. Align pump gasket in the same manner and place pump cover onto carburetor.

Fig. TP3-1—Engine model and type number is stamped into blower housing base.

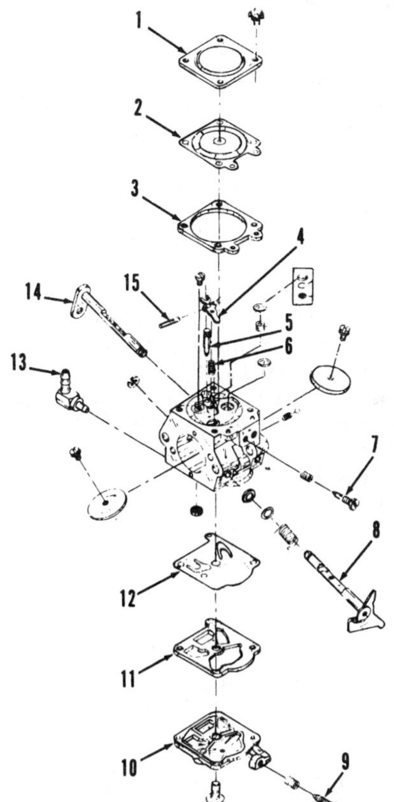

Fig. TP3-2—Exploded view of diaphragm type carburetor used on TC200 engine.

1. Cover
2. Diaphragm
3. Gasket
4. Metering lever
5. Inlet needle valve
6. Spring
7. Idle mixture needle
8. Throttle shaft
9. Idle speed screw
10. Cover
11. Gasket
12. Diaphragm
13. Fuel inlet
14. Choke shaft
15. Pin

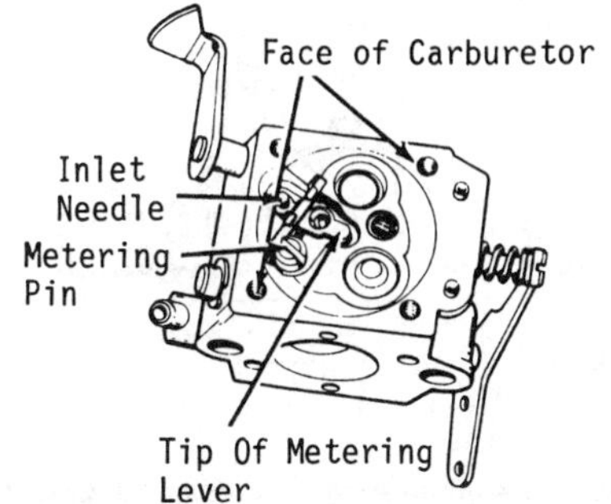

Fig. TP3-3—Tip of metering valve lever should be 0.060-0.070 inch (1.52-1.78 mm) from the face of carburetor body.

Numbers on throttle plate should face to the outside when throttle is closed. Apply a small amount of Loctite grade "A" to fuel inlet before installation.

IGNITION SYSTEM. All Model TC200 engines are equipped with a solid state ignition module located outside the flywheel. Correct air gap between laminations of module and magnets of flywheel is 0.012 inch (0.30 mm). Use Tecumseh gage part number 670297.

GOVERNOR ADJUSTMENT. Model TC200 engine is equipped with an air vane type governor. Refer to CRANKSHAFT under REPAIRS section for adjustment procedure.

LUBRICATION. Engine is lubricated by mixing gasoline with a good quality two-stroke, air-cooled engine oil. Manufacturer recommends a fuel:oil ratio of 24:1.

CARBON. Muffler and exhaust ports should be cleaned after every 50 hours of operation if engine is operated continuously at full load. If operated at light or medium load, the cleaning interval can be extended to 100 hours.

REPAIRS

TIGHTENING TORQUES. Recommended tightening torque specifications are as follows:

Crankcase cover to crankcase	70-100 in.-lbs. (8-11 N·m)
Cylinder to crankcase	60-75 in.-lbs. (7-8 N·m)
Carburetor	20-32 in.-lbs. (2.3-3.6 N·m)
Flywheel nut	180-240 in.-lbs. (20-27 N·m)
Ignition module	30-40 in.-lbs. (3.4-4.5 N·m)
Starter retainer screw	45-55 in.-lbs. (5.1-6.2 N·m)

CRANKSHAFT. To remove crankshaft, drain fuel tank, remove tank strap and disconnect fuel line at carburetor. Disconnect and remove spark plug. Remove the three screws retaining blower housing and rewind starter assembly and remove housing. Remove the two screws retaining ignition module. Use strap wrench to hold flywheel. Use flywheel puller tool (670299) to remove flywheel. Remove air cleaner assembly with carburetor, spacer, gaskets and screen. Mark and remove governor link from carburetor throttle lever. Remove the three 5/16 inch cap screws, then separate blower housing base from crankcase. Attach engine holder tool (670300) with the three blower housing base screws. Place tool in a bench vise. Remove muffler springs using tool fabricated from a 12 inch piece of heavy wire with a ¼ inch hook made on one end. Remove the four cylinder retaining nuts, then pull cylinder off squarely and in line with piston. Use caution so rod does not bend. Install seal protector tool (670206) at magneto end of crankshaft and seal protector tool (670263) at pto end of crankshaft. Remove crankcase cover screws, then carefully separate crankcase cover from crankcase. Rotate crankshaft to top dead center and withdraw crankshaft through crankcase cover opening while sliding connecting rod off crankpin and over crankshaft. Refer to Fig. TP3-4. Use care not to lose any of the 23 crankpin needle bearings which will be loose. Flanged side of connecting rod (Fig. TP3-5) must be toward pto side of engine after installation. Handle connecting rod carefully to avoid bending.

Standard crankpin journal diameter is 0.5985-0.5990 inch (15.202-15.215 mm). Standard crankshaft pto side main bearing journal diameter is 0.6248-0.6253 inch (15.870-15.880 mm). Standard crankshaft magneto side main bearing journal diameter is 0.4998-0.5003 inch (12.69-12.71 mm). Crankshaft end play should be 0.004-0.012 inch (0.10-0.30 mm).

To install crankshaft, clean mating surfaces of crankcase, cylinder and crankcase cover. Avoid scarring or burring mating surfaces.

Crankshaft main bearing in crankcase of early model engines did not have a retaining ring as shown in Fig. TP3-6. Retaining ring was installed as a running change in late model engines. To install new caged bearing in crankcase, place bearing on installation tool (670302) with the numbered side of bearing away from tool. Press bearing into crankcase until tool is flush with crankcase housing. Install retaining ring (as equipped). Place seal for magneto side onto seal installation tool (670301) so metal case of seal enters tool first. Press seal in until tool is flush with crankcase. Use the same procedure to install bearing and seal in crankcase cover using bearing installation tool (670304) and seal installation tool (670303).

New crankpin needle bearings are on bearing strips. Heavy grease may be used to retain old bearings on crankpin journal as required. During reassembly, connecting rod must not be forced onto crankpin journal as rod failure or bending will result. Apply Loctite 515 to mating surfaces of crankcase during reassembly and use seal protectors when installing lip seals over ends of crankshaft. When installing cylinder over piston, install a wooden block with a slot cut out for connecting rod under piston to provide support and prevent connecting rod damage. Exhaust ports in cylinder are on the same side of engine as muffler resting boss. Make certain cylinder is correctly positioned, stagger ring end gaps and compress rings using a suitable ring compressor which can be removed after cylinder is installed over piston. Install cylinder and push cylinder onto crankcase studs to expose 1-2 threads of studs. Install the four nuts onto exposed threads of studs, then push cylinder further down to capture nuts on studs. Tighten nuts in a crisscross pattern to specified torque.

Install muffler using fabricated tool to install springs. Install blower housing

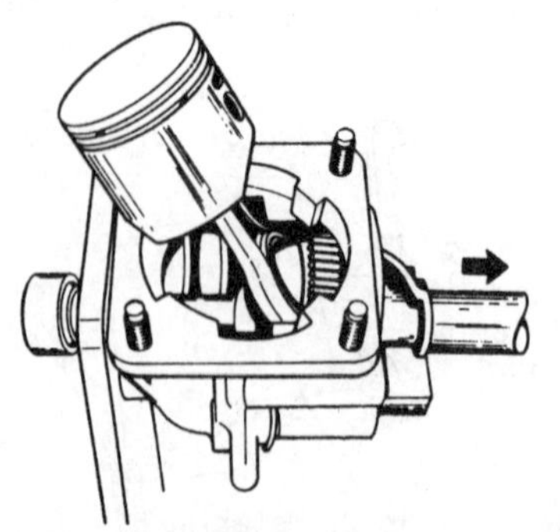

Fig. TP3-4—Connecting rod must be carefully worked over crankpin during crankshaft removal. Do not lose the 23 loose crankpin needle bearings.

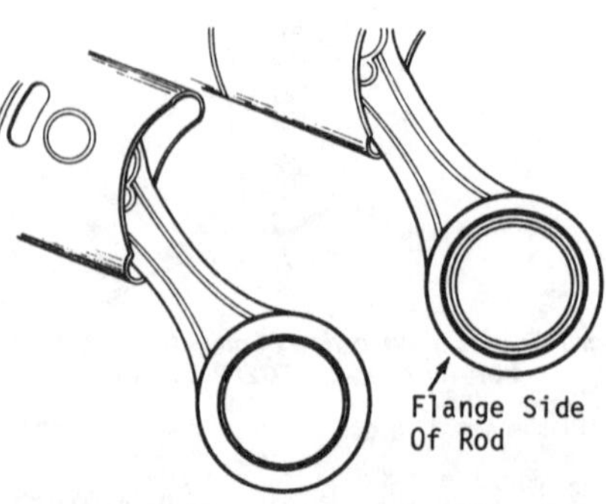

Fig. TP3-5—Flanged side of connecting rod must face pto side of engine after installation.

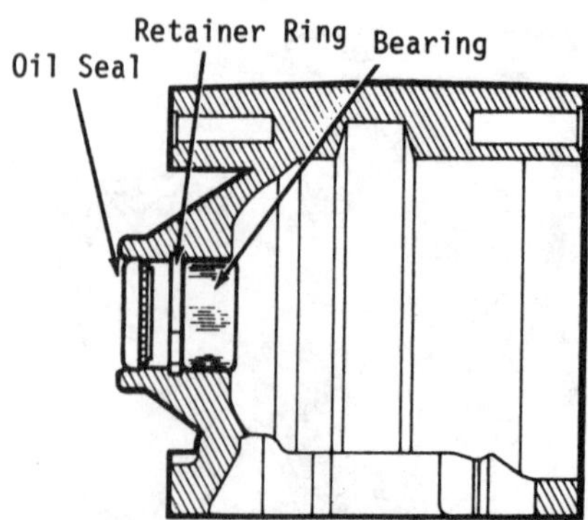

Fig. TP3-6—Early Model TC200 engines did not have retaining ring shown. Retaining ring was installed as a running change in late model engines.

base and tighten the three screws to specified torque.

Refer to Fig. TP3-7 to install governor air vane assembly. Speed adjustment lever is held in place by inserting screw into the blower housing base. Long end of governor spring hooks into the notch on neck of air vane. Short end hooks into the hole in speed adjustment lever. To decrease governed speed of engine, bend speed adjusting lever towards spark plug end of engine. To increase governed speed of engine, bend lever in the opposite direction. Throttle link is inserted into hole in the neck of the air vane and the hole closest to the throttle shaft in throttle plate.

Install carburetor, spacer, gaskets, screen and air cleaner body on engine. Tighten screws to specified torque. Install and adjust ignition module. Install blower housing/rewind starter assembly and tighten screws to specified torque. Install fuel tank.

PISTON, RINGS AND CONNECTING ROD. Standard piston diameter is 1.4327-1.4340 inch (36.39-36.42 mm). Standard width of both ring grooves is 0.050-0.051 inch (1.27-1.29 mm). Standard piston ring width is 0.46-0.47 inch (11.7-11.9 mm). Standard ring end gap is 0.004-0.014 inch (0.10-0.36 mm).

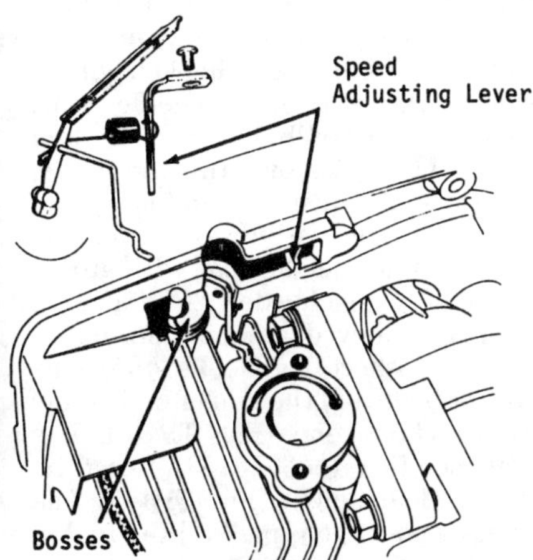

Fig. TP3-7—View of air vane governor assembly used on Model TC200 engines. Refer to text.

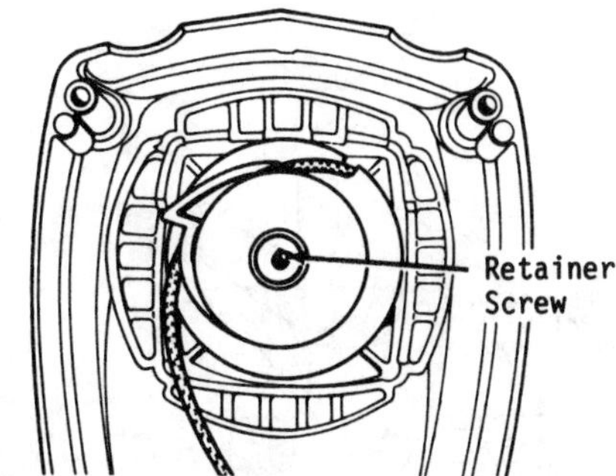

Fig. TP3-8—View showing rewind starter retaining screw.

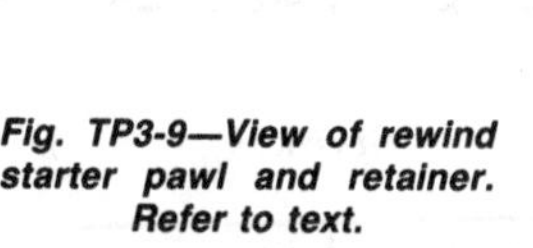

Fig. TP3-9—View of rewind starter pawl and retainer. Refer to text.

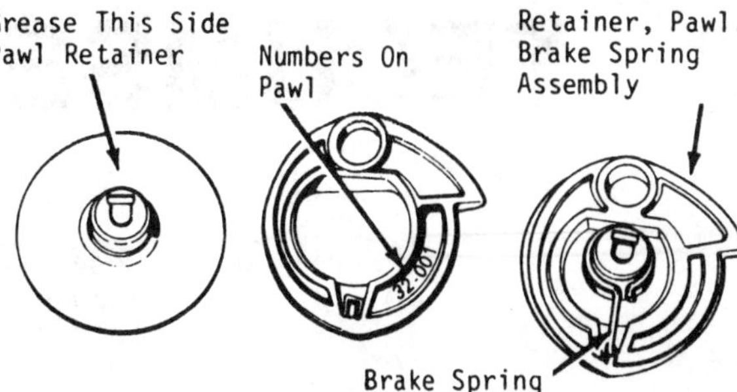

CYLINDER. Cylinder must be smooth and free of scratches or flaking. Clean carbon carefully as necessary. Standard bore size is 1.4375 inches (36.513 mm).

TECUMSEH SPECIAL TOOLS. Tecumseh special tools are available to aid in engine disassembly and reassembly are listed by use and tool part number.

FLYWHEEL PULLER 670299
AIR GAP GAGE 670297
ENGINE HOLDER.......... 670300
SEAL PROTECTOR
(MAG. END).............. 670206
SEAL PROTECTOR
(PTO END) 670263
SEAL INSTALLER
(MAG END) 670301
SEAL INSTALLER
(PTO END)............... 670303
BEARING INSTALLER
(MAG END) 670302
BEARING INSTALLER
(PTO END) 670304

REWIND STARTER. The rewind starter assembly is incorporated into blower housing. Blower housing design varies according to engine model and specification number. To release rewind spring tension, remove staple in starter handle and slowly let spring tension release by winding rope onto rope sheave. Remove the 5/16 inch retainer screw (Fig. TP3-8). Remove pawl retainer and pawl (Fig. TP3-9) and extract starter pulley. Use caution not to pull rewind spring out of housing at this time. Uncoiling spring can be very dangerous. If rewind spring is damaged or weak, use caution when removing spring from housing.

To reassemble, grease center post of housing and portion of housing where rewind spring will rest. Grip rewind spring firmly with needlenose pliers ahead of spring tail. Insert spring and hook tail into housing as shown in Fig. TP3-10. Make certain spring is seated in housing before removing needlenose pliers from spring. Grease top of spring. Insert starter rope into starter pulley and tie a left handed knot in end of rope. With neck of starter pulley up, wind starter rope in a counterclockwise rotation. Place end of rope in notch of pulley and place pulley in housing. Press down on pulley and rotate until pulley attaches to rewind spring. Refer to Fig. TP3-11. Lubricate pawl retainer with grease and place the pawl, numbers up, onto retainer. Place brake spring on center of retainer with tab locating into pawl Fig. TP3-9). Tab on pawl retainer must align with notch in center post of housing and locating hole in pawl must mesh with boss on starter pulley (Fig. TP3-12). Install retainer screw (Fig. TP3-8) and tighten to specified torque. Use starter rope to wind spring a minimum of 2 turns counterclockwise and a maximum of 3 turns. Feed starter rope through starter grommet and secure starter handle using a left-hand knot.

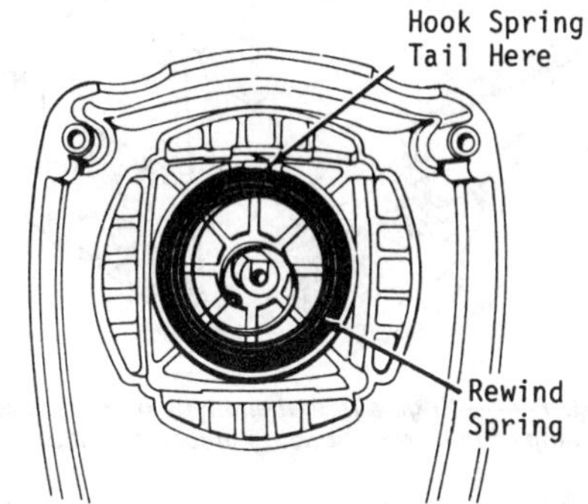

Fig. TP3-10—View of rewind spring and housing.

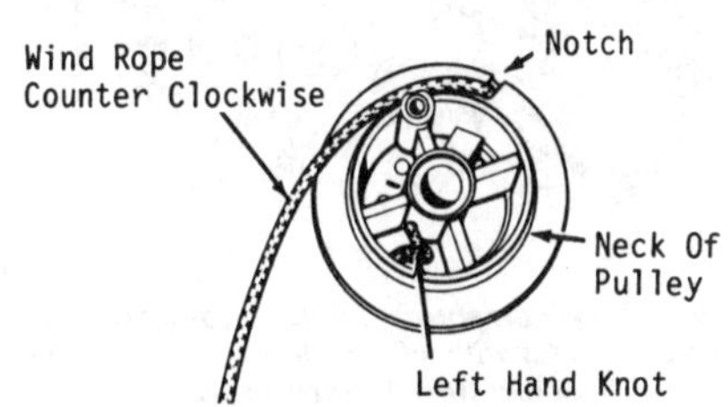

Fig. TP3-11—View showing rewind starter rope as shown and as outlined in text.

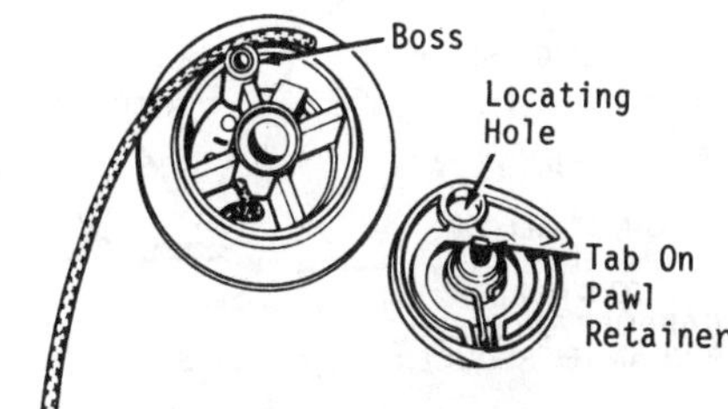

Fig. TP3-12—Boss must engage locating hole on pawl retainer. Refer to text.

TECUMSEH 2-STROKE

Model	Bore	Stroke	Displacement
TVS840, TVXL840	2.44 in. (62 mm)	1.81 in. (46 mm)	8.46 cu. in. (138 cc)

ENGINE IDENTIFICATION

Both models are vertical crankshaft, two-stroke, single-cylinder, air-cooled engines. Model TVS840 is equipped with standard piston rings and Model TVXL840 is equipped with one half keystone piston rings and one standard piston ring. All models are equipped with a float type carburetor.

Engine model and type numbers are stamped in blower housing as shown in Fig. TP4-1. Always furnish engine model and type numbers when ordering parts.

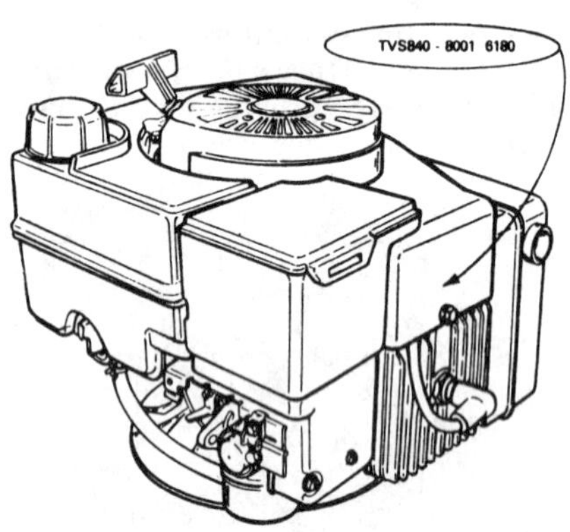

Fig. TP4-1—Engine model and type numbers are stamped in blower housing in location shown.

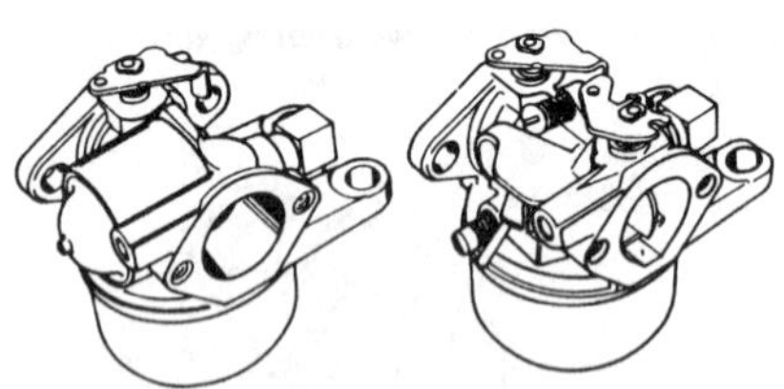

Fig. TP4-2—View showing fixed jet carburetor (left) and carburetor with adjustable low speed mixture screw (right). Refer to text.

MAINTENANCE

SPARK PLUG. Recommended spark plug for both models is a Champion CJ8Y. Specified spark plug electrode gap is 0.030 inch (0.76 mm).

CARBURETOR. Both models are equipped with a float type carburetor. Carburetor may be equipped with a fixed high speed jet and an adjustable low speed mixture screw or fixed high speed and low speed mixture jets. Refer to Fig. TP4-2.

Initial adjustment of low speed mixture screw on carburetor models so equipped, is one turn open from a lightly seated position. Make final adjustment with engine running and at operating temperature. Place throttle control in slow speed position and turn low speed mixture screw in a clockwise direction until engine falters. Note this position, then turn low speed mixture screw in a counterclockwise direction until engine starts to sputter or rpm decreases. Note this position, then turn low speed mixture screw to a position half-way between the two noted positions.

To disassemble carburetor equipped with low speed mixture screw, remove float bowl retaining bolt, float bowl, float hinge pin, float and fuel inlet needle and spring. Use a wire hook and remove fuel inlet needle seat. Use a 1/8 inch chisel to remove Welch plug (Fig. TP4-3). Do not attempt to drill Welch plug. Remove low speed mixture screw. Remove throttle and choke plates. Remove throttle and choke shafts.

When reassembling carburetor, install the new fuel inlet seat into carburetor body with grooved side of seat facing down (Fig. TP4-4). Install a new Welch plug and secure by tapping crown of Welch plug with a hammer and punch. Seal Welch plug with fingernail polish or other suitable sealer. When installing dampening spring on float, the short leg hooks onto carburetor body and the longer end points toward choke end (Fig. TP4-5). Use Tecumseh float tool 670253A to set the correct float height (Fig. TP4-6). The gage is a go-no go type. Pull the tool in a 90 degree direction to the hinge pin. The toe of the float, end opposite the hinge, must be under the first step and can touch the second step without gap. Carefully bend the float tab holding the fuel inlet needle to obtain correct height.

To disassemble carburetor equipped with fixed high speed and low speed jets, note that the fuel bowl retaining bolt stamped as shown in Fig. TP4-7 is manufactured with left-hand threads. Remove float bowl retaining bolt, float bowl, float hinge pin, float, fuel inlet needle and spring. Use a hooked wire to remove the fuel inlet needle seat. Remove primer bulb retaining ring and primer bulb. Remove throttle plate, throttle shaft, dust seal and spring.

To reassemble carburetor, install the new fuel inlet needle seat with grooved side of seat facing down (Fig. TP4-4). When installing dampening spring on float, the short leg hooks onto carburetor body and the longer end points toward choke end (Fig. TP4-5). Use Tecumseh float tool 670253A to set the correct float height (Fig. TP4-6). The gage is a go-no go type. Pull the tool in a 90 degree direction to the hinge pin. The toe of the float, end opposite the hinge, must be under the first step and can touch the second step without gap.

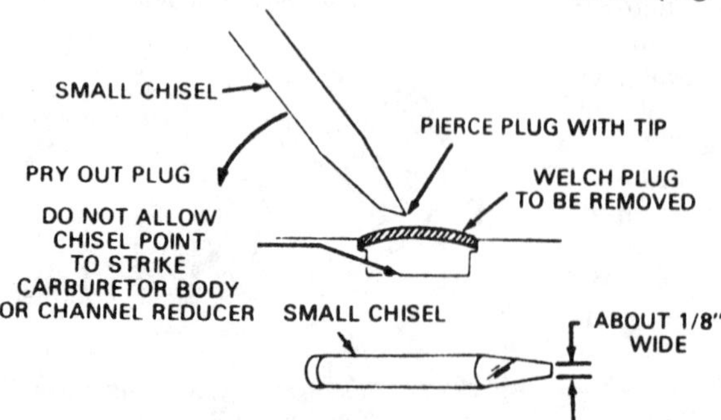

Fig. TP4-3—Use a 1/8 inch chisel to remove carburetor Welch plug.

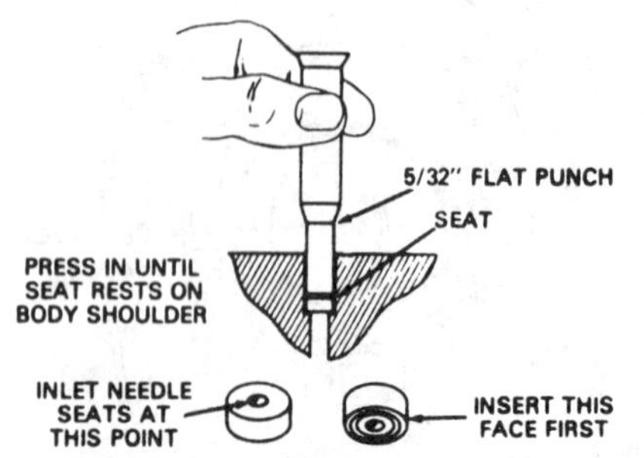

Fig. TP4-4—Fuel inlet seat must be installed in carburetor body with grooved side down.

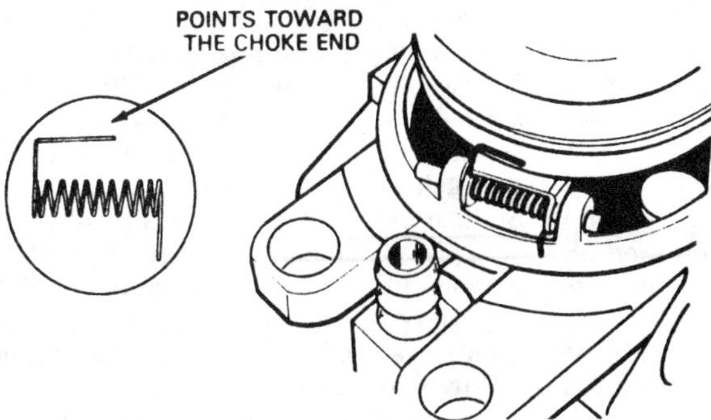

Fig. TP4-5—Install fuel inlet needle spring as shown. Refer to text.

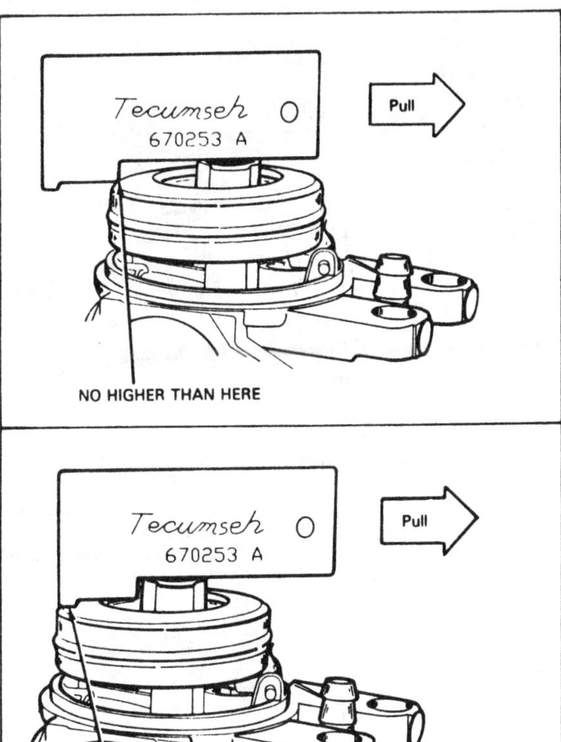

Fig. TP4-6—Tecumseh go-no go gage 670253 should be used to set correct float height. Refer to text.

Fig. TP4-7—Float bowl retaining bolt marked as shown is manufactured with left-hand threads.

Carefully bend the float tab holding the fuel inlet needle to obtain correct height. Place primer bulb and retaining ring in a 3/4 inch deep socket and install assembly into primer bulb recess in carburetor body until retaining ring is seated.

GOVERNOR. Both models are equipped with a mechanical type governor located at lower side of crankcase. All governor linkage positions should be marked prior to disassembly. To adjust governor, first place the throttle at wide-open position. Then loosen holding screw (Fig. TP4-8) and turn slotted governor shaft clockwise as far as it will go without forcing it. Tighten holding screw to secure adjustment.

IGNITION SYSTEM. Standard ignition system on all models is a solid-state electronic system which does not have breaker points. There is no scheduled maintenance.

To test for spark, remove spark plug cable from spark plug. Insert metal conductor into cable end and hold conductor 1/8 inch (3 mm) from cylinder shroud. Turn engine over and observe spark. A weak spark or no spark indicates a defective ignition coil. Also check for broken, loose or shorted wiring or a faulty spark plug.

Air gap between solid-state module and flywheel should be 0.0125 inch (0.318 mm).

CARBON. Carbon and other combustion deposits should be cleaned from exhaust port and EGR tube after every 75-100 hours of normal operation. Before cleaning the ports, remove the muffler and position the piston at bottom dead center. Clean ports using a pointed 3/8 inch (9.5 mm) wooden dowel rod. Refer also to EGR tube under REPAIRS section for EGR tube cleaning or removal.

LUBRICATION. Both models are lubricated by mixing regular unleaded gasoline with a good quality two-stroke air-cooled engine oil rated SAE 30 or SAE 40 at a 32:1 ratio. Always mix fuel in a separate container and add only mixed fuel to engine fuel tank.

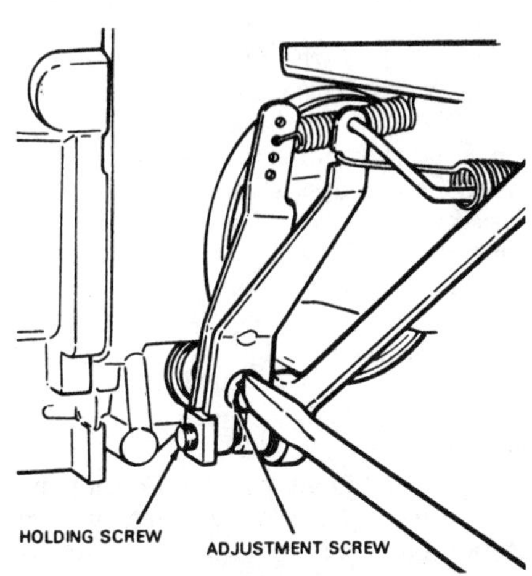

Fig. TP4-8—View of external governor linkage used on both models. Refer to text for adjustment procedure.

GENERAL MAINTENANCE. Check and tighten all loose bolts, nuts or clamps prior to each day of operation. Check for fuel leakage and repair if necessary.

Clean dust, dirt, grease or any foreign material from cylinder head and cylinder block cooling fins after every 100 hours of operation or more frequent if needed. Inspect fins for damage and repair if necessary.

REPAIRS

TIGHTENING TORQUES. Recommended tightening torques are as follows:

Spark plug	16-20 ft.-lbs. (22-27 N·m)
Crankcase cover	10-17 ft.-lbs. (13-23 N·m)
Adapter plate-to-cylinder	13-18 ft.-lbs. (18-24 N·m)
Carburetor	10-12 ft.-lbs. (13-16 N·m)
Flywheel nut	34-36 ft.-lbs. (46-49 N·m)

CYLINDER AND CRANKCASE. The cylinder is an integral part of one crankcase half. To disassemble engine, drain all fuel and remove fuel tank. Remove air cleaner assembly and muffler. Remove the four screws retaining blower housing to engine and remove housing assembly. Remove solid-state ignition module. Use strap wrench 670305 to hold flywheel, then remove flywheel retaining nut. Pull brake lever away from return spring as far as it will go and place an alignment pin through the lever hole so it catches on the shroud (Fig. TP4-18). This will hold the pad off the flywheel. Use flywheel puller 670306 to remove the flywheel. Remove the three screws retaining the speed control to the cylinder block. Mark all linkage connections and disconnect all governor and carburetor linkage. Remove the three screws from the pto mounting adapter. Lightly tap adapter to separate from crankcase. Remove carburetor, then remove intake elbow and reed valve assembly. Remove brake assembly. Remove the four crankcase bolts (Fig. TP4-9) and carefully remove crankcase cover from cylinder and crankcase assembly. Carefully lift crankshaft and pull piston assembly out of cylinder. Slide oil seal, retainer washer, ball bearing and governor slide ring off of crankshaft. Catch the governor flyweight balls as governor slide ring is removed. Remove the oil seal and ball bearing from magneto end of crankshaft. Slide the piston and connecting rod off toward the magneto end of crankshaft. The 31 needle bearings in connecting rod are loose and will fall out during connecting rod removal.

Clean and inspect all parts. Standard cylinder bore diameter is 2.437-2.438 inches (61.89-61.93 mm).

To reassemble engine, use heavy grease to retain connecting rod needle bearings (new bearings are retained on a strip) and position piston and connecting rod on crankshaft as shown in Fig. TP4-10. Make certain flanged side of connecting rod (Fig. TP4-11) is toward magneto end of crankshaft, then work connecting rod and piston assembly onto crankshaft. Install the three governor flyweight balls into the crankshaft. Install governor slide ring onto crankshaft so flat portion of ring covers the flyweight balls. Install ball bearing type main bearing, retainer washer and the large oil seal as shown in Fig. TP4-12. Install ball bearing type main bearing and smaller oil seal onto magneto end of crankshaft as shown in Fig. TP4-13. Stagger ring end gaps and carefully install piston and connecting rod assembly into cylinder and crankcase assembly. Tapered edge at bottom of cylinder will compress rings. Position governor follower arm so it rides on the pto side of the governor slide ring as shown in

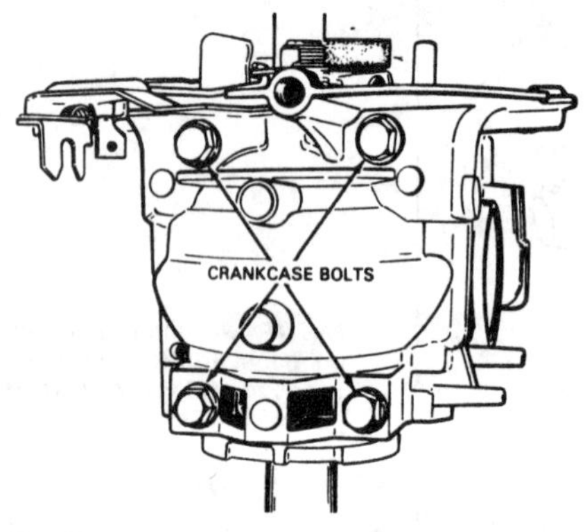

Fig. TP4-9—View showing location of the four crankcase cover bolts.

Fig. TP4-10—Piston and connecting rod assembly are installed by working connecting rod over crankshaft as shown.

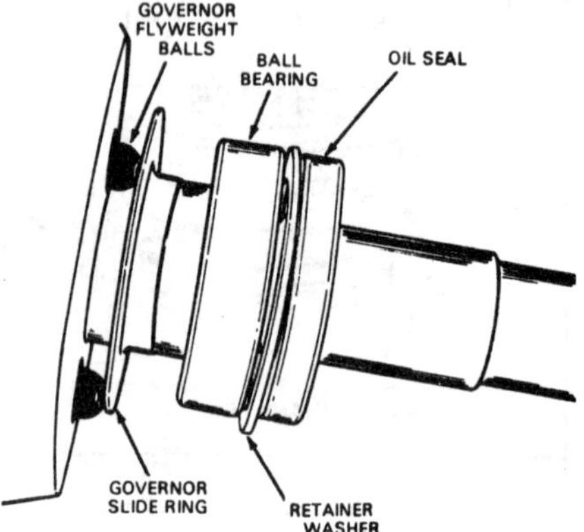

Fig. TP4-12—Install governor flyweight balls, governor slide ring, ball bearing, retainer washer and oil seal onto pto side of crankshaft following arrangement shown.

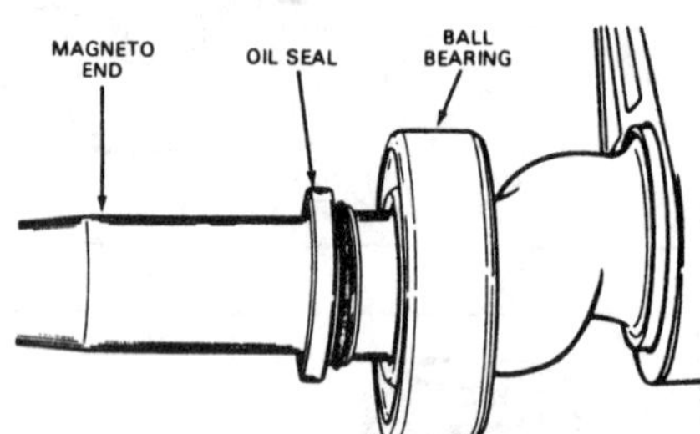

Fig. TP4-13—Install ball bearing and oil seal onto magneto side of crankshaft as shown.

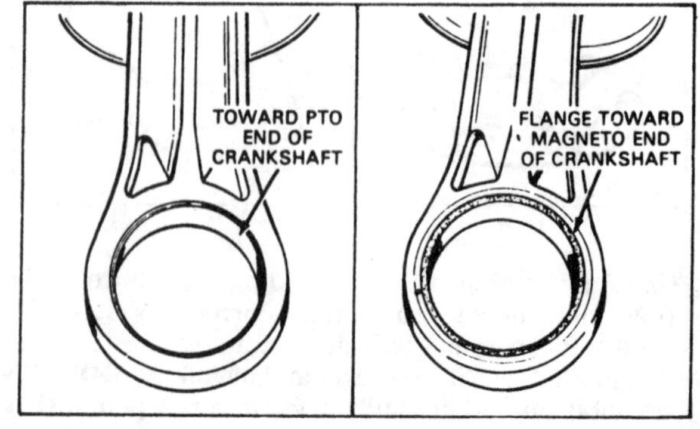

Fig. TP4-11—Connecting rod must be installed on crankshaft so flanged side of connecting rod is toward magneto end of crankshaft.

Fig. TP4-14. Apply a silicon type sealer onto crankcase and crankcase cover mating surfaces and install crankcase cover. Crankcase cover and crankcase must be aligned at pto end of crankcase as shown in Fig. TP4-15. Use a straightedge to make certain crankcase halves are aligned, then tighten the four crankcase cover bolts to specified torque. Refer to Fig. TP4-16 and install brake assembly. Install new "O" rings on intake elbow and reed assembly prior to installation. Position one fiber washer on each carburetor post prior to mounting carburetor on cylinder block (Fig. TP4-17). Float bowl of carburetor must be facing pto side of engine. Install the pto mounting adapter so the holes line up. Gently tap adapter until it is seated, then install the retaining screws. Install all governor and carburetor linkage. Disengage brake and install an alignment pin as shown in Fig. TP4-18. Install flywheel and tighten flywheel retaining nut to specified torque. Install solid-state ignition module with a 0.0125 inch (0.318 mm) gap between module and flywheel. Tighten module retaining screws to 30-40 in.-lbs. (3.4-4.5 N·m). Remove brake alignment pin. Install blower housing and rewind starter assembly. Install gasket and muffler, air cleaner assembly and fuel tank.

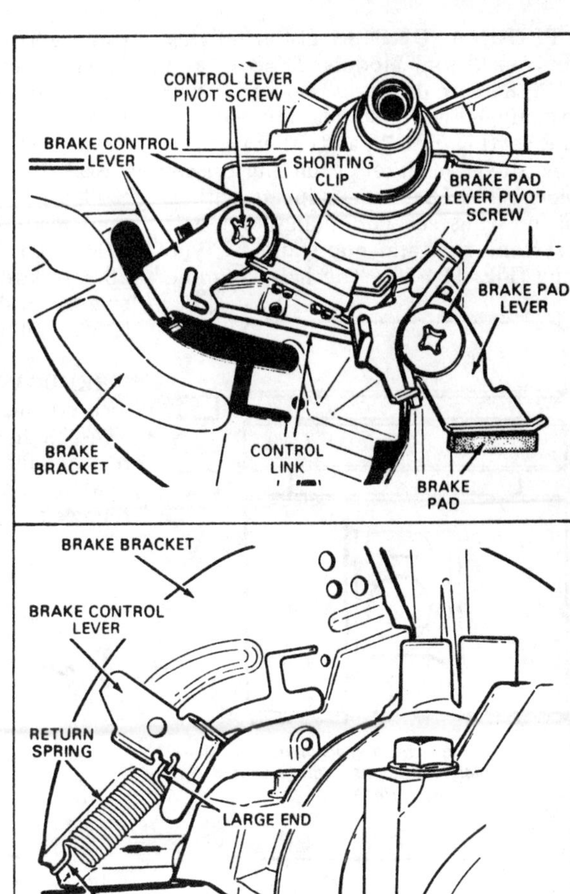

Fig. TP4-16—View showing assembled position of flywheel brake assembly.

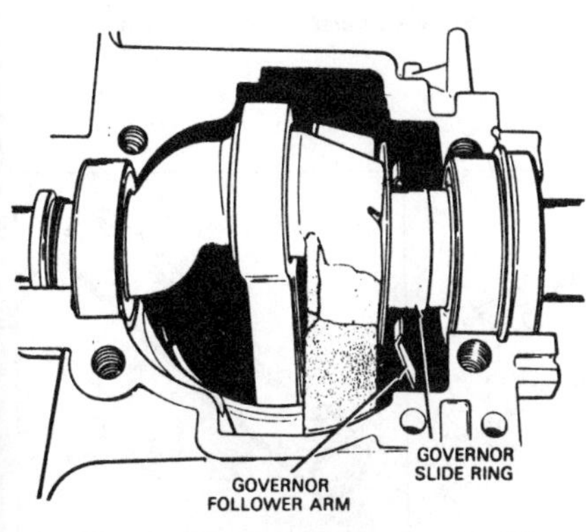

Fig. TP4-14—Governor follower arm must be on pto side of governor slide ring as shown after crankshaft installation in crankcase.

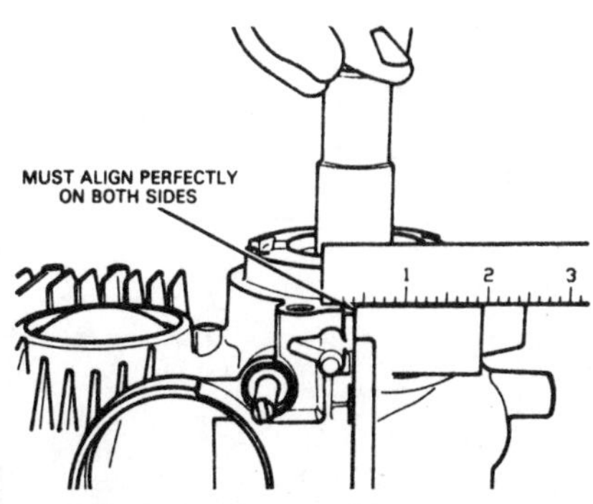

Fig. TP4-15—When joining crankcase halves, crankcase cover and crankcase must be perfectly aligned. Refer to text.

CRANKSHAFT AND CONNECTING ROD. Refer to previous CYLINDER AND CRANKCASE paragraphs for crankshaft and connecting rod removal procedure.

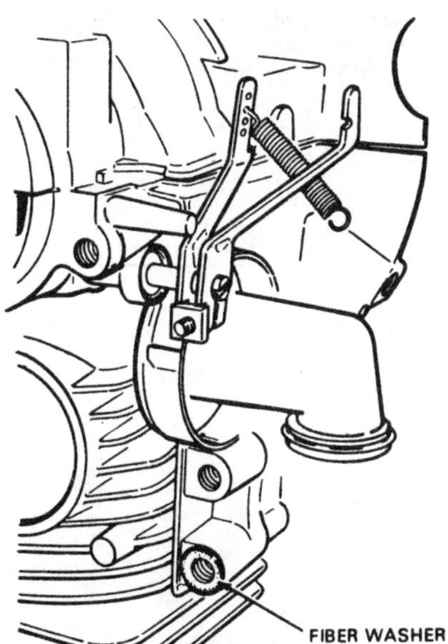

Fig. TP4-17—One fiber washer must be installed on each carburetor post prior to mounting carburetor on cylinder block.

Crankshaft main bearing journal diameter at pto side should be 0.9833-0.9838 inch (24.976-24.989 mm). Crankshaft main bearing journal diameter at magneto side should be 0.7864-0.7869 inch (19.975-19.987 mm). Crankpin journal diameter should be 0.9710-0.9715 inch (24.663-24.676 mm). Crankshaft end play should be 0.0004-0.0244 inch (0.010-0.620 mm).

Connecting rod and piston assembly must be installed on crankshaft as shown in Fig. TP4-11.

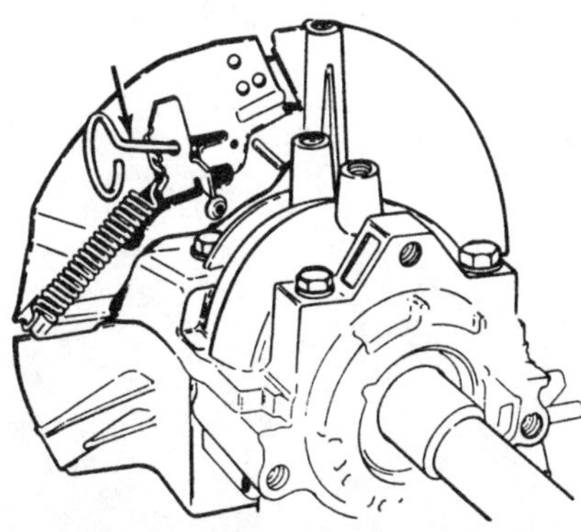

Fig. TP4-18—Disengage flywheel brake by extending brake lever as shown then install an alignment pin through hole in lever to retain position.

PISTON AND RINGS. Piston and ring sets used on Models TVS840 and TVXL840 are different. Model TVS840 is equipped with a standard type piston ring set (Fig. TP4-19) and either ring can be installed in either piston ring groove. Model TVXL840 is equipped with a piston ring set containing one half keystone ring and one standard type ring (Fig. TP4-20). The half keystone ring and piston ring groove are tapered. The half keystone type ring must be installed in the upper piston ring groove with the long beveled side of ring toward top of piston (Fig. TP4-20).

Standard piston diameter is 2.4325-2.4330 inches (61.785-61.798 mm). Piston ring groove width should be 0.0645-0.0655 inch (1.638-1.664 mm) for all standard type ring grooves. Piston ring width should be 0.0615-0.0625 inch (1.562-1.588 mm).

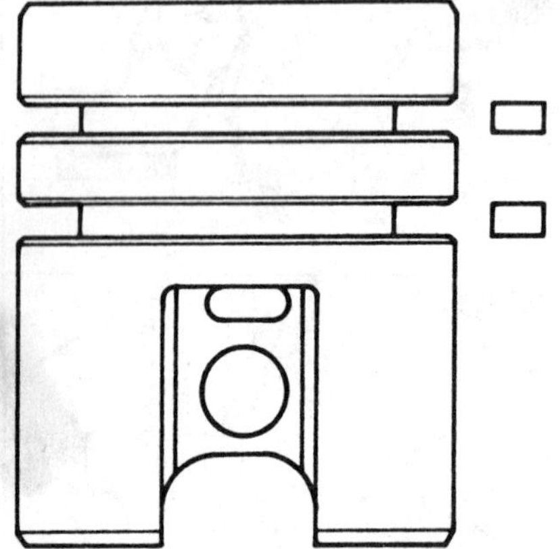

Fig. TP4-19—View showing piston used on Model TVS840 which is equipped with standard type rings.

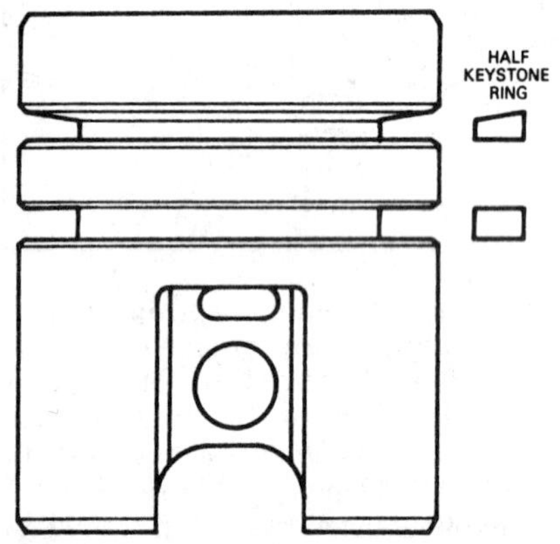

Fig. TP4-20—View showing piston used on Model TVXL840 which is equipped with one half keystone type ring and one standard type ring. Note correct position of each ring.

REED VALVE. The reed plate is located on the intake elbow (Fig. TP4-21). If clearance between end of reed and intake elbow is 0.010 inch (0.25 mm) or more, intake elbow must be renewed.

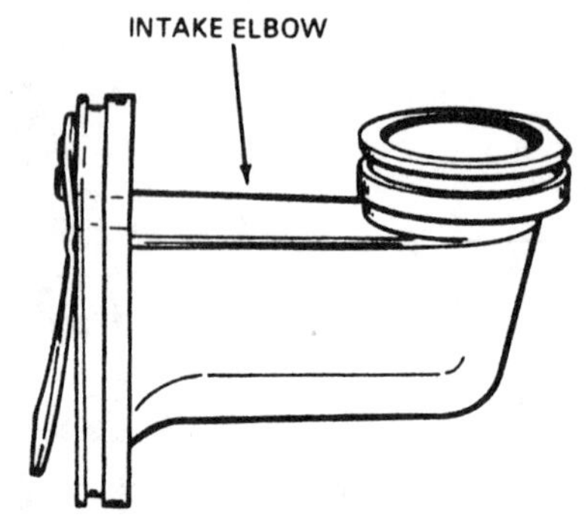

Fig. TP4-21—Reed valve is an integral part of the intake elbow. Refer to text.

EGR TUBE. Both models are equipped with an EGR (exhaust gas recirculation) tube located near the cylinder exhaust port. A controlled amount of exhaust gas is drawn into the crankcase to aid movement of the air-fuel mixture to the combustion chamber. This tube should be cleaned when cleaning combustion deposits from exhaust ports.

To remove the EGR tube, measure distance EGR tube protrudes from block, then clamp tube with vise grips and rotate in a clockwise direction. This will collapse the tube and allow removal (Fig. TP4-22). Install new tube with seam in tube facing 45 degrees to the exhaust port as shown in Fig. TP4-23 and to the same depth as old tube. Tecumseh tool 670318 is available to aid installation of EGR tube to correct depth.

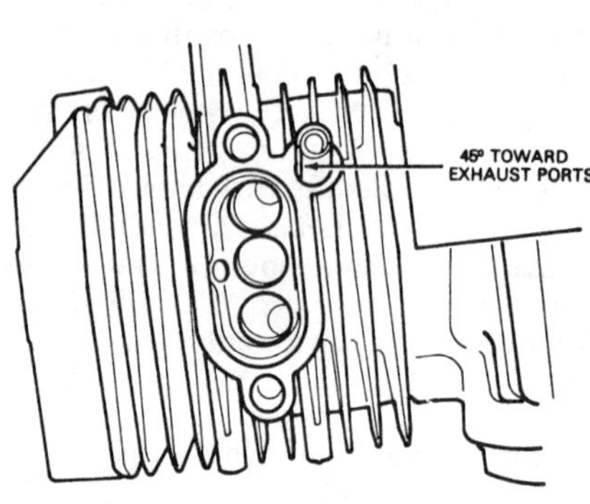

Fig. TP4-23—EGR tube must be installed with seal toward exhaust ports as shown.

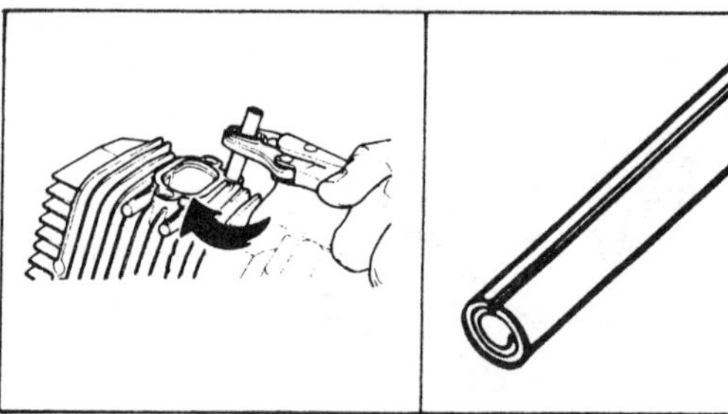

Fig. TP4-22—To remove the EGR tube, clamp vise grips onto tube then rotate in a clockwise direction to collapse tube.

TECUMSEH 4-STROKE

Model	Bore	Stroke	Displacement
LAV25, LAV30, TVS75, H25, H30	2.313 in. (58.74 mm)	1.844 in. (46.84 mm)	7.75 cu. in. (127 cc)
LV35, LAV35, TVS90, H35-prior 1983, ECH90	2.500 in. (63.50 mm)	1.844 in. (46.84 mm)	9.05 cu. in. (148 cc)
H35 after 1983	2.500 in. (63.50 mm)	1.938 in. (49.23 mm)	9.51 cu. in. (156 cc)
LAV40, TVS105, HS40, ECV105	2.625 in. (66.68 mm)	1.938 in. (49.23 mm)	10.49 cu. in. (172 cc)
TNT100, ECV100	2.625 in. (66.68 mm)	1.844 in. (46.84 mm)	9.98 cu. in. (164 mm)
V40, VH40, H40, HH40	2.500 in. (63.50 mm)	2.250 in. (57.15 mm)	11.04 cu. in. (181 cc)
ECV110	2.750 in. (69.85 mm)	1.938 in. (49.23 mm)	11.50 cu. in. (189 cc)
LAV50, ECV120, TNT120, TVS120, HS50	2.812 in. (71.43 mm)	1.938 in. (49.23 mm)	12.04 cu. in. (197 cc)
H50, HH50, V50, VH50, TVM125	2.812 in. (71.43 mm)	2.250 in. (57.15 mm)	12.18 cu. in. (229 cc)
V60, VH60, TVM140	2.625 in. (66.68 mm)	2.500 in. (63.50 mm)	13.53 cu. in. (222 cc)

ENGINE IDENTIFICATION

Engines must be identified by the complete model number, including the specification number in order to obtain correct repair parts. These numbers are located on the name plate and/or tags that are positioned as shown in Fig. T1 or T1A. It is important to transfer identification tags from the original engine to replacement short block assemblies so unit can be identified when servicing.

If selecting a replacement engine and model or type number of the old engine is not known, refer to chart in Fig. T1B and proceed as follows:

1. List the corresponding number which indicates the crankshaft position.
2. Determine the horsepower needed. (Two-stroke engines are indicated by 00 in the second and third digit positions).
3. Determine the primary features needed. (Refer to the Tecumseh Engines Specification Book No. 692531 for specific engine variations.
4. Refer to Fig. T1C for Tecumseh engine model number interpretation.

The number following the letter code is the horsepower or cubic inch displacement. The number following the model number is the specification number. The last three digits of the specification number indicate a variation to the basic engine specification.

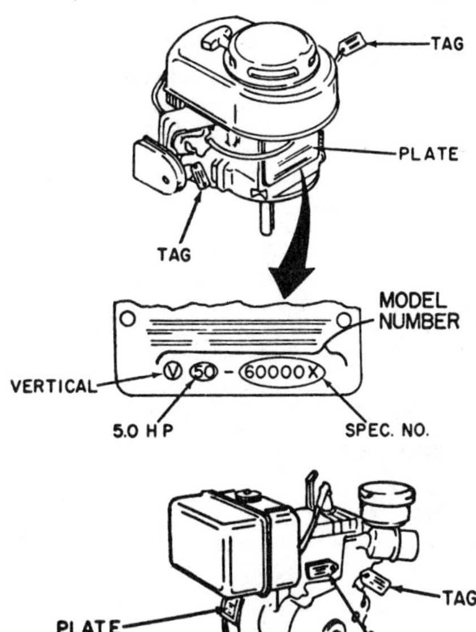

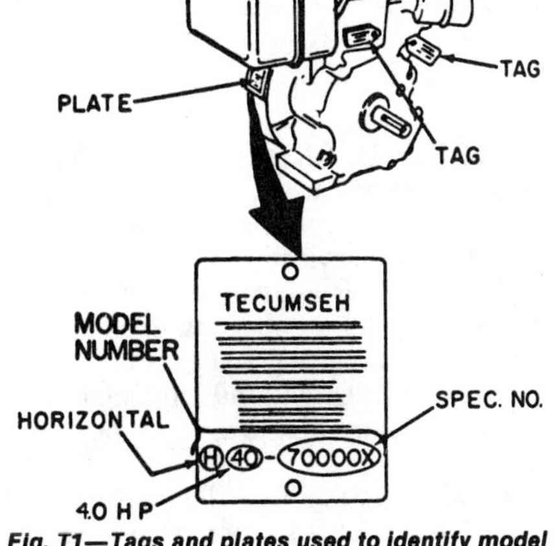

Fig. T1—Tags and plates used to identify model will most often be located in one of the positions shown.

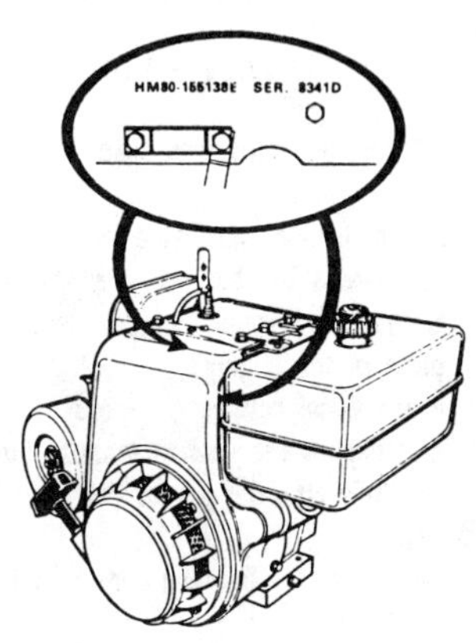

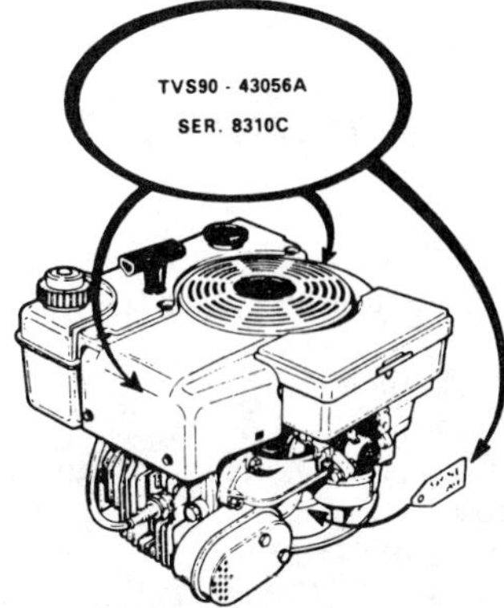

Fig. T1A—Locations of tags and plates used to identify later model engines.

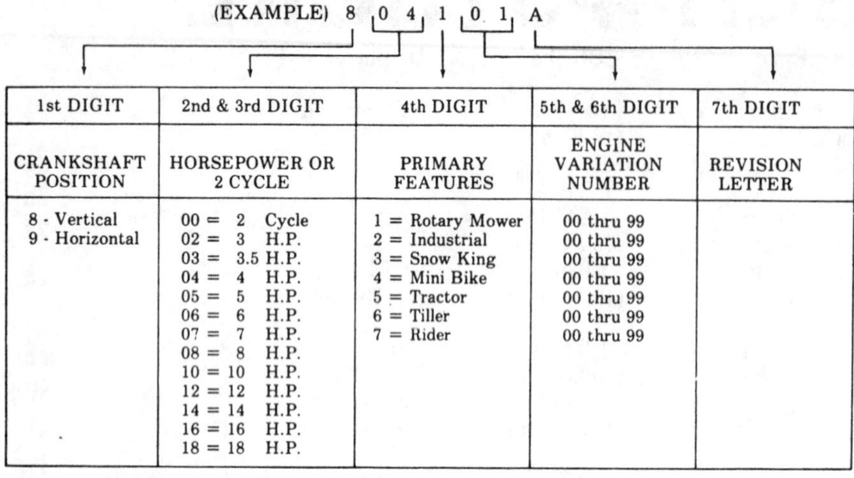

TECUMSEH NUMERICAL SERVICE NUMBER SYSTEM

(EXAMPLE) 8 0 4 1 0 1 A

1st DIGIT	2nd & 3rd DIGIT	4th DIGIT	5th & 6th DIGIT	7th DIGIT
CRANKSHAFT POSITION	HORSEPOWER OR 2 CYCLE	PRIMARY FEATURES	ENGINE VARIATION NUMBER	REVISION LETTER
8 - Vertical	00 = 2 Cycle	1 = Rotary Mower	00 thru 99	
9 - Horizontal	02 = 3 H.P.	2 = Industrial	00 thru 99	
	03 = 3.5 H.P.	3 = Snow King	00 thru 99	
	04 = 4 H.P.	4 = Mini Bike	00 thru 99	
	05 = 5 H.P.	5 = Tractor	00 thru 99	
	06 = 6 H.P.	6 = Tiller	00 thru 99	
	07 = 7 H.P.	7 = Rider	00 thru 99	
	08 = 8 H.P.			
	10 = 10 H.P.			
	12 = 12 H.P.			
	14 = 14 H.P.			
	16 = 16 H.P.			
	18 = 18 H.P.			

Fig. T1B—Reference chart used to select or identify replacement engines.

V - Vertical Shaft
LAV - Lightweight Aluminum Vertical
VM - Vertical Medium Frame
VH - Vertical Heavy Duty (Cast Iron)
TVS - Tecumseh Vertical Styled
TNT - Toro N' Tecumseh
ECV - Exclusive Craftsman Vertical
H - Horizontal Shaft
HS - Horizontal Small Frame
HM - Horizontal Medium Frame
HHM - Horizontal Heavy Duty (Cast Iron) Medium Frame
HH - Horizontal Heavy Duty (Cast Iron)
ECH - Exclusive Craftsman Horizontal

(EXAMPLE)

TVS90-43056A - is the model and specification number

TVS - Tecumseh Vertical Styled
90 - Indicates a 9 cubic inch displacement
43056A - is the specification number used for properly identifying the parts of the engine
8310C - is the serial number
8 - first digit is the year of manufacture (1978)
310 - indicates calendar day of that year (310th day or November 6, 1978)
C - represents the line and shift on which the engine was built at the factory.

Fig. T1C—Chart showing model number interpretation.

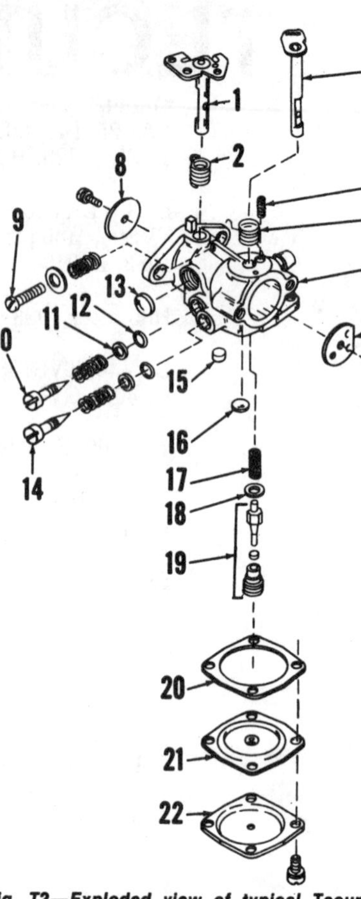

Fig. T2—Exploded view of typical Tecumseh diaphragm carburetor.

1. Throttle shaft
2. Return spring
3. Choke shaft
4. Choke stop spring
5. Return spring
6. Carburetor body
7. Choke plate
8. Throttle plate
9. Idle speed screw
10. Idle mixture screw
11. Washers
12. "O" rings
13. Welch plug
14. Main fuel mixture screw
15. Cup plug
16. Welch plug
17. Inlet needle spring
18. Gasket
19. Inlet needle & seat assy.
20. Gasket
21. Diaphragm
22. Cover

MAINTENANCE

SPARK PLUG. Spark plug recommendations are shown in the following chart.

14mm – 3/8 inch reach
Gasoline . . . . Champion J8
LP-Gas . . . . Champion J8
Kerosene . . . . Champion UJ12
18mm – 1/2 inch reach
Gasoline . . . . Champion D16 or MD16
Kerosene . . . . Champion D21
All 7/8 inch reach . . . . . . Champion W18

CARBURETOR. Several different carburetors are used on these engines. Refer to the appropriate paragraph for model being serviced.

Tecumseh Diaphragm Carburetor. Refer to model number stamped on the carburetor mounting flange and to Fig. T2 for identification and exploded view of Tecumseh diaphragm type carburetor.

Initial adjustment of idle mixture and main fuel mixture screws from a lightly seated position, is 1 turn open. Clockwise rotation leans mixture and counterclockwise rotation richens mixture.

Final adjustments are made with engine at operating temperature and running. Operate engine at rated speed and adjust main fuel mixture screw (14) for smoothest engine operation. Operate engine at idle speed and adjust idle mixture screw (10) for smoothest engine idle. If engine does not accelerate smoothly, slight adjustment of main fuel mixture screw may be required. Engine idle speed should be approximately 1800 rpm.

The fuel strainer in the fuel inlet fitting can be cleaned by reverse flushing with compressed air after the inlet needle and seat (19 – Fig. T2) are removed. The inlet needle seat fitting is metal with a neoprene seat, so the fitting (and enclosed seat) should be removed before carburetor is cleaned with a commercial solvent.The stamped line on carburetor throttle plate should be toward top of carburetor, parallel with throttle shaft

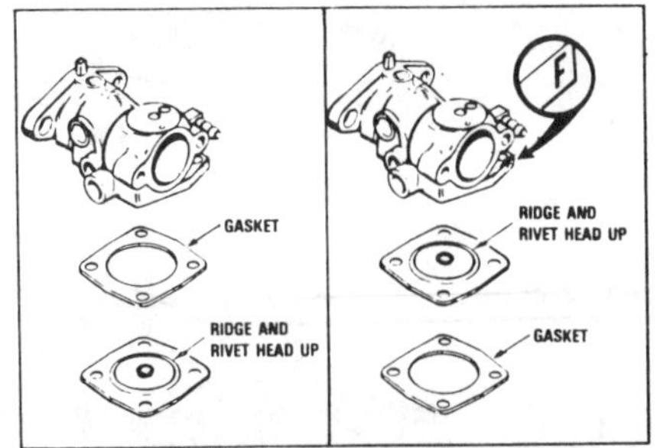

Fig. T2A—Illustration showing correct position of diaphragm on carburetors unmarked and marked with a "F".

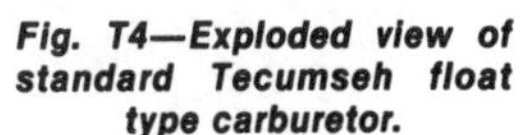

Fig. T4—Exploded view of standard Tecumseh float type carburetor.

1. Idle speed screw
2. Throttle plate
3. Return spring
4. Throttle shaft
5. Choke stop spring
6. Choke shaft
7. Return spring
8. Fuel inlet fitting
9. Carburetor body
10. Choke plate
11. Welch plug
12. Idle mixture needle
13. Spring
14. Washer
15. "O" ring
16. Ball plug
17. Welch plug
18. Pin
19. Cup plugs
20. Bowl gasket
21. Inlet needle seat
22. Inlet needle
23. Clip
24. Float shaft
25. Float
26. Drain stem
27. Gasket
28. Bowl
29. Gasket
30. Bowl retainer
31. "O" ring
32. Washer
33. Spring
34. Main fuel needle

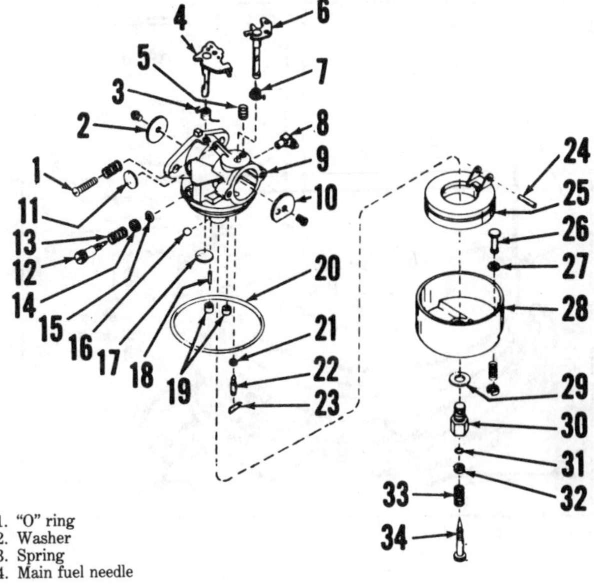

and facing OUTWARD as shown in Fig. T3. Flat side of choke plate should be toward the fuel inlet fitting side of carburetor. Mark on choke plate should be parallel to shaft and should face INWARD when choke is closed. Diaphragm (21–Fig. T2) should be installed with rounded head of center rivet up toward the inlet needle (19) regardless of size or placement of washers around the rivet.

On carburetor Models 0234-252, 265, 266, 269, 270, 271, 282, 293, 303, 322, 327, 333, 334, 344, 345, 348, 349, 350, 351, 352, 356, 368, 371, 374, 378, 379, 380, 404 and 405, or carburetors marked with a "F" as shown in Fig. T2A, gasket (20–Fig. T2) must be installed between diaphragm (21) and cover (22). All other models are assemblies as shown, with gasket between diaphragm and carburetor body.

Tecumseh Standard Float Carburetor. Refer to Fig. T4 for identification and exploded view of Tecumseh standard float type carburetor.

Initial adjustment of idle mixture and main fuel mixture screws from a lightly seated position, is 1 turn open. Clockwise rotation leans mixture and counterclockwise rotation richens mixture.

Final adjustments are made with engine at operating temperature and running. Operate engine at rated speed and adjust main fuel mixture screw (34) for smoothest engine operation. Operate engine at idle speed and adjust idle mixture screw for smoothest engine idle. If engine does not accelerate smoothly, slight adjustment of main fuel mixture screw may be required.

Carburetor must be disassembled and all neoprene or Viton rubber parts removed before carburetor is immersed in cleaning solvent. Do not attempt to reuse any expansion plugs. Install new plugs if any are removed for cleaning.

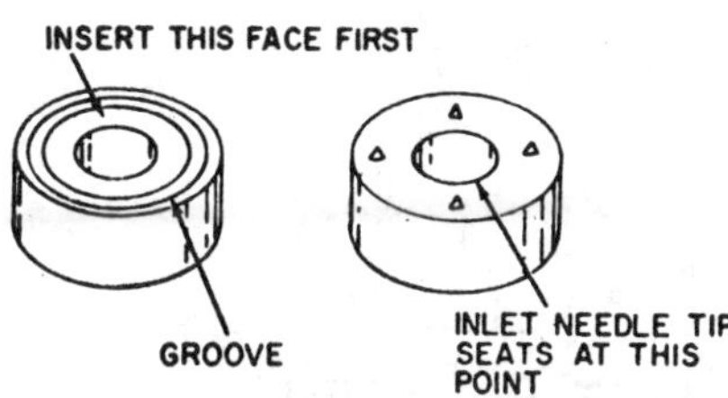

Fig. T7—The Viton seat used on some Tecumseh carburetors must be installed correctly to operate properly. All-metal needle is used with seat shown.

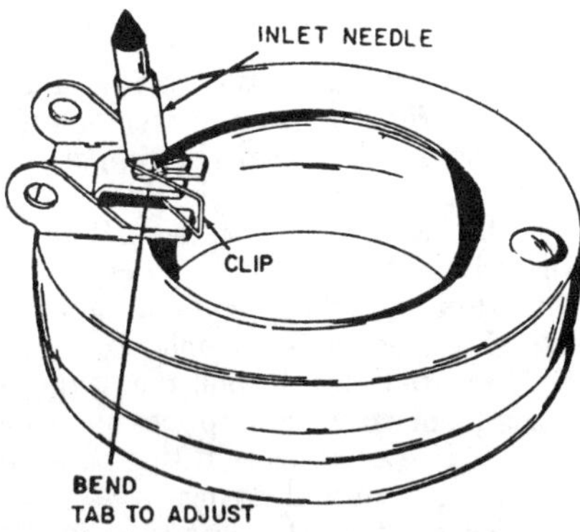

Fig. T5—View of float and fuel inlet valve needle. The valve needle shown is equipped with a resilient tip and a clip. Bend tab shown to adjust float height.

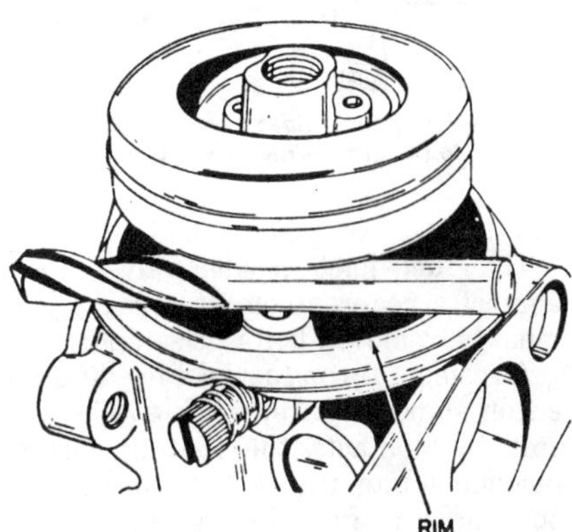

Fig. T8—Float height can be measured on some models by using a drill as shown. Refer to text for correct specifications.

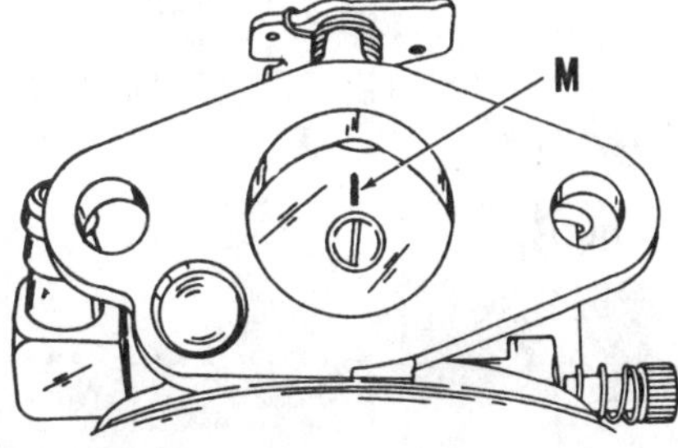

Fig. T3—The mark (M) on throttle plate should be parallel to the throttle shaft and outward as shown. Some models may also have mark at 3 o'clock position.

Fig. T6—A 10-24 or 10-32 tap is used to pull the brass seat fitting and fuel inlet valve seat from some carburetors. Use a close fitting flat punch to install new seat and fitting.

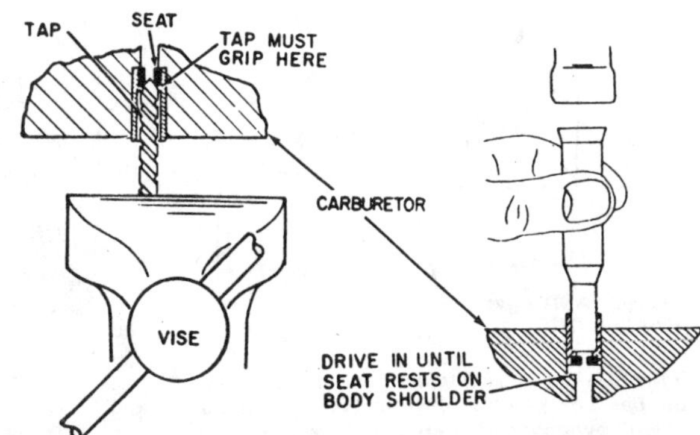

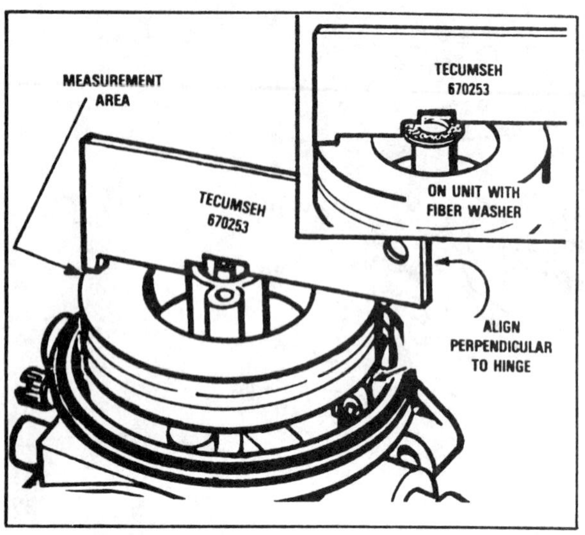

Fig. T8A—Float height can be set using Tecumseh float tool 670253 as shown.

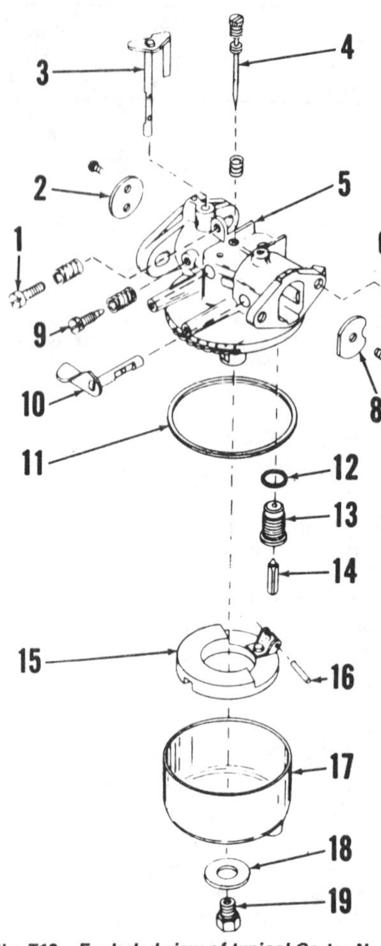

Fig. T13—Exploded view of typical Carter N carburetor.

1. Idle speed screw
2. Throttle plate
3. Throttle shaft
4. Main adjusting screw
5. Carburetor body
6. Choke shaft spring
7. Ball
8. Choke plate
9. Idle mixture screw
10. Choke shaft
11. Bowl gasket
12. Gasket
13. Inlet valve seat
14. Inlet valve
15. Float
16. Float shaft
17. Fuel bowl
18. Gasket
19. Bowl retainer

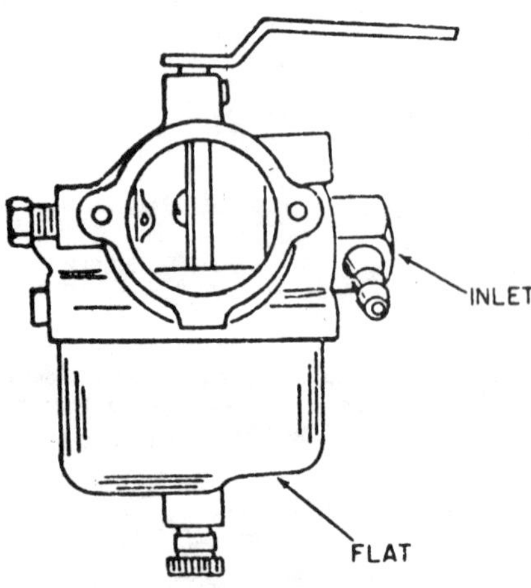

Fig. T9—Flat part of float bowl should be located under the fuel inlet fitting.

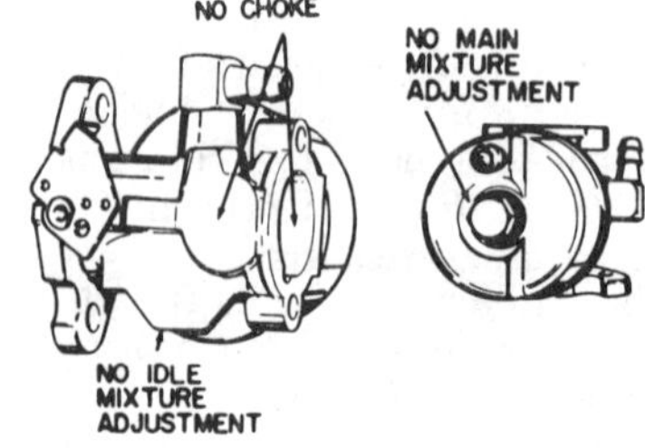

Fig. T11—Tecumseh "Automagic" carburetor is similar to standard float models, but can be identified by absence of choke and adjusting needles.

The fuel inlet needle valve closes against a neoprene or Viton seat which must be removed before cleaning.

A resilient tip on fuel inlet needle is used on some carburetors (Fig. T5). The soft tip contacts the seating surface machined into the carburetor body to shut off the fuel. Do not attempt to remove the inlet valve seat.

A Viton seat (21–Fig. T4) is used on some models. The rubber seat can be removed by blowing compressed air in from the fuel inlet fitting or by using a hooked wire. The grooved face of valve seat should be IN toward bottom of bore and the valve needle should seat on smooth side of the Viton seat. Refer to Fig. T7.

A Viton seat contained in a brass seat fitting is used on some models. Use a 10-24 or 10-32 tap to pull the seat and fitting from carburetor bore as shown in Fig. T6. Use a flat, close fitting punch to install new seat and fitting.

Install the throttle plate (2–Fig. T4) with the two stamped marks out and at 12 and 3 o'clock positions. The 12 o'clock line should be parallel with the throttle shaft and toward top of carburetor. Install choke plate (10) with flat side down toward bottom of carburetor. Float setting should be 0.200-0.220 inch (5.08-5.6

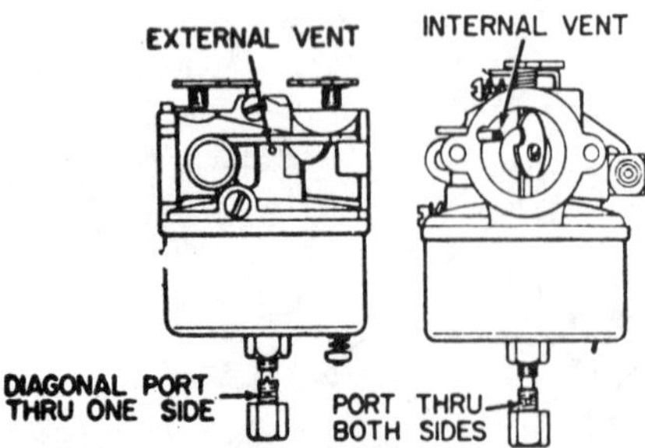

Fig. T10—The bowl retainer contains a drilled fuel passage which is different for carburetors with external and internal fuel bowl vent.

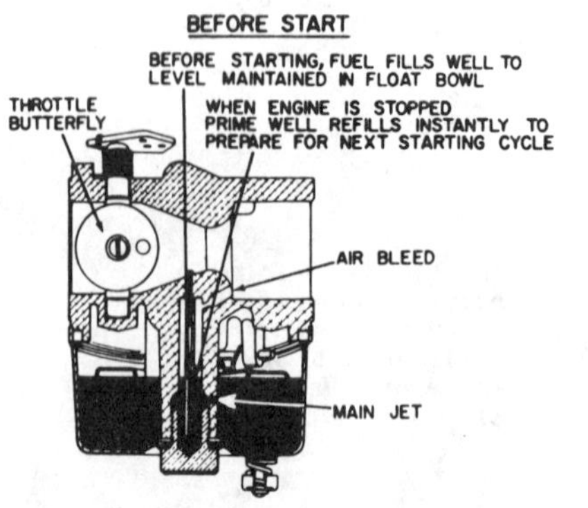

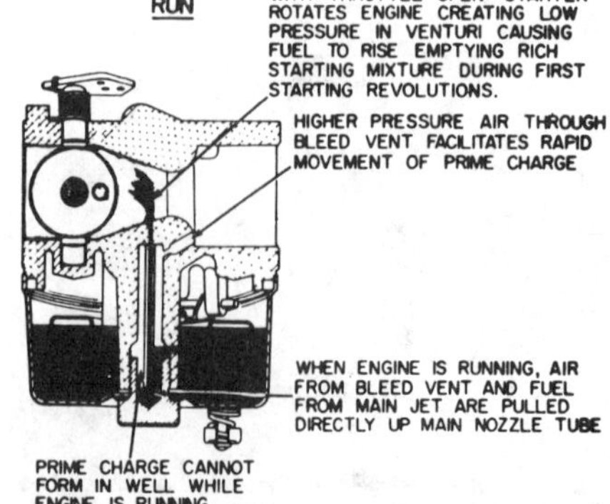

Fig. T12—The "Automagic" carburetor provides a rich starting mixture without using a choke plate. Mixture is changed by operating with a dirty air filter and by incorrect float setting.

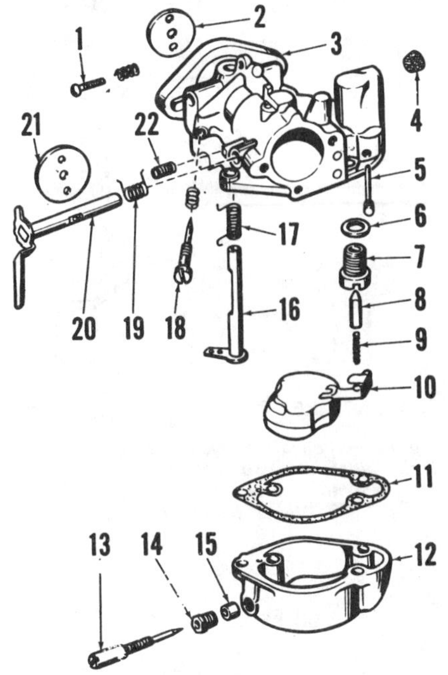

Fig. T14—Exploded view of Marvel-Schebler carburetor.

1. Idle speed screw
2. Throttle plate
3. Carburetor body
4. Fuel screen
5. Float
6. Gasket
7. Inlet valve seat
8. Inlet valve
9. Spring
10. Float
11. Gasket
12. Fuel bowl
13. Main adjusting screw
14. Retainer
15. Packing
16. Throttle shaft
17. Throttle spring
18. Idle mixture screw
19. Choke spring
20. Choke shaft
21. Choke ratchet spring

mm) and can be measured with a number 4 drill as shown in Fig. T8 or using Tecumseh float gage 670253 as shown in Fig. T8A. Remove float and bend tab at float hinge to change float setting. The fuel inlet fitting (8 – Fig. T4) is pressed into body on some models. Start fitting, then apply a light coat of Loctite to shank and press fitting into

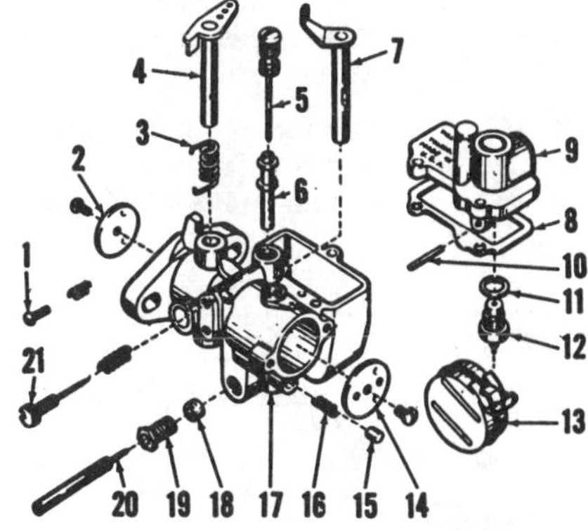

Fig. T15—Exploded view of typical Tillotson MT carburetor.

1. Idle speed screw
2. Throttle plate
3. Throttle spring
4. Throttle shaft
5. Idle tube
6. Main nozzle
7. Choke shaft
8. Gasket
9. Bowl cover
10. Float shaft
11. Gasket
12. Inlet needle & seat assy.
13. Float
14. Choke plate
15. Friction pin
16. Choke shaft spring
17. Carburetor body
18. Packing
19. Retainer
20. Main adjusting screw
21. Idle mixture screw

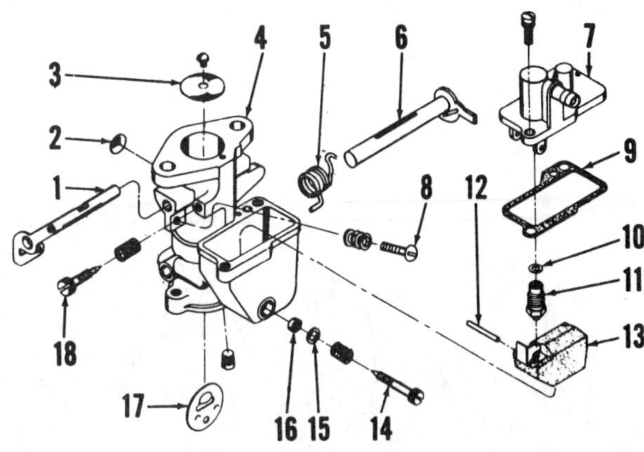

Fig. T16—Exploded view of typical Tillotson E carburetor.

1. Throttle shaft
2. Welch plug
3. Throttle plate
4. Carburetor body
5. Return spring
6. Choke shaft
7. Bowl cover
8. Idle speed stop screw
9. Bowl gasket
10. Gasket
11. Inlet needle & seat assy.
12. Float shaft
13. Float
14. Main fuel needle
15. Washer
16. Packing
17. Choke plate
18. Idle mixture screw

position. The flat on fuel bowl should be under the fuel inlet fitting. Refer to Fig. T9.

Be sure to use correct parts when servicing the carburetor. Some gaskets used as (20 – Fig. T4) are square section, while others are round. The bowl retainer (30) contains a drilled passage for fuel to the high speed metering needle (34). A diagonal port through one side of the bowl retainer is used on carburetors with external vent. The port is through both sides on models with internal vent. Refer to Fig. T10.

Tecumseh "Automagic" Float Carburetor. Refer to Fig. T11 and note carburetor can be identified by the absence of the idle mixture screw, choke and main mixture screw.

Float setting is 0.210 inch (5.33 mm) or float may be set with a number 4 drill as shown in Fig. T8 or Tecumseh float gage 670253 as shown in Fig. T8A. Air filter maintenance is important in order to obtain the correct fuel mixture. Refer to notes for servicing standard Tecumseh float carburetors. Refer to Fig. T12 for operating principles.

Carter Carburetor. Refer to Fig. T13 for identification and exploded view of Carter carburetor.

Initial adjustment of idle mixture screw and main fuel mixture screw from a lightly seated position is 1½ turns open for idle mixture screw and 1¾ turns open for main fuel mixture screw. Clockwise rotation leans mixture and counterclockwise rotation richens mixture.

Final adjustments are made with engine at operating temperature and running. Operate engine at rated speed and adjust main fuel mixture screw (4) for smoothest engine operation. Operate engine at idle speed and adjust idle mixture screw (9) for smoothest engine idle. If engine acceleration is not smooth, main fuel mixture screw may have to be adjusted slightly.

To check float level, invert carburetor body and float assembly. There should be 11/64 inch (4.37 mm) clearance between free side of float and machined surface of body casting. Bend float lever tang to provide correct measurement.

Marvel-Schebler. Refer to Fig. T14 for identification and exploded view of Marvel-Schebler Series AH carburetor.

Initial adjustment of idle mixture screw and main fuel mixture screw from a lightly seated position, is 1 turn open for idle mixture screw and 1¼ turn open for main fuel mixture screw. Clockwise rotation leans mixture and counterclockwise rotation richens mixture.

Final adjustments are made with engine at operating temperature and running. Operate engine at rated speed and adjust main fuel mixture screw (13) for smoothest engine operation. Operate engine at idle speed and adjust idle fuel mixture needle (18) for smoothest engine idle. Adjust idle speed screw (1) to obtain 1800 rpm.

To adjust float level, invert carburetor throttle body and float assembly. Float clearance should be 3/32 inch (2.38 mm) between free end of float and machined surface of carburetor body. Bend tang on float which contacts fuel inlet needle as necessary to obtain correct float level.

Tillotson Type "MT" Carburetor. Refer to Fig. T15 for identification and exploded view of Tillotson type MT carburetor.

Initial adjustment of idle mixture screw and main fuel mixture screw from

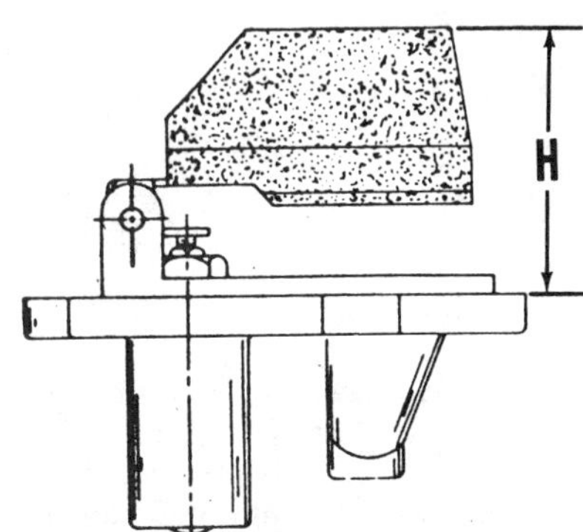

Fig. T17—Float height (H) should be measured as shown for Tillotson type "E" carburetors. Gasket should not be installed when measuring.

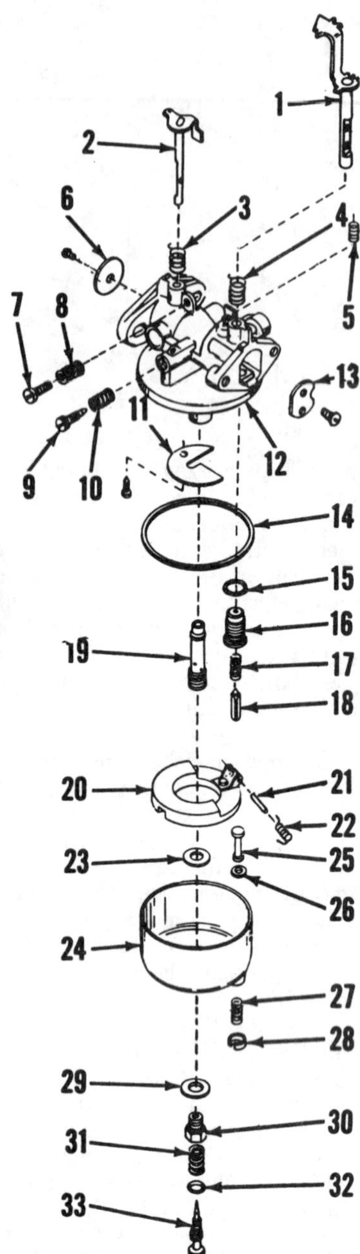

Fig. T18—Exploded view of Walbro LMG carburetor.

1. Choke shaft
2. Throttle shaft
3. Throttle return spring
4. Choke return spring
5. Choke stop spring
6. Throttle plate
7. Idle speed stop screw
8. Spring
9. Idle mixture screw
10. Spring
11. Baffle
12. Carburetor body
13. Choke plate
14. Bowl gasket
15. Gasket
16. Inlet valve seat
17. Spring
18. Inlet valve
19. Main nozzle
20. Float
21. Float shaft
22. Spring
23. Gasket
24. Bowl
25. Drain stem
26. Gasket
27. Spring
28. Retainer
29. Gasket
30. Bowl retainer
31. Spring
32. "O" ring
33. Main fuel screw

a lightly seated position, is ¾ turn open for idle mixture screw (21) and 1 turn open for main fuel mixture screw (20). Clockwise rotation leans fuel mixture and counterclockwise rotation richens fuel mixture. Final adjustments are made with engine at operating

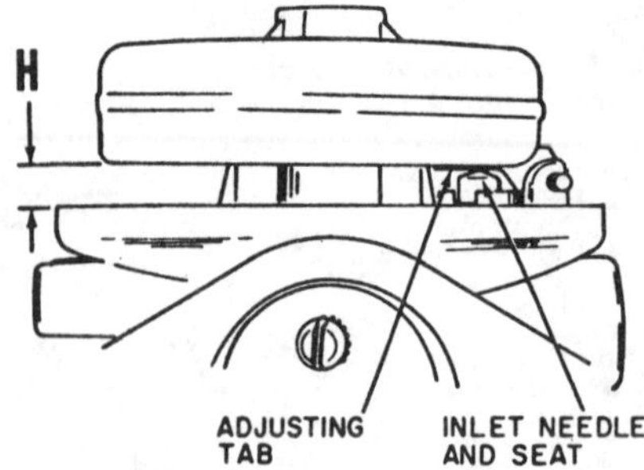

Fig. T19—Float height (H) should be measured as shown on Walbro float carburetors. Bend the adjusting tab to adjust height.

temperature and running. Operate engine at rated speed and adjust main fuel mixture screw for smoothest engine operation. Operate engine at idle speed and adjust idle mixture screw for smoothest engine idle. Adjust throttle stop screw (1) to obtain 1800 rpm. If engine acceleration is not smooth, main fuel mixture screw may have to be adjusted slightly.

To check float setting, invert carburetor throttle body and measure distance from top of float to carburetor body float bowl mating surface. Distance should be 1-13/32 inch (35.72 mm). Carefully bend float tang which contacts fuel inlet needle as necessary to obtain correct float level.

Tillotson Type "E" Carburetor. Refer to Fig. T16 for identification and exploded view of Tillotson type E carburetor.

Initial adjustment of idle mixture screw and main fuel mixture screw from a lightly seated position, is ¾ turn open for idle mixture screw (18) and 1 turn open for main fuel mixture screw (14). Clockwise rotation leans mixture and counterclockwise rotation richens mixture.

Final adjustments are made with engine at operating temperature and running. Operate engine at rated speed and adjust main fuel mixture screw for smoothest engine operation. Operate engine at idle speed and adjust idle mixture screw for smoothest engine idle. Adjust throttle stop screw (8) to obtain desired idle speed. If engine does not accelerate smoothly, it may be necessary to adjust main fuel mixture screw slightly.

To check float setting, invert carburetor throttle body and measure distance between top of float at free end to float bowl mating surface on carburetor. Distance should be 1-5/64 inch (27.38 mm). Refer to Fig. T17. Gasket should not be installed when measuring float height. Carefully bend float tang which contacts fuel inlet needle as necessary to obtain correct float level.

Walbro. Refer to Fig. T18 for identification and exploded view of Walbro carburetor.

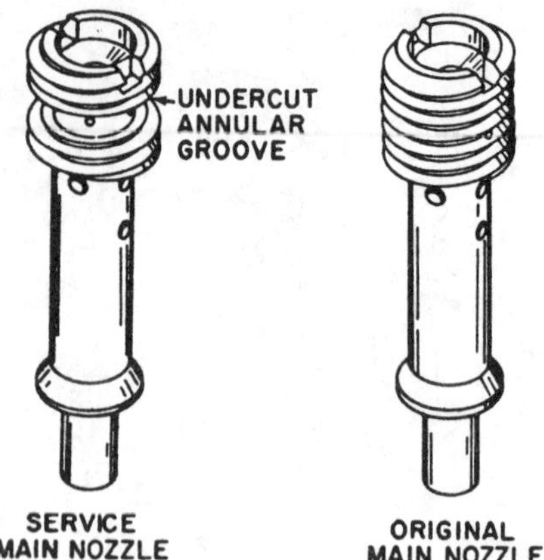

Fig. T20—The main nozzle originally installed is drilled after installation through hole in body. Service main nozzles are grooved so alignment is not necesasry.

Initial adjustment of idle mixture and main fuel mixture screws from a lightly seated position, is 1 turn open. Clockwise rotation leans fuel mixture and counterclockwise rotation richens fuel mixture.

Final adjustments are made with engine at operating temperature and running. Operate engine at rated speed and adjust main fuel mixture screw (33) for smoothest engine operation. Operate engine at idle speed and adjust idle mixture screw for smoothest engine idle. Adjust throttle stop screw (7) so engine idles at 1800 rpm.

To check float level, invert carburetor throttle body and measure clearance between free end of float and machined surface of carburetor. Clearance should be ⅛ inch (3.18 mm). Refer to Fig. T19. Carefully bend float tang which contacts fuel inlet needle as necessary to obtain correct float setting.

NOTE: If carburetor has been disassembled and main nozzle (19—Fig. T18) removed, do not reinstall the original

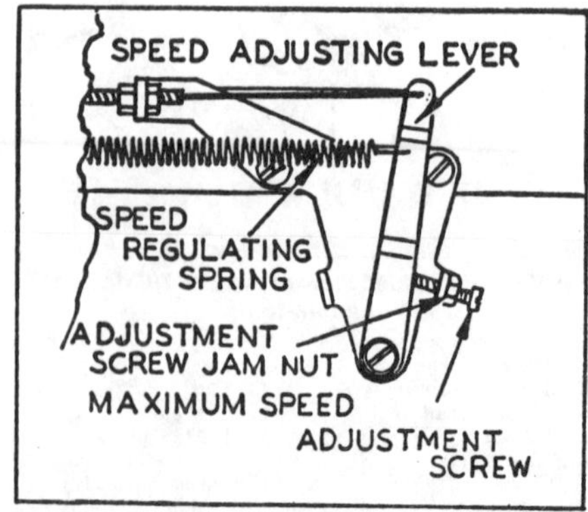

Fig. T21—Speed adjusting screw and lever. The no-load speed should not exceed 3600 rpm.

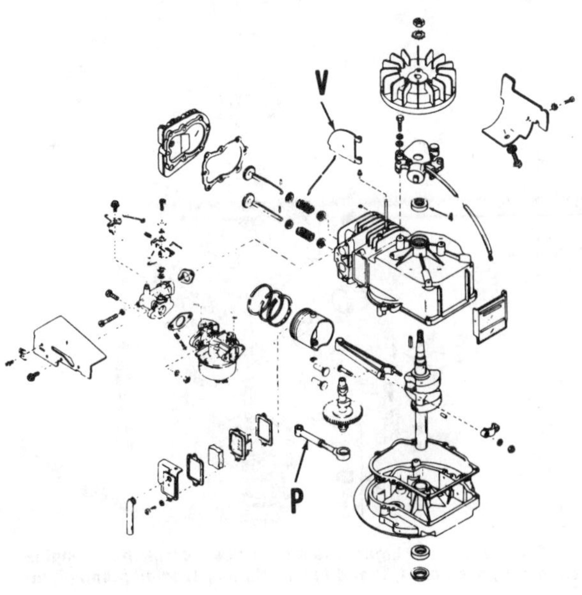

Fig. T21A—Exploded view of Light Frame vertical crankshaft engine. Model shown has plunger type oil pump (P) and air vane (V) governor.

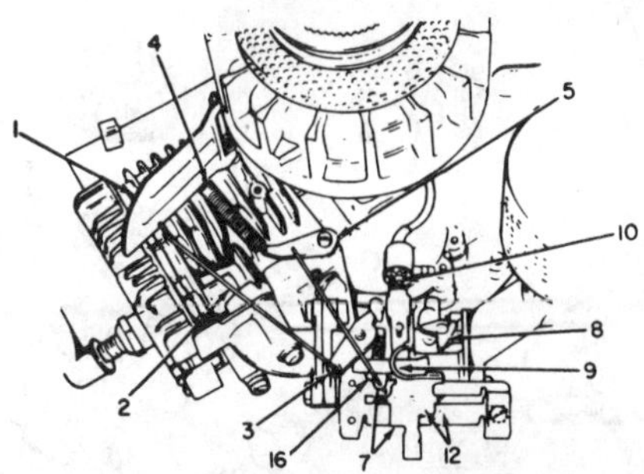

Fig. T24—View of one type of speed control with pneumatic governor. Refer also to Figs. T25, T26 and T27.

1. Air vane
2. Throttle control link
3. Carburetor throttle lever
4. Governor spring
5. Governor spring linkage
7. Speed control lever
8. Carburetor choke lever
9. Choke control
10. Stop switch
12. Alignment holes
16. Idle speed stop screw
18. High speed stop screw (Fig. T27)

equipment nozzle. Install a new service nozzle. Refer to Fig. T20 for differences between original and service nozzles.

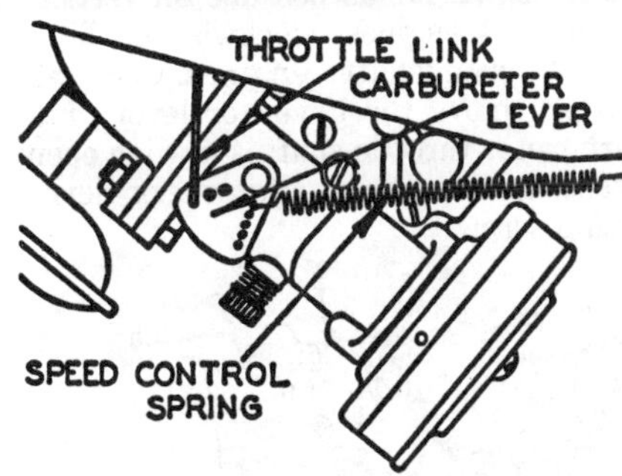

Fig. T22—Outer spring anchorage holes in carburetor throttle lever may be used to reduce speed fluctuations or surging.

PNEUMATIC GOVERNOR. Some engines are equipped with a pneumatic (air vane) type governor. On fixed linkage hookups, the recommended idle speed is 1800 rpm. Standard no-load speed is 3300 rpm. Operating speed range is 2600 to 3600 rpm. On engines not equipped with slide control (Fig. T23), obtain desired speed by varying the tension on governor speed regulating spring by moving speed adjusting lever (Fig. T21) in or out as required. If engine speed fluctuates due to governor hunting or surging, move carburetor end of governor spring into next outer hole on carburetor throttle arm shown in Fig. T22.

A too-lean mixture or friction in governor linkage will also cause hunting or unsteady operation.

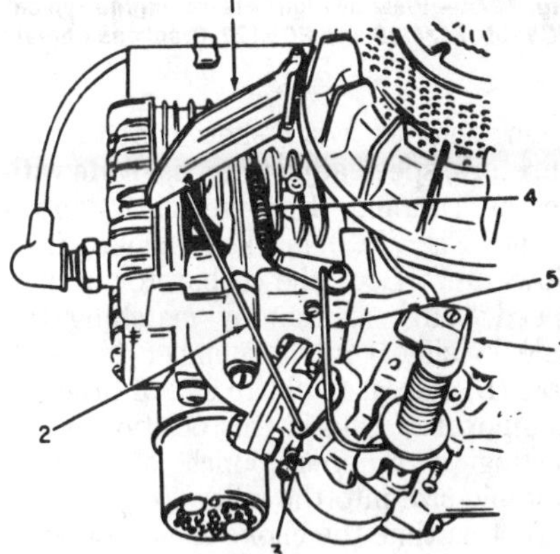

Fig. T25—Pneumatic governor linkage.

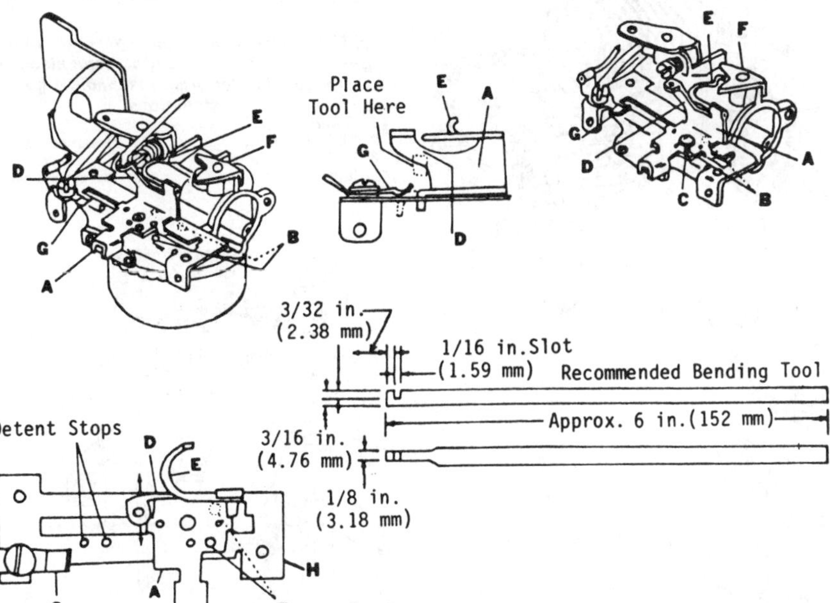

Fig. T23—When carburetor is equipped with slide control, governed speed is adjusted by bending slide arm at point "D". Bend arm outward from engine to increase speed.

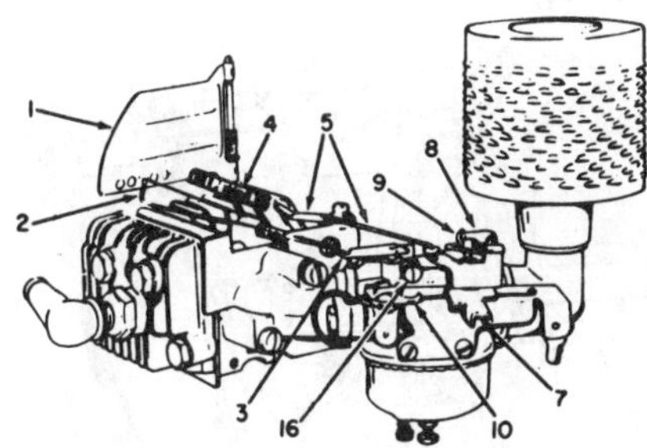

Fig. T26—Pneumatic governor linkage.

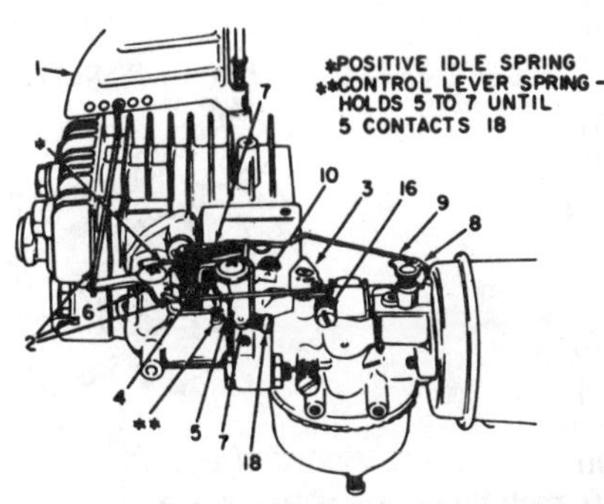

Fig. T27—Pneumatic governor linkage.

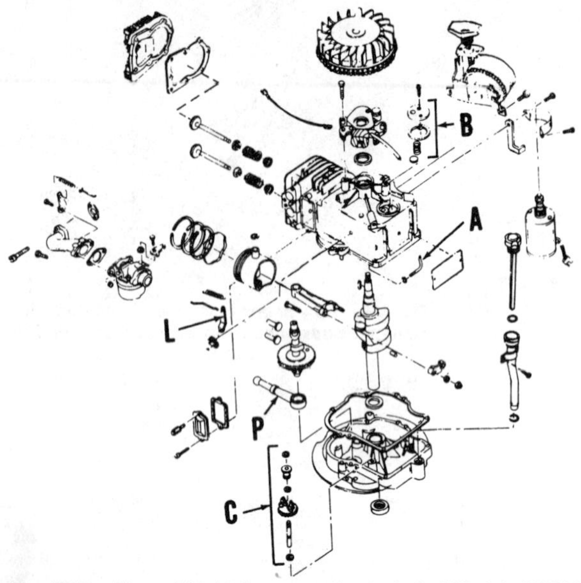

Fig. T27A—View of Light Frame engine typical of Models ECV100, ECV105, ECV110 and ECV120. Crankcase breather valve is shown at (B).

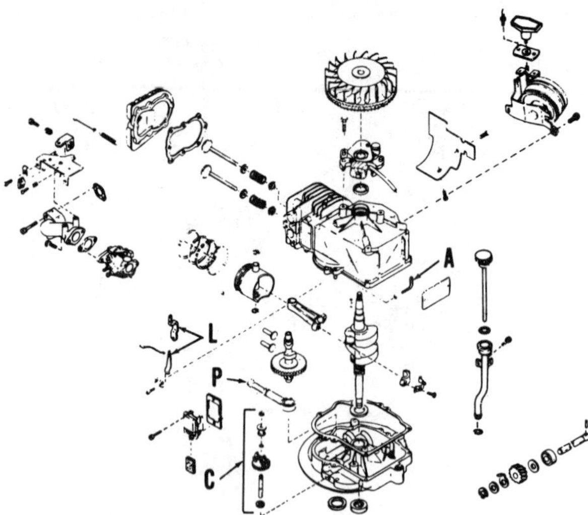

Fig. T28A—View of Light Frame vertical crankshaft engine. Mechanical governor (A, C and L) and plunger type oil pump (P) are shown.

On engines with slide control carburetors, speed adjustment is made with engine running. Remove slide control cover (Fig. T23) marked "choke, fast, slow, etc.". Lock the carburetor in high speed "run" position by matching the hole in slide control member (A) with hole (B) closest to choke end of bracket. Temporarily hold in this position by inserting a tapered punch or pin of suitable size into the hole.

At this time the choke should be wide open and the choke activating arm (E) should be clear of choke lever (F) by 1/64 inch (0.40 mm). Obtain clearance gap by bending arm (E). To increase engine speed, bend slide arm at point (D) outward from engine. To decrease speed, bend arm inward toward engine.

Make certain on models equipped with wiper grounding switch that wiper (G) touches slide (A) when control is moved to "STOP" position.

Refer to Figs. T24, T25, T26 and T27 for assembled views of pneumatic governor systems.

MECHANICAL GOVERNOR. Some engines are equipped with a mechanical (flyweight) type governor. To adjust the governor linkage, refer to Fig. T28 and loosen governor lever screw. Twist protruding end of governor shaft counterclockwise as far as possible on vertical crankshaft engines; clockwise on horizontal crankshaft engines. On all models, move the governor lever until carburetor throttle shaft is in wide open position, then tighten governor lever clamp screw.

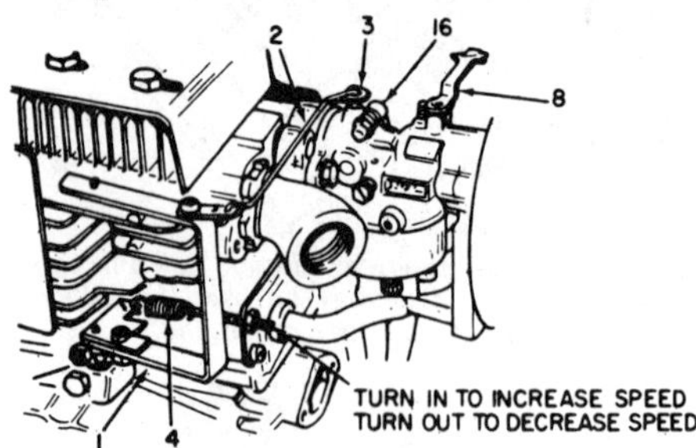

Fig. T29—View of mechanical governor with one type of constant speed control. Refer also to Fig. T30 through T43 for other mechanical governor installations.

1. Governor lever
2. Throttle control
3. Carburetor throttle lever
4. Governor spring
8. Choke lever
16. Idle speed stop screw

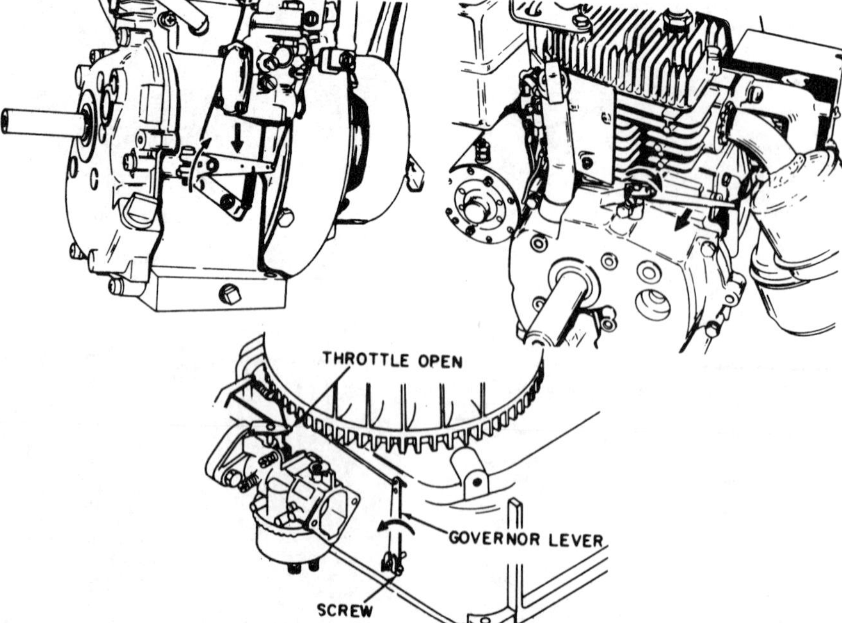

Fig. T28—Views showing location of mechanical governor lever and direction to turn when adjusting position on governor shaft.

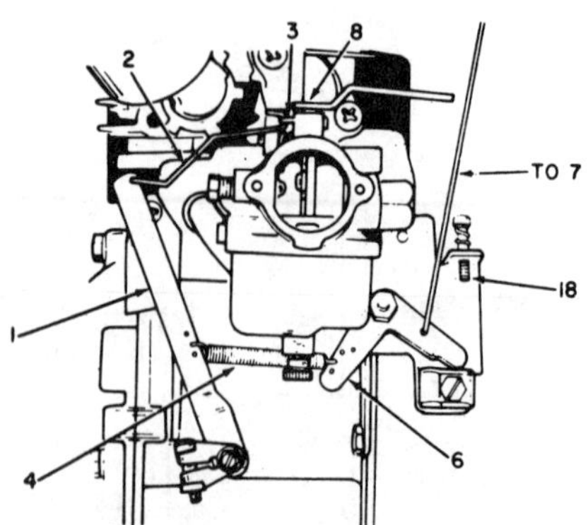

Fig. T30—Mechanical governor linkage. Refer to Fig. T29 for legend except for the following.

6. Bellcrank
7. Speed control lever
18. High speed stop screw

Binding or worn governor linkage will result in hunting or unsteady engine operation. An improperly adjusted carburetor will also cause a surging or hunting condition.

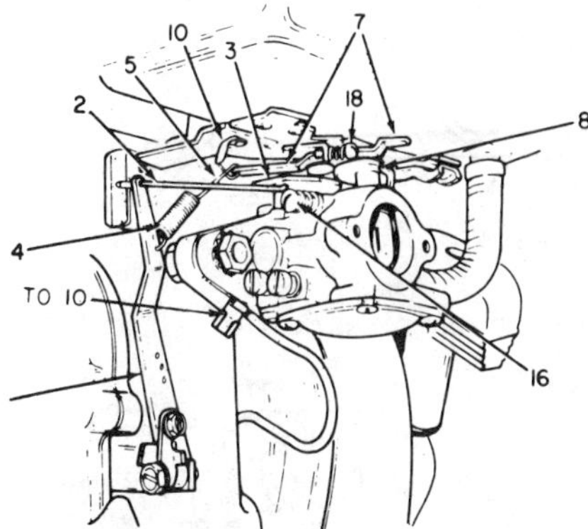

Fig. T31—Mechanical governor linkage.

1. Governor lever
2. Throttle control link
3. Carburetor throttle
4. Governor spring
5. Governor spring linkage
7. Speed control lever
8. Choke lever
10. Stop switch
16. Idle speed stop screw
18. High speed stop screw

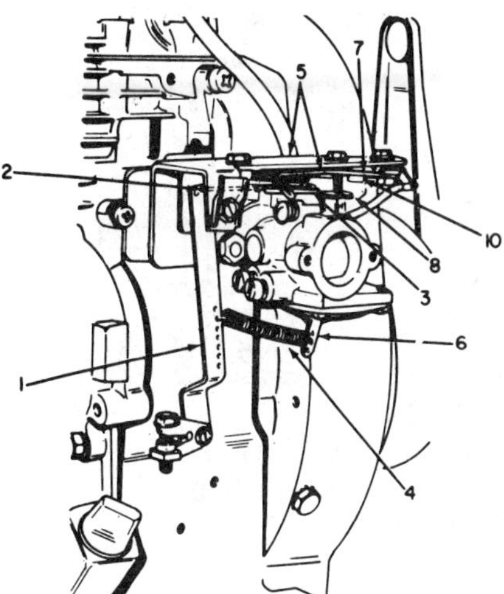

Fig. T32—Mechanical governor linkage. Refer to Fig. T31 for legend. Bellcrank is shown at (6).

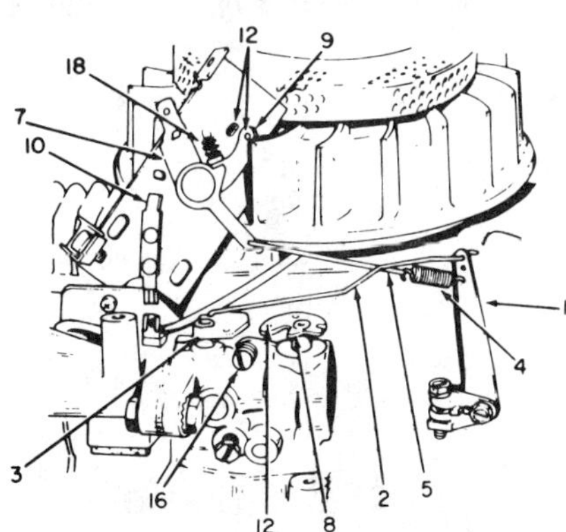

Fig. T33—Mechanical governor linkage. Control cover is raised to view underside.

1. Governor lever
2. Throttle control link
3. Carburetor throttle lever
4. Governor spring
5. Gover spring linkage
7. Speed control lever
8. Choke lever
9. Choke control
10. Stop switch
12. Alignment holes
16. Idle speed stop screw
18. High speed stop screw

Refer to Figs. T29 through T42 for views of typical mechanical governor speed control linkage installations. The governor gear shaft must be pressed into bore in cover until the correct amount of the shaft protrudes. Refer to illustration and chart in Fig. T44.

IGNITION SYSTEM. A magneto ignition system with breaker points or

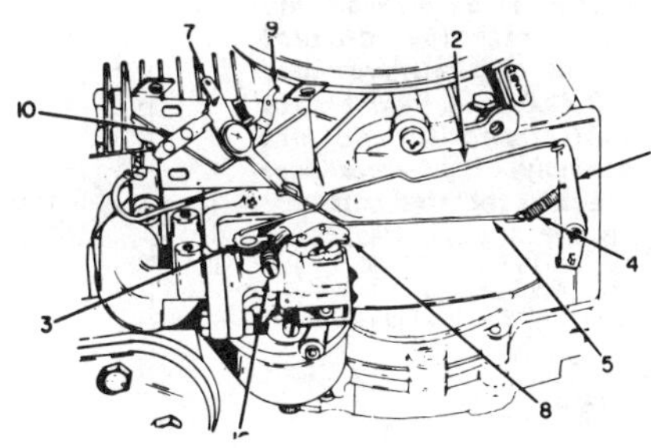

Fig. T34—Mechanical governor linkage with control cover raised. Refer to Fig. T33 for legend.

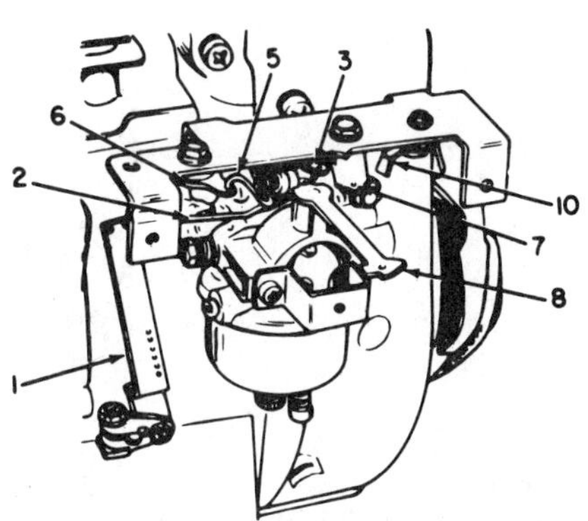

Fig. T35—Mechanical governor linkage. Governed speed of engine is increased by closing loop in linkage (5); decrease speed by spreading loop. Refer to Fig. T33 for legend.

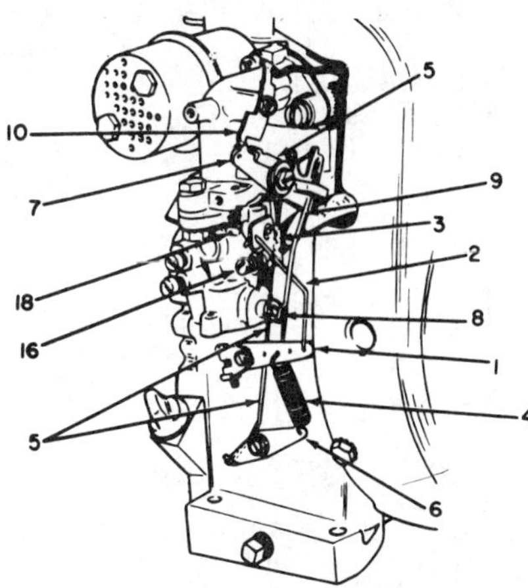

Fig. T36—Mechanical governor linkage.

1. Governor lever
2. Throttle control link
3. Throttle lever
4. Governor spring
5. Governor spring linkage
6. Bellcrank
7. Speed control lever
8. Choke lever
9. Choke control link
10. Stop switch
16. Idle speed stop screw
18. High speed stop screw

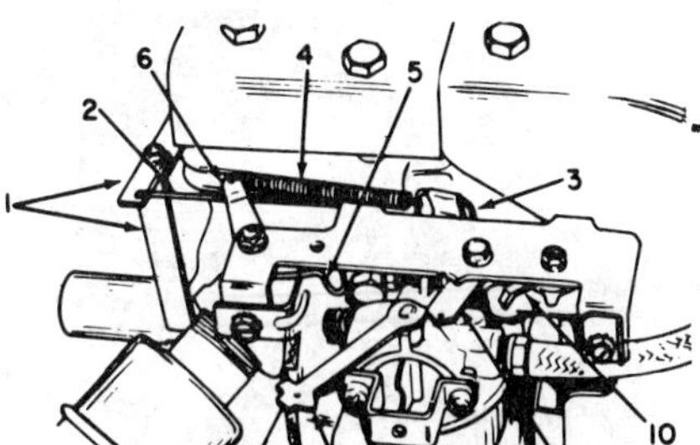

Fig. T37—View of control linkage used on some engines. To increase governed engine speed, close loop (5); to decrease speed, spread loop (5). Refer to Fig. T36 for legend.

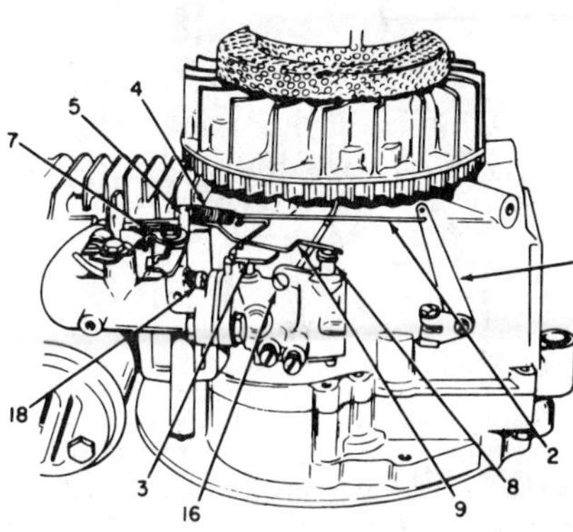

Fig. T38—View of mechanical governor control linkage. Governor spring (4) is hooked onto loop in link (2).

1. Governor lever
2. Throttle control link
3. Throttle lever
4. Governor spring
5. Governor spring linkage
7. Speed control lever
8. Choke lever
9. Choke control link
16. Idle speed stop screw
18. High speed stop screw

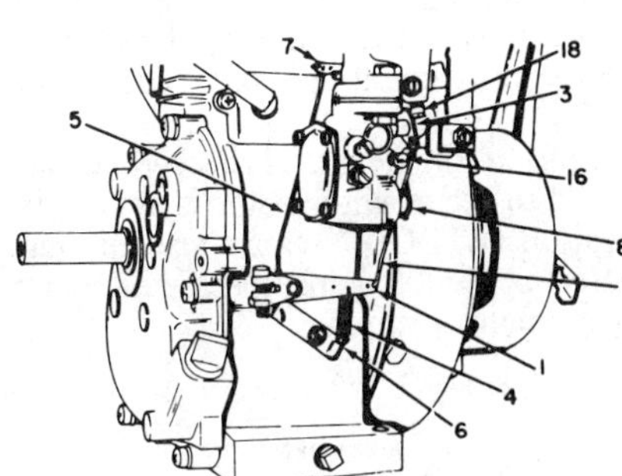

Fig. T39—Linkage for mechanical governor. Refer to have Fig. T38 for legend. Bellcrank is shown at (6).

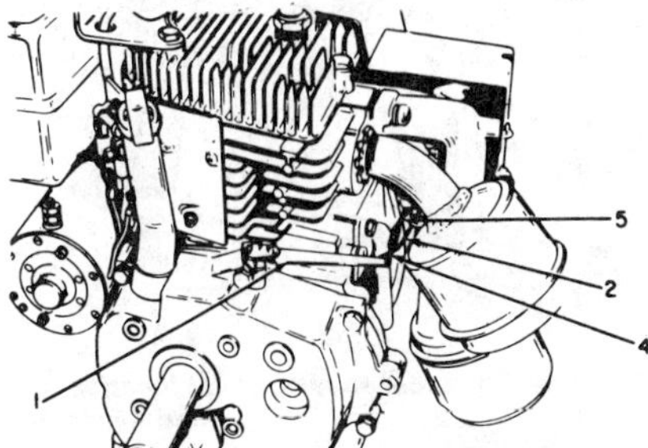

Fig. T40—Mechanical governor linkage. Refer to Fig. T38 for legend.

Fig. T41—Mechanical governor linkage. Refer to Fig. T38 for legend.

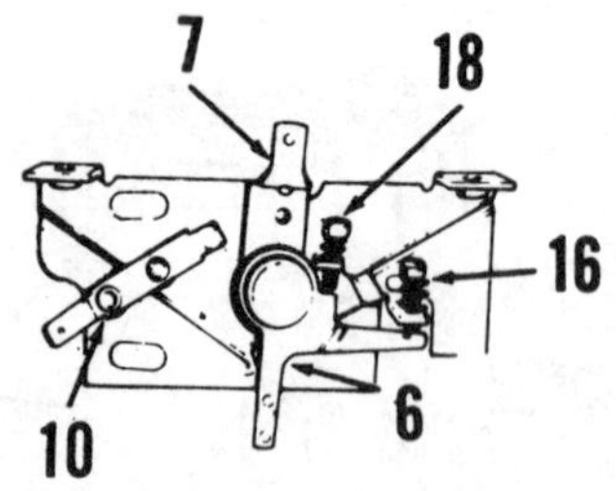

Fig. T42—Models with "Automatic" carburetor use control shown.

6. Governor spring bellcrank
7. Control lever
10. Idle speed stop screw
16. Idle speed stop
18. High speed stop screw

capacitor-discharge ignition (CDI) may have been used according to model and application. Refer to appropriate paragraph for model being serviced.

Breaker Point Ignition System. Breaker point gap at maximum opening should be 0.020 inch (0.51 mm) for all models. Marks are usually located on stator and mounting post to facilitate timing.

Ignition timing can be checked and adjusted to occur when piston is at specific location (BTDC) if marks are missing. Refer to the following specifications for recommended timing.

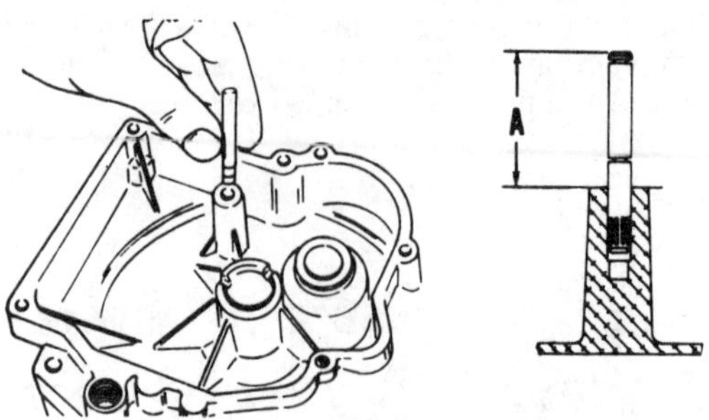

Fig. T44—The governor gear shaft must be pressed into bore until the correct amount of shaft protrudes (A, B, C, D or E). Refer also to illustrations for correct assembly of governor gear and associated parts.

B. 1-5/16 inches (33.338 mm)
C. 1-5/16 inches (33.338 mm)
D. 1-3/8 inches (34.925 mm)
E. 1-19/32 inches (40. 481 mm)

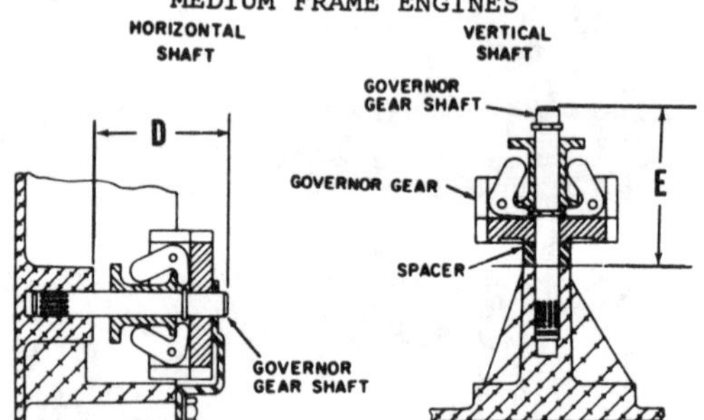

Models	Piston Position BTDC
HS40, LAV40, TVS105, ECV100, ECV105, ECV110, ECV120, TNT100, TNT120	0.035 in. (0.89 mm)
V40, VH40, LAV50, TVS120, H40, HH40, HS50	0.050 in. (1.27 mm)
LAV25, LAV30, LV35, LAV35, TVS90, H25, H30, H35, TVS75, ECH90	0.065 in. (1.65 mm)
V50, VH50, V60, VH60, H50, HH50, H60, HH60, TVM125, TVM140	0.080 in. (2.03 mm)

Some models may be equipped with the coil and laminations mounted outside the flywheel. Engines equipped

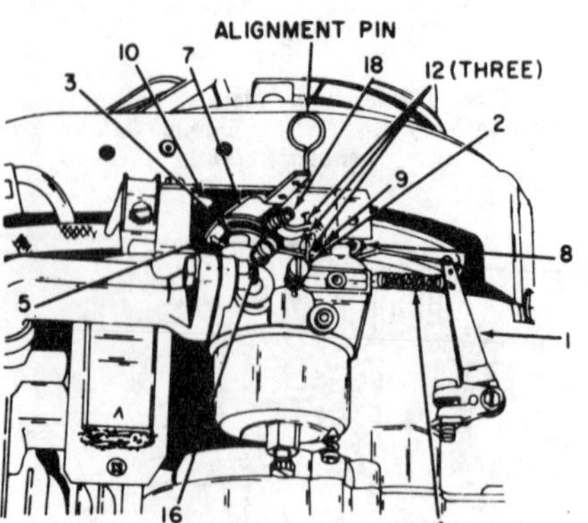

Fig. T43—On model shown, adjust location of cover so control lever is aligned with high speed slot and alignment holes are aligned.

1. Governor lever
2. Throttle control link
3. Throttle lever
4. Governor spring
5. Governor spring linkage
7. Control lever
8. Choke lever
9. Choke linkage
10. Stop switch
12. Alignment holes
16. Idle speed stop screw
18. High speed stop screw

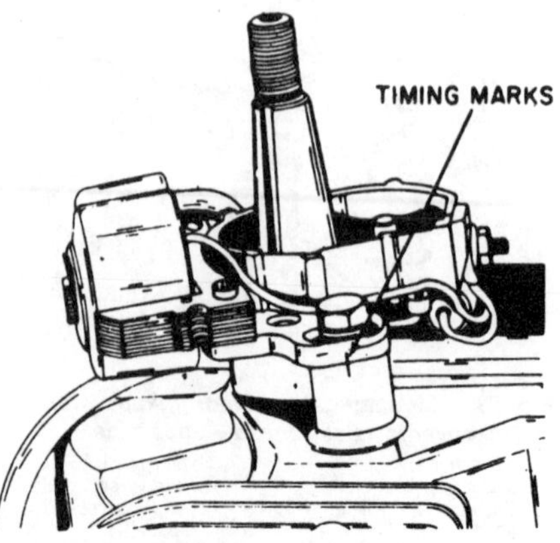

Fig. T45—Align timing marks as shown on magneto ignition system.

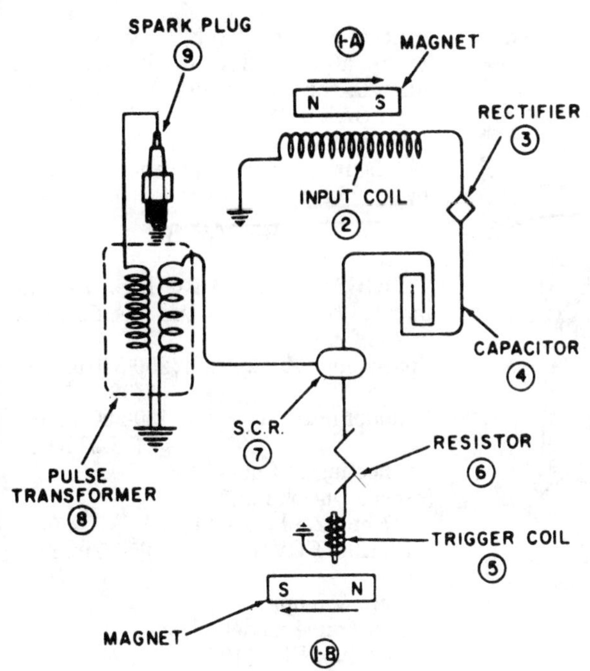

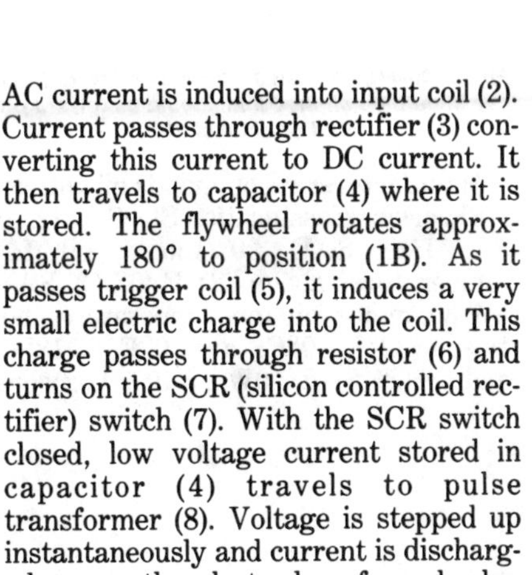

Fig. T46—Wiring diagram of solid state ignition system used on some models.

.015"

Fig. T47A—Set coil lamination air gap at 0.015 inch (0.38 mm) at points shown.

with breaker point ignition have the ignition points and condenser mounted under the flywheel with the coil and laminations mounted outside the flywheel. This system is identified by the round shape of the coil and a stamping "Grey Key" in the coil to identify the correct flywheel key.

The correct air gap setting between the flywheel magnets and the coil laminations is 0.015 inch (0.38 mm). Use Tecumseh gage 670259 or equivalent thickness sheet plastic to set gap as shown in Fig. T47A.

Solid State Ignition System. The Tecumseh solid state ignition system does not use ignition breaker points. The only moving part of the system is the rotating flywheel with the charging magnets. As the flywheel magnet passes position (1A–Fig. T46), a low voltage AC current is induced into input coil (2). Current passes through rectifier (3) converting this current to DC current. It then travels to capacitor (4) where it is stored. The flywheel rotates approximately 180° to position (1B). As it passes trigger coil (5), it induces a very small electric charge into the coil. This charge passes through resistor (6) and turns on the SCR (silicon controlled rectifier) switch (7). With the SCR switch closed, low voltage current stored in capacitor (4) travels to pulse transformer (8). Voltage is stepped up instantaneously and current is discharged across the electrodes of spark plug (9), producing a spark before top dead center.

Some units are equipped with a second trigger coil and resistor set to turn the SCR switch on at a lower rpm. This second trigger pin is closer to the flywheel and produces a spark at TDC for earlier starting. As engine rpm increases, the first (shorter) trigger pin picks up the small electric charge and turns the SCR switch on, firing the spark plug at TDC.

If system fails to produce a spark to the spark plug, first check high tension lead (Fig, T47). If condition of high tension lead is questionable, renew pulse transformer and high tension lead assembly. Check low tension lead and renew if insulation is faulty. The magneto charging coil, electronic triggering system and mounting plate are available only as an assembly. If necessary to renew this assembly, place unit in position on engine. Start retaining screws, turn mounting plate counterclockwise as far as possible, then tighten retaining screws to 5-7 ft.-lbs. (7-10 N·m).

Engines with solid state (CDI) ignition have all the ignition components sealed in a module and located outside the flywheel. There are no components

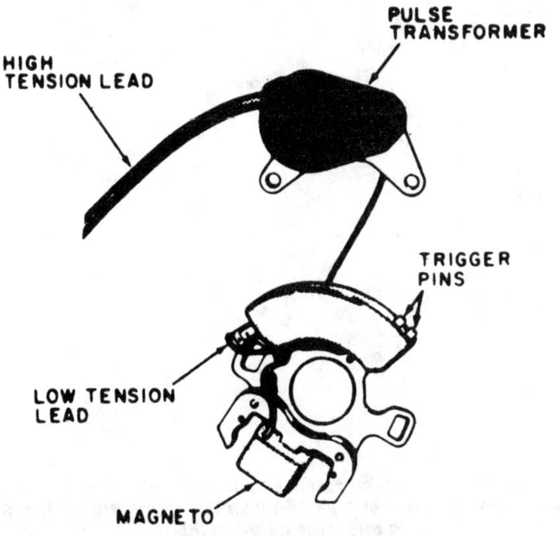

Fig. T47—Diagram of solid state ignition system used on some models.

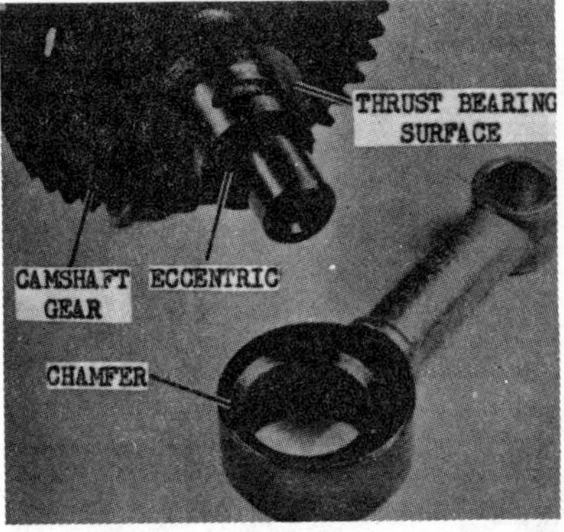

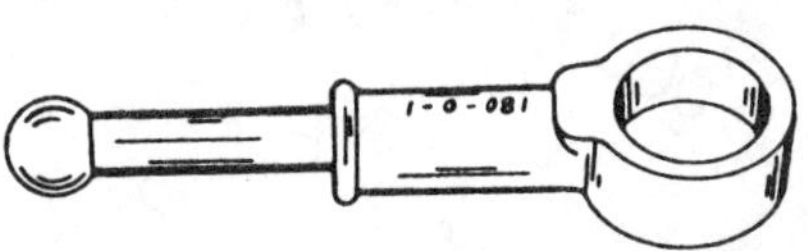

Fig. T48—Various types of barrel and plunger oil pumps have been used. Chamfered face of collar should be toward camshaft if drive collar has only one chamfered side. If drive collar has a float boss, the boss should be next to the engine lower cover, away from camshaft gear.

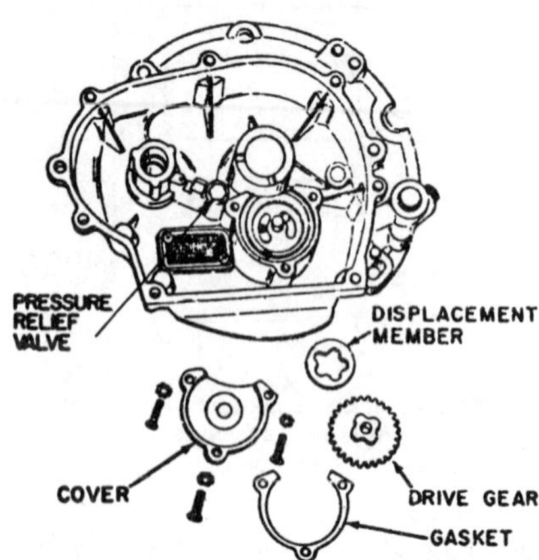

Fig. T49—Disassembled view of typical gear-driven rotor type oil pump.

under the flywheel except a spring clip to hold the flywheel key in position. This system is identified by the square shape module and a stamping "Gold Key" to identify the correct flywheel key.

The correct air gap setting between the flywheel magnets and the laminations on ignition module is 0.015 inch (0.38 mm). Use Tecumseh gage 670259 or equivalent thickness sheet plastic to set gap as shown in Fig. T47A.

LUBRICATION. Vertical crankshaft engines may be equipped with a barrel and plunger type oil pump or a gear driven rotor type oil pump. Horizontal crankshaft engines may be equipped with a gear-driven rotor type pump or with a dipper type oil slinger attached to the connecting rod.

Oil level should be checked after every five hours of operation: Maintain oil level at lower edge of filler plug or at "FULL" mark on dipstick.

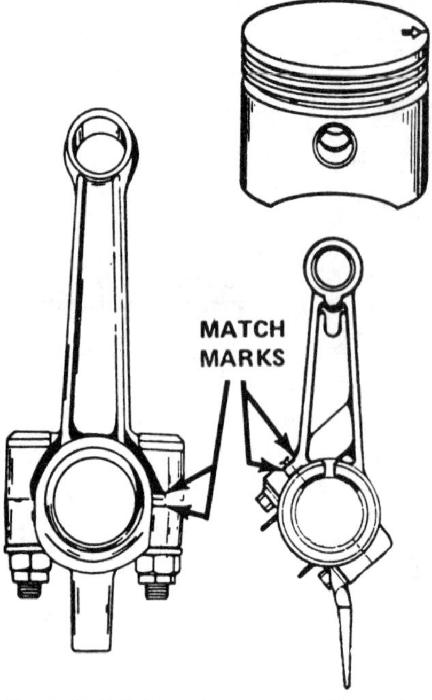

Fig. T50—Match marks on connecting rod and cap should be aligned and should be toward pto end of crankshaft.

Manufacturer recommends oil with an API service classification SC, SD, SE or SF. Use SAE 30 motor oil for temperatures above 32° F (0° C), SAE 5W-30 motor oil for temperatures between 32°F (0°C) and 0°F (−18°C) and a 90% SAE 10W oil-10% kerosene mixture or SAE 5W-30 for temperatures below 0° F (−18° C). Manufacturer explicity states: DO NOT USE SAE 10W-40 motor oil.

Oil should be changed after the first two hours of engine operation and after every 25 hours of operation thereafter.

REPAIRS

TIGHTENING TORQUES. Recommended tightening torque specifications are as follows:

Spark plug . . . . 250-360 in.-lbs. (28-41 N·m)
Cylinder head . . . . 160-200 in.-lbs. (18-23 N·m)
Connecting rod nuts (except Durlok nuts):
1.7 hp., 2.5 hp., 3.0 hp., ECH90, ECV100 . . . . . . 65-75 in.-lbs. (7-9 N·m)
4 hp. & 5 hp. light frame models, ECV105, ECV110, ECV120 . . . . 80-95 in.-lbs. (9-11 N·m)
5 hp. (3.7 kW) medium frame, 6 hp, (4.5 kW) . . . . 86-110 in.-lbs. (10-12 N·m)
Connecting rod bolts (Durlok):
5 hp. (3.7 kW) light frame . . . . 110-130 in.-lbs. (12-15 N·m)

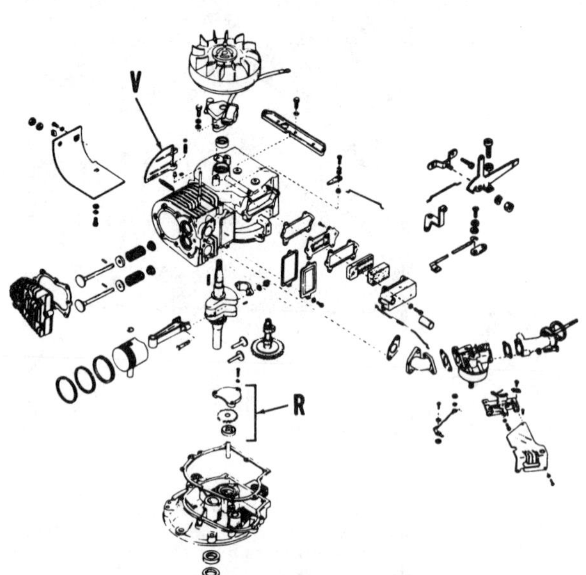

Fig. T49A—View of vertical crankshaft engine. Rotor type oil pump is shown at (R); governor air vane at (V).

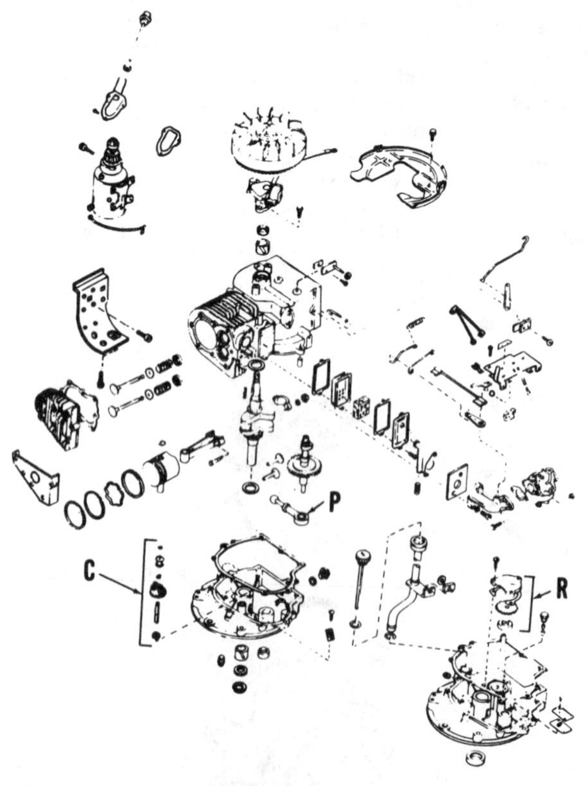

Fig. T50A—View of Medium Frame engine with vertical crankshaft. Some models use plunger type oil pump (P); others are equipped with rotor type oil pump (R).

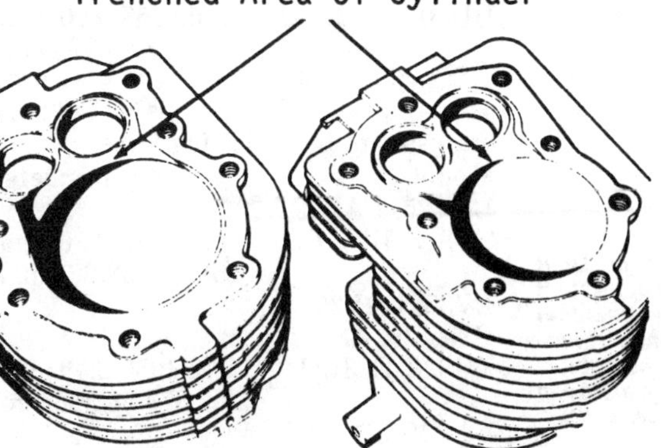

Fig. T51—Cylinder block on Models H50, HH50, V50, VH50, TVM125 and TVM140 has been "trenched" to improve fuel flow and power.

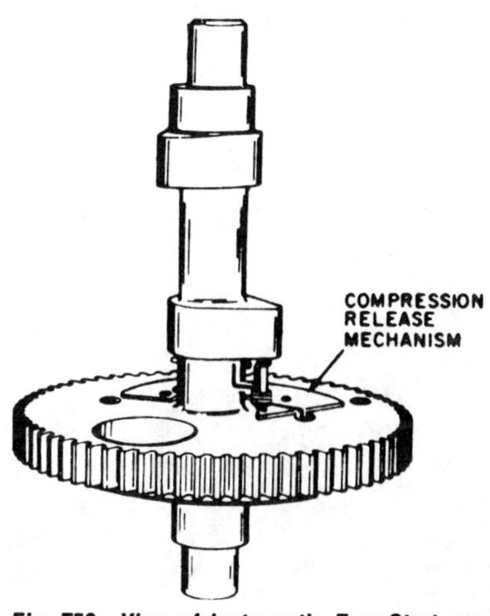

Fig. T53—View of Insta-matic Ezee-Start compression release camshaft.

5 hp. (3.7 kW)
medium frame, 6 hp.
(4.5 kW) . . . . 130-150 in.-lbs. (15-17 N·m)
All other models . . . . . . 95-110 in.-lbs. (11-12 N·m)

Flywheel:
Light frame . . . . 360-400 in.-lbs. (41-45 N·m)
Medium frame . . . . 430-480 in.-lbs. (49-54 N·m)
External ignition . . . . . 400-440 in.-lbs. (45-50 N·m)

Magneto stator . . . . 40-90 in.-lbs. (5-10 N·m)

Mounting flange . . . . 75-110 in.-lbs. (9-12 N·m)

Carburetor to intake pipe . . 48-72 in.-lbs. (5-8 N·m)

Intake pipe to cylinder . . . . 72-96 in.-lbs. (8-11 N·m)

Gear reduction
housing . . . . 100-144 in.-lbs. (11-16 N·m)

Gear reduction cover . . . . 75-110 in.-lbs. (9-12 N·m)

CONNECTING ROD. Piston and connecting rod assembly is removed from cylinder head end of engine. The aluminum alloy connecting rod rides directly on crankshaft crankpin.

Refer to the following table for standard crankpin journal diameter.

Model	Diameter
LAV25, LAV30, TVS75, H25, H30, LV35, LAV35, TVS90, H35-prior 1983, ECH90, TNT100, ECV100	0.8610-0.8615 in. (21.869-21.882 mm)
H35-after 1983, LAV40, TVS105, HS40, ECV105, ECV110, ECV120, TNT120, LAV50, HS50, TVS120	0.9995-1.0000 in. (25.390-25.400 mm)
V40, VH40, H40, HH40, H50, HH50, V50, VH50, TVM140, V60, VH60	1.0615-1.0620 in. (26.962-26.975 mm)

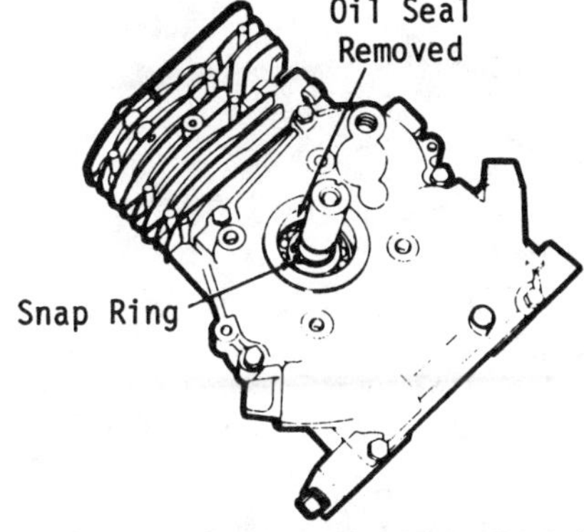

Fig. T52—On Models H30 through H50 it is necessary to remove snap ring under oil seal before removing crankcase cover.

Standard inside diameter for connecting rod crankpin bearing journal is shown in the following table.

Model	Diameter
LAV25, LAV30, TVS75, H25, H30, LV35, LAV35, TVS90, H35-prior 1983, ECH90, TNT100, ECV100	0.8620-0.8625 in. (21.895-21.908 mm)
H35-after 1983, LAV40, TVS105, HS40, ECV105, ECV110, ECV120, TNT120, LAV50, HS50, TVS120	1.0005-1.0010 in. (25.413-25.425 mm)
V40, VH40, H40, HH40, H50, HH50, V50, VH50, TVM140, V60, VH60	1.0630-1.0636 in. (27.000-27.013 mm)

Connecting rod bearing-to-crankpin journal clearance should be 0.0005-0.0015 inch (0.013-0.38 mm) for all models.

When installing connecting rod and piston assembly, align the match marks on connecting rod and cap as shown in

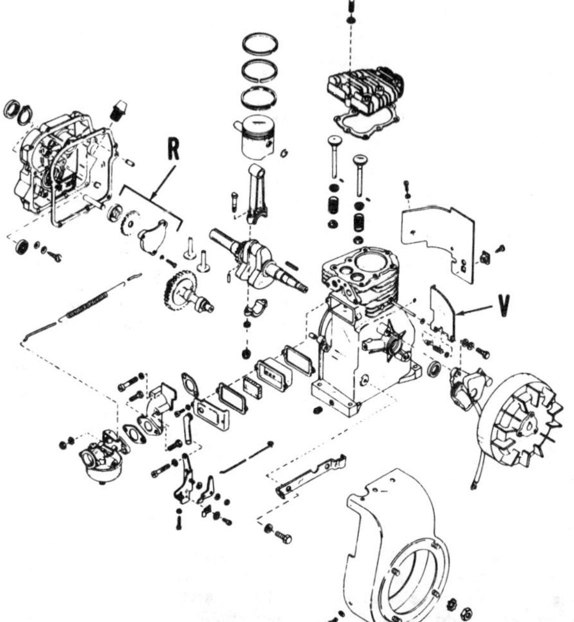

Fig. T52A—View of Light Frame engine with horizontal crankshaft. Air vane (V) type governor and rotor type oil pump (R) are used on model shown.

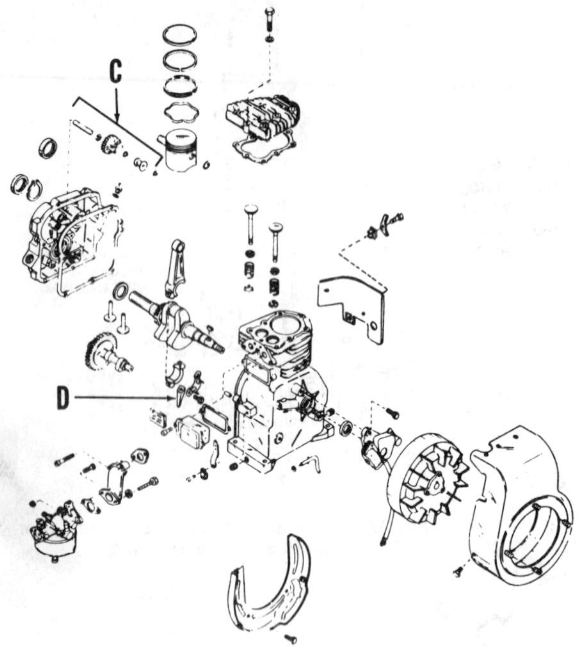

Fig. T53A—View of Light Frame horizontal crankshaft engine with mechanical governor and splash lubrication. Governor centrifugal weights are shown at (C) and lubrication dipper at (D).

Fig. T50. On some models, the piston pin hole is offset in piston and arrow on top of piston should be toward valves. On all engines, the match marks on connecting rod and cap must be toward power takeoff (pto) end of crankshaft. Lock plates, if so equipped, for connecting rod cap retaining screws should be renewed each time cap is removed.

PISTON, PIN AND RINGS. Aluminum alloy pistons are equipped with two compression rings and one oil control ring. Ring end gap for all models is 0.007-0.017 inch (0.18-0.43 mm).

Piston skirt-to-cylinder clearances are listed in the following table.

Model	Clearance
LAV25, LAV30, TVS75, H25, H30	0.0025-0.0040 in. (0.064-0.102 mm)
H50, HH50, VH50, TVM125, TVM140, V60	0.0035-0.005 in. (0.089-0.127 mm)
LV35, LAV35, TVS90, H35 (all), ECH90, LAV40, TVS105, HS40, TNT100, ECV100, ECV110, LAV50, TVS120, ECV120, TNT120, HS50, VH50, VH60	0.0045-0.0060 in. (0.114-0.152 mm)
ECV105	0.0050-0.0065 in. (0.127-0.165 mm)
V40, VH40, H40, HH40	0.0055-0.0070 in. (0.140-0.178 mm)

Standard piston diameters measured at piston skirt 90° from piston pin bore are listed in the following table.

Model	Diameter
LAV25, LAV30, TVS75, H25, H30	2.3090-2.3095 in. (58.649-58.661 mm)
LV35, LAV35, H35 (all), TVS90, ECH90	2.4950-2.4955 in. (63.373-63.386 mm)
LAV40, TVS105, HS40, TNT100, ECV100, EVC105	2.6200-2.6205 in. (66.548-66.560 mm)
H50, HH50, V50, VH50, TVM125, TVM140, V60, VH60	2.6210-2.6215 in.
V40, VH40, H40, HH40	2.4945-2.4950 in. (63.363-63.373 mm)
ECV110	2.7450-2.7455 in. (69.723-69.736 mm)
HS50, LAV50, TVS120, ECV120, TNT120	2.8070-2.8075 in. (71.298-71.311 mm)

Standard ring side clearance in ring grooves are shown in the following table.

Model	Clearance
LAV25, LAV30, TVS75, H25, H30	0.002-0.005 in. (0.05-0.13 mm)
LV35, LAV35, TVS90, H35 (all), V40, VH40, H40, HH40, ECH90	0.002-0.003 in. (0.05-0.08 mm)

Fig. T54—The camshaft and crankshaft must be correctly timed to ensure valves open at correct time. Different types of marks have been used, but marks should be aligned when assembling.

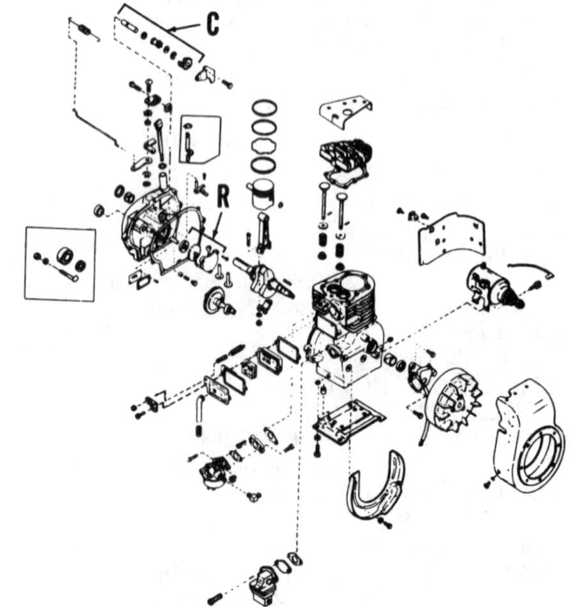

Fig. T55—View of Medium Frame horizontal crankshaft engine with mechanical governor (C). An oil dipper for splash lubrication is cast onto the connecting rod cap instead of using the rotor type oil pump (R).

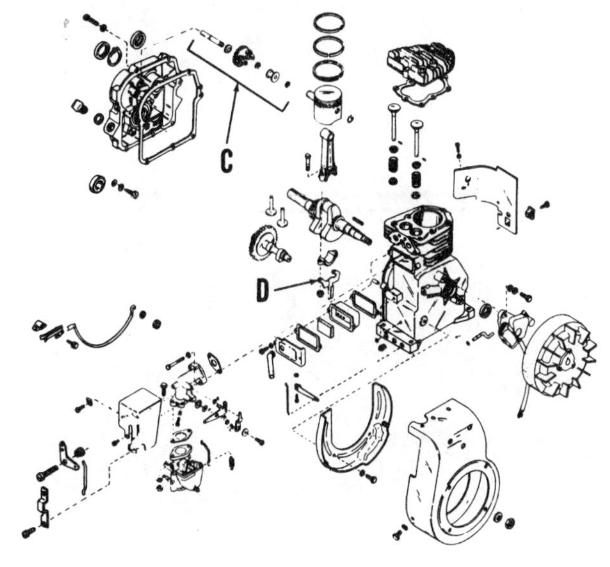

Fig. T56—View of Light Frame horizontal crankshaft engine with mechanical governor and splash lubrication. Notice that connecting rod cap on model shown is away from camshaft side of engine.

LAV40, TVS105,
HS40, ECV100,
ECV105, TNT100:
Compression rings . . . . 0.002-0.004 in.
(0.05-0.10 mm)
Oil control ring . . . . 0.001-0.004 in.
(0.03-0.10 mm)
ECV110:
Compression rings . . . . 0.002-0.004 in.
(0.05-0.10 mm)
Oil control ring . . . . 0.001-0.002 in.
(0.03-0.05 mm)
LAV50, HS50, TVS120:
Compression rings . . . . 0.003-0.004 in.
(0.08-0.10 mm)
Oil control ring . . . . 0.002-0.003 in.
(0.05-0.08 mm)
ECV120:
Compression rings . . . . 0.003-0.004 in.
(0.08-0.10 mm)
Oil control ring . . . . 0.001-0.002 in.
(0.03-0.05 mm)
TNT120:
Compression rings . . . . 0.003-0.004 in.
(0.08-0.10 mm)
Oil control ring . . . . 0.002-0.004 in.
(0.05-0.10 mm)
H50, HH50, V50,
VH50, TVM125,
TVM140, V60,
VH60 . . . . 0.002-0.004 in.
(0.05-0.10 mm)

Refer to CONNECTING ROD section for correct piston to connecting rod assembly and correct piston installation procedure. Piston pin should be a tight push fit in piston pin bore and connecting rod pin bore and is retained by snap rings at each end of piston pin bore. Install marked side of piston rings up and stagger ring end gaps equally around circumference of piston during installation.

CYLINDER AND CRANKCASE. Cylinder and crankcase are an integral casting on all models. Cylinder should be honed and fitted to nearest oversize for which piston and ring set are available if cylinder is scored, tapered or out-of-round more than 0.005 inch (0.13 mm).

Standard cylinder bore diameters are shown in the following table.

Model **Diameter**
LAV25, LAV30,
TVS75, H25, H30 . . . . 2.3125-2.3135 in.
(58.738-58.763 mm)
LV35, LAV35, H35 (all),
TVS90, ECH90, V40,
VH40, H40, HH40 . . . . 2.5000-2.5010 in.
(63.500-63.525 mm)
LAV40, TVS105,
HS40, ECV100,
TNT100, ECV105,
TVM140, V60,
VH60 . . . . 2.6250-2.6260 in.
(66.675-66.700 mm)
ECV110 . . . . 2.7500-2.7510 in.
(69.850-69.875 mm)
LAV50, TVS120,
H50, HH50, V50,
VH50, HS50,
ECV120, TNT120 . . 2.8120-2.8130 in.
(71.425-71.450 mm)

Refer to PISTON, PIN AND RINGS section for correct piston-to-cylinder block clearance.

Note also that cylinder block used on Models H50, HH50, V50, VH50, TVM 125 and TVM 140 has been trenched to improve fuel flow and power (Fig. T51).

Standard main bearing bore diameters for both main bearings, or as indicated otherwise, are shown in the following chart.

Model **Diameter**
LAV25, LAV30, TVS75,
H25, H30 (prior 1983),
LV35, LAV35, TVS90,
H35 (prior 1983),
ECH90, TNT100,
ECV100 . . . . 0.8755-0.8760 in.
(22.238-22.250 mm)
H35 (after 1983),
LAV40, TVS105,
HS40, V40, VH40,
H40, HH40, ECV110,
LAV50, TVS120,
HS50, H50 . . . . 1.005-1.0010 in.
(25.413-25.425 mm)
TNT120, HH50,
V50, VH50,
TVM125, TVM140,
V60, VH60 . . . . 0.9985-0.9990 in.
(25.362-25.375 mm)
H30 (after 1983):
Crankcase side . . . . . 1.0005-1.0010 in.
(25.413-25.425 mm)
Cover (flange) side . . 0.8755-0.8760 in.
(22.238-22.250 mm)
ECV105:
Crankcase side . . . . . 1.0005-1.0010 in.
(25.413-25.425 mm)
Cover (flange) side) . 1.2010-1.2020 in.
(30.505-30.531 mm)

A special tool kit is available from Tecumseh to ream the cylinder crankcase and cover (mounting flange) main bearing bores to install service bushing for some models.

CRANKSHAFT, MAIN BEARINGS AND SEALS. Crankshaft main bearing journals on some models ride directly in the aluminum alloy bores in the cylinder block and the crankcase cover (mounting flange). Other engines are originally equipped with renewable steel backed bronze bushings and some are originally equipped with a ball type main bearing at the pto end of crankshaft.

Refer to CONNECTING ROD section for standard crankshaft crankpin journal diameters. Standard diameters for crankshaft main bearing journals are shown in the following chart.

Model **Diameter**
LAV25, LAV30,
TVS75, H25, H30
(prior 1983)*,
LV35, TVS90,
H35 (prior 1983),
ECH90, ECV100,
LAV35 . . . . 0.8735-0.8740 in.
(22.187-22.200 mm)
H35 (after 1983),
H30 (after 1983)*,
LAV40, TVS105,
HS40, ECV105,
TNT100, V40,
VH40, H40, HH40,
ECV110, LAV50,
TVS120, HS50,
ECV120, TNT120,
V50, H50, HH50,
VH50, TVM125,
TVM140, V60,
VH60 . . . . 0.9985-0.9990 in.
(25.362-25.375 mm)

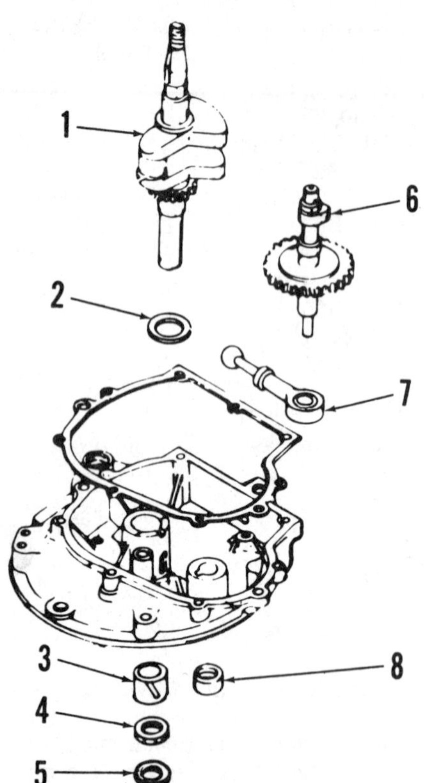

Fig. T61—An auxiliary pto shaft that turns at ½ the speed of the crankshaft is available by using a special extended camshaft. The hole in lower cover is sealed using lip type seal (8).

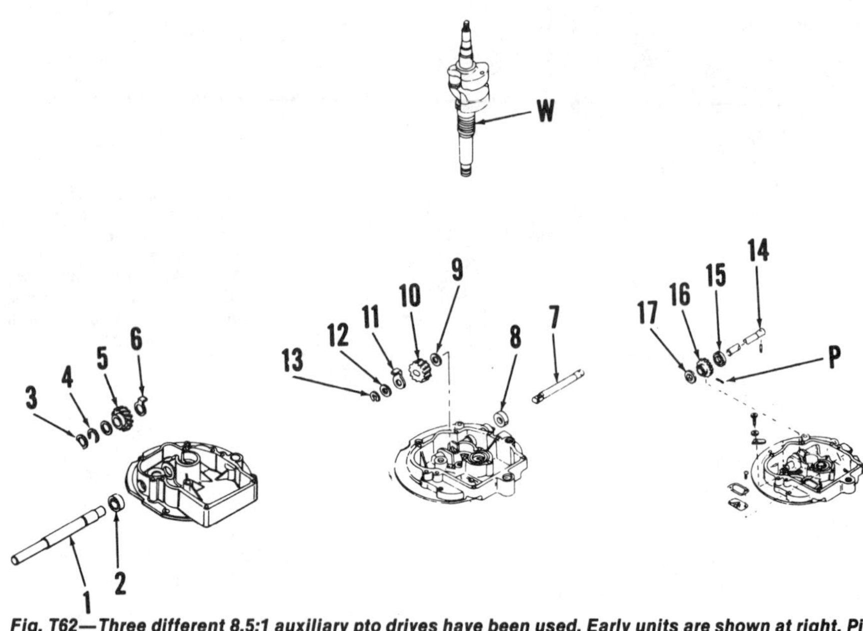

Fig. T62—Three different 8.5:1 auxiliary pto drives have been used. Early units are shown at right. Pin (P) attaches gear (16) to the pto shaft (14). On later units, the shaft is held in position by snap ring (4) or (13). Worm gear (W) on crankshaft drives the gear (5, 10 or 16) on all models.

1. Pto shaft
2. Seal
3. Washers (2)
4. Snap ring
5. Gear
6. Tang washer
7. Pto shaft
8. Seal
9. Thick washer
10. Gear
11. Tang washer
12. Washer
13. Snap ring
14. Pto shaft
15. Seal
16. Gear
17. Washer

*H30 models produced after 1983 have a main bearing journal diameter at flywheel side of 0.9985-0.9990 inch (25.362-25.375 mm) and a main bearing journal diameter at pto side of 0.8735-0.8740 inch (22.187-22.200 mm).

Clearance between crankshaft main bearing journals and main bearing bushings should be 0.0010-0.0025 inch (0.025-0.64 mm) for 0.8735-0.8740 inch (22.187-2200 mm) diameter journal or 0.0015-0.0025 inch (0.038-0.064 mm) for 0.9985-0.9990 inch (25.362-25.375 mm) journal.

Crankshaft end play for all models is 0.005-0.027 inch (0.13-0.69 mm).

On Models H30 through HS50, it is necessary to remove a snap ring (Fig. T52) which is located under oil seal. Note oil seal depth before removal of seal, remove seal and snap ring. Crankcase cover may not be removed. When installing cover, press new seal in to same depth as old seal before removal.

Ball bearing should be inspected and renewed if rough, loose or damaged. Bearing must be pressed on or off of crankshaft journal using a suitable press or puller.

Always note oil seal depth and direction before removing old seal from crankcase or cover. New seals must be pressed into seal bores to the same depth as old seal before removal on all models.

When installing crankshaft, align crankshaft and camshaft gear timing marks as shown in Fig. T54 except on certain engines manufactured for Craftsman. Refer to CRAFTSMAN engine section of this manual.

CAMSHAFT. The camshaft and camshaft gear are an integral part which rides on camshaft journals at each end of camshaft. Camshaft on some models also has a compression release mechanism mounted on camshaft gear which lifts exhaust valve at low cranking rpm to reduce compression and aid starting (Fig. T53).

Renew camshaft if lobes or journals are worn or scored. Spring on compression release mechanism should snap weight against camshaft. Compression release mechanism and camshaft are serviced as an assembly only.

Standard camshaft journal diameter is 0.6230-0.6235 inch (15.824-15.837 mm) for Models V40, VH40, H40, HH40, H50, HH50, V50, VH50, TVM125, TVM140, V60 and VH60 and 0.4975-0.4980 inch (12.637-12.649 mm) for all other models.
is longer on some models and that on models equipped with barrel and plunger type oil pump, the pump is operated by an eccentric on camshaft.

When installing camshaft, align crankshaft and camshaft timing marks as shown in Fig. T54 except on certain engines manufactured for Craftsman. Refer to CRAFTSMAN engine section of this manual.

OIL PUMP. Vertical crankshaft engines may be equipped with a barrel and plunger type oil pump or a gear-driven rotor type oil pump. Horizontal crankshaft engines may be equipped

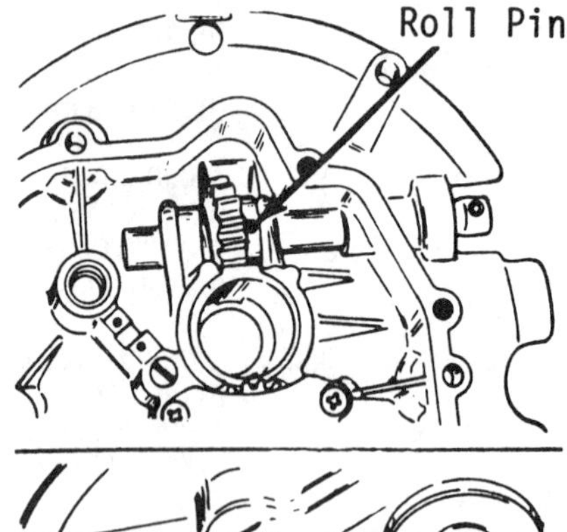

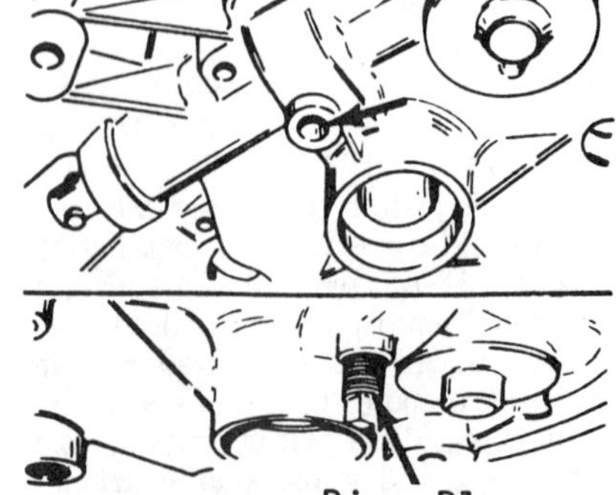

Fig. T63—The roll pin must be removed from early 8.5:1 auxiliary pto before the shaft and gear can be withdrawn. On some models (center) the boss is closed. The boss can be drilled (@ inch) and tapped to accept a 7-28 N.P.T. ‡ inch pipe plug as shown in lower view.

with a gear-driven rotor type pump or with a dipper type oil slinger attached to the connecting rod.

The barrel and plunger type oil pump is driven by an eccentric on the camshaft. Chamfered side of drive collar (Fig. T48) should be toward engine lower cover. Oil pumps may be equipped with two chamfered sides, one chamfered side or with flat boss as shown. Be sure installation is correct.

On engines equipped with gear-driven rotor oil pump, check drive gear and rotor for excessive wear or other damage. End clearance of rotor in pump body should be within limits of 0.006-0.007 inch (0.15-0.18 mm) and is controlled by cover gasket. Gaskets are available in a variety of thicknesses.

On all models with oil pump, be sure to prime pump during assembly to ensure immediate lubrication of engine.

VALVE SYSTEM. Valve tappet clearance (cold) is 0.010 inch (0.25 mm) for intake and exhaust valves for Models V50, VH50, H50, HH50, TVM125, TVM140 and V60. Valve tappet clearance for all other models is

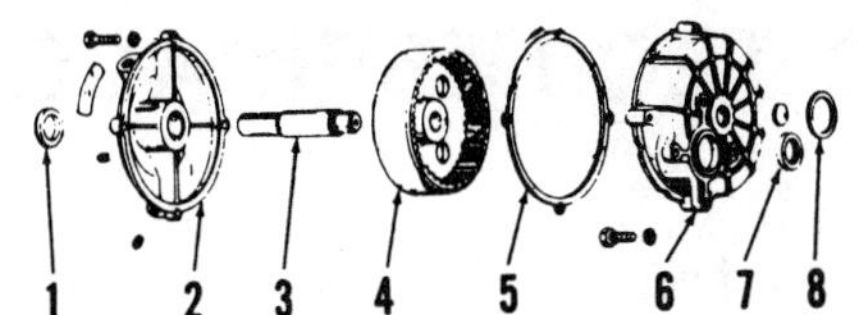

Fig. T64—Exploded view of the 6:1 gear reduction assembly. Housing (6) is bolted to cylinder block and the pinion gear is made onto end of crankshaft.

1. Seal
2. Cover
3. Output shaft
4. Gear
5. Gasket
6. Housing
7. Seal
8. Cork gasket

0.008 inch (0.03 mm) for intake and exhaust valves.

Valve face angle is 45° and valve seat angle is 46° except on very early production 2.25 horsepower (7.6 kW) models which have a 30° valve face and seat angle.

Valve seat width should be 0.035-0.045 inch (0.89-1.14 mm) for all models.

Valve stem guides are cast into cylinder block and are nonrenewable. If excessive clearance exists between valve stem and valve guide, guide should be reamed and new valve with oversize stem installed.

REDUCED SPEED PTO SHAFTS. A pto (power takeoff) shaft which rotates at 1/2 the speed of the crankshaft is available by extending the camshaft through the cover (lower mounting flange). Refer to Fig. T61. Except for the seal around the extended camshaft, service is similar to standard models.

A slow speed (8.5:1) auxiliary pto shaft is used on some vertical shaft engines (Fig. T62). A worm gear (W) on the crankshaft turns the pto gear and pto shaft. Several different versions of this unit have been used. A roll pin (Fig. T63) is used to hold the gear onto the shaft on early models. A pipe plug and a threaded boss are located as shown in center and lower views of some of these early models. The roll pin can be driven out of the gear and shaft through the threaded hole of models so equipped. Refer to Fig. T62 for order of assembly.

Disassembly and repair procedure for the 6:1 reduction will be evident after examination of the unit and reference to Fig. T64.

SERVICING TECUMSEH ACCESSORIES

12 VOLT STARTING AND CHARGING SYSTEMS

Engines with 3.5 horsepower (57 cc) or more may be equipped with a 12 volt direct current starting, generating and ignition unit system. The system includes the usual coil, condenser and breaker point for ignition, plus extra generating coils and flywheel magnets to generate alternating current. Also included is a silicon rectifier panel or regulator-rectifier for changing the generated alternating current to direct current, a series wound motor with a Bendix drive unit and a 12 volt wet cell storage battery.

Models LAV30, LAV35 and LAV40 may be equipped with a 12 volt starting motor with a right angle gear drive unit and a Nickel Cadmium battery pack. The "SAF-T-KEY" switch is located in the battery box cover. A battery charger which converts 110 volt ac house current to dc charging current is used to recharge the Nickel Cadmium battery pack.

Refer to the following paragraphs for service procedures on the units.

12 VOLT STARTER MOTOR (BENDIX DRIVE). Refer to Fig. TE1 for exploded view of 12 volt starter motor and Bendix drive unit used on some engines. This motor should not be operated continuously for more than 10 seconds. Allow starter motor to cool one minute between each 10 second cranking period.

To perform a no-load test, remove starter motor and use a fully charged 6 volt battery. Maximum current draw should not exceed 25 amperes at 6 volts. Minimum rpm is 6500.

When assembling a starter motor, use 0.003 and 0.010 inch spacer washers (14 and 15) as required to obtain an armature end play of 0.005-0.015 inch (0.13-0.38 mm). Tighten nut on end of armature shaft to 100 in.-lbs. (11 N·m). Tighten the through-bolts (30) to 30-34 in.-lbs. (3-4 N·m).

12 VOLT STARTER MOTOR (SAF-T-KEY TYPE). Refer to Fig. TE2 for exploded view of right angle drive gear unit and starter motor used on some Model LAV30, LAV35 and LAV40 models. Repair of this unit consists of renewing right angle gear drive parts or the starter motor assembly. Parts are not serviced separately for the motor (8). Bevel gears (9) are serviced only as a set.

When installing assembly on engine, adjust position of starter so there is a 1/16 inch (1.59 mm) clearance between crown of starter gear tooth and base of flywheel tooth.

NICKEL CADMIUM BATTERY. The Nickel Cadmium battery pack (Fig. TE3) is a compact power supply for the "SAF-T-KEY" starter motor. To test battery pack, measure open circuit voltage across black and red wires. If voltage is 14.0 or above, battery is good and may be recharged. Charge battery only with charger provided with the system. This charger (Fig. TE4) converts 110 volt AC house current to DC charging current. Battery should be fully charged after 14-16 hours charging time. A fully charged battery should test 15.5-18.0 volts.

A further test of battery pack can be made by connecting a 1.4 ohm resistor across black and red wires (battery installed) for two minutes. Battery voltage at the end of two minutes must be 9.0 volts minimum. If battery passes this test, recharge battery for service.

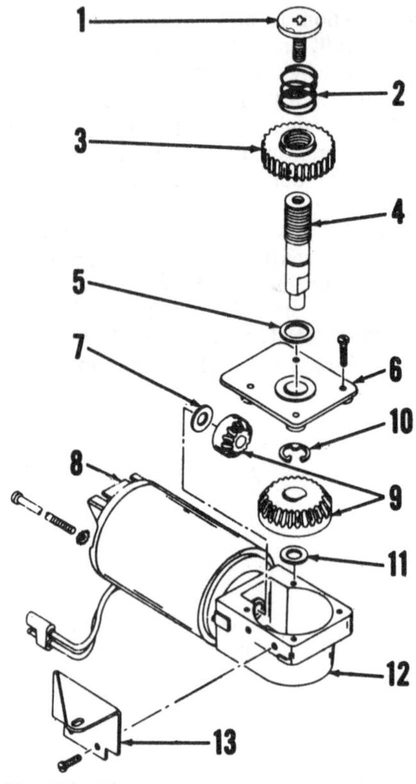

Fig. TE2—Twelve volt electric starter assembly used on some engines equipped with "SAF-T-KEY" Nickel Cadmium battery pack.

1. Screw L.H.
2. Spring
3. Starter gear
4. Shaft
5. Thrust washer
6. Cover
7. Shim washer
8. Starter motor
9. Gear set
10. Retaining ring
11. Shim washer
12. Gear housing
13. Mounting bracket

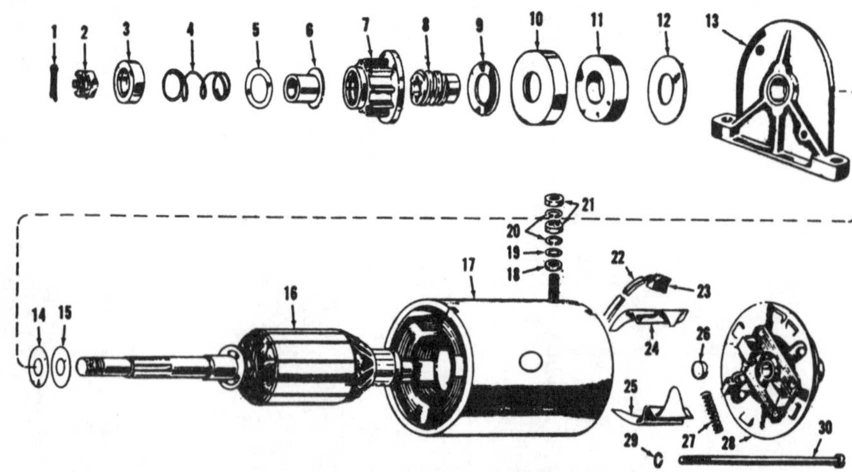

Fig. TE1—Twelve volt electric cranking motor used on some engines.

3. Pinion stop
4. Antidrift spring
5. Pinion washer
6. Antidrift sleeve
7. Pinion gear
8. Screw shaft
9. Thrust washer
10. Cushion
11. Rubber cushion
12. Thrust washer
13. Drive end cap
14. Spacer washer, 0.003 in.
15. Spacer washer, 0.010 in.
16. Armature
18. Insulation washer
22. Insulation tubing
23. Brush
24. Insulation
25. Insulation
26. Thrust spacer
28. Commutator end cap

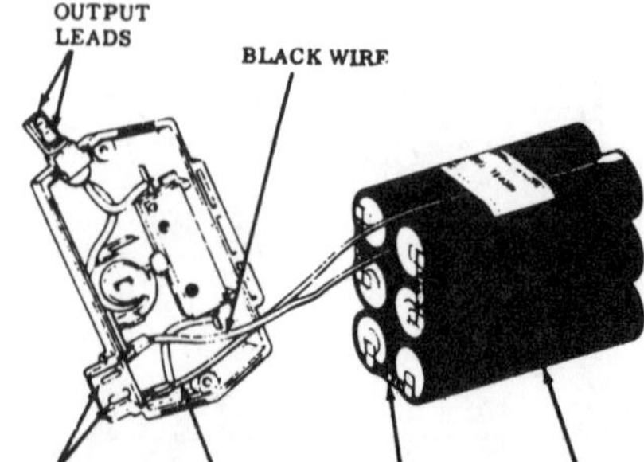

Fig. TE3—View of Nickel Cadmium battery pack connected to "SAF-T-KEY" starting switch.

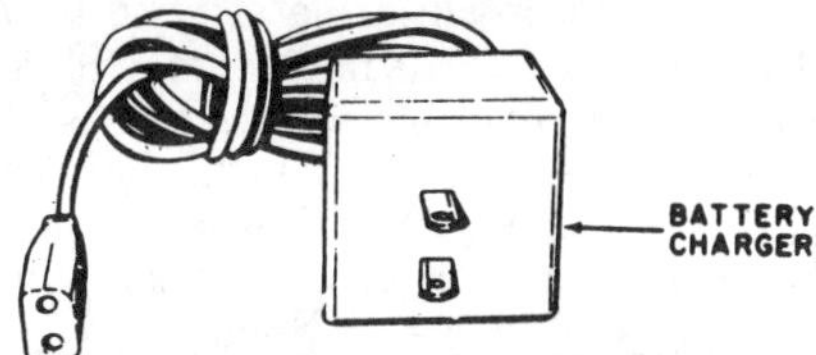

Fig. TE4—Recharge Nickel Cadmium battery pack only with the charger furnished with "SAF-T-KEY" starter system. Charger converts 110 volt ac to dc charging current.

ALTERNATOR CHARGING SYSTEMS. Flywheel alternators are used on some engines for the charging system. The generated alternating current is converted to direct current by two rectifiers on the rectifier panel (Figs. TE5 and TE6) or the regulator-rectifier (Fig. TE7).

The system shown in Fig. TE5 has a maximum charging output of about 3 amperes at 3600 rpm. No current regulator is used on this low output system. The rectifier panel includes two diodes (rectifiers) and a 6 ampere fuse for overload protection.

The system shown in Fig. TE6 has a maximum output of 7 amperes. To prevent overcharging the battery, a double pole switch is used in low output position to reduce the output to 3 amperes for charging the battery. Move switch to high output position (7 amperes) when using accessories.

The system shown in Fig. TE7 has a maximum output of 7 amperes and uses a solid state regulator-rectifier which converts the generated alternating current to direct current for charging the battery. The regulator-rectifier also allows only the required amount of current flow for existing battery conditions. When battery is fully charged, current output is decreased to prevent overcharging the battery.

Testing. On models equipped with rectifier panel (Figs. TE5 or TE6), remove rectifiers and test them with either a continuity light or an ohmmeter. Rectifiers should show current flow in one direction only. Alternator output can be checked using an induction ammeter over the positive lead wire to battery.

On models equipped with the regulator-rectifier (Fig. TE7), check the system as follows: Disconnect B+ lead and connect a dc voltmeter as shown in Fig. TE8. With engine running near full throttle, voltage should be 14.0-14.7 volts. If voltage is above 14.7 or below 14.0 but above 0, the regulator-rectifier is defective. If voltmeter reading is 0 the regulator-rectifier or alternator coils may be defective. To test the alternator coils, connect an ac voltmeter to the ac leads as shown in Fig. TE9. With engine running at near full throttle, check ac voltage. If voltage is less than 20.0 volts, alternator is defective.

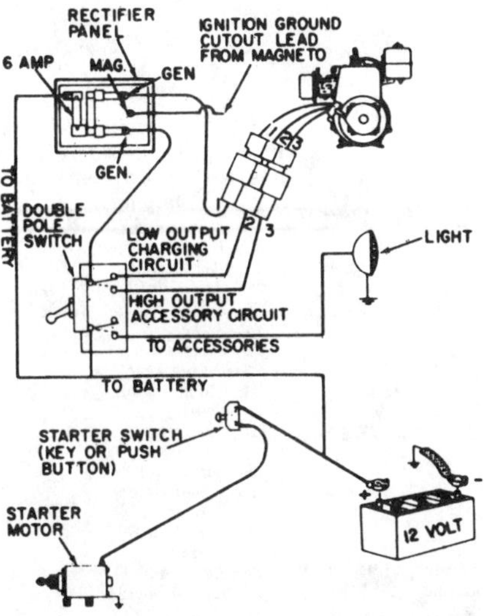

Fig. TE6—Wiring diagram of typical 7 ampere alternator and rectifier panel charging system. The double pole switch in one position reduces output to 3 amperes for charging or increases output to 7 amperes in other position to operate accessories.

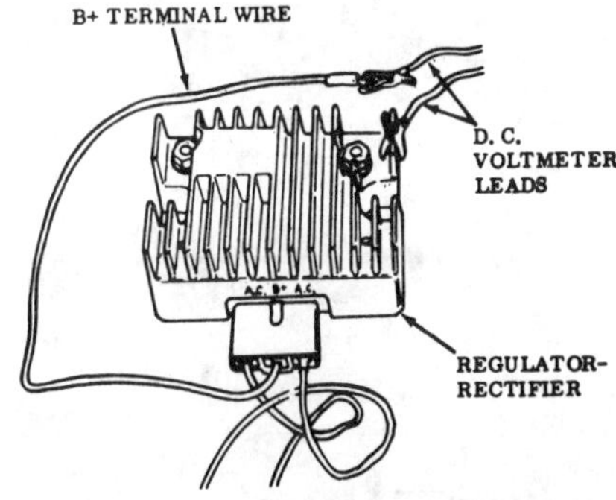

Fig. TE8—Connect dc voltmeter as shown when checking the regulator-rectifier.

110 VOLT ELECTRIC STARTER

110 VOLT AC-DC STARTER. Some vertical crankshaft engines are available with a 110 volt ac electric starting system as shown in Fig. TE11. When switch button (6) is depressed, switch (21) is turned on and the motor rotates driven gear (23). As more pressure is applied, housing (11) moves down compressing springs (19). At this time, clutch facing (24) is forced tight between driven gear (23) and cone (30) and engine is cranked.

Electric motor (3) is serviced only as an assembly. All other parts shown are available separately. Shims (18) are used to adjust mesh of drive gear (5) to driven gear (23).

110 VOLT AC-DC STARTER. Some engines may be equipped with a 110 volt ac-dc starting motor (Fig. TE12). The rectifier assembly used with this starter converts 110 volt ac house current to approximately 100 volts dc. A thyrector (8) is used as a surge protector for the rectifiers.

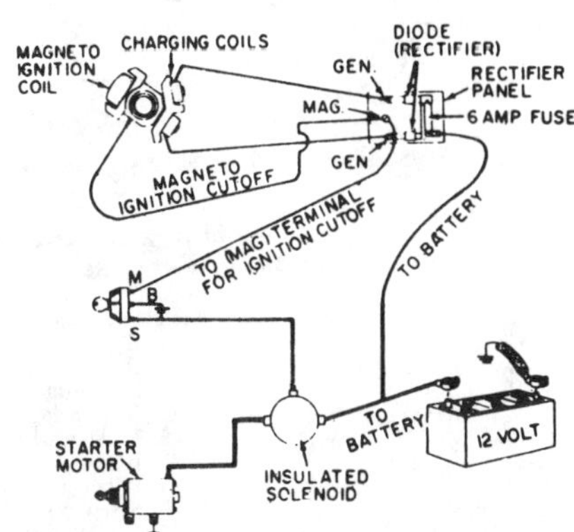

Fig. TE5—Wiring diagram of typical 3 ampere alternator and rectifier panel charging system.

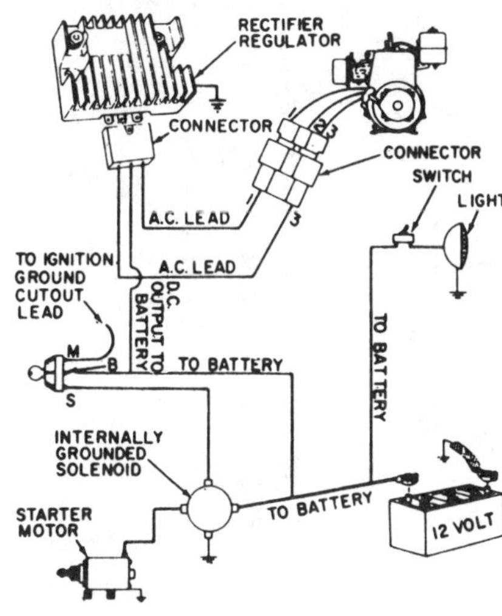

Fig. TE7—Wiring diagram of typical 7 or 10 ampere alternator and regulator-rectifier charging system.

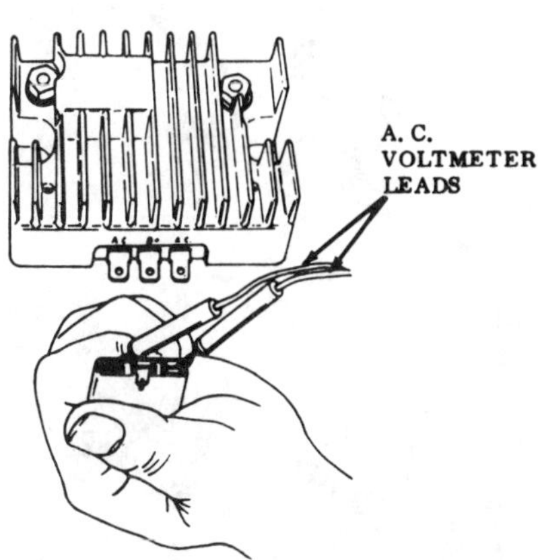

Fig. TE9—Connect ac voltmeter to ac leads as shown when checking alternator coils.

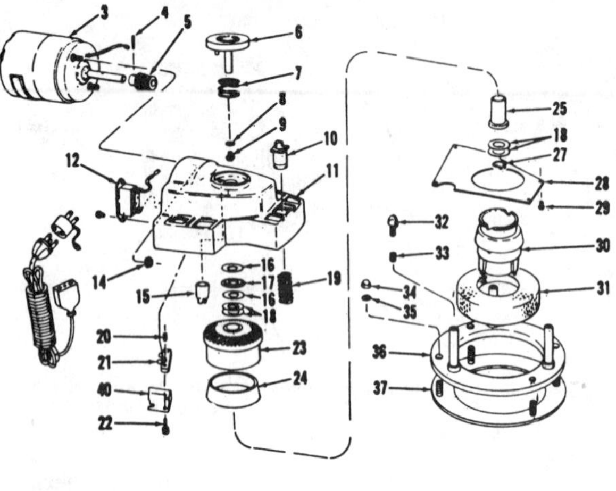

3. Motor
4. Roll pin
5. Pinion gear
6. Starter switch button
7. Switch return spring
8. Washer
9. Screw
10. Mounting bushings (3)
11. Starter housing
12. Receptacle
14. Nuts (2)
15. Pinion bearing

Fig. TE11—Exploded view of the 110 volt ac electric starter used on some engines.

16. Thrust races (2)
17. Thrust bearing
18. Thrust shims
19. Guide post springs (3)
20. Switch plunger spring
21. Switch
22. Screws (2)
23. Driven gear
24. Clutch facing
25. Bearing
27. Snap ring
28. Shield
29. Screws (4)
30. Lower cone
31. Screen
32. Cap studs (3)
33. Spring
34. Acorn nuts (4)
35. Lockwashers
36. Mounting ring
37. Reinforcing ring
40. Switch cover

To test starting motor, remove the dc output cable (9) from motor. Connect an ohmmeter between one of the flat receptacle terminals and motor housing. If a short is indicated, the motor is grounded and requires repair. Connect the ohmmeter between the two flat receptacle terminals. A resistance reading of 3-4 ohms must be obtained. If not, motor is faulty.

A no-load test can be performed with the starter removed. Do not operate the motor continuously for more than 15 seconds when testing. Maximum ampere draw should be 2 amperes and minimum rpm should be 8500.

Disassembly of starter motor is obvious after examination of the unit and reference to Fig. TE12. When reassembling, install shim washers (20) as necessary to obtain armature end play of 0.005-0.015 inch (0.13-0.38 mm). Shim washers are available in thicknesses of 0.005, 0.010 and 0.020 inch. Tighten elastic stop nut (32) to 100 in.-lbs. (11 N•m). Tighten nuts on the two through-bolts to 24-28 in.-lbs. (3 N•m).

To test the rectifier assembly, first use an ac voltmeter to check line voltage of the power supply, which should be approximately 115 volts. Connect the input cable (7) to the ac power supply. Connect a dc voltmeter to the two slotted terminals of the output cable (9). Move switch (3) to "ON" position and check the dc output voltage. Direct current output voltage should be a minimum of 100 volts. If a low voltage reading is obtained, rectifiers and/or thyrector could be faulty. Thyrector (8) can be checked after removal, by connecting the thyrector, a 7.5 watt ac light bulb and a 115 volt ac power supply in series. If light bulb glows, thyrector is faulty. Use an ohmmeter to check the rectifiers. Rectifiers must show continuity in one direction only.

WIND-UP STARTERS

RATCHET STARTER. On models equipped with the ratchet starter, refer to Fig. TE13 and move release lever to "RELEASE" position to remove tension

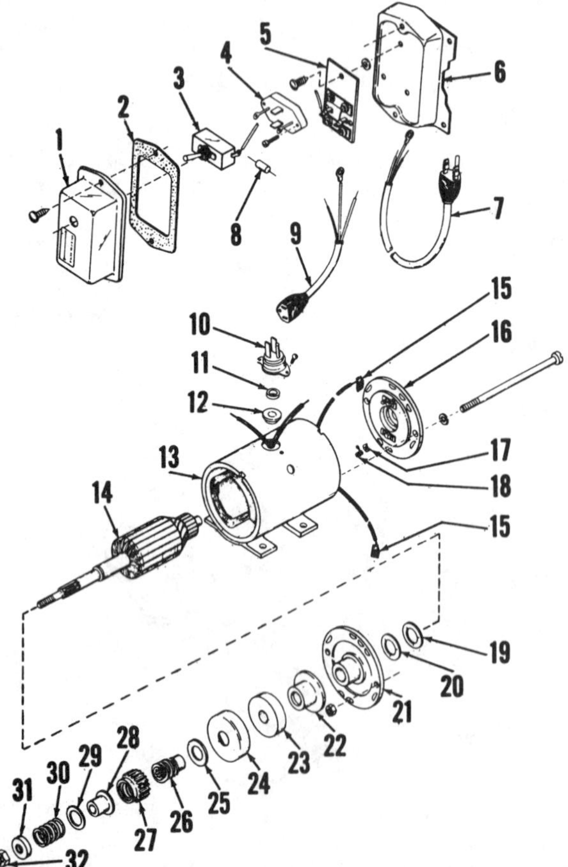

Fig. TE12—Exploded view of 110 volt ac-dc starter motor and rectifier assembly.

1. Cover
2. Gasket
3. Switch
4. Rectifier bridge
5. Rectifier base
6. Mounting base
7. AC input cable
8. Thyrector
9. DC output cable
10. Receptable
11. Washer
12. Insulator
13. Field coils & housing assy.
14. Armature
15. Brushes
16. End plate
17. Spring insulator
18. Brush spring
19. Nylon washer
20. Shim washers (0.005, 0.010 & 0.020 in.)
21. Drive end-plate
22. Thrust sleeve
23. Rubber cushion
24. Cup
25. Thrust washer
26. Screw shaft
27. Pinion gear
28. Spring sleeve
29. Washer
30. Antidrift spring
31. Pinion stop
32. Elastic stop nut

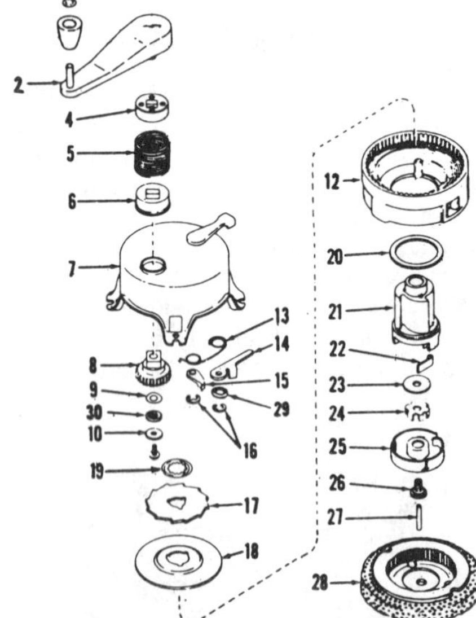

Fig. TE13—Exploded view of a ratchet starter assembly used on some engines.

2. Handle
4. Clutch
5. Clutch spring
6. Bearing
7. Housing
8. Wind gear
9. Wave washer
10. Clutch washer
12. Spring & housing
13. Release dog spring
14. Release dog
15. Lock dog
16. Dog pivot retainers
17. Release gear
18. Spring cover
19. Retaining ring
20. Hub washer
21. Starter hub
22. Starter dog
23. Brake washer
24. Brake
25. Retainer
26. Screw L.H.
27. Centering pin
28. Hub & screen
29. Spacer washers
30. Lockwasher

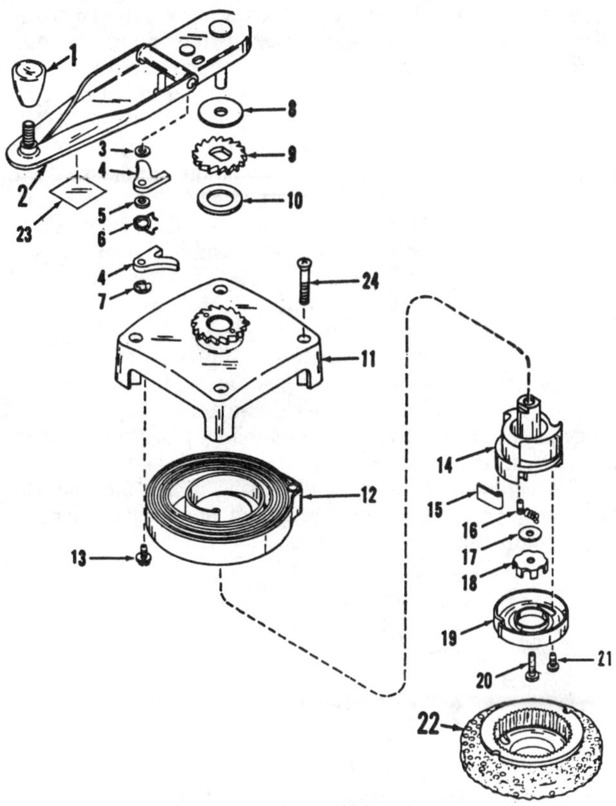

Fig. TE14—Components of impulse mechanical starter. When handle (2) is used to wind spring (12) the energy put into the spring cranks the engine.

1. Handle knob
2. Starter handle
3. Dog washer
4. Dog release
5. Dog spacer
6. Dog release spring
7. Pivot retainer
8. Ratchet
9. Ratchet
10. Ratchet spacer
11. Housing assy.
12. Spring & keeper
13. Keeper screw
14. Spring hub assy.
15. Starter dog
16. Spring & eyelet
17. Brake washer
18. Brake
19. Retainer
20. Screw
21. Screw
22. Hub & screen
24. Screw

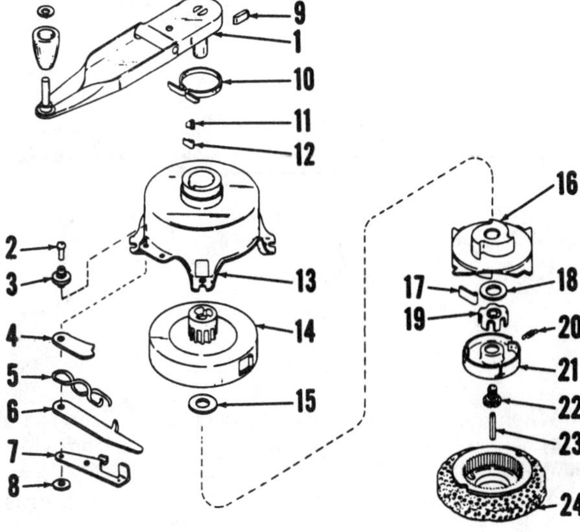

Fig. TE15—Exploded view of remote release impulse starter used on some models. Refer also to Fig. TE16.

1. Cranking handle
2. Rivet
3. Pivot stud
4. Release lever
5. Release spring
6. Trip lever
7. Trip release
8. Washer
9. Wind dog
10. Brake band
11. Lock dog spring
12. Lock dog
13. Housing
14. Spring & keeper
15. Bearing washer
16. Hub
17. Starter dog
18. Thrust washer
19. Brake
20. Spring
21. Retainer
22. Screw L.H.
23. Centering pin
24. Hub & screen

from main spring. Remove starter assembly from engine. Remove left-hand thread screw (26), retainer hub (25), brake (24), washer (23) and six starter dogs (22). Note position of starter dogs in hub (21). Remove hub (21), washer (20), spring and housing (12), spring cover (18), release gear (17) and retaining ring (19) as an assembly. Remove retaining ring, then carefully separate these parts.

CAUTION: Do not remove main spring from housing (12). The spring and housing are serviced only as an assembly.

Remove snap rings (16), spacer washers (29), release dog (14), lock dog (15) and spring (13). Winding gear (8), clutch (4), clutch spring (5), bearing (6) and crank handle (2) can be removed after first removing the retaining screw and washers (10, 30 and 9).

Reassembly procedure is the reverse of disassembly. Centering pin (27) must align screw (26) with crankshaft center hole.

IMPULSE STARTER. To overhaul the impulse starter (Fig. TE14), first release tension from main spring. Unbolt and remove starter assembly from engine. Remove retaining screw (21), retainer (19), spring and eyelet (16) and starter dogs (15). Remove screw (20), brake (18), washer (17), cranking handle (2), bearing (8), ratchet (9) and spacer (10). Withdraw spring hub (14) and if spring and keeper assembly (12) are to be renewed, remove keeper screw (13). Carefully remove spring and keeper assembly.

CAUTION: Do not remove spring from keeper.

Remove pivot retainer (7), dog releases (4), spring (6), washer (3) and spacer (5) from handle (2).

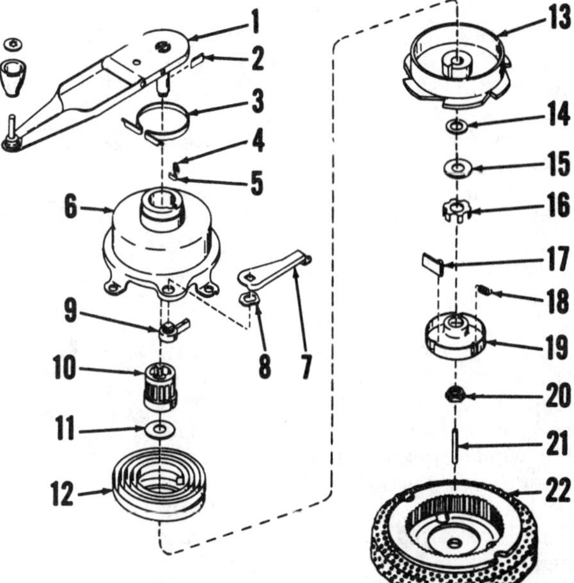

Fig. TE16—Exploded view of remote release impulse starter used on some models. Refer also to Fig. TE15.

1. Cranking handle
2. Wind dog
3. Brake band
4. Lock dog spring
5. Lock dog
6. Housing
7. Release lever
8. Wave washer
9. Release arm
10. Hub
11. Spacer washer
12. Power spring
13. Spring housing
14. Shim
15. Thrust washer
16. Brake
17. Starter dog
18. Spring
19. Retainer
20. Shoulder nut
21. Centering pin
22. Hub & screen

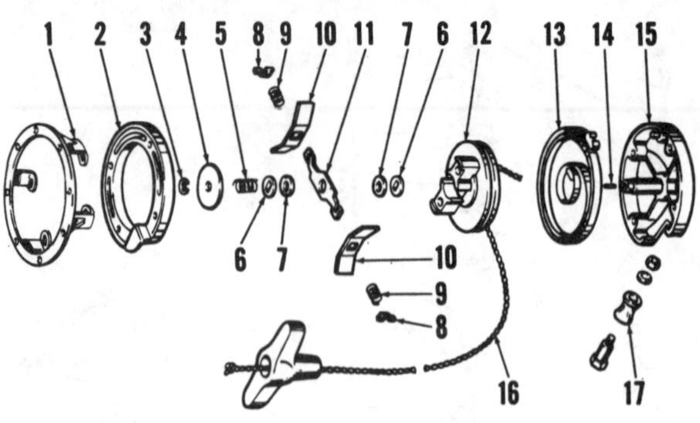

Fig. TE17—Exploded view of typical friction shoe rewind starter.
1. Mounting flange
2. Flange
3. Retaining ring
4. Washer
5. Spring
6. Slotted washer
7. Fiber washer
8. Spring retainer
9. Spring
10. Friction shoe
11. Actuating lever
12. Rotor
13. Rewind spring
14. Centering pin
15. Cover
16. Rope
17. Roller

When reassembling, install spring hub (14) before placing spring and keeper in housing (11). Install ratchet (9) so teeth are opposite ratchet teeth on housing.

REMOTE RELEASE IMPULSE STARTER (SURE LOCK). To disassemble the "Sure Lock" remote release impulse starter (Fig. TE15), first release tension from main spring. Unbolt and remove starter assembly from engine. Remove centering pin (23), then unscrew the left-hand threaded screw (22). Remove retainer (21), spring (20), brake (19), starter dog (17), thrust washer (18) and hub (16). Remove cranking handle (1), wind dog (9), brake band (10), lock dog spring (11) and lock dog (12). Carefully lift out spring and keeper assembly (14).

CAUTION: Do not attempt to remove spring from keeper. Spring and keeper are serviced only as an assembly.

Some models use pivot stud (3), release lever (4), release spring (5) and trip lever (6). Other models use rivet (2) and trip release (7). Removal of these parts is obvious after examination of the unit and reference to Fig. TE15.

Reassembly procedure is the reverse of disassembly. Press centering pin (23) into screw (22) about 1/3 of the pin length. Pin will align starter with center hole in crankshaft.

REMOTE RELEASE IMPULSE STARTER (SURE START). Some models are equipped with the "Sure Start" remote release impulse starter shown in Fig. TE16. To disassemble the starter, first move release lever to remove power spring tension. Unbolt and remove starter assembly from engine. Remove centering pin (21), shoulder nut (20), retainer (19), spring (18), starter dog (17), thrust washer (15) and shim (14). Remove cranking handle (1), wind dog (2) and brake band (3). Rotate spring housing (13) spring while pushing down on top of hub (10). Withdraw spring housing, power spring and hub from housing (6). Lift hub from spring.

This is the only wind-up type starter spring which may be removed from the spring housing (keeper). To remove spring, grasp inner end of spring with pliers and pull outward. Spring diameter will increase about 1/3 when removed. To install new power spring in spring housing, clamp spring housing in a vise. Anchor outer end of spring in housing. Install spacer washer (11) and hub (10), engaging inner end of spring in hub notch. Insert crankpin handle shaft in hub and while pressing down on cranking handle, rotate crank to wind spring into housing. When all of the spring is in the housing, allow cranking handle and hub to unwind.

To remove release lever (7) and release arm (9), file off peened-over sides of release arm. Remove lever, wave washer and arm.

When reassembling, place release arm on wood block, install wave washer and release lever. Using a flat face punch, peen over edges of new release arm to secure release lever. The balance of

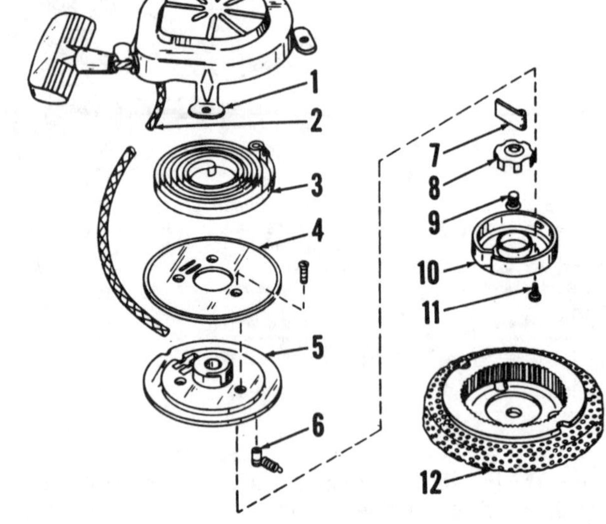

Fig. TE18—Exploded view of typical dog type rewind starter assembly. Some units of similar construction use three starter dogs (7).
1. Cover
2. Rope
3. Rewind spring
4. Pulley half
5. Pulley half & hub
6. Retainer spring
7. Starter dog
8. Brake
9. Brake screw
10. Retainer
11. Retainer screw
12. Hub & screen assy.

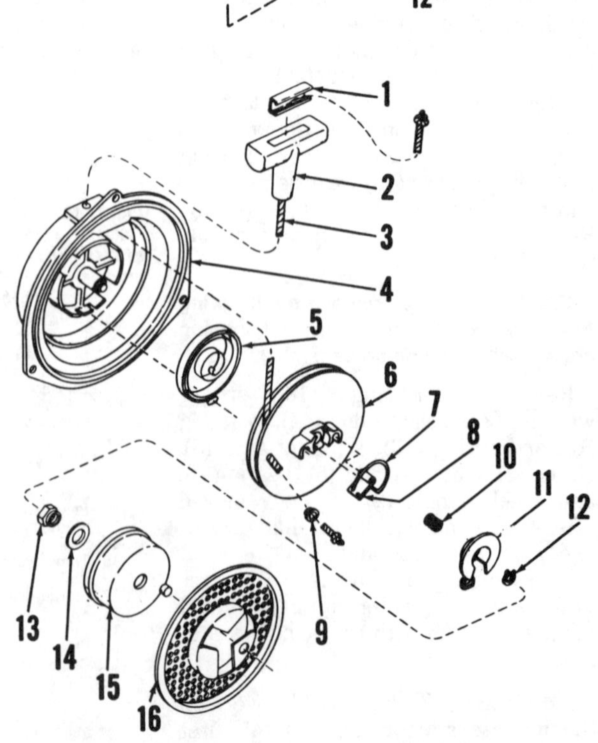

Fig. TE19—Exploded view of typical dog type rewind starter assembly used on some two-stroke engines.
1. Insert
2. Handle
3. Rope
4. Starter housing
5. Rewind spring
6. Pulley & hub assy.
7. Dog spring
8. Starter dog
9. Grommet
10. Brake spring
11. Brake
12. Retaining ring
13. Flywheel nut
14. Washer
15. Starter cup
16. Screen

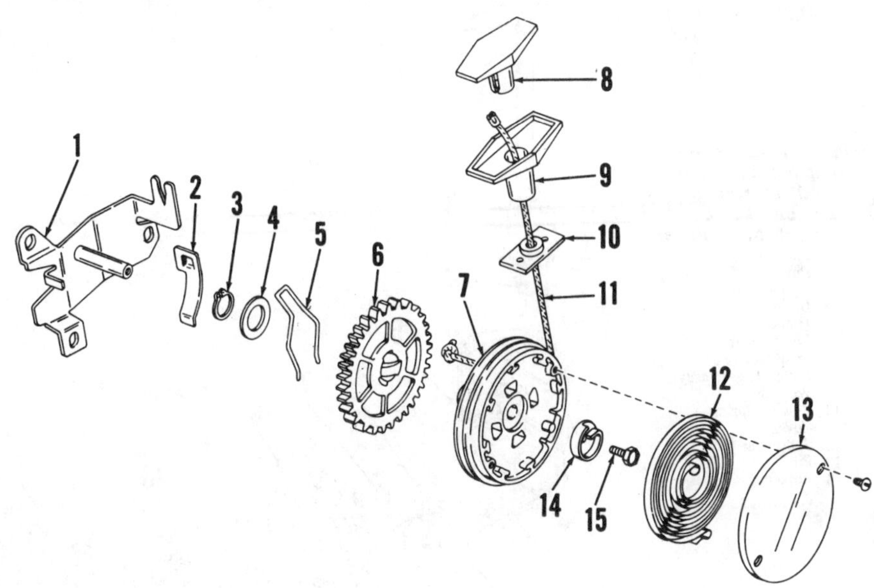

Fig. TE20—Exploded view of single-gear, side-mounted rewind starter used on some vertical crankshaft engines.

1. Mounting bracket
2. Rope clip
3. Snap ring
4. Washer
5. Brake
6. Gear
7. Pulley
8. Insert
9. Handle
10. Bracket
11. Rope
12. Rewind spring
13. Cover
14. Spring hub
15. Hub screw

reassembly is reverse of disassembly procedure. Install centering pin so end of pin extends below end of mounting legs. The pin will align the center of starter to center hole in end of crankshaft.

REWIND STARTERS

FRICTION SHOE TYPE. To disassemble the starter, refer to Fig. TE17 and hold starter rotor (12) securely with thumb and remove the four screws securing flanges (1 and 2) to cover (15). Remove flanges and release thumb pressure enough to allow spring to rotate pulley until spring (13) is unwound. Remove retaining ring (3), washer (4), spring (5), slotted washer (6) and fiber washer (7). Lift out friction shoe assembly (8, 9, 10 and 11), then remove second fiber washer and slotted washer. Withdraw rotor (12) with rope from cover and spring. Remove rewind spring from cover and unwind rope from rotor.

When reassembling, lubricate rewind spring, cover shaft and center bore in rotor with a light coat of Lubriplate or equivalent. Install rewind spring so windings are in same direction as removed spring. Install rope on rotor, then place rotor on cover shaft. Make certain inner and outer ends of spring are correctly hooked on cover and rotor. Preload the rewind spring by rotating the rotor two full turns. Hold rotor in preload position and install flanges (1 and 2). Check sharp end of friction shoes (10) and sharpen or renew as necessary. Install washers (6 and 7), friction shoe assembly, spring (5), washer (4) and retaining ring (3). Make certain friction shoe assembly is installed properly for correct starter rotation. If properly installed, sharp ends of friction shoes will extend when rope is pulled.

Remove brass centering pin (14) from cover shaft, straighten pin if necessary, then reinsert pin 1/3 of its length into cover shaft. When installing starter on engine, centering pin will align starter with center hole in end of crankshaft.

DOG TYPE (FOUR-STROKE) To disassemble the dog type starter, refer to Fig. TE18 and release preload tension of rewind spring by pulling starter rope until notch in pulley half (5) is aligned with rope hole in cover (1). Use thumb pressure to prevent pulley from rotating. Engage rope in notch of pulley and slowly release thumb pressure to allow spring to unwind. Remove retainer screw (11), retainer (10) and spring (6). Remove brake screw (9), brake (8) and starter dog (7). Carefully remove pulley assembly with rope from spring and cover. Note direction of spring winding and carefully remove spring from cover. Unbolt and separate pulley halves (4 and 5) and remove rope.

To reassemble, reverse the disassembly procedure. Then, preload rewind spring by aligning notch in pulley with rope hole in cover. Engage rope in notch and rotate pulley two full turns to properly preload the spring. Pull rope to fully extended position. Release handle and if spring is properly preloaded, the rope will fully rewind.

DOG TYPE (TWO-STROKE). Some engines are equipped with the dog type rewind starter shown in Fig. TE19. To disassemble the starter, first pull rope to fully extended position. Then, tie a slip knot in rope on outside of starter housing. Remove insert (1) and handle (2) from end of rope. Hold pulley with thumb pressure, untie slip knot and withdraw rope and grommet (9). Slowly release thumb pressure and allow pulley to rotate until spring is unwound. Remove retaining ring (12), brake (11), brake spring (10) and starter dog (8) with dog spring (7). Lift out pulley and hub assembly (6), then remove rewind spring (5).

Reassemble by reversing the disassembly procedure. When installing the rope, insert a suitable tool in the rope hole in pulley and rotate pulley six full turns until hole is aligned with rope hole in housing (4). Hold pulley in this position and install rope and grommet. Tie a slip knot in rope outside of housing and assemble handle, then insert on rope. Untie slip knot and allow rewind spring to wind the rope.

SIDE-MOUNTED TYPE (SINGLE GEAR). To disassemble the side-mounted starter (Fig. TE20), remove insert (8) and handle (9). Relieve spring tension by allowing spring cover (13) to rotate slowly. Rope will be drawn through bracket (10) and wrapped on pulley (7). Remove cover (13) and carefully remove rewind spring (12). Remove hub screw (15), spring hub (14), then withdraw pulley and gear assembly. Remove snap ring (3), washer (4), then withdraw pulley and gear assembly. Remove snap ring (3), washer (4), brake (5) and gear (6) from pulley.

Reassemble by reversing the disassembly procedure, keeping the following points in mind: Lubricate only the edges of rewind spring (12) and shaft on mounting bracket (1). Do not lubricate spiral in gear (6) or on pulley shaft. Rewind spring must be preloaded approximately 2½ turns.

When installing the starter assembly, adjust mounting bracket so side of teeth on gear (6) when in fully engaged position, has 1/16 inch (1.59 mm) clearance from base of flywheel gear teeth. Remove spark plug wire and test starter several times for gear engagement. A too-close adjustment could cause starter gear to hang up on flywheel gear when engine starts. This could damage the starter.

SIDE-MOUNTED TYPE (MULTIPLE GEAR). Some engines may be equipped with the side-mounted starter shown in Fig. TE21. To disassemble this type starter, unbolt cover (11) and remove items (7, 8, 9, 10, 11 and 14) as an assembly. Pull rope out about 8 inches (203 mm), hold pulley and gear assembly (8) with thumb pressure and engage rope in notch in pulley. Slowly release thumb pressure and allow spring to unwind. Remove snap ring (7), pulley and gear (8), washer (9) and rewind spring (10) from cover. Rope (14) and rewind spring can now be removed if necessary. Gears (3 and 6) can be removed after first removing snap ring (12), driving out pin (13) and withdrawing shaft (1). Remove brake (4).

To reassemble, reverse the disassembly procedure. Preload the rewind spring two full turns.

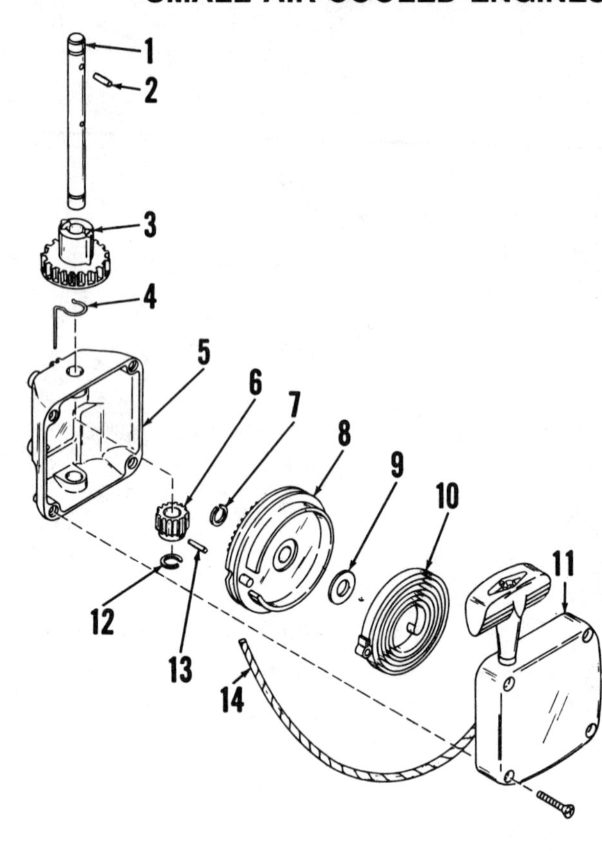

Fig. TE21—Exploded view of multiple-gear, side-mounted rewind starter assembly used on some vertical crankshaft engines.

1. Shaft
2. Pin
3. Upper gear
4. Brake
5. Housing
6. Lower gear
7. Snap ring
8. Pulley & gear assy.
9. Washer
10. Rewind spring
11. Cover
12. Snap ring
13. Pin
14. Rope

TECUMSEH CENTRAL PARTS DISTRIBUTORS

(Arranged Alphabetically by States)

These franchised firms carry extensive stocks of repair parts. Contact them for name of dealer in their area who will have replacement parts.

Charlie C. Jones Battery &
Electric Co., Inc.
Phone: (602) 272-5621
2440 West McDowell Road
P.O. Box 6654
Phoenix, Arizona 85005

Pacific Power Equipment Company
Phone: (415) 692-1094
1565 Adrian Road
Burlingame, California 94010

Pacific Power Equipment Company
Phone: (303) 371-4081
500 Oakland Street
Denver, Colorado 80239

Spencer Engine, Incorporated
Phone: (813) 253-6035
1114 West Cass Street
P.O. Box 2579
Tampa, Florida 33601

Sedco, Incorporated
Phone: (404) 925-4706
1414 Red Plum Road NW
Norcross, Georgia 30093

Small Engine Clinic
Phone: (808) 488-0711
98019 Kam Highway
Honolulu, Hawaii 96701

Industrial Engine & Parts
Phone: (312) 927-4100
1133 West Pershing Road
Chicago, Illinois 60609

Medart Engines & Parts of Kansas
Phone: (913) 888-8828
15500 West 109th Street
Lenexa, Kansas 66219

Grayson Company of Louisiana
Phone: (318) 222-3211
100 Fannin
P.O. Box 206
Shreveport, Louisiana 71102

W. J. Connell Company
Phone: (617) 543-3600
65 Green Street
Route 106
Foxboro, Massachusetts 02035

Central Power Distributor
Phone: (612) 633-5179
2010 W. County Rd. C.
Roseville, Minnesota 55113

Medart Engines & Parts
Phone: (314) 343-0505
100 Larkin William Industrial Ct.
Fenton, Missouri 63026

Original Equipment, Incorporated
Phone: (406) 245-3081
905 Second Avenue North
Box 2135
Billings, Montana 59103

E. J. Smith & Sons Company
Phone: (704) 394-3361
4250 Golf Acres Drive
P.O. Box 668887
Charlotte, North Carolina 28266

Uesco Warehouse, Incorporated
Phone: (701) 237-0424
715 25th Street North
P.O. Box 2904
Fargo, North Dakota 58108

Gardner Engine & Parts Distribution
Phone: (614) 488-7951
1150 Chesapeake Avenue
Columbus, Ohio 43212

Mico, Incorporated
Phone: (918) 627-1448
7450 East 46th Place
P.O. Box 470324
Tulsa, Oklahoma 74147

Power Equipment Systems
Phone: (503) 585-6120
465 Marion St.
P.O. Box 629
Salem, Oregon 97308

Sullivan Brothers, Incorporated
Phone: (215) 942-3686
Creek Road & Langoma Avenue
P.O. Box 140
Elverson, Pennsylvania 19520

Pitt Auto Electric Company
Phone: (412) 766-9112
2900 Stayton Street
Pittsburgh, Pennsylvania 15212

Locke Auto Electric Service,
Incorporated
Phone: (605) 336-2780
231 North Dakota Avenue
P.O. Box 1165
Sioux Falls, South Dakota 57101

Medart Engines & Parts of Memphis
Phone: (901) 774-6371
674 Walnut Street
Memphis, Tennessee 38126

Engine Warehouse, Incorporated
Phone: (713) 937-4000
7415 Empire Central Drive
Houston, Texas 77040

Frank Edwards Company
Phone: (801) 363-8851
100 South 300 West
P.O. Box 2158
Salt Lake City, Utah 84110

R B I Corporation
Phone: (804) 798-1541
101 Cedar Run Drive
Lake-Ridge Park
Ashland, Virginia 23005

CANADIAN DISTRIBUTORS

Suntester Equipment (Central), Ltd.
Phone: (403) 453-5791
13315 146th Street
Edmonton, Alberta, Canada T5L 4S8

Suntester Equipment (Central), Ltd.
Phone: (416) 624-6200
5466 Timberlea Boulevard
**Mississauga, Ontario, Canada
L4W 2T7**

Interprovincial Turf, Ltd.
Phone: (403) 227-3300
P.O. Box 2010
4404 42nd Avenue
Innisfail, Alberta T0M 1A0

Fallis Turf Equipment, Ltd.
Phone: (604) 277-1314
11951 Forge Place
**Richmond, British Columbia
V7A 4V9**

Consolidated Turf Equipment, Ltd.
Phone: (204) 633-7276
972 Powell Avenue
Winnipeg, Manitoba R3H 0H6

The Halifax Seed Company, Ltd.
Phone: (902) 454-0938
Box 8026, Station A
Halifax, Nova Scotia B3K 5L8

Ontario Turf Equipment Co., Ltd.
Phone: (519) 452-3540
50 Charterhouse Crescent
London, Ontario N5W 5V5

O.J. Company, Ltd.
Division of Otto Jangle Co., Ltd.
Phone: (514) 861-5379
294 Rang St. Paul
Sherrington, Quebec J0L 2N0

Brandt Industries, Ltd.
Phone: (306) 525-1314
705 Toronto Street
Regina, Saskatchewan S4R 8G1

WISCONSIN

TELEDYNE TOTAL POWER
3409 Democrat Road
Memphis, Tennessee 38118

Model	Bore	Stroke	Displacement
AA	2.25 in. (57.2 mm)	2.75 in. (69.9 mm)	10.9 cu. in. (180 cc)
AB, ABN, ABS	2.50 in. (63.5 mm)	2.75 in. (69.9 mm)	13.5 cu. in. (221 cc)
ACN, HACN	2.63 in. (66.7 mm)	2.75 in. (69.9 mm)	14.9 cu. in. (244 cc)

ENGINE IDENTIFICATION

Model HACN is a vertical crankshaft engine, while all other models in this section are horizontal crankshaft engines.

Engine model number, serial number and specification numbers are located on engine nameplate. Always furnish model number, serial number and specification number when ordering parts.

MAINTENANCE

SPARK PLUG. Recommended spark plug for all models is a Champion D16J, or equivalent. Spark plug electrode gap should be 0.030 in. (0.76 mm).

CARBURETOR. Engine may be equipped with one of a variety of carburetors of different manufacture. The following list shows engine and carburetor applications.

Engine Model	Carburetor
ABS, ACN	Marvel-Schebler VH53
AB, ABN	Marvel-Schebler VH63 or VH92
AA, AB, ABN	Stromberg OH 5/8
AB, ABN, ACN, HACN	Zenith 87B5

Note that engines of the same model may be equipped with different carburetors. For service procedure, refer to appropriate paragraph for model of carburetor being serviced.

Marvel-Schebler Carburetors. Initial adjustment of Marvel-Schebler carburetor for idle and main mixture screws from a lightly seated position is ½ turn open for idle mixture screw and 1½ turns open for main mixture screw. Refer to Fig. W1. Make final adjustments with engine at operating temperature and running. Adjust idle mixture screw to obtain smoothest idle and acceleration. Operate engine under load at full throttle and adjust main fuel mixture screw to obtain smoothest operation. As each adjustment affects the other, adjustment procedure may have to be repeated.

To check carburetor float setting, invert carburetor throttle body and float assembly. Measure distance between free end of float and body gasket surface. Distance should be ¼ inch (6.35 mm). Carefully bend float lever tang which contacts fuel inlet needle to obtain correct measurement.

Stromberg Carburetor. Initial adjustment of idle mixture and main fuel mixture screws from a lightly seated position, is ¾ turn open for idle mixture screw and 1 turn open for main fuel mixture screw. Make final adjustments with engine at operating temperature and running. Adjust idle mixture screw to obtain smoothest idle and acceleration. Operate engine under load at full throttle and adjust main fuel mixture screw to obtain smoothest operation. As each adjustment affects the other, adjustment procedure may have to be repeated.

Fig. W1—Typical Marvel-Schebler carburetor used on Wisconsin engines. Note location of fuel mixture adjustment needles.

To check float level, remove float chamber cover. Fuel level should be 17/32 inch (13.49 mm) below machined surface of float chamber flange. Carefully bend float lever close to float to obtain correct measurement.

Zenith Carburetor. Initial adjustment of idle mixture and main fuel mixture screws from a lightly seated position, is 1¼ turns open. Refer to Fig. W2. Make final adjustments with engine at operating temperature and running. Adjust idle mixture screw to obtain smoothest idle and acceleration. Operate engine under a load at full throttle and adjust main fuel mixture screw to obtain smoothest operation. As each adjustment affects the other, adjustment procedure may have to be repeated.

To check float level, invert carburetor throttle body and float assembly. Measure distance from float tops at their free ends to machined surface of throttle body. Distance should be 31/32 inch (24.61 mm). Carefully bend float arm close to each float body until correct measurement is obtained.

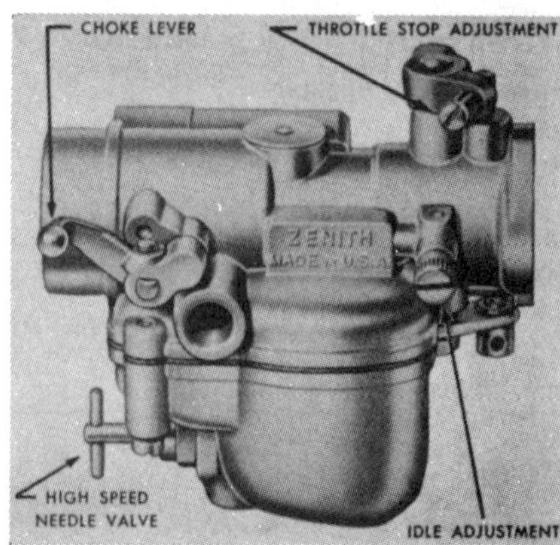

Fig. W2—Typical Zenith carburetor used on Wisconsin engines. Location of fuel mixture adjusting needles on Stromberg carburetors is similar.

GOVERNOR. A centrifugal flyweight type mechanical governor (Fig. W3) is used on all models. Governor weights (1) are hinged to the engine camshaft. Hardened pins (3) in weights transmit centrifugal force to governor sleeve (2). Movement of the governor sleeve operates the governor lever (6–Fig. W4) through a yoke.

Minor speed variations are obtained by changing governor spring tension through adjustment of screw (4). Major speed variations are obtained by repositioning hooked end of governor spring (5) in various holes of governor lever (6). Governor lever (6) has three adjusting holes with hole number one being nearest to pivot point of lever. Governed no-load and loaded speeds will vary by 100 to 350 rpm depending upon adjustment and placement of governor spring. Refer to the following table for governed speed adjustments.

Desired RPM Under Load	Spring Hole Number*	No-Load RPM
1600	1	1930
1700	1	2015
1800	1	2060
1900	1	2110
2000	1	2180
2100	1	2240
2200	1	2340
2300	1	2430
2400	1	2535
2500	2	2650
2600	2	2750
2700	2	2860
2800	2	2950
2900	3	3160
3000	3	3200
3100	3	3260
3200	3	3445
3300	3	3525
3400	3	3595
3500	3	3670
3600	3	3735

*Governor spring holes are numbered starting with hole closest to governor arm shaft as hole number 1, with hole number 3 being at outer end of arm. Spring PM-74 is used on all models.

Rod connecting carburetor throttle arm to governor bellcrank must be adjusted for proper length as follows: With engine stopped, disconnect carburetor rod from governor bellcrank or arm. Open carburetor throttle to wide open position. Move the governor bellcrank or arm towards carburetor as far as possible and adjust length of rod so bent end of rod is in register with hole in governor bellcrank or arm end. Then turn rod into swivel block on carburetor throttle arm two turns, insert bent end into hole in end of governor bellcrank or arm and secure with cotter pin.

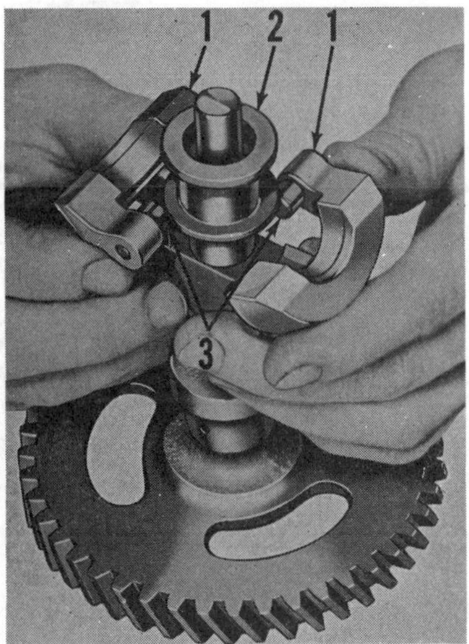

Fig. W3—Governor weights (1) are hinged to camshaft. Centrifugal force is transferred from weights to sleeve (2) through hardened pins (3). The motion of the sleeve controls the governor lever (6—Fig. W4).

MAGNETO. Engine may be equipped with either a Fairbanks-Morse or Wico magneto. Magneto breaker points and condenser are accessible by removing the end cover from the magneto. Set breaker point gap to 0.015 inch (0.38 mm).

To time magneto on all models, remove the flywheel shroud and inspection hole plug. Fig. W5 shows engine shroud cutaway to expose inspection hole. Turn engine until timing mark on cam gear is visible through inspection hole. Install magneto with timing marks on magneto gear and cam gear aligned as shown in Fig. W5. Reinstall inspection hole plug and flywheel shroud. Vane marked D-C on flywheel should be aligned with centerline mark (2). To check timing advance, whiten vane marked D-C with chalk or paint and use power timing light with engine running at operating speed. Timing should be advanced 17° on Models ACN and HACN and 28° on Models AA, AB, ABN and ABS. Marked vane will appear in line with running timing mark (3). When

Fig. W4—Fixed speed governor controls typical of Wisconsin engine models.

3. Bracket
4. Adjusting screw
5. Governor spring
6. Governor lever

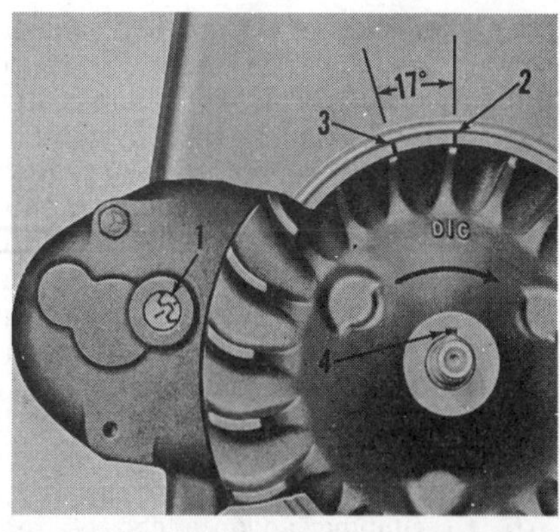

Fig. W5—Magneto timing on Wisconsin Models ACN and HACN. Running timing of other models is 28° BTDC. Models equipped with distributor do not have inspection hole or timing marks (1).

1. Magneto timing marks
2. Centerline mark
3. Running timing mark
4. Flywheel keyway

magnetos on all engine models are properly timed, the impulse coupling will snap just as the marked vane on the flywheel is in line with the centerline mark on the flywheel shroud when turning the engine slowly by hand.

On Model HACN, magneto is mounted in a suspended position below timing gear train and an oil drain tube is fitted in the magneto body to drain excess lubricating oil back to the engine crankcase.

DISTRIBUTOR. Models ABN and ACN equipped with an electric starter and generator or starter-generator combination have a battery ignition distributor. Prestolite IGW-4408 distributor is used on late production engines. Earlier engines were fitted with IGW 4405 or IGW-4179 distributors. Distributors rotate at crankshaft speed.

Breaker point contact gap is 0.020 inch (0.51 mm) and must be correctly adjusted before timing engine.

To time the distributor after removal from engine, remove screen from flywheel air intake opening. Turn crankshaft in normal direction of rotation by hand until piston can be felt to be rising on compression stroke at spark

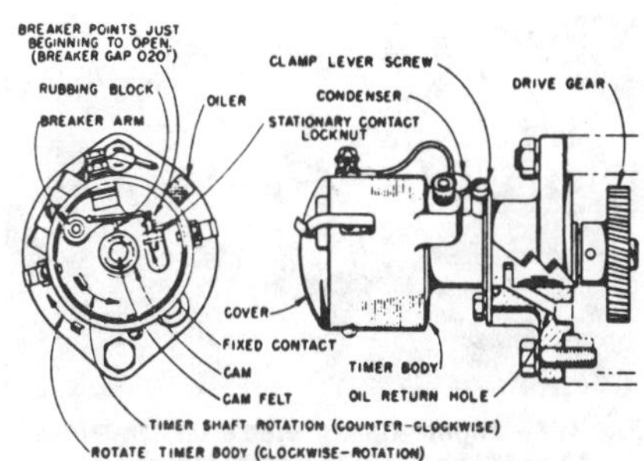

Fig. W6—Distributor used on models ABN and ACN, when equipped with battery ignition. Refer to text for timing procedure.

plug hole. Observe D-C marked vane on flywheel and align D-C mark with static timing mark (2 – Fig. W5) at top center of air shroud opening. Hold flywheel at this top dead center position. With distributor removed from engine, rotate distributor drive gear counterclockwise by hand until cam just starts to open the breaker points. Remount distributor on engine carefully so as not to disturb positions of either crankshaft or camshaft gear while engaging distributor drive gear. After distributor is securely remounted, loosen clamp lever screw (Fig. W6) so running advance can be set. Automatic advance designed into distributor flyweight mechanism accounts for 15° of the 17° required as running advance. Two additional degrees needed are set as initial advance on distributor body by turning it clockwise, points just starting to open. Two degrees is equivalent to 3/64 inch (1.191 mm) on circumference of distributor. Tighten clamp screw and use a timing light to check running advance. Note that engine must run at 1800 rpm or more for full automatic advance. If running advance does not provide 17° advance, adjust distributor as necessary and secure clamp screw.

LUBRICATION. Check oil level daily and maintain oil level at full mark on dipstick or at top of filler plug.

Manufacturer recommends oil with an API service classification SD or SE. Use SAE 30 oil when temperature is above 40°F (4°C); SAE 20 oil when temperature is between 40°F (4°C) and 5°F (–15°C) and SAE 10 oil when temperature is below 5°F (–15°C).

Engine oil should be changed after every 50 hours of operation. Crankcase capacity for Model ACN is 2 pints (0.95 L) and crankcase capacity for all other models is 1.75 pints (0.8 L).

Horizontal crankshaft models are splash lubricated by a dipper on connecting rod cap which swings through a constant level oil trough fitted in crankcase. Oil level in this trough is maintained by a cam-operated pump in the oil sump.

Fig. W7—Engine timing marks on Models AA, AB, ABN and HACN Wisconsin engines.

1. Magneto timing marks
2. Crankshaft gear
3. Camshaft timing
4. Cam gear
5. Magneto gear

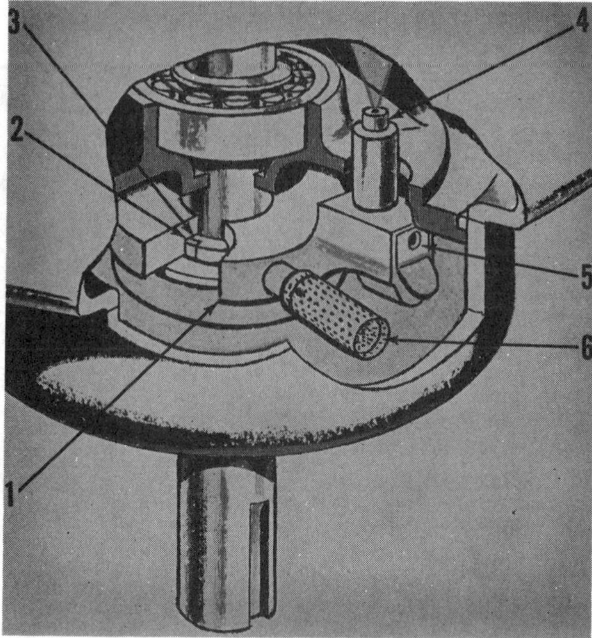

Fig. W8—Engine oil pump on Model HACN and other vertical shaft Wisconsin engines.

1. Pump body
2. Pump vane
3. Vane retainer
4. Oil spray jet
5. Relief valve
6. Suction strainer

Vertical crankshaft model engine (HACN) is spray lubricated by a vane type oil pump located at lower end of crankshaft. See Fig. W8. Oil is contained in adaptor base of engine. Oil pressure relief valve reed (5) will require renewal if bent or badly fitted to relief hole. Oil strainer screen (6) should be serviced whenever engine is disassembled for removal of sludge or contamination or for renewal if defective.

Models ABN and ACN equipped with ignition timer and battery instead of magneto require 3 to 5 drops of engine oil be applied to felt wick in cam sleeve and to shaft oiler after every 100 hours of operation.

REPAIRS

TIGHTENING TORQUES. Recommended tightening torques are as follows:

Spark plug	25-30 ft.-lbs. (34-40 N·m)
Head bolts	14-18 ft.-lbs. (19-24 N·m)
Connecting rod bolts	14-18 ft.-lbs. (19-24 N·m)
Crankcase bolts (except HACN)	6-8 ft.-lbs. (8-11 N·m)
Engine adapter base (HACN)	24-26 ft.-lbs (33-35 N·m)
Crankcase side cover (HACN)	6-8 ft.-lbs. (8-11 N·m)
Main bearing plates	14-18 ft.-lbs. (19-24 N·m)

CONNECTING ROD. Connecting rod and piston assembly are removed from top of cylinder bore after head is removed. Diameter of crankpin on crankshaft is 1.0 inch (25.4 mm) on all models.

Crankshaft end of connecting rod is babbitted and shim-adjusted. Connecting rod bearing clearance is 0.0007-0.002 inch (0.018-0.051 mm).

Side clearance is 0.004-0.010 inch (0.102-0.254 mm) for AA, AB, ABN and ABS models and 0.006-0.013 inch (0.152-0.330 mm) for ACN and HACN models.

Connecting rod end and cap are indexed for reassembly with oil hole in cap installed towards oil spray jet in Model HACN or towards oil pump body on all other models.

PISTON, PIN AND RINGS. Model AA and some earlier production AB models are fitted with three rings, one

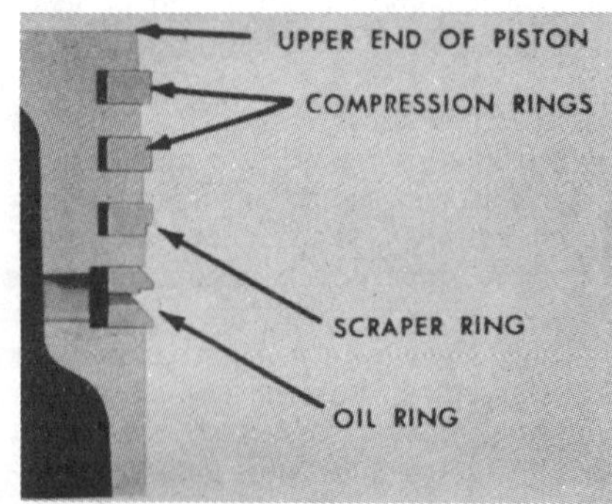

Fig. W9—Piston ring cross-section. Model AA and some production units of Model AB have only one compression ring. Later Model AB and Models ABN and ABS are as shown. Note wiping arrangement of scraper ring and oil control ring.

compression ring, one scraper ring and one oil control ring. Later AB models and all other models have a four ring piston with two compression rings, one scraper ring and one oil control ring (Fig. W9 and W10).

End gap for piston rings on all models is 0.012-0.022 inch (0.31-0.56 mm). On four ring pistons, stagger end gaps 90° around circumference of piston during assembly. On three ring pistons, stagger end gaps 120° around circumference of piston during assembly.

Side clearance for all piston rings of Models AA, AB, ABN and ABS is 0.002-0.003 inch (0.51-0.76 mm). Side clearance for Models ACN and HACN is; top ring, 0.002-0.0035 inch (0.051-0.089 mm); second and third rings, 0.001-0.0025 inch (0.025-0.064 mm); fourth ring, 0.0025-0.004 inch (0.064-0.101 mm). Pistons and rings are available in several oversizes as well as standard.

Piston skirt clearance in cylinder bore is 0.0045-0.0050 inch (0.114-0.127 mm) for Models AA, AB, ABN and ABS. For Models ACN and HACN operated at 3000 rpm and lower, piston skirt clearance is 0.0050-0.0055 inch (0.127-0.140 mm). If these models are operated at engine speeds exceeding 3000 rpm, skirt clearance should be increased to 0.0060-0.0065 inch (0.152-0.165 mm).

Floating piston pin is a light press fit in piston bosses with 0.0002-0.0008 inch (0.005-0.020 mm) clearance in connecting rod for all models. Piston pin is available in several oversizes as well as standard.

CYLINDER. When cylinder bore wear exceeds 0.005 inch (0.13 mm), rebore to nearest oversize for which piston and rings are available.

CRANKSHAFT. Shim-gaskets are fitted in combination between main bearing plate and crankcase to adjust crankshaft for end play of 0.002-0.004 inch (0.05-0.10 mm) in tapered roller bearings. Use dial indicator to measure end play.

Crankshaft and camshaft gear timing marks must be aligned during installation (Fig. W7).

CAMSHAFT. Hollow camshaft with drive gear revolves on a pin which is an interference fit in crankcase sidewalls. To remove, knock out expansion plug and drive out the pin and opposite plug working from flywheel side of the crankcase. Renew plugs during reassembly.

VALVE SYSTEM. Valve tappet clearance (cold) for Models AA, AB, ABN and ABS is 0.011-0.013 inch (0.28-0.33 mm) for both intake and exhaust valves. Valve tappet clearance (cold) for Models ACN and HACN is 0.008 inch (0.20 mm) for intake valve and 0.014 inch (0.36 mm) for exhaust valve.

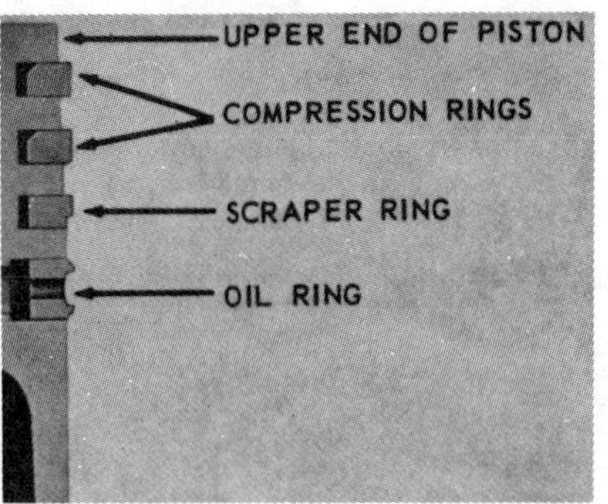

Fig. W10—Piston ring cross-section for Model ACN and HACN. Note position of inner chamfer (top) of compression rings and placement of scraper ring.

Valve tappet clearance on all models is adjusted by carefully grinding off the end of valve stem to increase clearance or by grinding seats deeper into block to reduce clearance.

Valve face and seat angles are 45°. Valve seat width should be 3/32 inch (2.38 mm) for all models.

Valve stem clearance in block bore or valve guides is 0.003-0.005 inch (0.08-0.13 mm) for all models. Renewable valve guides are available for Models ACN and HACN and should be installed if stem clearance reaches or exceeds 0.007 inch (0.18 mm). Models AA, AB, ABN and ABS compensate for stem wear by using new valves having stems which are 0.004 inch (0.10 mm) oversize.

SERVICING WISCONSIN ACCESSORIES

REWIND STARTER. Fig. W11 provides an exploded view of rewind starter of the type found on some model ADN engines. See Fig. W12 to identify parts. After starter is unbolted from engine shroud, remove retainer ring, retainer washer, brake spring, friction washer, friction shoe assembly and second friction washer as shown in Fig. W12. Hold the rope pulley in one hand, insert rope in pulley notch and allow rotor to rotate to unwind the recoil spring preload. Lift rotor from cover, shaft and recoil spring.

NOTE: Check the winding direction of recoil spring and rope to aid in reassembly.

Remove recoil from cover and unwind rope from rotor.

When reassembling the unit, lubricate recoil spring, cover shaft and its bore in rotor with grease. Install the rope on rotor. Install the rotor on the cover shaft engaging the recoil spring inner end hook. Preload the recoil spring four turns and install middle flange and mounting flange. Check friction shoe sharp ends and renew if necessary. Install friction washers, friction shoe assembly, brake spring and retainer ring. Make certain friction shoe assembly is installed properly for correct starter rotation. If properly install-

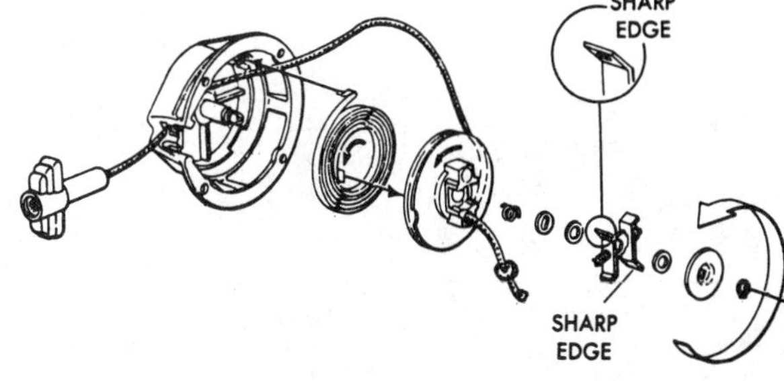

Fig. W11—Exploded view of rewind starter. See text.

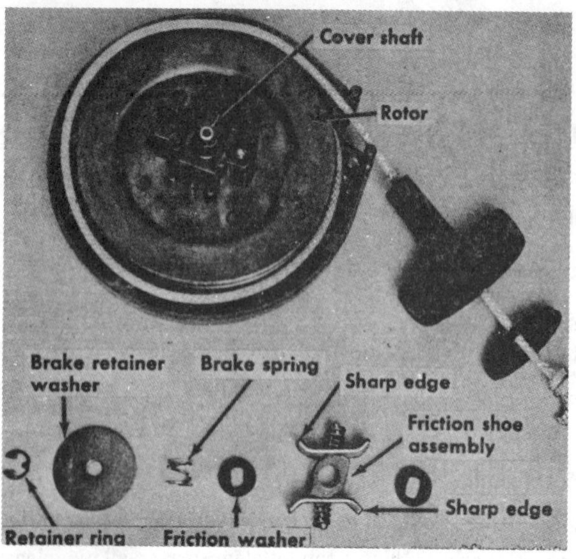

Fig. W12—View of starter with friction shoe assembly removed. Refer to text for service procedure.

ed, sharp ends of friction shoe plates will extend when rope is pulled.

Starter operation can be reversed by winding rope and recoil spring in opposite direction and turning the friction shoe assembly upside down.

FUEL PUMP. Fuel pump, part LP-42-B, is available as an option on some models. Should it be necessary to order a new crankcase assembly, be sure to specify if engine has a fuel pump so correct crankcase unit will be furnished.

To service fuel pump, disconnect fuel lines at pump and unbolt pump body from engine. Mark castings (11 and 12–Fig. W13) with a light file stroke or chisel mark at joining edges. Remove four assembly screws (15) and separate fuel head (12) from lower body (11). Invert fuel head and remove screw (16), followed by plate (10), valve springs and balls (3 and 6) and gasket (5). Parts (3, 5 and 6) are provided in rebuild kit. Install new valve parts exactly as old ones were positioned after thorough cleaning of fuel head casting (12). Secure valve assembly in place with securing screw (16).

Remove rocker arm spring (13) from lower (diaphragm) casting (11). Hold lower casting (11) in one hand with a thumb or finger against link (9). Press downward on metal center of diaphragm (2) to compress diaphragm spring (4), then turn diaphragm clockwise a quarter-turn to disengage diaphragm rod from link (9). Discard diaphragm (2) and spring (4). After lower casting is thoroughly cleaned, reverse removal procedure above to install new diaphragm spring (4) and diaphragm (2) from kit. Reinstall rocker arm spring (13).

Using a new mounting gasket, remount lower cast assembly in place on

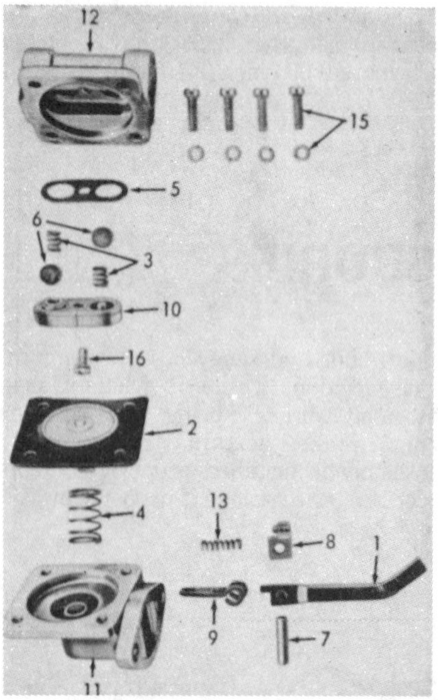

Fig. W13—Exploded view of fuel pump. Items marked (*) are furnished in repair kit for rebuilding pump.

1. Rocker arm
2. Diaphragm*
3. Valve spring*
4. Diaphragm spring*
5. Valve gasket*
6. Valve ball*
7. Rocker arm pin
8. Spring clip
9. Rocker arm link
10. Valve plate
11. Lower casting (mount)
12. Fuel head
13. Rocker arm spring
15. Assembly screws
16. Valve plate screw

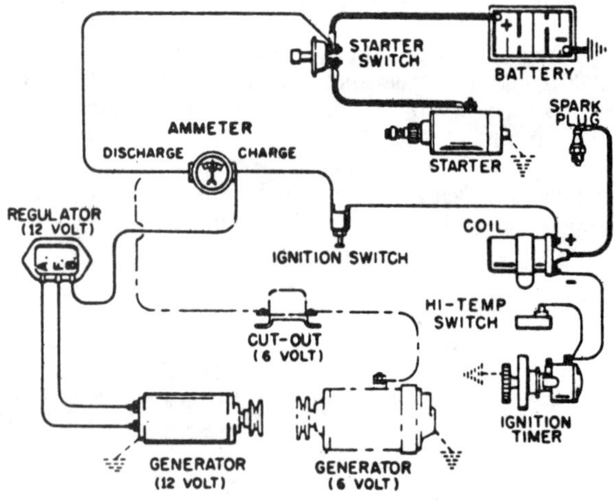

Fig. W14—Wiring circuit for models equipped with starter and generator. Twelve volt negative ground system is shown. Polarity is positive ground for all 6 volt systems and earlier 12 volt systems. Refer to text. Note high temperature switch in ignition circuit for overheating protection.

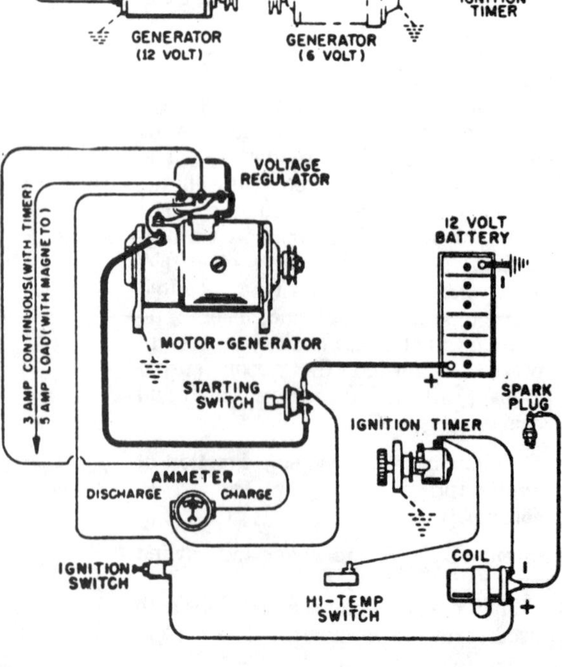
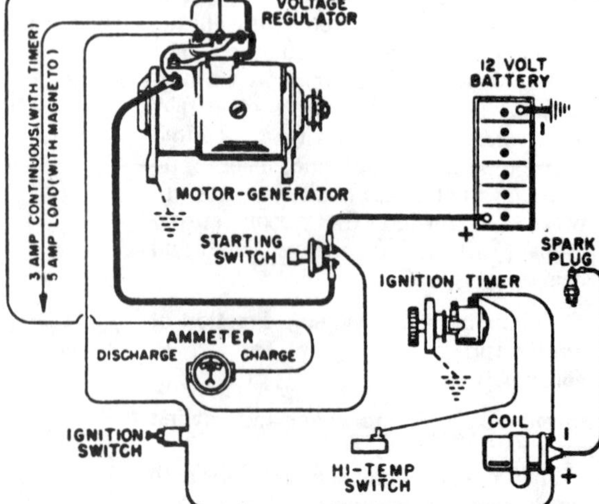

Fig. W15—Wiring layout for combination motor-generator. See text for service.

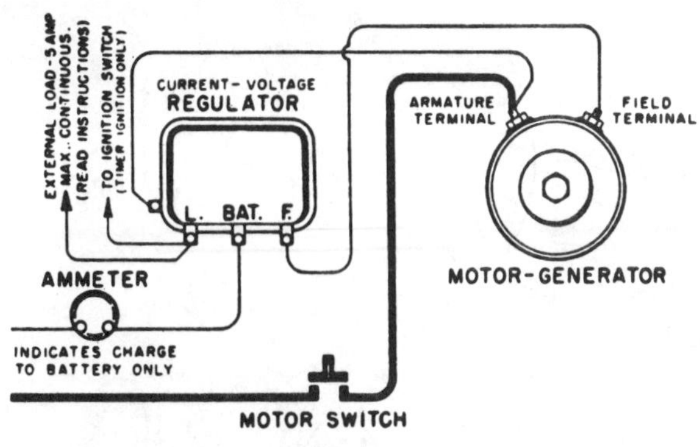

Fig. W16—Wiring connections for regulator used with motor-generator.

the engine. Secure with pump mounting screws.

Slowly crank engine by hand until diaphragm (2) lies flat on surface of lower casting (11). Place fuel head (12) back in position (aligning index marks) over new diaphragm and start threads of four assembly screws (15). Continue hand turning engine until diaphragm is pulled down in lower casting, then tighten assembly screws evenly and securely. Reconnect fuel lines.

ELECTRICAL SYSTEM. Models ABN and ACN may be equipped with battery ignition and distributor. When engine is so equipped, consult wiring diagrams of Fig. W14, W15 or W16 for circuit service.

All 6 volt systems are positive (+ ground. Twelve volt systems have been changed from positive (+) to negative (−) ground beginning with engine serial number 3985702, Model ACN.

Generator (6 Volt). Prestolite GAS-4103-1 generator is used. Rotation is clockwise, observed from drive end, with third brush control and cutout relay as shown in Fig. W14. Field coil draw is 3.4-3.8 amps at 5.0 volts with current output of 7.1 amps at 8.0 volts. Circuit breaker used with this generator, Prestolite CB-4008, has the following specifications: Circuit breaker contact gap is 0.015-0.045 inch (0.38-1.14 mm) with armature air gap of 0.010 inch (0.25 mm), with contacts closed, air gap is 0.030 inch (0.76 mm). Closing voltage is 6.5-7.3 volts and opening current is 0.5-2.5 amps.

Generator (12 Volt). Older, positive (+) ground units use Prestolite GJG-4007MP generator. Negative ground systems use GJG-4010M generator. Rotation is clockwise for either polarity. Field current draw is 1.7-1.9 amps at 10 volts. Motoring draw is 4-5 amps. Maximum current output is 10 amps (7000 rpm) at 15 volts.

Voltage-Current Regulator. A vibrating type regulator is used to control output of shunt-wound generator. Prestolite VBO-4201-Z1 is used in positive ground systems and VBO-4201-Y1 is used with negative ground generators. Cutout relay opens at 3-5 amps and closes at 12.6-13.6 volts. Current is controlled at 9-11 amps. Operating voltage is 14.3-14.8 volts at 110°F (43°C). Voltage regulator specifications call for the following armature air gaps: Circuit breaker, 0.025-0.027 inch (0.64-0.69 mm), voltage regulator and current regulator each at 0.048-0.052 inch (1.22-1.32 mm). Circuit breaker contact gap is 0.015 inch (0.38 mm) minimum.

Starter (6 Volt). Prestolite MAK-4008 starter, which may be found on older production models, is obsolete, though parts remain available. Rotation is counterclockwise as seen from drive end. Armature end play is 0.005-0.062 inch (0.13-0.20 mm). No-load test is 70 amps at 5.0 volts and 4700 rpm. Stalling torque is 2.5 ft.-lbs. (3.4 N·m) at 250 amps and 2.0 volts.

Present engine production uses starter MDH-4001M which rotates counterclockwise. Armature end play is 0.005 inch (0.13 mm). No-load current draw is 52 amps at 5.0 volts and 8100 rpm. Stall torque is 1.6 ft.-lbs. (3 N·m) at 230 amps and 2.0 volts.

Pinion position is 1-25/32 inch (45.24 mm) from face of mounting flange to edge of pinion with 1/16 inch (1.59 mm) tolerance.

Starter (12 Volt). Prestolite MDO-4002M starter is used. Rotation is counterclockwise. Armature end play is 0.005 inch (0.13 mm). No-load current draw is 38 amps at 10,000 rpm and 10 volts. Stall torque is 1.5 ft.-lbs. (1 N·m) at 170 amps and 4 volts. Later production models use Prestolite starter MGD-4102A, which has comparable specifications.

Pinion position is 1-25/32 inch (45.24 mm) from face of mounting flange to edge of pinion with 1/16 inch (1.59 mm) tolerance.

Motor-Generators. Starting motor-generator and regulator circuit diagram is shown in Fig. W15 and W16.

The 10 amp current output of generator function side of motor-generator is

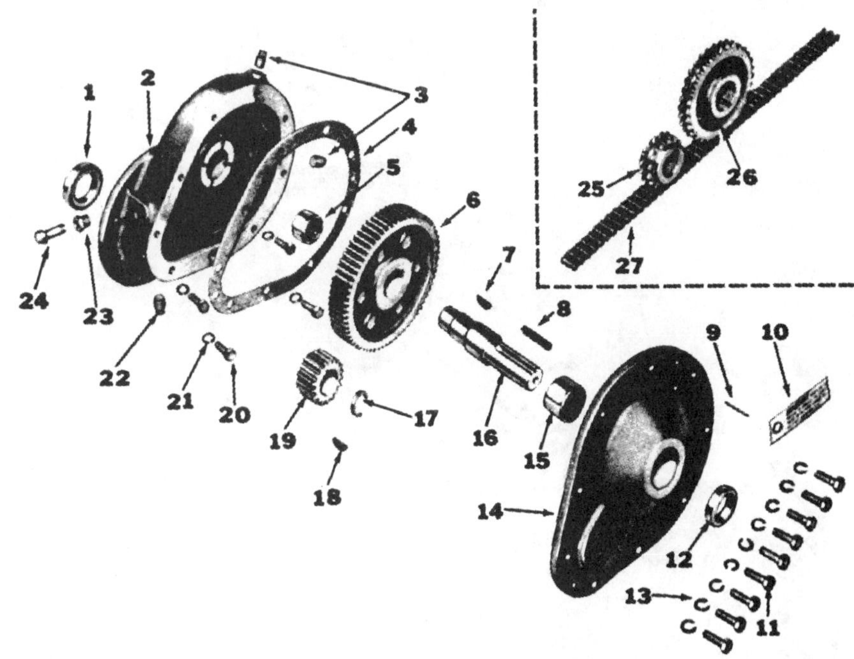

Fig. W17—Exploded view of typical reduction gear set used with Wisconsin engines. Inset shows chain-sprocket arrangement used in some models. Refer to text.

1. Crankshaft oil seal
2. Main housing & bearing plate
3. Pipe plug
4. Cover gasket
5. Inner bearing
6. Driven gear
7. Woodruff key
8. Key
9. Cover pin
10. Instruction tag
11. Cover screw
12. Takeoff shaft oil seal
13. Cover lockwashers
14. Housing cover
15. Outer bearing
16. Takeoff shaft
17. Drive gear retainer
18. Woodruff key-drive gear
19. Drive gear
20. Housing screw
21. Washer
22. Drain plug
23. Reducer
24. Breather
25. Drive sprocket
26. Driven sprocket
27. Drive chain

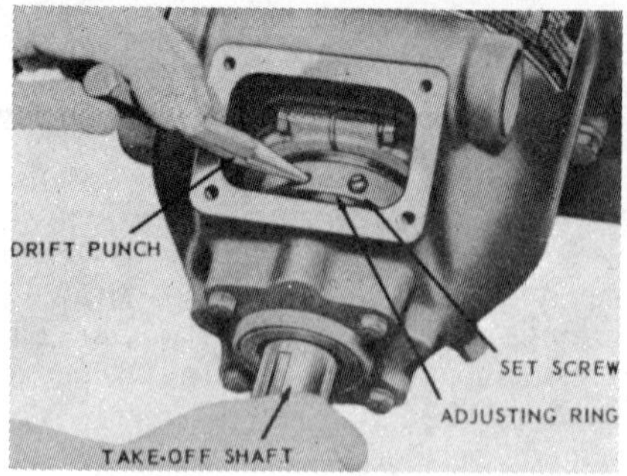

Fig. W18—Adjustment technique, Wisconsin multiple-disc clutch. See text for procedure.

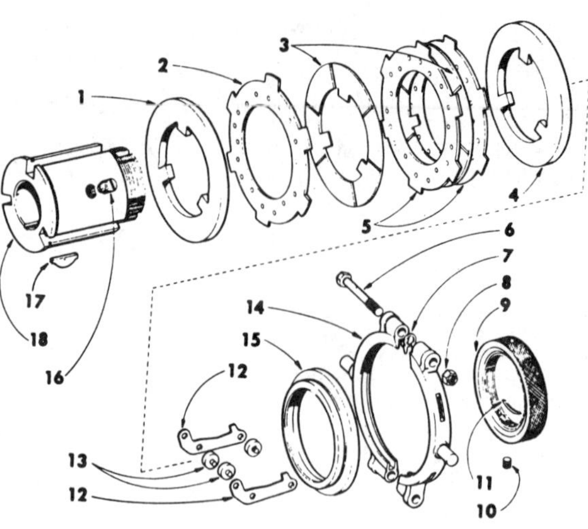

Fig. W20—Exploded view of clutch hub and disc assemblies. All parts are serviced. Collar assembly (14) may be die cast or bronze and key (17) is available in 3 styles. Furnish detailed information when ordering renewal parts.

1. Back clamping plate
2. Drive plate (sintered)
3. Driven plates (steel)
4. Clamping plate (front)
5. Drive plates (sintered)
6. Cap screw (2)
7. Shim (2)
8. Locknut (2)
9. Adjuster & lock
10. Adjuster set screw
11. Lock spring
12. Lever (6)
13. Lever rollers (9)
14. Collar assy.
15. Wedge sleeve
16. Hub set screw
17. Key
18. Hub

distributed separately at regulator terminals "L" and "BAT"–see Fig. W16. Current from "BAT" terminal is regulated to meet charging requirements for the battery. Ammeter shown in circuit indicates only battery charging rate, not the current draw for lights and accessories as in other systems. Measurement of this accessory load requires another ammeter be set in series with load on terminal marked "L". If engine has magneto ignition, up to 5 amps may be drawn from "L" terminal, however, if distributor ignition is used, accessory load must be limited to 3 amps as 2 amps must flow in coil-distributor circuit for satisfactory engine operation.

Motor-generator furnished after serial number 3985701 is negative ground (–) Delco-Remy 1101868, with Delco-Remy regulator 118984. Earlier production, positive (+) ground models are equipped with Delco-Remy motor-generator 1101972 and regulator 1118985.

These specifications apply to all models. Direction of rotation is clockwise when viewed from drive end. Field current draw is 1.43-1.54 amps at 12 volts, checked at 80° F (27° C). Cold output is 10 amps at 14 volts and 5450 rpm. No-load test draws an average 13 amps (maximum 18 amps) at 11 volts and 2500-3000 rpm.

Regulator cutout relay and point opening gaps are both 0.020 inch (0.51 mm). Closing range for cutout relay is 11.8-14.0 volts. When adjusting, set to 12.8 volts. Voltage regulator air gaps are 0.075 inch (1.91 mm) with voltage range of 13.6-14.5 volts. Setting should be adjusted to 14.0 volts.

REDUCTION UNITS. Gear drive and chain drive speed reduction units are available. Refer to Fig. W17 for exploded view. Reduction ratios offered are 2.92:1 and 2:1 with chain drive unit as shown in inset of Fig. W17. Spur gear model shown has a ratio of 3.266:1. Later model reduction units have spiral gears with reduction ratios of 3.25:1, 5.5:1 or 6:1. Special crankshafts are required on all reduction equipped engines. Complete engine and reduction model identification is needed when ordering parts or service material.

When mounted on engine, reduction housing and main bearing plate (2—Fig. W17) are fitted in place of regular bearing plate used on the basic engine. The same shim gaskets and service procedure for adjusting engine main bearing end play are used. Refer to CRANKSHAFT section for details.

Reduction units may be mounted on engine crankcase in any of four positions, depending on requirements. Note that interchangeable oil plugs (3 and 22) and breather reducer (23) will require shifting when reduction unit is mounted in positions other than as shown in Fig. W17. Design of housing is such that a drain hole will always be in position at low side and oil check plug will be at correct oil level with breather vent slightly higher with filler hole at top regardless of position of reduction housing in relation to engine block.

Reduction gears are lubricated by oil

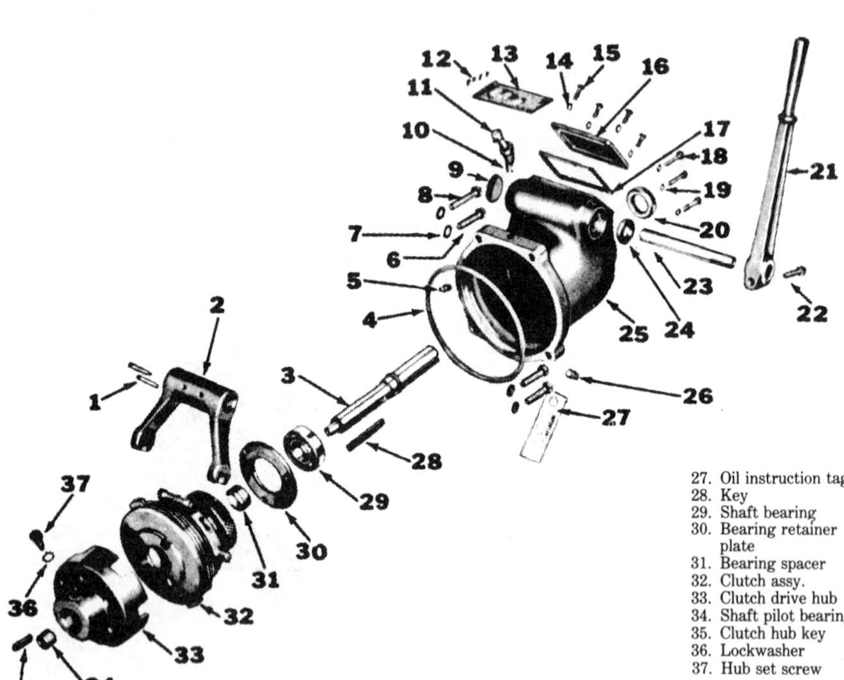

Fig. W19—Exploded view, engine side, of Wisconsin multiple-disc clutch and power takeoff assembly.

1. Roll pin (2)
2. Shifter yoke
3. Shaft
4. Cork seal
5. Oil level plug
6. Screw
7. Washer
8. Screw
9. Expansion plug
10. Elbow
11. Breather
12. Screw
13. Clutch instruction plate
14. Washer
15. Screw
16. Inspection hole cover
17. Cover gasket
18. Bearing retainer screw
19. Washer
20. Oil seal
21. Shift lever
22. Clamp screw
23. Shifter shaft
24. Shaft oil seal
25. Housing
26. Oil drain plug
27. Oil instruction tag
28. Key
29. Shaft bearing
30. Bearing retainer plate
31. Bearing spacer
32. Clutch assy.
33. Clutch drive hub
34. Shaft pilot bearing
35. Clutch hub key
36. Lockwasher
37. Hub set screw

of same type as used in engine crankcase. Gearcase capacity is approximately 1 pint (0.47 L); fill to check level plug opening. Oil should be changed at 500 hour intervals or more frequently for continuous operation. Check oil level daily.

CLUTCH. When equipped with a clutch and power takeoff assembly, engines require a special crankcase, crankshaft and main bearing plate with a special seal for the clutch housing. Complete model and part number identification is stamped on engine crankcase and must be furnished when ordering renewal parts.

Clutch is multiple-disc design which is manually operated and runs in oil. Service is limited to renewal of worn or defective parts, maintaining a "wet" oil level of approximately ½ pint (0.23 L) of same oil as is used in crankcase, and adjustment to overcome slippage due to normal wear of friction discs.

To adjust clutch, remove inspection plate (16–Fig. W19). Disengage clutch and rotate takeoff shaft by hand until set screw of knurled adjusting collar (Fig. W18) is visible at top. Loosen set screw and turn adjusting collar clockwise slightly, using a punch as in Fig. W18 while holding takeoff shaft from turning. Engage clutch. If in proper adjustment, clutch will engage with a slight snap. If necessary, continue to rotate collar until proper adjustment is made, then retighten set screw. Replace inspection cover and replenish clutch oil. Refer to Fig. W20 for exploded view of clutch components if renewal of defective parts becomes necessary.

WISCONSIN ROBIN

2-STROKE

Model	Bore	Stroke	Displacement
WT1-125V	2.2 in. (56 mm)	1.97 in. (50 mm)	7.49 cu. in. (123 cc)

ENGINE INFORMATION

Model WT1-125V is a two-stroke, single-cylinder, air-cooled vertical crankshaft engine. Engine model and specification numbers are located on the name plate. Engine serial number is stamped either on the crankcase or the engine's identification tag. The model, specification and serial numbers must be furnished when ordering parts.

MAINTENANCE

SPARK PLUG. Recommended spark plug is a Champion CJ8 or equivalent. Manufacturer recommends removing and cleaning spark plug and adjusting electrode gap after every 50 hours of normal operation. Specified spark plug electrode gap is 0.024 inch (0.6 mm).

CARBURETOR. Model WT1-125V engine is equipped with a float type carburetor (Fig. W25) with fixed low speed and main fuel jets. Plastic float (15) is nonadjustable. Refer to Fig. W25 for exploded view of carburetor during disassembly and reassembly.

GOVERNOR. Model WT1-125V is equipped with a centrifugal flyweight type governor. The governor plate, governor sleeve and governor yoke are located in the crankcase and lubricated by the oil mixed with the fuel. To adjust the external governor linkage, first make certain all linkage operates freely. Inspect linkage for worn or broken parts. Governor spring (1—Fig. W26) is normally installed in the center hole of governor lever (9) as shown. Loosen adjusting nut (7) and adjusting plate screw (5). Push adjusting plate (6) downward and tighten screw (5). Place speed control lever (2) in high speed position and tighten nut (7). Adjust throttle stop screw (22) to limit maximum engine speed to rpm specified by equipment manufacturer.

IGNITION SYSTEM. Model WT1-125V engine is equipped with a solid-state electronic ignition system. Specified air gap between ignition coil and flywheel is 0.016-0.020 inch (0.4-0.5 mm). To adjust, loosen retaining screws and move coil to obtain correct clearance. Tighten coil retaining screws to secure adjustment.

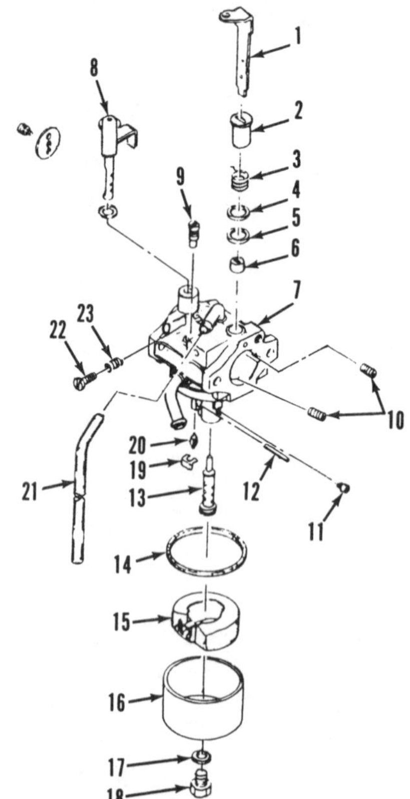

Fig. W25—Exploded view of the fixed jet float type carburetor used on Model WT1-125V engine.

1. Choke shaft
2. Ring
3. Spring
4. Cap
5. Seal
6. Ring
7. Body
8. Throttle shaft
9. Pilot jet
10. Air jets
11. Main jet
12. Pin
13. Nozzle
14. Gasket
15. Float
16. Float bowl
17. Gasket
18. Bolt
19. Clip
20. Fuel inlet needle
21. Tube
22. Throttle stop screw
23. Spring

To check ignition system, inspect all components for broken, frayed, loose or disconnected ignition wires. Make certain spark plug is clean and in good condition. Use a standard test plug to check for ignition spark. Ignition coil can be checked using a good ohmmeter. Connect one ohmmeter lead to the high tension wire (spark plug) and connect remaining ohmmeter lead to the iron coil laminations. Ohmmeter reading should be 13,000 ohms. If resistance reading is infinite, open winding in ignition unit, loose or broken spark plug connector or faulty high tension lead is indicated. If resistance reading is low, secondary coil is shorted.

LUBRICATION. The engine is lubricated by mixing oil with the fuel. Manufacturer recommends mixing a good quality two-stroke, air-cooled engine oil with gasoline at a 32:1 ratio. Use a separate container when mixing fuel and oil.

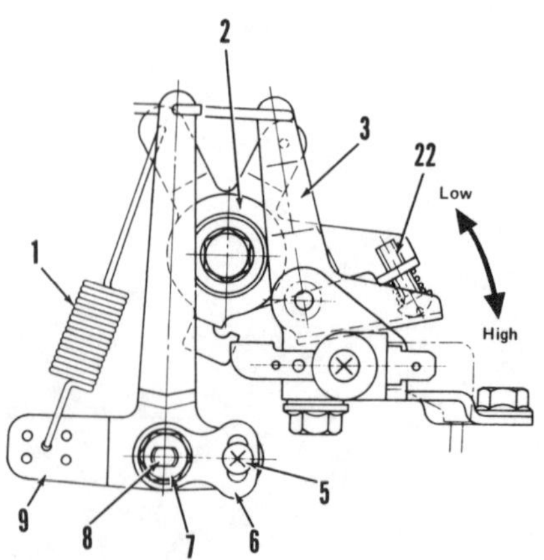

Fig. W26—View showing external governor linkage used on Model WT1-125V.

1. Governor spring
2. Speed control lever
3. Choke lever
5. Adjusting plate screw
6. Adjusting plate
7. Nut
8. Governor shaft
9. Governor lever
22. Throttle stop screw

CARBON. Manufacturer recommends removing cylinder head and cleaning carbon and combustion deposits from exhaust port, muffler and piston after every 500 hours of normal operation. Use a wooden scraper to remove combustion deposits to prevent damaging piston or cylinder.

GENERAL MAINTENANCE. Check and tighten all loose bolts, nuts or clamps prior to each day of operation. Check for fuel leakage and repair if necessary.

Clean dust, dirt, grease or any foreign material from cylinder head and cylinder block cooling fins daily. Inspect fins for damage and repair if necessary.

Clean spark plug and air cleaner after every 50 hours of normal operation. Clean fuel strainer and fuel tank after every 200 hours of normal operation.

REPAIRS

TIGHTENING TORQUES. Recommended tightening torques are as follows:

Spark plug	18-22 ft.-lbs. (24-30 N·m)
Flywheel nut	28-30 ft.-lbs. (38-40 N·m)
Cylinder nuts	13-16 ft.-lbs. (18-22 N·m)
Crankcase bolts	80-88 in.-lbs. (9-10 N·m)

CYLINDER AND CYLINDER HEAD. To remove cylinder and cylinder head, remove spark plug and the four head bolts (1 and 2—Fig. W27). Carefully remove cylinder head (3) from cylinder. Remove cylinder head gasket (4). Remove the four nuts (6), lockwasher (7) and flat washer (8). Carefully slide cylinder (5) off of piston and ring assembly. Remove gasket (9).

Clean carbon and combustion deposits from cylinder head and combustion chamber prior to reassembly. Inspect all parts for wear or damage. If cylinder head is warped 0.0004 inch (0.01 mm) or more, renew cylinder head. Standard cylinder bore diameter is 2.0900-2.0907 inches (56.000-56.019 mm). If cylinder bore diameter is 2.0959 inches (56.15 mm) or more, cylinder should be resized for the next size piston and ring set which is available. Oversize piston and ring sets are available in 2.215 inches (56.25 mm) and 2.224 inches (56.50 mm). If cylinder is 0.0004 inch (0.01 mm) or more out-of-round, or if cylinder taper exceeds 0.0006 inch (0.015 mm), cylinder must be resized.

When reassembling, install a new gasket (9) and make certain piston ring ends are around the locating pins in piston grooves. Carefully work cylinder down over piston. Install a new head gasket (4) with folded edge facing toward cylinder head. Install cylinder head and tighten head bolts following a crisscross pattern to torque specified in TIGHTENING TORQUES.

PISTON, PIN AND RINGS. To remove piston, refer to CYLINDER AND CYLINDER HEAD paragraphs to remove cylinder head and cylinder. Remove the two retaining rings (10—Fig. W28) and slide piston pin (11) out of piston pin bore and connecting rod pin bore. Remove needle bearing (7) as necessary. Remove piston rings (14 and 15).

Clean and inspect piston. Top piston ring is a chrome color and second ring is a darker color. Standard piston ring groove width is 0.0787 inch (2.0 mm). If ring groove width exceeds 0.0846 inch (2.15 mm), renew piston. Standard clearance between ring and ring groove is 0.0020-0.0035 inch (0.05-0.09 mm). If

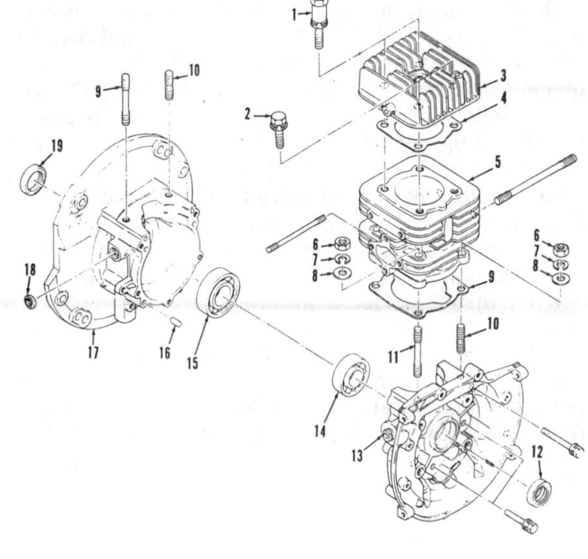

Fig. W27—Exploded view of WT1-125V engine crankcase, cylinder and cylinder head assembly.

1. Head bolt
2. Head bolt
3. Cylinder head
4. Gasket
5. Cylinder
6. Nut
7. Lockwasher
8. Flat washer
9. Gasket
10. Cylinder stud
11. Cylinder stud
12. Seal
13. Crankcase half
14. Main bearing
15. Main bearing
16. Dowel pin
17. Crankcase half
18. Governor seal
19. Seal

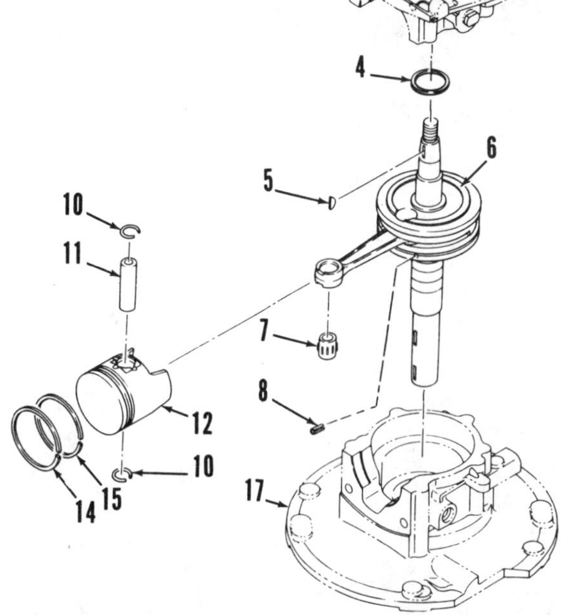

Fig. W28—Exploded view of connecting rod and crankshaft assembly.

1. Nut
2. Washer
4. Shim
5. Key
6. Connecting rod & crankshaft assy.
7. Needle bearing
8. Spring pin
10. Retaining rings
11. Piston pin
12. Piston
13. Crankcase half
14. Piston ring
15. Piston ring
17. Crankcase half

clearance exceeds 0.0059 inch (0.15 mm), renew piston and/or rings. Standard piston ring end gap is 0.004-0.010 inch (0.10-0.25 mm). If ring end gap exceeds 0.070 inch (1.52 mm), renew rings and/or recondition cylinder bore. Standard piston ring width is 0.0925 inch (2.35 mm). If ring width is 0.0921 inch (2.34 mm) or less, renew ring.

Standard piston pin bore diameter in piston is 0.4724 inch (12 mm). If piston pin bore diameter exceeds 0.4738 inch (12.035 mm), renew piston. Standard piston pin diameter is 0.4724 inch (12 mm). If piston pin diameter is 0.4712 inch (11.97 mm) or less, renew piston pin.

Standard piston diameter is 2.203 inches (55.96 mm). If piston diameter is 2.199 inches (55.85 mm) or less, renew piston. Standard clearance between piston and cylinder bore is 0.0016-0.0029 inch (0.041-0.074 mm). If clearance exceeds 0.0109 inch (0.277 mm), renew piston and/or recondition cylinder bore.

When installing piston on connecting rod, the "F" mark on top of piston should face toward flywheel side of engine after assembly. Always install new piston pin retainers and make certain piston ring ends are around the locating pins in piston grooves.

CRANKCASE, CRANKSHAFT AND CONNECTING ROD. To remove crankshaft and connecting rod assembly, remove cylinder head and cylinder as outlined in CYLINDER AND CYLINDER HEAD paragraphs. Remove piston and ring assembly. Remove the five bolts retaining upper and lower crankcase halves. Gently tap crankcase halves to separate. Remove crankshaft and connecting rod assembly. Crankshaft and connecting rod are serviced as an assembly only. Do not separate crankshaft and connecting rod. Remove retaining screws (1—Fig. W29), governor yoke (2), governor sleeve (3) and governor flyweight plate (4). It may be necessary to heat crankcase halves slightly to remove ball bearing type main bearings (14 and 15—Fig. W27).

Seals (12, 18 and 19) should be pressed into seal bores until seated. It may be necessary to heat crankcase slightly to install main bearings (14 and 15). Standard diameter for crankshaft main bearing journal is 0.79 inch (20.066 mm). If diameter is 0.7884 inch (20.025 mm) or less, renew crankshaft. Standard thrust clearance between crankshaft and crankcase is 0.0000-0.0079 inch (0.0-0.2 mm). If thrust clearance exceeds 0.04 inch (1.0 mm), vary thickness of shim (4—Fig. W28). Shims are available in 0.0039 inch (0.1 mm) and 0.0118 inch (0.3 mm) thicknesses. Standard crankshaft runout is 0.002 inch (0.05 mm). If crankshaft runout exceeds 0.005 inch (0.13 mm), renew crankshaft.

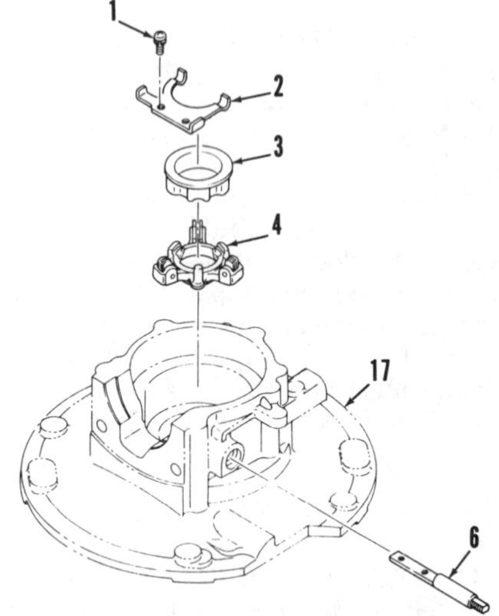

Fig. W29—Exploded view of governor assembly used on Model WT1-125V engine.

1. Screw
2. Governor yoke
3. Governor sleeve
4. Governor flyweight plate
6. Governor shaft
17. Crankcase half

WISCONSIN ROBIN

4-STROKE

Model	Bore	Stroke	Displacement
EY18W, EY18-3W	2.56 in. (65 mm)	2.16 in. (55 mm)	11.1 cu. in. (183 cc)

ENGINE INFORMATION

Wisconsin engines in this section are four-stroke, single-cylinder, air-cooled engines. Power rating is in relation to maximum engine rpm on all models. Engine model number is located on cooling shroud just above rewind starter.

MAINTENANCE

SPARK PLUG. Recommended spark plug for all models is a Champion L86 or equivalent. Spark plug electrode gap should be 0.020-0.025 inch (0.51-0.64 mm).

CARBURETOR. Refer to Fig. WR1 for exploded view of Mikuni carburetor used on both models.

Initial carburetor idle mixture screw (24—Fig. WR1) adjustment should be 1-1/2 turns open from a lightly seated position. Main fuel mixture is metered through main jet (15) and is nonadjustable.

Make final carburetor adjustments with engine at operating temperature and running. Adjust idle mixture screw to obtain smoothest idle operation and acceleration. Adjust idle stop screw (12) to obtain 1250 rpm at operating temperature.

To check or adjust float level, remove plug (22) and float bowl (20). Place carburetor body on end (on manifold flange) so float pin is in vertical position. Move float to close inlet needle valve.

NOTE: Needle valve is spring loaded. Float tab should just contact needle valve pin but should not compress spring.

Using a depth gage, measure distance between body flange and free end of float as shown in Fig. WR3. Distance should be 0.710-0.790 inch (18.03-20.27 mm). Carefully bend tab on float lever to obtain correct setting.

GOVERNOR. On Model EY18W, the mechanical flyweight governor is

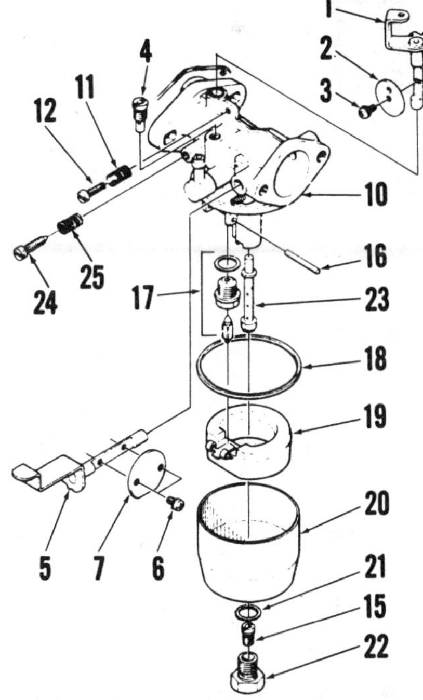

Fig. WR1—Exploded view of Mikuni carburetor used on Models EY18W and EY18-3W.

1. Throttle shaft
2. Throttle plate
3. Screw
4. Idle jet
5. Choke shaft
6. Screw
7. Choke plate
10. Carburetor body
11. Spring
12. Idle speed stop screw
15. Main jet
16. Float pin
17. Inlet needle & seat
18. Bowl gasket
19. Float
20. Float bowl
21. Gasket
22. Plug
23. Needle jet
24. Idle mixture screw
25. Spring

mounted on camshaft gear (Fig. WR5). Engine speed is controlled by the tension on governor spring (14). Before attempting to adjust governed speed, synchronize governor linkage by loosening clamp nut (16) and turning governor lever (15) counterclockwise until carburetor throttle plate is in wide-open position. Insert screwdriver in slot in end of governor lever shaft (1) and rotate shaft counterclockwise as far as possible. Tighten governor clamp nut.

To adjust for a particular loaded rpm, hook governor spring (14) to control lever (12) and governor lever (15). Start engine, loosen wing nut (9), move control lever (12) counterclockwise and adjust stop screw (13) until the required no-load speed is obtained. If engine is to operate at a fixed speed, tighten wing nut (9). For variable speed operation, do not tighten wing nut.

On Model EY18-3W, the centrifugal flyweight governor rotates on shaft (17—Fig. WR6) in the gear cover and is driven by the camshaft gear. Engine speed is controlled by the tension on governor spring (9). Before attempting to adjust governed speed, synchronize governor linkage by loosening wing nut (15) and turning control lever (11) clockwise until carburetor throttle valve is wide open. Lock control lever in this position by tightening wing nut. With clamp screw (21) loose, insert a screwdriver in slot at end of governor shaft (1) and turn shaft clockwise until you feel the vane end of shaft stop against the flyweight thrust sleeve (20), then tighten clamp bolt.

To adjust for a particular loaded rpm, hook governor spring (9) to control lever (11) and governor lever (3). Start the engine. Loosen wing nut (15), move control lever (11) clockwise and adjust stop screw (8) until the required no-load speed is obtained. If engine is to operate at a fixed speed, tighten wing nut (15). For variable speed operation, do not tighten wing nut.

For the following loaded engine speeds on Models EY18W and EY18-3W engines, adjust governor according to the following chart.

Loaded rpm	No-load rpm
1800	2370
2000	2515
2200	2665
2400	2815
2600	2975
2800	3140
3000	3310
3200	3485
3400	3670
3600	3855

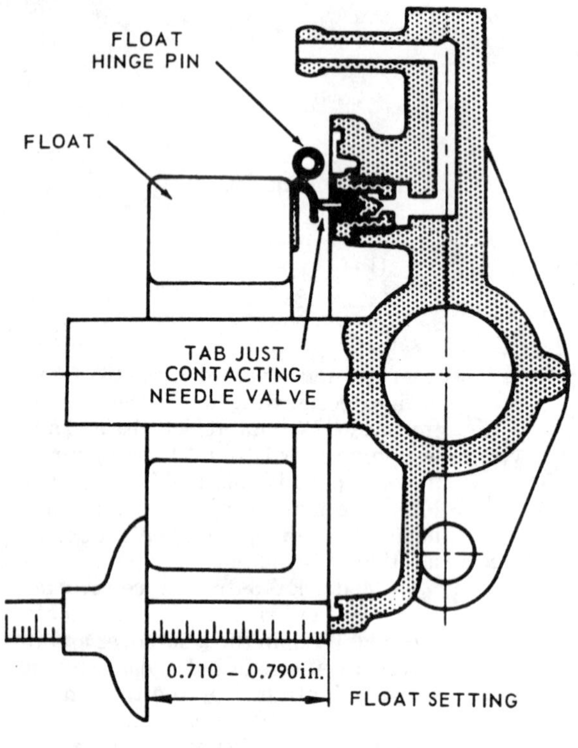

Fig. WR3—With the fuel bowl removed, stand carburetor on manifold flange and measure float setting as shown.

Fig. WR12—View of 23 degrees BTDC timing mark (M) on flywheel aligned with timing mark (D) on crankcase of Model EY18W engine.

IGNITION SYSTEM. Standard ignition system is a flywheel magneto type with breaker points and condenser. Optional solid-state ignition system is available for some models. Refer to appropriate following section for model being serviced.

Breaker-Point Ignition System. Breaker points, condenser and ignition coil are located behind flywheel. Initial breaker-point gap for both models should be 0.014 inch (0.36 mm).

To check and adjust engine timing, disconnect lead wire from shut-off switch. Connect test lead from a continuity light to lead wire and ground remaining test lead to engine. Slowly rotate flywheel in normal operating direction until light goes out. Immediately stop turning flywheel and check location of timing marks. Timing marks should align as shown in Figs. WR12 and WR13. If timing mark on flywheel is below timing mark on crankcase, breaker-point gap is too large. If flywheel marks are above marks on crankcase, breaker-point gap is too small. Carefully measure the distance necessary to align the two timing marks, then remove flywheel and breaker-point cover. Changing point gap 0.001 inch (0.03 mm) will change timing mark on flywheel 1/8 inch (3.18 mm). Reassemble and tighten flywheel retaining nut to torque specified under TIGHTENING TORQUES in REPAIRS section.

Solid-State Ignition System. Transistor type solid-state ignition system does not have breaker points. There is no scheduled maintenance.

To test for spark, remove spark plug cable from spark plug. Insert metal conductor into cable end, then use a suitable insulated tool and hold conductor 1/8 inch (3 mm) from cylinder shroud. Turn engine over and observe spark. A weak spark or no spark indicates a defective ignition coil. Also check for broken, loose or shorted wiring or a faulty spark plug.

LUBRICATION. Check engine oil level daily and maintain oil level at full mark on dipstick or at lower edge of filler plug.

Manufacturer recommends oil with an API service classification of SE or SF. Use SAE 30 oil when temperature is above 40° F (4° C); SAE 20 oil when temperature is between 15° F (-9° C) and 40° F (4° C) and SAE 10W-30 oil when temperature is below 15° F (-9° C).

Oil should be changed after every 50 hours of operation. Crankcase capacity for all models is 1-1/4 pints (600 cc).

An oil dipper mounted on the connecting rod cap provides splash type lubrication.

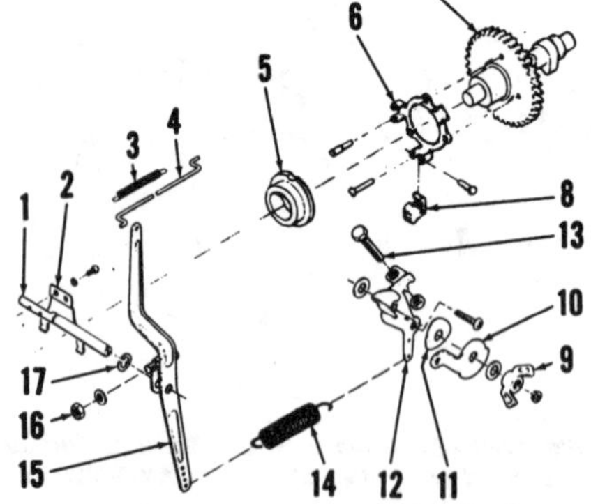

Fig. WR5—Exploded view of mechanical governor and linkage used on EY18W engine.

1. Governor lever shaft
2. Yoke
3. Link spring
4. Carburetor link
5. Thrust sleeve
6. Governor plate
7. Camshaft assy.
8. Flyweights (3)
9. Wing nut
10. Stop plate
11. Wave washer
12. Control lever
13. Speed stop screw
14. Governor spring
15. Governor lever
16. Clamp nut
17. Retaining ring

1. Governor lever shaft
2. Clip
3. Governor lever
4. Link spring
5. Carburetor link
6. Back plate
7. Cap screw
8. Stop screw
9. Governor spring
10. Friction washer
11. Control lever
12. Control wire lock screw
13. Stop plate
14. Washer
15. Wing nut
16. Clip
17. Shaft
18. Spacer
19. Governor gear
20. Thrust sleeve
21. Screw

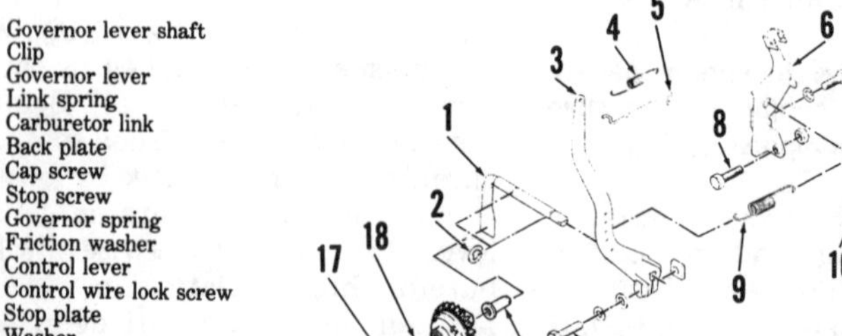

Fig. WR6—Exploded view of mechanical governor and linkage for Model EY18-3W engine.

Fig. WR13—View of 23 degrees BTDC timing mark (M) on flywheel aligned with timing mark (M) on crankcase of Model EY18-3W engine.

CRANKCASE BREATHER. Both models are equipped with a floating poppet type breather valve located in the breather plate behind the cover plate.

REPAIRS

TIGHTENING TORQUES. Recommended tightening torques are listed below:

Spark plug .... 27 ft.-lbs. (37 N·m)
Cylinder head bolts/nuts .. 22 ft.-lbs. (30 N·m)
Connecting rod bolts ..... 14 ft.-lbs. (19 N·m)
Gear cover cap screws .... 13 ft.-lbs. (18 N·m)
Flywheel nut .... 47 ft.-lbs. (64 N·m)

CYLINDER HEAD. Renew cylinder head if warpage exceeds 0.006 inch (0.15 mm) on all models. Always use a new head gasket when installing cylinder head. Tighten cylinder head nuts evenly in three stages; first to 12 ft.-lbs. (16 N·m), then to 18 ft.-lbs. (24 N·m), and finally to 22 ft.-lbs. (30 N·m).

CONNECTING ROD. Connecting rod and piston assembly are removed from cylinder head end after removal of cylinder head and gear cover. Check connecting rod against the following specifications:

Model EY18W

Connecting rod-to-crankpin clearance .... 0.0021-0.0031 in. (0.053-0.079 mm)
Maximum clearance .... 0.005 in. (0.13 mm)
Connecting rod side clearance ...... 0.0080-0.0235 in. (0.203-0.597 mm)
Connecting rod-to-piston pin clearance ...... 0.0004-0.0012 in. (0.010-0.031 mm)
Maximum clearance .... 0.003 in. (0.076 mm)

Model EY18-3W

Connecting rod-to-crankpin clearance ...... 0.0015-0.0025 in. (0.038-0.064 mm)
Maximum clearance .... 0.005 in. (0.13 mm)
Connecting rod side clearance .... 0.0039-0.118 in. (0.099-0.300 mm)
Maximum clearance .... 0.039 in. (0.99 mm)
Connecting rod-to-piston pin clearance ...... 0.0004-0.0012 in. (0.010-0.031 mm)
Maximum clearance .... 0.003 in. (0.076 mm)

When installing connecting rod and piston assembly, make certain match marks (cast ribs) on connecting rod and cap are together as shown in Fig. WR15. Install oil dipper with offset toward gear cover end of engine if engine is to be operated on a tilt toward take-off end. Mount dipper with offset toward flywheel end if operated on a tilt in that direction or with no tilt operation. Use a new lockplate and tighten connecting rod cap screws to specified torque.

PISTON, PIN AND RINGS. Both models are equipped with one compression ring, one scraper ring and one oil control ring. Install rings as shown in Fig. WR16. Check pistons, pins and rings against the following specifications:

Model EY18W

Ring end gap (all rings) .... 0.002-0.010 in. (0.05-0.25 mm)
Max. ring end gap (all rings) .... 0.040 in. (1.01 mm)
Ring side clearance (all rings) ...... 0.0005-0.0030 in. (0.013-0.076 mm)
Max. side clearance (all rings) .... 0.006 in. (0.15 mm)

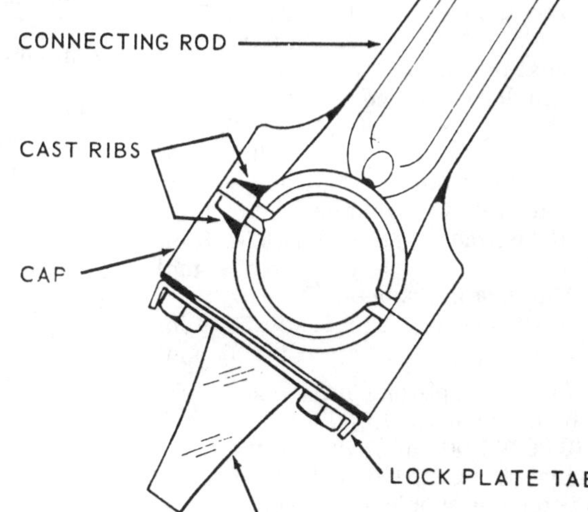

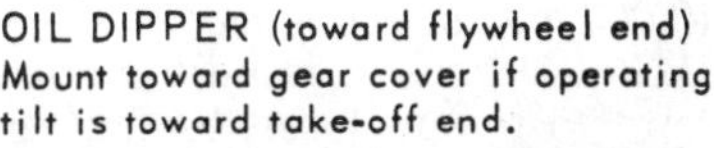

Fig. WR15—Connecting rod and cap must be installed with cast ribs together.

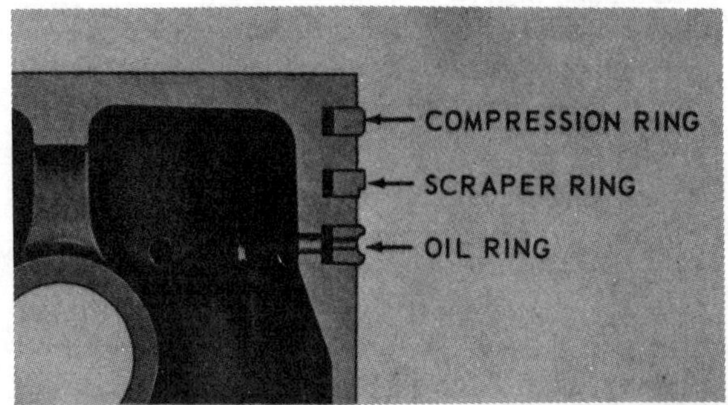

Fig. WR16—Install piston rings as shown. Stagger ring end gaps around circumference of piston.

Piston-to-cylinder clearance (thrust face) .... 0.0016-0.0032 in. (0.041-0.081 mm)
Max. piston-to-cylinder clearance (thrust face) .... 0.008 in. (0.20 mm)
Standard piston diameter (thrust face) .... 2.5567-2.5775 in. (64.940-64.961 mm)
Min. piston diameter (thrust face) .... 2.5535 in. (64.859 mm)

Model EY18-3W

Ring end gap (all rings) .... 0.002-0.010 in. (0.05-0.25 mm)
Max. ring end gap (all rings) .... 0.040 in. (1.01 mm)
Ring side clearance
Top ring .... 0.0020-0.0037 in. (0.051-0.094 mm)
Second ring .... 0.0016-0.0033 in. (0.041-0.084 mm)
Oil ring .... 0.0022-0.0040 in. (0.056-0.102 mm)
Max. ring side clearance (all rings)..... 0.006 in. (0.152 mm)
Piston-to-cylinder clearance (thrust face) .... 0.0012-0.0027 in. (0.031-0.069 mm)
Max. piston-to-cylinder clearance (thrust face) .... 0.008 in. (0.20 mm)
Standard piston diameter (thrust face) .... 2.5571-2.5578 in. (64.95-64.97 mm)
Min. piston diameter (thrust face) .... 2.5539 in. (64.87 mm)

Standard piston pin diameter for all models is 0.5509-0.5512 inch (13.992-14.00 mm) with a minimum diameter of 0.5496 inch (13.96 mm). Piston pin should be a 0.00035 inch (0.009 mm) interference fit to a 0.00039 inch (0.010 mm) loose fit in piston pin bore. Piston pin looseness must not exceed 0.0023 inch (0.06 mm). Standard piston pin clearance in connecting rod is 0.0004-0.0012 inch (0.010-0.029 mm). Piston pin clearance in connecting rod must not exceed 0.0047 inch (0.12 mm).

Piston and ring sets are available in oversizes as well as standard.

CYLINDER BLOCK. If cylinder wall is scored, or out-of-round more than 0.0004 inch (0.01 mm), or tapered more than 0.0006 inch (0.015 mm), the cylinder should be bored to nearest oversize for which piston and rings are available.

Standard cylinder bore diameter for both models is 2.5591-2.5599 inches (65.001-65.022 mm).

CAMSHAFT. Camshaft rides in bores in crankcase and gear cover. When removing camshaft assembly, lay engine on side to prevent tappets from falling out. If valve tappets are removed, identify them so they can be reinstalled in their original positions.

Camshaft journal diameter for Model EY18W engine should be 0.5889-0.5893 inch (14.958-14.968 mm) and for Model EY18-3W engine should be 0.6682-0.6687 inch (16.972-16.985 mm). If camshaft journal on either model is worn 0.0009 inch (0.023 mm) below minimum specified diameter, then camshaft must be renewed.

When reinstalling camshaft, lubricate camshaft bearing surfaces, lift valve tappets upward, then slide camshaft into position making certain the marked tooth on crankshaft gear is between the two marked teeth on camshaft gear.

CRANKSHAFT. The crankshaft on all models is supported in two ball bearings. Renew bearings if any indication of roughness, noise or excessive wear is noted. Crankshaft end play for all models should be 0.001-0.009 inch (0.025-0.229 mm). End play is controlled by an adjusting collar or shim located between crankshaft gear and gear cover main bearing. Three thicknesses of adjusting collars or shims are available for both models and are as follows:

EY18W
0.701-0.709 in.
0.709-0.717 in.
0.717-0.725 in.

EY18-3W
0.0215-0.0257 in.
0.0287-0.0343 in.
0.0362-0.0425 in.

To determine the correct thickness of adjusting collar or shim with gear cover removed, measure distance (A—Fig. WR17 or WR18) between machined surface of crankcase face and end of crankshaft gear. Measure distance (B) between machined surface of gear cover and end of main bearing. The compressed thickness of gear cover gasket (C) should be 0.007 inch (0.18 mm) for all models. Select adjusting shim or collar that is 0.001-0.009 inch (0.03-0.23 mm) less than the total of A, B and C. After reassembly, crankshaft end play can be checked with a dial indicator.

Check crankshaft against the following specifications:

Model EY18W

Standard crankpin diameter .... 1.0210-1.0215 in. (25.933-25.946 mm)
Min. crankpin diameter .... 1.019 in. (25.88 mm)
Connecting rod-to-crankpin clearance .... 0.0021-0.0031 in. (0.053-0.079 mm)
Max. connecting rod-to crankpin clearance .... 0.005 in. (0.13 mm)
Maximum crankpin taper & out-of-round .... 0.0002 in. (0.005 mm)

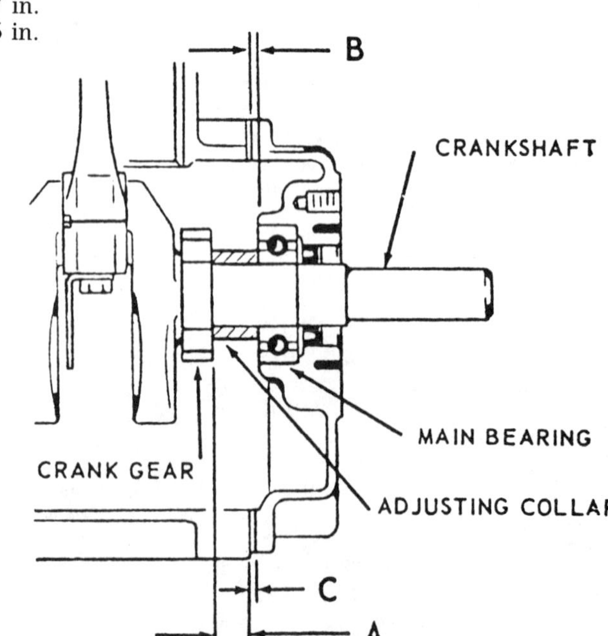

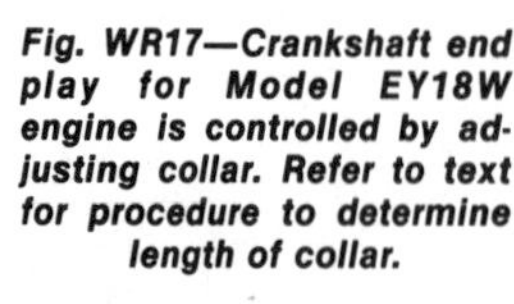
Fig. WR17—Crankshaft end play for Model EY18W engine is controlled by adjusting collar. Refer to text for procedure to determine length of collar.

Model EY18-3W

Standard crankpin
diameter . . . . 1.0217-1.0222 in.
(25.951-25.964 mm)
Min. crankpin
diameter . . . . 1.019 in.
(25.88 mm)

Connecting rod-to-crankpin
clearance . . . . . . 0.0015-0.0025 in.
(0.038-0.064 mm)
Max. connecting rod-to-crankpin clearance . . . . . 0.005 in.
(0.13 mm)

Maximum crankpin taper
& out-of-round . . . . 0.0002 in.
(0.005 mm)

Renew crankshaft if the above specifications are exceeded. When reassembling engine, make certain marked tooth on crankshaft gear is between the two marked teeth on camshaft gear. When renewing crankshaft oil seals, install seals with lips toward ball bearings.

VALVE SYSTEM. Valve tappet gap (cold) for all models should be 0.006-0.008 inch (0.15-0.20 mm) for intake and exhaust valves.

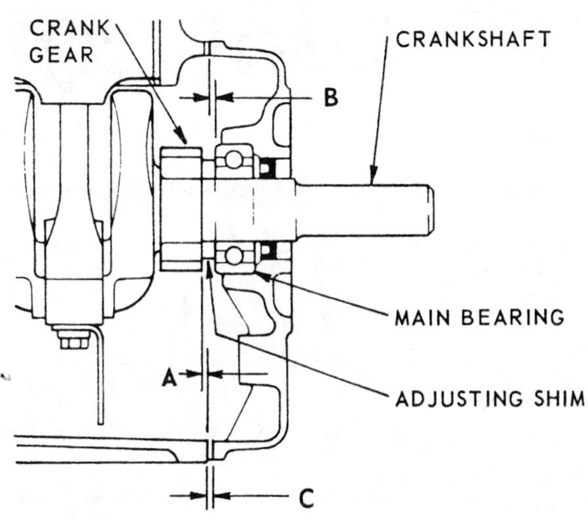

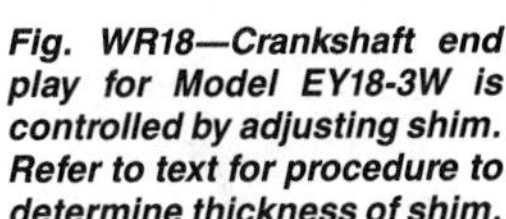
Fig. WR18—Crankshaft end play for Model EY18-3W is controlled by adjusting shim. Refer to text for procedure to determine thickness of shim.

Adjust valve tappet clearance as follows: To increase clearance, grind off end of valve stem. To reduce clearance, grind valve seat deeper or renew valve and/or valve tappet.

Valve face and seat angles are 45 degrees on all models and standard seat width should be 0.047-0.059 inch (1.2-1.5 mm). Maximum allowable seat width is 0.098 inch (2.5 mm).

Standard valve stem diameter for all models is 0.273-0.274 inch (6.93-6.96 mm) for intake and exhaust valves. Standard valve stem-to-guide clearance is 0.0016-0.0039 inch (0.041-0.099 mm) for intake and exhaust valves. Maximum stem clearance in valve guide should be 0.006 inch (0.15 mm) for intake and exhaust valves.

WISCONSIN ROBIN

4-STROKE

Model	Bore	Stroke	Displacement
W1-080	2.01 in. (51 mm)	1.50 in. (38 mm)	4.74 cu. in. (77.6 cc)
W1-145, W1-145V	2.48 in. (63 mm)	1.81 in. (46 mm)	8.73 cu. in. (143 cc)
W1-185, W1-185V	2.64 in. (67 mm)	2.05 in. (52 mm)	11.2 cu. in. (183 cc)
W1-230	2.68 in. (68 mm)	2.44 in. (62 mm)	13.7 cu. in. (225 cc)

ENGINE IDENTIFICATION

All models are four-stroke, air-cooled, single-cylinder gasoline engines. Models W1-145V and W1-185V are vertical crankshaft engines. All other models are horizontal crankshaft engines. On all models, the engine model and specification numbers are located on the name plate on flywheel shroud. The serial number is stamped on the crankcase base. Always furnish engine model, specification and serial numbers when ordering parts.

MAINTENANCE

SPARK PLUG. Recommended spark plug for all models except Model W1-230 is a Champion CJ14 or equivalent. Recommended spark plug for Model W1-230 is a Champion L92YC or equivalent. Specified spark plug electrode gap for all models is 0.025 inch (0.6 mm).

CARBURETOR. All models are equipped with a Mikuni float type carburetor. All models except Model W1-230 are equipped with carburetor shown in Fig. WR25. Carburetor has fixed low speed and high speed jets. Model W1-230 is equipped with carburetor shown in Fig. WR26. Carburetor has an adjustable low speed mixture screw and a fixed main fuel (high speed) jet.

Initial adjustment of low speed mixture screw (24—Fig. WR26) should be 1-3/8 turns open from a lightly seated position. Carburetor used on all other models does not have a fuel mixture adjustment screw. Main fuel mixture for all carburetors is metered through main jet (15—Fig. WR25 or WR26) and is nonadjustable.

Make final carburetor adjustment on Model W1-230 with engine at normal operating temperature and running. Adjust low speed mixture screw to obtain smoothest idle operation and acceleration. On all models, adjust idle stop screw (12) to obtain an idle speed of 1250 rpm at normal operating temperature.

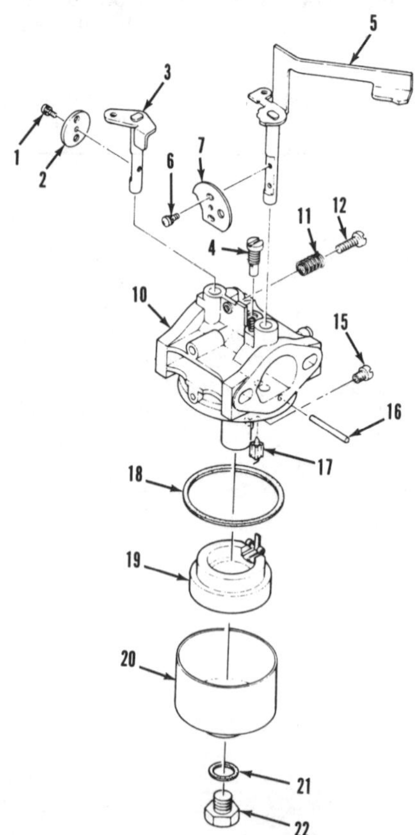

Fig. WR25—Exploded view of carburetor used on all models except Model W1-230.

1. Screw
2. Throttle plate
3. Throttle shaft
4. Pilot jet
5. Choke shaft
6. Screw
7. Choke plate
10. Carburetor body
11. Spring
12. Throttle stop screw
15. Main jet
16. Pin
17. Fuel inlet valve
18. Gasket
19. Float
20. Float bowl
21. Gasket
22. Plug

To check or adjust float level, remove bowl plug (22—Fig. WR25 or WR26) and fuel bowl (20). Place carburetor body on end (on manifold flange) so float pin is in vertical position. Move float to close inlet needle valve.

NOTE: Needle valve is spring loaded. Float tab should just contact needle valve pin but should not compress spring.

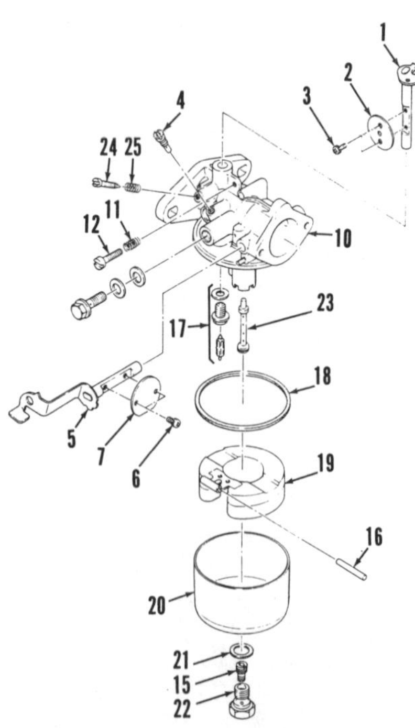

Fig. WR26—Exploded view of carburetor used on Model W1-230.

1. Throttle shaft
2. Throttle plate
3. Screw
4. Pilot screw
5. Choke shaft
6. Screw
7. Choke plate
10. Carburetor body
11. Spring
12. Throttle stop screw
15. Main jet
16. Pin
17. Fuel inlet valve assy.
18. Gasket
19. Float
20. Float bowl
21. Gasket
22. Plug
23. Nozzle
24. Low speed mixture screw
25. Spring

On Models W1-080 and W1-230, float (19) should be parallel with float bowl mating surface (Fig. WR27). On Models W1-145, W1-185, W1-145V and W1-185V, measure float setting as shown in Fig. WR28. Dimension "A" should be 0.492-0.571 inch (12.5-14.5 mm). On all models, carefully bend tab on float lever to obtain correct setting.

GOVERNOR. All models are equipped with a centrifugal flyweight type governor. Governor assembly is located in the gear cover and is driven by the camshaft gear on Models W1-080, W1-145, W1-185 and W1-230 (Fig. WR29). Governor assembly is located in the oil pan and is driven by the camshaft gear on Models W1-145V and W1-185V (Fig. WR30).

To adjust external governor linkage on all models, loosen clamp screw on governor lever (Fig. WR31). Turn speed control lever clockwise (toward high speed position) until throttle valve in carburetor is opened fully. Hold lever in this position. Insert screwdriver in slot at end of governor shaft. Turn clockwise as far as shaft can be turned. Tighten governor lever clamp screw. Two different governor springs are available to obtain a variety of engine speeds. Changing governor spring location in holes will also vary engine speed.

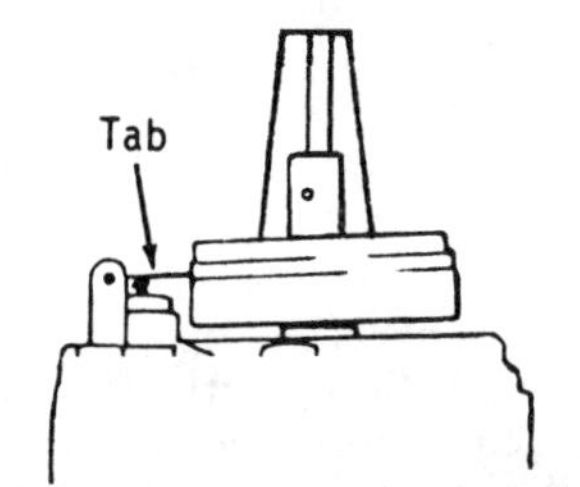

Fig. WR27—On Models W1-080 and W1-230, float should be parallel with float bowl mating surface.

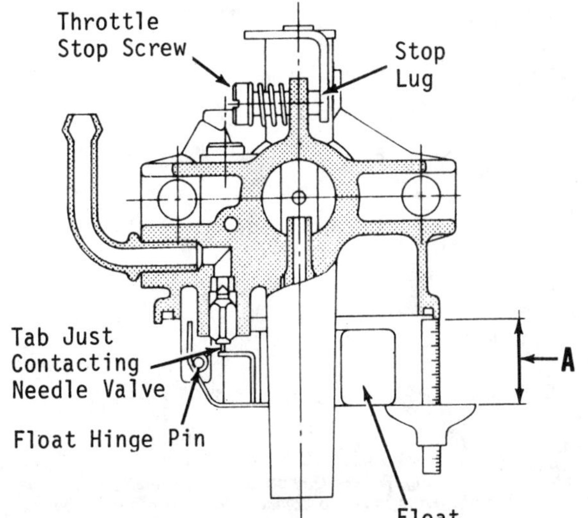

Fig. WR28—On Models W1-145, W1-185, W1-145V and W1-185V, refer to text for float dimension "A."

IGNITION SYSTEM. Early models are equipped with a breaker point type ignition system. Late models are equipped with a solid-state ignition system. Refer to the appropriate paragraphs for ignition type being serviced.

Breaker-Point Ignition System. Breaker points and condenser are located behind flywheel and the ignition coil is located outside flywheel. Initial breaker-point gap for all models is 0.014 inch (0.36 mm). Specified air gap between ignition coil and flywheel is 0.020 inch (0.5 mm).

To check and adjust engine timing, disconnect lead wire from shut-off switch. Connect test lead from a continuity light to lead wire and ground remaining test lead to engine. Slowly rotate flywheel in normal operating direction until light goes out. Immediately stop turning flywheel and check location of timing marks. Timing marks should align. If timing mark on flywheel is below timing mark on crankcase, breaker-point gap is too small. Carefully measure the distance necessary to align the two timing marks, then remove flywheel and breaker-point cover. Changing point gap 0.001 inch (0.03 mm) will change timing mark on flywheel approximately 1/8 inch (3.18 mm). Reassemble and tighten flywheel retaining nut to torque specified under TIGHTENING TORQUES in REPAIRS section.

Solid-State Ignition System. Transistor type solid-state ignition system does not have breaker points. There is no scheduled maintenance. Specified air gap between ignition coil and flywheel is 0.020 inch (0.5 mm).

To test for spark, remove spark plug lead from spark plug. Insert metal conductor into cable end, then use a suitable insulated tool and hold conductor 1/8 inch (3 mm) from cylinder shroud. Turn engine over and observe spark. A weak spark or no spark indicates a defective ignition coil. Also check for broken, loose or shorted wiring or a faulty spark plug.

LUBRICATION. Check engine oil level daily and maintain oil level at full mark on dipstick or at lower edge of filler plug.

Manufacturer recommends oil with an API service classification of SE or SF. Use SAE 30 oil when temperature is above 40° F (4° C), SAE 20 oil when temperature is between 15° F (-9° C) and SAE 10W-30 oil when temperature is below 15° F (-9° C).

Oil should be changed after every 50 hours of operation. Crankcase capacity is 0.85 pints (402 cc) for Model W1-080, 1.3 pints (615 cc) for Models W1-145, W1-185 and W1-185V, 1.2 pints (568 cc) for Model W1-145V and 1.37 pints (648 cc) for Model W1-230.

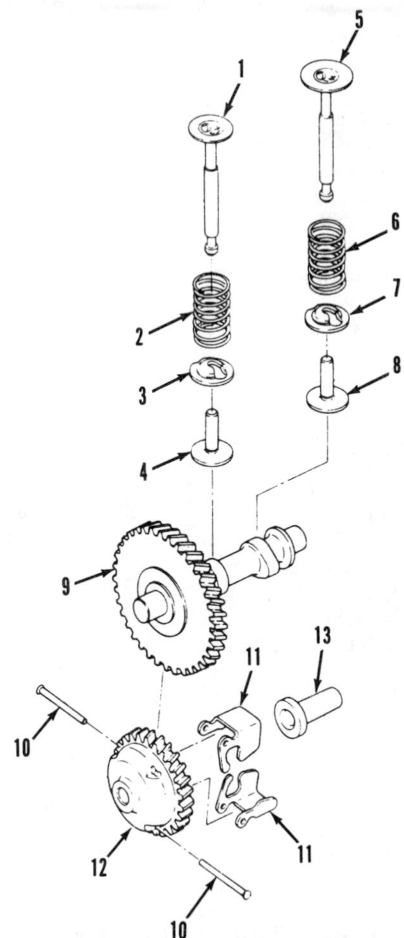

Fig. WR29—Exploded view of governor flyweight assembly used on horizontal crankshaft models.

1. Exhaust valve
2. Spring
3. Retainer
4. Tappet
5. Intake valve
6. Spring
7. Retainer
8. Tappet
9. Camshaft & gear
10. Pins
11. Flyweights
12. Governor gear
13. Sleeve

GENERAL MAINTENANCE. Check and tighten all bolts, nuts or clamps prior to each day of operation. Check for fuel or oil leakage and repair if necessary.

Clean dust, dirt, grease or any foreign material from cylinder head and cylinder block cooling fins after every 50 hours of normal operation. Inspect fins for damage and repair as necessary.

Cylinder head should be removed and carbon and other combustion deposits cleaned after every 500 hours of normal operation.

REPAIRS

TIGHTENING TORQUES. Recommended tightening torque specifications are as follows:

Spark plug:
- W1-080, W1-230 .... 17-18 ft.-lbs. (23-24 N·m)
- W1-145, W1-145V, W1-185, W1-185V ......9-11 ft.-lbs. (12-14 N·m)

Cylinder head bolts/nuts:
- W1-080 .... 7-8 ft.-lbs. (9-11 N·m)
- W1-145, W1-145V. W1-185, W1-185V .....14-17 ft.-lbs. (19-23 N·m)
- W1-230 .... 24-26 ft.-lbs. (33-35 N·m)

Connecting rod cap screws:
- W1-080 ....4-6 ft.-lbs. (6-8 N·m)
- W1-145, W1-145V .... 8 ft.-lbs. (10.5 N·m)
- W1-185, W1-185V, W1-230 .... 12-14 ft.-lbs. (16-19 N·m)

Crankcase cover/oil pan:
- W1-080, W1-145, W1-185 .... 6-7 ft.-lbs. (8-10 N·m)
- W1-145V, W1-185V .... 7 ft.-lbs. (9 N·m)
- W1-230 .... 12-14 ft.-lbs. (16-19 N·m)

Flywheel nut
- (all models) .... 44-47 ft.-lbs. (60-64 N·m)

CYLINDER HEAD. On all models, renew cylinder head if warpage exceeds 0.006 inch (0.15 mm). Always use a new head gasket when installing cylinder head. Tighten cylinder head bolts or nuts evenly and in stages until torque specified in TIGHTENING TORQUES is obtained.

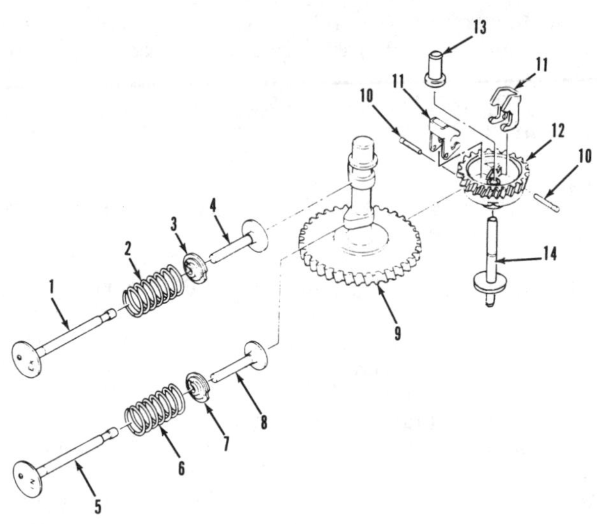

Fig. WR30—Exploded view of governor flyweight assembly used on vertical crankshaft models.

1. Exhaust valve
2. Spring
3. Retainer
4. Tappet
5. Intake valve
6. Spring
7. Retainer
8. Tappet
9. Camshaft & gear
10. Pins
11. Flyweights
12. Governor gear
13. Sleeve
14. Stem

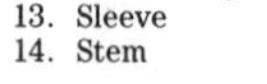

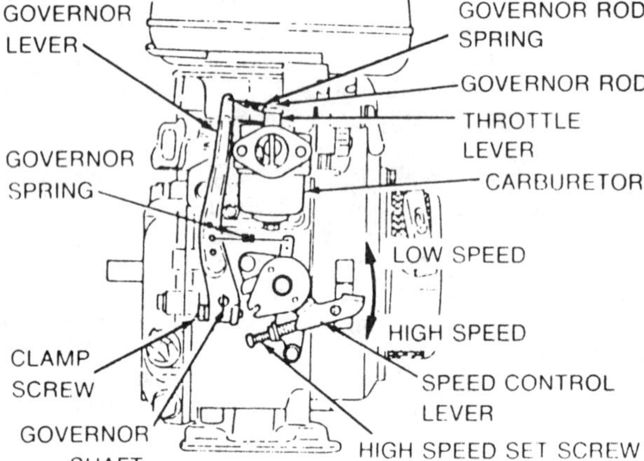

Fig. WR31—View of external governor linkage similar to linkage used on all models. Refer to text for adjustment procedure.

CONNECTING ROD. To remove piston and connecting rod assembly, remove all cooling shrouds. Remove cylinder head and crankcase cover or oil pan. Remove connecting rod retaining bolts and remove connecting rod and piston assembly from cylinder bore.

Connecting rod-to-crankpin clearance for all models except Model W1-145V and W1-185V should be 0.0015-0.0025 inch (0.038-0.064 mm). If clearance exceeds 0.008 inch (0.203 mm), renew bearing and/or crankshaft.

Connecting rod side clearance for all models should be 0.0039-0.0118 inch (0.099-0.300 mm). If side clearance exceeds 0.039 inch (0.99 mm), renew connecting rod and/or crankshaft.

Connecting rod-to-piston pin clearance for all models should be 0.0004-0.0012 inch (0.010-0.030 mm). If clearance exceeds 0.005 inch (0.13 mm), renew piston pin and/or connecting rod.

When installing connecting rod and piston assembly, make certain match marks (cast ribs) on connecting rod and cap are adjacent to each other as shown in Figs. WR32 and WR33. On horizontal crankshaft engines equipped with an oil dipper, install dipper with offset facing toward gear cover end of engine if engine is to be operated on a tilt toward pto side. Mount dipper with offset facing toward flywheel end if operated on level or tilt toward flywheel side. On all

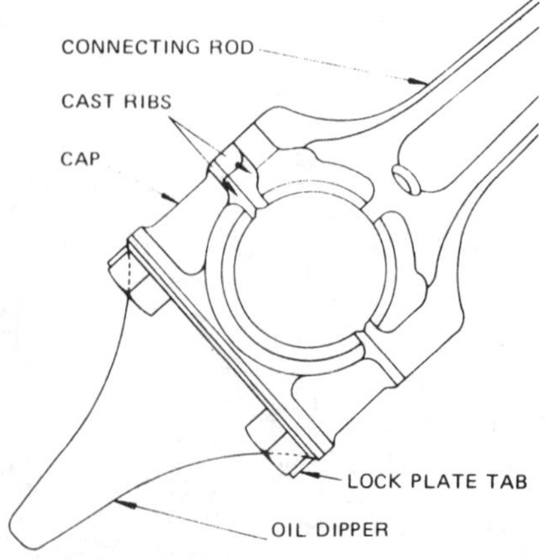

Fig. WR32—On horizontal crankshaft models, cast ribs must be adjacent to each other for correct assembly.

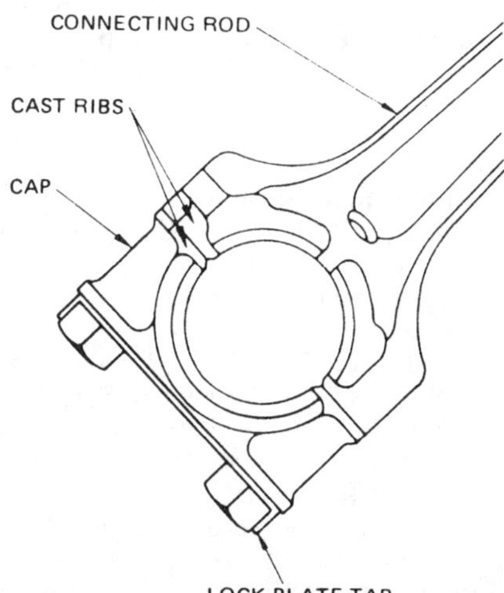

Fig. WR33—On vertical crankshaft models, cast ribs must be adjacent to each other for correct assembly.

models, use a new lockplate and tighten connecting rod cap screws to torque specified in TIGHTENING TORQUES.

PISTON, PIN AND RINGS. All models are equipped with one compression ring, one scraper ring and one oil control ring. Install rings as shown in Fig. WR34. Stagger ring end gaps at 90 degree intervals around piston.

Piston ring end gap for Models W1-080, W1-145, W1-145V and W1-185V should be 0.0080-0.0160 inch (0.203-0.407 mm). If ring end gap exceeds 0.059 inch (1.5 mm), renew rings and/or recondition cylinder bore. Piston ring end gap for Models W1-185 and W1-230 should be 0.002-0.010 inch (0.05-0.25 mm). If ring end gap exceeds 0.040 inch (1.01 mm), renew rings and/or recondition cylinder bore.

Piston ring side clearance in piston ring groove should be as follows:

Models W1-080, W1-145, W1-145V, W1-185V

Top ring	0.0035-0.0053 in. (0.089-0.135 mm)
Second ring	0.0023-0.0041 in. (0.058-0.104 mm)
Oil ring	0.0004-0.0025 in. (0.010-0.064 mm)

Models W1-185

Top ring	0.0019-0.0037 in. (0.048-0.094 mm)
Second ring	0.0004-0.0021 in. (0.010-0.053 mm)
Oil ring	0.0004-0.0025 in. (0.010-0.064 mm)

Model W1-230

Top ring	0.002-0.004 in. (0.050-0.102 mm)
Second ring	0.0015-0.0030 in. (0.038-0.076 mm)
Oil ring	0.0004-0.0025 in. (0.010-0.064 mm)

On all models, if piston ring side clearance exceeds 0.006 inch (0.152 mm), renew piston rings and/or piston.

Standard piston-to-cylinder clearance for Model W1-080 is 0.0003-0.0019 inch (0.008-0.048 mm). Standard piston-to-cylinder clearance for Models W1-145, W1-185, W1-145V and W1-185V is 0.0008-0.0023 inch (0.020-0.058 mm). Standard piston-to-cylinder clearance for Model W1-230 is 0.0012-0.0027 inch (0.030-0.069 mm).

Standard piston diameter is 2.006-2.007 inches (50.95-50.98 mm) for Model W1-080. If piston diameter is 2.0031 inches (50.88 mm) or less, renew piston. Standard piston diameter is 2.469-2.470 inches (62.71-62.74 mm) for Models W1-145 and W1-145V. If piston diameter is 2.4661 inches (62.64 mm) or less, renew piston. Standard piston diameter is 2.629-2.630 inches (66.78-66.80 mm) for Models W1-185 and W1-185V. If piston diameter is 2.6261 inches (66.70 mm) or less, renew piston. Standard piston diameter is 2.675-2.676 inches (67.95-67.97 mm) for Model W1-230. If piston diameter is 2.6272 inches (67.87 mm) or less, renew piston.

Standard piston pin diameter for Model W1-080 is 0.4327-0.4330 inch (10.990-11.000 mm). If piston pin diameter is 0.4325 inch (10.99 mm) or less, renew piston pin. Standard piston pin diameter for all other models is 0.5509-0.5512 inch (13.993-14.000 mm). If piston pin diameter is 0.5496 inch (13.96 mm) or less, renew piston pin.

On all models, piston pin should be 0.00035 inch (0.009 mm) interference fit to a 0.00039 inch (0.010 mm) loose fit in piston pin bore. Piston pin looseness must not exceed 0.0023 inch (0.06 mm). Standard piston pin clearance in connecting rod is 0.0004-0.0012 inch (0.010-0.029 mm). If piston pin clearance exceeds 0.0047 inch (0.12 mm), renew piston pin and/or connecting rod.

Piston and ring sets are available in oversizes as well as standard.

CYLINDER BLOCK. If cylinder wall is scored, or out-of-round more than 0.0004 inch (0.01 mm), or tapered more than 0.0006 inch (0.015 mm), the cylinder should be bored to nearest oversize for which piston and rings are available.

Standard cylinder bore diameter is 2.0079-2.0086 inches (51.000-51.018 mm) for Model W1-080. If diameter exceeds 2.0139 inches (51.15 mm), recondition cylinder. Standard cylinder bore diameter is 2.4803-2.4811 inches (63.000-63.020 mm) for Models W1-145 and W1-145V. If diameter exceeds 2.4862 inches (63.149 mm), recondition cylinder. Standard cylinder bore diameter is 2.6378-2.6385 inches (67.000-67.018 mm) for Models W1-185 and W1-185V. If diameter exceeds 2.6437 inches (67.15 mm), recondition cylinder bore. Standard cylinder bore diameter is 2.6770-2.6777 inches (68.000-68.014 mm) for Model W1-230. If diameter exceeds 2.683 inches (68.15 mm), recondition cylinder bore.

CAMSHAFT. On all models, camshaft rides in bores in crankcase and gear cover/oil pan. When removing camshaft, position engine to prevent tappets from falling free. If valve tappets are removed, identify them so they can be reinstalled in their original positions.

Standard camshaft journal diameter is 0.3932-0.3937 inch (9.987-10.000 mm) for journal at pto side and 0.3926-0.3937 inch (9.972-10.000 mm) for journal at flywheel side on Model W1-080. If either journal diameter is less than 0.39.18 inch (9.95 mm), renew camshaft. Standard camshaft journal diameter for flywheel and pto side is 0.5895-0.5899 inch (14.973-14.983 mm) for Models W1-145, W1-185, W1-145V and W1-185V. If diameter is less than 0.5890 inch (14.961 mm), renew camshaft. Standard camshaft journal diameter for flywheel and pto side is 0.6680-0.6688 inch (16.967-16.977 mm) for Model W1-230. If diameter is less than 0.6670 inch (16.942 mm), renew camshaft.

When reinstalling camshaft, lubricate camshaft bearing surfaces, lift valve tappets upward, then slide camshaft into position, making certain timing marks on camshaft and crankshaft gears are aligned.

CRANKSHAFT. The crankshaft on all models is supported in two ball bearing

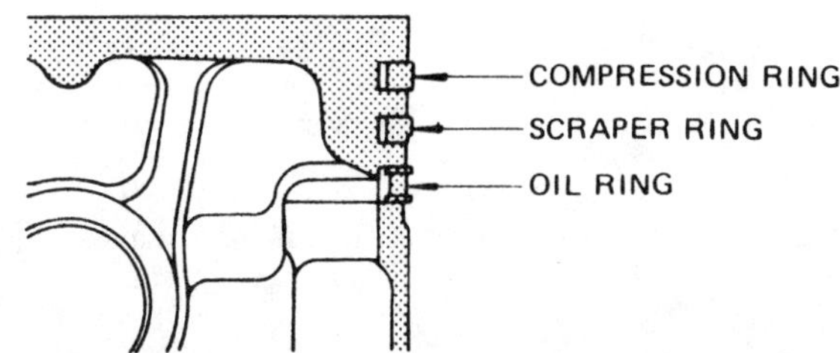

Fig. WR34—Install piston rings as shown. Stagger ring end gaps at 90 degree intervals around piston.

type main bearings. Renew bearings if any indication of roughness, noise or excessive wear is noted. Crankshaft end play should be 0.000-0.008 inch (0.00-0.20 mm) for all models. End play is controlled by an adjusting collar or shim (17—Fig. WR35) located between crankshaft gear and gear cover/oil pan main bearing. Three thicknesses of adjusting collars or shims are available for each model.

To determine the correct thickness of adjusting collar or shim with gear cover removed, measure distance (A—Fig. WR36) between machined surface of crankcase face and end of crankshaft gear. Measure distance (B) between machined surface of gear cover and end of main bearing. The compressed thickness of gear cover gasket (C) is 0.009 inch (0.23 mm). Select adjusting shim or collar that is 0.001-0.009 inch (0.03-0.23 mm) less than the total of A, B and C. After reassembly, crankshaft end play can be checked with a dial indicator.

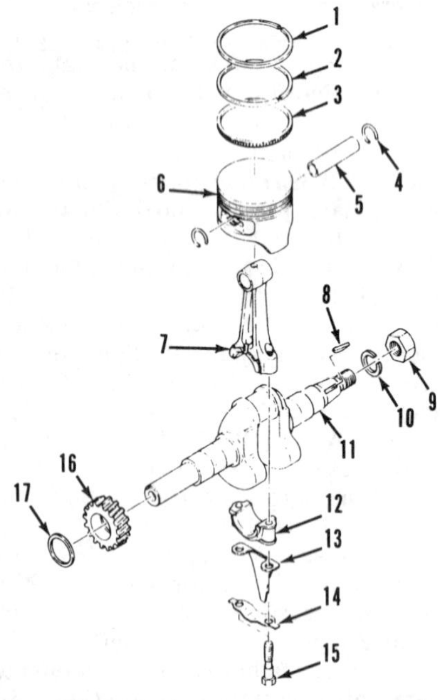

Fig. WR35—Exploded view of crankshaft, connecting rod and piston assembly on Model W1-185. All models are similar.

1. Compression ring
2. Scraper ring
3. Oil control ring
4. Retainer
5. Piston pin
6. Piston
7. Connecting rod
8. Key
9. Nut
10. Lockwasher
11. Crankshaft
12. Connecting rod cap
13. Slinger
14. Lockplate
15. Cap screw
16. Gear
17. Shim

Standard crankpin journal diameter is 0.7856-0.7860 inch (19.954-19-964 mm) for Model W1-080. If journal diameter is 0.781 inch (19.84 mm) or less, renew crankshaft. Standard crankpin journal diameter is 0.9429-0.9434 inch (23.950-23.962 mm) for Model W1-145. If journal diameter is 0.9390 inch (23.85 mm) or less, renew crankshaft. Standard crankpin journal diameter is 0.9436-0.9450 inch (23.967-24.00 mm) for Model W1-145V. If journal diameter is 0.9391 inch (23.85 mm) or less, renew crankshaft. Standard crankpin journal diameter is 1.0216-1.0220 inches (25.950-25.959 mm) for Model W1-185. If journal diameter is 1.0177 inches (25.85 mm) or less, renew crankshaft. Standard crankpin journal diameter is 1.021-1.023 inches (25.93-26.00 mm) for Models W1-185V and W1-230. If journal diameter is 1.017 inches (25.83 mm) or less, renew crankshaft.

On all models, clearance between crankpin journal and connecting rod should be 0.0015-0.0025 inch (0.038-0.064 mm). If clearance exceeds 0.008 inch (0.20 mm), renew connecting rod bearing and/or crankshaft. Maximum allowable crankpin taper and out-of-round is 0.0002 inch (0.005 mm).

When reassembling engine, make certain timing marks on camshaft and crankshaft gears are aligned after installation. When renewing crankshaft oil seals, install seals with lips toward ball bearings.

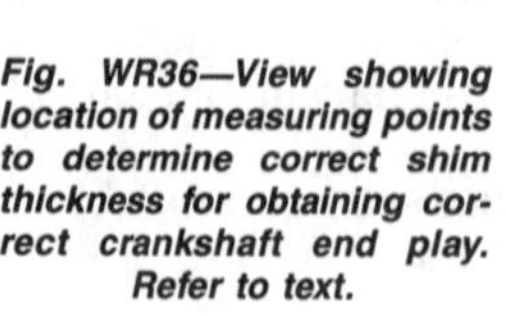

Fig. WR36—View showing location of measuring points to determine correct shim thickness for obtaining correct crankshaft end play. Refer to text.

VALVE SYSTEM. Valve tappet gap (cold) should be 0.003-0.005 inch (0.08-0.13 mm) for all models except Model W1-230. Valve tappet gap (cold) should be 0.0063-0.0079 inch (0.16-0.20 mm) for Model W1-230.

To increase clearance, grind off end of valve stem. To reduce clearance, grind valve seat deeper or renew valve and/or valve tappet.

On all models, valve face and seat angles should be 45 degrees. Standard seat width is 0.047-0.059 inch (1.2-1.5 mm). Maximum allowable seat width is 0.098 inch (2.5 mm).

Standard valve stem diameter is 0.2165 inch (5.5 mm) for Model W1-080. If diameter is 0.2105 inch (5.35 mm) or less, renew valve. Standard valve stem diameter is 0.256 inch (6.5 mm) for Models W1-145V and W1-185V. If diameter is 0.2501 inch (6.35 mm) or less, renew valve. Standard valve stem diameter for Models W1-145 and W1-185 is 0.2543 inch (6.460 mm) for intake valve stem and 0.2528 inch (6.421 mm) for exhaust valve. If diameter is 0.2500 inch (6.35 mm) or less, renew valve. Standard valve stem diameter for Model W1-230 is 0.276 inch (7 mm). If diameter is 0.270 inch (6.86 mm) or less, renew valve.

Standard valve stem-to-guide clearance for all models is 0.0010-0.0024 inch (0.025-0.061 mm) for intake valve stem and 0.0022-0.0039 inch (0.056-0.099 mm) for exhaust valve. Maximum allowable valve stem-to-guide clearance is 0.0118 inch (0.30 mm) for intake and exhaust valve stems.

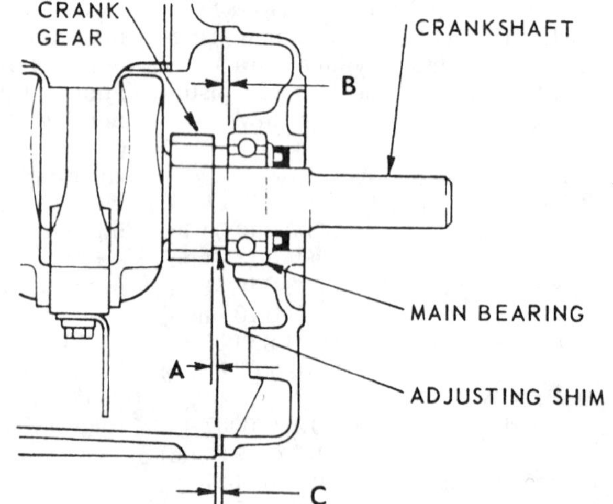

SERVICING WISCONSIN ROBIN ACCESSORIES

REWIND STARTER

Early Style

OVERHAUL. To disassemble the rewind starter, refer to Fig. WR60 and release spring tension by pulling rope handle until about 18 inches (457 mm) of rope extends from unit. Use thumb pressure against ratchet retainer to prevent reel from rewinding and place rope in notch in outer rim of reel. Release thumb pressure slightly and allow spring mechanism to slowly unwind. Twist loop of return spring and slip loop through slot in ratchet retainer. Refer to Fig. WR61 and remove nut, lockwasher, plain washer and ratchet retainer. Reel will completely unwind as these parts are removed. Remove compression spring, three ratchets and spring retainer washer. Slip fingers into two of the cavity openings in reel hub (Fig. WR62) and carefully lift reel from support shaft in housing.

CAUTION: Take extreme care that power spring remains in recess of housing. Do not remove spring unless new spring is to be installed.

If power spring escapes from housing form a 4½ inch (114 mm) wire ring and twist ends together securely. Starting with the outside loop, wind spring inside the ring in a counterclockwise direction.

NOTE: New power springs are secured in a similar wire ring for ease in assembly.

Place spring assembly over recess in housing so hook in outer loop of spring is over the tension tab in housing. Carefully press spring from wire ring and into recess of housing.

Using a new rope of same length and diameter as original, place rope in handle and tie a figure eight knot about 1½ inches (38 mm) from end. Pull knot into top of handle. Install other end of rope through guide bushing of housing and through hole in reel groove. Pull rope out through cavity opening and tie a slip knot about 2½ inches (64 mm) from end. Place slip knot around center bushing as shown in Fig. WR63 and pull knot tight. Stuff end of rope into reel cavity. Spread a film of light grease on power spring and support shaft. Wind rope ¼ turn clockwise in reel and place rope in notch on reel. Install reel on support shaft and rotate reel counterclockwise until tang on reel engages hook on inner loop of power spring. Place outer flange of housing in a vise and use finger pressure to keep reel in housing. Hook a loop of rope in the reel notch and preload power spring by turning reel 7 full turns counterclockwise. Remove rope from notch and allow reel to slowly turn clockwise as rope winds on pulley and handle returns to guide bushing on housing.

Install spring retainer washer (Fig. WR61), cup side up, and place compression spring into cupped washer. Install return spring with bent end hooked into hole of reel hub. Place the three ratchets in position so they fit the contour of the recesses. Mount ratchet retainer so loop end of return spring extends through slot. Rotate retainer slightly clockwise until ends of slots just begin to engage the three ratchets. Press down on retainer, install flat washer, lockwasher and nut, then tighten nut securely.

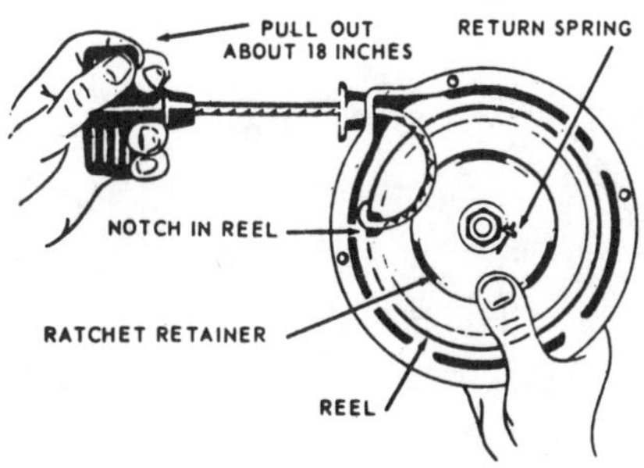

Fig. WR60—View showing method of releasing spring tension on rewind starter assembly.

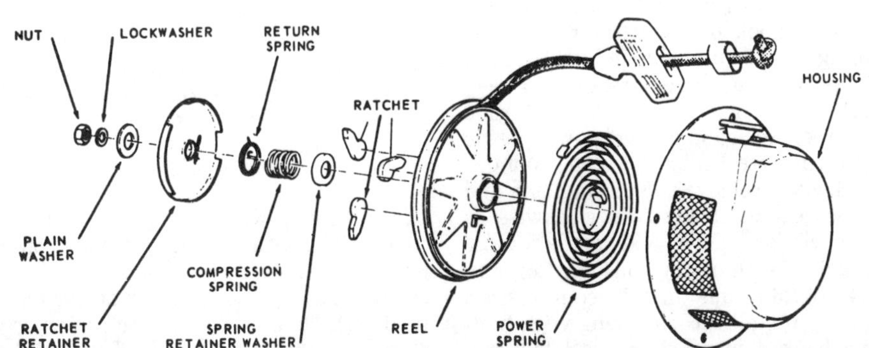

Fig. WR61—Exploded view of rewind starter assembly used on Models EY18W and EY18-3W engines.

New Style

OVERHAUL. To disassemble the rewind starter, refer to Fig. WR64 and release spring tension by pulling rope handle until about 14 inches (356 mm) of rope extends from unit. Use thumb pressure against ratchet retainer to prevent reel from rewinding and place rope in notch in outer rim of reel. Release thumb pressure slightly and allow spring mechanism to slowly unwind. Twist loop of return spring and slip loop through slot in ratchet retainer. Refer to Fig. WR65 and remove clip, thrust washer and ratchet retainer. Reel will completely unwind as these parts are removed. Remove compression spring

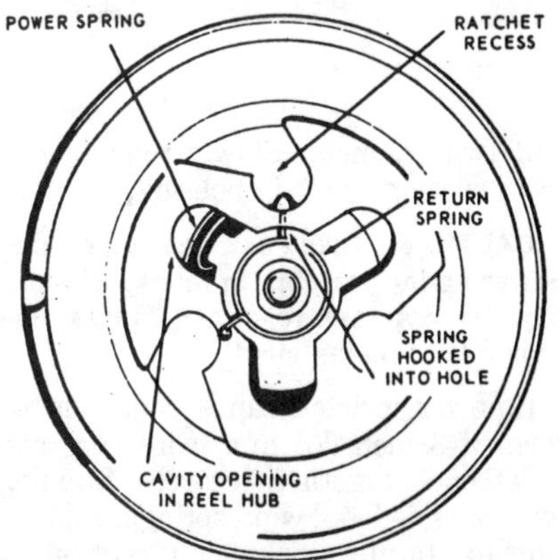

Fig. WR62—Use fingers in reel hub cavities to lift reel from support shaft.

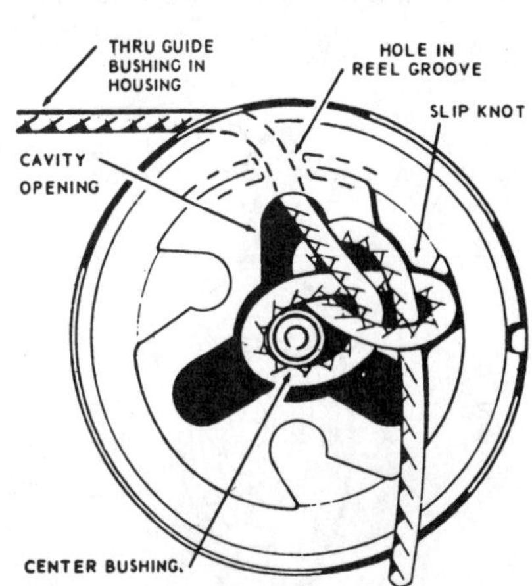

Fig. WR63—Install rope through guide bushing end hole in reel groove, then tie slip knot around center bushing.

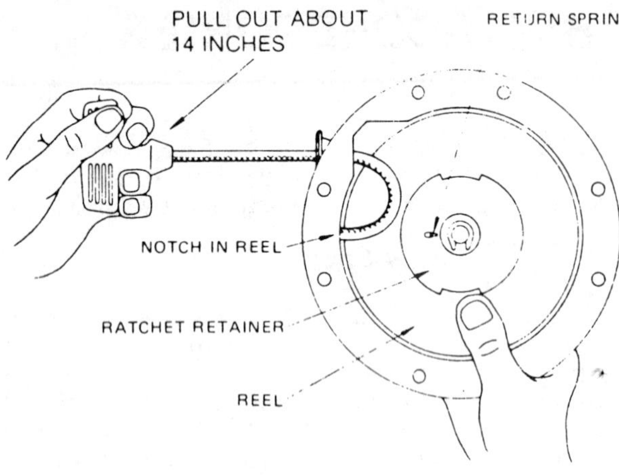

Fig. WR64—View showing method of releasing spring tension on rewind starter assembly.

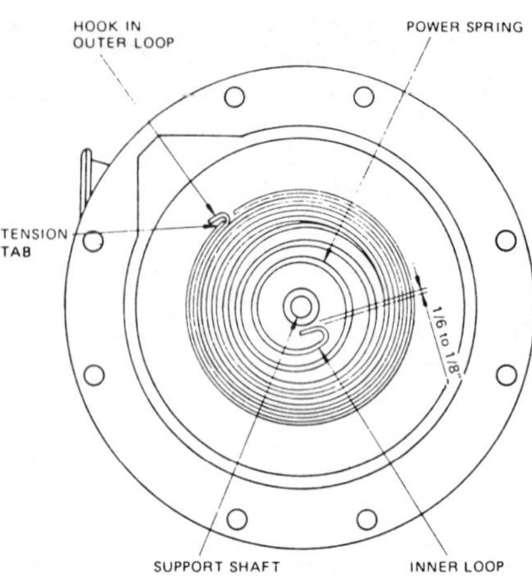

Fig. WR67—Reel must be installed with hook in outer loop of power spring engaged on tension tab and inner loop of spring spaced as shown from suupport shaft.

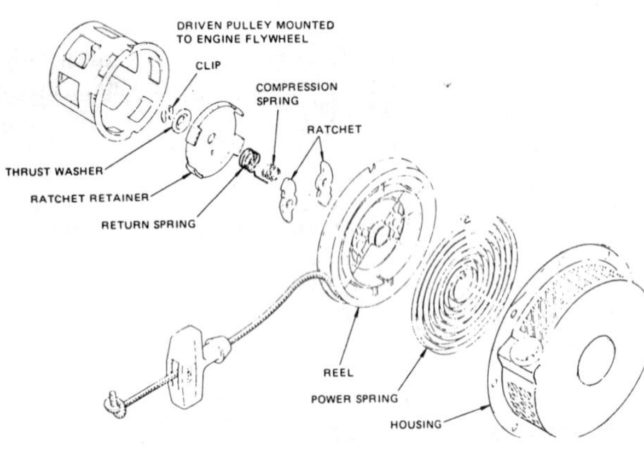

Fig. WR65—Exploded view of rewind starter assembly used on Model W1-145 engine.

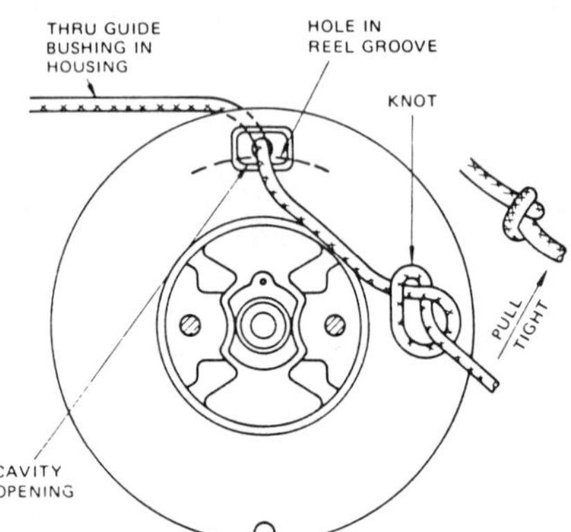

Fig. WR68—Install rope through guide bushing and hole in reel groove, then tie knot as shown.

and two ratchets. Slowly remove reel from support shaft in housing.

CAUTION: Take extreme care that power spring remains in recess of housing. Do not remove spring unless new spring is to be installed.

If power spring escapes from housing, form a 3 inch (76 mm) wire ring and twist ends together securely. Starting with outside loop, wind spring inside the ring in counterclockwise direction as shown in Fig. WR66.

NOTE: New power springs are secured in a similar wire ring for ease in assembly.

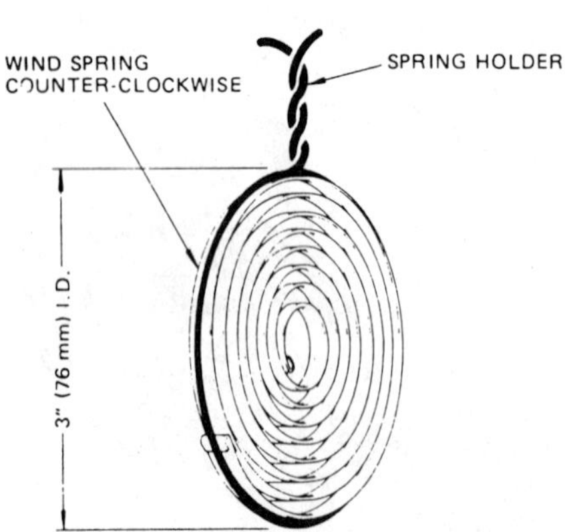

Fig. WR66—Illustration showing fabricated spring holder used to rewind power spring.

Place new spring assembly over recess in housing so hook in outer loop of spring is over the tension tab in the housing. Carefully press spring from wire ring and into recess of housing. See Fig. WR67.

Using a new rope of same length and diameter as original, place rope in handle and tie a figure eight knot about 1 inch (25 mm) from end. Pull knot into top of handle. Install other end of rope through guide bushing of housing and through hole in reel groove. Pull rope out through cavity opening and tie a knot about 1 inch (25 mm) from end. Tie knot as illustrated in Fig. WR68 and stuff knot into cavity opening. Wind rope 2½ turns clockwise on reel, then lock rope in notch of reel pulley. Install reel on support shaft and rotate reel counterclockwise until tang on reel engages hook on inner loop of power spring. Place outer flange of housing in a vise and use finger pressure to keep reel in housing. Hook a loop of rope into reel notch and preload power spring by turning reel four full turns counterclockwise. Remove rope from notch and allow reel to slowly turn clockwise as rope winds on pulley and handle returns to guide bushing on housing. Refer to Fig. WR65 and install compression spring. Install return spring with bent end hooked into hole of reel hub and looped end toward outside. Install the

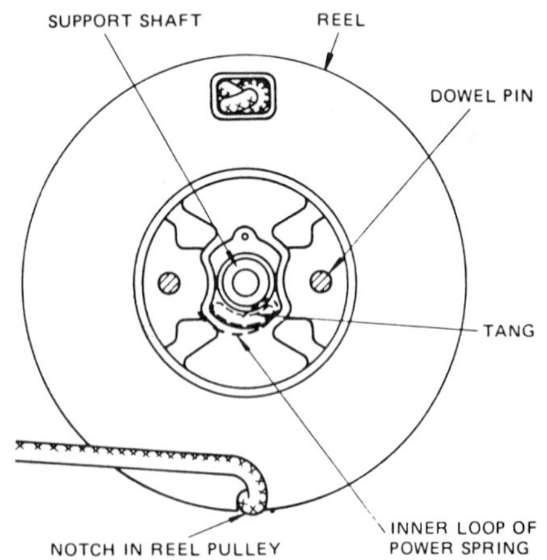

Fig. WR69—When installing reel in housing, engage inner loop of power spring on tang of reel. Refer to text.

two ratchets, with tips pointed in a counterclockwise direction, into the contours of the hub recesses. Mount ratchet retainer so loop end of return spring extends through slot. Rotate retainer slightly clockwise until ends of slots just begin to engage the ratchets. Press down on retainer, install washer and clip.

GEAR REDUCTION UNIT

Gear-driven speed reduction units are available on Model EY18-3W engines. See Fig. WR70 for exploded view of reduction unit. Reduction ratio is 6:1 Special crankshafts are required on reduction models. Takeoff end of crankshaft is machined to serve as the drive gear. Reduction unit is mounted directly to engine gear cover. Reduction unit may be mounted on engine in any of four positions, depending on requirements. Note that cover (9) will require shifting to maintain position of level plug (1) and fill plug (2) when unit is mounted in positions other than shown in Fig. WR70. Reduction gears are lubricated by same type of oil used in engine crankcase. Gearcase capacity is approximately 3/8 pint (0.18 L), filled to level plug opening. Oil change interval should not exceed 500 hours of operating time and more frequently in heavy continuous duty. Maintain level by regular periodic checks.

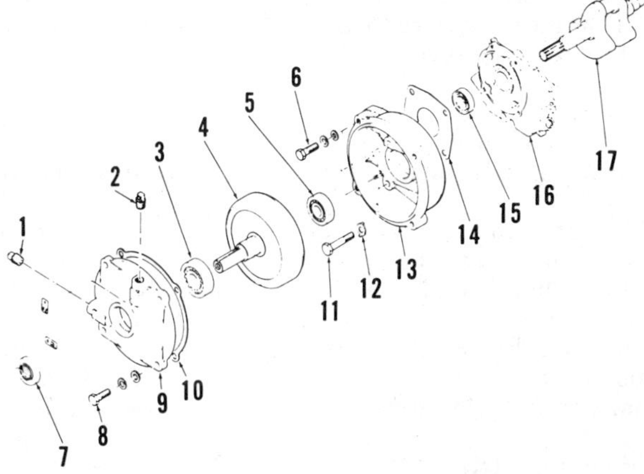

Fig. WR70—Exploded view of gear reduction unit used on Model EY18-3W engine.

1. Level plug
2. Fill plug
3. Bearing
4. Drive shaft/gear
5. Bearing
6. Bolt
7. Seal
8. Bolt
9. Cover
10. Gasket
11. Bolt
12. Lock tab
13. Housing
14. Gasket
15. Seal
16. Engine gear cover
17. Crankshaft with gear

TELEDYNE TOTAL POWER CENTRAL PARTS DISTRIBUTORS

(Arranged Alphabetically by States)

These franchised firms carry extensive stocks of repair parts. Contact them for name of dealer in their aira who will have replacement parts.

Joe H. Brady & Assoc.
308 South 31st Street
Birmingham, Alabama 35233

Parts Service Company, Inc.
301 Columbus Street
Montgomery, Alabama 36104

Keeling Company
P.O. Box 15310
North Little Rock, Arkansas 72231

Southwest Products Corp.
2949 N. 30th Avenue
Phoenix, Arizona 85017

Lanco Engine Services, Inc.
664 Marsat Court, Suite D
Chula Vista, California 92011

E. E. Richter & Son
P.O. Box 8186
Emeryville, California 94662

Lanco Engine Services, Inc.
13610 Gramercy Avenue
Gardena, California 90249

Central Equipment Company
4477 Garfield Street
Denver, Colorado 80216

Highway Equipment & Supply Co.
5366 Highway Avenue
Jacksonville, Florida 32205

P.H. Neff & Sons, Inc.
5295 N.W. 79th Avenue
Miami, Florida 33166

Highway Equipment & Supply Co.
1016 West Church Street
Orlando, Florida 32805

Highway Equipment & Supply Co.
6015 U.S. Highway 301 North
Tampa, Florida 33610

Georgia Engine Sales & Service
5715 B Oak Brook Parkway
Norcross, Georgia 30093

Lanco Engine Services Hawaii, Inc.
3219 Valena Street
Honolulu, Hawaii 96819

Teledyne Total Power
1230 North Skyline Drive
Idaho Falls, Idaho 83402

Wilson Engine Power, Inc.
5727 N.E. 16th Street
Des Moines, Iowa 50313

Harley Industries, Inc.
P.O. Box 336
Wichita, Kansas 67201

Wilder Motor & Equipment Co., Inc.
4022 Produce Road
Louisville, Kentucky 40218

William F. Surgi Equipment Co.
221 Laitram Lane
Harahan, Louisiana 70123

Diesel Engine Sales & Service
199 Turnpike Street
Stroughton, Massachusetts 02072

Engine Supply of Novi, Inc.
44455 Grand River
Novi, Michigan 48050

Teledyne Total Power
4401 85th Avenue North
Brooklyn Park, Minnesota 55443

Northern Engine & Supply Inc.
Hoover Road
Virginia, Minnesota 55792

Allied Construction Equipment Co.
4015 Forest Park Avenue
St. Louis, Missouri 63108

King-McIver Sales, Inc.
6375 New Burnt Poplar Road
Greensboro, North Carolina 27419

Northern Engine & Supply Co.
2710 3rd Avenue, North
Fargo, North Dakota 58102

John Reiner Co.
145 Commerce Road
Carlstadt, New Jersey 07072

Central Motive Power
3740 Princeton Drive, N.E.
Albuquerque, New Mexico 87107

John Reiner & Co., Inc.
6681 Moore Road
Syracuse, New York 13211

Cincinnati Engine & Parts Co., Inc.
2863 Stanton Avenue
Cincinnati, Ohio 45206

Cincinnati Engine Parts Co.
1051 Lucas Road
Mansfield, Ohio 44905

Harley Industries, Inc.
5408 South 103rd East Avenue
Tulsa, Oklahoma 74147

Lucky Distributing
8111 NE Columbia Blvd.
Portland, Oregon 97218

Joseph L. Pinto, Inc.
719 East Baltimore Pike
East Lansdowne, Pennsylvania 19050

Teledyne Total Power
1061 Main Street
N. Huntington, Pennsylvania 15642

Wilder Motor & Equipment
1219 Rosewood Drive
Columbia, South Carolina 29201

Southern Supply Company
366 North Royal Street
Jackson, Tennessee 38301

RCH Distributors, Inc.
P.O. Box 161272
Memphis, Tennessee 38186

Wilder Motor & Equipment Co.
301 15th Avenue North
Nashville, Tennessee 37203

Harley Industries
9226 Premier Row
Dallas, Texas 75247

Harley Industries, Inc.
12227 L FM 529 Spencer
Houston, Texas 77041

Teledyne Total Power
1127 International Parkway
Fredericksburg, Virginia 22405

Engine Sales & Service Co.
601 Ohio Avenue
Charleston, West Virginia 25302

Teledyne Total Power
2244 West Bluemound Road
Waukesha, Wisconsin 53187

CANADIAN DISTRIBUTORS

Mandem Inc.
5925 83rd Street
Edmonton, Alberta T6E 4

Pacific Engines & Equipment
1391 William Street
Vancouver, British Columbia V5L 2

Mandem Inc.
21 Murray Park Road
Winnipeg, Manitoba R3J 3

Mandem Inc.
481 Edinburgh Drive
Moncton, New Brunswick E1E 4

Mandem Inc.
3 Bestobell Road
Toronto, Ontario M8W 4

Mandem Inc.
8550 Delmeade Road
Montreal, Quebec H4T 1

United Continental Engines
8550 Delmeade Road
Montreal, Quebec H4T 1

Mandem Inc.
1250 St. John Street
Regina, Saskatchewan S4R 1

MAINTENANCE LOG

NOTES